The Human Eye

Structure and Function

Clyde W. Oyster

The University of Alabama at Birmingham

Sinauer Associates, Inc. • Publishers
Sunderland, Massachusetts

On the cover:
Leonardo da Vinci. A study of the proportions of the eye and eyelid. Detail from Turin 15574. Royal Library, Turin. By permission of the Italian Minister for Cultural Resources and Activities.

Part openers:
From *Abbildungen des menschlichen Auges* by Samuel Thomas Soemmerring, 1801.

The Human Eye: Structure and Function

 For information or to order, address: Sinauer Associates, Inc., P.O. Box 407, Sunderland, Massachusetts 01375-0407 U.S.A. Fax: 413-549-1118.
Internet: publish@sinauer.com, www.sinauer.com.

Library of Congress Cataloging-in-Publication Data

Oyster, Clyde W., 1940–
The human eye : structure and function / Clyde W. Oyster.
p. cm.
Includes bibliographical references and index.
ISBN 0-87893-645-9
1. Eye—physiology. 2. Eye—Anatomy. I. Title.
[DNLM: 1. Eye—physiology. WW 103 098h 1999]
QP475.097 1999
612.8'4—dc21
DNLM/DLC 99-25402
for Library of Congress CIP

Printed in U.S.A.

9 8 7 6 5 4 3 2 1

[*To Ellen*]

Through whose eyes
I have seen so much
that I might not
have seen without her

Brief Contents

Contents

PART TWO
Components of the Eye

Preface

This book is about both structure and function of the human eye

A wing would be a most mystifying structure if one did not know that birds flew. One might observe that it could be extended a considerable distance, that it had a smooth covering of feathers, and that strength and lightness were prominent features of its construction. These are important facts, but by themselves they do not tell us that birds fly. Yet without knowing this, and without understanding something of the principles of flight, a more detailed examination of the wing itself would probably be unrewarding.

■ Horace B. Barlow, *Sensory Communication*

Once upon a time, a group of young men and women gathered in Berkeley's eternal springtime and discovered that they shared, as one of them recently wrote, an "enthralling delusion: We were going to explain the brain." We were aided and encouraged in this endeavor by the Berkeley faculty members who were our graduate and postdoctoral advisors (I won't name them all—they know who they are). One of the lessons we learned from them is embodied in the accompanying quote—namely, that understanding a thing well cannot be done by looking at it from just one point of view. Structure is unlikely to be meaningful unless function is taken into account. The lesson applies just as well to the eye as it does to the wing of a dove.

I was one of Horace Barlow's graduate students in those heady times, and although I cannot think as well as he does (few people can), I would be a poor pupil indeed if I did not try to reflect some of the principles that made his work so original and so important. Thus I am obliged by his example to write not only about the structure of the eye but also about its function—that is, about what it does and how it works.

So this book is not about the anatomy of the eye, though that is its major emphasis, and it is not about the physiology and biochemistry of the eye, though those topics are considered. As vision scientists try to better understand the eye, they use tools from anatomy, biochemistry, physics, physiology, psychology, and other disciplines. By drawing on all of them as need be, I have tried to tell a more complete story than could be told from the perspective of any one of them. As I point out in the Acknowledgments, I have had help in trying to tell this story properly.

The text is in two parts, introduced by a Prologue. The Prologue makes the points that the human eye is just one sort of many that have evolved over several hundred million years, and that there is more to vision than our experience of it. Part One, which comprises Chapters 1 through 7, deals with the initial stages of ocular development, the eye as an optical system, and the other systems—vascular, muscular, innervational—that are necessary for the eye's normal functioning. Much of this material is ultimately concerned with linkages or connections between the eye and the rest of the body or between the eye and the external world. Part Two, Chapters 8 through 16, is devoted to the structure and function of the component parts of the eye. The concerns here are mostly local—that is, the specific roles played by individual tissues in terms of the eye's overall performance. But because tissues can interact even though they are physically separate, the different parts of the eye exert influences on one another; these interactions are also considered. Finally, eyes normally change throughout life, and these changes are outlined in a brief Epilogue. Like the rest of the body, our eyes grow, mature, and age. Structure and function at one age are not necessarily typical of structure and function at others.

Eyes do not always work as they should

In writing about how the eye works, I could not avoid the obvious fact that some eyes work better than others. There are many ways eyes can fail and many degrees of failure—some minor, some catastrophic. These are the problems that concern optometrists and ophthalmologists, the people whose business it is to recognize and treat ocular disorders. Because much of their diagnosis depends on evaluation of the eye's structure, it is important to address the anatomical basis of abnormal function whenever it is known. Clinicians, like anatomists, are studying the structure of the eye, but with different methods and different views; thus, I have included the clinical views of ocular structure wherever possible. And since one way to treat some disorders is surgery, which is essentially a controlled alteration of the eye's anatomy, the elements of the more common ocular surgeries are considered.

Ocular pathology is a vast subject, however, and this is not a clinical text. Therefore, I have concentrated on common problems that are directly related to structural anomalies, to anatomical variation, and to surgical alterations. Thus, I have said much about glaucoma, cataracts, and corneal refractive surgery, for example, but little about infectious diseases.

What we understand about structure depends on how we study it

People tend to think of anatomy as a collection of cut-and-dried facts, stacked like firewood and ready for consumption or use. I would like to dispel that notion; facts about the anatomy of the eye are no more substantial and no less elusive than are facts in other disciplines. The difficulties are of two kinds. One is that there is no such thing as the anatomy of the eye; eyes vary in structural detail, and when we say that the structure is such and such, we are really saying that this is the most common, the most studied, or the most probable of the variant forms. Variation may be disconcerting because it forces us to hedge our facts round with qualifications, transforming seemingly hard facts into soft and fuzzy facts, but variation is real; in some cases, knowing that variation exists may explain what would otherwise be incomprehensible.

The other difficulty with anatomical facts is that all methods for anatomical study, including dissection and direct observation, provide limited, restricted views of the eye's structure. Because of these limitations, anatomical facts sometimes have a shadowy "now you see it, now you don't" quality about them. Thus many illustrations and descriptions of ocular structure are interpretations of the evidence, educated guesses based on methods that reveal structural details selectively and incompletely. In short, the quality of the facts depends critically on the evidence underlying them, on the methods by which the evidence was obtained, and on how often the observations have been repeated and confirmed. For this reason, another part of the story about the eye's structure and function deals with some of these methods and their limitations.

These methodological discussions are interspersed throughout the text as "boxes" that deal with experimental techniques, problems in interpretation, and some analytical methods. Their collective message is that our understanding of the eye changes as we find new ways to visualize it. And since surgeons are in the business of altering the anatomy of the eye on the basis of their interpretations of the structure, I have also included several discussions of surgical principles.

Efforts to understand the eye began more than 2000 years ago—we're still working on it

The human eye has several histories: an evolutionary history dealing with the emergence of our species and our eyes over geological time, a developmental history that unfolds within an individual's lifetime, and an intellectual history, extending back at least to the classical age of Greece, that documents human efforts to understand the eye.

Much of the intellectual history of the eye is a visual history composed of drawings that reveal what was understood about the eye at any given time. As luck would have it, the building next door to mine houses the Reynolds Historical Library, a superb collection of books about science and medicine in which I could examine original volumes dating back a thousand years or so. My paging through many old and some recent books resulted in a series of "vignettes" that sketch the development of understanding about the eye through brief glimpses of people and ideas. The main theme running through the series is how the eye was depicted and what the representations say about the understanding of structure. This theme, and my very rusty Latin, led me to concentrate on authors who illustrated their work. But there are exceptions; some scientists who are too important to ignore preferred words to drawings.

The vignettes are not scholarly analyses or characterizations; I'm not a trained historian, and I have depended on biographies written by others for my information about people's lives. I can look knowledgeably at drawings of the eye, however, and my comments about what people understood come from my examination of the primary sources, with a few exceptions that were not available to me. That element of the vignettes is my scholarly contribution. My sources of information are collected in the Historical References section.

Some readers may regard these historical asides as irrelevant clutter. If so, skip over them, but there is a good reason for reading them someday, if not now. Simply put, one can't judge the originality of an idea without knowing the history of the subject. (That conception of the importance of knowing history, incidentally, is not original with me.) Genuinely original ideas are rare, and there is no way to know what is new unless one knows what is old.

This book draws on and points to other literature about the eye

The literature relating to the structure and function of the eye is formidably large. To give some idea of what I mean by "large," the 1998 meeting of the Association for Research in Vision and Ophthalmology (ARVO) had 5210 presentations, the majority of which will eventually be published in one of the scientific or clinical journals. A quick subject search of *Index Medicus* for the period from 1995 to August 1998 produced 3963 publications under the heading "retina," 3087 under "cornea," and 676 under "extraocular muscles." There are other categories for the other parts of the eye and related structures, for the visual pathways in the brain, for visual functions, and so on—all adding up, on average, to hundreds of pages about the eye and vision published every day. And all of these topics are the subjects of books.

I have read only a small fraction of this literature. Although I know how to search for the good stuff, my selective reading is unlikely to have included everything of importance. And I may not have interpreted correctly everything that I did read. It follows that this book should not be one's sole source of information about the eye.

I have recommended other things to read in the Additional Reading and References sections that conclude each chapter. In round numbers, there are references to 100 books and 600 journal articles. Many of the references are cited because they provided the basis for what I have to say on a subject and, more importantly, for what is illustrated in the figures. Some of the articles contradict statements I make in the text, and interested readers are free to make their own judgments.

The references have been filtered in several ways to arrive at the final list. As much as possible, I selected articles specific to the human eye, my second choice being eyes of other primates. And I have emphasized references that struck me as particularly useful, insightful, or readable. When I could find appropriate book chapters or review articles, I have cited them in preference to the primary literature containing original scientific or clinical observations on which the reviews were based. My reason for doing so is simply that the technical jargon in the primary literature can be a significant barrier to anyone who is new to a specialty.

Two books have been cited repeatedly. My copy of *Histology of the Human Eye*, by Michael J. Hogan, Jorge A. Alvarado, and Joan P. Weddell, has been worn to tatters by constant use. The book was published in 1971, but the photomicrographs and illustrations are superb and timeless. In particular, the three-dimensional renditions of ocular structure by Joan Weddell are the standard by which all others should be judged; her drawings inspired more than a few of the illustrations in this book.

Molecular Biology of the Cell, by Bruce Alberts, Dennis Bray, Julian Lewis, Martin Raff, Keith Roberts, and James D. Watson, now in its third edition, has been a constant, reassuring companion throughout my writing. The book has only a few sections devoted specifically to the eye, but it deals with all the stuff of which eyes are made—molecules, cells, and tissues—at a level of sophistication and clarity that no book on the eye has ever offered. Anyone familiar with *MBC* will recognize my debt to the authors in certain matters of style and organization.

This book is meant to be read

I tend to think of this book as a series of related short stories or essays about the eye. Although the series is constructed such that later chapters build on earlier ones, I have included liberal cross-referencing and a glossary to accommodate readers who prefer to dip into books here and there. These features should also assist teachers who might like to use this book as a text, but whose sense of logical development leads them to a different sequence of presentation.

One of my goals in writing about the human eye was to make what we know about this marvelous organ accessible to people who are not already specialists in vision. To this end, I have tried to use technical language sparingly and to define the inevitable specialized terms as I use them. I have also made very few assumptions about what the reader already knows. Numerous scientific disciplines interact in studying the eye and vision, and the specialized vocabularies and knowledge bases that facilitate communication among anatomists, for example, may impede communication between anatomists and biochemists. For this reason, I have not assumed expertise in any particular discipline on the reader's part.

My sense of storytelling has prompted a leisurely pace with diversions, digressions, and some dawdling to admire the view. It's a good story as far as it goes—it's not complete—and I see no need to hurry it along; in any case, there is far less meandering in the telling of the story than there was in its development. For those of the "just give me the facts, Jack" persuasion ("Will that be on the board examinations?"), I suggest rereading my earlier comments about "facts," and if that fails to cure the compulsion, finding another book. This one is meant to be read, and the illustrations are meant to be studied. It is meant to bring deeper understanding of the eye and its role in vision, while eliciting in the reader a sensation more akin to pleasure than to pain.

Acknowledgments

I get by with a little help from my friends,
Oh, I get high with a little help from my friends.
John Lennon and Paul McCartney

I had more than a little help from my friends in writing this book, but to avoid any misunderstanding of their contributions, John and Paul's phrase "get high" should be construed figuratively. I am particularly indebted to the following people, all of whom read and commented on at least one chapter of the draft manuscript: Larry J. Alexander, John F. Amos, Jimmy D. Bartlett, Thomas F. Freddo, Michael F. Land, Robert E. Marc, John C. Morrison, Jane M. Olver, Roswell R. Pfister, Jacob G. Sivak, Daniel J. Sullivan, Richard C. Van Sluyters, and Gerald Westheimer.

Their sometimes extensive critiques eliminated numerous errors, large and small, and helped bring order to some very muddled text; the book is far better than it would have been without them. I didn't always follow their suggestions, however, and they are not responsible for errors of fact or interpretation that remain. Those are my doing.

Dan Sullivan read the entire manuscript while he was an optometry student and offered important insights from a student's perspective, telling me frankly what worked for him and what didn't. And critiques of the historical vignettes by Charlotte G. Borst, then Director of the Reynolds Historical Library, provided slightly painful but extremely valuable lessons in how *not* to write about the past. Of my scientific reviewers, the burden fell heaviest on Robert Marc, who managed to find something of value in 200 pages of manuscript and more than 100 figures and told me how to recast them. Those pages are now Chapters 13 thorugh 16, dealing with the retina and optic nerve. Only a good friend would have done it.

Peter Farley, my patient, resourceful editor at Sinauer Associates, has been an encouraging, reassuring voice on the telephone and a correspondent by e-mail for more years than I care to mention. He and the project editor, Paula Noonan, orchestrated all the people and operations that contributed to the making of this book. Working with them has been a genuine pleasure.

Anatomy demands illustrations, a lot of them, and I am grateful to Nancy Haver and to the folks at Precision Graphics for doing so well with the pitiful sketches and jottings they sometimes had to work with. The illustrations are at least as important as the text that accompanies them.

Leo Semes did the ultrasound and gonioscopy images of my eyes, Kate Henning plotted my corneal topography, and Kim Washington, clinic photographer, took the external and fundus photos of my eyes, all at the UAB School of Optometry clinic. Tim L. Pennycuff, archivist, and Faye Harkins, curatorial assistant, arranged for illustrations from rare books in the Reynolds Historical Library to be photographed.

Bob Rodieck kindly allowed me to read much of the manuscript for his book *The First Steps in Seeing* before it was published, and it helped considerably in writing the material on phototransduction. I borrowed several figures in almost identical form from his book. Tim Kraft tutored me in the workings of photoreceptors and helped calculate the relative sensitivities of rods and cones; Denis Baylor confirmed our calcualations.

This book is dedicted to Ellen Shizuko Takahashi, my longtime academic colleague, research collaborator, traveling companion, best friend, dance partner, and spouse. She helped in all her roles. And I appreciate her willingness, at one time in her life, to be photographed in her bath.

Clyde W. Oyster
Birmingham, Alabama

Prologue

A Brief History of Eyes

Were there no example in the world of contrivance except that of the eye, it would be alone sufficient to support the conclusion which we draw from it, as to the necessity of an intelligent Creator.

■ William Paley, *Natural Theology*

To suppose that the eye, with all its inimitable contrivances for adjusting the focus to different distances, for admitting different amounts of light, and for the correction of spherical and chromatic aberration, could have been formed by natural selection, seems, I freely confess, absurd in the highest degree.

■ Charles Darwin, *The Origin of Species*

The Antiquity of Eyes and Vision

Thinking about the eye gave Charles Darwin "a cold shudder"

The human eye was William Paley's premier example of natural objects that could have come into being only through the offices of a supreme designer. Paley reasoned that if a complex mechanical contrivance like a watch requires a watchmaker, something as wonderful as the eye, which is beyond the wit and ability of man to create, must have an eyemaker. A half century later, Charles Darwin had to confront and try to circumvent this appealing argument from design when he wrote *On the Origin of Species.* It was not an easy job, and in discussing the eye under the heading "Organs of Extreme Perfection and Complication," Darwin began with the quotation opposite, in which he admitted the apparent absurdity of the eye's evolving without plan or design. It strains imagination to think that something as complex and intricate as the human eye could have arisen through a mindless process like natural selection, which is little more than a sieve that passively separates existing structures that offer an advantage for survival from those that do not, favoring transmission of the former to successive generations.

Darwin's great idea about the workings of evolution ran counter to the commonsense notion that complex things, be they mechanical or biological, have a purpose and must have been designed for that purpose. And in one sense, this notion is correct; neither a Rolex nor an eye is likely to have been created from nothing by chance, by the right pieces just happening to fall together in just the right way. But complex things do evolve, acquiring their structural and functional complexity in stages. Modern timepieces evolved from rather crude and humble origins, and there are many transitional forms along the way (and some obvious failures as well). The same is true of eyes. The difference between the timepiece and the eye is the agent of evolution; for the timepiece, intelligent, purposeful design was clearly a factor, and not long before Paley wrote, substantial prizes had been offered as incentives to the designers of chronometers.

But in Darwin's conception, nature and natural selection directed the eye's evolution, and there was no clear-cut goal, no plan, and no particular purpose other than survival of the organisms bearing the eyes. In trying to persuade people to consider an alternative view of the way the world works, Darwin had no choice but to confront Paley's argument on Paley's turf—if natural selection could not account for the evolution of a "perfect" and complicated eye, the theory would be dead in its tracks. A substantial case had to be made for the existence of various working, useful eyes that could represent, in retrospect, the evolutionary lineages of modern, complex eyes.

Here is Darwin's argument.

> Reason tells me, that if numerous gradations from a simple and imperfect eye to one complex and perfect can be shown to exist, each grade being useful to its possessor, as is certainly the case; if further, the eye ever varies and the variations be inherited, as is likewise certainly the case; and if such variations should be useful to any animal under changing conditions of life, then the difficulty of believing that a perfect and complex eye could be formed by natural selection, though insuperable by our imagination, should not be considered as subversive of the theory.*

Darwin went on to show that the eyes of living animals vary considerably and that they range from exceedingly simple to highly complex. And since the owners of all these eyes use them for one purpose or another, they are "useful" ipso facto. Eyes of a particular type also vary, and some of the variations can be passed along from parents to offspring. Although most variations will either be less good than the original or will be neutral in effect, some will be improvements that will be transmitted selectively because the "changing conditions of life" give the owners of the improved eyes a better chance to survive and reproduce. And there you have it—the necessary conditions for natural selection to do the job are in place and natural selection cannot be ruled out as the eye's designer.

In marshaling his evidence, Darwin was working at a tremendous disadvantage. He did not know the full variety of forms that eyes can assume—several have been recognized only within the last few decades—and much of the detailed structural variation among eyes of similar type was yet to be studied. Darwin had no knowledge about when or how many times eyes and their key components had originated and did not know the mechanism through which good variants—the improvements—could be passed to successive generations. Gregor Mendel was still cultivating his pea plants, so there could be no discussion of genes, DNA, mutations, or any of the other important elements of molecular genetics. And Darwin erred, as his friend Thomas Huxley pointed out, in thinking that evolution could be described as a long, gradual series of almost infinitesimal changes. Most species change hardly at all during their tenure as discrete species (a phenomenon called stasis by the evolutionary biologists), and the formation of new species can be rapid (on a geological timescale) and dramatic. But Darwin was aware of the difficulties and fully cognizant of the challenge posed by the eye and Paley's discussion of it. Thus he wrote to an American colleague (Asa Gray): "The eye to this day gives me a cold shudder, but when I think of the known fine gradations, my reason tells me I ought to conquer the cold shudder."

The history of the eye is embedded in the history of animals and molecules

Although I will mention an example of soft-tissue preservation below, most of what we know about ancient animals and their relationships to living species is based on bones, teeth, and shells. Soft tissues rarely fossilize, and we will never

*The passages from Darwin have been quoted elsewhere numerous times, but usually in slightly different form. There were five editions of *On the Origin of Species* after the first (1859), and I have used one of the later editions; Darwin's modifications of the original text make his points more clearly and forcefully.

have a fossil record of the eye's evolution. We learn more all the time about our hominid ancestors from their bones, and sometimes their footprints, but their eyes are irretrievably gone.

Fortunately, there are other ways to reconstruct the history of eyes. First, we can estimate the ages of some of the key molecules from which eyes are constructed. This analysis establishes the earliest possible time for the appearance of eyes, for they could not have been made until their molecular components were made. Second, the family tree of the animal kingdom, which continues to be elaborated and refined, provides a way to see where eyes appear in the pattern of branching. From this evidence, we can ask if eyes appeared once early in the family tree and thus appear on all the branches leading to living species or if eyes evolved independently at different times in different parts of the tree. Finally, we can bring the differences among eyes of living species to bear on the problem. Because eyes come in diverse and recognizable forms, this diversity can be superimposed on the family tree to establish the lineage or lineages of eyes.

In the end, Darwin's fear that the eye would be the downfall of his theory was groundless. Eyes are more ancient and more varied than he ever imagined, and the path from primitive to sophisticated eyes is, in retrospect, free of obstacles. As the evolutionary theorist Richard Dawkins has pointed out, it is impossible to imagine a transition from a rudimentary to a sophisticated eye occurring in a single step; the odds of it happening as a matter of chance are too small to calculate. But the eyes of living animals have been millions of years in the making. Changes in their structure were accepted or rejected by natural selection, and the accretion of the accepted changes, all of which were useful or they would not have been accepted, has produced the eyes familiar to us.

But many living animals do not have highly sophisticated eyes. Vision appears to be so useful, even in its most rudimentary forms, that eyes of living animals can be arranged in a sequence from primitive to complex because many species have not needed to upgrade their eyes. Although this sequence should not be construed as representing the steps taken by evolution, it does indicate that perfection is not so much in the eye of the beholder as in the eye of the possessor. Eyes like ours may be "perfect and complicated," but less elaborate forms are useful to their owners even without the potential for vision offered by our eyes. As a result, the vast majority of animals have eyes of some sort, and eyes have been around since animals first appeared as forms of life.

As Darwin suggested, our eyes are all the more marvelous when we can think about them (and ourselves, for that matter) as the most recent products of winnowing by millions of years of dispassionate, disinterested quality control. My cat and the cockroaches he used to chase could think the same, of course; they and their very different kinds of eyes are no less wonderful.

Several of the eye's critical molecules are ancient

By comparing DNA from different species, molecular geneticists can specify how much of the genetic material is shared. Closely related species have much of their DNA in common, while distant relatives share less of their DNA sequences. By this criterion, our closest living relatives are chimpanzees, with whom we share 98.4% of our DNA; gorillas are more distant relatives, for we share less, 97.7%, of our DNA with them.*

*The method most often used is called DNA hybridization. This and other molecular methods for establishing relationships among animals have been the subject of heated debate because the molecular evidence is sometimes at odds with older classifications based on analyses of morphology. The dust has mostly settled, and the prevailing point of view seems to be that while no method is foolproof, the molecular techniques can be quite powerful and should be applied whenever possible to supplement the traditional ways in which degrees of relationship among species are specified. See, for example, *Patterns in Evolution,* by Roger Lewin.

The difference between human and chimpanzee DNA has increased steadily since the two lines diverged from their common ancestor. If the underlying rate of mutation has been constant and if the rate can be estimated, it should be possible to extrapolate back in time to the point of common ancestry. This task is not particularly easy, because the assumption of a constant rate of change is difficult to test—constancy is likely to be a rough approximation—and estimation of the rate is also problematic. Any conclusions therefore leave room for argument, but current data indicate that the human line diverged from the chimpanzee line about 7 million years ago and from the gorilla line almost 10 million years ago. For more distant primate relatives, like the Old World monkeys, we must look back more than 30 million years for a common ancestor.

We can use this method of establishing relationships among species for individual molecules by comparing DNA or RNA sequences responsible for the molecule's construction. In the eye—any eye—the most important molecule is the photopigment that absorbs light and makes vision possible; all photopigments have a similar structure. Chapter 13 will explore the details; here we need know only that the light-absorbing part of the pigment is usually a derivative of vitamin A, and this small molecule is coupled to a large protein called an **opsin**. When the photopigment absorbs light, the chromophore bound to the opsin changes its shape, thereby initiating a series of changes in the opsin that "activate" the photopigment and pass its activation along to other molecules in an enzymatic cascade that signals the initial light absorption. Opsins vary in their primary structure—the sequence of amino acids in the protein—but their three-dimensional structure is always similar, with seven helical regions that fold back and forth within cell membranes.

Not only are opsins from different species similar in structure, but the genetic instructions for their construction are also similar. This resemblance among opsins from diverse organisms—humans and flies, for example—says that these proteins are highly **conserved**, meaning that over the course of evolution, the genetic specifications for the opsins have diverged much less than the genome of the animals in which the molecules are found. The human line parted company with the line leading to flies more than 500 million years ago. But when the appropriate DNA sequences are compared, the data give a still older age for opsin—600 million years or so. In short, the molecule that all animals use to detect light has been present and in use throughout the history of the animal kingdom.

Although opsin is ancient, we cannot conclude that it originally formed to detect light. The opsins are part of a protein family called **G protein–linked receptors** used by cells to interact with their environment and with other cells (see Chapters 5, 7, 11, and 13). All G protein–linked receptors have the same structure of seven folds, and the value of opsin as a detector of light may have been an accidental discovery made by the ancestors of animals. If so, when multicellular animals first developed photosensitive cells and the cells aggregated to form the first eyes, opsins were probably already at hand for use in these primitive light-detecting organs.

Most eyes have a lens that either concentrates or focuses light in an image. All these lenses are made of proteins, some of which are uniquely lens proteins. Unlike the opsins in photopigments, however, lens proteins come in more varieties and there are significant differences in lens composition among major animal groups. For example, the lens proteins of vertebrates differ from those of arthropods. The main proteins in human lenses are a group called **crystallins** (see Chapter 12), of which there are several types. The crystallins that have been studied are at least as old as the vertebrates, almost 500 million years of age. Similarly, the proteins in arthropod lens cones are as old as the phylum itself, roughly 600 million years of age. The differences between arthropod and vertebrate lens proteins are consistent with the independent origins of arthropod lens cones and of the lenses in vertebrate eyes.

Lens proteins appear to have been used for other things before being pressed into service when eyes formed lenses. Several of the lens proteins are enzymes with different uses in cells, and some lens proteins may have precursors in bacterial proteins, which pushes their age back to near the beginning of life. Materials from which lenses could be constructed appear to have been at hand when such construction began.

Eyes were invented by multicellular animals almost 600 million years ago

Putting a date on the first appearance of eyes depends on what one means by "eye." If the term refers to a multicellular organ, even if it has just a few cells, then by definition, eyes could not form before there were multicellular animals. But many unicellular organisms—those members of the kingdom Protista that are neither plants nor animals—can detect light by using aggregations of pigment molecules, and they use the information to modify their motility or metabolic activity. One of the familiar living examples, probably known to anyone who has taken a biology class, is the protozoan *Euglena,* which has an eyespot near its motile flagellum (Figure 1). Some living protists are very like their ancestral forms embedded in ancient sedimentary rocks, and this similarity suggests that the ability to detect light and modify behavior in response to light has been around for a very long time. Animals arose from one of these unicellular creatures, perhaps from one already specialized for a primitive kind of vision.

An eye is a collection of cells specialized for light detection by containing photosensitive pigment along with a means of limiting the direction of incident light that will strike the photosensitive cells. This definition says nothing about image formation, lenses, eye movements, or any of the other features we associate with our own eyes, but it does recognize the simplest form of functional and anatomical specialization—namely, detection of light. Everything else can be built up from this simple beginning, and some animals appear to have had eyes almost from the beginning of the animal kingdom.

Animals were scarce 600 million years ago, in the geological era called the Precambrian. There are very few fossil remains from that time (though more keep turning up, usually making the news), and most evidence of their presence is indirect, such as small tunnels in rock that could be ancient worm burrowings. But 50 million years or so later, which is just a moment in geological time, fossilized bits and pieces of animals abound, suggesting that a great burst of evo-

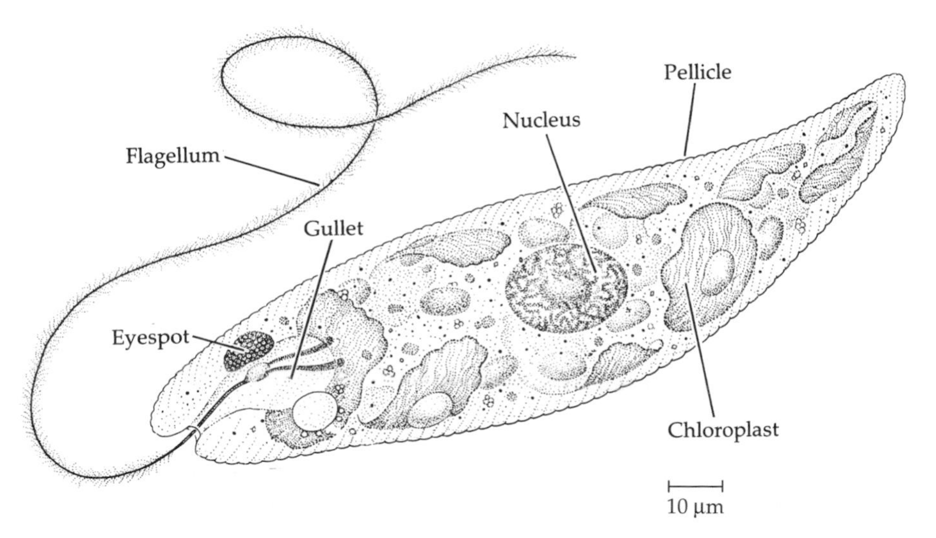

Figure 1
The Protozoan *Euglena*
Euglena is aquatic, commonly found in pond and creek water and biology lab aquaria. It moves by beating of the flagellum and rotates around its long axis as it moves. When the long axis is not aligned with the direction of incident light, the eyespot is periodically shaded by the base of the flagellum as the organism rotates. Shading the eyespot alters the flagellum beat, reorienting the long axis until rotation no longer shades the eyespot. This process tends to make *Euglena* move toward the light source. (From Brusca and Brusca 1990.)

lutionary creativity occurred in the 50-million-year interval. This surge of new life is called the "Cambrian explosion." Animals became abundant.

The first direct evidence for the early origin of eyes comes from fossils that are about 530 million years old, a time shortly after the Cambrian explosion; they were found on a mountainside in British Columbia in a deposit known as the Burgess Shale. The Burgess Shale fossils are extraordinarily important because they were once soft-bodied creatures, many of them lacking shells and other hard parts that fossilize easily. Their preservation is little short of miraculous (as are the delicate microdissection methods used to reconstruct three-dimensional structure from these flattened fossils), and they are one of the few known repositories of early soft-bodied fauna.

Not all of the Burgess animals had eyes, or at least they are not recognizable as such, but some did, and the reconstructed organs look superficially like eyes of living crustaceans, particularly shrimp and crabs. (In the published accounts, the authors note that details like the facets in most arthropod eyes were usually not seen. Gross features—location, size, and hemispheric shape—are responsible for the designation of these structures as eyes.) Some of the eyes were mounted on small stalks and may have been movable; others appear to be on or a part of the body surface. One animal, *Opabinia,* had five eyes: two lateral pairs and a single medial eye; at least one of the lateral pairs had stalks and could have been movable. And some trilobitelike animals in the Burgess Shale had faceted eyes much like those of later fossil trilobites. (Trilobites have been extinct for about 200 million years, but they have left a rich fossil record in which their compound eyes show clearly. Although they were arthropods, their exact relationship to living groups is a matter of debate. Their extinction notwithstanding, they were a successful group of animals, consisting of about 4000 widely distributed species that persisted for about 300 million years.)

Although the presence of eyes on some of the Burgess animals indicates that eyes have been around for a very long time, it is unlikely that these were the first eyes; they seem much too large and (potentially) well developed to be brand-new inventions. The best we can do is put the origin of eyes somewhere between the beginning of the Cambrian explosion, about 600 million years ago, and the death of the Burgess animals, some 530 million years ago. But if the Burgess animals are a fair representation of the Cambrian fauna, there is no evidence for any precursors to our eyes; the Burgess eyes we know about point to arthropods, not to vertebrates.

Eyes arose not once, but numerous times, in different animal groups

The lack of anything resembling a vertebrate eye among the Burgess animals is another reason for thinking that not all eyes are equally ancient and, more generally, that eyes evolved more than once. There is certainly reason to suspect that arthropod and vertebrate eyes evolved independently and that the arthropod type of eye is the more ancient form. But the rarity of fossils like the Burgess Shale animals means that when and where eyes appeared cannot be determined directly from hard evidence. We must extrapolate back in time along the lineages of living animals.

The history of animals is a highly branched tree whose trunk is the ancestor of all animals; the major branches represent the highest taxonomic level of description, the phyla, with progressively smaller branches and twigs for the finer levels of classification—class, order, family, genus, and species. If two diverging branches both represent animals with eyes, it is reasonable to assume that eyes were a feature of their common ancestor—that is, of the larger branch from which they arose. Because humans and chimpanzees have very similar eyes, for example, it is a safe bet that we inherited them from our common ancestor. (And because humans and chimps have similar color vision, our common ancestor probably had it also.)

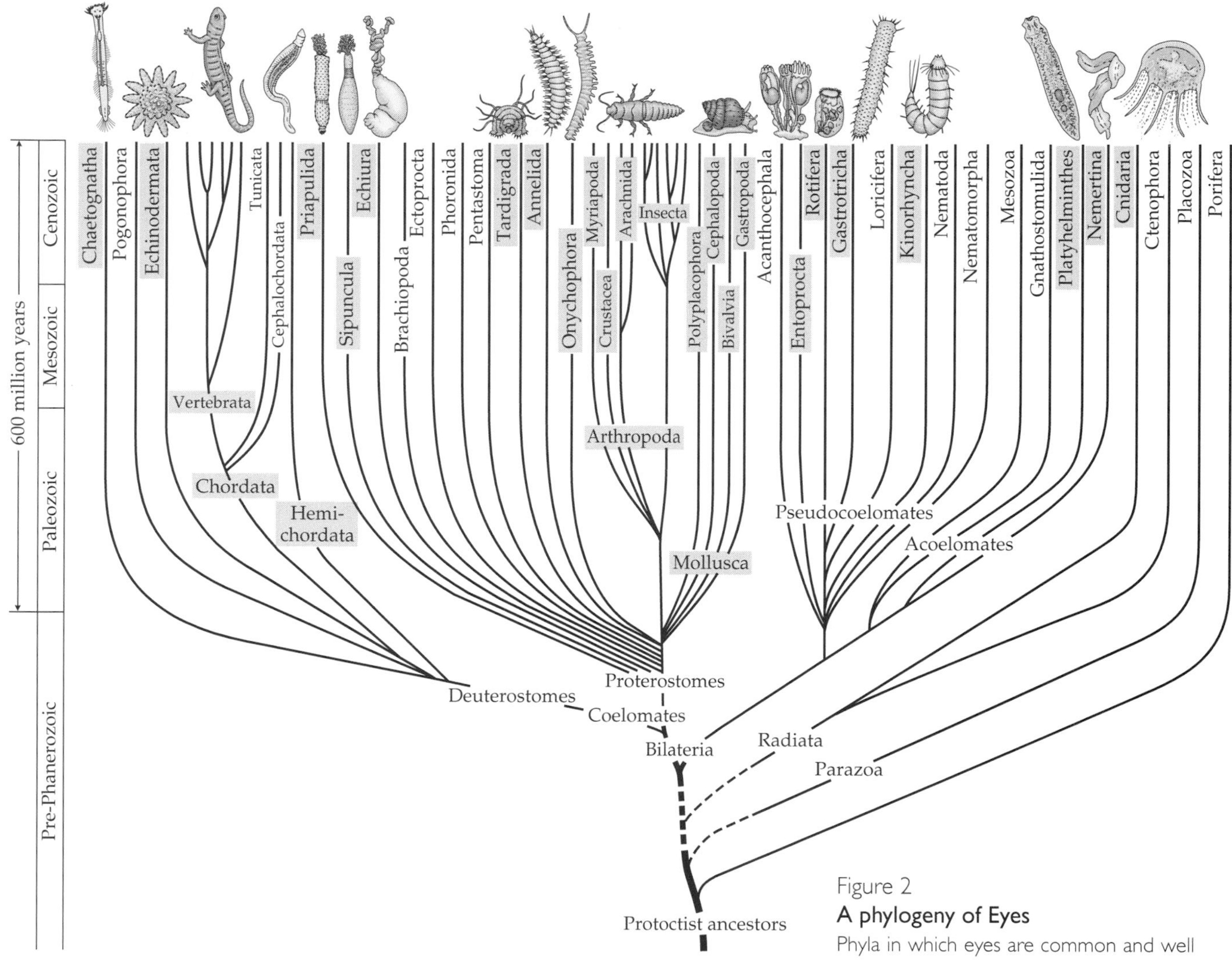

Figure 2
A phylogeny of Eyes
Phyla in which eyes are common and well developed are shaded. The appearance of eyes in a lineage where they were absent indicates de novo evolution of the eyes. Gaps in the record caused by eyeless creatures suggest that eyes evolved more than once. (After Margulis and Schwartz 1988.)

If, on the other hand, one of two branches has eyed forms within it and the other does not, there are two possible conclusions. The ancestral animals may have had eyes that were lost in one of the daughter branches. The more likely possibility, however, is that eyes evolved de novo in one daughter branch of an eyeless ancestor. The more that eyes are scattered among the branches of the tree without connecting links between them, the more we must think of eyes evolving independently more than once.

This kind of retrospective analysis has been done at different levels of taxonomic classification and therefore at different levels of detail, but it is easiest (and still instructive) at the coarsest scale, that of animal phyla. Figure 2 shows the 33 phyla of the animal kingdom as a family tree going back in time; our phylum, Chordata, is the fourth from the left.* Most of these phyla are probably recognizable only to card-carrying zoologists (I am not one of them), but they should be familiar when some of the members are indicated by their common names. Just to the left of the Chordata, for example, are the Echinodermata (sea stars), while the Arthropoda

*This is but one of numerous versions of the phylogenetic tree of animals; it is based on Margulis and Schwartz 1988. Trees differ because of the criteria used to construct them. Some of the more recent trees based on cladistics, which looks not at morphological similarities but at the presence or absence of shared characteristics, such as eyes or fingernails, arrange the phyla differently. For the simple kind of analysis done here, however, the point can be made with any of them.

occupy the center of the diagram, with the subphyla containing insects, crustaceans, spiders, and millipedes. Next to them are the Mollusca, with octopus and squid, snails, and my namesake, oysters, among the bivalves. Less familiar groups include Cnidaria (jellyfish and coral), Porifera (sponges), Annelida (earthworms and leeches), Nematoda (parasitic roundworms), Platyhelminthes (flatworms), and so on. The inset drawings indicate the basic body plan for animals in each phylum.

The phyla in which eyes are widely distributed among orders and families are shaded in Figure 2; eyes may be present in the other phyla, but are not widely distributed or not present throughout the life cycle (some animals may have eyes only as larvae). Roughly two-thirds of the phyla have some members with eyes, and in all but a few of these, eyes are common throughout the phylum. But 14 phyla either lack eyes or have very rudimentary eyes distributed sporadically among their members. Moreover, there are gaps among the branches—places where diverging phyla differ in their possession of eyes.

Look, for example, at the phyla Cnidaria and Ctenophora near the right-hand side of the figure. Eyes are common among the motile cnidarians, like jellyfish. Eyes are missing, however, among the sessile members of the phylum (corals, sea anemones, sea fans) and among members of the neighboring phylum, the ctenophores (another sort of jellyfish). It is likely, first, that the parent branch from which these phyla arose, the Radiata, had no eyes and, second, that the eyes of the jellyfish in the phylum Cnidaria were their own, independent inventions.

Closer to home, in the phylum Chordata, our subphylum (Vertebrata) has eyes in all orders, families, genera, and species. Yet members of the closest subphyla, Tunicata and Cephalochordata, either do not have eyes or have such a primitive form that it is difficult to imagine them being ancestral to vertebrate eyes. (The most common tunicates are the sea squirts; they begin as tiny, motile larvae that become considerably larger, sessile adults. Cephalochordates comprise about two dozen species of small, fishlike creatures called lancelets. The best-known genus is *Amphioxus,* which may be similar to the ancestral vertebrates.) This evidence suggests that the early chordates lacked the large eyes of later vertebrates, which seem to have been invented anew.

There are other gaps in the tree that can be used to generate the same message: Eyes evolved numerous times in the course of evolution. When the tree is examined at a finer scale, using orders or families, and when detailed differences in eye and photoreceptor structure are considered, eyes appear to have evolved at least 40 and perhaps as many as 65 times within the animal kingdom. (New data on the Pax6 gene and its association with eye formation in animals as diverse as humans and insects have been interpreted to mean that eyes evolved only once. Other conclusions are possible, however, and the issue is far from settled.) The most ancient, unbroken line of eyes is probably that occupied by the phyla in the center of Figure 2, the line producing the molluscs, arthropods, annelids, and onychophorans (velvet worms—imagine a worm with a lot of short, chubby legs along the body and a pair of antennae on the head, and you have the picture). Arthropod eyes were probably the first, appearing almost 600 million years ago.

Eyes are most common in groups of motile animals living in lighted environments

The six major phyla in which most members have eyes that are elaborated beyond a primitive stage are the Chordata, Annelida, Onychophora, Arthropoda, Mollusca, and Cnidaria. Together, they constitute about 96% of known living species. (Since insects are included, that isn't surprising. The insect subphylum Uniramia has almost 900,000 known species, which represents more than 80% of *all* known species; there are about 50,000 vertebrate species.) Do these sighted creatures share characteristics that sightless or marginally sighted members of other phyla do not have?

The division between sighted and sightless seems to depend primarily on availability of light and degree of motility. To take an extreme case, a sessile (fixed, immobile) animal living in deep ocean water is unlikely to have eyes. Sunlight does not penetrate much beyond a few hundred meters of water depth, so there is not much to see save the occasional bioluminescent organism; where there is no light, there can be no vision. Add to this the animal's immobility, which prevents it from relocating to a lighted environment, and any advantage from vision becomes vanishingly small. Parasitic animals that spend their lives inside other organisms and burrowing animals that normally live their lives underground are in a similar situation, their motility notwithstanding. Eyes don't help in the dark.

Sessile animals living in shallow water—on coral reefs, for example—are another group for whom vision is relatively unimportant, despite the presence of light in their environment. Many of these animals feed by filtering water or, in the case of sea anemones, by sensing and grasping small animals that come within reach of their tentacles. The dominant senses here are tactile and chemical ("taste"); sensing distant objects, whether as prey or possible predators, is much less important than sensing the local environment. Vision, which is a distance sense par excellence, confers no special advantage.

As always, there are exceptions. Some animals, for example, are motile as juveniles living in shallow, sunlit water and retain their eyes when they take up their adult, sedentary lifestyle; others discard their larval eyes during metamorphosis to the adult form. Some sessile tube worms have small eyes on their tentacles that are part of a protective reflex; decreasing illumination, like a shadow, signals the animals to withdraw into their tubes. And many of the tube worms' relatives do not have eyes, but can still sense and react to the general level of illumination in their environment by using photosensitive cells scattered over their body surface. Among other things, they can use this sensing of diffuse light to synchronize their activities with the solar day. But scattered photosensitive cells are not eyes, and the resulting "sensation" is not vision.

For the most part, eyes are associated with running, crawling, hopping, swimming, or flying through a world where light, and variations in light, can be used to make motion purposeful. The purposes have to do with food, shelter, and sex; eyes will be found wherever they can be used to discover what is where. (Although we use our eyes and vision for many different tasks, we probably do not escape the foregoing generalization; the relationship of our daily activities and visual behaviors to one or more of the three basic imperatives is merely less direct and stereotyped than in other species.)

The Diversity and Distribution of Eyes

The first step in vision is an eye that can sense the direction of incident light

A single photoreceptor cell or even a small group of them on the body surface is useless for anything other than detecting the presence and amount of light (Figure 3*a*). Light incident from the right strikes the same photoreceptors with the same intensity as light incident from the left; the directions of incidence are therefore indistinguishable. Moreover, light passing through the animal's body cannot be distingushed from light striking the photoreceptors directly. Thus, variations in light intensity cannot be used to tell the animal *where* something is in its environment.

But vision is largely about where things are located, and the key to specifying "where" is having an arrangement of photosensitive elements that are affected differently as the direction of incident light changes. The first requirement is an array of photoreceptors; having more than one receptor means that different

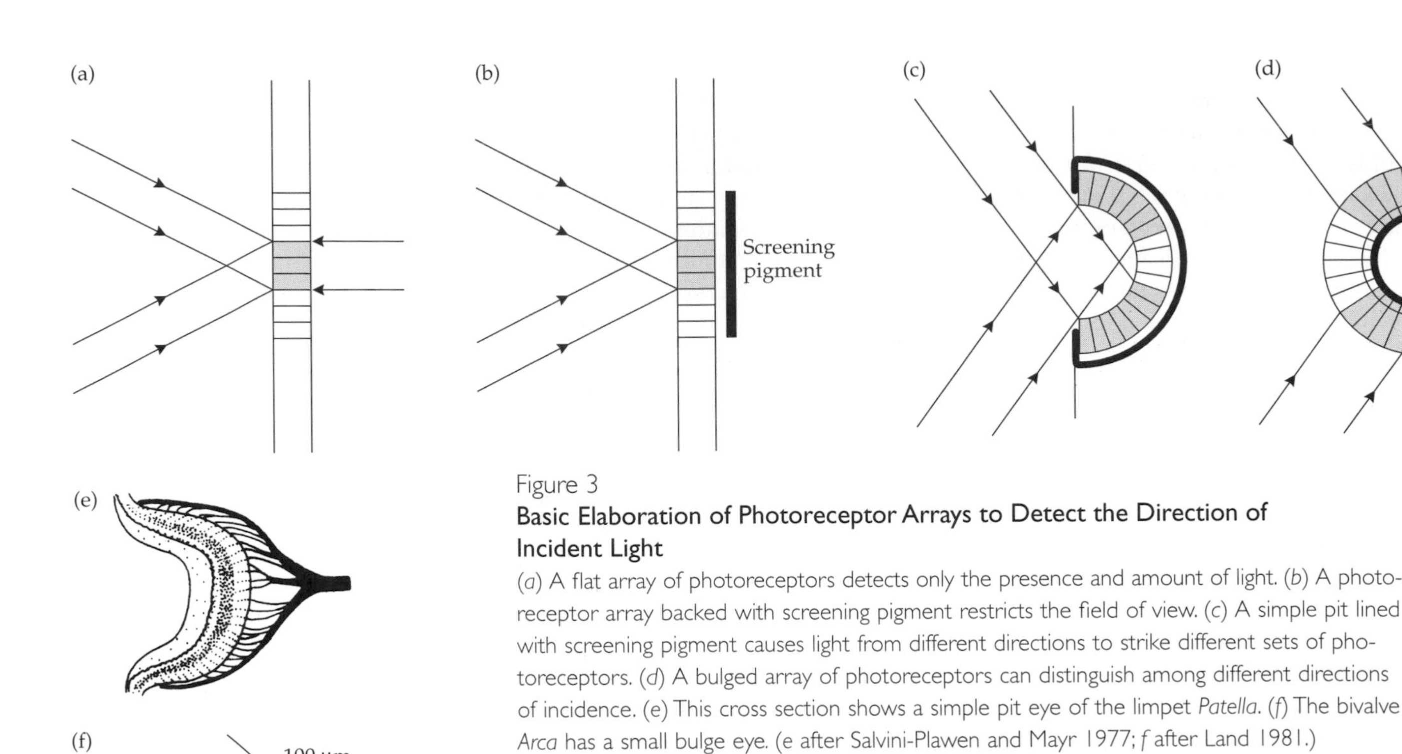

Figure 3
Basic Elaboration of Photoreceptor Arrays to Detect the Direction of Incident Light
(*a*) A flat array of photoreceptors detects only the presence and amount of light. (*b*) A photoreceptor array backed with screening pigment restricts the field of view. (*c*) A simple pit lined with screening pigment causes light from different directions to strike different sets of photoreceptors. (*d*) A bulged array of photoreceptors can distinguish among different directions of incidence. (e) This cross section shows a simple pit eye of the limpet *Patella*. (*f*) The bivalve *Arca* has a small bulge eye. (e after Salvini-Plawen and Mayr 1977; *f* after Land 1981.)

receptors can be influenced by light from different directions. This array, which constitutes a primitive **retina**, is potentially sensitive to light from any direction, even light passing through the animal's body, so it is lined on the inner side with light-absorbing pigment, commonly melanin (Figure 3*b*). This simple step reduces the retina's field of view from a sphere (light coming from any direction) to a hemisphere (light coming from the retina's side of the animal). If an animal had two of these elementary eyes on either side of its body, it could potentially distinguish between light coming from the left and light coming from the right.

This restriction of the field of view may be useful in a crude way, but any information about location is still extremely coarse. Refining the direction sense requires additional restrictions, restrictions that make different photoreceptors look at different parts of the field of view. This restriction can be accomplished in one of two ways. In the first method, the retina and its screening pigment are displaced from the plane of the body surface into a shallow pit or depression (Figure 3*c*). The rim of the pit restricts the field of view so that light coming from above, for example, is incident on photoreceptors lining the lower half of the pit but is prevented from striking photoreceptors on the upper half. An eye like this cannot resolve spatial detail, but the simple change from a flat array to a cupped array of photoreceptors means that light from different directions will affect different sets of photoreceptors. Darkening produced by a predator looming in front could be distinguished from darkening produced by a predator sneaking up from behind. It may not be much in the way of vision, relatively speaking, but the information could certainly be useful in deciding which way to flee.

Another way to restrict what different photoreceptors see is by arranging them as a bulge from the body surface (Figure 3*d*). Here, light coming from above is incident on photoreceptors lining the upper side of the bulge, but screening pigment prevents the light from passing through the eye to reach photoreceptors on the lower side. Either system, pit or bulge, will therefore do the job, creating an association between receptor location and direction of incident light. And at this simple level of organization, there is little to recommend one system over the other. The pit solution has some advantages, but they will dominate only when these elementary eyes are elaborated.

Although the pit and the bulge probably represent the earliest forms of eyes, they are present in living animals. Figure 3*e* is a simple pitlike eye from a limpet, and Figure 3*f* is a simple bulge eye from a bivalve mollusc. Neither of these creatures is highly visual and their rudimentary eyes are probably little changed from their earliest evolution.

At least ten types of eyes can be distinguished by differences in their optical systems

Eyes vary enormously in their detailed structure, but these variations are based on a smaller number of themes, which are the optical strategies that eyes use to collect and image light. It is fair to say that nature employs almost every optical trick available, including a few that are extremely difficult to reproduce in glass or plastic. The number of themes, which I have put at ten, is arbitrary; what I call a theme may be a variation to someone else or vice versa (and I have deliberately omitted a rare one). Ten is a nice number, however. More detailed examples of the different eyes in operation will be found in the "bestiary" that concludes this chapter.

The ten eye types are shown schematically in Figure 4, organized according to their probable origins as pits or bulges. The pit-derived eyes, as a group, are called **simple eyes**, referring to the presence of a single optical system (although the system may have several elements) and multiple photoreceptors (see Figure 4*a*). The group of bulge-derived eyes are **compound eyes**; they all have multiple optical systems (see Figure 4*b*). The simplest eye types are the shallow cup eye and the basic bulge eye. As already mentioned, these optical themes are two ways to gain some sense of direction without resorting to an optical system.

A deeply cupped eye with a small opening to the interior eye chamber is a **pinhole eye**. The optical theme is the restricted aperture, which must now be considered a key element in the eye's optical function. The aperture gives the eye a coarse image of the external world on the retina, but in so doing, the amount of light entering the eye is reduced. The classic example of a pinhole eye is that of *Nautilus*, but as we will consider later, it is not a particularly good eye. Pinhole eyes with a transparent medium—a kind of primitive lens and vitreous—in the eye offer more possibilities for concentrating and imaging light.

Covering the surface of the eye and incorporating a lens within it produces a **simple refracting eye**. It is shown in Figure 4*a* in two forms: One operates in air, where the outer surface (cornea) is a major refractive element; the other operates in water, where the cornea is optically ineffective and most refraction is done by the lens. Simple refracting eyes come in a wide range of sizes and shapes, the vast majority of which form images on the retina. All vertebrate eyes fall into this category, as do those of several invertebrate phyla. Examples of animals that have simple refracting eyes and are discussed in the bestiary later in this chapter are octopi, goldfish, pigeons, jumping spiders, and humans.

Many simple refracting eyes have reflecting surfaces behind their retinas, but these mirrors (**tapeta**) play no role in image formation. In **simple reflecting eyes**, however, the mirror is an essential part of the optical pathway. The mirror both focuses light and folds the optical pathway. Simple reflecting eyes are not common, probably because it is difficult to construct large mirrored surfaces with high optical quality, and a mirror as the sole optical element is known in only a few species. The best-known example of a simple reflecting eye is that of the scallop.

All compound eyes have multiple optical systems, by definition, but they can be subdivided by the degree to which the optical systems are independent of one another. In **apposition eyes**, the optical systems are totally independent, each system forming its own small image. Apposition eyes have small clusters of pho-

(a) Simple eyes

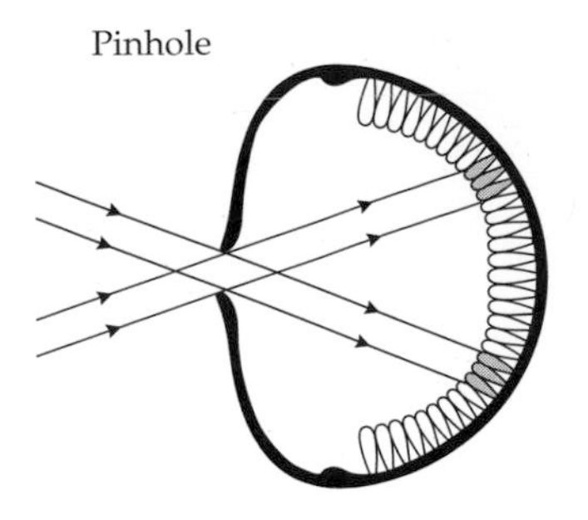

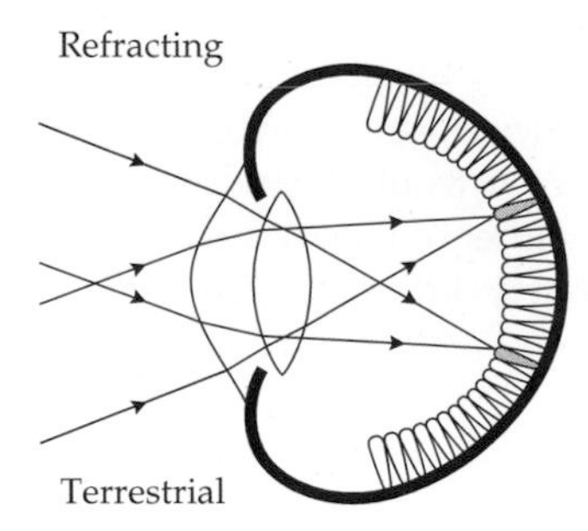

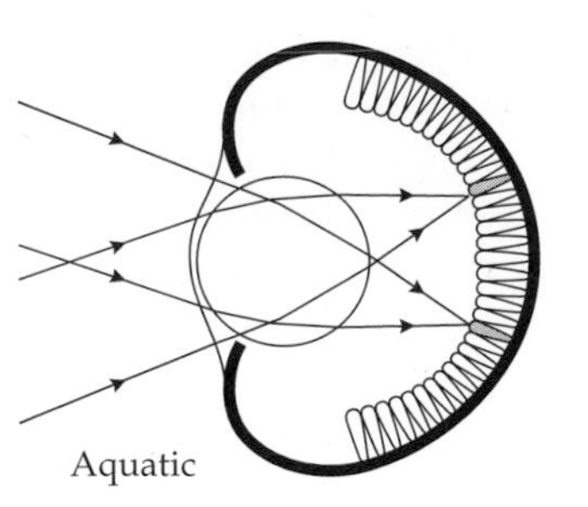

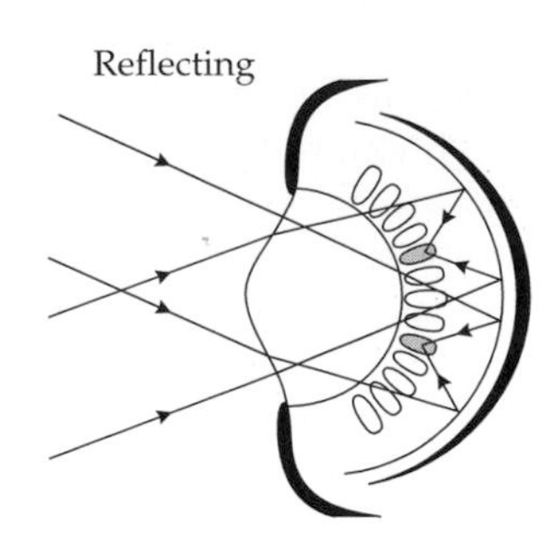

(b) Compound eyes

Apposition eyes (Independent optical systems)

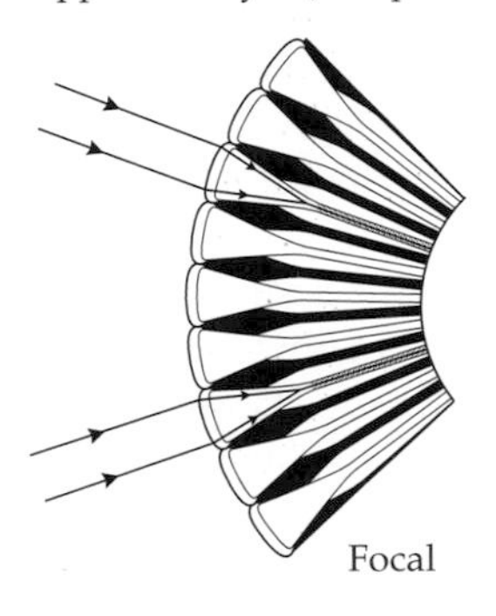

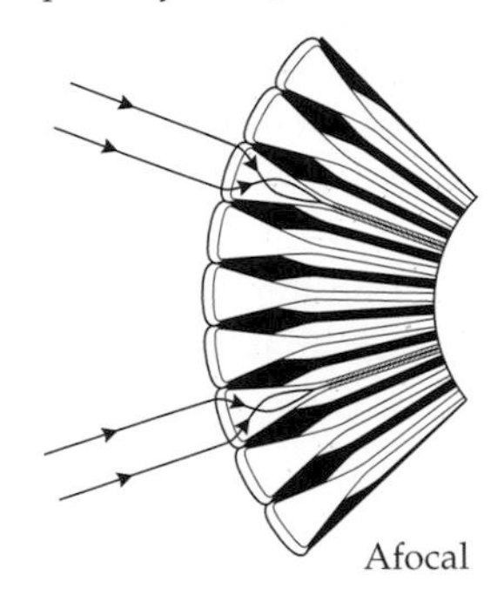

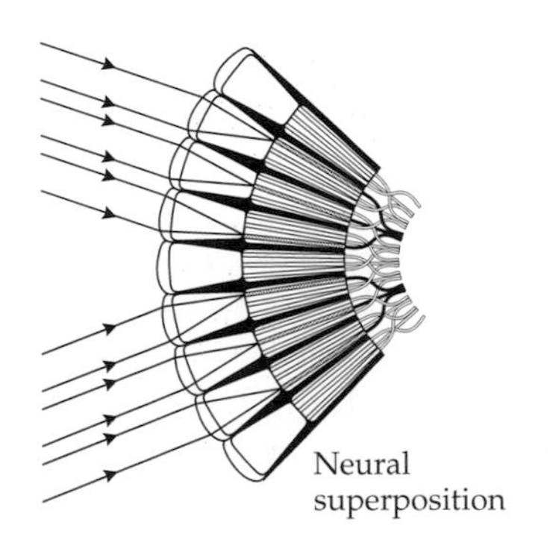

Optical superposition eyes (cooperative optical systems)

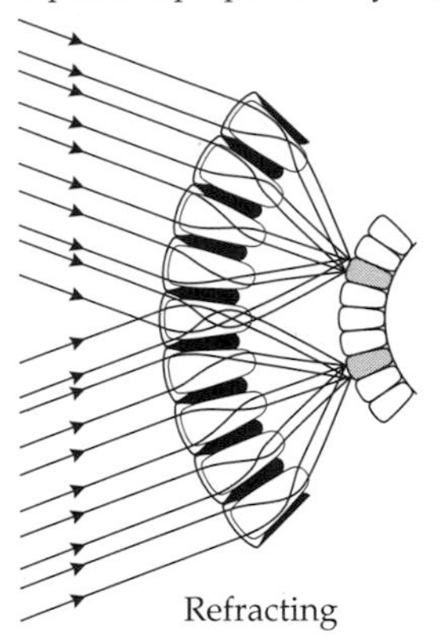

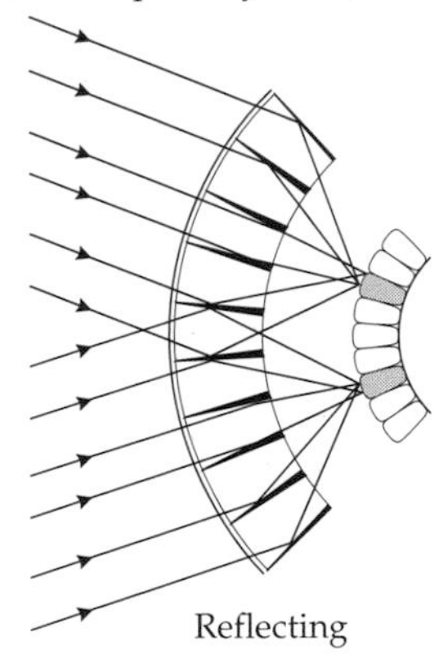

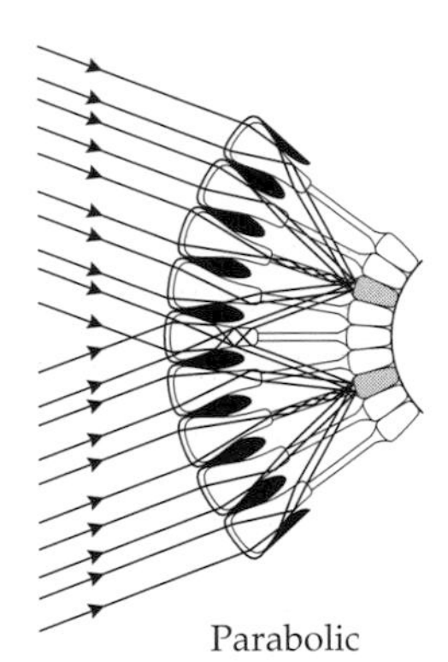

Figure 4
Ten Types of Eyes
(*a*) Simple eyes are pit or cuplike eyes with a single optical system. The basic cup is the simplest theme, with no optical elements. In a *pinhole* eye, the aperture is small enough to be optically effective—that is, to produce a small blur circle for a point source of light. *Refracting* eyes can produce sharp images on the array of photoreceptors. Most of the refraction is done by the cornea in the *terrestrial* form of these eyes, but by the lens in the *aquatic* version. In a *reflecting* simple eye, a mirror behind the retina reflects and focuses light. (*b*) Compound eyes have multiple optical systems; the prototype is a small bulge of photoreceptors with no optical elements. In apposition eyes, the optical systems are independent. The optical systems form images on the photoreceptors in *focal apposition* eyes, but there are no focused images in *afocal apposition* eyes. Images from different optical systems are combined by the neural circuitry in *neural superposition* eyes. The optical systems work together in superposition eyes so that light from different optical systems falls on a single photoreceptor element. Light is concentrated by refraction in *refracting superposition* eyes, by reflection in *reflecting superposition* eyes, or by a combination of both in *parabolic superposition* eyes.

toreceptors, generally from five to eight in number, that typically fuse along a central axis to form a photosensitive **rhabdom** (see Figure 15 for more detail). Each rhabdom receives an image from its own optical system; the optical system may have a separate cornea and lens, or these may be fused into a single structure. Whatever the case, each optical system–photoreceptor combination is an **ommatidium**.

Three types of eyes use the apposition theme. **Focal** (or **simple**) **apposition eyes** have ommatidia that bring light to a focus on the photoreceptors; that is, each ommatidium produces a small image of the part of the world it sees. Refraction is done largely by the cornea in terrestrial species (see the description of the honeybee in the ocular bestiary later in this chapter) or by the lens in aquatic species (for example, the horseshoe crab).

Afocal apposition eyes are very similar in appearance to focal apposition eyes, but the optical elements in the ommatidia do not produce focused images;

instead, they work like small telescopes in which parallel rays of light entering the system are still parallel when they exit the system. As far as the photoreceptors are concerned, whether the light reaching them is parallel or converging to a focus is immaterial (the photoreceptors have no way to detect the direction of incidence of the light rays), and afocal systems are at least as good as focal ones. Butterflies have eyes of this type.

Some eyes use the apposition of independent optical systems but have the ability to distinguish among the different parts of the image within an ommatidium. Because the fields of view of neighboring ommatidia overlap to some extent, any given point in the visual world is imaged and detected by several photoreceptors in different ommatidia. The potential confusion of a single point being imaged several times is solved by the neural wiring, which brings the photoreceptor signals for the separate images together at the next level of the nervous system. This combining of image signals is called superposition, and because it is being done by the neural wiring, such an eye is called a **neural superposition eye**. Note, however, that the *optical* theme is still that of focal apposition. The best-known example of a neural superposition eye is that of the common housefly.

When the optical systems in compound eyes are not independent—that is, when they operate in concert to produce an image—the eyes are called **optical superposition eyes**. These eyes are characterized by a distinct separation between the photoreceptors and the elements of the optical systems. Most commonly, the imaging is done by refraction in a **refracting superposition eye**. Because each photoreceptor or photoreceptor cluster receives light from numerous lenses, these eyes can be quite sensitive, able to detect small amounts of light. The example in the bestiary is the firefly; other nocturnal insects, like moths, have refracting superposition eyes, as do many deep-sea crustaceans.

What can be done by refraction can also be done by internal reflection in **reflecting superposition eyes**. Crayfish have this sort of eye, and the somewhat tricky optics are discussed for this animal in the bestiary. Finally, some optical systems—in crab eyes, for example—have been shown recently to use both refraction and reflection in generating the image superposition. Because of the shape of the special lens, these are called **parabolic superposition eyes**. Reflecting and parabolic superposition eyes are potentially more sensitive than apposition eyes.

All animals have been confronted during their evolution with the same physics of light; the ways in which light interacts with matter, the laws of reflection and refraction, and the need for transparent media have always been present. Different solutions have been adopted, as shown by the variety of eye types in existence, but they are not of equal potential in terms of the quality of visual performance they can support. Compound eyes, for example, are inherently limited in performance in comparison to simple eyes, yet they are the oldest eyes and they dominate numerically: Many more species and individual animals have compound eyes than have simple eyes

This numerical dominance suggests that factors other than optical excellence bear on the type of eye adopted and the degree to which it is elaborated. Optical excellence may carry a high price in terms of some biological coinage and may not represent good value for a particular organism.

Vertebrates always have simple eyes, but invertebrates can have compound eyes, simple eyes, or both

Figure 5 shows the animal kingdom as a disc with segments for the different phyla; the size of each segment represents the number of species in each phylum as a fraction of all known species. (The word "known" is significant; recent estimates place the total number of species well above the number that are currently known. Insects probably occupy even more of the animal kingdom than depicted here.)

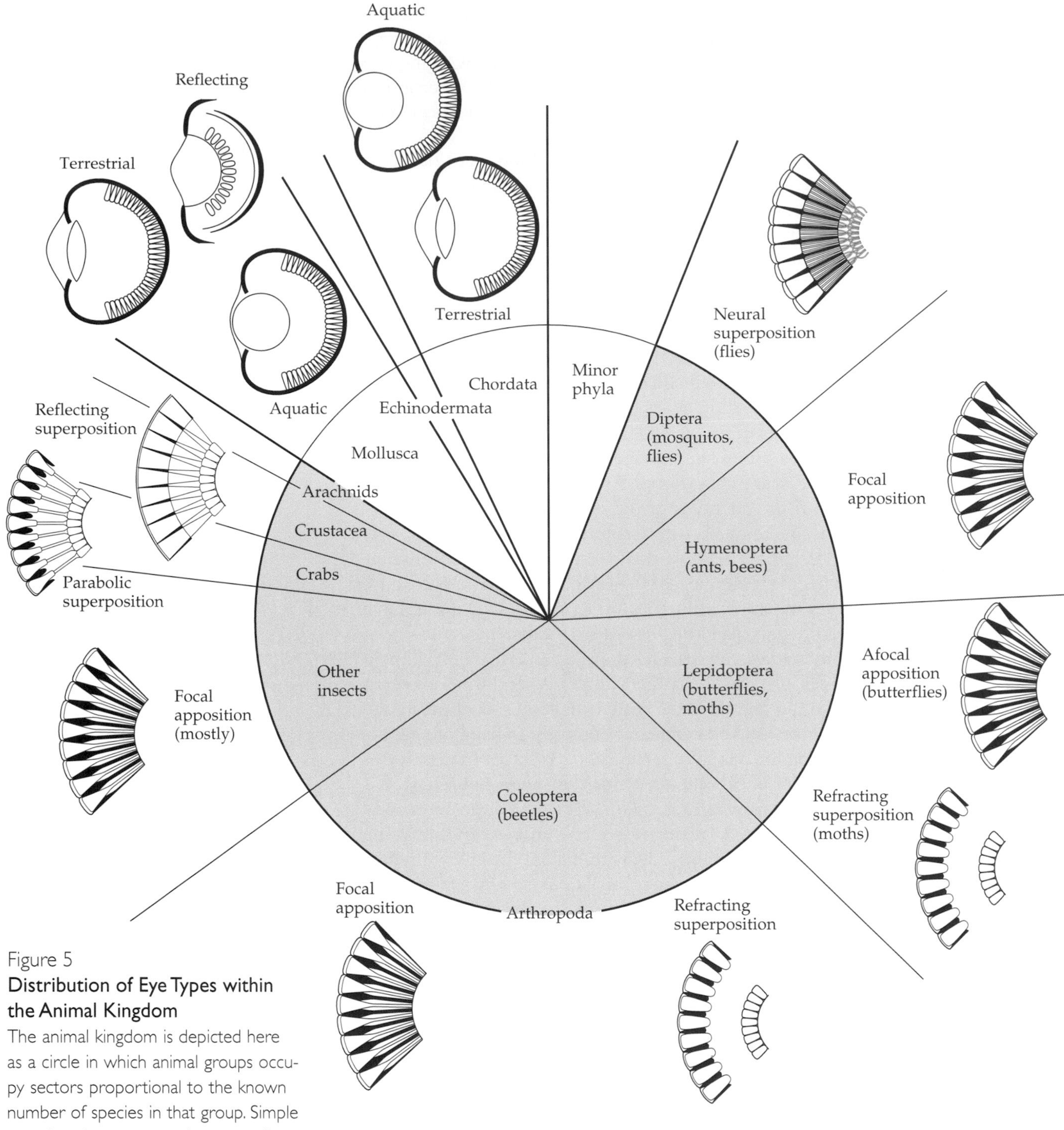

Figure 5
Distribution of Eye Types within the Animal Kingdom
The animal kingdom is depicted here as a circle in which animal groups occupy sectors proportional to the known number of species in that group. Simple eyes found among vertebrates, molluscs, arachnids, and a few minor phyla account for about 20% of the total. Almost 80% of the animal kingdom consists of species with compound eyes, all of which are arthropods. Several types of compound eyes are restricted to particular orders or suborders. (After Ali 1984.)

Arthropod species far outnumber any other phylum, and the vast majority of arthropods have compound eyes; the most notable exceptions are arachnids—spiders and scorpions—which have multiple simple eyes. In terms of numbers, therefore, compound eyes are the predominant form of eye in the animal kingdom. Arthropods are invertebrates, of course, but one cannot generalize about eye type from arthropods to other invertebrates. Almost all molluscs, which form the second largest invertebrate phylum, have simple eyes, and some can rival vertebrate simple eyes in complexity and performance (see the discussion of the octopus in the ocular bestiary later in this chapter). The simple eyes in other invertebrate phyla are less elaborate, but they are simple eyes all the same.

Thus, if we take the large division between vertebrates and invertebrates, the only definitive statement is that vertebrates have simple eyes and *only* simple eyes. Invertebrate eyes encompass the entire spectrum of eye types; all the themes and variations of both simple and compound eyes are represented. We can make one clean division among invertebrates, however. Insects *always* have compound eyes of one sort or another at some part of their life cycle, usually as adults.

Assignments become messier at this point. Beginning with arthropods and their compound eyes, two eye types are found in the order Lepidoptera; only butterflies have *afocal* apposition eyes, while moths have *refracting* superposition eyes (but this type is also found elsewhere—in fireflies in the order Coleoptera and in some crustaceans, for example). The other types of superposition eyes are assigned to crustaceans: Crayfish, shrimp, lobsters, and other relatives have *reflecting* superposition eyes; some "true" crabs have *parabolic* superposition eyes. Simple (focal) apposition eyes are the rule in the order Hymenoptera—ants, bees, wasps—but they are distributed without any clear pattern among the other orders of insects and to arthropods outside the Insecta—to many crustaceans and the horsehoe crab (*Limulus*), for example. Some of the "true" flies in the order Diptera have neural superposition eyes, but not all of them; mosquitoes, among others, have focal apposition eyes. The order Hemiptera, the "true bugs," contains a mixture of eye types, but most are focal apposition eyes, and the largest insect order, the Coleoptera (beetles), contains a mélange of simple apposition eyes and refracting superposition eyes.

If there is any message in this distribution of compound eye types, the most likely is that focal apposition eyes are the ancestral form. Their wide distribution among the arthropods hints that focal apposition eyes are parental to the rest; the sharp restrictions of some types to particular groups of organisms—like afocal apposition to butterflies—suggest later derivation. Lines of descent are conjectural, but it is not difficult to imagine an optical change converting a focal apposition eye to an afocal one, nor is it difficult to imagine separating rhabdoms from optical elements to convert an afocal apposition eye to a superposition eye.

As for simple eyes, the first point is that there are two distinct groups, which are based not on optical type but on lineage. Vertebrate and invertebrate simple eyes may appear similar, but they evolved independently of one another. Thus, the similarities indicate convergence on the same optical solution, while the differences in detail, like the structure of the retina, speak to the different starting points on these two evolutionary tracks. Vertebrate simple eyes are all variations on the theme of refraction, with modifications to meet the demands of resolution, sensitivity, changing focus of the eye, and seeing in air rather than in water. Substantial, consistent differences among the eyes of different vertebrate families allow us to talk about avian eyes, reptilian eyes, or primate eyes, but these differences, however interesting, are variations on a single theme.

As for invertebrates, many arthropods have simple eyes in addition to their compound eyes, but these simple eyes rarely form images. As mentioned earlier, however, some arthropods—particularly some spiders and scorpions—have superb simple eyes. And among these animals, the simple eyes vary in quality: Some are relatively poor in terms of sensitivity or resolution; others, like those of jumping spiders (see below) are extremely good. Some worms (Annelida) have small simple eyes, for example, as do some jellyfish (Cnidaria) and the velvet worms (Onychophora).

The phylum Mollusca is the richest repository of simple eyes in all their complexity and variety. Snails and some shellfish (Gastropoda) have eyes ranging from small pits (limpets), to vitreous-filled pinhole eyes (abalones), to refracting simple eyes (terrestrial snails and many sea snails). Some bivalves have numerous image-forming eyes, and at least one—the scallop—has a set of reflecting simple eyes. The best-developed eyes are those of the cephalopods (with the exception of *Nautilus*, which has a pinhole eye). Octopi, squid, and cuttlefish have refracting simple eyes that rival the best of the aquatic simple eyes among vertebrates—that is, those of fish.

As a result, the eyes of molluscs can be arranged as a sequence of possible steps along the evolutionary path to complexity; the transitional forms are all there and obviously useful to the living animals who possess them. Had the full story of mollusc eyes been available to Charles Darwin, he might not have gotten a cold shudder.

Paths and Obstacles to Perfection

Simple eyes improve as they become larger

Almost any change in a rudimentary pit eye is an improvement. Deepening the pit, restricting the aperture, covering the surface with a transparent layer (a cornea), and forming a region of higher refractive index in the anterior part of the eye (a lens) are all simple changes that lead to a better-defined retinal image. It makes no difference if these changes are made singly or in combination, nor is the sequence of change important; an altered eye will be a better eye in terms of its image-forming ability, and there is no obvious impediment to evolving a basic refracting simple eye. But while these changes might produce an eye suitable for a snail, this eye is a far cry from the apparent perfection and complexity of the human eye. Is it possible for the eye to change from basic to perfect in small stages and yet be useful to its owner, though less than perfect, at every stage?

The main criterion for performance in a diurnal eye is its visual acuity, its ability to resolve spatial detail. The finer the detail the eye can detect, the better it will be at providing the visual system with information it needs to decide what is where. Rudimentary eyes are not very good at this, but the way in which evolutionary improvements could be made is straightforward. It is a matter of producing large images that can be examined in detail by many photoreceptors.

The route to larger images is through larger eyes and larger retinas (Figure 6). Although the angular size of the image does not change as eye size increases, the *linear* size of the image does increase; it follows directly that more photoreceptors will be included within the image. An increase in the number of photoreceptors examining the image means that it can be resolved at a higher level of detail. In essence, that is all there is to it; some qualifications remain to be considered, but evolutionary changes in the direction of larger eyes can hardly fail to produce better eyes in the process.

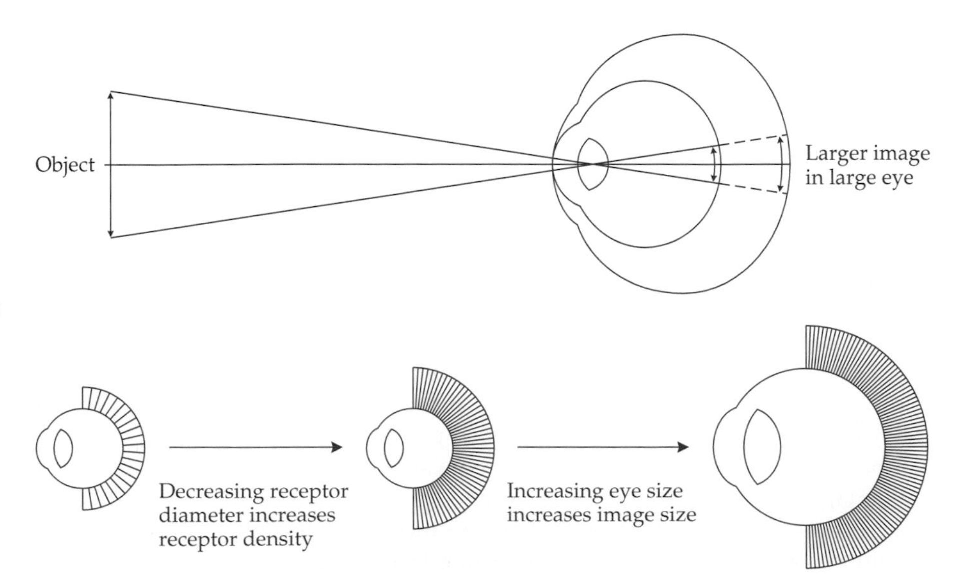

Figure 6
Elements of Construction for High Resolution in Simple Eyes
(*Top*) Although images in small and large eyes may have the same angular subtense, the image in the large eye will have larger linear dimensions. (*Bottom*) For a small eye, resolution can be increased if the receptors are thinner and more tightly packed. Any additional increase in resolution requires that the eye be larger, thus increasing the absolute size of the image.

How large the eye can be depends on the size of the animal; a snail with its simple eyes expanded to human eye size would be a bizarre creature and not one likely to evolve (Dawkins [1996] illustrates this hypothetical creature). But even small eyes gain resolution if they have thinner photoreceptors. The more photoreceptors that can be used to examine an image of a given size, the better the resolution of spatial detail. Raptorial birds achieve their exceptional visual acuity by having eyes that are unusually large relative to the size of the animal, as well as by having very thin, densely packed photoreceptors.

There is a limit on how thin photoreceptors can be made in the interest of high spatial resolution. When the diameter of a cylindrical photoreceptor becomes smaller than about 1 μm, light no longer funnels down the long axis of the cylinder. Instead, some of the light escapes from the photoreceptor and passes into neighboring photoreceptors. As a result of this phenomenon, photoreceptors are no longer optically independent of one another. And if the photoreceptors are not optically independent, a difference in light intensity between neighboring photoreceptors will be smeared out over both of them, thus limiting the resolution of fine detail.

Even though photoreceptor diameter limits the resolving capacity of small eyes, an animal will always do better visually with eyes that are as large as its overall size permits. For vertebrate eyes, this translates to an increase in eye size with body size as shown in Figure 7, where body size is indicated by body weight. Tiny animals, like shrews or bats, have very small eyes, while the largest vertebrate eye belongs to the largest animal, the baleen whale. Birds' eyes, which have been marked with a different symbol in Figure 7, tend to have eyes that are large relative to their weight. Their large eyes are consistent with their high visual acuity.

Large eyes are also the route to high sensitivity under dim light conditions (note the large eyes of owls in Figure 7). Large eyes can have large apertures (pupils), which increase the amount of light admitted to the eye, and the large images can be examined by more photoreceptors. In this case, however, signals from the photoreceptors are not kept separate, as they are in high-acuity eyes, but are pooled to increase the probability that a few photon absorptions will be detected. The larger and brighter the retinal image can be, the more effective the pooling strategy becomes, so the selection pressure on nocturnal eyes pushes them to become larger, just as it does, for a different reason, in diurnal eyes.

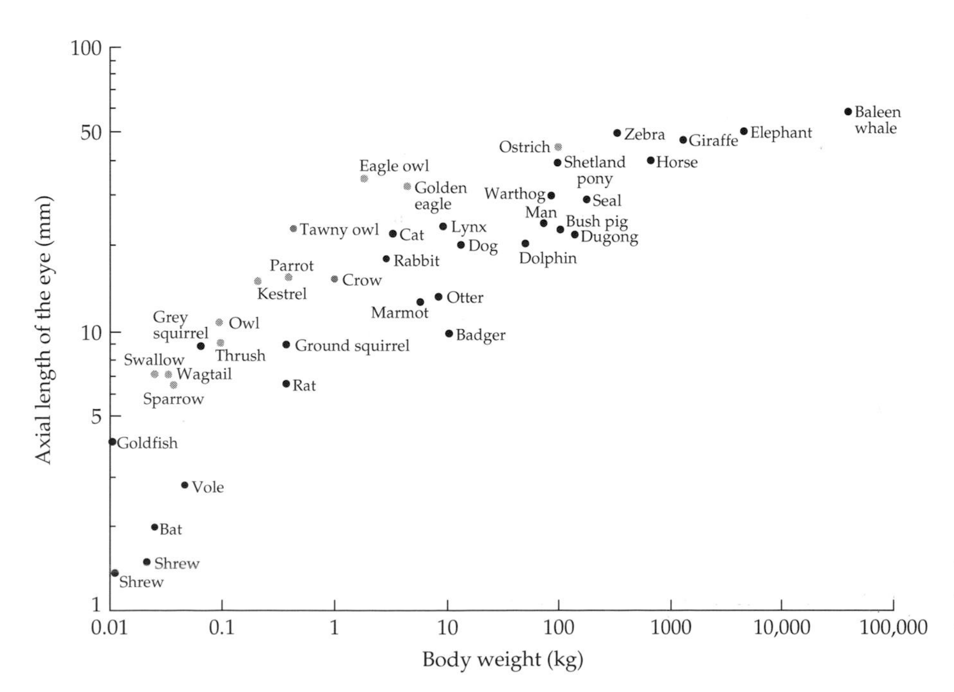

Figure 7
Eye Size and Body Weight
Eye size increases with body weight, but there is no single, simple relationship. Birds typically have eyes that are relatively large for their body weight. (After Hughes 1977, with modifications.)

But, as the saying goes, there's no such thing as a free lunch. Increased size means that more cells are required to construct the eye, individual components of the eye must increase in size in a correlated way, and ancillary systems, like extraocular muscles and their control systems, may be required to take advantage of the eye's improved performance. These "extras" may be too costly for some animals, particularly small ones, but they do not appear to be serious obstacles to the evolution of the eye.

More cells devoted to the construction of eyes should not be a problem for an animal that is evolving in the general direction of becoming generally larger; a few million more cells to increase eye size should not be much of a burden. And while we don't understand the process exactly, the retina is a key player in telling the rest of the eye how much to grow during pre- and postnatal development (see Chapter 1 and the Epilogue); this mechanism for correlating growth of the components in an individual eye could do the same thing as eyes evolved. As for the evolution of ancillary systems to move the eye, to change the aperture size, or to alter the optical efficiency of the cornea and lens, none seems especially difficult. Nature has done these things routinely in various ways, suggesting that these ancillary systems are relatively easy to come by. In the end, it appears that increasing the size of an eye is a natural and easy step along the path to "perfection and complication."

Elaborate simple eyes may have evolved rapidly

Although the steps in the evolution of eyes are lost to us, the process can be simulated in computer graphics. Beginning with a flat array of photoreceptors backed by a layer of screening pigment, the computer is given instructions about some changes it can make in the primitive eye, and a criterion for deciding if the changes improve the eye's performance. As discussed earlier, the essential criterion for an improvement that would be viewed positively by natural selection is spatial resolution; any change that improves resolution, however marginally, is more acceptable than changes that are neutral or negative in their effects on resolution. The changes to be evaluated are (1) making the flat retina into a pit, (2) deepening the pit, (3) wrapping the pigment around the front of the pit to create an aperture, (4) increasing the refractive index in the anterior part of the pit to make a lens, and (5) matching the photoreceptor size and packing to the size of the finest detail in the eye's image.

Figure 8 shows what happens after the computer is turned loose to do its thing. The original flat array of photoreceptors and screening pigment is at the top; the final eye, which the computer reached in 1829 steps, is at the bottom. The final eye is an aquatic type (the cornea was assumed to be optically ineffective) very similar in appearance to octopus and goldfish eyes, as we'll see in the ocular bestiary later in the chapter. Overall eye size has been normalized in the calculations, so the final eye is the same size as the original photoreceptor array, but the sequence incorporates a very substantial increase in size, even if it is not apparent.

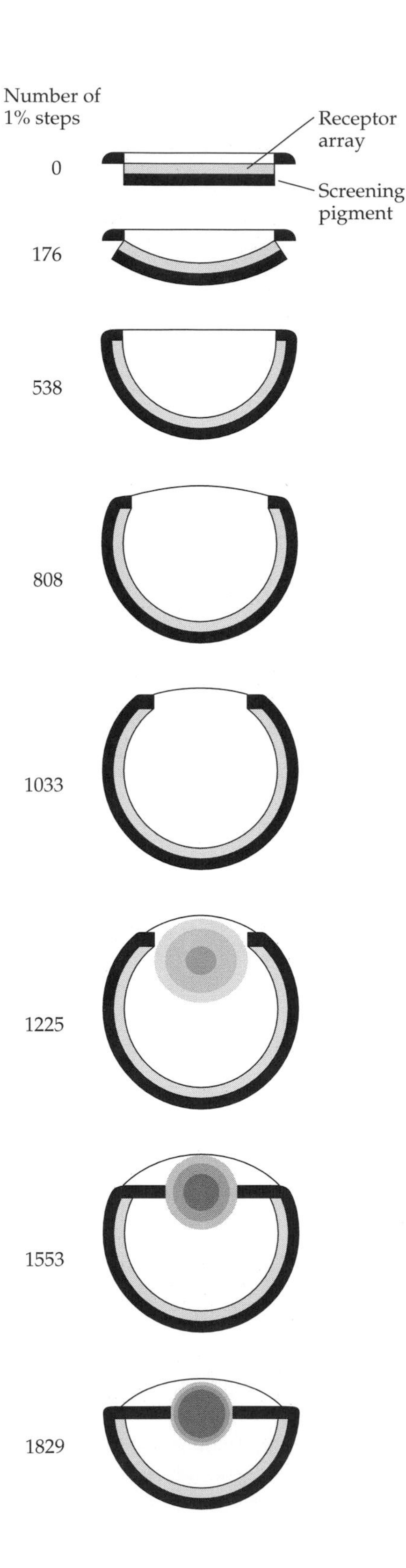

Figure 8
Possible Steps in the Evolution of a Simple Eye
At the beginning (*top*), the eye is modeled as a flat array of photoreceptors with a transparent layer in front of them and a layer of screening pigment behind them. If structure is allowed to change by 1% at each step, using improved spatial resolution as the criterion for accepting or rejecting a change, an optically sophisticated simple eye is produced in fewer than 2000 steps. The lens first seen here at step 1225 has a gradient refractive index. (After Nilsson and Pelger 1994.)

The sequence, from top to bottom, shows a gradual deepening of the pit, the formation and narrowing of an aperture, and the development of a gradient index lens. All these steps, and the intermediate ones that are not illustrated, represent improvements because spatial resolution increased at each step. Thus if this were a real evolutionary sequence, we could say without hesitation that each eye would be useful to its owner and would be better than its predecessors. Optical perfection is attainable in fewer than 2000 useful steps!

These steps represent 1% changes in structure from one step to the next. The number of generations of animals required to implement these changes is 363,992. (I am omitting the underlying calculations.) Four hundred thousand generations, to use a round number, is not a lot, particularly when we consider that all this would have taken place in relatively small aquatic animals for which the interval between generations is only a year or so. This entire evolutionary sequence appears to be able to unfold in 500,000 years, a mere 1% of the time that animals have been in existance.

Just because eyes *could* have evolved in a half million years doesn't mean they did evolve that quickly, however. Eyes belong to animals that have many other parts, and there is no reason to think that eyes evolved independently of the rest of the organism. But whenever improved vision was an advantage, eyes could evolve quickly to exploit that advantage.

Compound eyes have inherent optical limitations in their performance

The image in compound eyes is a composite image, a mosaic of all the small images generated by individual ommatidia (Figure 9). Even so, it would seem that an increase in the size of the eye, along with a compensatory decrease in the size of the ommatidia, should produce the same effect as an increase in the size of a simple eye—that is, a larger image examined in more detail by numerous ommatidia and increased spatial resolution. But it doesn't work; the amount of detail depends not on the overall image size but on the resolving capacity of the ommatidia that are producing the small component images. And the resolving capacity of an ommatidium is related to its diameter; the smaller it is, the poorer its resolving capacity.

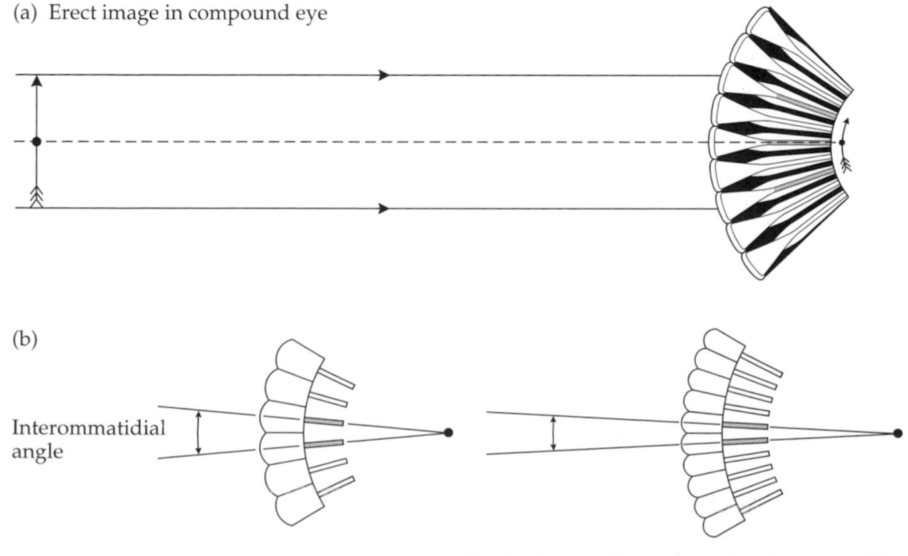

Figure 9
Optics and the Resolution Limit in Compound Eyes
(*a*) Compound eyes generate an erect representation of the world that is a composite of all the small images formed by the individual ommatidia. (*b*) Spatial resolution is proportional to the angle subtended by the optical axes of adjacent ommatidia (the interommatidial angle). Because of diffraction by the ommatidial apertures, high spatial resolution (small interommatidial angles) requires very large eyes with many large ommatidia.

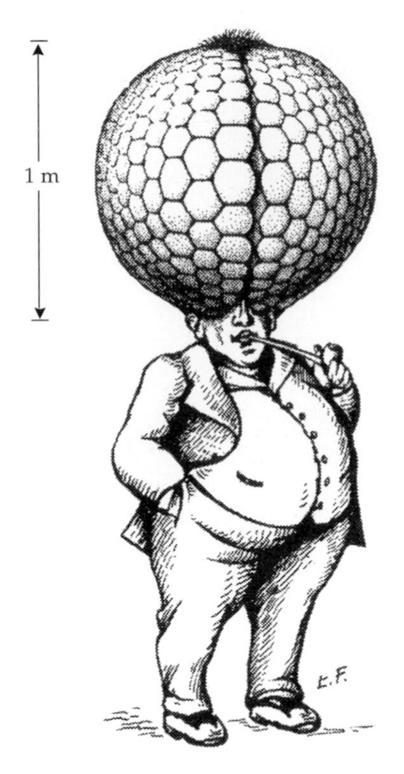

Figure 10
The Problem with Compound Eyes
Since resolution in compound eyes can be increased only by increasing the size of the ommatidia and the overall size of the eye, substituting our simple eyes with compound eyes of equivalent resolution would produce a grotesque being. In the drawing, each facet represents 10,000 ommatidia, and the large central facets near the midline correspond to the foveas in our retinas. (From Kirschfeld 1976.)

The problem is diffraction at the apertures of the ommatidia. Light interacts with the edges of the aperture much as ocean waves are altered as they pass through the opening in a breakwater; the waves spread out as they pass the opening, and the smaller the opening, the more the waves diverge. In a similar way, diffraction at the aperture in any optical system produces a smearing or spreading of the image that causes a distant point of light, like a star, to be imaged as a small, circular disc of light, and the smaller the aperture, the more the smearing and the larger the disc of light. As details like small point objects are smeared out, the system cannot resolve them as details.

Diffraction is a minor problem in our eyes, where the minimum aperture diameter is about 2 mm, but it is a major consideration in compound eyes, where the apertures of the ommatidia may be 100 times smaller. Resolution is inversely related to aperture size, and a compound eye with ommatidia 20 μm in diameter will have a resolution at least 100 times poorer than a simple eye with a 2 mm aperture. There is no escape from this limitation.

Good resolution in compound eyes therefore requires ommatidia that are as large as possible in terms of their absolute size, but still small in terms of their field of view; a small field of view is necessary to see fine detail. But if a compound eye were to have ommatidia 1 mm in diameter, to circumvent the diffraction limit, and fields of view of a fraction of a degree, for good resolution, the eye would be enormous; Figure 10 shows how we might look with compound eyes whose performance matched our simple eyes. The demands are impossible.*

Although compound eyes are limited in their resolving capacity, they can be made more sensitive. This task, however, requires a particular optical theme; the small apertures in focal apposition eyes limit their light-gathering capacity along with limiting their resolution. A better solution for compound eyes is optical superposition (Figure 11). Using numerous facets of the eye to form an image on a single rhabdom has an effect similar to enlarging the pupil in a simple eye; as far as the single rhabdom is concerned, more light is being admitted. In some cases, the sensitivity of an optical superposition eye exceeds that of an apposition eye by a thousandfold (though not all of this gain comes from the optical superposition—there are other factors). Thus, optical superposition eyes are associated with nocturnal habitats for terrestrial species or deep water for aquatic species.

No animal with small compound eyes can have high spatial resolution, but in terms of sheer numbers of animals and species, compound eyes are nature's dominant solution to the problem of seeing. And given the optical limitations of these eyes, that is a curious result. Or is it? With the exception of some crustaceans, animals with compound eyes are relatively small, and their lifestyle does not require seeing small objects at large distances. Consider, for example, a large, predatory insect like a praying mantis, which has a pair of large focal apposition eyes. The mantis catches prey within a long arm's reach; at this distance, a small fly will be large in terms of visual angle and well within the resolution limit of the mantis's eye. Turning the predator–prey relationship around, the animals most likely to eat the mantis are considerably larger than the mantis. It should be able to see them at distances large enough to allow time for evasive strategies.

Most animals with compound eyes—the insects and crustaceans—seem to have this kind of visual world in which small things are important close up and

*Doubling the size of the ommatidia while maintaining their small fields of view means doubling the number of ommatidia as well, and this can be done only by increasing the size of the entire eye. To increase spatial resolution by a factor of four, a compound eye must be increased 16-fold in its linear dimensions and 256-fold in area. That is too much to ask, particularly if the animal is small.

large things are important only at large distances. The limited resolution of their compound eyes meets the need easily. And in meeting that need, the small compound eye provides a very large visual field such that an animal with two compond eyes sees the entire world around it. This type of vision can be accomplished with small simple eyes, but not quite so easily.

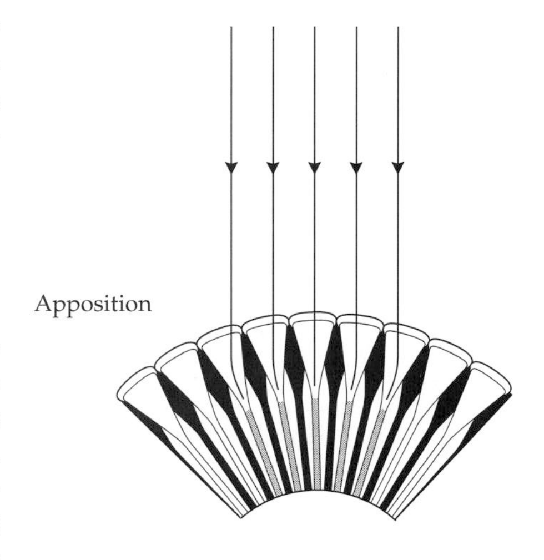

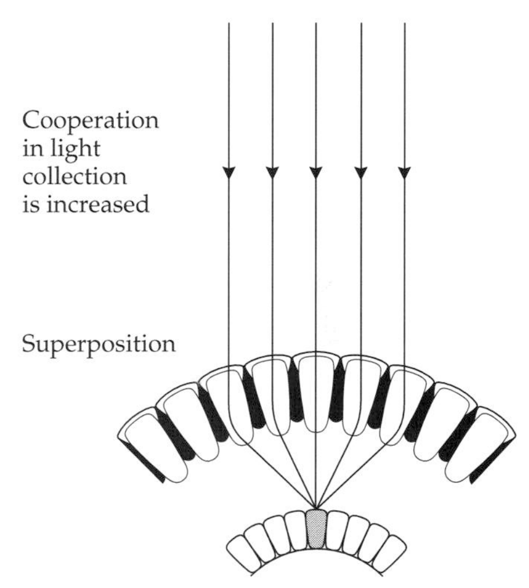

Figure 11
Increasing Sensitivity in Compound Eyes
Increasing the size of the rhabdoms can increase sensitivity in compound eyes, but significant gains require a change in the eye's optical properties. Ommatidia must work in concert to collect light for any one rhabdom, which, as shown, is essentially a change from an apposition to a superposition eye.

An Ocular Bestiary: Fourteen Eyes and Their Animals

The rest of this chapter is devoted to specific animals and their eyes, selected to illustrate each major eye type in detail, using species whose eyes and vision have been reasonably well studied. Half of the animals represented have one of the types of compound eye (honeybee, horseshoe crab, butterfly, housefly, firefly, crayfish, crab). Simple eyes are represented by a pinhole eye (*Nautilus*), a reflecting simple eye (scallop), and several examples of refracting simple eyes that illustrate evolutionary convergence (octopus, goldfish), visual excellence (pigeon), variation within this eye type (jumping spider), and relationships to primate eyes (human).

1. Compound eye—focal apposition, terrestrial variety: Honeybee (*Apis mellifica*)

PHYLUM: Arthropoda

CLASS: Insecta

ORDER: Hymenoptera

FAMILY: Apidae

Bees have compound eyes of the focal apposition type (Figure 12), which is the most common type of compound eye. The cornea in each ommatidium is the primary refracting element, and the jellylike lens cone behind it is mainly a transparent optical path for light imaged on the rhabdom. The tip of the rhabdom lies at the primary focal point of the corneal refracting surface; this focal length is about 60 μm. Light entering the tip of the rhabdom is guided down through the thin cylinder in which the photopigment lies by the higher refractive index of the rhabdom compared to the surrounding cytoplasm; the rhabdom has some features of optical fibers used for data transmission. Since the average diameter of each ommatidium is about 25 μm, the system is severely diffraction limited. Bees can resolve spatial detail subtending just under 1° of visual angle; humans do about 60 to 100 times better.

Bees communicate with one another mainly by tactile and chemical senses. Their famous "waggle dance," with which they transmit information about the direction and distance to a food source, is often performed in the dark. In fact, vision is of little use for anything done inside the hive, and the hive is where a worker typically spends all but the last few weeks of her life. Outside, however, vision is used for orientation, locating and recognizing food sources, and, by the drones, for their mating flight with the future queen. In several of these operations, bees reveal visual abilities that we do not possess.

The bees' focal apposition eyes, and most other types of compound eyes, can detect the plane of light polarization. In ordinary unpolarized light, the electric vectors of the incident light can have any angle with respect to the direction at which the light is traveling, but light is said to be polarized if the electric vectors are restricted to a narrow range of angles—for example, parallel to the horizon. One way this happens is by reflection, as when sunlight is reflected from the surface of water; polarized sunglasses are meant to block this horizontally polarized

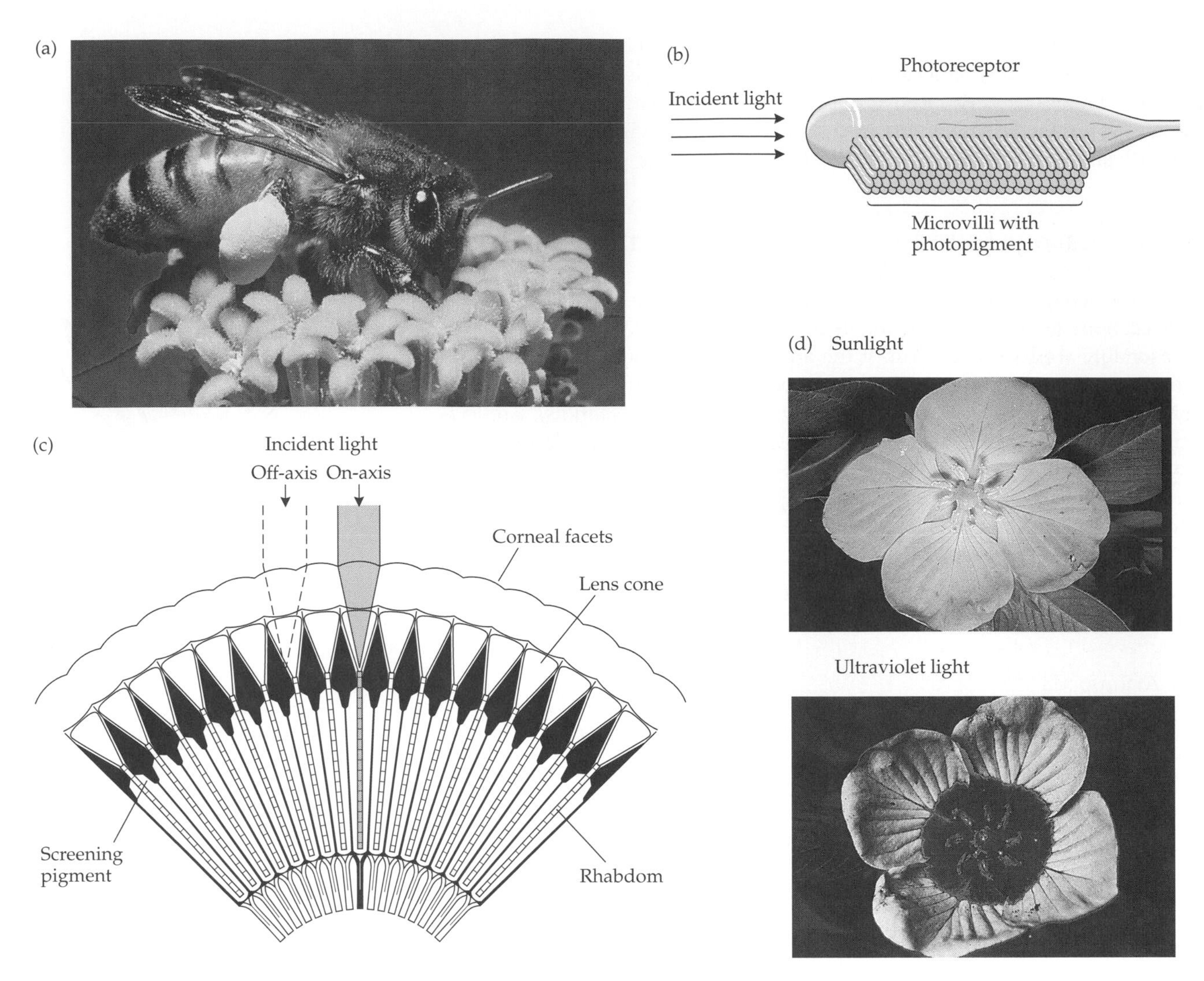

Figure 12
A Focal Apposition Eye—Terrestrial Variety: Honeybee
(a) The eye of the honeybee is elliptical, with the long axis near vertical, having tightly packed hexagonal facets. *(b)* Photoreceptors—retinula cells—are of the brush type; the microvilli from adjacent photoreceptors interlace to form the rhabdom. *(c)* A section through the eye shows the corneal facets overlying cone-shaped "lenses" that funnel light to long, cylindrical rhabdoms behind the apices of the lens cones. Each rhabdom is surrounded by the retinula cells, whose bristles mingle to form the rhabdom, and each ommatidium is separated from its neighbors by pigment-containing cells; the screening pigment is particularly obvious around the lens cones. Light striking an ommatidium along its optical axis will be focused on the underlying rhabdom (large arrow), but light incident at small angles off the axis (small arrow) will not reach the rhabdom. *(d)* Many flowers appear different in ultraviolet light, with markings not visible to us under natural illumination. Bees can see these patterns, which often converge on the nectar- and pollen-bearing parts of the flower. (*a* by Thomas C. Boyden, © 1998/Dembinsky Photo Associates; *c* after Nilsson 1989; *d* courtesy of Thomas Eisner.)

reflected light. Also, sky light is normally polarized to some extent, the amount of polarization increasing as a function of angular distance from the sun. Photographers know about this phenomenon and use polaroid filters to darken the sky for more dramatic photographs, but we can't see the change in polarization directly. Bees *can* see it, and they use the information to establish their directions of flight relative to the sun.

This ability to detect the plane of light polarization resides in the structure of the photoreceptors in compound eyes, which look somewhat like miniature brushes (see Figure 12*b*). The photopigments are in the microvilli—the bristles on the brush—and most of the photopigment molecules are parallel to the long axes of the microvilli. Because of this arrangement, the photopigment in a receptor is preferentially sensitive to light whose plane of polarization is parallel to the microvilli, and the sensitivity declines as the polarization angle deviates from alignment with the microvilli.

We associate bees with brightly colored flowers, and it is no surprise to find that bees have color vision. The surprise is that their color vision was disputed until first demonstrated in the 1930s by the great student of bees, Karl von Frisch (1886–1982; Nobel laureate 1973). Bees do not see reds as well as humans do, but they do much better at the detection of ultraviolet (UV) light. (We can see ultraviolet light if the intensity is high enough, but the optical media in our eyes absorb strongly in the UV region of the spectrum, thereby reducing its transmission to the retina.) In bees, however, one of the three photopigments is tuned for strong absorption in the UV range.

One way that bees use their ultraviolet sensitivity is in the recognition of flowers; as Figure 12*d* shows, some flowers reveal striking patterns under ultraviolet illumination that are invisible to humans. Some of the patterns are almost like direction signs, pointing the way to the nectar source. Because the bees can see the differences in UV light reflections from the flower petals, they can see these special patterns and have been shown to use them in selecting their food sources. The flower-bearing plants provide these signals and go to the trouble of producing nectar because pollen brushed onto the bees as they feed is carried from one plant to another. This mutual benefit—food for the bees and pollination for the plants—underlies the coevolution of bees and flowers.

II. Compound eye—focal apposition, aquatic variety: Horseshoe crab (*Limulus polyphemus*)

PHYLUM: Arthropoda
CLASS: Chelicerata
SUBCLASS: Merostomata
ORDER: Xiphosura
FAMILY: Limulidae

Horseshoe crabs get their name from the shape of their shell, which looks like a horseshoe from which their long tail (telson) protrudes (Figure 13). They are often referred to as living fossils, partly because they look primitive, as if they have been around for a very long time (which they have), and partly because the other members of their subclass are extinct. Horseshoe crabs also remind people of the extinct trilobites to which they are distantly related, but *Limulus* is more closely related to living spiders, scorpions, and ticks.

Limulus has five eyes: a small pair of ocelli near the midline, a pair of large lateral eyes (Figure 13*b* and *c*), and a small midline ventral eye (they also have photoreceptors on their tails that play a role in setting the animal's internal clock). The large lateral pair of eyes, the ones of interest here, are focal apposition eyes, like those of honeybees, but they are modified for aquatic vision in which the lens is the primary refractive element. As Figure 13*d* shows, the lenses in the ommatidia are not separate from the cornea, and they protrude inward as small cones with flat tips, each cone having a center-to-edge gradient of refractive index. The index gradient affects light–dark fringes passing through a thin section of the cornea and lens cone. The fringes are displaced by the difference in refractive index between the section and its surrounding medium, but the dis-

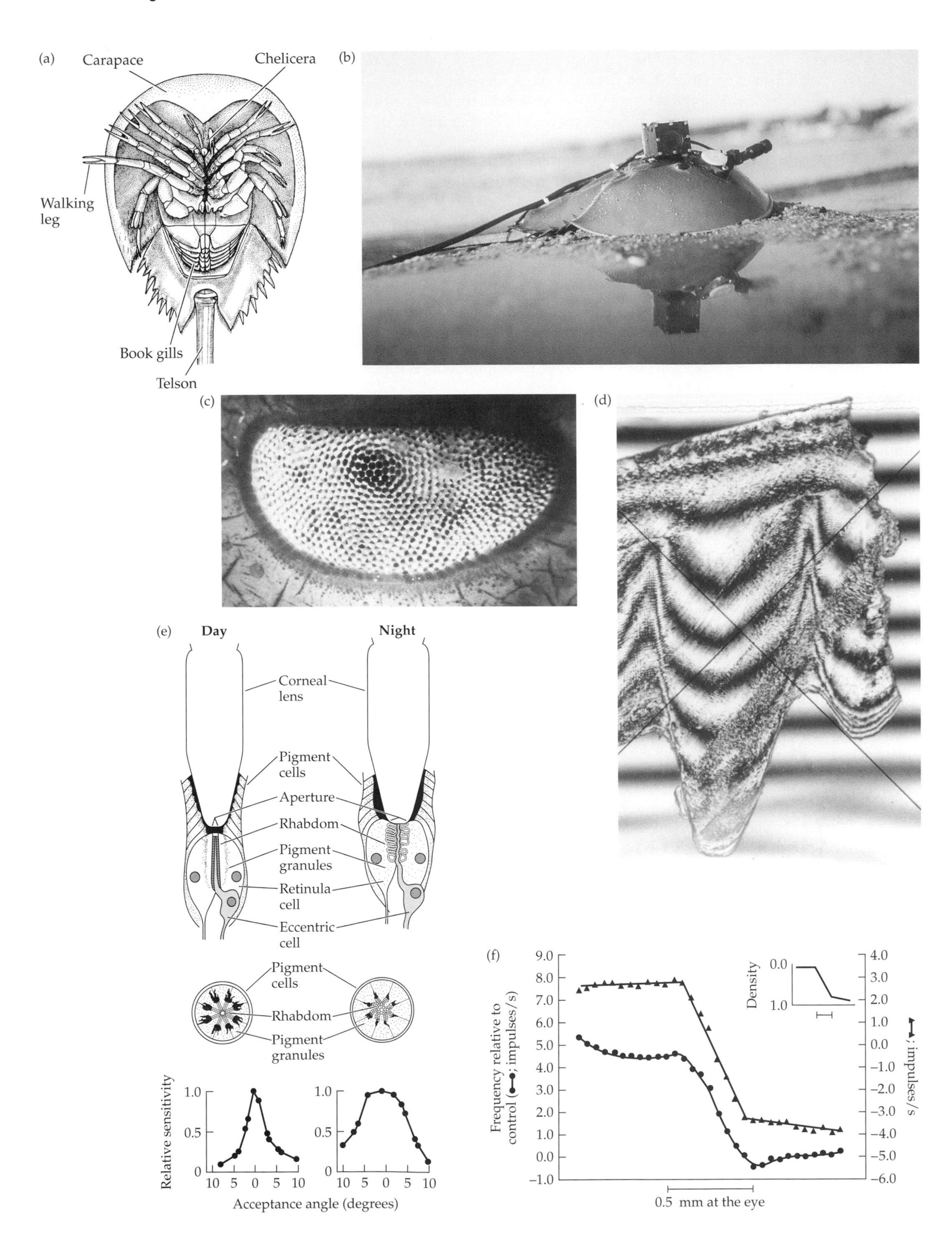
(a)
Carapace
Chelicera
Walking leg
Book gills
Telson
(b)
(c)
(d)
(e)
Day
Night
Corneal lens
Pigment cells
Aperture
Rhabdom
Pigment granules
Retinula cell
Eccentric cell
Pigment cells
Rhabdom
Pigment granules
Relative sensitivity
1.0
0.5
0
10 5 0 5 10
Acceptance angle (degrees)
(f)
Frequency relative to control (●—●; impulses/s)
▲—▲; impulses/s
Density
0.0
1.0
0.5 mm at the eye
9.0
8.0
7.0
6.0
5.0
4.0
3.0
2.0
1.0
0.0
–1.0
4.0
3.0
2.0
1.0
0.0
–1.0
–2.0
–3.0
–4.0
–5.0
–6.0

Figure 13
A Focal Apposition Eye—Aquatic Variety: Horseshoe Crab
(*a*) A view of *Limulus* from below shows the animal's legs, tail, and the horseshoe shape of its shell. (*b*) This *Limulus* is wearing a miniature video camera ("CrabCam") aligned with the visual axis of one ommatidium from which electrical signals were being recorded. (c) One of the lateral eyes is shown in this enlarged view. (*d*) When alternating light and dark stripes (a grating) are projected through a thin section of the cornea and lens cone of *Limulus*, the parallel grating stripes are markedly distorted within the lens cone, thereby indicating the refractive index gradient within the lens. (e) The ommatidial structure in *Limulus* varies diurnally. At night the rhabdom becomes shorter and fatter, and the screening pigment migrates away from the rhabdom. As a result, the angle from which the ommatidium collects light becomes larger at night. (*f*) When a step of light is moved across a single ommatidium in the *Limulus* eye, the eccentric cell discharge follows the change in light intensity exactly (upper curve). When neighboring ommatidia are also allowed to see the intensity step, however, the eccentric cell discharge from the test ommatidium changes, exaggerating the changes in intensity at the top and bottom of the step (lower curve). (*a* from Brusca and Brusca, 1990; *b* courtesy of Robert Barlow; *c* by Larry Wasserman, courtesy of Robert Barlow; *d* from Land 1979; e after Barlow, Chamberlain, and Levinson 1980; *f* after Ratliff and Hartline 1959.)

placement is greatest along the axis of the lens cone, indicating that the refractive index is greater here than elsewhere. This difference gives the lens cone focusing properties that it would not have if the refractive index were uniform throughout; as a result, light is focused onto the tip of the rhabdom.

Compared to those of the honeybee, the ommatidia in the eye of *Limulus* are large in absolute terms, so they are not so severely diffraction limited. The ommatidia are large relative to the size of the eye, however, subtending a visual angle of about 7.7°. (The comparable value for the honeybee's eye is just under 1°.) Therefore, the ability of the *Limulus* eye to resolve spatial detail is seven to eight times poorer than that of the honeybee. But *Limulus* normally goes about its business under dim light conditions, and sensitivity is probably more important than high spatial resolution.

Along the New England coast, horseshoe crabs gather in large numbers in shallow water along the beaches on moonlit evenings in May and June. Males and females pair off, making nests by scraping out depressions in the sand near the high-tide mark, where the females lay eggs that are externally fertilized by the males. Unpaired males often release sperm around mating pairs or into the nest after the pair have gone. After this orgy, they return to deeper water, leaving the eggs in the sand to their fate. Enough young seem to make it through the larval stages to keep the species going, but the vast majority of the eggs never become horseshoe crabs.

The nuptial behavior of the horseshoe crabs is guided partly by vision (but a variety of factors determine the timing of the event). When concrete castings of old *Limulus* shells are placed along the shore with other concrete objects of equal volume (spheres or cubes), the horeshoe crabs can tell the difference; they prefer the concrete shells as potential mates to the spheres and cubes, thereby demonstrating a capacity for discriminating visual form. (Admittedly, confusing a concrete casting of a shell with a real *Limulus* is not strikingly intelligent behavior, but that is another problem.)

Although animals can often be classified as diurnal or nocturnal depending on the time they are most active, most are capable of working with intermediate levels of light at dawn or dusk, and their eyes usually have mechanisms for altering their sensitivity. In *Limulus,* the structure of the ommatidia is modified in a cyclic pattern throughout a 24-hour day (see Figure 13*e*); pigment migration and changes in the rhabdom give the ommatidia wider fields of view and greater sensitivity at night than during daylight. This cycle, which is a circadian rhythm, is centrally controlled; a central "clock" sends neural signals to the eye, where the release of a neuromodulator brings about redistribution of the screening pigment and change in the retinula cell shape. Pigment migration, changes in cell shape, and cell movement are common strategies in both compound and simple eyes for adjusting the eye's sensitivity, although the linkage to a central clock seen in *Limulus* is not as well established in other species.

For all its oddity, *Limulus* occupies a significant place in the history of vision reseach, largely because of an anatomical feature that makes the eye amenable to study. Each ommatidium has a large "eccentric cell," whose axon is the main pathway for neural signals being sent from the eye to the brain. The action potentials transmitted by eccentric cell axons were some of the first signals from an eye to be recorded and studied, initially by Haldan Keffer Hartline (1903–1983; Nobel laureate, 1967), with subsequent contributions from his collaborators and students. A particularly interesting result is shown in Figure 13*f*; the figure compares the light intensity across a light–dark border (an edge) to the neural discharge from the eccentric cell in an ommatidium across which the edge was moved. The neural representation of the edge in the action potentials from the eccentric cell is an exaggerated version of the original light–dark discontinuity; as it is sometimes stated, the edge has been "enhanced." Although the ommatidia in the *Limulus* eye are optically independent, inhibitory interactions among neighboring eccentric cells produce the edge-enhancement effect.

A similar form of lateral inhibition, with a similar contrast-enhancing effect, is present in all vertebrate retinas that have been studied; the phenomenon is probably ubiquitous (see Chapter 14). Under the right circumstances, we can see the result of these retinal interactions in our own eyes when a light-to-dark transition appears to be highlighted by adjacent thin strips of a whiter white and a blacker black. These unexpected bright and dark bands are "Mach bands," named for the physicist Ernst Mach (1838–1916), who first described and studied them (see Ratliff 1965). Thus we have a connection between the sonic boom produced by an airplane flying faster than the speed of sound—Mach 1—and the decidely unspeedy horseshoe crab.

III. Compound eye—afocal apposition: Monarch butterfly (*Danaus plexippus*)

PHYLUM: Arthropoda

CLASS: Insecta

ORDER: Lepidoptera

FAMILY: Danaidae

At several points in an insect's life the animal changes considerably from one stage to the next. In almost 90% of the species, the transformation—the metamorphosis—is so dramatic that the stages look like totally different animals. Butterflies and moths are particularly striking examples. A monarch butterfly begins as a larva that we call a caterpillar (Figure 14); the elongated body is segmented, with pairs of legs on half or more of the segments, and the head has chewing mouthparts, antennae, and small aggregations of photoreceptors (ocelli) for eyes. The monarch caterpillar spends most of its time eating the leaves of the milkweed plants on which eggs were originally laid. When mature, the caterpillar hangs down from a milkweed branch and begins the process of transformation by secreting a shell called a chrysalis. Within the chrysalis, the body tissues are broken down and rearranged—the animal at this stage is a pupa—to create a different creature. The result, a monarch butterfly, has brightly colored wings, six legs on its thorax, mouthparts designed to suck nectar, and a nifty pair of compound eyes.

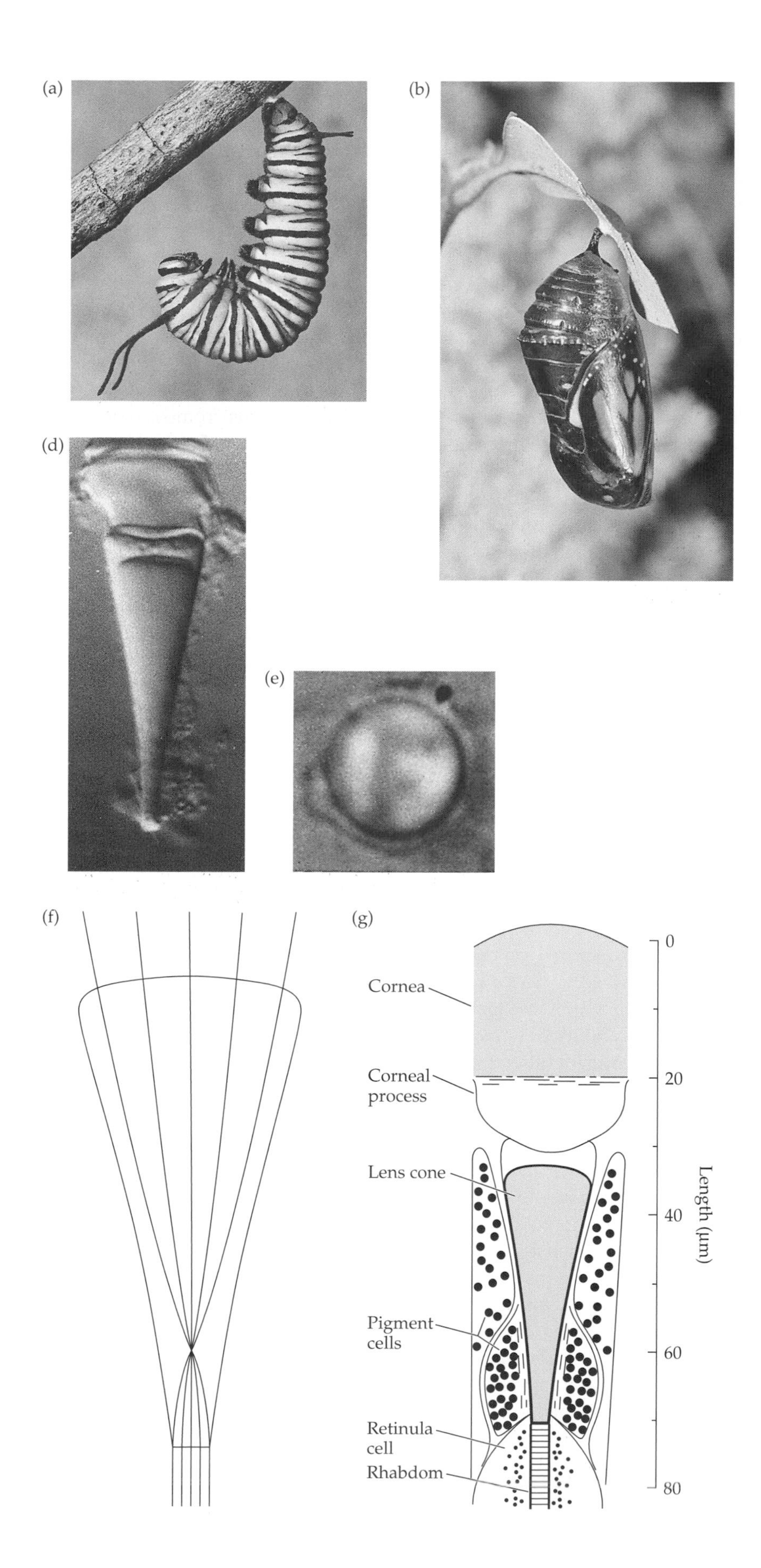

Figure 14
An Afocal Apposition Eye: Monarch Butterfly
This insect progresses from a caterpillar (*a*), to the pupal stage in its chrysalis (*b*), to the newly emerged butterfly (*c*). (*d*) A microscopic view from the side shows the corneal facet and lens cone through which an image—the letter *F*—formed by the optical system can be seen (e). (*f*) Light is focused within the lens cone and exits as parallel rays. Because the light rays are parallel both at the corneal surface and when they leave the lens cone, the system is *afocal,* as are terrestrial telescopes. (*g*) In this diagram of the anterior part of an ommatidium, a corneal process and a space between the cornea and the lens can be seen; these features differ from focal apposition eyes and are responsible for the afocal operating characteristics of the butterfly's ommatidia. (*a* by Skip Moody, ©1994/Dembinsky Photo Associates; *b* by William J. Weber/Visuals Unlimited; *c* by Gustav W. Verderber/Visuals Unlimited; *d, e* from Nilsson 1989.)

Butterflies have apposition eyes, quite similar to those of bees, with the exception of their image-forming properties. In these *afocal* apposition eyes, parallel rays of light entering the optical system leave the system as a narrower bundle of parallel rays—each ommatidium is like a miniature terrestrial telescope (see Figure 14*g*). Light is focused by the cornea at a point in the posterior part of the lens cone, and refraction *by the lens cone* makes the incident ray bundle parallel again before it reaches the rhabdom. Where the lens cone in a bee's eye makes no refractive contribution, the lens cone in the butterfly's eye is a key optical element. In its overall effect, this is not an image-forming optical system like that found in the focal apposition eyes of bees; instead, the optical system funnels incident light into the rhabdom.

In an eye like the human eye, an unfocused ray bundle is disastrous because the defocusing prevents any resolution of fine detail. This is of no consequence in the butterfly eye, however, because the resolution is set by the size and acceptance angle of the ommatidium and not by the size of photoreceptors that are sampling different parts of a detailed image. As long as the incident light reaches the rhabdom, it makes no difference whether it is focused or not—what the rhabdom sees is the same in either case.

The advantage offered by the afocal system appears to be improved spatial resolution. The acceptance angles of the ommatidia are more sharply tuned—that is, narrower—so resolution more nearly approaches the limit set by diffraction at the small aperture. For the monarch butterfly, the resolution limit will be slightly better than that of the honeybee. Good spatial resolution is useful at this stage of the animal's life, when it must fly from one food source to another, recognizing the flowers and the nectar-containing parts of them (in the larval stage, it had only to crawl around on a plant and chew leaves). The butterfly is also the sexual stage of life; potential mates must be recognized (although a large part of the recognition process is chemical communication) and chased through the air. And good vision is useful for any flying creature; flying into unseen obstructions wastes time and is potentially damaging. Butterflies don't fly very fast, but they sometimes navigate very long distances. Monarch butterflies migrate each autumn from as far as Canada to a winter home in Mexico, traveling thousands of kilometers at a pace of 6 km per hour. It is not clear how they manage this feat, but vision undoubtedly plays a role.

The bright colors and bold patterns on butterfly wings imply that butterflies have good vision and use it to recognize members of their own species. But there is another visual consideration: how the butterfly is seen by other animals. Color and pattern may be meant as much to confuse or inform predators as to inform other butterflies.

Butterflies are potential food for large birds, all of which have excellent color vision, but monarchs are distasteful and toxic because the milkweed they ate as larvae is poisonous. The monarchs' bright orange and black patterns are a way of advertising their unpalatability. A young blue jay will make only one attempt to eat a monarch; the ensuing bout of vomiting teaches a lasting lesson. This taste test, of course, means the demise of one monarch, but it also means that one more jay will leave all other monarchs strictly alone. The effectiveness of the strategy is evidenced by mimicry. The smaller viceroy butterfly (*Limenitis archippus*) has evolved a color and pattern almost identical to that of the monarch. Viceroys are apparently quite tasty, but they try conceal the fact by masquerading as monarchs. The deception works only on birds that have previously encountered a monarch butterfly—uninitiated birds will gobble up viceroys quite happily—but at least some potential predators remain on the sidelines.

IV. Compound eye—apposition with neural superposition: Housefly (*Musca domestica*)

PHYLUM: Arthropoda

CLASS: Insecta

ORDER: Diptera (true flies)

FAMILY: Muscidae

The aerial acrobatics of houseflies are recognized in the name *muscae volitantes* ("flying flies") given to the small dark spots we sometimes see flitting through our visual field when we look around at uniformly illuminated surfaces, like the pages of a book. The spots are shadows cast on the retina by small floaters, cellular debris, moving around in our eye's vitreous chamber as our eyes move. Houseflies, of course, have no vitreous. Their eyes are compound eyes (Figure 15), optically of the focal apposition type, but with an important variation that has led to their classification as neural superposition eyes.

Unlike most apposition eyes, those in flies do not have a fused rhabdom. Instead, screening pigment in the fly's ommatidia keeps the retinula cells apart (see Figure 15*b*). Since the optical system is similar to that of the honeybee, a small image is formed at the tip of the rhabdom. Different parts of the image will fall on different retinula cells, with the result that each retinula cell views a slightly different part of the world through the optical system. And because the fields of view of neighboring ommatidia overlap somewhat, retinula cells in neighboring ommatidia see the same point in visual space. This multiple imaging could be a source of great confusion; how could the fly make sense of a world in which every resolvable point in space is imaged a half dozen times?

The trick is to reunite the separate images by feeding the signals from all the different retinula cells that view the same point in space onto a common site at the next level of the nervous system. This task requires a highly specific pattern of wiring (see Figure 15*c*), which has been demonstrated quite convincingly in flies and other species with eyes of this type. The recombining of separate images is superposition, and the fact that it is done by the wiring makes it neural superposition.

Neural superposition doesn't improve the eye's resolution, which is still limited by the size of the ommatidial apertures; some flies have about the same resolving capacity as bees, on the order of 1° or so, but the housefly has a resolving capacity of about 2.5°. Superposition makes the eye more sensitive, however; in houseflies, cells in seven ommatidia are combined in the neural image, thereby increasing sensitivity about sevenfold. This increased sensitivity means that the flies can go about their business slightly earlier in the day than bees and other insects that have simple apposition eyes can, and they can keep at it slightly later. Where there is competition with other species for food, the small temporal expansion of their niche may be advantageous. But the possibility that the flies can see better than their predators at dawn and dusk may be even more important.

Ommatidia in compound eyes are rarely uniform in size over the entire ocular surface. Eyes often have zones or streaks in which the ommatidia are larger and therefore have better spatial resolution. (Similar high-resolution "streaks" are also present in vertebrate retinas; they are usually horizontal and probably have to do with the importance of the horizon as a visual reference.) Diurnal flying insects in particular often have a zone of higher resolution—akin to the fovea or area centralis in some vertebrate retinas—pointing forward in their

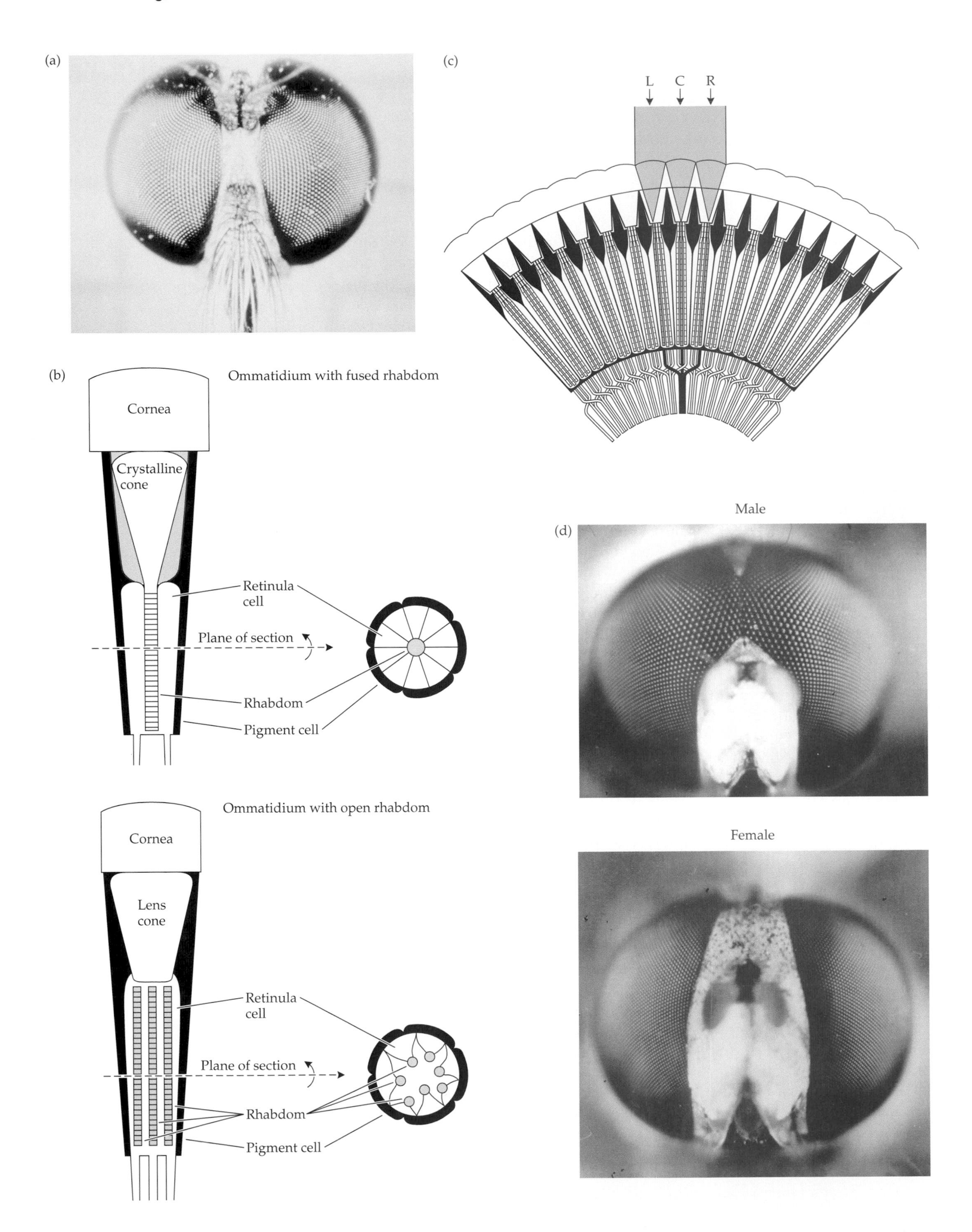
(a)
(c)
L
C
R
(b)
Ommatidium with fused rhabdom
Cornea
Crystalline cone
Retinula cell
Plane of section
Rhabdom
Pigment cell
Ommatidium with open rhabdom
Cornea
Lens cone
Retinula cell
Plane of section
Rhabdom
Pigment cell
Male
(d)
Female

◀ Figure 15
A Neural Superposition Eye: Housefly
(*a*) The many facets in the eyes show clearly when viewed from the front. The facets in the eyes are hexagonal, although the exact shape cannot be seen here. (*b*) Optically, the fly's eye is an apposition eye, but with an open rhabdom; unlike in other apposition eyes, parts of the image can be seen by different photoreceptors within an ommatidium. (*c*) Light entering along the axis of an ommatidium will fall on a central rhabdom (C), while light just off the axis will fall on other rhabdoms (L and R, for example). When the light is coming from a distant point as it is here, the point will be imaged on three different rhabdoms in three different, but neighboring, ommatidia. The neural circuitry connecting the ommatidia to the central nervous system combines the signals from the different retinula cells as shown for the three illuminated here. This combining of signals—neural superposition—avoids the confusion of multiple imaging and increases the eye's sensitivity. (*d*) Facets in the fly's eye are not always the same size and may differ between males and females. In the species shown here, the male's eyes are larger, extending over the top of the head, and have a frontal region of enlarged facets. (*a* courtesy of Michael F. Land; *b* and *c* after Nilsson 1989; *d* from Land 1989.)

flight direction; this is true of bees, butterflies, and houseflies. The larger facets near the front of the eye are visible in the housefly eye in Figure 15*a*, but show more clearly in another species (Figure 15*d*). The strategy is the same in vertebrate eyes, in which resolution in some parts of the visual field is sacrificed for better resolution in others (I use the plural because there may be more than one direction of importance and therefore more than one region of specialization in the eye).

Sexual dimorphism—structural differences between males and females—does not extend to the eyes in vertebrates, but it is common in insect eyes; the forward-pointing high-resolution zone in fly eyes is an example. The zone in the male fly's eye points upward more than in the female's eye, and the male's eye facets are larger for better resolution. When males pursue females, they chase by keeping the female imaged in the eye's acute zone (*Cherchez la femme!*), using shifts in the image position as error signals to correct their flight path. Males also have retinula cells within the ommatidia of the high-resolution zone that are not found in females, and they have a special pattern of connectivity to the central nervous system. For these reasons, this part of the male fly's eye has been dubbed "the love spot."

V. Compound eye—refracting superposition: Firefly (*Photuris* spp.)

PHYLUM: Arthropoda
CLASS: Insecta
ORDER: Coleoptera
FAMILY: Lampyridae

At one time in my life I often spent summer evenings chasing and catching the beetles I knew as "lightning bugs" (fireflies) and putting them in a bottle; the hope, never realized to my satisfaction, was to have a kind of biological flashlight to carry around. In some parts of the tropics, people carry fireflies in bottles as lanterns to light their way at night, and women use bunches of them wrapped in gauze as glowing ornaments for their hair (according to Brusca and Brusca [1990]). There must be some trick that I never figured out.

Fireflies have eyes with afocal optical systems, similar to those in the eyes of butterflies, but with a significant difference: The ommatidia are not optically independent, and they cooperate in bringing light to a particular rhabdom

(Figure 16). These are refracting superposition eyes. The feature distinguishing superposition eyes from apposition eyes is the large separation between the optical elements, the cornea and lens cone, and the rhabdoms. The space is called the clear zone and is filled with a transparent material, making it roughly akin to the vitreous in a simple eye. In the firefly's eye, however, the clear zone is not completely homogeneous; fine crystalline threads run radially from the lens cones to the rhabdoms.

Refracting superposition eyes vary in the amount of refraction by the corneal surface and in the refractive index gradients within the cornea or lens cone, but the refractive index gradients are required to make the overall system work like a small telescope constructed with two distinct lenses; parallel light striking the cornea is still parallel when it leaves the lens cone. This system is not greatly different from the optics in the butterfly eye, but in this case the parallel bundles of light must cross the clear zone to reach the rhabdoms. Of more importance, light from a single point in space must come through several optical elements to impinge on a single rhabdom. Making this happen requires nearly perfect spherical geometry. The optical elements are part of a spherical surface, and the rhab-

(a)

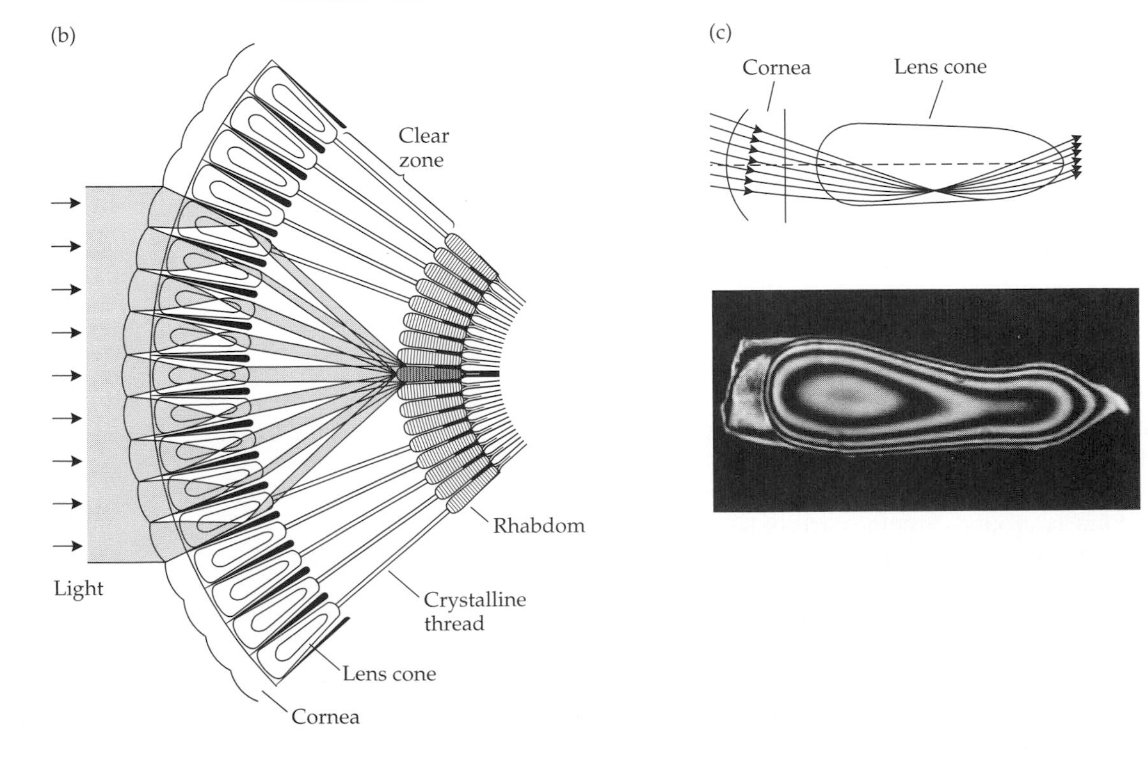

Figure 16
A Refracting Superposition Eye: Firefly
(*a*) This firefly is signaling. (*b*) A section through the firefly's eye shows a separation between the optical elements (cornea and lens cones) and the rhabdoms of the photoreceptors. The space—the clear zone—is typical of optical superposition eyes. In firefly eyes, crystalline threads cross the clear zone between the lens cones and the rhabdoms; some have been omitted for clarity. The central rhabdom is shown as receiving light passing through nine optical systems. (c) Light rays follow curved paths through a lens cone, with rays exiting roughly in parallel. Curved ray paths are common in gradient index systems, and the existence of the index gradient is shown by the interference pattern formed by a section of lens cone in the photograph. (*a* courtesy of Thomas Eisner; *b* and *c* after Nilsson 1989; photograph in *c* courtesy of D.-E. Nilsson.)

doms are oriented along radii of the sphere, about halfway between the center of curvature and the surface.

In Figure 16*b*, nine ommatidia in a row superimpose their images on a single rhabdom, but there can be considerably more than this; considering the eye in three dimensions, there will be a circular cluster of a hundred or more ommatidia whose ray bundles are superimposed. The obvious advantage of the superposition eye compared to an apposition eye is a gain in sensitivity, which is just what one would want for an eye operating in a dimly lit environment—the firefly's world.

The light-emitting organs of fireflies consist of a cluster of cells called photocytes within which molecules (luciferins) break down in the presence of an enzyme (luciferase); light is part of the energy released when the luciferin breaks down. The photocytes lie just below the transparent epidermis and are lined on the inner side by a layer of reflective cells. Light is emitted normally in pulses; the pulse frequency and duration are species-specific and are used for signaling between males and females. Males signal as they fly around and look for a light pulse in response from a stationary female; the duration between the male's signal and the female's response is also characteristic for a species.

Part of the fireflies' need for a sensitive eye is related to detecting these signals and responses that are a prelude to mating. For the males in particular, the ability to see obstacles as they are flying at night is important. They are unlikely to be injured by crashing into obstacles (which they do), but the crash slows their search for a responsive female and the competition is often heavy. More important still is the ability to see the female as the male approaches after she has responded. Things may not be as they seem.

A meadow on a summer's evening will often be the haunt of several species of fireflies. The males fly over the meadow, signaling to prospective mates, while the females sit on the sidelines, from where they respond to the appropriate signal. Some of these females are deceptive; they can make not only the response that is characteristic of their species but also the response that is appropriate to the female of another species. A male from the other species who is duped by this mimicry may approach expecting sex, only to end up as Jezebel's dinner. The game is not entirely one-sided, however; males have learned to approach cautiously, and visually inspect the lady from a close but safe distance. And there is some evidence (disputed) that males at risk sometimes mimic other species in which the females are predatory, perhaps to make these dangerous females reveal themselves.

VI. Compound eye—reflecting superposition: Crayfish

PHYLUM: Arthropoda

SUBPHYLUM: Crustacea

CLASS: Malacostra

ORDER: Decapoda

INFRAORDER: Astacidae

FAMILY: Cambaridae

Crayfish are closely related to lobsters, differing largely in habitat (crayfish usually live in fresh water, while lobsters are marine) and construction of the carapace (lobsters incorporate more calcium carbonate and their shells are considerably harder than those of crayfish).

Many decapods, particularly crayfish, lobsters, and shrimp, have optical superposition eyes that use reflection to bring an object seen by many ommatidia together on a single rhabdom (Figure 17). In general, the eye's structure is much like that of a refracting superposition eye (in the firefly, for example), particularly in the obvious separation between the optical elements of the ommatidia and the rhabdoms. But in the aquatic eye of the crayfish, the cornea is of little use in refraction, leaving it up to the rest of the optical system to form an image, as in the *Limulus* eye. Here, however, the optical components lack any refractive index gradient that could underlie refraction, and recognizing that omission was part of the key to realizing how these eyes work. If light entering an optical element is redirected, and the redirection is not done by refraction, then what about mirrors?

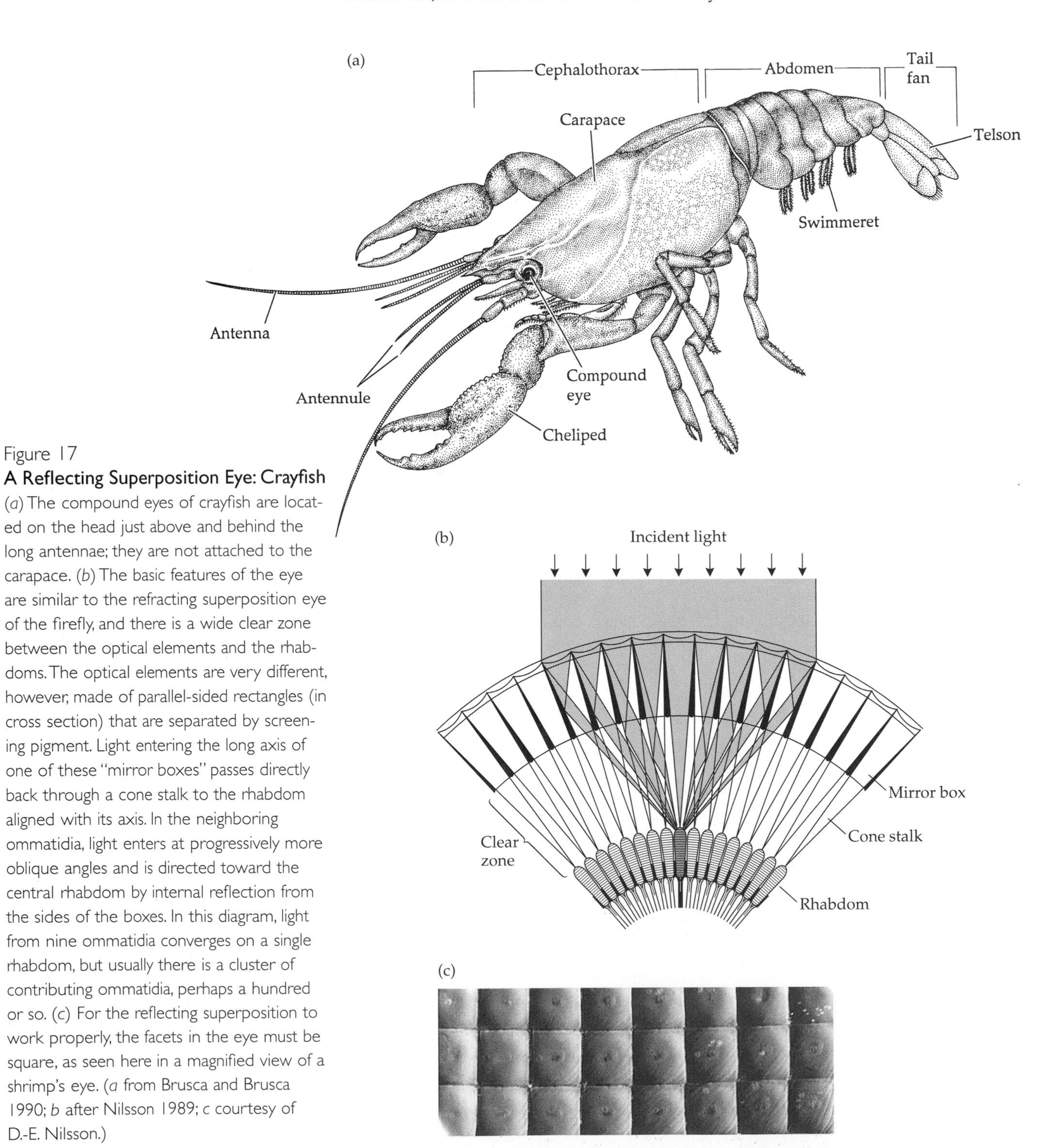

Figure 17
A Reflecting Superposition Eye: Crayfish
(*a*) The compound eyes of crayfish are located on the head just above and behind the long antennae; they are not attached to the carapace. (*b*) The basic features of the eye are similar to the refracting superposition eye of the firefly, and there is a wide clear zone between the optical elements and the rhabdoms. The optical elements are very different, however, made of parallel-sided rectangles (in cross section) that are separated by screening pigment. Light entering the long axis of one of these "mirror boxes" passes directly back through a cone stalk to the rhabdom aligned with its axis. In the neighboring ommatidia, light enters at progressively more oblique angles and is directed toward the central rhabdom by internal reflection from the sides of the boxes. In this diagram, light from nine ommatidia converges on a single rhabdom, but usually there is a cluster of contributing ommatidia, perhaps a hundred or so. (*c*) For the reflecting superposition to work properly, the facets in the eye must be square, as seen here in a magnified view of a shrimp's eye. (*a* from Brusca and Brusca 1990; *b* after Nilsson 1989; *c* courtesy of D.-E. Nilsson.)

The principal of reflection in a plane is extremely simple—the angle of reflected light rays equals the angle of incident light rays—and in one dimension it is easy enough to see how these eyes are working. As Figure 17*b* shows, the optical elements in the eye have nearly parallel sides that are reflective, either because of internal reflection (the refractive index of the element is higher than the index of the surroundings) or because the element has a reflective coating. A light ray entering the system parallel to the optic axis will pass through without reflection or refraction and will strike the rhabdom directly behind the optical element. When the ray is oblique, however, it will encounter a reflecting surface before it can leave the element and therefore be redirected toward a neighboring rhabdom. As indicated, the greater the obliquity of incidence, the greater the amount of redirection to more distant rhabdoms. As a result, any particular rhabdom receives light directly from its own optical element and redirected light from an array of neighboring elements. Thus, there is superposition in this particular plane.

Applying the simple law of reflection in a three-dimensional configuration is surprisingly tricky; for superposition by reflection to work in an array of optical elements, the elements must have a particular shape—namely, a square cross section. Only this geometry will permit superposition by a circular cluster of optical elements onto a single rhabdom. The facets in a crayfish eye are square (see Figure 17*c*). They are strikingly different from the hexagonal or roughly circular facets in other compound eyes.

These reflecting superposition eyes offer the same advantage as those that use refraction: greater sensitivity than that of nonsuperposition eyes. For crayfish dwelling in turbid water or for lobsters and shrimps living at dimly illuminated ocean depths, the increased sensitivity should be quite useful. Sensitivity is the more important consideration, and spatial resolution for the crayfish eye is not particularly good; the facets are about 50 μm square with an angular separation of about 2.3°.

Crustaceans have difficulty growing larger. Their exoskeleton is essentially dead tissue that is not added to once it is formed, except for the incorporation of minerals to harden it. If the animal is to grow, it must get rid of the old exoskeleton; then it can expand and create a new, larger exoskeleton. This process of molting is called *ecdysis,* from a Greek word meaning "to strip off." (Thus, a fancy word for a nightclub stripper is "ecdysiast.") Ecdysis is under hormonal control, and when the time comes—it occurs more frequently in young, rapidly growing crustaceans—the inner parts of the exoskeleton are dissolved to be incorporated into the new exoskeleton, a large split is created between the thorax and tail, and the crustacean wriggles out of the remnants of its old exoskeleton. The new exoskeleton is soft, and the animal must enlarge by taking on water, stretch the new exoskeleton to fit its larger size, and then remineralize the exoskeleton so that it becomes hard again. The animal is quite vulnerable to predators at this stage. (Crustaceans are not the only animals to do this: Ecdysis also occurs among insects; the exoskeletons of cicadas that have molted are commonly seen clinging to tree branches after the cicada has departed.)

Large crayfish have larger eyes than small crayfish; it would be interesting to know if the eyes increase in size only during the molt, and if the size increases as the result of the addition of ommatidia, an increase in the size of existing ommatidia, or a combination of both. For reflecting superposition eyes to work properly, the alignment of optical elements and rhabdoms must be extremely precise if each point in the world is to be imaged on a single rhabdom by hundreds of optical elements. Knowing how these eyes grow larger while maintaining their optical precision may provide some clues to their evolution—that is, to how the requisite optical alignments were established in the first place.

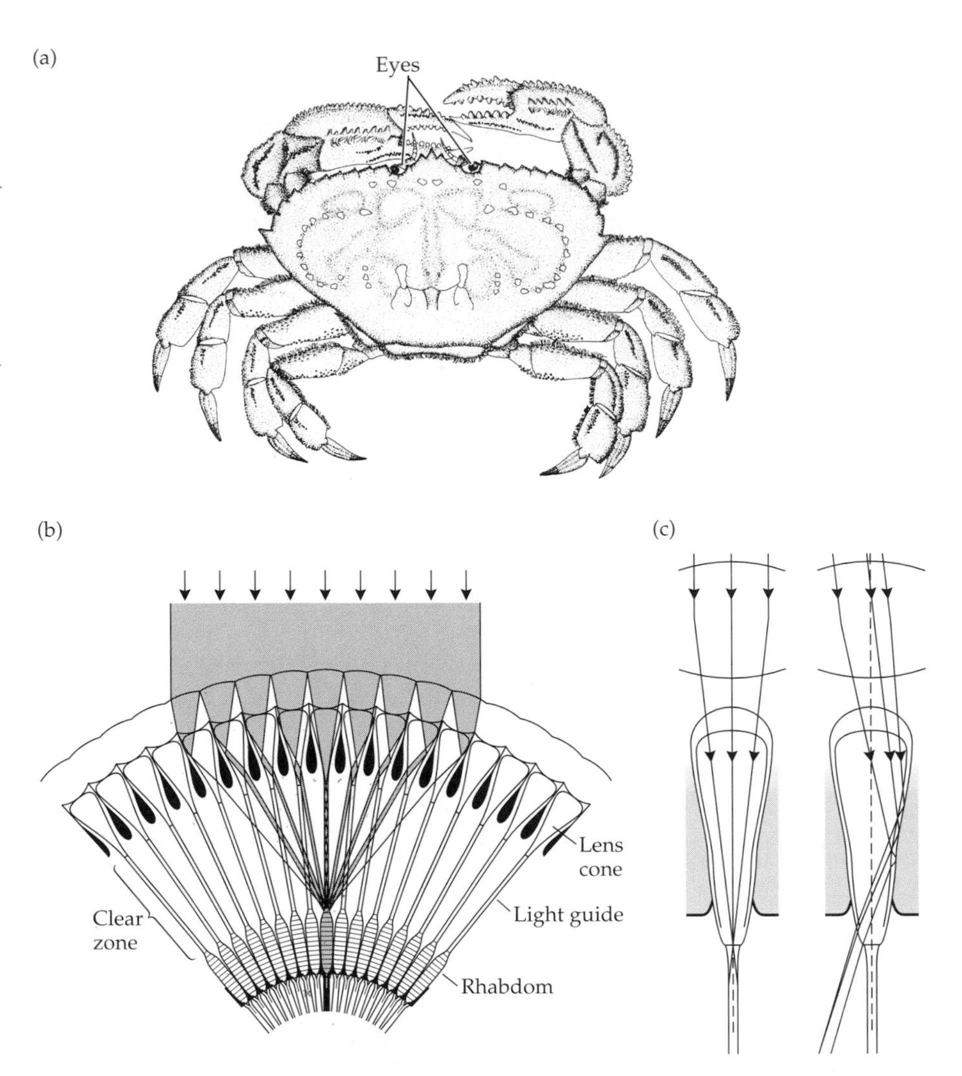

Figure 18
A Parabolic Superposition Eye: "True" Crabs
(*a*) Arrows identify the eyes on this crab. (*b*) A cross section of a crab eye is very similar to the refracting superposition eye of a firefly, except for the unusual shape of the lens cones. The cooperative redirecting of light by several optical elements to a single rhabdom is, in effect, like that in other superposition eyes. Light entering along the axis of an ommatidium is brought to a focus at the apex of the lens cone and channeled by the light guide directly to the underlying rhabdom. The light is focused by refraction in the optical system. Light entering off the axis undergoes refraction *and* internal reflection; the reflection redirects the light to a neighboring rhabdom. (*c*) The unusual shape of the lens cones leads to the characterization of this system as parabolic superposition. (*a* from Brusca and Brusca 1990; *b* and *c* after Nilsson 1989.)

VII. Compound eye—parabolic superposition: Crabs

PHYLUM: Arthropoda
SUBPHYLUM: Crustacea
CLASS: Malacostra
ORDER: Decapoda
INFRAORDER: Brachyura

The brachyurans are the "true crabs" with which most of us are familiar; thus one might expect them to have eyes like those of other decapods, such as shrimp and lobsters. In general, the eyes are similar: They are optical superposition eyes with the telltale clear zone between the optical elements and the photoreceptors (Figure 18). In some of the brachyuran crabs, however, the optical facets are neither square, as they are in reflecting superposition eyes, nor are the lenses of the gradient index type that would be necessary for an aquatic refracting superposition eye. Yet the eyes can be shown experimentally to perform optical superposition.

These eyes appear to combine elements of refraction and reflection in their ommatidia, and the name "parabolic" superposition comes from the distinctive shape of the lens cylinders in these eyes (see Figure 18*c*). Light entering along the ommatidial axis is focused by a corneal "lens" at the posterior tip of the lens cone or cylinder and conveyed by internal reflection in a light guide directly back to the corresponding rhabdom. Oblique ray bundles strike the sides of the lens cone, undergoing internal reflection and recollimation, as in an afocal system, and are redirected to neighboring rhabdoms. As in all superposition eyes, a particular rhabdom receives light from numerous optical elements.

This optical combination of refraction and reflection is a relatively new discovery, and the full range of animals to which it applies is probably yet to be determined. And there seems to be some variation in the relative contributions of refraction and reflection. For example, the xanthid or mud crabs, which include the delicious Florida stone crabs, have square facets in their eyes, suggesting that reflection may be a more important consideration than refraction. In others, the facets are circular or nearly so, and these may depend less on reflection. There is undoubtedly more to be discovered about this optical theme and the range of animals that use it.

In many crabs, the eyes—whatever the optical type—are mounted on movable stalks. Eye movements (or more accurately, eyestalk movements) can be elicited by appropriate visual stimuli, and by the use of progressively more detailed spatial patterns the eye movement responses can be taken as an indicator that the eye can resolve the detail to the point at which eye movements cease; that is, one can measure a crab's visual acuity. These eye movements keep the retinal image of the visual world stationary while the animal changes its posture or as it moves around. Most animals, including humans, have similar reflex eye movements, but because in crabs fewer neurons are involved and the reflexes are not overlaid with other eye movement patterns, crabs give us an opportunity to understand these behavioral patterns at their simplest.

VIII. Simple eye—pinhole: *Nautilus*

PHYLUM: Mollusca

CLASS: Cephalopoda

SUBCLASS: Nautiloidae

Pinhole eyes may have evolved numerous times, but this type of eye has been retained in its purest form, filled with seawater rather than with a secreted vitreous, only by the few species of *Nautilus*. Most people are probably unfamiliar with *Nautilus* as an animal, knowing it only by the beautiful spiral shell and perhaps an insipid poem by Tennyson that schoolchildren were once required to memorize. But the shell houses a fairly sizable animal (Figure 19) that is closely related to squid, octopus, and cuttlefish (these other cephalopods lack any external shell). *Nautilus* is not quite as predatory as its relatives, but it is a reef scavenger, feeding on small fish and crustaceans.

The *Nautilus* eye is almost half the size of the human eye—therefore about 1 cm in diameter—and its aperture varies from about 0.5 mm to perhaps 3 mm. The eye chamber is lined with photoreceptors and supporting cells that constitute the retina, but no cornea covers the eye, nor is there any lens or vitreous. By forgoing any refracting surfaces to focus or concentrate light and by allowing the eye chamber to be filled with seawater, the pinhole eye is restricted to poor resolution (the blur circle can be no smaller than the aperture; see Figure 19) and poor sensitivity (restricting the rays entering the eye makes the image very dim). The minimum angle of resolution is about 8°, a value that is bettered by many humbler creatures with either simple or compound eyes.

The puzzling thing about the *Nautilus* eye is not that it evolved—the transformation from a pit to a pinhole eye is simply a matter of making the pit deeper and the aperture smaller—but that it has been retained, for it is not a very good eye and it could easily be made better. Moreover, *Nautilus* has been around for a very long time, perhaps 500 million years or so, which is plenty of time to

(a)

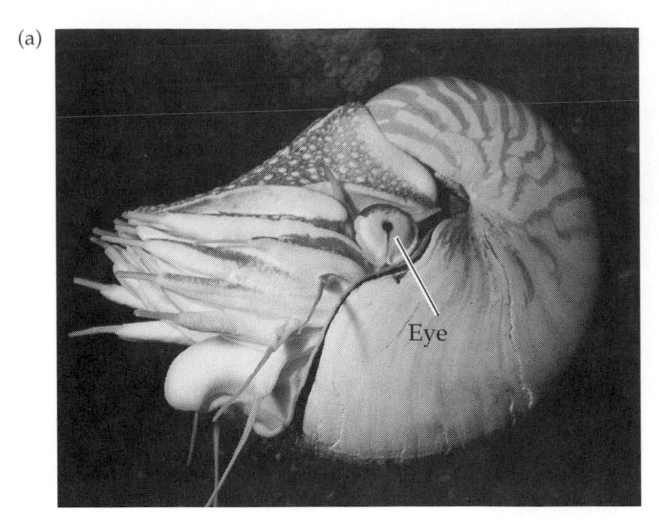

(b)

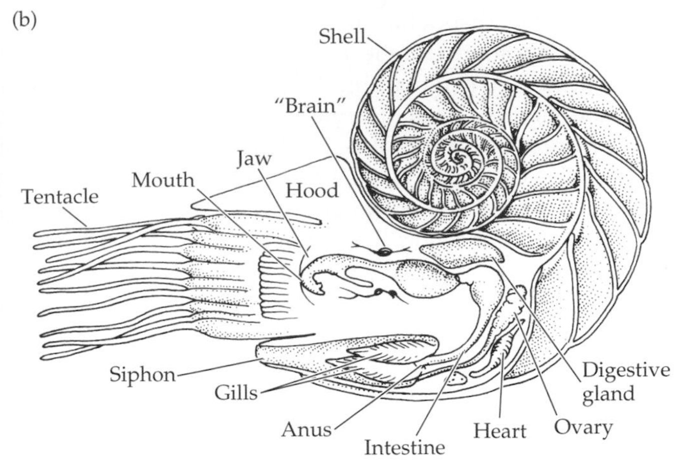

(c)

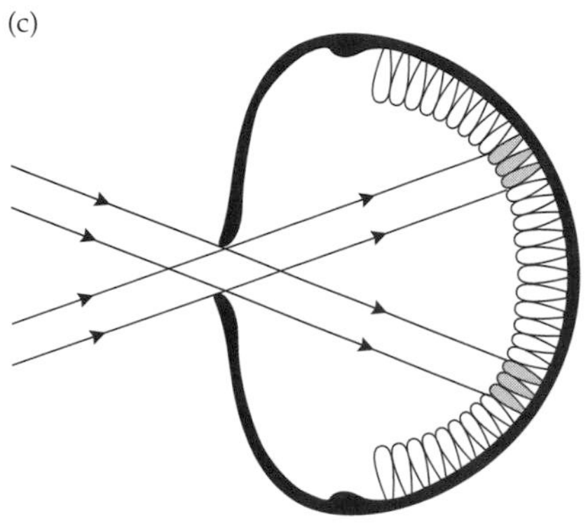

(d)

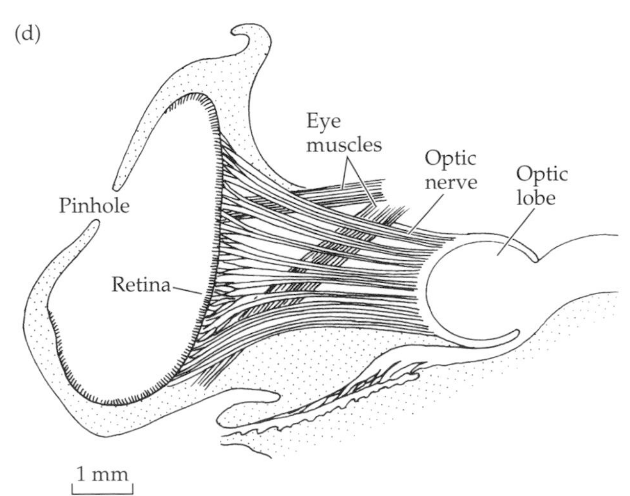

Figure 19
A Simple Pinhole Eye: *Nautilus*
(*a*) The aperture of the pinhole eye is the dark spot at the arrow as the animal looks out from its shell. (*b*) A sectional drawing of a *Nautilus* shows the animal living in the most recently built addition to its spiral shell. The empty chambers are pumped free of water to give the animal buoyancy, and the propulsive siphon is directed forward, so the animal swims backward. (*c*) The optics of a pinhole eye are extremely simple. The "image" on the retina is the same size as the aperture, so a well-defined "point" image requires a pinhole-sized aperture. The smaller the aperture, however, the dimmer the image will be. (*d*) A cross section through the *Nautilus* eye shows the pinhole—here, about 1 mm in diameter—and the retina lining the back of the chamber filled with seawater. Nerve fibers from the receptors form bundles that collectively constitute the optic nerve. Some muscles appear to be in a position to move the eye, but very little is known about their operation. (*a* by David J. Wrobel/Biological Photo Service; *b* after Hickman, Roberts, and Hickman 1988; *c* after Nilsson 1990; *d* after Young 1964.)

improve its eye. If eyes were automobiles, *Nautilus* would be chasing prey with a Model T Ford while its cephalopod relatives hunted in Ferraris. So if there is an embarrassing question about natural selection and the evolution of eyes, it is why in the world this clunky eye has not been improved for millions of years.

The question remains one of nature's little mysteries. The *Nautilus* eye may have evolved along a path leading to an optical cul-de-sac (literally) from which there was no alternative path to a better eye. Or perhaps a Model T eye suits the lifestyle of *Nautilus* and the only problem is our perception that nature should have done better. After all, *Nautilus* is a survivor, and a 500-million-year lineage is difficult to argue with.

IX. Simple eye—refracting, aquatic variety: *Octopus*

PHYLUM: Mollusca
CLASS: Cephalopoda
ORDER: Octopoda

Octopus is a predator specializing in crabs, lobsters, small fish, and bivalves. As bottom dwellers, octopi hide in rocks, crevices in reefs, and discarded shells, boots, or cans (for the small animals), and dart out to grab their prey with their arms. Crabs and other crustaceans are paralyzed by a secretion of the salivary gland and dismembered before being chewed by the rasping mouthparts and

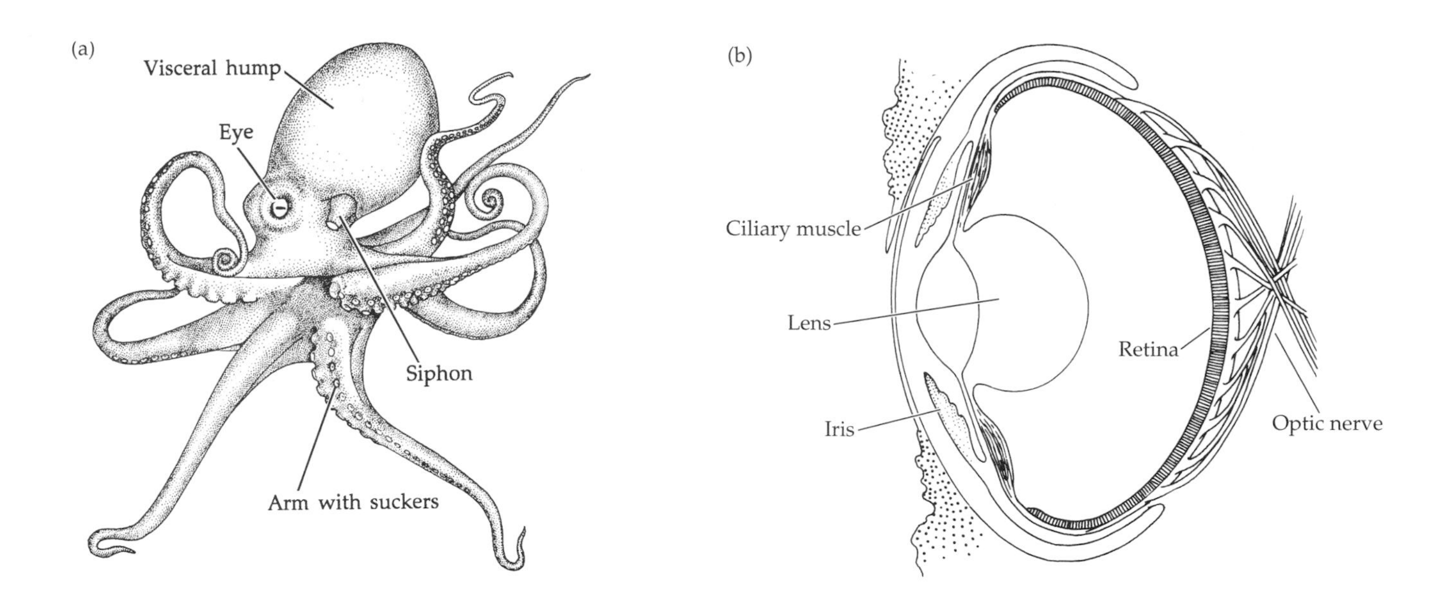

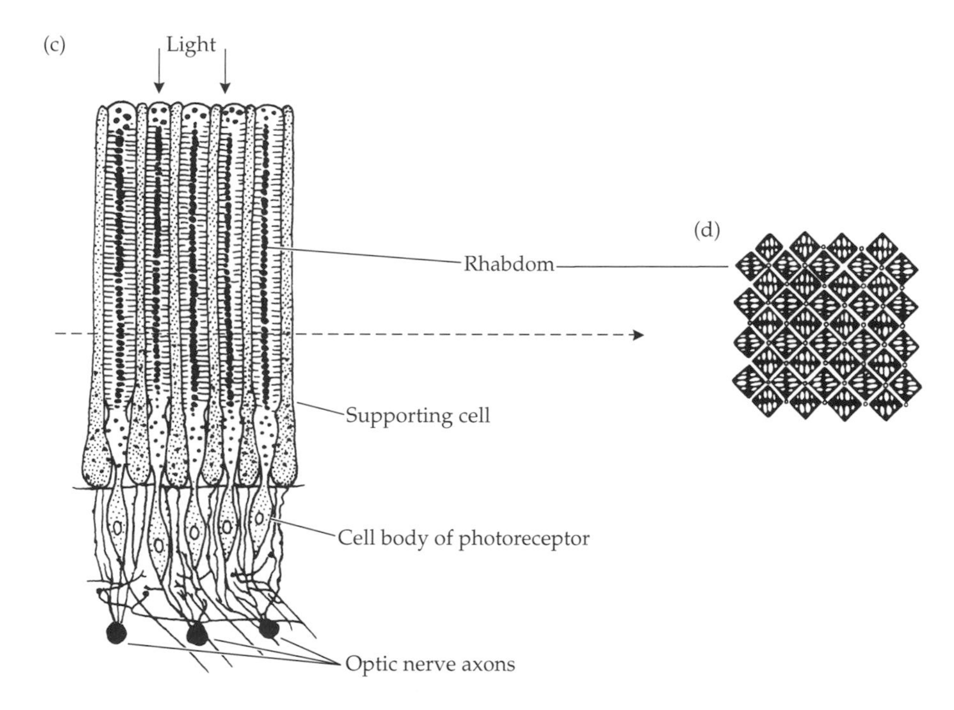

Figure 20
A Refracting Simple Eye—Aquatic Variety: *Octopus*
(*a*) The octopus eye is relatively large and has a horizontal slit pupil. (*b*) A section through the octopus eye shows its remarkable similarity to vertebrate simple eyes of the aquatic variety. The large spherical lens has associated muscle to move it back and forth, the iris contains muscle for changing the width of the pupil, and the retina with its numerous photoreceptors has bundles of nerve fibers that exit the back of the eye to form the optic nerve. The eye's extraocular muscles are not shown here. (*c*) A section through the retina shows long photoreceptors separated by pigmented supporting cells. The tips of the photoreceptors are next to the membrane surrounding the vitreous chamber of the eye. (*d*) A transverse section through the photoreceptors at the plane indicated shows that the photoreceptors fit tightly together in a square array. The bristles on adjacent rows of receptors are at right angles to one another. (*a* from Brusca and Brusca 1990; *b* after Wells 1966; *c* and *d* after Young 1964.)

swallowed. *Octopus* removes the soft parts of the bivalves from their shells and swallows them whole. *Octopus* can move very rapidly, propelling itself by a jet of water expelled from its siphon. The ability to move quickly is important not only for catching prey but also to avoid being a meal for a large fish or seal. Rapid movement and the need to recognize small creatures that may be well camouflaged add up to a need for good vision. The eyes of *Octopus* are the most highly developed of all invertebrate simple eyes.

The octopus eye is quite large (Figure 20). It has a nearly spherical lens, a large vitreous chamber, and a well-developed retina. The eye has musculature to move the lens and change the eye's focus, extraocular muscles to move the eye, and a rectangular pupil whose aperture can be varied in size. The lens is the gradient index type that affords high power with little or no spherical aberration, as in fish lenses (see the next section). In its gross morphology, the octopus eye, like those of squid and cuttlefish, is remarkably like the human eye, despite the otherwise enormous differences between humans and cephalopods.

The retina in the octopus eye looks nothing like the human retina, however; it has more in common with the retina of arthropods. The octopus retina is everted (see Figure 20*c*), with the photoreceptors at the inner surface pointing toward the incident light, and the photoreceptors are the "brush" variety, with photopigment-containing microvilli protruding from opposite sides of the cell. Alternating rows of photoreceptors are oriented at right angles to one another, creating a square packing of the photoreceptor cells (see Figure 20*d*) within which the rows of microvilli alternate between horizontal and vertical orientations. The photoreceptors are backed up by a layer of screening pigment and a complex of nerve cells that constitutes the rest of the retina. Long axons from the cells collect together in bundles that exit the back of the eye as an optic nerve (see Figure 20*b*).

The highly regular arrangement of the photoreceptor microvilli suggests polarization sensitivity of the sort discussed earlier for honeybees. This has been tested—*Octopus* is a good subject for behavioral experiments—and shown to be correct; octopi can easily detect the difference between horizontally and vertically polarized light. *Octopus* also exhibits excellent form discrimination, with the ability not only to discriminate complex shapes and different orientations of objects, but also to ignore size in making its judgments about similarity of shape. In terms of discriminating fine spatial detail, *Octopus* does nearly as well as humans do, in large part because of the small size and high packing density of its photoreceptors.

It is less clear if *Octopus* has color vision; the balance of contradictory results indicate that this capability is unlikely. Octopi do change color by regulating the size of chromatophores in the skin, but the colors are muted and the visually significant change is in the pattern of mottling of the skin. Some authors have linked these changes to the animals' emotional state. Such a relationship may or may not be true, but the idea that an octopus can indicate states that we would call happy, sad, or affectionate is appealing—I've always thought they are soulful creatures.

X. Simple eye—refracting, aquatic variety: Goldfish

PHYLUM: Chordata

SUBPHYLUM: Vertebrata

CLASS: Osteichthyes (bony fishes)

UPRAORDER: Teleostei

ORDER: Cypriniformes

The vast majority of living fish are teleosts (modern bony fish), whose origins are much more recent than other bony fish like sturgeons (Chondrostei), lungfish (Dipneusti), or coelacanths (Crossopterygii). Sharks and rays (cartilaginous fish) are more ancient still. Goldfish are teleosts, closely related to carp, and they are not always gold; the French refer to them as *poisson rouge* ("red fish"), which is the way Henri Matisse painted them.

All fish have refracting simple eyes in which almost all of the refraction is done by the lens. As seen in a section through a goldfish eye (Figure 21), the lens is spherical, which is the first indication of high dioptric power. What one cannot see, however, is the significant variation of refractive index, which is highest and fairly constant in the central lens core (nucleus) and declines outward to its lowest value at the lens circumference. This index gradient not only increases the lens power (relative to a uniform index lens), but also makes the lens work properly. A spherical, uniform index lens will not form a good image, because of spherical aberration, but the gradient index lens is almost aberration free (see

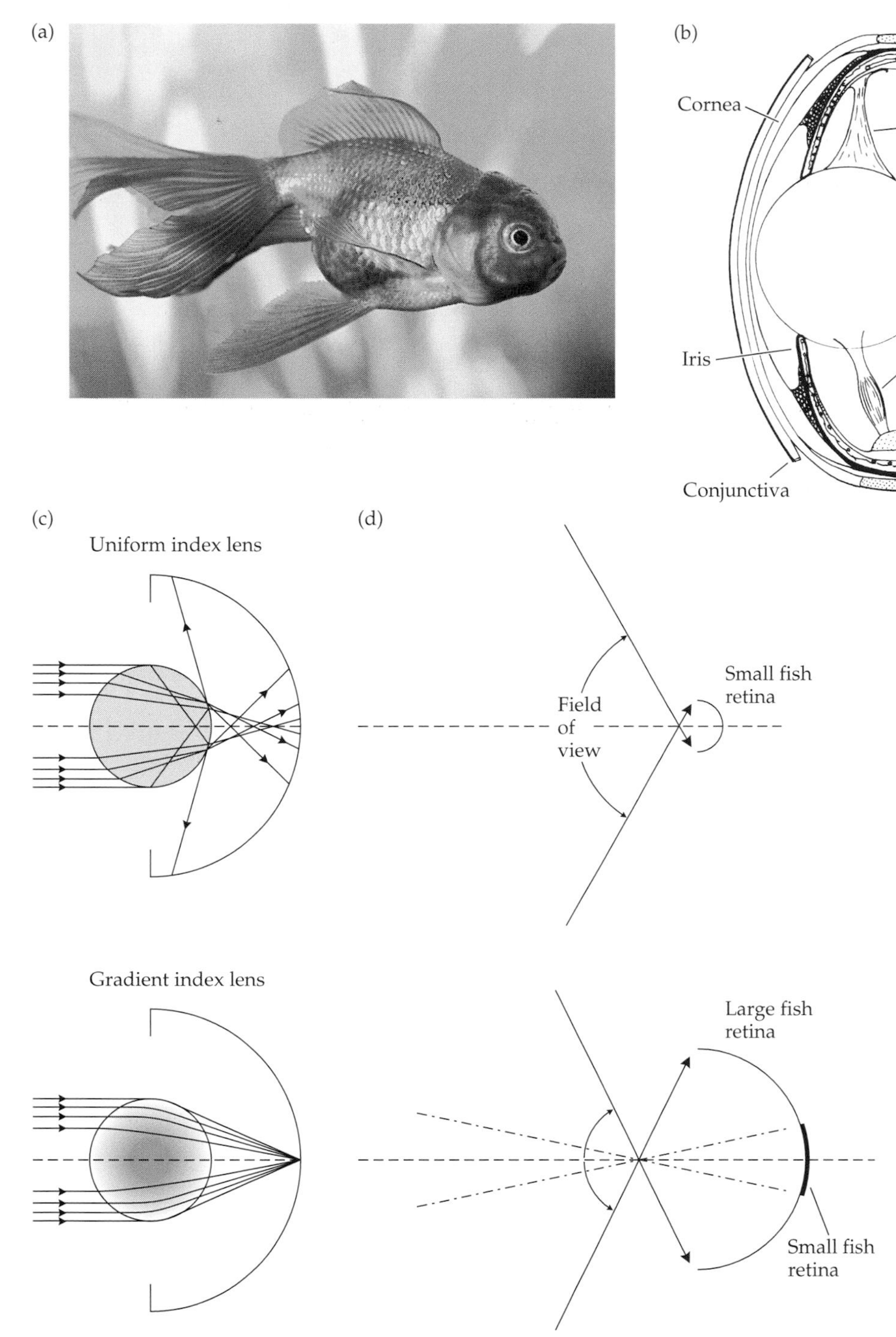

Figure 21

A Refracting Simple Eye—Aquatic Variety: Goldfish

(*a*) The goldfish eye is relatively large and has a circular pupil. (*b*) A section through a typical teleost eye, like that of goldfish, shows the spherical lens bulging through the pupil, a suspensory ligament and the retractor lentis muscle that allow the lens to be moved back and forth, and a fairly thick retina, from which nerve axons leave bundled together as a single optic nerve. The extraocular muscles are not shown. (*c*) A spherical lens with a uniform refractive index has an enormous amount of spherical aberration and is incapable of forming a decent image. A gradient index lens, of the type normally found in fish eyes, is free of aberrations, allowing both peripheral and central rays to focus in the same plane. (*d*) A small goldfish eye has about the same field of view as it does after the fish has grown considerably larger. But because the original small retina is now surrounded by retina added as the eye grew, it "sees" a much smaller part of the overall visual field. This implies that the wiring of the retina into the central nervous system is plastic, capable of changing as the animal grows. (*a* by Bill Kamin/Visuals Unlimited; *b* after Walls 1942.)

Figure 21*c*). The gradient index lens also has greater optical power than a uniform index lens with the same radius of curvature. The index gradient appears to be a direct consequence of the way the lens grows, which is from the inside out by the continual addition of new layers of cells around the existing lens. The concentration of proteins, and therefore the refractive index, gradually increases in the older, central part of the lens as new layers are added.

The human lens and those in other terrestrial vertebrates have similar index gradients, even though the surfaces are not spherical and the lens contributes less to focusing than the fish lens does. The persistence of the index gradient is probably a reminder that our distant ancestors were fish and that their neat solution to the aberration problem was too good to give up.

The spherical gradient index lens is one of several features that make the goldfish eye similar to the octopus eye. Both have ciliary muscles to move the lens and change the eye's focus, variable pupils, extraocular muscles, and well-developed retinas with high photoreceptor densities. The goldfish and octopus eyes are also comparable in size (for animals of comparable body weight). These points of similarity constitute a classic example of **convergent evolution**, wherein similar structures have evolved independently of one another. And there is no question about the independence of the lineages. Goldfish are vertebrates, octopi are molluscs, and the lines diverged 500 million or so years ago, long before vertebrates and their eyes appeared. The similarity of the eyes reflects the shared optical problems that had to be solved and the limited options available to a simple eye in solving them.

There are differences in goldfish and octopus eyes, however; the most striking is the organization of the retinas, which clearly reveals their independent origins. The goldfish retina is a typical vertebrate retina in which the photoreceptors (there are both rods and cones) line the posterior or outer surface of the retina, with their photopigment-containing parts pointing away from the incident light. But, as noted earlier, the octopus retina is constructed on the standard invertebrate plan; the photoreceptors form rhabdomlike arrangements, and they are on the inner surface of the retina, pointing toward the incident light. Thus the retinas are fundamentally different, but since the problem of forming a high-quality retinal image is the same, these very different animals hit on the same optical solution to the problem.

One of the interesting things about goldfish is their prodigious growth; they range from a few centimeters in length as juveniles to the huge 40- or 50-cm-long fish (*koi*) that swim in Japanese garden ponds. Given a chance, little goldfish with little eyes grow up to be large goldfish with large eyes. In mammals, eyes are relatively large at birth, and while there is subsequent enlargement, the increase in eye size is only 50% or so of the size at birth; human eyes at birth, for example, are about two-thirds their adult size. Although this enlargement must involve the addition of tissue to the immature eye, no new cells are added to the human retina after birth; it grows mainly by passive enlargement. In goldfish, the eye enlargement from birth to maturity is much greater, perhaps a tenfold increase, and the retina grows too, adding new neurons at the periphery of the retina throughout the life of the animal.

Vertebrate retinas contain five major classes of neurons, with numerous cell types or subclasses, and in any goldfish retina all these cell types are present at all stages of development, from birth to maturity. This opportunity to see all the stages of neuronal development at the same time has made fish retinas the objects of intense study. It is possible not only to study the development of characteristic neuronal geometries and mechanisms underlying their formation but also to intervene, altering the retinal environment in specified ways, and thereby find out which aspects of cellular development are predetermined and which are dependent on the neural environment in which development takes place. Both genetic and epigenetic factors are involved, but much of the final structure of retinal neurons is not specified in advance and depends on interactions among cells as they grow and mature. In general, epigenetic factors are more important in large nervous systems, like those of fish, than in small ones, like those of insects.

Goldfish have good vision, but the continual addition of neurons to the retina means that the visual system is being constantly reorganized. The problem, basically, is that a small eye sees much the same thing as does a large eye, in the sense that their fields of view are almost identical. The peripheral retina in a small eye sees objects at the edges of the visual field, but after the eye has enlarged, the original peripheral retina has been shifted toward the center as new retina is added at the periphery (see Figure 21*d*). Unlike the human retina, in which an image on a particular bit of retina is always associated with an object located in a specific direction, the directional specification by bits of goldfish retinas must constantly change, thus implying a constant, ongoing reorganization within the central nervous system. How this reorganization is accomplished is another story, but it is relevant to us because our visual systems show various kinds of plasticity early in development that become hardwired later. Knowing how the human brain came to be organized as it is and why it cannot repair its neurons after they have been damaged may depend on understanding how animals such as goldfish maintain their neural flexibility throughout life.

XI. Simple eye—refracting, terrestrial variety: Pigeon (*Columba livia*)

PHYLUM: Chordata

SUBPHYLUM: Vertebrata

CLASS: Aves

ORDER: Columbiformes

FAMILY: Columbidae

I have been a bird-watcher as long as I can remember—my mother claimed that my first spoken word was "birdie"—and birds have always defined for me a timeless sense of place. Venice, for example, is pigeons. No visitor to the Piazza San Marco can ignore the thousands of fearless, freeloading pigeons, underfoot, overhead, sometimes in your hair, filling that glorious space with bustling activity, blowing feathers, and pigeon droppings that are deposited with complete impartiality—birds are quite egalitarian in this respect.

There are around 250 species of pigeons and doves in the family Columbidae, nine of which are resident in North America, but the domestic pigeon, or rock dove, is by far the most conspicuous. Pigeons are highly social, with well-defined behaviors for regulating their social ordering and relationships; their courtship displays go on continuously, in and out of breeding season, perhaps contributing to the lifelong bonding of pairs. And pigeons are uninhibited, making them one of the easiest birds in which to study behavior; whatever they do, they will do in public, usually where you can sit with a cup of coffee and watch. Their navigational abilities are legendary, and they have been much studied by scientists trying to figure out what sort of compasses they use for finding direction. (The answer seems to be that they can use any information available: sun position, star positions, the earth's magnetic field, wind direction, local terrain—you name it.) Like most birds, much of pigeons' behavior is associated with excellent vision, vision that can be superior to that of humans.

A cross section of a pigeon's eye is shown in Figure 22*b*. It is a simple refracting eye constructed for acute, diurnal vision. Note that the eye is not spherical; although it is often described as flattened, as if the poles were pushed closer with a bulging at the equator, that description does not capture the most important feature. The posterior part of the eye, where the retina lies, is roughly a hemisphere, and its curvature is everywhere convex. The anterior part of the eye is quite different, however. Although the cornea is convex and highly curved, the rest of the anterior segment is concave, giving the eye as a whole a sort of bell-shaped appearance, a shape that is even more exaggerated in hawks and eagles. Because of the fluid pressure within, eyes should naturally have a roughly spherical shape; in birds' eyes, the concavity of the anterior segment is maintained by bony plates (ossicles) within the sclera that resist the outward pressure of the internal fluid. The overall result of this shape is a large eye with a small aperture and a wide field of view.

The structure projecting from the retina into the vitreous is the **pecten** (see Figure 22*b*, right), a highly vascularized, densely pigmented tissue found in all

(a)

(b)

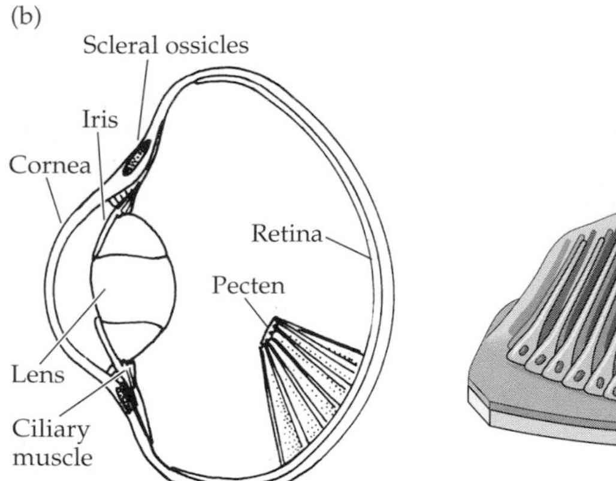

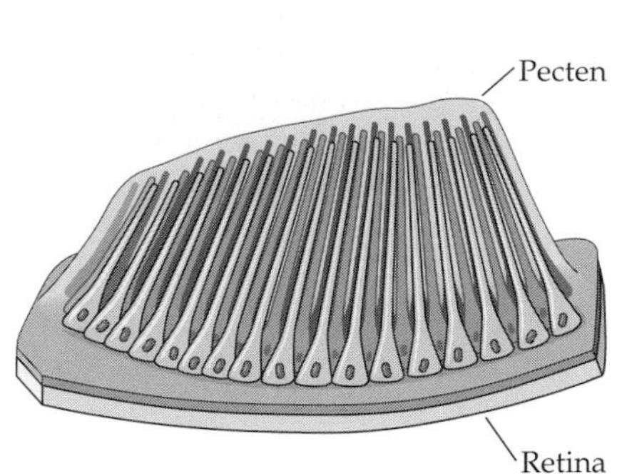

(c)

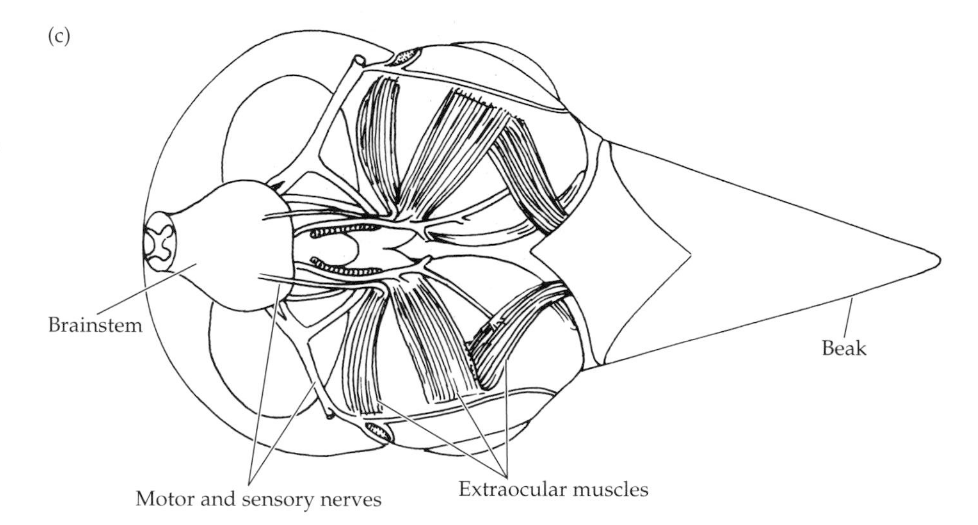

Figure 22
A Refracting Simple Eye—Terrestrial Variety: Pigeon
(*a*) The rock dove, or domestic pigeon, is the universal urban bird. (*b, left*) A section through the pigeon's eye shows its distinctly aspheric shape that is retained by the presence of bony ossicles (plates) in the anterior sclera. The relatively flat lens is surrounded by a heavy belt of muscle, the contraction and relaxation of which can produce very large changes in lens power. The pecten is seen here as a dark projection into the vitreous from the optic nerve head. A three-dimensional drawing of the pecten (*right*) shows it to be an elongated ridge of tissue with fluted sides; its dark color is due to heavy melanin pigmentation. The interior of the pecten is filled with blood vessels that are thought to be analogous to the central retinal artery and vein in mammalian eyes. In addition to having a nutritive function, the pecten is an intraocular sunshade. (*c*) When the eyes are shown in situ, the very large size of birds' eyes is obvious; they occupy most of the volume of the head, and the two eyes almost touch at the midline, with only a thin plate of bone (not illustrated) to separate them. Each eye, like that of humans, has six muscles that rotate the eye within the orbit. (This is a drawing of a sparrow, not a pigeon.)
(*a* by Len Rue, Jr./Visuals Unlimited; *b* and *c* after Walls 1942.)

bird eyes, most elaborately in diurnal birds. (A similar structure, the conus papillaris, is characteristic of diurnal lizards.) The function of the pecten has long been a matter of debate, and the only point of consensus is a nutritive function; unlike mammalian retinas, bird retinas are not extensively vascularized, and it is thought that the blood vessels in the pecten supply the inner retina by diffusion. Optical explanations for the pecten abound, but only one, which has gone through several cycles of acceptance and rejection, seems plausible and important.

When the pecten is mapped onto the pigeon's visual field, it lies above the horizon and above the area of highest visual acuity in the retina. Because the pecten is elongated, extending across a sizable portion of the retina, it is located *between* the area centralis and the image of the sun on the inferior retina. This position is important because the sun's image is very bright, so bright that it acts as a secondary light source within the eye from which light reflects to other parts of the retina. This scattered light interferes with the resolution of fine spatial detail by reducing image contrast. But the pecten is in a position to absorb stray light with its dense melanin pigment, thereby keeping the sun's image from scattering onto the part of the retina most concerned with acute vision. The pecten is oriented so that it does not obstruct the eye's field of view. In short, the pecten operates as an interocular sunshade, and its prominence in diurnal birds is obviously advantageous.

Birds' eyes are invariably large relative to the head. As Figure 22*c* shows, the eyes almost touch; they are separated by only a thin layer of bone. The visual axes of the eyes are directed laterally such that the angle between the axes is about 160°. (The comparable angle in an animal like a rabbit is almost 180°. Some birds have more frontally directed eyes; in hawks, the visual axes diverge by only 20° or so.) Thus the visual fields of the two eyes have only a small region in front where the fields overlap and where binocular vision is possible. The pigeon's total field of view, however, is very large, almost 360°; that is, pigeons can see behind and in front without moving either head or eyes. If you have ever seen a falcon strike at a pigeon in flight, the desirability of panoramic vision for the pigeon is obvious.

Eyes with high visual acuity often have a fovea in the retina, a place where the inner layers of the retina have been moved aside so that nothing lies between the photoreceptors and the incoming light. Foveas are especially well developed in bird retinas, and some species have two foveas in each retina, one looking out to the side and the other looking straight ahead. Pigeons have only a single side-looking fovea, but another retinal region with high cone density looks anteriorly and falls within the region where the visual fields of the two eyes overlap; this is the bird's "pecking field," so called because it sees the visual world just in front of the pigeon's beak.

As a general rule, the ability of birds' eyes to resolve fine spatial detail is as good and often better than that of human eyes. Birds also have better color discrimination, in the sense that they can discriminate color over a broader range of the visible spectrum. Both of these superiorities can be attributed to the photoreceptors. The receptors are predominantly cones (in diurnal species) that are very long and thin, allowing them to be packed together at high density. Where we may use 150,000 cones to sample a small part of the retinal image, a pigeon can employ twice as many. The higher the sampling density, the better the eye's visual acuity. The enhanced color discrimination comes from the use of four, rather than three, cone photopigments combined with color filters (oil droplets) in the cones.

The combination of a large eye and large retinal image with high-density photoreceptor sampling by wavelength-selective photoreceptors produces an eye superbly adapted for a visually dependent lifestyle. The price is a loss of flexibility; diurnal birds cannot operate as well as we can in dimly lit environments, which is one reason they disappear at sunset. Night belongs to owls, nighthawks, and whippoorwills, whose large eyes have special adaptations for seeing in dim light.

XII. Simple eye—refracting, terrestrial variety: Jumping spiders (*Metaphidippus* spp.)

PHYLUM: Arthropoda
CLASS: Chelicerata
SUBCLASS: Arachnida
ORDER: Araneae
FAMILY: Salticidae

The spiders with which most of us are familiar are the silk-spinning varieties that build webs of varying geometry in which to entangle their prey. Spiders typically have eight simple eyes, each directed to a different part of the total visual field. But most of these common spiders are not highly visual; many are nocturnal and their dominant sense is tactile. The spider usually lurks in the center or near the edge of the web, using vibrations transmitted through the strands of silk to detect prey caught in the web, to locate the prey, and to estimate its size. Male and female spiders often communicate from a distance by tapping signals on the web strands.

Figure 23 ▶
A Refracting Simple Eye—Terrestrial Variety: Jumping Spiders
(*a*) In this photograph, four of the spider's eight eyes are clearly visible along the front of its head. (*b*) A horizontal section through the spider's head shows all eight eyes in section, along with the fields of view (dotted lines) based on the sizes of the retinas (darkly shaded areas) in the different eyes. The large pair of tubular anteromedian eyes have the smallest fields of view, for which the spider compensates by moving the optical tubes and the retinas. (*c*) These drawings of the anteromedian eye positions show both eyes looking to the right as a version eye movement (top), looking approximately straight ahead in fixation (middle), and a divergence of the optic axes with the left eye looking more to the left than the right eye (bottom). These eye movements can be seen in juvenile spiders, which are almost transparent. (*d*) An enlarged view of the retina in the right anteromedian eye shows the photoreceptors with their long photopigment-containing processes radiating in toward the retina's "fovea." The tips of the photoreceptors form four distinct layers (labeled 1 through 4). This layering seems to be a compensation for chromatic aberration wherein red, green, and blue light focuses on layers 1, 2, and 3, respectively. (*a* courtesy of Michael F. Land; *b* through *d* after Land 1969.)

Some spiders are active, diurnal hunters, however, using vision to detect their prey and to guide their attack. Jumping spiders fall into this category (wolf spiders are another group of hunters), and their frontally directed eyes are quite impressive (Figure 23*a*). Relative to the size of the spider, the corneas are huge, almost a millimeter in diameter in some species, and this fact alone suggests that these eyes are meant for serious business; their spatial resolution is not likely to be limited by diffraction. Behavioral estimates indicate that jumping spiders can resolve spatial details of about 5′ of visual angle (human spatial resolution is about 0.5′ of arc).

Figure 23*b* is sectional drawing through the head of a jumping spider which shows all the eyes and their fields of view in the horizontal plane, with the exception of the small posteromedian eyes. The eyes of interest here are the principal eyes (anteromedian), which look directly in front of the animal. They are almost tubular, indicating a long focal length for the optical system, and their fields of view are quite restricted, as suggested by the small extent of retina at the back of the eye. At this level of detail, the design is reminiscent of the central visual area in the human eye—that is, the fovea and its associated optical path—which we use to examine different parts of our visual world at high resolution.

The jumping spider has a problem, however, because the eye's optical system is not movable; the cornea–lens system is rigidly attached to the exoskeleton. Thus, it would seem that the spider must move its body if its principal eyes are to be brought to bear on different objects of interest. But unlike the tubular or bell-shaped eyes of birds, which are limited in mobility by the surrounding bone, there is nothing to prevent the jumping spider from moving the retina and ocular tube to examine different parts of the visual field. And this is what the spider does, using small strips of muscle attached to the eyes; several different eye positions are shown in Figure 23*c*.

In moving its retinas, the jumping spider exhibits several movement patterns that are similar to human eye movements. The retinas may make large, rapid changes of position (saccades—see Chapter 4) of up to 25° or so in magnitude, in which the movements in the two eyes may be in the same direction (version eye movements) or in different directions (divergence or convergence). There are also torsional eye movements, in which the ocular tube twists around its anterior–posterior axis, and small horizontal scanning movements when the eyes examine a new object. The similarity extends to the eye muscles; as in vertebrate eyes, six muscles on each eye are used to generate the movements.

The retinas in the eyes of jumping spiders are elongated vertical strips of photoreceptors, about 20 μm wide by 200 μm high. The narrow horizontal dimen-

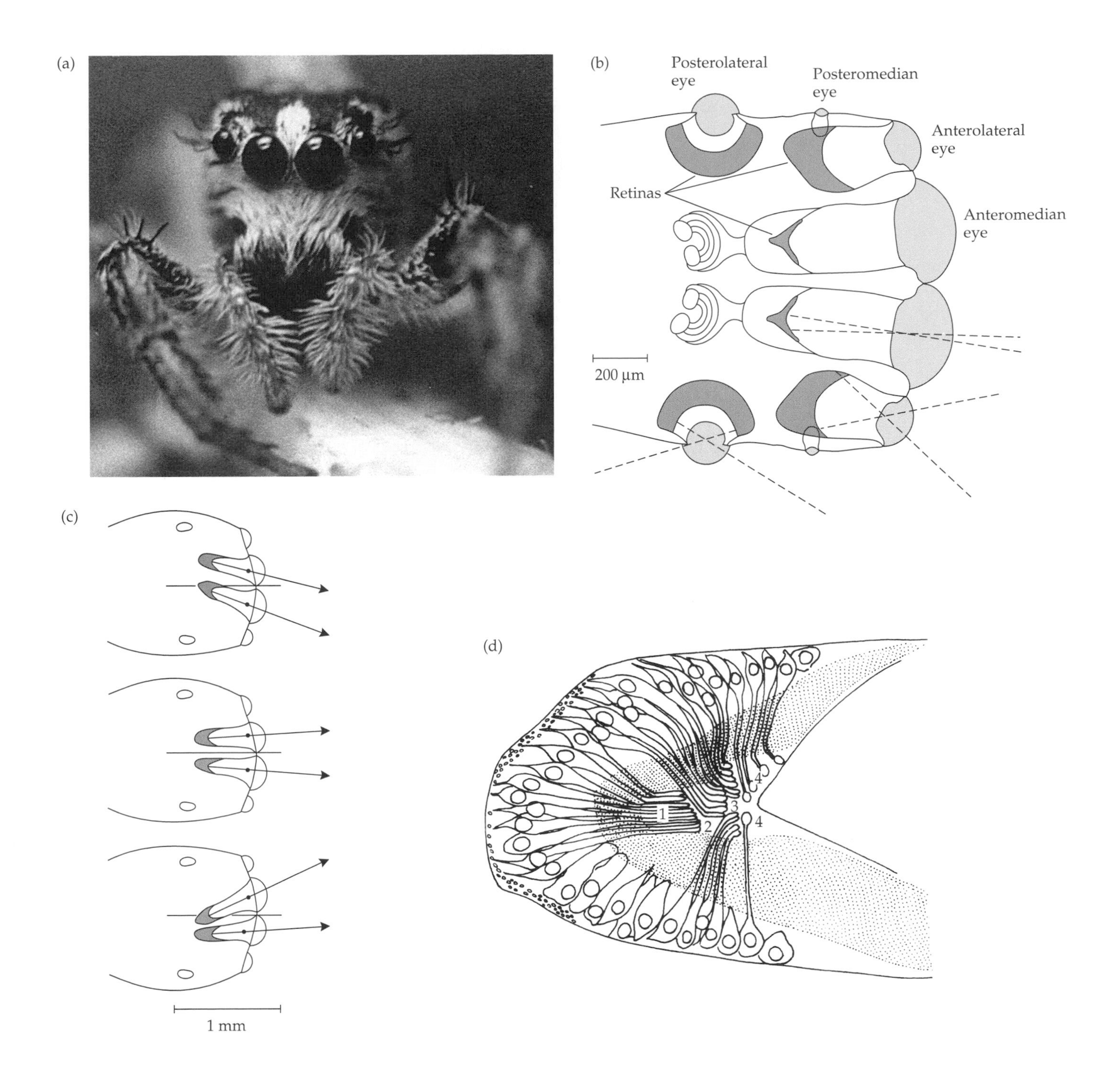

sion is the feature that makes the horizontal movements of the retina necessary. Perhaps the most curious feature, however, is the existence of four layers of photoreceptors in each retinal strip (see Figure 23*d*). The two deepest layers, those most posterior in the eye, are nearly congruent, and their photoreceptors cover the entire retinal strip. Photoreceptors in the two superficial layers, however, are more restricted in their distribution, occupying small regions near the center of the retinal strip.

This unusual retinal structure appears to compensate for the eye's longitudinal chromatic aberration (blue light undergoes more refraction than red light, and light of different wavelengths is focused in different planes). Different photoreceptors are placed in different layers according to their peak wavelength sensitivities, red more posterior than blue. (Human eyes also have chromatic aberration, but it is relatively small and usually not important. Humans are myopic to blue light, however; I remember a mildly scandalous nude scene in the 1960s musical *Hair* that was bathed in dim blue light and disappointingly out of focus.)

Large as jumping-spider eyes are, they are not the largest among spiders. This award goes to a group of nocturnal hunters in the genus *Dinopis* that ambush prey by dropping a netlike web on them from above. Their main eyes are the posteromedian eyes (the smallest eyes in jumping spiders), in which the corneal diameter can be around 1.5 mm. In these animals, however, the large eye size is an adaptation for light gathering and high sensitivity rather than for high spatial resolution. *Dinopis* eyes are about 2000 times more sensitive than the eyes of the diurnal jumping spider.

XIII. Simple eye—reflecting: Scallops

PHYLUM: Mollusca
CLASS: Bivalvia
ORDER: Mytioida
FAMILY: Pectinidae

Scallops are probably best known by one of their parts—the large edible muscle that links the two shell halves together in the living animal. The shape of the shell is distinctive, however, and in the guise of *coquille Saint-Jacques* ("the seashell of St. James"), the scallop shell was the emblem of pilgrims traveling to Santiago de Compostela in northern Spain, where James's headless body was interred. And in Greek mythology, Venus arose from a scallop shell, as depicted in Sandro Botticelli's *Birth of Venus.*

These creatures of legend and gastronomic delight live in shallow to deep water, depending on the species, where they embed the hinged side of their shells in the sand. When the shells are open, small tentacles around the shell rims are used to catch small prey that drift into the shell opening. The small, glistening discs scattered among the tentacles in Figure 24*a* are eyes, a few of which are shown at higher magnification in Figure 24*b*. Each animal typically has 50 to 100 eyes. Like other bivalves, scallops are food for predators, particularly sea stars, and they will close their shells in response either to decreased illumination or to motion, both of which may indicate the presence of a predator. In a more vigorous escape reaction, scallops rhythmically clap their shells together, propelling themselves out of harm's way on bursts of expelled water.

The eyes of scallops are simple eyes with several interesting modifications. As shown in the cross-sectional view (see Figure 24*c*), each eye has a large lens of unusual shape; two retinas, one lying in front of the other; and a reflecting surface—a mirror, or **tapetum**—behind the posterior retina. Light refracted by the lens is reflected back by the tapetum to a focus on the anterior retina. In common with vertebrate retinas, but unlike most retinas in invertebrate simple eyes, the nerve fibers leave from the front surfaces of the retinas; that is, they are inverted retinas.

A more detailed view of the photoreceptors in the two retinas (see Figure 24*d*) shows the inversion in both retinas, but also shows that the receptors in the two retinas are structurally different. The photopigment-containing processes in the posterior retina form a fan-shaped array that is directed posteriorly, while those in the anterior retina are parallel and directed anteriorly. The two sets of photoreceptors also respond differently. Light on the anterior photoreceptors pro-

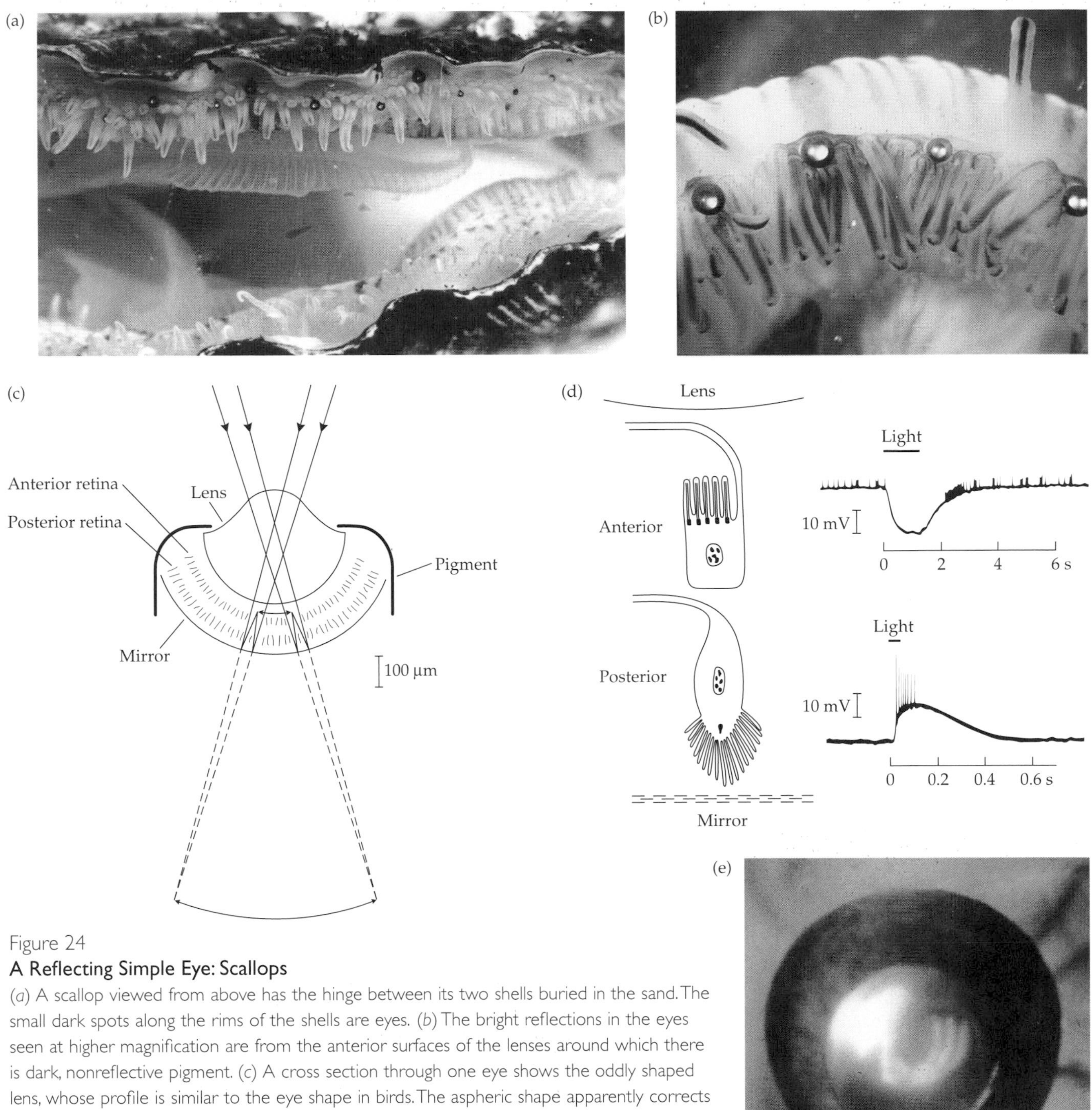

Figure 24

A Reflecting Simple Eye: Scallops

(*a*) A scallop viewed from above has the hinge between its two shells buried in the sand. The small dark spots along the rims of the shells are eyes. (*b*) The bright reflections in the eyes seen at higher magnification are from the anterior surfaces of the lenses around which there is dark, nonreflective pigment. (*c*) A cross section through one eye shows the oddly shaped lens, whose profile is similar to the eye shape in birds. The aspheric shape apparently corrects for aberration. Light passing through the lens must traverse two layers of retinal tissue before striking a mirror, from which it is reflected and focused. The plane of focus coincides with the distal (anteriormost) retinal layer. If the mirror were not present, the image would lie well behind the eye, as shown. (*d*) A schematic view of two photoreceptors in the different retinal layers shows the different structural features of the receptors; those in the distal retina are "brushes," while the proximal retinal receptors are "fans." The recording to the right of each cell shows its response to a brief flash of light; distal receptors hyperpolarize, like vertebrate photoreceptors, while proximal receptors depolarize, like most invertebrate photoreceptors. (*e*) The eyes do form images, as shown by this photograph of one eye in which an image of Michael Land's hand can be seen. (*a* and *b* courtesy of Michael F. Land; *c* after Land 1981; *d* and *e* after Land 1984.)

duces a more negative intracellular electric potential (hyperpolarization), a response that is similar to that of vertebrate photoreceptors, although the underlying ionic currents producing the responses are different (see Chapter 13). The posterior photoreceptors, however, become more positive (depolarize) when illuminated, and this is the typical pattern in invertebrate photoreceptors. Because receptor signaling is done by depolarization, the posterior receptors are signaling *increases* in light intensity (ON) while the anterior receptors are signaling *decreases* in intensity (OFF).

The scallop's behavioral sensitivity to small amounts of object movement requires a well-formed image. At high magnification, one can see the image formed in the eye, in this case the photographer's hand (Figure 24*e*). This image is in the plane of the anterior retina where the photoreceptors will respond to movement of parts of the image of only 2°. Posterior receptors will not resolve details this small and seem to be more sensitive to overall changes in illumination.

Most of the focusing of light in the scallop's eye is done by its mirror, which is the reason for calling it a reflecting simple eye. Although this is a respectable way to form an image—all the best optical telescopes use it—spherical mirrors produce considerable aberration. In telescopes we solve this problem by making the mirrors parabolic, but the scallop's mirror is spherical. The image produced by the scallop eye would be quite poor if it were not for the lens. The distinctly aspheric shape of the lens appears to compensate for aberrations that would be introduced by the reflecting mirror, so the lens–mirror combination produces a relatively aberration-free image.

Mirrorlike tapeta of the sort found in scallop eyes are very common, but they rarely contribute anything to image formation. In cats and dogs, for example, the tapeta increase the eyes' sensitivity under dim light conditions by reflecting light that has escaped absorption so that the photoreceptors have another shot at it. But while these eyes would not be optically impaired if the tapetum were absent, the scallop eye would not operate properly without its mirror.

The one known instance of a totally reflecting simple eye of any size is in a deep-sea crustacean, *Gigantocypris,* whose eyes are parabolic reflecting surfaces with the photoreceptors in the center of the paraboloid. Like an astronomical reflecting telescope operated in its prime focus configuration, these eyes have enormous light-gathering capacity, which is consistent with the animal's habitat at ocean depths where very little light penetrates.

XIV. Simple eye—refracting, terrestrial variety: Humans (*Homo sapiens*)

PHYLUM: Chordata

CLASS: Mammalia

INFRACLASS: Eutheria (placental mammals)

ORDER: Primates

FAMILY: Hominidae

Human eyes are inherited from primate ancestors, and the differences between human eyes and those of our closest relatives (chimpanzees and gorillas) are almost negligible. Indeed, there is not much difference between our eyes and those of more distantly related Old World monkeys, from which the human line departed about 30 million years ago. Thus when we read about our hominid ancestors, such as "Lucy" and other australopithecines, we can assume that she and her kin saw the world 4 or 5 million years ago much as we do now.

Primates have been around for about 60 million years and our family, the Hominidae, for perhaps 20 million years. The earliest primates were small creatures, probably insectivores, and probably nocturnal. They began to flourish after the dinosaurs went into decline. The path from the earliest primates to humans is not known, or at least not agreed upon, but once primates were no longer restricted by the presence of dinosaurs, they appear to have expanded rapidly into new territories and ecologic niches. One expansion made them diurnal and arboreal, and what was probably a nocturnal or crepuscular (twilight) eye would have had to develop specializations for good spatial acuity under bright light conditions. (We have here a version of the old "chicken or egg" problem—which came first? Did an arboreal, diurnal lifestyle drive the development of high-acuity eyes, or did the ocular developments permit the move into the trees and the sunshine? We don't know, of course, but it is likely that the shift in lifestyle and the changes in the eye went hand in hand, with adaptations in one enhancing the value of adaptations in the other.)

Somewhere along the way, primates acquired the capacity for color discrimination (if the earliest primates were indeed nocturnal, their retinas likely were rod dominated, and any cones may have had only one photopigment). An interesting case has been made for the coevolution of trichromatic color vision in primates, and colored fruit. Many primate species are still largely fruit eaters, and the basic idea is similar to the coevolution of flowers and bees. Primates aren't in the pollination business, but they can spread a tree's seeds by spitting out seeds or fruit pits as they eat the fruit or by swallowing the seeds and distributing them by defecation. Colored fruit attracts the animals if they have color vision, and the mutual advantages to tree and primate encourage one to develop color and the other to develop the ability to see color.

Humans come from a line of primates that descended from the trees, developed a bipedal form of locomotion, and turned their manipulative and visual skills (and larger brains) to the production of tools and weapons. And as the cave paintings in France and Spain will testify, *Homo sapiens* used vision, almost from the beginning of the species, to create visual art.

Our eyes are versatile simple eyes of the terrestrial kind. The cornea is the main refractive component, providing about two-thirds of the eye's total optical power (most of the rest is contributed by the lens). We do not have quite the visual acuity of diurnal birds nor their range of color discrimination, but our eyes are better suited to a wide range of lighting conditions; where birds are specialists, we are generalists. Our fish ancestry shows in our gradient index lens, as mentioned earlier, but we use a totally different system for using the lens to change the eye's focus. Fish deal with changes in object distance by moving the lens forward and backward along the eye's optic axis, as do some terrestrial mammals, while our lens remains stationary, changing optical power by changing shape (curvature).

The versatility of the human eye is well suited to an animal who survives not by strength, or stamina, or adaptation to a restricted niche, but by wit and the power of wit to transcend the limitations of biology. Our combination of eyes and brain—our vision—has been one of the driving forces of our rapid cultural evolution.

References and Additional Reading

The Evolution of Eyes

Dawkins R. 1996. The forty-fold path to enlightenment. Chapter 5, pp. 138–197, in *Climbing Mount Improbable.* WW Norton, New York.

Goldsmith TH. 1990. Optimization, constraint, and history in the evolution of eyes. *Q. Rev. Biol.* 65: 281–322.

Halder G, Callearts P, and Gehring WJ. 1995. New perspectives on eye evolution. *Curr. Opin. Genet. Dev.* 5: 602–609.

Land MF and Fernald RD. 1992. The evolution of eyes. *Annu. Rev. Neurosci.* 15: 1–29.

Lewin R. 1996. *Patterns in Evolution: The New Molecular View.* Scientific American Library, New York.

Margulis L and Schwartz KV. 1988. *Five Kingdoms: An Illustrated Guide to the Phyla of Life on Earth,* 2nd Ed. WH Freeman, San Francisco.

Mollon JD. 1989. "Tho' she kneel'd in that place where they grew. . .": The uses and origins of primate colour vision. *J. Exp. Biol.* 146: 21–38.

Morris SC. 1998. *The Crucible of Creation: The Burgess Shale and the Rise of Animals.* Oxford University Press, Oxford.

Nilsson D-E and Pelger S. 1994. A pessimistic estimate of the time required for an eye to evolve. *Proc. R. Soc. Lond. B* 256: 53–58.

Salvini-Plawen L von and Mayr E. 1977. On the evolution of photoreceptors and eyes. *Evol. Biol.* 10: 207–263. (The notion that eyes evolved between 40 and 65 times comes from this important paper. The language is meant for specialists, and it is difficult reading.)

Zuker CS. 1994. On the evolution of eyes: Would you like it simple or compound? *Science* 265: 742–743.

The Diversity of Eyes

Ali MA. 1984. Prologue. Pp. 1–7 in *Photoreception and Vision in Invertebrates,* Ali MA, ed. Plenum Press, New York.

Barlow HB. 1952. The size of ommatidia in apposition eyes. *J. Exp. Biol.* 29: 667–674.

Charman WN. 1991. The vertebrate dioptric apparatus. Chapter 5, pp. 82–117, in *Vision and Visual Dysfunction,* Vol. 2, *Evolution of the Eye and Visual System,* Cronly-Dillon JR and Gregory RL, eds. CRC Press, Boca Raton, FL.

Dusenbery DB. 1992. *Sensory Ecology.* WH Freeman and Company, New York.

Hughes A. 1977. The topography of vision in mammals of contrasting life style: Comparative optics and retinal organisation. Chapter 11, pp. 613–756, in *The Visual System in Vertebrates, Handbook of Sensory Physiology,* Vol. VII/5, Cresitelli F, ed. Springer-Verlag, Berlin.

Kirschfeld K. 1976. The resolution of lens and compound eyes. Pp. 354–370 in *Neural Principles in Vision,* Zettler F and Weiler R, eds. Springer-Verlag, Berlin.

Land MF. 1981. Optics and vision in invertebrates. Chapter 4, pp. 471–592, in *Vision in Invertebrates, Handbook of Sensory Physiology,* Vol. VII/6B, Autrum H, ed. Springer-Verlag, Berlin.

Land MF. 1989. Variations in the structure and design of compound eyes. Chapter 5, pp. 90–111, in *Facets of Vision,* Stavenga DG and Hardie RC, eds. Springer-Verlag, Berlin.

Land MF. 1991. Optics of the eyes of the animal kingdom. Chapter 6, pp. 118–135, in *Vision and Visual Dysfunction,* Vol. 2, *Evolution of the Eye and Visual System,* Cronley-Dillon JR and Gregory RL, eds. CRC Press, Boca Raton, FL.

Land MF. 1997. Visual acuity in insects. *Annu. Rev. Entomol.* 42: 147–177.

Nilsson D-E. 1989. Optics and evolution of the compound eye. Chapter 3, pp. 30–73 in *Facets of Vision,* Stavenga DG and Hardie RC, eds. Springer-Verlag, Berlin.

Nilsson D-E. 1990. From cornea to retinal image in invertebrate eyes. *Trends Neurosci.* 13: 55–64.

Walls GL. 1942. *The Vertebrate Eye and Its Adaptive Radiation.* Cranbrook Institute of Science, Bloomfield Hills, MI. Reprinted by Hafner Publishing, New York, 1963.

Eyes and Their Animals

Barlow HB and Ostwald TJ. 1972. Pecten of the pigeon's eye as an inter-ocular shade. *Nature* 236: 88–90.

Barlow RB, Chamberlain SC, and Levinson JZ. 1980. *Limulus* brain modulates the structure and function of the lateral eyes. *Science* 210: 1037–1039.

Barlow RB, Ireland LC, and Kass L. 1982. Vision has a role in *Limulus* mating behavior. *Nature* 296: 65–66.

Brusca RC and Brusca GJ. 1990. *Invertebrates.* Sinauer Associates, Sunderland, MA.

Case JF. 1984. Vision in mating behavior of fireflies. Chapter 9, pp. 195–222, in *Insect Communication,* Lewis T, ed. Academic Press, London.

Eisner T, Silberglied RE, Aneshansley D, Carrel JE, and Howland HC. 1969. Ultraviolet videoviewing: The television camera as insect eye. *Science* 166: 1172–1174.

Fernald RD. 1990. Teleost vision: Seeing while growing. *J. Exp. Zool.* Suppl. 5: 167–180.

Fernald RD and Wright SE. 1983. Maintenance of optical quality during crystalline lens growth. *Nature* 301: 618–620.

Gould JL and Gould CG. 1988. *The Honey Bee.* Scientific American Library, New York.

Herzog ED and Barlow RB. 1992. The *Limulus*-eye view of the world. *Vis. Neurosci.* 9: 571–580.

Hickman CP, Roberts LS, and Hickman FM. 1988. *Integrated Principles of Zoology,* 8th Ed. Times/Mirror/Mosby College Publishing, St. Louis.

Horridge GA. 1969. The eye of the firefly *Photuris. Proc. R. Soc. Lond. B* 171: 445–463.

Land MF. 1965. Image formation by a concave reflector in the eye of the scallop, *Pecten maximus*. *J. Physiol.* 179: 138–153.

Land MF. 1969. Structure of the retinae of the principal eyes of jumping spiders (Salticidae: Dendryphantinae) in relation to visual optics. *J. Exp. Biol.* 51: 443–470.

Land MF. 1978. Animal eyes with mirror optics. *Sci. Am.* 239 (6): 126–134.

Land MF. 1979. The optical mechanism of the eye of *Limulus*. *Nature* 280: 396–397.

Land MF. 1984. Molluscs. Pp. 699–725 in *Photoreception and Vision in Invertebrates*, Ali MA, ed. Plenum Press, New York.

Lloyd JE. 1981. Mimicry in the sexual signals of fireflies. *Sci. Am.* 222 (7): 138–145.

Meyer DB. 1977. The avian eye and its adaptations. Chapter 10, pp. 549–611, in *The Visual System in Vertebrates, Handbook of Sensory Physiology*, Vol. VII/5, Cresitelli F, ed. Springer-Verlag, Berlin.

Muntz WRA and Raj U. 1984. On the visual system of *Nautilus pompilius*. *J. Exp. Biol.* 109: 253–263.

Nilsson D-E, Land MF, and Howard J. 1984. Afocal apposition optics in butterfly eyes. *Nature* 312: 561–563.

Passaglia C, Dodge F, Herzog E, Jackson S, and Barlow R. 1997. Deciphering a neural code for vision. *Proc. Natl. Acad. Sci. USA* 94: 12649–12654.

Ratliff F. 1965. *Mach Bands: Quantitative Studies on Neural Networks in the Retina*. Holden-Day, San Francisco.

Ratliff F and Hartline HK. 1959. The response of *Limulus* optic nerve fibers to patterns of illumination on the receptor mosaic. *J. Gen. Physiol.* 42: 1241–1255.

Vogt K. 1980. Die Spiegeloptik des Flusskrebsauges. The optical system of the crayfish eye. *J. Comp. Physiol.* 135: 1–19.

Wells MJ. 1966. Cephalopod sense organs. Pp. 523–545 in *Physiology of Mollusca*, Vol. 2, Wilbur KM and Yonge CM, eds. Academic Press, New York.

Young JZ. 1964. *A Model of the Brain*. Clarendon, Oxford.

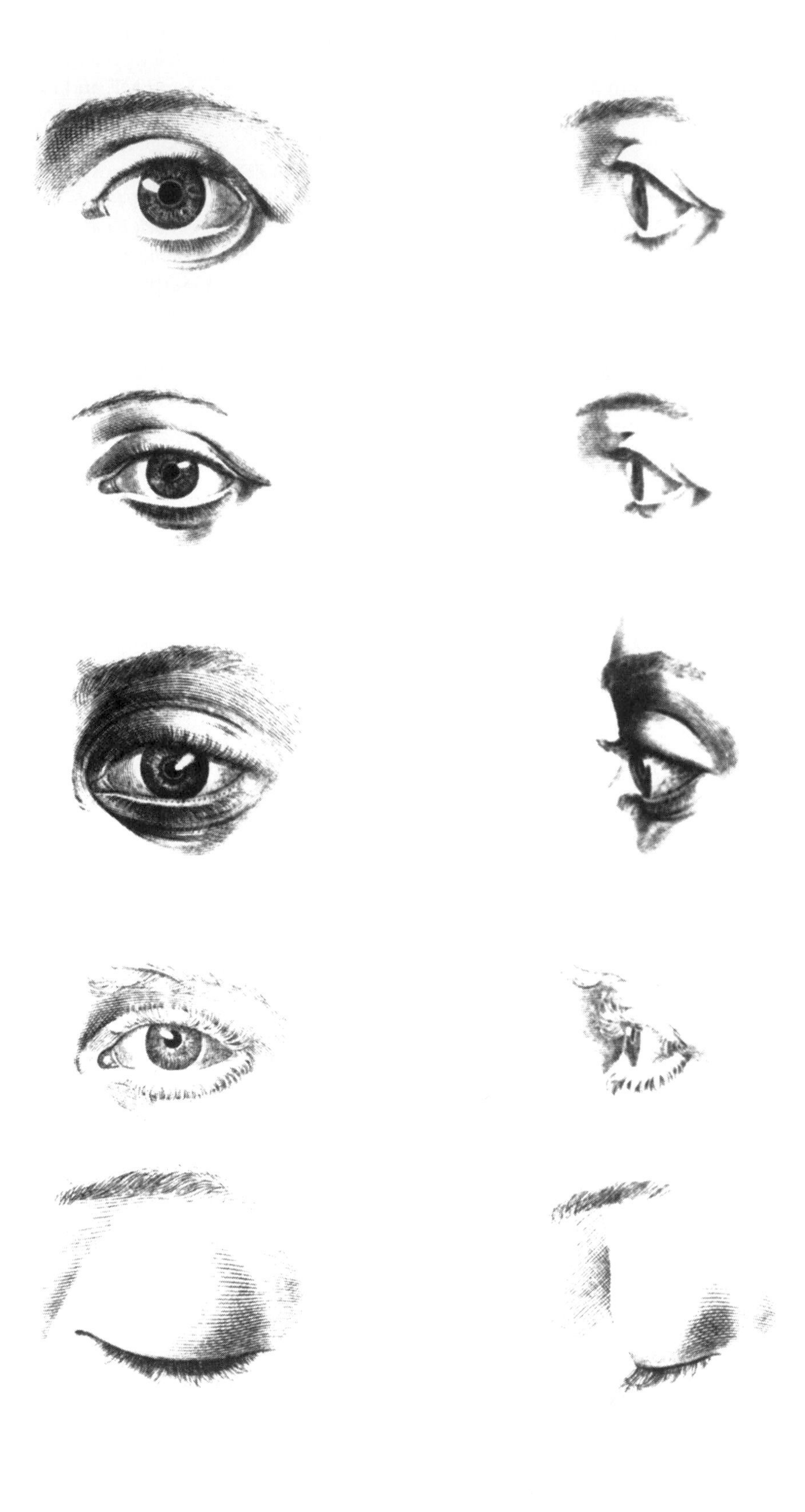

Part One

Ocular Systems

All roads lead to the eye…

■ Paul Klee

Chapter 1

Formation of the Human Eye

To begin personally on a confessional note, I was at one time, at my outset, a single cell . . . I do not remember this, but I know that I began dividing. I have probably never worked so hard, and never again with such skill and certainty. At a certain stage, very young, a matter of hours of youth, I sorted myself out and became a system of cells, each labeled for what it was to become—brain cells, limbs, liver, the lot—all of them signalling to each other, calculating their territories, laying me out . . . I didn't plan on this when it was going on, but my cells, with a better memory, did.

■ Lewis Thomas, *The Fragile Species*

Like the human body, the human eye has inauspicious beginnings; it appears a few weeks after fertilization of the ovum and formation of the zygote as almost insignificant changes in a small group of cells on the anterior end of the developing nervous system. In just over a month, these small beginnings produce a miniature eye in which the basic elements of the adult organ are easily recognizable. At term, the infant's eye is a small version of the adult eye (about two-thirds its final size); all that remains to be completed is the final development of some photoreceptors.

Every eye has a history, and because no two histories are ever the same, no two eyes are identical in all details. Most of the time, the differences between two eyes are small and insignificant, largely because all eyes develop and grow according to the same set of rules and are subject to many of the same influences; this is the way development is supposed to work. But the genetic instructions that guide development cannot anticipate every possible contingency. Developmental variations are inevitable, and they are often recorded as structural variations in mature eyes. Some of the variation is normal—that is, expected—and some is outside the normal range and therefore abnormal. Recognizing what is abnormal and understanding how abnormalities arise require knowing something about the expected course of ocular development.

Embryogenesis is a complicated business, involving myriad events whose timing and sequence are crucial to the successful development of the embryo. Many of the critical features of embryogenesis, particularly the mechanisms responsible for the occurrence of particular events at specific times, are not well understood and are the subjects of intense investigation. It is clear, however, that proper development of the eye depends to some extent on events prior to the eye's first appearance. Thus the story begins when each of us was a single cell.

Some Developmental Strategies and Operations

Embryogenesis begins with cell proliferation, cell movement, and changes in cell shape

In operational terms, cells have a limited repertoire of possible transformations. Perhaps the most basic of these is **proliferation**, an increase in cell number by mitosis and division. Many of the changes during the early stages of embryogenesis are proliferative. The most obvious consequence of cell proliferation is an increase in tissue volume, but because proliferation may vary either by rate, waxing and waning over time, or by spatial location, affecting only specific populations of cells, a uniform expansion of the embryo's size is not the only possible result. Different parts of the embryo can expand at different rates and at different times, making the possible consequences of this simple increase in cell number both varied and profound.

Cells can also change their shape and move around; since cell movement normally involves changes in shape, these changes are often lumped together as cell **motility**. Like cell proliferation, cell motility can vary with time and with spatial location in the embryo, again opening up a vast range of possible outcomes. In general, however, the movements of cells in a particular population affect the spatial organization of the embryo, while changes in cell shape, without movement, are related to the configuration of specific parts of the embryo.

Proliferation and motility account for much of the structural change in the early development of the embryo and of the eye, as we will see.

Specialized tissues are formed by collections of cells that have become specialized themselves

Individual cells contain the entire genome of the organism, and therefore have the potential to become any one of the roughly 200 classes of cells in the body. At some time and place, however, embryonic cells make a commitment to a particular developmental path; this **determination** means that the cell's future options have become limited. It may mean, for example, that the cell is destined to become a neuron, and regardless of its intrinsic genetic potential it will never become a fibroblast.

The next step is **differentiation**, which makes the commitment explicit; differentiation means that the cell begins to manufacture the proteins and intracellular organelles necessary for the lifestyle to which it has been committed. Also, the cell usually acquires a characteristic shape and structure that provide observable evidence of differentiation. Differentiation may follow determination immediately, but there can be a considerable delay; a cell may not differentiate for weeks or months after determination. Determination and differentiation are usually accompanied by proliferation and changes in shape of the differentiated cells, and these processes ultimately underlie the characteristics of the different tissues in the body and the eye. The specific triggers for determination and differentiation at particular times and locations are generally unknown, but there is considerable evidence that interactions among cells are necessary for these processes to occur.

Proliferation, movement, and differentiation in a cell group may require communication with other cells

Statements that particular cells proliferate, move, or differentiate at a specific time are statements about *what* happens; there are still underlying questions about *why* or *how* these changes occur. One general explanation for a change at a particular time and place is that the event is built into the genetic instructions that guide embryogenesis. This may well be the case for small organisms having several thousands of cells, for which hundreds of thousands of instructions may be enough to completely specify the outcome, but it is unlikely to work when millions or billions of cells are involved. A set of comparably detailed instructions would be astronomically large. Thus changes at particular times and places depend in part on the occurrence of previous events—changes are contingent—and on communication among developing cells. The communication process is called **induction**.

We know induction occurs in embryogenesis because of experimental manipulations in which developing cell groups are removed or transplanted at specific stages of development. In some species, removing the primitive eye prevents development of the lens, for example, while transplanting the primitive eye to another location in the embryo elicits lens formation in an abnormal location. Signals from the first part of the eye to form are controlling some of the subsequent developmental events.

Some forms of cellular communication, like synaptic interactions among neurons, are well understood, but the interactions responsible for induction are less clear. There are several possible modes of communication, but induction appears to be largely chemical in nature. The signals need to act on particular cells, implying that the communicating molecules are specific for the target cells, probably as a result of a lock-and-key relationship between the induction molecules and receptors on the target cells. This restriction also implies that the communicating cell groups are close together, since proximity will help keep the effect local and require only a small signal. Some of the molecules used for inductive signals are known, but most are not. Knowing what the various inducers are and how they are switched on and off (or how their receptors are switched on and off) will provide a much deeper understanding of embryogenesis. For the time being, however, we are limited to talking about induction as an operation whose mechanism remains to be clarified in detail.

Induction is an extremely important process. Because of it, the sequence and timing of developmental events can be regulated by the local environment, events can be synchronized without the use of a master clock, and more generally, development can be coordinated. What happens at a particular place and time is contingent on what has gone before. Induction plays a significant role in the development of the eye, and knowing that the phenomenon occurs will be of great value in our appreciation of the sequence of structural changes and our recognition of situations in which induction has occurred or failed.

Embryonic Events before the Eyes Appear

The blastocyst forms during the first week of embryogenesis

The first stages of embryogenesis are purely proliferative, transforming the fertilized egg—the zygote—from a single cell into a small cluster of seemingly identical cells. The human zygote is quite small, about 100 µm in diameter, and the initial process of creating more cells does not add volume; instead the proliferation is referred to as **cleavage**, and the zygote is divided into smaller cells. Each cleavage doubles the cell number, and by the time the cell cluster enters the uterus from the fallopian tube, it contains about 16 cells and is called the **morula**. The morula is enclosed by the glycoprotein shell—the zona pellucida—that surrounded the zygote from the outset and within which cleavage has occurred.

The morula divides several more times, and the resulting tightly packed cell cluster transforms within a day or so into a ball of cells with a hollow interior—the **blastocyst** (Figure 1.1). The internal cavity is the **blastocoel**. Between the morula and blastocyst stages the cells become compacted and tight anchoring junctions form between them, more cell divisions follow, and finally a major structural rearrangement takes place. By the time the morula contains 32 cells, it is a solid cluster with internal and external cells. The external cells have a sodium pump that moves Na^+ ions in toward the center of the cluster, which brings water in by osmotic pressure, causing the cluster to expand like a small blister; this process—called cavitation—forms the blastocoel. In cross section, as in Figure 1.1, the blastocyst has an exterior shell of cells surrounding the blastocoel, with a small dense cell cluster remaining on one side. The cell cluster is the **inner**

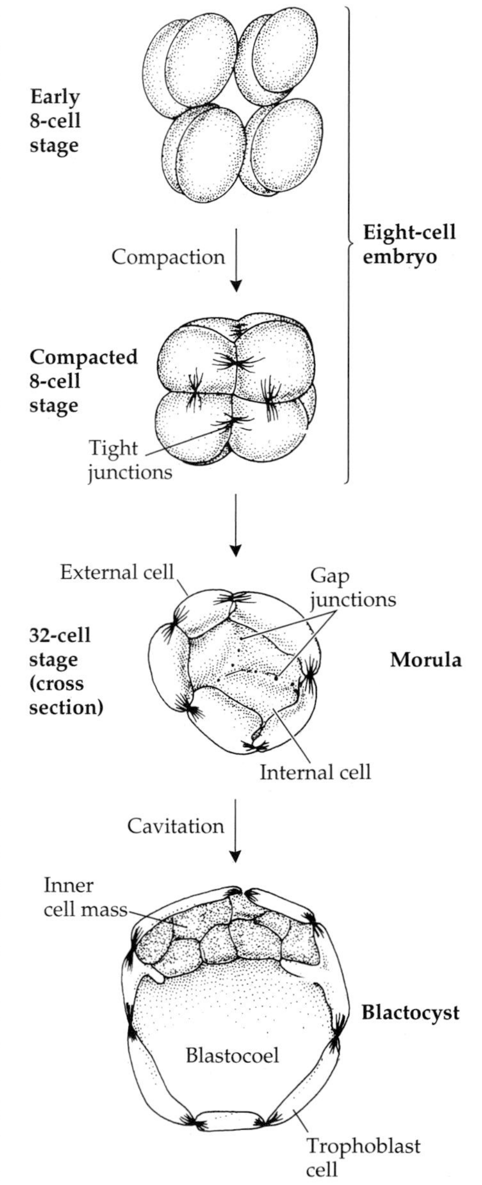

Figure 1.1
Formation of the Blastocyst
The arrows indicate a temporal path from the eight-cell embryo (after the first three cell divisions) to the blastocyst, a process that occupies the first week of embryogenesis.

cell mass and consists of the cells that were at the interior of the morula. The cells around the outside form the **trophoblast**.

The inner cell mass and the trophoblast have different fates and different roles in mammalian embryogenesis, and this division of the blastocyst into two parts is the first obvious step of differentiation. The inner cell mass will become the embryo per se, while the trophoblast cells will contribute to the **chorion**, the embryonic part of the placenta; the trophoblast cells are responsible for the proper implantation of the blastocyst into the uterine wall.

The second embryonic week is devoted largely to blastocyst implantation and the beginnings of some important structural changes in the inner cell mass. Implantation involves an interaction of the trophoblast with the tissue of the uterine wall (the endometrium); the trophoblast cells grow and invade the uterine tissue, thus anchoring the blastocyst to the uterus and beginning the initial stages of forming the placenta, which will bring the maternal and fetal circulatory systems into close contact and exchange (Figure 1.2). The development of the placenta continues for quite some time, and it is critically important if embryogenesis is to proceed normally; it is supporting structure, however—the embryo forms from the inner cell mass.

The inner cell mass becomes the gastrula, which is divided into different germinal tissues

As the trophoblast cells are invading the endometrium to implant the blastocyst, the cells of the inner cell mass proliferate and reorganize to produce a geometrically distinct structure, the **gastrula**. As cells in the inner mass proliferate, they begin to form a flat plate of tissue that separates from the adjacent trophoblast, thereby creating a small cavity. As growth and expansion continue, the cavity will become the amnion surrounding the embryo.

There is another step of differentiation at this point. The small plate formed by the inner cell mass has two layers of cells, the **epiblast** and the **hypoblast** (see Figure 1.2*c*). The cells of the hypoblast migrate along the inside of the blastocoel to become the lining for the yolk sac represented by the expansion of the blastocoel. As a result, the hypoblast cells, like those of the trophoblast, are extraembryonic and do not contribute directly to the embryo. The embryo forms from the cells of the epiblast.

The epiblast continues to expand and flatten, but as it does so, it develops a line—the **primitive streak**—that defines its long axis, running from one edge in toward the center of the plate of epiblastic cells. As proliferation continues, cells along the primitive streak migrate down and then laterally, eventually making a three-layered disc or plate of cells (Figure 1.3).

To this point, gastrulation has included much continued proliferation, movement, and changes in cell shape. The embryo now has polarity; the primitive streak is the axis of bilateral symmetry and it is now possible to distinguish the caudal (tail) end from the rostral (head) end of the embryo. The rostral–caudal axis is an important gradient along which different events take place depending on their relative position along the axis. In addition to these processes, cellular determination has divided the embryo into different tissues.

The three layers of cells in the gastrula are the primary **germ tissues**. The cell migrations creating the layers have been accompanied by changes in the nature of the cells, and from this time onward, they and their progeny will proceed along different developmental paths. The upper layer of cells closest to the amnion is now called the **ectoderm**, which will give rise to the nervous system, the epidermis and other epithelial layers, and the major parts of the eye. The middle layer of cells, the **mesoderm**, will become blood vessels, muscle, part of the urogenital system, bone, and connective tissue; the orbital bones, the extraocular

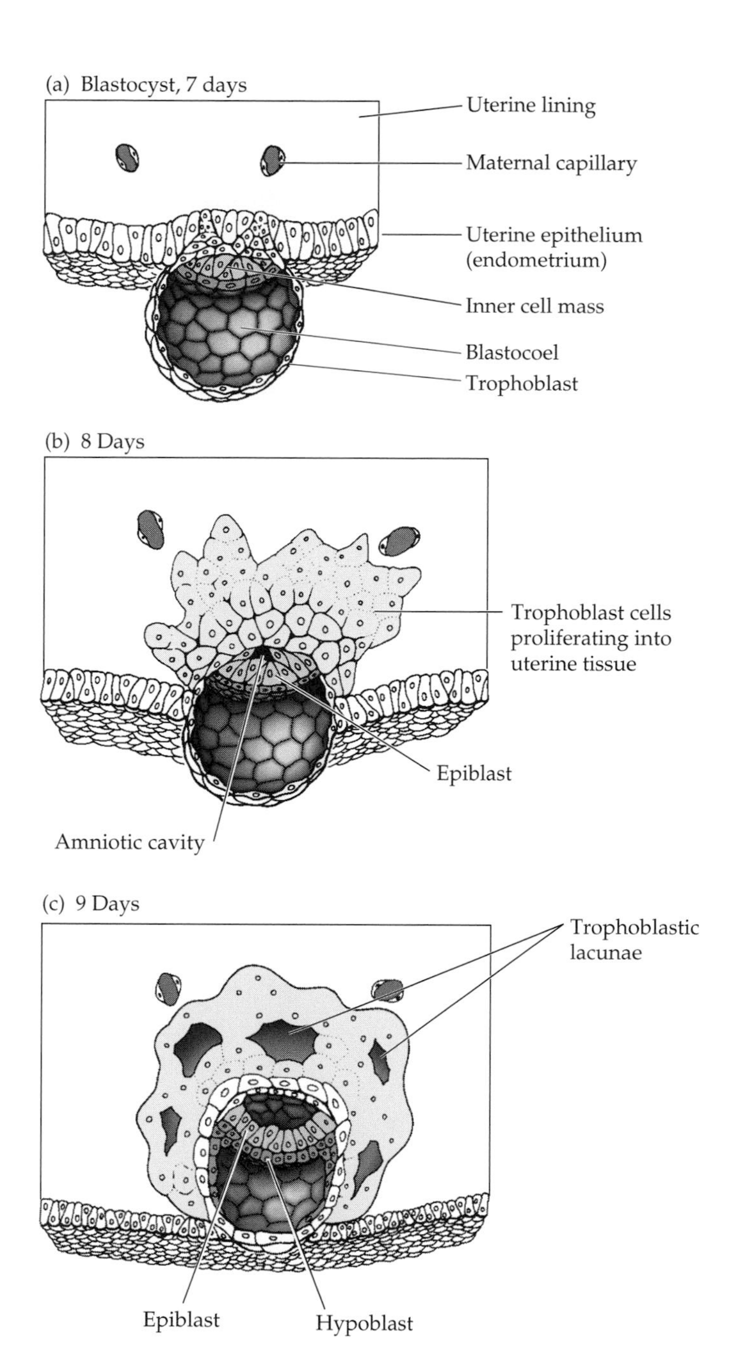

Figure 1.2
Implantation of the Blastocyst
Trophoblast cells adjacent to the inner cell mass proliferate and invade the epithelium (endometrium) lining the uterus. As this process continues during the second week of embryogenesis, the blastocyst becomes embeddeded in the uterine wall, uterine blood vessels move into the area, and the inner cell mass of the blastocyst begins the process of gastrulation by differentiation and cell migration. (From Gilbert 1997.)

muscles, and the blood vessels associated with the eye are mesodermal derivatives. Finally, the inner layer of cells next to the blastocoel and developing yolk sac is **endoderm**, from which the linings of the gut, the respiratory system, and related organs will form; it plays no direct role in formation of the eye and orbit.

Although the ectodermal cells appear to be alike at this stage, they are already destined for different structures. It may not be possible to specify precisely which cells will end up where, but it is clear that there are separate groups of cells on either side of the midline that will contribute to the primitive eye, specifically to the retina, from one group, and to the lens, from another. As embryogenesis proceeds, these groups of cells will become widely separated for a time and then come back together, at which point their interaction will trigger the development of the lens. It appears, however, that the presumptive retinal and presumptive lens cell groups are already interacting in this gastrula stage where

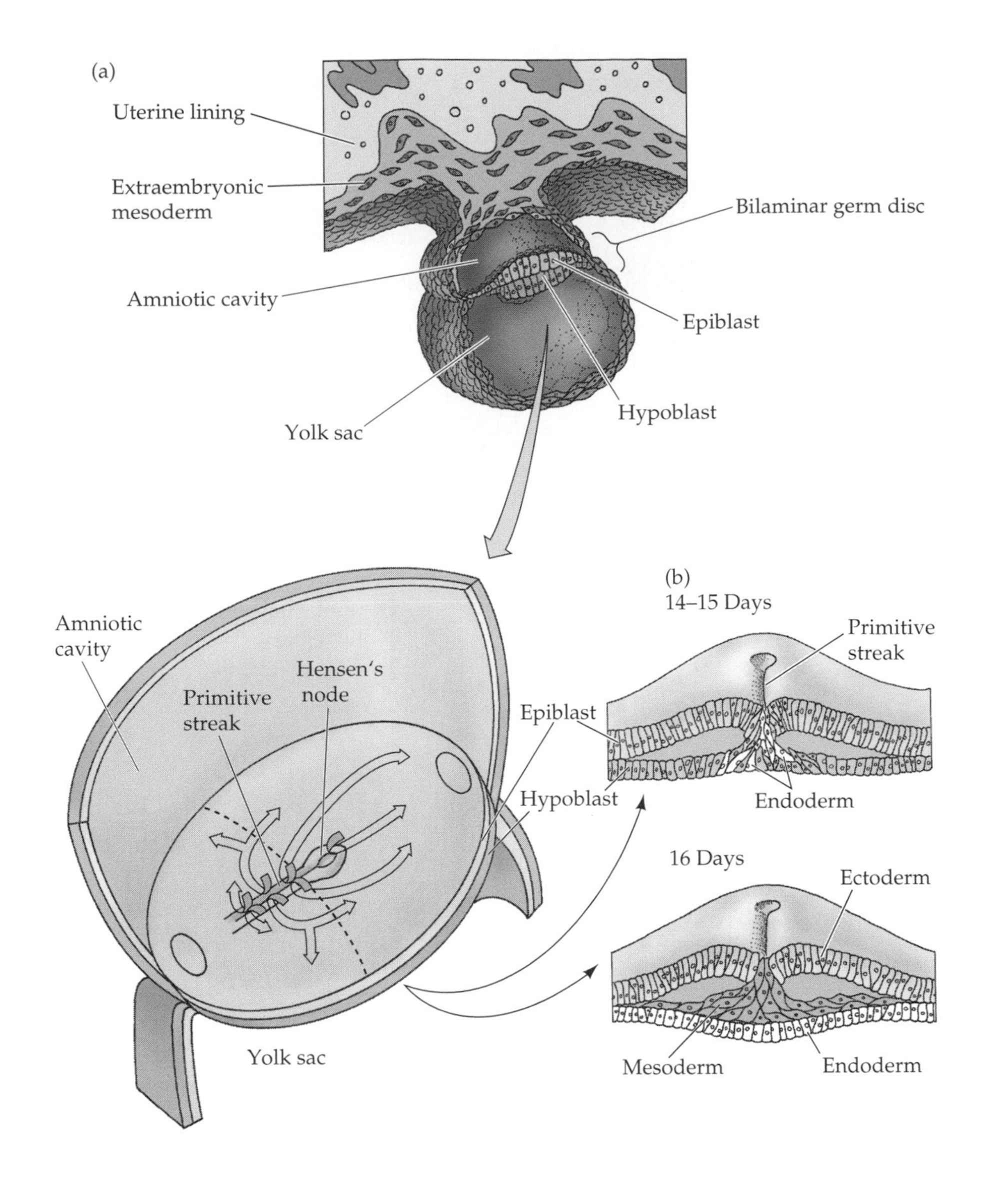

Figure 1.3
Gastrulation and Formation of the Primary Germinal Tissues
(*a*) The late-stage gastrula is a flat plate of cells dividing the embryo into two cavities, the amniotic cavity and the yolk sac. (*b*) A line, the primitive streak, begins to extend along the amniotic surface of the plate, and the migration of cells from the streak into the plate between epiblast and hypoblast creates a three-layered embryonic disc. (From Gilbert 1997.)

they are close together, the retinal group in effect priming or conditioning the lens group to react appropriately at their later meeting.

Neurulation begins the development of the nervous system

Gastrulation is followed by **neurulation**. Beginning first at the rostral end and proceeding caudally along the axis of the primitive streak, the ectoderm thickens on either side of the axis, thus forming parallel **neural folds** separated by a **neural groove** (Figure 1.4). The groove lies just above an elongated column of cells, the **notochord**, that has formed in the underlying mesoderm. The neural groove deepens, invaginates, closes, and pinches off from the superficial layer of ectoderm, forming a tube of ectodermal cells, the **neural tube**, that eventually runs from the rostral to the caudal end of the embryo.

When the neural tube closes and separates from the surface, other cells along the line of closure separate from both the surface and the neural tube and move into separate clusters along either side of the midline. These cells constitute the **neural crest**. In time they migrate away from the neural tube to form structures peripheral to the central nervous system (Figure 1.5). Among other things, neural crest cells will become the ganglia of the autonomic nervous system; most of the

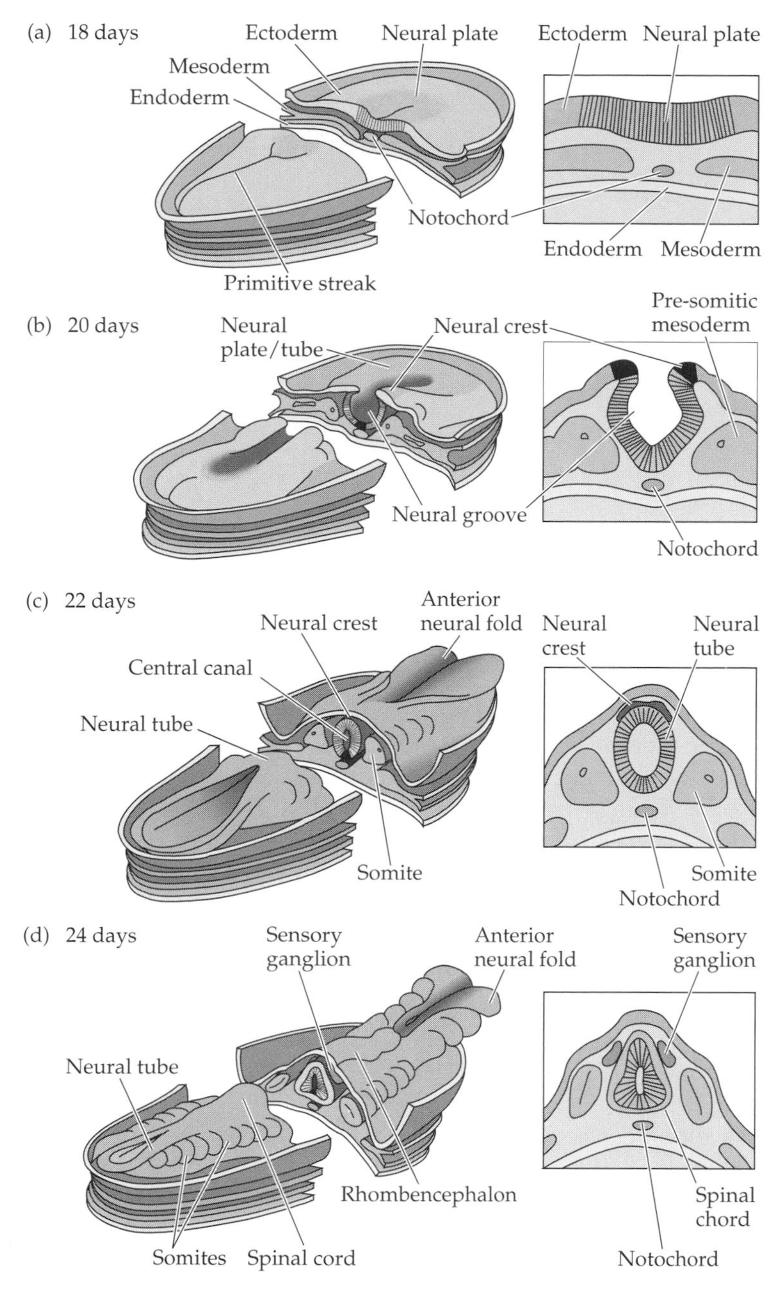

Figure 1.4
Formation of the Neural Tube
(a) Neurulation begins with formation of the neural plate, with its flanking, parallel neural folds, along the line of the primitive streak. *(b)* The plate invaginates to form the neural tube and continues until the tube separates from the surface ectoderm; other cells adjacent to the neural tube form the neural crest. *(c)* The expanding rostral end of the tube will become the brain—it closes last *(d)*. (From Purves et al. 1997.)

body's sensory nerves, including those associated with the eye; the fibroblasts in the cornea, sclera, and iris stroma; the endothelium lining the eye's anterior chamber; and the myoblasts that form the ciliary muscle. The brain, spinal cord, oculomotor nerves, retina, iris muscles, and the eye's internal epithelial layers develop from the neural tube.

Neurulation is an extremely important step in vertebrate embryogenesis, and several models have been proposed for the the mechanisms underlying the sequence of changes from neural plate to neural tube. In general, these proposals have to do with the changes in cell shape that cause a flat plate of cells to invaginate and roll up into a tube. At least some of the changes that occur during neurulation are similar to events that occur as the primitive eye is forming; the difference is that neurulation occurs along a line, thus forming a tube, while the formation of the eye involves invagination around a point, forming a spherical vesicle. Some of the possible mechanisms will be considered later in the context of the eye.

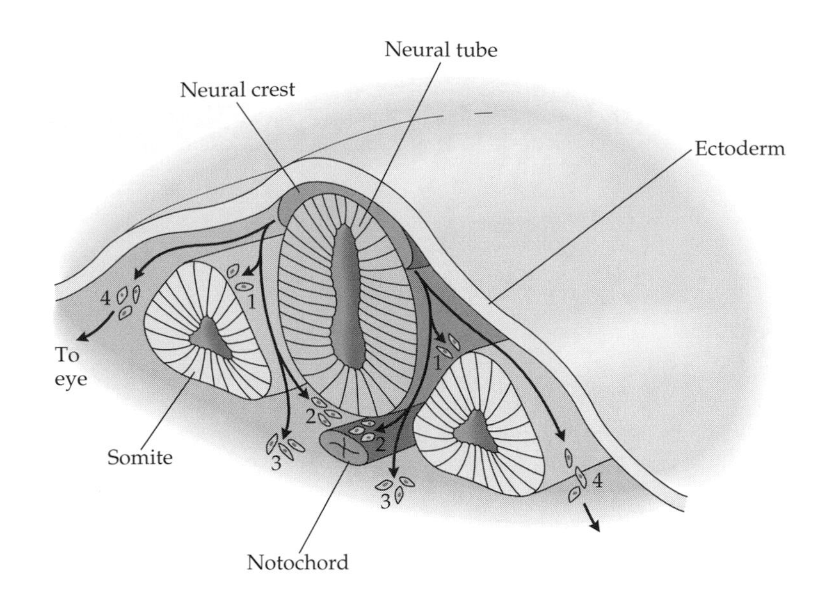

Figure 1.5
Migration of Neural Crest Cells
Cells from the neural crest migrate peripherally to contribute to different structures. Pathways 1 and 2 contain cells that will become sensory or autonomic ganglia. Cells in pathway 3 go to the adrenal gland. Cells in pathway 4 contribute to the eye, among other things.

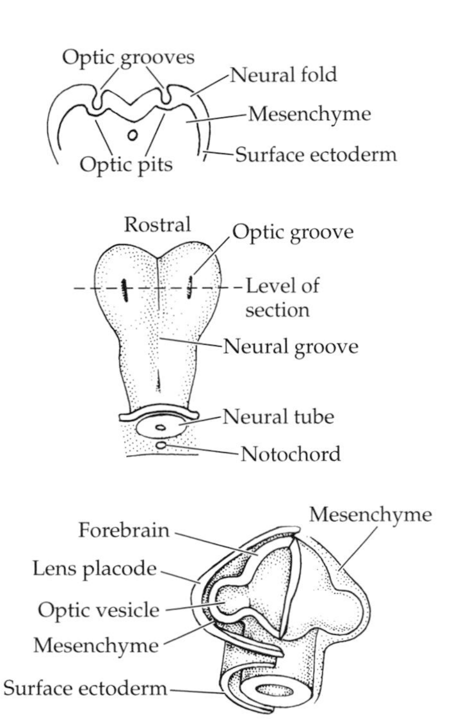

Figure 1.6
Formation of the Optic Vesicle
The eyes begin as optic pits, two small bumps on the rostral end of the neural tube. By the time the neural tube closes, the optic pits have expanded into small spheres—the optic vesicles—attached to the sides of the neural tube by short connecting tubes—the optic stalks. Continued cell proliferation results in an expansion of the optic vesicle and elongation of the optic stalk until the vesicle has grown outward sufficiently that it comes in contact with the surface ectoderm, at which time the lens begins to form.

The differentiation into primary germ tissues during gastrulation has been supplemented during neurulation. The neural tube, which is now inside the embryo and not connected to the surface, is of ectodermal origin, but the tissue is now called **neural ectoderm** to distinguish it from the exterior, **surface ectoderm**. The neural ectoderm will produce all the neurons, glial cells, and epithelia of the brain and spinal cord (including the retina and optic nerve of the eye), while the surface ectoderm will become primarily the exterior layers of the skin, including the corneal epithelium, conjunctiva, eyelid skin, and associated glands. The lens of the eye and much of the lacrimal drainage system are also derived from surface ectoderm.

Complete closure of the ends of the neural tube, which is the final act of neurulation, occurs around the end of week 4 of embryogenesis.

Formation of the Primitive Eye

Ocular development begins in the primitive forebrain

As the neural tube is closing at the rostral end, two small bumps appear on the neural tube, one on either side. These **optic pits** are the beginnings of the eyes. From this point on, events are bilaterally symmetric if all goes well, and two primitive eyes will form. As a matter of convenience, however, the subsequent discussion will refer to only one eye of the symmetric pair.

Cell proliferation at the optic pit produces a progressively larger bulge on the side of the neural tube. The bulge enlarges, as if it were a balloon inflating, to become a spherical **optic vesicle** connected to the neural tube by a short, cylindrical **optic stalk** (Figure 1.6). Both the optic vesicle and the optic stalk are made of a layer of epithelial cells, meaning that they are hollow; the vesicle is a balloon, and the stalk is a tube. Continued cell proliferation results in a steadily enlarging optic vesicle, but expansion is eventually limited by the presence of the outer surface ectoderm, which is expanding less rapidly. Contact or close proximity of the optic vesicle and surface ectoderm is another significant step, one that leads directly to significant changes within these ectodermal layers.

The optic vesicle induces formation of the lens

This stage of the eye's formation is of particular interest because it demonstrates so clearly the contingent nature of ocular development. In amphibians, transplanting a growing optic vesicle to another part of the embryo causes the surface ectoderm overlying the optic vesicle in its new location to form a lens, even though the location on the embryo may be quite inappropriate for an eye. In the absence of the optic vesicle, the region of surface ectoderm that would have formed a lens under normal circumstances does not make the necessary transformation. In short, the surface ectoderm alters to form a lens by induction from the optic vesicle; if there is no vesicle, there will be no lens.

The interpretation of this classic experimental observation has had to be modified over the years to account for differences among species (it doesn't always work) and to incorporate the notion that several stages of induction may be involved. As mentioned earlier, interactions between presumptive retina and presumptive lens tissues in the gastrula stage might be an initial sensitization of the presumptive lens tissue, predisposing it to a later definitive interaction.

Other tissues may contribute to development of the lens when the presumptive lens tissue encounters them as the embryo changes its shape during gastrulation and neurulation (Figure 1.7). The early endoderm is one possibility (which would mean that the endoderm has an indirect contribution to ocular development), and the presumptive cardiac mesoderm is another. In other words, the final step of lens induction may work only when the surface ectoderm that gives rise to the lens has been primed by an appropriate sequence of encounters with other tissues. When the surface ectoderm is properly prepared, the inductive signal produced by the optic vesicle prompts lens formation.

Elaboration of the Primitive Eye

The optic cup and the lens form from different germinal tissues by changes in cell shape

Once the optic vesicle has grown large enough to lie close to the surface ectoderm, the two apposing surfaces thicken, partly by elongation of the cuboidal ectodermal cells and partly by proliferation; the thickened region of the surface ectoderm is the **lens placode**. Next, both surfaces buckle inward, producing two invaginations—one in the surface ectoderm, which changes the lens placode to a **lens vesicle**, and one in the underlying optic vesicle, which becomes an **optic cup** (Figure 1.8). As mentioned earlier, these buckling, invaginating changes are reminiscent of the events during neurulation (and gastrulation in other species); the underlying mechanism may be similar in all cases. If so, the following description is one of the simpler versions of the story as it applies to the eye.

Cells are not amorphous blobs of goo; they have an internal cytoskeleton constructed of microtubules, which gives them some rigidity, and the cytoskeleton is attached to fibrils of actin, a contractile protein that acts as muscle does to change the arrangement of the skeleton. Furthermore, the actin fibrils in many cells, particularly epithelial cells, are organized as bands or belts that wrap around the cell just inside the cell membrane. (This feature is present very early in embryogenesis; cells in the blastocyst have both actin belts and structural junctions between cells.) Tightening such an actin belt will force a rearrangement of the cytoskeleton and a corresponding change in cell shape (Figure 1.9).

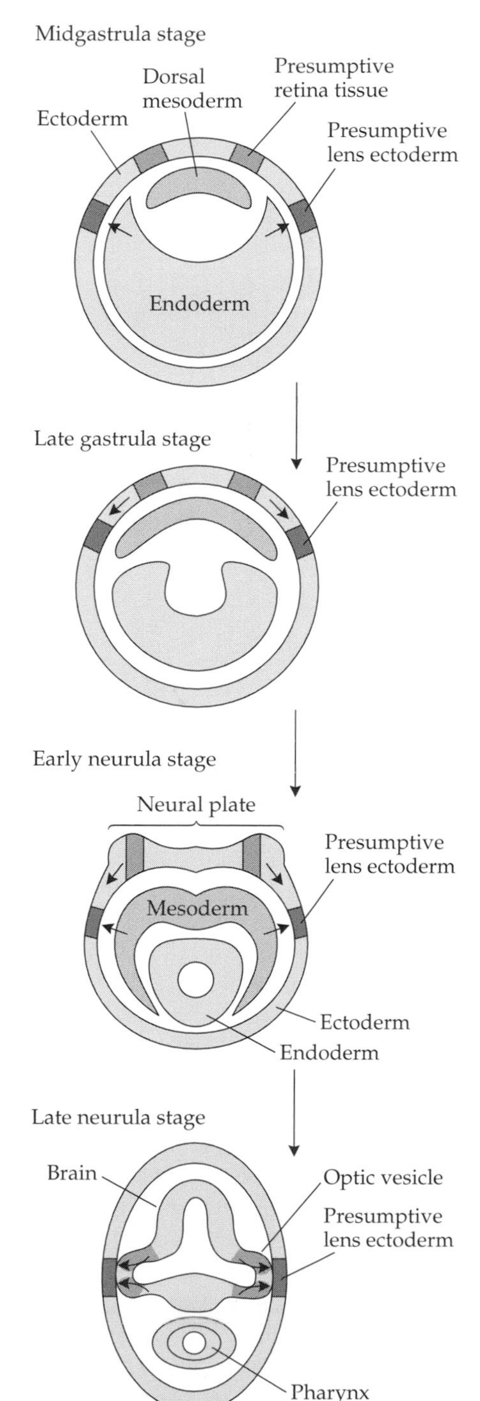

Figure 1.7
Sequence of Possible Inductive Influences on the Presumptive Lens
As the embryo changes from the midgastrula stage through the various steps of neurulation, the tissue that will eventually become the lens comes close to several other tissues in a particular sequence. Presumptive retina tissue is nearby until the neural tube starts to form, at which time endoderm and then mesoderm are in proximity and may exert influences that prepare the presumptive lens tissue for later signals from the optic vesicle. Arrows indicate lens-inducing signals.

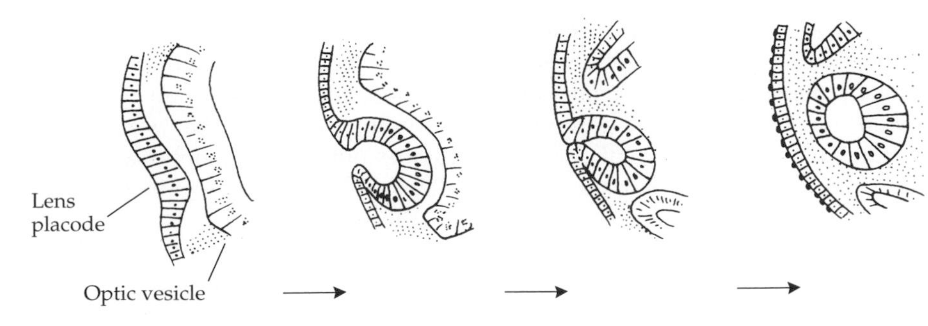

Figure 1.8
Formation of the Lens Vesicle
The lens placode buckles inward, producing a pronounced depression in the exterior surface. Cell shape changes, cell proliferation, and cell migration produce a spherical lens vesicle connected to the surface ectoderm by a small annulus of cells. Shortly thereafter, the vesicle separates completely from the surface ectoderm. Throughout this sequence, the invagination of the lens placode is accompanied by simultaneous invagination of the optic vesicle.

This mechanism for altering the shape of a single cell is not sufficient to explain a major alteration in the shape of an epithelial layer consisting of many cells. But if many small, individual shape changes can be coordinated, a large change in the surface configuration can be produced. This kind of coordination does occur, because the actin belts in neighboring epithelial cells are directly opposite one another (see Figure 1.9) and the intervening cell membranes are bound together by extracellular glycoproteins. This is one of several kinds of anchoring junctions made by cells with their neighbors or with their basement membranes.

Because cells are bound together in this way, they are not independent of their neighbors, and the forces developed when a few cells change shape affect a larger group of cells. Thus when individual cells become more wedge shaped, with their apices outward, the surface in which they reside buckles inward. In addition, whatever the signal for actin belt tightening may be, it is unlikely to be confined to a single cell; more probably, a group of cells will be affected to varying degrees depending on their proximity to the signal source.

Thus, one action of the signal for lens formation may be triggering contraction of actin fibrils in a group of target cells so that they invaginate. (This mechanism implies that the direction of buckling depends on the position of the actin belts within the target cells; in this example, the belts are closer to the *outer* surface and the resultant buckling is *inward.*) Another possible action is embodied in the "tractor" model of invagination, which invokes the movement of adhesive molecules within cells, causing some of them to change their shape and to shift inward relative to the others. Since the cells are anchored to one another, the moving cells act as miniature tractors, pulling the surface into an invaginated configuration.

In any event, the invaginations that change the lens placode to a lens vesicle and the optic vesicle to an optic cup are aided by local cell proliferation. This

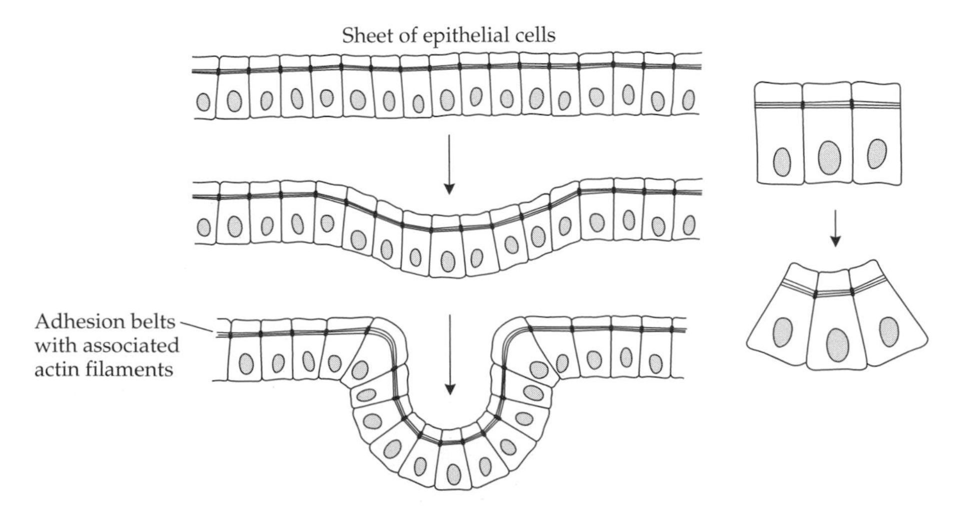

Figure 1.9
Change in Cell Shape as a Mechanism for Invagination
One model for invagination invokes change in cell shape. Columnar epithelial cells are attached to one another by anchoring junctions. Contraction of the protein belt should force a change in the cell's configuration, making it become more wedge shaped in cross section. Because this cell is attached to its neighbors and because the signal for contraction is unlikely to be confined to a single cell, there will be similar shape changes in a number of cells in the layer, making the layer change from flat to indented. The overall effect has been likened to pulling a drawstring to create a bag from a layer of cloth.

combination of changes in cell shape and continuous addition of new cells produces very rapid and dramatic deepening of the invaginations. The lens vesicle is nearly complete within a day or so after the initial appearance of the lens placode.

The optic cup is initially asymmetric, with a deep groove on its inferior surface

While the primitive lens is forming, the optic cup also continues to invaginate more deeply. This invagination is quite asymmetric, however; it proceeds much more rapidly on the inferior surface of the optic cup. As a result, the optic cup is initially not cup shaped at all, but instead has a deep groove on the inferior side that extends from the front of the cup back into the optic stalk. This groove is the **choroidal fissure** (Figure 1.10). The choroidal fissure forms because some parts of the invaginating optic cup have much more cell proliferation than others; this differential proliferation rate may be difficult to visualize, but the resulting shape of the optic cup is similar to the shape produced by pressing into one side of a balloon with the rigid edge of one hand. The asymmetric deformation that results is quite different from the radially symmetric deformation produced by pressing into the balloon with a closed fist.

Formation of the choroidal fissure is accompanied by growth of a primitive blood vessel that enters the fissure along the underside of the optic stalk and extends forward toward the rim of the cup and the lens. This vessel will become first the **hyaloid artery**, which will supply the developing lens, and later the **central retinal artery**.

Closure of the choroidal fissure completes the optic cup

After the hyaloid vessels have grown into the fissure and have reached the vicinity of the lens, the fissure begins to close by a new wave of cell proliferation in the optic cup. The edges of the fissure fuse as they come in contact. Viewed in cross section (Figure 1.11), closure of the fissure produces two abutting layers of cells from which the inner and outer layers of the cup extend to either side. This two-layered partition must be removed so that the inner and outer layers are continuous (see Figure 1.11)—a task that can be accomplished either by cell death or by cell rearrangement.

Under the first possibility, death and disintegration of the cells removes the partition, leaving inner and outer cell layers whose continuity can be established by the formation of anchoring junctions. The second possibility involves the breakdown of anchoring junctions along a perpendicular line passing through the center of the partition; cells above the line would be incorporated into the inner layer of the cup, and those below the line into the outer layer. The mechanism is not known, but either death or rearrangement is likely to be triggered by mutual induction from the parts of the optic cup on either side of the choroidal fissure.

When fusion is complete, the optic cup becomes a two-layered hemisphere whose outer layer is continuous with the optic stalk. The hyaloid vessels have been trapped within the optic cup and optic stalk by the closure of the choroidal fissure.

The lens vesicle forms in synchrony with the optic cup

Because of the induction by the optic vesicle, the initial invagination of the lens placode to form the lens vesicle coincides with the invagination of the optic vesicle itself. These invaginations proceed together, so that when the lens vesicle has become nearly spherical, though still attached to the surface ectoderm, the choroidal fissure has begun to close. The invagination of the lens vesicle from the surface ectoderm culminates when the near-spherical vesicle pinches off from the

Direct view of optic cup with lens and surface ectoderm removed

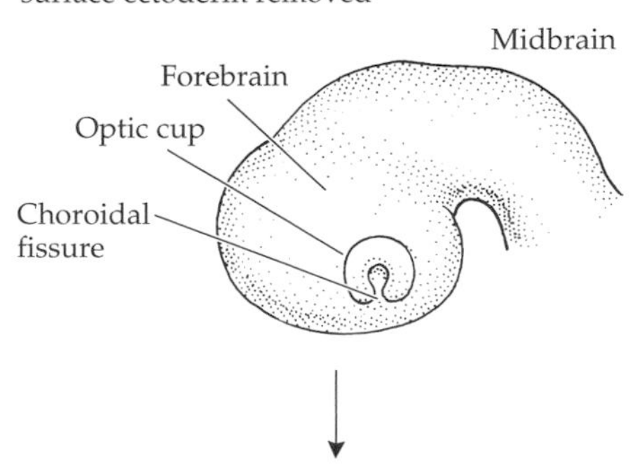

Section through optic cup prior to separation of lens vesicle from the surface

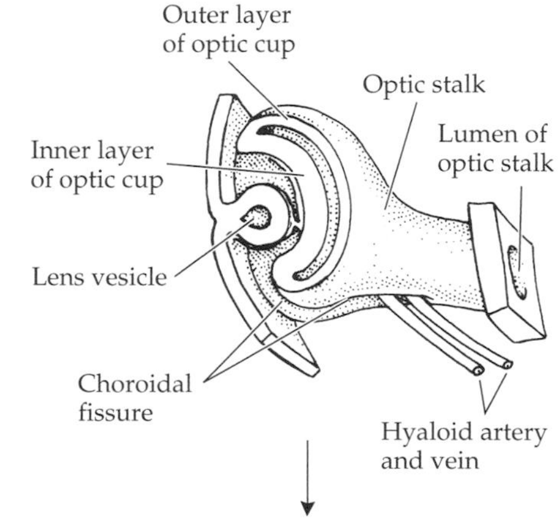

Later section through optic cup prior to closure of choroidal fissure

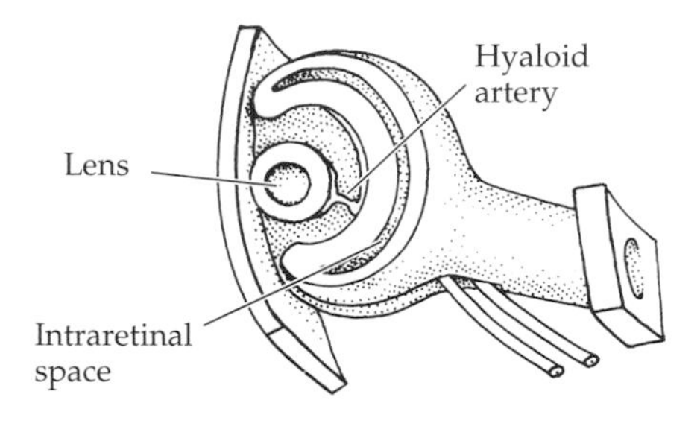

Figure 1.10
Formation of the Optic Cup
The optic vesicle invaginates in synchrony with the invagination of the lens placode. The invagination progresses much more rapidly on the inferior side of the vesicle, and the developing optic cup has a deep groove—the choroidal fissure—along its inferior side. A blood vessel grows forward along the base of the fissure toward the developing lens vesicle; this is the hyaloid artery.

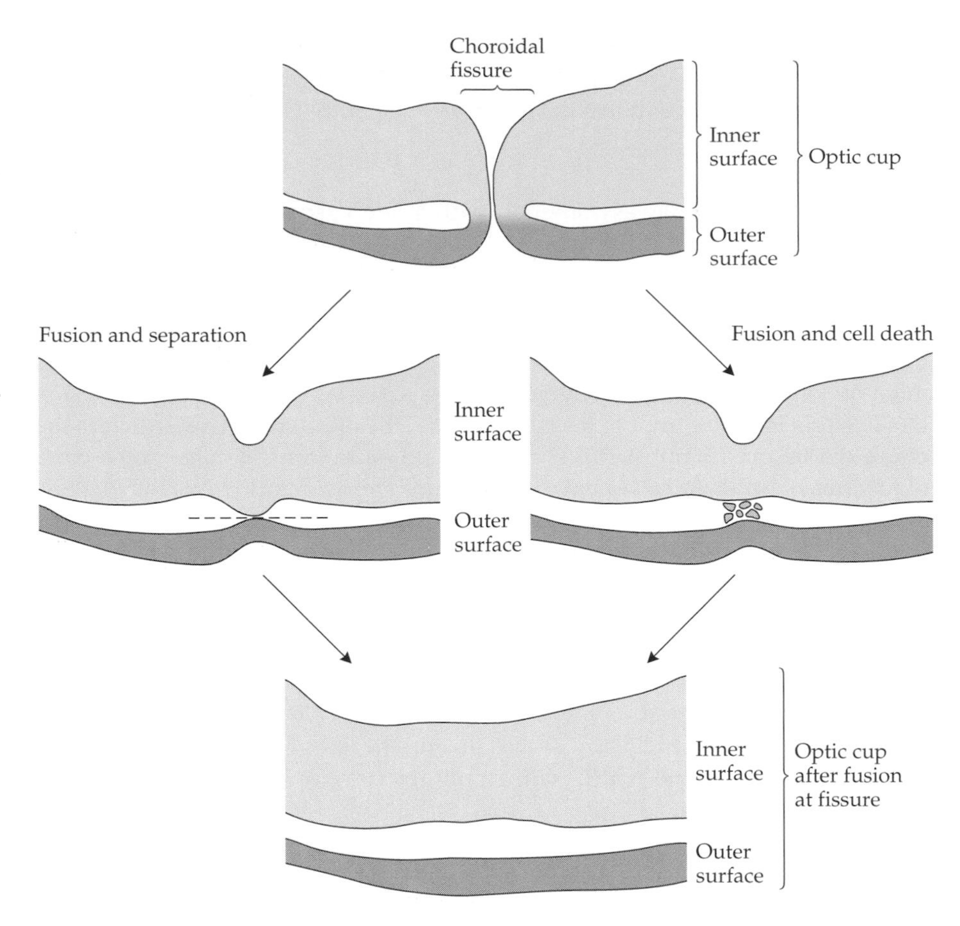

Figure 1.11
Fusion and Obliteration of the Choroidal Fissure
The vertical section through the optic cup perpendicular to the line of the choroidal fissure shows the lips of the fissure as they have just touched. Removal of the small vertical partition created when the margins of the fissure come together can be accomplished either by splitting the partition or by cell death. Either strategy should produce a seamless two-layered joining of the optic cup.

surface to become an independent structure that can now be called the lens (Figure 1.12). In a short time, the choroidal fissure will close, and the hyaloid artery will begin to develop a highly branched network of vessels around the lens.

The primitive lens is a spherical layer of epithelial cells enclosing a hollow interior. As a first step toward a complete lens, the hollow interior of the lens is filled. The epithelial cells closest to the optic cup—that is, those on the posterior surface—elongate and grow into the lumen toward the anterior surface; when they reach the anterior surface, the lumen is obliterated. These posterior epithelial cells have now made their contribution to lens growth; they eventually lose their nuclei and subsequently do not divide. Given the proximity of the posterior surface of the lens to the optic cup, this step of cell elongation is probably induced by the optic cup.

The primitive lens is the first ocular structure to exhibit cell differentiation

The cells that fill the interior of the lens exhibit several properties that are characteristic of lens cells, the first of which is their propensity for elongation when given the opportunity or the signal to do so; all the future growth of the lens involves a similar process of cell elongation (see Chapter 12). In addition, the lens cells begin to synthesize their characteristic proteins, the crystallins. This step marks the transition from determination, which began with the formation of the lens placode, to differentiation, wherein the cells are now constrained to a specific developmental future (in some species, the crystallins begin to be synthesized in the lens placode). The cells in the remaining epithelial layer on the anterior surface are determined, but have yet to differentiate; their differentiation will produce the future growth of the lens.

For the most part, the cells of the optic cup, the optic stalk, and the surface ectoderm from which the lens separated remain undifferentiated, as do most of the cells in the mesoderm surrounding the optic cup. An obvious exception is in the outer layer of the optic cup, where pigment can now be seen in some cells as they begin to synthesize melanin; they will become the retina's pigmented epithelium. The only other differentiated tissue will be the endothelial cells forming the walls of the primitive blood vessels, but almost all of these are transient and will be replaced as development proceeds. Thus, the lens is the first of the permanent ocular tissues to reveal its adult character.

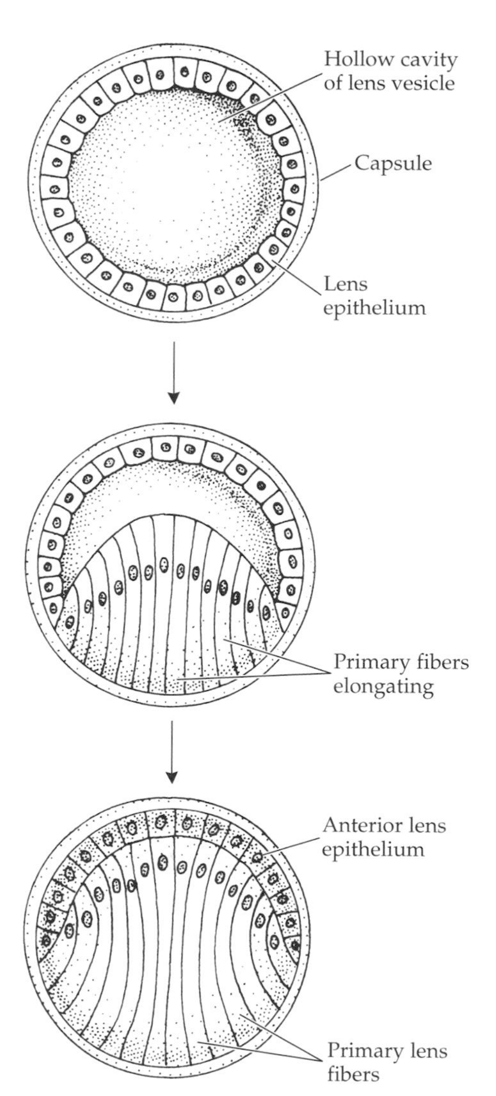

Figure 1.12
Formation of the Primitive Lens
Continued invagination of the lens vesicle eventually produces an independent, hollow sphere. The interior of the lens vesicle is filled by the cells closest to the optic cup; they elongate toward the anterior surface, thereby filling the lumen and making the lens a solid structure. All subsequent growth of the lens will come from the epithelial cells that remain on the anterior surface.

All future growth of the lens comes from the early lens cells, some of which are "immortal" stem cells

The first few hundred epithelial cells in the lens vesicle are the progenitors of the adult lens; some of them fill the hollow center of the lens vesicle, while the remaining cells are the source of future lens growth. All new lens cells will be the descendants of these pioneers, with no cellular contribution from cells in surrounding ocular tissues. This process will be described more fully in Chapter 12; its steps include differentiation of the epithelial cells around the circumference of the lens, elongation, and extension over the previous cells to form a new lens shell. Lens shells are added throughout life.

For this process to work, some of the cells in the anterior epithelium must in a sense be immortal. They proliferate, both to provide new epithelial cells and new cells for the lens substance, but cell division must provide not only new cells but also exact replicas of the parent cells. In this way, the parent cell never disappears, never dies; in the form of an identical descendant, it is always present. These cells are **stem cells**; they are not at the end of a differentiation pathway, they can divide indefinitely, and at least one of their daughters does not differentiate. When these conditions are met, stem cells are immortal, and they serve as a source of new cells for tissues that are constantly being replaced, like many epithelial tissues, or are continually being added to, like the lens.

Differentiated lens cells, or lens fibers as they are called, are quite different from most other cells in the eye. Once they form from the lens epithelium, lens fibers are permanent; they never divide, they are never destroyed in the normal course of events, and they cannot be replaced. This characteristic of permanence is shared only with retinal neurons. In addition, most of the crystallin proteins of the lens fibers are unique, existing nowhere else in the body. Thus the lens is a most unusual tissue, with its own set of peculiar properties—some of considerable clinical importance—that are established almost from the beginning of the lens's existence as a structural entity.

The precursors of the future retina, optic nerve, lens, and cornea are present by the sixth week of gestation

The lens differentiates and the choroidal fissure closes about 5 to 6 weeks into gestation. Although there is a great deal yet to be added, the basic form of the eye has been established and several major components are in place (Figure 1.13). The optic cup and the optic stalk represent the beginnings of the retina and the optic nerve, respectively; in the future, the outer rim of the cup will contribute epithelial layers to the iris and ciliary body, as well as the iris muscles. As we have seen, the lens is present, with all the cells it needs for its future growth. The overlying surface ectoderm from which the lens vesicle separated represents the future corneal epithelium. All of these ectodermally derived components are embedded in relatively undifferentiated mesoderm, which will eventually give rise to the outer coats of the eye, the extraocular muscles, the orbit, orbital blood vessels, and so on.

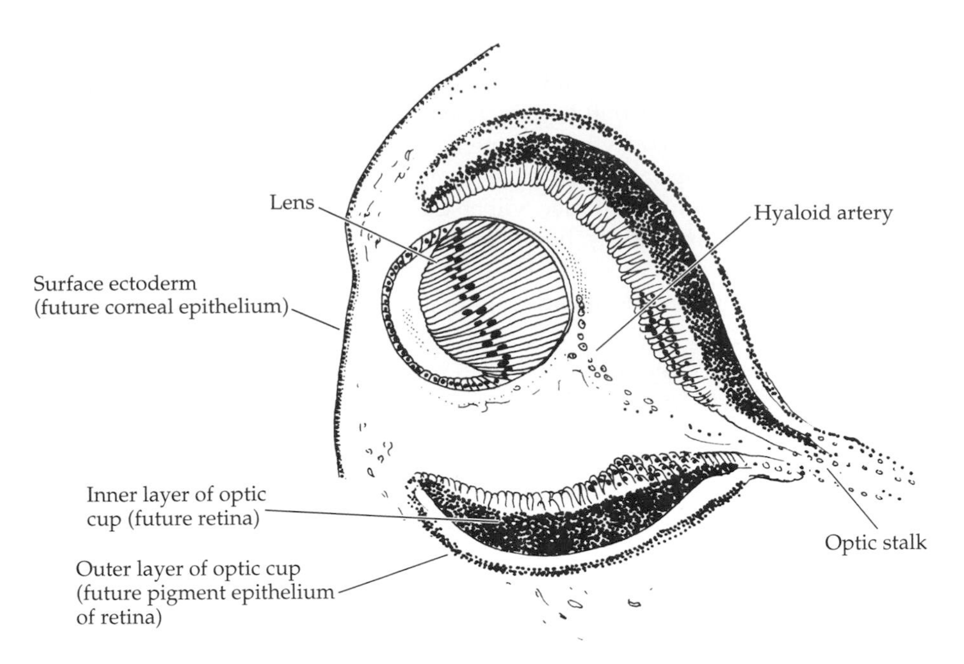

Figure 1.13
The Eye after Five Weeks of Gestation
This vertical section through an eye at 5 weeks shows the eye embedded in mesoderm that has just begun, or will begin within a week, to differentiate to form the sclera, the extraocular muscles, and the cartilage templates of the orbital bones. Most of the optic cup and optic stalk will form the retina and optic nerve; the rim of the cup, however, will grow forward to become the epithelial layers of the ciliary body and iris, and the iris muscles.

The first 5 or 6 weeks of gestation are a period in which the cellular processes underlying ocular development are mainly cell proliferation, movement, and changes of shape. There has been some determination—cells in the optic cup will become neurons, glial cells, or epithelial cells, for example—but with the exception of the lens, there has been very little differentiation to begin formation of the adult tissues. A great deal of differentiation is yet to come, however, and subsequent stages of ocular development cannot be discussed simply as changes in shape and form of epithelial layers. Instead, the characteristics of differentiated cells will have to be considered in terms of their contributions to different ocular structures; these changes will be discussed in successive chapters as part of the detailed considerations of the various parts of the eye and its adnexa.

In general, the eye develops from inside to outside

The retina, now represented by the optic cup, is the innermost of three coats of tissue in the mature eye; and the lens, which may be thought of as the innermost of all ocular structures, has begun its unique developmental course. Events to come will add structures to these basic elements. The uveal tract will be next, followed in order by the sclera and the cornea, the extraocular muscles, and the bones of the orbit. This inward-to-outward sequence, however, best describes the first appearances of tissue differentiation, not the sequence in which structures complete their development; the cornea and sclera, for example, are structurally complete at birth (although they will continue to increase in size for some years), but the retina is not yet fully formed.

The importance of the inside-to-outside generalization resides in the possibility that the innermost structures—the optic cup in particular—are not simply a template on which the rest of the eye is molded, but that they control the development of external structures during embryogenesis and continue to exert some influence well into the postnatal period. In other words, the developing nervous system, represented here by the optic cup, may dictate the development of its future sensory input from the eye. Although this is a difficult hypothesis to test, there are indications that it contains an element of truth; recent studies on the development of myopia, for example, implicate the retina as the prime controller of the correlated growth of the eye's refractive components (curvatures of the cornea and lens, anterior chamber depth, axial length of the eye). Failures of correlation result in refractive error (see Chapter 2). In addition, some developmental anomalies are most easily interpreted as inductive failures wherein defects in external structures reflect and perpetuate an anomaly in the optic cup.

Failures of Early Development

"If anything can go wrong, it will"

This version of Murphy's Law is a wry commentary on life that applies not only to its original context—the assembly and maintenance of aircraft—but also to the assembly of an animal and its eyes. The development of any organism is extremely complicated, and with so many opportunities for things to go wrong, it is remarkable that development succeeds as often it does (although it may be less often than we think—some estimates suggest that fewer than half of all fertilizations result in successful embryos). Of the embryos that survive to term, approximately 5% have abnormalities, some of which are quite serious.

Embryological errors during the early phases of ocular development—that is, during the first 4 to 6 weeks of gestation—usually have profound consequences, affecting not just the eye but other parts of the embryo. There has been so little differentiation, for example, that improper formation of intercellular anchoring junctions is unlikely to be confined to the budding eye and will involve much of the developing nervous system. Errors of this type are so catastrophic that the embryo rarely survives to term. And when the embryo does survive, the ocular defects are often associated with, and very much secondary to, other severe anomalies.

One or both eyes may fail to develop completely

Of the few isolated ocular defects attributable to errors very early in gestation, an extreme and very rare example is initial formation of the optic vesicle followed by its atrophy. The result is the complete absence of an eye, a condition called **degenerative anophthalmos**. More commonly, one or both eyes are significantly smaller than normal, perhaps no larger than a pea. This condition, called **microphthalmos**, may not be an early error, however; histological examination of microphthalmic eyes often reveals the presence of differentiated tissues, showing that development must have proceeded for at least a few months. However, this phenomenon reinforces the point made earlier about the importance of induction as a driving force in development. We cannot be certain, but the small size of the microphthalmic eye is probably due to failure of the optic cup to enlarge properly; the cup could, and did, induce differentiation within the surrounding mesoderm, but because the cup in its manifestation as developing retina did not enlarge, the surrounding tissues remained correspondingly small. Also, microphthalmic eyes are associated with very small bony orbits, the proportions of which are commensurate with the small eyes.

Clinical management of microphthalmos involves the use of plastic orbital inserts that cause the orbits to enlarge as the child grows, so that the head will be shaped normally and prosthetic eyes can be implanted for a cosmetically normal appearance. The orbital inserts provide an example of mechanical, as opposed to chemical or molecular, induction; bone is quite responsive to mechanical forces, and in lieu of a normally growing eye it will respond to successively larger implants with continual growth and enlargement.

Congenital absence of the lens may be an early developmental failure

Given the sequence of events during the early stages of ocular development, other possible defects are easily imagined; improper formation of the lens vesi-

Vignette 1.1

The Eye of Mann

IDA CAROLINE MANN (1893–1983) was born the year that the Spanish histologist Santiago Ramón y Cajal published his great work on the structure of retinal neurons (see Vignette 14.2). That coincidence is surely meaningless, but in the same way that retinal anatomists invariably pay homage to Ramón y Cajal, it is impossible to discuss the development of the human eye without referring to Ida Mann. Although she was not the first person to study aspects of ocular development—Aristotle had commented on it 2000 years earlier—she was the first to give a complete account of human ocular development based on her own observations, observations that have proved to be accurate and insightful.

Serious study of ocular development must be done at a microscopic scale; thus the subject could not be pursued effectively until the appropriate histological methods had been developed, in the latter part of the nineteenth century. And since much of the technical work had been done by German anatomists, it is not surprising that some of them were among the first to use the methods to study ocular development. Their work was there for Ida Mann to build on when she began to study the eye.

The first of Ida Mann's books, *The Development of the Human Eye,* was published in 1928 (with revised editions in 1950 and 1964); it represents almost 10 years of study and writing. Her book was the most complete study of human ocular development that had been done, and it has had enormous influence. Even now, most illustrations of the eye at various stages of development derive their inspiration, if not all their details, from her drawings. Several of these drawings are shown in Figures 1 and 2.

These illustrations are special. Ida Mann drew them herself because, as she says, she took pleasure in doing it. This personal touch is important; Ida Mann speaks to us in the most direct way possible about what she saw and understood. Approaching science this closely is increasingly rare as scientists are forced to work with layers of technicians, students, and computers between them and their objects of study and between them and their audience. And for the same reason, Mann's work has unusual authority and integrity; we know it is genuine because she did it herself. Rendering all those drawings undoubtedly gave Ida Mann a far deeper understanding than she would have had otherwise, and we are the beneficiaries of her efforts.

Figure 1 contains two related drawings, which are among the simplest of the more than 200 in her book. This illustration is included here because it shows what the anatomist is trying to do. The upper part is a three-dimensional representation of the optic cup and the lens vesicle at about 4 weeks of age. At this stage of development, the lens vesicle has not yet separated from the surface ectoderm and the choroidal fissure has not closed, but the hyaloid artery is within the optic cup. The sections in the lower part of the figure are the raw data—that is, the developing eye as the anatomist saw it—and the task was transforming these sections into the three-dimensional representation of the eye shown in the upper part of the figure. The relationship between the tissue slices and the complete structure is fairly clear in this case because the developing eye is geometrically simple, but reconstructions of this type can be very difficult.

Figure 2 is a much more detailed longitudinal section through the eye at about 10 weeks of age. The cornea is present, the anterior chamber has been defined by the developing iris, the retinal layers have appeared near the posterior pole, and two extraocular

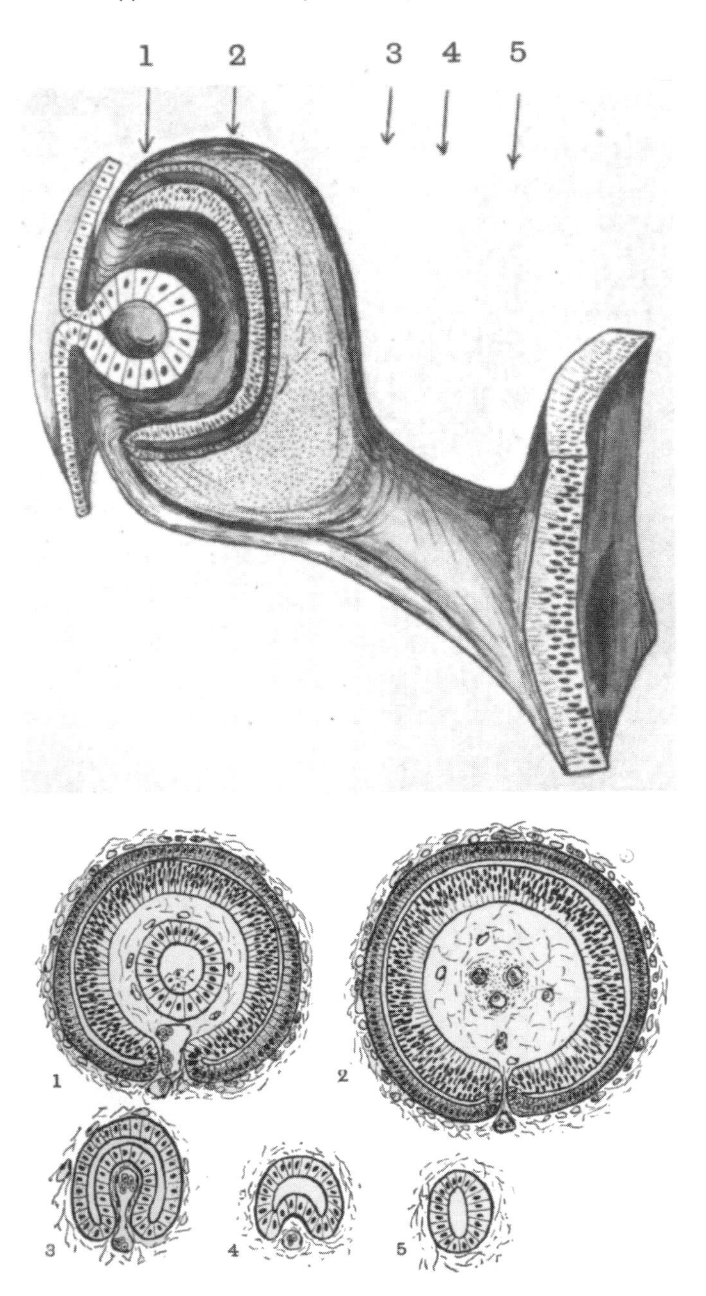

Figure 1

The Eye at Four Weeks of Gestation

The upper part of the figure shows a three-dimensional reconstruction of the optic stalk, the optic cup, and the lens vesicle at 4 weeks of gestation. The numbers and arrows correspond to the planes of the transverse sections in the lower part of the figure. (From Mann 1928.)

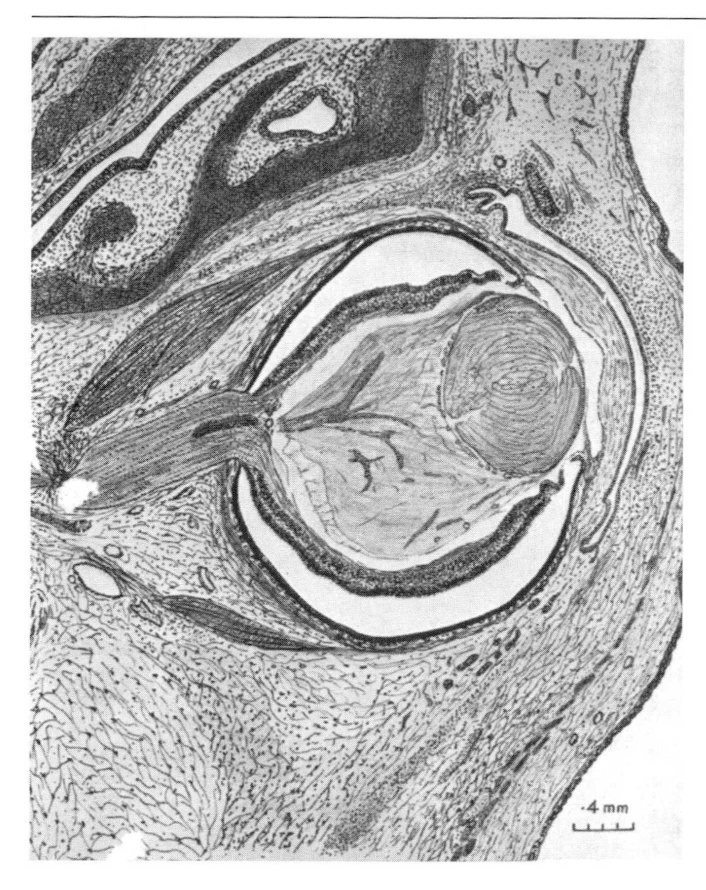

Figure 2
The Eye at Ten Weeks of Gestation
In this horizontal section, the medial side of the eye is up and the lateral side is down. The original is a drawing, not a photograph; at this level of detail, it represents several days of steady, painstaking work. (From Mann 1928.)

muscles (the medial and lateral recti) are complete from origins to insertions. We are seeing the developing eye as Ida Mann saw it, and she used her artistic sensibility and scientific judgment to render significant details with great clarity. Even in well-stained tissue, some of these details would have low contrast and they would be less obvious in a photograph. To put it another way, this is a subjective rendering of the developing eye, a view filtered through the eyes and mind of Ida Mann. (A photograph wouldn't necessarily be better; the choice of subject is a matter of judgment, and images can be altered when they are printed. In fact, the ease with which images can be undetectably manipulated by computer graphics these days makes them suspect as objective or neutral records. It is a question of who has proved to be trustworthy. You can trust Ida Mann.)

While working on the development of the eye, Mann also studied medicine and ophthalmology. But even after she began a clinical career, she continued to investigate ocular development, this time in terms of development gone awry. Her second book, *Developmental Abnormalities of the Eye,* was published in 1937 (with a revised edition in 1957). It is here that we find a careful, critical discussion of colobomas and the association of typical colobomas with improper closure of the choroidal fissure. (Mann is careful to say that the idea did not originate with her, but she is the one who marshaled a convincing array of evidence in support of the hypothesis.) This book, like her first, is a classic, and both are still worth reading; she wrote clearly and gracefully and told her stories well.

Mann became chief surgeon at Moorfields Eye Hospital in London, the only woman to be chief surgeon in the hospital's then 150-year history, and in 1945 she became the first professor of ophthalmology at Oxford. In 1946, she and Antoinette Pirie published a book titled *The Science of Seeing,* which was meant to give laypeople a better understanding of vision and visual disorders. (The book addresses such questions as the number of things that can produce blindness and what can be done about them.)

Ida Mann married in 1944 when she was 51 years old. Five years later, she resigned her academic positions to move with her husband to Australia in search of a better climate for his failing health. He lived only a few more years, but she remained in Perth after his death and became involved with efforts to control the extremely high prevalence of trachoma among the aborigines. (Trachoma is an infection of the conjunctiva—the tissue covering the anterior part of the eye and the inner surface of the eyelids—by the organism *Chlamydia trachomatis.* It can lead to corneal ulceration and blindness. At that time, about 3% of Caucasians and more than 50% of aborigines in Western Australia had the infection.)

Mann's experience with trachoma and other eye diseases not commonly encountered in Europe and North America led to work in other countries for the World Health Organization, and more books resulted. She wrote several travel books, using her married name (Caroline Gye), and in 1966 published another book about the eye, called *Culture, Race, Climate and Eye Disease.* This book, like the earlier ones, broke new ground, this time in a field of study she called "geographical ophthalmology"; it is a fascinating account of the variety and distribution of vision problems that affect us, enlivened by Mann's insightful comments on the cultural diversity of our species.

Most of the many awards and honors that Ida Mann received for her work are listed in her obituary in the *Australian Journal of Ophthalmology* (volume 12, pages 95–96, 1984). There, several of her former colleagues and friends concluded their account of her life and work as follows: "Ida Mann was not only an ophthalmologist of great wisdom and a scientist of great merit but she was also a woman of grace and compassion. We were fortunate to have her living amongst us for so long."

cle with resultant absence of the lens (**aphakia**) is one of the more obvious. Congenital aphakia does occur, but it seems not to be a failure of primary induction of the lens vesicle. Instead, most observations on aphakic eyes report the presence of lens remnants, suggesting that the aphakia was produced by a failure after differentiation, resulting in degeneration and atrophy.

The difficulty here, as is often the case with congenital anomalies of the eye, is the limited opportunity for histological examination of affected eyes. Moreover, as discussed earlier, the lens cells differentiate very early, so the presence of seemingly mature fragments may not be particularly meaningful. Thus, there is disagreement in the literature concerning the presence and interpretation of lens fragments in congenitally aphakic eyes; we cannot rule out the possibility of a primary failure of the lens vesicle in some cases or, in others, a later failure of lens induction by the optic cup.

Incomplete closure of the choroidal fissure can produce segmental defects in the adult eye

A more prevalent group of structural defects can be attributed to improper closure of the choroidal fissure in the growing optic cup. If the optic cup serves as a template, or induces, the proper differentiation and development of tissues external to it, it follows that a defect in the cup itself may be perpetuated as a spatially corresponding defect in more external structures. The manifest defects are **colobomas**, and they appear as incomplete, improperly formed, or depigmented segments in such structures as the iris, lens, ciliary body, choroid, and retina, either singly or in combination. An iridial coloboma, with a piece of the iris missing, and a choroidal coloboma are shown in Figure 1.14. The choroidal coloboma appears very white because all of the melanin pigment and blood vessels normally found in the choroid are absent; the white region is the inner surface of the sclera (see Figure 1.14*b*).

A "typical" coloboma will always be situated in the inferior aspect of the eye, a location corresponding to the original position of the choroidal fissure. An isolated iridial coloboma would therefore be an indication that only the rim of the optic cup failed to close completely (the rim of the cup forms the iris epithelium and muscles), while an isolated choroidal coloboma would suggest that while the rim of the cup closed, part of the choroidal fissure nearer the optic stalk did not.

Atypical colobomas are located somewhere other than the inferior part of the eye, and since their location does not correspond to the normal position of the choroidal fissure, their etiology is unclear. If one assumes, however, that atypical colobomas also represent failures of induction, their occurrence may indicate that the choroidal fissure is not the only place where the optic cup may develop improperly; the cup might develop cracks or breaks in places where it is normally continuous, for example. Alternatively, the optic cup sometimes rotates around its anterior–posterior axis, thereby placing the choroidal fissure in an unusual and unexpected position. Such rotation would mean that typical and atypical colobomas both reflect a failed closure of the choroidal fissure.

Figure 1.14
Typical Colobomas
(a) In this iridial coloboma, a segment of the iris is missing. *(b)* This choroidal coloboma appears as a large circular area just below the optic nerve head; the choroidal pigment is absent and the white area is the inner surface of the sclera. (*a* courtesy of John F. Amos; *b* from Alexander 1994.)

(a)

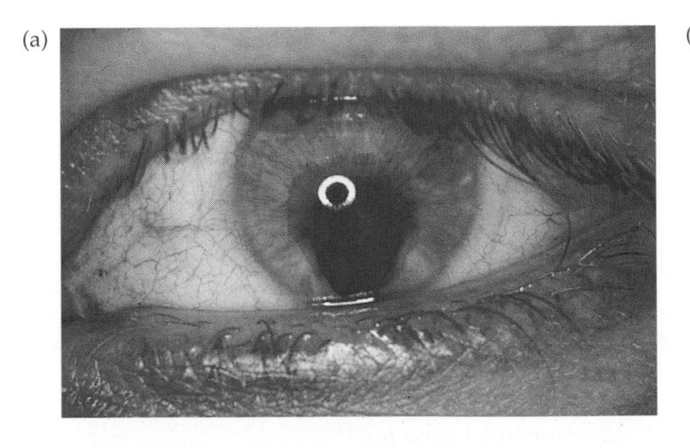

(b)

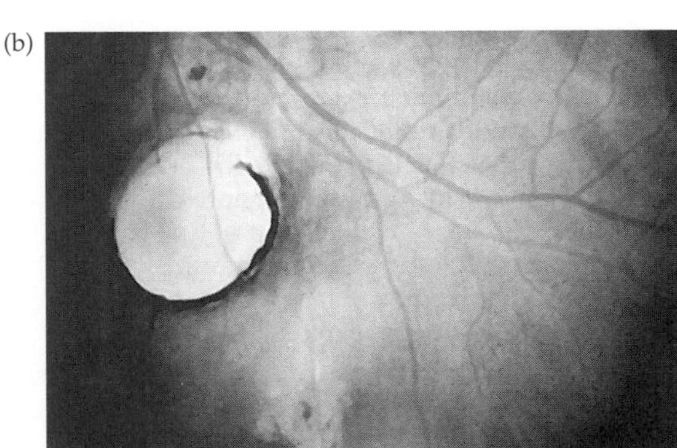

References and Additional Reading

Developmental Strategies and Operations

Alberts B, Bray D, Lewis J, Raff M, Roberts K, and Watson JD. 1994. Cellular mechanisms of development. Chapter 21, pp. 1037–1137, in *Molecular Biology of the Cell*, 3rd Ed. Garland, New York.

Beebe DC. 1994. Homeobox genes and vertebrate eye development. *Invest. Ophthalmol. Vis. Sci.* 35: 2897–2900.

Edelman GM. 1988. *Topobiology.* Basic Books, New York.

Maclean N and Hall BK. 1987. *Cell Commitment and Differentiation.* Cambridge University Press, Cambridge.

Embryonic Events before the Eyes Appear

Gilbert SF. 1997. *Developmental Biology,* 5th Ed. Sinauer Associates, Sunderland, MA. (Chapters 5, 6, and 7.)

Moore KL. 1988. *The Developing Human: Clinically Oriented Embryology,* 4th Ed. WB Saunders, Philadelphia.

Purves D et al. 1997. *Neuroscience.* Sinauer Associates, Sunderland, MA.

Purves D and Lichtman JW. 1985. Early events in neural development. Chapter 1, pp. 3–23, in *Principles of Neural Development.* Sinauer Associates, Sunderland, MA.

Formation of the Primitive Eye

Barischek YR. 1992. Embryology of the eye and its adnexae. *Dev. Ophthalmol.* 24: 1–142.

Cook CS, Ozanics V, and Jakobiec FA. 1991. Prenatal development of the eye and its adnexa. Chapter 2, pp. 1–93, in *Duane's Foundations of Clinical Ophthalmology,* Vol. 1. Tasman W and Jaeger EA, eds. JB Lippincott, Philadelphia.

Duke-Elder S and Cook C. 1963. The general scheme of development. Chapter 2, pp. 19–48, in *Normal and Abnormal Development: Embryology,* Vol. III, Part 1, *System of Ophthalmology,* Duke-Elder S, ed. Henry Kimpton, London.

Grainger RM, Henry JJ, Saha MS, and Servetnick M. 1992. Recent progress on the mechanism of embryonic lens formation. *Eye* 6: 117–122.

Hilfer SR and Yang J-JW. 1980. Accumulation of CPC-precipitable material at apical cell surfaces during formation of the optic cup. *Anat. Rec.* 197: 423–433.

Mann I. 1964. *The Development of the Human Eye.* Grune and Stratton, New York.

Oliver G and Gruss P. 1997. Current views on eye development. *Trends Neurosci.* 20: 415–421.

Piatigorsky J. 1981. Lens differentiation in vertebrates. A review of cellular and molecular features. *Differentiation* 19: 134–153.

Reichenbach A and Pritz-Hohmeier S. 1995. Normal and disturbed early development of the eye anlagen. *Prog. Ret. Res.* 14: 1–46.

Saha M, Spann CL, and Grainger RM. 1989. Embryonic lens induction: More than meets the optic vesicle. *Cell. Diff. Devel.* 28: 153–172.

Wakely J. 1976. Scanning electron microscopy of lens placode invagination in the chick embryo. *Exp. Eye Res.* 22: 647–651.

Weale RA. 1982. *A Biography of the Eye.* HK Lewis, London. (Chapter 2.)

Failures of Early Development

Alexander L.J. 1994. *Primary Care of the Posterior Segment*, 2nd Ed. Appleton & Lange, Norwalk, CT.

Mann I. 1957. *Developmental Abnormalities of the Eye.* JB Lippincott, Philadelphia.

Chapter 2

Ocular Geometry and Topography

How far can the best eyes see a common house-fly? A hundred yards? It is quite impossible. Very well: eyes that cannot see a house-fly that is a hundred yards away cannot see an ordinary nail-head at that distance, for the size of the two objects is the same. . . .

"Never mind a new nail; I can see that, though the paint is gone, and what I can see I can hit at a hundred yards, though it were only a mosquito's eye. . . ." The rifle cracked, the bullet sped its way, and the head of the nail was buried in the wood, covered by the piece of flattened lead.

There, you see, is a man who could hunt flies with a rifle.

■ Mark Twain, *Fenimore Cooper's Literary Offenses*

Before Sam Clemens became Mark Twain, he had been, among other things, a riverboat pilot, a placer miner, and a newspaper reporter, occupations in which success is related to keen observation. He knew perfectly well from his own experience that neither flies nor nail heads can be seen from a distance of 100 yards. James Fenimore Cooper, on the other hand, was not an acute observer of the world around him. And because of his impoverished perception, Cooper's frontier romances were often silly, contrived, and implausible (and badly written, as Twain gleefully pointed out). The accuracy of a Kentucky rifle in a frontiersman's hands is legendary, but only a marksman invented by Cooper would boast that he could see an object we know to be invisible and then hit it with a rifle ball.

It isn't possible to hunt flies with a rifle, even if there were a reason for trying, but how badly did Cooper err? How big would the target have to be at a 100 yards for a marksman just to see it and make a remarkable shot? If the best eyes cannot see a housefly 100 yards away, at what distance *can* a fly be seen? (I recall a samurai movie in which the character played by Toshiro Mifune snatched flies out of the air with his chopsticks. That's believable, though surely very difficult.) What is the difference between the best eyes and all the others? What are the limits of human vision, and to what extent is the eye a limiting factor? In short, what is it about the eye's optics that affects how well we see fine detail?

Elements of Ocular Structure

The human eye is a simple eye

All vertebrate eyes are simple eyes with a single optical system that forms a single image. (Simple eyes may form only one image, but the image-forming properties may vary depending on the part of the optical system that is utilized. Thus, the "four-eyed fish" of South and Central America, *Anableps anableps,* uses one part of each eye's optical system for aerial vision and another for vision below the waterline.) The single image is examined by many photoreceptors, and the more photoreceptors that examine an image of given size, the finer will be the detail extracted from the image.

The advantage of simple eyes is that they can have apertures large enough to avoid the diffraction limits of compound eyes (see the Prologue) and to have large images for the photoreceptors to sample. The human eye is a fairly representative vertebrate eye; its macroscopic anatomy is quite simple, in the everyday sense of the word. The eye is a fluid-filled chamber enclosed by three coats, or layers, of tissue. It is an optical system made of leather, water, and jelly, as Gordon Walls once put it.

The outermost of the three coats of the eye consists of cornea, limbus, and sclera

The transparent **cornea** makes up about 16% of the eye's outer coat; the white, opaque **sclera** accounts for most of the rest (Figure 2.1). Both tissues consist mainly of densely woven collagen fibers, which make the tissues rigid, resistant to penetration, and able to protect the more delicate inner layers. The cornea is the eye's principal refractive element, and it owes its transparency to the regularity of its structural organization.

The **limbus** is the region of transition from cornea to sclera; it is an annulus of tissue about 1.5 mm wide around the cornea. The limbus is of interest because it contains specialized structures for removing fluid (the aqueous humor) from the eye. Since improper drainage of aqueous humor may be associated with glaucoma, which is a potentially blinding disorder, discussions of the limbal structure and function inevitably center on the problem of aqueous drainage.

The middle coat—the uveal tract—includes the iris, ciliary body, and choroid

Most of the blood vessels of the eye and most of its melanin pigment are located in the **uveal tract**. (The Latin word *uvae* means "grapes"; the posterior surface of the iris and the folds of the ciliary body apparently reminded early investigators of grapes.) The uveal tract is not structurally homogeneous; it contains three distinct, but continuous, structures: from front to back, the **iris**, the **ciliary body**, and the **choroid** (see Figure 2.1).

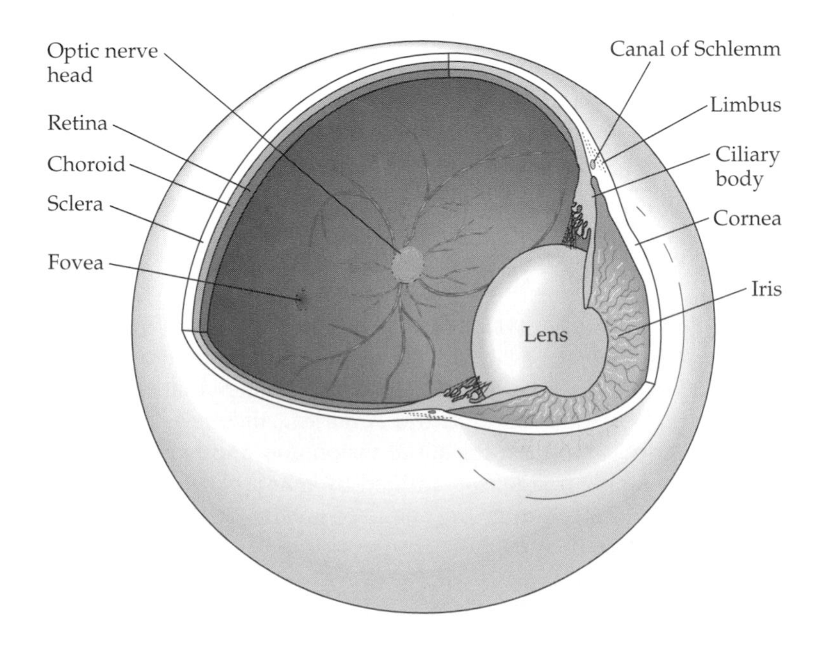

Figure 2.1
Major Structures and Regions of the Eye
The right eye, viewed from the lateral side and slightly above, with the entire lateral-superior quadrant removed.

The iris, visible through the cornea, is the part of the uveal tract most accessible to direct inspection. It is not a complete layer, however; the hole in its center is the **pupil**. Because of its melanin pigment, the iris is opaque and light can enter the posterior portion of the eye only though the pupil; thus, the iris is an aperture stop for the eye's optical system. Two muscles in the iris, the **sphincter** and the **dilator**, allow the size of the pupil to be varied, thereby regulating the amount of light entering the eye and modulating the image-degrading effects of diffraction and aberration, which are discussed later in the chapter.

The ciliary body is adjacent to and continuous with the iris. It can be visualized as a ring, or annulus, of muscle encircling the anterior portion of the eye just inside the sclera, with a set of highly vascular folds, the **ciliary processes**, on the inner side of the muscular ring. The musculature is part of the system for altering the refractive power of the lens (**accommodation**), while the ciliary processes, collectively, are the site at which the **aqueous humor** (also known simply as aqueous) forms; the watery aqueous is the fluid that fills the anterior portion of the eye.

The remainder of the uveal tract, which is well over half of it, consists of the choroid, the main components of which are blood vessels, including an extensive capillary bed (the **choriocapillaris**), and dense melanin pigment. The melanin pigment serves the same role as the matte black interior of a camera: It absorbs and thus eliminates scattered light that might otherwise degrade the image. Most of the choroidal blood vessels supply or drain the choriocapillaris, which lies on the inner side of the choroid and is the blood supply for the photoreceptors in the retina.

The eye's innermost coat—the retina—communicates with the brain via the optic nerve

If the eye were a camera, the **retina** would be the photosensitive film, but this analogy very much understates the function of the retina. Perhaps a better way to think of it is as something like a computer that receives inputs from 100 million photodetectors that are sampling the pattern of light and dark in the image formed by the eye's optical system. The retina processes the information it receives and transmits it to the brain via the million or so neurons of the **optic nerve**, which exits through the sclera on the posterior-nasal aspect of the eye.

Both functionally and anatomically, the retina is by far the most complex structure in the eye. Although the retina contains only three major classes of cells—neurons, glial cells, and epithelial cells—the neurons can be divided into about 50 to 100 anatomically, physiologically, or histochemically distinguishable cell types. The possible ways to interconnect these cell types are more numerous still, and since the way in which cells are connected is thought to be related to what the retina does, understanding the connectivity is a major concern.

In general, the job of the retina is, first, to detect light in the retinal image and, second, to inform the brain about the features of the image that can be used to construct a mental image of the external objects that were imaged on the retina. But the retina's art is abstract, and it does not send the brain a snapshot of the retinal image. There is no point-by-point representation of the image. Understanding the retina will require a discussion about what is being abstracted from the retinal image and how it is done.

For clinicians, the concern is not so much how the retina works, but the kinds of things that can interfere with its normal operation and how they can be prevented or alleviated. Since good vision requires an in-focus retinal image, the list of possible problems begins with the operation of the eye as an optical system, and continues with almost anything that affects the milieu of the retina: intraocular pressure, integrity of the vascular systems, and so on.

Vignette 2.1
The Medieval Eye

SHIPS THAT DOCKED IN THE ANCIENT CITY of Alexandria, Egypt, were routinely searched for manuscripts to be copied or confiscated as additions to the Alexandrian library, which was meant to contain every book ever written. This goal was probably never achieved, but the library's collection has been estimated at almost a million papyrus scrolls (books in those days), and many of them were the sole copy. The final demise of the library of Alexandria around A.D. 500 was a loss of incalculable magnitude; our intellectual heritage was decimated.

Part of the loss may have been *The Book of the Eye,* a text written by Herophilus (circa 335–circa 280 B.C.), who was regarded as the greatest of the Greek anatomists. Its existence is inferred from other writings that survived. From other authors who seem to have seen such a book, we know that Herophilus understood and described the anatomy of the eye in some detail and that he discovered and described the nerves of the eye, among other things. Some contemporary scholars believe that Herophilus used illustrations in his book, which makes its loss all the greater; they would have been the first known anatomical illustrations of the eye. More important still, Herophilus would have been writing about things he had observed; he dissected the human body and was one of the last to do so for nearly 15 centuries. (The possibility, hotly debated, that Herophilus sometimes practiced human vivisection is too appalling to contemplate for long.)

Much of what we know about the work of Herophilus and other Greek anatomists comes from the writings of Galen (A.D. 129–circa 199), who was the principal inheritor and exponent of the knowledge acquired by the Greek philosophers. (All fields of intellectual inquiry, including what we now call science, were philosophy as far as the ancients were concerned.) Galen was extremely well educated, widely read in the classical literature, extensively traveled, and the dominant figure in anatomy and medicine until the Renaissance. His enormous influence came from his erudition, his strongly held opinions, his skill as a polemicist, and the enormous volume of his writings (then, and later, he was referred to as a "windbag").

Because human dissection was not permitted in Galen's time, his detailed descriptions of ocular dissections refer primarily to bovine and simian eyes. He was thoroughly familiar with the work of Herophilus, however, and whatever preconceptions Galen had about the structure of the human eye probably derived from this source. In any event, he was certainly influenced by the earlier Greek writers on vision and the function of the eye, and it was their ideas about vision, expressed in Galen's words, that was to influence thinking about the eye for almost a millennium. (Vignette 3.2 has more to say about Galen's influence.)

If we did not have some idea about the structure of the eye, it would be difficult to reconstruct from Galen's description; there are no pictures, the meanings of words are obscure, and translators do not always agree about Galen's terminology, as they point out in extensive footnotes. But some essential features are clear: Galen understood the three-layered structure of the eye, the relationship of the retina to the optic nerve, the difference between the aqueous and vitreous chambers, and the existence of oculorotary muscles (although he assumed that humans also have a retractor bulbi muscle like that of some other animals). It is also clear that Galen viewed the lens as the seat of vision.

The first illustration of Galen's ideas, or at least the first we know of, was done by an Arab physician, Hunain ibn Ishak, in the latter part of the ninth century A.D. This representation of the eye (Figure 1) is highly symbolic, but some of the key structural features of the eye are present and consistent with Galen's descriptions. The eye has three major coats—the sclera and cornea, the choroid, and the retina in our terms—with a thin outer layer that could correspond to the episclera and Tenon's capsule (see Chapter 3). The drawing includes a lens, the fluid-filled aqueous and vitreous chambers, a pupil, and an optic nerve. But while the major elements are present, their sizes, shapes, and positions differ from our modern version. The optic nerve, for example, exits at the posterior pole and runs straight back from the eye, instead of running medially, and the lens is depicted as a sphere located in the center of the eye. And just in front of the lens is a structure with no anatomical counterpart; it represents the visual spirit and reveals the drawing's Greek inspiration.

Vision was the most difficult sense for the ancients to understand. While the wind on one's face offered clues to the nature of sound and its medium of transmission over long distances, there were no comparable hints about light. What it was, how it was transmitted, and how it might interact with the eye were mysteries about which one could only try to deduce solutions. Plato and his followers, who influenced Galen, treated vision as a sense analogous

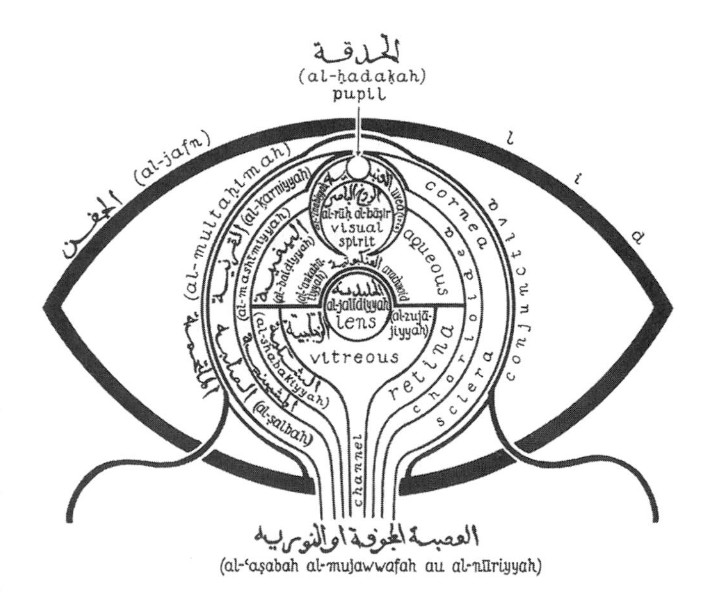

Figure 1
The Human Eye a Thousand Years Ago
A cross section through the eye is shown framed by the aperture of the lids as viewed from the front. Note the central, spherical lens, the hollow optic nerve that is continuous with the vitreous chamber, and the structure representing the "visual spirit" just in front of the lens. (From Polyak 1941.)

to touch; it was as if rigid, invisible rays of a vital spirit emanated from the eye and vibrated when they encountered an external object. (Plato's student Aristotle disagreed and believed that light, whatever it was, came from objects to the eye.) They believed that the vibrations were concentrated by the lens—hence its central location—and transmitted to the brain by fluid in the eye and the optic nerve. Thus, the visual spirit in this drawing is the source of the emanating rays through which the eye sensed the environment.

The drawing in Figure 2 was done a century later, around A.D. 1000, and it marks an immense change in thinking about the eye; there is no longer any representation of a visual spirit. This drawing was done by the great mathematician Ibn Al-Haytham (965–1039), generally known by the Latin version of his name, Alhazen. Like Euclid before him, Alhazen had a geometer's love for optics, and his investigations convinced him that rays of light enter the eye from external objects, not the other way around. (Alhazen neither knew nor cared about the *nature* of light, which made much of the philosophers' considerations irrelevant.) By eliminating the visual spirit, Alhazen demysticized the eye, making its operation subject to the same optical principles as any glass lens. Where the eye was once a matter of wonder and speculation, he made the eye an object that could be studied and possibly understood.

Alhazen's drawing is not explicit about anatomical structure, but he shows the optically important proportions and curvatures of the cornea and sclera quite accurately. His major failing is the lens; he placed it more nearly in its correct location than Ibn Ishak had, but it is still depicted as a large circle, presumably a cross section of a sphere. (Note that the pupil, represented by the small circle at the front of the lens, and the optic nerve head, the circle at the back of the lens, are drawn as they would be seen when viewed from the front of the eye; they have been rotated 90° relative to the plane of the main drawing. This stylistic representation of features in their most characteristic aspect, with different vantage points for different features, is quite ancient; it is particularly obvious in Egyptian art. Thus the large circle representing the lens may also be a frontal view.)

Alhazen understood image formation in terms of the camera obscura (pinhole camera), and his drawing suggests that he believed the image to be formed at the back surface of the lens. (The quantitative law of refraction was first derived by the Dutch astronomer and mathematician Willebrord Snell about 600 years later.) This location is wrong, but it is consistent with the powerful spherical lens that his drawing may be indicating, and it places the emphasis where it should be—on the image-forming properties of the eye. The ocular image would interact with the optic nerve, which Alhazen shows running from the back of the lens to the point at which it joins the nerve from the other eye at the optic chiasm. (The representation of the optic chiasm is also unprecedented.) Alhazen's depiction of the optic nerve within the eye is incorrect, but it is consistent with the prevailing view of the optic nerve as the sensory element of the eye. Because of its obvious blood vessels, the retina was thought to be mainly nutritive in function (Galen again).

In the late medieval period, Alhazen's work was translated into Latin and became the basis of several thirteenth-century treatises on optics, including the *Opus Majus* of Roger Bacon (circa 1220–1292), the *Perspectiva Communis* of John Pecham (circa 1230–1292), and the *Perspectiva* of Witelo (or Vitellio; circa 1230 or 1235–after 1275), all of which included diagrams dealing with the optics of the eye. The importance of these works is not that they added much new, but that they were excellent syntheses and brought optical knowledge to a much wider audience. Bacon and Pecham hedged on Alhazen's dismissal of the visual spirit, but Witelo adopted Alhazen's position unreservedly, and this was the view that was to influence the scholars who made the next major advances. But those advances had to await the Renaissance.

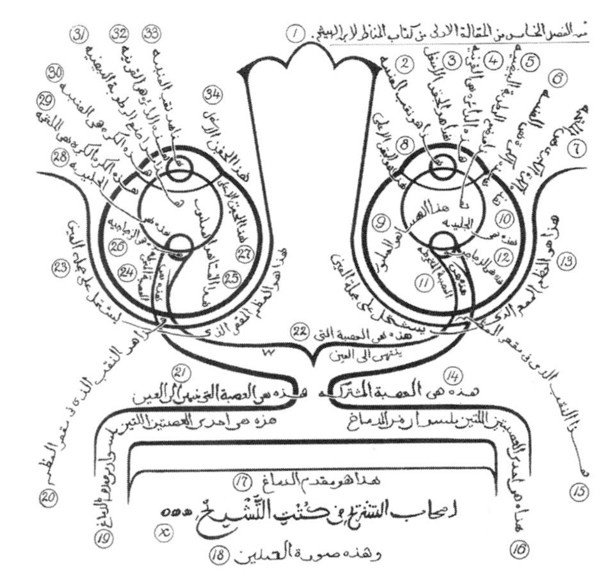

Figure 2
The Eye of Alhazen
This horizontal section through the midlevel of the orbits shows the optic nerves joining behind the eyes at the optic chiasm. The relative curvatures and proportions of the cornea and sclera are fairly accurate, but other features, like the lens and optic nerve head, are not. This eye has no "visual spirit" from which rays might leave the eye to sense external objects. (From Polyak 1941.)

Most of the volume of the eye is fluid or gel

To complete this brief overview of ocular structure, we must consider the **lens**, which lies about a third of the way between the front and the back of the eye; it is held in place by a suspensory ligament, the **zonule**, that attaches to the ciliary body. Functionally, the lens contributes about a third of the eye's total refractive power for distance (unaccommodated) vision, with the capacity, in young persons, for another 12 diopters (D) or so during accommodation.

Most of the volume of the eye is optically and structurally empty. The **anterior chamber**, between the cornea and the iris, and the **posterior chamber**, between the iris and lens, are filled with aqueous; the **vitreous chamber** behind the lens is filled with the gel-like **vitreous humor**. Despite their paucity of well-defined structure, however, both the aqueous and the vitreous have significant clinical problems associated with them. As mentioned earlier, one form of glaucoma is associated with an imbalance between aqueous formation and drainage. And the vitreous, for all its apparent stability, can change sufficiently to predispose the retina to detachment, thus separating the photoreceptors from their blood supply such that they could be lost.

Image Quality and Visual Performance

Images of point sources are always small discs of light whose size is a measure of optical quality

The retinal image produced by refraction at the cornea and lens is a map of the external world on the surface of the retina or, more precisely, on the retinal layer containing the photoreceptors. It is a small, inverted replica of the pattern of light emitted by or reflected from external objects, and every point in space is mapped by the eye's optics to a corresponding point on the retina. The replication of the external world in the retinal image is not perfect, however; points in space, like stars, are not points in the image. Instead, they are small discs of light whose size is a measure of image quality. The smaller the disc image of a point source, the better the image.

The light in the image of a single star is distributed around a maximum, rather like a Gaussian, or normal, distribution is, and it can be characterized by the size of the maximum and the degree of spread around it. A small amount of spread is associated with a more pointlike image and better resolution of fine detail. Two stars that are very close together in the sky provide a test case for resolution of detail; the question is whether we can see two stars, or just one. The image of two nearby stars consists of two partly overlapping distribution functions whose peaks are the centers of the stars' images (Figure 2.2*a*). There is a deep trough between the peaks where there is less light, and this difference between the amount of light in the peaks and the trough allows us to perceive two separate objects. The trough disappears if the stars are closer together, however (Figure 2.2*b*), because light from the individual images adds to fill up the trough. The overall image has a single broad plateau and offers little hint that the light came from two individual objects. Detail present in the world cannot be extracted from the retinal image; the detail is below the resolution threshold.

Two stars that should be resolvable at a given separation may not be resolvable if their images are out of focus. In this case, the spread of the image distribution functions has been increased by blur (Figure 2.2*c*). Where there would have been a trough in the overall light distribution for in-focus images, the defocusing has spread the light in the individual images and filled the trough. The only way to resolve these blurred images is to move them farther apart.

Since the spread of light in the image of a single point determines how close it can be to another point image (with the same spread) and still be seen as separate, this minimum separation can be used as a criterion for visual performance. Called the **Rayleigh criterion** (Figure 2.2*d*), it states that the centers of the images must be separated by half their spread if they are to be seen as two objects. For *focused* images, this criterion reduces to a statement of the minimum angle of separation ($\varnothing_{min}$), in degrees, as follows:

$$\varnothing_{min} = \frac{180}{\pi} \times \frac{1.22\lambda}{d}$$

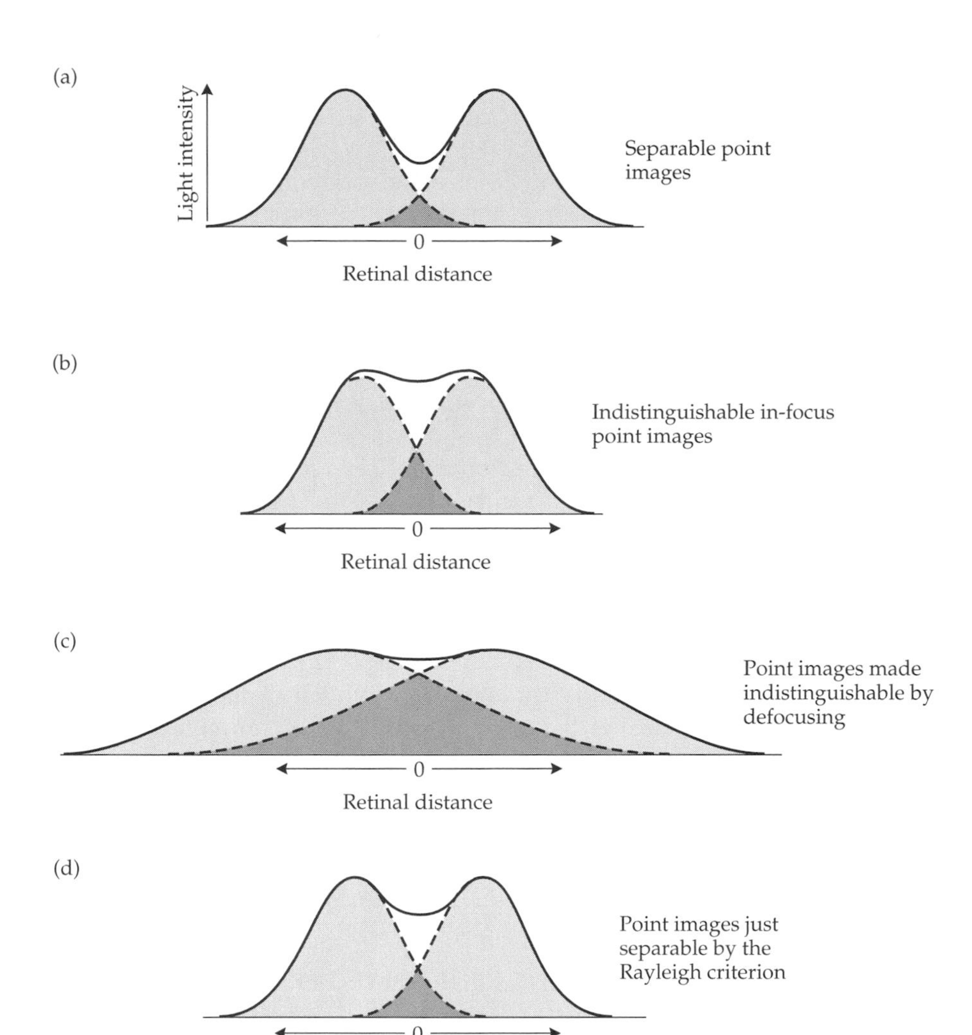

Figure 2.2
Retinal Images of Stars
Each curve shows the amount of light falling on the retina at various points along a line through the centers of the images of two very small point sources of light, such as stars. (*a*) There is some overlap of the two images, but the peaks at the centers of the images are quite distinct. (*b*) When the stars are very close together, their images overlap extensively, forming a flat-topped intensity distribution in which the original point images are lost. (*c*) Points that are separated as in part *a*, but that have more image spread introduced by defocusing, cannot be distinguished as individual points. (*d*) Point images separated by half the image spread have a shallow trough between the peaks that allows them to be just barely distinguished as separate points. This is the separation specified by the Rayleigh criterion.

where λ is the wavelength of light and d is the diameter of the pupil, both in millimeters. Green light in the middle of the visual spectrum (555 nm) and a 2 mm pupil give a minimum angle of separation of 1.16′ arc.

The Rayleigh criterion tells us how far apart two points or two parallel lines must be placed if they are to be seen as separate—that is, to be resolved. Moreover, it indicates that the major limiting factor is an anatomical variable—namely, pupil size. Finally, and most significantly, it is a statement about the visual system's best performance; anything that increases image spread will increase the minimum angle of resolution, which means that just resolvable objects must be farther apart than the optimum distance. And since the surest way to increase image spread is to defocus the image—to blur it—we must have in-focus imagery for the visual system to operate at its best.

The amount of smear or spread in the image of a point source is related to the range of spatial frequencies transmitted by the optical system

Notions of frequency and mixtures of frequencies are probably most familiar in the acoustic domain, where pure tones are sinusoidal amplitude variations at a single temporal frequency (middle A at concert pitch is 440 Hz) and other sounds are combinations of different pure tones at different amplitude (middle A played on a violin and a trumpet are distinguishable because of other frequencies [harmonics] combined with the fundamental frequency).

This acoustic analogy is applicable to vision. The optical equivalent of a pure tone is a pattern of alternating light and dark bands—a grating—in which the variation (or modulation) is sinusoidal (Figure 2.3). Combinations of different sinusoidal variations will produce light-intensity variations that are not sinusoidal. In fact, a mathematical theorem states that any complex object or image can be treated as a mixture of pure sinusoidal variations that differ from one another in their frequency, amplitude, and phase (the relative position of the beginning of a frequency cycle). Among other things, this statement means that any pattern with abrupt discontinuities between light and dark will contain high frequencies (they are necessary for the abrupt transition), while a broad, uniformly illuminated patch of light will have a spatial frequency of zero, or very close to zero.

Once again, stars are wonderful test objects. They are infinitesimally small points of light that can be thought of as mixtures of spatial frequencies from zero to infinity, all with the same amplitude. Thus a perfect optical *image* of a star should also contain all spatial frequencies at constant amplitude. This is never possible, however; the imaging process always removes some of the high spatial frequencies. Removing high spatial frequencies makes the image less like a point of light and more like a small disc.

The measure of the amplitude variation (or modulation depth) of light is **contrast**. Contrast is defined as the average modulation depth divided by the average amount of light in the pattern (see Figure 2.3). If the amount of light at the peak of the modulation is L_{max} and the amount of light at the modulation trough is L_{min}, then

$$\text{Percent contrast} = \frac{(L_{max} - L_{min})}{(L_{max} + L_{min})} \times 100$$

L stands for **luminance**, which is related to the number of light quanta per unit area and time in the light *source*. At 100% contrast, $L_{min} = 0$.

Since both visual objects (sources) and their images can be characterized by the frequencies and contrasts of their component sinusoids, we can evaluate an

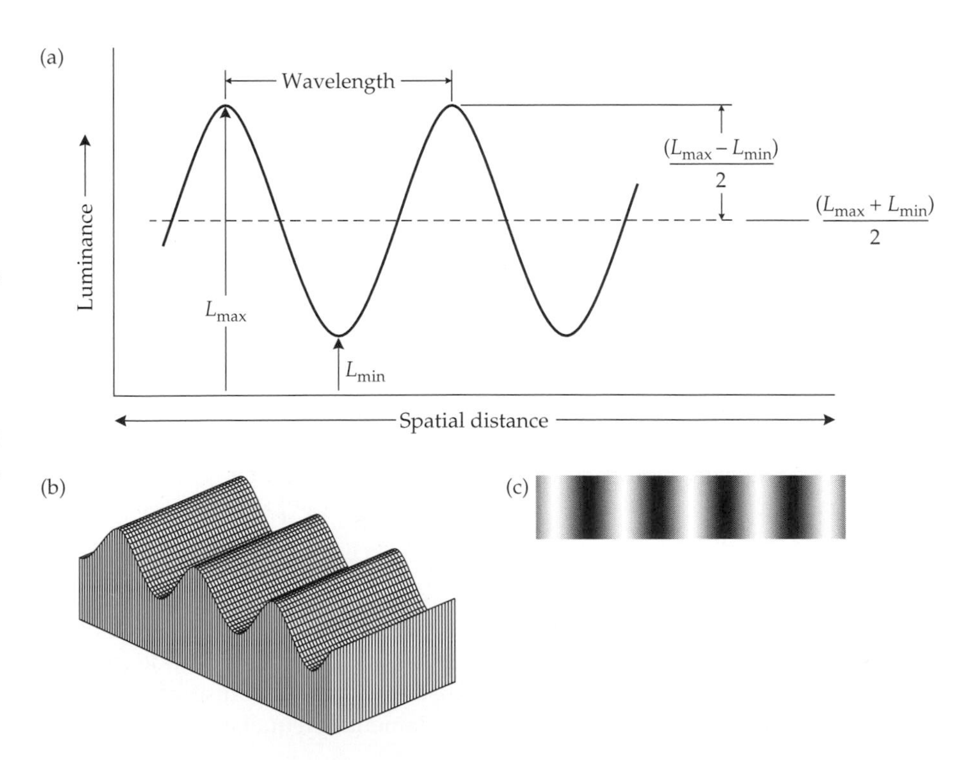

Figure 2.3
Sinusoidal Gratings and Spatial Frequency
In sinusoidal gratings, the luminance varies periodically according to the equation $L = a \sin(bx) + c$, where L is luminance, a is the amplitude of the modulation, x is the distance across the retina (bx is the wavelength or cycle width), and c is the value of L when $x = 0$ (it is called the phase of the grating). (*a*) A plot of the luminance modulation in such a grating. (*b*) Its three-dimensional structure. (*c*) The appearance of a sinusoidal grating to an observer; unlike a square-wave grating, which consists of alternating black and white stripes, the sinusoidal grating has blurry stripes. The average luminance of the grating is calculated as $(L_{max} + L_{min})/2$, and the average modulation depth is calculated as $(L_{max} - L_{min})/2$; contrast of the grating is the average modulation depth divided by the average luminance.

optical system by determining how faithfully it transmits the frequencies contained in the source to the image it forms. Faithfulness, in this context, is measured by the contrast in the image relative to the contrast in the object. If, for example, an object grating of a particular spatial frequency has 100% contrast and the image also has 100% contrast, transmission is perfect. At the other extreme, if 100% contrast in the object is reduced to 0% in the image (no variation at all between the original light and dark grating stripes), then transmission has failed completely and this component of the object will not be contained in the image.

This is the basis for measuring the **modulation transfer function** of an optical system, which shows the ratio of image contrast to object contrast over a wide range of spatial frequencies from zero (uniform illumination) to the high-frequency limit of the optical system. A spatial frequency of zero corresponds to an extended, uniformly illuminated object; all we are measuring is the intensity of light in the image relative to the object. For optical systems with transparent media, like the normal eye, the losses in transmission will be only a few percent, occurring mainly by reflectance at the surfaces of the optical elements. This, however, is the best the system can do; once the target contains some variation—that is, when spatial frequencies are greater than zero—the ratio of image contrast to object contrast will be less than the maximum at zero frequency.

The important issue is now the spatial frequency at which the ratio of image contrast to object contrast becomes zero. The higher this value, the better the image quality, because the spread of light in the image of a point source of light is inversely related to the spatial frequencies transmitted by the optical system: The higher the spatial frequencies transmitted, the less the image spread, and the closer two point images can be located and still be resolved. (If the system could transmit infinitely high spatial frequencies, the image would be an infinitesimally small point with no spread at all. This is a theoretical limit, however, one that cannot be achieved in real optical systems with finite aperture sizes.)

The modulation transfer function for a human eye with a 2 mm pupil is shown in Figure 2.4.* At zero spatial frequency, the curve has been set to a value of 1, and the contrast for all other spatial frequencies has been plotted relative to this maximum. As spatial frequency increases, the curve declines, meaning that the image contrast is decreasing relative to the object contrast. At about 60 cycles/degree, the curve goes to zero; at this point, there is no difference between peaks and troughs in the image and the sinusoidal variation in the object is no longer being transmitted by the eye's optics. This is an absolute limit, called the **cutoff frequency**. It is determined solely by pupil (or aperture) size and by the wavelength of light emitted from the source, as follows:

$$\text{Cutoff frequency} = \frac{\pi}{180} \times \frac{d}{\lambda}$$

in cycles/degree, where d is pupil diameter and λ is the wavelength of light, both in millimeters.

For a 2 mm pupil and green light (555 nm), the calculated cutoff frequency is 1.05 cycles/minute of arc (63 cycles/degree), or, expressed as angular subtense, 0.95′ arc/cycle. (This value is similar to, but smaller than, the Rayleigh criterion value calculated earlier, but one can be derived from the other. The Rayleigh value is the reciprocal of the cutoff frequency, multiplied by 1.22.) The cutoff frequency determined experimentally (55 cycles/degree) is lower than the theoret-

*This modulation transfer function was not measured directly by determining the ratio of image contrast to object contrast for gratings of different spatial frequencies, but was derived instead from the measured light spread in the image of a very thin streak of light—that is, the line spread function. The modulation transfer function and the line spread function are mathematically related.

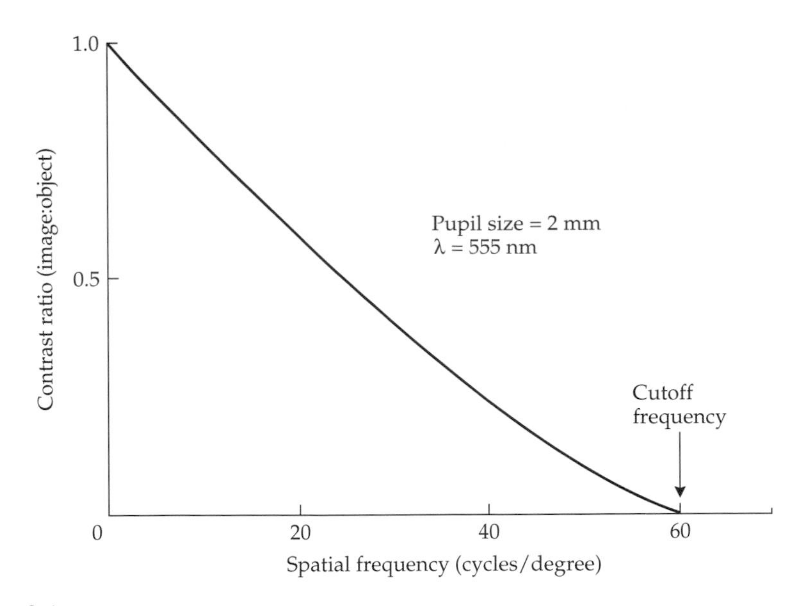

Figure 2.4
The Modulation Transfer Function for the Human Eye
The modulation transfer function shows the contrast in the image relative to the contrast in the object over the complete range of spatial frequencies transmitted by the optics. The ratio of image contrast to object contrast is 1.0 for zero spatial frequency (a uniformly illuminated object), but it declines with increasing spatial frequency until the contrast ratio is zero at the cutoff frequency. The shape of the modulation transfer function is affected by aberrations and defocusing, but the cutoff frequency is a constant—that is, an absolute limit—determined by the pupil size and the color of the light. In theory, the light distribution of the retinal image for any object can be determined from the modulation transfer function. (After Westheimer 1964.)

ical value because of an unavoidable imprecision in the original measurement of the line spread function.

The modulation transfer function of the eye is a measure of the quality of its image formation; later in this chapter we will consider how it is affected by aberrations and defocusing. The cutoff frequency gives an absolute limit for the ability of the visual system to resolve fine detail; high spatial frequencies not transmitted by the optics cannot be seen.

The contrast sensitivity function specifies how well different spatial frequencies are seen by the visual system

Other interesting questions concern how closely the visual system as a whole—that is, the combination of eye and brain—approaches the limit imposed by the optical system and how well the resolving capacity of the visual system is expressed by clinical measures of visual acuity. As it happens, the resolving power of the whole visual system falls somewhat short of the optical limit under most circumstances; that is, the ultimate limit on our ability to see fine detail is biological, not optical. The anatomical limit is imposed by photoreceptor size and spacing (see Chapter 15).

Although the modulation transfer function is a measure of optical performance, an analogous method can be used to determine the maximum spatial frequency detectable by the visual system as a whole. The targets are sinusoidal gratings whose contrast can be varied. For each of numerous spatial frequencies from low to high, the task is to determine the minimum grating contrast for which the pattern can be reliably detected by the observer (or, the contrast below which the grating cannot be detected). This value is a contrast threshold, and its reciprocal is a contrast sensitivity. If a grating can be detected at low contrast, the sensitivity is high. But if a grating cannot be detected at maximum contrast (100%), the threshold is effectively infinite and the sensitivity is zero, which is another way of saying that the grating is invisible.

It makes no difference whether the data are plotted as contrast thresholds or as contrast sensitivities as functions of spatial frequency; they are equivalent. The most common method, however, is the use of contrast sensitivity, probably because of the ease of interpretation; when the curve declines, sensitivity and visual performance are decreasing. These plots of contrast sensitivity for different spatial frequencies are called **contrast sensitivity functions** (Figure 2.5). This function superficially resembles the eye's modulation transfer function, which is shown by a shaded line in the figure, but the curves mean very different things; the modulation transfer function depicts properties of the optical system, while the contrast sensitivity function depends on the optics, the neural processing, and the perceptual apparatus.

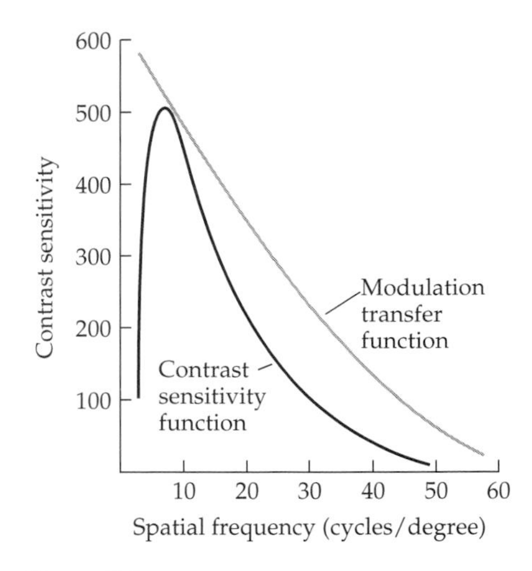

Figure 2.5
A Contrast Sensitivity Function for the Visual System
The solid curve is a contrast sensitivity function showing that the visual system is most sensitive to spatial frequencies around 5 or 6 cycles/degree. The sensitivity is somewhat reduced at lower frequencies and markedly reduced at higher frequencies; at around 50 cycles/degree, the sensitivity becomes too low (the contrast threshold is too high) to measure. The dotted curve shows contrast sensitivity determined solely by the eye's optics and its modulation transfer function; the visual system as a whole falls somewhat short of the performance of the optical system alone, indicating that the limitation on visual acuity and spatial resolution is biological, not optical. (After Woodhouse and Barlow 1982.)

The contrast sensitivity function in Figure 2.5 was measured with moderate pupil size, low photopic illumination, and corrected refractive error. Thus, it is a "typical" contrast sensitivity function. Sensitivity peaks around 5 cycles/degree and falls off at lower and higher spatial frequencies. The decline at higher frequencies is the more pronounced, and by extrapolating the curve down to zero sensitivity, we obtain a spatial frequency value that represents the maximum spatial frequency detectable by the visual system; here, the maximum frequency is about 50 cycles/degree. (This upper limit is not the same as the optical cutoff frequency, which is an absolute limit imposed by the pupil size. The maximum frequency for contrast sensitivity is highly variable.) The decline in sensitivity below 5 cycles/degree can be attributed to the effects of neural processing on low spatial frequencies.

Contrast sensitivity functions have been used extensively in experimental studies of visual performance; their value is the relative purity of the resolution task (clinical measures of visual acuity—see Box 2.1—are easily influenced by extraneous variables, like recognition, learning, or criterion shift) and the ability to detect effects on low or intermediate spatial frequencies, where some visual deficits may be particularly manifest. The possibility that specific pathologies are detectable in their early stages because characteristic, and therefore diagnostic, changes in contrast sensitivity functions appear has generated considerable interest in their clinical use.

We can see flies when their images subtend about one minute of visual angle

Visual angles can be specified in either image space or object space (Figure 2.6). To make the angles equal, their apices should be the eye's nodal points, but as a matter of practicality, the object space angle is usually specified with its apex in the plane of the pupil, and the image space angle is assumed to be identical. This is not quite accurate, but the error is generally too small to worry about.

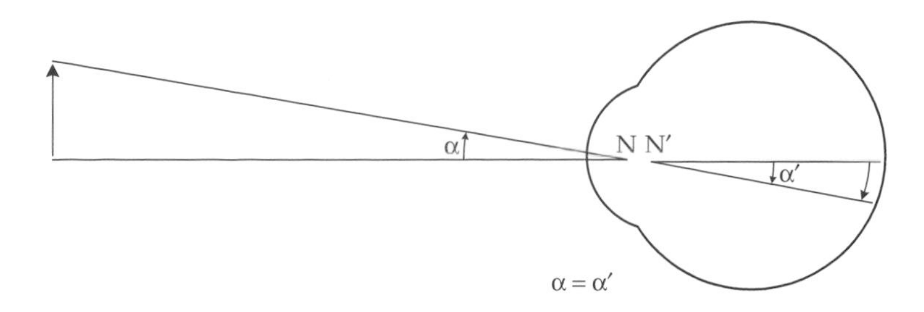

Figure 2.6
Visual Angles in Object Space and Image Space
The visual angle in object space is the angular subtense (α) of a target at the primary nodal point of the eye (N). The angle subtended by the target's image (α′) relative to the secondary nodal point (N′) is equal to the object space angle.

[Box 2.1] Evaluating Visual Resolution

VISUAL ACUITY IS THE MAIN CLINICAL CRITERION for assessing visual performance; it is determined for every patient, usually several times under different circumstances. Usually the patient's task is to read lines of alphabet letters that diminish in size progressively from line to line. When the patient can no longer read the letters, the resolution threshold has been crossed.

Snellen letters, which are named for their designer, the Dutch ophthalmologist Herman Snellen (1834–1908), have a specific construction (Figure 1): The width of the lines that form the letters is one-fifth the overall letter height. Letters whose height subtends a visual angle of 5′ of arc at a distance of 20 feet (approximately 6 m) are the norm; the Snellen ratio 20/20 (6/6) for normal visual acuity means that at a test distance of 20 feet, or 6 m (the numerator in the ratio), the patient can reliably identify letters subtending 5′ of arc at 20 feet, or 6 m (the denominator in the ratio).* If the patient can read only the larger letters subtending 5′ of arc at 40 feet, the acuity is 20/40 (6/12). Some definitions of legal blindness use 20/200 (6/60) acuity as the criterion; in this case, letters that can just be recognized at 20 feet are so large that they would subtend 5′ of arc at 200 feet (the 20/200 letter is 10 times larger in linear dimensions than the 20/20 letter).

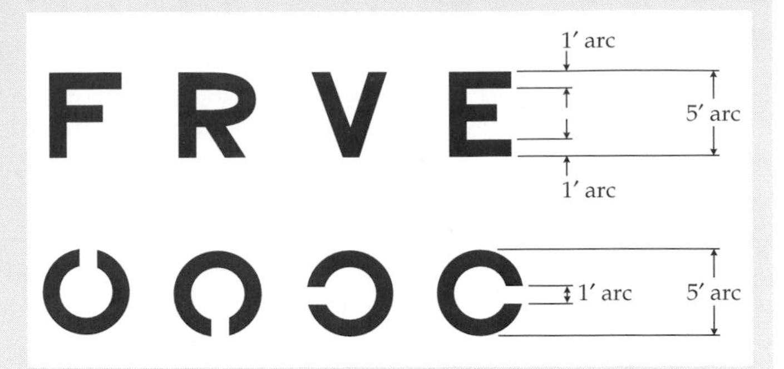

Figure 1
Letters for Determining Visual Acuity
The letters for testing visual acuity are constructed of simple lines whose width is one-fifth the height of the letters. At a testing distance of 20 feet (6 m), the standard letter height is 3/8 inch (8.73 mm), which subtends a visual angle of 5′ arc at the eye. Landolt *C*s (the second of the two lines) are constructed as circles with a gap that is equal to the line width and is one-fifth the diameter of the circle. The patient's task is to specify the orientation of the gap, whether up, left, right, or down.

Despite its clinical utility, Snellen acuity is not a direct measure of the visual system's ability to resolve fine detail, in part because of extra information contained in a Snellen letter. A blurred letter *T*, for example, will have a different overall shape (and different spatial frequency content) than an equally blurred letter *D*. Patients can often use this difference to guess letters correctly even when the detail of the letters is no longer visible. Since a correct guess will make the acuity seem better than it really is, correct identification of, say, more than half of the letters in a line is normally demanded.

Using the same letter in different orientations is a way of eliminating overall letter shape as a recognition factor; the Landolt *C*, named after the French ophthalmologist Edmond Landolt (1846–1926), is one example (see Figure 1). Since there are normally four possible orientations of the letter, one can be correct 25% of the time by guessing, so correct responses at least 50% of the time are required. The so-called tumbling *E*s often used with preschool children are another variation.

Measures of visual acuity are affected not only by optical and pathological factors but also by external considerations, such as the nature of the acuity targets. Acuity depends on target contrast, for example, and low-contrast targets, such as those generated by projection, may give poorer acuity values than well-illuminated printed targets with high contrast. But target contrast is not standardized for clinical use, nor is the illumination level in examination rooms. As a result, acuity measurements in clinics are not as meaningful as one might like; comparisons of acuity measures made in different places, at different times, and by different clinicians may vary because of differences in illumination, not because of any intrinsic factors that are of interest.

Much of the time, however, visual acuity is being used to measure the success of a refractive correction, with the expectation that corrected visual acuity in a normal eye will be 20/20 (6/6), and precision of the measurement is not of great concern; whether or not the true value is a bit more or a bit less is of no great consequence. In some cases, however, accuracy may be important if acuities need to be monitored over time or if the acuity measurement may itself be of diagnostic significance. A simple, effective approach to better accuracy is to plot the results of acuity determinations as **frequency-of-seeing curves**. These curves plot the percentage of letters of a given size that are identified correctly as a function of letter size expressed as their Snellen ratio (Figure 2).

In a normal case, for example, none (0%) of the 20/15 (6/4.5) letters might be identifiable, while all (100%) of the 20/30 (6/9) letters might be identified correctly. For intermediate sizes, percentages of the letters identified correctly might be on the order of 25% and 75%. In this manufactured example, a requirement for 50% correct would yield an acuity of about 20/22.5 (6/6.75). In normal individuals, the transition between 0% correct and 100% is usually quite abrupt, passing through only one or two intermediate letter sizes.

In **amblyopia**, which is reduced visual acuity not attributable to refractive error or pathology, the transition from 0% to 100% is much more gradual, encompassing perhaps three or four intermediate letter sizes. The frequency-of-seeing curve is necessary for establishing a meaningful acuity measure for an amblyopic eye, and the gradual rise of the curve is a diagnostic sign (normal eyes with

*Although metric units are standard for optical calculations and for the ophthalmic industry, clinicians in the United States use an outmoded visual acuity specification. Metric specifications are used elsewhere, so the normal corrected acuity in most of the world is 6/6, not 20/20. Why the outdated practice continues is unknown, but it is followed here, with the metric equivalents in parentheses.

uncorrected refractive error will have reduced visual acuity, but the transition between 0% and 100% correct will still be abrupt).

The contrast sensitivity functions discussed in the text have moved into clinical settings as automated testing instruments have been developed. They generally are not used to determine a high spatial frequency cutoff directly, however; the curve determined by measurements made at lower frequencies must be extrapolated to determine a high-frequency cutoff. But this is not always done; the most common procedure is to compare the contrast thresholds at lower spatial frequencies to a set of normal limits determined by the instrument designer. Different patterns of deviation from the expected values may provide clues to the causal factors, which may be optical or pathological in nature.

Contrast sensitivity functions characterize the performance of the visual system more completely than conventional measures of visual acuity and therefore offer a potentially richer source of information about the effects of visual deficits. For this reason, some new acuity charts have been developed and introduced during the past decade or so that incorporate contrast as a variable element (Figure 3). These newer acuity charts differ in design—some using letters of variable contrast, others use grating patterns—but the intent is to provide clinicians with the kind of information available from contrast sensitivity functions without the attendant technical and practical limitations. If clinical trials show these methods to have significant diagnostic advantages, they will undoubtedly come to be used routinely.

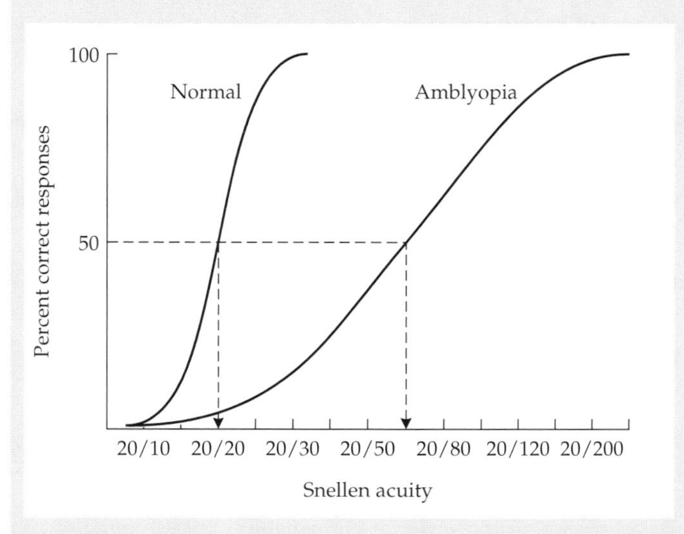

Figure 2
Acuity Determined from Frequency-of-Seeing Curves
Each line of an acuity chart has up to a dozen letters or symbols to be identified. None of the letters can be identified correctly when they are below the acuity threshold, but all of them can be identified correctly with larger letter sizes. Intermediate letter sizes produce correct identifications between these extremes, and the plot of the percentage of correct identifications as a function of letter size is a frequency-of-seeing curve. The acuity threshold is often taken to be the letter size at which 50% of the letters were correctly identified. For normals, the transition from 0% to 100% correct is abrupt, but individuals with amblyopia typically have an extended range of letter sizes between 0% and 100% correct.

Figure 3
Acuity Charts with Variable Contrast
The letters on a single page of the Pelli-Robson chart are all the same size, but their contrast decreases from top to bottom. By using different pages with different letter sizes, correct identification is a function of both letter size and contrast. The added dimension of contrast has the potential to make the test more sensitive to visual dysfunction and to discriminate among reduced visual acuities arising from different sources. (After Pelli, Robson, and Wilkins 1988.)

The Rayleigh criterion and the cutoff frequencies for the modulation transfer function and contrast sensitivity function indicate that we can resolve point objects when they are separated by about 1′ of arc in visual angle. (The reciprocal of the cutoff frequencies—that is, degrees/cycle—is used here. Thus 60 cycles/degree is 1/60°/cycle, or 1′ of arc.) Neither the two-point resolution task embodied in the Rayleigh criterion nor the grating detection task in measuring contrast sensitivity cutoff frequency is the same as detecting a single light or dark object—like a fly—but the thresholds are not very different. Thus most people can see small objects at high contrast when the object subtends about 1′ of visual angle.

Specifying the threshold in angular terms incorporates objects of different sizes; a bowling ball can be considerably farther away than a fly and still subtend 1′ of visual angle. And what is the greatest distance at which a fly can be seen? Assuming the fly is 4 mm long, the answer is 13.8 m (0.004 m/tan 1′). The 100-yard (91.4 m) distance at which James Fenimore Cooper's marksman supposedly saw and hit a target the same size is about seven times farther away. He might have been able to see the fly if it had been about the size of a Ping-Pong ball (32 mm).

The Anatomy of Image Formation

The quality of a focused image is affected by pupil size, curvatures of optical surfaces, and homogeneity of the optical media

Point objects are spread into small discs as focused images in any optical system, no matter how good the system is; the eye, astronomical telescopes, and fine cameras are all subject to the effect, the most fundamental source of which is diffraction by the aperture. **Diffraction** is the classic example of light having a wavelike character (although it can be explained without treating light as waves). As concentric wave fronts encounter the aperture, part of the wave front is blocked by the opaque material surrounding the aperture and part passes through the aperture. As the wave fronts pass through the aperture, they spread out, as if they were bent by the edges of the aperture. This spreading increases as the aperture becomes smaller, so very small apertures are almost like secondary light sources from which light radiates out in all directions to fill the hemispherical space behind the aperture.

This unavoidable image spread from diffraction is seen in the modulation transfer function as a loss of high spatial frequency content in the image relative to the object, and as noted earlier, the cutoff frequency of the system is directly related to pupil size: The larger the pupil, the higher the cutoff frequency, and the less spread in the image.

Calculated values for the cutoff frequencies of the eye at different pupil sizes are given in Table 2.1, along with their approximate equivalents in the clinical Snellen notation. Note that a 2 mm pupil diameter, which is near the minimum of the normal pupil size range, is associated with a cutoff frequency of 63 cycles/degree and a maximum visual acuity of 20/10 (visual acuity notation is discussed in Box 2.1). Since the best visual acuities are normally in the order of 20/10 to 20/15, diffraction from the smallest normal pupil size seems unlikely to be a limiting factor for visual acuity. Larger pupil sizes, with their larger cutoff frequencies, will certainly not generate diffraction limitations on visual acuity.

Table 2.1

Diffraction Limits for an Aberration-Free Eye

Entrance pupil diameter (mm)	Cutoff frequency (cycles/degree)	Approximate Snellen acuity	
0.5	15.6	20/40	(6/12)
1.0	31.2	20/20	(6/6)
1.5	47.4	20/13	(6/4)
2.0	63.0	20/10	(6/3)
2.5	78.6	20/8	(6/2.4)
3.0	94.2	20/6	(6/1.8)

Source: Westheimer 1964.

The curves in Figure 2.7 are the modulation transfer functions calculated for different pupil sizes, assuming the eye has no aberrations of any kind. Contrast is attenuated progressively as the spatial frequency increases, meaning that the finite pupil size imposes not only an absolute limit on the system—the cutoff frequency—but also relative limits. And the effect is less severe for larger pupil sizes; *in the absence of aberrations,* larger pupil sizes with their higher cutoff frequencies are associated with less attenuation below the cutoff frequency (Figure 2.7*a*). If these curves are replotted so that the cutoff frequency is unity, whatever its absolute value, and other spatial frequencies are plotted as a fraction of the cutoff frequency, the curves collapse into a single curve that represents the performance of an eye limited only by diffraction (Figure 2.7*b*). If the spatial frequency is a specific fraction of the cutoff frequency, there will be a specific amount of contrast attenuation, regardless of pupil diameter. But since larger pupils are associated with larger cutoff frequencies, it follows that the eye should operate with the pupil as large as possible—theoretically.

Although pupil size is the only anatomical variable in the equations for the Rayleigh criterion or for cutoff frequency, it is not the only anatomical component affecting image quality. Aberrations are introduced by the particular shape of the eye's optical surfaces. If the cornea and lens had perfectly spherical surfaces, for example, the eye would exhibit considerable **spherical aberration**, in which light entering the eye farthest from the optic axis focuses in front of the paraxial rays. The dioptric difference between these two extremes is a measure of the amount of spherical aberration. Since larger pupil sizes admit more of the peripherally entering rays, spherical aberration and retinal image degradation should increase as pupil size increases; this consequence is an argument for keeping pupil size as small as possible—just the opposite of the conclusion we reached when considering diffraction.

Spherical aberration is reduced by aspherical surfaces whose radius of curvature increases from center to periphery. Measurements of the eye's spherical aberration are not particularly simple and have not been made very often, but there are two consistent results: First, the eye has some spherical aberration, on the order of 0.5 to 1.5 D, but this aberration is considerably less than one would expect if the eye lacked a form of aberration correction. Second, the major factors operating to minimize the eye's spherical aberration are peripheral flattening of the cornea and a refractive index gradient in the lens, where the index decreases from center to periphery. Most results indicate that the spherical aberration of the lens is negative in sign, thereby making the total spherical aberration of the eye less than the spherical aberration of the cornea.

The total effect of spherical and other aberrations on the eye's optical performance can be seen in the experimentally determined modulation transfer functions for the eye (Figure 2.8). The eye never achieves the theoretically optimum performance, and the deviation from optimum increases with increasing pupil size. The simple conclusion, therefore, is that the image-degrading effects of aberrations far outweigh any limitations that might be imposed by diffraction; image quality will be better with small pupils than with large pupils. Small pupil sizes also produce an increased depth of focus, which means that there is less precision required in the position of the optical image relative to the retina. (The retinal image formed through a small pupil will be dimmer, however.)

In discussions of image quality, the optical media are assumed to be transparent and free of local variations in refractive index (like bubbles) or opacities. If present, any of these inhomogeneities can adversely affect the modulation transfer function and produce further spread and distortion in the retinal image. In extreme cases, an optical inhomogeneity may produce two images of a single object (**monocular diplopia**). In general, extensive opacities in the cornea or lens require surgical intervention to restore some semblance of normal image quality.

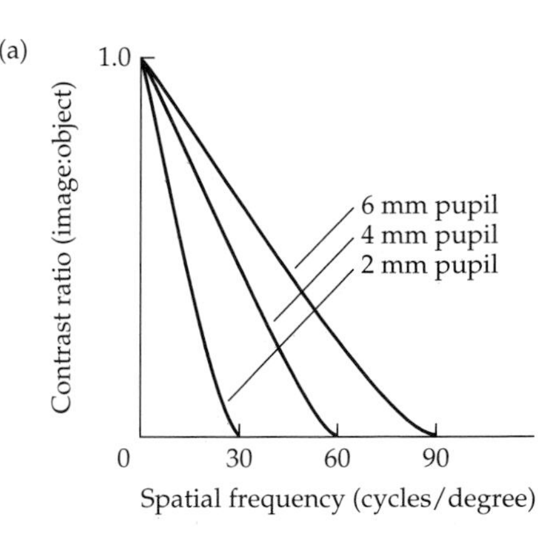

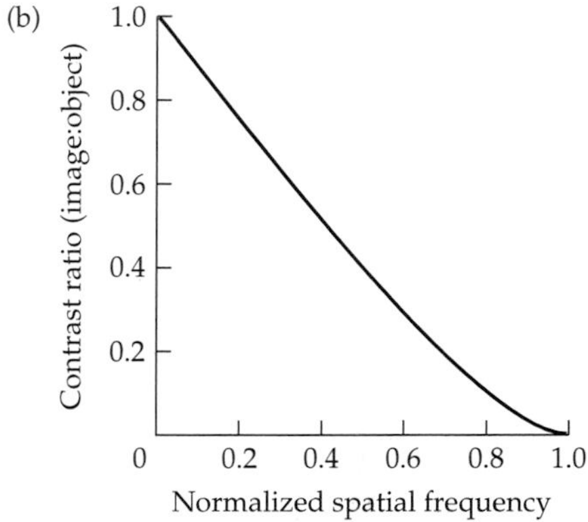

Figure 2.7
Modulation Transfer Functions for an Aberration-Free Eye
(*a*) As the pupil size becomes smaller, the cutoff frequency decreases, and this decrease corresponds to increased light spread in the image. (*b*) When spatial frequency is plotted in relative terms, as a fraction of the cutoff frequency, the curves for different pupil sizes collapse into a single curve that provides complete information about contrast attenuation as a function of spatial frequency for the optical system. (After Westheimer 1964.)

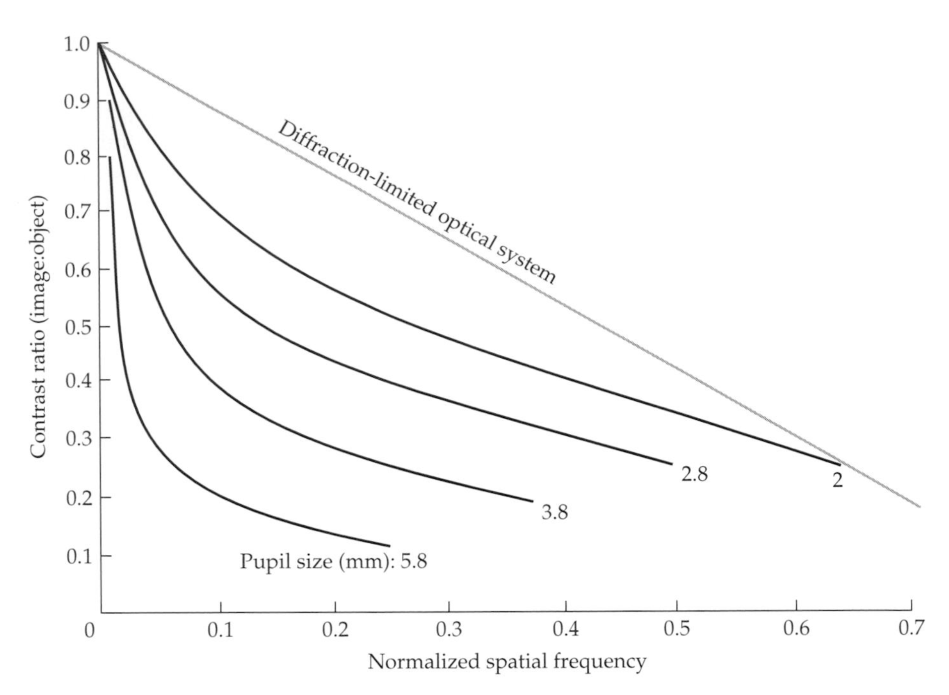

Figure 2.8
Modulation Transfer Functions for the Human Eye
Modulation transfer functions for a real eye with different pupil sizes are compared to the theoretical curve for an aberration-free eye (a diffraction-limited optical system). As pupil size increases, the curves deviate farther from the ideal and exhibit severe contrast attenuation at large pupil sizes. This increased attenuation with larger pupils is due to the eye's aberrations. (After Campbell and Green 1965.)

Image quality in a particular eye is difficult to assess directly, and since it is an inherent property of the eye's optical anatomy, it is difficult to improve. Image quality is quite easy to reduce, however; corneal refractive surgery (see Chapter 8), for example, always reduces image quality to some extent, even as it corrects refractive error. In practice, ensuring that the eye is working at its inherent limit—doing its best—means ensuring in-focus imagery, because even small amounts of defocusing have profound effects on image quality.

Defocusing produces large changes in the modulation transfer function

Modulation transfer functions can be used to assess the effects of defocusing (Figure 2.9). In general, the experimental and theoretical pairs of curves in Figure 2.9 are similar, but their divergences illustrate the fact that a 2 mm pupil does not eliminate all aberration effects. More important, however, is the severe loss of contrast exhibited with progressively larger amounts of defocusing. For the 3.5 D defocusing in the bottom pair of curves, contrast has been reduced by 90% or more for all spatial frequencies greater than one-tenth of the cutoff frequency (in other words, for all spatial frequencies above 5 or 6 cycles/degree). This is a severe deficit; a patient with this amount of defocusing would have a Snellen acuity around 20/400 (6/120).

To some extent, the reduction in the modulation transfer functions with defocusing is like the effect of increasing pupil size with in-focus imagery (see Figure 2.8). The modulation transfer functions for 2.5 D defocusing and for an in-focus image with a 5.8 mm pupil diameter are roughly equivalent, for example. The

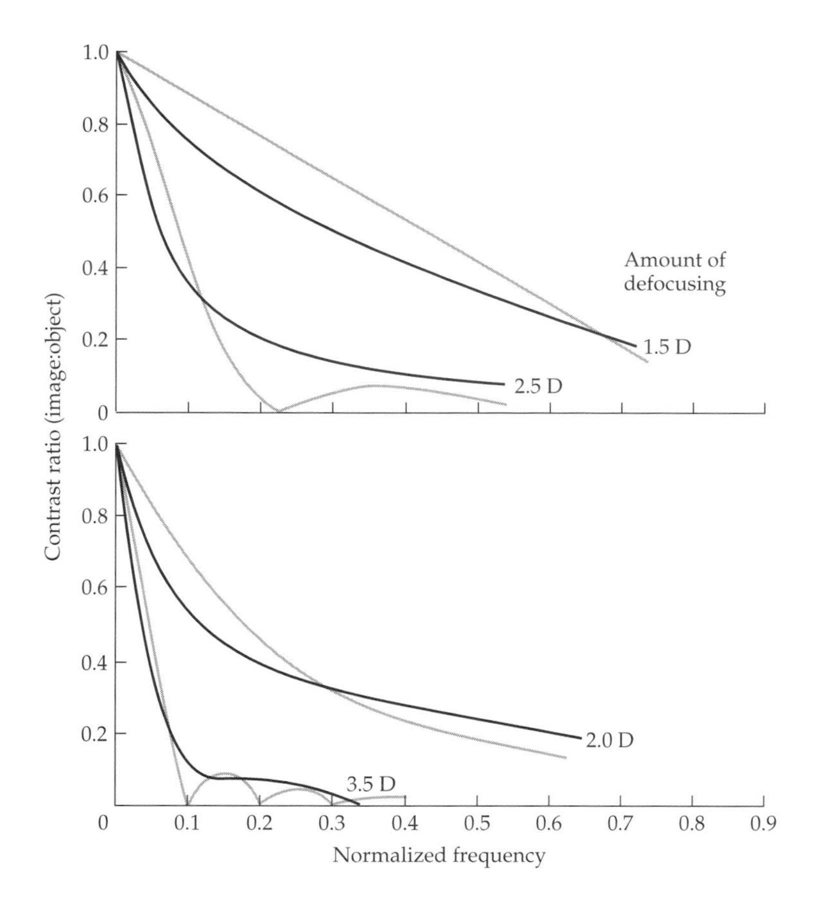

Figure 2.9
Modulation Transfer Functions with Defocusing
The effect of defocusing (with a constant pupil size) is similar to the effect of increasing pupil size with an in-focus image. The contrast attenuation for 3.5 D defocusing is severe at higher spatial frequencies and will reduce visual acuity to around 20/400 (6/120). The dashed curves are theoretical functions for an idealized diffraction-limited eye; the solid curves represent experimental results. (After Campbell and Green 1965.)

practical difference is that pupil diameters of 5.8 mm with daytime illumination are not to be expected, whereas refractive errors of 2.5 D are relatively common. The refractive error is the more serious impediment to optimal performance by the optical system.

The major anatomical factors that determine the refractive power of the eye are the curvatures of the cornea and lens and the depth of the anterior chamber

The anatomical variables that affect image quality are also relevant to refractive error and image defocusing. For the image of a distant object to be in focus on the retina, the refractive power of the eye must match, in a sense, the axial length of the eye (or vice versa). If the refractive power is too large relative to the axial length, the optical image will be formed in front of the retina and the retinal image will be out of focus; this is **myopia** (Figure 2.10*a*). Conversely, if the refractive power is too weak relative to the axial length, the optical image will lie behind the retina (or at least it *would*, if the retina were not in the way). Again, the retinal image is blurred; this condition is **hyperopia** (Figure 2.10*b*). When all the components are properly matched and the optical image coincides with the retinal plane, the refractive error is zero; this is **emmetropia** (Figure 2.10*c*).

Optically, the eye is a two-element system, made up of the cornea and the lens, both of which are plus lenses. The overall dioptric power of the system is given by the equation

$$F_{\text{total}} = F_{\text{cornea}} + F_{\text{lens}} - \left(\frac{t}{n}\right)(F_{\text{cornea}} \times F_{\text{lens}})$$

where t is the distance between the posterior surface of the cornea and the anterior surface of the lens (the **anterior chamber depth**, which is normally about 3.5

(a) Myopia

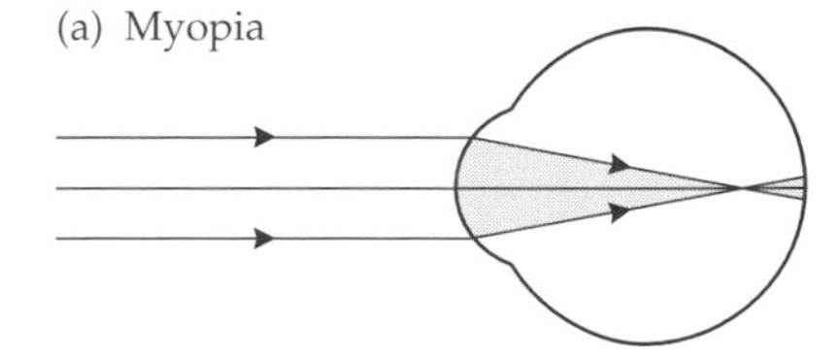

(b) Hyperopia

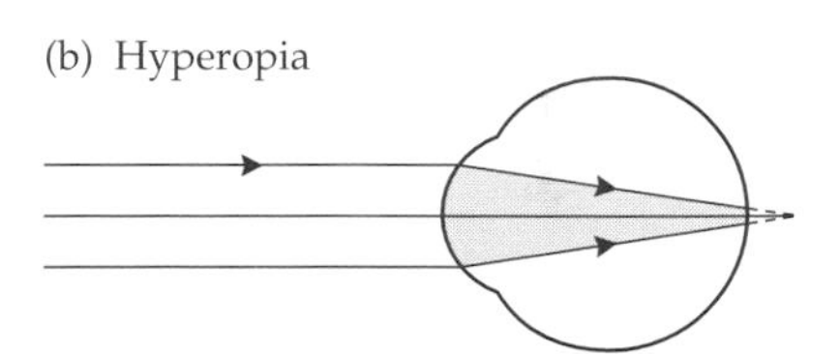

(c) Emmetropia

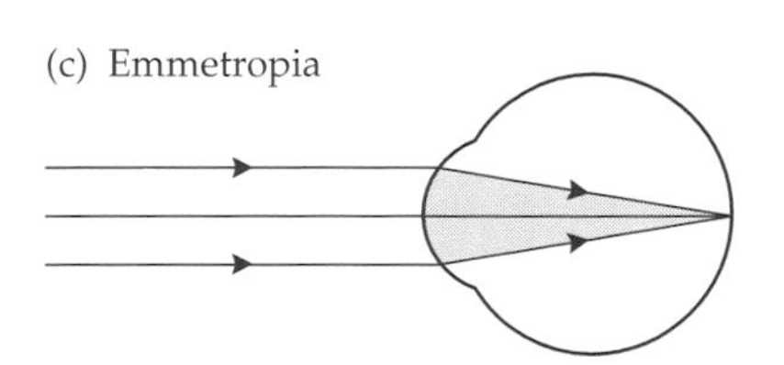

Figure 2.10
Image Location in Myopia, Hyperopia, and Emmetropia
The image plane lies in front of the retina in myopic eyes (*a*) and behind the retina in hyperopic eyes (*b*). The image plane shifts by about 100 μm for each diopter of refractive error. Emmetropic eyes (*c*), which have no refractive error, form images of distant objects on the retina, in the plane of the photoreceptors.

mm), and n is the refractive index of the aqueous humor. With refractive powers constant, a shallower anterior chamber increases the total power of the system, while a deeper chamber decreases overall power.

Most of the refraction of light occurs at the anterior surface of the cornea; it is the only air-to-tissue interface in the system, so the change in refractive index is greater here than anywhere else. Thus, the cornea's anterior radius of curvature is the major factor in the refractive status of the eye.* An average value for the anterior radius of curvature is around 8 mm, which translates to a power of +47 D (the anterior surface power equals the difference between the cornea's refractive index [about 1.376] and the refractive index of air [1.000], all divided by the radius of curvature in meters). Combining the powers of the anterior and posterior surfaces and allowing for their 0.5 mm separation gives a total corneal power of +41.1 D.

The overall power of the lens is about half that of the cornea; therefore the lens contributes about one-third of the eye's total power in its unaccommodated condition, which is the situation when we are viewing distant objects. Average radii of curvature for the unaccommodated adult lens are around 10 mm for the anterior surface and –6 mm for the posterior surface. (These are the values commonly used in schematic eyes for calculation purposes. More recent measurements give higher central radii of curvature; see Chapter 12.) Combining these measurements with a thickness around 4 mm gives a total unaccommodated lens power of about +20 D (assuming the lens has a refractive index of 1.413). When the lens increases in power with maximum accommodation, the anterior radius of curvature is about 5 mm and the posterior radius is about –5 mm. The combined effect increases lens power to around +30 D.

The optical properties of the lens are less well understood than those of the cornea. Measurements of its surface curvatures are more difficult, and although the refractive index is known to vary from the center to the periphery of the lens, the details of the variation are not known. Calculating lens power with a surrounding index of 1.336 for the aqueous and vitreous and a single average value for the lens (for example, 1.413, as given) underestimates lens power. The lens is really a gradient index system, and it needs to be treated as a set of nested shells (like an onion) whose indices change in small steps from the central to the peripheral shells. When an index gradient is incorporated into the calculations, the total lens power increases by about +2 D over calculations with a single, uniform index.

When the cornea and lens are considered as a combined optical system, these values produce a total power for the eye of around +57 D; another often cited value around +60 D is based on a slightly smaller corneal radius of curvature. The precise value is not important, since no model is totally accurate; the main point is that the dioptric power of the "typical" unaccommodated, emmetropic eye is around +60 D, give or take a bit. Of this +60 D, the cornea contributes about two-thirds and the lens about one-third.

In simple myopia or hyperopia, the defocused image of a point source is a blur circle, the size of which is proportional to the amount of refractive error. In **astigmatism**, however, there is no single plane of focus (Figure 2.11). A point source usually does not have a circular image, because the plane of focus varies from one meridian to the next. For example, if the target to be imaged were a small cross (+), astigmatism might cause the image of the vertical limb (|) to be focused in one plane (just in front of the retina, perhaps), while the image of the horizontal limb (–) might be focused closer to the lens—that is, farther from the retina. Midway between these two planes, the limbs of the cross would be equally out of focus, creating what is called the circle of least confusion.

*The posterior surface of the cornea has a smaller radius of curvature than the anterior surface, but the refractive indices of the cornea and the aqueous are so similar that the interface has relatively little refractive power. At a calculated value of about –6 D, its contribution to the cornea's refractive power is relatively small, though not negligible.

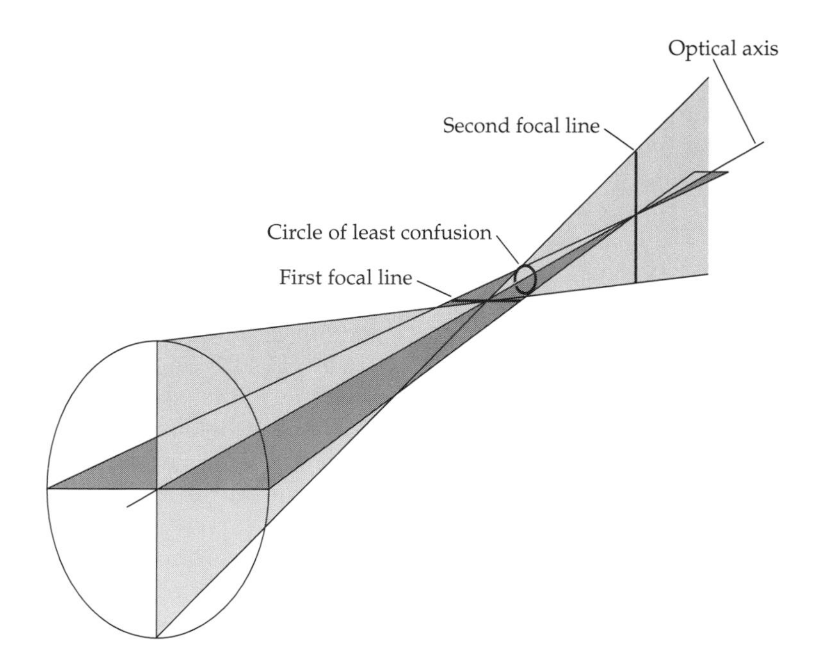

Figure 2.11
Astigmatism
There is no single plane of focus for an astigmatic optical system. Instead, the meridians of greatest and least power have focal planes defining a range of focus with a blurred "circle of least confusion" midway between the extremes.

As it happens, the image of the horizontal limb of the cross is formed by light rays passing through the vertical meridian of the eye's optical system, while the rays through the horizontal meridian form the image of the vertical limb; thus, the optical power in the two meridians must be different, with more positive power in the vertical meridian in this example. The reverse situation is also possible, with the greater power in the horizontal meridian. Also, the meridians of greatest and least power may be something other than strictly horizontal and vertical (**oblique astigmatism**), or the meridians may not be orthogonal (**irregular astigmatism**).

Astigmatism could be produced by a tilting or skewing of the cornea or lens with respect to the anterior–posterior axis of the eye or by a lack of radial symmetry in one or more optical surfaces. The most common cause appears to be asymmetry of the anterior corneal surface.

Schematic eyes are approximations of the eye's optically relevant anatomy

The first Nobel prize for studies of the eye and vision was awarded in 1911. It went to Allvar Gullstrand, a Swedish ophthalmologist, for his work on the optics of the eye.* Gullstrand invented the slit lamp—a now standard piece of clinical apparatus—and made significant improvements in Hermann von Helmholtz's ophthalmoscope, but his prize-winning work is encapsulated in several schematic eyes that still bear his name.

Individual eyes vary considerably, as we will see, and optically important data may not be readily available for a particular eye; thus much of what we understand about the optical function of the human eye is based on model eyes that approximate an average or typical eye to some degree. A simple approximation is called a reduced eye, in which the cornea and lens have been replaced by a single optical surface having the equivalent power of the cornea–lens combination (Figure 2.12). A closer approach to reality might have a separate cornea and lens, but each represented by a single surface instead of two. In its ultimate form, the

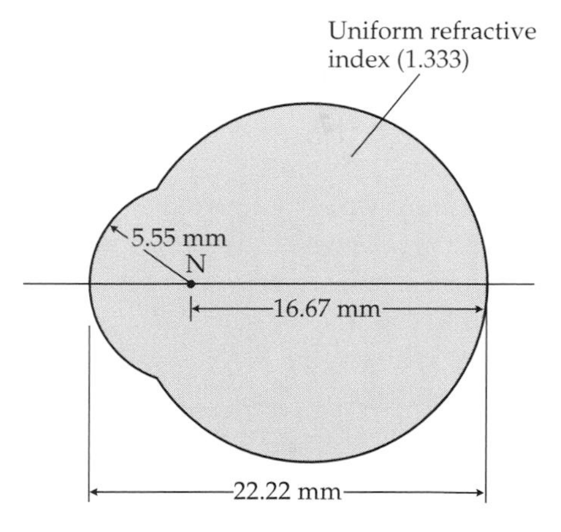

Figure 2.12
A Reduced Eye
The cornea–lens combination of the real eye can be represented in several simplified forms. This one is a single surface at which the refractive index changes and whose power is equivalent to the eye's total power. The eye has a single nodal point (N). The corneal radius of curvature and the axial length are unrealistically small.

*Vision scientists have been honored twice more since Gullstrand's time. The prize in physiology or medicine went to Ragnar Granit, H. K. Hartline, and George Wald in 1967 and to David Hubel and Torsten Wiesel in 1981 (shared with neuroscientist Roger Sperry).

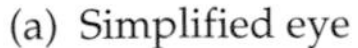

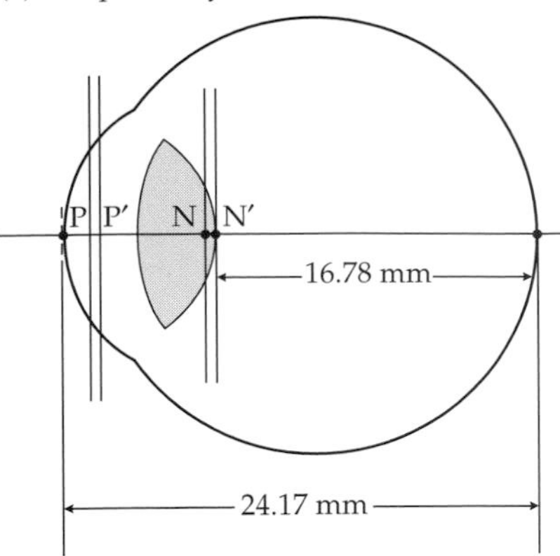

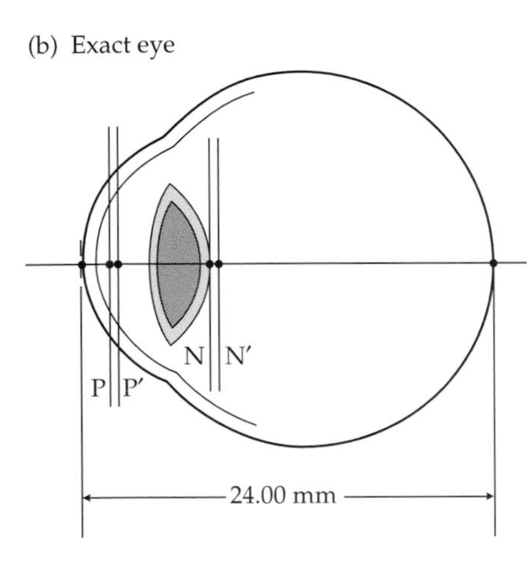

Figure 2.13
Gullstrand's Schematic Eyes
(*a*) The number 2 simplified eye. For many purposes, the approximations introduced by ignoring some optically relevant features are not significant and are outweighed by easier calculations. The simplified eye is often what people mean when they refer to the optical properties of "the eye." (*b*) The exact eye. This cross-sectional drawing of the eye shows, to scale, the curvatures and dimensions of the eye, along with the locations of the principal planes (P and P′) and nodal points (N and N′). All the optical surfaces have constant radii of curvature; that is, they are spherical surfaces. The corneal radius of curvature (7.70 mm) is realistic and the lens is treated as having two refractive indices, the higher one in the center of the lens.

model eye is an attempt to represent the optical components of the biological eye as accurately as possible, with curvatures, thicknesses, and separations that mimic the anatomy of the eye.

Two of Gullstrand's schematic eyes are shown in Figure 2.13. His "simplified eye" has a single corneal surface and a lens with a single refractive index (Figure 2.13*a*); the "exact eye" includes both corneal surfaces and a lens with different central and peripheral refractive indices (Figure 2.13*b*). The differences are small, but significant. The simplified eye has higher dioptric power, a flatter cornea, and a thicker lens, for example. The differences between the schematic eyes mean that they will give somewhat different answers to optical calculations; the exact eye will be required for some purposes, while the simplified eye may be adequate for others. It depends on the degree of accuracy required. But when we refer to "the eye" in the context of its optical characteristics, we are referring to one of these eyes. These are the principal planes we use as a standard in discussing the focal length or refractive power of "the eye," and these are the nodal points we use for calculating image size or angular magnification.

The exact eye, despite its name, exhibits several departures from reality; the constant radii of curvature of the corneal and lenticular surfaces are the most obvious. The lack of an aperture stop (the pupil) is another, although it can be added quite easily. These features mean that the exact eye will exhibit severe spherical aberration (among others), quite unlike the real eye, and it is not useful for evaluating the physical optics of the eye; it is a model of the eye's geometric optics. The exact eye is also slightly hyperopic, a feature that disappears when more precise gradient index modeling of the lens is incorporated into the model.

For most calculations involving the effects of changing curvatures or axial length to simulate refractive error and its correction, the simplified eye is used. For example, when we refer to the angular subtense of an object, it is the angle subtended at the primary nodal point; the angular subtense of the image—subtended at the secondary nodal point—will be the same, by the definition of nodal points. To calculate the object's angular subtense, therefore, we need to know the size of the object and its distance from the primary nodal point, which will be the distance from the cornea *plus* 7.1 mm. An object that subtends 1° at the primary nodal point will have a retinal size equal to tan 1° × 16.78 mm (the posterior nodal distance)—that is, approximately 0.3 mm, or 300 μm. Thus for small angles, the standard form of the relation between angular subtense and retinal image size is 1′ arc = 5 μm on the retina.*

The all-inclusive term for refractive errors is **ametropia**. Schematic eyes have been useful in evaluating the optical consequences of correcting different categories of ametropia. At one time, a distinction was made between *refractive* ametropia, in which the axial length is "normal" but the eye's refractive power is not, and *axial* ametropia, in which the refractive power is normal but the axial length is unusually long or short. (Although the dichotomy between axial and refractive seems plausible and is certainly easy to think about, there are wide ranges of variation in axial length and corneal curvature associated with the same amount of ametropia. The variation in axial length will be considered later. Ametropias attributable solely to axial length or solely to optical power are probably quite rare.)

Modeling a refractive ametropia is quite simple; all the curvatures and locations of principal points or nodal points remain the same as in, say, Gullstrand's simplified eye, and only the axial length is varied. Modeling axial ametropia, on

*The conversion factor for large angles requires calculating the size of a circular arc subtended by a given angle with a radius of 16.78 mm. In this case, size equals 16.78 mm multiplied by the angle in *radians*. For a 1° angle (1° = π/180 radians), this calculation yields a size that is almost identical to the small-angle approximation, differing by less than 0.001 mm. For a 10° angle, however, the calculated sizes differ by about 3%, and the difference increases very rapidly for still larger angles.

the other hand, requires changing the curvatures of surfaces and then recalculating the positions of the principal planes and nodal points, while leaving the axial length fixed at 24 mm. This done, equivalent amounts of ametropia in the two systems can be compared in terms of the sizes of the optical images, the blurred retinal image sizes, the amount of anatomical change required to produce equal amounts of defocusing, and so on. More important, however, is the ability to evaluate the effects of correcting the ametropia with spectacle lenses, contact lenses, or—in an aphakic model with the lens removed—intraocular lens transplants.

Eye Shape and Size

Vertebrate eyes vary considerably in shape

Vertebrate eyes vary from distinctly flattened (oblate) spheroids (in some fish, for example) to tubular, almost bell-shaped eyes (in some birds). Thus, to say that the human eye is basically spherical is not altogether trivial; in this, as in other anatomical respects, the human eye is somewhere in the middle of the observed range of variation.

Near-sphericity facilitates movement of the eye. The eye rests in a pocket of orbital fat and connective tissue (see Chapter 3), and we can think of it as the ball in a ball-and-socket joint; the ball is free to rotate in any direction around a fixed center of rotation (which, ideally, is the ball's center of curvature). This freedom of movement is a direct consequence of the spherical shape. In animals whose eyes are decidedly aspherical, movements of the eyes in some directions are quite restricted; in owls, for example, the eyes are nearly immobile around the horizontal and vertical axes of rotation; the animals compensate by having a large range of head movement. Since the sclera of the human eye is not perfectly spherical, it can be shown experimentally that the center of rotation is not a point fixed exactly at the center of curvature of the sclera. The deviation is quite small over the normal range of eye movements, however, and for most purposes it can be treated as negligible if the ball-and-socket model is used as a basis for discussing movements of the eyes.

Both the cornea and the sclera are aspheric

The greater curvature (smaller radius of curvature) of the cornea relative to the sclera is the eye's main departure from sphericity. In an idealized horizontal section of the eye, the sclera would be a circle having a radius of about 12 mm with a segment of an 8 mm radius circle added at the front to represent the cornea. This combination would make the anterior–posterior length of the eye longer than the scleral diameter. In fact, a horizontal section of the eye is almost completely enclosed by a 12 mm radius circle, as shown in Figure 2.14; in other words, the eye's major diameters—horizontal, vertical, and anterior–posterior—are approximately equal.

The curvature of the cornea is not constant, flattening from center to periphery, and the scleral radius of curvature also varies. The anterior sclera has a somewhat *smaller* radius that brings the scleral cross section inside the circle, and then it flattens (the radius of curvature increases) near the cornea, making a significant departure anteriorly from the idealized circular cross section. As a result, the bulge of the cornea does not extend much beyond the circle. There is also a characteristic distortion of the sclera on the posterior-nasal aspect where the optic nerve exits the eye.

Although adult eyes can vary significantly in overall size, as we will see, normal eyes adhere to this basic shape; distortions are abnormal. Both the cornea and the sclera can develop large areas of structural weakness and thinning of the tissue that result in outward bulging in the affected region. In the sclera, these

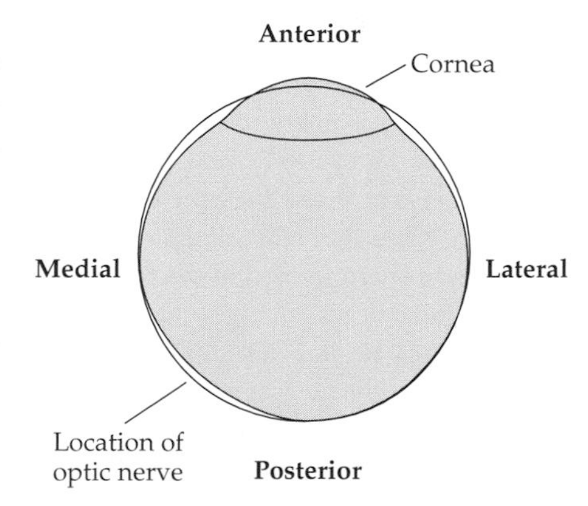

Figure 2.14
A Horizontal Section of the Human Eye
If the eye were perfectly spherical, any section that included the center of curvature would be circular. This illustration shows the departure of the eye's cross section from a circle that best fits the sclera. The sclera is flattened around the optic nerve and around the cornea.

Vignette 2.2
Fundamentum Opticum

AT THE BEGINNING of the seventeenth century, no one understood how the eye worked as an optical system. Refraction of light at curved surfaces could be dealt with only if the problem was reduced to chief rays that struck the surface perpendicularly, a gross oversimplification. There was disagreement about which of the eye's components was sensitive to light (most people thought it to be the lens), and the prevailing anatomical authority, Vesalius (see Vignette 3.2), placed the lens near the center of the eye, where its true optical role could not be appreciated.

But important reinterpretations were in progress. With no evidence other than a conviction that the image in the eye should be large, Felix Platter (1526–1614) designated the retina as the light-sensitive surface in the eye. The retina seemed to be the only structure large enough to capture a large image. Then Fabricius (see Vignette 12.1) described the true size, shape, and position of the lens in 1600. This was an important step, and Fabricius knew it, but he was unable to evaluate the optical consequences of his work; that would require someone who understood refraction and could do the mathematics of optical ray tracing.

The person for the job was Johannes Kepler (1571–1630). Although he had entered the University of Tübingen with the intention of becoming a Lutheran clergyman, Kepler studied astronomy and mathematics, becoming first a teacher of mathematics in Graz, Austria, and later succeeding the Danish astronomer Tycho Brahe as imperial mathematician at Prague. Since astronomy was then a highly visual discipline (most observations are made by instruments these days), Kepler was concerned about how the eye worked and how its workings might affect astronomical observations. Accordingly, he attempted to evaluate the image-forming properties of the eye.

The problem in doing so was the absence of a general law of refraction—what we now call Snell's law was several decades away—and an inability to deal with aspherical surfaces. Thus, the surfaces of the eye were assumed to be spherical, and the bending of light rays at these surfaces was treated by an approximation. Even so, Kepler got most of the story right; his calculations showed an *inverted* image formed in the plane of the retina. Kepler showed no diagrams of his conclusions, but they were illustrated later by René Descartes in *La Dioptrique* (Figure 1). Descartes is also responsible for the familiar version of Snell's law, using sine or cosine functions; Snell used cosecants.

In the diagram, cones of light rays from each of three points in space are admitted to the eye through the pupil, and each cone is focused on a corresponding retinal point after refraction by the cornea and lens. The image is inverted; the left-hand point in space (V) is to the right on the retina (R), while the right-hand point in space (Y) is to the left on the retina (T). The apparent inversion of the retinal image was bothersome; a century earlier, Leonardo da Vinci had assumed that the orientation of the retinal image must correspond to the way we see the world, and it is difficult to see why this correspondence is not a necessary one. Kepler didn't want to believe the image was inverted, and he went to considerable pains to find something wrong with his calculations. But he couldn't do it and had to attribute the idea that the image turns right side up again before being perceived to a nonoptical feature of the visual process.

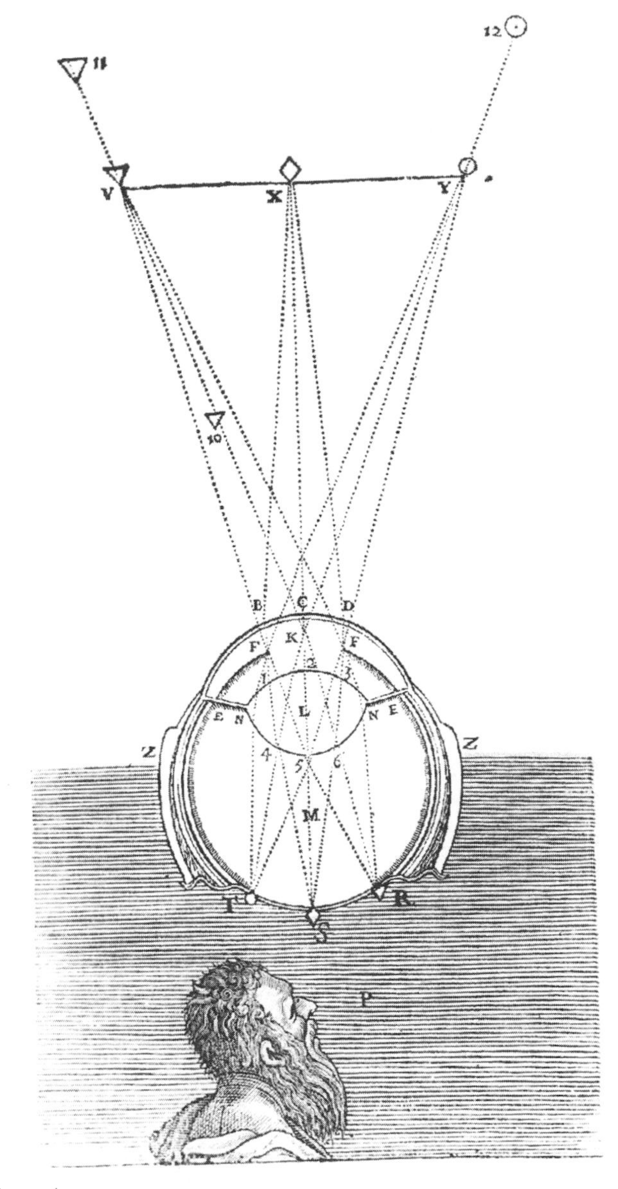

Figure 1
The Eye Forms an Inverted Image on the Retina
René Descartes's version of image formation is similar to that described by Kepler. Rays emanating from points in space are focused as points on the retina. In the process, however, the image is reversed (right to left in this illustration) with respect to the object. (From Descartes 1637.)

Figure 1 hints at the way Kepler could have observed the retinal image inversion. The man at the bottom of the drawing appears to be looking at the image from behind the eye. As it happens, this device is not altogether fanciful; it was used by the Jesuit as-

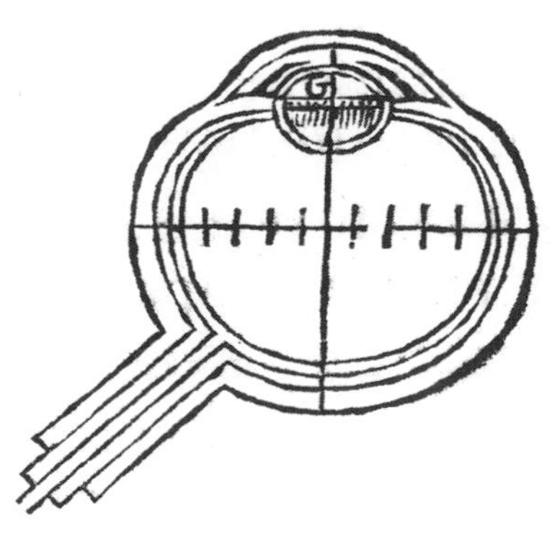

Figure 2
Visual Optics Comes of Age
This horizontal section of the human eye is the first good approximation of the eye's optical anatomy. Although the lens is too far forward and the cornea is too large, the drawing is sufficiently accurate that it can be used for reasonable calculations about the optical properties of the eye. (From Scheiner 1619.)

tronomer Christoph Scheiner (1573–1650) to demonstrate directly that Kepler was correct.

Among other things, Scheiner invented the pantograph, a device used to convert drawings from one scale to another, and he discovered sunspots independently of Galileo. As a good Catholic, Scheiner reacted in a predictable way to the Lutheran Kepler's concept of a heliocentric solar system and attempted—unsuccessfully—to keep the sun revolving around Earth. Like Kepler, however, Scheiner was interested in vision, and his studies of the eye were published in 1619 as *Oculus Hoc Est: Fundamentum Opticum*. Figure 2 is taken from Scheiner's book; it is the first cross-sectional drawing of the eye that appears to be accurate from a modern perspective. (It is better than Descartes's version in Figure 1, which was published almost 20 years later.)

Here we see the lens in its correct position, with differing curvatures for the anterior and posterior surfaces and an indication of the attachment of the zonule at the lens equator. The optic nerve, for the first time, is shown exiting the eye on the nasal side of the midline, not directly from the posterior pole. The central core of the optic nerve is continuous with the retina, as it should be, and even the anterior sclera's change in curvature is depicted accurately. One can quibble with the too shallow anterior chamber and the unrealistic forward projection of the iris, but all in all, it is a remarkably good illustration.

Kepler didn't test his prediction of an inverted image focused in the plane of the retina. Scheiner did, however, with a wonderfully simple and direct experiment. The trick is simply finding a way to see the retinal image in situ. Scheiner did it on an isolated eye by removing a flap of sclera and choroid near the eye's posterior pole, thus creating a small window through which he could see the image formed by a distant object aligned with the eye's optic axis. Voilà! The image was upside down and in focus, as Kepler had predicted, and as Figure 1 illustrates. The ghost of Leonardo must have been wondering why *he* didn't think of that. (This experiment was repeated later by Descartes, which is why he could draw Figure 1 without being fanciful. If one uses the eye of an albino rabbit, the inverted image of a bright object can be seen through the sclera and choroid without dissection.)

Scheiner also left us with a method for determining an eye's refractive status; the method utilizes the Scheiner disc, which is an opaque disc pierced with two pinholes, each about 0.5 mm in diameter and separated by about 3 mm. When the Scheiner disc is placed just in front of the eye and centered with respect to the pupil, light from a distant *point* source will enter the eye as two discrete pencils of light, passing through the pupil on either side of the optic axis. If the eye is emmetropic, the beams will converge on a single point on the retina, and one will perceive a single point of light (Figure 3). If the eye is ametropic (whether myopic or hyperopic), two points will be seen. Their separation is a function of the magnitude of the refractive error, which we can measure by adding supplementary lenses or moving the light source until the two points become one. Some modern optometers (or refractometers), with all their high-tech sophistication, are memorials to Christoph Scheiner; his simple, elegant principle can be found in the heart of these devices.

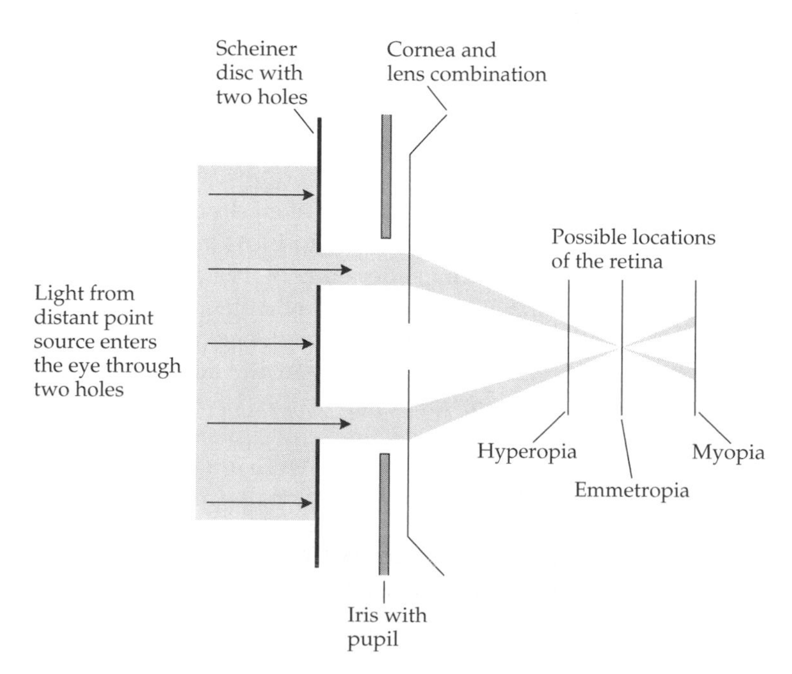

Figure 3
The Scheiner Principle
Viewing a distant point source of light through an opaque disc containing two small pinholes will reveal the eye's refractive status. If the eye is emmetropic, only a single point will be seen; if a refractive error exists, two points will be seen, and the separation of the two points is proportional to the magnitude of the ametropia.

bulges are **staphylomas**, distinguished by their general location as anterior or posterior staphylomas. The converse situation, where the sclera bulges inward, often indicates the presence of an orbital tumor that is pressing on the eye.

The most common form of corneal bulging is called **keratoconus**, so named for the almost conical shape the weakened tissue is forced to assume by the intraocular pressure on its posterior surface. Keratoconus is associated with very high myopia and astigmatism (see Chapter 8).

On average, the adult human eye measures twenty-four millimeters in all dimensions

Large eyes are advantageous because their long focal length means that the image size will be large in its linear dimensions. When an image is projected on an array of photoreceptors, an increase in the size of the image will simultaneously increase the number of photoreceptors on which the image projects; light–dark variations contained within the area of a single photoreceptor when the image is small can be scanned by numerous photoreceptors, and therefore resolved, when the image is enlarged.

A potential disadvantage is the fact that the amount of light admitted to the eye is directly related to pupil size, and unless the eye's aperture increases along with the eye size, the image in a large eye will be dimmer, which may limit the system's performance under low light conditions. The variable pupil in the human eye therefore increases in size in dim illumination and becomes smaller in bright light.

The average diameter of the adult human eye is about 24 mm, measured vertically, horizontally, or from anterior to posterior. This number appeared in a comparative perspective with other vertebrate eye sizes in Figure 7 in the Prologue, in which mammalian eyes were shown to range in diameter from about 1 mm (shrew) to nearly 120 mm (blue whale). In general, eye sizes in different species correlate most strongly with body weight. Although there are some obvious exceptions—birds have larger eyes than would be expected for their body weight—human eyes fit the relationship well and are therefore neither unusually large nor unusually small.

Most of the data on human eye size prior to the twentieth century were obtained by direct measurement with calipers on eyes removed postmortem; the average value for the diameters of the eye was just over 24 mm. There are a variety of potential problems with this method of measuring eye size; more meaningful results come from in vivo measurements of ocular diameters. The most extensive data on eye size—specifically, axial length—were obtained by an interesting use of X rays.

X rays are sufficiently close to light in the spectrum of electromagnetic radiation to be absorbed by the retinal photopigments and produce a visual sensation. A thin, vertical sheet of X rays passing through the center of the eye from the side will intersect the retina along a circular cross section, and the resulting visual perception is that of a circle of light. As the X-ray sheet is moved back toward the posterior pole of the eye, the perceived circle becomes smaller and smaller until the X rays are at the point of tangency with the retina; the perception is now a single point of "light," and this point marks the posteriormost extent of the retina. Since the position of the anterior pole of the eye, the corneal apex, can be determined optically with the same piece of apparatus, the distance between the locations of the optical and X-ray beams is the anterior–posterior (**axial**) length of the eye.

Solve Stenström used the X-ray method to measure axial length in 1000 adult eyes, in which other optical properties, including refractive error, were also measured. His axial length data are plotted in Figure 2.15 as a relative frequency dis-

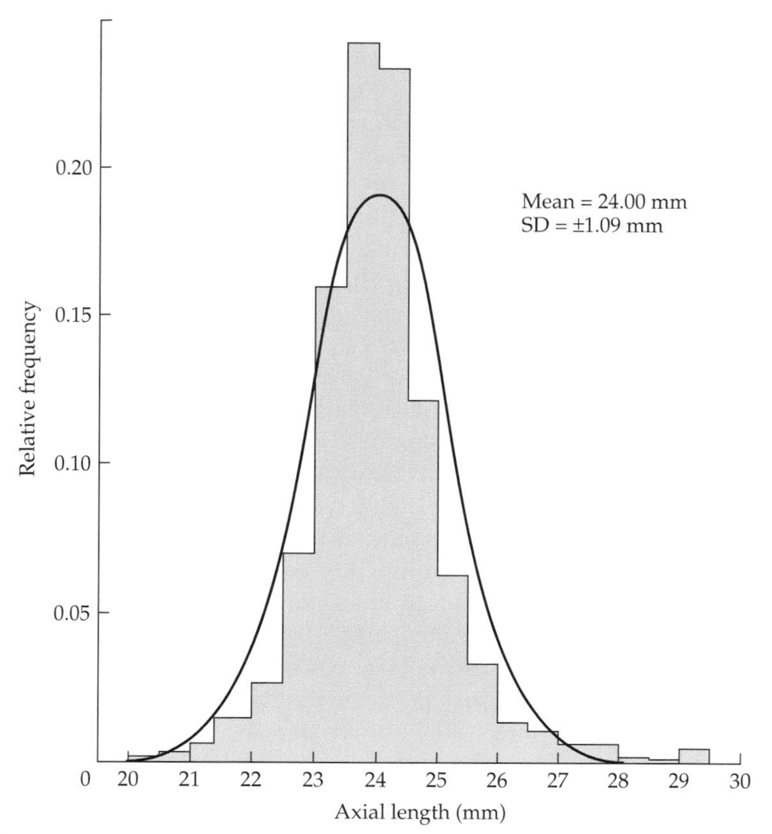

Figure 2.15
The Axial Lengths of a Thousand Eyes
The most common and the average axial lengths of the eye are both about 24 mm, but they range from 20 to 30 mm. The smooth curve is a normal (Gaussian) probability function that best fits the data. Thus we can say that the eye's average axial length is 24 mm, with a standard deviation of ±1.0 mm. (Replotted from Stenström 1946.)

tribution.* The most common axial length corresponds to the peak of the distribution (the **mode** of the distribution); it is about 24 mm. Since the distribution is fairly symmetric around the mode, it follows that the arithmetic average (the **mean**) of all the axial lengths will also be close to 24 mm. This, then, is the basis for the statement that the average axial length of the adult human eye is 24 mm; because the measurement method does not include the thickness of the choroid and sclera, the older direct measurements are slightly larger, and our value of 24 mm is the average *optical* length of the eye, not the average *anatomical* length.

Axial lengths and other anatomical features vary among individuals, but most eyes are emmetropic

Although 24 mm may be the most common axial length, there is considerable variation among individuals; the values range from 20 to almost 30 mm. The standard deviation is fairly small, however; just over 1 mm. Thus, about two-thirds of the eyes (±1 standard deviation) have axial lengths between 23 and 25

*The 1000 eyes were the right eyes of 1000 individuals. Results using both eyes from 500 individuals would differ because of the tendency for a pair of eyes to be similar. The individuals were of both sexes, but none of the data differed significantly between males and females. All individuals were Caucasians, and the data may not be fully applicable to other racial groups.

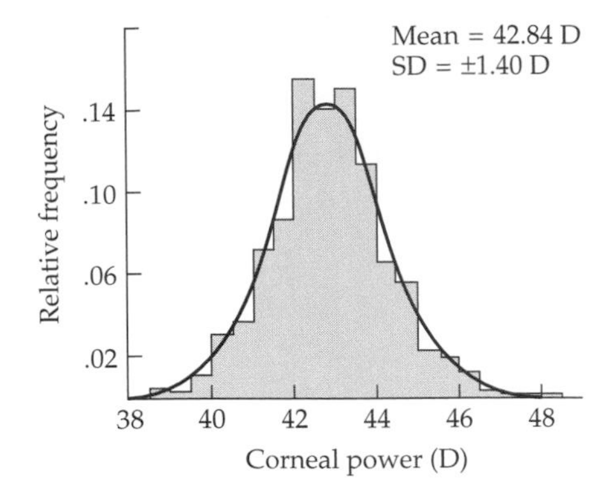

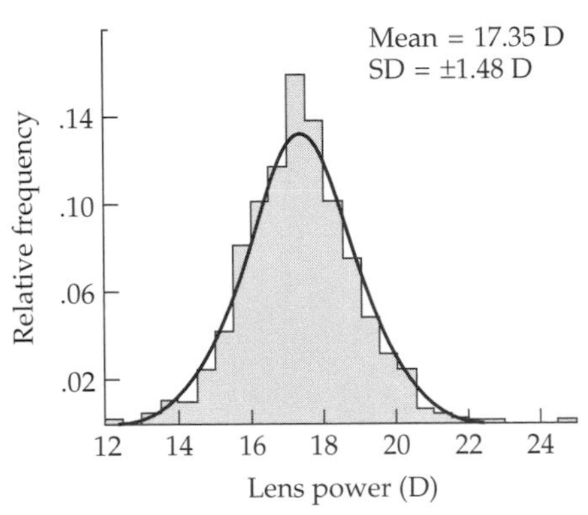

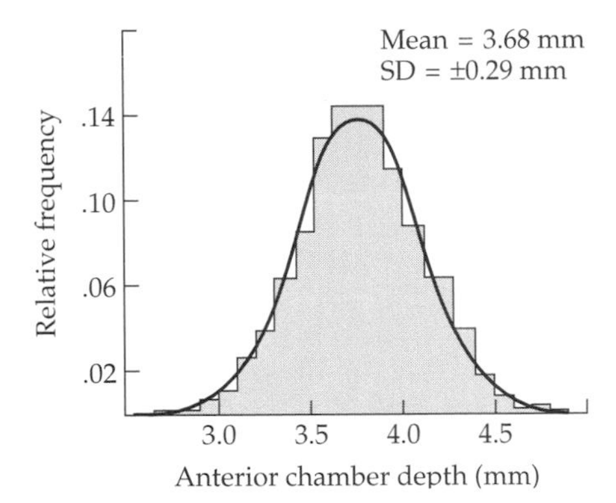

Figure 2.16
Corneal and Lens Powers and Anterior Chamber Depth in a Thousand Eyes
In addition to axial length, the other optically relevant parameters of an eye are its corneal power, lens power, and anterior chamber depth (the distance separating the cornea and the lens). All three parameters are normally distributed, and the mean and standard deviation for each data set are indicated. The data are from the same eyes that are used for the axial length distribution in Figure 2.15. (Replotted from Stenström 1946.)

mm. The actual data are well described by a normal, or Gaussian, distribution and can therefore be characterized by its mean and standard deviation.

Other optically relevant anatomical variables show the same kind of variability (Figure 2.16). Corneal power has a mean value of about 43 D (SD = 1.40 D), the mean anterior chamber depth is 3.68 mm (SD = 0.29 mm), and the mean lens power is 17.35 D (SD = 1.48 D).* Again, the different data sets are well described by normal distribution functions.

The total optical power in a particular eye is given by the combination of corneal and lens powers and the anterior chamber depth. Since all of these characteristics are variable, it would be reasonable to expect comparable variation in the optical power of eyes. And since refractive error is a function of the optical power relative to axial length, which is another variable element, refractive error might also be distributed as a normal (Gaussian) variable in the population.

But look at the distribution of refractive errors for these eyes (Figure 2.17). Unlike the distributions of the anatomical variables, the refractive error distribution is skewed—there are more high myopias than high hyperopias—and the peak of the distribution at zero refractive error (emmetropia) is unusually pronounced. A normal distribution function having the same mean and variance (the smooth curve in Figure 2.17) does not describe the data at all well.

At least two important conclusions can be drawn from these data. First, most eyes—certainly more than expected—are emmetropic or nearly so, and there must be some constraint on variability in any particular eye. For example, if an eye with an axial length longer than the mean value is to be emmetropic, the long axial length must be compensated for by a corneal power less than the mean value. This suggests, as a second conclusion, that the anatomical elements in a particular eye are not independent of one another; instead, they are correlated, and correlated in such a way that there is a strong tendency toward emmetropia. Ametropia implies a failure of correlation.

Most refractive error is related to relatively large or small axial lengths

With appropriate statistical methods, it is possible to specify the strength of the correlation between pairs of variables or among any number of variables. To some extent, these relationships are apparent in plots of relevant variables against one another. One of the more revealing of such plots is shown in Figure 2.18*a*, in which refractive error has been plotted against axial length; if the correlation between refractive error and axial length were perfect, the data points would form a straight line. The data obviously do not lie on a line, but there is a clear trend showing increasing myopia (or decreasing hyperopia) as axial length increases. The coefficient of correlation (a statistical measure of the strength of association between the variables) is around 0.76, which is respectably high (1 is the maximum value). Similar plots of corneal power or anterior chamber depth versus refractive error exhibit more scatter and have lower correlation coefficients.

If a mechanism to correlate the anatomical variables of an eye so that emmetropia is achieved were at work, one would also expect to see appropriate correlations between the major anatomical elements; one pair is illustrated in Figure 2.18*b*. Here, corneal radius of curvature has been plotted against axial length, showing a modest relation between them: As axial length increases, corneal radius tends to increase (corneal power decreases), which is in the correct direction to minimize refractive error. It also makes sense that an eye with a larger scleral radius of curvature should have a larger corneal radius of curvature. The correlation between corneal radius and axial length is not as good as

*Because of the way it was calculated, lens power is about 3 D lower than the true lens power.

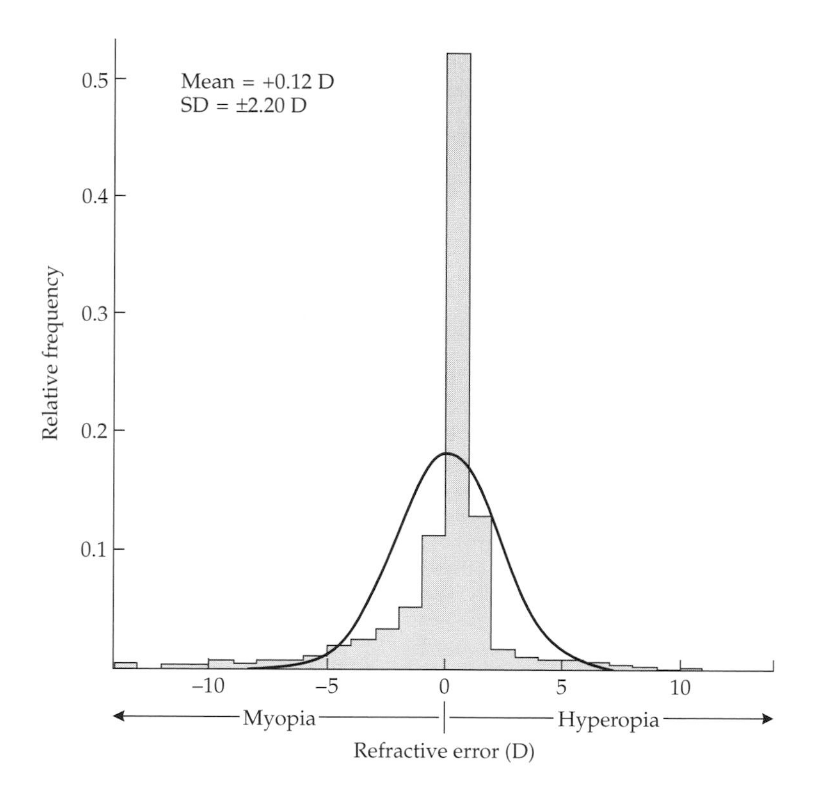

Figure 2.17
Refractive Errors in a Thousand Eyes
The refractive errors in the eyes in which axial lengths and other parameters were measured form a distribution that cannot be described by a normal probability function. There are more eyes with near zero refractive error (emmetropic) and more myopic eyes than expected, and fewer hyperopic eyes than expected. By far the most common refractive error is none. The optical parameters for a particular eye appear to be highly dependent on one another; if they were independent, the refractive error distribution would be normal. (Replotted from Stenström 1946.)

the correlation between refractive error and axial length, however. These relationships imply that axial length is the major determinant of refractive error, which results from a failure of the normal correlation between axial length and other anatomical elements.

Other sorts of analysis, including partial coefficients of correlation (which can show the association between axial length and refractive error with corneal power factored out) or multiple regression (which gives weights to the different variables in terms of their relative contributions), make the story clear. Not only is axial length the primary determinant of refractive error, it accounts for almost 50% of refractive error. Corneal power contributes somewhat less than 20%, and the anterior chamber depth about 5%. The remainder, just over 20%, must be attributed to unknown factors, of which the lens is certainly one (a point confirmed in other, later investigations).

The tendency for myopic eyes to be larger than average and hyperopic eyes to be smaller than average suggests that growth has gone on too long in some eyes, making them large and myopic because the other anatomical elements could not keep pace to compensate, while it has stopped too soon in others, making them short and hyperopic, presumably because compensatory growth in other elements did not have an opportunity to be fully expressed. The retina may be the master controller of normally correlated growth.

The Eye's Axes and Planes of Reference

The eyes rotate around nearly fixed points

If the eye were truly a ball in a ball-and-socket joint, its center of rotation would be a fixed point in space coinciding with the sclera's center of curvature. But when the eye rotates horizontally or vertically, it shifts up or down or to one side or the other; rotation is accompanied by translation, and the center of rotation must be moving as the eye moves. However, the movement of the center of rotation within the normally occurring range of purely vertical or purely horizontal

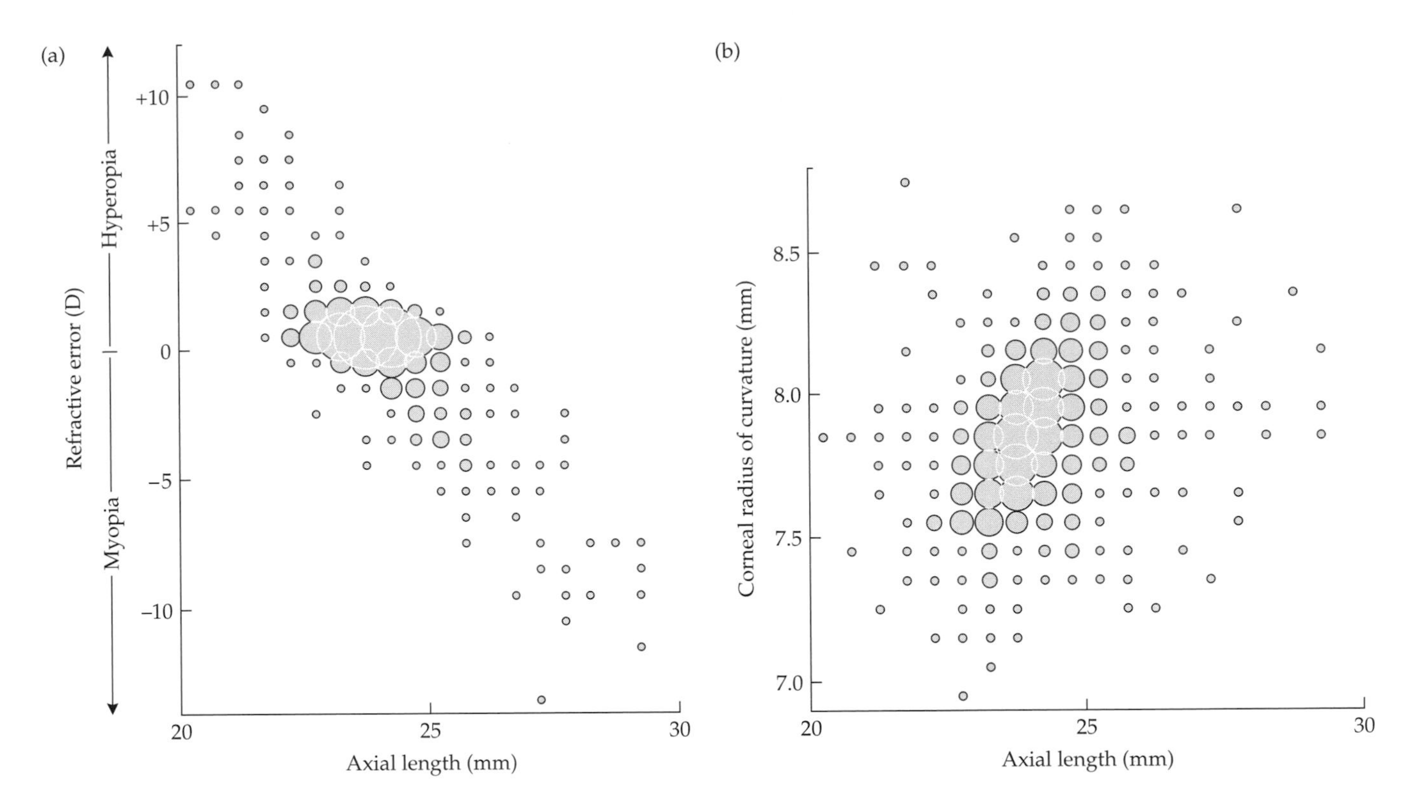

Figure 2.18
Correlations between Optical Parameters
(*a*) Refractive error as a function of axial length. If different refractive errors were due solely to different axial lengths, a plot of the two variables would be a straight line. Although the scatterplot shows a linear relationship, many points do not fall on the line. Axial length is a major component of refractive error, but it is not the only component (note, for example, that the most common axial length—24 mm—can be associated with both emmetropic and myopic eyes). The area of the circles is proportional to the number of observations lying near the center of the circle. (*b*) Corneal radius of curvature as a function of axial length. Corneal radius tends to increase (that is, corneal power tends to decrease) as axial length increases. Although this correlation is in the correct direction to minimize refractive error (decreased corneal power could compensate for a long axial length), the wide scattering of the data shows that correlation is weak. (Replotted from Stenström 1946.)

eye movements (roughly ±20° from the straight-ahead position of gaze) is only a few hundred micrometers, and there are very few situations in which this small deviation is important.

Although the center of rotation is relatively fixed in position in the orbit, its location within the eye is somewhat inconvenient because it does not coincide with the eye's optical or visual axes. On average, the center of rotation lies in the horizontal plane that bisects the eye, 13 mm behind the corneal apex and 0.5 mm nasal to the line of sight. There is individual variation, of course, but for practical purposes, the center of rotation is taken to be 13.5 mm behind the corneal apex on the line of sight (or, in terms of Gullstrand's simplified eye, just over 6 mm behind the posterior nodal point). Thus, when we refer to an eye movement of, say, 20°, we are referring to an angular rotation around this stationary, somewhat imaginary, point lying 13.5 mm behind the corneal apex.

Like the head, the eyes have three sets of orthogonal reference planes

The three reference planes for the head are the sagittal planes that pass vertically through the head from front to back, with the midsagittal plane bisecting the head along the sagittal suture of the skull; the coronal (or transverse) planes that cut through the head vertically from side to side; and the horizontal planes that slice horizontally, front to back and side to side. These sets of planes are mutually perpendicular—that is, orthogonal. If the centers of rotation of the two eyes are positioned with perfect symmetry, they will both lie in the same horizontal and coronal planes at the intersections with two sagittal planes lying about 30 mm on either side of the midsagittal plane. Moreover, if the eyes are in the correct position, the head's reference planes will coincide with the eye's sagittal, coronal, and horizontal planes.

The "correct position" is called the **straightforward position of gaze**, and although it has a precise definition, it means basically that the anterior–posterior

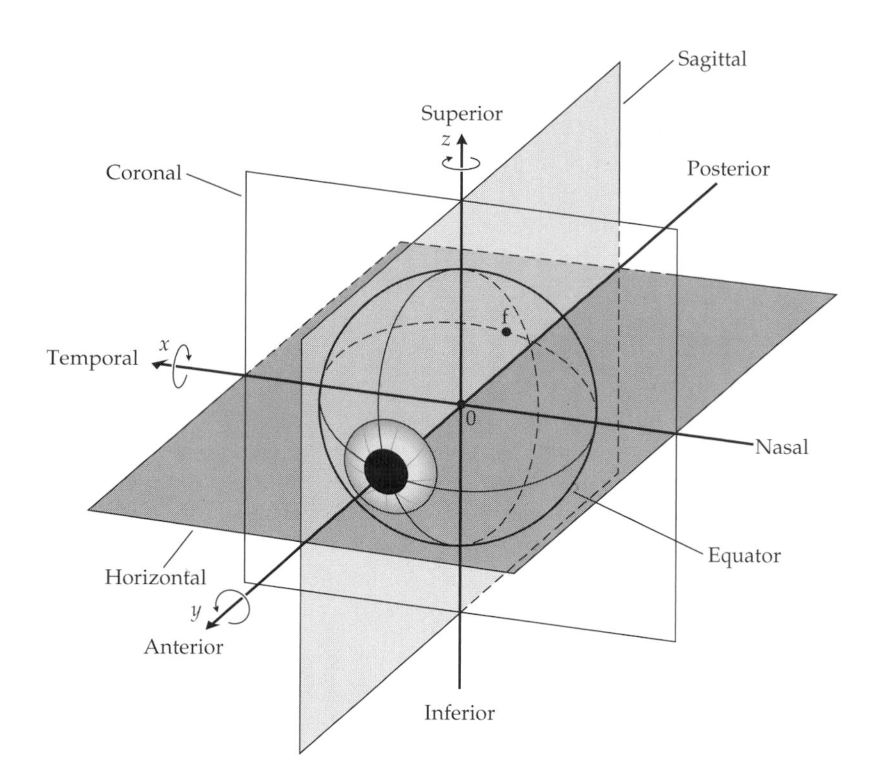

Figure 2.19
Reference Planes for the Head and Eyes The head is transected by three orthogonal sets of reference planes: sagittal, coronal, and horizontal. When the eyes are in their straightforward position of gaze, their corresponding reference planes are parallel to those of the head. The eye's reference planes intersect, in principle, at the center of curvature of the sclera, which is near the eye's center of rotation. The fovea (f) is temporal to the anterior–posterior axis.

axes of the eyes are parallel, as when we look at a very distant object, and perpendicular to the coronal plane of the head held erect. Under these conditions, each eye may be said to be in the **primary position of gaze**. (Again, there is a technical definition of primary position, but it is not easily determined except under special circumstances.)

With the eyes in the straightforward position of gaze, the sagittal, coronal, and horizontal planes of the eye are parallel to the corresponding planes of the head that intersect at the eye's center of rotation (Figure 2.19). The coronal plane intersects the eyeball just over halfway between the anterior and posterior poles, thus dividing the eye into anterior and posterior halves (the anterior "half" will be slightly larger); the line of intersection with the eye is the **equator**. The equator is an imaginary circle whose exact location is difficult to establish even in an isolated eye, and the division of the eye into anterior and posterior halves is therefore an approximation. More commonly, the subdivision in everyday use refers to the eye's **anterior segment**, which is the part of the eye anterior to the choroid—it includes the ciliary body, lens, iris, cornea, and part of the sclera—while the rest of the eye (choroid, retina, optic nerve, and the remainder of the sclera) forms the **posterior segment**. The equator lies *within* the posterior segment.

A vertical division of the eye by a sagittal plane creates medial (nasal) and lateral (temporal) halves. Again, this division is an imaginary one, and care is needed in specifying how the division was made. If the eye were in the straightforward position of gaze, the division would be relative to the line of sight (which we'll discuss shortly), and the dividing plane should bisect the fovea, making the fovea the reference point relative to which temporal and nasal would be specified on the retina. If the division were made relative to the pupillary axis, however, the plane should pass through the anatomical posterior pole; thus the fovea would be in the temporal retina, relatively speaking. In other words, when we use the nasal–temporal terminology, it is sometimes necessary to be explicit about how the division was made and what axis or landmark was used as the center of reference.

A horizontal plane containing the center of rotation will divide the eye into superior and inferior halves; since this plane contains both the line of sight and the pupillary axis, it can be determined empirically, and it will bisect both the fovea and the posterior pole. Of the several divisions we have made, it comes the closest to producing mirror-symmetric halves.

The practical consequence of these theoretical constructions is a way of talking about locations. The vertical (sagittal), transverse (coronal), and horizontal planes cut the ocular sphere into eight segments that can be used to specify approximate location; the insertion of the superior oblique muscle, for example, is in the posterior-superior-lateral segment of the globe. Also, the iris is anterior to the lens, the fovea is temporal to the optic nerve head, and so on. There will be some inevitable points of confusion—the anterior epithelium of the iris is not on the anterior surface of the iris, but it *is* anterior to the posterior epithelium—but remembering that these designations are made relative to a coordinate system means that the designations are consistent and sensible.

Since the reference planes of the eye contain the center of rotation, we can use them to specify the major *axes* of rotation. Rotation of the eye in the horizontal plane, for example, can be regarded as a rotation about an axis that is perpendicular to the plane; in this case, horizontal rotation occurs around a *vertical* axis (see Figure 2.19). Vertical rotations are made around a horizontal axis, and so-called torsional rotations are made around the anterior–posterior axis.

The pupillary axis is a measure of the eye's optical axis

Although the eye's anterior–posterior axis has been referred to repeatedly, it is difficult to specify in vivo. We cannot use the anatomical anterior–posterior axis, because its determination requires direct measurement on the isolated eyeball. Two other anterior–posterior axes are easily located, however, and they are therefore the most useful axes. The first of these, the **pupillary axis**, is defined as the line perpendicular to the cornea at its apex that passes through the center of the eye's entrance pupil (the entrance pupil is the image of the real pupil as viewed through the cornea; it lies slightly in front of the real pupil, as shown in Chapter 10). You can locate the pupillary axis by holding a penlight below your observing eye, while moving from side to side relative to the subject's stationary eye (Figure 2.20). When the bright first Purkinje image of the penlight appears to be centered in the entrance pupil, it is on the pupillary axis.

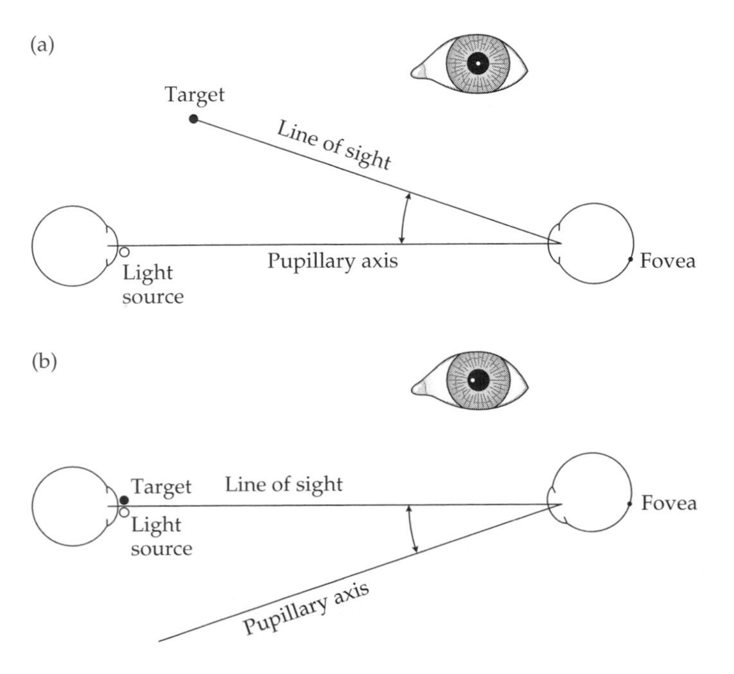

Figure 2.20
Estimating Angle Kappa
(*a*) A small point source of light has been positioned so that its first Purkinje image is centered in the entrance pupil. By definition, the light source lies on the pupillary axis. (*b*) Here the patient is looking directly at the light source, which places the light on the visual axis. Under this condition, the first Purkinje image is displaced nasally from the center of the entrance pupil. Normally, the displacement should be about 0.5 mm and should be the same in both eyes. The 0.5 mm displacement corresponds to the approximately 5° magnitude of angle kappa (see text for explanation).

If the eye's optical elements were perfectly aligned, as they are in schematic eyes, the pupillary axis would be the optical axis of the cornea and its extension would be the optical axis of the lens. It is very unlikely, however, that this ideal case is ever realized. The cornea may not be radially symmetric with respect to the pupillary axis, for example, and the lens may be tilted or skewed slightly vertically or horizontally; any tilt of the lens will deviate its optical axis away from alignment with the pupillary axis. Moreover, the pupil may not be perfectly centered with respect to the cornea, and if it weren't, the pupillary axis would not be aligned with the true optic axis of the cornea (the pupil is typically off center nasally by about 0.5 mm). These variations are usually small enough to be ignored for practical purposes. Thus, the pupillary axis is our best estimate of the location of the eye's optic axis; if extended through the eye, it should exit very near the anatomical posterior pole.

The line of sight differs from the pupillary axis by the angle kappa

Another major axis, the **line of sight**, is the line between an external object of regard and the center of the entrance pupil. Since normal subjects always use the center of the fovea to look at (regard) a small object, the line of sight should correspond very closely to the **visual axis**, which is the projection of the foveal center into object space through the eye's nodal points. Note, however, that the line of sight and the visual axis cannot coincide exactly, since they pass through points—the center of the entrance pupil and the anterior nodal point—that are several millimeters apart. The angular difference between the line of sight and the visual axis is only about 10′ of arc, however.

There is a significantly larger difference between the pupillary axis and the line of sight. The pupillary axis is determined optically, and it is our most accessible correlate of the eye's anatomical anterior–posterior axis and of the central axis of the eye's optical system. The line of sight, on the other hand, is determined by the location of the point on the retina being used for fixation relative to the eye's anatomical posterior pole; the fovea is the normal retinal fixation point, and it is a few degrees temporal to the posterior pole. Therefore, remembering the inversion by the eye's optics, the line of sight will be deviated a few degrees *nasally* from the pupillary axis (Figure 2.21). The correct designation of the angle formed by the intersection of the line of sight and the pupillary axis at the center of the entrance pupil is angle lambda; in clinical parlance, however, it is invariably referred to as **angle kappa**. (Technically, angle kappa is the angle formed by the intersection of the pupillary and *visual* axes. Since the difference between angles kappa and lambda is only a few minutes of arc, however, the technical distinction is rarely important.)

The angles kappa in the two eyes should have the same magnitude

Clinical measurements of angle kappa have two basic uses. First, an angle kappa outside the normal range of values may indicate that a nonfoveal point on the retina is being used for monocular fixation; this condition is known as **eccentric**

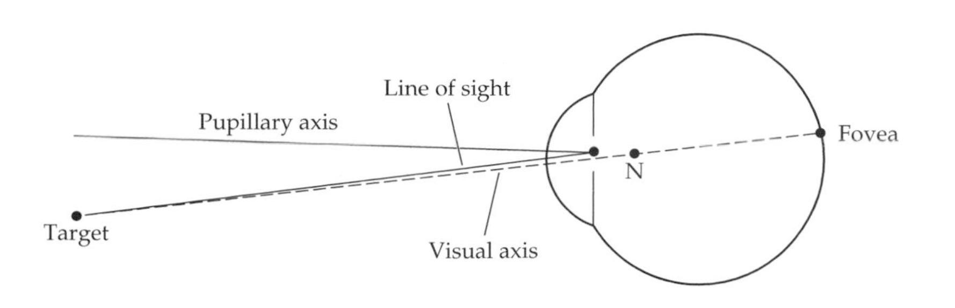

Figure 2.21
The Eye's Optical Axes
The pupillary axis, the line of sight, and the visual axis are imaginary lines related to the direction in which the eye is pointed. The visual axis cannot be located empirically, because it depends on knowledge of the location of the primary nodal point (N). The pupillary axis and the line of sight can be located if the individual can maintain steady fixation on a small target. Angle kappa, in clinical terminology, is the angle between the pupillary axis and the line of sight.

fixation, but there are much better methods for detecting the presence and determining the magnitude of the eccentricity. Second, different angles kappa in the two eyes under binocular conditions can indicate the presence of an oculomotor deviation called **strabismus** (or heterotropia), in which the two eyes do not simultaneously use both foveas for fixation. The size of the difference between the angles kappa can be used as a rough estimate of the magnitude of the strabismus.

If we return to the designation of the straightforward position of gaze or, roughly speaking, primary position, it should be clear that the line of sight and the pupillary axis cannot simultaneously be directed straight ahead. The straightforward position of gaze is correctly defined in terms of the lines of sight, however, and any use of a fixation target in a clinical or experimental setting indicates that the line of sight is the primary reference axis, even if that is not stated explicitly. In some situations, the experimental subject cannot be counted on to fixate a target with the fovea—for example, some experimental animals may not have foveate retinas, and human infants have foveas that are not fully developed—and here the objectively determined pupillary axis is the reference of choice.

Eye position is specified by the direction of the line of sight in a coordinate system whose origin lies at the eye's center of rotation

Because the center of rotation of the eye can be regarded (with a few caveats mentioned earlier) as a fixed point in space, it is a useful point of origin for either spherical or rectilinear coordinate systems. The most common use of a spherical coordinate system is in specification of eye position during or after eye movement, relative to primary position. This subject, the kinematics of the eye, is very old, and there are several ways the coordinate systems can be constructed; it is also a highly technical subject that will not be pursued here, except to say that it is critical whenever the eye's rotational position must be specified accurately.

A practical consequence of the various coordinate systems is their use in ophthalmic instruments that require a specification of eye position; perimeters used in plotting visual fields and visual field defects use the Listing coordinate system, for example, while binocular instruments, such as troposcopes or synoptophores, use the Fick system of coordinates. Although this practice means that a particular direction of gaze will be specified by different numbers in different types of instruments, the coordinate systems are mathematically interchangeable.

For strictly anatomical purposes, three-dimensional rectilinear coordinates have been used to determine the absolute and relative locations of structures in the eye and the orbital cavity. The average spatial locations—that is, the x, y, and z coordinates—of the origins and insertions of the extraocular muscles, for example, were worked out more than a century ago, and until very recently they were the primary data for analyses and descriptions of extraocular muscle actions (see Chapter 4). Determining the spatial location of ocular structures by direct measurement, which was the original method, has been supplanted by methods such as magnetic resonance imaging or laser scanning. A contemporary use of spatial location data, however it is measured, is the production of three-dimensional computer reconstructions of the eye and orbit. As the level of detail incorporated in the computer data base increases, one should be able to take a dynamic tour through the eye, able to examine any region or structure from different perspectives and at different levels of magnification. This development has obvious potential both for visualizing ocular anatomy and for planning surgical procedures.

References and Additional Reading

Image Quality and Visual Performance

Campbell FW and Green DG. 1965. Optical and retinal factors affecting visual resolution. *J. Physiol.* 181: 576–593.

Campbell FW and Gubisch RW. 1966. Optical quality of the human eye. *J. Physiol.* 186: 558–578.

Pelli DG, Robson JG, and Wilkins AJ. 1988. The design of a new letter chart for measuring contrast sensitivity. *Clin. Vision Sci.* 2: 187–199.

Wandell BA. 1995. Image formation. Chapter 2, pp. 13–43, in *Foundations of Vision.* Sinauer Associates, Sunderland, MA.

Westheimer G. 1964. Pupil size and visual resolution. *Vision Res.* 4: 39–45.

Westheimer G. 1992. Visual acuity. Chapter 17, pp. 531–547, in *Adler's Physiology of the Eye,* 9th Ed., Hart WM, ed. Mosby Year Book, St. Louis.

Westheimer G and Campbell FW. 1962. Light distribution in the image formed by the living human eye. *J. Opt. Soc. Am.* 52: 1040–1045.

Woodhouse JM and Barlow HB. 1982. Spatial and temporal resolution and analysis. Chapter 8, pp. 133–164, in *The Senses* (Cambridge Texts in the Physiological Sciences 3), Barlow HB and Mollon JD, eds. Cambridge University Press, Cambridge.

The Anatomy of Image Formation

Bennett AG and Rabbetts RB. 1989. *Clinical Visual Optics,* 2nd Ed. Butterworths, London. (Chapters 2, 3, 4, 12, and 20.)

Charman WN. 1991. The vertebrate dioptric apparatus. Chapter 5, pp. 82–117, in *Vision and Visual Dysfunction,* Vol. 2, *Evolution of the Eye and Visual System,* Cronly-Dillon JR and Gregory RL, eds. CRC Press, Boca Raton, FL.

Charman WN and Jennings JAM. 1976. The optical quality of the monochromatic retinal image as a function of focus. *Brit. J. Physiol. Optics* 31: 119–134.

Emsley HH. 1969. The schematic eye—unaided and aided. Chapter X, pp. 336–403, in *Visual Optics,* Vol. 1, *Optics of Vision,* 5th Ed. Hatton Press Ltd., London.

Johnson CA. 1994. Evaluation of visual function. Chapter 17, pp. 1–20, in *Duane's Foundations of Clinical Ophthalmology,* Vol. 2, Tasman W and Jaeger EA, eds. JB Lippincott, Philadelphia.

Miller D. 1991. Optics of the normal eye. Chapter 7, pp. 7.2–7.28, in *Optics and Refraction: A User-Friendly Guide,* Vol. 1, *Textbook of Ophthalmology,* Podos SM and Yanoff M, eds. Gower Medical Publishing, New York.

Tomlinson A, Hemenger RP, and Garriott R. 1993. Method for estimating the spheric aberration of the human crystalline lens *in vivo. Invest. Ophthalmol. Vis. Sci.* 34: 621–629.

Westheimer G. 1972. Optical properties of vertebrate eyes. Chapter 11, pp. 449–482, in *Handbook of Sensory Physiology,* Vol. VII/2, *Physiology of Photoreceptor Organs,* Fuortes MGF, ed. Springer-Verlag, Berlin.

Eye Shape and Size

Hirsch MJ and Weymouth FW. 1947. Notes on ametropia—A further analysis of Stenström's data. *Am. J. Optom.* 24: 601–608.

Hughes A. 1977. The topography of vision in mammals of contrasting life style: Comparative optics and retinal organization. Chapter 11, pp. 613–756, in *Handbook of Sensory Physiology,* Vol. VII/5, *The Visual System in Vertebrates,* Cresitelli F, ed. Springer-Verlag, Berlin.

Laird IK. 1991. Anisometropia. Chapter 10, pp. 174–198, in *Refractive Anomalies,* Grosvenor T and Flom MC, eds. Butterworth-Heinemann, Boston.

Rushton RH. 1938. The clinical measurement of the axial length of the living eye. *Trans. Ophthalmol. Soc. U.K.* 58: 136–140.

Sorsby A, Benjamin B, Davey JB, Sheridan M, and Tanner JM. 1957. *Emmetropia and Its Aberrations.* Medical Research Council Special Report Series, No. 293. Her Majesty's Stationery Office, London.

Stenström S. 1946. Untersuchungen über die Variation und Kovariation der optischen Elemente des menschlichen Auges. *Acta Opthal. (Copenh.)* Suppl. 26. Translation by Woolf D. 1948. Investigation of the variation and the correlation of the optical elements of human eyes. *Am. J. Optom.* 25: 218–232, 286–299, 340–350, 388–397, 438–449, 496–504.

Walls GL. 1942. *The Vertebrate Eye and Its Adaptive Radiation.* Cranbrook Institute of Science, Bloomfield Hills, MI. Reprinted by Hafner Publishing, New York, 1963.

The Eye's Reference Planes and Axes

Alpern M. 1969. Specification of the direction of regard. Chapter 2, pp. 5–12, in *The Eye,* Vol. 3, *Muscular Mechanisms,* 2nd Ed. Davson H, ed. Academic Press, New York.

Fry GA and Hill WW. 1962. The center of rotation of the eye. *Am. J. Optom.* 39: 581–595.

Chapter 3

The Orbit

We must always remember that our bone is not only a living, but a highly plastic structure: the little trabeculae are constantly being formed and deformed, demolished and formed anew.

■ D'Arcy Wentworth Thompson, *On Growth and Form*

Bones carry in their shape, or morphology, the signature of soft tissues with which they were embedded during life—tissues including muscles, ligaments, tendons, arteries, nerves, veins, and organs. . . . The evolutionary history of humanity itself may be read from the fossil record—a record comprising mostly teeth and bones.

■ Tim D. White, *Human Osteology*

Our bones must be not only strong and rigid to serve their purposes of mechanical support and protection (in the case of the skull), but also responsive and changeable enough to grow for 20 years or so postnatally, to mold to the changing shapes and sizes of surrounding tissues, and to repair themselves when damaged. Bones do all these things routinely, and the bones surrounding the eyes are no exception. They protect while growing and remodeling.

Animals vary widely in their dependence on vision and in the hazards to which their eyes are exposed in the course of their daily lives. Like many of our primate relatives, we humans are social, adventurous, sometimes quarrelsome, visually dependent creatures whose activities can have traumatic consequences. In short, our eyes are important, but vulnerable, and because they are exposed to danger, they need protection. That protection comes from the bones that surround the eyes; these bony shelters within which the eyes reside are the **orbits.**

The more completely the eye is surrounded with bone, however, the more restricted is its field of view. As a rough generalization, animals that are preyed upon often have incomplete or "open" orbits from which their laterally directed eyes protrude, thus affording a panoramic field of view. Predators are more likely to have frontal eyes with closed orbits that restrict the field of view somewhat but offer greater protection from injury. In terms of orbital structure, we fall in the predatory class.

The Bony Orbit

The orbits are roughly pyramidal

Using the orbit as a mold for a plaster casting produces a pear-shaped lump of plaster about 5 cm long, whose stem end is the apex of the orbit and whose broad base represents the external opening of the orbital cavity. A geometric approximation of this shape is a somewhat lopsided pyramid, or tetrahedron. This approximation makes it easy to think of the orbit as having four sides: medial and lateral walls, a ceiling, and a floor (Figure 3.1*a*).

When the orbits are viewed from above (Figure 3.1*b*), some relationships between the two orbits are apparent. The medial walls of the orbits are nearly parallel, about 25 mm apart in adults; the lateral walls, if extended back until they intersected, would be perpendicular. Or to put it another way, if the medial and lateral walls intersected, the angle of intersection would be about 45°.

The orbital pyramid is truncated; the opening at the apex of the orbit is the **optic foramen**, through which the optic nerve and the ophthalmic artery pass.

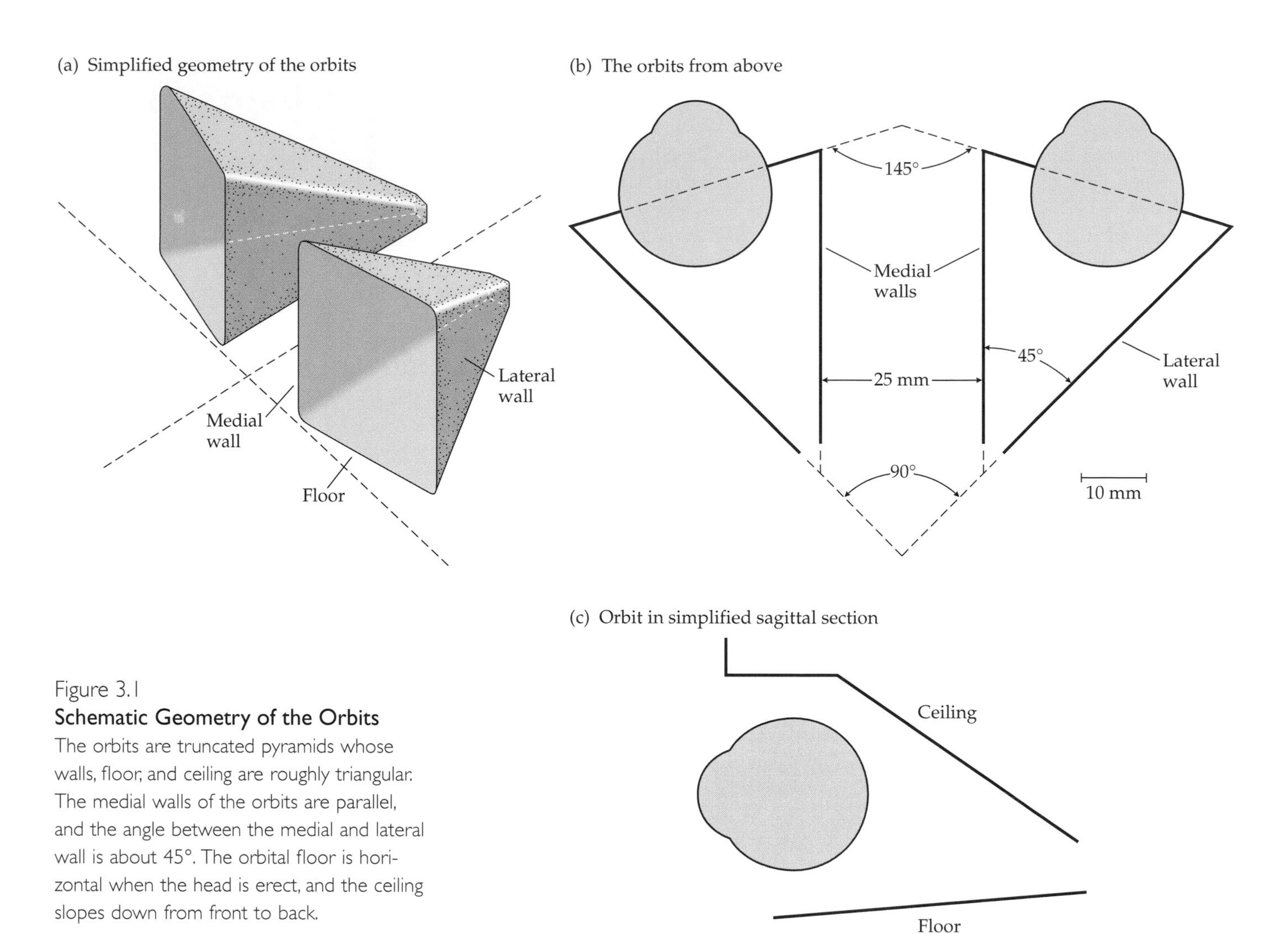

Figure 3.1
Schematic Geometry of the Orbits
The orbits are truncated pyramids whose walls, floor, and ceiling are roughly triangular. The medial walls of the orbits are parallel, and the angle between the medial and lateral wall is about 45°. The orbital floor is horizontal when the head is erect, and the ceiling slopes down from front to back.

The floor of the orbit is roughly triangular (as is the ceiling, which has been removed in Figure 3.1*b*), and the side of the triangle bounded by the lateral wall does not extend as far forward as the medial side does; thus the eyes are somewhat more exposed laterally, but the lateral field of view is correspondingly more extensive.

A view from the medial side (Figure 3.1*c*) shows that the floor of the orbit is nearly horizontal, while the ceiling slopes down significantly from front to back. Thus the medial and lateral walls are also roughly triangular. The top and side views in Figure 3.1, though quite schematic, show the normal positions of the eyes in the orbits. The center of rotation of the eye is just behind the rim of the orbit on the lateral side, and the prominent forward extension of the orbital ceiling affords special protection above. In addition, the anterior–posterior axes of the eyes are fairly well centered with respect to the orbital margins, both vertically and horizontally.

The large bones of the face form the orbital margin and much of the orbit's roof, floor, and lateral wall

The seven bones from which the orbit is constructed have complex shapes that fit together like pieces of a three-dimensional jigsaw puzzle to form the roughly pyramidal geometry described in the previous section. The fit is not perfect, however, because of various gaps and holes through which blood vessels and nerves enter and exit the orbital cavity. In addition, individual bones are not always confined to a single side of the orbit; thus the junctions of the orbital walls are not as sharp or well-defined as the drawings in Figure 3.1 imply.

By far the strongest part of the orbit is its anterior rim or margin, which is formed by three of the large bones of the face: the **frontal**, **zygomatic**, and **maxillary** bones. The maxillary bone contributes to the inferior and medial parts of the orbital rim, the frontal forms the superior and lateral rim, and the zygomatic is part of the inferior and lateral rim (Figure 3.2). All three bones are unusually thick at the orbital margin, and the sutures along which they join are extensive. Both features—thickness and extent of suture—make the orbital margin very resistant to fracture, so much so that a heavy blow to the orbital margin may leave the margin intact, but cause the bone to fracture at another location, as we'll discuss shortly.

Viewed from the front, the orbital margin in adults is roughly rectangular, with the horizontal dimension (approximately 39 mm) larger than the vertical (approximately 35 mm). The first of its most prominent landmarks is the **supraorbital notch**, which lies almost in the center of the superior orbital margin. Sometimes the notch is bridged with bone to become a foramen, but when it remains a notch, it is easy to feel through the overlying skin and muscle in vivo. The notch (or foramen) carries the supraorbital artery, vein, and nerve. The **infraorbital foramen** is almost directly below the supraorbital notch, lying in the maxillary bone about 4 mm down from the inferior orbital margin. It carries the infraorbital artery, vein, and nerve.

The lateral and medial edges of the orbital margin typically have small protrusions of bone called tubercles. The **lateral orbital tubercle** (also called Whitnall's tubercle) is on the frontal process of the zygomatic bone, just below

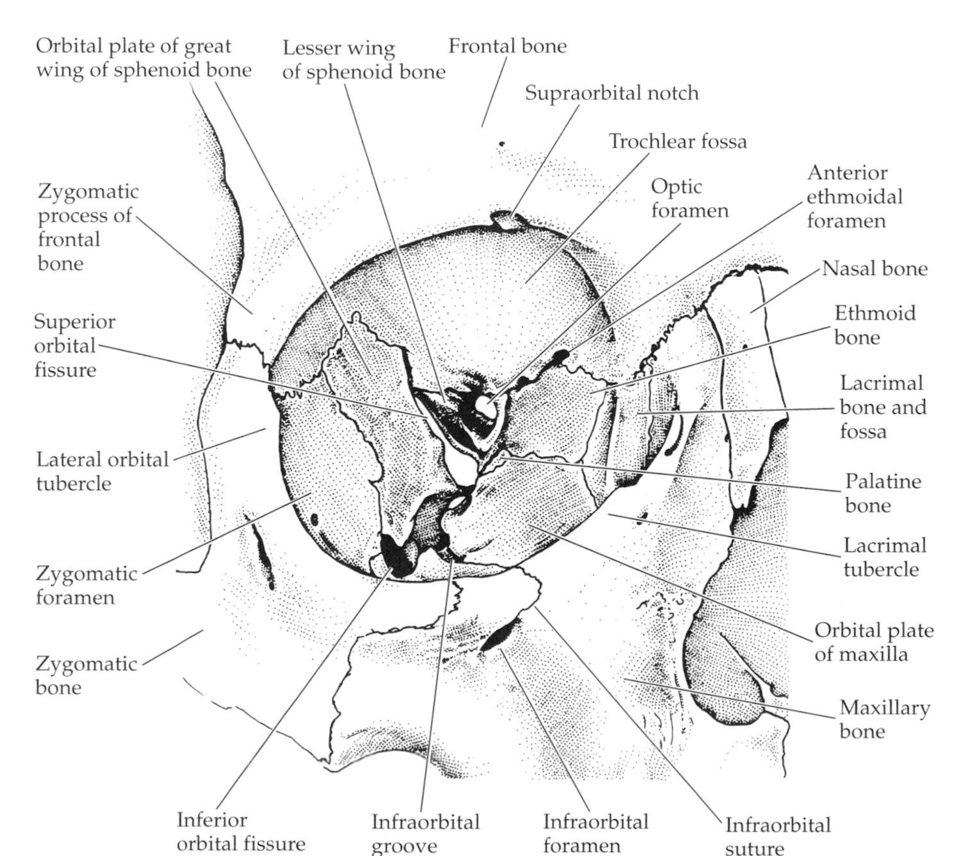

Figure 3.2
The Right Orbit from the Front
This view of the orbit is from a slightly lateral perspective, so the optic foramen is centered in the border created by the orbital margin; from this vantage point, both lateral and medial walls are visible. All seven bones contributing to the orbit are included, as are all the foramina.

the suture between the zygomatic and frontal bones. It marks the attachment site of the lateral portion of the levator muscle's tendon and several ligaments. The lateral orbital tubercle is not always well developed; the more obvious feature is the notchlike discontinuity at the zygomatic–frontal suture just above the lateral tubercle; it can be felt easily through the skin. The **lacrimal tubercle** on the medial margin is somewhat lower than the lateral tubercle, and it is continuous with the lower end of the anterior lacrimal crest on the frontal process of the maxillary bone. Again, it is an attachment site for ligaments, but it is not always very prominent.

The maxillary, frontal, and zygomatic bones contribute not only to the face and orbital margins, but also to the walls, roof, and floor of the orbit. Plates of bone extend back from the orbital margin; the orbital plate of the frontal bone forms most of the orbital ceiling (Figure 3.3*a*), the orbital plate of the maxillary bone forms most of the floor of the orbit (Figure 3.3*b*), and the plate of the zygomatic forms part of the lateral wall (Figure 3.3*c*). These bony plates are relatively thin and can be broken with corresponding ease; their major function, therefore, is less protection than support; they provide rigid anchoring points for orbital muscles and ligaments and separate the orbital contents from the surrounding sinuses and other structures.

The sphenoid bone fills the apex of the orbital pyramid and contributes to the lateral and medial walls

The other major bone of the orbit is the **sphenoid**; in essence, this bone closes the orbit at the back. The sphenoid is a block of bone at the midline of the skull from which a pair of bony wings extend on each side; the upper wings are smaller (**lesser**) than the lower (**greater**) wings. The central block of bone, the body of the sphenoid, has a large sinus within it, and it is penetrated on either side of the midline by the **optic canals**, through which the optic nerves pass; in other words, the optic foramina are the entrances to the optic canals. The upper surface of the sphenoid, which is saddle shaped, is called the **sella turcica** (literally "Turkish saddle"); it forms the pituitary fossa that partly surrounds the pituitary gland.

The lesser wing of the sphenoid forms the part of the orbital ceiling near the orbital apex that is not filled by the orbital plate of the frontal bone (see Figure 3.3*a*). Similarly, the portion of the lateral wall that is not filled by the zygomatic is completed by the greater wing of the sphenoid; it is the major component of the lateral wall. The large gap between the greater and lesser wings of the sphenoid is the **superior orbital fissure** (see Figures 3.2 and 3.3*a*); it separates the orbital ceiling from the lateral wall.

Thus, the construction of the ceiling and lateral wall of the orbit is quite simple; the ceiling is formed by the orbital plate of the frontal bone with a small contribution from the sphenoid posteriorly, and the lateral wall consists of the

Figure 3.3 ▶
The Ceiling, Floor, Lateral Wall, and Medial Wall of the Right Orbit
(*a*) The orbital ceiling is formed by the orbital plate of the frontal bone; it contains the lacrimal fossa (for the lacrimal gland) and the trochlear fossa. (*b*) The orbital floor is formed mostly by the orbital plate of the maxillary; the infraorbital fissure separates the floor from the lateral wall. (*c*) The lateral wall is a combination of the zygomatic bone in front, and the greater wing of the sphenoid posteriorly; in the posterior orbit, the lateral wall is separated from the ceiling and the orbital apex by the superior orbital fissure. (*d*) The medial wall consists of the frontal process of the maxillary, and the lacrimal, ethmoid, and sphenoid bones; the deep nasolacrimal fossa is formed where the maxillary and lacrimal bones abut each other. The optic foramen is high on the medial wall at the orbital apex.

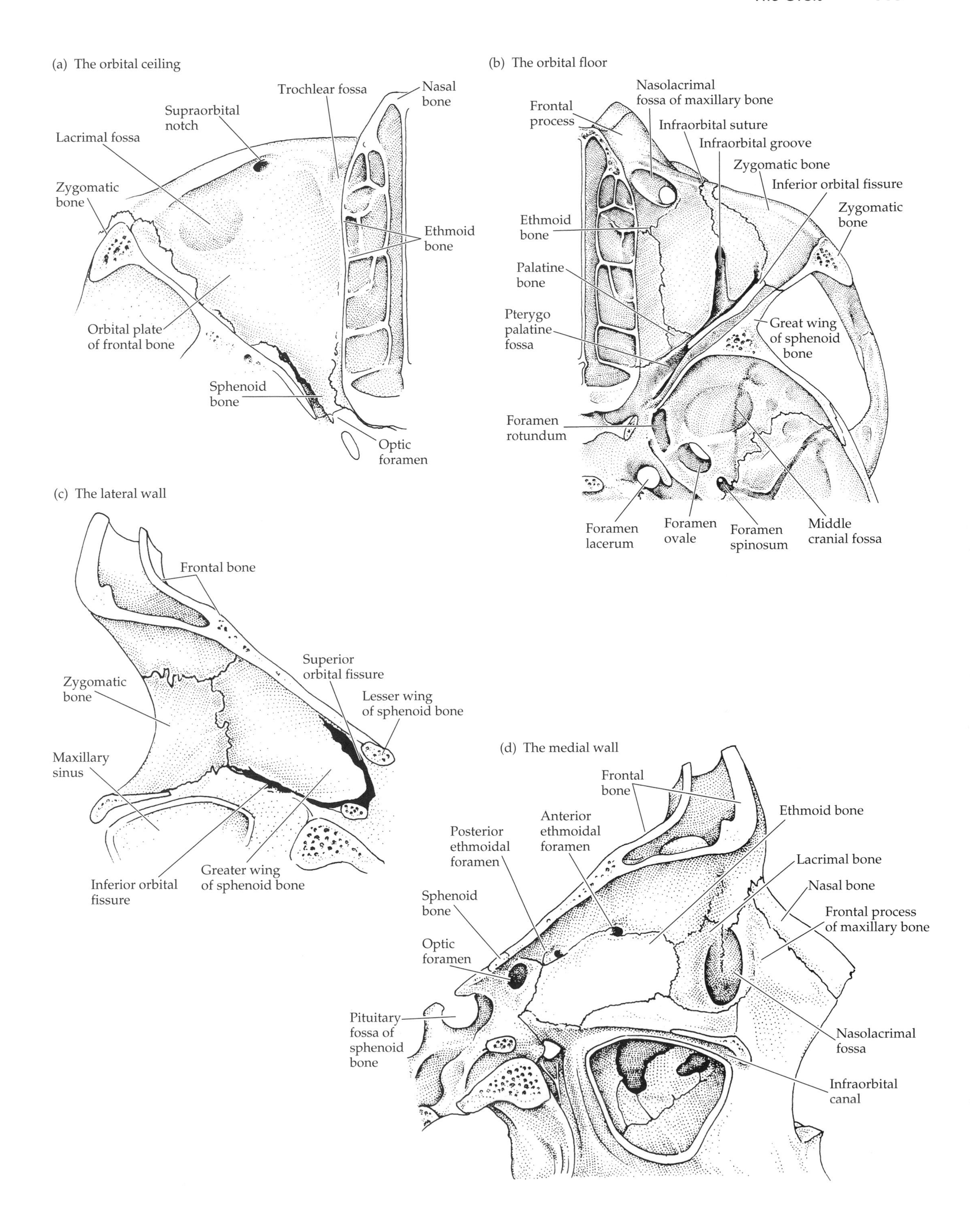
(a) The orbital ceiling
Trochlear fossa
Nasal bone
Supraorbital notch
Lacrimal fossa
Zygomatic bone
Ethmoid bone
Orbital plate of frontal bone
Sphenoid bone
Optic foramen
(b) The orbital floor
Nasolacrimal fossa of maxillary bone
Frontal process
Infraorbital suture
Infraorbital groove
Zygomatic bone
Inferior orbital fissure
Zygomatic bone
Ethmoid bone
Palatine bone
Pterygo palatine fossa
Great wing of sphenoid bone
Foramen rotundum
Foramen lacerum
Foramen ovale
Foramen spinosum
Middle cranial fossa
(c) The lateral wall
Frontal bone
Superior orbital fissure
Zygomatic bone
Lesser wing of sphenoid bone
Maxillary sinus
Inferior orbital fissure
Greater wing of sphenoid bone
(d) The medial wall
Frontal bone
Anterior ethmoidal foramen
Ethmoid bone
Posterior ethmoidal foramen
Lacrimal bone
Nasal bone
Sphenoid bone
Frontal process of maxillary bone
Optic foramen
Pituitary fossa of sphenoid bone
Nasolacrimal fossa
Infraorbital canal

Vignette 3.1

Sovereign of the Visible World

Dispel from your mind the thought that an understanding of the human body in every aspect of its structure can be given in words; for the more thoroughly you describe, the more you will confuse: it is therefore necessary to draw as well as describe.

Leonardo da Vinci (*Notebooks*)

THE WORD "GENIUS" has been devalued by overuse, but one of the many pleasures of wandering the streets of Florence is knowing that the person to whom the designation best applies once trod the same path. As Vasari wrote,

> *[t]he greatest gifts often rain down upon human bodies through celestial influences as a natural process, and sometimes in a supernatural fashion a single body is lavishly supplied with such beauty, grace, and ability that wherever the individual turns, each of his actions is so divine that he leaves behind all other men and clearly makes himself known as a genius endowed by God (which he is) rather than created by human artifice. Men saw this in Leonardo da Vinci, who displayed great physical beauty . . . , a more than infinite grace in every action, and an ability so fit and so vast that wherever his mind turned to difficult tasks, he resolved them completely with ease.* (*The Lives of the Artists*, 1568)

Leonardo da Vinci (1452–1519) was not a Florentine by birth, though the family home in Vinci was nearby and was under Florentine influence, but he spent much of his youth and young adulthood in the city, where he trained in the studio of Andrea del Verrocchio. His artistic genius was recognized early, and the acclaim has never ceased. Some art historians point to one of Leonardo's later works, *The Virgin of the Rocks,* as the first modern painting in the history of art, one containing a natural and realistic scene of elegance, grace, and technical virtuosity that produces, quite unexpectedly, vague unease and subtle perplexity; Leonardo painted for the viewer's mind, not just the viewer's eye, and that is a thoroughly modern point of view. Another, better-known enigma, his *Mona Lisa,* was described by Andre Malraux as "the subtlest homage genius has ever paid to a human face."

Like other Renaissance artists, Leonardo believed that representing the body in paint or marble required knowledge of what lies beneath the skin. Life and vitality dwell in muscles, arteries, and guts. But where Michelangelo and others had mainly artistic concerns, Leonardo was engaged in a grand quest to understand nature: plants, animals, rocks, rivers, the cosmos, and, of course, human beings. To understand humankind, he felt it necessary to understand the structure and function of the human body. Although the term had yet to be coined, Leonardo was a "scientist." And in science as in art, this self-educated, illegitimate son of a small-town notary was a genius.

In Leonardo's time, dissection of the human body was forbidden, and even though the secular authorities were unlikely to pursue rumors of surreptitious dissections, accessible corpses were rare, dissections were necessarily clandestine, and the risk was significant. (Also, it must have been very unpleasant; there were no preservatives, so one worked for days and nights in the stench of a decaying, fly-blown body.) But for 30 years or so, Leonardo dissected whenever he had the chance, and the results of his efforts are contained in his anatomical notebooks, most of which are now in the royal library at Windsor Castle. Among other things, Leonardo was the first to understand the compound curvature of the spine, the existence of the frontal sinuses, the presence of the ventricles in the brain (he made wax castings of them), and the structure of the valves in the heart. In the process of recording these observations, Leonardo anticipated many of the now standard devices of anatomical illustration; different vantage points, sections in different planes, and overlays.

Vision and the eye were central to Leonardo's thinking—so much so that his goal in his life and work was "knowing how to see" (*saper vedere*). He wrote eloquently on the importance of vision, of knowing how to see, and he knew a great deal about the process. He knew about the gradual onset of presbyopia and wore spectacles for near work after the age of 50, he knew about light and dark adaptation, and he proved to his own satisfaction that vision was not an emanation from the eye to external objects, but was instead the result of light entering the eye and interacting with it. He worked out a design for an astronomical telescope long before Galileo, but never constructed it, and he was, of course, a master of the painterly problems of color mixing and perspective. But some of his conclusions about the eye, most notably his belief that the retinal image is erect, were quite wrong.

Leonardo erred, ironically, because he misunderstood the anatomy of the eye; in fact, he probably never dissected a human eye. Compared to his drawings of structures that he had carefully dissected, such as the valves of the heart, his drawings of the eye are crude sketches of the optical elements and probably reflect his interpretation of what he had read, combined with what he had learned from some dissections of fish eyes (as suggested by his drawing of a large spherical lens). Combine a too powerful, improperly positioned lens with a rather imperfect understanding of refraction and one ends up misunderstanding image formation by the eye.

The magnitude of Leonardo's accomplishments is most evident in the structures he was able to study carefully. One example is in a drawing of the eyes and cranial nerves viewed from above (Figure 1); on the left, the eyes and the nerves have been isolated from surrounding structures, but they are illustrated in situ on the right. The optic nerves, chiasm, and tracts are quite obvious, with the olfactory nerves running over them. The other nerves are the superior division of the oculomotor (III) and the ophthalmic division of the trigeminal (V_1). This drawing is the earliest known anatomical illustration of the optic chiasm and the main cranial nerves. It is also quite accurate.

In Leonardo's illustration of the orbit and its surroundings (Figure 2), the orbit is shown intact on the right and partly cut away on the left. In the cutaway version, the superior and inferior orbital fissures are obvious, as are the frontal, ethmoidal, and maxillary sinuses around the orbit (the purpose of the cutaway drawing was to show the relationship of the orbit and the sinuses). The right-hand drawing gives a very clear impression of the strength of the orbital

The title of this vignette was taken from *The Creators,* by Daniel J. Boorstin (Random House, New York, 1992).

margin, while also including details such as the lateral orbital tubercle and the posterior lacrimal crest.

If these drawings seem unpolished in comparison to modern anatomical illustrations, remember that these were Leonardo's working drawings—they are primary data, first representations of his observations, not figures for publication. Even so, their quality is superb, and they anticipate almost every device used by contemporary illustrators. For example, the drawings in Figure 1 do not show all the anatomical structures that would have been visible; instead, it includes just the eyes, their nerves, and the bones of the skull. The drawing is an abstraction, made to isolate one set of related structures from everything else so that one is not overwhelmed with obscuring details. Contemporary practice is the same in showing, for example, nerves, arteries, and veins in separate figures. The cutaway drawing in Figure 2 is another of Leonardo's innovations. As for the artistry, Leonardo imbues his drawings with an almost reverential dignity, a quality rarely repeated in anatomical illustrations.

The most remarkable feature of these drawings is that they have no known precedent; they are completely the products of Leonardo's originality; nothing like them had ever been seen before. But did anyone see these drawings at the time? Leonardo intended to publish his anatomical work—a dozen or so illustrations had been done in publishable form—but most of it remained in his voluminous notebooks, which were scattered after his death and rarely seen again until the nineteenth century. As a result, Leonardo's place in the history of science is uncertain. No matter how original and unprecedented his work may have been, if no one knew of the work for 400 years, he could not have had much influence on subsequent developments in science. With the exception of the handful of paintings that have been admired ever since he did them, the full extent of his genius may have been of little consequence, however much we can appreciate it in retrospect through his notebooks.

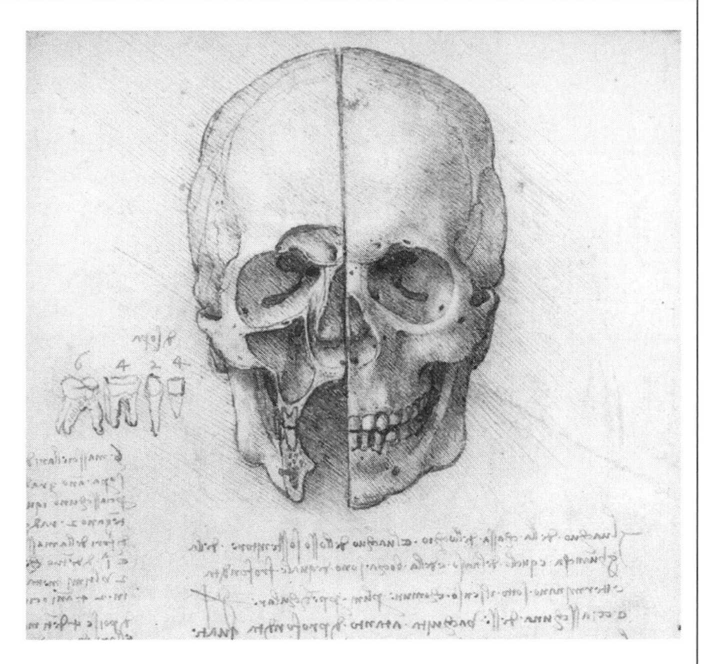

Figure 2
Views of the Skull from the Front
Leonardo's inventiveness in illustrating anatomy is evident in the cutaway drawing on the left; the frontal, ethmoidal, and maxillary sinuses show very clearly. (From The Royal Collection ©1999 Her Majesty Queen Elizabeth II.)

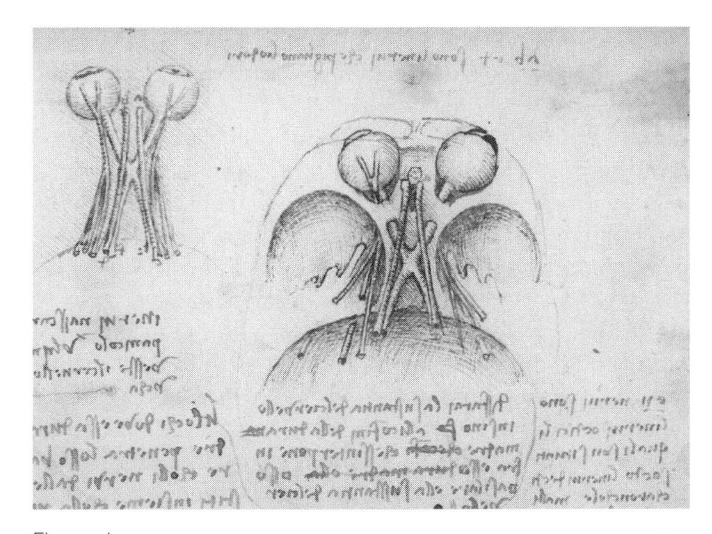

Figure 1
Views of the Eyes and Cranial Nerves from Above
Leonardo drew the eyes and nerves in isolation on the left and in situ on the right. His comments are written in Italian from right to left, with left–right reversal of all the letters. He was left-handed and found that this form of writing allowed him to hold his pen naturally and not smudge the ink as he wrote. (From The Royal Collection ©1999 Her Majesty Queen Elizabeth II.)

But one fact needs to be explained: We know from the historical records that a systematic study of anatomy, recorded in the visual language of illustrations, did not exist before Leonardo. After Leonardo, however, anatomical illustration blossomed and flourished. Is that coincidence, or was Leonardo's work the germinative element? The question is impossible to answer because any evidence linking Leonardo to later work is gone, and may not have existed. Ideas, not facts, are the most important elements of scientific progress, and Leonardo was working out ideas about how to study anatomy and how to record one's observations.

But good ideas are hard to confine; they circulate widely, rapidly, and mostly by word of mouth, leaving no trace of their path from mind to mind. The anatomical details Leonardo discovered were important, but they were far less significant than his methods for discovering and recording them. And we know from diaries of the time that Leonardo's drawings were known to exist and that some people had seen them; the ideas, if not the details, were almost certainly in the intellectual air, like whiffs of smoke from a distant fire. The odor would be novel and revelatory, quite enough for other people to light a few fires of their own. It may have happened.

In terms of direct, documentable evidence, Andreas Vesalius (see Vignette 3.2) is rightly credited with being the founder of anatomy as a systematic body of knowledge. But Leonardo was there before him, developing a visual language for the subject and telling posterity how to see. Vesalius may have smelled the smoke from Leonardo's fire.

orbital process of the zygomatic anteriorly, with most of the posterior part filled by the orbital face of the sphenoid's greater wing. The orbital floor is also uncomplicated; most of the floor is the orbital plate of the maxillary bone, with a bit of the anterior-lateral edge filled by the zygomatic and the apex at the back filled by a very small process from the underlying **palatine** bone (see Figure 3.3*b*). The palatine, the fifth of the orbital bones mentioned thus far, makes the smallest contribution to the orbital structure.

The lacrimal and ethmoid bones complete the medial wall between the maxillary bone in front and the sphenoid in back

The orbit's medial wall (Figure 3.3*d*) includes four bones: the maxillary anteriorly and the sphenoid posteriorly, with the **ethmoid** and **lacrimal** bones filling the gap between them. The small lacrimal bone abuts the frontal process of the maxillary bone, which has a pronounced groove just behind the orbital margin. A corresponding groove in the lacrimal bone creates a deep **nasolacrimal fossa**, where the bones fit together, with well-defined anterior and posterior **lacrimal crests** (ridges) in the maxillary and the lacrimal bones, respectively. The fossa continues down through the maxillary bone as the **nasolacrimal canal**. Most of the medial wall posterior to the lacrimal bone is the orbital plate of the ethmoid bone. This plate is so thin, however, that it is called the *lamina papyracea* ("sheet of paper"). It is all that separates the orbital cavity from the ethmoidal sinuses.

Three major and several minor foramina permit blood vessels and nerves to enter or exit the orbital cavity

None of the orbital surfaces are perfectly flat, continuous plates of bone; they all have holes or gaps for the passage of other structures, or places where the bone has formed bumps or depressions to accommodate the orbital contents.

The orbital ceiling has no foramina except for the supraorbital notch at the anterior margin, but there are two fairly well defined depressions in the orbital plate of the frontal bone; both are anterior, just behind the orbital margin (see Figure 3.3*a*). On the lateral side, a very shallow depression in the orbital plate, the **lacrimal fossa**, marks the location of the lacrimal gland above and lateral to the eye. The other landmark on the orbital plate of the frontal bone is on the anterior-medial side; it is a tiny pit in the bone—the **trochlear fossa**—that marks the location of the **trochlea**, which is the cartilaginous U-shaped pulley for the tendon of the superior oblique muscle (see Chapter 4). Ossification of the trochlea or the ligaments holding it in place sometimes fills in the fossa and leaves behind a small bony protrusion called the *spina trochlearis*.

The lateral wall of the orbit lacks significant landmarks, with the exception of two small foramina in the orbital process of the zygomatic bone through which small branches of the main zygomatic nerve pass (the zygomatico-temporal and zygomatico-facial—the foramina have the same names as the nerves). There are large gaps between the lateral wall and the adjacent ceiling and floor of the orbit. The lesser and greater wings of the sphenoid form the posterior parts of the ceiling and lateral wall of the orbit, respectively, leaving the prominent superior orbital fissure as the space between the wings. Most of the nerves and the major veins of the eye and orbit pass through the superior orbital fissure (the exceptions are the optic and zygomatic nerves and the ophthalmic artery).

The posterior gap between the lateral wall and the orbital floor is the **inferior orbital fissure** (see Figure 3.3*b*). It extends forward in the floor of the orbit as the **infraorbital groove**, then burrows beneath the floor as a canal, eventually emerging on the front of the maxillary bone as the infraorbital foramen (see Figure 3.2). The floor just above the infraorbital canal has a prominent **infraorbital suture** in the bone, which is often the site of fracture. The infraorbital groove, canal, and foramen are associated with the infraorbital artery, vein, and nerve (see Chapters 5 and 6), which enter through the inferior orbital fissure along with the zygomatic nerve and veins communicating with the pterygoid plexus.

On the medial wall of the orbit, the major landmarks are the nasolacrimal fossa of the lacrimal sac and the nasolacrimal canal, mentioned earlier, as well as the anterior and posterior **ethmoidal foramina** for the ethmoidal arteries, veins, and nerves (see Figure 3.3*d*). The list of orbital landmarks is completed by the optic foramen, which is located high on the medial wall at the apex of the orbit (see Figures 3.2 and 3.3*d*). The optic nerve and ophthalmic artery run through the optic foramen.

In summary, the three major foramina—the superior and inferior orbital fissures, and the optic foramen—are in the posterior half of the orbit, and most of the blood vessels and nerves running to and from the orbit pass through these openings. Minor foramina are situated in the medial wall (the ethmoidal foramina) and the lateral wall (the zygomatic foramina). The prominent nasolacrimal fossa and canal on the anterior-medial wall are associated with the lacrimal sac and the nasolacrimal duct (see Chapter 7).

Blowout fractures of the orbit are a consequence of the relative weakness of the orbital plates

The relatively heavy construction of the orbital margin makes it quite resistant to blunt trauma, but it can be broken; breakage of one or more of the bones forming the margin is called a **comminution fracture**. But trauma to the facial aspect of the maxillary, frontal, or zygomatic bones often produces fracture in part of the bone away from the site of impact. In particular, the orbital plates may rupture, a situation called **blowout fracture**. Blowout fractures of the orbital plate of the maxillary bone—that is, of the orbital floor—are the most common. Most blowout fractures occur when an object larger than the orbital opening strikes the orbital margin; baseballs are common culprits. The mechanism responsible for the fracture is often said to be a compression of the eye and the orbital contents such that the force is transmitted to the relatively weak orbital plate of the maxillary bone, causing it to fracture; in this sense, the orbit blows *out*.

This explanation comes from a study that attempted to simulate the conditions for blowout fracture by the following method: A hurling ball was placed over the closed lids of the orbit of a cadaver, and the ball was struck sharply with a hammer. A cracking sound was interpreted as fracturing bone. The results of this experiment were, first, the orbital floor could be fractured without causing a break in the orbital margin, and, second, if the eye and orbital contents were removed prior to striking the hurling ball, fracture of the orbital floor was always accompanied by fracture of the orbital margin. The authors therefore concluded that the presence of the eye and orbital contents are necessary for a blowout fracture to occur and that the force producing the fracture must be transmitted through these soft tissues.

These results have been disputed and contradicted several times, but whatever the results may be, the experiment is fundamentally flawed: The geometry of the orbit is quite variable, so it is impossible to guarantee that the experimental conditions are the same from one trial to the next. Since no definitive experimental answer is possible, the only recourse is some basic physics, which suggests that the forces developed by blunt impact will be more effectively transmitted by rigid bone than by compressible soft tissues. Thus, the shock wave traveling through the bone from the site of impact is the most likely cause of fracture at a weak point in the bone. In addition, the initial direction of fracture may be into rather than out from the orbital cavity. Whatever the case, the result is a blowout fracture.

In a blowout fracture of the maxillary bone, the orbital floor ruptures—most commonly along the infraorbital suture—and the broken pieces of the orbital plate on either side of the fracture probably separate transiently like the opening of a fault during an earthquake. As the shock wave subsides, the fault may partly close again, but its momentary opening may allow the tissue above to drop down and be caught by the pieces of orbital plate. The connective tissue surrounding the inferior rectus and inferior oblique muscles (see Chapter 4) is especially likely to be trapped in this way, thus restricting eye movements, and the eye on the affected side may now be lower in the orbit than the other eye because the broken orbital floor no longer provides any support. Another possible consequence is pressure on the infraorbital nerve that produces anesthesia of the lower lid and cheek. In addition, the eye itself and the lacrimal and ethmoid bones may be damaged.

Abnormal positioning of the eye relative to the orbital margin may indicate local or systemic pathology

The exact configuration of the orbital margins and the positions of the eyes relative to them vary considerably among individuals, but some of the variation is important. **Exophthalmos** refers to an abnormal protrusion of the eye, or *proptosis;* it is measured as the position of the cornea's anterior pole relative to the notch in the lateral orbital margin. The converse, **enophthalmos**, is an abnormal recession or sinking back of the eye into the orbit. Characterizing these conditions is not always clearcut, since what may seem exophthalmic for one person may be quite normal for another, and racial differences in the structure of the brow and upper eyelid compound the problem. For a particular person, however, changes in the fore and aft positions of the eyes (or just one eye) may have diagnostic significance.

Bilateral or unilateral exophthalmos, for example, is often associated with thyroid disorders, such as Graves' disease; hypertrophy of the extraocular muscles and of other orbital tissues pushes the eyes forward from the normal position. Unilateral exophthalmos may also be produced by an orbital tumor. Enophthalmos is a common consequence of extreme malnutrition, but it is also associated with lesions that affect the sympathetic nervous system. Vertical or horizontal displacements of the eye by tumors or trauma are also possible; as noted in the previous section, a drop in vertical position of one eye can be associated with a blowout fracture.

Normally, the orbits are very symmetric with respect to the midline of the head, and they are in the same horizontal plane. On occasion, however, one orbit and therefore one eye may be farther from the midline than the other. This asymmetry lacks significant consequences, except that it should be recognized and taken into account when one is prescribing refractive corrections using spectacle lenses (unless specified otherwise, the optical centers of spectacle lenses will be symmetric about the midline). One orbit and one eye can be vertically displaced as well. The problem in this case is that the two eyes must be in the same horizontal plane if they are to work together normally. An individual with a vertically displaced eye will compensate with a characteristic head tilt to one side or the other that brings the two eyes into horizontal alignment. If forced to hold the head erect—for example, during a routine eye examination—the individual may report double vision.

See Box 3.1 for methods used to examine the eye and orbit in vivo.

Infection and tumors can enter the orbital cavity through the large sinuses that surround the orbit

The orbital plates separate and isolate the eye from surrounding structures and tissues, but because the plates are relatively thin, they are not impregnable. Problems originating outside the orbit can affect the eye. Both the medial wall and the orbital floor are adjacent to large sinuses in the ethmoid and the maxillary bones, respectively (Figure 3.3*b* and *d*). Thus infections in the sinuses, particularly the ethmoidal sinuses that lie next to the thin lamina papyracea, can be transmitted to the orbit; the most common consequence is an infection of orbital tissues—but generally not the eye itself—called **orbital cellulitis**. (About 60% of the cases of orbital cellulitus are said to be secondary to a sinusitis.) Large tumors in the sinuses may produce degeneration of the intervening orbital plate, thereby invading the orbital cavity directly, rather than by metastasis.

The same considerations apply to the orbital apex, which lies near the large sinus within the sphenoid bone (Figure 3.3*c*), and to the orbital ceiling, whose anterior part is adjacent to the frontal sinus in the frontal bone (Figure 3.3*c* and *d*). Most of the orbital ceiling, however, separates the orbital cavity from the overlying frontal lobe of the brain; fractures of the orbital ceiling may therefore be associated with intracranial hemorrhage or infection. The proximity of the orbital ceiling to the frontal lobe also explains why the earliest attempts to cure various psychiatric disorders by prefrontal lobotomies employed a surgical approach through the orbital ceiling.

Only the lateral wall of the orbit is free of association with any large sinuses. Here, the major exterior structures are muscles of the face and jaw that run between the lateral wall of the orbit and the heavy zygomatic arch (which, incidentally, affords even more protection on the lateral side). Near the back of the orbit, the lateral wall separates the orbital cavity from the tip of the brain's temporal lobe.

Thus the walls, ceiling, and floor of the orbit are well away from the exterior of the head, and only on the lateral side, which is the most exposed to trauma, do the structures external to the orbit provide additional protection. The concerns about orbital relations are quite different for the ceiling, floor, and medial wall, where invasion is more likely to originate from pathology than from trauma.

Connections between the Eye and the Orbit

Connective tissue lines the interior surface of the orbit

The earlier diagrams showing the position of the eye in the orbit (see Figures 3.2 and 3.3) did not illustrate the structures that hold the eye in this position or form the socket within which the eye rotates. Much of this support system is connective tissue that directly or indirectly connects the eye and other structures to the orbital bones (or, to be more precise, to the *lining* of the bones).

Bones are covered with a layer of tough connective tissue, the **periosteum**, that adheres tightly to the bone in places. The orbital bones are no exception, but the periosteal lining of the orbit is referred to specifically as the **periorbita**. The entire inner surface of the orbital bones is covered by the periorbita, which is continuous at the orbital margin with the external periosteum of the facial bones; the periorbita also fuses at the optic foramen with the dura mater surrounding the optic nerve. Part of the structural support for the eye consists of bands of connective tissue that fuse with the periorbita at one end and with connective tissue surrounding the eye at the other.

Connective tissue surrounds the eye and extraocular muscles

The sclera is covered with a thin layer of connective tissue containing small blood vessels that is called the **episclera**. The episclera covers all of the eyeball except the cornea. In turn, the episclera is itself covered by another layer of connective tissue that tends to fuse with it; this outer layer is **Tenon's capsule** (Figures 3.4 and 3.5). Tenon's capsule has the same relationship to the eye and its extraocular muscles as a glove has to one's hand and fingers; if the eye is the

Box 3.1 Visualizing the Orbit and Its Contents in Vivo

THE SAME TRICK USED by echolocating bats to reconstruct space and locate objects within it—like insects—is used to probe and reconstruct the normally invisible contents of the orbit. Ultrasound is high-frequency sound above the range of human hearing—that is, above 20 kHz (1 kHz = 1000 cycles/s)—but like any other sound waves, ultrasound can be conducted by media that are optically opaque and it will reflect from interfaces where there is a change in the acoustic conduction index. Therefore, the basic principle of ultrasonography is to send a pulse of ultrasound into the body and to detect the echoes that come back from the different surfaces encountered by the pulse; the more time that is required for an echo to come back to the detector, the deeper, or farther away, the reflecting surface lies. Diagnostic ultrasound has frequencies of 5 MHz and greater; the spatial resolution increases as the frequency increases.

In its simplest form, the ultrasound picture is a record of the time between the emission of an ultrasound pulse and the sequence of echoes that it produces; this form is called the A-scan mode. Figure 1*a* shows the detector's response to an ultrasound pulse sent into the eye along its anterior–posterior axis. In this case, the acoustic and optical surfaces are nearly identical, and the record shows echoes from the cornea, the surfaces of the lens, and several closely spaced echoes from the retina, choroid, and sclera. If we know how much the speed of sound is slowed down relative to air, the differences in time between echoes can be translated to differences in distance. Thus, A-scan ultrasonography can be used to measure the axial length of the eye in vivo, for example, and in a specialized form it is used routinely to measure corneal thickness.

B-scan ultrasonography produces a picture of one planar section through the eye and orbit. In this case, we are producing not just a single sound path through the tissue from which echoes are recorded, but multiple paths, each with its own pattern of echoes. The paths originate from a common point in front of the eye, which is the location of the ultrasound source, and radiate from there, giving the final picture a fanlike configuration. The echoes from each path are shown as bright dots on a video screen, producing the kind of image shown in Figure 1*b*.

The advantage of ultrasonography as a diagnostic tool is its ease and simplicity; the method is not invasive, it can be done quickly, and there is no discomfort to the patient. It can provide evidence for or confirmation of the presence of tumors, foreign objects, and retinal detachment, among other things. For some purposes, ultrasonography is the method of choice for probing ocular and orbital structures in vivo. What ultrasonography does not do well, however, is provide good information about exact location, and because of the attenuation of echoes from structures well away from the source, its usefulness for probing the posterior part of the orbit is limited. These limitations are largely overcome by the more precise and detailed pictures obtained from computed tomography or magnetic resonance imaging.

Conventional X rays have important uses, but they are not particularly good for visualizing structures in isolation. Since everything between the X-ray source and the film plane casts its X-ray shadow on the film to greater or lesser degrees, the X-ray picture superimposes the shadows of spatially separate, but overlaid structures, sometimes making interpretation of the images rather difficult. What is often more helpful is an accurate spatial reconstruction of the body's deep tissues and organs, which is what computed tomography and magnetic resonance imaging provide.

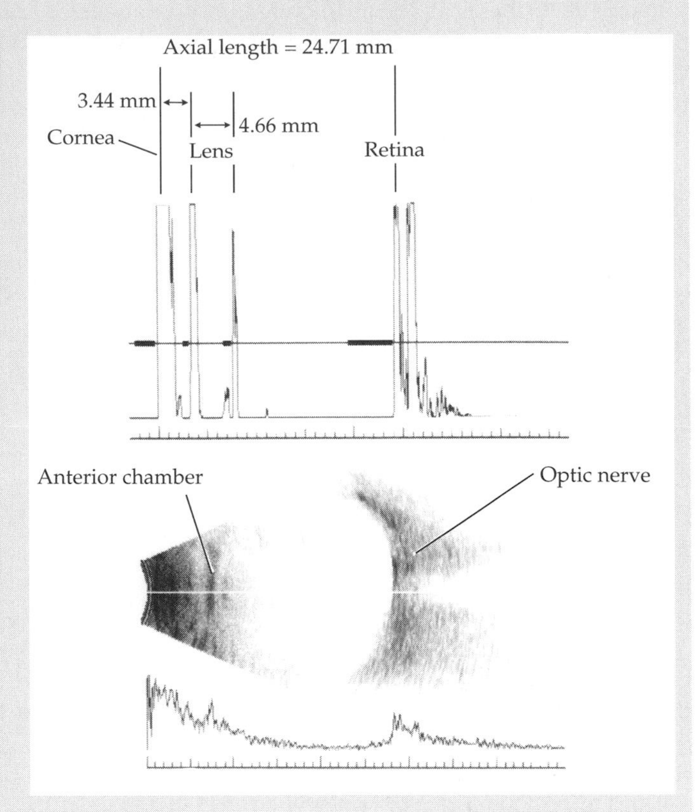

Figure 1
Ultrasonography of the Eye and Orbital Cavity
(*a*) This A-scan image shows the sequence of echoes recorded when an ultrasound pulse is sent into the eye along the anterior–posterior axis. Well-defined echoes return, as indicated, from the major ocular surfaces and from the tissue just behind the eye. (*b*) This B-scan image, which combines many A-scans to form an image of a horizontal acoustic section through the eye, outlines the whole interior surface of the eye, as well as the corneal and lenticular surfaces, though not as clearly.

Computed tomography (CT) is an X-ray method in which the X-ray attenuation at a particular point in the body is computed from many X-ray beams passing through the point from different directions in a plane. When the attenuation factors for all resolvable points in the plane are displayed as variations in light intensity, the result is like a section through the body where the tissues with the greatest X-ray attenuation, such as bone, are bright while watery tissues transparent to X rays are dark. CT scans use many parallel beams from many different directions, and the matrix of points within the tissue is therefore quite large, perhaps up to

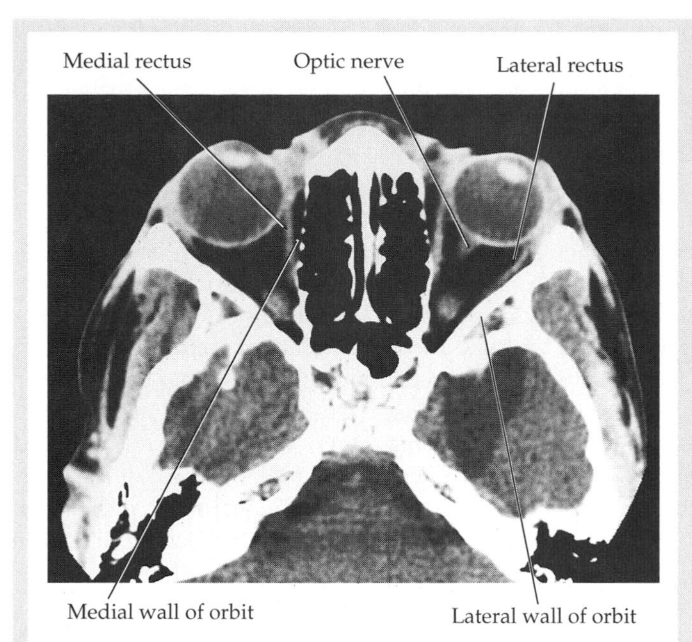

Figure 2
Horizontal CT Scan through the Eyes and Orbits
This CT scan shows the bony structures of the orbit as having the highest density, but the eyes and the lenses within them, the optic nerves, and the horizontal rectus muscles are also visible. One of the patient's eyes habitually rotated laterally, and the eyes clearly have misaligned visual axes. (Courtesy of Katherine Niemann.)

100,000. For a section through the head, the resolution would be around 1 mm or less.

A horizontal CT scan through the head at the midlevel of the orbits is shown in Figure 2. The spatial resolution in this image is somewhat less than 1 mm, and we can easily see the medial and lateral walls of the orbits, the eyes (including the lenses), the medial and lateral rectus muscles, and the optic nerves. The bony structures show up best, soft tissues less well. Blood vessels and highly vascularized neoplasms can be made visible by intravenous injection of radiopaque substances into the bloodstream.

Magnetic resonance imaging (MRI) produces similar pictures, but with some important differences that are based on the nature of the technique. MRI takes advantage of the weakly magnetic character of atomic nuclei that have odd-numbered atomic weights. Hydrogen, with its single proton, is the most common of these atoms in the body, and it is therefore the prime target of the method. When a strong external magnetic field is generated around the body, these nuclei tend to align themselves with the field. These atomic nuclei will also absorb radio waves at a particular frequency—the *nuclear resonance frequency*—and brief pulses of radio waves can temporarily knock the nuclei out of alignment with the magnetic field. When the radio pulse stops, the perturbed atoms will return to alignment within a characteristic period of time (the *relaxation time*), giving up energy in the form of detectable radio wave frequencies as they do so. The relaxation time depends both on the atomic species and on the environment in which it is located; hydrogen, for example, has a shorter relaxation time in fat than in water. In essence, it is the radio wave frequencies emitted by the relaxing atomic nuclei that are recorded and used to spatially reconstruct the tissue in terms of the intensity of the frequencies.

The MRI picture in Figure 3 shows the same structures as those in the CT scan in Figure 2, but the images look rather different. For example, the bone that is so evident in the CT image is hardly visible by MRI; bone has very low water content, and therefore very little free hydrogen, so it emits only a weak signal. There is very good definition of soft tissues, and the extraocular muscles, the optic nerves, and even the chiasm are quite obvious. In general, the ability of MRI to show soft tissue so well is what makes it the method of choice for many purposes; localizing nonmetallic foreign bodies in the orbit is an obvious example. MRI should not be used if there is any reason to suspect a metal object in the orbit (or anywhere else for that matter). The strong magnetic field could easily displace the object, producing additional damage (fortunately, dental amalgams are not influenced by the magnetic fields).

MRI has several variations. There are two modes of nuclear relaxation, called T_1 and T_2, that differ in the contrast with which various tissues can be imaged. Therefore, depending on the tissue or body region of interest, one or the other mode may be preferable; the modes may also be used in combination for other sorts of contrast enhancement. The method can be modified to emphasize water movement, a configuration called nuclear magnetic angiography (NMA), which provides a very detailed picture of blood vessels without requiring that anything be injected into the bloodstream. In addition, atoms other than hydrogen are accessible to the method. Of these atoms, resonance imaging of sodium produces strong signals from the cerebrospinal fluid and vitreous; sodium resonance is said to be helpful in the diagnosis of occlusive stroke and in the detection of certain types of neoplasms.

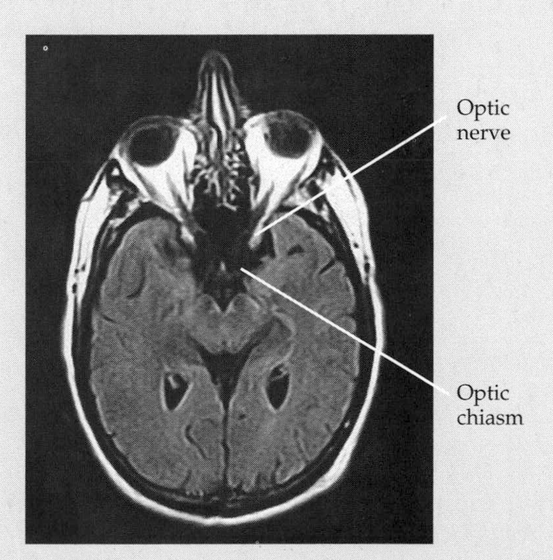

Figure 3
MRI of the Eyes and Orbits
Unlike the CT scan (see Figure 2), this MRI picture does not show bone well. Imaging of soft tissues is quite good, however, and in this image the optic nerves can be followed through the optic canals to the optic chiasm (which was obscured by the sphenoid in the CT scan). (Courtesy of Mark Swanson.)

hand and the extraocular muscles are the fingers, Tenon's capsule is the glove. It covers all of the eyeball except the cornea, the limbus, and the sclera anterior to the insertions of the rectus muscles; in other words, Tenon's capsule begins about 5 mm from the corneal margin.

Check ligaments connect Tenon's capsule to the periorbita

The bands of connective tissue that run between Tenon's capsule and the periorbita are called **check ligaments**. Their anatomy is quite variable and descriptions of them are correspondingly disparate, but most authors agree that major check ligaments are associated with the medial and lateral rectus muscles (see Figures 3.4 and 3.5). The connective tissue of Tenon's capsule that sheathes the medial and lateral recti on the side of the muscle closest to the orbital wall gives rise to connective tissue bands—the check ligaments—that run out and forward from the muscles to fuse with the periorbita. For the check ligament of the lateral rectus, the periorbital insertion is on the orbital margin near the lateral orbital tubercle. The check ligament from the medial rectus runs to the lacrimal bone behind the posterior lacrimal crest.

The medial and lateral check ligaments are continuous with Tenon's capsule surrounding the extraocular muscles and the eye, and the portion of Tenon's capsule on the underside of the eye is said to be thickened (or to have additional connective tissue strands joining it to the medial and lateral check ligaments); as a result, one can imagine the eye resting in a hammock or sling of connective tissue that is continuous, beginning at one side of the orbit and passing under the eye to connect to the other side of the orbit (see Figure 3.5). This sling is the **suspensory ligament** of the eye or, in older terminology, the *ligament of Lockwood*. Thus, the suspensory ligament is at least partly responsible for the proper vertical positioning of the eye in the orbit.

Other bands of connective tissue can be found on both the superior and inferior aspects of the eye, but they are less well defined and less consistent in location than the medial and lateral check ligaments. There is, however, a connection between the fascial sheath of the superior rectus and the fascial sheath of the levator muscle just above it (see Figure 3.5); elevation of the eye by contraction of the superior rectus is associated with elevation of the upper eyelid, and the mechanical connection between the superior rectus and levator muscles is part-

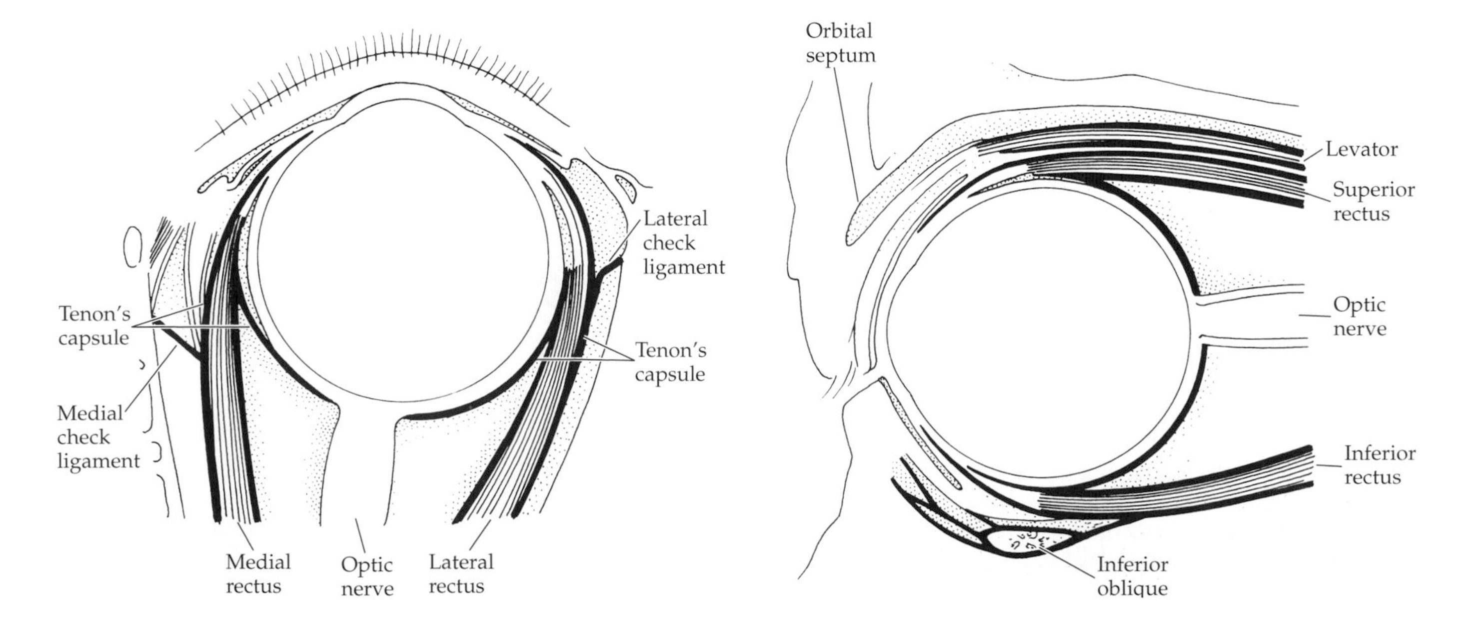

Figure 3.4
Check Ligaments, Tenon's Capsule, and Other Connective Tissue
(*a*) This horizontal section shows the medial and lateral check ligaments running between the fascia around the medial and lateral rectus muscles to the periorbita on the orbital walls. (*b*) This vertical section shows the septum orbitale, connections between the superior rectus and levator muscles, and connections between the inferior rectus and oblique muscles to the lower eyelid.

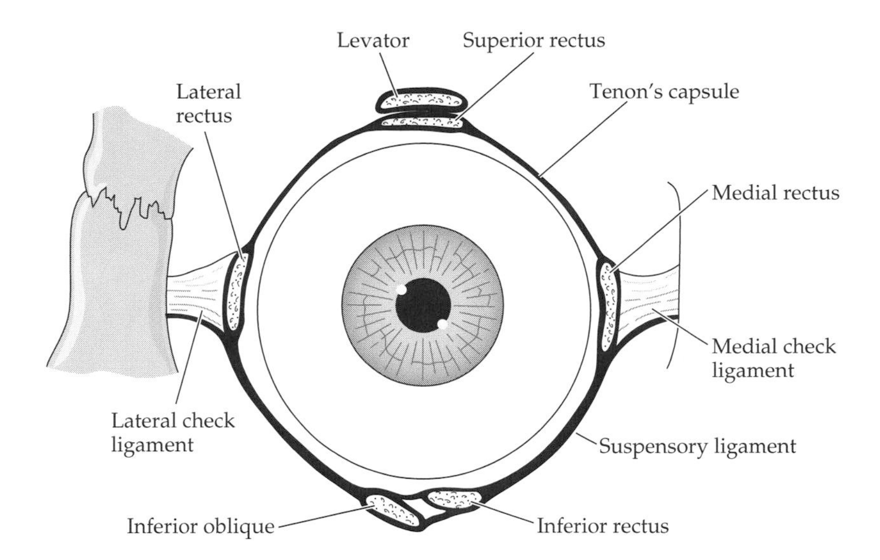

Figure 3.5
Check Ligaments, Tenon's Capsule, and the Suspensory Ligament
The lateral and medial check ligaments are continuous with the portion of Tenon's capsule that is on the underside of the eye, creating a sling—the suspensory ligament—on which the eye rests.

ly responsible (the levator muscle inserts into the upper eyelid; see Chapter 7). An additional connective tissue sling attached to the periorbita of the orbital ceiling supports the levator muscle; it is called the **superior transverse ligament** or *Whitnall's ligament* (see Chapter 7).

Similarly, fascial connections are sometimes said to exist between the inferior muscles (inferior rectus and oblique) and the orbital floor; although these connections undoubtedly exist, there is very little consistency among individuals. Connective tissue connections between the inferior muscles and the lower eyelid are more commonly reported, and they would explain the slight depression of the lower lid as the eye looks downward.

The eye's four rectus muscles run through slings or pulleys consisting of connective tissue and smooth muscle fibers (Figure 3.6). These pulleys affect the actions of the rectus muscles in moving the eyes (see Chapter 4).

All structures in the orbital cavity are lined and interconnected with connective tissue

The three-dimensional structure of all the interconnected sheets and bands of connective tissue in the orbit is complicated. In general, however, it is realistic to think of all the major structures in the orbit—the eye, extraocular muscles, nerves, and blood vessels—as being sheathed with connective tissue, with numerous interconnections among these components. Returning to the analogy of the hand and glove for the eye and Tenon's capsule, respectively, the fingers of the glove have webbing between them and other nearby structures. Moreover, connective tissue fuses with the periorbita at sites other than the check ligaments. A particularly important location is the orbital apex, where the connective tissue sheaths of the rectus muscles merge with the periorbita and the dura mater of the optic nerve. As the muscles contract and the eye rotates, the optic nerve flexes, resulting in mechanical pulling and tugging on the connective tissue attachments among the muscles, nerve, and periorbita; infection of the optic nerve may therefore manifest itself as pain during normal movements of the eyes.

Abnormal development of the connective tissue may affect movement of the eyes

Although the term "check ligament" implies that these bands of connective tissue check or restrict rotation of the eye in a particular direction, neither the check ligaments nor the other bands of connective tissue appear to come into play

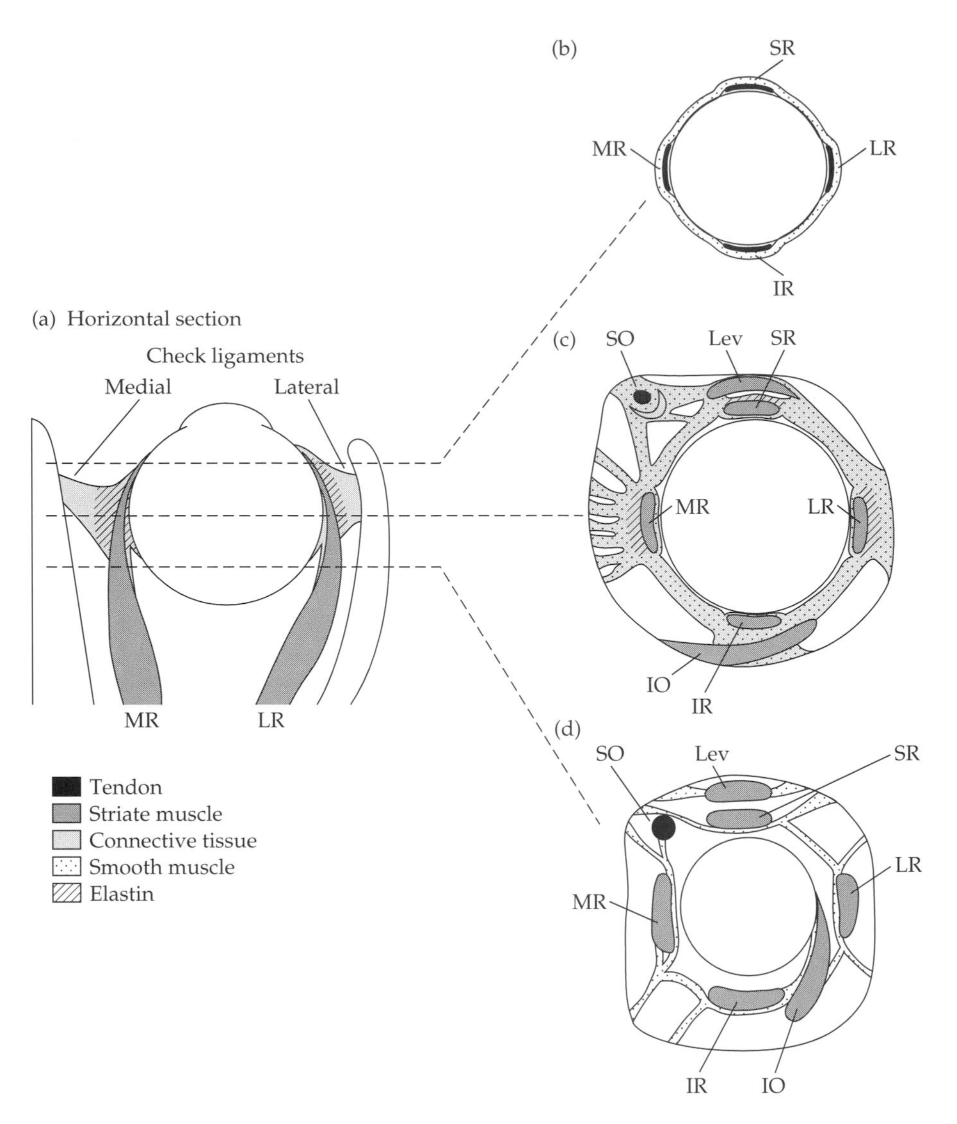

Figure 3.6
Rectus Muscle Pulleys and Slings
(a) A horizontal section through the eye shows the medial and lateral check ligaments and three planes of transverse section through the eye (dashed lines). The check ligaments contain not only collagen but also smooth muscle cells and elastin. *(b)* A transverse section near the anterior edge of the check ligaments cuts through the rectus muscle tendons, all of which are covered with connective tissue. *(c)* In a section near the eye's equator, the rectus muscles are completely surrounded with connective tissue that also contains smooth muscle cells and elastin. These sleeves of tissue around the rectus muscles are their pulleys. *(d)* Farther back, a section shows the muscles supported by slings of connective tissue. Both ends of the slings attach to the orbital bones. (After Demer, Miller, and Poukens 1996.)

within the normal range of eye movements. Genuine restrictions of ocular rotation, however, may result from abnormal development of check ligaments or other fascial connections, a situation that requires surgical intervention. And clearly the connective tissue connections associated with the eye and the extraocular muscles—that is, Tenon's capsule and the rectus muscle pulleys—have a great deal to do with the actions of the muscles when they contract (see Chapter 4). Whether to remove the fascial connections or to leave them intact as much as possible is therefore one of the numerous variables affecting surgery on the extraocular muscles.

Fat fills the spaces in the orbital cavity that are not occupied by other structures

Although the various fascial connections, particularly the suspensory ligament, support the eye, the normal position of the eye in the orbit is maintained by pillows of fat (see Figures 7.4 and 7.12). In general, any space in the orbit that is not occupied by the eye, muscles, blood vessels, and so on, is filled by fat that cushions, supports, and separates these structures. Depletion of orbital fat, as in extreme malnutrition, is therefore accompanied by a drop in vertical eye position and by enophthalmos.

The septum orbitale prevents herniation of orbital fat into the eyelids

Fatty tissue in the orbital cavity has pockets of fat cells surrounded by sheets of connective tissue that are continuous with the rest of the orbital connective tissue, but the fat still has a tendency to ooze around under the influence of pressure or gravity. (And if one wants to assign a function to much of the orbital connective tissue, this is a reasonable one—it helps keep the fatty tissue in place.) The fat near the front of the orbit above and below the eye is kept within the orbit by yet another sheet of connective tissue, the **septum orbitale**. This sheet is attached to the periorbita all around the orbital margin (see Figure 3.4*b*) and enters the eyelids to fuse with the rigid tarsal plates (see Chapter 7). When the septum orbitale is breached, fat can move freely into the lids, sometimes making normal movements and closure of the lids are impossible.

Development of the Orbital Bones

Many bones form first as cartilage templates

The first step in the formation of most bone in the body is the differentiation of **chondrocytes** from the mesoderm; these cells, which are responsible for producing cartilage, proliferate and secrete a matrix consisting of proteoglycans and collagen. For large bones, there are often several initial sites of chondrocyte activity (foci) that will eventually meet and fuse as the individual foci expand by continual production of the matrix. This combination of chondrocytes and their matrix is **cartilage**, and the first stage of bone formation is the generation of a cartilaginous replica or template that will become bone by a process of replacement.

Cartilage is plastic; it can be deformed by mechanical force, and perhaps more importantly, it can respond readily to inductive influences from other developing tissues. That is, cartilaginous templates of the bones are affected by what is going on around them, and the complex geometry of the adult bones is more a result of their developmental environment than of any inherent predilection for a particular shape.

As the foci of cartilage expand by the addition of cells and matrix, the most active regions of growth are at the periphery of the expanding cartilage, meaning that the regions of the original foci are, in a sense, older then the active regions at the periphery. The transition from cartilage to bone also begins first at the growth foci, and the periphery is the last to make the transition. Chondrocytes in the mature regions of the cartilage expand and die, leaving large cavities and channels in the matrix that are invaded by **osteoblasts**, which are the bone-forming cells. Like chondrocytes, osteoblasts secrete a matrix, but one composed of a different type of collagen that is converted to bone by the deposition of calcium phosphate, which gives bone its characteristic rigidity. As the osteoblasts surround themselves with matrix, they mature, losing much of their secretory activity, and become mature **osteocytes**. The continuation of this replacement process transforms the original piece of cartilage into a bone surrounded by a thin layer of cartilage that is responsible for future growth of the bone.

Most orbital bones do not have cartilaginous templates

Most of the bones contributing to the orbit are not preformed in cartilage; the preceding description of bone formation applies only to the ethmoid bone and to the body and lesser wings of the sphenoid. All the rest (frontal, maxillary, zygomatic, lacrimal, palatine, and the greater wings of the sphenoid) are **dermal** bones formed by foci of osteoblasts within sheets of connective tissue that have

Vignette 3.2
The Anatomy of Vesalius

ONE OF THE MOST INFLUENTIAL, most imitated, and most blatantly plagiarized books in the history of science may well be *De humani corporis fabrica* ("Structure of the Human Body") by Andreas Vesalius (1514–1564). Published in 1543, the *Fabrica* (as this work is known colloquially) contains more than 200 large prints depicting the anatomy of the human body; it is the accomplishment that Leonardo da Vinci began but did not finish.

Vesalius was born in Brussels and educated there and in Paris. Anatomy as Vesalius studied it as a medical student in Paris had very little to do with the human body; most dissections were done on other animals, and they were not done often. When a corpse was available, the professor usually read aloud the appropriate passages from Galen, while an underling employed for the purpose attempted to find and demonstrate the parts mentioned, usually with limited success. Students were supposedly gentlemen and were not to dirty their hands. Vesalius, to his lasting credit and fame, rebelled.

Vesalius was hardly out of his teens when he realized that something was very wrong; either Galen was an unreliable guide, or his teachers did not understand Galen. Or perhaps both. In any event, Vesalius took the same path that Leonardo had started along some 50 years before; he found bodies (sometimes by robbing graves) and taught himself anatomy by doing his own dissections. He soon acquired a unique body of detailed, firsthand knowledge, became a professor at the University of Padua when he was 23, and published the *Fabrica* about 5 years later.

We do not know what prompted Vesalius to cast his knowledge of human anatomy in the form of illustrations. Perhaps the idea was in the Italian air. Not long after Leonardo's death, the Italian physician Berengario da Carpi (1460–1530) published two illustrated anatomical textbooks, and some of the illustrations are similar in detail to some of Leonardo's drawings; da Carpi may have seen Leonardo's work. Vesalius was undoubtedly aware of these books. As a professor, Vesalius had to teach, and he had already experienced the limitations of verbal descriptions of anatomy and dissection procedures. Perhaps that experience was enough to convince him that illustrations were needed. Or, knowing nothing about the details of Leonardo's earlier investigations, he may have simply realized that he was in a position to do something unique and unprecedented. Whatever the motivation, the science of anatomy begins with Vesalius.

The *Fabrica* contains several plates devoted to the eye and associated structures; the one reproduced here is the first known illustration of a dissected eye, and it reveals the dissection method. Vesalius first cut through the sclera at the equator, then peeled it away from both the anterior and posterior segments, then cut through the choroid (and probably the retina). When these parts had been peeled away, he was left with the lens in a large blob of vitreous. For some purposes, this is a good technique; in particular, it shows the continuity of the structures in the uveal tract (iris, ciliary body, and choroid) very clearly and makes it easy to see the highly vascular nature of the choroid. What it fails to do, however, is retain the lens in its proper relationship to other structures; Vesalius shows the lens near the center of the vitreous chamber and, by so doing, perpetuates an old error.

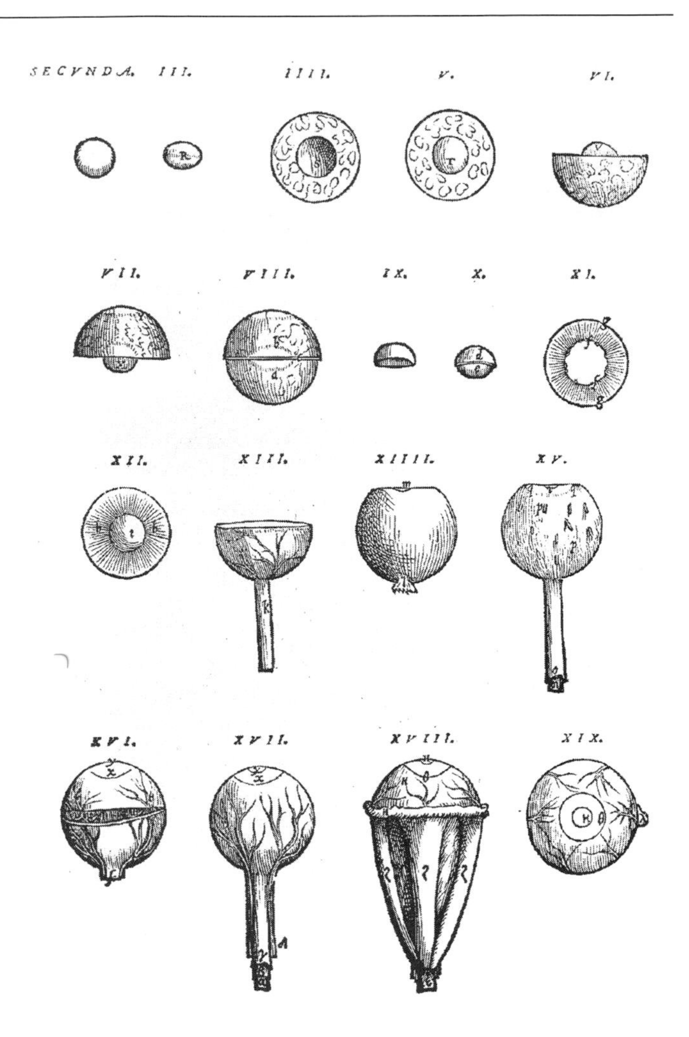

The Eye Dissected
Parts *secunda* through X show the lens in isolation and in relation to the vitreous body. XI and XII are views of the ciliary body and iris from behind. Parts XIII, and XVI through XIX show views of the sclera with blood vessels. Part XV shows the choroid with blood vessels and the optic nerve. Part XVIII is the first drawing of the rectus muscles; the cuff around the equator of the eye is probably the conjunctiva. (From Vesalius 1543.)

The drawing labeled XVIII is the first to illustrate the extraocular muscles. The rectus muscles are the best part of the drawing, although Vesalius did not seem to recognize that the superior rectus (or the optic nerve, for that matter) runs medially as it goes from the eye to the apex of the orbit. The obliques are not shown, and the tissue that Vesalius included within the muscle cone formed by the recti is a retractor bulbi muscle that is not found in humans. (The retractor bulbi in other animals acts to pull the eye back into the orbit, presumably for protection.) One can only guess that

Vesalius included the retractor bulbi because he saw it so frequently in other animals and persuaded himself that it was present on the human eye.

Despite their importance in the history of understanding the human eye, these illustrations were not the best of Vesalius's efforts; the basis of his fame lies elsewhere, particularly in the so-called musclemen in which the skeletal musculature is rendered with great skill and artistry (and with the rather surreal placement of the dissected figures in Venetian landscapes). Vesalius was regarded as a competent draftsman, and part of the appeal of the ocular drawings shown here is that they probably came from his own hand. For the more important and complicated illustrations, however, a great deal more than competent draftsmanship was required; they were done by an accomplished artist, most likely someone from the nearby studio of the painter Titian in Venice.

The *Fabrica* was of monumental importance, but Vesalius did not understand all of human anatomy at a single stroke, nor did he do away with Galenic anatomy easily. Galen's authority had reigned for 1000 years, and people who had built reputations on their knowledge of Galen were not about to let some young smart aleck kick the props out from under them. Vesalius was subjected to intense, public vilification in highly personal terms, most notably by one of his former teachers. Others, however, responded more favorably and constructively; both Fallopius and Eustachius (see Vignette 4.1) used the *Fabrica* as the starting point for their own investigations, allowing them to correct, clarify, and supplement Vesalius. The other reaction to Vesalius was imitation, sometimes with acknowledgment of his inspiration, but more often without it.

All of the drawings for the *Fabrica* were transferred to large blocks of fine-grained pear wood that were cut and incised so that the illustrations could be made as relief prints. In general, woodblock prints are not as delicate or detailed as prints from metal engravings, nor are the blocks as durable (although the *Icones Anatomicae,* which is available in many medical libraries, was printed in 1935 from Vesalius's original blocks). Therefore, a relatively quick and easy way to "improve" on Vesalius was to redo the illustrations from the *Fabrica* as metal engravings with, of course, a few minor changes that could be called corrections (on occasion, the corrections were simply to change the background landscapes from Venice to somewhere else). In any event, once one has seen the *Fabrica,* many of the anatomical illustrations over the next 100 years or so look like old, familiar friends. (Vesalius did not find the imitations published in his lifetime to be either amusing or flattering.)

Vesalius did not have many years to enjoy his fame. To avoid being brought before the inquisition on charges of body snatching and desecration, he offered and was permitted to make a pilgrimage to Jerusalem as penance. But while there, he heard that his former professorship at Padua, which he had left shortly after publishing the *Fabrica,* was again vacant, because of the untimely death of Fallopius; unfortunately, Vesalius badly misjudged the seaworthiness of the ship he took in his haste to return. When the ship was delayed by storms, its provisions turned out to be bad and inadequate. Vesalius was put ashore, desperately ill, to die on the small island of Zante (Zákinthos), off the Ionian coast of Greece.

already differentiated in the mesoderm. In other words, the steps of cartilage formation and replacement are omitted. The subsequent growth and development of the dermal bones is much like those of the other bones except that the dermal bones do not retain regions of cartilage at which postnatal growth occurs. Instead, the dermal bones have regions around their edges, at the sutures between bones, that are not fully ossified at birth and have the potential for new bone deposition for almost 20 years. The large bones of the skull at birth, for example, have extensive areas of soft tissue between them, which are the *fontanelles*. Most of the fontanelle regions ossify fairly quickly, but the capacity for the addition of new bone at the edges of the skull bones persists for years.

The difference in the formation of the dermal bones and the endochondral bones, such as the sphenoid, ethmoid, occipital, and parts of the temporal bone, reflects their different origins in our vertebrate ancestors. The endochondral bones were part of the early braincase, and this heritage is reflected by the fact that all the cranial nerves and the major arteries exit the base of the brain through one of the endochondral bones: the optic nerve and ophthalmic artery through the sphenoid, the branches of the trigeminal through the sphenoid, and the spinal cord through the occipital, for example. The dermal bones were not part of the early braincase or vertebral column but were instead bony head plates to which muscles of mastication were attached; the muscles attached to the inner surfaces of the dermal plates and ran between the plates and the braincase. The association between the dermal bones and muscles of mastication has been retained, but the muscles now attach to the outer side of the bones, and the inner surfaces of the bones lie next to the brain.

Thus the first signs of bone formation are somewhat different for the endochondral and dermal bones. Development of the sphenoid and ethmoid begins with the differentiation of chondrocytes in relatively avascular regions of the mesoderm, whereas development of the other orbital bones begins with the direct formation of osteoblasts in regions of high vascularity. Despite these differences, however, the histology of the adult orbital bones is essentially the same.

The orbital plates begin to form during the sixth week of gestation

Very little is known about the details of formation of the orbital bone except that the process is ongoing, starting about week 6 in gestation and continuing for years. The beginnings of bone formation at 6 weeks are not dramatic, except to histologists; these first steps are differentiation in the mesoderm surrounding the eye cup and the first appearance of chondrocytes or osteoblasts at various locations. The cartilaginous templates require a month or so to form, after which the cartilage is gradually replaced by bone.

At birth, the orbital bones have roughly the same configurations as the adult bones, but they are smaller, their sutures have not ossified, and they will continue to grow. There are some differences, to be sure; the infant's orbital margin is more nearly circular, the orbital fissures and foramina are relatively large, the optic canal through the sphenoid bone is very short, and the floor of the orbit may be incomplete (the persistence of the infraorbital suture is a sign of late closure). The dimensions of the infant's orbits, like those of the eyes, are roughly three-quarters of their adult values. The subsequent postnatal growth of the eye, its associated structures, and the bony orbit are very well matched, so the size of the orbit scales perfectly to match the size of its contents.

The capacity of bone for growth, repair, and remodeling lasts many years

For all its rigidity, bone has a great capacity for remodeling itself in response to various extrinsic factors. In fact, the structure continually breaks down and re-forms. Controlled breakdown is the responsibility of the **osteoclasts**, cells that are always present and can erode the existing bone matrix. A change in the mechanical stress on a bone or in chemical signals from adjacent tissues can send the osteoclasts into action, causing old bone to be removed in one place by the osteo*clasts* and new bone to be added in another place by the osteo*blasts*. The result is a change in the conformation of the bone. For the bones of the skull, particularly the dermal bones, it is easy to imagine growth and expansion at their edges accompanied by a continual process of removal on the inner sides of the bones, addition of bone to the outer sides, and thus a gradual enlargement of the skull and the orbits.

The eyes and orbits rotate from lateral to frontal positions during development

When the primitive eyes first form, they are on the sides of the embryo's head and their optic axes are directed laterally. At birth, however, the eyes have moved to their normal frontal positions; the eyes and their developing orbital bones have therefore rotated almost 90° from their original positions on the head.

Eye and orbital position rotate in most primates, most carnivores, and raptorial birds as they develop. In primates, however, the rotation coincides with a more rapid and relatively greater expansion of the brain and skull bones behind the primitive eyes than of the brain and bones in front of them. Thus one can think of the eyes being pushed to frontal positions by the rapid development of a very large brain, but this is an evolutionary push, not a mechanical one, and neither the mechanism nor the reason is known. We do know that frontal eyes have evolved independently several times and that large brains are not a guarantee of frontal eyes; dolphins, for example, have very large brains and lateral eyes.

Most developmental anomalies of the orbital bones are associated with anomalies of the facial bones

Since the major contributors to the orbit are the bones of the face, problems with development of the orbital bones are rarely confined to the orbit. Most developmental problems with the orbital bones are associated with anomalous ossification affecting the bones of the face and skull generally, a category of problems referred to as **craniofacial dysostosis**. Crouzon's disease, for example, involves abnormal enlargement of the sphenoid, among other things, with the possibility of blindness; optic atrophy (atrophy of the optic nerve) may result if abnormal growth of the sphenoid constricts the nerve within the optic canal (this is an example of bone failing to respond to its environment). Management of this sort of anomaly requires attention from several different surgical disciplines.

References and Additional Reading

The Bony Orbit

Dallow RL. 1986. Ultrasonography of the eye and orbit. Chapter 4, pp. 55–69, in *Diagnostic Imaging in Ophthalmology,* Gonzalez CF, Becker MH, and Flanagan JC, eds. Springer-Verlag, New York.

Doxanas MT and Anderson RL. 1984. *Clinical Orbital Anatomy.* Williams & Wilkins, Baltimore.

Fisher YL, Nogueira F, and Salles D. 1998. Diagnostic ophthalmic ultrasonography. Chapter 108 in *Duane's Foundations of Clinical Ophthamology*, Vol. 2. Tasman W and Jaeger EA, eds. JB Lippincott, Philadelphia.

Garrity JA and Forbes GS. 1991. Computed tomography of the orbit. Chapter 24 in *Duane's Clinical Ophthalmology,* Vol. 2, Tasman W and Jaeger EA, eds. JB Lippincott, Philadelphia.

Kronish JW and Dortzbach RK. 1991. Magnetic resonance imaging of the orbit. Chapter 25 in *Duane's Clinical Ophthalmology,* Vol. 2, Tasman W and Jaeger EA, eds. JB Lippincott, Philadelphia.

Linberg JV, Orcutt JC, and Van Dyk HJL. 1993. Orbital surgery. Chapter 14 in *Duane's Clinical Ophthalmology,* Vol. 5, Tasman W and Jaeger EA, eds. JB Lippincott, Philadelphia.

Meyer DR. 1996. Orbital fractures. Chapter 48 in *Duane's Clinical Ophthalmology*, Vol. 2, Tasman W and Jaeger EA, eds. JB Lippincott, Philadelphia.

Rootman J. 1988. *Diseases of the Orbit.* JB Lippincott, Philadelphia.

Sherman DD and Lemke BN. 1992. Orbital anatomy and its clinical implications. Chapter 21 in *Duane's Clinical Ophthalmology,* Vol. 2, Tasman W and Jaeger EA, eds. JB Lippincott, Philadelphia.

Smith B and Regan WF. 1957. Blowout fracture of the orbit: Mechanism and correction of internal orbital fracture. *Am. J. Ophthalmol.* 44: 733–739.

Warwick R. 1976. The bony orbit and paranasal sinuses. Chapter 1, pp. 1–29, in *Woolf's Anatomy of the Eye and Orbit,* 7th Ed. WB Saunders, Philadelphia.

White TD. 1991. Skull. Chapter 4, pp. 37–99, in *Human Osteology.* Academic Press, San Diego.

Whitnall SE. 1932. Osteology. Part 1, pp. 1–108, in *The Anatomy of the Human Orbit and Accessory Organs of Vision.* Oxford University Press, London. Reprinted by Robert E. Kreiger Publishing, Huntington, NY, 1979.

Zide BM and Jelks GW. 1985. Bones, vessels, and nerves. Chapter 1, pp. 1–12, in *Surgical Anatomy of the Orbit.* Raven, New York.

Connections between the Eye and the Orbit

Demer JL, Miller JM, and Poukens V. 1996. Surgical implications of the rectus extraocular muscle pulleys. *J. Pediatr. Ophthalmol. Strabismus* 33: 208–218.

Demer JL, Miller JM, Poukens V, Vinters HV, and Glasgow BJ. 1995. Evidence for fibromuscular pulleys of the recti extraocular muscles. *Invest. Ophthalmol. Vis. Sci.* 36: 1125–1136.

Koorneef L. 1982. Orbital connective tissue. Chapter 32 in *Duane's Biomedical Foundations of Ophthalmology,* Vol. 1, Tasman W and Jaeger EA, eds. JB Lippincott, Philadelphia.

Development of the Orbital Bones

Alberts B, Bray D, Lewis J, Raff M, Roberts K, and Watson JD. 1994. Fibroblasts and their transformations: The connective tissue cell family. Chapter 22, pp. 1179–1186, in *Molecular Biology of the Cell,* 3rd Ed. Garland, New York.

de Haan AB and Willekens BL. 1975. Embryology of the orbital walls: A preliminary report. Pp. 57–64 in *Orbital Disorders* (Modern problems in ophthalmology, vol. 14), Bleeker GM, Gaston JB, Kronenberg B, and Lyle TK, eds. S. Karger, Basel, Switzerland.

Mann I. 1950. The orbit and its contents. Chapter 7, pp. 256–274, in *The Development of the Human Eye.* Grune & Stratton, New York.

[Chapter 4]

The Extraocular Muscles

I have had a growing feeling over the years that many ophthalmologists would be quite happy if our eyes did not move at all. Think how this would simplify things. How easy it would be to refract children or map a visual field. . . . But eye movements must have something to do with vision even if we are not sure just what.

■ David A. Robinson, comments on receiving the Proctor medal, 1987

David Robinson's tongue-in-cheek comment contains a serious point—a reminder that while it is possible to see without moving our eyes, the position of our eyes in the orbits is always changing in the course of everyday seeing, whether or not we are aware of it. Vision is possible without eye movements, but the way we use vision to explore our world is possible only because our eyes are movable. It is not so much that vision affects eye movements and eye movements affect vision, but that the two are inextricably bound up in one another.

All vertebrates have movable eyes, and with some variation in anatomical arrangement, all use six muscles organized as three agonist–antagonist pairs to move the eyes. (Invertebrates that have hard exoskeletons, such as arthropods, often have eyes that cannot move, because they are rigidly attached to the exoskeleton. Some skirt this restriction by having eyes on movable stalks. Others, such as jumping spiders, have movable retinas that partly compensate for the immovability of the optical system; see the Prologue.) The extraocular muscles are the effectors, and their motor nerves are the final common path by which several sets of interacting neural circuits control eye position. In a sense, the muscles are simply transducers, converting electrical signals into mechanical force and doing so in a very stereotypical fashion.

If muscles are simply devices for applying force to the eye, differences in the patterns of eye movement among species, the variety of stimuli that can elicit eye movements, and failures of normal operation are most likely to be related to the sensory systems and the neural circuits that compute command signals for eye movements rather than to the muscles themselves. But as we will see, the extraocular muscles have some unusual anatomical features that may be related to special demands placed on the muscles by different types of eye movement.

Patterns of Eye Movement

Our eyes are always moving, and some motion is necessary for vision

Even when we are looking fixedly at something and we think our eyes are motionless, they are moving. This movement is very small and requires a very sensitive system for monitoring eye position to detect it. There is a high-frequency tremor (**micronystagmus**) of a few minutes of arc in amplitude on which is superimposed a low-frequency drift, interrupted now and then by a rapid flick of the eyes of about 10 minutes of arc (Figure 4.1). Under these conditions, which we would characterize as steady fixation, the retinal image on the fovea is constantly jittering around in exactly the same pattern as the eye movement. (In terms of the cone photoreceptors in the fovea, the tremor sweeps the image across one or two cone diameters and the rapid flicks sweep it across 20 or so.)

With a special optical configuration, the effect of this micronystagmus on retinal image motion can be eliminated so that the image is absolutely fixed in position on the retina, creating what is called a stabilized image. When an image is stabilized, vision fades and the target imaged on the retina becomes less and less visible, eventually disappearing altogether. The ever-present small eye movements are apparently necessary for vision. (Afterimages produced by a camera flash are a familiar sort of stabilized image. They disappear in a minute or so because they are stabilized, but their effect can be shown to persist for up to 30 minutes.)

As far as effective vision is concerned, these small tremors and flicks are an irreducible minimum—the smallest amount of eye and retinal image movement that can occur without vision being compromised. But there is another extreme; too much retinal image movement does not give the visual system the time it needs to extract information from the retinal image, and vision suffers. The range between too little and too much image motion is fairly small, and eye movements appear to have the job of minimizing retinal image motion, keeping it in the range where vision works best.

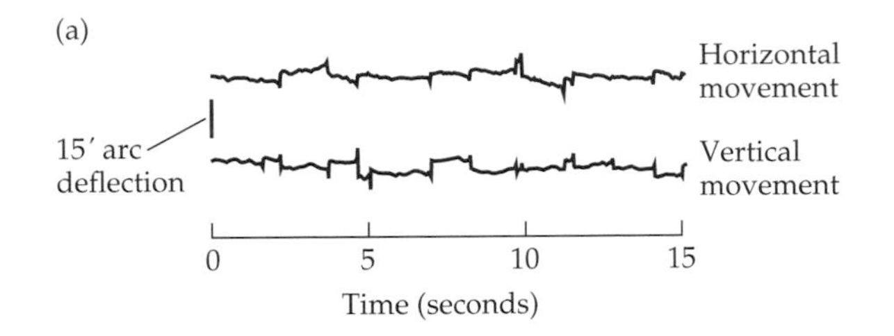

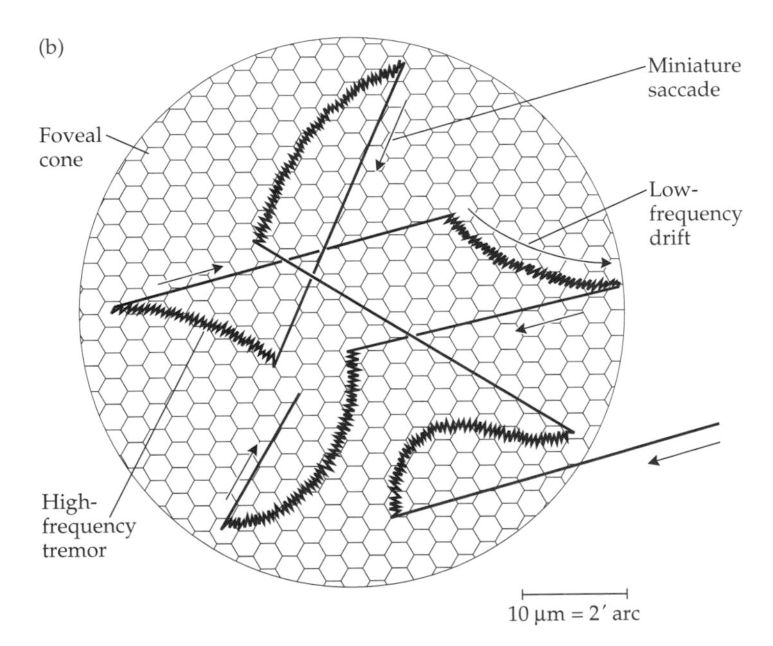

Figure 4.1
Miniature Eye Movements during Steady Fixation
(*a*) Horizontal and vertical eye position as a function of time. Eye position has three components: a high-frequency tremor, low-frequency drifts, and miniature rapid eye movements (saccades). The total range of movement is about 20′ of arc. (*b*) Eye position relative to the central bouquet of foveal cone photoreceptors during a 10 s interval. The array contains several hundred cones. (*a* after Steinman et al. 1973; *b* after Pritchard 1961.)

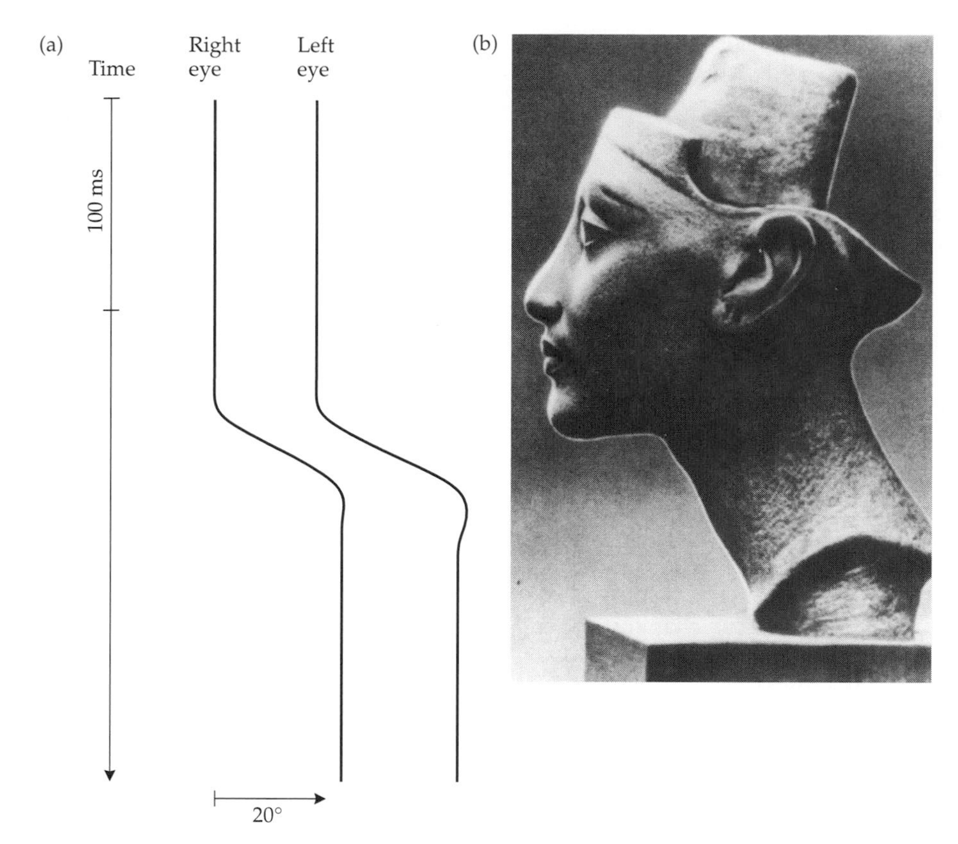

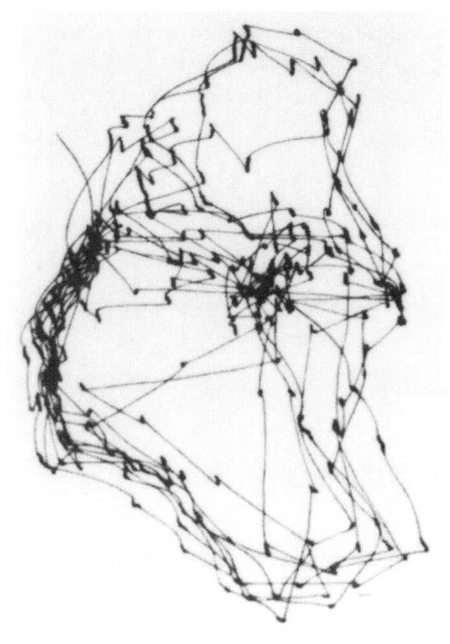

Figure 4.2
Saccades and Looking
(*a*) These traces are records of eye positions oriented as if we were looking at the person making the eye movements. About 140 ms after presentation of the stimulus, both eyes move rapidly to the subject's left, taking up new positions after about 40 ms. The saccades in the two eyes are very similar but not exactly matched. (*b*) Saccades are used to examine different parts of our visual world. A continuous 2-minute record of eye position (right) during examination of a photograph of a bust of Nefertiti (left) shows saccades as thin lines connecting points of fixation where the eye paused to examine details. (*a* after Westheimer 1954; *b* from Yarbus 1967.)

Large, rapid eye movements are used for looking around, for placing retinal images of interest on the fovea

The human retina has a small region at the center of its fovea where cone photoreceptors are thinnest and packed most tightly together (see Chapter 15). This region is less than 1° of visual angle in diameter, and it gives us the best resolution of fine detail. When we look at something, the goal is to image that thing at the center of the fovea; we could do this by moving our head and body and keeping our eyes stationary, but eye movements are much more efficient.

Saccades, or **saccadic eye movements**, can vary considerably in magnitude, from less than 1° to more than 20° as the situation demands. Their peak velocity during the movement depends on the size of the saccade, but it can reach 500°/s for large saccades. Because the velocity is so high, the movements are brief, rarely taking more than 50 ms (milliseconds) from start to finish. Saccades always occur in both eyes simultaneously, and the movements are well matched in size, speed, and direction. The 20° saccades in Figure 4.2*a* occurred about 140 ms after the stimulus appeared and took about 40 ms to complete. The slight overshoot and small correction to achieve the new eye position are common. Under normal circumstances, we rarely make saccades larger than 15° or so; if a movement larger than this is required, it is done as a combination of eye and head movement.

Saccades are voluntary eye movements in the sense that they are optional; we can look at something or not, as we choose. Once initiated, however, execution of the movement is automatic. And during the eye movement, when the retinal image is sweeping rapidly across the retina, vision is automatically suppressed.

We use saccades for looking around and examining our environment (see Figure 4.2). What we are doing is catching or acquiring particular images with our foveas so that we can examine them better. To examine the images better, we need to hold them on the fovea of our eyes so that they are not moving too much. After saccades catch the image, other eye movements hold it.

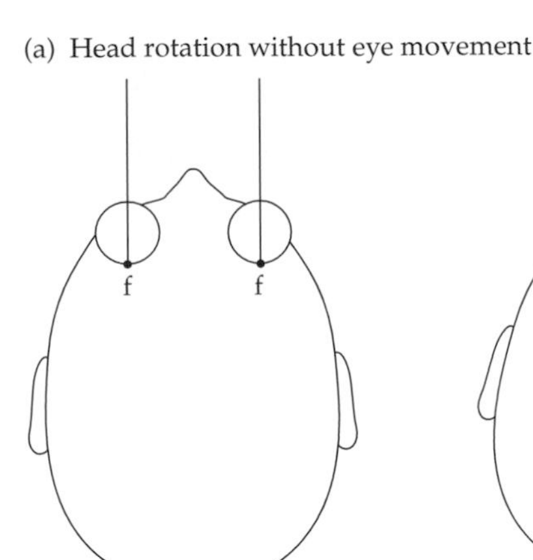

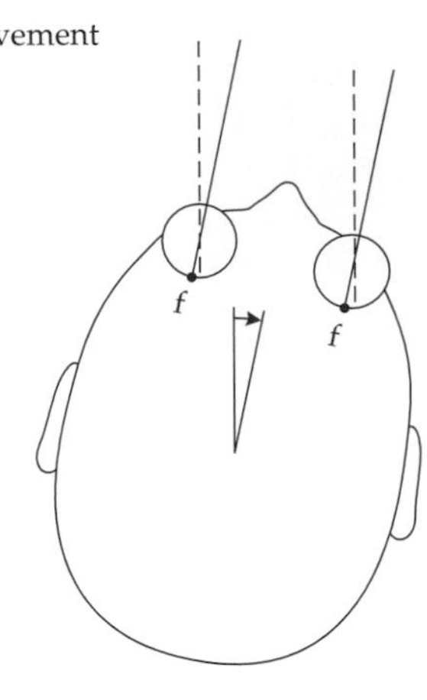

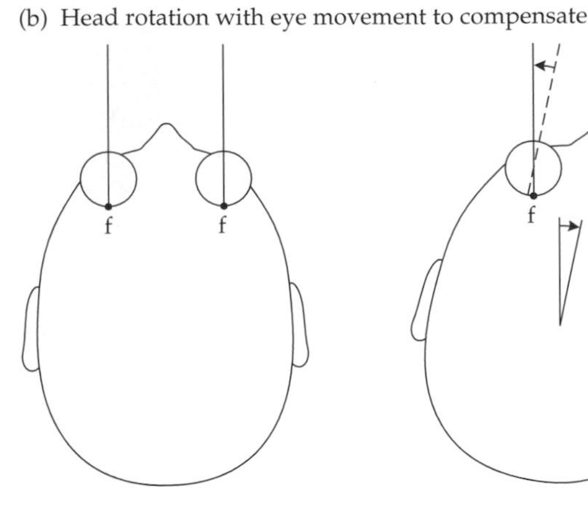

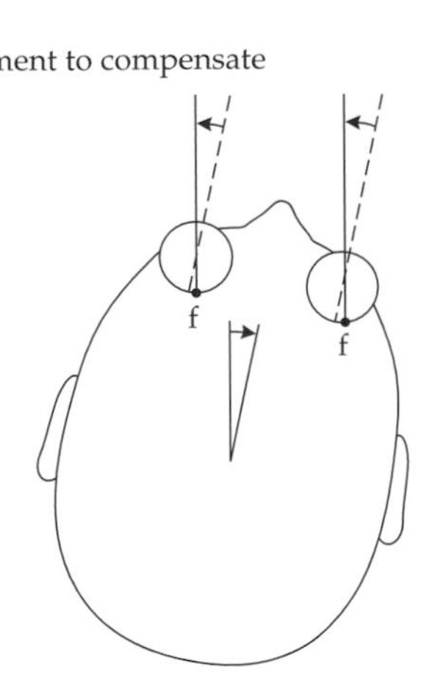

Figure 4.3
Eye Movement Opposite to Head Rotation Minimizes Image Movement
(*a*) A head rotation without eye movement produces retinal image movement. An object that was imaged on the foveas (f) before the head rotation now lies on an extrafoveal point (right). (*b*) If the eyes move opposite to the head rotation by exactly the same amount, an object imaged on the foveas does not move off the foveas (right).

Slow eye movements are used to track or follow movement and to compensate for changes in head and body position

If the eyes were not moving and were fixed in position in the orbits, any change in head or body position would produce a corresponding change in retinal image position; the image would be swept across the retina by an amount equal to the head or body rotation. If, however, the body movement were accompanied by an eye movement of the same velocity, but in the *opposite* direction, the retinal image would remain stationary (Figure 4.3). Compensatory eye movements of this type are generated by receptors in the inner ear (in the semicircular canals and otolith organs) that detect acceleration and change with respect to gravity; these **vestibulo-ocular reflexes** are probably the most primitive form of eye position control. As a simple example of the reflexes at work, tilting one's head toward the right shoulder will produce a clockwise torsion of the eyes that is opposite to the head tilt and will tend to keep the eyes in their original orientation.

Using signals from the inner ear to minimize retinal image motion is indirect and not very precise, although some animals whose visual requirements are not too stringent may get along quite nicely with no other means for eye position control. A better strategy is *visual* detection of the direction and velocity of retinal image motion when it first occurs, using this information to generate compensatory eye movements that bring the retinal image motion back to zero (or below the threshold for detecting it).

Three eye movement control systems employ this strategy. The first, which is common to all animals that have movable eyes, is the **optokinetic reflex**, for which the optimal stimulus is movement of all, or a large portion, of the retinal image. In nature, this stimulus arises whenever an animal slowly rotates its head without moving its body; the response is a rotation of the eyes opposite to the direction of head rotation. In a laboratory setting, the animal or human subject is stationary inside a rotating cylinder that has a pattern on its interior surface. In this case, the eyes will track the slow rotation of the cylinder, interrupted periodically by quick eye movements in the opposite direction. The alternating slow and quick phases in opposite directions constitute a **nystagmus**, and the entire pattern of reflex movements is an optokinetic nystagmus. Use of a rotating stimulus isolates the optokinetic reflex; in the natural case of head turn, the eye movement may be influenced by both visual and vestibular stimuli.

A second category of visually guided reflexes consists of the **smooth pursuit eye movements**, in which the eyes track a small object that is moving relative to a stationary background. Pursuit eye movements are similar to the slow component of the optokinetic reflex, but they represent a situation in which the optokinetic reflex is subordinated; when the eyes track the small object, thus reducing its retinal image motion to zero, it follows that all the rest of the retinal image—the background—must be in motion, which is the appropriate stimulus for the optokinetic system. Pursuit movements are found only in species that have a fovea or its functional equivalent, so the pursuit movement is an indication of the fovea's importance and the need to minimize motion of the *foveal* portion of the retinal image above all else.

The **fixation reflex** is the third position control system that probably uses information about the direction and velocity of retinal image motion (the notion of a separate fixation reflex is fairly recent, and its workings are still under investigation). Although fixation might be considered a limiting case of pursuit in which the object whose retinal image motion is to be minimized is *not* moving relative to the background, several lines of evidence suggest that the pursuit reflex is switched on and off for the specific task of tracking *moving* objects. Anecdotal evidence of situations in which fixation is normal but pursuit is abnormal, and vice versa, also implies that the systems are separate.

After a saccade has captured a retinal image, all these reflexes help hold the image on the fovea, either by compensating for postural changes or by detecting "slip" of a moving or stationary image off of the fovea. The smooth pursuit reflex is the only one that is voluntary—we can decide whether or not to track an object with our eyes and decide which of a number of objects to track. The resulting eye movement, however, is matched by the workings of the reflex arc to the object movement. Like saccades, these eye movements occur in both eyes, matched in size, speed, and direction.

These compensatory and image-holding movements are normally slow, with velocities rarely greater than 20°/s. Under appropriate circumstances, however, the smooth pursuit system can match target velocities up to about 90°/s.

Eye movement velocities may vary by a factor of 10^5

The difference in velocity between the slowest following movements and the fastest saccades is from about 0.05 to 500°/s. This 10,000-fold difference must be reflected in muscle contraction velocities. Taking the eye's radius of curvature to be about 12 mm, 1° on the surface of the eye subtends about 0.2 mm, and each 1 mm change in muscle length corresponds to about 5° of eye rotation. From this, the calculated range of muscle contraction velocities will be on the order of 0.01 to 100 mm/s. There are few, if any, other muscles in the body whose normal functioning requires them to operate over such an enormous range of contraction velocities.

Because of this unusually large velocity range, it is convenient—and probably physiologically appropriate—to consider the difference between slow and fast movements to be fundamental. That is, there are rapid eye movements (saccades) and slow eye movements, which include everything else (including vergence eye movements, discussed in the next section). Both saccades and slow movements can be elicited by either visual or nonvisual stimuli, both can be modified by expectation and learning, and both can be gated on or off by attention and selection. These are general similarities, however; significant differences in movement velocity and innervational patterns remain.

Since the eyes have overlapping fields of view, their movements must be coordinated

Thus far, the eye movements we have discussed are **versions**—that is, movements of the eyes in the same direction (the *same* direction means that both eyes simultaneously rotate right, left, up, down, clockwise, or counterclockwise—all directions are referenced to the individual, not to the observer).* It is not part of the definition, but the eyes are also expected to move with approximately the same angular velocity and magnitude of rotation, as shown in Figure 4.2*a*.

To make version eye movements, the active muscles of the two eyes must be linked or yoked in a particular way. A version to the right, for example, requires simultaneous contraction of the right lateral rectus and the left medial rectus; a downward version requires simultaneous contraction of both inferior recti. These muscles are described in the section "Control of Eye Position," and the neural circuitry underlying versional yoking will be considered later in the chapter.

Version refers to a binocular situation in which both eyes are open and seeing, but versional yoking of muscles in the two eyes persists when one eye is covered; if the left eye is covered, for example, a rightward movement of the right eye will be accompanied by a rightward movement of the left eye. In clinical terms, eye movements under this monocular condition are no longer versions, but **ductions**. With the left eye covered, the rightward movement of the right eye is an abduction. Use of the term "ductions" for eye movements under monocular conditions recognizes explicitly that we do not know exactly what the covered eye is doing.

If we viewed objects only at a certain distance that never changed, and if the versional yoking between the two eyes were perfect, we could get by with version eye movements and nothing else. These conditions do not hold, however. Since we are frontal-eyed animals with foveate retinas, we need to align the two eyes so that the foveas are directed, very precisely, toward the same object in space, which for objects at different distances requires that the angle between the visual axes be changed (Figure 4.4). That is, there is now a need to minimize foveal image motion in both eyes at the same time, requiring that the eyes sometimes move in opposite directions.

Eye movements in opposite directions are **vergences**. In the horizontal plane, rotation of both eyes toward the midline is **convergence**; rotation of both eyes laterally is **divergence**. Vergences may also occur in the vertical plane (one eye elevates while the other depresses) or around the anterior–posterior axis (both eyes intort—**encyclovergence**—or both extort—**excyclovergence**).

Vergence stimuli—blur or disparity—are relatively complex and require processing at the cortical level. Vergences have longer latencies than other types of movement have, and they are typically slow movements; large disparity-driven vergences have maximum velocities of around 30°/s. Under normal circumstances, changes in eye position often have both version and vergence components.

f f

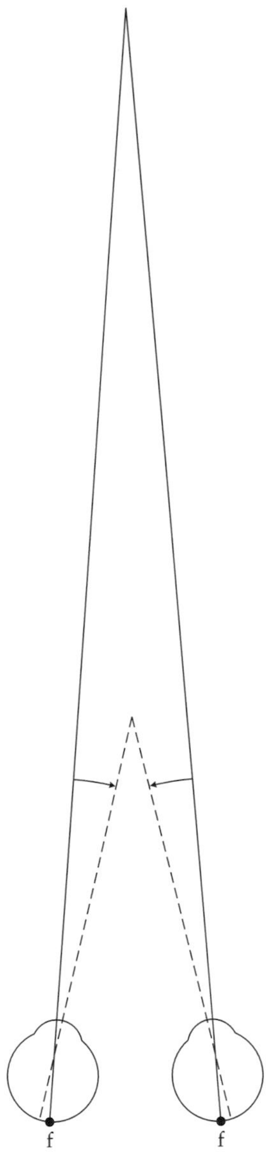

Figure 4.4
Vergences Are Required to Look at Objects at Different Distances
When a distant object is imaged on both foveas, a nearer object will be imaged temporal to the foveas. If the nearer object is to be imaged on the foveas, the eyes must converge; that is, the right eye must rotate to the left and the left eye must rotate to the right.

Slow movements of the eyes in opposite directions help keep corresponding images on the foveas in both retinas simultaneously

There are two broad functional categories of vergence movements that differ in the visual stimuli that activate them. **Accommodative vergence** is elicited by retinal image *blur*, which is the stimulus for accommodation. The goal of the accommodative system is to minimize image blur, and as the lens changes shape to refocus the image, there will be a concomitant change in the vergence position of the eyes. In accommodating from a distant to a near object, for example, lens power increases and the eyes converge (also the pupil constricts); this combination of accommodation, convergence, and pupil constriction is referred to as the *near triad* (see Chapters 10 and 11).

*The clinical terminology is as follows: A version to the right is *dextroversion*, to the left is *levoversion*, up is *sursumversion* or *supraversion*, down is *deorsumversion* or *infraversion*, clockwise around the anterior–posterior axis is *levocycloversion*, and counterclockwise is *dextrocycloversion*. But since the statements "rightward version" or "upward version" are accurate and unambiguous, only purists insist on Latin terms.

Accommodation and accommodative vergence can be elicited by image blur in only one eye (with the other covered), and if everything worked perfectly, the amount of convergence would precisely align both eyes at the dioptric distance for which the eye accommodated. But this perfection is rarely achieved, and one cannot expect accommodative vergence to provide anything better than approximate alignment.

The requisite precision is achieved by **fusional vergence**, whose stimulus (disparity) is binocular and related to the amount of foveal misalignment. If, for example, an object were imaged slightly temporal to the foveas, this disparity would elicit a *con*vergence (see Figure 4.4.) The threshold disparity for fusional vergences is very small, measured in seconds of arc, and the system is capable of aligning the foveal centers with this degree of precision.

Vergences are slow eye movements, working best at velocities lower than 5°/s. Under everyday conditions of seeing, where shifts of visual attention from one distance to another are often rapid, other eye movements, such as saccades, are superimposed on the vergences.

Strabismus is a misalignment of the two visual axes under binocular viewing conditions

When the systems for controlling eye position work as they should, the two eyes are not only coordinated in their movements, but also aligned so that a visual target of interest is imaged simultaneously on the two foveas. Failure to achieve this sort of alignment is a **strabismus** (or **heterotropia**).

A strabismus has several significant variables in addition to the amount of misalignment (the **angle of strabismus**). Either eye may deviate from alignment, the deviation may be in any direction, and the deviation may be present all the time or only under certain conditions; all of these features are taken into account in the descriptive terminology. Thus, an individual whose right eye deviates outward (laterally) has a right **exotropia**. If the deviation is present at all times and all fixation distances, it is a **constant** right exotropia. Similarly, an inward deviation of the left eye is a constant left **esotropia**. If the left eye is sometimes the deviating eye while the right eye deviates at other times, the strabismus is **alternating**. Strabismus that appears only at certain times, fixation distances, or gaze positions is **intermittent**; strabismus in which the angle of deviation changes is **noncomitant** (when the deviation angle is invariant, the strabismus is **comitant**). Most strabismus is misalignment in the horizontal plane, but there can be vertical deviations—they are **hypertropias**. Box 4.1 discusses a basic method for detecting and measuring strabismus.

Control of Eye Position

The extraocular muscles are arranged as three reciprocally innervated agonist–antagonist pairs

The **superior rectus** and **inferior rectus** muscles attach to the eye directly above and below the margin of the cornea, respectively, and their major actions when they contract are vertical rotations of the eye around its horizontal axis (Figure 4.5*a*). Their actions are directly antagonistic; the superior rectus produces an upward rotation of the eye's anterior pole (**elevation**) and the inferior rectus produces a downward rotation (**depression**). The muscles are **reciprocally innervated**; when one contracts to rotate the eye, the other relaxes. The reciprocal innervation is organized centrally by interconnections between extraocular motor nuclei (which is discussed later).

The muscle pair responsible for horizontal rotations around the eye's vertical axis are the **medial rectus** and the **lateral rectus**, which attach, respectively, on

Box 4.1 Detecting Ocular Misalignment

You can observe a lot by watching.
Lawrence P. "Yogi" Berra

ALTHOUGH GROSS MISALIGNMENTS OF the eyes in the horizontal plane are easy to see, a constant strabismus of moderate amount is surprisingly difficult to notice by direct inspection. But it takes no more than a handheld occluder to make even very small deviations quite obvious. The trick is to elicit eye movements that reveal the presence of a deviation—movement is readily seen.

To test for strabismus, we need to establish that the eyes are misaligned under binocular conditions, which we do using the **cover–uncover test**. The patient is given a small target on which to fixate—it may be either a distant or a near target—and instructed to keep looking at the target with whichever eye is available. Each eye, in turn, is covered, then uncovered, thus briefly interrupting binocularity and then reestablishing it. Attention is directed to the eye that is *not* being covered and uncovered. If no strabismus is present, covering an eye will have no effect; the uncovered eye will simply maintain fixation.

But consider the example in Figure 1, which shows a right esotropia. Covering and uncovering the right eye elicits no movement of the left eye; it maintains fixation throughout. When the left eye is covered, however, the right eye *ab*ducts to take up fixation, indicating that it had been deviated nasally under binocular conditions—this is the evidence for esotropia. When the left eye is uncovered, there are two possible results. The first is no movement of either eye, suggesting that the strabismus is alternating, and the left eye has now become the deviating eye (this can be confirmed by covering the left eye again to see if the right eye abducts to take up fixation). The other possibility is that both eyes move to the left when the left eye is uncovered, with the left eye taking up fixation again while the right eye returns to its deviated position. This result indicates a *constant* right esotropia.

The cover–uncover test has some refinements, but even this basic form provides a great deal of information; we know a strabismus is present, we have established that it is a unilateral deviation of the right eye, and the size of the eye movement gives us a rough idea about the size of the deviation—small, medium, or large. If the strabismus were present at different test distances, we would have even more evidence that the strabismus is constant, not intermittent.

The amount of the deviation (the angle of strabismus) can be determined using ophthalmic prisms and the **alternate cover test**.* Alternate cover means that the cover is moved to cover first one eye and then the other, permitting no opportunity for binocular vision. Each time the cover is shifted from one eye to the other, the eye being uncovered moves to take up fixation. The

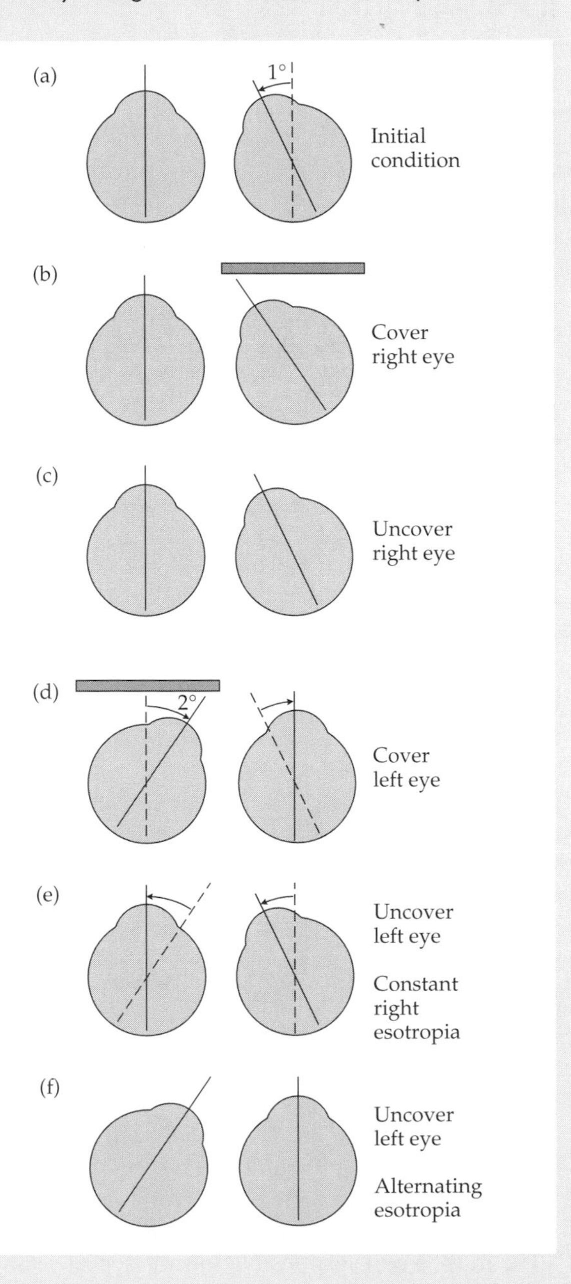

Figure 1
The Cover–Uncover Test for Detecting Strabismus
Covering (*b*), then uncovering (*c*) the patient's right eye produces no movement of the left eye. Covering the left eye (*d*), however, elicits an abduction of the right eye to take up fixation and an adduction of the left eye behind the cover. If the strabismus is constant, uncovering the left eye will prompt a leftward version to the original condition where the left eye is fixating and the right eye is deviated (*e*). If there were no movements when the left eye was uncovered, the right eye has taken up fixation, the left eye is deviated, and the strabismus is alternating (*f*). Note that the primary angle of deviation (1°) is smaller than the secondary angle of deviation (2°), indicating that greater than normal innervation was required for the right lateral rectus to abduct the right eye.

goal here is to find the amount of ophthalmic prism that will eliminate any movement. For the right esotropia in the example, the prism should be placed in front of the right eye with its base–apex axis horizontal and the base on the temporal side (*base out*). When the amount of prism matches the angle of deviation, the eyes will no longer have to abduct to take up fixation, and no movement will be observed; if the prism is too small, one will still see some abduction, and if too large, the movement will change to adduction. For experienced observers, this method of estimating the strabismus angle is accurate to about 2 Δ (about 1.2°).

In the cover–uncover test, we watched the eye not being covered and uncovered; no movement meant no strabismus, but even in the absence of strabismus one might see movement of the *covered* eye as the cover is removed. This movement indicates the presence of a **heterophoria**, the most common form of ocular misalignment. The important distinction is that strabismus (or heterotropia) exists in the presence of stimuli to binocular vision, while phorias are manifest *only* when binocularity is disrupted. (Phorias are often called "latent" oculomotor deviations, as if they were lurking in the shrubbery, waiting to leap out at the first opportunity, but there is nothing latent about them. Phorias are *manifest* deviations that appear when binocular vision is made impossible. Under binocular conditions, phorias do not exist, are not defined, and cannot be discussed.)

When one eye is covered, stimuli to fusional vergence are eliminated and the covered eye will drift to a position determined solely by the accommodative vergence system; removing the cover reestablishes fusional vergence control and the now uncovered eye will move back to its normal binocular position (which in the absence of strabismus should be perfect alignment with the other eye). The direction and magnitude of a phoria represent the demand required of the fusional vergence system for bifixation; its magnitude is determined with prisms and the alternate cover test.

*The amount of misalignment is an angular measure; it can be specified in degrees, but the usual unit of measure is the **prism diopter**. One prism diopter (Δ) corresponds to a 1 cm misalignment of the visual axes at a distance of 1 m, 2 prism diopters to 2 cm displacement at the same distance, and so on. The relationship between prism diopters and the angle subtended in degrees is: $\Delta = 100 \tan \alpha°$. Thus, an angle of 45° (tan 45 = 1) is equivalent to 100 Δ.

the medial and lateral sides of the eye (Figure 4.5*b*). Contraction of the medial rectus rotates the anterior pole toward the midline (**adduction**), and the lateral rectus rotates the eye laterally (**abduction**) (the Latin prefixes *ab-*, "away from," and *ad-*, "toward," refer to direction relative to the midline of the head; thus *ad*duction is toward the midline, *ab*duction is away from the midline). Again, these actions are directly antagonistic, and the muscles are reciprocally innervated.*

If the eye had only rectus muscles, the innervational commands for gaze position and movement could be a simple rectilinear coordinate system, with horizontal position being a function of relative innervation to the medial and lateral recti and vertical position dictated by the relative amount of innervation to the superior and inferior recti. The eye has another axis of rotation, however, and it is necessary to consider rotations around the anterior–posterior axis (these are **torsions**).

The actions of the **superior oblique** and **inferior oblique** are not particularly obvious, and they will be considered in more detail later. For the moment, the obliques may be thought of as the muscles responsible for torsional eye rotations, but as we will consider later, torsional movements are not their major actions. If one thinks of the eye as having a clock face superimposed on it—viewed from the front—an **intorsion** is a rotation of of the 12 o'clock position toward the midline (6 o'clock rotates away from the midline); the reverse—rotation of 12 o'clock away from the nose—is **extorsion**. The muscle best arranged to produce intorsion is the superior oblique; extorsion is done by the inferior oblique (Figure 4.5*c*).

The eyes are stationary when the opposing forces exerted by the extraocular muscles are in balance

The eye can rotate freely in its pocket of orbital fat, limited only by the extraocular muscles and the various fascial connections of the eye to the orbit; any force

*Early Greek or Latin names for the extraocular muscles referred to the expressiveness of eye movements. For example, the lateral rectus was at one time named for its role in sidewise glances of flirtation, the superior rectus for looking upward to beseech divine guidance or intervention, and the inferior rectus for looking down in disdain. The current names indicate the location on the eye and the direction in which the muscles extend to reach the eye—directly (in a straight line—rectus) or obliquely.

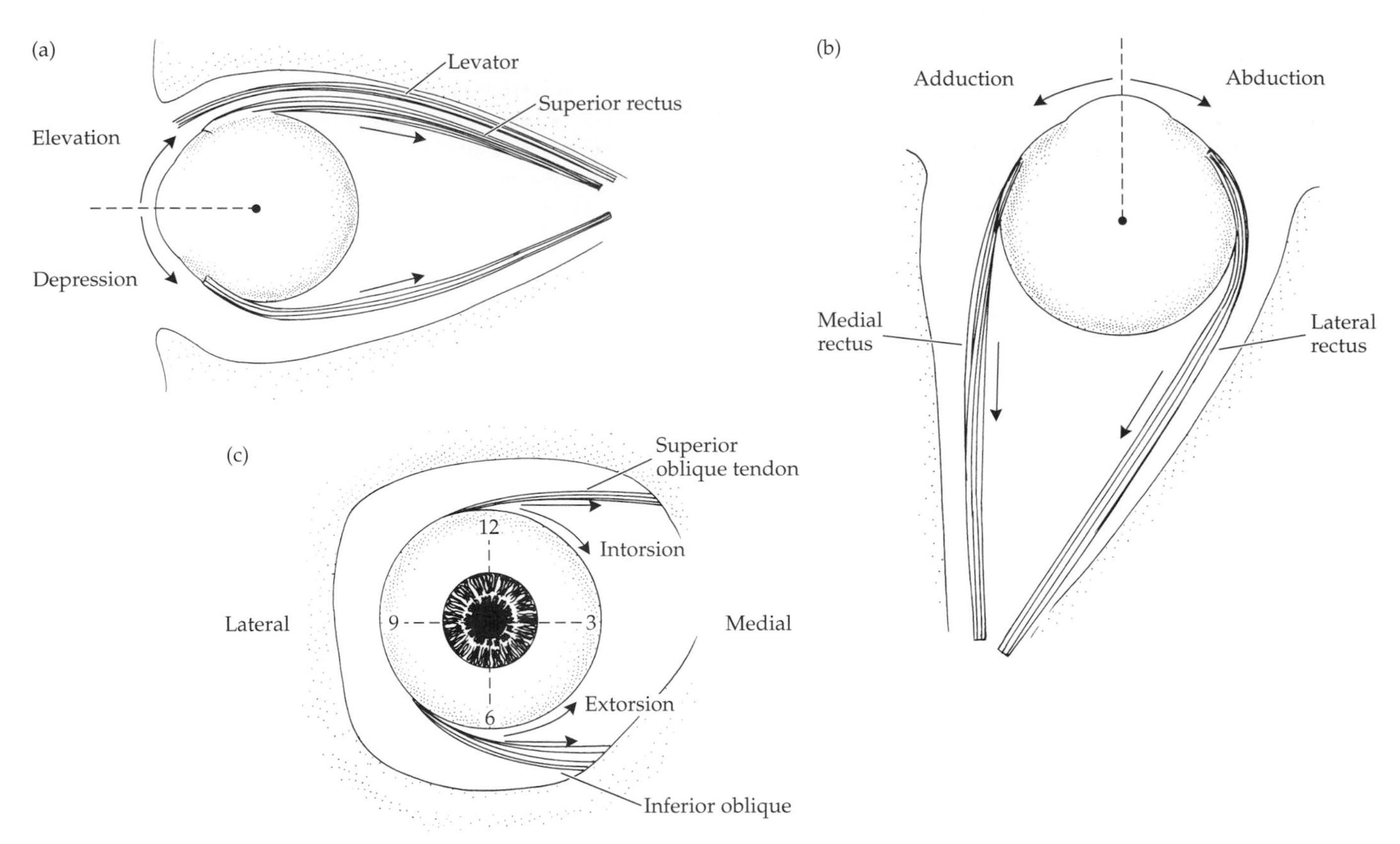

Figure 4.5
The Extraocular Muscle Pairs
(*a*) Viewed from the side, the superior rectus runs above the eye and the inferior rectus below it. The main action of these muscles is to rotate the eye around a horizontal axis, which is perpendicular to the plane of section near the center of the eye (dot), such that the eye elevates when the superior rectus contracts and depresses when the inferior rectus contracts. (*b*) The medial and lateral recti, viewed from above, rotate the eye around a vertical axis (dot). The medial rectus rotates the eye toward the nose (adducts) and the lateral rectus rotates the eye away from the nose (abducts). (*c*) The superior and inferior obliques are shown running to the eye from the medial wall of the orbit. Part of their action is rotation of the eye around its anterior–posterior axis. The superior oblique will produce intorsion (12 o'clock rotates nasally), the inferior oblique extorsion (6 o'clock rotates nasally).

applied to the eye from any source will produce rotation. But the extraocular muscles are never completely relaxed. They have a normal state of tonus, or tonic level of contraction, and even when relaxed the muscle tissue has some inherent elasticity, like a rubber band; as a result, the muscles and associated tissues are *always* exerting force on the eye. Thus, for the eye to be stationary, whatever the direction of gaze, the forces exerted on the eye by the six extraocular muscles and other orbital tissues must be in equilibrium, with the force exerted by any one muscle perfectly counterbalanced by an opposing force.

When the eye is stationary in an adducted gaze position, for example, force is applied to the eye by the contracted, shortened medial rectus. This force will be counterbalanced by the elasticity of the relaxed, but stretched, lateral rectus. The eye will remain stationary as long as these active and passive forces are in balance—that is, of equal magnitude and opposite direction. For each of the seemingly infinite number of stationary positions of gaze, there will be a unique set of active and passive forces in the six muscles that produce the required state of equilibrium.

Imbalanced forces produce eye rotations

Any muscle contraction disrupts the equilibrium of the stationary eye. The eye rotates in the direction of the imbalancing force and continues rotating until an equilibrium can be reestablished in a new position of gaze. If the lateral rectus were to contract, the eye would rotate laterally (abduct) as a result of this new force and of the simultaneous relaxation of the medial rectus produced by reciprocal innervation of the muscle pair. At some point, however, the active force in the lateral rectus would be counterbalanced by the passive stretch developed in the medial rectus, establishing a new equilibrium in a new, abducted position of gaze.

What makes this essentially mechanical problem interesting is considering what must happen if the new equilibrium position of the eye is to be specified in advance of the rotation—that is, if the magnitude of the rotation is to be predictable and controllable. For this to happen, the contractile force developed in the lateral rectus must be related directly and precisely to the difference between present and anticipated eye positions. But this task requires considering the passive force developed in the medial rectus as it lengthens, because this counterbalancing passive force will determine where the eye comes to rest. The neural circuitry responsible for innervation to the muscles must in some sense know about the active and passive properties of muscles and other orbital tissues if it is to produce the appropriate neural commands to move the eye to a particular, specified position of gaze.

Muscle force is related to muscle length

Force developed by a muscle and applied to the eye can be either active, as a consequence of muscle contraction, or passive, produced by the inherent elasticity of the muscle as it is stretched. It is therefore reasonable to expect a particular relationship between the length of a muscle and the force applied by the muscle to the eye. Force should increase as the muscle contracts and shortens from its resting length (the length the muscle assumes when it receives no innervational command and is left to its own devices). Also, because of the inherent elasticity of the muscle, force should increase when it is stretched beyond its resting length. A complete picture of force versus length should therefore be a U-shaped function, with the minimum force at the bottom of the U corresponding to the resting length of the muscle. And since the muscle will change length for different eye positions, the curve should also describe how muscle force changes with eye position.

Knowing the approximate form of the force–length curve is good, but not good enough; to alter muscle mechanics surgically or to make predictions about the operation of the oculomotor system, we need to know more about the force magnitudes and how quickly these forces change as eye position changes. In other words, the forces must be measured, preferably in intact, normally functioning human extraocular muscles. Extremely smali strain gauges, which measure force, implanted in muscles of patients undergoing strabismus surgery have allowed us to measure the forces that develop as the patient looks from one position of gaze to another. Figure 4.6 summarizes some of the results.

With the eye in primary position, the medial rectus exerted about 11 g of force, reflecting its tonic state of contraction. As the muscle lengthened during abduction (in which the prime mover would be the *lateral* rectus), the force developed in the muscle first decreased to a minimum at about 15° of abduction and then increased as the muscle was stretched.

When the eye *ad*ducts and moves into the field of action of the medial rectus, the force increases slowly, then more rapidly and linearly at a rate of around 8 g

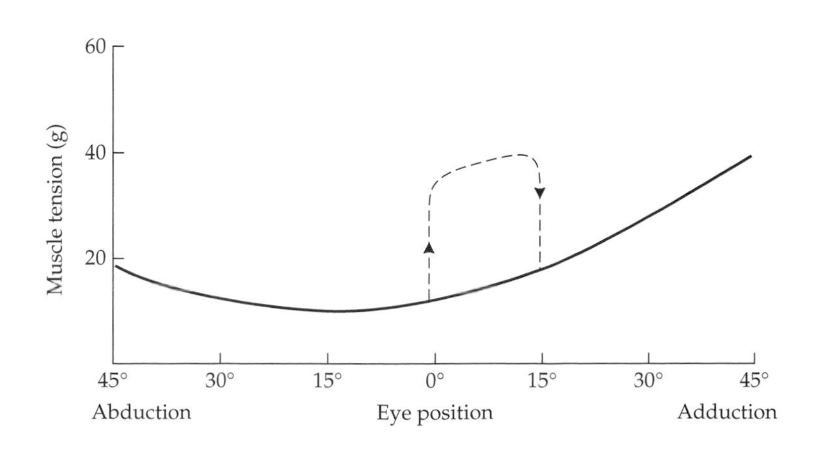

Figure 4.6
Force Developed in an Extraocular Muscle throughout the Muscle's Field of Action
The curve is the force at equilibrium applied to the eye by the medial rectus from adducted to abducted gaze; force was measured with an implanted strain gauge in an otherwise normal muscle. The increased force in abducted gaze is passive, produced by the inherent elasticity of the muscle as it lengthens when relaxed; increased force in adducted gaze is active, produced by muscle contraction. The dashed line shows the force required to adduct the eye by 15°. (After Collins et al. 1975.)

of force for every 5% increase in muscle length. (On average, the medial rectus and its tendon together are about 45 mm long; a 5% change in length is just over 2 mm, corresponding to an eye rotation of about 10°.) Note that the forces required to *move* the eye from one position to another are different from the equilibrium conditions that *maintain* a particular position of gaze. The dashed line with arrows in Figure 4.6 shows the rise and decline of force as the eye moves from primary position to 15° of adduction. The force increased by almost 20 g to move the eye, but maintaining it in the new position required only 5 or 6 g of additional force over that exerted in primary position.

To maintain the eye in more extreme positions of adducted gaze, the medial rectus must apply more force to the eye, as shown by the upward trend in the force curve for larger amounts of adduction. More force is required because the relaxed lateral rectus is being stretched passively as the eye is adducted and its passively developed force will increase; it is this passive component that the medial rectus must counterbalance to maintain the adducted gaze position.

Most of the force that maintains eye position is passive force

Each of the extraocular muscles has a force–length curve, and it is instructive to consider the force–length curves for both members of an agonist–antagonist pair. Based on the discussion so far, the opposing forces in the two muscles should balance out at all positions of gaze to maintain the eye in a fixed position. Figure 4.7 shows the force–length curves for the lateral and medial recti. The curves are identical by design. I have represented the lateral rectus by reversing the medial rectus curve left to right (so that muscle contraction is in the abducted direction), flipping it upside down (so that the forces in the two muscles are opposite in direction), and translating the curve so that minimum force is at a somewhat abducted position of gaze (corresponding to the minimum for the medial rectus at an abducted position). The justification for this graphical manipulation resides in the basic similarity of the experimental force–length curves for the lateral and medial recti.

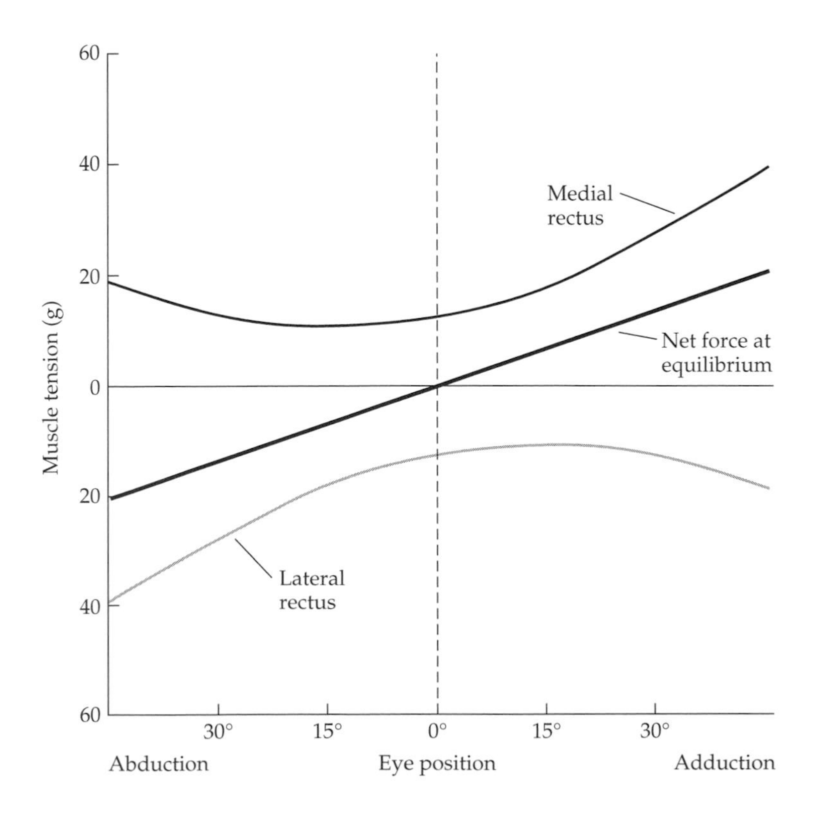

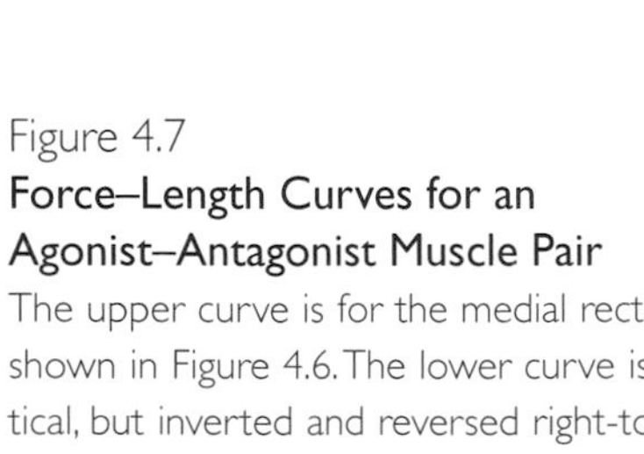

Figure 4.7
Force–Length Curves for an Agonist–Antagonist Muscle Pair
The upper curve is for the medial rectus as shown in Figure 4.6. The lower curve is identical, but inverted and reversed right-to-left; it represents force developed in the lateral rectus. The difference of the two curves, shown by the straight line, is the net force applied to the eye by the muscle pair at equilibrium for the different gaze positions. Net force is zero only for the primary position of gaze; at other positions, there must be a passive force, equal in magnitude but opposite in direction to the net force, that holds the eye in position.

In the primary position of gaze, the forces in the two muscles are equal in magnitude but opposite in direction. This is the expected equilibrium condition in which the forces cancel and the net force is zero. For all adducted gaze positions, however, the passive force in the stretched lateral rectus is smaller than the force in the medial rectus, and for all abducted positions of gaze, the medial rectus force is the smaller. If we take the difference between the two curves, the result is a straight line running through zero at primary position but having nonzero values everywhere else. And this result seems to belie everything said so far; except at primary position, there is a force imbalance because the force developed in the active muscle to maintain eye position is greater than the passive force in its stretched antagonist.

In fact, there is nothing wrong with our original premise, which stated that the eye is stationary when all the opposing forces are in balance. Figure 4.7 says simply that not all of the relevant forces are generated by just the two muscles in an agonist–antagonist pair; there are four other muscles attached to the eye and while they probably do not contribute *actively* to maintaining the eye in any of the adducted or abducted gaze positions, they will have tonic states of contraction and their own elasticity. And there are numerous fascial connections represented by Tenon's capsule, check ligaments, and so on (see Chapter 3), that will exert passive forces on the eye. Thus, the apparent force imbalance in Figure 4.7 is a measure of other passive forces applied to the eye by other muscles and orbital tissues. The direction of these passive forces *opposes* the agonist muscle in maintaining eye position; it has to develop more force than one might expect on the basis of the passive force in the antagonist alone.

There are two basic conclusions to be drawn from this analysis; first, the relationship between muscle force and muscle length is nonlinear, and second, it is never realistic to consider events in one muscle, or even the two muscles of an opposing pair, in isolation; other tissues and other muscles also contribute to the force balance equation. Both considerations affect extraocular muscle surgery (see Box 4.2); the nonlinearity makes it more difficult to predict the effect of a particular surgical procedure, and the interplay of forces from all the muscles may require that an operation to correct the eye's alignment involve more than one muscle.

Equilibrium muscle lengths and forces for different gaze positions are functions of the innervational command

The neural circuitry that controls eye position and movement is responsible for apportioning innervational commands to the motor nerves to move the eye in a particular direction. Any change in position around any of the rotational axes will be associated with a change in the innervation to, and the forces developed by, at least two of the extraocular muscles. It appears that a particular level of innervation to a muscle always produces the same state of contraction or relaxation and a particular set of innervational commands to the six muscles always produces the same eye position relative to all three axes of rotation.* And this is true regardless of any previous innervational patterns. To put it another way, the history of eye movements is not important; the nervous system does not have to remember and take account of past eye positions when setting the current innervation levels.

*This is a restatement of Donder's law, which says that for any given direction of gaze, the eye always has the same orientation with respect to its three cardinal axes of rotation. Listing's law is a stronger statement, specifying that the orientation of the eye is *as if* it rotated around an axis that is perpendicular to the plane containing the line of sight for primary position and the line of sight for the new gaze position.

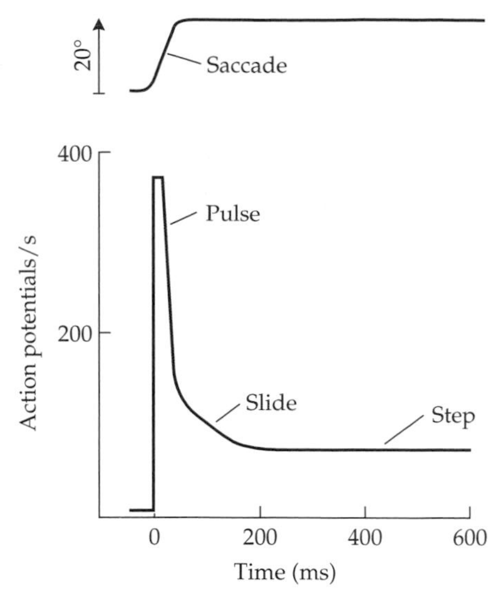

Figure 4.8
Discharge of Action Potentials by Extraocular Motor Neurons during Saccades
The upper trace is eye position before, during, and after a 20° saccade. The action potential frequency of the motor neuron (lower trace) has an initial high-frequency burst of activity (the "pulse") followed by a slow decline in frequency (the "slide") to a steady level of firing (the "step"). The step is higher than the firing level before the eye movement. In general, the amplitude of the pulse is correlated with eye movement velocity, and the magnitude of the step is related to the difference between initial and final eye positions. (After Leigh and Zee 1991.)

Different patterns of innervation are required for fast and slow eye movements

When the change from one eye position to another is very slow, the innervation to the extraocular muscles also changes slowly from one level to another. Muscle length corresponds to a certain steady rate of nerve impulses in its motor nerve, so the mean firing rate will increase gradually as the eye moves into the muscle's field of action or it will decrease gradually as the eye moves out of the field of action. The mean firing rate of a motor neuron corresponds to a certain muscle length, the length corresponds to a particular eye position, and there is little more to be said. But the gradual changes of mean firing rate that occur for slow eye movements will not work for saccades.

Although the eye does not weigh much (around 7.5 g in adults), its mass produces inertia that resists movement. To overcome the inertia, the eye needs a swift kick to get it moving and up to speed; this kick is the innervational command (Figure 4.8). Prior to a 20° saccade, a medial rectus neuron fires action potentials at a steady rate commensurate with the original eye position. Just before the eye moves, the firing rate increases abruptly to a very high level, and then drops back somewhat less quickly to a new steady firing level that is higher than the original steady firing rate. This pattern is a *pulse-slide-step;* the pulse is the high-frequency burst of neural activity that initiates the eye movement, the step is the difference between the initial and final steady-state firing levels, and the slide is the transition between pulse and step.

The height of the pulse (the maximum action potential frequency) dictates eye movement velocity during the saccade; the size of the step is related to the angular magnitude of the eye rotation—that is, to the angular difference between the original and the new eye positions. The innervation slide begins after the eye has come to rest and probably represents a decline in muscle force that is compensating for a corresponding but opposing slow change in passive force.

Extraocular motor neurons are located in three interconnected nuclei in the brainstem

Neurons with axons innervating the superior oblique and the lateral rectus occupy cranial nerve nuclei IV (trochlear) and VI (abducens), respectively; the remaining muscles are innervated by neurons in the oculomotor nucleus (III). The locations of these nuclei in the brainstem are shown in Figure 5.20. The neurons for the superior, medial, and inferior recti and the inferior oblique occupy separate, slightly overlapping regions in the oculomotor nucleus.

Not all of the cell bodies in these nuclei have axons going to one of the extraocular muscles. Some of them are interneurons that connect to other sets of motor neurons and are responsible for yoking of the active muscles during version eye movements. The abducens nucleus, whose primary neurons send axons to the lateral rectus, has many interneurons that cross the midline and extend up the brainstem in the **medial longitudinal fasciculus** to the contralateral oculomotor nucleus, where they terminate on neurons innervating the medial rectus muscle (see Figure 4.10). For version eye movements, the interneurons are activated along with the primary lateral rectus neurons and have similar discharge characteristics. Since they will activate the contralateral medial rectus neurons when the ipsilateral lateral rectus neurons are active, a version eye movement results. A smaller population of interneurons in the medial rectus portions of the oculomotor nucleus operate in the same way, making excitatory contacts on contralateral abducens neurons.

Because so many nuclear interconnections are running together in the medial longitudinal fasciculus, lesions in this part of the brainstem can severely disrupt normal eye movements. A unilateral lesion produces **internuclear**

ophthalmoplegia, which involves deficient adduction toward the affected side, among other things. Most of the time, vergence eye movements are not affected (they may be, however, if the lesion is close to the oculomotor nerve nucleus, since vergence control neurons can be found in the area).

Interneuron links between the vertical rectus muscles and between the obliques are less well documented, but they are likely to serve a similar yoking operation. Only the trochlear nucleus appears to lack interneurons.

Motor commands are the result of interactions between visual and nonvisual inputs to the motor control centers

The motor neurons are the final common path in the eye position control system, and their neural activity is highly stereotyped; the signals conveyed by the motor neurons to the extraocular muscles are always related to eye movement velocity and eye position. These signals, however, can be elicited by a variety of stimuli, both visual and nonvisual.

Figure 4.9 illustrates a way of thinking about the problem of controlling eye position. The eye receives an input from the external world, represented by the retinal image. Two sets of visual information extracted from the image are sent to the central nervous system; one, at a relatively low level, has to do with basic information about image location on the retina and the velocity of retinal image motion; the second—going to the box labeled "visual cortical areas" on Figure 4.9—is the main visual information stream from which decisions can be made about the identity and importance of particular elements in the overall retinal image. Visual cortical areas should be thought of as operations and interactions throughout the primary and association visual cortical areas. Both the position and velocity information and an input from the visual cortical areas impinge on the large box labeled "relative spatial localization." This operation is also distributed in several locations, including prefrontal motor cortex, cerebellum, and superior colliculi.

The box labeled "relative spatial localization" on Figure 4.9 also receives inputs from a variety of other sources: the vestibular system, auditory system, proprioceptors, the tactile system, and so on. Thus eye position is subject to a variety of influences and may be affected not only by what is seen, but also by

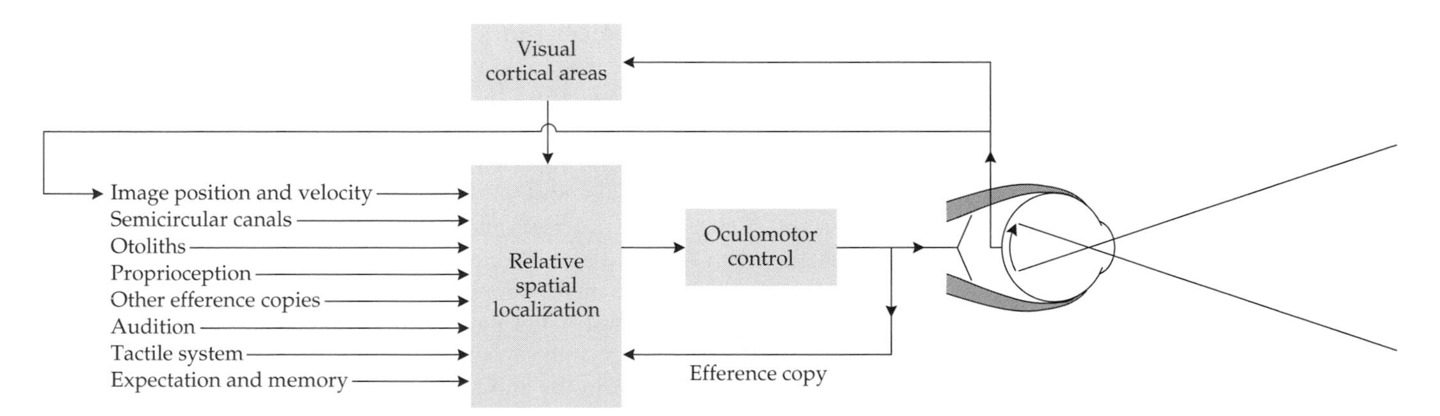

Figure 4.9
An Overview of the Eye Position Control System
The signals from motor neurons to the extraocular muscles are the result of numerous influences, both visual and nonvisual. The consensus is determined by combining inputs from the retina, the visual areas of the cortex, and other sensory systems, and signals concerning the current status of oculomotor commands to the extraocular muscles (efference copy). Most of oculomotor control is in the brainstem; relative spatial localization is distributed among cortex, superior colliculi, and cerebellum. (After Collewijn 1989.)

such things as body position and movement, the location of sound sources, or the location of unseen objects that have touched one's body.

The single output from relative spatial localizaton goes to the box labeled "oculomotor control," thereby implying that all the various inputs have been utilized in arriving at a judgment about where the eye is *now* and where it should be *next*. The visual inputs may dominate under some circumstances and the nonvisual inputs under others, but they all have their say. Oculomotor control translates this consensus into appropriate neural signals to the extraocular muscles, and a copy of these signals is sent back to the relative spatial localization box.

A copy of the innervational command is used to verify the system's operation

The label "efference copy" on Figure 4.9 refers to information about the oculomotor command signal that is sent back to the relative spatial localization box so that the *intended* oculomotor command can be compared to the *actual* command. This comparison is a way of ensuring that the motor control system is doing what it is supposed to do and, if not, adjusting things appropriately.

The effect of efference copy can be observed in the following way. Note that when your eyes move to look from one place to another, the visual world remains stationary. Objects that are truly stationary *appear* stationary in spite of eye and retinal image motion. Now, observe that when you move one eye by jiggling it with your finger placed on the lateral side of the eye, there is a great deal of apparent motion of your visual world. Although the eye and the retinal image moved in both cases, the different perception is due to the operation of efference copy when the eyes were moved by normal contraction of the extraocular muscles and the absence of efference copy when the eyes were passively moved.

In the first case, the neural commands to move the eyes were sent back as efference copy for comparison; the brain *knew* that the eyes would move, it *expected* retinal image motion as a consequence, and it *attributed* the retinal image motion to eye movement and not to movement of real-world objects. When the eyes were moved by jiggling with a finger, however, no commands were sent to the extraocular muscles and there was no efference copy. The brain had no choice but to *assume* that the eyes were stationary and that the retinal image motion was produced by the movement of objects in the world. The result was apparent movement. Efference copy provides the brain with a continuous record of what the eyes are supposed to be doing; without it, our visual world would be a very confusing place.

Individuals who have a partially paralyzed extraocular muscle often exhibit what is called **past-pointing**. When asked to look in the direction of action of the paretic muscle and to indicate with their hand where they are looking (the situation is arranged so that they cannot see the hand), these individuals will indicate a direction with the hand that is much more rotated than the eye. The most likely explanation is that the innervational command required to elicit contraction from the weak muscle is stronger than normal, and so is the efference copy signal; the brain therefore thinks the eye has rotated more than it really has, and this produces the past-pointing by the unseen hand. Some studies have concluded that proprioceptive signals from sensory organs in the muscles (discussed later in the chapter) may also contribute to past-pointing.

Extraocular motor neurons receive inputs from premotor areas in the brainstem to generate appropriate signals for saccadic eye movements

Several locations in the brain are known to be involved in saccadic eye movements, including the primary visual cortex in the occipital lobe, a motor area in the frontal motor cortex, and the superior colliculi. In particular, signals from the colliculi and the frontal motor areas impinge on brainstem premotor neurons that modify the input signals before telling the motor neurons what to do. The brainstem circuitry is a part of the "oculomotor control" box in Figure 4.9.

Figure 4.10 is a schematic diagram of the brainstem premotor circuitry for saccadic eye movements, with the various collections of cells in their approximate anatomical locations. The eye movement is a version to the right, so the active muscles are the right lateral rectus and the left medial rectus.

This premotor stage of saccade initiation has two sets of signals from the superior colliculus. One is a position signal that activates cells called *long-lead burst neurons.* The name comes from their activity; as shown, they fire a burst of action potentials that precede not only the saccade but also any other premotor neuron activity.

The other input is a trigger signal to *omnipause neurons;* these cells are constantly active *except* during a saccade. As shown, they are inactivated when the long-lead burst neurons are firing. The omnipause neurons make inhibitory connections with several other groups of cells in the circuit, and their role can be thought of as gating the premotor circuitry. That is, the next stage of the saccadic circuit is rendered quiescent by the steady inhibition from the omnipause neurons until there is to be a saccade. Then, the omnipause neurons are turned off, thereby opening the gate to the other circuit components.

The long-lead burst neurons activate a group of cells called *excitatory burst neurons.* One contact for the excitatory burst neuron is to the ipsilateral abducens neurons and the abducens interneurons; the input will generate the pulse in the discharge of the motor neuron (see Figure 4.8).

A second contact is with cells in a *neural integrator.* Cells in the integrator are called *tonic neurons* or *eye position–related neurons.* A neural integrator is required in the oculomotor system because the signals being used to generate the eye movements, of whatever type, are related to eye velocity, while the outputs from the motor neurons must be related to eye position (the step in the discharge pattern; see Figure 4.8). Velocity is the time derivative of position, so going *from* velocity *to* position requires integration. The same neural integrators are thought to be used for all eye movements in the horizontal plane; the integrators for vertical eye movements are separate. As Figure 4.10 shows, the output from the cells in the integrator goes to both the abducens neurons and the abducens interneurons, where it will generate the step in their discharge.

The third contact made by excitatory burst neurons is to *inhibitory burst neurons.* Their outputs are inhibitory to the contralateral neural integrator and to the contralateral abducens neurons and interneurons going to the *ipsilateral* medial rectus neurons. This is the connectivity underlying reciprocal innervation.

There are several uncertainties and omissions in this scheme. The relationship between the long-lead burst neurons and the omnipause neurons is a matter for some disagreement, for example, and there is no indication that the workings of this system can be affected by the cerebellum, as it can. But the basic features of motor neuron activity during saccades—pulse, step, reciprocal innervation, versional yoking—are readily explained by this circuit. The major uncertainties may well be the nature and precise target of the triggering and position signals from the superior colliculus and frontal eye fields.

The pathways for smooth pursuit movements and for vergences go through the cerebellum, but vergences have a separate control center near the oculomotor nucleus

Since information from the vestibular system goes to the cerebellum, it is not surprising that eye movements elicited by the vestibular system have the cerebellum as part of the reflex pathway. But the same can be said for visually driven

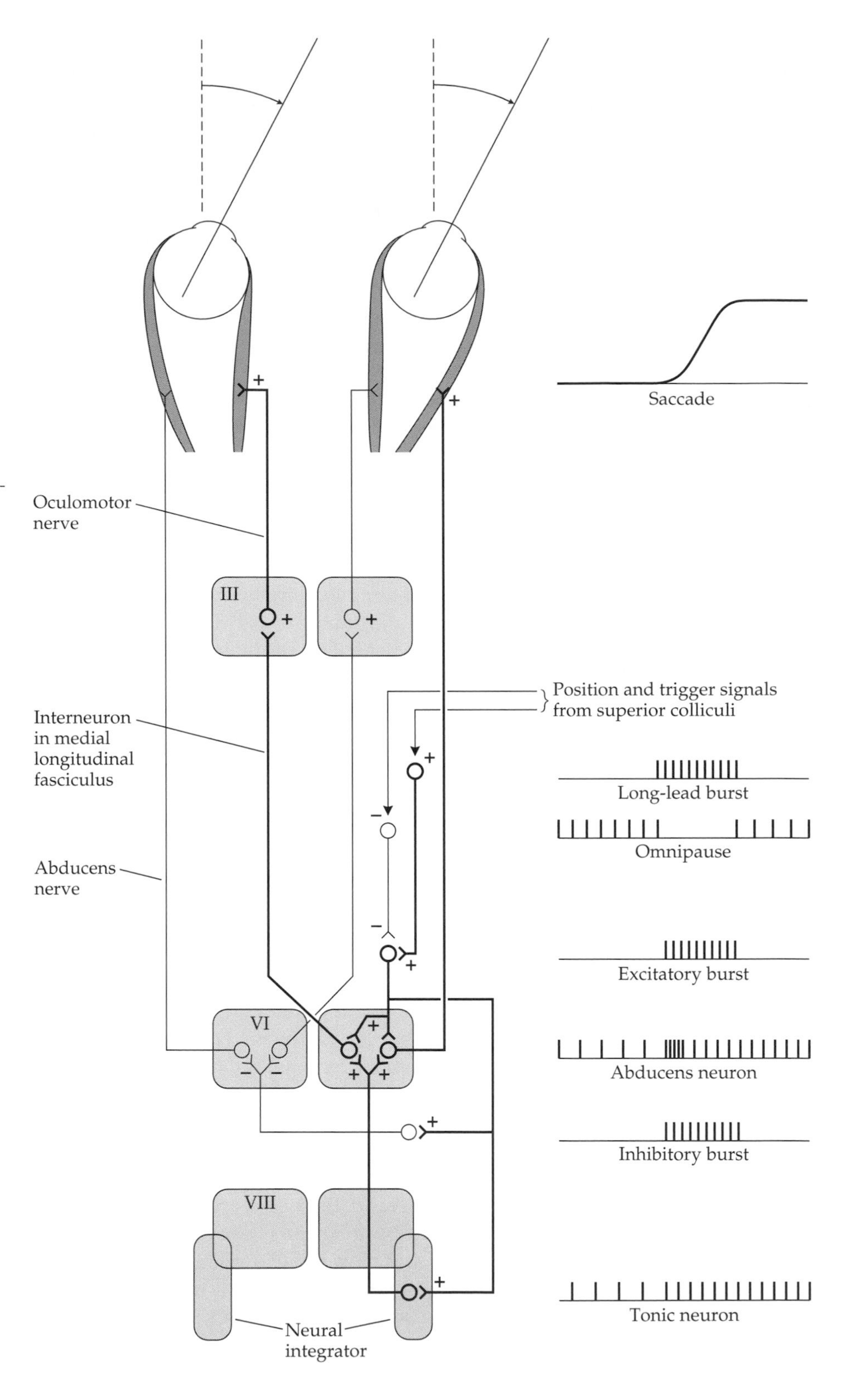

Figure 4.10
Brainstem Circuitry for Saccadic Eye Movements
This somewhat simplified schematic shows the basic connectivity by which both eyes make a saccade to the right. The input responsible for the pulse in the motor neuron discharge comes from excitatory burst neurons that activate abducens neurons and abducens interneurons going to the contralateral medial rectus neurons. The excitatory burst neuron activity is also integrated and sent to the abducens neurons and interneurons to become the step component of the motor command. Activation of inhibitory burst neurons reduces the motor signals to the antagonist muscles. (+ is excitation and – is inhibition.)

movements, such as smooth pursuit movements or optokinetic nystagmus. The reason, presumably, is that head or body position can change during slow eye movements, and since the cerebellum is in the business of coordinating movement, it needs to have a role in modulating slow, visually driven eye movements. The cerebellum has less to do with saccades, because they are so rapid; there is little that can be done to modify a saccade during its brief duration.

Signals from the frontal motor eye field and parts of the association visual cortex (medial temporal and medial superior temporal visual areas) descend to nuclei in the pons and then to the flocculus in the cerebellum (Figure 4.11). Purkinje cells in the flocculus are active during smooth pursuit movements—their activity is correlated to movement velocity—and microstimulation of these cells will elicit smooth pursuit movements. Lesions in the flocculus will produce smooth pursuit deficits.

Since the signals from the cerebellum are related to velocity, they must be integrated to produce a signal that specifies eye position. The smooth pursuit system is thought to use the same neural integrators used by the saccadic system. The integrated signal probably goes directly to the appropriate motor neurons. For smooth pursuit movements in the horizontal plane, the targets would be abducens motor neurons and abducens interneurons going to the contralateral medial rectus motor neurons (see Figure 4.11*a*).

An area near the oculomotor nucleus in monkeys contains cells whose activity is related to the vergence angle of the eyes. These *vergence cells* or *near response cells* send signals to medial rectus motor neurons and perhaps to cells in the Edinger–Westphal nuclei that are controlling accommodation (see Chapter 5).

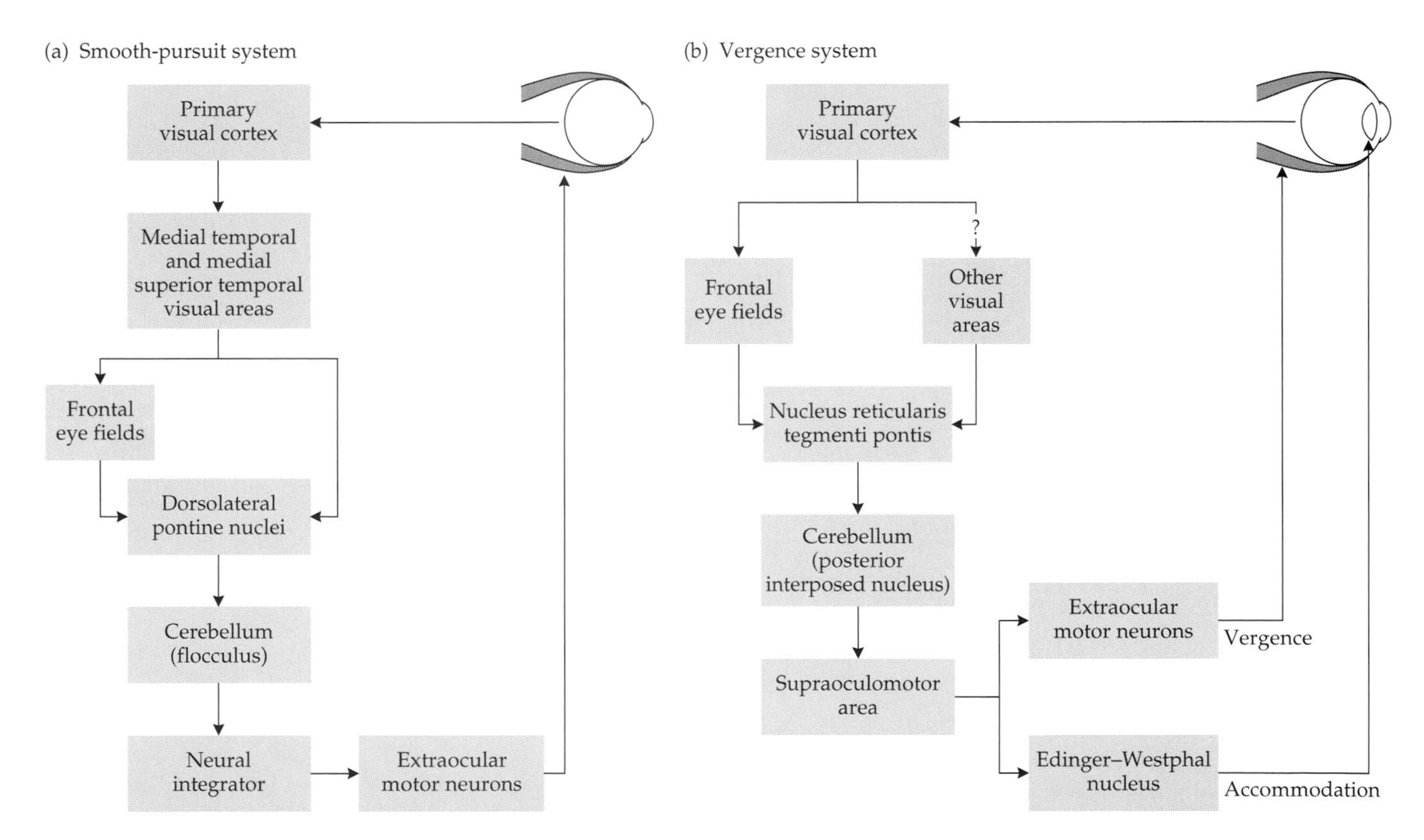

Figure 4.11
Elements of Smooth Pursuit and Vergence Pathways
The afferent pathways for smooth pursuit and vergence involve the primary visual cortex, other visual areas, and the frontal eye fields. Both pathways descend to different parts of the pons before going to different parts of the cerebellum. (*a*) In the smooth pursuit pathway, the signals from the cerebellum go to the neural integrator and then to the appropriate extraocular motor neurons. (*b*) In the vergence pathway, cerebellar signals go to the cluster of vergence neurons in the supraoculomotor area and then to the appropriate extraocular motor neurons for vergence and perhaps to the Edinger–Westphal nucleus for accommodation. (*b* after Gamlin et al. 1996.)

Box 4.2 Changing the Effects of Extraocular Muscle Contraction

IN PRINCIPLE, strabismus surgery should be a straightforward problem in mechanics; if one eye deviates in a certain direction by a certain amount, correction should be a matter of weakening the primary agonist muscle or strengthening the primary antagonist by the appropriate amounts. For example, in a right esotropia the right lateral rectus is presumably weak, allowing the normal tonus of the medial rectus to rotate the eye medially. To restore the eye to its normal straight-ahead position of gaze, the lateral rectus must made be more effective, or the medial rectus must be made less effective, or both.

A muscle can be made more effective by shortening. One method of shortening a muscle is to detach it at its insertion, cut a piece from the tendon (and muscle if necessary), and reattach it (**resection**) (see part *a* of Figure 1). Cutting a piece out of the belly of the muscle and rejoining the cut ends (**myectomy**) also shortens the muscle. Third, suturing a tuck in the muscle shortens it (**tucking**) (part *b* of the figure). Of these three basic methods, tucking has some obvious advantages; in particular, it is reversible, and when combined with adjustable sutures (an option in several of these procedures), it allows the result to be modified without additional surgery. For a variety of reasons, however, resection or myectomy may be preferred in specific situations.

Two methods for weakening a muscle are making cuts in the muscle so that it will elongate (**marginal myotomy**) (part *c* of the figure) or relocating its insertion to a less advantageous position (**recession**) (part *d*). Recession and resection of an antagonist–agonist pair is common.

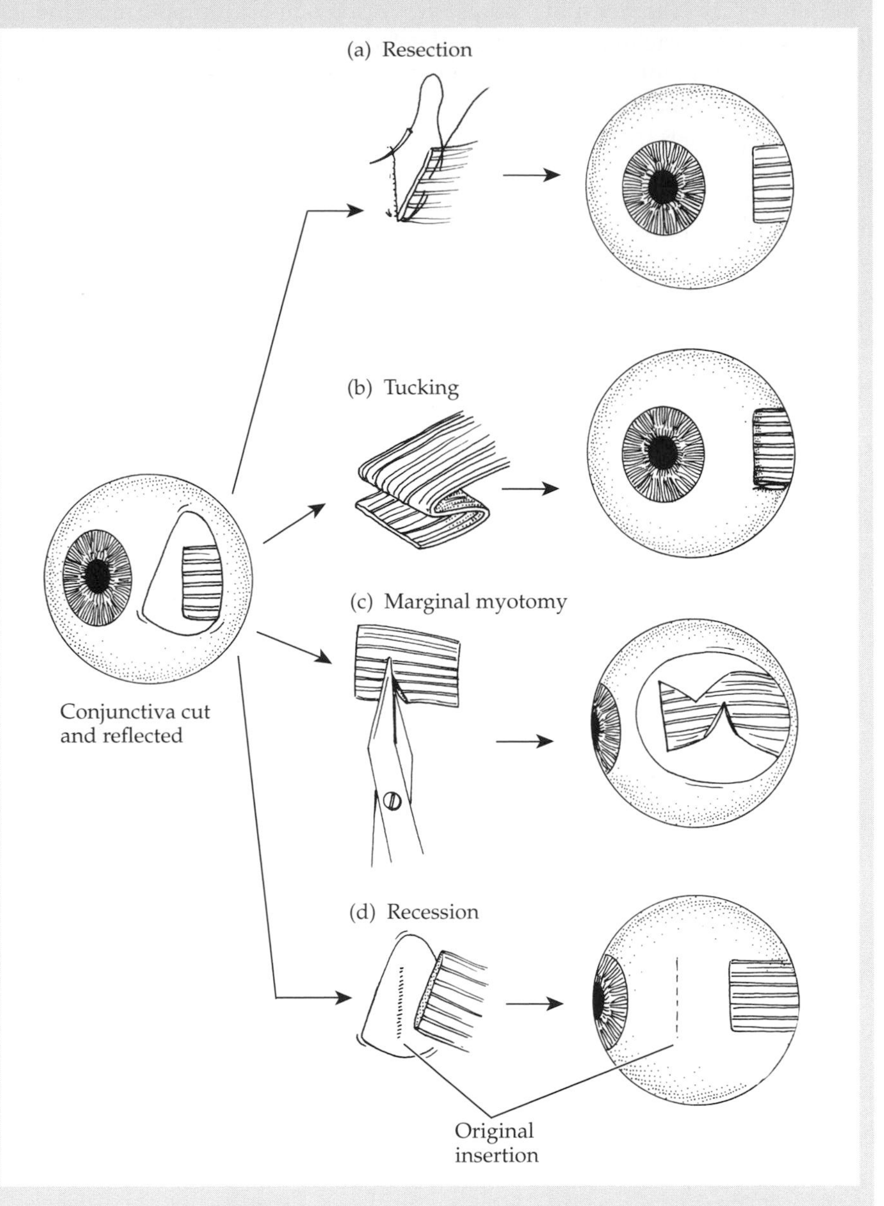

Figure 1
Basic Forms of Extraocular Muscle Surgery
(*a*) In resection, the muscle is detached from its insertion, a piece of the tendon is removed to shorten it, and it is then reattached to its insertion. (*b*) In tucking, the muscle is folded like the letter *Z*, and the fold is held in place by sutures. (*c*) Marginal myotomy makes a muscle less effective by cutting some of the muscle fibers and allowing the muscle to elongate. (*d*) For the rectus muscles, a recession means moving the muscle's insertion posteriorly on the globe.

Another type of extraocular muscle surgery is **transposition**, which refers to gross relocation of the insertion of a muscle in order to radically change its action (recession is also a movement of a muscle's insertion, but the new insertion is in line with the original one—the muscle action does not change and the muscle is simply rendered less effective). Several insertions may be moved, and sometimes muscle insertions are interchanged. There are numerous variations on this theme, but transpositions are less common than other procedures, and they are often performed as a last resort.

Although these surgical procedures may seem straightforward, the simplicity is illusory, and numerous uncertainties can affect the results. For example, knowing how much a muscle should be shortened or lengthened is complicated by the nonlinear relationship between muscle length change and the magnitude of the strabismus. Is it better to shorten the weaker muscle of the agonist–antagonist pair, to weaken the stronger muscle, or to combine these procedures? Should the surgery be confined to just one muscle pair, or does the interaction among muscles demand that all six be included? Should the operation include only the deviated eye, or both eyes? What should be done about the connective tissue sheaths and connections between muscles that affect muscle actions—cut them away, or leave them intact? Does a 5 mm recession always produce the same result, or does surgical technique influence the outcome? Will the surgical procedure compromise the blood supply to the anterior segment of the eye (see Chapter 6)? These and other questions do not always have clear-cut answers, which means that strabismus surgery is not simple mechanics; favorable outcomes are critically dependent on the surgeon's judgment and experience.

In addition, strabismus may not be a muscular problem. Strabismus is often—perhaps most of the time—associated with sensory anomalies that may have contributed to the strabismus, may be secondary to it, or may impede proper diagnosis. Mechanical realignment of the eyes will not necessarily eliminate an associated sensory deficit; in some instances, the sensory anomaly is such that a strabismus will reappear after an initially successful realignment of the eyes. Thus for adults, strabismus surgery is usually elective surgery with the goal of realigning the eyes for the sake of appearance; reestablishing normal binocular vision is impossible if the strabismus is long-standing. Infants and children with strabismus, however, are a different matter. The visual system is relatively plastic for some years postnatally, and early alignment of the eyes appears to be critical for establishing and maintaining normal binocular vision. The question with strabismus in infancy and childhood concerns the best way to align the eyes if strabismus develops at a very young age.

Patients with strabismus are normally evaluated first for nonsurgical correction. Some forms of strabismus, for example, can be eliminated by correction of the refractive error or by supplemental corrections for near vision; in others, ophthalmic prisms may compensate for the deviation. Orthoptics, which is vision training directed toward goals such as expanding the working range of the vergence control system or reducing the common suppression of vision in a deviated eye, is also a useful alternative to surgery in some cases and almost always a supplement to surgery in others. (Conventional orthoptic procedures are difficult or impossible with very young children, a limitation that has led to many forms of games, play activities, and exercises included under the rubrics of vision, perceptual, or behavioral training. It is difficult to evaluate these treatment strategies, and claims about their value or efficacy are largely anecdotal.) Another alternative to surgery (discussed in the text) is injection of botulinum toxin into a muscle or muscles; the toxin renders the muscle less effective by reducing neurotransmitter release. The effect is long-lasting, but ultimately reversible, and it may sometimes be the method of choice in very young children.

Current efforts to better characterize the actions of the extraocular muscles are directed toward making strabismus surgery less of an art form; the hope is to predict the outcome of any proposed alteration of the muscles. A computer-based model of the extraocular muscle system discussed in the text can incorporate data from diagnostic tests to evaluate the problem, make recommendations about the type of correction that is required, and evaluate alternative procedures. Although the preliminary reports are encouraging, some things, such as the formation of scar tissue and other adhesions as the tissue heals, may never be predictable. The surgeon's artistry will probably always be a key ingredient in the correction of strabismus.

Many details of the vergence pathways are not known, but as Figure 4.11*b* indicates, at least one source of input to the vergence neurons is the posterior interposed nucleus in the cerebellum, whose cells have vergence-linked activity. This part of the cerebellum receives inputs from the cerebellar cortex and from a precerebellar nucleus (nucleus reticularis tegmenti pontis), which has inputs from the frontal eye fields and probably other visual areas as well. This schema is too sketchy to explain how vergence commands are generated, but it does suggest the separateness of the vergence and smooth pursuit pathways, even though both pass through the cerebellum.

Extraocular Muscle Structure and Contractile Properties

Muscle fibers are the units from which muscles are constructed

Muscles are composed of thin, elongated, multinucleate cells called muscle fibers; they are multinucleate because each fiber forms embryologically by the fusion of numerous smaller cells whose individual nuclei are retained. Within a muscle, the fibers are collected in bundles, each bundle surrounded by a connective tissue sheath, the **perimysium** (Figure 4.12*a*). The muscle itself consists of many of these bundles, and the entire collection is also wrapped by connective tissue (the **epimysium**). The epimysium of the extraocular muscles is continuous with and generally considered to be a part of Tenon's capsule (see Chapter 3).

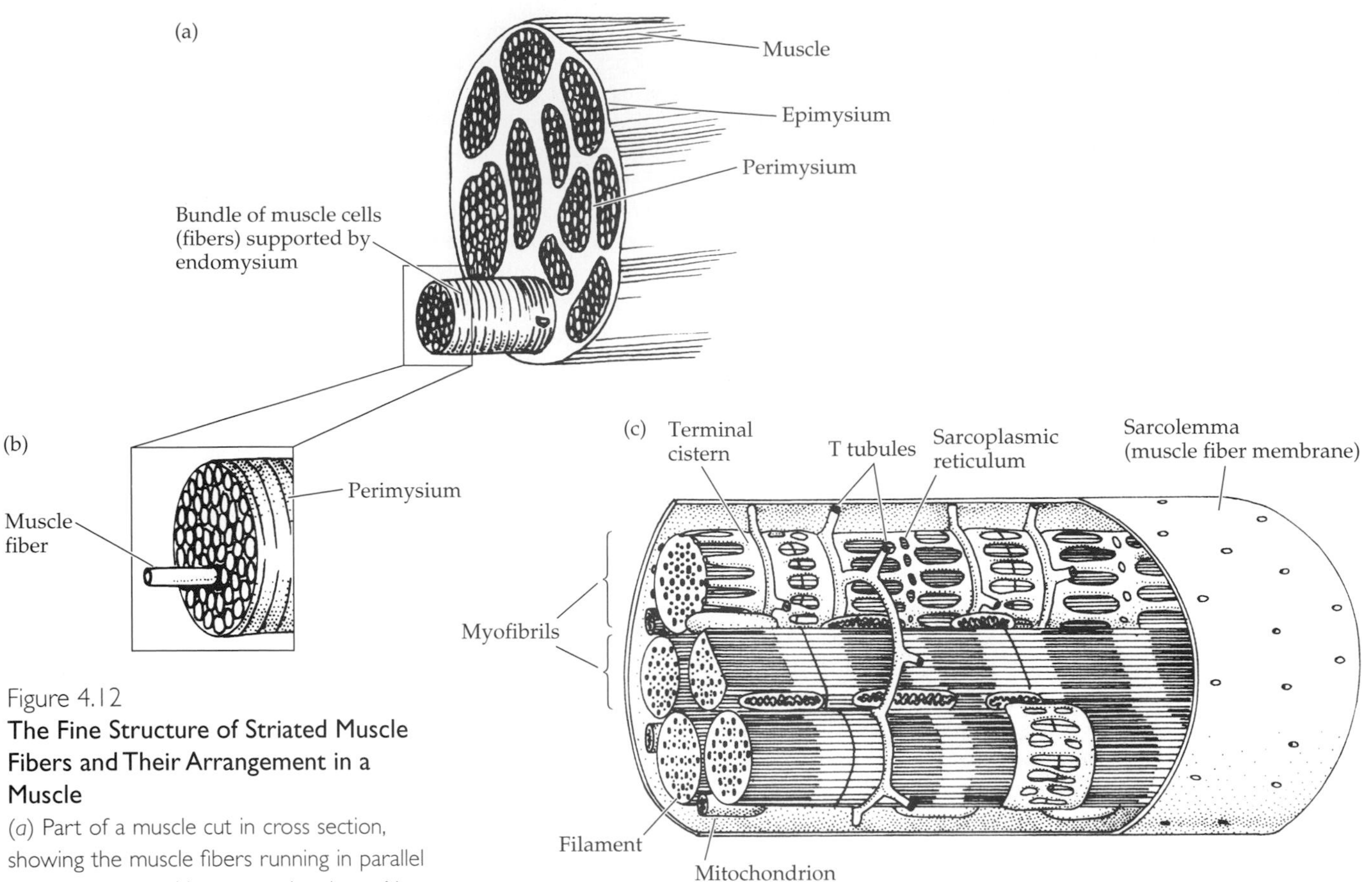

Figure 4.12
The Fine Structure of Striated Muscle Fibers and Their Arrangement in a Muscle
(*a*) Part of a muscle cut in cross section, showing the muscle fibers running in parallel bundles separated by connective tissue (the perimysium). (*b*) A muscle fiber bundle, bound by the perimysium. (*c*) A muscle fiber consists mainly of parallel myofibrils, which are bundles of contractile proteins, surrounded by the sarcoplasmic reticulum and interspersed with T tubules.

The muscle fiber's membrane, the **sarcolemma**, has a characteristic structure. At regular intervals, the membrane invaginates to form a tunnel or tube running from one side of the fiber to the other; these tunnels are the **T tubules** (Figure 4.12*c*). Because of them, no part of the cell's interior cytoplasm can be very far from the sarcolemma and the extracellular space. An irregular complex of membranes in the cytoplasm, the **sarcoplasmic reticulum**, extends between neighboring T tubules and surrounds all the bundles of contractile proteins (**myofibrils**) within the cell. The sarcoplasmic reticulum is a reservoir of intracellular calcium that is released when the sarcolemma, including the T tubule system, is depolarized by the action of the neurotransmitter released from the motor nerve terminals contacting the muscle fiber. The released calcium triggers a series of reactions that produce contraction and shortening of the muscle fiber.

Striated muscle fibers have a parallel arrangement of contractile proteins that interleave to cause contraction

Most of the mass of a muscle fiber consists of myofibrils, the structure and arrangement of which give striated muscle its characteristic banded appearance. The major proteins in the myofibrils are **actin** and **myosin**. Molecules of actin and myosin join to form long actin and myosin filaments; actin filaments are referred to as "thin" filaments because the long molecular coils are thinner than the "thick" myosin filaments. The thick and thin filaments have a highly ordered arrangement (Figure 4.13) that is stabilized and maintained by other structural proteins (not shown in the illustration). The two classes of filaments interleave with one another, and varying the amount of interleaving produces changes in muscle length.

Muscle contraction can be explained by a sliding filament model. The actin and myosin filaments move relative to one another by a kind of molecular ratchet (Figure 4.14) in which the heads of the myosin molecules bend, dissociate from the adjacent actin filament, straighten, and reattach to the actin in a new location. Each repetition of this cycle constitutes only a small ratcheting movement, about 12 nm in extent, but the cycle is very fast (less than 5 ms) and the filaments can move past one another at rates up to about 15 μm/s. Since the ratcheting action is occurring simultaneously throughout the myofibril system in the muscle fiber, the overall rate of contraction can be several hundred times faster, measured in millimeters per second.

Striated muscle fibers differ in structural, histochemical, and contractile properties

Although the foregoing description applies to all striated muscle fibers, they differ in structural detail and contractile properties; three or four muscle fiber types can be distingushed on the basis of differences in histochemical staining properties, numbers of mitochondria, extent and density of the sarcoplasmic reticulum, speed of contraction, and resistance to fatigue produced by chronic stimulation, among others. To summarize an involved story, the simpler threefold classification includes two types of **fast-twitch** fibers, one "red" and one "white." The term "twitch" refers to their property of rapid all-or-none contraction; that is, twitch fibers either contract maximally or they do not contract at all. Red fast-twitch fibers have high mitochondrial density and oxidative metabolism, white fibers low. The third muscle fiber type is also "red," but it is called **slow-twitch** because of its relatively lower contraction velocity.

Skeletal muscles vary in their proportions of fast-twitch and slow-twitch fibers. In general, muscles responsible for very fast movement of a joint, such as the muscles of the fingers, have a high proportion of fast-twitch fibers. Large

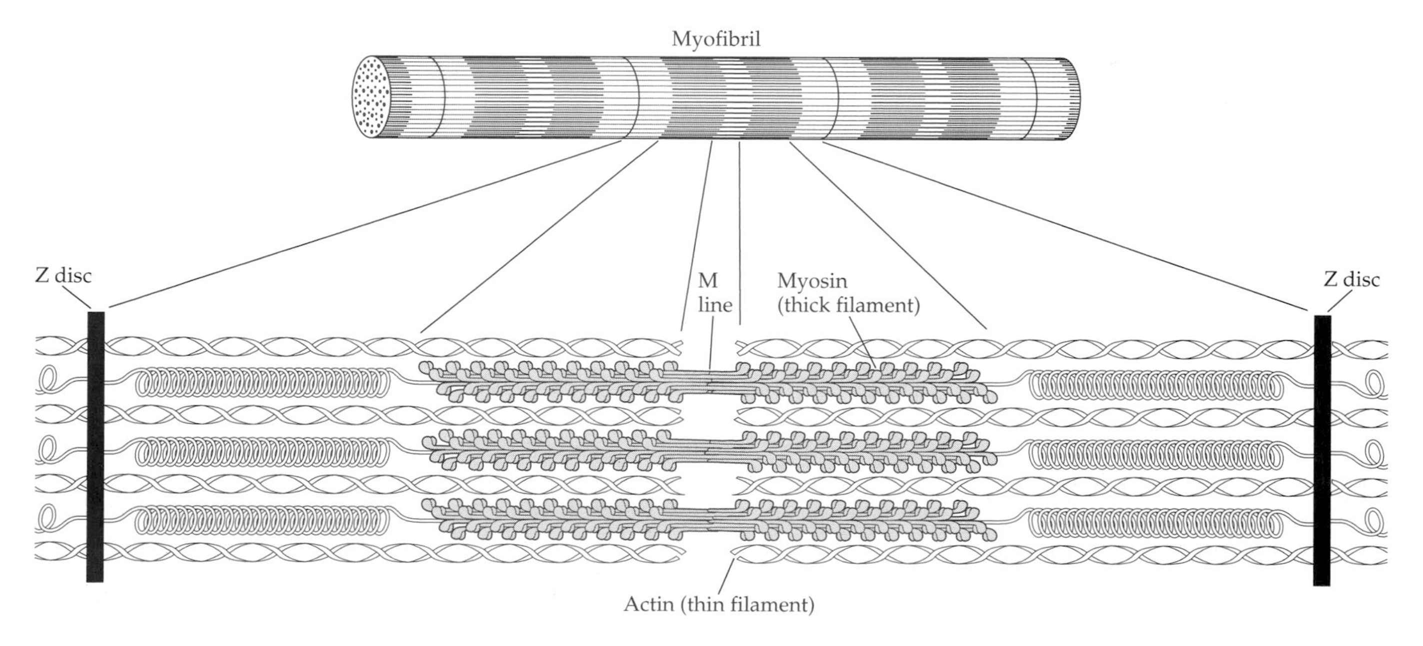

Figure 4.13
Actin and Myosin in a Myofibril
Contractile proteins form long, thin strands running parallel to one another. The actin strands are thinner than the myosin strands, and alternating bundles of actin fibrils interleave with myosin fibril bundles. Muscle contraction is produced by an increase in the amount of interleaving.

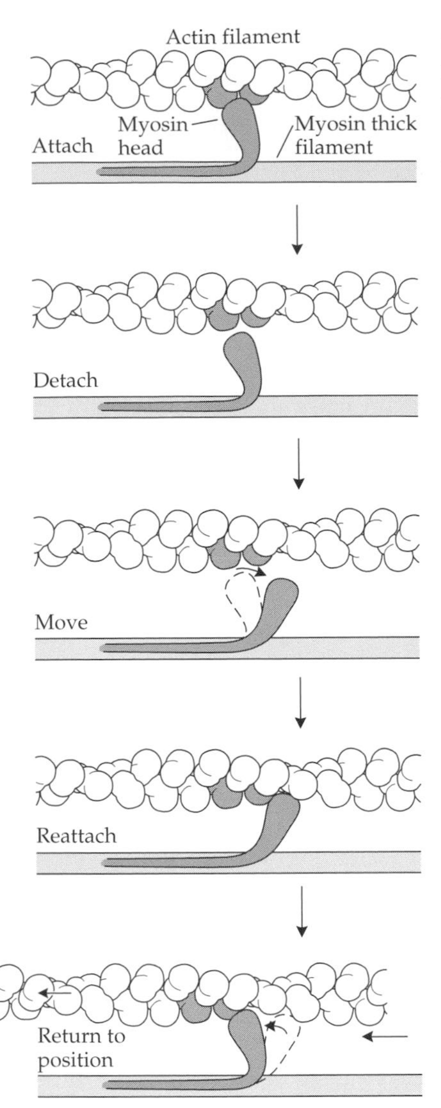

Figure 4.14
The Ratchet Model for Interleaving and Contraction
The small protruding heads of myosin fibrils are movable. To increase the interleaving with adjacent actin fibrils, the myosin heads detach from the actin, swing back and reattach, and return to their original position, carrying the actin fibril along with them. Numerous rapid repetitions of this sequence throughout a myofibril will cause it to shorten very quickly. Reversing the sequence decreases interleaving and allows the myofibril to lengthen. (After Alberts et al. 1994.)

muscles whose primary task is maintaining posture are more likely to be dominated by slow-twitch fibers. All three types are also found in the extraocular muscles.

The extraocular muscles contain muscle fiber types not found in skeletal muscles

Extraocular muscle fibers are striated, and in many respects they are diminutive versions of skeletal muscle fibers. Mean cross-sectional diameters of extraocular muscle fibers are less than 20 μm (the range is from around 10 to 50 μm), whereas skeletal muscle fibers are typically more than 100 μm in diameter.

Striated muscle fibers are normally innervated by a single nerve process (a branch of an axon) that terminates in a broad expanded structure called a **motor end plate** (also *terminaison en plaque*). The motor end plate fits into a shallow depression in the muscle fiber (Figure 4.15*a*), and here the neurotransmitter is released from the axon terminal to activate the muscle fiber. Muscle fibers receiving input from motor end plates are called *singly innervated fibers* because there is normally just one end plate region per muscle fiber (where there are several end plates, they come from the same axon, and because they are close together, the muscle fiber is still considered to be singly innervated). Fast- and slow-twitch fibers in both skeletal muscle and extraocular muscle are singly innervated fibers.

In addition to the singly innervated fibers, extraocular muscles contain another group of muscle fibers that are not contacted by conventional motor end plates and that receive numerous contacts from the motor nerve axons. These fibers are referred to as *multiply innervated.* The multiple axon terminals are small, round swellings at the ends of the axonal processes. They are called **en grappe endings** (or *terminaisons en grappe*) because they occur in clusters (*grappes* in French) (Figure 4.15*b*). These nerve endings are not seen in skeletal muscles, but they are very similar to nerve endings in smooth muscles. Thus, some extraocular muscle fibers are similar to smooth muscle fibers elsewhere in the body, even though they are technically striated muscle fibers.

The singly and multiply innervated fibers can be partically separated by size. Singly innervated fibers have a larger mean diameter, which has led to their being called "thick" fibers. The smaller-diameter multiply innervated fibers are "thin" fibers. The diameter ranges for the two fiber groups overlap somewhat.

On a strictly anatomical basis, the extraocular muscles have two broad classes of muscle fibers: the singly innervated thick fibers, which are like those found in other skeletal muscles, and the multiply innervated thin fibers, which are unique to the extraocular muscles. (The levator muscle does not contain any thin, multiply innervated fibers, and it is therefore much more like the body's other striated muscles.) Since both of these groups can be subdivided into several types on the basis of histochemical features, there are five or six fiber types in the extraocular muscles. The meaning of this five- or sixfold subdivision is unclear, but the division into thick and thin fibers seems to be functionally significant.

Thick and thin extraocular muscle fibers differ in their contractile properties

The singly innervated thick fibers are twitch fibers like their counterparts in skeletal muscles. They contract very rapidly, although fast-twitch fibers are faster than slow-twitch fibers, and they exhibit all-or-none contractions. Thus, when the nerve impulses arriving at the motor end plate are sufficient to activate the muscle fiber, the fiber contracts rapidly to the full extent of which it is capable.

We have no direct evidence about the contractile properties of the multiply innervated thin fibers in human extraocular muscles, but they seem to be similar to the multiply innervated fibers in amphibians and birds that are known to exhib-

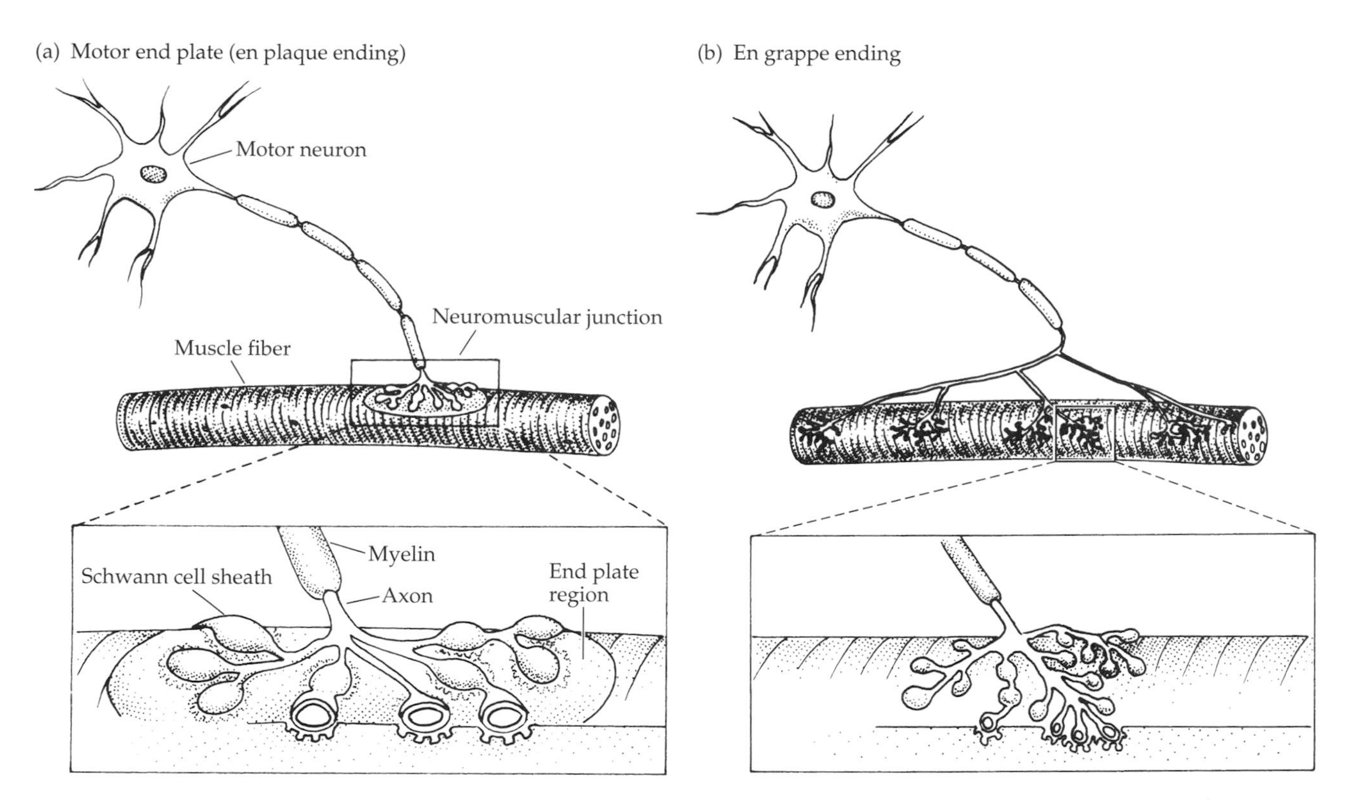

Figure 4.15
Motor Nerve Endings on Extraocular Muscle Fibers
(*a*) A conventional motor end plate on a thick fiber. There is usually one motor end plate per fiber. (*b*) An en grappe ending on a thin fiber. There are many small terminal endings on the fiber, but all come from the same nerve axon.

it relatively slow, graded contractions. Muscle fibers of this type are called **tonic** fibers; instead of contracting in an all-or-none fashion, their contraction is graded, increasing monotonically as the strength of the neural input increases. If the thin fibers in extraocular muscles have tonic contractile properties (and circumstantial evidence suggests they do), they are likely to provide a mechanism for very fine control of muscle contraction and, therefore, fine control of eye movements.

Muscles that have only twitch fibers can increase muscle contraction only by recruiting progressively more and more fibers to aid in the contraction. One can imagine the contractile force increasing in small steps, each step representing the contraction of another group of muscle fibers that has been recruited. Tonic fibers, on the other hand, allow the contraction to be graded not only by recruitment of more fibers but also by grading of the contraction of the fibers themselves. Contraction can be graded in a continuous fashion, such that the size of the steps becomes extremely small and the number of steps becomes very large. This mode of contraction suggests that thin fibers are particularly suitable for small, slow eye movements, while thick fibers, with their fast, twitchlike contractions, would be most useful for saccades.

Different muscle fiber types are not randomly distributed within the muscles

Thick and thin fibers are segregated to a considerable degree. Cross sections of the rectus muscles (Figure 4.16) show a core of thick fibers lying close to the eye (in the global part of the muscle) surrounded by a sheath of thin fibers on the outer (orbital) surface of the muscle. In the obliques, the sheath of thin fibers completely surrounds the core of thick fibers. The boundary between thick fiber core and thin fiber sheath is not absolute; some thin fibers are found in the global portion of the muscle, and some thick fibers are found in the orbital portion. Even so, the majority of fibers in the two groups are segregated from each other.

It is not clear why the muscle fibers should be separated in this way. One possibility is that the separation represents nothing more than a developmental gra-

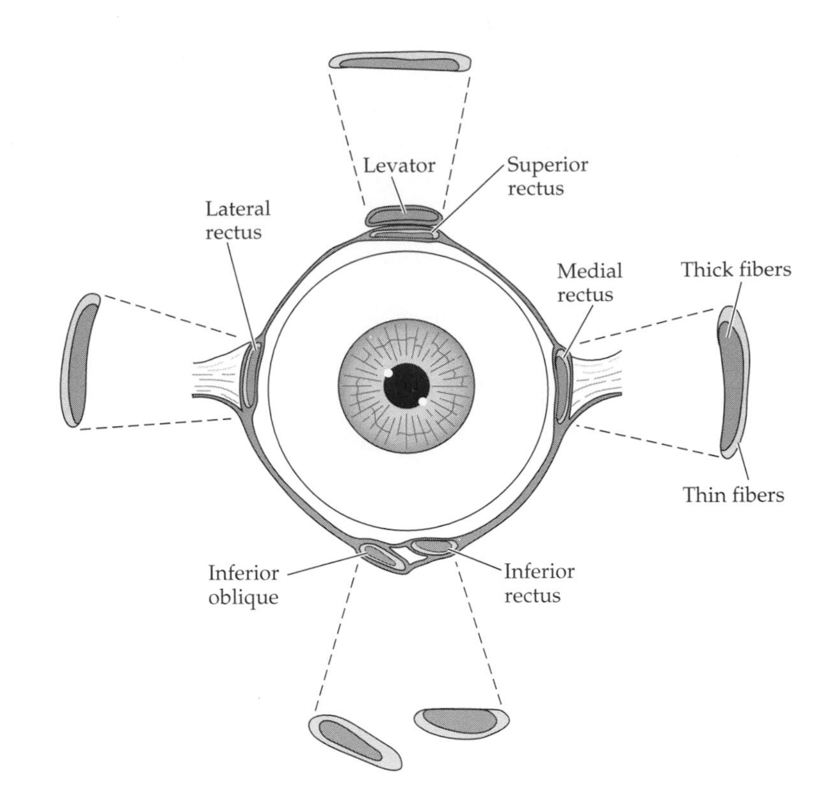

Figure 4.16
Distribution of Thick and Thin Fibers in the Extraocular Muscles
In the rectus muscles, shown here cut in cross section near the equator of the eye, thick fibers dominate in a region of the muscle closest to the globe (shaded). The outer parts of the muscles are dominated by thin fibers. The inferior oblique, shown below the inferior rectus, has an inner core of thick fibers (shaded) completely surrounded by a sheath of thin fibers; the belly of the superior oblique (not shown) is similar. Since the levator has no thin fibers, it has no subdivision. The dividing lines between thin fiber and thick fiber regions are not as sharply defined as they are in the illustration.

dient, wherein thick fibers form first, generating the muscle core, and the thin fibers are added later outside the core of thick fibers (see the section on development at the end of this chapter). But if development explains how the fibers come to be distributed as they do, it says nothing about why the thin fibers should be present in the first place. We can only speculate, but the presence and distribution of the thin fibers may contribute to the smooth, continuous development of force in a slowly contracting extraocular muscle.

Different muscle fiber types may receive different innervational commands

The existence of different muscle fiber groups in extraocular muscle has been known for many years, and there has been much speculation about how the unusual structural properties of the extraocular muscles might be related to the functionally different types of eye movements. An earlier hypothesis, for example, associated thick fibers with version movements and thin fibers with vergences. For a variety of reasons, this conjecture no longer seems tenable, and a more likely association—if one exists—is that of thin fibers with slow movements and of thick fibers with saccades.

This idea has some intuitive appeal, and the preceding discussion has already hinted at it, but some experimental results pose considerable difficulties. Recordings of neural activity from single oculomotor neurons in alert, normally behaving monkeys, for example, indicate that *all* extraocular motor neurons are active during *all* types of eye movement. In other words, a given motor neuron will be active whenever the eye moves into its field of action, regardless of whether the eye movement is a version or a vergence, fast or slow. If the signals transmitted via the motor neurons are not separated by type of eye movement, there can be no assignment of specific eye movement commands to specific groups of extraocular muscle fibers. At face value, the experimental results mean that the unusual structural features of the extraocular muscles have no special functional significance.

But motor neurons going to a particular muscle have different thresholds of activity, meaning that some begin to fire nerve impulses for very small eye rotations, while others remain inactive until the required eye rotation is fairly large. Low-threshold neurons generally have quite gradual increases in activity as a function of eye rotation; high-threshold neurons, once activated, increase their activity at a much faster rate (Figure 4.17). This distinction between low- and high-threshold motor neurons may provide the functional dichotomy that allows us to justifiably think about different functional roles for the different classes of muscle fibers.

Specifically, if the thin tonic fibers were activated by low-threshold motor neurons and the thick twitch fibers by high-threshold motor neurons, we would have a functional subdivision between small eye movements, most of which are slow movements, via the activation of tonic fibers, and large eye movements, invariably saccades, via the activation of twitch fibers. A hint that this subdivision exists comes from studies using very fine needle electrodes to record electrical activity of muscle fibers in the global and orbital portions of human extraocular muscles during eye movements: The patterns of activity were different and were consistent with a "division of labor" in which orbital (presumably thin, tonic) fibers contributed mostly to small, slow fixational eye movements and global (presumably thick, twitch) fibers were primarily responsible for saccades.

Another factor could be the response properties of the muscle fibers themselves. For example, saccades are initiated by high-frequency bursts of neural activity—the pulse. These high-frequency bursts of activity may be quite inappropriate and ineffective when applied to tonic fibers, with their slow contractile properties, and possibly different cell membrane properties. In other words, signals meant to elicit saccades might impinge on tonic fibers, but these signals might have little or no effect. Conversely, slow changes in neural activity, of the sort required for slow eye movements, might not be effective on the presumably high-threshold twitch fibers. If so, we would have another reason for associating slow eye movements with tonic fibers and for associating saccades, especially the large ones, with the twitch fibers.

Extraocular muscles have very small motor units

The extraocular muscles contain between 15,000 and 35,000 muscle fibers each, most of which are thought to run the entire length of the muscle, and these fibers

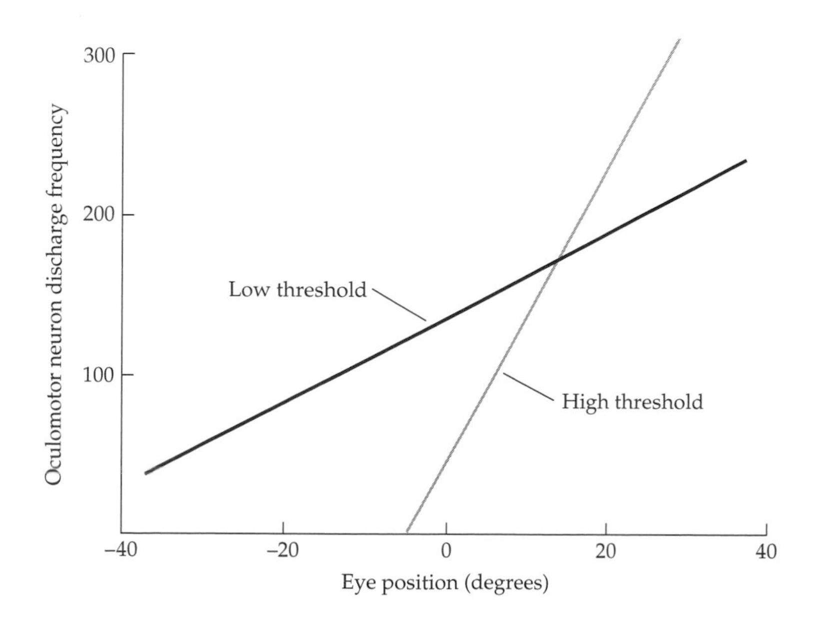

Figure 4.17
Activation Thresholds and Firing Rate Increases in Extraocular Motor Neurons as a Function of Eye Position The graph shows action potential frequency as a function of eye positions progressing into motor neuron fields of action. The activity of the low-threshold motor neuron is associated with a very small change in eye position; once activated, the firing increases slowly with increasing eye position. The high-threshold neuron does not begin to fire until the eye is well into the muscle's field of action, but its firing increases rapidly with additional change of eye position. (After Robinson and Keller 1972.)

are innervated by several thousand motor nerve axons (Table 4.1). The number of muscle fibers is always larger than the number of nerve axons, and their ratio (number of muscle fibers/number of motor nerve axons) gives the average size of a **motor unit**, which Charles Sherrington (see Vignette 5.1) defined as the number of muscle fibers innervated by a single nerve axon.

This definition assumes that a muscle fiber receives input from only one axon, which is certainly true for the singly innervated twitch fibers, and that all the muscle fibers in a motor unit have the same threshold of activation by the nerve impulses arriving from the motor nerve axon—this is less certain, but a reasonable assumption. The importance of the motor unit is that it represents the smallest functional element of muscle contraction for the twitch fibers in that it corresponds to the activation of a single nerve axon. A muscle with many small motor units should exhibit more finely graded contraction than a muscle with relatively few, large motor units.

The numbers in Table 4.1 are the sort used to calculate the size of extraocular muscle motor units, but note that the numbers come from different sources and they are not very consistent; depending on the numbers one chooses, the calculated motor unit sizes vary significantly. The numbers of motor nerve axons are suspect in part because the motor nerves are carrying sensory axons from sensory organs in the muscles and, in the oculomotor nerve, around 1500 parasympathetic axons destined for the ciliary ganglion (see Chapter 5). Including these nonmotor axons in the numbers used to calculate motor unit size will make the motor units seem smaller than they really are. The best numbers are probably those for the lateral rectus and abducens nerve and the superior oblique and trochlear nerve; these numbers yield motor unit sizes between 12 and 14.

If we assume that these numbers are slight underestimates that apply to all the extraocular muscles, we arrive at a value of 15 for the motor unit size in human extraocular muscles. This value is similar to the numbers obtained by more direct methods in other species, and whatever the true motor unit size is, it is certainly quite small in comparison to motor units in the skeletal muscles, which rarely have fewer than 100 muscle fibers per motor unit and may contain several thousand.

With an average extraocular motor unit size of 15, the calculated *number of motor units* in the extraocular muscles is quite large, ranging from around 100 in the obliques to over 200 in the lateral rectus. This number implies that there can be at least 100 small steps of contraction between the resting length of a muscle and its fully contracted length. If we assume, generously, that a muscle can contract to half its resting length (which is around 40 mm for most of the muscles), the contraction steps will be *no larger* than 0.2 mm, which corresponds to about

Table 4.1

Muscle Fibers, Nerve Fibers, and Motor Units

Muscle	Number of muscle fibers	Number of nerve fibers	Size of motor unit
Medial rectus	29,000		7.7
Lateral rectus	35,000	2000–2500 (VI)	14–17.5
Superior rectus	20,000		5.3
Inferior rectus	26,000		6.9
Superior oblique	15,000	1100–1200 (IV)	12–13.6
Inferior oblique	18,000		4.8

Source: Eggers 1993 (numbers rounded up to the nearest thousand); original numbers from various sources.

1° of eye rotation. In short, the small motor unit size in extraocular muscles is consistent with the small, relatively constant load they operate against and the very small contractions they often make.

Acetylcholine at the neuromuscular junctions depolarizes the cell membrane by opening sodium channels

All extraocular muscle fibers, whether thick or thin, receive inputs from motor nerve axons that use **acetylcholine** as the neurotransmitter. The action of the neurotransmitter at the conventional motor end plate on the thick fibers is summarized in Figure 4.18. Action potentials arriving at the nerve terminal depolarize it, and this depolarization opens Ca^{2+} channels in the terminal membrane, allowing Ca^{2+} to flow into the terminal. The increase in Ca^{2+} causes synaptic vesicles to fuse with the terminal membrane at the synaptic cleft and to release acetylcholine into the cleft. Acetylcholine acts on the muscle fiber by binding to acetylcholine receptors in the sarcolemma. The structure of the receptors is such that only acetylcholine and a few other molecules of similar configuration will fit the receptor site well enough to bind with it.

Acetylcholine receptors are associated with membrane **channels** that regulate the movement of sodium ions (Na^+) across the sarcolemma. (They are ligand-gated channels, and acetylcholine is the ligand.) Binding of acetylcholine to the receptor sites opens the Na^+ channels at the neuromuscular junction, permitting a large influx of Na^+, which depolarizes the membrane. The local depolarization spreads rapidly across the entire membrane because other Na^+ channels are voltage sensitive; as one region depolarizes, the depolarization causes neighboring channels to open, and the depolarization quickly propagates. The process is quickly terminated, however; the binding of acetylcholine at the receptor sites is transient, and free acetylcholine within the synaptic cleft is broken down by acetylcholinesterase (a degradative enzyme) into acetate and choline. The choline is recycled by a membrane transporter that returns it to the nerve terminal.

The spread of depolarization along the sarcolemma may differ among muscle fiber types, producing different contractile properties

The action of acetylcholine will be the same at the receptor sites of twitch fibers that have large motor end plates and on those of thin fibers where acetylcholine is released from the small en grappe terminals. Studies on tonic muscle fibers in other species, however, indicate that the sarcolemma of the tonic fibers responds differently, possibly because the voltage-sensitive Na^+ channels are not widely distributed across the membrane. The low density of voltage-sensitive channels implies that only a very strong stimulus will depolarize the entire membrane and elicit a complete contraction of the muscle fiber. Furthermore, this low density suggests that graded neural inputs, producing different amounts of acetylcholine release, will produce different degrees of membrane depolarization and, therefore, different amounts of contraction. But the membrane properties of human extraocular tonic fibers are not known.

We do know, however, that the effects of pharmacological agents on the extraocular muscles are consistent with the assumption that the thin muscle fibers have tonic contraction properties. For example, agents that mimic the effect of acetylcholine will produce very sustained contraction of the extraocular muscles, but do not have the same effect in skeletal muscles. The difference is attributed to the presence of thin fibers in extraocular muscles and the absence of thin fibers in skeletal muscle.

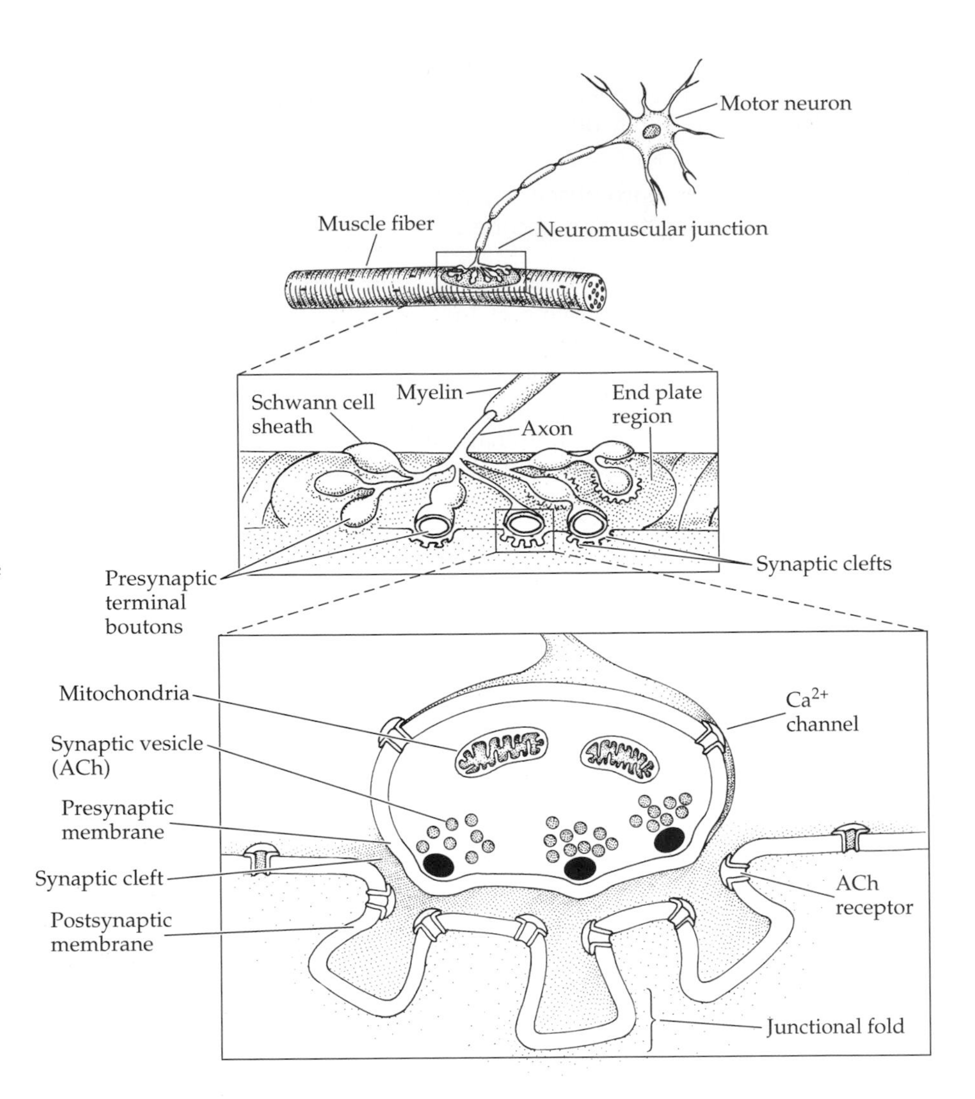

Figure 4.18

Activation of a Muscle Fiber by Release of Acetylcholine from a Motor Neuron Nerve Terminal

Action potentials transmitted down the axon depolarize the terminal, permitting an influx of Ca^{2+} through voltage-gated Ca^{2+} channels. Calcium promotes the mobilization and fusion of acetylcholine-filled vesicles with the terminal membrane. The acetylcholine (ACh) is released into the synaptic cleft between the end plate and the sarcolemma of the muscle fiber. Acetylcholine molecules bind transiently to acetylcholine receptors in the sarcolemma, causing the receptors to change their configuration so that Na^+ passes into the cell more readily. Nearby voltage-sensitive Na^+ channels are thereby activated, and a wave of Na^+ influx sweeps across the sarcolemma. Unbound acetylcholine is broken down to acetate and choline by acetylcholinesterase in the synaptic cleft.

Extraocular muscles exhibit high sensitivity to agents that mimic or block the action of acetylcholine

The agents most commonly used to enhance the action of acetylcholine are anticholinesterases. They act at the neuromuscular junction by disabling the acetylcholinesterase that normally breaks down free, unbound acetylcholine. Removing or reducing the amount of acetylcholinesterase means that acetylcholine released from the nerve terminal will have a prolonged, continuous effect, since there is no longer anything present to destroy it and stop its action. Administration of an anticholinesterase might seem a useful therapy for a weakened muscle, and it is, but the duration of the effect is fairly short, so it cannot be used as a long-term solution to muscle weakness.

A variety of agents interfere with the normal action of acetylcholine; most of them act by competing with acetylcholine for receptor sites—curare is a classic example. As long as the competitor occupies the receptor sites, the acetylcholine cannot interact with the muscle fiber membrane and therefore cannot induce normal muscle contraction. Some of these competitive blockers are considered to be neurotoxins, such as those found in snake venom, and they may have very prolonged effects at the neuromuscular junctions.

The effect of acetylcholine can also be reduced by interfering with its release from the nerve terminal. Agents that reduce the Ca^{2+}-mediated mobilization and membrane fusion of synaptic vesicles will limit the amount of acetylcholine that is released. Botulinum A toxin, which will be discussed shortly, operates in this fashion.

Extraocular muscles often exhibit early symptoms of myasthenia gravis

Because of their high sensitivity to the effects of the naturally occurring neurotransmitter and its degradative enzymes, the extraocular muscles and the levator muscle of the upper eyelid (see Chapter 7) are often the first striated muscles to show the effects of **myasthenia gravis**, an autoimmune disease in which the immune system makes antibodies that destroy the acetylcholine receptors. The early symptoms are drooping upper eyelids (**ptosis**), strabismus, which is sensed by the patient as double vision (**diplopia**), and abnormal eye movements.

Ptosis and eye position abnormalities can occur for numerous reasons; part of the differential diagnosis of myasthenia gravis involves the administration of anticholinesterases. In response to the anticholinesterase and the prolonged action of the acetylcholine at whatever receptor sites remain, the ptosis will be reduced or eliminated and the eye movement patterns will return to normal. These effects are short-lived, but sufficient to establish that the problem is at the neuromuscular junction. If the ptosis and altered eye movements were due to other problems, either with the muscles themselves or with the innervational signals to the muscles, the anticholinesterase would not alleviate the symptoms.

There is more to the diagnosis of myasthenia gravis than this, but the ocular response to anticholinesterases is one of the more critical observations. Unfortunately, there is no cure or effective long-term symptomatic relief, and the disease may progress to involve many other muscles of the body.

Neurotoxins that interfere with acetylcholine action can be used to alleviate strabismus and blepharospasm

Most types of extraocular muscle surgery are irreversible, and since obtaining a desired result is not particularly straightforward (see Box 4.2), reversible, nonsurgical alternatives are always worth exploring. One approach uses injections of a bacterial toxin, botulinum A toxin, into the extraocular muscles. As mentioned earlier, botulinum toxin interferes with mobilization and membrane fusion of synaptic vesicles in the motor nerve terminals and thereby reduces the amount of acetylcholine that would normally be released by a given level of terminal depolarization. (Synaptic vesicle fusion is mediated by Ca^{2+}, and the botulinum toxin blocks the normal action of Ca^{2+}. The mechanism is not fully understood, but the blockage is intracellular and not a blockage of the Ca^{2+} channels in the terminal membrane.) In an esotropia, for example, injecting the toxin into the medial rectus reduces the adduction produced by tonic innervation of the muscle, thereby allowing the lateral rectus to rotate the eye outward into alignment with the fellow eye.

This procedure may seem risky, but the effects of the agent are quite localized, and any toxin not readily incorporated into nerve terminals is quickly broken down. The toxin is quite stable, therefore long-lasting in effect, and the change in muscle action it produces may last months. This method has the advantage that it is reversible with the passage of time, and the effect can be graded over several injections to achieve the desired result. At present, botulinum toxin is most often used to alleviate strabismus in children, in part because the effect is reversible and also because the procedure does not require a general anesthetic.

Although its original use was in strabismus treatment, botulinum toxin is probably more commonly used in the treatment of **blepharospasm**, a chronic spasm of the orbicularis muscle in the eyelids (see Chapter 7) that produces

Vignette 4.1

Locating the Extraocular Muscles

PEOPLE HAVE KNOWN for a long time that the eye has muscles to move it. The Greeks knew about them, and so did Galen. But what, exactly, did they know? After all, Vesalius so many years later showed only the rectus muscles and included a retractor bulbi muscle that the human eye does not possess (see Vignette 3.2).

The oblique muscles were a problem, and it is not clear that anyone prior to the mid-sixteenth century knew they existed. In retrospect, however, it is easy to see why the superior oblique might have been overlooked; the belly of the muscle is outside the rectus muscle cone, and its long, flat tendon could easily have been mistaken for and confused with Tenon's capsule and other connective tissue around the eye. Because the inferior oblique has almost no tendon, it does not pose the same difficulty as the superior oblique, but if the inferior oblique were severed from its origin in the course of dissection, it could be mistaken for part of the presumptive retractor bulbi muscle. Without advance knowledge of what to expect, the muscle arrangement is tricky.

Two of Vesalius's contemporaries figured out how many extraocular muscles there are and provided accurate descriptions of their relationships to the eye. The first was Fallopius (Gabriele Falloppio, 1523–1562), who was a student of Vesalius, but turned out to be a poor surgeon nonetheless; too many of his patients died, so he turned to medicine. He eventually assumed the chair of anatomy at Padua—the position held earlier by Vesalius—in 1551. His best-known anatomical work concerned the female reproductive system; he coined the term "vagina," and his name was given to the fallopian tubes, but he was also the first to give an accurate description of the extraocular muscles. This description required that he correct and modify Vesalius, who he greatly admired, and he did so with a kindness that was rare at the time.

The *Observationes Anatomicae,* published in 1561, is the only book that is indisputably Fallopius's own work, but it contains no illustrations. Fallopius expressed himself clearly, however, and there is no difficulty with his descriptions. In his view, there were seven muscles in the orbit (he discussed the orbicularis in the eyelids but correctly did not count it among the orbital muscles); the orbital muscles were the four recti, the two obliques, and the levator. Fallopius stated flatly that the retractor bulbi did not exist in humans, he pointed out that the levator was a muscle separate from the superior rectus (Vesalius thought they were a single muscle), and he described and named the trochlea, through which the superior oblique tendon passes. He recognized the obliques as an opposing muscle pair meant to produce "circular" (torsional) movements of the eyes. But most of these highly original observations went unrecognized until much later, perhaps because of the lack of illustrations.

Someone else did do the illustrations, however. Eustachius (Bartolommeo Eustachio, 1520–1574) was another of the sixteenth century's great anatomists—his name is given to the eustachian tubes—and was also the physician to several popes. Eustachius was a formidable scholar; he knew not only Latin and Greek, but also Hebrew and Arabic, and he read widely in all these languages. Naturally, he was well versed in Galen's anatomy, and his initial reaction to the *Fabrica* of Vesalius (see Vignette 3.2) was quite hostile. His opposition moderated in time as he realized from his own dissections how much there was in Galen to be corrected and that there was room to correct Vesalius as well. His new observations were recorded with an artist's assistance in a series of 47 copperplate engravings that were to be his lasting statement about human anatomy.

Several of these plates illustrate the eye; Figure 1 shows a plate featuring the extraocular muscles. Drawings II, III, and IV show the

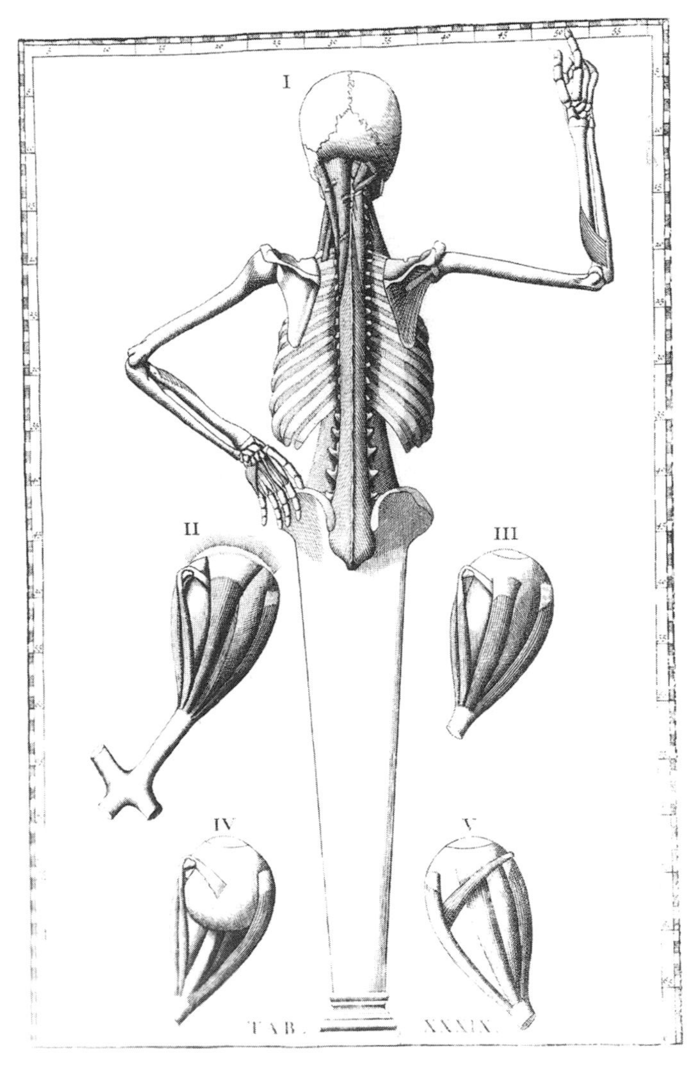

Figure 1
The Extraocular Muscles According to Eustachius
These drawings were the first to show the correct number and location of the extraocular muscles. They are on Tabula XXXIX in the *Tabulae Anatomicae Clarissimi Viri,* 1714, from which they were photographed for this illustration (Reynolds Historical Library, University of Alabama at Birmingham). (A facsimile edition [*Edition Medicina Rara*] is also available.)

eye from above, with the levator removed in III and the superior rectus removed in IV. The superior oblique, complete with tendon, shows clearly in drawing IV; the insertion is almost directly under the superior rectus, which is better than Fallopius's less accurate statement that the insertion is between those of the superior and lateral recti. The inferior oblique is illustrated in drawing V, from below the eye; it is shown, correctly, to the outside of the inferior rectus. These drawings are particularly noteworthy for showing the muscles as they are positioned in situ so that their effects on the eye as they contract can be appreciated. This important feature was often ignored by later anatomists and is sometimes ignored even now—a few modern anatomy texts fail to show the extraocular muscles in their true and meaningful positions.

Some years ago, I spent part of an afternoon wandering along a short stretch of the Via Flaminia, one of the ancient Roman roads, looking at the stones rutted by chariot and wagon wheels, musing about the Caesars, Roman legions, and invaders from the north who followed this path to victory or defeat. Four hundred years before me, the aging, ill Eustachius had also followed the Via Flaminia, responding to a summons from his pope that he could not refuse; his journey ended when he died beside the road. At his death, the illustration shown here and the rest of Eustachius's anatomical plates had not been published and they were lost for a century and a half; they finally appeared in 1714 as the *Tabulae Anatomicae Clarissimi Viri*, published under the auspices of Clement XI. His service to the papacy was finally rewarded.

forcible, involuntary lid closure. Reducing the amount of acetylcholine released by the motor nerve terminals reduces the state of orbicularis contraction.

Sensory Endings in Extraocular Muscles and Tendons

Skeletal muscles have two major types of sensory organs

Since the motor commands sent to a given muscle must vary depending on the muscle's current state of contraction, the neural systems that coordinate and regulate muscle actions require continually updated information about the status of the muscles. For the body's skeletal muscles, most of this information is provided by two sets of sensory receptors: **muscle spindles**, which report muscle length, and **Golgi tendon organs**, which report muscle tension. The role of these sensory receptors in regulating muscle contraction is quite important and well understood for skeletal muscles, but the extraocular muscles are a different story. For a long time, muscle spindles and tendon organs seemed to be absent from extraocular muscles. Then they were observed in some species, but thought to be absent in humans. Now they are known to be present in human extraocular muscles and their tendons, but there are still questions about why they are there and what they do.

Golgi tendon organs in skeletal muscle tendons are enclosed in a long, slender capsule within which are braided, coiled strands of collagen fibers (Figure 4.19*a*); the collagen coils are continuous with a few strands of muscle fiber that enter one end of the capsule. The sensory nerve fibers have highly branched terminal arrays that interweave extensively with the collagen coils. When the muscle is under tension, the collagen coils are stretched, putting mechanical pressure on the nerve fiber terminal branches within the coils. This pressure elicits nerve impulses in the nerve fibers. As the muscle develops more tension, there is more stretching within the tendon organ, and a higher frequency of action potentials transmitted centrally by the sensory nerve fibers (group Ib afferents), which typically contact inhibitory interneurons. Tendon organs are most sensitive to tension in the muscle that is due to muscle contraction.

Muscle spindles are also enclosed in a capsule (Figure 4.19*b*), but they are located within the muscles, not the tendons. The capsule contains around ten muscle fibers of at least two kinds: "Bag" fibers have nuclei aggregated in the central part (the nuclear bag) of the spindle; "chain" fibers have nuclei dispersed throughout their length. Sensory fibers of one class (group Ia afferents) have terminal branches that wrap around the central region of both bag and chain fibers; those of another class (group II afferents) wrap primarily around the more peripheral part of the chain fibers.

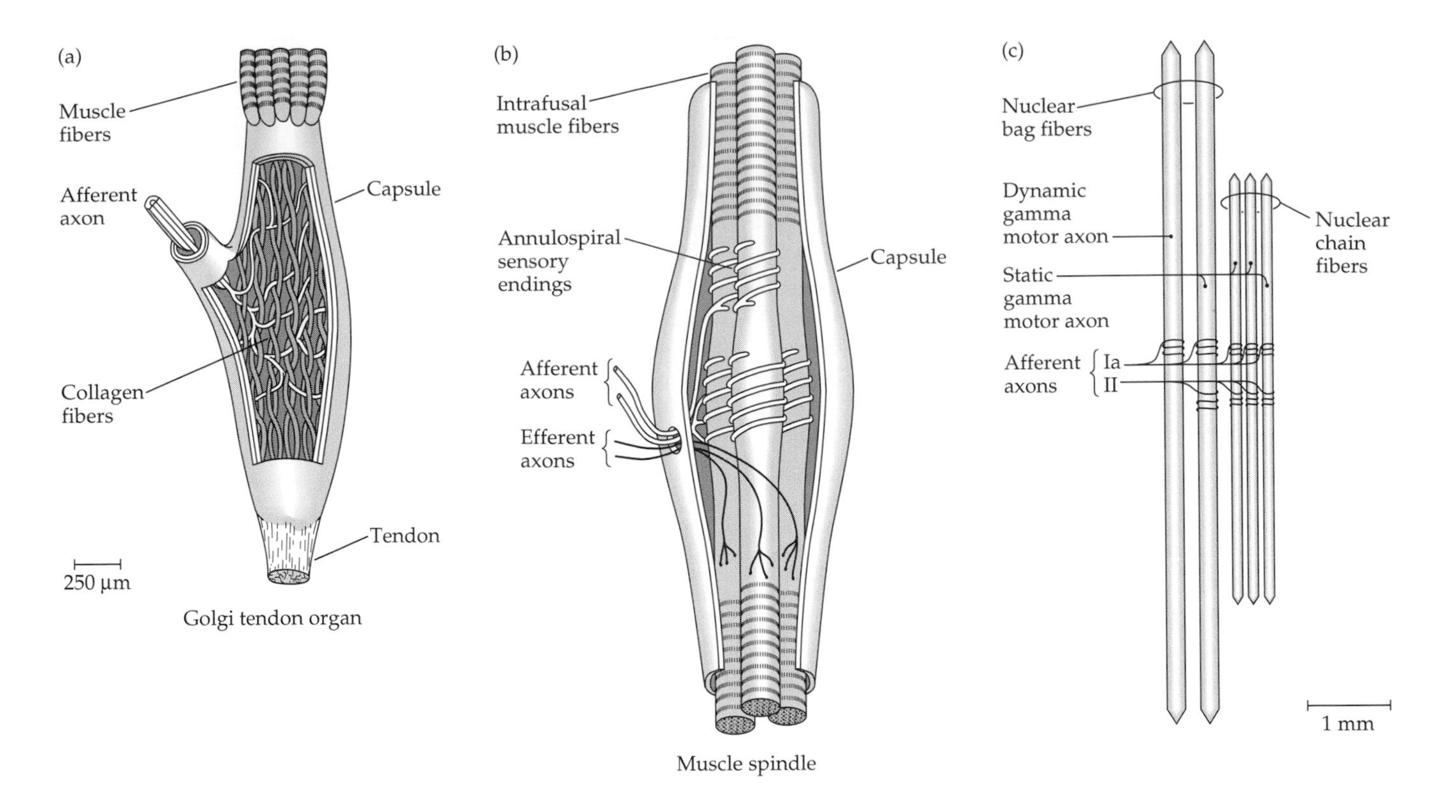

Figure 4.19
Structure of Muscles Spindles and Golgi Tendon Organs
(*a*) Golgi tendon organs consist of a capsule attached to a group of muscle fibers at one end and collagen of a tendon at the other. Inside the capsule, branches of a sensory (afferent) axon weave in and out among strands of collagen from the tendon. Contraction of the muscle fibers stretches the collagen in the capsule and elicits nerve impulses from the sensory fibers. (*b*) Muscle spindles contain specialized muscle fibers that are bundled together within a capsule. Two types of sensory nerve fibers that spiral around the muscle fibers are activated when the muscle fibers are stretched. (*c*) There are also two types of gamma-efferent motor nerves to the muscle fibers in the spindle. (After Purves et al. 1997.)

Both the bag and the chain muscle fibers receive inputs from motor nerve axons that originate centrally (the gamma efferents) (see Figure 4.19*c*). The generation of nerve impulses in the sensory nerve fibers requires stretching of the muscle fibers within the spindle (the muscle fibers of the spindle are **intrafusal**, while those in the muscle are **extrafusal**). This stretch may arise either from stretching of the muscle by an increased load or from contraction of the intrafusal fibers as a result of input from the gamma-efferent neurons. The sensory afferents from the muscle spindles make excitatory contacts to the primary (alpha) motor neurons going to the extrafusal muscle fibers.

Human extraocular muscles have anatomically degenerate sensory organs and exhibit no stretch reflexes

A simple way to stretch the intrafusal fibers is to stretch the entire muscle passively. When the muscle is stretched, the sensory nerve fibers in the spindles will be activated and their signals will elicit activity in the alpha motor neurons innervating the muscle's extrafusal fibers. Thus, the muscle will contract. This is a stretch reflex, and the most common example is the reflex contraction of the quadriceps muscle elicited by a sharp tap on the patellar tendon. (This stimulus will also produce activity from the sensory fibers in the tendon organs, but it is considerably less than that from the spindle afferents.) There is more to the muscle spindle story than this, but the association of muscle spindles and stretch reflexes is the key point here.

Since human extraocular muscles were once thought to possess no muscle spindles, the absence of an extraocular muscle stretch reflex would not be surprising. And when the same procedures used to elicit stretch reflexes from skeletal muscles are used with human extraocular muscles in the course of strabismus surgery, no stretch reflexes can be elicited. But muscle spindles *are* present in the extraocular muscles. Relative to the organs in skeletal muscles, they are not numerous, ranging from about 4 or 5 in the inferior oblique to around 35 in the inferior rectus. They are also structurally degenerate; that is, the spindles are small, their capsules are not well defined, there are fewer intrafusal fibers than skeletal muscle spindles, and the differentiation of bag and chain fibers is less obvious. They do have both sensory and motor nerve axons, however.

Reports on tendon organs associated with human extraocular muscles are sketchy, but they seem to be present—they certainly are in other primates—and they are anatomically degenerate in much the same ways as the muscle spindles are.

The obvious question concerns the role of these sensory organs, the muscle spindles in particular. Since the muscles don't have stretch reflexes, what are the spindles doing? Part of the answer may be that extraocular muscles have no need for stretch reflexes; unlike their skeletal counterparts, extraocular muscles do not experience large, sudden changes of load in the course of their normal operation, and there is no obvious reason for having a mechanism to compensate for such loading changes. This lack of changing load could explain the degeneracy of the muscle spindles in both number and structure; spindles are present because they are a natural constituent of striated muscles, but since their major function is unnecessary, there is no pressure to retain the system to its full extent. A similar argument holds for tendon organs, which are normally affected by increased muscle contraction after external loading; when the load never changes, there is not much need for tendon organs.

Passive extraocular muscle stretch may produce bradycardia

Although extraocular muscles do not have a stretch reflex, passive stretching of extraocular muscles may have an interesting and disturbing result. In some individuals, extraocular muscle stretch slows the heartbeat, and in some cases, the heart may stop completely. This phenomenon, well known to strabismus surgeons, is the **oculocardiac reflex**.

The anatomical pathway for this reflex is not known; some authors attribute the oculocardiac reflex not only to the extraocular muscles but also to sensory nerves within the eye. The fact that the reflex can be elicited by pulling on a detached extraocular muscle, however, suggests that sensory organs in muscle are primarily responsible. The oculocardiac reflex has no obvious functional role, and it probably represents aberrant neural wiring in the brainstem between the part of the trigeminal nucleus to which the muscle spindle afferents are going and the nucleus of the vagus nerve, whose axons innervate the heart. The miswiring presumably is a developmental anomaly in some individuals.

Sensory endings in extraocular muscles probably do not convey information about eye position

However degenerate in structure and number the sensory organs in the extraocular muscles may be, they presumably do something, perhaps serving a role that is normal in skeletal muscle, but of minor importance. One possibility, much studied, discussed, and debated, concerns proprioceptive feedback from the extraocular muscle spindles—that is, information from the muscle's sensory organs about the amount of eye rotation and the position of the eye within the orbit. The analogy with the skeletal system is that in addition to the reflex interactions between sensory organs and muscles, we derive information about limb position from receptors in the muscles and joints. Perhaps something similar accounts for eye position information.

Experiments designed to test for the presence of proprioceptive information from the extraocular muscles show some effect on perception of visual direction, such as in past-pointing, and some influence on ocular alignment. At present, the evidence suggests that the contribution of spindles or tendon organs to ocular proprioception is both subtle and modest.

Sensory signals from the extraocular muscles may be involved in motor learning, motor plasticity, and development

In other species, it has been shown that sensory signals from the extraocular muscles reach the cerebellum, which is involved in motor learning and plasticity. Therefore, one possibility is that sensory information about extraocular muscle length and tension is used to adapt motor innervation patterns to changing circumstances. Such a role could be particularly important during the period of growth and development when the sizes and masses of the eye and the extraocular muscles are slowly changing. In addition, the planes of action of the extraocular muscles are highly correlated with the planes of the semicircular canals in the vestibular apparatus; it is possible, but again untested, that this correlation is established during development by the interaction of signals from the muscles and the semicircular canals in the cerebellum.

Thus, the main contributions of the sensory organs in the extraocular muscles may be made as the oculomotor system develops and grows. In adults, therefore, the extraocular muscle spindles and tendon organs may have little or nothing to do with the operation of the motor system. It would be interesting to know if their anatomical degeneracy were related to age.

Actions of the Extraocular Muscles

All of the extraocular muscles except the inferior oblique have their anatomical origins at the apex of the orbit

A roughly elliptical ring of tendon at the apex of the orbit, the **common tendon of Zinn**, encloses the optic foramen and crosses part of the superior orbital fissure (Figure 4.20); the common tendon is the shared origin of the four rectus muscles, each of which attach to a different part of it.

The origins of the superior and inferior recti are on the superior and inferior parts, respectively, of the common tendon lateral to the optic foramen, while the medial rectus originates on the medial side of the foramen, and the lateral rectus originates on the lateral part of the tendon that bridges the superior orbital fissure. All nerves to the rectus muscles and to the eye (two divisions of the oculo-

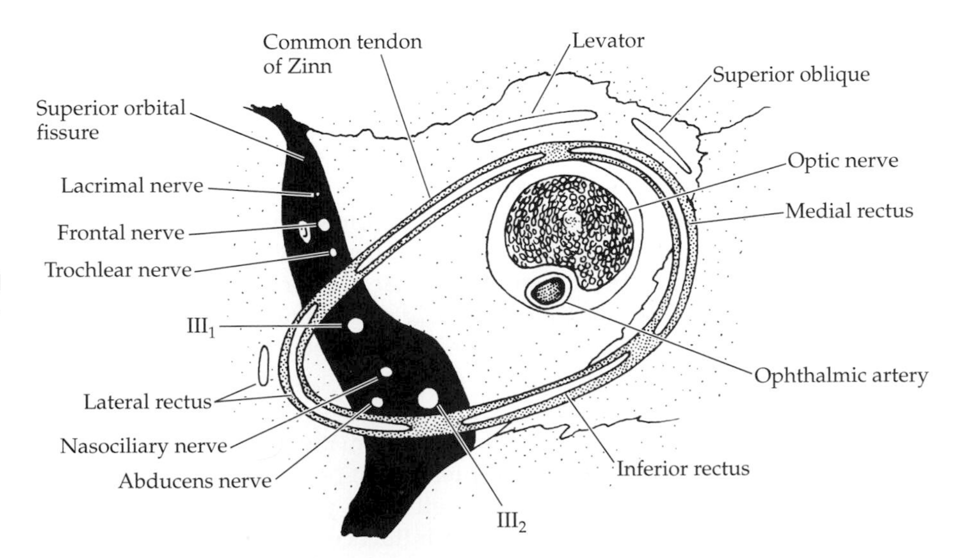

Figure 4.20
Anatomical Origins of the Extraocular Muscles at and around the Common Tendon of Zinn
The common tendon of Zinn is an elliptical band of tendon that encloses the optic foramen and bridges the lower portion of the superior orbital fissure; the four rectus muscles arise directly from the tendon in the locations shown. The lateral rectus normally has two sites of origin, a major one on the common tendon and a minor one just lateral to it. The origin of the medial rectus is higher than the origin of the lateral rectus, and the origin of the inferior rectus is slightly medial to that of the superior rectus. The levator and the superior oblique originate from the sphenoid bone just above (levator) and medial to (superior oblique) the optic foramen.

motor nerve [III_1 and III_2], the abducens nerve, and the nasociliary nerve) pass through the superior orbital fissure within the the common tendon. The origin of the superior oblique is above and medial to the optic foramen on the sphenoid bone just outside the common tendon, and the levator muscle going to the upper lid originates directly above the optic foramen, again on the sphenoid bone.

The anatomical origin of the inferior oblique and the functional origin of the superior oblique are anterior and medial in the orbit

As Figure 4.21 shows, the inferior oblique has its origin near the floor of the orbit on the anterior part of the medial wall, specifically on the orbital plate of the maxilla, just below the nasolacrimal fossa. This point of attachment is almost directly beneath the trochlea on the frontal bone, which is the *functional* origin of the superior oblique.

The arrangement of the superior oblique is unusual, and it is the only one of the extraocular muscles whose anatomical and functional origins are obviously in different places. From its anatomical origin at the orbital apex, the superior oblique runs forward along the upper medial side of the orbit. Before reaching the trochlea, the muscular portion ends, and the extension of the muscle is a long tendon that runs through the trochlea and turns out and back to run to its insertion on the eye. When the muscle contracts, the portion of the muscle between the orbital apex and the trochlea is what shortens. This contraction, however, pulls on the tendon running through the trochlea and thus pulls on the eye. The force exerted on the eye is directed along a line between the tendon's insertion on the eye and the trochlea. Thus the trochlea is the functional origin of the superior oblique.

The functional and anatomical origins are identical for the inferior oblique, whose course from origin to insertion on the eye is roughly parallel to the path of the superior oblique tendon running from the trochlea to the eye. The approximate parallelism of the paths of the inferior oblique muscle and the superior oblique tendon make them an agonist–antagonist pair.

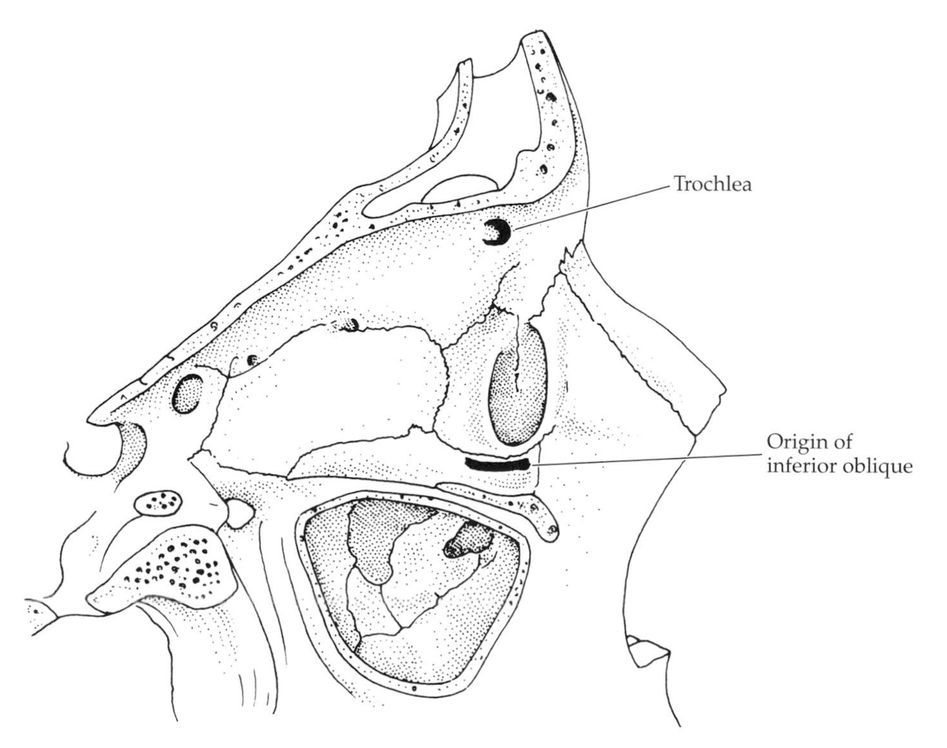

Figure 4.21
The Functional Origin of the Superior Oblique (the Trochlea) and the Anatomical Origin of the Inferior Oblique
This view of the medial wall of the orbit shows the anatomical origin of the inferior oblique on the orbital plate of the maxilla just lateral to the nasolacrimal fossa. This site lies directly below the trochlea, through which the tendon of the superior oblique passes, on the anterior medial edge of the orbital portion of the frontal bone.

The four rectus muscles are arranged as horizontal and vertical pairs, all inserting onto the anterior portion of the globe

All extraocular muscles have tendons, and the insertions on the eye are the lines along which the collagen fibers of the tendons fuse and interweave with the collagen of the sclera. From the front (Figure 4.22), only the insertions of the rectus muscles are visible; the insertions of the superior and inferior recti are directly above and below the cornea, respectively, centered on the sagittal plane, and those of the medial and lateral recti are medial and lateral to the cornea, respectively, centered on the horizontal plane.

The average distances of the rectus muscle insertions from the cornea margin vary. The medial rectus insertion is closest (5.5 mm), followed in sequence by the inferior rectus (6.7 mm), the lateral rectus (6.9 mm), and the most distant, the superior rectus (7.3 mm). The distance between the insertion and the cornea for any particular muscle normally varies among individuals by ±0.5 mm.

Beginning with the medial rectus and proceeding clockwise, an imaginary line joining the insertions spirals outward; this line is called the **insertion spiral of Tillaux**. Despite its undisputed existence, however, the insertion spiral has no known physiological significance.

The horizontal recti rotate the eye in the horizontal plane around a vertical axis

When the eye is in primary position, the anatomical arrangement of the horizontal recti is very simple and the primary actions of the medial and lateral recti are easily deduced. As Figure 4.23 shows, the lines joining the midpoints of the muscle origins and insertions are nearly horizontal and are very close to the horizontal plane through the eye. Actually, the origin of the lateral rectus is slightly lower than the medial rectus origin (see Figure 4.20) and the insertion is slightly higher (see Figure 4.22); these differences are illustrated in Figure 4.23. Thus the muscles are not precisely coplanar and they cannot be perfectly antagonistic, but these small differences are unlikely to be significant. Contraction of either muscle produces a force directed posteriorly in the horizontal plane along the line

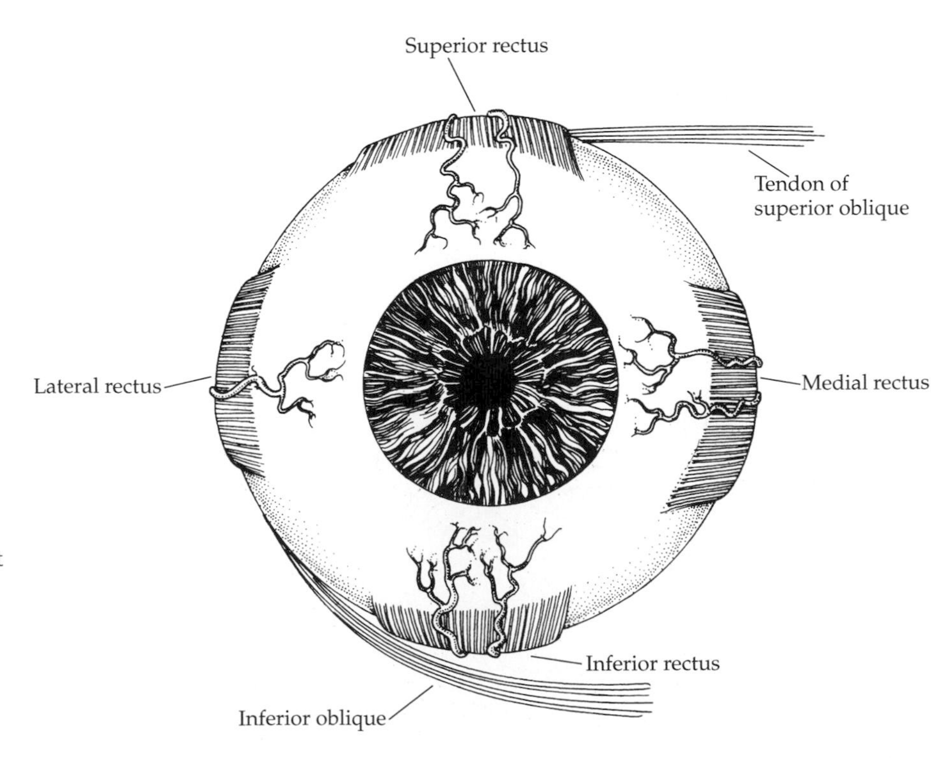

Figure 4.22
Insertions of the Rectus Muscles on the Eye
The medial rectus insertion is closest to the corneal margin, and the superior rectus insertion is farthest away. The midpoint of the lateral rectus insertion is slightly higher than that of the medial rectus, and the superior rectus insertion is slightly more temporal than the inferior rectus insertion. The distances from the insertions to the corneal margin vary among individuals by ±0.5 mm.

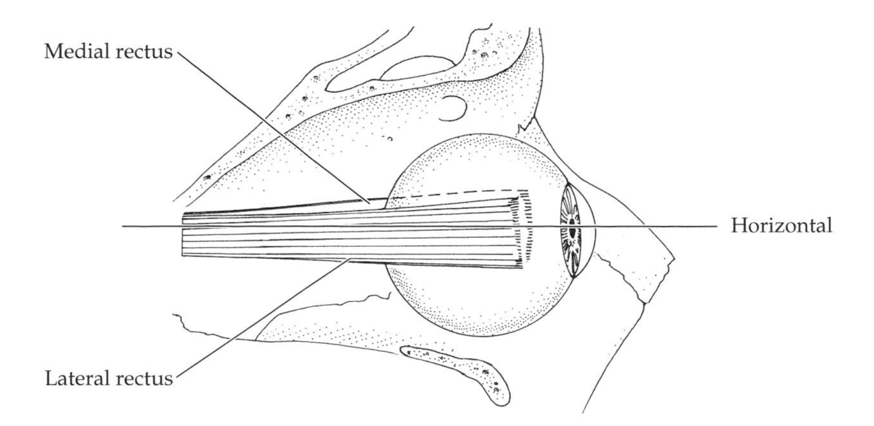

Figure 4.23
The Recti Viewed from the Lateral Side
The paths followed by the horizontal recti from their origins to their insertions are shown here in relation to the horizontal plane of the eye. The medial rectus is almost perfectly bisected by the horizontal plane, but the lateral rectus is lower at its origin than at its insertion. The superior rectus runs up from its origin to the superior part of the globe, and the inferior rectus runs down to its inferior point of insertion.

between the insertion and the muscle's origin, thus producing a horizontal rotation of the eye around its *vertical* axis.

Contraction of the medial rectus rotates the eye's anterior pole toward the medial side of the orbit (adduction). Contraction of the lateral rectus rotates the eye in the opposite direction—laterally (abduction). Although these muscles are a functional pair, they are not matched in all respects. The medial rectus has an anatomical insertion closer to the cornea (already noted), its tendon is wider and shorter, and it is appreciably larger in bulk (though not in terms of the number of muscle fibers; see Table 4.1) and therefore capable of developing greater force during contraction. Since the muscles are reciprocally innervated, these differences do not come into play under normal circumstances; the lateral rectus does not have to overpower the medial rectus. When the effectiveness of muscles needs to be altered, however, these factors may be important; very often, the desired surgical outcome is achieved more easily by altering the stronger muscle of the pair, which in this case is the medial rectus.

The vertical recti are responsible for upward and downward rotations of the eye

The superior rectus runs from its origin at the orbital apex along the top of the eye to its insertion, and the inferior rectus follows a similar course below the eye, as we saw in Figure 4.5*a*. Since the superior rectus exerts force on the upper side of the eye, the eye responds by rotating upward around the horizontal axis. Contraction of the inferior oblique produces a downward rotation around the horizontal axis.

The insertions of the superior and inferior recti are vertically aligned (see Figure 4.22), and the anatomical origins are nearly in vertical alignment as well (the midpoint of the superior rectus insertion is slightly more lateral; see Figure 4.20). Thus, the two muscles follow nearly the same path to the eye, one above and one below, and form another symmetric agonist–antagonist pair. If rotations produced by contractions of the vertical recti are to be purely vertical, however, the muscles should be aligned with the vertical plane containing the eye's anterior–posterior axis, and they are not.

Viewing the paths followed by the vertical recti to the eye from above (Figure 4.24), we can see that the origins of the muscles are medial to the midline of the eye, and they run forward and laterally from their origins to their insertions. As a result, the muscle axes are not aligned with the eye's sagittal plane. This misalignment complicates matters; it means that the contractile force cannot be confined to just one of the eye's three axes of rotation as it would be if the muscles ran directly back in the sagittal plane. Instead, the orientation of the vertical recti implies some rotational components around the vertical and anterior–posterior axes as well; we will return to this problem later.

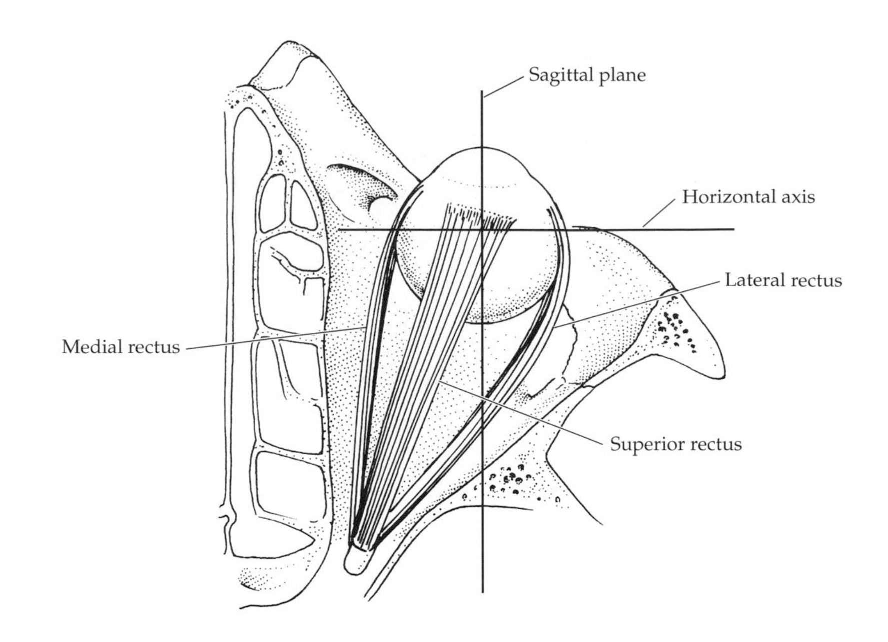

Figure 4.24
The Recti Viewed from Above
The medial rectus runs almost parallel to the medial wall of the orbit; the lateral rectus runs laterally from its origin to its insertion. The superior and inferior recti both have lateral courses from origins to insertions; they are not quite aligned, because the inferior rectus origin is slightly medial to the superior rectus insertion.

The recti define a muscle cone within the orbital cavity that contains most of the ocular blood vessels and nerves

The muscles originating at the orbital apex run forward in the orbit—the rectus muscles to the eye directly, the superior oblique indirectly by way of the trochlea, and the levator to the upper eyelid. As Figure 4.25 shows, the rectus muscles define a cone that widens toward the eye; the nerves and arteries to the eye and the rectus muscles are found within the muscle cone, having entered its apex through the optic foramen (optic nerve and ophthalmic artery) and the lower part of the superior orbital fissure (abducens, nasociliary, and oculomotor nerves; see Figure 4.20). All other nerves and blood vessels in the orbit enter and exit outside the muscle cone.

Since all the nerves to or from the eye and the extraocular muscles run within the muscle cone, the target zone for anesthetic injections (retrobulbar block; see Chapter 5) is roughly the center of the muscle cone.

The oblique muscles constitute a third functional pair, inserting onto the posterior portion of the eye

The insertions of the obliques are most easily seen when the eye is viewed from behind (Figure 4.26). The insertion of the superior oblique tendon lies, for the

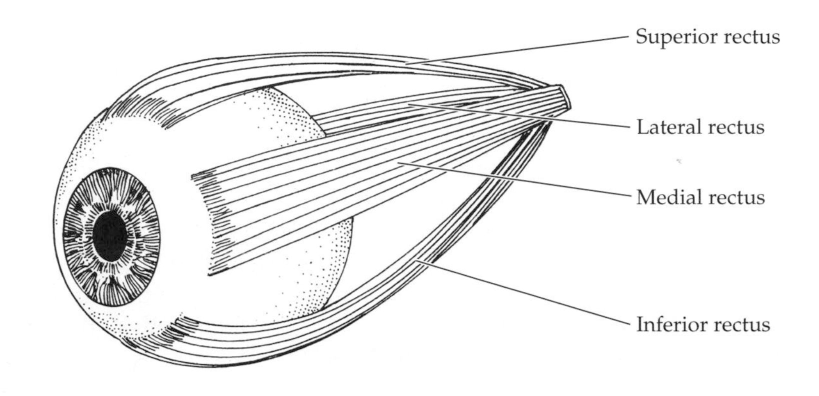

Figure 4.25
The Muscle Cone
The rectus muscles are close together at the apex of the orbit, but they become separated more as they run forward to their insertion on the eye. They define a cone with the eye as its base and the common tendon of Zinn as its apex. All nerve and blood vessels to the eye and the rectus muscles lie within the muscle cone.

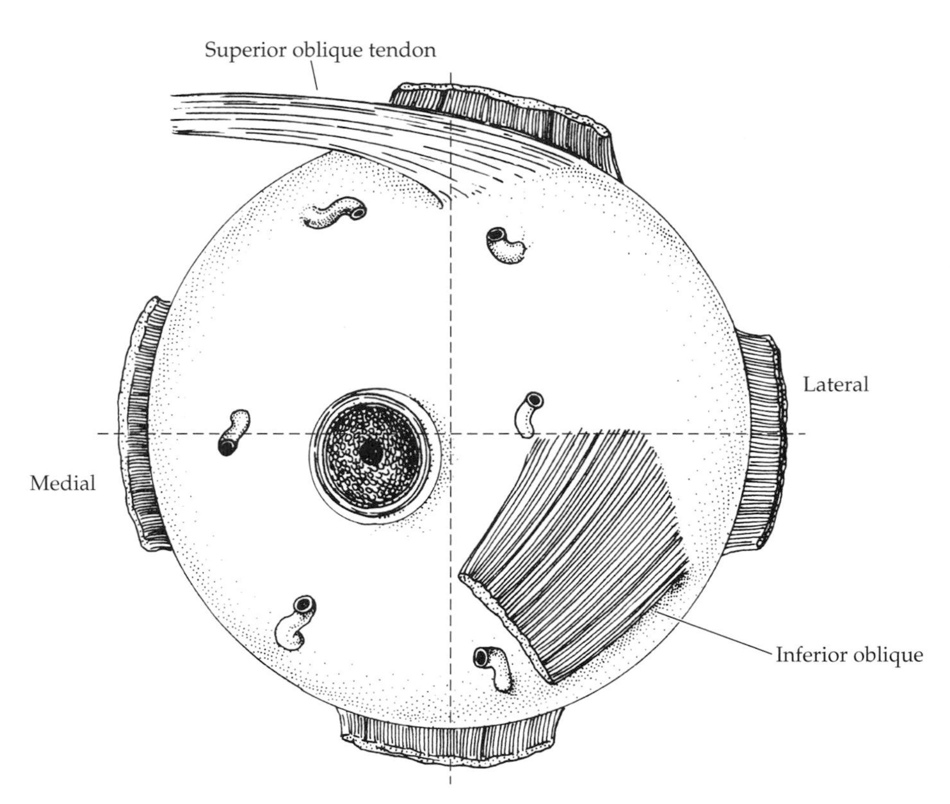

Figure 4.26
Insertions of the Oblique Muscles
The anatomical insertions of the obliques are quite asymmetric. The inferior oblique insertion is more lateral and posterior, and closer to the horizontal plane. The medial edge of the inferior oblique insertion is near the location of the fovea.

most part, within the lateral posterior quadrant on the upper half of the eye. One might expect the inferior oblique to have a symmetric insertion on the underside of the eye, but it is more laterally and posteriorly placed than the superior oblique insertion, and it is much closer to the horizontal plane.

The placement of the inferior oblique insertion makes surgical access relatively difficult, and its proximity to the fovea means that any surgery in this area requires great care. The inferior oblique tendon is also very short (1 mm). For all these reasons, extraocular muscle surgeons generally avoid cutting and reattaching the inferior oblique tendon. Note that the tendon of the superior oblique passes under the superior rectus to reach its insertion; that is, the superior oblique tendon lies between the superior rectus and the eyeball. The inferior oblique, however, approaches its insertion *below* the inferior rectus; thus the inferior rectus lies between the inferior oblique and the eye (see Figure 4.22).

Figure 4.27 shows the obliques viewed from above. Both of these muscles run laterally and posteriorly from their origins (the trochlea for the superior oblique) to their insertions on the posterior aspect of the eye. Unlike the pairs of rectus muscles, the obliques do not follow paths in the same plane; because the inferior oblique runs more posteriorly than does the superior oblique, their agonist–antagonist pairing cannot be as precise as that of the other muscle pairs.

When the obliques contract, the lines of force applied to the eye are not perpendicular to any of the eye's principal axes of rotation. Contraction of the superior oblique, for example, produces a downward rotation of the eye because the force is being applied to the superior aspect of the globe, but downward rotation is not the only effect; rotations around the other two axes can also be expected, and similar considerations apply to the inferior oblique. In fact, the actions of the obliques are sufficiently difficult to compute that authors differ in their assignments of the primary actions of these muscles, some arguing for elevation and depression, others for extorsion and intorsion. These differences of opinion reflect the fact that the detailed actions of the extraocular muscles in general, and of the obliques in particular, are very difficult to specify.

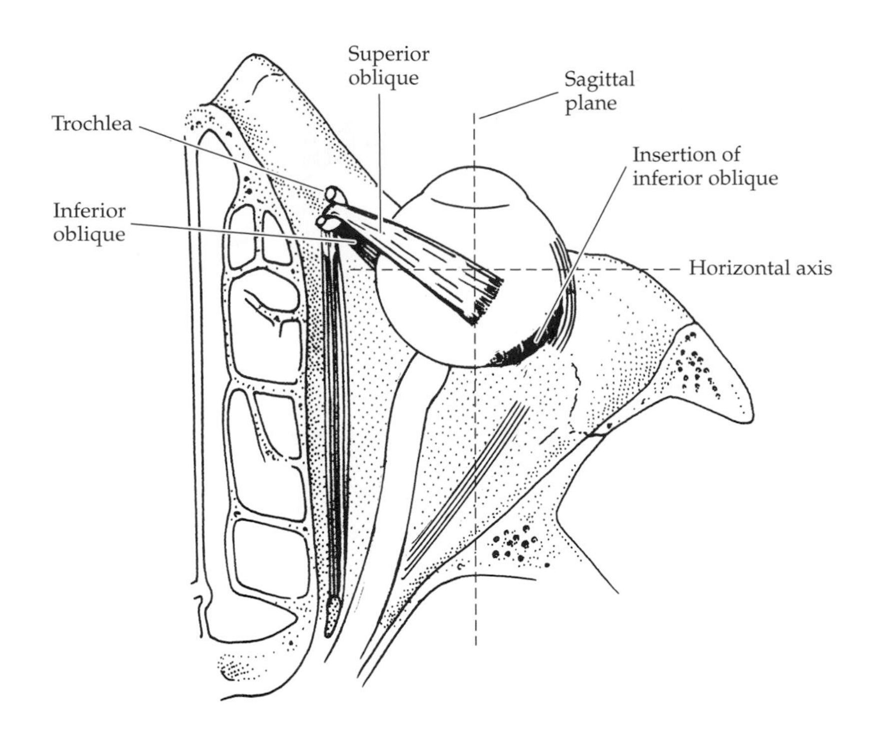

Figure 4.27
The Obliques Viewed from Above
After the tendon of the superior oblique passes through the trochlea, it runs laterally and back to its insertion on the globe. The asymmetric locations of the insertions of the obliques means they are much less aligned than the rectus muscle pairs are.

Extraocular muscle actions cannot be measured directly

In theory, the direction in which the eye rotates when an extraocular muscle contracts should be dictated by the geometry of the origin and insertion of the muscle. Since the force developed in a contracting muscle should be applied along a line between the centers of origin and insertion, and the eye rotates around a fixed center of rotation, knowing the locations of the origins and insertions in orbital space should make the problem of muscle action soluble. And once this problem has been solved, we have a basis on which to evaluate improper muscle actions and correct them.

Figuring out muscle actions by computation might seem unnecessarily indirect, but it is very difficult to approach the problem experimentally. To do so, one would need to elicit contraction of each muscle in isolation (the whole muscle, not just a part of it), accompanied by appropriate relaxation of its antagonist. This task has not been done, and there is no obvious method for doing it. In short, we are stuck with the problem of trying to *deduce* the actions of the extraocular muscles.

The classic description of action of the extraocular muscles is based on the geometry of their origins and insertions

Nineteenth-century investigators, most notably Volkmann, made extensive, careful measurements of the locations of extraocular muscle origins and insertions relative to a fixed coordinate system in the orbit with the eye in primary position. These measurements provide the basic information about the spatial relationships between origins and insertions that are required to work out the extraocular muscle actions. The anatomical insertions, however, are not necessarily the places where the forces that develop in contracting muscles act on the eye. Instead, the force is applied at the point where the muscle first contacts the eye (Figure 4.28); this is the **physiological insertion** of the muscle. With the exception of extreme rotations, the physiological insertion does *not* coincide with the anatomical insertion. Physiological insertion points can be derived in various ways; one method is an accurate diagram, as Figure 4.28 implies. The desired

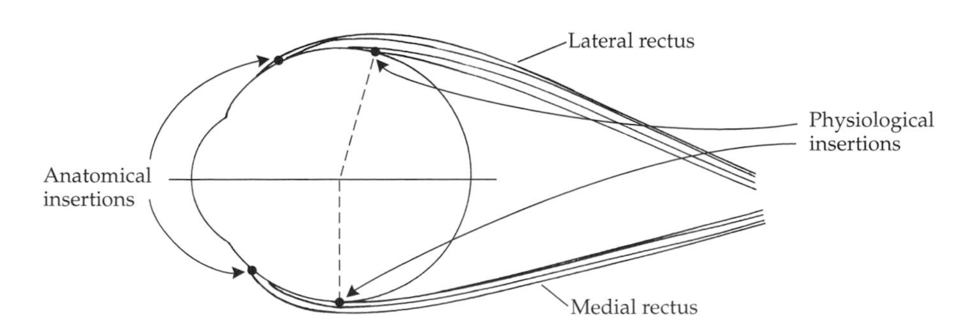

Figure 4.28
Physiological Insertions of the Extraocular Muscles
The physiological insertion of an extraocular muscle is defined as the point where the muscle first contacts the globe on the way to its anatomical insertion; it is the point where muscle force acts on the eye.

points are where the muscle is tangent to the circular cross section of the sclera. The extent to which the physiological insertions move on the eye as the eye rotates will be an important consideration in the final analysis of muscle action.

The traditional approach to solving the muscle action problem treated the eye as a sphere, rotating around a fixed point at its center, to which strings were attached, each string representing one of the extraocular muscles as it ran between its origin and insertion. The question to be solved by calculation was the eye rotation produced when each of the strings was pulled in turn. One result of this sort of analysis is summarized by the **Hering diagram** (Figure 4.29).

The center point of the diagram represents the anterior pole of the eye when the eye is in primary position, and the six lines running out from the center are the trajectories along which the anterior pole would move if each of the six muscles were to contract in isolation. The medial and lateral recti, as expected, are shown rotating the eye in the horizontal plane (either adduction or abduction). The trajectories for the vertical recti run up (superior rectus) and down (inferior rectus), showing their respective actions of elevation and depression, but they deviate from true vertical in the direction of adduction. This deviation indicates that the vertical recti produce rotation not only around the eye's horizontal axis (up or down) but also around the vertical axis, and both muscles are adductors.

The anterior pole trajectories have small striations on them that represent the horizontal axis of the eye in primary position, and these striations deviate from horizontal as the trajectories curve away from true vertical. They are showing intorsion produced by contraction of the superior rectus and extorsion produced by contraction of the inferior rectus.

The trajectories representing contraction of the obliques show pronounced torsional actions, as well as abduction by both muscles, elevation by the inferior oblique, and depression by the superior oblique.

The Hering diagram has a firm basis in anatomical fact, but the derivation of muscle actions involves several unrealistic assumptions. In particular, the notion of a single muscle acting alone, unaided and unaffected by any other muscles, is quite unjustified. At minimum, one must consider what happens in an agonist–antagonist pair, and for some directions of eye rotation, especially elevation and depression, normal movements may involve cooperation between the vertical recti and the obliques. In fact, the question about the action of a single muscle is an inappropriate question for which the answer, if obtainable, would have very little relevance to normal eye movements and patterns of extraocular muscle contraction.

Another problem is the possibility that muscle actions change as the eye moves to other positions of gaze; muscle origins are fixed in position, but the insertions move in three-dimensional space as the eye rotates, thereby changing the geometric relationship between origins and insertions from which the muscle actions are derived. And since individuals may exhibit strabismus in some

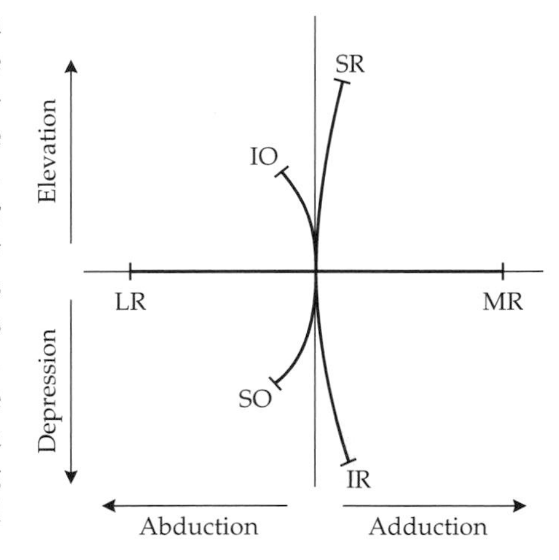

Figure 4.29
Calculated Actions of Isolated Muscles in Primary Position of Gaze: The Hering Diagram
The lines in this diagram are the hypothetical paths followed by the eye's anterior pole for each of the extraocular muscles acting in isolation. The trajectories corresponding to contraction of the medial (MR) and lateral (LR) recti are pure adduction and pure abduction, respectively. The major actions of the superior (SR) and inferior (IR) recti are elevation and depression, respectively, with elements of adduction and some torsion for both muscles. The superior (SO) and inferior (IO) obliques have strong torsional components, along with elevation (IO) and depression (SO).

positions of gaze and not in others, there are compelling clinical reasons for knowing more about variation in muscle actions with different positions of gaze.

Boeder diagrams attempt to describe the actions of the extraocular muscles completely

Figures 4.30 and 4.31 show calculations of extraocular muscle actions that attempt to describe muscle actions realistically and in a clinically useful way. They are included here not because they are totally accurate—they aren't—but because they have been so influential in our thinking about extraocular muscle actions.

Secondary positions of gaze are all the abducted, adducted, elevated, and depressed positions of gaze beginning from primary position—that is, all points along the horizontal and vertical lines that intersect primary position. Just as muscle actions were calculated for primary position, they can be calculated for secondary gaze positions (Figure 4.30).

Among the salient features are the near verticality of the trajectories for the vertical recti in abducted gaze and the increasing verticality, with reduced curvature, for the obliques in adducted gaze. The diagram says, in effect, that the vertical recti have purely vertical actions in *ab*ducted gaze where they have become the prime elevator and depressor of the eye (the obliques have lost most of their vertical components in abducted gaze). Similarly, the obliques have predominantly vertical actions in *ad*ducted gaze, contributing much more to elevation and depression than the vertical recti do. Incidentally, the actions of the obliques are noticeably asymmetric, because of the asymmetry of their insertions on the eye and the differences in their lines of action that were mentioned earlier.

When the eye is elevated or depressed from primary position, the diagram shows that the horizontal recti are no longer simple muscles of adduction and

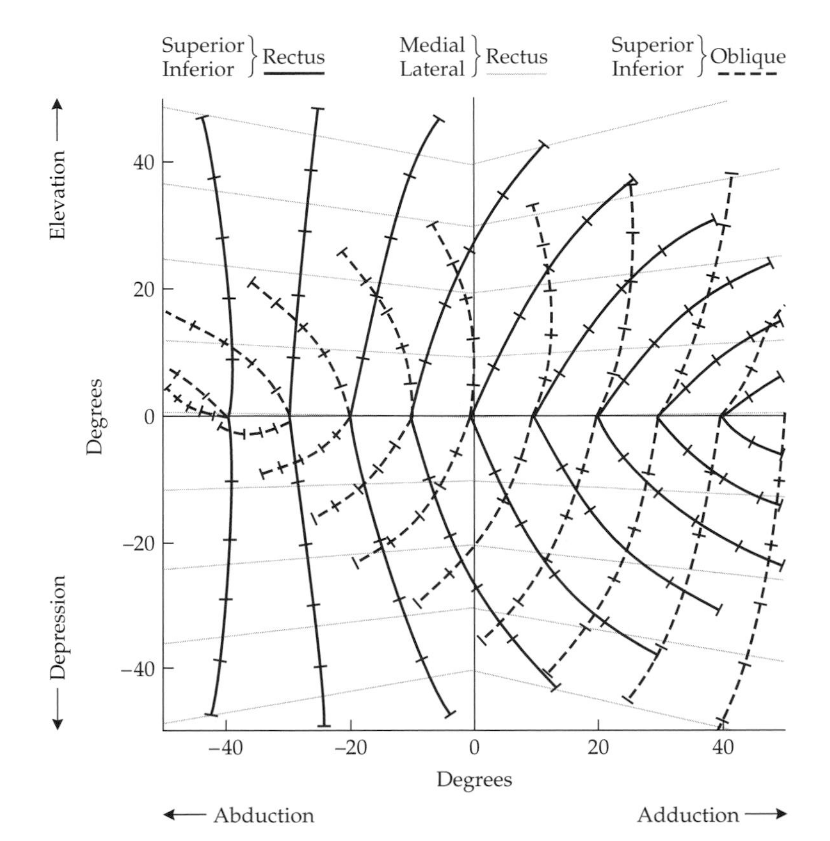

Figure 4.30
Hypothetical Changes in Muscle Actions with Different Positions of Gaze
In extreme abducted gaze, the vertical recti appear to have almost purely vertical actions, becoming more and more torsional muscles as one moves into adducted gaze positions. The obliques seem to do just the opposite; their vertical actions dominate in adducted gaze and their torsional actions dominate in abducted gaze. The medial and lateral recti exhibit vertical components in the actions in elevation and depression. (After Boeder 1961.)

abduction; they seem to acquire vertical actions as well, helping elevate the eye in elevated gaze positions and depressing the eye in depressed gaze positions.

Boeder's diagram provides a rationale for the **confrontation test**, as well as some of its more sophisticated variants, that is used clinically to check extraocular muscle actions. One compares elevation and depression of the patient's eyes in extreme abducted and adducted positions of gaze. An inability or deficit in elevation or depression when the eye is abducted implies an underaction of one of the vertical recti; a deficit in adducted gaze implicates one of the obliques. The test, in other words, is meant to isolate as much as possible the two sets of muscles contributing to elevation and depression; inability to abduct or adduct directs attention to one of the horizontal recti. The confrontation test is simple, but for reasons that will be made apparent, it can be misleading.

Boeder's calculations assumed that any muscle could be treated independently, which is probably unrealistic. He later took up the problem of cooperation and synergism among extraocular muscles, however, and tried to characterize extraocular muscle actions more realistically. The essential ideas are, first, that the forces applied to the eye by the six extraocular muscles must be in equilibrium for any given position of gaze and, second, that muscle force is inversely proportional to muscle length. One can calculate muscle lengths for any position of gaze on strictly geometric grounds; for each position the distance from origin to insertion must change along with the changing geometry, and this distance is the required muscle length. Assuming muscle length and muscle force to be linearly related, and the linearity constant to be the same for all muscles, plots of muscle lengths for different gaze positions should show directly the contribution made by each muscle to each gaze position.

The results of this analysis require six diagrams, one for each muscle (Figure 4.31). Each diagram contains two families of curved lines; the thin lines are isolength curves showing all the different positions of gaze that the eye could assume with the muscle having a specific length. The curve labeled 0, for example, is the resting length of the muscle, which in this case refers to the length of the muscle when the eye is in primary position (note that all the 0 curves go through the origin). Positive numbers on the curves indicate the amount (in millimeters) the muscle has lengthened from primary position; negative numbers are the amount of shortening as the muscle contracts from its resting state. Using the lateral rectus as an example, each millimeter of shortening corresponds to successively greater amounts of abduction, as one would expect, while the muscle elongates for adducted positions of gaze.

The heavy lines are the family of curves orthogonal to the isolength curves. They represent the expected change in eye position as the length of the muscle changes—that is the muscle's action. For example, shortening and lengthening of the lateral rectus when the eye is in the horizontal plane are associated with pure abduction and adduction, respectively. When the eye is elevated or depressed, however, the lines representing lateral rectus action curve from the adducted to abducted gaze positions, meaning that the lateral rectus has taken on a component of elevation in elevated positions of gaze, and a component of depression in depressed positions of gaze.

In interpreting the action curves for the other muscles, it is instructive to compare them to the original diagram (see Figure 4.30). As before, the vertical recti are almost pure elevator and depressor in abducted gaze, with their vertical actions reduced in adducted gaze. The obliques show more torsion in abduction, and more elevation or depression in adduction. But where the superior oblique appears to be almost a pure depressor at 20 to 30° of adduction, the inferior oblique has a relatively small action of elevation and a strong component of adduction. The effect of the asymmetric insertions of these muscles is even more obvious than before.

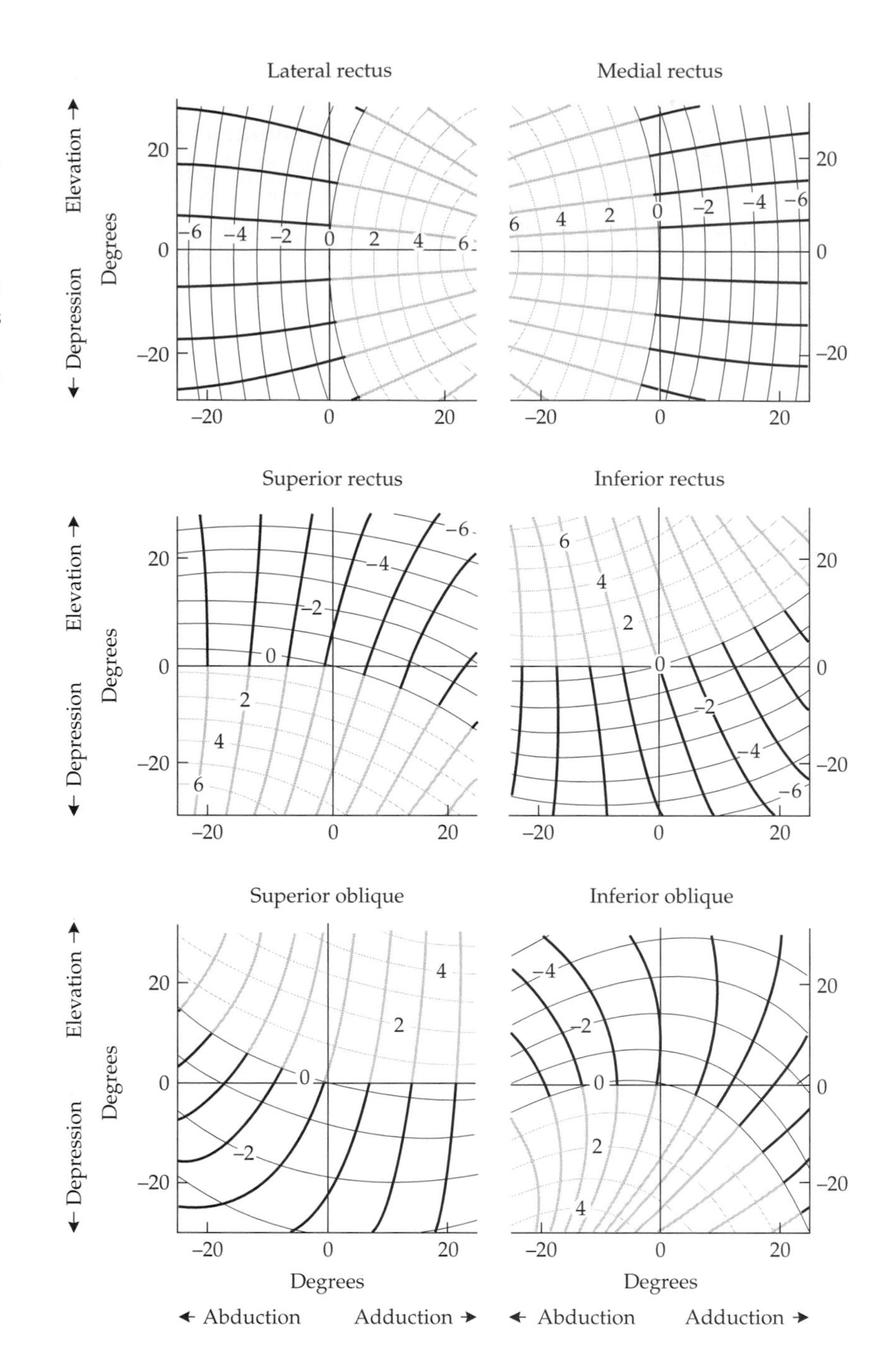

Figure 4.31
Actions of Agonist–Antagonist Muscle Pairs
These diagrams, one for each of the six muscles, have taken the passive force in the antagonistic muscle into consideration when considering the action of the agonist. The heavy lines in each diagram represent muscle action, and the light lines are isolength curves (see text for details). Some of the major features of the simpler analysis (see Figure 4.30) remain, in particular the strongly vertical actions of the obliques in adducted gaze and of the vertical recti in abducted gaze. (After Boeder 1962.)

The presence of Tenon's capsule and muscle pulleys invalidates the geometric model of extraocular muscle actions

To make the analysis of muscle actions more accessible and dynamic, David A. Robinson attempted to incorporate all the appropriate calculations in a computer simulation of the eye and extraocular muscles. The simulation was meant to give an accurate representation of all the anatomical information about muscle origins and insertions, the normal elasticity and tonus of the extraocular muscles, the passive forces from check ligaments and other tissues that might resist ocular rotation, and so on. In mechanical terms, this simulation was rather like a ball rotating in a socket, but with elastic bands—not strings—attached to the ball in place of the extraocular muscles. When first turned loose on the problem, the computer promptly predicted that the appropriate resting position of the eye was looking directly backward toward the apex of the orbit!

Something was badly wrong, and since Robinson's model was not qualitatively different from Boeder's, something had to have been wrong with all prior analyses of extraocular muscle forces and actions. In retrospect, there were at least two basic faults. First, as we saw at the outset, the force developed in a muscle is *not* linearly related to its length (see Figure 4.6). Second, all the previous analyses had assumed that as the eye moves around, the muscles slip freely over the surface of the globe, but such freedom of movement means that the physiological insertions of the muscles would not be fixed in position *on the eye*. That is, the points where force is applied to the eye by a muscle will move around as the eye moves.

In fact, the extraocular muscles have several kinds of anatomical restrictions on their freedom of movement. One is Tenon's capsule and associated connective tissue, which were discussed in Chapter 3. Tenon's capsule and other fascial elements are very much like a corset for the extraocular muscles; the muscles are bound to the globe and to one another, thus preventing much change in the positions of the physiological insertions as the eye moves.

One effect of the binding by Tenon's capsule is illustrated in Figure 4.32 for the medial rectus. When the eye elevates and the muscle is free to slip across the surface of the eye, its path from origin to anatomical insertion will be straight and the physiological insertion will shift up onto the superior aspect of the eye (Figure 4.32*a*). This effect suggests that contraction of the medial rectus in elevated positions of gaze will contribute to further elevation of the eye, which is precisely what Boeder's diagrams indicate (see Figure 4.30). When the medial rectus is bound to the globe, however, the physiological insertion moves hardly at all (Figure 4.32*b*). The path of the muscle from origin to anatomical insertion is no longer straight, but this is not important, since the force will be applied at the *physiological* insertion. In this case, when the medial rectus contracts, the force

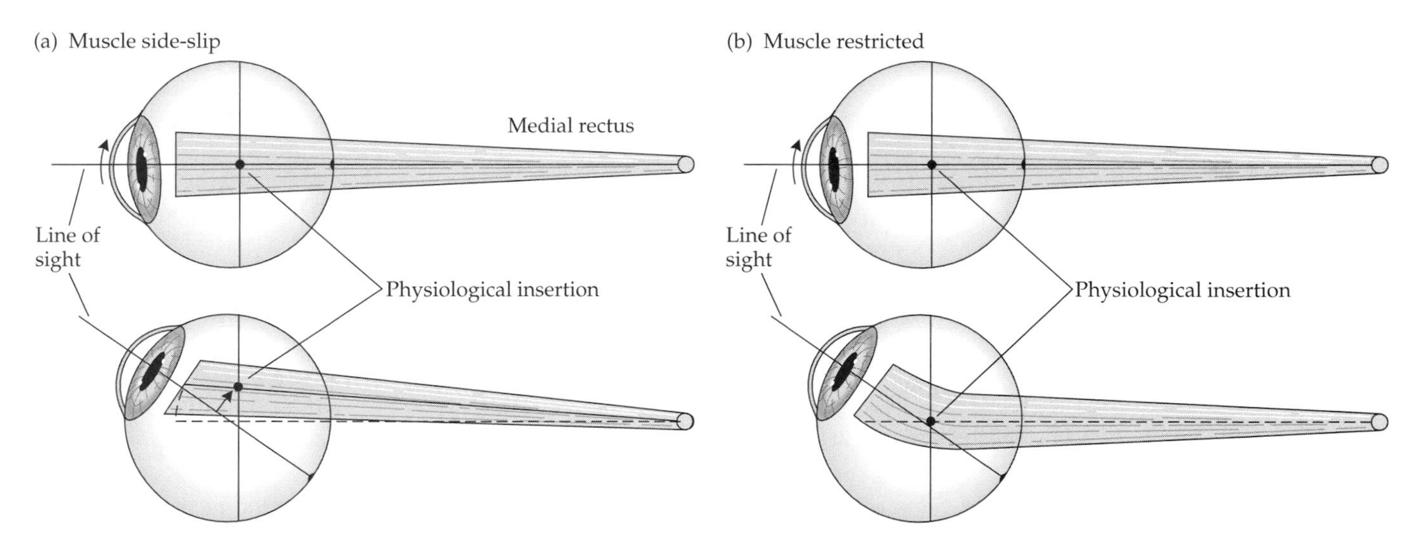

Figure 4.32
Possible Movement of Physiological Insertions and Change of Muscle Action Due to Muscle Side-Slip
(*a*) If the medial rectus could slip freely over the surface of the eye, the physiological insertion would move above the horizontal plane as the eye was elevated. This movement should produce a component of elevation for the medial rectus when the eye is elevated, as indicated in Figure 4.30. (*b*) When muscle side-slip cannot occur, because of mechanical restrictions (Tenon's capsule, muscle pulleys), the physiological insertion remains in the horizontal plane even though the eye is elevated. The action of the medial rectus should therefore remain pure adduction regardless of the elevated gaze position of the eye.

will continue to be in the horizontal plane, and despite the elevated eye position, the postulated action will be the same as when the eye was in primary position—namely, pure adduction.

The fascial connections also mean that the extraocular muscles are mechanically coupled to one another. This sort of coupling was noted in Chapter 3 for the superior rectus and the levator, and this particular linkage has long been recognized and understood. What had not been recognized was the possibility of mechanical interaction among the other muscles, but it happens and must be taken into account. In short, there is no way for the actions of the vertical recti, for example, to be isolated from the state of contraction in the horizontal recti.

Other connective tissue restrictions are created by the pulleys, or by connective tissue and smooth muscle slings, through which the rectus muscles run on their way to the eye (see Chapter 3). The pulleys restrict the muscles to particular paths through the orbit, from which they cannot deviate, as the eye moves. (Although the pulleys are said to contain some smooth muscle, the implication that the tightness or locations of the pulleys can be varied has not been established.) The pulleys are analogous to the trochlea of the superior oblique in that they are the functional origins of the rectus muscles; they do not represent such an abrupt change in the direction of muscle force, however. These restrictions on the muscles must also be taken into account when the muscle actions are being modeled.

"There is no simple way to describe the action of these muscles on the eye!"

This admonition from Miller and Robinson (1984) is well founded: the actions of the extraocular muscles defy simple description because there are so many interacting variables that must be considered and represented. Yet it is important that we have a believable description that is as accurate as we can make it, simple or not; the diagnosis and treatment of extraocular muscle anomalies demands it. When mechanical interactions between muscles and restrictions on muscle movements are taken into account, the picture of extraocular muscle actions changes, as Figure 4.33 shows. These plots suggest that some important revisions to the classic descriptions of extraocular muscle actions are required.

For example, the lines of action for the vertical recti have very little curvature and little deviation from vertical, which means that the subsidiary actions discussed earlier are nonexistent (with the exception of a curious little glitch in extreme adduction). In other words, the superior rectus elevates the eye and the inferior rectus depresses the eye, and these actions show very little change with eye position. In particular, the vertical recti are quite effective in both adducted and abducted positions of gaze.

The medial and lateral recti remain the simplest muscle pair, showing actions of pure adduction or abduction for any horizontal gaze position. They do depart from pure adduction and abduction when the eye is elevated or depressed, but note that the effects are just the opposite of those shown earlier; when the eye is elevated and adducted, for example, contraction of the medial rectus tends to

Figure 4.33 ▶
Extraocular Muscle Actions When Mechanical Restrictions and Linkages between Muscles Are Taken into Account
In these diagrams, the light lines are isoforce curves (see text for details) and the dark lines represent muscle actions. These action curves were constructed (by eye) as an orthogonal set to the original isoforce curves generated by a computer model of the extraocular muscle system. Muscle actions predicted by this analysis are different from and to some extent simpler than those in Figure 4.31, particularly for the obliques, which now have strong components of elevation or depression in all gaze positions. (After Robinson 1975.)

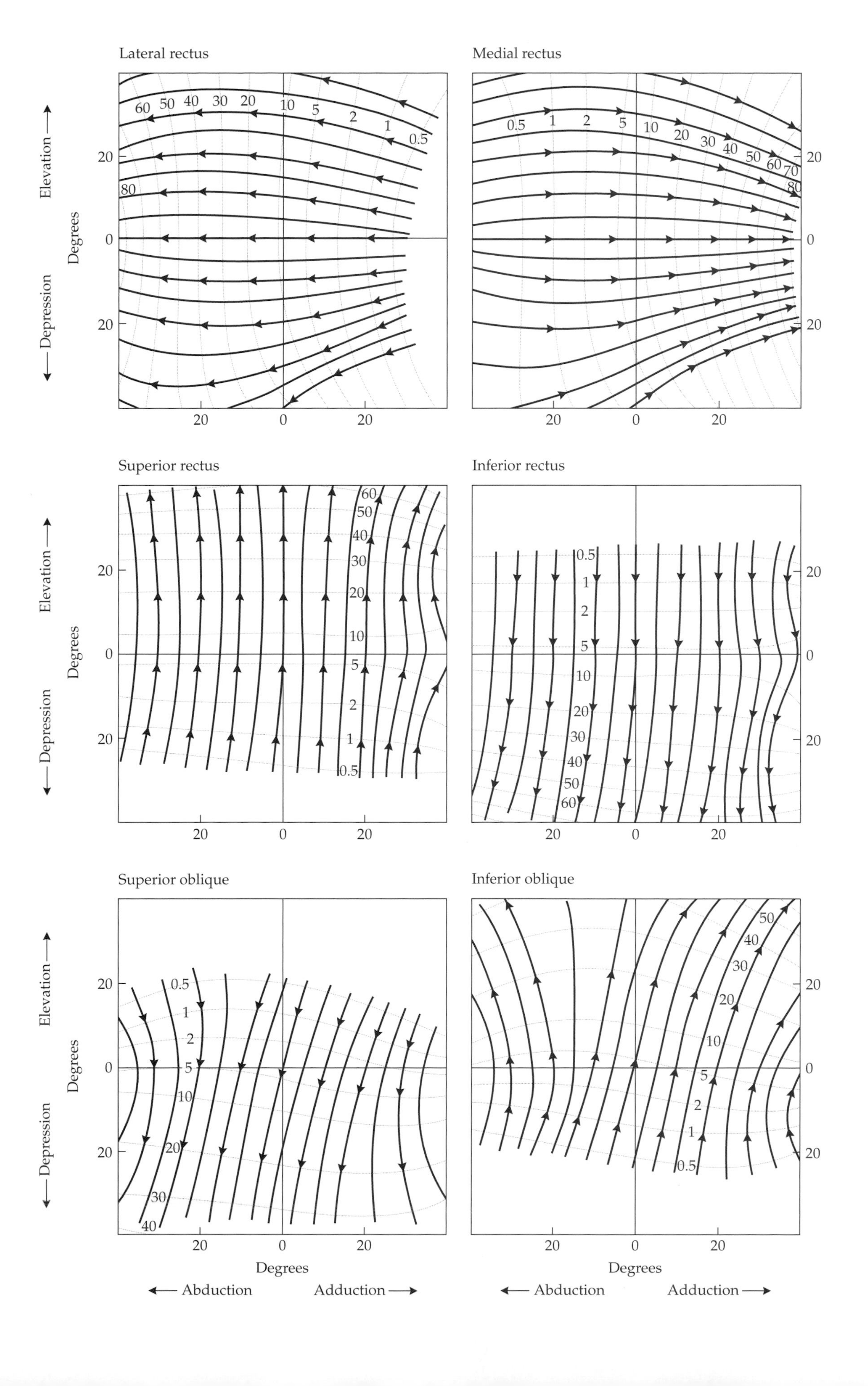

Lateral rectus
Medial rectus
Superior rectus
Inferior rectus
Superior oblique
Inferior oblique
Elevation
Depression
Degrees
Abduction
Adduction

depress the eye, not elevate it. This effect is called bridling; when a bridled horse raises its head, tugging on the reins pulls the head back down, rather than raising it even more. This effect is a consequence of the binding of the muscle to the eye by Tenon's capsule (see Figure 4.32) and of the restrictions imposed by the muscle pulley. The lateral rectus is also a depressor in upward gaze, and both muscles are elevators when the eye is depressed.

Perhaps the most striking trajectories are those for the obliques. Their dominant actions now appear to be elevation and depression over most of the range of gaze positions. The torsional components are still present, but they are much reduced; they are most pronounced at the extremes of adduction and abduction. In addition, the torsional components change direction depending on whether the eye is elevated or depressed, abducted or adducted. Thus while the obliques are the muscles primarily responsible for torsional movements, their main actions are elevation and depression of the eye. Finally, note that the obliques differ over most gaze positions in the horizontal component of their actions; the inferior oblique usually, but not always, adducts, while the superior oblique usually, but not always, abducts.

Although the detailed actions of the extraocular muscles cannot be described simply, some simplifying summary statements about how they move the eye are possible. For the rectus muscles, it is fair to say that the superior rectus elevates, the inferior rectus depresses, the medial rectus adducts, and the lateral rectus abducts. And since the primary actions of the recti vary so little with gaze position and their subsidiary actions are so small, these statements require almost no qualifiers. The obliques are more complicated, but it is true that the superior oblique depresses the eye, and this component of its action is directly antagonistic—in all fields of gaze—to the elevation produced by the inferior oblique. And the obliques are the only muscles that have significant torsional components to their actions; the direction of torsion varies with gaze position, but the two muscles are always torsional antagonists.

The bad news for clinicians is the similarity in the actions of the vertical recti and the obliques. Elevation and depression of the eye from adducted and abducted positions of gaze do not separate the actions of the vertical recti and the obliques as earlier analyses indicated. Therefore, a confrontation test can be used to check for deficits in vertical eye movements, but the test cannot reliably specify which of the vertically acting muscles is deficient. More sophisticated tests are required to make this distinction.

A realistic model of the extraocular muscle system is important for the diagnosis and treatment of muscle paresis

All descriptions of extraocular muscle actions, whether a computer simulation or a simple table listing primary, secondary, and tertiary muscle actions, are models of the extraocular muscle system, differing in their assumptions and the anatomical features they take into account. There is nothing unusual about modeling; we do it all the time in an informal way. But formal models are commonly based in mathematics so that relationships among variables can be expressed quantitatively. Although this mathematical approach often makes the models more difficult to understand intuitively, there is an overriding advantage: Formal models can make quantitative predictions, so they are testable in ways that informal models are not.

To understand the oculomotor system and control of eye position by the extraocular muscles, we need computer models because they represent the only practical way to deal with so many interacting variables. By making the variables that affect muscle action explicit and expressing them in quantitative form as functions or numerical parameters, we can assess the effects of change in one or more of the variables. Muscles can be made to overact or underact, their insertions on the eye can be relocated, the mechanical linkage between muscles can be altered, and so on. It is possible to see predictions about the amount and type of strabismus that would result from paralysis of one muscle or another.

It is possible to simulate not only the effects of muscle paresis but also the effects of surgical correction. Figure 4.34*a*, for example, shows the simulated set of eye positions for what is called the tight lateral rectus syndrome. In this syndrome, the lateral rectus behaves as if it were much less elastic than normal; the result is exotropia and difficulty adducting the eye because of the inelasticity of the lateral rectus. These features show clearly in the simulation. After a simulated 7 mm recession (a weakening; see Box 4.2) of the lateral rectus (Figure 4.34*b*), the exotropia has been reduced almost to zero and the range of adduction has been increased considerably (for the patient to whom these simulations were compared, the required surgery was recession of the lateral rectus in combination with resection of the medial rectus).

The computer model that generated Figure 4.34 has been refined in the past 20 years or so as new and better data about the mechanical properties of the extraocular muscles have become available. The model is unlikely to supplant the need for judgment and experience by extraocular muscle surgeons, but the value of the model is substantial. The hard fact about the extraocular muscles and eye movements is that the system is *not* simple; it is possible to understand, in a general way, how the extraocular muscles work and what they do, and this level of understanding has been the point of this chapter, but the many interacting variables make it difficult to understand and deal with the system in a detailed, quantitative fashion. But computers, properly instructed, do this sort of job very well.

Development of the Extraocular Muscles

Each muscle develops from several foci in the mesoderm surrounding the optic cup

All of the extraocular muscles are of mesodermal origin. Their development can be described as occurring in half a dozen stages, but for our purposes it is enough to think of three: (1) differentiation of myoblasts from the mesoderm, (2) proliferation with additional differentiation, and (3) fusion of the primitive muscle cells to form multinucleate muscle fibers. Development seems to occur simultaneously all along the course of the future muscle; that is, each muscle forms from several foci that expand and eventually join. Thus, the muscle ori-

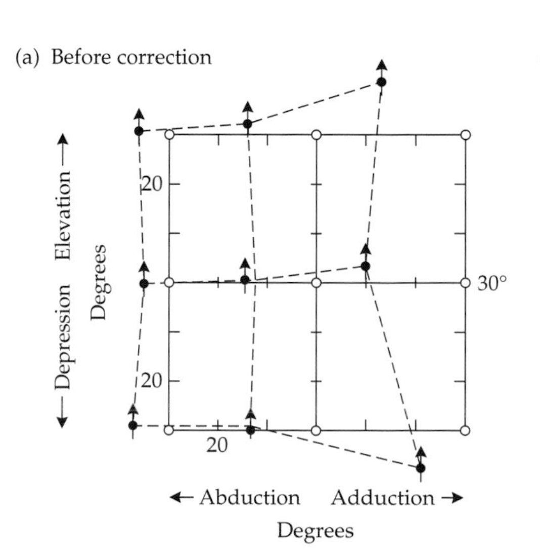

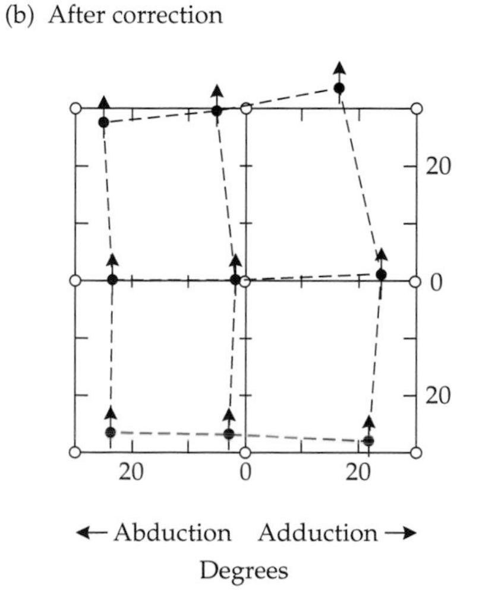

Figure 4.34
Simulated Defect and Simulated Correction
(*a*) This simulation of tight lateral rectus syndrome shows the problem patients have with adduction and the exaggerated elevation and depression in adducted gaze positions. The square grid with open circles shows the normal fixation pattern. (*b*) The results of a simulated correction corresponding to a 7 mm recession (see Box 4.2) of the lateral rectus. The improvement in adduction is quite obvious. (After Robinson 1975.)

Vignette 4.2
In the Service of the Eye

WITH SOME BEADS, SEQUINS, AND FEATHERS, the grotesque masks over the faces in Figure 1 could belong to revelers at carnival in Venice or Mardi Gras in New Orleans, but actually the masks are an early method for treating strabismus. The intent was to force the deviated eye to be used in alignment with its fellow eye if it were going to be used at all; thus, for an esotropia, the aperture in the mask was located so that the deviated eye would have to be rotated outward into normal alignment if it were to see anything, and for an exotropia, the deviated eye would have to rotate inward. It is unlikely that this technique ever converted a constant strabismus to normal eye alignment, but the man who is often regarded as one the pioneers of ophthalmology should be given credit for trying to solve the problem in a sensible way. (Though in his multivolume history of ophthalmology, Julius Hirschberg suggests that similar strabismus masks had first been used almost 2,000 years earlier by the Greeks.)

Georg Bartisch (1535–1606) was born in a small town near Dresden, where he later settled and lived most of his life. He had little formal education, though he was literate, and he was apprenticed to a barber-surgeon in his early teens. In those days, surgery was a lowly profession separate from medicine and despised by physicians; its practitioners were much like itinerant peddlers, hawking their surgical wares as they wandered from place to place, bleeding their patients, sewing up wounds, and generally wreaking bloody havoc in the hope of cures. If we take Bartisch at his own word, however, he was an exception. He carefully listed all the personal qualities that a surgeon should possess (a daunting list) and described at length the things that must be done to ensure maximum benefit to the patient in a variety of surgical procedures. Although many of the surgical procedures seem horrific, since they were done without anesthesia, Bartisch comes across as a man genuinely concerned with his patients' welfare; he rails, for example, at fly-by-night surgeons who leave town immediately and offer no postoperative care.

All this and more is contained in a book rather grandly titled ΟΦΘΑΛΜΟΔΟΥΛΕΙΑ, *Das ist Augendienst,* first published in 1583. (The Greek title can be rendered as *Ophthalmodouleia,* that is, ophthalmology, and *Augendienst* is roughly "service to the eye".) Despite the title, it is fairly clear that Bartisch knew neither Latin nor Greek, and *Das ist Augendienst* was written in German. Bartisch describes cataract surgery, instruments designed for special purposes (such as exenteration of the orbital contents with a single swipe of a special blade), a variety of herbal remedies, and other treatments that were probably his own inventions. He also shows us, in a novel way, what he understood about the anatomy of the eye.

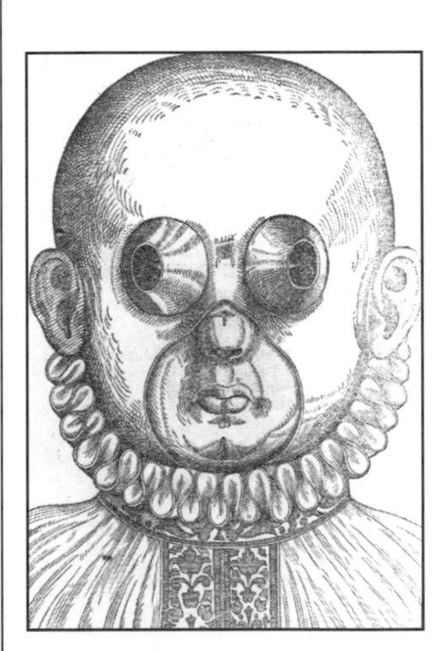

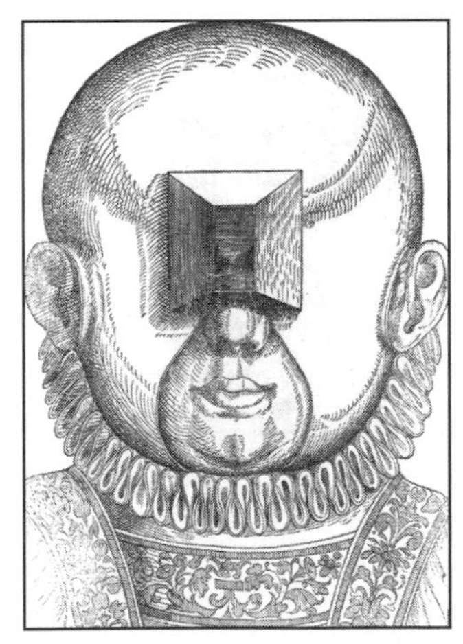

Figure 1
Strabismus Correction Masks
The mask on the left was used to treat esotropia, the mask on the right exotropia. The basic idea is still with us in various forms of partial occlusion, but now we use spectacles, not full facial masks. (From Bartisch 1583.)

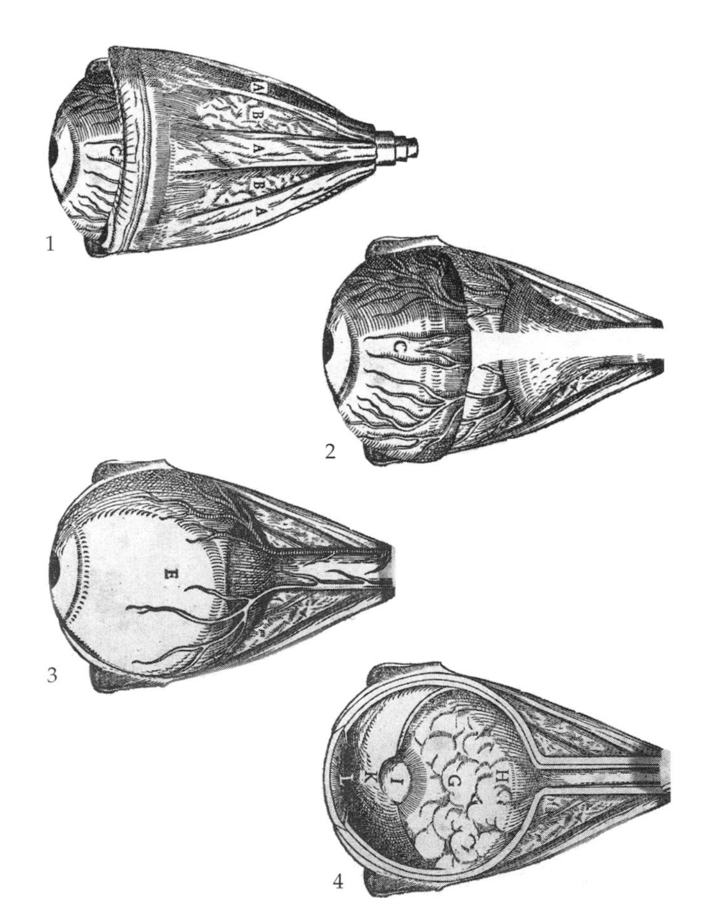

Figure 2
Visual Dissection of Ocular Structure according to Bartisch
One of the overlays created by Bartisch is shown here as successive layers are folded back to reveal the underlying structures; there are six overlays in the original. The successive drawings revealed by turning back the overlays have been rotated 90°, and other drawings of the eye surrounding the overlay have not been included here. (From Bartisch 1583.)

After many pages of testimonials to his skills, Bartisch began his mainly clinical treatise with a discussion of anatomy; Figure 2 shows his illustration of the eye. His innovation was depicting the eye's three-dimensional structure in a number of overlaid drawings; the first view is external, showing the sclera, but that could be folded back to reveal the choroid underneath, and so on to reveal the deeper structures as each overlay was folded back in succession. This illustrative approach meant, of course, that each copy of the book required assembly of the figure by pasting the overlays in position (there were not a great many copies, however). Here, we see the drawing as each overlay is folded back in turn. (Another drawing by Bartisch showed a view of the head from above, in which one could remove the top of the skull to see the brain beneath, and then remove additional layers down to the base of the skull.)

Although the way in which Bartisch illustrated ocular anatomy was novel, the anatomy itself was derivative, as one can see by comparing this figure with the drawings in Vignette 3.2. What Bartisch knew about the structure of the eye he got from Vesalius; even some of the small details have been copied faithfully. The position of the lens in the eye is particularly revealing; the lens is well back of its true position, indicating that in spite of his careful descriptions of the operation for "couching" cataracts, Bartisch did not understand the relationship between cataracts and the lens. But he did not claim to be a student of anatomy, and the correct depiction of the lens by Fabricius (see Vignette 12.1) was yet to come.

Bartisch's use of overlays to convey an impression of three-dimensional structure has been employed more recently. *The Human Eye in Anatomical Transparencies*, published in 1943 by Bausch & Lomb Optical Company, both acknowledges and pays homage to Bartisch's ingenuity. In addition to an excellent text and a historical appendix, this wonderful book contains eight doubled-sided transparent pages, so arranged that one can visually dissect the eye and the orbit from front to back or from one side to the other. (My copy of the third printing in 1945 has its original price marked on the flyleaf—$6.50, only a few dollars more than the price of a hardback novel at the time. I doubt that B&L made money on this venture.) This method for visually dissecting three-dimensional structure is very powerful—so powerful that computer-based displays often employ it to great effect. But only the technology is new—the critical idea is more than 400 years old.

gins near the developing orbital bones and the muscle insertions adjacent to the developing eye form at the same time.

The extraocular muscles appear after the optic cup, but before the orbital bones

In human embryos, the first signs of developing extraocular muscles are clusters of myoblasts appearing in the mesoderm around the optic cup. The few studies that have been made of human extraocular muscle development agree that these developmental foci are detectable by prenatal day 30. There is less agreement, however, about the number of foci and the timing of their appearance for different muscles. The foci either appear almost simultaneously at day 30 or in sequence between days 27 (lateral rectus) and 33 (inferior rectus and oblique); some muscles may have just one or a few foci. Whatever the case, pioneering motor nerve axons reach the developmental foci within a week of their appearance. The levator muscle begins to form later, its first foci appearing around day 46.

During the next month or so, the myoblasts grow and fuse with one another, and the primitive muscles begin to show their characteristic striation. They also acquire an immature form of innervation from motor nerve axons, in which the developing muscle fibers are each contacted by several different axons. This polyneuronal innervation is quite different from the final pattern, in which each muscle fiber receives input from a single axon.

Different muscle fiber types form late in gestation and continue to develop postnatally

Distinctions among different types of extraocular muscle fibers cannot be made until very late in gestation, and the dominant muscle fiber group consists of the thick, singly innervated fibers. In fact, there is very little evidence that thin, multiply innervated fibers are present at birth, suggesting that development of the muscles continues well into the postnatal period. The type of muscle fiber may be determined by the type of innervation (single motor end plates or multiple en grappe terminals) it receives, and the late appearance of thin fibers in the extraocular muscles may be associated with late, prolonged development of the neural inputs.

In other species, more motor neurons are produced during development than are present in the adult; a wave of cell death in the late prenatal and early postnatal periods reduces the overabundance of neurons to adult levels. And as already mentioned, most of the innervation is polyneuronal. Thus, the reorganization of the motor innervation involves the death of supernumerary axons, the withdrawal of axon collaterals, and possibly the establishment of en grappe terminals with subsequent formation of the thin muscle fiber system.

Most developmental anomalies are associated with the connective tissue of the muscles or with their innervation

Extraocular muscles rarely fail to develop, but instances of their absence have been documented histologically. A more common observation is the opposite situation: the existence of extra muscles or extra slips of muscle with separate insertions on the eye. Most of the developmental abnormalities, however, are associated with the connective tissue sheaths and the check ligaments of the muscles.

The usual explanation of **Brown's syndrome**, for example, is a fascial sheath around the superior oblique tendon that is too short. Since the sheath is attached to Tenon's capsule at one end and the trochlea at the other, and the distance from the superior oblique insertion to the trochlea normally *increases* during adduction, the superior oblique cannot depress the eye properly in adducted gaze.

On occasion, one or more muscles may be completely or partially formed from connective tissue rather than from muscle tissue. This condition is assumed to be a developmental failure, not a subsequent replacement of muscle tissue by connective tissue. The lateral rectus seems particularly prone to this anomaly, which is one explanation for **Duane's retraction syndrome**: Since most of the lateral rectus is connective tissue, the affected eye cannot be abducted. In addition, the connective tissue is inelastic and effectively prevents adduction of the eye when the medial rectus contracts. The medial rectus contraction pulls the eye back into the orbit; hence the term "retraction syndrome."

It is not clear that all cases of Duane's retraction syndrome are due to fibrotic lateral recti. Another possible explanation is anomalous innervation such that the lateral and medial recti contract together during attempted adduction. This explanation implies that the normal innervation of the lateral rectus by the abducens nerve has been replaced or is dominated by innervation from the oculomotor nerve.

The oculomotor system is not fully operational at birth

Eye movements in newborn infants are poorly coordinated, and their following eye movements are very inaccurate; a year or so is required for these eye movement patterns to approximate those of adults. Several factors contribute to the immaturity of the oculomotor system, including incomplete structural development in the fovea, in the motor control pathways, and in the extraocular muscles.

Any visual function requiring signals from foveal cones is deficient at birth because the foveal cones are not fully developed (see Chapter 16). As a result, infants have not only very poor visual acuity, but also deficiencies in foveal fixation and in pursuit movements that require the fovea. Other eye movements that do not require a mature fovea, such as optokinetic nystagmus, are much more like adult eye movements. Even in this case, however, there are marked discrepancies between the velocity and magnitude of movements in the two eyes, which is one of the indicators that the neural circuitry responsible for such things as the versional yoking of the eyes is incomplete. In addition, the late-occurring cell death in the oculomotor nuclei and their associated parts of the brainstem indicates that the extraoculomotor system has yet to be fully organized.

The extraocular muscles are immature at birth in several respects, the most obvious being their small size; the muscles in the newborn have roughly two-thirds the adult length and cross-sectional area. Most of the subsequent growth of the extraocular muscles is by hypertrophy—that is, by an increase in muscle fiber size rather than by an increase in the number of muscle fibers (hyperplasia), and the muscles grow in length and bulk for some years, not stopping until the bones of the face and orbit cease to grow. Another immature feature at birth is the near absence of thin, multiply innervated fibers in the extraocular muscles. If the thin fiber system is primarily responsible for slow eye movements, as discussed earlier, we have yet another anatomical factor that underlies the immaturity of slow eye movements in infants.

References and Additional Reading

Patterns of Eye Movement

Alpern M. 1969. Types of movement. Chapter 5, pp. 65–174, in *The Eye,* Vol. 3, *Muscular Mechanisms,* 2nd Ed., Davson H, ed. Academic Press, New York.

Collewijn H, Erkelens CJ, and Steinman RM. 1988. Binocular co-ordination of human horizontal saccadic eye movements. *J. Physiol.* 404: 157–182.

Collewijn H, Van der Mark F, and Jansen TC. 1975. Precise recordings of human eye movements. *Vision Res.* 15: 447–450.

Coppola D and Purves D. 1996. The extraordinarily rapid disappearance of entoptic images. *Proc. Natl. Acad. Sci. USA* 93: 8001–8004.

Erkelens CJ, Van der Steen J, Steinman RM, and Collewijn H. 1989. Ocular vergence under natural conditions. I. Continuous changes of target distance along the median plane. *Proc. R. Soc. Lond. B* 236: 417–440.

Goldberg ME, Eggers HM, and Gouras P. 1991. The ocular motor system. Chapter 43, pp. 660–678, in *Principles of Neural Science,* 3rd Ed., Kandel ER, Schwartz JH, and Jessell TM, eds. Elsevier, New York.

Kowler E. 1991. The stability of gaze and its implications for vision. Chapter 4, pp. 71–92, in *Vision and Visual Dysfunction,* Vol. 8, *Eye Movements,* Carpenter RHS, ed. CRC Press, Boca Raton, FL.

Land MF and Furneaux S. 1997. The knowledge base of the oculomotor system. *Philos. Trans. R. Soc. Lond. B* 352: 1231–1239.

Meyer CH, Lasker AG, and Robinson DA. 1985. The upper limit of human smooth pursuit velocity. *Vision Res.* 25: 561–563.

Pola J and Wyatt HJ. 1991. Smooth pursuit: Response characteristics, stimuli and mechanisms. Chapter 6, pp. 138–156, in *Vision and Visual Dysfunction,* Vol. 8, *Eye Movements,* Carpenter RHS, ed. CRC Press, Boca Raton, FL.

Pritchard RM. 1961. Stabilized images on the retina. *Sci. Am.* 204 (6): 72–78.

Purves D, Augustine GJ, Fitzpatrick D, Katz LC, LaMantia A-S, and McNamara JO. 1997. Eye movements and sensory-motor integration. Chapter 19, pp. 361–374, in *Neuroscience.* Sinauer Associates, Sunderland, MA.

Rashbass C. 1961. The relationship between saccadic and smooth tracking eye movements. *J. Physiol.* 159: 326–338.

Rashbass C and Westheimer G. 1961. Disjunctive eye movements. *J. Physiol.* 159: 339–360.

Steinman RW, Cushman WB, and Martins AJ. 1982. The precision of gaze. *Hum. Neurobiol.* 1: 97–109.

Steinman RM, Haddad GM, Skavenski AA, and Wyman D. 1973. Miniature eye movements. *Science* 181: 810–819.

Walls GL. 1962. The evolutionary history of eye movements. *Vision Res.* 2: 69–80.

Westheimer G. 1954. Mechanism of saccadic eye movements. *AMA Arch. Ophthalmol.* 52: 710–724.

Yarbus AL. 1967. *Eye Movements and Vision,* B Haigh, transl. Plenum, New York.

Control of Eye Position

Büttner-Ennever JA. 1981. Anatomy of medial rectus subgroups in the oculomotor nucleus of the monkey. Pp. 247–252 in *Progress in Oculomotor Research. Developments in Neuroscience,* Vol. 12, Fuchs AF and Becker W, eds. Elsevier/North Holland, New York.

Cannon SC and Robinson DA. 1987. Loss of the neural integrator of the oculomotor system from brainstem lesions in monkey. *J. Neurophysiol.* 57: 1383–1409.

Carpenter RHS. 1988. *Movements of the Eyes.* Pion, London.

Clendaniel RA and Mays LE. 1994. Characteristics of antidromically identified oculomotor internuclear neurons during vergence and versional eye movements. *J. Neurophysiol.* 71: 1111–1127.

Collewijn H. 1987. The physiology of eye movements. Pp. 19–35 in *Eye Movement Disorders,* Sanders EACM, de Keizer RJW, and Zee DS, eds. Martinus Nijhoff/Dr W. Junk, Dordrecht, Netherlands.

Collewijn H. 1989. The vestibulo-ocular reflex: Is it an independent subsystem? *Rev. Neurol. (Paris)* 145: 502–512.
Collins CC. 1975. The human oculomotor control system. Pp. 145–180 in *Basic Mechanisms of Ocular Motility and Their Clinical Implications,* Lennerstrand G and Bach-y-Rita P, eds. Pergamon, Oxford.
Collins CC, Scott AB, and O'Meara DM. 1975. Muscle tension during unrestrained human eye movements. *J. Physiol.* 245: 351–369.
Crawford JD, Cadera W, and Vilis T. 1991. Generation of torsional and vertical eye position signals by the interstitial nucleus of Cajal. *Science* 252: 1551–1553.
Fuchs AF, Kaneko CRS, and Scudder CA. 1985. Brainstem control of saccadic eye movements. *Annu. Rev. Neurosci.* 8: 307–337.
Gamlin PDR, Yoon K, and Zhang H. 1996. The role of cerebro-ponto-cerebellar pathways in the control of vergence eye movements. *Eye* 10: 167–171.
Guthrie BL, Porter JD, and Sparks DL. 1983. Corollary discharge provides accurate eye position information to the oculomotor system. *Science* 221: 1193–1195.
Keller EL. 1991. The brainstem. Chapter 9, pp. 200–223, in *Vision and Visual Dysfunction,* Vol. 8, *Eye Movements,* Carpenter RHS, ed. CRC Press, Boca Raton, FL.
Leigh RJ and Zee DS. 1991. The properties and neural substrate of eye movements. Part I, pp. 1–290, in *The Neurology of Eye Movements,* 2nd Ed. FA Davis, Philadelphia.
Lisberger SG, Morris EJ, and Tychsen L. 1987. Visual motion processing and sensory-motor integration for smooth pursuit eye movements. *Annu. Rev. Neurosci.* 10: 97–129.
Mays LE and Gamlin PDR. 1995. Neuronal circuitry controlling the near response. *Curr. Opin. Neurobiol.* 5: 763–768.
Miller JM and Robins D. 1992. Extraocular muscle forces in alert monkey. *Vision Res.* 32: 1099–1113.
Moschovakis AK, Scudder CA, and Highstein SM. 1996. The microscopic anatomy and physiology of the mammalian saccadic system. *Prog. Neurobiol.* 50: 133–254. (This is a detailed, encyclopedic review of the subcortical circuitry for saccadic eye movements.)
Porter JD, Guthrie BL, and Sparks DL. 1983. Innervation of monkey extraocular muscles: Localization of sensory and motor neurons by retrograde transport of horseradish peroxidase. *J. Comp. Neurol.* 218: 208–219.
Robinson DA. 1981. The use of control systems analysis in the neurophysiology of eye movements. *Annu. Rev. Neurosci.* 4: 463–503.
Robinson DA. 1987. The windfalls of technology in the oculomotor system. *Invest. Ophthalmol. Vis. Sci.* 28: 1912–1924.
Robinson DA, O'Meara DM, Scott AB, and Collins CC. 1969. Mechanical components of human eye movements. *J. Appl. Physiol.* 26: 548–553.
Rodieck RW. 1988. Looking. Chapter 13, pp. 294–325, in *The First Steps in Seeing*. Sinauer Associates, Sunderland, MA.
Scott AB. 1979. Ocular motility. Chapter 23 in *Duane's Foundations of Clinical Ophthalmology,* Vol. 2, Tasman W and Jaeger EA, eds. JB Lippincott, Philadelphia.
Sparks DL and Mays LE. 1990. Signal transformations required for the generation of saccadic eye movements. *Annu. Rev. Neurosci.* 13: 309–336.
Stone LS and Lisberger SG. 1990. Visual responses of Purkinje cells in the cerebellar flocculus during smooth pursuit eye movements. I. Simple spikes. *J. Neurophysiol.* 63: 1241–1261.
Westheimer G and Blair SM. 1973. Oculomotor defects in cerebellectomized monkeys. *Invest. Ophthalmol.* 12: 618–621.
Zhang Y, Mays LE, and Gamlin PDR. 1992. Characteristics of near response cells projecting to the oculomotor nucleus. *J. Neurophysiol.* 67: 944–960.

Extraocular Muscle Structure and Contractile Properties

Alberts B, Bray D, Lewis J, Raff M, Roberts K, and Watson JD. 1994. Muscle. Chapter 11, pp. 847–858, in *Molecular Biology of the Cell,* 3rd Ed. Garland, New York.
Burke RE. 1981. Motor units: Anatomy, physiology, and functional organization. Chapter 10, pp. 345–422, in *Handbook of Physiology, Section 1: The Nervous System,* Vol. 2, *Motor Control, Part 1,* Brooks VB, ed. American Physiological Society, Bethesda, MD.
Eggers HM. 1993. Functional anatomy of the extraocular muscles. Chapter 31 in *Duane's Foundations of Clinical Ophthalmology,* Vol. 1, Tasman W and Jaeger EA, eds. Lippincott-Raven, Philadelphia.
Finer JT, Simmons RM, and Spudich JA. 1994. Single myosin molecule mechanics: Piconewton forces and nanometer steps. *Nature* 368: 113–119.
Oda K. 1986. Motor innervation and acetylcholine distribution of human extraocular muscle fibers. *J. Neurol. Sci.* 74: 125–133.
Porter JD, Baker RS, Ragusa RJ, and Brueckner JK. 1995. Extraocular muscles: Basic and clinic aspects of structure and function. *Surv. Ophthalmol.* 39: 451–484.
Porter JD, Burns LA, and McMahon EJ. 1989. Denervation of primate extraocular muscle: A unique pattern of structural alterations. *Invest. Ophthalmol. Vis. Sci.* 30: 1894–1908.
Robinson DA and Keller EL. 1972. The behavior of eye movement motoneurons in the alert monkey. *Bibl. Ophthal.* 82 (Cerebral Control of Eye Movements and Motion Perception): 7–16, S. Karger, Basel.
Scott AB. 1981. Botulinum toxin injection of the eye muscles to correct strabismus. *Trans. Am. Ophthalmol. Soc.* 79: 734–770.
Scott AB and Collins CC. 1973. Division of labor in human extraocular muscles. *Arch. Ophthalmol.* 90: 319–322.
Sellin LC. 1985. The pharmacological mechanism of botulism. *Trends Pharmacol. Sci.* 6: 80–82.
Spencer RF and McNeer KW. 1987. Botulinum toxin paralysis of adult monkey extraocular muscles: Structural alterations in orbital, singly innervated muscle fibers. *Arch. Ophthalmol.* 105: 1703–1711.
Spencer RF and McNeer KW. 1991. The periphery: Extraocular muscles and motor neurons. Chapter 8, pp. 175–199, in *Vision and Visual Dysfunction,* Vol. 8, *Eye Movements,* Carpenter RHS, ed. CRC Press, Boca Raton, FL.

Spencer RF and Porter JD. 1988. Structural organization of the extraocular muscles. Chapter 2, pp. 33–79, in *Reviews of Oculomotor Research,* Vol. 2, *Neuroanatomy of the Oculomotor System,* Büttner-Ennever JA, ed. Elsevier, Amsterdam.

Squire JM. 1986. *Muscle: Design, Diversity and Disease.* Benjamin/Cummings, Menlo Park, CA.

Sensory Endings in Extraocular Muscles and Tendons

Gauthier GM, Nommay D, and Vercher J-L. 1990. The role of ocular muscle proprioception in visual localization of targets. *Science* 249: 58–61.

Gordon J and Ghez C. 1991. Muscle receptors and spinal reflexes: The stretch reflex. Chapter 37, pp. 564–580, in *Principles of Neural Science,* 3rd Ed., Kandel ER, Schwartz JH, and Jessell TM, eds. Elsevier, New York.

Keller EL and Robinson DA. 1971. The absence of stretch reflex in extraocular muscles of the monkey. *J. Neurophysiol.* 34: 908–919.

Lewis RF, Zee DS, Gaymard BM, and Guthrie BL. 1994. Extraocular muscle proprioception functions in the control of ocular alignment and eye movement conjugacy. *J. Neurophysiol.* 72: 1028–1031.

Lukas JR, Aigner M, Blumer R, Heinzl H, and Mayr R. 1994. Number and distribution of neuromuscular spindles in human extraocular muscles. *Invest. Ophthalmol. Vis. Sci.* 35: 4317–4327, with numerous corrections in *Invest. Ophthalmol. Vis. Sci.* 36: 521–522, 1995.

Porter JD. 1986. Brainstem terminations of extraocular muscle primary afferent neurons in the monkey. *J. Comp. Neurol.* 247: 133–143.

Purves D, Augustine GJ, Fitzpatrick D, Katz LC, LaMantia A-S, and McNamara JO. 1997. Spinal cord circuits and motor control. Chapter 15, pp. 291–309, in *Neuroscience.* Sinauer Associates, Sunderland, MA.

Ruskell GL. 1979. The incidence and variety of Golgi tendon organs in extraocular muscles of the rhesus monkey. *J. Neurocytol.* 8: 639–653.

Ruskell GL. 1984. Spiral nerve endings in human extraocular muscles terminate in motor end plates. *J. Anat.* 139: 33–43.

Ruskell GL. 1989. The fine structure of human extraocular muscle spindles and their potential proprioceptive capacity. *J. Anat.* 167: 199–214.

Actions of the Extraocular Muscles

Boeder P. 1961. The co-operation of extraocular muscles. *Am. J. Ophthalmol.* 51: 469–481.

Boeder P. 1962. Co-operative action of extraocular muscles. *Brit. J. Ophthalmol.* 46: 397–403.

Clark RA, Miller JM, and Demer JL. 1997. Location and stability of rectus muscle pulleys: Muscle paths as a function of gaze. *Invest. Ophthalmol. Vis. Sci.* 38: 227–240.

Demer JL, Miller JM, and Poukens V. 1996. Surgical implications of the rectus extraocular muscle pulleys. *J. Pediatr. Ophthal. Strabimus* 33: 208–218

Demer JL, Miller JM, Poukens V, Vinters HV, and Glasgow BJ. 1995. Evidence for fibromuscular pulleys of the recti extraocular muscles. *Invest. Ophthalmol. Vis. Sci.* 36: 1125–1136.

Diamond GR and Eggers HM. 1993. *Strabismus and Pediatric Ophthalmology,* Vol. 5, *Textbook of Ophthalmology,* Podos SM and Yanoff M, eds. Mosby, St. Louis.

Ettl A, Kramer J, Daxer A, and Koornneef L. 1997. High-resolution magnetic resonance imaging of the normal extraocular musculature. *Eye* 11: 793–797.

Helveston EM. 1993. *Surgical Management of Strabismus,* 4th Ed. Mosby, St. Louis.

Koorneef L. 1982. Orbital connective tissue. Chapter 32 in *Duane's Foundations of Clinical Ophthalmology,* Vol. 1, Tasman W and Jaeger EA, eds. Lippincott-Raven, Philadelphia.

Miller JM. 1989. Functional anatomy of normal human extraocular muscles. *Vision Res.* 29: 223–240.

Miller JM and Robinson DA. 1984. A model of the mechanics of binocular alignment. *Comput. Biomed. Res.* 17: 436–470.

Porter JD, Poukens V, Baker RS, and Demer JL. 1996. Structure-function correlations in the human medial rectus extraocular muscle pulleys. *Invest. Ophthalmol. Vis. Sci.* 37: 468–472.

Robinson DA. 1975. A quantitative analysis of extraocular muscle cooperation and squint. *Invest. Ophthalmol. Vis. Sci.* 14: 801–825.

Simonsz HJ, van Minderhout HM, and Spekreijse H. 1997. Sixty strabismus cases operated with the Computerized Strabismus Model 1.0: When does it benefit, when not? *Strabismus* 5: 203–214.

Development of the Extraocular Muscles

Freedman HL and Kushner BJ. 1997. Congenital ocular aberrant innervation—New concepts. *J. Pediatr. Ophthalmol. Strabismus* 34: 10–16.

Gilbert PW. 1957. The origin and development of the human extrinsic ocular muscles. *Contrib. Embryol.* 36: 61–78.

Miller NR, Kiel SM, Green WR, and Clark AW. 1982. Unilateral Duane's retraction syndrome (Type 1). *Arch. Ophthal.* 100: 1468–1472.

Porter JD and Baker RS. 1998. Anatomy and embryology of the ocular motor system. Chapter 25, pp. 1043–1099, in *Walsh & Hoyt's Clinical Neuro-Ophthalmology,* Vol. 1, 5th Ed., Miller NR and Newman NJ, eds. Williams & Wilkins, Baltimore.

Sevel D. 1981. A reappraisal of the origin of human extraocular muscles. *Ophthalmology* 88: 1330–1338.

Sevel D. 1986. The origins and insertions of the extraocular muscles: Development, histologic features, and clinical significance. *Trans. Am. Ophthalmol. Soc.* 24: 488–526.

Chapter 5

The Nerves of the Eye and Orbit

Life is a question of nerves and fibers, and slowly built-up cells in which thought hides itself and passion has its dreams.

■ Oscar Wilde, *The Picture of Dorian Gray*

Contemporary theories of perception, cognition, intelligence, and other mental attributes . . . tend to regard the fact that the brain resides in a body as a superfluous detail. [But it may be] . . . that a first step in deciphering the brain is to understand the way it serves the needs of the body it tenants.

■ Dale Purves, *Body and Brain*

Elements of Neural Organization

The brain deals with information about the external world and the body

The nervous system receives information from the external world and from the body, evaluates the information, makes decisions, and sends commands to a variety of effectors, mostly muscles and glands. The eyes are particularly important in this process; most of what we know about the world is based on information derived through seeing.

"Information" is a key word in any discussion of the nervous system. Information can be defined technically and sometimes specified quantitatively (as in bits or bytes), but for present purposes, information is news. The amount of information, or newsworthiness, is related to the rarity of an event's occurrence; the less likely it is that something will happen, the more informative its occurrence is. ("Dog bites man" is commonplace and therefore not informative; "man bites dog" is unexpected and therefore newsworthy.) For sensory systems, most information is conveyed by change, such as a change in air pressure for the auditory system or a change in mechanical pressure for the somatosensory system. For vision, changes in intensity, wavelength, or location of light over space and time are the main pieces of information the brain uses to reconstruct the external world. The less predictable the changes, the more informative they are.

Peripheral nerves connect the brain to "the body it tenants" and to the world outside the body. In discussing them, the significant issues are the kinds of information the nerves convey, where the information is coming from and where it is going, and the consequences of disrupting the flow of information.

Neurons are the anatomical elements of neural systems

A **neuron** is a single nerve cell. Neurons vary enormously in their morphology (see Figure 5.1) and in their modes of operation; they may use different neurotransmitters, some may conduct pulses of electrical activity where others do not, and so on. Their common feature, however, is that they are the building blocks, the basic elements, from which the nervous system is constructed. A "typical" neuron—for example, a spinal motor neuron (Figure 5.1*a*)—is bipolar: It receives inputs at one end from sensory receptors or other neurons, and provides outputs at the other end to neurons, muscles, or glands. Inputs are usually delivered to small branches of the cell, the **dendrites**, which are collectively referred to as the dendritic *arbor,* while the outputs are from another group of branches arising from a single long process called the **axon**. The nerve cell body (the **soma**, plural somata) contains the cell's apparatus for metabolism and protein synthesis and lies somewhere between the two ends of the cell.

Neurons are specialized to conduct electrical activity, which usually takes the form of brief changes in the electrical polarity of the cell's membrane that can be

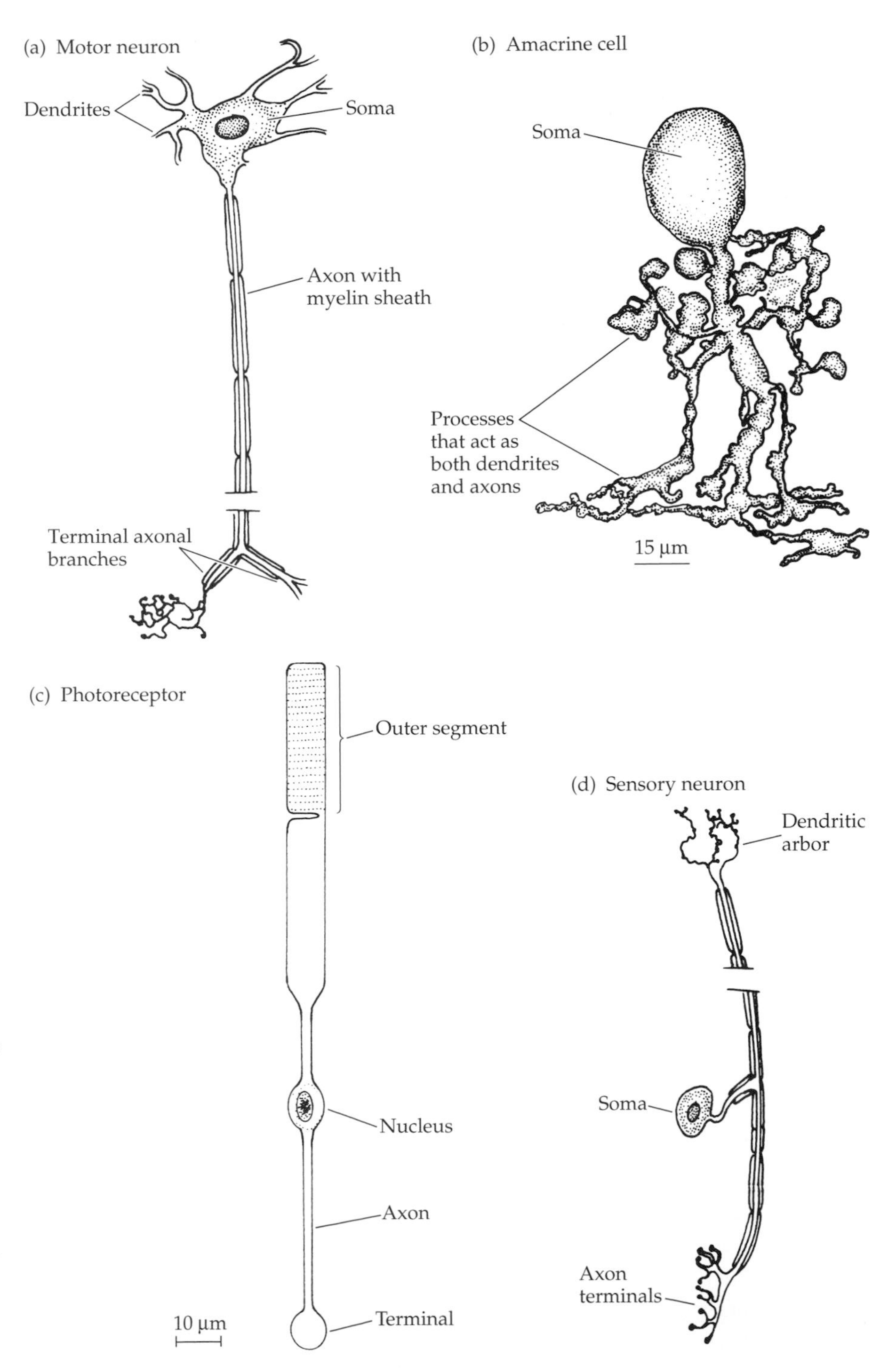

Figure 5.1
Variation in Neuronal Morphology
(*a*) A motor neuron with dendrites arising from the soma and a long axon—shown here as interrupted—with its terminal axonal branches. (*b*) A retinal amacrine cell, which is an atypical neuron; the processes arising from the soma operate simultaneously as dendrites and axons, both receiving inputs and making outputs. (*c*) A rod photoreceptor has no dendrites and does not conduct action potentials, but neurotransmitter molecules are released from the terminal. (*d*) In this sensory neuron from the trigeminal nerve, the soma is much closer to the axon terminals than to the dendritic arbor. The site of action potential generation is near the sensory endings, making the long process extending peripherally from the soma appear to be more like axon than like dendrite.

propagated from one end of the cell to the other without the amplitude being diminished. The frequency of these **action potentials** is related to the strength of the inputs to the cell; the information conveyed by the action potentials, however, depends on where the inputs come from and where the outputs go. In other words, the *meaning* of a burst of action potentials of a particular frequency is quite different in a motor neuron from that, say, in a neuron in the optic nerve.

Some neurons are not bipolar—that is, there is no clear distinction between dendrites and axon (a retinal amacrine cell; Figure 5.1*b*); some do not conduct action potentials (a photoreceptor; Figure 5.1*c*); and some have their cell bodies much closer to the output end than to the input end (a trigeminal nerve neuron; Figure 5.1*d*). Nonetheless, all neurons convey information in the form of electrical activity.

Neural circuits consist of neurons linked mostly by unidirectional chemical synapses

The primary mode of communication among neurons, and between neurons and various effectors, is chemical in nature. At sites of close apposition between two neurons, or between neuron and effector, called **synapses**, axon terminals of one cell release a chemical **neurotransmitter** that interacts with chemical **receptors** on the dendrites of the second neuron or on cells of a muscle or gland (Figure 5.2*a* and *b*). The cell that releases the neurotransmitter is the **presynaptic** cell; the cell on which the neurotransmitter acts is the **postsynaptic** cell. Perhaps a half dozen substances are used by neurons as neurotransmitters; others whose actions on neurons are less direct are called **neuromodulators**. Whatever the case, the effect is either to **excite** (increase) or **inhibit** (decrease) the electrical activity in the target cell.

Chemical synapses are unidirectional: Information flows from the presynaptic to the postsynaptic cell, but transmission back across a particular synapse in the opposite direction is impossible. A postsynaptic cell can feed back onto a presynaptic cell, but such transmission requires another synapse whose directionality is opposite the first. These **recurrent** connections are very common in the retina and central nervous system. The unidirectionality of synapses allows us to ignore the fact that the electrical activity in a neuron, wherever it is generated in the cell, will spread throughout the cell, invading both axon and dendrites. The electrical activity will evoke neurotransmitter release only where the neurotrans-

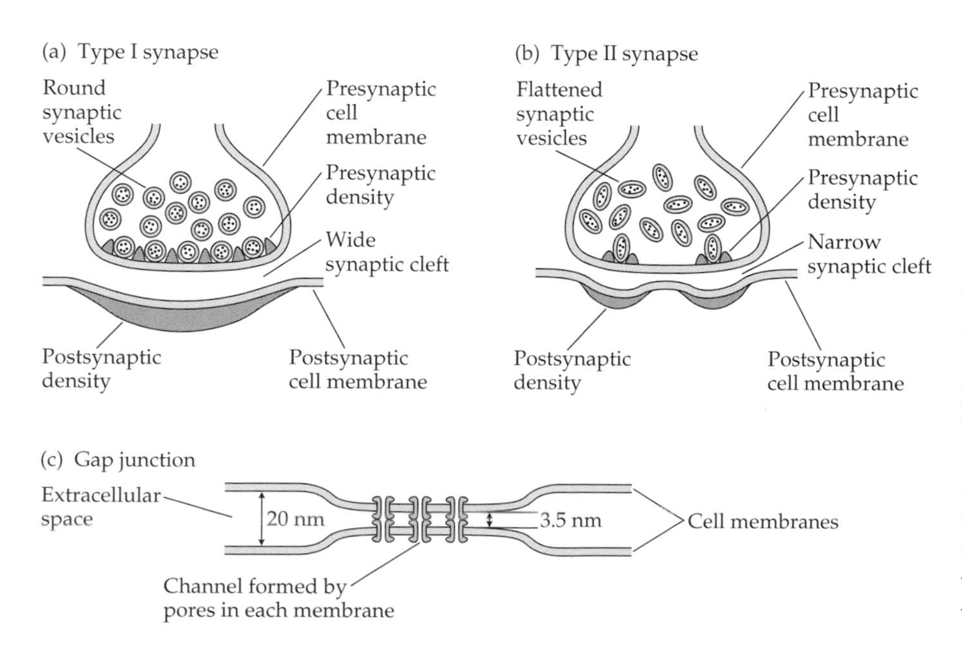

Figure 5.2
Sites of Communication between Neurons
(*a*) A conventional chemical synapse as reconstructed from electron microscopic sections. There is a wide synaptic cleft into which the contents (neurotransmitter molecules) of the vesicles on the presynaptic side are released. The membranes on both sides of the cleft are electron dense, most notably on the postsynaptic side. (*b*) A chemical synapse, generally thought to release an inhibitory neurotransmitter. The cleft is relatively narrow, the regions of pre- and postsynaptic membrane density are smaller, and the vesicles are markedly elliptical. (*c*) At gap junctions, the cell membranes are pressed tightly together, leaving almost no space between them. This arrangement places ion channels in the two membranes directly in contact with one another and current can flow in either direction. Conductance through gap junctions varies with changes in the pore size.

mitter has been accumulated in the cell adjacent to the synapses, which are typically at the axon terminals. At sites where the cell receives inputs, no neurotransmitter is available for release and the electrical activity cannot be communicated to the presynaptic cell. (The qualifier "chemical" in reference to synapses has been used deliberately; gap junctions [Figure 5.2*c*] between neurons or muscle fibers can be sites of *bi*directional *electrical* communication. Gap junctions are common in the retina and in some of the intrinsic muscles of the eye.)

The direction of neural information flow distinguishes between sensory and motor nerves

The unidirectionality of chemical synapses means that it is proper to talk about a direction of information flow; an optic nerve fiber conducts signals *from* the eye *to* the brain, but not the other way around. Information can flow from brain to eye only by another set of nerve fibers. Thus, nerves that conduct information to the brain are **sensory**, while nerves that carry signals from the brain to the peripheral muscles and glands are **motor**. In many cases, however, the designations "motor" or "sensory" nerve are indications of the dominant group of nerve fibers (rather than all the fibers); many nerves contain both sensory and motor axons. Even the optic nerve in some species contains a high proportion of fibers that convey signals from the brain to the retina—the *centrifugal fibers*—presumably allowing the brain to modify whatever the retina is doing. (Evidence for centrifugal fibers in the human optic nerve is very weak; they probably do not exist.)

Motor outputs are divided anatomically and functionally into somatic and autonomic systems

The neural systems of which the peripheral nerves are part, either as input or output, can be categorized meaningfully in several ways. The most sizable distinction is between the **autonomic nervous system** and everything else, which for want of a better term is called the **somatic**, or **voluntary**, **nervous system**. All of the spinal and cranial nerves are associated with the somatic system (which includes the heart and diaphragm as voluntary muscles).

The autonomic system is both anatomically and functionally distinctive; the output of the autonomic system is to the body's smooth muscles and glands; its inputs come primarily (though not exclusively) from the body's **enteroceptors**, whose job is monitoring the status of internal conditions in the body's organs. Whereas the cell bodies of motor neurons in the somatic nervous system are contained within the brain and spinal cord, the cell bodies of autonomic neurons lie in ganglia that are separate and sometimes quite distant from the central nervous system.

The autonomic system is subdivided into the sympathetic and parasympathetic systems

The autonomic system has two parts, the **sympathetic nervous system** and the **parasympathetic nervous system**. Again, the distinction between them is part anatomic and part functional. Anatomically, the systems differ in the location of the nerve ganglia: The somata of the sympathetic neurons lie in two long chains of ganglia on either side of the vertebral column; the parasympathetic ganglia are scattered throughout the body in close proximity to the target organs of the parasympathetic neurons.

Functionally, the sympathetic system uses norepinephrine (noradrenaline) as a neurotransmitter, but the parasympathetic system uses acetylcholine (there may be exceptions, as will be noted later in this chapter). The two systems of neurons often innervate the same targets, but usually with antagonistic effects.

One or both parts of the autonomic system innervate the eye's intrinsic muscles, its vascular smooth muscle, some smooth muscle in the eyelids, and the glands.

The Optic Nerve and the Flow of Visual Information

In the optic nerve, the location of axons from retinal ganglion cells corresponds to their location on the retina

The **optic nerve** (cranial nerve II) is the eye's main sensory nerve. Its structure in and near the eye will be discussed in Chapter 16; the issue here is the paths the axons follow to the brain and their central destinations—that is, the routing and destinations of information sent out from the retina.

The human optic nerve contains about 1 million nerve fibers, all of which are axons of the ganglion cells that form the innermost cellular layer of the retina. Ganglion cell axons from all parts of the retina converge to the optic nerve head, which is about 3 mm nasal to the posterior pole of the eye and about 1 mm below the horizontal plane (Figure 5.3*a*). The axon bundle exits the eye as the optic nerve, runs nasal and posterior in the orbit to the optic foramen, through which it passes into the optic canal in the sphenoid bone. After passing through the sphenoid bone, the optic nerves from each eye join at the optic chiasm (which will be discussed shortly).

The axons enter the nerve at the nerve head in a fairly orderly and consistent way. To a large extent, the initial location of axons in the nerve corresponds to the location of the parent ganglion cells in the retina (Figure 5.3*b*). Ganglion cells in the superior retina have axons in the superior half of the nerve, and the axons in the inferior part of the nerve are from ganglion cells in the inferior retina. Ganglion

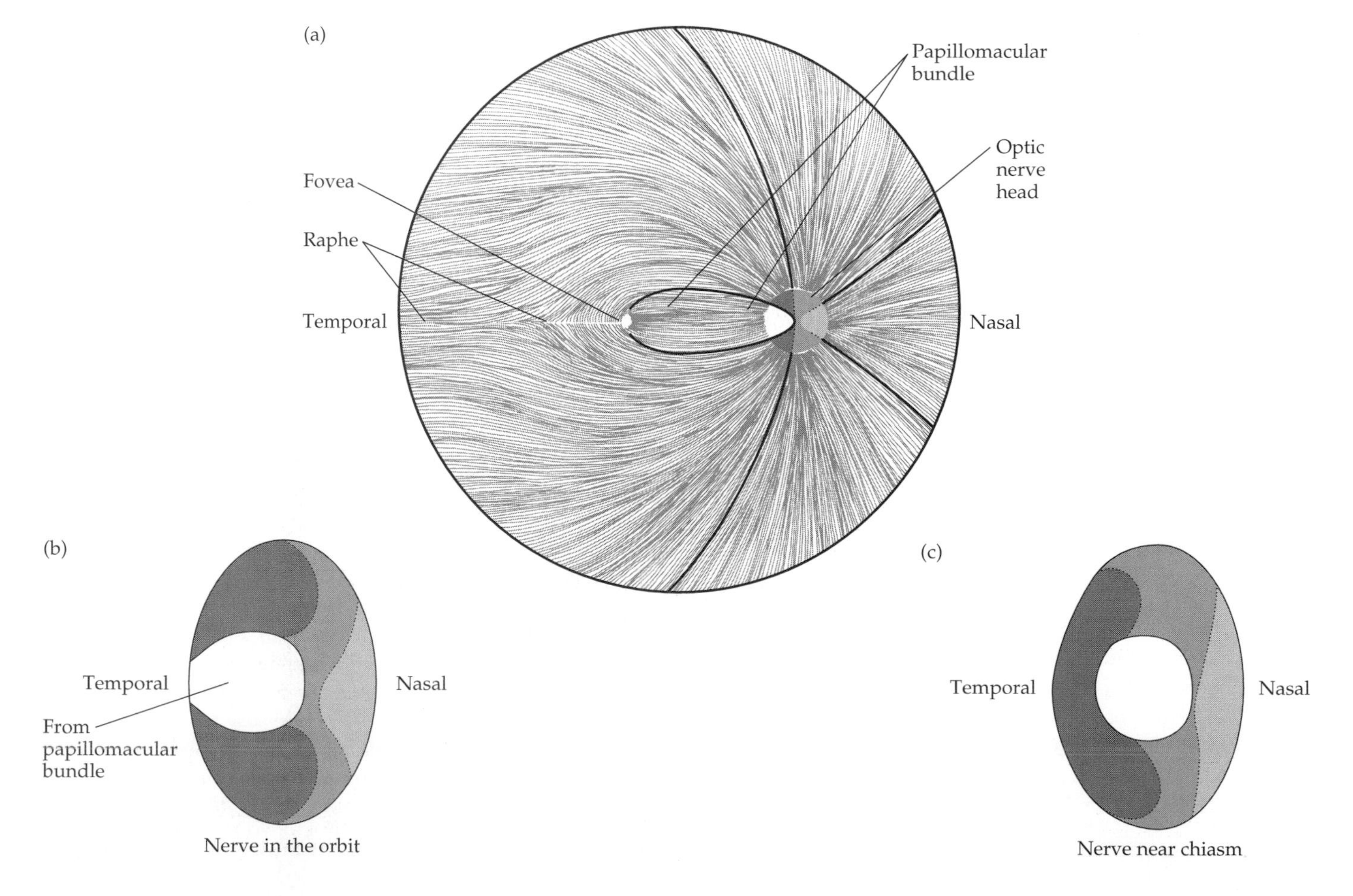

Figure 5.3
Ganglion Cell Axons Converging to Form the Optic Nerve
(*a*) The ganglion cell axons that constitute the optic nerve run across the retina and converge in a systematic way; axons from foveal ganglion cells (the papillomacular bundle) lie in a wedge-shaped region on the lateral side of the nerve head, with axons from the superior-temporal retina above them and axons from the inferior-temporal retina below them. Axons from retina regions medial to the nerve head occupy the medial side of the nerve. (*b*) This section through the nerve behind the eye within the orbit shows the axons from the papillomacular bundle shifting toward the center of the nerve. (*c*) Near the optic chiasm, the axons have shifted even more, but there is still considerable order. The foveal axons are now quite central in the nerve and the progressively more peripheral annuli of axons in the nerve correspond to more peripheral locations in the retina. (After Naito 1989.)

Vignette 5.1

The Integrative Action of the Nervous System

DISCUSSIONS OF NEURONS, NERVES, and the nervous system often require words like proprioception, nociception, recruitment, motor unit, receptive field, reciprocal innervation, final common path, and synapse. These words were first used by Charles Scott Sherrington (1857–1952) to describe basic properties of the nervous system, and they can be found in his book *The Integrative Action of the Nervous System*, after which this vignette is titled. Although based on work done almost a century ago (the book was first published in 1906 and reissued in 1947), it is still worth reading; facts and details have changed, but Sherrington stressed principles of neural organization, and it is testimony to his brilliance that the principles he elucidated have survived.

Sherrington's accomplishments are particularly remarkable in view of the enormous limitations he confronted in trying to study the nervous system: The first recordings of electrical activity in nerves had only just been made, the status of neurons as anatomically discrete elements was still a matter of debate, and none of the apparatus taken for granted by neuroscientists of today existed. Sherrington dealt with these obstacles by simplification. He focused on spinal reflexes in which the stimuli could be tightly controlled and the responses could be easily monitored, often using experimental animals in which the influences of the central nervous system were eliminated by surgical sectioning of the spinal cord at a high cervical level; these animals were also "decerebrate" preparations—that is, their entire cerebrum was also removed, a task that is anything but simple.

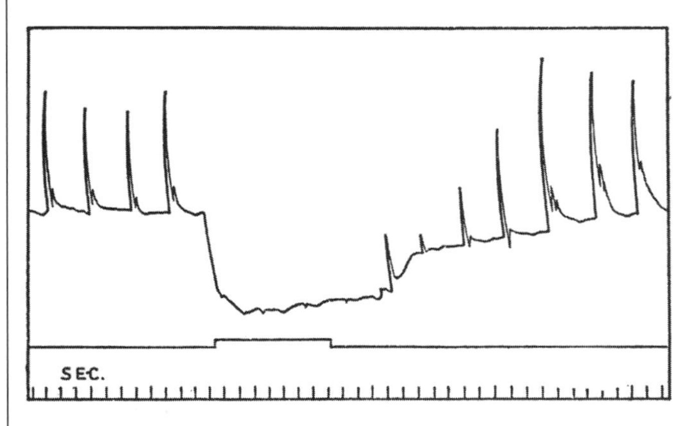

Figure 1
Reciprocal Innervation
The upper trace shows contractions of the knee extensor as the patellar tendon reflex is elicited. The step on the middle trace represents the duration of an electrical stimulus applied to the nerve going to the antagonist flexor muscle; stimulating the flexor's nerve inhibits the reflex contraction in the extensor because of the reciprocal innervation. (From Sherrington 1906.)

Preparing to study spinal reflexes required a long and tedious study of the spinal nerves, working out where the various nerves went and what structures they innervated. In the process, Sherrington was able to clarify the important distinction between the dorsal roots and ganglia, which contain sensory neurons, and the ventral roots, which convey axons of spinal motor neurons. Then came the "simple" experiments, in which natural or electrical stimuli could be used to elicit the contraction of particular muscles. Figure 1, for example, shows recordings of tension developed in the extensor muscle of the knee. The rhythmic contractions produced by striking the patellar tendon are interrupted by electrical stimulation of the nerve going to the antagonist muscle, which is a flexor of the knee. The result demonstrates both inhibition and reciprocal innervation.

One of Sherrington's students was John Carew Eccles (1903–1997), who was awarded the Nobel prize in physiology or medicine in 1963 (with Alan Hodgkin and Andrew Huxley), in part for his demonstration of inhibition on spinal motor neurons. Sherrington could not demonstrate inhibition directly—it was inferred from the results—but his student was able to accomplish the feat some 50 years later. And the experiments were similar in logic if not in technique. Where Sherrington used muscle contraction as his indicator of neural excitation and inhibition, Eccles was able to use the membrane potential of the motor neuron.

Although Sherrington was the first to demonstrate reciprocal innervation clearly, he credited his predecessors with the original idea, a behavior said by his biographers to be typical. In this instance, he credited René Descartes (1596–1650) and used illustrations from two editions of Descartes's *Tractatus De Homine* to make the point. One of them is included here as Figure 2; it shows a drawing of the eye and extraocular muscles with a system of nerves and valves to control the flow of a nerve fluid that Descartes (and almost everyone else) thought to be responsible for the neural activation of muscles. Descartes's conception is wrong in several respects, in particular his idea that the reciprocal innervation was a peripheral mechanism requiring interconnection of muscles within the orbit, but there is no question that his general understanding of reciprocal innervation was correct.

Descartes's writings, in their original editions, were part of Sherrington's library. His early schooling included a strong background in classics, literature, and languages (he read Latin and was fluent in French and German), all of which combined to make him a lifelong writer of poetry (some of which was published), an ardent bibliophile, and a serious collector of rare books. He is said to have given more incunabula (books printed before 1501) to the British Museum than any other individual donor.

Sherrington's immense erudition and continual curiosity made him a difficult lecturer; it was much too easy to stray from a story, posing questions, speculating about answers, and expressing wonderment about the brain and humankind and nature. Although other physiologists and his graduate students could find pleasure and profit in the changing levels of this open-ended discourse, undergraduates and lay audiences were less likely to appreciate a tale with no ending and little obvious plot. But there are other ways to teach. Sherrington had an enormous influence throughout the world with a book called *Mammalian Physiology: A Course of Practical*

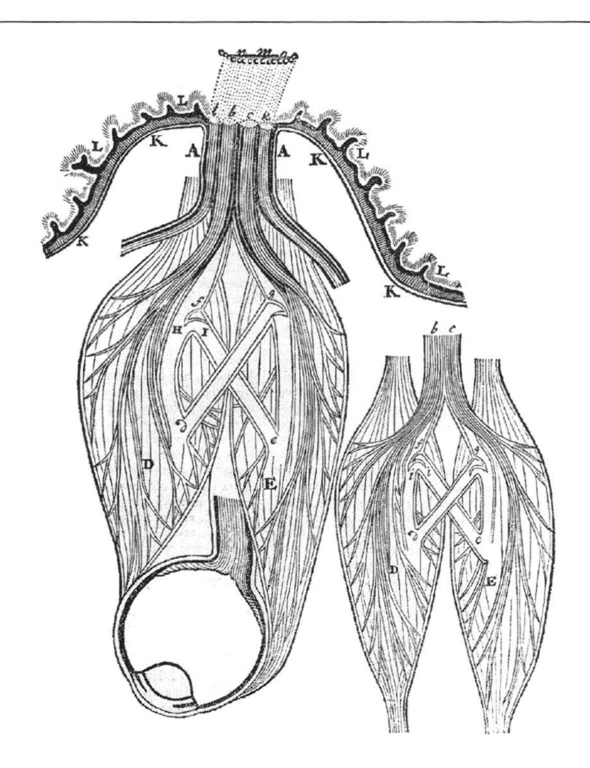

Figure 2
Reciprocal Innervation in Extraocular Muscles According to René Descartes
The eye is drawn in situ on the left, with an isolated agonist–antagonist muscle pair on the right. The open channels are the nerves; the nerve to each muscle also has a connection to its antagonist. When fluid flows through the nerve to activate one muscle (on the left), the connection to the antagonist (on the right) prevents it from contracting. (From Descartes 1677.)

Exercises (first published in 1919, with a revised edition in 1929). I vividly recall repeating some of the experiments in my physiology classes in 1960, even using the same kind of soot-blackened kymograph drums to record changes in muscle tension. (The original version of Figure 1 was a kymograph record; after the recording pen had traced its way through the soot, the paper was sprayed with varnish to keep the soot from smudging.)

Sherrington received the Nobel prize in physiology or medicine in 1932, when he was 75 years old and about to retire from his Oxford professorship. The award was shared with his younger friend, Edgar Douglas Adrian (1889–1977), the Cambridge physiologist who was the first to record action potentials from optic nerve axons—among other things. Retirement for Sherrington meant the continuation of a voluminous and affectionate correspondence with his friends and former students and the writing of more books. *Man on His Nature* was first published in 1941, with a second much-revised edition in 1951, and *The Endeavour of Jean Fernel* in 1946 (Fernel was a sixteenth-century French physician, physiologist, and anatomist; he is credited with coining the term "pathology"). Sherrington's life ended peacefully at the age of 94, as he sat by his fireplace reading a book—in French.

cells that are nasal to the optic nerve head have axons in the medial (nasal) side of the nerve, while the axons from temporal ganglion cells are on the lateral (temporal) side of the nerve. Axons coming from ganglion cells in the vicinity of the fovea occupy a large wedge of the nerve on its lateral side (see also Figure 16.24).

As the axons travel back toward the optic chiasm, they retain much of their topographic order, although axons from neighboring retinal regions intermingle somewhat. (This description is based on recent studies of axonal order in the optic nerve of the Japanese macaque monkey. The few data on humans indicate that the two species are similar.) The parts of the nerve occupied by axons from different parts of the retina change; close to the chiasm (Figure 5.3*c*), the foveal axons have moved into the center of the nerve, and their original lateral position has been filled by axons coming from more temporal retinal locations. Axons from the nasal retina are still medial in the nerve, but they have spread more toward its midline.

For clinicians, the consistent ordering of axons in the nerve means that localized lesions or infections may produce visual field defects with characteristic size and location, thus providing clues to the site and nature of the lesion.

The optic nerve is flexible and its path from the eye to the optic foramen is curved; both features permit the eye to rotate freely without the hindrance that would be produced if the nerve were stiff and straight. Orbital fat and connective tissue attachments support the nerve in the orbit. These connective tissue attachments between the dura mater around the nerve and other elements of the orbital fascia, particularly the epimysium (Tenon's capsule) of the rectus muscles, are densest where the nerve leaves the orbit through the apex of the muscle cone at the optic foramen. When the nerve is infected (a condition called **optic neuritis**), contraction of the muscles during eye movements can be painful because of the pulling and tugging on the nerve through the fascial connections. (Unspecialized sensory nerve terminals are scattered throughout the orbital connective tissue.)

Axons from the two optic nerves are redistributed in the optic chiasm

After the two optic nerves pass through their respective optic canals and exit the sphenoid bone, they merge as the **optic chiasm**, from which the bilateral **optic tracts** arise, each tract containing fibers going to their central destinations (Figure 5.4). In humans, the optic chiasm is a site for redistribution of the optic nerve fibers from the two eyes; axons from the nasal hemiretinas cross to the opposite side of the brain (the *contralateral* projection), while temporal retinal axons remain on the side from which they originated—that is, the *ipsilateral* side (Figure 5.5). (The directions nasal and temporal are relative to the center of the fovea.) The crossing of axons from one side to the other is called **decussation**; since only the nasal axons cross in humans, the decussation is partial, but it can be nearly complete in other species. Albino animals and Siamese cats have anomalous patterns of decussation; structural abnormalities in the lateral geniculate nuclei of human albinos indicate the presence of a similar anomaly.

Not all of the crossing axons take the most direct and shortest route from one side to the other. Some cross the midline and turn anteriorly, as if they were going into the contralateral optic nerve, then finally loop back to join the temporal retinal fibers of the other eye that are heading for the optic tract. Other nasal axons remain on the ipsilateral side as if they were trying to enter the ipsilateral optic tract, then change direction to cross the midline and go into the contralateral tract. Between these extremes are axons that run directly through the center of the chiasm. As we will consider later in this chapter, the dispersion and direction changes made by decussating nasal axons probably reflect the problems encountered and decisions made by pathfinding axons during development.

In any case, nasal and temporal retinal axons are separated and distributed within the chiasm in a characteristic way (see Figure 5.5). At the midline, only the decussating nasal axons are present. At the anterior edge between the optic nerves, most axons are from the *inferior* nasal retinas; at the posterior edge (between the tracts) most are from the *superior* nasal retinas. At the lateral sides

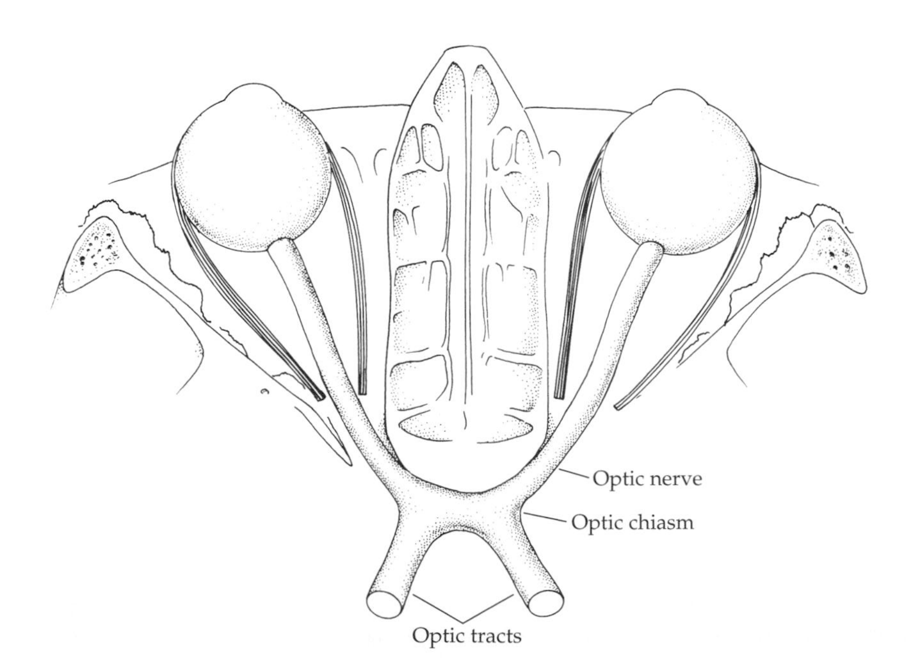

Figure 5.4
The Optic Nerves from the Eyes to the Optic Chiasm
The optic nerves run in a gently curved path medially from the eyes to the optica foramina, where they pass through the optic canals in the sphenoid bone; as they exit on the upper side of the sphenoid, they merge to form the optic chiasm. The optic tracts run back to the brain from the chiasm.

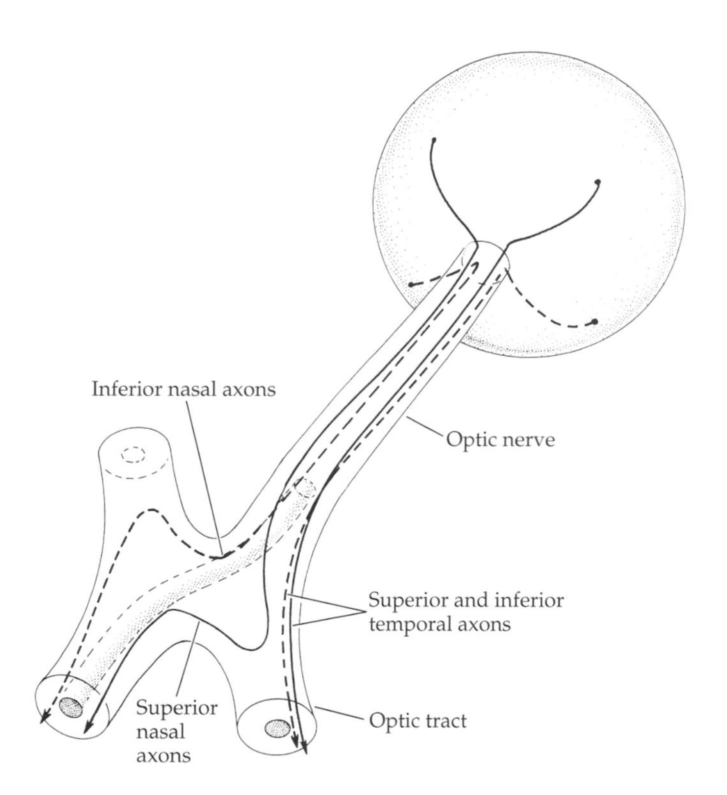

Figure 5.5
Axon Paths through the Optic Chiasm
Nasal retinal axons cross in the optic chiasm to the contralateral side; temporal retinal axons remain ipsilateral. Inferior-nasal axons tend to cross the midline at the anterior edge of the chiasm, while superior-nasal fibers cross posteriorly. The axons running along the sides of the chiasm are a mixture of ipsilateral temporal and contralateral nasal axons, but only those from the right retina are shown here. The shaded region shows the path of axons coming from ganglion cells near the fovea.

of the chiasm, temporal and nasal axons intermingle. Thus, lesions that are localized to one part or another of the chiasm necessarily affect different sets of axons and therefore produce different defects in the visual fields of the eyes.*

The decussation of axons in the chiasm is imperfect

If a vertical line were drawn through the center of the fovea to separate nasal from temporal, and if *all* nasal axons crossed at the chiasm, severing one optic tract would produce a region of blindness that bisected the fovea (or in anatomical terms, there would be a line separating ganglion cells on the nasal side of the *contralateral* retina whose axons had been severed from those on the temporal side that were unaffected). This is not what happens, however; when the experiment is done, the line dividing nasal and temporal hemiretinas is rather fuzzy. A vertical strip extends above and below the fovea within which ganglion cells may send their axons either ipsilaterally or contralaterally (Figure 5.6); this region is the **area of nasotemporal overlap**. The strip widens toward the retinal periphery, suggesting that the division between nasal and temporal is even less precise away from the fovea.

Although the overlap region is narrowest at the fovea, there are clearly ganglion cells on the temporal side of the fovea that send their axons contralaterally and, conversely, nasal ganglion cells that send their axons ipsilaterally. One important consequence of this arrangement is that the entire fovea and a bit of the region around it has some represention in both optic tracts and therefore on both sides of the visual cortex. This arrangement has implications for stereopsis

*Each eye has a visual field, the region of space within which objects can be seen when the eye is stationary and fixation with the fovea is maintained on a stationary target. The size of the monocular visual fields is specified in angular terms: Relative to the line of sight, the monocular visual fields extend about 90° laterally, about 60° medially (the nose gets in the way and limits the medial extent), about 45° superiorly (limited by the brow), and about 60° inferiorly. When both eyes are used, the binocular field extends about 180° horizontally, and the central 120° is the binocular area that is seen simultaneously by both eyes.

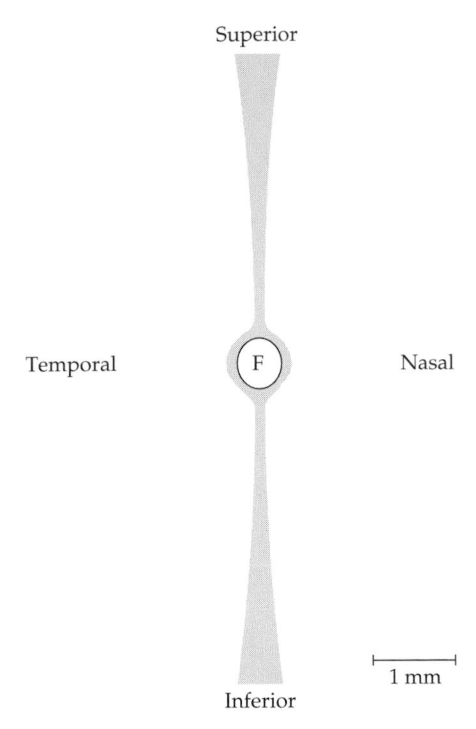

Figure 5.6
Nasotemporal Overlap of Crossed and Uncrossed Ganglion Cell Axons
A schematic view of the retina, showing the region (shaded) in which ganglion cell axons may project to either the contralateral or the ipsilateral side. Ganglion cells in a ring around the fovea (F) and in a vertical strip above and below the fovea may have either contralateral or ipsilateral projections. (After Fukuda et al. 1989.)

(a binocular system for depth perception), but it also means that a lesion on one side of the brain cannot completely remove axons coming from half of the fovea. In effect, the fovea is doubly represented and to some extent protected from the full impact of a unilateral lesion.

Spatial ordering of axons changes in the optic tracts

Most discussions of the human optic tract imply or assume that retinal axons from the two eyes course through the tract in an orderly fashion and that the order is **retinotopic** (that is, spatial relationships in the retina are preserved elsewhere). Such an arrangement suggests that a small bundle of axons in a particular part of the tract will be from corresponding parts of the two retinas and therefore correspond to the same region of the visual field; from one bundle of axons to the next there should be a systematic change in retinal and visual field regions. If this is true, a small tract lesion at a particular site should produce corresponding visual field defects in both of the monocular visual fields and different individuals with identical lesions (assuming such a situation is possible) should have identical visual field defects.

This view has been challenged in recent years by studies that use specific axonal labeling methods (see Box 5.2) to see how axons from particular sets of ganglion cells run through the nerve and tract. This work, particularly as it applies to primates, is still in progress, but it is already clear that the order in the tract is not strictly retinotopic; for example, neighboring axons may originate from ganglion cells in widely separated retinal regions, but they are likely to be of the same functional type. After axons pass through the chiasm, they are rearranged within the tract so that axons carrying the same sort of visual information cluster together, regardless of where they originate in the retina. Thus the tract may contain not just one continuous representation of a visual hemifield, but several separate representations.

The idea that tract axons should be segregated by cell type rather than by retinal location makes sense in terms of where they are going; their target, the lateral geniculate nucleus, has different, separate regions not only for inputs from different classes of retinal ganglion cells but also for inputs from the two eyes (see the next section). It is not surprising that this separation should be reflected in the tract; it is like freeway traffic distributing itself in different lanes in anticipation of an imminent exit or of one farther down the road. (Also like real freeway traffic, there are lane weavers and those that cut across traffic at the last minute to make their exit; traffic may be generally orderly, but there can be local disorder.) Unfortunately, this rearrangement makes matters more difficult for clinical neurology. It is difficult to predict what sort of visual field defect will result from a small optic tract lesion or to specify the location of a lesion based on the visual field defect.

In the lateral geniculate nuclei, which are primary targets of axons in the optic tracts, inputs from the two eyes are separated into different layers

Axons in the optic nerve can go to any of several destinations (Figure 5.7). About 90% of the optic nerve and tract axons are part of the primary visual pathway, terminating in ("projecting to") the **lateral geniculate nuclei** (**LGN**s) on either side of the midbrain. Coronal slices through one LGN shows distinct layers (Figure 5.8). Although some layers merge at the anterior and posterior ends, most of the LGN has six layers; they are numbered 1 through 6, beginning at the bottom (the ventral side) of the nucleus. Axons coming from the two eyes go to different layers. Layers 1, 4, and 6 receive inputs from the contralateral eye; layers 2, 3, and 5, ipsilateral inputs.

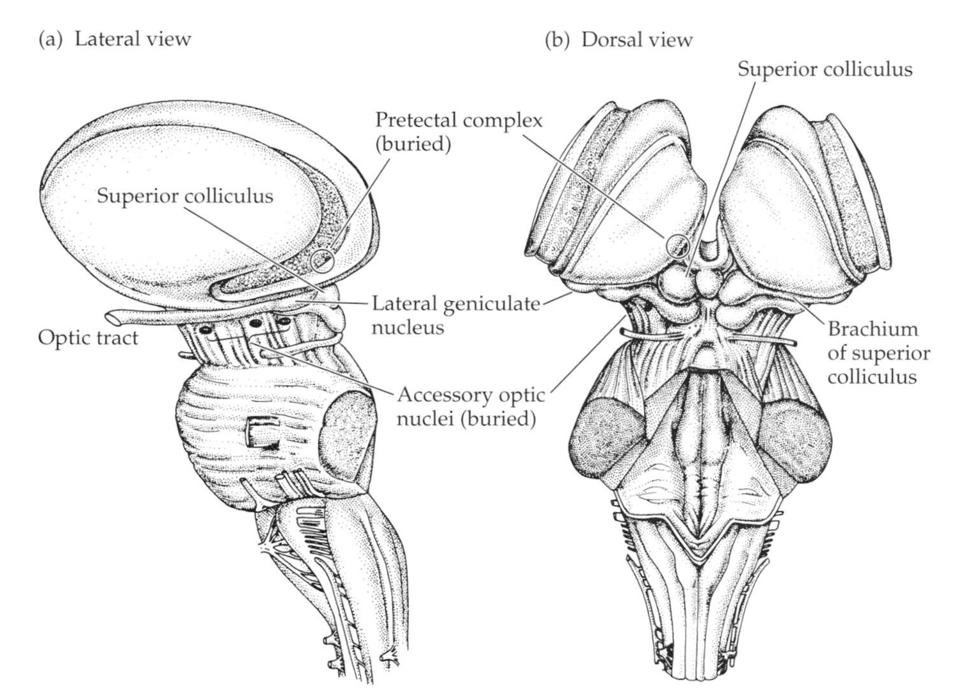

Figure 5.7
Brain Regions Innervated by Retinal Ganglion Cell Axons: The Direct Retinal Projections
The brainstem and midbrain, shown from the side (*a*) and in a posterior view (*b*), illustrating the locations of the various structures innervated by retinal ganglion cell axons. The major projection is to the lateral geniculate nuclei, whose cells send axons to the primary visual cortex (not shown). A secondary projection goes to the superior colliculi via the brachia of the superior colliculi. Small bundles of axons run to the pretectal nuclei lying just in front of the colliculi and to the accessory optic system, whose three terminal nuclei lie along the sides of the brainstem. Axons may leave the upper surface of the optic chiasm to innervate the suprachiasmatic nucleus in the hypothalamus (not shown). (After Martin 1989.)

The left LGN receives axons from the left temporal retina and the right nasal retina; it is therefore concerned with the right half of the visual world (the right LGN sees the left half). But since the axons from one eye go to three different layers, it appears that the each eye's hemifield is represented three times and that each half of the visual world is represented six times in the LGN. To some extent, these multiple representations of the visual world are a functional segregation of retinal inputs, wherein the retinal axons terminating in the different layers are from different classes of retinal ganglion cells.

Cells in the lateral geniculate nuclei are themselves segregated to some extent by size. Layers 1 and 2 are the large-cell (*magno*cellular) layers—one each for inputs from each eye; layers 3 through 6 are the small-cell (*parvo*cellular) layers—two each for inputs from each eye. In general, retinal axons from the large parasol ganglion cells (see Chapter 14) terminate in the large-cell layers, while axons from the smaller midget ganglion cells and from other ganglion cells terminate in the small-cell layers. One part of the functional segregation occurring here concerns color vision: Parasol ganglion cells do not convey wavelength-specific information, so the LGN cells receiving inputs in layers 1 and 2 cannot be part of our color vision system.

The axons of the LGN neurons continue the primary visual pathway as the **optic radiations** leading to the primary visual cortex in the occipital lobe.

Axons terminating in the lateral geniculate nuclei are spatially ordered

One of the important features of the retinogeniculate projection (the path from retina to LGN) is its retinotopic organization: Ganglion cells that are neighbors in the retina will have their terminals as neighbors in the LGN, regardless of how much their axons may have wandered apart in the nerve or tract. To put it another way, retinal ganglion cell axons are not free to terminate at random locations in the LGN; instead, neurons at a particular location in the nucleus will always be activated by visual stimuli at a particular location in the visual field, and movement of the stimulus from one location to another will be accom-

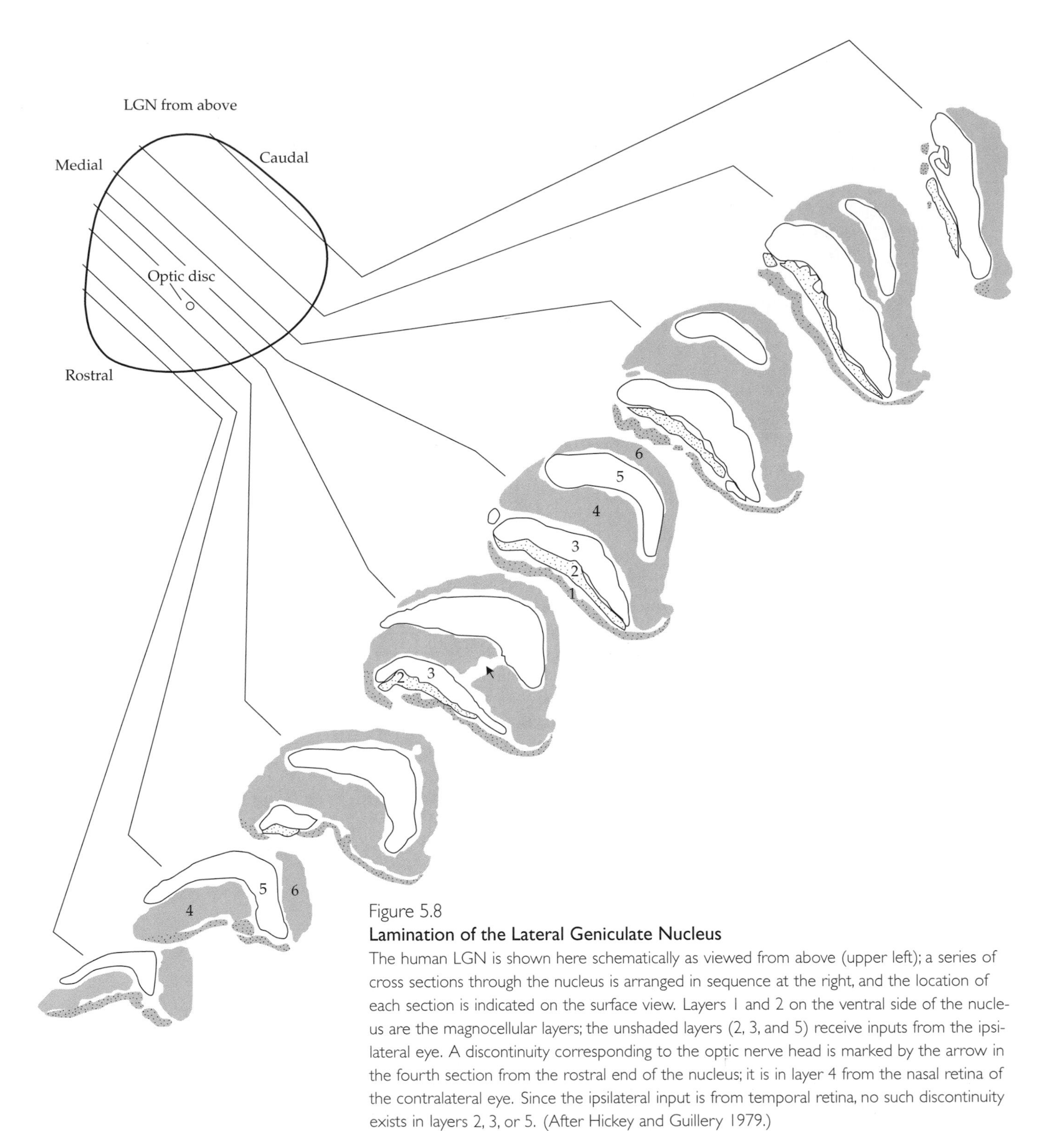

Figure 5.8
Lamination of the Lateral Geniculate Nucleus
The human LGN is shown here schematically as viewed from above (upper left); a series of cross sections through the nucleus is arranged in sequence at the right, and the location of each section is indicated on the surface view. Layers 1 and 2 on the ventral side of the nucleus are the magnocellular layers; the unshaded layers (2, 3, and 5) receive inputs from the ipsilateral eye. A discontinuity corresponding to the optic nerve head is marked by the arrow in the fourth section from the rostral end of the nucleus; it is in layer 4 from the nasal retina of the contralateral eye. Since the ipsilateral input is from temporal retina, no such discontinuity exists in layers 2, 3, or 5. (After Hickey and Guillery 1979.)

panied by a corresponding shift in the location of activated neurons in the LGN. It follows that destruction of a particular part of the LGN will always produce a **scotoma** (an area of relative or complete blindness) in the same part of the visual field.

Although the mapping of the retina onto the LGN retains spatial order, the maps are distorted (Figure 5.9); some parts of the visual field and therefore some parts of the retina are represented by more axons and therefore occupy more of

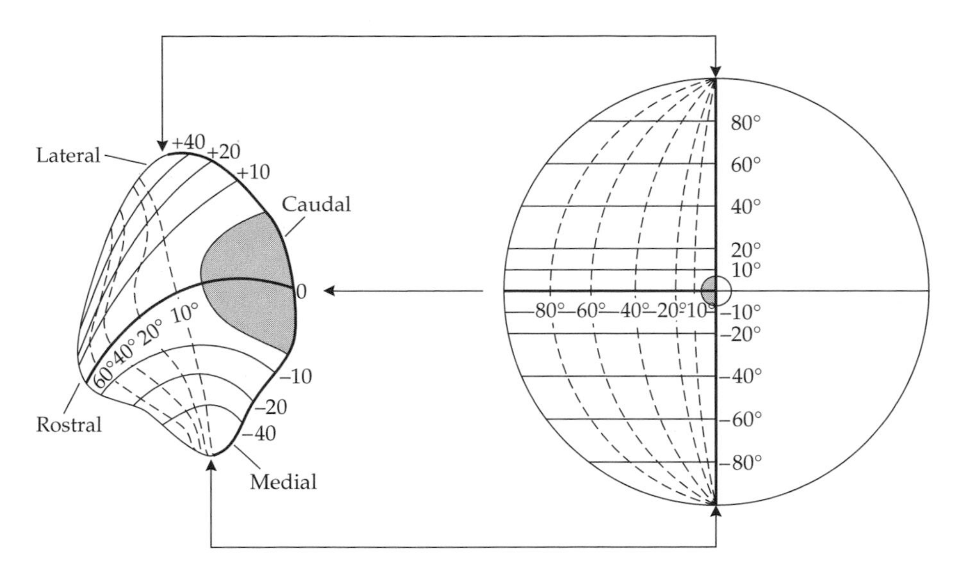

Figure 5.9
Retinotopic Mapping onto the Lateral Geniculate Nucleus
The dorsal surface of the right LGN in rhesus monkey is shown with the lines of latitude and longitude from the left visual hemifield mapped onto it. The central 5° of the visual field (shaded) projects to a relatively large area at the caudal end of the LGN and the peripheral parts of the visual field are compressed. The vertical meridian in the visual field lies along the caudal border of the LGN; the horizontal meridian runs along the rostral–caudal axis of the nucleus. (After Malpeli and Baker 1975.)

the map's surface area. The fovea and the retina immediately around it are represented by a large area at the posterior (caudal) end of the nucleus, and the horizon runs along the anterior–posterior axis. Thus the superior visual field (inferior retina) is lateral in the LGN, and the inferior field (superior retina) is medial.

The fovea and its corresponding part of the visual field are expanded relative to more peripheral retina and visual field. This enlargement of the area devoted to inputs from the central retina persists in the projection from the LGN to the primary visual cortex, where it has been called "cortical magnification." What this means, simply, is that while the central 5° of the visual field is less than 3% of the total retinal area, this central 5° occupies well over half of the surface area of the LGN and the primary visual cortex. Given that roughly half of all retinal ganglion cells are found in the central region (see Chapter 15), however, it should come as no surprise that their terminations are spread over a proportionately large part of any target nucleus and, by direct extension, the primary visual cortex. In any case, the distortion or magnification is very much like drawing a picture on a sheet of rubber, and then stretching the sheet; the picture will be distorted, but all of the relative spatial relationships between parts of the picture will be retained. For the retinogeniculate projection, the important point is not just that the map is distorted, but that the distortion is always the same.

Because each hemiretina has been mapped three times in different layers, there are several maps. The maps in the different layers are aligned (Figure 5.10). If a pin were inserted into the LGN perpendicular to the surface, it would contact retinal afferents corresponding to a particular location in visual space. As the pin was inserted deeper, passing from one layer to the next, the retinal afferents would correspond to the same location in space, regardless of which eye the afferents came from. Or to turn the emphasis around, a particular point in space will be imaged at specific points on the retinas in both eyes and then as up to six points along a straight line or narrow column through the LGN. This implicit columnar organization in the LGN is transferred in a more explicit way to the primary visual cortex.

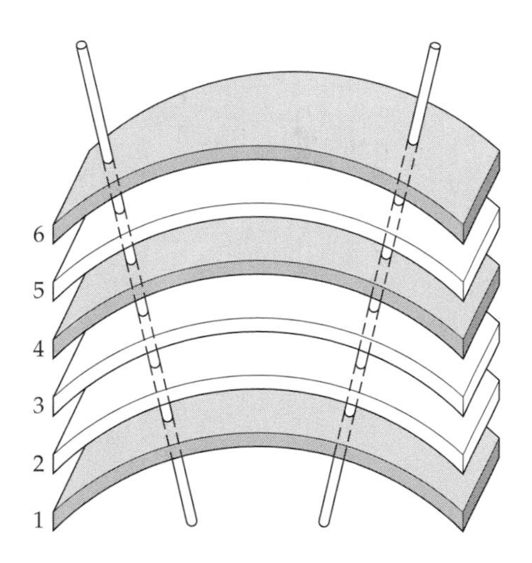

Figure 5.10
The Maps to the Different LGN Layers Are in Register
Specific points in the visual world are represented as lines through the LGN. Since the lines are not interrupted from one layer to the next, the visual field maps in the different layers are aligned with one another. The contralateral layers are shaded.

Some axons leave the optic tracts for other destinations

Roughly 10% of retinal ganglion cell axons do not terminate in the lateral geniculate nuclei, but most of the methods for studying these subsidiary projections require experimental techniques that cannot be used with humans. As a result, the details of these projections are not fully understood, and much of what we think is going on in the human brain is assumed by analogy to other primates that have been studied experimentally.

A subsidiary projection may arise as a small group of retinal axons parts company with the other axons in the optic tract and continues to another destination. If subsidiary projections form in this way, some retinal ganglion cells do not contribute to the primary visual pathway and are not directly involved in visual perception. In some species, several subsidiary retinal projections appear to form in this way, with the signals conveyed by certain classes of retinal ganglion cells going only to secondary targets and therefore being used for other purposes (e.g., control of reflex eye movements).

Alternatively, axons in the optic tract may branch, with a branch going to the LGN and a second branch (the **axon collateral**) going to another target. In this case, the information streams in the primary and secondary projections would be the same, although the uses made of the information could be quite different. There is evidence, again in other species, that some subsidiary projections are formed by axon collaterals. It is not clear, however, that all classes of ganglion cell axons have collateral projections.

Axons terminating in the superior colliculi form discontinuous retinotopic maps

The largest of the secondary pathways is a bundle of axons called the **brachium of the superior colliculus** (see Figure 5.7). As the name implies, most of the axons in these bilateral brachia go to the superior colliculi, where they synapse on cells in the superficial layers (like the LGN, the superior colliculus is a layered structure). Most of these axons are probably axon collaterals. They create a retinotopic mapping of the contralateral half of the binocular visual field onto the colliculus, wherein the left half of the visual field is mapped onto the right superior colliculus, and vice versa. This arrangement, so far, is much like the projections to the LGNs.

In the superior colliculi, however, the axons from the two eyes are not segregated in different layers. Instead, the inputs from the two eyes are interlaced, much like interleaving the fingers of two hands. The regions occupied by axon terminals from *one* eye are discontinuous, forming a complex arrangement of stripes, clumps, and islands of terminating axons (Figure 5.11). The spaces between the stripes belonging to one eye are occupied by terminals from the other eye, thus forming a complementary set of stripes.

The portion of the colliculus occupied by axon terminals coming from ganglion cells in the central retina is considerably smaller than in the LGN, and it was thought at one time that the fovea was not represented in the superior colliculi at all, but axons from ganglion cells in the central retina do terminate in the anterior part of the colliculus. The foveal representation in the colliculi, however, is not expanded as it is in the lateral geniculate nuclei.

The superior colliculi receive a variety of inputs other than those from the retinas, including projections from the visual cortex, the eye fields in the frontal cortex, and perhaps as many as 40 subcortical parts of the brain. Thus the superior colliculi are sites of interaction between direct retinal information, visual information that has undergone further processing in the cortex, and information from other sensory systems. Much of this interaction has to do with controlling saccadic eye movements (see Chapter 4).

Axons forming the afferent part of the pupillary light reflex pathway terminate in the pretectal nuclear complex

In rhesus monkeys, some retinal ganglion cell axons leave the brachia of the superior colliculi and go to the complex of **pretectal nuclei** near the superior colliculi.

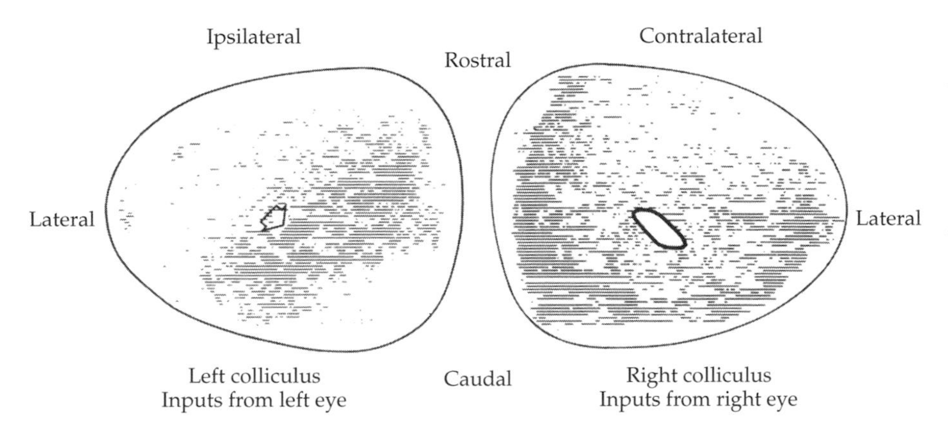

Figure 5.11
Mapping of Retinal Inputs onto the Superior Colliculi
These schematic views of the surfaces of the superior colliculi in rhesus monkey show the locations of inputs from the left retina. Thus the map of the left colliculus is the ipsilateral projection, and the map of the right colliculus is the contralateral projection. If one map is flipped over onto the other, the projections overlap only slightly. Open spaces in the maps correspond to inputs from the right retina. The space outlined in the contralateral projection (right) corresponds to the optic nerve head in the nasal retina; the comparable region in the ipsilateral projection (arrow) receives retinal input because it corresponds to the temporal retina. (After Pollack and Hickey 1979.)

The situation is presumably similar in humans. The pretectal nuclei are not particularly easy to define anatomically, and description and nomenclature have been revised several times. Currently, however, the complex is described as five separate cell clusters lying in the midbrain just anterior to the superior colliculi; they are the **nucleus of the optic tract**, the **pretectal olivary nucleus**, and the **anterior**, **medial**, and **posterior pretectal nuclei**. The nucleus of the optic tract and the pretectal olivary nucleus receive dense input from retinal ganglion cell axons, the medial and posterior pretectal nuclei receive less dense input, and the anterior pretectal nucleus may receive no input at all.

The nucleus of the optic tract is the most studied of the pretectal nuclei. In monkeys and other mammals, such as cats and rabbits, the retinal information relayed through the nucleus of the optic tract is used for eye movement control, specifically for the slow phase of optokinetic nystagmus (see Chapter 4).

Neurons in the pretectal olivary nuclei are involved in pupil control. The cells exhibit activity levels that are related to the level of retinal illumination, which is a necessary feature for setting pupil size, and the olivary pretectal neurons appear to project to the Edinger–Westphal nuclei (as we'll see later in the chapter), which are part of the pupillary light reflex pathway. The identity of the retinal ganglion cells providing the inputs to the olivary pretectal nuclei is not known.

No specific functions have been proposed for the other pretectal nuclei.

Retinal inputs to the accessory optic system may help coordinate eye and head movement

Axons leaving the collicular brachia as part of the accessory optic system run along the side of the brainstem to three small interconnected accessory nuclei: the **dorsal**, **lateral**, and **medial terminal nuclei** (see Figure 5.7). This axon bundle, which was originally called the transpeduncular tract, is observable in humans, but again, most of what we know about it is derived from studies in monkey.

All three of the terminal nuclei receive direct retinal inputs, although in comparison to other retinal projections, like those to the geniculate nuclei or the superior colliculi, the inputs are very small; probably only a few thousand axons are involved. Nonetheless, the accessory optic system may be quite important; the neurons of the terminal nuclei connect to the vestibular nuclei and to the cerebellum. The system as a whole is thought to be involved in coordinating eye and head movements.

Retinal axons may provide inputs to a biological clock

The **suprachiasmatic nucleus** is a small cell cluster in the hypothalamus just above the optic chiasm. In some species, this nucleus receives direct input from retinal axons. In humans, stains for degenerated myelin (Box 5.1) show degen-

Box 5.1 Tracing Neural Pathways: Degeneration and Myelin Staining

SOME KNOWLEDGE ABOUT NERVES and innervation is quite ancient. Greek anatomists, for example, knew about the larger peripheral nerves and correctly assumed that a nerve entering a muscle was somehow associated with the muscle's contraction. In fact, most information that can be gleaned from careful dissection and direct inspection has been available for several centuries. But identifying a nerve and the peripheral structure associated with it is only part of the story; in particular, what happens to the nerve when it enters the central nervous system? It is now difficult to isolate and follow. Something other than dissection is required to deal with this problem.

Many neurons in sensory and motor nerves and tracts within the brain and spinal cord have axons sheathed in myelin laid down by glial cells—Schwann cells in peripheral nerves and oligodendroglia in the central nervous system. Dyes that selectively stain myelin are therefore useful in preparing atlases of the brain, for example; nuclei containing cell bodies and axon terminals generally lack myelin and show up clearly among fiber tracts whose myelin has made them densely stained. And the bundles of axons are easy to see. Figure 1 shows myelin staining in the optic nerves, chiasm, and tracts in an individual who lost one eye many years before death; all the myelinated fibers from the enucleated eye had degenerated long before, leaving the myelinated axons from the remaining eye clearly visible and allowing their paths through the chiasm to be followed. The partial decussation, in which some axons cross the midline at the chiasm while others don't, is obvious.

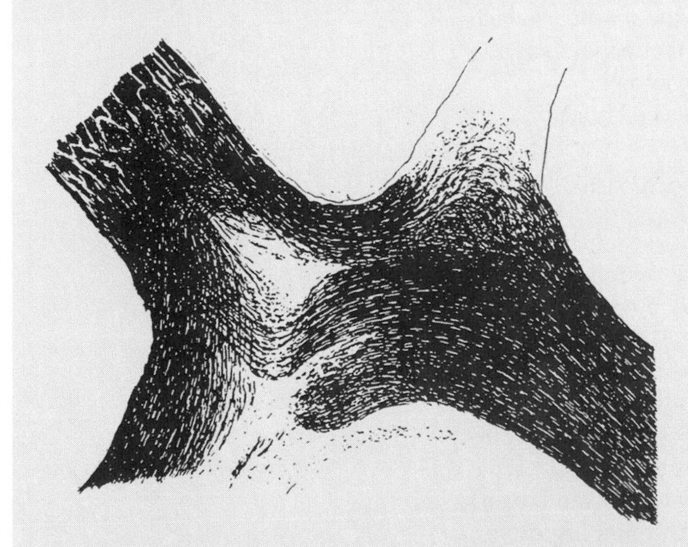

Figure 1
Myelin Staining
This section through the optic chiasm, stained for myelin, shows that all optic nerve fibers from one eye had degenerated as a result of enucleation of the eye 24 years prior to death. The paths followed by the decussating and nondecussating axons are obvious. (From Polyak 1957).

But the axon paths in Figure 1 are clear only because axons from one eye are missing. Had this not been the case, it would be impossible to distinguish axons coming from the different retinas. And this is the problem with myelin staining: All axons with myelin sheaths are stained, and particular groups of axons cannot be isolated. The way around this difficulty is to examine not the normal myelin, but myelin that is degenerating.

When neurons die and degenerate after they have been damaged, their myelin sheaths also degenerate, leaving behind characteristic traces of debris. Making this degeneration selective is a matter of controlling the damage done to the nervous system; specific nerves or portions of nerves are severed or small lesions are placed in specific parts of the brain. In time, the results of this experimental damage can be seen in the pattern of degenerating myelin. One of the earliest tract-tracing methods (the Marchi technique) labels degenerating myelin fragments but does not stain normal myelin; thus the dark stain in the processed brain tissue will show where the degenerating axons traveled.

Unfortunately, some axons are not myelinated, and the myelin sheaths on other axons do not extend all the way to the ends of the axon. Retinal ganglion cell axons and the fibers in the long ciliary nerves are two examples—myelin on the ganglion cell axons does not extend inside the eye, nor does the myelin on the long ciliary nerve fibers extend into the cornea. Therefore, a myelin degeneration stain would not be able to trace these axons in their entirety.

Other stains, however, are selective for the degeneration products within the axons and their parent neurons, and the presence or absence of myelin thus is not a consideration. In general, these stains are called silver stains, since silver nitrate or other silver compounds are essential components of these often complex mixtures of reagents. Most of the chemical reactions that make these stains work are not known in detail, but the important result is the precipitation of silver onto degenerating neurofilaments in the damaged axons and cells. (The methods often carry the names of their developers—Nauta, Bodian, Fink-Heimer, among others.) One of the difficulties here is that not all neurons have the same degree of neuro- or microfibrillar content and some may not be stained by one or another method.

One by-product of neuronal degeneration does not require special staining procedures. It is called *transneuronal atrophy* because cells receiving inputs from axons in a nerve will die and degenerate after the input nerve axons die. The phenomenon is particularly obvious in the lateral geniculate nuclei after one optic nerve has been severed. In Figure 2, the parts of the LGN that were deprived of input from optic tract axons stain very lightly because most of the cells have atrophied and the intracellular components normally stained are no longer present. In this case, layers 2, 3, and 5 have atrophied because the eye on the same side had been enucleated.

Because these methods require carefully controlled lesions or tract sections, they are not applicable to studies of the human cen-

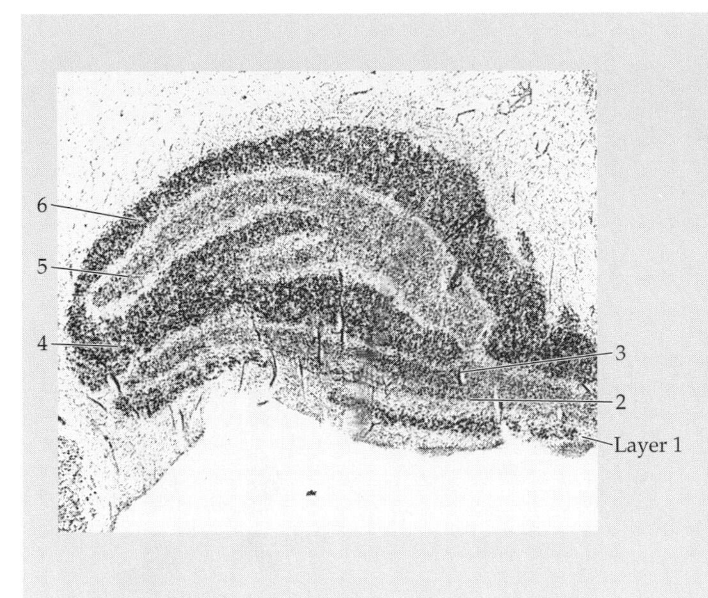

Figure 2
Transneuronal Atrophy in the Lateral Geniculate Nucleus
Layers 2, 3, and 5 are lighter than the other layers because the cells within them are less numerous and smaller. The eye ipsilateral to this LGN section had been enucleated many years before death; this demonstrates the ipsilateral projection to layers 2, 3, and 5. (From Horton and Hedley-White 1984.)

tral nervous system except in cases of well-defined surgical lesions, such as the ocular enucleations that preceded the results in Figures 1 and 2. There are also limitations imposed by the sensitivity of the techniques (small degenerating tracts may escape observation), by their technical difficulty (most silver and gold stains are somewhat capricious; the same histological protocol may produce variable results for no apparent reason), and by their dependence on the degeneration produced by well-defined lesions or nerve tract sections. Newer types of labels, however, exploit properties possessed by all normal neurons to define the paths they follow (see Box 5.2).

eration in the suprachiasmatic nucleus of individuals who lost one eye prior to death. Another nearby nucleus, the paraventricular nucleus, also exhibits degeneration, suggesting that it also receives retinal input. The route by which retinal axons reach these nuclei is not clear; they may enter the hypothalamus directly from the upper surface of the optic chiasm.

This pathway is thought to be involved in regulation and resetting of a biological clock responsible for diurnal (**circadian**) rhythms in blood pressure, intraocular pressure, photoreceptor outer segment shedding (see Chapter 13), and other daily variations. (The term "biological clock" is a convenient way to summarize evidence pointing to *something* in the brain that provides an inherent rhythmicity to the body's functions. What the clock consists of, whether it is to be found in a single location, and how it works are all unanswered questions.)

Interest in the existence and operation of this pathway in humans is related to the phenomena of jet lag and seasonal affective disorder (the SAD syndrome), a form of clinical depression commonly experienced by people living where there are just a few hours of winter daylight. In both cases, therapy may involve periods of exposure to bright light, which may alter melatonin levels and help reset a biological clock that has been upset by the rapid change of time zones in the case of jet lag, or by too little daylight in the case of seasonal affective disorder. Given the many unknowns, the therapeutic rationale is essentially a guess about how the system operates.

Lesions of the optic nerves and tracts produce defects in the visual fields

The route followed by retinal ganglion cell axons through the orbit and along the base of the brain is not free of hazards. Transmission of visual information can be affected by vascular problems, such as arterial blockages or aneurysms, or by expanding tumors in nearby structures. Whatever the nature of the lesion, if it interferes with the conduction of signals through the visual pathway, the result will be a visual defect—a scotoma—in which vision has been reduced, perhaps lost completely, in part of the visual field of one or both eyes.

Figure 5.12 shows a schematic overview of the eyes and visual pathways, along with the monocular visual fields as they would appear after lesions at various places along the primary visual pathway. The most obvious consequence of

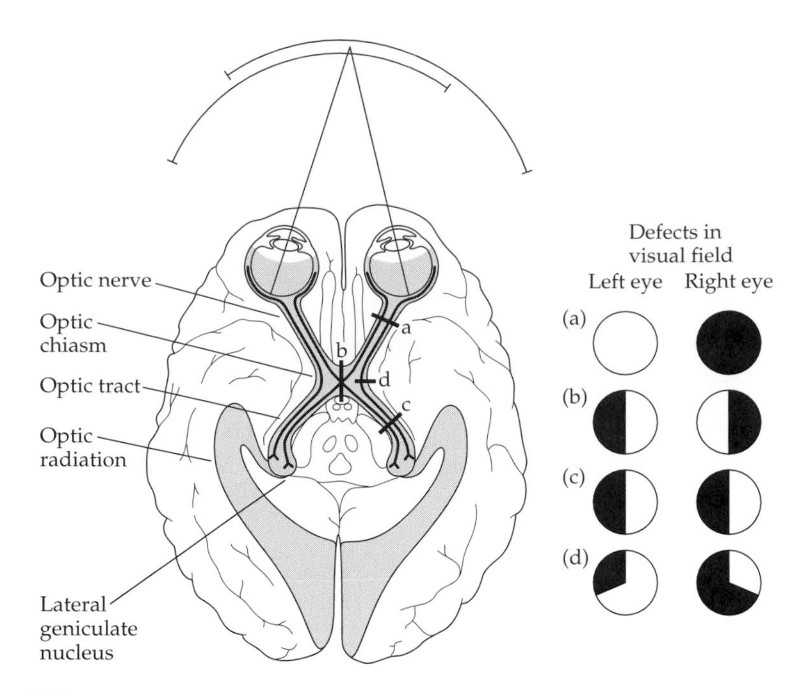

Figure 5.12
Visual Field Defects Produced by Sectioning along the Retinogeniculate Pathway
(*a*) A section through the right optic nerve between the eye and the optic chiasm will produce a complete loss of the right eye's visual field. Splitting the optic chiasm along the midline (*b*) will cut the crossing nasal retinal axons and produce hemianopias in the temporal visual fields of both eyes. Sectioning the right optic tract (*c*) will affect right temporal and left nasal axons, producing blindness in the left halves of both visual fields. A cut along the right side of the optic chiasm (*d*) has less predictable results, but may affect more than half of the right eye's visual field and at least the superior-temporal quadrant of the left eye's field.

the anatomy is the fact that any lesion between the chiasm and the eye will affect the visual field of only one eye (Figure 5.12*a*), whereas lesions of the chiasm or the optic tracts will affect both monocular fields. This result is a consequence of the partial decussation within the chiasm and the mingling of axons from both eyes in the optic tracts.

If one optic tract were completely severed, as Figure 5.12*c* implies, the scotomas would be **hemianopias**, in which half of each monocular field is blind. For the section of the *right* optic tract illustrated here, the severed axons would be those from the right temporal retina (producing a hemianopia in the *nasal* hemifield of the right eye) and those from the left nasal retina (producing a hemianopia in the *temporal* hemifield of the left eye). The combined result would be blindness in the entire *left* half of the binocular visual field.

Most discussions of hemianopias leave the impression, and often show it explicitly, that the hemianopia is complete and splits the foveal region in half. But given the earlier evidence that both halves of the fovea project to both sides of the brain, this kind of foveal splitting as the result of a lesion would not be expected. Unfortunately, we do not know exactly what happens to the field defect in the region of the fovea; a convincing demonstration of foveal splitting would require methods for plotting the scotoma and monitoring eye movements that are much more sensitive than those in standard use. In short, we do not know if a hemianopia splits or spares the foveal region. (Cortical lesions often produce hemianopias with "macular sparing," meaning that the scotoma does not include the fovea and the region around it. The usual, probably correct explanation for the sparing is that very few lesions are large enough to affect all of the extensive cortical area devoted to the fovea.)

Most lesions in the optic tract do not affect all the axons in the tract. As a result, the defects rarely include the entire hemifield in either eye, and the monocular defects often differ in size and location. Even a partial lesion will involve both eyes, however; although the axon terminals from the two eyes are segregated in different layers in the lateral geniculate nuclei, they are intermingled within the tracts and even small lesions will affect axons from both eyes to some extent.

The effects of lesions at the chiasm are complicated because of the way the decussating nasal retinal axons are dispersed within the chiasm. In the simplest case, the chiasm is split along its midline into right and left halves (see Figure 5.12*b*). If this were done, the only axons affected would be the crossing nasal retinal axons; since axons from both eyes would be affected, the result would be a temporal hemianopia in the visual field of each eye—in other words, bitemporal hemianopia. Smaller lesions along the midline confined to either the anterior or the posterior edge of the chiasm would affect only part of the crossing nasal fibers, and the field defects would not extend through an entire hemifield. An anterior lesion along the midline, for example, would interfere with *inferior* nasal axons from both eyes (see Figure 5.5), producing scotomas in the *superior temporal* quadrants of the visual fields. These scotomas—called **quadrantanopias**—are superior and bitemporal. Inferior bitemporal quadrantanopias would be produced by lesions affecting the axons at the posterior edge of the chiasm.

Lesions along the sides of the chiasm should affect both ipsilateral temporal and nasal retinal axons and, to some extent, depending on the size of the lesion, contralateral nasal axons. The lesion shown in Figure 5.12*d* will cut all of the ipsilateral temporal retinal axons, some of the ipsilateral *superior* nasal axons (they have yet to cross the midline), and some of the contralateral *inferior* nasal axons that have already crossed. The expected result is a scotoma in the visual field of the right eye that includes the entire *nasal* hemifield plus part of the *inferior temporal* field. The defect for the left eye will be a quadrantanopia that is superior and temporal. Note that this situation, a lesion along one side of the chiasm, produces the most asymmetry between the defects in the two visual fields; the defect in the field of the ipsilateral eye is much larger.

As before, the effects of real lesions in real patients are rarely this neat, but the differing consequences of lesions in different locations can still be observed. In Figure 5.13, for example, the visual field defects were the result of a pituitary tumor. Since the pituitary is a midline structure that lies directly below the chiasm, its expansion would be expected to affect the chiasm along its midline

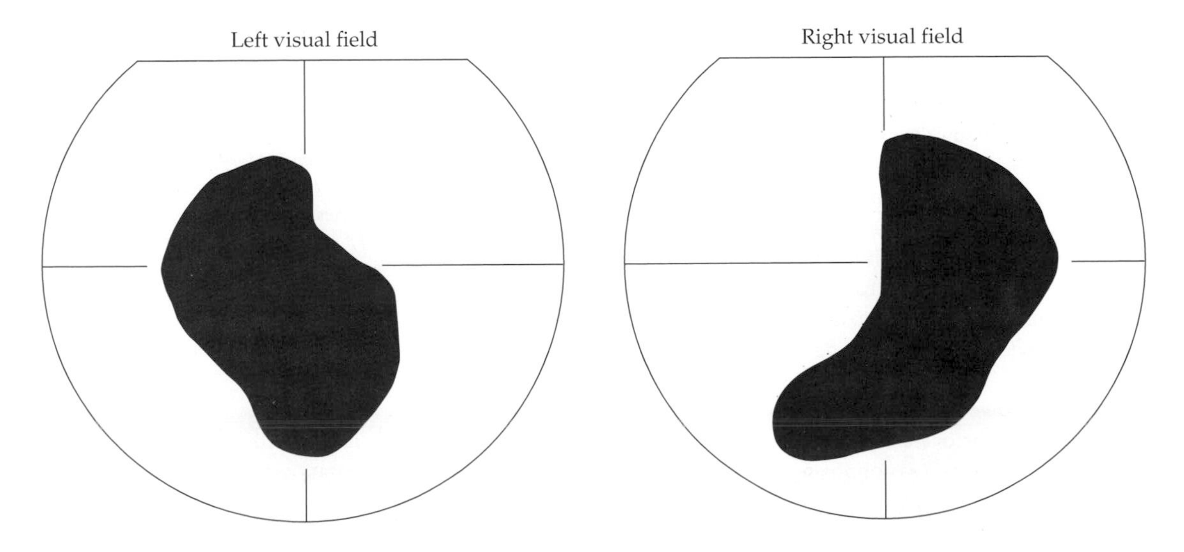

Figure 5.13
Visual Field Defects from Pathology
These bitemporal scotomas resulted from a pituitary adenoma that affected the chiasm along its midline.

Vignette 5.2

Seeing One World with Two Eyes: The Problem of Decussation

LATE ON A JUNE AFTERNOON IN 1886, the bodies of two men were pulled from Lake Starnberg (Starnberger See) not far from Munich. They were Ludwig II, the "mad king" of Bavaria, and his personal psychiatrist, Johann Bernhard Aloys von Gudden (1824–1886). As events were reconstructed, the two men had gone for their second of two walks that day, part of the therapy for Ludwig's madness, but with no attendant following behind, as was the usual practice. Ludwig is thought to have rushed into the lake as if to commit suicide, whereupon Gudden followed to prevent it, but he was overpowered and drowned by the king, who then drowned himself.

Ironically, Gudden's advocacy of more humane treatment for the mentally ill was his downfall. He ran a psychiatric institute and believed that the common practice of isolation and physical restraint in asylums (under incredibly filthy and degrading conditions) was of no therapeutic value and that patients benefited from normal human contact. Patients were not to be totally removed from society; long walks and conversations, with their psychiatrist if no one else, would aid their return to normality. Hence the walks and conversations with King Ludwig, and Gudden's untimely death.

Gudden was also a neuroanatomist who made many original observations on the terminations and origins of cranial nerves and brain fiber tracts. One such tract is what we now call the accessory optic system; Gudden saw its transpeduncular tract on the surface of the midbrain and found one of its three terminal nuclei. Another tract that appears to connect the two sides of the brain just behind the optic chiasm, also more developed in animals other than humans, is known as Gudden's commissure (or supraoptic decussation); although it lies very close to the chiasm, this tract is not known to carry retinal afferents and may have nothing to do with vision.

One of Gudden's basic anatomical methods was a paradigm of selective destruction. For example, he would sever a single peripheral or cranial nerve in a very young animal and allow the animal to survive to maturity—waiting years in some cases. Because the lesion was made only on one side of the body, the other side served as a normal control, first for behavior and then for anatomy. By allowing the animals to survive for a long time, when Gudden finally sectioned the brain and nerves he was able to observe the atrophy and myelin degeneration that occurs in the brain after nerves have been damaged. This method produced the first clear evidence in primates that some fibers from each optic nerve cross in the chiasm while others do not—that is, evidence for partial decussation (see Figure 1 in Box 5.1).

The question about the paths followed by optic nerve axons at the chiasm was an old one, raised long before axons were understood in the modern sense, and there were two opposing views. Some held that the chiasm was simply a place where the optic nerves adhered to one another with no merging or fusing of the two nerves; this meant that the left and right eyes were connected to the left and right sides of the brain, respectively. The opposing view held that the nerves crossed at the chiasm—that is, there was complete decussation—and the two eyes were connected to opposite sides of the brain.

For people who thought about it, the fact that we have two eyes but only one, seamless perception of the world was a problem. Since a large portion of the visual world is seen simultaneously by both eyes, signals from the two eyes must be combined in some way. But if the two eyes communicate independently to the two sides of the brain, either by no decussation or by complete decussation, there must be a mechanism deep in the brain for bringing the signals from the two eyes together. The solution imagined by the great French philosopher, mathematician, and physiologist René Descartes (1596–1650) in Figure 1 shows the optic chiasm to be a place where the optic nerves come in contact with one another with no mingling of axons and no decussation. In Descartes's view, both eyes communicate with the ipsilateral sides of the brain, specifically with the ventricles. From the fluid-filled ventricles, however, the "impressions" from the two eyes are conveyed to a common site, the pineal gland, where they combine to produce a kind of cyclopean eye. This scheme would also work if there were complete decussation at the chiasm.

As vague and erroneous as Descartes's anatomical conception was, he was absolutely correct on one important issue about which he had no information at all, and that is the retinotopic mapping into the brain. In Figure 1, an object is mapped onto the retinas in both eyes by their optical systems, and this point-by-point mapping is preserved in the visual "impressions" that reach the brain. Moreover, the impressions from the eyes are combined so that the separate points in the two retinas receiving the image of a particular point on the object (the point of the arrow, for example) are mapped precisely onto one another. In modern terms, this is normal retinal correspondence.

The notion that signals from the two eyes could be brought into register peripherally by partial decussation at the optic chiasm seems to have been first enunciated by Isaac Newton (1642–1727); the suggestion appears in his treatise on optics as "query" number 15 in Part I of Book Three:

> *Are not the Species of Objects seen with both Eyes united where the optick Nerves meet before they come into the Brain, the Fibres on the right side of both Nerves uniting there, and after union going thence into the Brain in the Nerve which is on the right side of the Head, and the Fibres on the left side of both Nerves uniting in the same place, and after union going into the Brain in the nerve which is on the left side of the Head, and these two Nerves meeting in the Brain in such a manner that their Fibres make but one entire Species or Picture, half of which on the right side of the Sensorium comes from the right side of both Eyes through the right side of both optick Nerves to the place where the nerves meet, and from thence on the right side of the Head into the Brain, and the other half on the left side of the Sensorium comes in like manner from the left side of both Eyes.*

Newton did not illustrate his idea, which makes Figure 2 all the more remarkable; this clear representation of crossing by nasal reti-

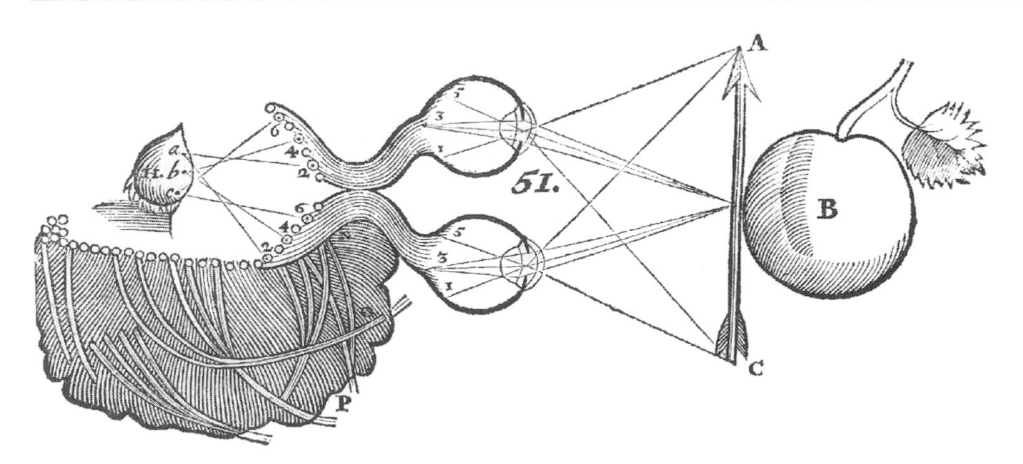

Figure 1
Combining Inputs from the Two Eyes without Decussation: Descartes's View
René Descartes drew the optic nerves and tracts as if there were no crossing whatsoever at the chiasm. He managed to create a unified perception of the world by combining the inputs from the two eyes at the pineal gland, but the means of combination is not specified. Notions about retinotopic mapping and retinal correspondence are explicit in the drawing. (From Descartes 1637.)

nal axons is from *Le Mechanisme, ou le Nouveau Traité de l'Anatomie du Globe de l'Oeil, &c.* by John Taylor, published in 1738. (Taylor published his book in several languages, including English, but I have seen only the French edition.)

John Taylor (1703–1772) was English by birth, and he is most charitably described as an itinerant oculist, meaning that he traveled a lot, performing surgery as he went. He was also a tireless and extravagant promoter for whom no amount of self-aggrandizement was excessive; his claims to the French title *chevalier* (knight) and to a Cambridge education were both spurious, for example. Worse still, he seems to have combined his talents as huckster and con artist with a very good mind—a dangerous man. Johann Sebastian Bach died blind after several operations by Taylor, who was unscrupulous enough to name Bach as a famous patient successfully treated. (Taylor was ordered out of Germany shortly after his treatment of Bach, apparently because of another bad incident in another city.)

But the man was clever, easily clever enough to conceal the source of his inspiration for Figure 2. He cites no evidence for his anatomy other than the notion that it must be this way, but his summary statements on the subject are clear, succinct, and correct: "The left part and the right part of each eye communicate with each other," and "half of the fibers of one optic nerve change place with half of the fibers of the other at the place where they meet in the head."*

Taylor says nothing about where the idea of partial decussation may have originated; given what we know of his character, however, that omission, which implies that the idea was his since not attributed to anyone else, is not surprising. He probably got the idea from reading Newton and was quite happy to pass Newton's genius off as his own. In any event, he got it right, but it took Gudden and another century to produce the evidence.

*"*Le côté gauche & le côté droit de chaque oeil communiquent l'un avec l'autre*" and "*la moitié des fibres d'un nerf optique change de place avec la moitié des fibres de l'autre, à l'endroit où ils se rencontrent dans la tête.*"

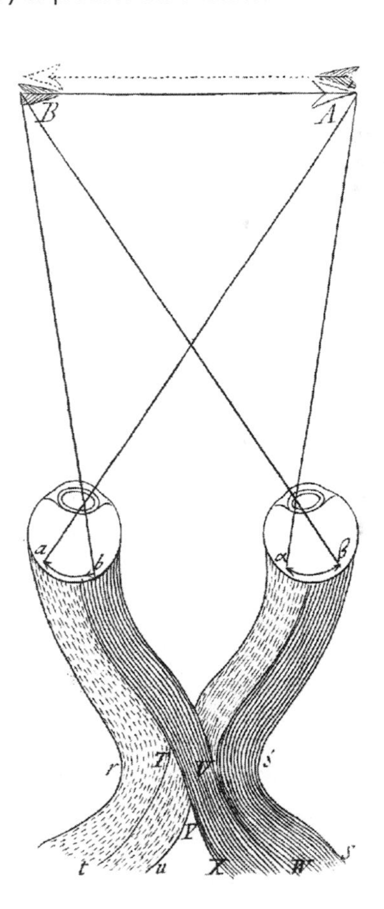

Figure 2
The First Illustration of Partial Decussation
This drawing clearly shows nasal retinal axons crossing in the chiasm while the temporal axons remain on the ipsilateral side. The nasal and temporal axons remain separate in the optic tracts, however, which leaves open the question about how the corresponding hemifields in the two eyes are ultimately combined. (From Taylor 1738.)

where the decussation occurs. The visual fields, as expected, show the bitemporal hemianopias, but they are neither complete nor perfectly symmetrical.

Lesions in the secondary visual pathways can be observed only as motor deficits

Retinal axons that leave the optic tract for destinations other than the LGN do not contribute directly to visual perception. This divergence of retinal afferents has two interesting consequences. First, it is possible to interrupt the primary visual pathway, particularly at levels past the LGNs, so as to produce complete blindness while certain visual behaviors, such as the pupillary light reflexes, remain intact and normal. Second, brain lesions that involve one or more of the subsidiary pathways without affecting the primary visual pathway will not produce visual field defects.

Thus, abnormal pupillary light reflexes, accommodative deficiencies, and altered eye movement patterns with normal, unimpaired visual fields direct one's attention to the subsidiary visual pathways and their target nuclei—that is, to the midbrain.

The Trigeminal Nerve: Signals for Touch and Pain

Two of the three trigeminal divisions carry signals from the eye and surrounding tissues

The trigeminal nerve (cranial nerve V) differs from other cranial nerves by collecting the cell bodies of its neurons in a large ganglion, the **gasserian** (**semilunar**) **ganglion**, lying outside the brainstem. The ganglion is connected to the brainstem by two bundles of nerve fibers: its **sensory** and **motor roots** (of which the sensory root is much larger). Nerve fibers from the periphery enter the ganglion through one of three branches: the opththalmic (V_1), the maxillary (V_2), and the mandibular (V_3) divisions. All the sensory nerves of the eye (other than the optic nerve) are associated with the ophthalmic division. The trigeminal system can be thought of as a treelike structure with a trunk—the gasserian ganglion and its connections to the brain stem—from which the three major divisions (V_1, V_2, V_3) radiate, each bearing still more, smaller branches (Figure 5.14).

The **ophthalmic division** (V_1), which will be our main concern, has three major branches, the nasociliary, frontal, and lacrimal nerves. The **nasociliary** nerve is the only one of these three to innervate the eye, and it has dependent branches to the nasal sinuses (the anterior and posterior ethmoidal nerves), to the skin of the nose and brow (the infratrochlear nerve), and to the eye (the long ciliary nerves and the sensory root of the ciliary ganglion), which we will discuss shortly. The **frontal** nerve has two dependent branches—the supratrochlear and supraorbital nerves—both of which go to the skin of the brow and forehead. The **lacrimal** nerve has only a single branch, the communicating branch, which is the route used by parasympathetic fibers going to the lacrimal gland via the lacrimal nerve; these parasympathetic fibers leave the lacrimal nerve as it passes through the lacrimal gland, and most lacrimal nerve fibers go out to the skin of the face.

Branches of the **maxillary division** (V_2) extend to the teeth and gums in the upper jaw, the lips and facial skin, and the mucosa of the maxillary sinus. The major branch, the **zygomatic** nerve, enters the orbit, and its dependent **infraorbital** nerve runs beneath the orbital floor to innervate the skin of the lower eyelid and the face. Other small branches, the zygomatico-facial and zygomatico-temporal nerves, exit the orbit through small foramina in the zygomatic bone to go out to the skin of the cheek.

Finally, the **mandibular division** (V_3) has numerous branches, which are not shown in Figure 5.14, because none of them pass through the orbit and they have

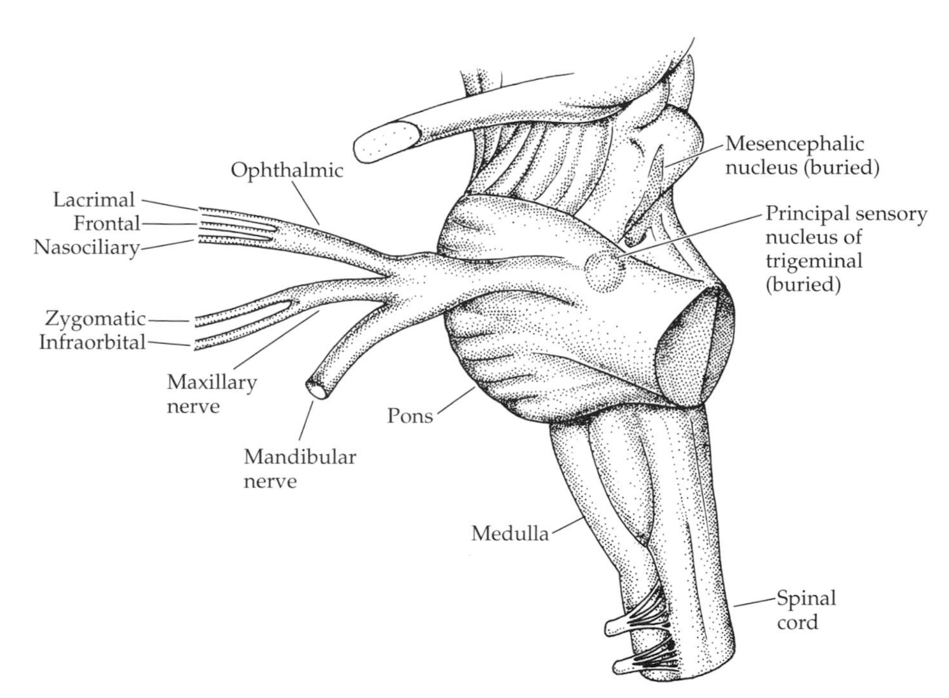

Figure 5.14
Schematic Overview of the Trigeminal System
The large sensory root and small motor root of the trigeminal come out through the pons to the large gasserian ganglion, from which the major divisions of the trigeminal split off. The branches of V_1 (ophthalmic) and V_2 (maxillary) that innervate the eye or pass through the orbit are illustrated schematically; all the somatosensory signals of the eye are conveyed by the nasociliary nerve of V_1. (After Martin 1996.)

nothing to do with the eye and its associated structures. As its name suggests, the mandibular division innervates the teeth, gums, and skin of the lower jaw.

All somatosensory information from the eye is conveyed by the nasociliary nerve to the ophthalmic division of the trigeminal

Somatosensory fibers originating in the tissues of the eye leave the eye through two sets of nerves: the long and short ciliary nerves, which we will discuss in more detail in the sections that follow. The two **long ciliary nerves** exit the posterior sclera several millimeters on either side of the optic nerve, and they run medially to join the nasociliary nerve. Several **short ciliary nerves**, usually 10 to 15, exit the sclera all around the optic nerve. They are mixed nerves, containing sensory, sympathetic, and parasympathetic fibers, and they all run from the eye to the **ciliary ganglion** that lies on the lateral side of the optic nerve about 1 cm behind the eye. The sensory fibers in the short ciliary nerves pass through the ganglion and reach the nasociliary nerve as the sensory root of the ciliary ganglion (see Figures 5.15 and 5.16).

Sensory nerve fibers from the cornea, conjunctiva, limbus, and anterior sclera join to form the long ciliary nerves

Sensory nerve endings are found in the cornea, the uveal tract, the conjunctiva overlying the anterior portion of the eye, and to some extent, the sclera. There have been occasional, unconfirmed reports of specialized sensory endings in the ocular tissues, but they are invariably bare nerve endings—that is, delicate terminal branches embedded in the tissues. In the general classification of sensory receptors they are **nociceptors**, activated by noxious or potentially noxious stimuli. Such stimuli include mechanical pressure, which is perceived as pain, and substances that are released by damaged cells or inflammatory reactions (histamine, potassium, serotonin, prostaglandins, and so on). The mechanisms by which these stimuli elicit action potentials in the nerve fibers are not fully understood.

The vast majority of the nerve endings are in the cornea, where they end in the outermost layer (the epithelium), although some are found within the underlying stroma (see Chapter 8). The innervation density of the cornea is extremely high, perhaps higher than anywhere else on the surface of the body, and it is exquisitely sensitive to mechanical pressure on its surface; small bits of dust and grit on the corneal surface can produce considerable discomfort. A cluster of these nerve endings is associated with an individual nerve fiber that runs radially out of the cornea, joining with other fibers as it goes.

Once the bundles of fibers are outside the cornea, in the limbus and sclera, the individual nerve fibers acquire a myelin sheath and go to either the medial or the lateral side of the eye, where the fiber bundles merge to form the long ciliary nerves. Throughout their course in the limbus and sclera, the fiber bundles are joined by other fibers from the overlying episclera and conjunctiva, the iris, the sclera, and the ciliary body. The long ciliary nerves run posteriorly in the sclera to a point just behind the ciliary body, where they pass inward to the choroid; they continue back in the choroid until they reach positions near the optic nerve, where they exit the eye through the sclera (Figure 5.15).

After leaving the eye, the long ciliary nerves run above and more or less parallel to the optic nerve for about 10 mm, where they merge with the nasociliary nerve, continuing through the ophthalmic division to the gasserian ganglion, where their cell bodies are located. The axonal portions of the cells go through the sensory root of the ganglion and terminate on cells in the main sensory nucleus of the trigeminal in the brainstem.

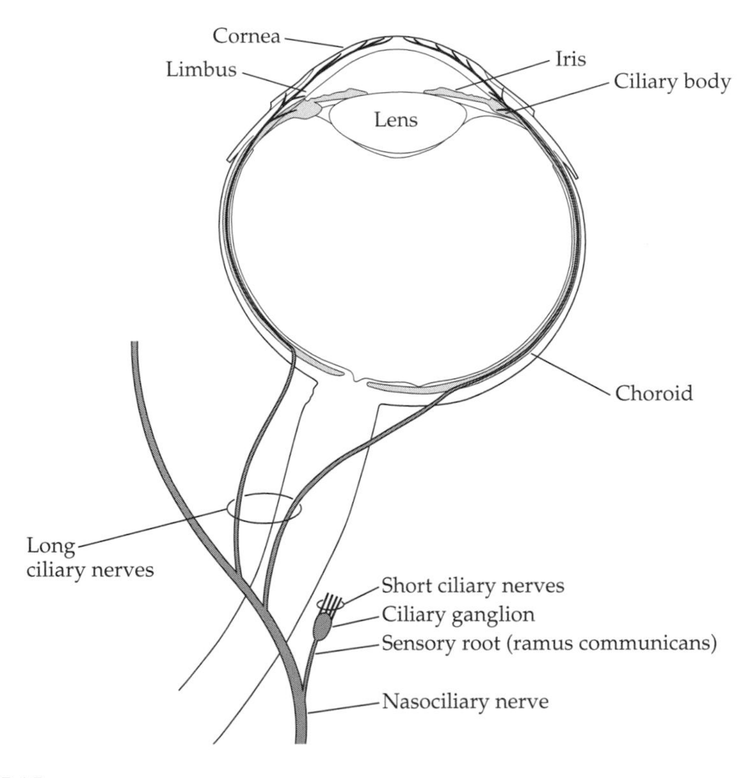

Figure 5.15
The Long Ciliary Nerves Join the Nasociliary Nerve
In this schematic section through the eye, nerve fibers from the cornea, limbus, iris, and ciliary body join on either side of the eye to run within the choroid as two long ciliary nerves. These nerves exit the sclera on either side of the optic nerve, running back and medially to join the nasociliary nerve. The majority of fibers in the long ciliary nerves are from the cornea.

Stimulation of corneal or conjunctival nerve endings elicits sensations of touch or pain, a blink reflex, and reflex lacrimation

The cornea is extremely sensitive to mechanical pressure, about ten times more sensitive than the fingertips. The sensation produced by stimulation of the corneal nerve fibers is almost always one of pain; the exception is stimuli just at threshold, which are perceived as touch. (Sensations of heat or cold are debatable, but warm stimuli seem to produce a painful sensation rather than a feeling of warmth, while cool stimuli may be perceived initially as sensations of touch, becoming painful as the stimulus is made colder.) Disagreeable as it is, the sense of pain is important; it provides a clear message that something is wrong and needs attention.

Stimulation of the corneal nerve fibers can trigger reflex blinking and reflex tear production (lacrimation), both of which may dislodge the object that stimulated the cornea and triggered the reflexes. The efferent parts of these reflex arcs are the motor innervation via the facial nerve (VII) to the orbicularis muscle of the eyelids (see Chapter 7) for reflex blinking and the parasympathetic innervation of the lacrimal gland for increased tear production (discussed later in the chapter). Since the corneal nerve fibers terminate in the brainstem, there must be interneurons in the brainstem that relay information from the brainstem nuclei of the trigeminal to the facial nerve nucleus proper and to its so-called lacrimal nucleus. The details of these intermediate parts of the reflex pathways are not known, however.

Other sensory fibers from the eye are conveyed by the short ciliary nerves and the sensory root of the ciliary ganglion

Many of the sensory fibers in the iris, ciliary body, and choroid do not join the long ciliary nerves. Instead, they form small bundles in the choroid, exiting the eye as part of the dozen or so short ciliary nerves that surround the optic nerve (Figure 5.16). The short ciliary nerves are connected to the small ciliary ganglion, a parasympathetic ganglion that will be discussed in more detail later in the chapter. The main point here is that the sensory fibers in the short ciliary nerves run directly through the ganglion without synapsing, and they exit as a single fiber bundle called the **sensory root** (of the ciliary ganglion), or **ramus communicans**. The sensory root joins the nasociliary nerve, and the fibers ultimately terminate in the brainstem along with those of the long ciliary nerves.

The long and short ciliary nerves differ in the parts of the eye they innervate, although there is some overlap, and in the density and specificity of their innervation. In general, the short ciliary nerves probably have fewer fibers, whose terminal endings are more widely distributed (neither the precise number nor the terminal distribution is known, however). As a result, pain elicited by stimulation of these sensory fibers is diffuse; it is very difficult to tell where in the eye the sensation is coming from, and usually the sensation is of dull rather than sharp pain.

Most other branches of the ophthalmic nerve carry somatosensory fibers from the skin of the eyelids and face

As far as the eye itself is concerned, the description of its somatosensory innervation is complete. All other branches of the trigeminal nerve as a whole, of its ophthalmic division, and of the nasociliary nerve convey sensory innervation from structures outside the eye. Many of these nerve fibers, particularly those innervating the facial skin, are associated with more specialized sensory endings that mediate the sensations of touch, heat, and cold, as well as pain.

The nasociliary nerve loses its identity where it is joined by the **anterior ethmoidal** nerve; the forward extension is the **infratrochlear** nerve (or to put it

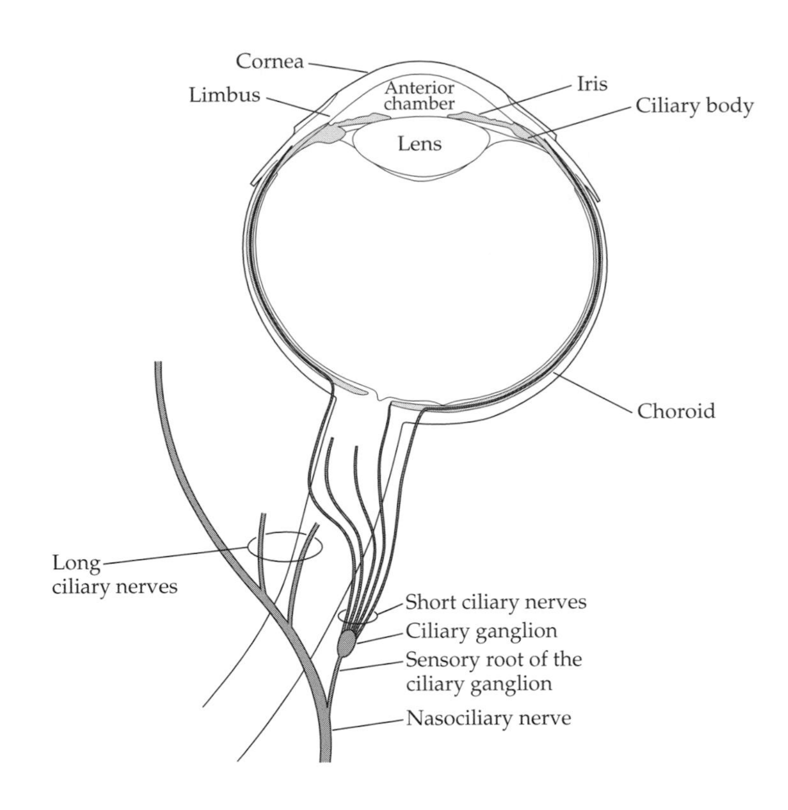

Figure 5.16
The Short Ciliary Nerves and the Ciliary Ganglion
Nerve fibers originating at free, unspecialized terminals throughout the uveal tract join together as 10 to 15 small nerve bundles, the short ciliary nerves, that exit the sclera around the optic nerve. They run to the ciliary ganglion on the lateral side of the optic nerve, about 10 mm behind the eye. Sensory fibers pass through the ganglion into its sensory root (ramus communicans), which joins the nasociliary nerve near the superior orbital fissure.

Figure 5.17
Branches of V_1 within the Orbit
The major subdivisions of V_1 pass through the superior orbital fissure as the nasociliary, frontal, and lacrimal nerves. The nasociliary gives rise to the ramus communicans (sensory root of the ciliary ganglion), the long ciliary nerve, the posterior and anterior ethmoidal nerves, and the infratrochlear nerve. The frontal has two branches, the supraorbital and supratrochlear nerves, but the lacrimal has only one (the communicating branch from the zygomatic temporal nerve). The nasociliary system is medial in the orbit, the frontal is superior, and the lacrimal is superior-lateral.

another way, the nasociliary is formed by the merging of the anterior ethmoidal and infratrochlear nerves) (Figure 5.17). As its name suggests, the infratrochlear nerve passes just below the trochlea; it carries sensory fibers from the lacrimal sac, above which it passes, the plica semilunaris and lacrimal caruncle in the medial canthus (see Chapter 7), the skin of the upper medial portion of the eyelids, and the upper part of the nose and the brow. Figure 5.18 shows the area of sensory innervation of the skin.

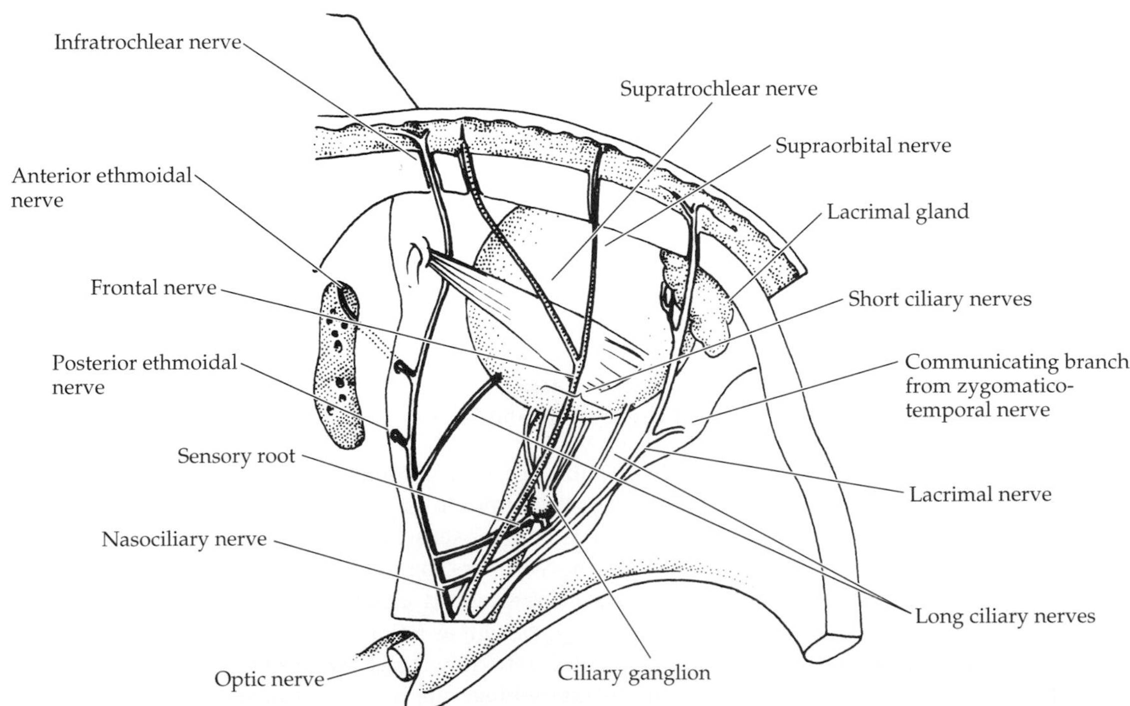

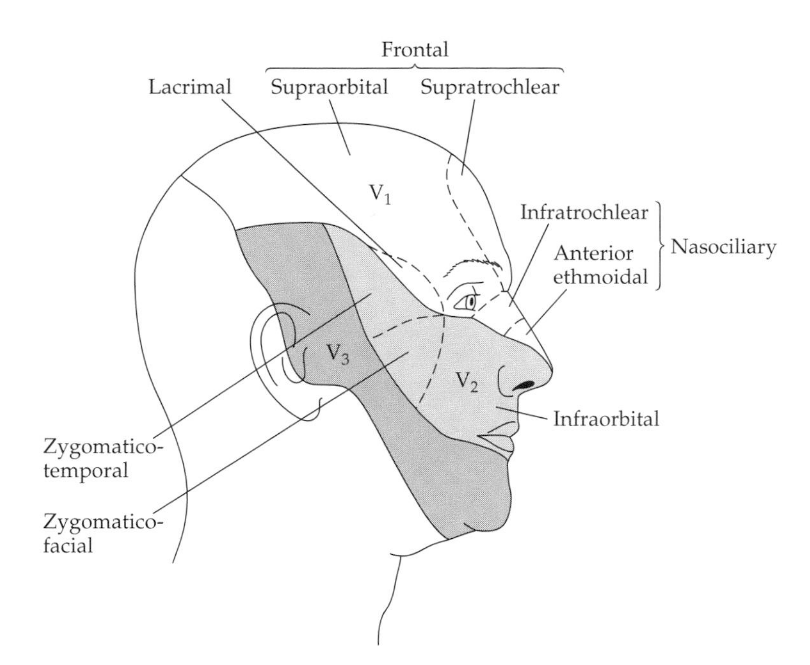

Figure 5.18
Innervation Areas on the Face for the Trigeminal Nerve
The ophthalmic nerve (V_1) innervates the skin along the side of the nose (infraorbital and anterior ethmoidal nerves from the nasociliary nerve), lateral to the orbit (lacrimal), and of the forehad and scalp (supraorbital and supratrochlear branches of the frontal nerve). The maxillary nerve (V_2) innervates the skin of the upper jaw and lower eyelid (infraorbital nerve) and the lateral side of the face (zygomatico-temporal and zygomatico-facial nerves). The lower jaw and skin of the face back to the ear is innervated by the mandibular nerve (V_3).

The anterior ethmoidal nerve consists of fibers that innervate the nasal mucosa and some that innervate the skin at the tip of the nose. One other branch of the nasociliary nerve, the **posterior ethmoidal** nerve, innervates the nasal mucosa and joins the nasociliary just after it enters the orbit through the posterior ethmoidal foramen.

The skin on the upper eyelid and the forehead is innervated by the **supratrochlear** and **supraorbital** nerves, the two branches of the frontal nerve. Figure 5.18 shows their regions of innervation. The supratrochlear nerve runs along the upper medial wall of the orbit above the trochlea; the supraorbital enters the orbit via the supraorbital notch, or foramen, in the orbital margin of the frontal bone. The nerves join about halfway back in the orbit above the levator, thus forming the frontal nerve, which continues posteriorly to the superior orbital fissure, where it leaves the orbit in company with the nasociliary and lacrimal nerves.

Facial skin on the lateral side of the orbit is innervated by the lacrimal nerve (see Figure 5.18). The nerve runs directly back through the orbit along the upper medial wall to its exit at the superior orbital fissure. It has only a single branch, the **communicating branch** (of the lacrimal nerve), which carries the postganglionic parasympathetic fibers that go to the lacrimal gland (discussed later in the chapter).

A few branches of the maxillary nerve pass through the orbit from the facial skin and the maxillary sinus

The remainder of the skin below and lateral to the orbit sends sensory fibers to the brain via the maxillary division of the trigeminal. As Figure 5.19 shows, the facial skin directly below the orbit is innervated by the infraorbital nerve, which runs through the infraorbital foramen in the maxilla and the infraorbital canal, below the orbital floor, and the infraorbital groove within the orbital floor. (The infraorbital also has branches that innervate the mucosa of the maxillary sinus just below the orbit.) Skin overlying the cheekbone sends fibers through two small zygomatic foramina, the **zygomatico-temporal** and **zygomatico-facial**, that emerge on the lower lateral wall of the orbit and join to form the zygomatic nerve. Parasympathetic fibers bound for the lacrimal gland run in the zygomatico-temporal nerve for a short distance, leaving via the communicating

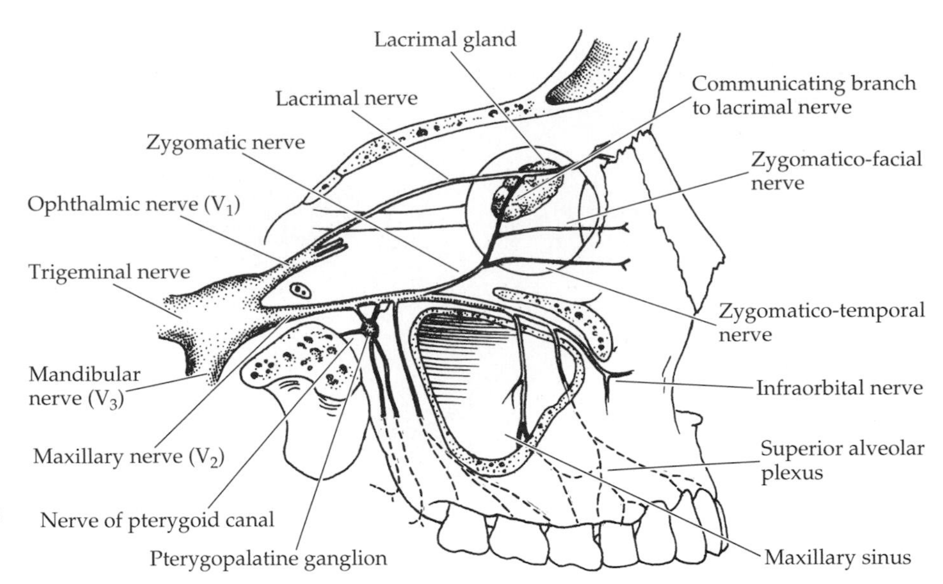

Figure 5.19
Branches of V_2 within the Orbit
The maxillary nerve (V_2) branches in the inferior orbital fissure into the zygomatic and infraorbital nerves. The infraorbital nerve runs through the infraorbital groove and canal below the floor of the orbit, emerging onto the face through the infraorbital foramen. The zygomatic nerve has two branches, the zygomtico-temporal and zygomatico-facial nerves, that exit the orbit through formina in the zygomatic bone. A communicating branch from the zygomatico-temporal to the lacrimal nerve carries parasympathetic axons arising from cells in the pterygopalatine ganglion; the communicating branch goes to the lacrimal nerve.

branch to reach the lacrimal nerve (that is, the communicating branch joins the zygomatico-temporal and lacrimal nerves).

The zygomatic and infraorbital nerves merge within the inferior orbital fissure to form the maxillary division of the trigeminal, which is joined along its course to the gasserian ganglion by several other small nerves; of particular interest here are the small ganglionic branches from the **pterygopalatine ganglion** (also called the **sphenopalatine**, or **Meckel's**, **ganglion**) that carry parasympathetic fibers for the lacrimal gland.

Lesions in the branching hierarchy of the ophthalmic nerve produce anesthesia that helps identify the lesion site

Since the ocular and orbital branches of the trigeminal system are conveying sensory information, lesions or disruptions of these nerves produce anesthesia in the innervated region. Diagnoses of the probable lesion site depend on knowing how the nerves are distributed (see Figure 5.18) and on being able to determine the regions in which sensation has been lost. For example, corneal anesthesia can result from lesions affecting the whole ophthalmic division of the trigeminal, the nasociliary nerve, the long ciliary nerves, or from local destruction of the sensory terminals within the cornea. Although all these cases are characterized by corneal anesthesia, the difference that indicates the location of the lesion is the other regions involved.

If the anesthesia is confined to the cornea, the problem is either local—within the cornea itself—or somewhere along the path of the long ciliary nerves before they join the nasociliary. A lesion of the nasociliary, however, anesthetizes not only the cornea but also the skin on the nose that is innervated by the infratrochlear and anterior ethmoidal nerves. (It would also involve the other sensory fibers from the eye that reach the nasociliary via the sensory root of the ciliary ganglion, but there is no way to test for loss of sensation in the uveal tract.) And the closer the lesion is to the origin of the trigeminal system at the brainstem, the more extensive is the region of anesthesia. A lesion that removed the entire ophthalmic division would affect not only the eye, but also the skin of the nose, the entire brow (frontal nerve), and the facial skin lateral to the orbit (lacrimal nerve). A still more central lesion could include the fibers in the maxillary division as well, thereby reducing sensitivity on the cheek and below the orbit.

For surgery the eye is anesthesized by **retrobulbar block**, an injection of anesthetic behind the eye. A long, gently curved needle is used to approach from the lateral side of the eye; the needle is passed between the lateral and inferior rectus muscles to make the injection within the muscle cone near the ciliary ganglion. All nerves within the muscle cone are affected, including those innervating the extraocular muscles.

Viral infection of the trigeminal system can produce severe corneal damage

Lesions that disrupt neural transmission are not the only disorders affecting sensory nerves. A familiar disorder is infection of the nerves by members of the herpes family of viruses. In the most common, relatively minor form, the viral infection produces small, irritating blebs in the skin innervated by the infected nerve; a certain amount of local tissue destruction normally heals without consequence. Viral infections of the trigeminal system, however, can involve the cornea; what might be small blebs in the skin can be significant ulcerations in the cornea that destroy its transparency and optical quality and provide a route for further secondary infection. These viral infections demand early recognition and aggressive treatment. Unfortunately, successful treatment may not be enough to prevent some loss of corneal sensitivity and to reduce the possibility that reinfection will not be recognized as quickly.

The Extraocular Motor Nerves

The three cranial nerves that innervate the extraocular muscles contain axons from clusters of cells in the brainstem

The six extraocular muscles (see Chapter 4) are supplied by three cranial nerves: the **oculomotor** (III) to the superior rectus, medial rectus, inferior rectus, and inferior oblique; the **trochlear** (IV) to the superior oblique; and the **abducens** (VI) to the lateral rectus. The oculomotor nerve also innervates the levator muscle of the upper lid.

The clusters of cell bodies in the brainstem from which the axons going to the extraocular muscles arise are the nuclei of the motor nerves. They are very obvious in histological sections, but the cell bodies in these oculomotor nuclei have extensive dendritic arbors that extend well beyond the cluster of cell bodies that we define as the nucleus. Since the inputs to the cells often impinge on the peripheral parts of the dendrites, it is more realistic to think of a "nucleus" as the core or center of a larger functional entity. Since the cell axons usually come from the somata, however, the nucleus is the site of the cell outputs.

Cells in different parts of the oculomotor nerve nucleus innervate the levator, the superior and inferior recti, the medial rectus, and the inferior oblique

The paired oculomotor nerves arise from large nuclei just below the rostral end of the aqueduct of Sylvius (Figure 5.20). Since these nuclei contain neurons whose axons are destined for five different muscles, they are more complicated than nuclei whose axons have a single destination. The description that follows is based on experiments in which single extraocular muscles in macaque monkey were injected with fluorescent or radioactive substances that are taken up by the axon terminals and transported back to the cell bodies from which the axons arise (see Box 5.2).

Cells that innervate a particular muscle are consistently observed to be most densely clustered in a particular part of the nucleus; they are never scattered at

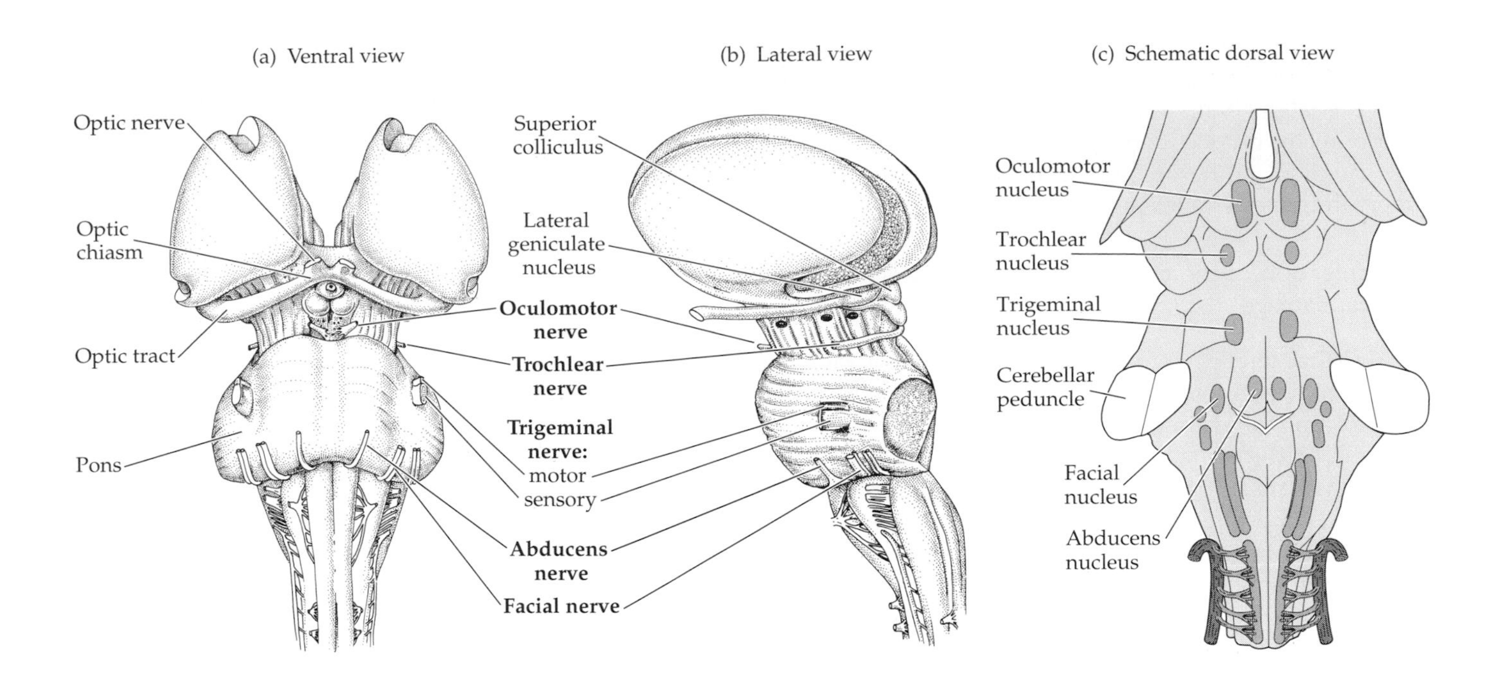

Figure 5.20
Cranial Nerves and Nerve Nuclei for the Extraocular Muscles and Eyelids
The extraocular motor nerves—oculomotor (III), trochlear (IV), and abducens (VI)—the facial nerve (VII), and the trigeminal nerve (V) are shown in relation to other structures when viewed from the ventral (*a*) and lateral (*b*) sides of the brainstem. (*c*) The locations of important nuclei shown in a schematic dorsal view of the brainstem.

random. As Figure 5.21 shows, cells for the different muscles are represented in a rostral to caudal sequence, although the regions are displaced vertically with respect to one another and thus overlap to some extent; roughly, the sequence is medial rectus (most rostral), inferior rectus, superior rectus, inferior oblique, and levator (most caudal). In effect, the axons destined for different muscles arise from cells that seem to form subnuclei within the main nucleus.

There are several complications, however. First, the cells for the different muscles are not completely separate; cells sending axons to the medial rectus mingle somewhat with those sending axons to the inferior rectus, for example, and such mingling occurs between other sets of cells as well. In addition, the "subnuclei" are defined by the locations of the cell bodies; since the cells have extensive dendritic trees, the real overlap will be more extensive. Finally, the representation of the medial rectus neurons is discontinuous; they form three separate clusters.

Since the oculomotor nuclei are fairly large, the fact that there is a certain amount of consistent spatial ordering of cells within the nucleus raises the possibility that lesions in the vicinity of the nuclei can have different effects, depending on their location. A rostral lesion, for example, is more likely to affect the neurons innervating the medial recti, while a caudal lesion is more likely to affect the levator and the muscles responsible for upward gaze (superior rectus and inferior oblique). In addition, a midline lesion is more likely to affect the medial recti than is a lesion on the lateral sides of the nucleus. But these are statements about probability, not certainty; the complex spatial organization of the oculomotor nucleus precludes unambiguous statements. Perhaps the most important generalization is that small lesions in the vicinity of the oculomotor nucleus are unlikely to affect all the muscles innervated by the oculomotor neurons.

One aspect of the organization of the oculomotor nuclei is much clearer. Injections of radioactive tracers into the levator, medial rectus, inferior rectus, and inferior oblique associated with the *right* eye label neurons in the *right* oculomotor nucleus; that is, most of the cells in the oculomotor nucleus send their axons into the ipsilateral oculomotor nerve and to the muscles on the ipsilateral side. The neurons innervating the superior recti are the exception; their axons cross the midline between the oculomotor nuclei to pass through the contralat-

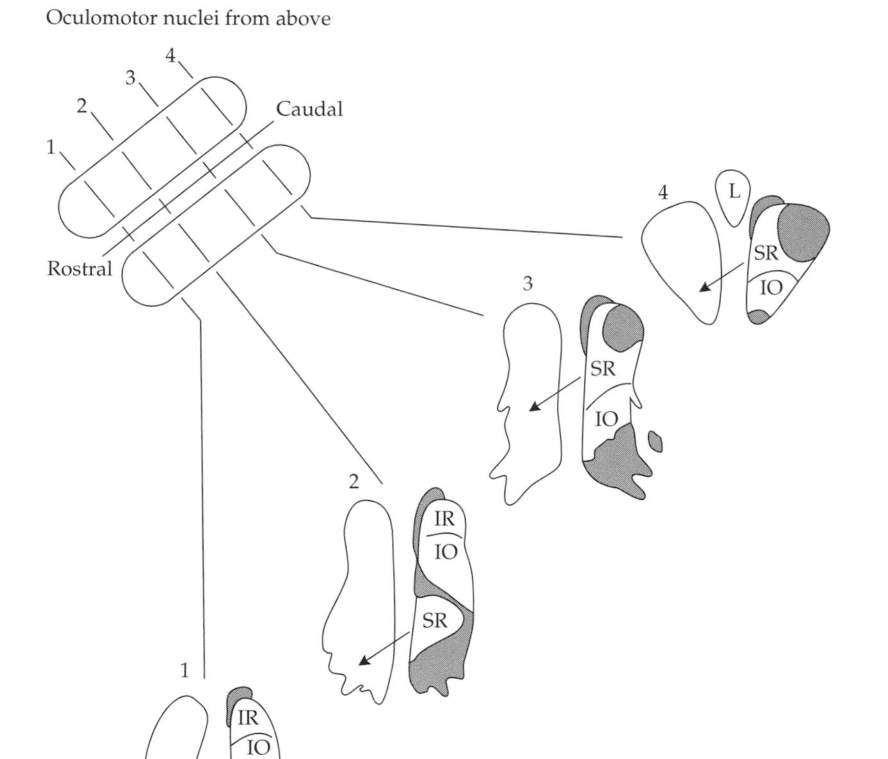

Figure 5.21
Extraocular Muscle Representations in the Oculomotor Nucleus
The neurons innervating the five muscles supplied by the oculomotor nerve are partially segregated in the nucleus of the nerve. Four sections through the paired nuclei are shown here; the approximate locations are indicated at upper left.. Neurons innervating the ipsilateral medial rectus are indicated with dark stippling and regions containing neurons for the other muscles—superior rectus (SR), inferior rectus (IR), inferior oblique (IO), and levator (L)—are labeled. The three-dimensional geometry of the nucleus is complicated; medial rectus neurons run in a region along the ventral side of the nucleus, but they are also found dorsally, in separate locations in the caudal part. Superior rectus neurons are not found in the rostralmost part, but they occupy a large part of the ventral-medial nucleus near the caudal end. Inferior rectus neurons are not found in the caudal part of the nucleus, but inferior oblique neurons extend throughout. The boundaries between muscle representations are not sharp, and there is always some overlap. The arrows show that the superior rectus neurons send axons to the opposite oculomotor nerve. (After Büttner-Ennever and Akert 1981.)

eral nucleus and enter the contralateral oculomotor nerve. Thus the right superior rectus is innervated by axons originating in the *left* oculomotor nucleus, and vice versa. Note, however, that severing one oculomotor nerve will still affect only the muscles on the ipsilateral eye.

Axons destined for different muscles run together in the oculomotor nerve until it exits the cavernous sinus just behind the orbit

The axons from different parts of the oculomotor nucleus form small bundles that coalesce as they run from the nucleus to the ventral side of the brainstem, emerging as several small rootlets that join to form the oculomotor nerve. The nerve passes forward, running through the cavernous sinus on its lateral side and from there to the superior orbital fissure. Just before reaching the fissure, the nerve usually divides, thus entering the apex of the muscle cone (i.e., within the annular tendon of Zinn) as the **superior** and **inferior muscular branches** of the oculomotor nerve (Figure 5.22). Typically, the superior branch innervates the levator and superior rectus, while the inferior branch innervates the medial and inferior recti and the inferior oblique.

The superior branch runs forward along the underside of the superior rectus until it is about one-third of the way along the muscle; at this point the branch may divide—one division going into the superior rectus and the other passing around and up to the levator. More commonly, the single branch enters the superior rectus, and any fibers destined for the levator pass directly through to emerge on the upper surface of the superior rectus.

The inferior muscular branch runs forward, below and lateral to the optic nerve, sending branches, in passing, under the optic nerve to the medial rectus

Box 5.2 Tracing Neuronal Connections: Axonal Transport Methods

SINCE BEING OBSERVED by Greek anatomists, nerves were thought to communicate between the brain and the peripheral structures by fluid flowing through them. As late as 1765, a medical student at the University of Edinburgh made careful notes about the evidence that demonstrated this fact (the lecturer was Alexander Monro, secundus, author of two volumes on the structure and function of the nervous system). For example, Monro claimed that if a very fine silk thread was tied tightly around a small peripheral nerve, the muscle innervated by the nerve could no longer be made to contract. This effect showed that fluid flows within the nerve fibers, and blocking the fluid flow also blocked the ability of the nerve to activate its target.* In addition, Monro might have added, van Leeuwenhoek (see Vignette 8.1) had observed small droplets of fluid oozing from the ends of cut nerves.

It is true that tying off a small motor nerve in the manner described causes the nerve to fail such that reflex contraction cannot be elicited; moreover, as Ramón y Cajal (see Vignette 14.2) and others observed many years later, there will be swelling on the side

*The lecture notes were made by William Withering (1741–1799), who became a surgeon in Edinburgh. They are part of the collection of the Reynolds Historical Library, University of Alabama at Birmingham, cataloged under the author's name, and titled *Anatomical and Physiological Observations. Collected from the Lectures of Doct^r Hunter & Doct^r Monro.*

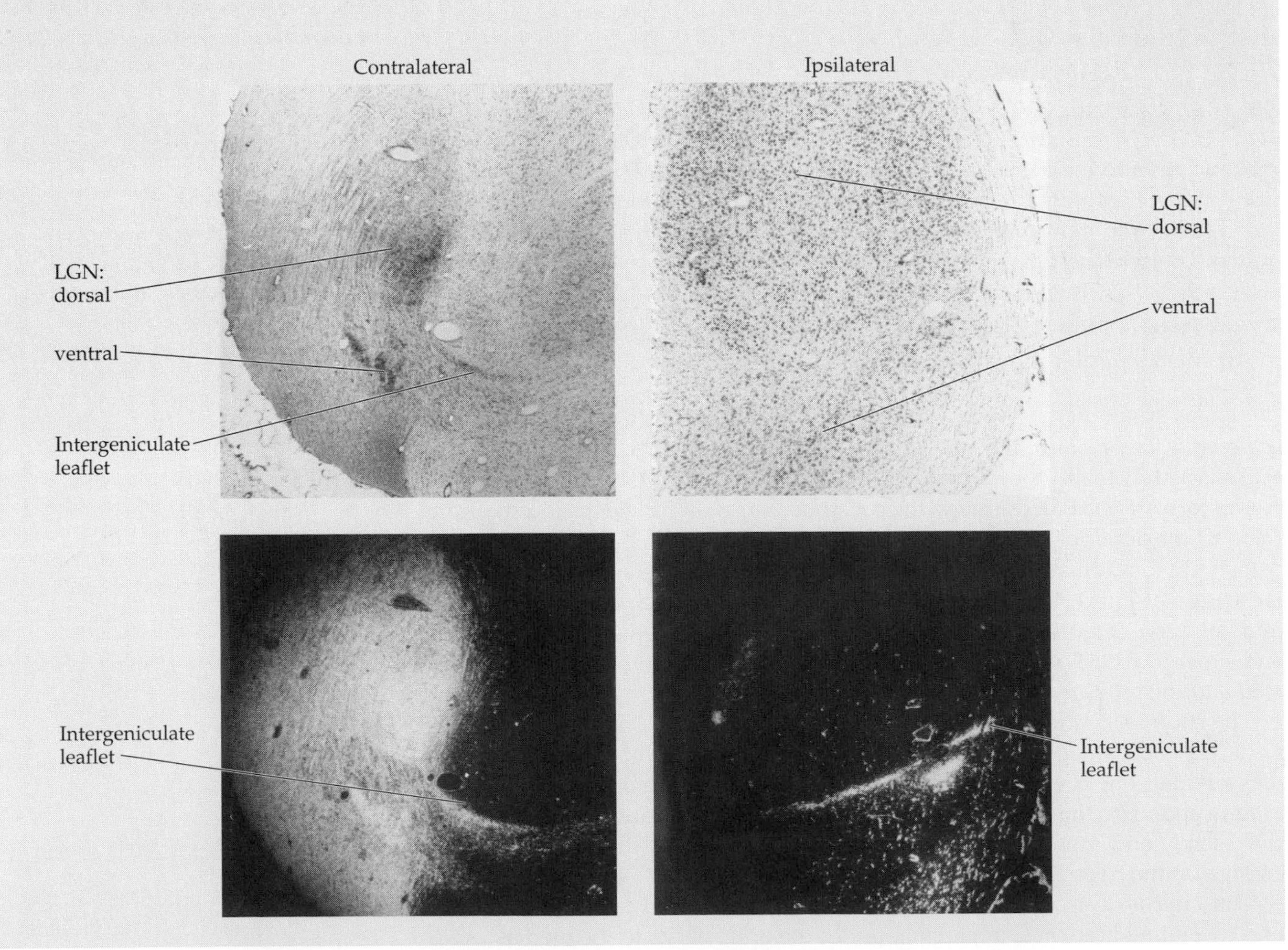

Figure 1
Autoradiography: Orthograde Transport
The upper panels are sections through the lateral geniculate nuclei in rabbit made 24 hours after tritiated leucine was injected into the vitreous chamber of one eye. The dorsal and ventral LGN are dark on the side contralateral to the injection because of black dots in the emulsion coating the sections; the dots are produced by beta particles emitted by tritiated leucine that was transported to the ganglion cell axon terminals. The individual dots are too small to be visible here, but they can be seen and counted at higher magnification. The lower panels are dark-field photomicrographs of the same sections. All of the dots in the emulsion produced by beta emission are now seen as bright spots because they are reflecting light in this type of illumination. This part of the rabbit's dorsal lateral geniculate nucleus receives no ipsilateral retinal input, as shown by the absence of bright dots. The ventral LGN and a structure called the intergeniculate leaflet have both contralateral and ipsilateral retinal inputs, however.

of the ligature closest to the spinal cord. But while Monro was right about fluid flowing within the nerve's axons, his argument that the fluid has anything to do with neurotransmission was incorrect; conduction fails because the ligature interferes with the propagation of action potentials. (Electricity was a new and interesting phenomenon in the late eighteenth century—witness Ben Franklin flying his kites—and there were already suggestions that it had something to do with nerve action. It was not well understood, however, and Monro assumed that if electricity were involved in nerve conduction, it would not be affected by a piece of thread tied around the nerve.) The fluid flow in the nerve fibers is now called axonal or axoplasmic flow.

There is considerable intracellular traffic between nerve cell bodies and their axon terminals; in general, this movement of proteins and other materials is **axoplasmic transport,** referred to as **orthograde transport** when the transport is from the cell body to the axon terminal, and **retrograde transport** when transport is from the terminal to the cell body. Depending on the substances transported, transport systems may be either fast or slow.

Axoplasmic transport becomes a servant of anatomy when detectable tags or labels are attached to the molecules being transported. For example, the amino acids proline and leucine are taken up by nerve cell bodies and incorporated in proteins that are transported to the nerve terminals. These amino acids can be made radioactive by synthesizing them with tritium (^{3}H) instead of ordinary hydrogen (tritium emits beta particles—electrons—as it decays). When the tritiated amino acids are experimentally placed in the vicinity of nerve cell bodies, they become incorporated in proteins by the cells and are transported in the usual way. Now, however, the proteins are tagged, and we can see where they go within the cells. Their locations are revealed when the beta particles interact with a photographic emulsion laid over a histological section, appearing as small black dots in the emulsion for normal bright-field microscopy or, more dramatically, as small, glowing points of light in dark-field illumination. Figure 1 shows both bright-field and dark-field pictures of the lateral geniculate nucleus after [^{3}H]leucine injected into the vitreous chamber of one eye had been taken up by ganglion cells and transported through their axons.

Autoradiography, which uses radioactive labels like tritium, is just one of numerous methods that exploit axonal transport. Another popular method uses an enzyme, horseradish peroxidase (HRP), that is taken up by axon terminals and by damaged axons, after which it undergoes both orthograde and retrograde transport. HRP can be seen in tissue sections because it will oxidize certain compounds (chromagens) in the presence of hydrogen peroxide and produce a very dark reaction product. Alternatively, the molecules of HRP can be tagged with a radioactive or fluorescent compound. Figure 2 shows an example of retrograde transport of HRP into retinal ganglion cells. This example was taken from an HRP-labeled retina used to show the region of nasotemporal overlap in the retina (see Figure 5.6).

None of these methods are foolproof. When a tagged compound is injected into a brain nucleus, for example, there is always some diffusion away from the injection site and therefore some uncertainty about the size of the region exposed to the tagged compound. This diffusion raises an inevitable question about the specificity of the label: Were all labeled cells within the nucleus of interest or was there diffusion into nearby cell groups? And some molecules, like HRP, are taken up by axons passing through a nucleus if they are damaged by the injection needle, potentially causing transport and labeling of cells that have no functional connection with the nucleus of interest. In addition, some tritiated amino acids can be made to cross synapses from axon terminals to the cells they innervate, a phenomenon called transneuronal transport. When this happens, the pattern of radioactivity will not show precisely where the labeled cells terminate.* The phenomenon can sometimes be used to advantage, however, to show several synaptically linked parts of a pathway simultaneously.

With proper controls, axonal transport methods are very powerful. Their value for human anatomy, however, is by analogy with results from other primates. But this situation may change. Some synthetic compounds, including one called Dye-I, do not require the transport systems in living cells and will flow through long-dead cells in tissue that has been treated with formaldehyde. The hope, yet to be realized, is that human brain tissue obtained postmortem can be studied by methods similar to those used so effectively in other species.

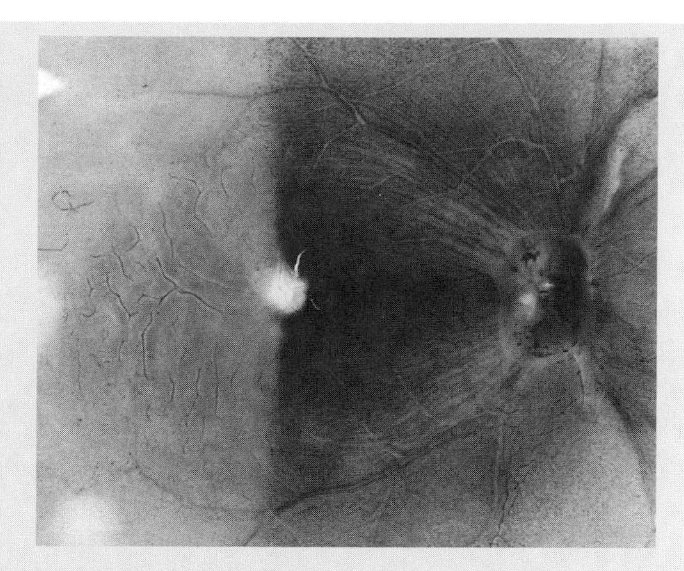

Figure 2
Retrograde Transport of Horseradish Peroxidase
This photograph shows a monkey retina after injection of HRP into the contralateral optic tract. The background is darker at the right, in the nasal retina, because most of the dark HRP reaction product is confined to ganglion cells whose axons cross in the chiasm to the contralateral side. At higher magnification, it would be possible to see details of the dendritic branching of the cells. (From Fukuda et al. 1989.)

*It's a problem. I coauthored a manuscript that was rejected for publication because a reviewer thought the labeling pattern had resulted from transneuronal transport and did not show where retinal axons were terminating, as we claimed. We were correct, as it turns out, but later studies by other investigators using several different amino acids were required to make the point unequivocally.

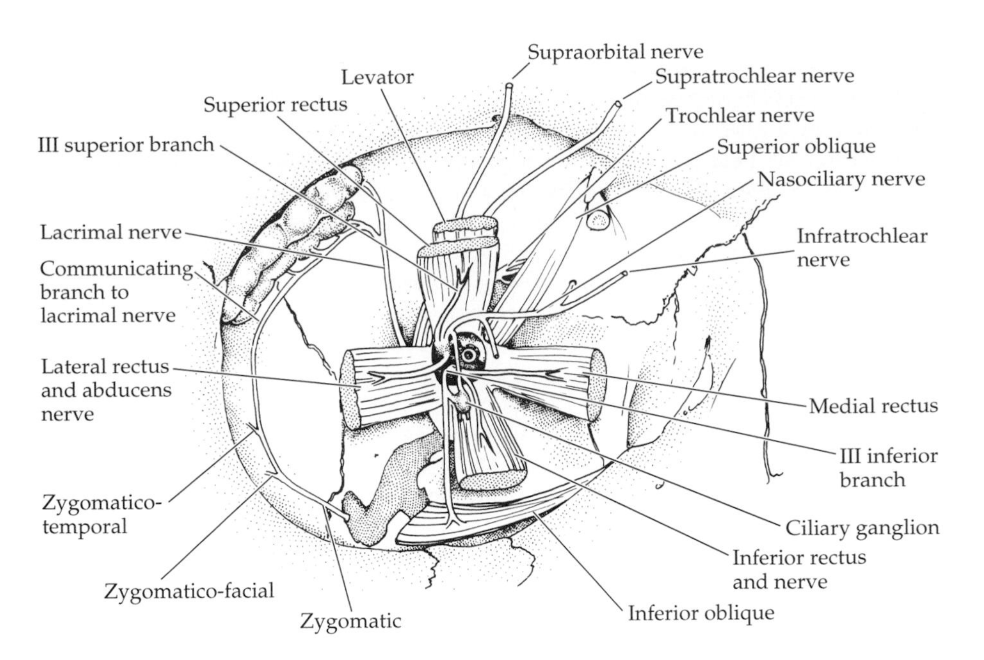

Figure 5.22
Orbital Branches of the Oculomotor Nerve
This view of the orbit from the front shows that nerve III (the oculomotor) has two branches entering the muscle cone through the superior orbital fissure; the superior branch typically supplies the superior rectus and levator, while the inferior branch sends subsidiary branches to the medial rectus, inferior rectus, and inferior oblique. Parasympathetic fibers headed for the ciliary ganglion exit the inferior division as the motor root of the ciliary ganglion. There is considerable variation in the details of branching to the individual muscles.

and down to the inferior rectus; these subsidiary branches leave the inferior division about one-third of the way forward in the orbit. After the fibers for the recti exit the inferior branch, the remaining fiber bundle continues forward to the eye, where it supplies the inferior oblique.

The oculomotor nerve contains parasympathetic fibers bound for the ciliary ganglion

A small bundle of axons leaves the inferior oculomotor branch going to the inferior oblique and runs upward to the ciliary ganglion; this is the **motor root** (of the ciliary ganglion), which carries preganglionic parasympathetic fibers to the ganglion. These axons come from cells in the **Edinger–Westphal nucleus**, which lies in the brainstem near the ipsilateral oculomotor nucleus. The axons from the Edinger–Westphal nuclei join the oculomotor nerves before they exit the brainstem.

Cells in the trochlear nerve nucleus innervate the contralateral superior oblique

The next pair of cranial nerves, the trochlear nerves (IV), originate in nuclei just caudal to the large oculomotor nuclei (see Figure 5.20*c*). The axon bundles from the trochlear neurons, however, do not exit on the base of the brainstem; instead they run dorsally on either side of the aqueduct of Sylvius toward the inferior colliculi, where the axon bundles cross the midline to the opposite side (decussate) and emerge on the dorsal surface just behind the inferior colliculi. The axon bundles, now the trochlear nerves, wrap around the sides of the brainstem and then run forward parallel to and below the oculomotor nerves. They pass through the cavernous sinus on the lateral sides en route to the superior orbital fissure, where they enter the orbits *outside* the muscle cone of the recti. After passing through the superior orbital fissure, the trochlear nerve turns medially, crossing over and above the levator near the back of the orbit to reach the superior oblique, which it enters about one-third of the way forward along the muscle.

The superior obliques share with the superior recti the property of contralateral innervation (and since these are the only contralaterally innervated muscles, the fact is easy to remember; superior—rectus or oblique—means contralateral innervation). The trochlear nerves, with just over 1000 axons each, are the smallest of the nerves going to the extraocular muscles (see Table 4.1). Since they exit on the dorsal side of the brainstem, they are also the longest of the nerves. Their small size and extra length make the trochlear nerves particularly vulnerable to trauma.

Abducens nerve cells innervate the ipsilateral lateral rectus

The nuclei of the abducens nerves (VI), lying just below the cerebellum, are much farther down in the brainstem than the trochlear nuclei (see Figure 5.20*c*). Abducens axons exit on the base of the brainstem at the posterior edge of the pons and run forward below the oculomotor and trochlear nerves to the cavernous sinus, where they are located medially and inferiorly near the internal carotids, and then to the superior orbital fissure. The abducens nerve enters the muscle cone within the annulus of Zinn and runs along the inner side of the lateral rectus to enter the muscle.

The abducens nerve, like the trochlear nerve, innervates a single muscle, so the effects of lesions anywhere along the course of the nerve are confined to the lateral rectus (assuming, of course, that the lesion does not affect other nerves—see below). Since the abducens is the motor nerve most laterally placed in the orbit, it is the most commonly involved in trauma to the side of the head.

All of the oculomotor nerves pass through the cavernous sinus on their way to the orbit

The foregoing descriptions of the paths of the motor nerves to their target muscles only hint at their clinically relevant anatomy, which has to do with their relationships to other structures along their course from the brainstem to the orbit. It should be clear, however, that there are not many places where a lesion could affect all the extraocular motor nerves: Within the orbit, where the oculomotor nerve has branched and the nerves are fanning out to the different muscles, one would expect problems with only one or a few muscles; near the brainstem, the nerves are fairly well separated and are not likely to be affected together. Between the brainstem and the orbit, however, the nerves run in parallel, and they all pass through or within the walls of the **cavernous sinus**, which is a large venous sinus of variable structure lying alongside the sphenoid bone (there are sinuses on both sides of the sphenoid).

Figure 5.23 shows a schematic cross section through the cavernous sinuses and the sphenoid bone; although the sinuses are normally bilaterally symmetrical, the two sides are illustrated somewhat differently to show some of the range of anatomical variability. The important generalizations are, first, the presence of the carotid artery passing through the sinus alongside the sphenoid bone and, second, the presence of all the motor and sensory nerves to the eye (except the optic nerve). The nerves passing through the sinus typically have a vertical sequence; from dorsal to ventral, the usual pattern is III, IV, VI, V_1, and V_2 (see Figure 5.23).

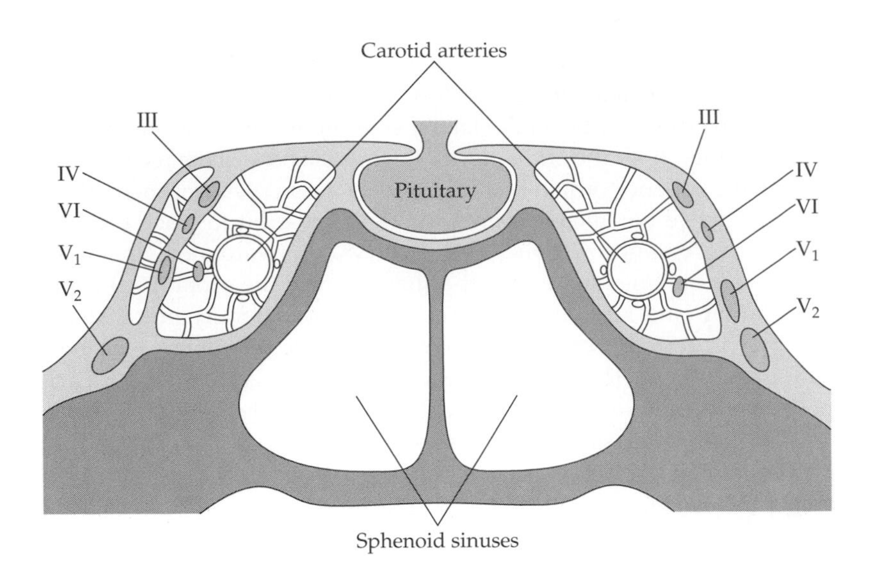

Figure 5.23
The Cavernous Sinuses and the Orbital Nerves
The nerves for the extraocular muscles and two branches of the trigeminal nerve run through the cavernous sinus in close proximity to one another. Their positions in the sinus vary, and the two sides of the drawing are different to show some of the variation. The most common arrangement is shown at right.

Since the nerves are a few millimeters apart as they pass through the sinus, they could be affected separately by small lesions. But the problems that occur here—carotid aneurysm, thrombosis, meningioma—tend to be large or to become large, and typically all the nerves are involved. The clinical signs are paralysis of *all* the extraocular muscles on the ipsilateral side (**external ophthalmoplegia**) and anesthesia of the skin around the orbit. In addition, because of the parasympathetic fibers in the oculomotor nerve, part of the eye's parasympathetic innervation will be affected.

The extraocular motor nerves probably contain sensory axons from muscle spindles and tendon organs

Although we have discussed nerves III, IV, and VI as motor nerves, they are not strictly motor nerves; the oculomotor nerve contains parasympathetic fibers going from the Edinger–Westphal nucleus to the eye and contains sensory fibers originating in the muscle spindles and tendon organs in the muscles (see Chapter 4). Or, to be precise, we *assume* there are sensory fibers in the human extraocular motor nerves, as there are in other primates. In monkeys, the sensory fibers run back in the nerves or nerve branches supplying the muscles to a point where the nerves pass close to the gasserian ganglion of the trigeminal nerve. Here, the sensory fibers leave the extraocular motor nerves and pass by ill-defined routes into the sensory root of the trigeminal and then forward in the brainstem to their cell bodies in the mesencephalic nucleus.

This arrangement means that lesions to the extraocular motor nerves will affect both sensory and motor fibers; only near or within the brainstem is there any possibility of affecting motor or sensory fibers independently. But in light of the discussion in Chapter 4 about the uncertain role of the muscle spindle and tendon organ systems in extraocular muscles, there is no way to detect an independent disruption of the sensory fiber pathway if it were to happen. The situation here is quite different from that of the mixed peripheral nerves, where lesions can produce both paralysis and anesthesia.

Innervation of the Muscles of the Eyelids

Three sets of muscles are associated with the eyelids

The structure and arrangement of the muscles associated with the eyelids are discussed in Chapter 7. For present purposes, it is sufficient to know that there are two striated muscles and two sets of smooth muscles in the lids. Lids are closed—that is, the palpebral aperture is narrowed—by contraction of the **orbicularis oculi**, a large striated muscle within both upper and lower eyelids. The upper eyelid can be raised by contraction of the **levator palpebrae superioris**, whose tendon inserts into the upper lid; its innervation, by the oculomotor nerve, was considered earlier in the chapter. In addition, the upper lid can be raised and the lower lid depressed by the **superior** and **inferior tarsal muscles**, respectively (they are also called the **palpebral muscles of Müller**); the tarsal muscles, as we'll see shortly, are smooth muscle innervated by the sympathetic nervous system.

The orbicularis is innervated by the facial nerve

The **facial** nerve (VII) nucleus lies near that of the abducens (see Figure 5.20*c*), and the axons of the facial nucleus cells emerge from the ventral side of the brainstem near the lower border of the pons. The nerve runs laterally toward the internal auditory meatus, with one subdivision remaining in the temporal bone to supply the middle ear and other structures, and the other exiting the skull via the stylomastoid foramen. Once outside the skull, the facial nerve branches extensively to supply the muscles of the face and jaw (Figure 5.24). The orbicularis is supplied by two branches, or sets of branches, called the **upper** and **lower zygomatic branches** (of the facial nerve).

Lesions that irritate the facial nerve and produce unusual levels of activity in the nerve fibers going to the orbicularis muscle can produce **blepharospasm** (see Chapter 7), an involuntary, forcible lid closure. Failure of transmission to the orbicularis and muscle paralysis may produce an inability to close the lids and an eversion (**ectropion**) of the lower lid (see Chapter 7).

The superior and inferior tarsal muscles are innervated by the sympathetic system

The superior tarsal muscle, which can raise the upper lid, is innervated by the sympathetic system, but the route by which sympathetic fibers reach the tarsal

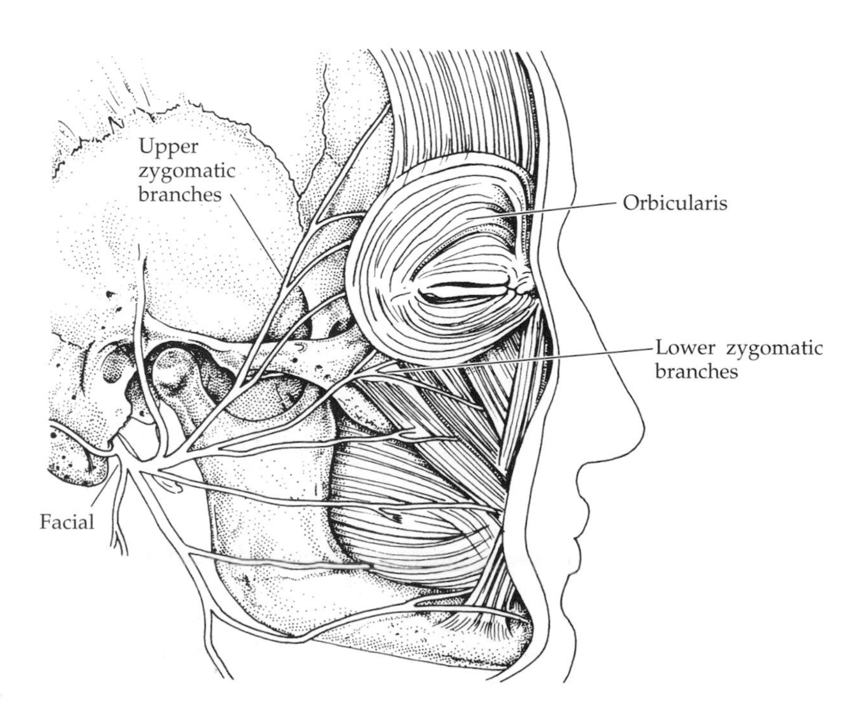

Figure 5.24
Branching and Distribution of the Facial Nerve
After exiting the skull thorugh the stylomastoid foramen, the facial nerve (VII) branches extensively to supply muscles of the face and jaw. Its upper and lower zygomatic branches innervate the orbicularis muscle of the eyelids.

muscle is not well understood. The sympathetic fibers are thought to arise from cells in the superior cervical ganglion along with fibers going to the intrinsic muscles of the eye and to follow the same path to the sympathetic plexus around the carotid artery. But from the carotid plexus to the tarsal muscle in the lid, the path seems genuinely unknown; the most common statement is that sympathetic fibers bound for the superior tarsal muscle enter the superior division of the oculomotor nerve, running in the branch going to the levator. If so, the fibers presumably continue through the levator, exiting at the anterior end of the muscle to innervate the superior tarsal muscle.

A recent study of innervation to the superior tarsal muscle in cynomolgous monkey found that after horseradish peroxidase was injected into the superior tarsal muscle and time was allowed for retrograde transport (see Box 5.2), cell bodies were labeled in the ipsilateral superior cervical ganglion *and* the ipsilateral pterygopalatine ganglion. Since the pterygopalatine ganglion is a parasympathetic ganglion, these results suggest that the superior tarsal muscle receives both sympathetic and parasympathetic innervation. The function of the parasympathetic innervation is unclear, but presumably it would be antagonistic to the sympathetic innervation, which elicits contraction of the muscle.

Innervation of the inferior tarsal muscle is equally uncertain. In this case, the sympathetic fibers may enter the inferior division of the oculomotor nerve, running in the branch supplying the inferior oblique, which is the extraocular muscle lying closest to the inferior tarsal muscle. From here, the fibers could continue forward to the inferior tarsal muscle. The possibility of parasympathetic innervation of the inferior tarsal muscle has not been investigated, but the situation is likely to be the same as that of the superior tarsal muscle.

Sympathetic fibers to the human tarsal muscles have been reported in *all* of the extraocular motor nerves and in the major branches of the ophthalmic division of the trigeminal nerve (nasociliary, frontal, and lacrimal nerves). These sympathetic fibers supposedly join these nerves as they enter the orbit through the superior orbital fissure. Unfortunately, the experimental method does *not* guarantee that the labeled fibers are sympathetic or that they are destined for the tarsal muscles. At present, the most that can be said is that the details of the innervation to the tarsal muscles are not known.

Ptosis may result from either oculomotor or sympathetic lesions

The upper eyelid is held in position by the tonus in the levator and the superior tarsal muscles. It follows from this that either a lesion to the oculomotor nerve, affecting innervation of the levator, or a sympathetic system lesion, affecting the superior tarsal muscle, will result in ptosis (drooping upper eyelid). Ptosis is an important diagnostic sign, but the ptosis itself is not a problem unless it is severe enough to interfere with vision when the eye is in primary position. In that case, the ptosis can be reduced surgically.

Autonomic Innervation of Smooth Muscle within the Eye

The superior cervical ganglion is the source of most sympathetic innervation to the eye

The sympathetic system related to the eye is illustrated schematically in Figure 5.25. Sympathetic fibers destined for the eye originate at cell bodies in the **superior cervical ganglion**, the uppermost ganglion of the sympathetic trunk; the superior cervical ganglia lie in the upper part of the neck, just below the base of the skull. Fibers from the ganglia run upward as the **internal carotid nerves** along with the internal carotid arteries to the level of the cavernous sinuses, where the nerves break up into numerous small, interweaving bundles—the

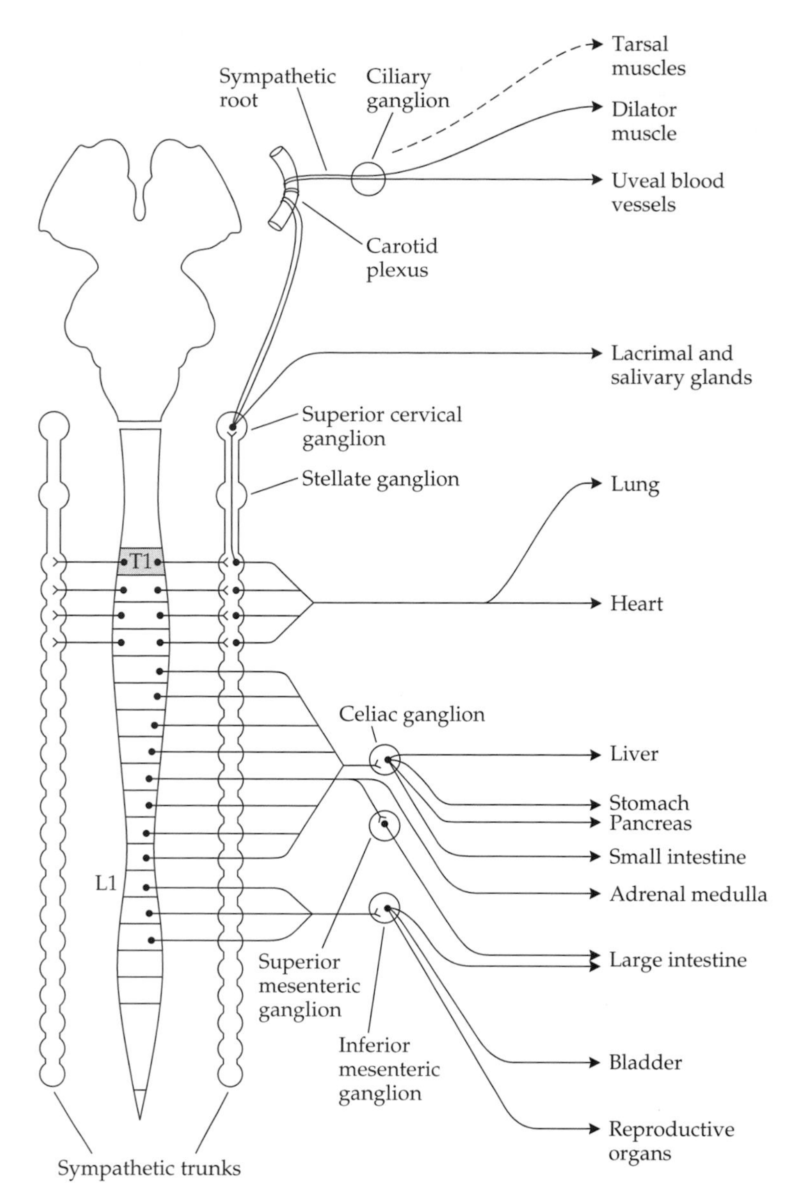

Figure 5.25
The Sympathetic Trunks and Sympathetic Innervation to the Eye and Other Organs
The sympathetic trunks and associated ganglia lie on either side of the spinal cord; the sympathetic innervation to the eye and other organs is distributed through the sympathetic nerve trunks and ganglia. The sympathetic pathway to the eye originates in the brainstem, from which axons descend the spinal cord to synapse on cells at the first thoracic segment (T1). Axons from the T1 cells travel up through the sympathetic trunk to the superior cervical ganglion, where they synapse on the cells that innervate targets in the eye and orbit. Some sympathetic axons may leave the carotid plexus and run in the oculomotor nerve to the tarsal muscles in the eyelids (dashed lines); this pathway is not well characterized, however. (After Martin 1996.)

sympathetic, or **carotid**, **plexus**—around the carotid arteries. From here, the classic version of the story holds that one of the bundles of sympathetic fibers, which will become the **sympathetic root of the ciliary ganglion**, runs along with the ophthalmic artery as it branches from the internal carotid, thereby entering the orbit via the optic foramen.

An alternative version states that several small sympathetic fiber bundles enter the orbit through the superior orbital fissure in conjunction with the nasociliary nerve and enter the eye via the long ciliary nerves. Very little evidence supports such a sympathetic pathway to the eye *in primates.* It is certainly true that other species often used experimentally do not have a sympathetic root to the ciliary ganglion; the sympathetic fibers reach the eye by a different route. What seems to have happened, however, is that this alternative pathway was assumed to apply to humans and other primates. If nothing else, this situation illustrates the potential problems in generalizing from one species to another.*

It would not be surprising to find that the long ciliary nerves contain some sympathetic nerve fibers (there are some indications of sympathetic innervation in the cornea, for example); the autonomic system in general seems to be flexible about the routes it takes to its targets. On the basis of the evidence in hand, however, it would be surprising if these fibers were anything more than a very small part of the eye's sympathetic innervation. Thus, with this qualification, the primary pathway for sympathetic fibers to the human eye will be taken to be the sympathetic root of the ciliary ganglion.

Sympathetic fibers enter the eye in the short ciliary nerves

After passing through the optic foramen, the sympathetic fibers run alongside the optic nerve to the ciliary ganglion. The fibers do *not* make synaptic connections in the ciliary ganglion. Instead, as they pass though the ganglion, the original single bundle of sympathetic fibers that formed the sympathetic root divides into smaller bundles that enter the dozen or so short ciliary nerves. The short ciliary nerves pass through the sclera to the choroid, where the fibers disperse into a sparse network of fibers as they proceed to their target structures.

Sympathetic innervation of the dilator muscle acts at alpha-adrenergic receptors to dilate the pupils

It is very difficult to follow sympathetic nerve fibers through the eye to their targets. As a result, statements about the innervation are largely based on pharmacology—the way in which the muscle is affected by drugs that mimic or block the normal sympathetic transmitter (norepinephrine)—or on immunohistochemical localization of receptor sites or neurotransmitter-related enzymes (see Box 15.1).

The dilator muscle of the iris increases pupil diameter—dilates the pupil—when it contracts, an action known by several lines of evidence to be the result of sympathetic innervation. Cutting the sympathetic pathway in the neck (cervical sympathectomy), for example, eliminates the normal tonus in the dilator muscle, and the unopposed action of the sphincter muscle produces miosis (pupil constriction). Drugs that mimic the sympathetic neurotransmitter action produce **mydriasis** (pupillary dilation); drugs that block sympathetic transmission produce miosis. The nerve terminals in the dilator contain norepinephrine, and the appropriate receptor sites (alpha-adrenergic) are found on the dilator muscle fibers.

Cervical sympathectomy and pharmacological agents affecting the sympathetic system have no consistent effects on accommodation, indicating that the ciliary muscle is not a target of the sympathetic innervation (although it has been suggested more than once).

*As near as I can tell, this version of the sympathetic pathway comes from a single diagram in the clinical continuing education literature in which there may have been an inadvertent substitution of "long" for "short" ciliary nerve. The diagram was later used in another widely referenced paper that focused on the *pre*orbital portions of the sympathetic pathway. Subsequent authors seem to have accepted the diagram as correct in its entirety without pursuing its sources.

The arterioles in the uveal tract receive sympathetic innervation that produces vasoconstriction

Most of the arterioles in the uveal tract, particularly in the choroid, appear to receive sympathetic innervation that acts at alpha-adrenergic receptors on their smooth muscle fibers. Although this aspect of the anatomy is clear, it is more difficult to say exactly what the sympathetic innervation is doing.

In general, when the arterial and venous pressures on either side of a vascular bed are constant, blood flow will be inversely proportional to resistance in the vascular bed; if the resistance increases, the flow will be reduced. The resistance is increased by vasoconstriction. For the choroidal circulation, activation of the sympathetic system reduces blood flow, suggesting that sympathetic innervation to the arterioles is producing contraction of the smooth muscle and vasoconstriction.

What this sympathetic vasoconstriction means functionally is not very clear, but it may help maintain constant blood flow under circumstances (hard physical labor, stress, or excitement) that usually elevate systemic arterial pressure. Unless counteracted, the elevated arterial pressure would increase choroidal blood flow, possibly to levels that would break down the normal blood–tissue barriers. This relationship suggests that the sympathetic vasoconstriction in the choroidal vasculature plays a compensatory role by preventing an undesirable increase in choroidal blood flow in the face of increased arterial pressure (see Chapter 6).

Horner's syndrome is the result of a central lesion in the sympathetic pathway

The classic case of an isolated sympathetic lesion affecting the eye is **Horner's syndrome**, which is typically characterized by ptosis, miosis, and anhidrosis (lack of sweating) of the face and neck on the affected side. Other signs may be reduced intraocular pressure, vasodilation of conjunctival blood vessels, and enophthalmos (sinking of the eye into the orbit). Both ptosis and miosis result from the interrupted innervation of the superior tarsal muscle and the dilator, respectively; anhidrosis results from the lack of normal sympathetic innervation of the sweat glands in the skin. The other signs, which may be transient if present at all, are related to the lack of normal sympathetic innervation of the vascular system.

The lesion site in a case of Horner's syndrome may be well away from the eye. The postganglionic neurons lie in the superior cervical ganglion in the neck, and the preganglionic fibers from the spinal cord are lower still, exiting the spinal cord at the first thoracic segment. Thus the lesion may lie anywhere along the pathway in the upper thorax and neck up to the point where fibers to the facial sweat glands part company with fibers to the eye and lids. This point is about where the common carotid bifurcates into the internal and external carotids; fibers to the sweat glands follow the external rather than the internal carotid.

Parasympathetic fibers entering the eye originate in the ciliary or the pterygopalatine ganglion

Parasympathetic innervation of the ocular smooth muscle is somewhat better defined than the sympathetic innervation; most of the parasympathetic inputs are to the sphincter of the iris and to the ciliary muscle, with some innervation of the choroidal vasculature as well. The parasympathetic fibers that innervate the sphincter and the ciliary muscle reach the eye via the short ciliary nerves, having originated from the cell bodies that constitute the ciliary ganglion (Figure 5.26). Thus, they are *post*ganglionic parasympathetic fibers; the *preganglionic* fibers originate in the Edinger–Westphal nucleus and reach the ciliary ganglion by way of the oculomotor nerve and the motor root of the ciliary ganglion.

An ongoing debate revolves around a parasympathetic pathway to the ciliary muscle that may either bypass the ciliary ganglion or pass through it without a

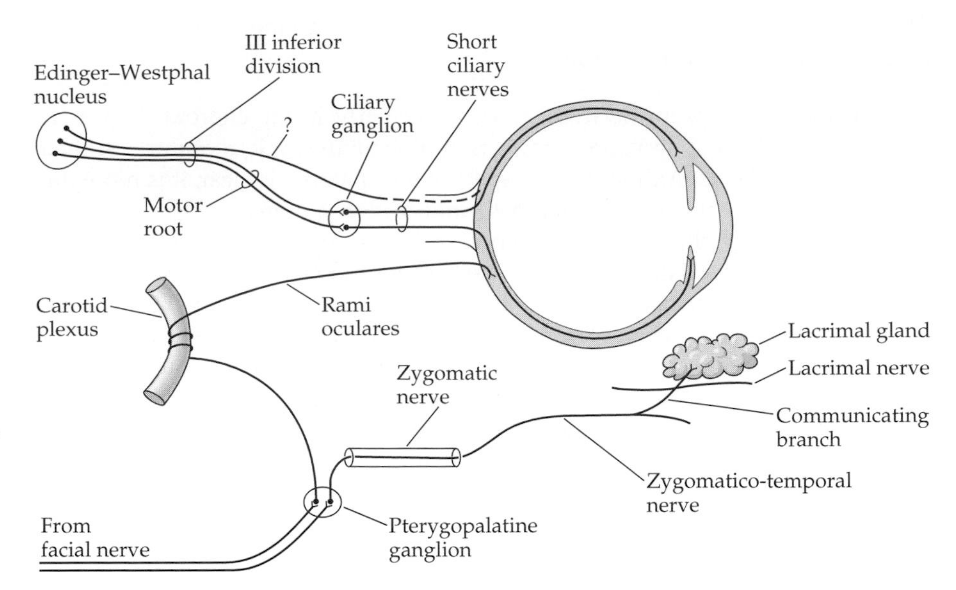

Figure 5.26
Parasympathetic Innervation of the Eye and Orbit
Postganglionic parasympathetic fibers to the eye arise from two sources: the ciliary ganglion within the orbit, whose preganglionic inputs come from the Edinger–Westphal nucleus through the oculomotor nerve, and the pterygopalatine ganglion just below the orbit, whose preganglionic input arrives through deep branches of the facial nerve. Some axons from the Edinger–Westphal nucleus may bypass the ciliary ganglion ("?"). Axons from ciliary ganglion cells innervate the intraocular musculature. Axons from pterygopalatine ganglion cells innervate the choroidal vasculature via the rami oculares (see Figure 5.27) and the lacrimal gland via the zygomatico-temporal and the communicating branch to the lacrimal gland.

synapse. HRP (horseradish peroxidase) injected in the vicinity of the ciliary body, for example, produces labeled cell bodies near the oculomotor nucleus, confirming the existence of some brainstem cells that do not synapse before reaching the eye. But there is no way of knowing if the labeled cells were terminating in the ciliary muscle or were merely passing through the area on their way to another place; fibers of passage damaged by the injection needle could take up HRP. The evidence from the past 25 years is suggestive, but like the HRP study, it is not decisive. I suspect there is more to this story, and I agree with Gerald Westheimer (one of the people who initiated this debate) that "there is something special about the innervation of the ciliary muscle." At the moment, however, the nature of that special something is not clear.

In addition to the foregoing oculomotor portion of the parasympathetic input to the eye, another portion associated with the facial nerve appears to bypass the ciliary ganglion and go not to the sphincter and ciliary muscles, but to the choroidal vasculature. The plexus of small autonomic nerves around the carotid artery behind the orbit has small branches that do not enter the orbit with the sympathetic root to the ciliary ganglion, but pass instead through the superior orbital fissure, eventually entering the eye along with—but separate from—the short ciliary nerves (Figure 5.27). These **rami oculares** appear to be *post*ganglionic parasympathetic fibers originating in the pterygopalatine ganglion, and their terminals can be found in the walls of the choroidal arterioles.

Axons from cells in the ciliary ganglion innervate the sphincter and the ciliary muscle

Although the postganglionic fibers are not easily traced once they are in the choroid, they clearly target the sphincter and ciliary muscle: Both structures show immunohistochemical labeling associated with cholinergic inputs, both muscles are activated by cholinergic drugs (producing accommodation and miosis), and both are inactivated by acetylcholine blockers (resulting in accommodative paralysis and mydriasis). The effects of lesions in the oculomotor nerve, ciliary ganglion, and short ciliary nerves confirm the anatomical pathway that the parasympathetic innervation follows from the brainstem to the eye. There appears to be some cholinergic innervation of the dilator muscle as well; its action is unclear, but it is probably antagonistic to the functionally dominant sympathetic innervation.

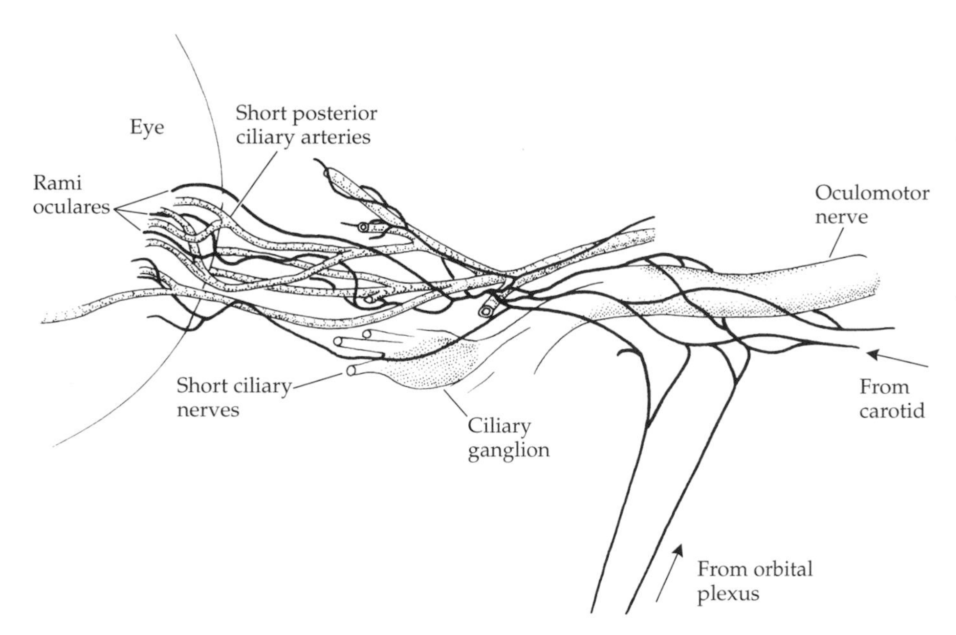

Figure 5.27
Rami Oculares
Parasympathetic axons from the pterygopalatine ganglion form a plexus in the orbit and are part of the carotid plexus. Small bundles of them enter the superior orbital fissure and run to the eye along with the short ciliary nerves and short posterior ciliary arteries; these are the rami oculares. They do not pass through the ciliary ganglion. (After Ruskell 1970.)

Axons from the pterygopalatine ganglion cells innervate vascular smooth muscle in the choroid

Preganglionic fibers to the pterygopalatine ganglion run in the facial nerve, and several studies have demonstrated that facial nerve stimulation can alter choroidal blood flow and intraocular pressure; the effect on the blood vessels appears to be vasodilation. Although there is no question that this innervational system exists—the anatomical evidence for the pathway via the rami oculares is convincing—questions remain about the nature of the system and how it operates. One problem is the difficulty in demonstrating any cholinergic innervation of the choroidal vasculature by immunohistochemical methods. It is possible, however, that the effect of this innervational system on the vascular muscle is mediated by release of one of the modulatory neuropeptides that affect smooth muscle, specifically VIP (vasoactive intestinal polypeptide). The extent to which this parasympathetic innervation of the uveal vasculature is involved in the regulation of blood flow is not known; the best one can do is note that the potential for regulation is present.

Accommodation and pupillary light reflexes share efferent pathways from the Edinger–Westphal nuclei to the eyes; pupillary reflexes are mediated by retinal signals reaching the Edinger–Westphal nuclei through the pretectal complex

The input to the ciliary ganglion, the *pre*ganglionic fibers, are from cells in the Edinger–Westphal nucleus, an aggregation of cells near the oculomotor nucleus on both sides of the brain. From the Edinger–Westphal nucleus, the fibers join the oculomotor nerve, running in its inferior division and then in the branch to the inferior oblique. A discrete bundle of fibers leaves the inferior oblique branch to go to the ciliary ganglion; this is the motor root of the ciliary ganglion (Figure 5.28). The preganglionic fibers synapse on cells in the ciliary ganglion whose axons enter the eye in the short ciliary nerves and run through the choroid to the ciliary muscle and the iris sphincter.

The pupillary light reflexes are pupil constriction when light is shined in either eye. Constriction of the pupil in the eye receiving the light stimulus is the **direct reflex**; constriction of the pupil in the fellow eye is the **consensual reflex**. The pupil constriction is mediated by parasympathetic innervation to the sphinc-

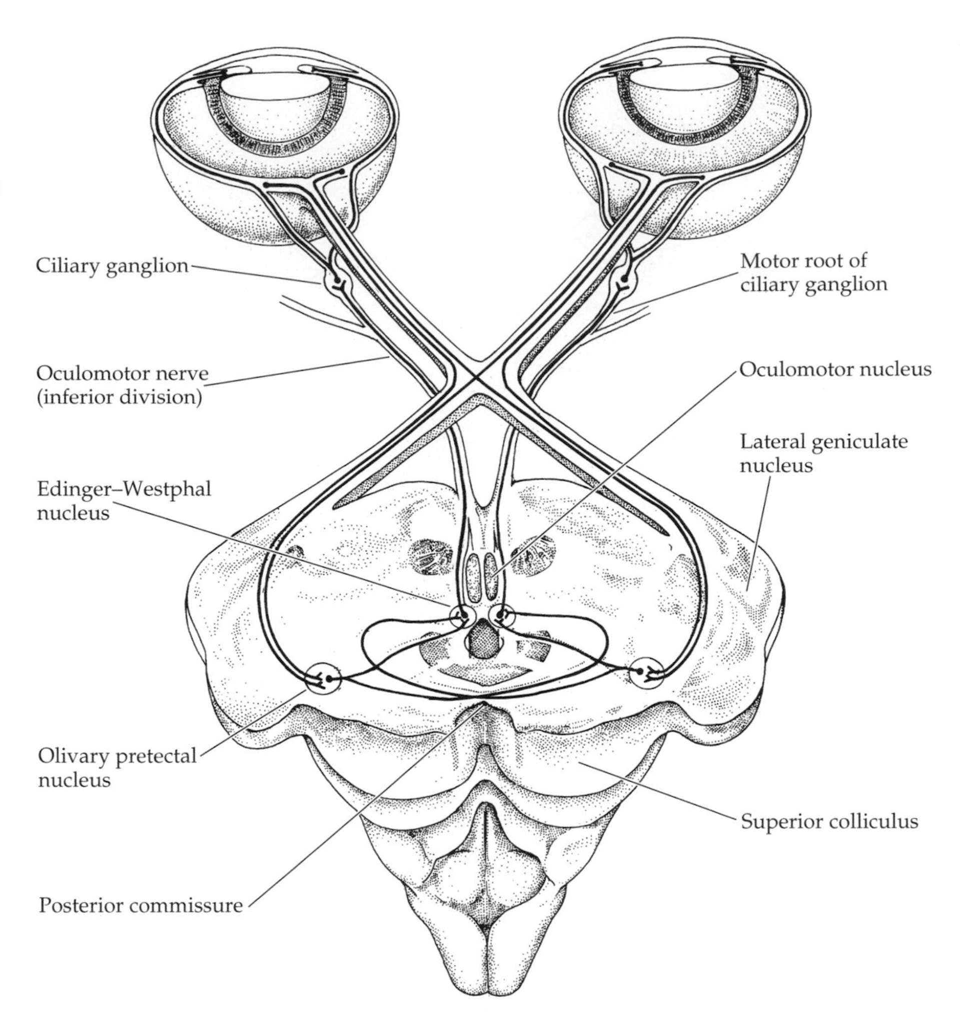

Figure 5.28
Elements of the Pupillary Reflex Pathway
Retinal ganglion cell axons make direct inputs to the olivary pretectal nucleus within the pretectal complex in the midbrain. Signals from the olivary pretectal cells go to both Edinger–Westphal nuclei. The efferent pathways from the Edinger–Westphal nuclei go to the ciliary ganglia and then to the sphincter muscles. (After Kourouyan and Horton 1997.)

ter. The afferent part of the reflex arc involves optic nerve fibers, probably from a subset of retinal ganglion cells whose activity is a function of the average amount of light reaching the retina. As discussed earlier, these axons leave the optic tract and run to the olivary pretectal nuclei, where they synapse on pretectal cells. The olivary pretectal nucleus cells, in turn, are thought to send signals to *both* Edinger–Westphal nuclei, accounting for the presence of both direct and consensual reflexes. The path from the olivary pretectal nucleus to the contralateral Edinger–Westphal nucleus is through the posterior commissure (see Figure 5.28).

A comparable diagram for accommodation cannot be drawn, except for the pathway from the Edinger–Westphal nuclei to the ciliary muscle, which is the same. (But recall the qualification noted earlier: Some neurons may bypass the ciliary ganglion.) The afferent part of the accommodative pathway certainly arises in the retina and travels in the primary visual pathway to primary and association areas of the visual cortex. But the route from here back to the Edinger–Westphal nuclei is mainly a matter for speculation.

Deficient pupillary reflexes may be associated with midbrain lesions

The abnormality of pupillary light reflexes that is most commonly attributed to a lesion in the efferent part of the reflex arc is **Argyll Robertson pupil**, which is typically characterized by a somewhat constricted pupil (or pupils) with no direct or consensual reflexes, but retention of the miosis that normally accompanies accommodation (see Chapters 10 and 11). It is usually bilateral.

The persistence of accommodative miosis implies that the pathway from the Edinger–Westphal nucleus to the eye is intact and that the sphincter muscles are normal. There is no reason to suspect an afferent defect, and since both direct and consensual reflexes are affected, the lesion site is most likely somewhere between the olivary pretectal nuclei and the Edinger–Westphal nuclei. The midbrain lesions found in patients with Argyll Robertson pupil were so often syphilitic that this syndrome was once regarded as a diagnostic sign of neurosyphilis. But there must be other causes, for while advanced syphilis is no longer common, Argyll Robertson pupil is still around.

The parasympathetic pathway from the Edinger–Westphal nucleus to the eye carries both pupillomotor and accommodative fibers, so any deficiency in one system or the other is expected to be associated either with lesions before the Edinger–Westphal nucleus or with a local problem with the target muscle. Argyll Robertson pupil exhibits the pupillary defect without accommodative involvement, and in theory, the opposite situation might also occur. With the exception of the normal loss of accommodation in presbyopia, however, there is little evidence for isolated neurological deficits of accommodation. The reason, probably, is that the accommodative control neurons providing input to the Edinger–Westphal nuclei are located close to the oculomotor nuclei, and any lesion in this region will involve both intra- and extraocular motor systems.

Innervation of the Lacrimal Gland

Axons from cells in the pterygopalatine ganglion reach the lacrimal gland via the zygomatic and lacrimal nerves

The lacrimal gland receives postganglionic parasympathetic fibers from cells in the pterygopalatine (sphenopalatine) ganglion, which lies in the pterygopalatine fossa of the sphenoid bone below the posterior part of the orbital floor. These axons leave the ganglion through several small branches that join the maxillary division of the trigeminal nerve, passing through the maxillary nerve to the zygomatic nerve and then to the zygomatico-temporal branch of the zygomatic (Figure 5.29). From here, the fibers leave via the communicating branch to join the lacrimal nerve, which they leave as it passes above the lacrimal gland.

These parasympathetic axons terminate in arrays around the secretory cells of the gland (see Chapter 7). Secretion from the lacrimal gland cells appears to be affected by the release from the parasympathetic axon terminals not only of acetylcholine, but also of one or more neuromodulatory peptides (in particular, VIP and substance P). Immunohistochemical studies also indicate the presence of sympathetic inputs to the lacrimal gland, in particular to the smooth muscle in the the small arterioles. The route by which these sympathetic fibers reach the gland is unclear, and the effect of sympathetic stimulation on tear production is still a matter of debate and investigation.

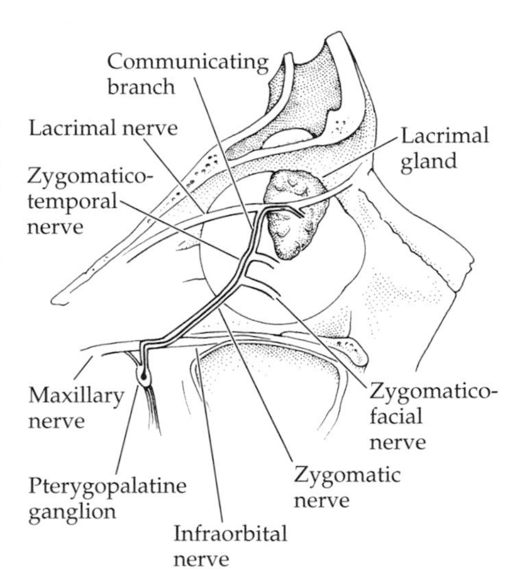

Figure 5.29
Innervation of the Lacrimal Gland
Parasympathetic axons from cells in the pterygopalatine ganglion reach the lacrimal gland through the zygomatic nerve, the zygomatico-temporal nerve, the communicating branch to the lacrimal nerve, and the lacrimal nerve.

The efferent pathway for lacrimal innervation begins in the facial nerve nucleus

Preganglionic fibers to the pterygopalatine ganglion appear to originate in cell bodies within a portion of the facial nerve nucleus, with axons running in some internal branches of the facial nerve—namely, the greater petrosal, deep petrosal, and pterygoid canal nerves.

This parasympathetic pathway to the lacrimal gland is the efferent side of a reflex arc originating at the sensory endings in the cornea and conjunctiva; activation of the cornea's sensory fibers can elicit tear production by the lacrimal gland. There is, presumably, an intermediate connection between the termination of the trigeminal fibers from the cornea in the brainstem and the preganglionic parasympathetic cells in the facial nucleus, but the details of this connection are not known. It is clear, however, that other inputs impinge on the parasympathetic pathway such that lacrimal gland stimulation—and excess tear production—can be associated with pain, emotional distress, or very intense light stimuli.

In addition, tear production by the secretory cells of the lacrimal gland can be influenced by hormones and other modulators that may be produced and circulated either locally or systemically. Thus, while the innervation to the lacrimal gland is quite important, it would be a mistake to attribute all changes in tear production to changes in innervational level.

Basal tear production may require tonic innervation of the lacrimal gland

Lesions in either the afferent or efferent portions of the lacrimal reflex pathway prevent reflex weeping, but the efferent lesions may be more significant because they eliminate all tonic innervation of the lacrimal gland. If a certain level of activity in the parasympathetic secretory fibers is necessary for the normal basal rate of tear production—and the "if" is significant—then a reduced innervation level could be a causative factor in the clinical syndrome of dry eye (see Chapter 7). Reduced tear production is not necessarily neurogenic, however; it may be due to a problem within the lacrimal gland itself. As usual, the diagnostic problem involves additional information about other associated symptoms.

Some Issues in Neural Development

The only parts of the ocular innervation for which we have extensive developmental information are the retina and optic nerve (see Chapter 16). Briefly, the retina begins to differentiate at about 5 weeks into gestation; ganglion cell axons appear at about 6 weeks and enter the central nervous system a week or so later. The optic nerve is not complete until the axons are myelinated, which occurs shortly before birth, and the development of foveal cones continues well into the postnatal period.

What is missing from these brief statements about the timing of innervation is any indication of how the developing axons or dendrites arrive at their target structures. And from a functional perspective, this is an extremely important issue; to work properly, the nervous system must be wired in a very specific way that is consistent and repeatable. Abducens neurons must find the lateral rectus, and not a different muscle; optic tract axons must end up in the lateral geniculate nucleus or another target and do so with the correct retinotopic order. How is this specificity accomplished?

The question, which is tantamount to asking how the nervous system assembles itself, does not have a simple answer, but the problem is being studied intensively, and it is possible to discuss some of the strategies the nervous system uses to assemble itself in the proper way.

Specialized growth cones guide the extension of axons and dendrites

Neurons do not start out looking like neurons. Once they differentiate, however, they begin to grow the processes that will become the axons and dendrites of the mature cell. Growth originates at the cell body, and the developing processes are extensions of the cell body.

Each developing process has a specialized region at its tip, called a **growth cone** (Figure 5.30). Growth cones have a cluster of small filopodia with some special adhesive properties that allow them to crawl across an appropriate substrate. New cell membrane is continually being synthesized just behind the growth cone as it crawls along, and the path of the growth cone is marked by the ever-lengthening process trailing behind it. The problem each growth cone faces is finding its way to the region where it will receive synaptic inputs or, if it is an axon, to the target cells to which it will be presynaptic.

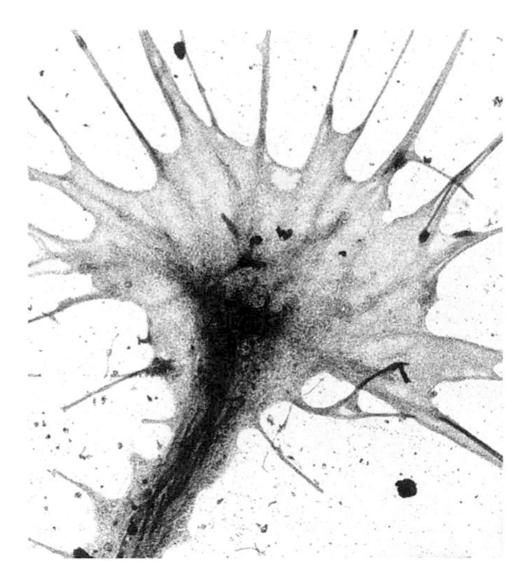

Figure 5.30
A Growth Cone
This photograph shows the growth cone as a filamentous expansion at the end of a growing neuronal process. As long as the culture plate has the appropriately adhesive substrate, the growth cone can crawl over the surface, leaving a trail of newly formed dendrite or axon behind it. (From Letourneau 1979.)

Pathfinding by growth cones depends on recognition of local direction signs

Growth cones on developing retinal ganglion cell axons must grow across the retina to the optic stalk, enter the stalk, grow to the site of the optic chiasm, decide whether to go to the ipsilateral or the contralateral side, enter the optic tract, and finally decide which of the several target nuclei to enter. Developing neurons in the gasserian ganglion face the same kind of problem, but in the opposite direction (their axons will be very short, but the dendrites must reach out to the periphery, seeking out the surface ectoderm that they will innervate).

"Seeking" and "deciding" are loaded words, implying that the growth cones have built-in knowledge about where to go and what to do. They don't, of course, but they do interact with their environment, and the end result of this interaction gives the appearance of directed search and knowledgeable decision. The essence of this pathfinding problem is identifying the environmental features that tell a growth cone whether it is on course.

The question applies mainly to the first cells—the "pioneer" neurons—that sprout axons and dendrites. Once they have found their destinations, later-developing neurons can follow the trail blazed by the pioneers. But this lesser task performed by later-developing cells hints at one environmental factor that may assist the pioneers: Specifically, structural or mechanical restrictions may confine or guide the pioneering axons along particular paths (a phenomenon called stereotropism). For example, the ganglion cell axons that will form the optic nerve have only one obvious exit from the developing eye, and that is the optic stalk. As axons grow into the optic stalk, the cellular components of the stalk degenerate ahead of them, leaving a clear path for the axons to follow from the retina to the site of the optic chiasm.

But this mechanical guidance is only part of the story, and probably a small part; many nerves form with no such preexisting structure to guide them. Moreover, the optic nerve axons have no mechanical guide after they reach the chiasm; there are two possible exits from the chiasm (three if the other optic nerve is included) but no structural barrier to force temporal retinal axons ipsilaterally and nasal retinal axons contralaterally. Other influences must be at work.

At present, the best we can do is identify the possible influences, and there are at least three. First, neurons growing in cell culture extend their processes along particular paths depending on the adhesiveness of the available substrates; the growth cones follow the spatial pattern of a substrate to which they adhere well, while avoiding less adhesive regions. If neurons can do this in vitro, it is reasonable to think they can do it in vivo; the only requirement is a properly adhesive substrate. One candidate for an in vivo adhesive substrate is the protein laminin, which is found on the surfaces of developing **neuroglial cells**, the supporting cells found throughout the nervous system. Growth cones find the glial laminin to be adhesive, and since the glial cells are scattered throughout the system, they may be like adhesive stepping-stones that guide the growth cones along particular paths. But while a particular type of laminin may guide growth cones, a deeper question remains: How does laminin appear in the appropriate places at the appropriate times?

Two other possible guidance mechanisms are gradients of electrical current flow and concentration gradients of a protein called **nerve growth factor**; neuronal processes growing in culture are sensitive to both. The difficulty here is that neither influence alone can explain all the features of neural pathfinding. For example, a nerve growth factor gradient centered on the optic stalk might well explain the convergence of ganglion cell axons to this site, but it cannot explain the rerouting of axons at the optic chiasm. In fact, none of the proposed guidance mechanisms, nor any combination thereof, explains the pattern of decussation at the chiasm. In short, several mechanisms for pathfinding by pioneer neurons are plausible, all of which are probably used, but it is difficult to be specific about how a particular neuron or subset of neurons reaches its destination.

Target recognition and acquisition may require specific markers produced by the target cells

However they do it, developing neurons make their way to the vicinity of their targets, which may be central nuclei or peripheral muscles. Now the question is, How do they recognize the target?

In this case, an explanation based on a neurotropic substance like nerve growth factor may be sufficient. Cells in a central target nucleus, for example, may produce a tropic molecule to which approaching axons are sensitive and toward which they will grow. The only requirement is that production of the tropic substance be temporally correlated with arrival of the axons, but this correlation may be a matter of induction: Production could be triggered by the approach of the axons themselves. In any event, nerve growth factor and other tropic factors are known to exist in different places in the developing embryo, thus giving the hypothesis some basic experimental support.

Tropic factors may also explain the innervation of multiple targets by axons running in the same nerve or tract. For example, pioneering optic tract axons may bypass the lateral geniculate nucleus in favor of the superior colliculus because the LGN's tropic factor has not been switched on; later axons may arrive when the LGN tropic factor is being expressed, however, thereby going to the LGN rather than to the colliculus. Alternatively, the sequence could be reversed, with the pioneers innervating the LGN and the stragglers going to the colliculus. Or the LGN and the colliculus might have different tropic factors to which subsets of retinal axons have different sensitivities.

Many early neurons are eliminated as mature patterns of connectivity are established

After neurons recognize and acquire their targets, synaptic connections must be made, and made in such a way that the final pattern of connectivity is precise. But an obvious feature of neuronal connectivity is the initial imprecision of the process; in the developing nervous system, there are many more neurons than in the mature system, and they target their termination sites rather haphazardly.

The brain seems to use a conservative strategy to assemble itself: It produces more neurons than it needs (better too many than too few) and allows the neurons to compete for the right to survive (the losers die). An early overproduction of neurons is well established for most parts of the nervous system, as is massive cell death late in gestation (even postnatally in some species). The retina and the cranial nerve nuclei all exhibit cell overproduction and subsequent cell loss. Although these observations imply a competitive situation, there must be some rules to the game. What determines which cells will live and which will die?

The general answer is that neurons are competing for synaptic space. The situation is like a demolition derby with 10,000 automobiles converging on a parking lot that has only 4000 parking spaces: The first to arrive have a definite advantage, but the rules of the game permit bumping, jostling, and outright displacement, and early arrival is no guarantee that the parking space will be retained.

Neurons can be jostled and displaced because newly established synapses are tenuous; they become more firmly established over time, apparently because of a combination of trophic factors and increasingly correlated electrical activity in the axons and their target cells. (Note that tro*p*ic factors are involved in guidance, while tro*ph*ic factors are involved in maintenance. Tropic factors may also be trophic, however.) The more synapses an axon can make on a target cell and the more effective the individual synapses become, the more the electrical activity in the pre- and postsynaptic cells will be correlated. This increased correlation may be reinforced by eliciting trophic factor release, which helps strengthen the synaptic connectivity, thus increasing the electrical correlation, and so on.

An axon that has managed to make only a few contacts with postsynaptic targets is in trouble; the connections will not be reinforced and may lose their space. If the axon has not managed to establish strong synapses on any target cell, the axon and its parent cell body may degenerate and die.

Competition of this type occurs in the developing visual system. Under normal circumstances, retinal axons from the two eyes intermingle in the early stages of development and segregate later to define the mature pattern of layering in the geniculate nuclei or the interleaving pattern of stripes in the colliculi. The competitive balance between the two eyes can be altered by reducing or eliminating of the activity in the developing optic nerve and tract of one eye, thereby giving the axons of the other eye more opportunity to acquire synaptic space. Methods for altering this balance include sectioning one optic nerve, blocking the activity of retinal ganglion cells with a neurotoxin, or in a subtler paradigm, covering one eye so that it must develop in the absence of any light stimuli (this last approach should alter the normal pattern of electrical activity in the ganglion cell axons). In all cases, the deprived eye loses territory in the target nuclei; its terminals occupy less area in the superior colliculi, for example, and its geniculate layers will be considerably smaller than normal.

Adult connectivity patterns are not always complete at birth, and postnatal development is subject to modification

Considerably more could be said about neural development, and even more needs to be discovered. The important message, however, is that all the elements of neural development—pathfinding, target recognition and acquisition, synapse establishment and maintenance—are interactive, contingent processes. At all stages of development, growing neurons interact with local features of their environment: They interact with adhesive proteins, chemical and electrical gradients, and other types of cells; they interact with the target cells; and they interact with each other. The result is a system that appears to be stereotypical and hard-wired, but is actually quite susceptible to modification; potentially, anything that affects the normal interaction between neurons and their environment will generate a developmental abnormality. Moreover, the period of susceptibility may be quite prolonged, extending into the postnatal years.

The periods of susceptibility are **critical periods**. Critical periods vary in duration and in time of occurrence for different parts of the nervous system and for different stages of the developmental process. The critical period for pathfinding, for example, occurs early and ends when a target has been acquired. The critical period for synapse establishment and maintenance begins later and may persist a long time. In fact, some capacity for synapse modification may persist throughout life.

Ocular albinism is associated with a pathfinding error in the development of optic nerve axons

Ocular albinism is associated with abnormal layering of the lateral geniculate nuclei that can be explained by a very early misrouting of axons at the chiasm and an anomalous pattern of decussation. Albinos often have disturbances of their eye movements, usually nystagmus, and in albino rabbits, the projection of retinal axons into the pretectal nuclei is abnormal. The same may be true in humans. In any event, there is circumstantial evidence for connectivity errors in the albino's visual system that become evident as altered eye movements. The apparent relationship between deficient melanin synthesis and axonal pathfinding errors is both intriguing and mysterious; it highlights the dependence of neural development on a normal chemical environment.

Anomalous innervation of the extraocular muscles may be the result of pathfinding or target recognition errors

Pathfinding and target recognition errors should produce anomalous innervation, such that an extraocular motor nerve innervates the wrong muscle, for example. A possible example in the oculomotor system is Duane's retraction syndrome (see Chapter 4), in which attempted abduction of the eye causes it to retract into the orbit. The syndrome may occur for several reasons, but one of them may be anomalous innervation of the lateral rectus by axons meant for the medial rectus; in this case, attempted medial gaze, which is normally the sole responsibility of the medial rectus, would produce co-contraction of the medial and lateral recti, thereby pulling the eye back into the orbit. By way of evidence, at least one patient with Duane's retraction syndrome is known to have lacked abducens nuclei and nerves, with innervation supplied to the lateral recti by an oculomotor nerve branch.

An example of axons that seem to have temporarily lost their way occurs in the long ciliary nerves. Instead of running directly forward between the ciliary body and the sclera, the nerve sometimes turns out through the sclera, emerges on the surface in the episclera, and then turns back along the way it came until it regains the normal path to the cornea. The resulting **nerve loop of Axenfeld** is associated with a branch of the anterior ciliary artery that penetrates the sclera near the anterior extent of the ciliary body.

How these nerve loops form is unknown, but the pioneering axons appear to be transiently distracted by the arterial branch, recognizing their error when they emerged in quite inappropriate surroundings. Although the error is corrected, later neurons slavishly follow the pioneers, thereby demonstrating the importance of interactions among developing neurons of similar type.

Some forms of amblyopia may be related to problems with postnatal establishment and maintenance of synaptic connections

Gross errors of innervation are not common in isolation, although they often occur in conjunction with other more widespread developmental anomalies. Problems that may be associated with competition and synapse maintenance, for which the critical period extends postnatally, may be more prevalent.

Amblyopia, which is characterized by reduced visual acuity that is not accounted for by refractive error or pathology, may be one of the better-known examples. Amblyopia occurring in conjunction with strabismus or anisometropia (unequal refractive errors in the two eyes) is particularly suspect; either the strabismus or the anisometropia could upset the normal competitive balance between the axonal projections from the two eyes, resulting in reduced neural input from one eye and, potentially, reduced visual acuity. This line of argument has considerable experimental support from studies in cats and primates, and it is now common practice to correct strabismus (and anisometropia) at very young ages. The intent is to reestablish a normal competitive situation *before* the end of the critical period, thus forestalling the appearance of amblyopia. Amblyopia that persists after the age of 6 to 8 years is relatively intractable, and the greater the deficit, the less likely the lost acuity can ever be fully restored.

Innervation of the extraocular muscles begins early in gestation, sensory innervation much later

Extraocular motor nerves enter the developing muscles around gestational weeks 5 to 6, and the characteristic motor end plates and sensory endings in the muscles can be distinguished about 6 weeks later, around the end of the first trimester. Less is known about the sensory innervation via the trigeminal system, except that it lags the motor innervation; fibers of the long ciliary nerves appear at the margin of the cornea at around 26 weeks, invading the cornea and its epithelium shortly thereafter. These examples provide some justification for a rough generalization that the innervation of a particular structure begins soon after the tissue has begun to differentiate; the tissues that differentiate earliest are the first to be innervated.

Postnatal Neuron Growth and Regeneration

Most postnatal neuron growth is interstitial growth

Before developing neurons reach and acquire their target structures, most of the new membrane that forms is added just behind the growth cones on the axons or dendrites. It is much like a spider spinning silk; as the growth cone crawls along, it leaves a trail of newly formed membrane. Once targets have been acquired and the processes have no need to extend farther, the growth cones disappear and subsequent increases in size of the neuron must be by another mechanism, called **interstitial** growth, in which new cell membrane is created within the preexisting structure. Figure 5.31 illustrates these two modes of growth.

Many neurons, particularly those in the peripheral nerves, retain a capacity for interstitial growth for many years. As the body grows and the neuron's soma and the muscle it innervates become more distant from one another, new membrane must be continually added between the soma and the axon terminal. Thus a spinal motor neuron innervating a leg muscle, for example, may have an axon

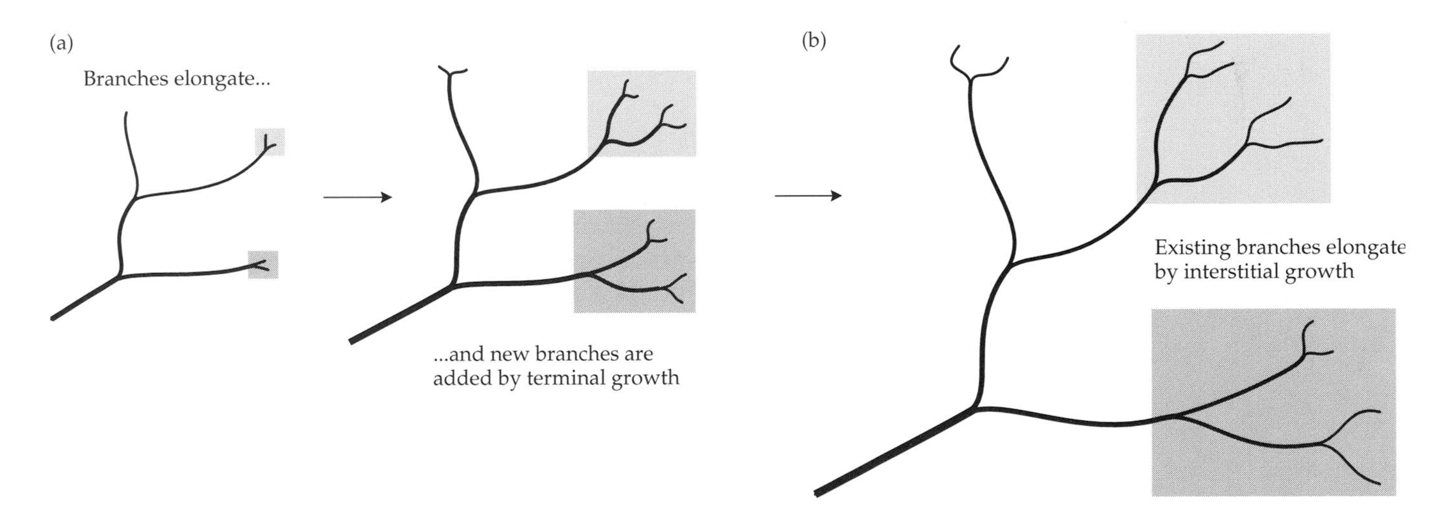

Figure 5.31
Terminal and Interstitial Modes of Neuron Growth
(a) Much of the initial growth of neurons occurs at the ends of developing processes, where new cell membrane forms just behind the advancing growth cones. *(b)* Later, after cells have acquired their targets, growth is interstitial, adding membrane in the middle of processes already formed. Interstitial growth allows neurons to expand as needed when growth of the body generally increases the distance between the cell body and its target structure.

a millimeter or so long in the embryo, 15 or 20 cm long at birth, and perhaps a meter long in the adult. Dendritic arbors can grow in the same way. In the brain, neurons can also increase in size by growing interstitially, but since the brain reaches adult size earlier than the rest of the body, the growth in central neurons does not go on as long.

Neurons do not undergo mitosis postnatally

Once neurons differentiate and begin to grow, they are said to withdraw from the cell cycle, meaning that they can no longer reproduce themselves. This inability to reproduce provides a rationale for the nervous system's massive overproduction of neurons in the embryo and the waves of neuronal cell death late in gestation. It is better to err on the side of overproduction, subsequently eliminating unneeded cells, than to have too few neurons with no way to make more of them.

Thus we are born with all the neurons we will ever have; any that die of natural causes or those destroyed by trauma or disease cannot be replaced. But the fact that trauma and disease do occur raises an important question: What happens when a neuron is damaged?

Spinal neurons in peripheral nerves can regenerate after being damaged

For many nonmammalian species, the severing of neurons between their cell bodies and axon terminals (axotomy) is not terribly serious over the long term; not only axons but whole limbs can be regenerated. In mammalian nervous systems, however, the consequences of axotomy are more severe. When the axon of a spinal motor neuron is severed, for example, the terminal segment that has been cut off will degenerate in a few weeks, and if an entire nerve has been severed, the target muscle that has lost its innervation will also change in characteristic ways. But depending on circumstances (how close the cut was to the cell body, the amount of damage to the myelin sheaths of the axons—cutting is worse than crushing, for example), the damaged neurons change in characteristic ways in response to the trauma. If they survive, they can regenerate their axon terminals and follow the path of the original, now degenerated, axons back to the target structure. To some extent, the target can be reinnervated with new, normally functioning synapses.

Reinnervation, however, is not the same as the original pattern; for a muscle, the motor units will be larger, there is likely to be polyneuronal innervation, and some of the original neurons may not regenerate. So while function can be reestablished to some extent, it is unlikely to be as good as the original. But damage to the peripheral nerves is not a hopeless proposition; both sensory and motor innervation can be at least partially restored.

Central nervous system neurons do not regenerate following major damage

Axotomy of the cranial nerves or sectioning of tracts in the brain is irreversible. If the optic nerve is sectioned, for example, there may be some sprouting of processes from the cut ends of the axons. But although these sprouts look like an attempt at regeneration, eventually the terminal ends of the axons degenerate and the parent ganglion cells in the retina die.

It appears that central and cranial nerve neurons do not regenerate not because they can't, but because they aren't allowed to. If the optic nerve is severed experimentally and a graft of peripheral nerve is inserted next to the stump of the optic nerve, the ganglion cells do not die and their axons happily regenerate into the peripheral nerve implant. The implant can be used as a bridge to lead the regenerating axons back to their original target nucleus in the brain, where some functional reconnections can be made.

Experiments of this type suggest that something in the central nervous system environment differs from the peripheral environment. Something in the central nervous system either prevents regeneration or does not provide the necessary stimulus for it. Some of the critical elements appear to be various nerve growth factors and glial cell proteins; they play an important role in the early development of the nervous system and are retained in the peripheral nervous system, but they are missing in the adult brain.

At first glance, this situation seems to be the wrong way around. The central nervous system and its nerves are so important that here, if any place, is where the capacity for regeneration should be retained and encouraged. But the connectivity in the brain is very specific, and establishing this connectivity in the first place is a highly contingent process. Contingency is so important, in fact, that it is probably impossible to duplicate in detail without going through the whole process of development all over again.

From this perspective, the inability of central neurons to regenerate, or the fact that they are discouraged from doing so, is a statement about the impossibility of the task: The developmental history of a neuron cannot be re-created because the conditions during development cannot be re-created. In the peripheral nervous system, contingency and specificity are less important during development, so the task of reinnervation is less demanding. It is not necessary that regenerating motor neurons reinnervate precisely the muscle fibers they originally supplied for the muscle to function, but scrambling the spatial topography of retinal inputs to the lateral geniculate nucleus would produce utter visual confusion.

Corneal nerve endings will regenerate following local damage

Nerve fibers in the cornea may be severed or damaged by trauma, infection, or surgery, but if the damage is confined to the cornea, reinnervation is possible. The immediate response of a corneal nerve fiber to damage is degeneration. The severed or damaged terminal portion degenerates and disappears, and the cut end of the main nerve fiber degenerates to some extent as well, but it does not degenerate completely; the degeneration rarely goes farther than the margin of the cornea.

In time, the denervated region will be resupplied, first by sprouting of new branches from adjacent, undamaged fibers and later by regeneration of the damaged fibers. How quickly this happens depends on the size of the affected region, but even a large area, like that produced by keratoplasty (see Chapter 8), may be fully resupplied in a few months.

Since the cornea is innervated by a branch of the trigeminal system, the fact that its nerve endings can regenerate may seem to belie the earlier statement that cranial nerves and tracts cannot regenerate. In this case, however, the damage is quite peripheral and involves only a small part of any individual neuron. The problem is very different from severing the nasociliary nerve, for example, which cuts closer to the parent cell bodies and leaves a long, complicated peripheral path along which regeneration would have to follow; regeneration of the nasociliary nerve does not occur. Damage in the cornea proper leaves the nerve fibers basically intact and in place, with only a small amount of fiber to regenerate and no stringent demands on the specificity of the reinnervation in the corneal epithelium.

Neuronal degeneration can affect other, undamaged neurons

Severing the optic nerve provides another example of the consequences of damage to central neurons. In this case, not only do the optic tract axons and their parent ganglion cells degenerate, but there are also obvious changes in the lateral geniculate neurons that the axons innervate. Specifically, the LGN neurons also degenerate, a phenomenon called **transneuronal atrophy**. It is as if the LGN neurons, now deprived of their inputs, have no particular reason to hang around any longer, so they die. The atrophy is quite evident in histological sections, such as that shown in Figure 2 of Box 5.2.

The degree of transneuronal atrophy varies in different parts of the brain, and the varying susceptibility to deafferentation probably has to do with the number of sources of input to a particular group of cells. LGN neurons are highly susceptible because retinal ganglion cells axons are the dominant source of input to the lateral geniculate nuclei; if that input is removed—along with the neurotrophic factors that are probably associated with it—the cells are deprived of anything to sustain them. In other nuclei, however, the cells may have enough different sources of input that removing just one of them does not remove enough of the factors necessary for cell maintenance to produce ill effects.

Peripheral neurons and their targets are generally much less sensitive to denervation. Muscles, for example, will atrophy from disuse if they are not reinnervated, but they do not deteriorate altogether. Similarly, severing the descending spinal tracts will not produce transneuronal atrophy in the spinal motor neurons; the motor neurons have other inputs, and they are operating in an environment where factors for neuron growth and maintenance normally are present.

References and Additional Reading

Elements of Neural Organization

Bennett MVL, Barrio LC, Bargiello TA, Spray DC, Hertzberg E, and Sáez JC. 1991. Gap junctions: New tools, new answers, new questions. *Neuron* 6: 305–320.

Martin JH. 1996. Introduction to the central nervous system. Chapter 1, pp. 1–32, in *Neuroanatomy: Text and Atlas*, 2nd Ed. Appleton & Lange, Stamford, CT.

Purves D, Augustine GJ, Fitzpatrick D, Katz LC, LaMantia A-S, and McNamara JO. 1997. The organization of the nervous system, Chapter 1, pp. 1–34, and Synaptic transmission, Chapter 5, pp. 85–98, in *Neuroscience,* Sinauer Associates, Sunderland, MA.

Swanson LW, Lufkin T, and Colman DR. 1999. Organization of nervous systems. Chapter 2, pp. 9–37, in *Fundamental Neuroscience*, Zigmond MJ, Bloom FE, Landis SC, Roberts JL, and Squire LR, eds. Academic Press, San Diego.

Wilson-Pauwels L, Akesson EJ, and Stewart PA. 1988. *Cranial Nerves: Anatomy and Clinical Comments*. B.C. Decker, Inc., Toronto.

The Optic Nerve and the Flow of Visual Information

Clarke RJ and Gamlin PDR. 1995. The role of the pretectum in the pupillary light reflex. Chapter 12, pp. 149–159, in *Basic and Clinical Perspectives in Vision Research,* Robbins JG, Djamgoz MBA, and Taylor A, eds. Plenum, New York.

Cooper HM and Magnin M. 1986. A common mammalian plan of accessory optic system organization revealed in all primates. *Nature* 324: 457–459.

Cowey A and Perry VH. 1980. The projection of the fovea to the supeior colliculus in rhesus monkeys. *Neuroscience* 5: 53–61.

Fukuda Y, Sawai H, Watanabe M, Wakakuwa K, and Morigiwa K. 1989. Nasotemporal overlap of crossed and uncrossed retinal ganglion cell projections in the Japanese monkey (*Macaca fuscata*). *J. Neurosci.* 9: 2353–2373.

Gamlin PDR, Zhang H, and Clarke RJ. 1995. Luminance neurons in the pretectal olivary nucleus mediate the pupillary light reflex in the rhesus monkey. *Exp. Brain Res.* 106: 177–180.

Guillery RW, Okoro AN, and Witkop CJ. 1975. Abnormal visual pathways in the brain of a human albino. *Brain Res.* 96: 373–377.

Hickey TL and Guillery RW. 1979. Variability of laminar patterns in the human lateral geniculate nucleus. *J. Comp. Neurol.* 183: 221–246.

Hoffmann KP and Distler C. 1989. Quantitative analysis of visual receptive fields of neurons in nucleus of the optic tract and dorsal terminal nucleus of the accessory optic tract in macaque monkey. *J. Neurophysiol.* 62: 416–428.

Horton JC and Hedley-White ET. 1984. Mapping of cytochrome oxidase patches and ocular dominance columns in human visual cortex. *Phil. Trans. R. Soc. Lond. B* 304: 255–272.

Kaas JH. 1997. Topographic maps are fundamental to sensory processing. *Brain Res. Bull.* 44: 107–112.

Malpeli JG and Baker FH. 1975. The representation of the visual field in the lateral geniculate nucleus of *Macaca mulatta*. *J. Comp. Neurol.* 161: 569–594.

Martin JH. 1996. The visual system. Chapter 6, pp. 161–198, in *Neuroanatomy: Text and Atlas*, 2nd Ed. Appleton & Lange, New York.

Naito J. 1989. Retinogeniculate projection fibers in the monkey optic nerve: A demonstration of the fiber pathways by retrograde axonal transport of WGA-HRP. *J. Comp. Neurol.* 284: 174–186.

Pollack JG and Hickey TL. 1979. The distribution of retino-collicular axon terminals in Rhesus monkey. *J. Comp. Neurol.* 185: 587–602.

Polyak S. 1957. Optic nerves, chiasma, tract and subcortical centers. Chapter VI, pp. 288–389, in *The Vertebrate Visual System*. University of Chicago Press, Chicago.

Reese BE. 1993. Clinical implications of the fibre order in the optic pathway of primates. *Neurol. Res.* 15: 83–86.

Reese BE and Baker GE. 1992. Changes in fiber organization within the chiasmatic region of mammals. *Visual Neurosci.* 9: 527–533.

Schiller PH and Malpeli JG. 1978. Functional specificity of lateral geniculate nucleus laminae of the rhesus monkey. *J. Neurophysiol.* 41: 788–797.

Simpson JI. 1984. The accessory optic system. *Ann. Rev. Neurosci.* 7: 13–41.

Sparks DL. 1986. Translation of sensory signals into commands for control of saccadic eye movements: Role of primate superior collliculus. *Physiol. Rev.* 66: 118–171.

Warwick R. 1976. The visual pathway. Chapter 7, pp. 325–395, in *Eugene Wolff's Anatomy of the Eye and Orbit*, 7th Ed. WB Saunders, Philadelphia.

Wässle H, Grünert U, Martin PR, and Boycott BB. 1990. Retinal ganglion cell density and cortical magnification factor in the primate. *Vision Res.* 30: 1897–1911.

The Trigeminal Nerve: Signals for Touch and Pain

Burton H. 1992. Somatic sensations from the eye. Chapter 4, pp. 71–100, in *Adler's Physiology of the Eye*, 9th Ed., Hart WM, ed. Mosby Year Book, St. Louis.

Dodd J and Kelly JP. 1991. Trigeminal system. Chapter 45, pp. 701–710, in *Principles of Neural Science*, 3rd Ed., Kandel ER, Schwartz JH, and Jessell TM, eds. Elsevier, New York.

Kraus EE and Smith CH. 1997. Trigeminal nerve. Chapter 36 in *Duane's Foundations of Clinical Ophthalmology*, Vol. 1, Tasman E and Jaeger EA, eds. Lippincott-Raven, Philadelphia.

Liu GT. 1998. Anatomy and physiology of the trigeminal nerve. Chapter 33, pp. 1595–1648, in *Walsh & Hoyt's Clinical Neuro-Ophthalmology*, Vol. 1, 5th Ed., Miller NR and Newman NJ, eds. Williams & Wilkins, Baltimore.

Porter JD. 1986. Brainstem terminations of extraocular muscle primary afferent neurons in the monkey. *J. Comp. Neurol.* 247: 133–143.

Ruskell GL. 1985. Innervation of the conjunctiva. *Trans. Ophthalmol. Soc. U.K.* 104: 390–395.

Warwick R. 1976. The orbital nerves. Chapter 6, pp. 275–324, in *Eugene Wolff's Anatomy of the Eye and Orbit*, 7th Ed. WB Saunders, Philadelphia.

The Extraocular Motor Nerves

Büttner-Ennever JA and Akert K. 1981. Medial rectus subgroups of the oculomotor nucleus and their abducens internuclear input in the monkey. *J. Comp. Neurol.* 197: 17–27.

Kourouyan HD and Horton JC. 1997. Transneuronal retinal input to the primate Edinger-Westphal nucleus. *J. Comp. Neurol.* 381: 68–80.

Martin JH. 1996. The somatic and visceral motor functions of the cranial nerve. Chapter 13, pp. 383–416, in *Neuroanatomy: Text and Atlas*, 2nd Ed. Appleton & Lange, New York.

Sacks JG. 1983. Peripheral innervation of extraocular muscles. *Am. J. Ophthalmol.* 95: 520–527.

Swanson MW. 1990. Neuroanatomy of the cavernous sinus and clinical correlations. *Optom. Vis. Sci.* 67: 891–897.

Umansky F and Nathan H. 1982. The lateral wall of the cavernous sinus. *J. Neurosurg.* 56: 228–234.

Innervation of the Muscles of the Eyelids

Kraus EE and Smith CH. 1996. Facial nerve. Chapter 37 in *Duane's Foundations of Clinical Ophthalmology*, Vol. 1, Tasman W and Jaeger EA, eds. Lippincott-Raven, Philadelphia.

Manson PN, Lazarus RB, Morgan R, and Iliff N. 1986. Pathways of sympathetic innervation to the superior and inferior (Müller) tarsal muscles. *Plast. Reconstr. Surg.* 78: 33–40.

Porter JD, Burns LA, and May PJ. 1989. Morphological substrate for eyelid movements: Innervation and structure of primate levator palpebrae superioris and orbicularis oculi muscles. *J. Comp. Neurol.* 287: 64–81.

Autonomic Innervation of Smooth Muscle within the Eye and Innervation of the Lacrimal Gland

Akert K, Glicksman MA, Lang W, Grob P, and Huber A. 1980. The Edinger–Westphal nucleus in the monkey. A retrograde tracer study. *Brain Res.* 184: 491–498.

Brodal P. 1998. Peripheral autonomic nervous system. Chapter 17, pp. 493–527, in *The Central Nervous System: Structure and Function*, 2nd Ed. Oxford University Press, New York.

Burde RM. 1988. Direct parasympathetic pathway to the eye: Revisited. *Brain Res.* 463: 158–162.

Flügel C, Tamm ER, Mayer B, and Lütjen-Drecoll E. 1994. Species differences in choroidal vasodilative innervation: Evidence for specific intrinsic nitrergic and VIP-positive neurons in the human eye. *Invest. Ophthalmol. Vis. Sci.* 35: 592–599.

Jaeger RJ and Benevento LA. 1980. A horseradish peroxidase study of the innervation of the internal structures of the eye: Evidence for a direct pathway. *Invest. Ophthalmol. Vis. Sci.* 19: 575–583.

Kourouyan HD and Horton JC. 1997. Transneuronal retinal input to the primate Edinger–Westphal nucleus. *J. Comp. Neurol.* 381: 68–80.

Maloney WF, Younge BR, and Moyer NJ. 1980. Evaluation of the causes and accuracy of pharmacologic localization in Horner's syndrome. *Am. J. Ophthalmol.* 90: 394–402.

Ruskell GL. 1970. An ocular parasympathetic pathway of facial nerve origin and its influence on intraocular pressure. *Exp. Eye Res.* 10: 319–330.

Ruskell GL. 1990. Accommodation and the nerve pathway to the ciliary muscle: A review. *Ophthalmic Physiol. Opt.* 10: 239–242.

Sinnreich Z and Nathan H. 1981. The ciliary ganglion in man. *Anat. Anz.* 150: 287–287.

Van der Werf F, Baljet B, Prins M, Timmerman A, and Otto JA. 1993. Innervation of the superior tarsal (Müller's) muscle in the cynomolgous monkey: A retrograde tracing study. *Invest. Ophthalmol. Vis. Sci.* 34: 2333–2340.

Westheimer G and Blair SM. 1973. The parasympathetic pathway to internal eye muscles. *Invest. Ophthalmol.* 12: 193–197.

Some Issues in Neural Development, Postnatal Neuron Growth, and Regeneration

Alberts B, Bray D, Lewis J, Raff M, Roberts K, and Watson JD. 1994. Neural development. Chapter 21, pp. 1119–1130, in *Molecular Biology of the Cell,* 3rd Ed. Garland, New York.

Collewijn H, Apkarian P, and Spekreijse H. 1985. The oculomotor behavior of human albinos. *Brain* 108: 1–28.

Daw NW. 1995. *Visual Development.* Plenum Press, New York.

Goldman CS. 1996. Mechanisms and molecules that control growth cone guidance. *Annu. Rev. Neurosci.* 19: 341–377.

Goodman CS and Tessier-Lavigne M. 1997. Molecular mechanisms of axon guidance and target recognition. Chapter 4, pp. 108–178, in *Molecular and Cellular Approaches to Neural Devleopment.* Cowan WM, Jessel TM, and Zipursky SL, eds. Oxford University Press, New York.

Guillery RW. 1996. Why do albino and other hypopigmented mutants lack normal binocular vision, and what else is abnormal in their central retinal pathways? *Eye* 10: 217–221.

Jeffery G. 1997. The albino retina: An abnormality that provides insight into normal retinal development. *Trends Neurosci.* 20: 165–169.

Letourneau PC. 1979. Cell substratum adhesion of neurite growth cones, and its role in neurite elongation. *Exp. Cell Res.* 124: 127–138.

Miller NR, Kiel SM, Green WR, and Clark AW. 1982. Unilateral Duane's retraction syndrome (Type 1). *Arch. Ophthalmol.* 100: 1468–1472.

Purves D, Augustine GJ, Fitzpatrick D, Katz LC, LaMantia A-S, and McNamara JO. 1997. Construction of neural circuits. Chapter 21, pp. 395–417, in *Neuroscience.* Sinauer Associates, Sunderland, MA.

Purves D and Lichtman JW. 1985. *Principles of Neural Development.* Sinauer Associates, Sunderland MA.

Rozsa AJ, Guss RB, and Beuerman RW. 1983. Neural remodeling following experimental surgery of the rabbit cornea. *Invest. Ophthalmol. Vis. Sci.* 24: 1033–1051.

Stirling RV and Dunlop SA. 1995. The dance of the growth cones—Where to next? *Trends Neurosci.* 18: 111–115.

Vidal-Sanz M, Bray GM, Villegas-Perez MP, Thanos S, and Aguayo AJ. 1987. Axonal regeneration and synapse formation in the superior colliculus by retinal ganglion cells in the adult rat. *J. Neurosci.* 7: 2894–2909.

Chapter 6

Blood Supply and Drainage

Blood has always been regarded as a form of natural wealth: a rich liquid asset settled on each individual as a birthright, a priceless deposit which can neither be spent nor accumulated, only lost or dispersed through injury or illness. There are no plutocrats, only paupers. . . . Unlike the wealth of a miser, which accumulates without it doing any useful work, the value of blood can be exploited only if it is kept ceaselessly on the move. It is useless when stationary, but it is beyond price as long as it visits and revisits every part of the body.

■ Jonathan Miller, *The Body in Question*

Although some cells in the eye are exceptions to the rule, very few cells in the body are more than 100 μm distant from a capillary through which blood is flowing. Capillaries are everywhere because cells require a steady supply of substances essential to their special functions and continued viability; components for constructing amino acids and proteins, oxygen, simple sugars, and various elements in ionic form (calcium, potassium, and so on) are some of the major ones. And in the course of their operation, cells generate products for secretion—by-products that they cannot retain for reasons of space or toxicity. These by-products must be distributed or discarded. Supply, dispersal, and disposal are handled by the cardiovascular system.

Assuming the heart and lungs do their job of pumping and oxygenating the blood, the crucial parts of the system for cells and tissues are the capillary beds, the arrays of small vessels between arteries and veins along which the interchanges between blood and cells take place. For interchanges between blood and cells to occur as they should, it is essential that the structure of the capillaries not be compromised and that blood flows through the capillaries at a rate appropriate for the needs of the tissue.

Distributing Blood to Tissues

Arteries control blood flow through capillary beds, and veins regulate blood volume

Blood leaving the heart through the aorta is never out of contact with a layer of endothelial cells until it returns to the heart through the superior and inferior venae cavae. Capillaries, where most of the interchanges between blood and tissue take place, are short tubes of endothelium with no external layers other than the endothelium's basement membrane. Arteries and veins are larger-diameter endothelial tubes to which additional layers have been added. Both arteries and veins have three layers in their walls: an inner *tunica intima* that includes the endothelium, a middle *tunica media* with smooth muscle and elastic tissue, and an outer *tunica adventitia* that is mostly connective tissue (Figure 6.1). The tunica media is typically much thicker in arteries than in veins of comparable size.

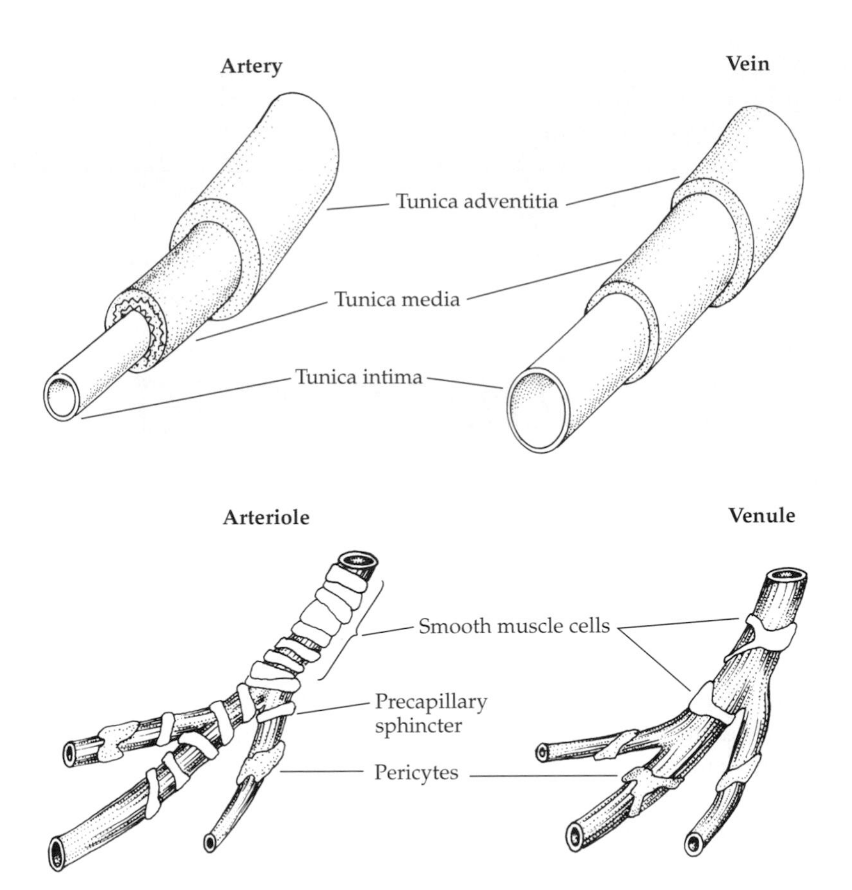

Figure 6.1
Structure of Arteries and Veins
Both arteries and veins have three layers, or tunics, in the vessel wall. The middle layer, tunica media, contains smooth muscle and is considerably more developed in arteries than in veins. Arterioles and venules are tubes of endothelium wrapped in smooth muscle cells and pericytes. (After Rhodin 1980.)

Arterioles and venules (small arteries and veins) do not have complete muscular and connective tissue layers around them.

Because of the extra layers of tissue surrounding the endothelium in the arteries and veins, neither fluid nor solutes can move easily out of these vessels into the surrounding tissues; diffusion of dissolved gases is also limited. Veins are conduits for returning oxygen-depleted blood to the heart for pumping to the lungs. But veins are distensible, and the volume of venous blood increases as they expand. The smooth muscle of veins allows some constriction, however, which increases venous return to the heart and reduces the total venous blood volume. Thus the venous system is a kind of reservoir of blood that can be drawn on at greater or lesser rates depending on need. At any given time, about two-thirds of the total blood volume is in the veins.

Arteries are also conduits for blood, but their heavy layer of smooth muscle allows the diameter of the lumen to be varied considerably; smooth muscle contraction constricts the lumen. The volume of blood flowing into a capillary bed decreases when the precapillary arterioles constrict and increases when the vessels dilate (assuming constant arterial pressure). In short, the arterial system not only brings blood to tissues but also regulates the rate of blood flow through them, particularly by relaxation or contraction of the precapillary sphincters (see Figure 6.1).

Blood flow through capillary beds can be controlled locally or systemically

Interchanges between blood in capillaries and cells outside require a certain amount of time; although the distance between cells and blood is usually less than 100 μm, the interchange is primarily by diffusion and cannot be instantaneous. (Diffusion time is proportional to distance squared; glucose can diffuse 10 μm in 50 ms, but it takes 5 s to diffuse 100 μm.) Therefore, the capillary blood should be moving slowly enough for diffusion to occur, but not so slowly that important diffusible substances are depleted before blood finishes traversing the capillaries. Most tissues have an optimum rate of blood flow, within the broad range between too slow and too fast, that is commensurate with their particular metabolic requirements. The requirements may change, however, requiring a change in the flow rate; an active muscle needs more oxygen and must eliminate more lactic acid than an inactive muscle does, for example, and blood flow through the skin must vary to increase or reduce heat loss from the body.

The rate of flow through capillaries is determined largely by the status of the precapillary arterioles supplying the capillary bed. Contraction of the smooth muscle in the walls of the arterioles reduces their diameter, increases the resistance to blood flow, and reduces the flow rate; relaxation of the smooth muscle has the opposite effects and increases blood flow. In the absence of any limit to total blood volume, the simplest strategy for regulating flow would be based on local demands; that is, the flow into a particular tissue would be regulated by the tissue itself. Local regulation does occur, but since blood volume is limited and would be insufficient if all tissues were simultaneously demanding maximum flow rates, central control establishes priorities about where needs are greatest at any given time.

Figure 6.2 shows the local and extrinsic factors that influence muscle tonus in the precapillary arterioles. Local control is possible because vascular smooth muscle always maintains some tonic contraction, even in the absence of innervation. Tonic contraction is due to electrical instability of the muscle cells' membrane potential; some cells are so unstable that their membrane potentials oscillate, making them rudimentary pacemakers for all the other smooth muscle cells to which they are electrically coupled. Vasomotor tone is controlled locally either by influences on these local pacemaker cells or by direct effects on all the smooth muscle cells in the local vasculature.

One form of control is an intrinsic response of the smooth muscle to changes in arterial pressure; it is called the **myogenic response**. As pressure rises, the increased force on the vessel walls causes them to expand. Expansion stretches the muscle cells and elicits increased activity in the pacemaker cells; increased pacemaker activity produces muscle contraction, vasoconstriction, and a reduction of flow to the original level. Conversely, reduced vessel pressure is a stimulus for vasodilation to offset what would otherwise be a reduction of blood flow. This ability to maintain normal flow rates in the face of changes in systemic blood pressure is called **autoregulation**. When blood pressure rises during exercise, for example, tissues whose metabolic needs are unchanged use autoregulation to maintain their normal flow rates.

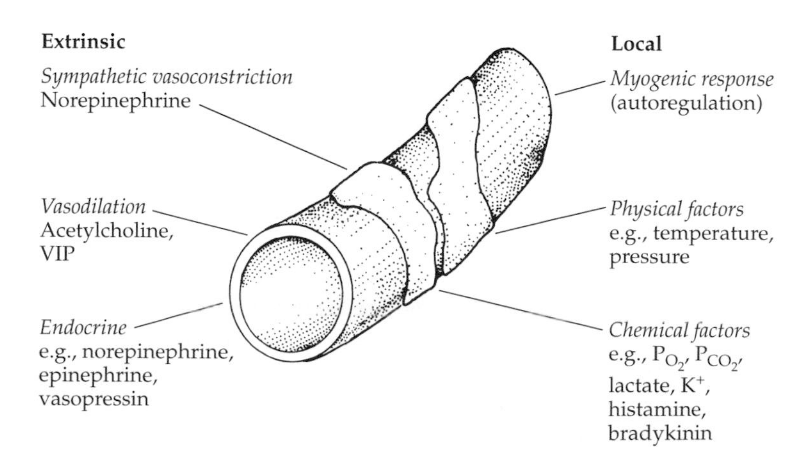

Figure 6.2
Control of Arterial Diameter and Blood Flow
Numerous extrinsic and local factors can affect vascular smooth muscle and therefore affect the rate of blood flow through arterioles and the capillary beds they supply. (After Levick 1995.)

Vascular smooth muscle is also sensitive to other local conditions, such as the partial pressures of oxygen (P_{O_2}) and carbon dioxide (P_{CO_2}). Decreased P_{O_2} and increased P_{CO_2} produce vasodilation, which increases flow, increases the oxygen supply, and carries off the carbon dioxide. A variety of metabolites, local concentrations of potassium (K^+) and hydrogen (H^+) ions, adenosine, and other substances are also known to be vasoactive; as their concentrations increase, the usual response is vasodilation (decreased smooth muscle activity) and increased blood flow. Both local temperature and external pressure also affect smooth muscle activity.

Because of the number of factors that can act locally to influence vasomotor activity and the uncertainty about their relative strengths and possible interactions, the mechanisms for local regulation of the microcirculation are not fully understood. But there is clearly considerable variation among precapillary arterial systems in local regulatory ability. In some systems, local regulation is quite important and may control blood flow not only through the whole capillary bed but also through very small sections of it. Other systems exhibit almost no local regulation. In the eye, the extremes are evident in the dual supply to the retina. The choroidal blood flow, which supplies the retinal photoreceptors, is normally high and not locally regulated. The blood supply to the inner retina, however, is more like that of the brain in being almost entirely regulated by local tissue needs.

Central control of the cardiovascular system is done by the autonomic nervous system. The parasympathetic system acts mainly through the vagus nerve to control heart rate, but some vessels receive parasympathetic innervation, and the release of either acetylcholine or vasoactive intestinal polypeptide (VIP) will produce vasodilation.

The sympathetic system acts directly on vascular smooth muscle. In most cases, sympathetic innervation produces vasoconstriction as norepinephrine released from sympathetic terminals interacts with alpha-adrenergic receptors on the smooth muscle cells. Modulation of the sympathetic innervation can thus increase or decrease blood flow by relaxation or contraction of the smooth muscle. Some vascular smooth muscle has beta-adrenergic receptors, which raises the possibility of vaso*dilation* mediated by sympathetic innervation; if used at all, however, this appears to be a minor route for vascular control.

Vascular smooth muscle is innervated to change local blood flow either to redistribute arterial blood or to assist autoregulation in compensating for the local effects of changes in systemic blood pressure. During vigorous exercise, for example, the heart rate and mean arterial blood pressure increase, and more blood is shifted from the venous to the arterial system, primarily to supply more blood to the active muscles and to the skin for thermoregulation. The need for increased blood flow in the muscles and skin is met in part by the increased amount of arterial blood, but also by redistribution; since there is normally no need for the gut and kidneys to be active during exercise, their blood flow will be reduced automatically to very low levels (conflicts may arise, however, if one exercises vigorously just after eating a heavy meal).

There is still a problem for other organs, however: The increased blood pressure and arterial blood volume mean that they will experience increased blood flow whether they need it or not. Where increased flow might be unacceptable, sympathetic innervation counteracts the increased flow by producing vasoconstriction. The choroidal circulation of the eye is subject to this kind of compensatory regulation, but other tissues, like the retina and brain, appear to cope with changes in systemic blood pressure by other locally acting regulatory mechanisms.

In addition to innervational effects, the presence of different receptor types on vascular smooth muscle means that endogenous epinephrine and norepinephrine released by the adrenal gland can affect the blood vessels, as can hormones, such as angiotensin and vasopressin, and substances released locally from cells in inflamed tissues, such as histamine and bradykinin.

Capillary beds in a tissue may be independent or interconnected

There are several different arrangements for supplying blood to a tissue (Figure 6.3). The supply may be either **segmented**, in which the regions supplied by different arterioles are spatially separate, or **overlapping**, in which different arterioles supply the same region. And in either of these arrangements, the arterioles may lack anastomoses, or shunts (i.e., they may be **end-arterial**), or they may be interconnected by shunts before they arrive at the capillary bed (i.e., they may be **redundant**). The term "collateral circulation," often encountered in the clinical literature, appears to refer to any form of dual supply—that is, to redundancy, overlap, or both.

As Figure 6.3 shows, the venous drainage frequently mimics the pattern of the arterial supply, whether segmented or overlapping, anastomotic or end-arterial. A capillary bed can have a segmented, end-arterial supply and redundant drainage, however. In other words, the nature of the arterial supply and the venous drainage are often the same, but they need not be.

Any of these arrangements is suitable for getting blood into capillaries, but they have different functional implications, best illustrated by the extreme cases of segmented end-arterial systems and overlapping anastomotic systems. The critical feature of a segmented end-arterial system is the lack of alternative paths by which blood can reach a particular part of the capillary bed. If a branch at any level of the arterial hierarchy is blocked, there is no other way to supply the tissue fed by the dependent arterioles. Thus tissues with segmented end-arterial supply are vulnerable to arterial branch occlusion; the supply to the inner retina is a classic example. If there is any advantage to an end-arterial system, it is probably the ease with which blood flow can be controlled at the local level.

In an anastomotic overlapping system, the shunts provide alternative routes for flow in the precapillary portion of the supply, and the overlap in the fields of supply adds to the possibilities. Such redundant systems are probably less amenable to local control of blood flow because several precapillary arterioles must be altered to effect any significant change. The obvious advantage, however, is the system's invulnerability to local blockages in the supply. If one arterial branch is occluded, blood can easily take another path to supply the capillary bed. Much of the supply to the anterior segment of the eye is highly redundant.

Similar considerations apply on the drainage side, except that it is difficult to think of any good reason for a segmented drainage that lacks shunts. This is the situation for the drainage of the inner retina, however, and venous branch occlu-

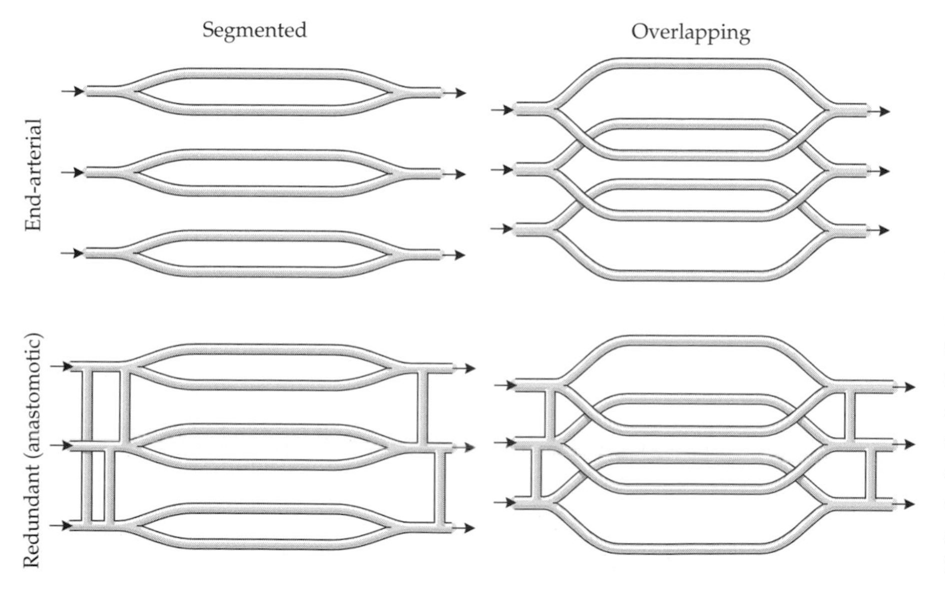

Figure 6.3
Patterns of Supply and Drainage of Capillary Beds
The pattern of supply and drainage of a capillary bed determines its ability to compensate for arterial or venous occlusion. Segmented, end-arterial systems lack any collateral circulation and are especially vulnerable to blockage in supply or drainage. At the other extreme, blood can reach a region within a capillary bed by various routes in overlapping, anastomotic systems; thus they have great capacity for compensation.

sion always compromises drainage to part of the retina, producing edema and potential tissue damage. All other drainage systems in the eye are anastomotic at some level and do not share the retina's vulnerability to venous occlusion.

The interchange between blood and cells depends partly on the structure of the vascular endothelium

Although the entire vascular system is lined with endothelial cells, the endothelium is not identical throughout the system. Some capillaries are constructed to be very restrictive, creating selective barriers between blood and tissues, while others allow free passage of large molecules to and from the blood. But most of the vascular endothelium is a single layer of broad, flattened cells that fit together tightly, leaving only a little space for the passage of fluid or solutes between the cells (Figure 6.4). The endothelium is always surrounded by a basement membrane, whose thickness may vary from one type of capillary to another. A typical capillary has a space of around 15 nm between endothelial cells; though small, this intercellular cleft is wide enough that fluid and small molecules can pass between cells in going from the plasma to the extracellular space. The cells contain small pinocytotic vesicles, which are thought to be involved in the transfer of large molecules between the inside and the outside of the capillary wall.

Most capillaries are continuous, meaning that the endothelium is unbroken except for the intercellular clefts. Within this group, however, considerable variation in structure affects capillary permeability. Some have numerous places in which the endothelial cells have become very thin, reduced to nothing but two layers of cell membrane; these are **fenestrated** capillaries, with "closed" fenestrations (see Figure 6.4*a*). At the fenestrations, the diffusion distance between plasma and the extracellular space is only about 0.1 μm, and these fenestrated capillaries are much more permeable than other continuous capillaries.

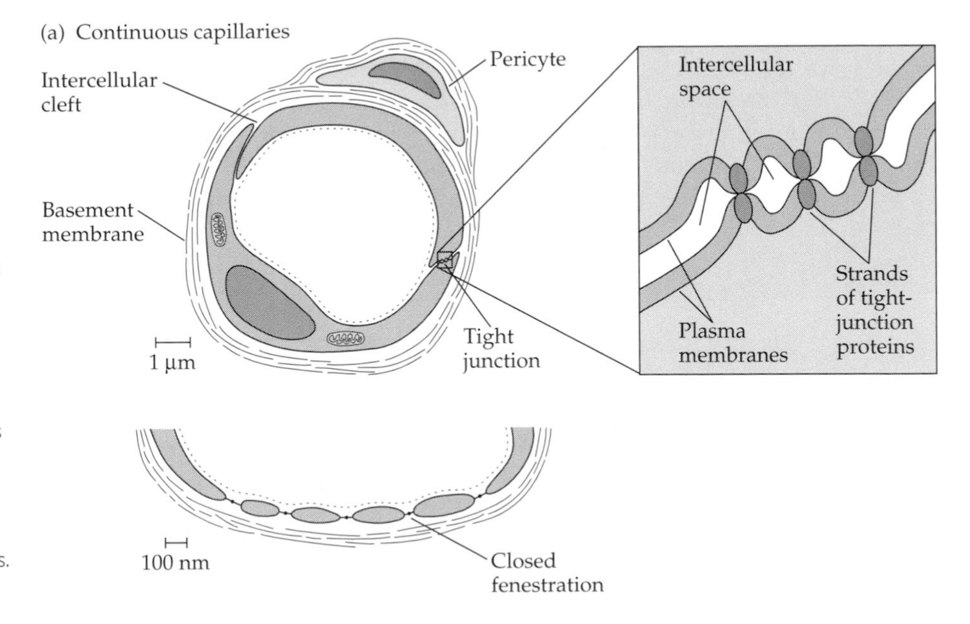

Figure 6.4
Structure of Capillary Endothelial Walls
The ease with which molecules can move across the endothelial walls of capillaries varies considerably depending on the structure of the endothelial cells. (*a*) Most capillaries are continuous, with no holes or pores in their endothelial walls. Barrier-type capillaries (top) have considerable overlap between neighboring endothelial cells and highly restrictive tight junctions between cells. Other continuous capillaries may have regions of extreme endothelial thinning—closed fenestrations—that make them quite leaky (bottom). Most capillaries lie between these extremes. (*b*) Discontinuous capillaries have pores in the endothelial walls, permitting almost unrestricted movement of large molecules and some cells in and out of the capillaries.

Where the interchange between blood and tissue must be highly selective, the capillary endothelium is very tightly constructed; the space between cells is smaller, and there are extensive occluding (tight) junctions between adjacent cells. Although endothelial cells are normally connected with tight junctions, these junctions in typical capillaries are local and widely spaced. In capillaries with a barrier function, however, the tight junctions are extensive and numerous, thereby imposing considerable restriction on the movement of molecules through the intercellular space. Interchange between blood and tissue is controlled, and the traffic in large lipophobic molecules, such as proteins, depends on the presence of intracellular transport systems.

Discontinuous capillaries (see Figure 6.4*b*) have small pores in the endothelium that make them highly permeable, even to large molecules; white blood cells can also squeeze through these openings. Discontinuous capillaries are common in the liver, spleen, and bone marrow, but they are not found in the eye.

All of the eye's capillaries are continuous, but they span the range from fenestrated to highly restrictive. Capillaries with closed fenestrations are found in the choroid and the ciliary processes; restrictive barrier capillaries are characteristic of the retina, where they constitute a blood–retina barrier, and the iris, where they form a blood–aqueous barrier.

Capillary endothelium is renewable, and capillary beds can change

Only a few cell types in the body are permanent—neurons, lens cells, and cardiac muscle cells are the main ones. The rest continually die and are replaced at varying rates. Vascular endothelial cells are in the nonpermanent category, with cell lifetimes in the order of months to years. As individual cells die, they are replaced, and the endothelium can renew and maintain itself. The entire endothelium will be replaced over a period of years. (By contrast, the turnover time for corneal epithelial cells is slightly less than 2 weeks.)

Vascular endothelium maintains itself by cell migration and replication; if a region is damaged, for example, cells on the edge of the damaged region move in to fill the gap, becoming thinner and flatter in the process. Next, local cell proliferation replaces missing cells until the normal number of cells is reestablished. Normal cell death is handled in a similar way, on a smaller scale.

Extensive damage to small blood vessels and capillary beds is a more serious problem. Vessels may be damaged so badly that their structure cannot be rebuilt, and the only solution is construction of a new set of vessels. Fortunately, the same capacities for proliferation and migration can be exploited in this construction, which mimics the way in which blood vessels develop and grow.

Under the right conditions, endothelial cells can breach their basement membrane and send out pseudopodia that become small sprouts, or bulges, from the cell (Figure 6.5). A sprout, which is a protrusion of the cytoplasm surrounded by cell membrane, can grow for some distance and become fairly large, incorporating a large fraction of the cell's cytoplasm. This cytoplasmic extension appears to contain the cell's machinery for cell division, so when the cell divides, the daughter cell is an extension of the original cytoplasmic sprout. Next, the cells change shape. They form vacuoles, and as the vacuoles enlarge, the cells rearrange themselves, creating a lumen that is continuous with the lumen in the original vessel. This is the beginning of a new capillary branch, and reiterating this sequence of migration, division, vacuolation, and rearrangement eventually produces a completely new capillary bed.

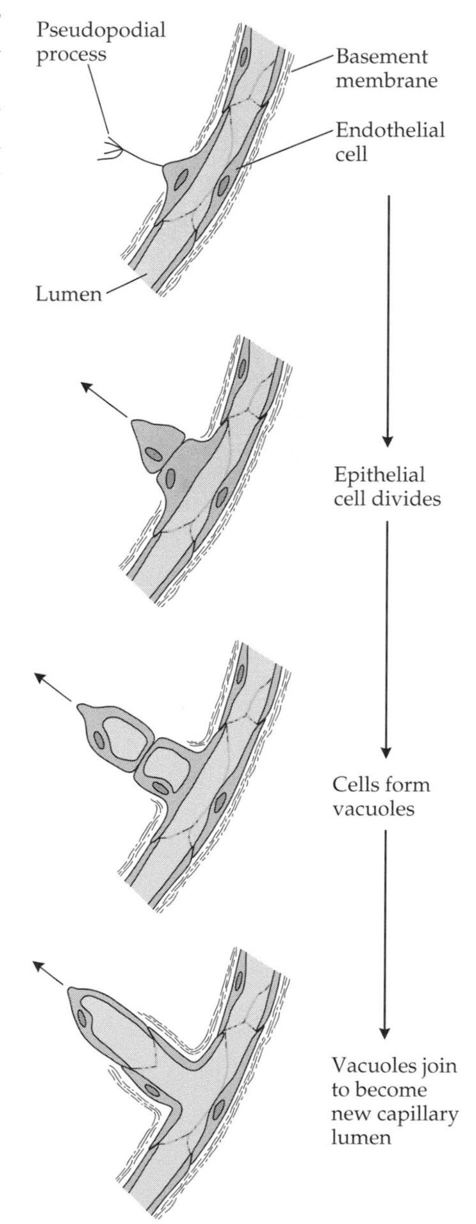

Figure 6.5
Formation of New Capillaries
Capillary beds grow by a process involving cell division, migration, and vacuole formation that creates a sprout from an existing capillary wall. (After Alberts et al. 1994.)

Neovascularization is a response to altered functional demands

Formation of new capillaries in a mature vascular system is called **neovascularization**. The stimulus for the formation of new capillaries is a change in nearby

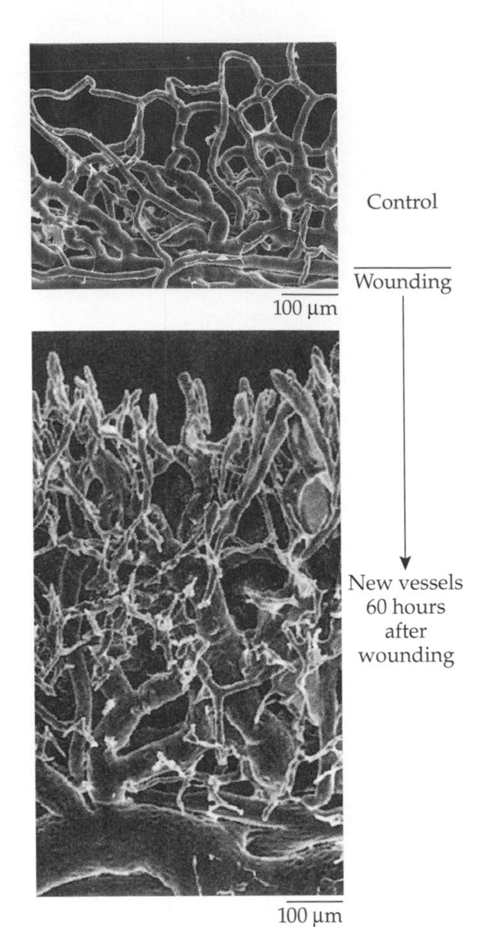

Figure 6.6
Neovascularization
Many new capillaries can be formed in a short time. The neovascularization here was a response to chemical damage, but many other stimuli are appropriate. (From Burger, Chandler, and Klintworth 1983.)

tissue, which may release growth-inducing factors when not adequately supplied with oxygen or when the tissue cells are damaged. Neovascularization can be quite rapid; in the example shown in Figure 6.6, an extensive set of new capillaries at the margin of the cornea developed within 3 days of the time when nearby tissue was damaged. In cases like this, neovascularization is clearly beneficial, aiding in rebuilding of the tissue by increasing the supply of oxygen and nutrients and removing cellular debris more efficiently. Once the damage has been repaired, the parts of the new capillary bed that are no longer required can be closed down and removed.

Neovascularization is necessary if tissues are to be repaired, maintained, and supplied as they grow. Normally, the vascular response is appropriate to the need; new capillaries do not proliferate indiscriminately, and neovascularization ceases when new tissue has been supplied or when damaged tissue has been resupplied. Neovascularization can get out of control, however, and be quite detrimental, particularly when the new capillaries form in places where they do not normally exist. The growth of tumors, for example, is often accompanied by extensive neovascularization, without which the tumor cells could not proliferate so rapidly. The tumor cells produce angiogenic factors to which the vascular system responds as if the signal were coming from normal cells; the more the tumor grows, the more angiogenesis it stimulates, thereby recruiting the normal vascular response for inimical purposes. But tumors are not the only inducers of inappropriate vascular proliferation; the normal responses to local oxygen deprivation or tissue damage may sometimes produce neovascularization that seems massively out of proportion to the damage.

Structurally weakened capillaries may be prone to excessive neovascularization

Although neovascularization can occur anywhere in the eye's capillary beds, the most damaging forms are associated with the capillaries in the inner retina. The details of retinal circulation will be considered later in the chapter; the important fact for the moment is that retinal capillaries normally grow *into* the retina from the larger vessels running over its surface; they never grow into the adjacent vitreous. When the retinal circulation is compromised, however, the result can be extensive neovascularization, with new vessels growing into the vitreous. These vascular tufts also induce or encourage the formation of connective tissue, producing a loss of transparency and forming adhesions that may lead to retinal detachment.

The interaction of factors that elicit neovascularization is not fully understood, but retinal neovascularization is clearly triggered by abnormalities in the circulation. Diabetes is one of the major culprits; venous branch occlusion, in which one of the venules draining part of the retina has become blocked, is another. In both cases, though for different reasons, the capillaries become leaky—suggesting that the capillaries have lost the tight junctions among their endothelial walls—and there is local tissue hypoxia, which is one of the stimuli for the formation of new capillaries. To make matters worse, angiogenic inhibitors normally present in the vitreous may be turned off, thereby permitting abnormal vascular proliferation into the vitreous.

In any case, the microcirculation has been structurally altered and the retina is signaling distress; as it is supposed to do, the vascular system does what it can to resupply the tissue. Unfortunately, the effort will fail because the venous obstruction and reduced blood flow in part of the retina or the damage to the microcirculation in diabetes cannot be overcome by the formation of new capillaries. And since the problems cannot be overcome, the stimuli for neovascularization persist and the capillary proliferation continues unabated. New capillaries are always leaky, so there is continual hemorrhage, clotting, scar formation, and other undesirable sequelae. Clinical management of extensive neovascularization and its complications is not simple, which is one of the reasons for emphasizing the earliest possible detection and medical control of diabetes.

The Ophthalmic Artery and Ophthalmic Veins

The ophthalmic artery distributes blood to the eye and its surroundings

The **ophthalmic arteries** branch from the internal carotid arteries at the level of the cavernous sinus. Each ophthalmic artery enters the orbit through the optic canal and foramen in company with the optic nerve, after which it crosses from lateral to medial above the optic nerve to run forward just above the medial rectus (Figure 6.7). At various points along the way, branches supply the eye and its associated structures, and with one exception (the infraorbital artery), all of the arteries found within the orbit are branches of the ophthalmic artery. There is considerable variability in the way branches arise from the ophthalmic artery within the orbit; Figure 6.7 shows the most commonly observed pattern and two variants.

Figure 6.8 gives a schematic overview of the ophthalmic artery system and the tissues supplied by its branches.

Within the orbit, the eye and the extraocular muscles are supplied by the *central retinal artery* to the inner retina; by the *medial* and *lateral posterior ciliary arter-*

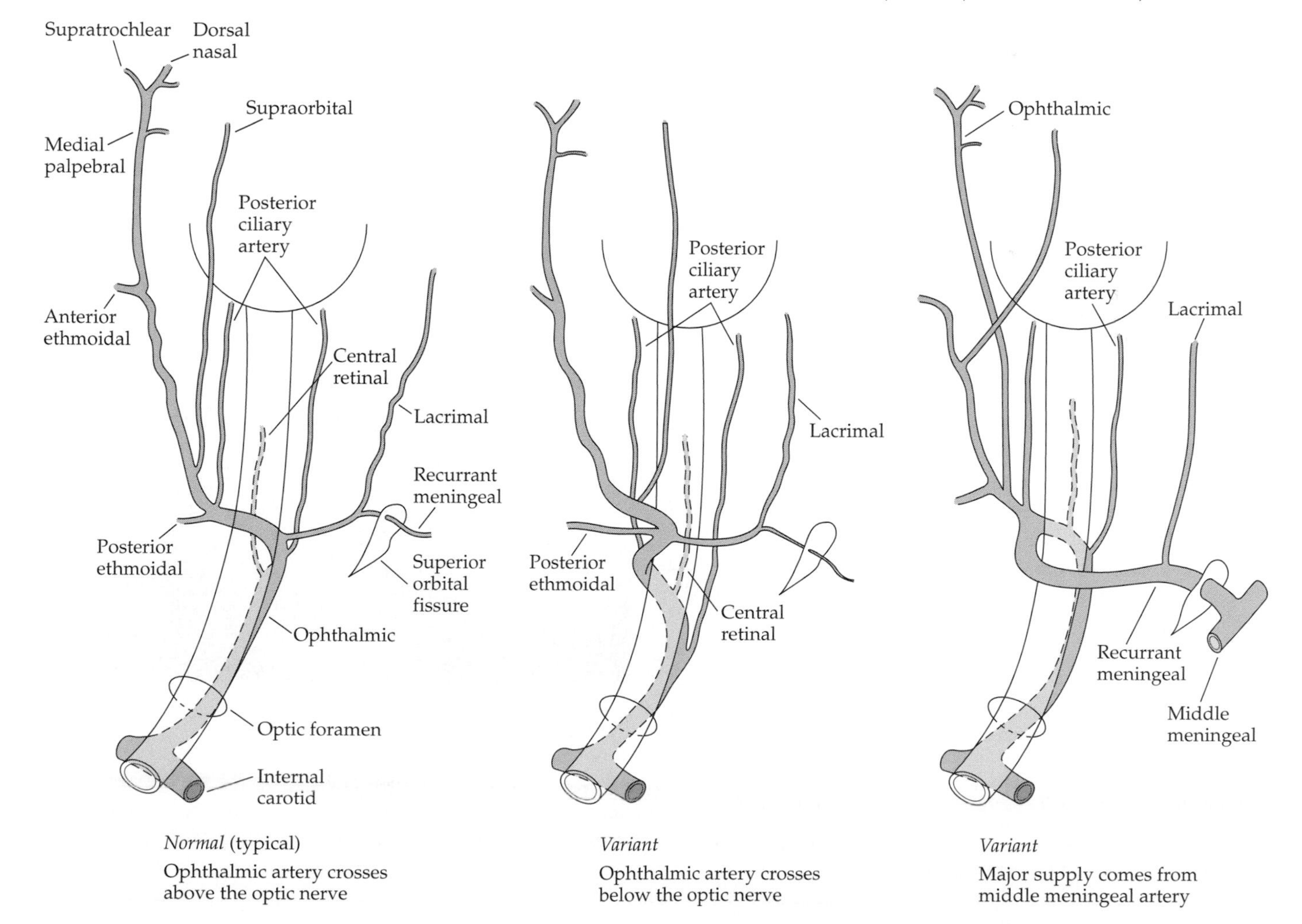

Figure 6.7
The Ophthalmic Artery and Its Branches in the Orbit
The ophthalmic artery branches from the internal carotid artery and enters the orbit through the optic foramen, along with the optic nerve. In the typical arrangement, it crosses above the nerve from lateral to medial and runs forward in the orbit above the medial rectus. Some large branches leave the orbit (the ethmoidal, lacrimal, supraorbital, supratrochlear, and dorsal nasal arteries); the others supply the eye and extraocular muscles. (The branches to the muscles are not included.) There are many variations of this arrangement; two are shown here. (After Hayreh and Dass 1962.)

Vignette 6.1

Circulation of the Blood

THE UNIVERSITY OF CAMBRIDGE in the late sixteenth century was becoming one of the world's great universities, but the education it provided was not uniformly excellent; even by the standards of the times, the preparation of young men to become physicians and surgeons was poor. That situation began to change when one of the smaller and more obscure colleges, Gonville, acquired a new master, a man named John Caius. Caius had studied in Padua, where he learned anatomy from Andreas Vesalius (see Vignette 3.2) and lived in the same house as that great anatomist. Caius remained a confirmed follower of Galen (see also Vignette 2.1), but he came to believe that dissection and observation were critical for anatomical learning and training, and that the methodology of Vesalius, if not his interpretations, should become part of the education offered at Cambridge. As a result, a provision was made that twice a year, the body of a criminal executed in the city of Cambridge would be publicly dissected by a teacher of anatomy.

The opportunity to observe two dissections a year is scarcely a medical education, but it was a start, and Gonville College (now known as Gonville and Caius College*) also offered scholarships specifically for aspiring physicians. One of these scholarships, made available by a bequest from Matthew Parker, the archbishop of Canterbury, was for young men born in Kent and educated at the King's school, which was associated with Canterbury. The scholarship specifically recognized the deficiencies of Cambridge medical education by providing an opportunity to supplement the years at Cambridge with 2 additional years at the universities of Paris, Montpellier, Bologna, or Padua.

In 1593, the newest Parker scholar accepted his annual stipend of 3 pounds and 8 pence; 400 years ago, this sum was thought to be quite sufficient to keep a scholar housed (badly) and fed (poorly) for a year. The 15-year-old scholar was William Harvey (1578–1657). His education at Cambridge is not well documented, but it consisted primarily of the *trivium*—rhetoric (including readings of the classical authors), logic, and grammar—for the first 3 years, followed by the *quadrivium*—music, arithmetic, geometry, and astronomy; these subjects had been the core of university education for almost 500 years. The medical part of his training normally would have been the observation of an occasional dissection and a few lectures here and there, but mostly reading and discussion of works by Galen and Hippocrates. The influence of Dr. Caius was evident, however, in Harvey's decision to complete his medical training at Padua.

*Although it is one of the oldest of the Cambridge colleges, Gonville and Caius is probably not as well known as others, such as Trinity or King's. It plays a role, however, in the film *Chariots of Fire*, which tells the story of the Olympic sprinter Harold Abrahams and his great competitor and teammate, Eric Liddell, "the Flying Scotsman." Abrahams was a student at Gonville and Caius when he won the gold medal in the 100 m dash in the 1924 Paris games. Gonville and Caius can also claim Francis Crick, codiscoverer of the double helical structure of DNA, as a former student.

When Harvey arrived in Padua in 1600, the professor of anatomy and surgery was Fabricius ab Aquapendente (see Vignette 12.1), the fourth in a line of distinguished anatomists; Vesalius, Columbo, and Fallopius (see Vignette 4.1) were his predecessors. The great dissecting theater, which can still be seen at Padua (Figure 1), had recently been completed. Like the university in Bologna, from which it derived, Padua was a university governed largely by the students, and Harvey was the representative for the English students throughout his stay in Padua. (Because of a common language—Latin—that all students of the time could read, speak, and understand, the ancient universities were quite international in character.)

Harvey's later conclusion that blood circulates in the body had its origins in Padua. His teacher, Fabricius, discovered the valves in peripheral veins, and Harvey probably first saw them while peering down from one of the dissecting theater galleries. The problem, however, was recognizing the function performed by the valves. Fabricius realized that the valves were like little gates, but he subscribed to the prevailing idea that venous blood flows away from the heart just as arterial blood does, and this erroneus belief led him to think about valves controlling the *rate* of flow rather than the *direction* of flow.

Harvey seems to have been skeptical about this interpretation almost from the beginning, and a great deal of his work after leaving Padua for medical practice in London was directed at determining the direction of venous blood flow. The sole illustration in his great work *Exercitatio Anatomica de Motu Cordis et Sanguinis in Animalibus* (1628), included here as Figure 2, shows the now familiar demonstration of valves in a superficial vein *preventing* flow of venous blood away from the heart. He also described the valves in considerable detail, noting that the cusps of the valves were oriented so that the valves would close readily to prevent backflow in the veins. There were a host of other problems as well: the correct sequence of contraction and relaxation of the atria and ventricles, the recog-

Figure 1 The Dissecting Theater at Padua
This model is a replication of the dissecting theater where Harvey first saw the valves in the veins of the limbs. The theater was designed by Fabricius, who oversaw its construction. It was recently renovated and named in his honor. (Courtesy of Stefano Piermarocchi.)

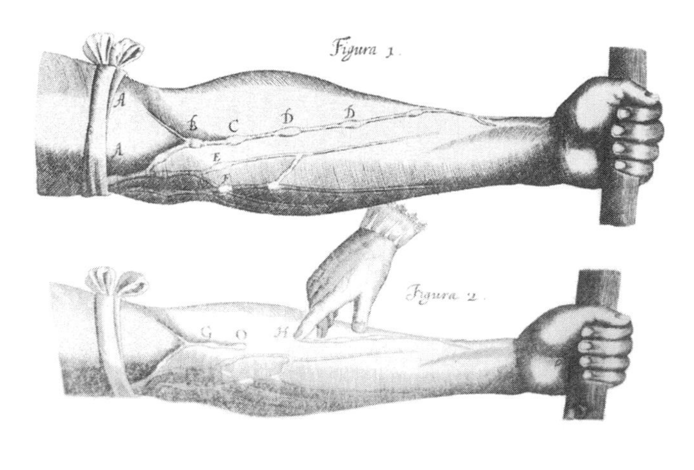

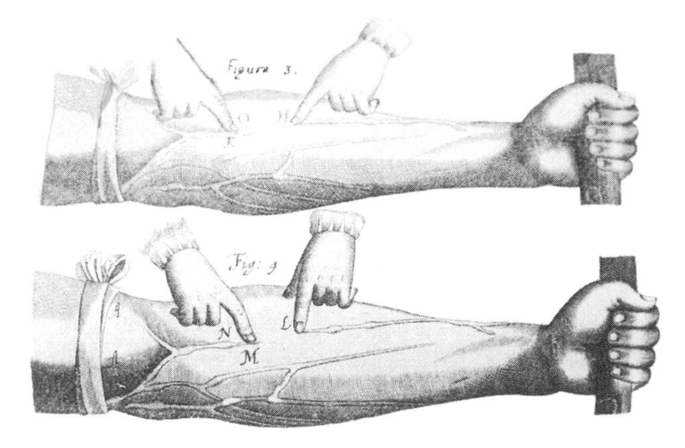

Figure 2 Valves in Peripheral Veins Prevent Blood Flow Away from the Heart
A ligature around the arm produces distension of the veins that makes the valves prominent (Figura 1). Blood cannot flow back through a valve to fill a section of vein that has been emptied (Figura 2). Blood cannot be pushed back through a valve (Figura 3), but a section of vein emptied below a valve (Fig. 4) will be refilled when pressure is released. (From Harvey 1628.)

nition that the right and left ventricles pumped blood to separate arterial systems, the idea that left ventricular contraction was followed by arterial expansion, and so on.

The most difficult problem in describing a pattern of circulation from heart to arteries, from arteries to veins, and through veins back to the heart, was the lack of any observable connection between arteries and veins. Capillaries were unknown and would remain so until the invention of the microscope (see Vignette 8.1). This missing link was a weak point in Harvey's argument because he had to infer the existence of something like capillaries while being unable to demonstrate their presence. Part of his argument was quantitative: He calculated the volume of blood pumped by the heart per unit time and, from the surprisingly large number he derived, pointed out that the extremities would swell enormously with accumulated fluid if there were not some means to remove the fluid from the tissue at a correspondingly high rate. This finding implied direct connectivity between the arterial and venous sides of the system.

Not until some years after Harvey's death was his great, inspired idea about the circulation of the blood universally accepted, and even more years passed before its full import was understood. He was right, of course, and it is this discovery for which he is primarily remembered. But he also pursued a busy career as a physician of considerable importance, and he wrote another important book, *Exercitationes de Generatione Animalium* (1651), concerning the problems of conception and embryogenesis. And for some of us, one of his greatest virtues may have been his preference for coffee over tea; he is said to be the first influential Englishman to praise the virtues of this beverage, and toward the end of his life coffeehouses began to flourish in London.

ies to the uveal tract and optic nerve; and by branches called *muscular arteries* to the extraocular muscles and, by extensions from the rectus muscles, to the anterior segment of the eye. The lacrimal sac and the lacrimal part of the orbicularis muscle (see Chapter 7) are supplied by the *dorsal nasal artery.* The lacrimal gland is supplied by the *lacrimal artery.*

Outside the orbit, the ethmoidal sinuses are supplied by anterior and posterior ethmoidal arteries. The eyelids are supplied by *medial* and *lateral palpebral arteries,* which are branches of the dorsal nasal artery (or directly from the ophthalmic artery) and the lacrimal artery, respectively. Skin and muscle of the upper brow are supplied by the *supratrochlear, supraorbital,* and *lacrimal arteries.* Meninges in the cranial fossa are supplied by the recurrent meningeal artery.

Blood supplied to tissues by the ophthalmic artery is drained to the cavernous sinus by the ophthalmic veins

Two large veins in the orbit collect blood from the eye and other stuctures and carry it out of the orbit through the superior orbital fissure to the cavernous sinus. As their names imply, the **superior** and **inferior ophthalmic veins** run in the upper and lower parts of the orbital cavity. Blood from the inferior part of the

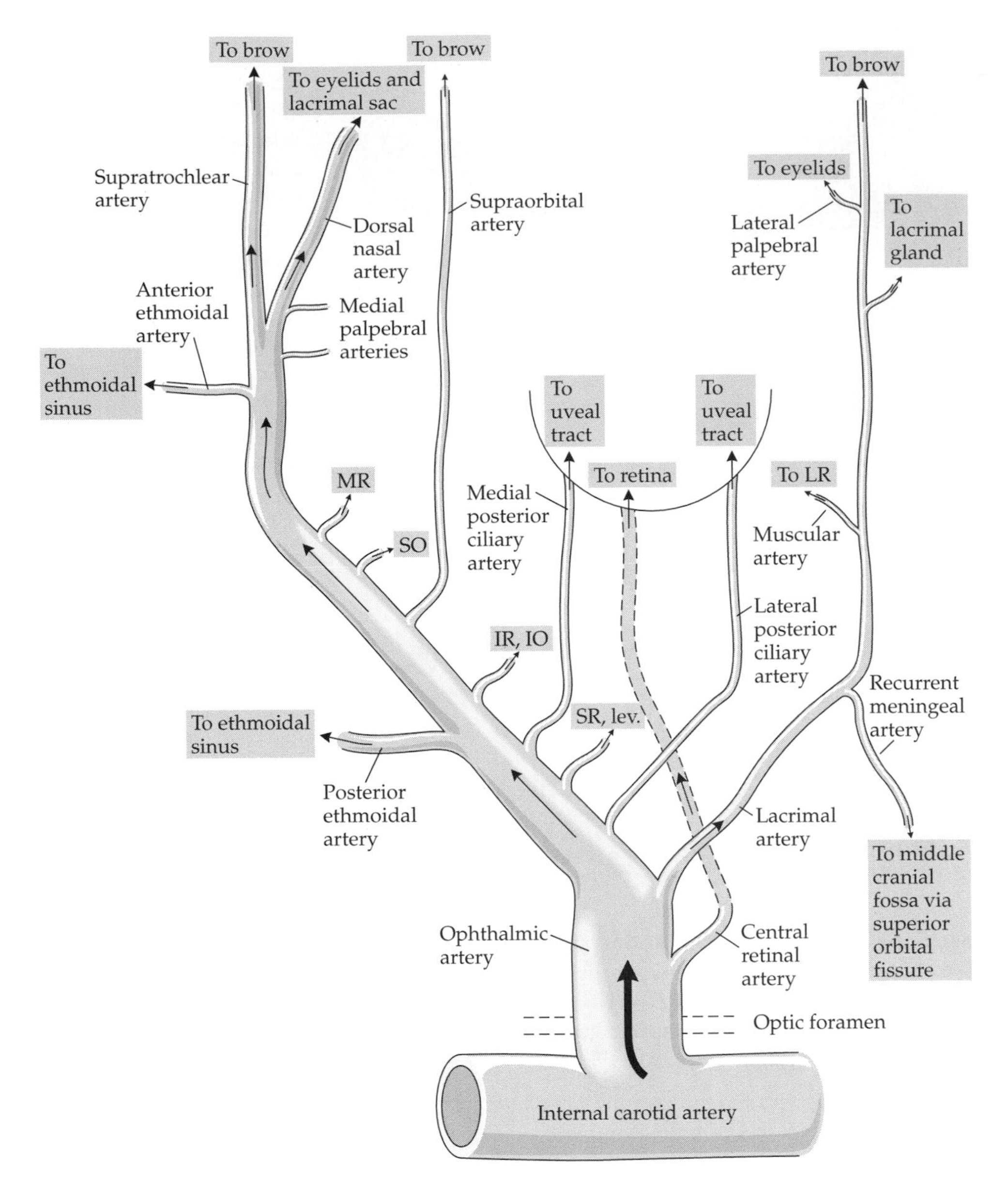

Figure 6.8
Distribution of Blood by the Ophthalmic Artery System
Blood is supplied to the eye by the central retinal artery, the posterior ciliary arteries, and the rectus muscle branches that will become anterior ciliary arteries. All other arteries in the orbit are passing through to supply structures outside the orbit. (MR, medial rectus; SO, superior oblique; IR, inferior rectus; IO, inferior oblique; SR, superior rectus; LR, lateral rectus; lev, levator.)

eye and the inferior muscles normally drains into the inferior ophthalmic vein; the rest of the blood is drained by the superior ophthalmic vein (Figures 6.9 and 6.10). This is a rough generalization, however. The venous branching pattern is so variable that the descriptions and names in the literature are not always consistent, and it is difficult to find a typical pattern. Here, I have emphasized the parallelism between arteries and veins to describe the normal situation.

Within the orbit, the eye and extraocular muscles are drained by the *central retinal vein* from the inner retina and optic nerve, the *vortex veins* (*vena vorticosa*) from the uveal tract, and the *muscular veins* of which those from the rectus muscles receive blood from the anterior segment of the eye.

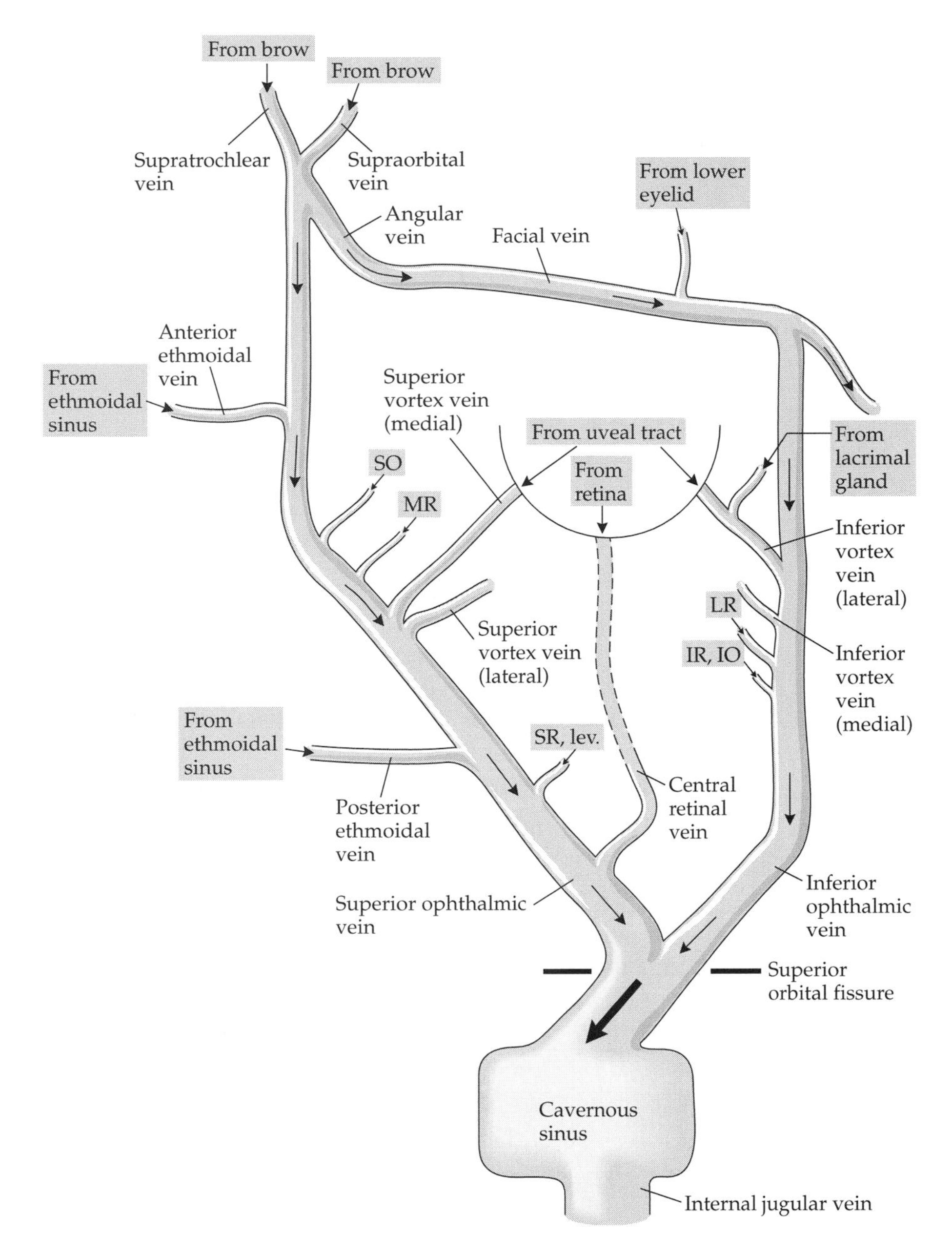

Figure 6.9
The Ophthalmic Veins and Their Tributaries
Most orbital blood flows to the cavernous sinus behind the orbit through the superior and inferior ophthalmic veins, which receive blood from the eye and elsewhere. The ophthalmic veins have anastomoses with veins of the face, in particular with the angular and facial veins. (MR, medial rectus; SO, superior oblique; IR, inferior rectus; IO, inferior oblique; SR, superior rectus; LR, lateral rectus; lev, levator.)

The lacrimal gland is drained by several *lacrimal veins* that are usually small and variable in number. They typically drain into one of the vortex veins.

Outside the orbit, the ethmoidal sinuses are drained by anterior and posterior ethmoidal veins. The eyelids and the skin and muscle around the orbit are drained by veins of the face, primarily the *angular* and *facial veins*. Two veins that are named the same as two arteries running through the orbit, the *supraorbital* and *supratrochlear veins*, do not enter the orbit as discrete entities; they merge at the orbital margin to become the superior ophthalmic vein. The *infraorbital vein*

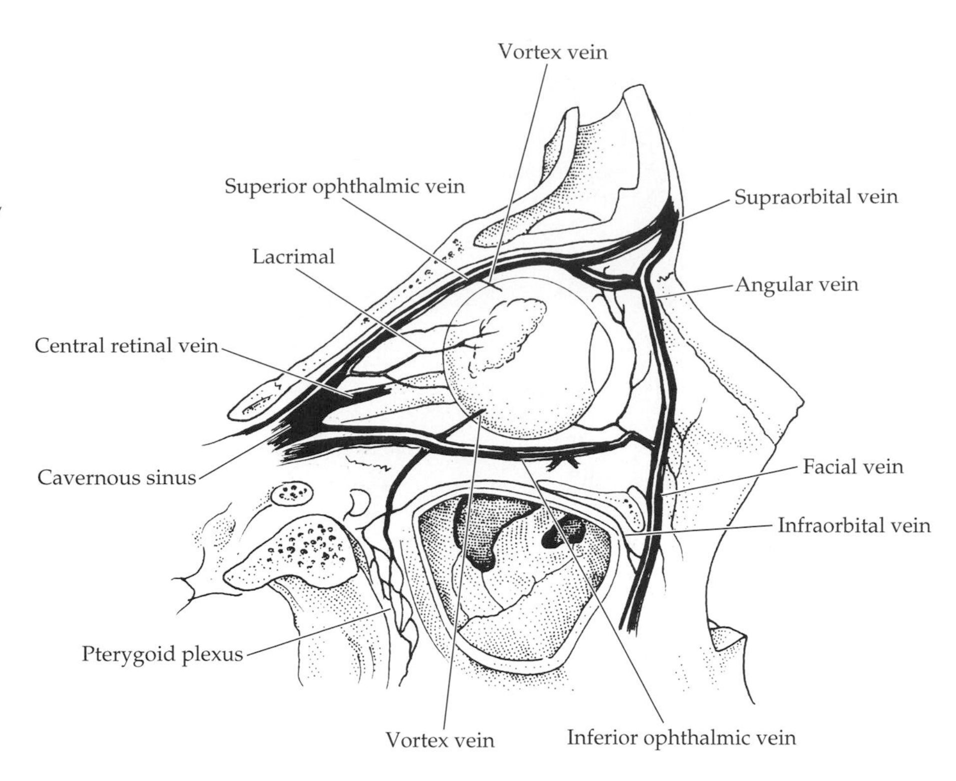

Figure 6.10
Drainage of Blood by the Ophthalmic Veins
The main veins draining the eye are the vortex veins, the central retinal vein, and the episcleral veins (not shown), all of which flow into the ophthalmic veins.

is the only vein from the face that enters the orbit, running below the orbital floor in parallel with the *infraorbital artery* (the infraorbital artery is not part of the ophthalmic artery system; it arises from the maxillary artery).

Supply and Drainage of the Eye

Muscular arteries supply both the extraocular muscles and the anterior segment of the eye

Typically, the ophthalmic artery has one or more small branches supplying each of the extraocular muscles. Two of these **muscular arteries** enter the belly of each muscle on its ocular side, about halfway between the muscle's origin and its tendon (there is usually only one branch to the lateral rectus muscle). After entering the muscle, the arteries divide into numerous small arterioles that run toward the ends of the muscle, branching into capillaries as they go. A parallel system of venules into which the capillaries drain produces a ladderlike pattern of blood vessels, with the capillary rungs running perpendicular to the long axis of the muscle fibers.

In the levator and the oblique muscles, the muscular arteries end at the capillary beds in the muscles. In the rectus muscles, however, the arterioles supplying the capillary beds are side branches and the muscular arteries continue through the muscles to emerge on their orbital surfaces several millimeters posterior to the muscle tendons (Figure 6.11). As these arteries run forward over the muscle tendons onto the eye, their name changes; they are now the **anterior ciliary arteries**. They will supply the conjunctiva, cornea, limbus, ciliary muscle, and iris.

Small venules in each muscle coalesce to form one or two small veins (**muscular veins**) that drain the muscle. These veins are highly variable in location and the degree to which they merge with one another and with the anterior ciliary or vortex veins, but they ultimately join either the superior or the inferior ophthalmic veins.

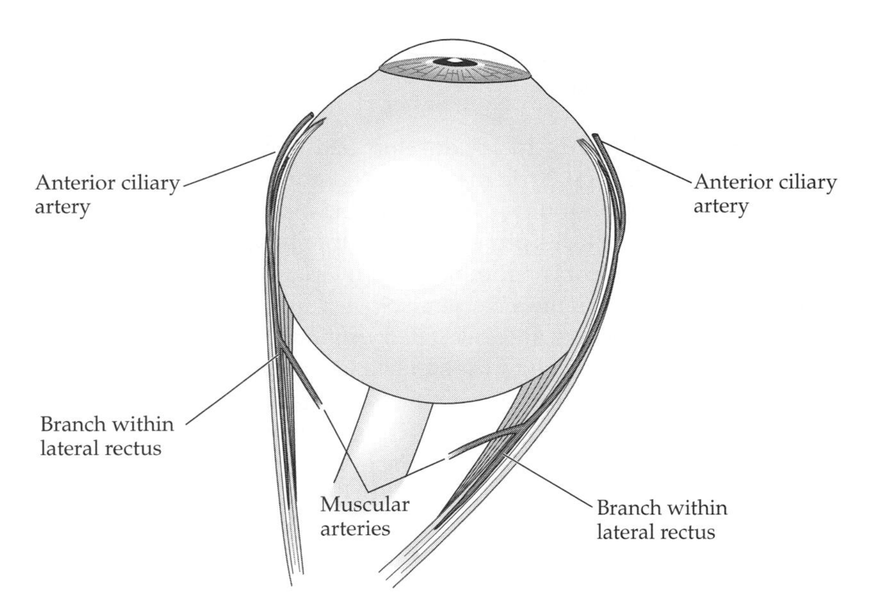

Figure 6.11
Muscular Arteries to the Rectus Muscles Become Anterior Ciliary Arteries
As shown here for the medial and lateral recti, the arteries to the rectus muscles supply capillary beds within the muscles and continue forward, outside the muscles, as the anterior ciliary arteries.

The anterior ciliary arteries contribute to the episcleral and intramuscular arterial circles

There are typically seven anterior ciliary arteries, one for the lateral rectus muscle and two for each of the other rectus muscles. Shortly after the anterior ciliary arteries pass over the rectus muscle tendons, the arteries bifurcate; one set of branches remains on the surface of the eye while the other penetrates the sclera to reach the ciliary body (Figure 6.12). The branches continuing forward in the episcleral tissue are the **episcleral arteries**. They will form the **episcleral arterial circle** and supply the corneal arcades and the conjunctiva (see the next section).

The anterior ciliary artery branches going into the eye are the **major perforating arteries**. These arteries branch again as they reach the ciliary body, and anastomoses among these branches (and possibly branches from other arteries) form the **intramuscular arterial circle**. As the name suggests, the intramuscular circle lies in the ciliary muscle, just posterior to the junction between the ciliary body and the iris.

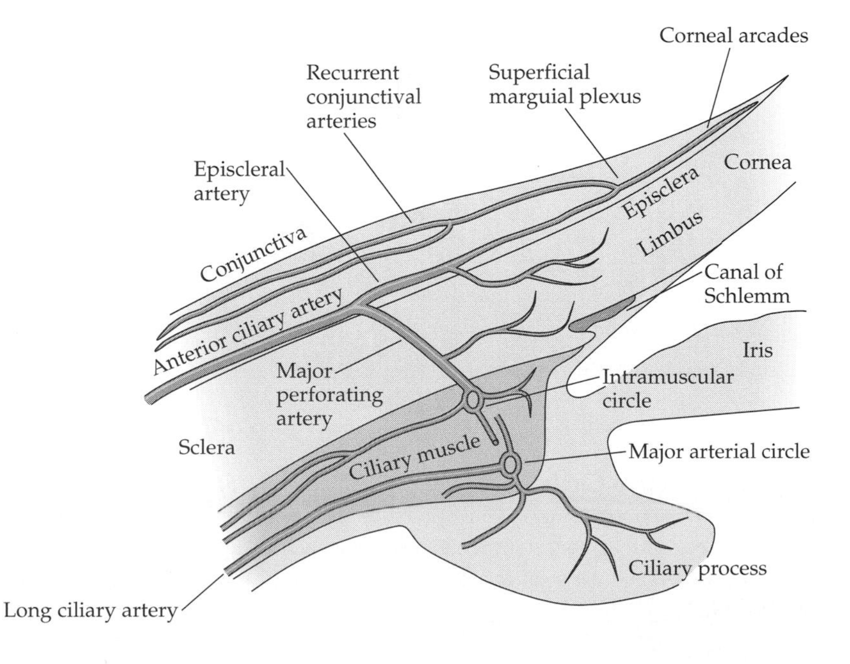

Figure 6.12
Anterior Ciliary Arteries Have Both Superficial and Deep Branches
After passing over the rectus muscle tendons, the anterior ciliary arteries branch. One set of branches, the episcleral arteries, remains on the surface of the eye; the other set, the major perforating branches, goes through the sclera to the ciliary body.

The conjunctiva and corneal arcades are supplied by branches from the episcleral arterial circle and drained by the episcleral and anterior ciliary veins

The extent to which the shunts develop between episcleral artery branches varies, and the episcleral circle may not be complete; in Figure 6.13, it is shown with two small discontinuities, one superior and one inferior. But even with these breaks in the circle, its presence means that this superficial part of the anterior ciliary artery system is redundant and overlapping.

Most branches from the episcleral circle run forward toward the cornea, branching as they go; they are now called **conjunctival arteries**. Further anastomoses near the cornea produce the **superficial marginal plexus**. The first of two distinguishable sets of branches from the marginal plexus are small arteries running directly *away from* the cornea in the conjunctival stroma. Because of the change in direction (away from the cornea instead of toward it), these branches are called **recurrent** (in this case conjunctival) **arteries**. They are the most superficial of all the arteries in the anterior ciliary artery system.

The second set of branches from the superficial marginal plexus consists of small arterioles that run radially in toward the margin of the cornea. These small arteries, their capillaries, and the small veins leading away from the cornea are the **corneal arcades**; they run within the palisades of Vogt (see Chapters 8 and 9) in the limbal conjunctiva. The cornea is normally avascular, which means that only cells in the peripheral cornea are supplied by the corneal arcades. Cells in the central cornea are about 5 mm away from the arcades, and they rely on the tear film covering the corneal surface for oxygen and other substances.

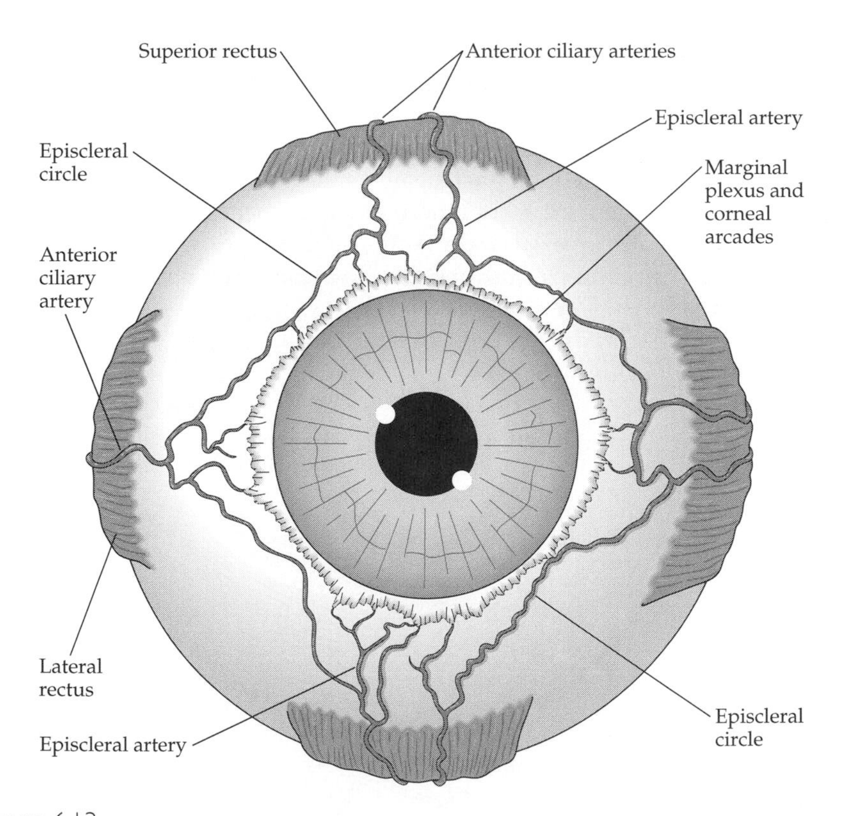

Figure 6.13
The Episcleral Arterial Circle
Anastomoses among branches of the episcleral arteries form a nearly complete arterial circle. Branches from the episcleral arterial circle supply the limbal and bulbar conjunctiva that overlies the blood vessels and the arcades around the margin of the cornea. (After Morrison and Van Buskirk 1983.)

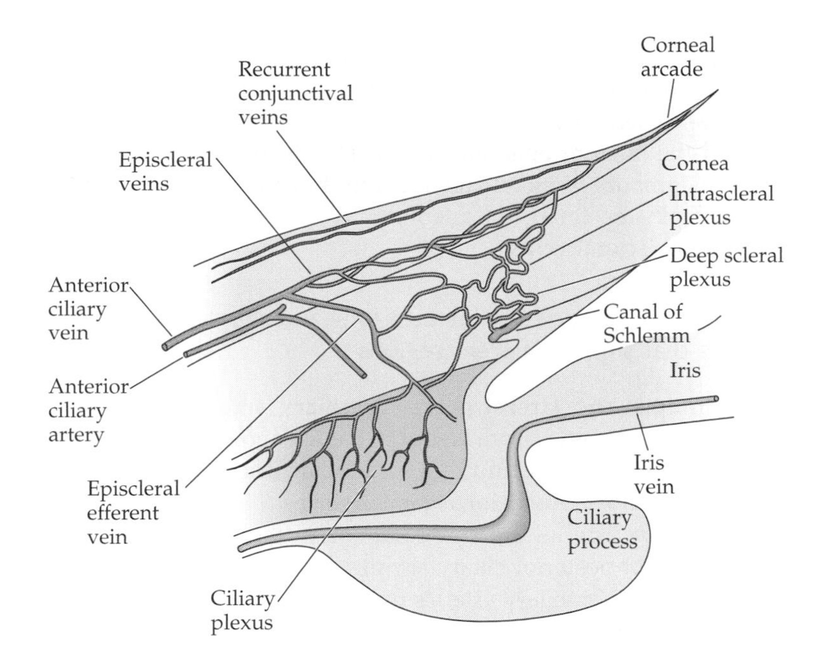

Figure 6.14
Drainage of the Limbus and Conjunctiva
The episcleral efferent veins receive blood from the ciliary muscle (the ciliary plexus), the limbus (the intrascleral plexus), the conjunctiva, and the corneal arcades. The drainage of the ciliary muscle probably overlaps the drainage field of veins going to the choroid from the ciliary processes and the iris. (After Hogan, Alvarado, and Weddell 1971.)

The system of episcleral veins drains the conjunctiva, corneal arcades, and limbus

The number of **episcleral veins*** varies, but there are usually at least four, and like the anterior ciliary arteries, they are associated with the four rectus muscles. Unlike the anterior ciliary arteries, however, the episcleral veins are not necessarily extensions of corresponding muscular veins (although they may be). They may run to the nearest of the ophthalmic veins or to the nearest of the vortex veins (which are discussed later). The episcleral veins carry aqueous humor leaving the eye at the limbus.

The episcleral veins also collect blood from the superficial **conjunctival veins** draining the conjunctiva and corneal arcades as well as from veins coming from a venous plexus (the **ciliary plexus**) in the ciliary body and veins in the limbal stroma. The veins from the ciliary plexus may, but most often do not, run along with the major perforating arteries; they are more often illustrated than they are named, but the term **episcleral efferent veins** has been used in the past and is used here (Figure 6.14). Unlike the arterial supply, there are no consistently observable venous circles, so there are no venous counterparts to either the episcleral or the intramuscular arterial circles. Nevertheless, the anterior venous system is heavily anastomotic.

Thus, the surface of the anterior segment is drained by small veins that are part of the corneal arcades and recurrent conjunctival veins that merge at the superficial marginal plexus. The marginal plexus is drained by conjunctival veins that are joined by small veins from the limbus to form the episcleral veins. The episcleral veins are joined by the efferent veins coming from the ciliary plexus.

The principle difference between the arterial and venous systems in this region is found in the limbus; although relatively few episcleral artery branches extend into the limbal stroma, numerous small veins come from the limbal stroma to join the episcleral veins. The arterial supply and venous drainage appear to be mismatched because many of the veins do not drain capillary beds in the limbal stroma but instead carry aqueous humor from the **canal of Schlemm** (see Chapter 9). The venous plexuses in the limbal stroma are the **deep scleral**

*The episcleral veins run parallel not only to the episcleral arteries in the episclera but also to the anterior ciliary arteries. The name of the episcleral veins does not change, however, so there are no anterior ciliary veins in the conventional nomenclature.

plexus, which lies closest to the canal of Schlemm, and the **intrascleral plexus**, which is more superficial (see Figure 6.14). There are probably anastomoses between the deep scleral plexus and the ciliary plexus and between the intrascleral plexus and the efferent episcleral veins. The deep and intrascleral venous plexuses carry a mixture of blood and aqueous, but an occasional **aqueous vein** arises directly from the canal of Schlemm and carries aqueous unmixed with blood out to the episcleral veins.

The posterior ciliary arteries divide into long and short posterior ciliary arteries that supply different regions

Typically, the **medial** and **lateral posterior ciliary arteries** branch from the ophthalmic artery as it passes from lateral to medial above the optic nerve. Each of these arteries branches several times, producing one **long posterior ciliary artery** and several **short posterior ciliary arteries** (Figure 6.15). In effect, the long posterior ciliary arteries are continuations of the parent posterior ciliary arteries, and the short posterior ciliary arteries are subsidiary branches.

The two long posterior ciliary arteries (medial and lateral) run to the posterior surface of the eye and pass through the sclera. The entry points of the arteries are close to the horizontal plane bisecting the eye, about 3.5 to 4.0 mm nasal or temporal to the optic nerve. Once through the sclera, the long posterior ciliary arteries run forward in the choroid toward the ciliary body. In passing through the choroid, the arteries have very few branches and thus do not contribute extensively to the choroidal circulation.

Although the medial and lateral long posterior ciliary arteries have very few branches in the choroid until they are well forward of the equator, there are some branches, most from the *lateral* long posterior ciliary artery. These recurrent branches arise near the equator and run back toward the region of the fovea,

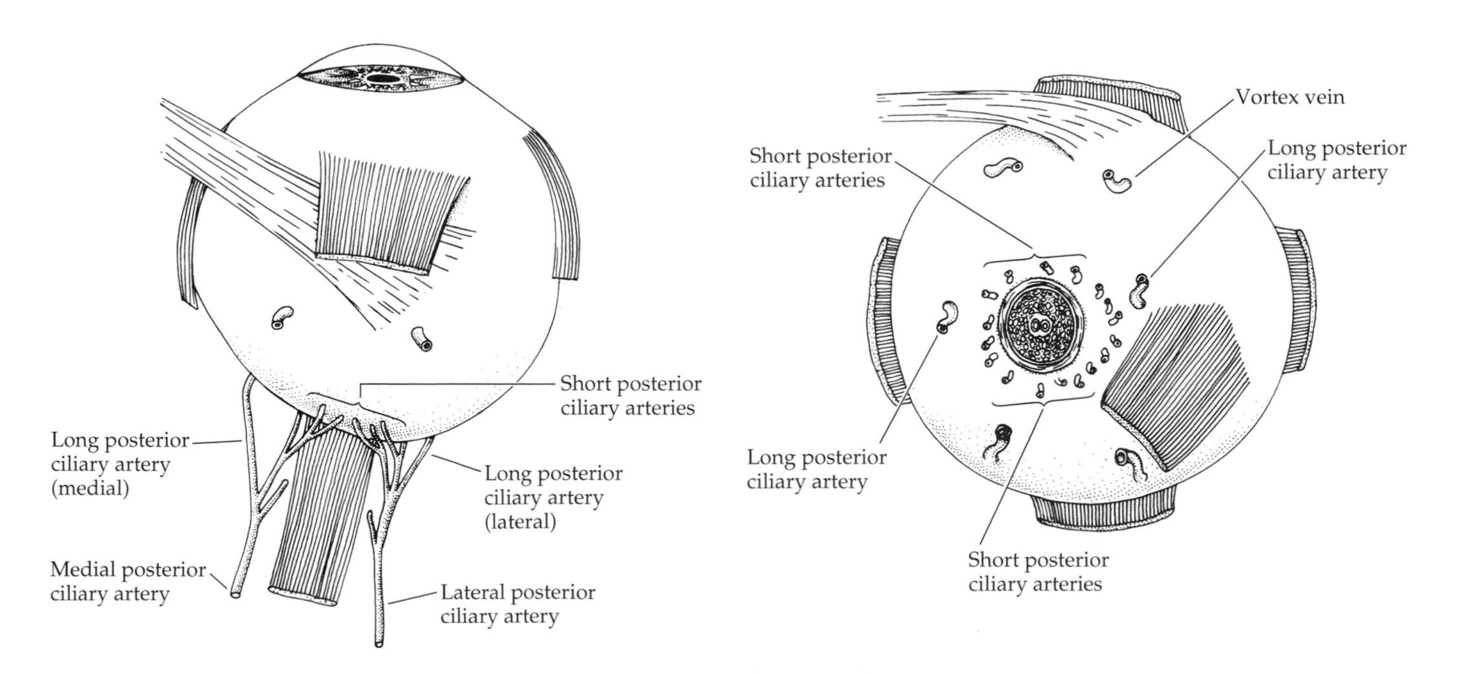

Figure 6.15
The Posterior Ciliary Artery System
The posterior ciliary artery system typically arises as two branches, the medial and lateral posterior ciliary arteries, from the ophthalmic artery. Small branches penetrating the sclera around the optic nerve are the short posterior ciliary arteries; the main arteries continue as the medial and lateral long posterior ciliary arteries. (After Hogan, Alvarado, and Weddell 1971.)

where they appear to contribute to the choriocapillaris. Since the choriocapillaris in this region is also supplied by the short posterior ciliary artery system, the presence of the recurrent branches from the lateral long posterior ciliary artery represents an alternative, redundant source of supply for the choriocapillaris just behind the fovea.

A ring of about 20 short posterior ciliary arteries enters the sclera around the optic nerve (see Figure 6.15). Most of them pass into the choroid, where they branch extensively to supply the choriocapillaris, which is the blood supply for the retinal photoreceptors. Some branches form an arterial circle in the sclera around the optic nerve and contribute to the nerve's blood supply.

The intramuscular and major arterial circles are formed in the ciliary body by branches from the anterior and long posterior ciliary arteries

Blood reaches the anterior uveal tract from two sources: the half dozen major perforating arteries from the anterior ciliary artery system, and the two long posterior ciliary arteries. Subsequent branches of the major perforating arteries and the long posterior ciliary arteries within the ciliary body then create two incomplete arterial circles: The **major arterial circle** lies near ciliary processes, and the **intramuscular arterial circle** is in the outer part of the ciliary muscle.

The formation of the arterial circles in the ciliary body has been described in two ways.* According to one scheme, the arterial circles arise primarily from different sources. The intramuscular circle, in this view, is formed by anastomoses among branches of the major perforating arteries; the major arterial circle is formed in a similar way by branches of the long posterior ciliary arteries (Figure 6.16*a*). There is said to be some contribution of the major perforating artery to the major arterial circle, but it is small and largely in the superior and inferior parts of the circle where branches of the long posterior ciliary arteries do not reach. The intramuscular circle is also incomplete, and there may be only a small (or no) contribution from the lateral major perforating artery. Branches from the intramuscular circle supply all but the inner portion of the ciliary muscle, the anterior choroid, and the iris. Branches from the major arterial circle go to the iris and provide the main supply to the ciliary processes. Thus the arterial circles are largely, though not wholly, independent, and their main region of overlapping supply is the iris, where both contribute to the **minor arterial circle**.

Another scheme has the intramuscular circle receiving contributions from both the major perforating arteries and the long posterior ciliary arteries; the circle is formed by their intermingling branches and by anastomoses among them. The major arterial circle is formed by branches from the long posterior ciliary arteries *and* by branches from the intramuscular circle (Figure 6.16*b*). Although the notion that branches from the intramuscular circle supply much of the ciliary muscle and that branches from the major arterial circle go to the ciliary processes still applies in this scheme, the interpretation about the *source* of blood is different. Because both arterial circles are fed by both arterial sources (major perforating branches and long posterior ciliary arteries), this scheme is one of overlapping circulation not only in the iris, but in the entire anterior uveal tract.

Both vascular schemes have two sets of branches arising from the intramuscular circle in common. Some branches run directly into the iris toward the pupil, where they join with other arteries from the major arterial circle to form the

*Vascular castings (see Box 6.1) of the anterior segment are very complex and probably quite variable. As a result, different investigators have seen different things in their dissections and scanning electron micrographs, leading to different interpretations. Although the people who do this work prefer their own descriptions, there is currently no way for an outsider to justify choosing one description over the other. A consensus will emerge, but it is probably a few years down the line.

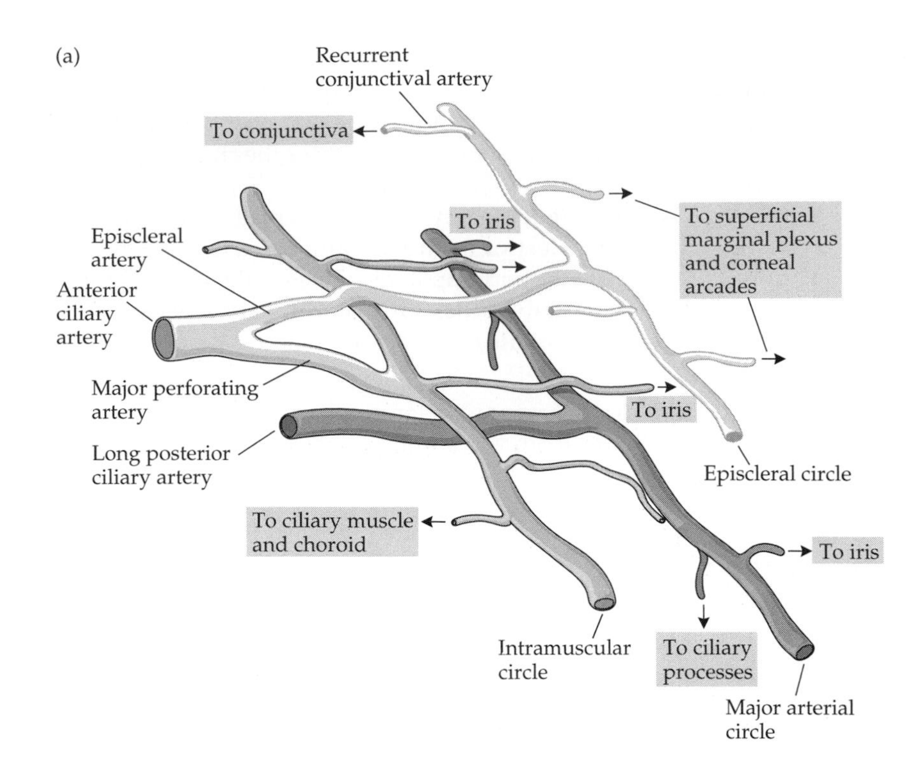

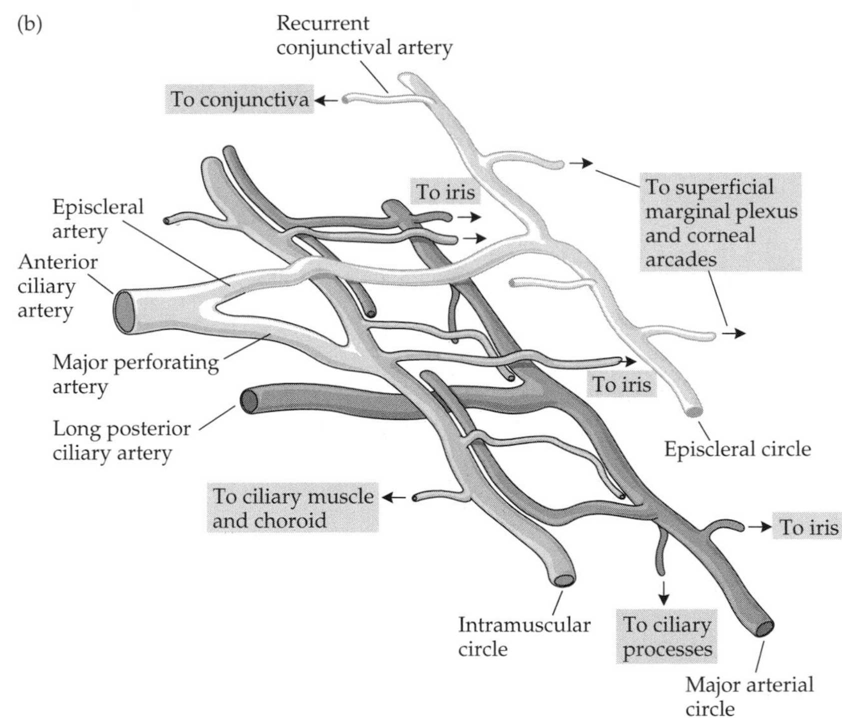

Figure 6.16
The Intramuscular and Major Arterial Circles
Two schemes for the formation of arterial circles. (*a*) The intramuscular circle shown here is formed almost solely by major perforating branches from the anterior ciliary arteries; the major arterial circle is formed mainly by branches from the long posterior ciliary arteries. According to this scheme, the regions supplied by the anterior and long posterior ciliary artery systems are relatively independent. (*b*) Both the anterior and the long posterior ciliary artery systems contribute significantly to both the intramuscular and major arterial circles. In this arrangement, there is considerable collateral supply in the anterior segment. (*a* after Funk and Rohen 1990; *b* after Morrison et al. 1996.)

minor arterial circle of the iris (see Figure 6.16). These arteries and their corresponding veins in the iris do not have a particular name, but they do have some special features in their construction. Specifically, their endothelium is of the barrier type, and the arteries have a sheath that is thought to prevent kinks in the arteries as the iris tissue is compressed when the pupil dilates (see Chapter 10).

A set of recurrent branches from the intramuscular circle runs posteriorly through the ciliary muscle to the choroid. These branches supply capillary beds within the ciliary muscle (particularly the meridional and radial portions; see Chapter 11) and contribute to the supply of the capillary beds in the anterior choroid. It is not clear if this arterial supply to the anterior choroid is an independent source or whether there are anastomoses with the posterior ciliary arteries. The region of the ciliary muscle supplied by branches from the intramuscular circle appears to include all but the innermost circular portion—that is, all but the part of the ciliary muscle adjacent to the ciliary processes.

In both schemes, branches from the intramuscular circle extend into the region of the major arterial circle. In the first case, there are relatively few of these branches, a condition that maintains the relative independence of the two arterial circles. In the second case, however, there are many more such branches, and after reaching the region of the major arterial circle, they turn to join and contribute to it. In some places, anastomoses connect these branches of the intramuscular circle and the branches of the long posterior ciliary artery.

One set of branches from the major arterial circle goes into the iris and runs radially toward the pupil. These branches join with similar branches from the intramuscular circle to form the minor arterial circle. Since all the arteries examined in the iris have tightly constructed, barrier-type endothelium, this specialization must apply to the branches from the major arterial circle as well as those from the intramuscular circle.

Many branches from the major arterial circle enter the ciliary processes (of which there are about 80) and branch extensively to form dense aggregations of small arterioles, capillaries, and venules (Figure 6.17). In fact, the ciliary processes are essentially masses of intertwining blood vessels covered by two layers of epithelium. Or one might say that the vascular anatomy is like an arterial fountain spraying blood against the epithelium. The epithelial layers that separate blood from the aqueous-filled posterior chamber are largely responsible for the transformation from blood to aqueous humor (see Chapter 11).

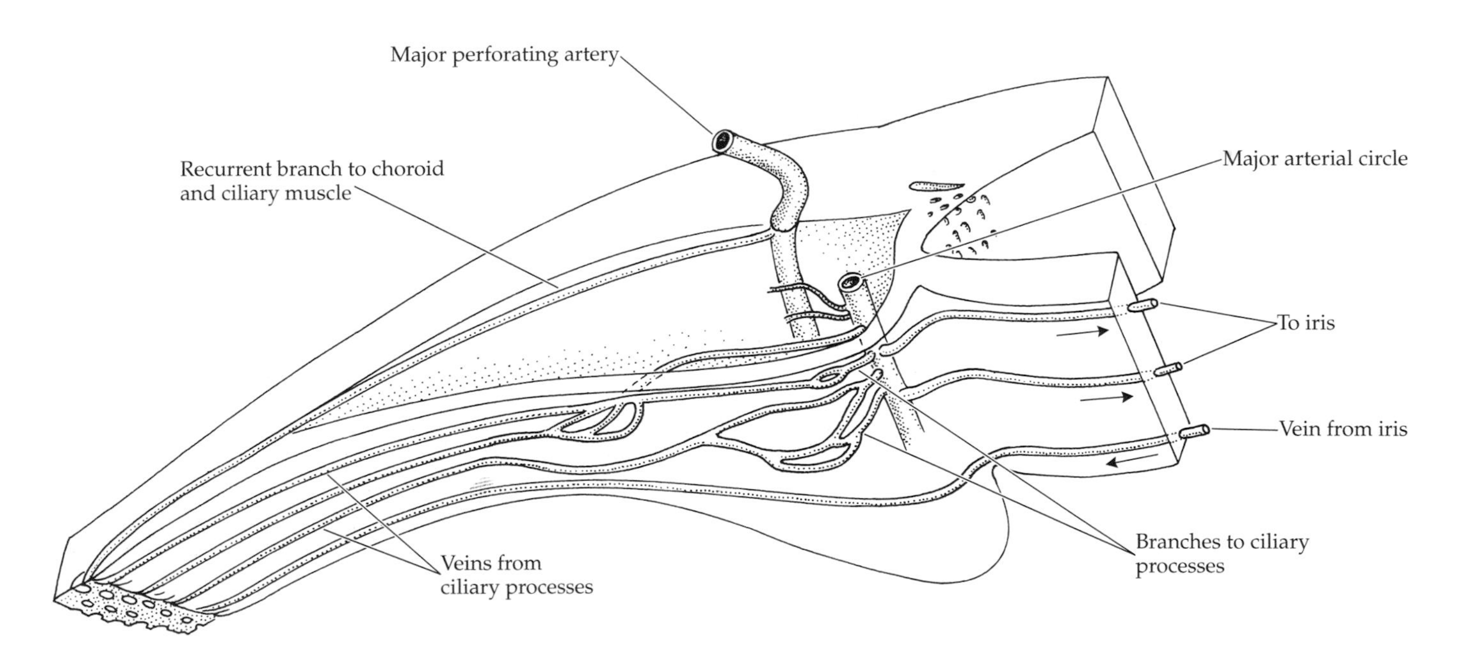

Figure 6.17
Blood Supply to the Ciliary Processes by the Major Arterial Circle
The capillaries in the ciliary processes are supplied by branches from the major arterial circle. These capillary beds are drained by veins that join those from the iris that run back to the choroid and the vortex veins. (After Funk and Rohen 1990.)

Castings (a technique discussed in Box 6.1) of the vessels in the ciliary processes show several distinct clusters of capillaries. It has been suggested that these capillary clusters are functionally different, since they are supplied by separate sets of arterioles whose responses to vasoactive substances may differ. Thus, if blood flow through the ciliary processes is related to the rate of aqueous production, pharmacologic alteration of the precapillary arterioles might offer a way to alter the rate of aqueous outflow. At present, however, the significance of the capillary clustering has not been established.

The capillaries in the ciliary processes are structurally different from those fed by other branches from the major arterial circle and different from any other capillaries in the anterior segment of the eye. They are large in diameter, often larger than the arterioles that supply them, and thus have an increased surface area for diffusion. Moreover, the capillaries have closed fenestrations, increasing their permeability. These characteristics are consistent with the process of aqueous formation from the capillary blood.

The blood in the ciliary processes is collected by large venules that run directly back along the inner surface of the ciliary body in parallel with veins coming from the iris to the choroid, where they join the system of choroidal veins going to the vortex veins (see Figure 6.17).

Blood supply to the anterior segment is redundant, but there is some segmentation in supply to the iris

The anterior segment appears to be well protected against any interruption of blood supply, but this invulnerability depends critically on the completeness of the arterial circles and the degree to which their fields of supply overlap. As indicated earlier, however, neither the intramuscular circle nor the major arterial circle is complete. One indication of this incompleteness is the persistent reporting of reduced blood flow (ischemia) in parts of the iris following strabismus surgery on one of the vertical rectus muscles in which one or more anterior ciliary arteries was tied off.

Figure 6.18 offers an explanation for this observation. Branches from both the intramuscular and the major arterial circles supply the iris, and any gaps in supply from one system are normally filled by the other. Two places are likely to lack a dual supply, however; these are parts of the superior and inferior quadrants into which branches from the long posterior ciliary arteries do not extend. Here the supply is solely from the intramuscular circle, which is part of the anterior

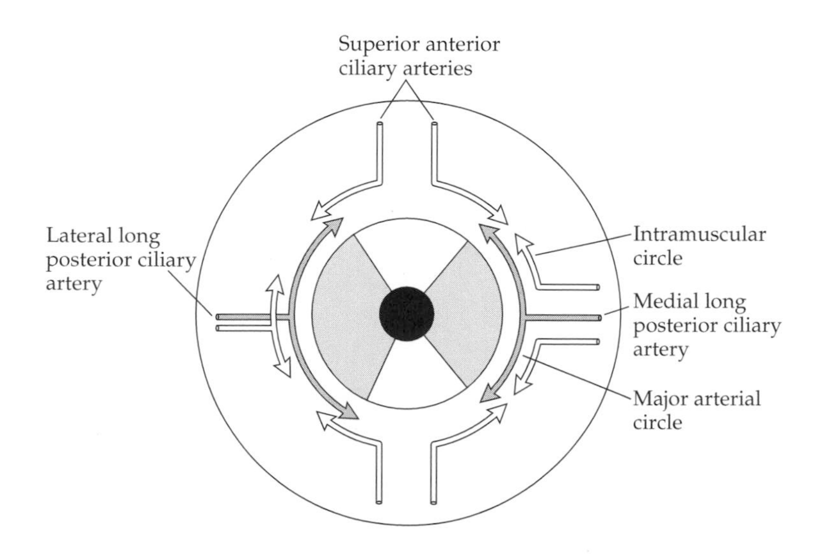

Figure 6.18
Overlap and Segmentation in the Blood Supply to the Iris
Because the major arterial circle is incomplete superiorly and inferiorly, the long posterior ciliary arteries contribute little to the blood supply to the superior and inferior segments of the iris (unshaded regions). These portions of the iris may experience reduced blood supply following surgery to the superior and inferior rectus muscles that affects their anterior ciliary arteries. The medial and lateral sectors of the iris supplied by branches from the major arterial circle (shaded regions) have a dual blood supply because of branches into the iris from the intramuscular circle. The two sources of blood supply do not completely overlap in the lateral sector, however. (After Olver and Lee 1989.)

ciliary artery system. It follows that tying off the superior or inferior anterior ciliary arteries will reduce blood flow to the superior and inferior sectors of the iris.

The real situation is not as clear-cut as the schematic in Figure 6.18 implies. Individual variation means that tying off anterior ciliary arteries will compromise circulation less in some cases than in others. This variation suggests, however, that surgery on the medial and lateral recti will rarely cause a problem, while surgery on the vertical recti runs a higher risk of compromising circulation.

The short posterior ciliary arteries terminate in the choriocapillaris, which supplies the retinal photoreceptors

After the short posterior ciliary arteries pass through the sclera and enter the choroid, they branch extensively, and almost all the arteries in the choroid posterior to the equator are dependent branches from the short posterior ciliary arteries (as already mentioned, the possible exceptions are recurrent branches from the lateral long posterior ciliary artery on the temporal side). Branches from the long posterior ciliary arteries and recurrent branches from the intramuscular circle can be found near the ciliary body.

Branches from each of the short posterior ciliary arteries are distributed to a different portion of the choroid, as Figure 6.19 indicates. In general, this arrangement constitutes segmentation in the arterial supply, but it is not clear if the segment dividing lines are firm or if the borders of the segments overlap. There are shunts among dependent arterial branches within a segment, but it is not known if the main short posterior ciliary arteries have anastomoses. Long-lasting reductions of blood flow in one of the segments following experimental blockage of a single artery suggests that high-level shunts are rare or nonexistent.

Figure 6.20 shows a small portion of the choroidal vasculature, as viewed looking inward toward the center of the eye. This picture shows a vascular casting (see Box 6.1), in which arterioles and venules can be distinguished by different impressions left in the plastic by the arterial and venous endothelia. Most of the vessels shown are venules, but arterioles are also visible. The capillary bed

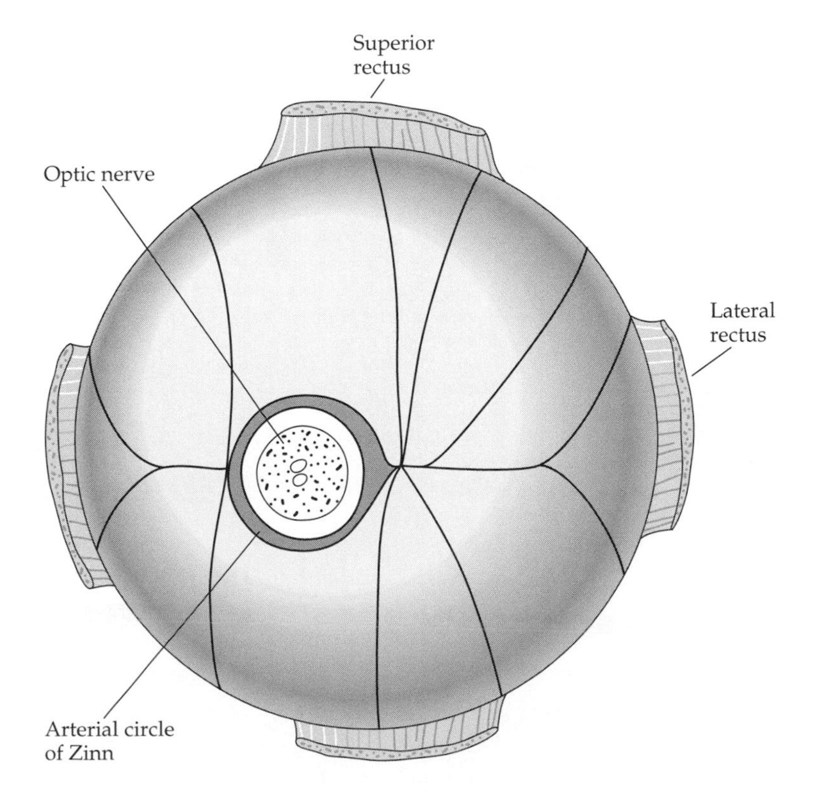

Figure 6.19
Segmentation of Supply from the Short Posterior Ciliary Arteries to the Choroid
Each of the short posterior ciliary arteries supplies a portion of the choroid with little or no overlap between neighboring regions supplied by different arteries. Eleven such regions are shown here as they map onto the back of the eye. Several regions abut lateral to the optic nerve, suggesting that the part of the choroid supplying the foveal photoreceptors may be relatively well protected against disruption of its blood supply. (After Olver 1990.)

[Box 6.1] Tracing Hidden Blood Vessels: Vascular Casting

STUDYING THE VASCULAR SYSTEM in any tissue or organ is difficult because the small vessels are embedded in the tissue; they cannot be seen directly, and exposing them by dissection may not be possible. This is particularly true of the eye, in which the vessels are small and numerous, ramifying in a dense, bewildering maze. Moreover, the locations of blood vessels and the patterns of supply and drainage are highly variable, much more so than nerves and innervational patterns, leading to sometimes contradictory descriptions of the eye's blood vessels. The range of variation has yet to be fully documented and consensus on some issues has yet to be achieved, but hard evidence—literally—about the organization of the blood vessels within the eye is now available.

The remarkable object in Figure 1 is a vascular casting. It is a complete replica of all the blood vessels in the eye, including the capillary beds. The casting was made by injecting a plastic resin and its polymerizing agent into a few of the larger arteries that supply the eye until the resin appeared in the veins. After the resin hardened, the eye was immersed in a caustic solution (potassium hydroxide) for 1 to 2 days to destroy all the soft tissue. Finally, the plastic was washed in distilled water to remove residual bits of tissue, leaving only the casting of all the blood vessels perfused by the resin.

Injecting fluids that will solidify into the blood vessels is not a new idea; in principle, it can be traced to Leonardo da Vinci's castings of the cerebral ventricles in wax. For small, delicate structures like blood vessels, however, wax is too fragile to be a practical casting material, and until recently, the material most commonly used was latex, often colored—red for arteries and blue for veins. The few pictures we have that show the complete structure of canal of Schlemm in the limbus are latex castings made by Ashton (see Chapter 9). Latex castings lack rigidity, however, so spatial relationships among blood vessels are not well retained, and latex does not easily penetrate the all-important capillary beds. Usually, the latex has to be injected at a pressure that ruptures the delicate capillary walls.

The plastics now in use are initially fluid enough to be injected at pressures within a normal physiological range (less than 175 mm of mercury) and to flow through the capillaries without rupturing them. The hardening of the plastic is controlled by temperature and by concentration of the polymerizing agent mixed with the resin.

Vascular castings show the paths and relationships of all the blood vessels in the eye. The ramification of major vessels or the anastomotic arterial circles within the eye can be traced with low magnification and microdissection, as discussed in the text. But the wealth of detail in the castings is best seen by sputtering them with gold so that they are electron dense and suitable for scanning electron microscopy. The castings retain microscopic differences between the endothelial cells in arterioles and venules, and these differences serve as signatures, making it possible to determine whether a particular piece of the casting is from the lumen of an artery or a vein (Figure 2). For this reason, the capillary beds and their patterns of supply and drainage can be examined at the level of the smallest vessels without ambiguity.

It would be nice to think that a vascular casting accurately represented *the* ocular vasculature, complete and definitive, but as usual, nothing is that simple. For example, how do we know a casting is complete? There is no standard against which it can be com-

Figure 1
A Vascular Casting of the Human Eye
This vascular casting of the left eye is seen from below before any gold sputtering for electron microscopy; the lateral rectus is uppermost and the lacrimal gland is in the upper right corner. The photo reveals that a substantial part of the eye's tissues is occupied by myriad small blood vessels. (From Olver and McCartney 1989.)

(a) Artery

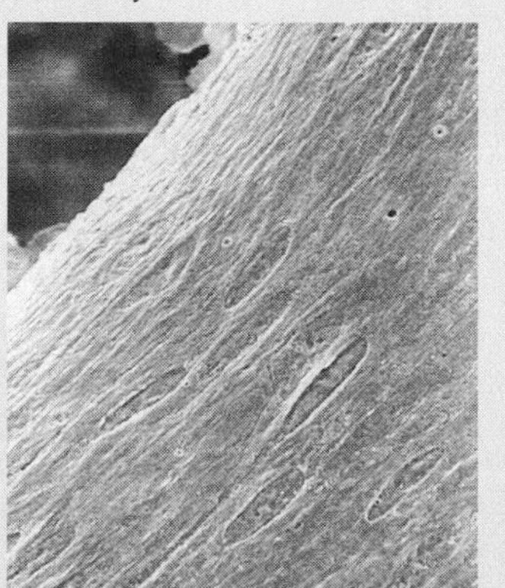

(b) Vein

Figure 2
Signatures of Arterial and Venous Endothelium
An arterial casting (*a*) has elongated furrows left by the endothelial nuclei in the arterial wall. The appearance is quite different from the rounded nuclear impressions in a venous casting (*b*). (From Olver and McCartney 1989.)

pared. Completeness can be judged only from multiple castings that exhibit the same level of detail and seem to be similar. And even when the casting seems to be complete, questions have been raised about the accuracy with which the casting represents the *functional* properties of the system. The latter point addresses the possibility that the resin injection pressure forces open arterial branches that are normally closed in vivo by sphincters, thus making the vascular anatomy appear more anastomotic and interconnected than the normal pattern of blood flow. (But since the normal pattern of blood flow is imperfectly understood, it is difficult to specify features of the vascular castings that are clearly abnormal.) Finally, there is the variability problem: It becomes not so much a matter of specifying the true pattern of vascularization, but of recognizing the most common or representative pattern and delimiting the range of variation.

For these and other reasons, a definitive description of any part of the eye's internal vasculature requires that numerous castings be examined in detail, preferably by several investigators. The final story will emerge as results accumulate over time. This process is under way, but it is not yet complete, as indicated in the text. Thanks to the vascular casting method, we have some exquisite representations of the vasculature of individual eyes, but we have yet to reach consensus on all the details within the vasculature of the "typical" human eye.

that is the target of the arterioles and the source of the blood in the veins is in the background, partially obscured by the overlying large vessels; known as the **choriocapillaris**, this capillary bed lines the entire inner surface of the choroid. Although the choriocapillaris is separated from the photoreceptors by an intervening membrane (Bruch's membrane; see Chapter 11) and the retinal pigment epithelium, the distance is less than 20 μm, and the photoreceptors depend on the choriocapillaris for supply. Details of this anatomical interface between photoreceptors and choriocapillaris are discussed in Chapters 11 and 13.

The choriocapillaris is a single layer of capillaries that are interconnected to form a completely anastomotic system in which there are potentially many routes along which blood could flow between any two points in the capillary bed

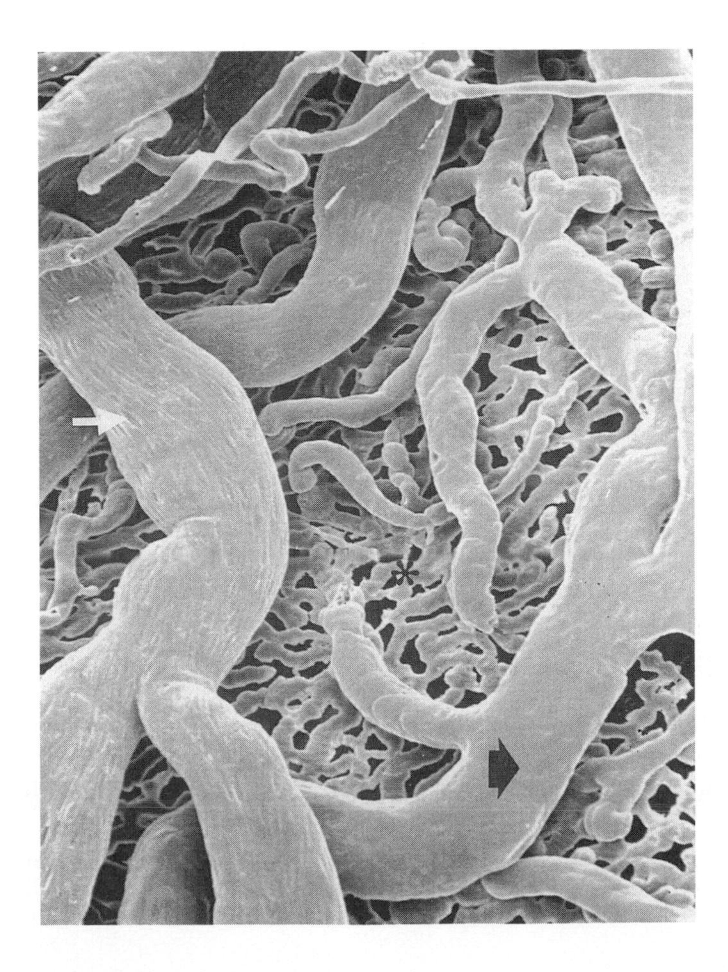

Figure 6.20
The Choroidal Blood Vessels
This vascular casting of the choroid, as viewed from the scleral side by a scanning electron microscope, shows it to consist of intertwined arterioles (e.g., white arrow) and venules (e.g., black arrow) that supply and drain the choriocapillaris (asterisk). (From Olver 1990.)

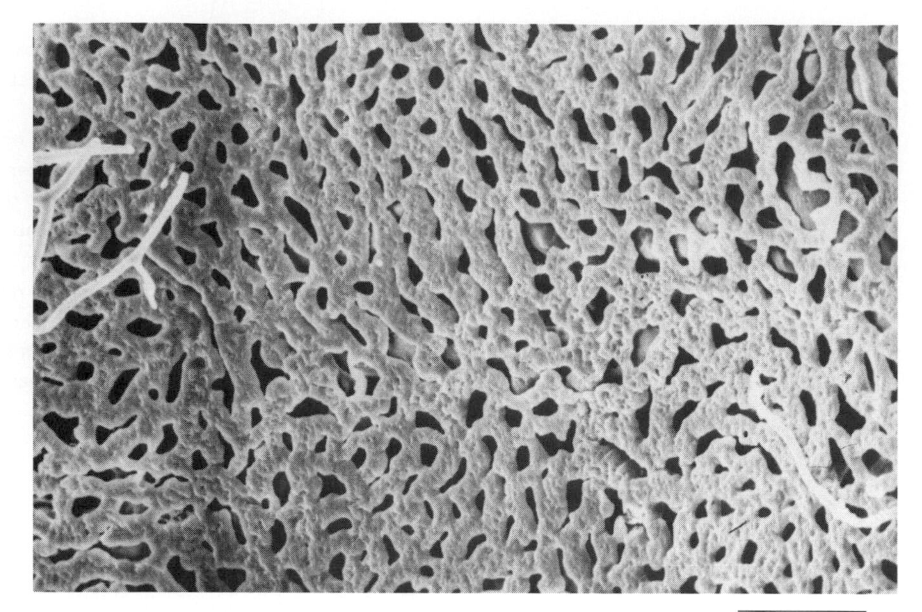

Figure 6.21
The Choriocapillaris
Viewed from the retinal side, the choriocapillaris is a dense meshwork of capillaries that are extensively anastomotic. The small vessels at the left belong to the retinal circulation. (From Olver 1990.)

(Figure 6.21). These capillaries, like those in the ciliary processes, are large and have closed fenestrations.

Studies of the choroidal circulation with fluorescein angiography (see Box 16.1) indicate that the choroidal circulation is segmented; blockage of supply to a region cannot be made up by blood from other sources. But the anatomy of the choriocapillaris shows no evidence of segmentation; if an artery supplying one part of the choriocapillaris were blocked, it would seem that blood could flow into the deprived region from surrounding parts of the choriocapillaris.

Although blood *could* flow along many paths through the choriocapillaris, flow must be from points of high pressure to low pressure; without a pressure differential, there would be no flow. The highest pressure points are the arterioles supplying the choriocapillaris; the lowest are the venules draining it. Thus blood flows through the choriocapillaris from arterioles to venules; the number and locations of arterioles and venules are critical in establishing the pattern of flow.

In regions of the choriocapillaris where venules outnumber arterioles, the system behaves *as if* it were segmented and unable to compensate for arterial blockage. Where arterioles outnumber venules, the system does not exhibit segmentation; arterial blockages are easily overcome, but venous blockages are not, leading to edema in the tissue. Because the relative number of arterioles and venules varies across the choriocapillaris, different regions have different patterns of blood flow (see Chapter 11); arterial blockage and loss of blood supply are the problem in some regions, while venous obstruction and edema are the concern in others.

The short posterior ciliary arteries contribute to the supply of the optic nerve through the circle of Zinn

Some of the short posterior ciliary arteries closest to the optic nerve where it exits the eye send small branches into the sclera around the nerve, where they branch and join to form the arterial **circle of Zinn*** (Figure 6.22). The completeness and

*The historically accurate name is the circle of Haller and Zinn. More recently, the neutral term "perioptic nerve arteriolar anastomoses (PONAA)" has been introduced. By using the term "circle of Zinn," I have ignored accuracy and neutrality in favor of simplicity.

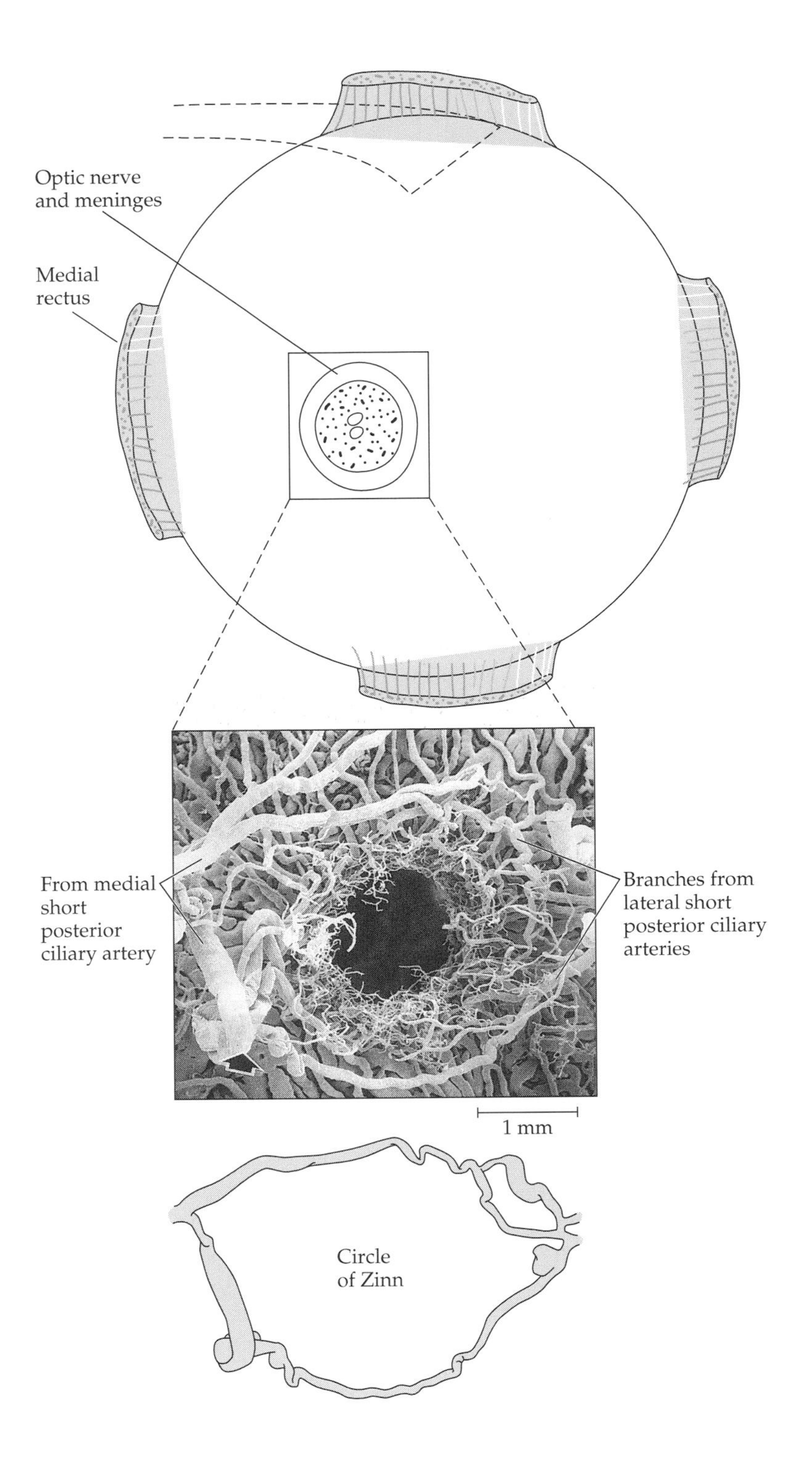

Figure 6.22
The Circle of Zinn
The arterial circle of Zinn lies in the sclera around the optic nerve. It is shown here as a vascular cast. Branches from lateral and medial short posterior ciliary arteries form anastomoses above and below the nerve head. (Photograph from Olver, Spalton, and McCartney 1990.)

even the existence of the circle of Zinn has been questioned many times, but recent vascular castings indicate that it is always present, and it is complete in three out of four eyes.

Many small arterioles branch off the circle of Zinn, some entering the region where the optic nerve penetrates the sclera (the **lamina cribrosa**; see also Chapter 16). Others enter the optic nerve head in front of the lamina cribrosa; still others run posteriorly along the outside of the optic nerve to join the plexus of small vessels in the pia mater around the nerve (Figure 6.23). The plexus in the pia mater is also derived from the short posterior ciliary arteries, but by branches running directly to the pial plexus without first going to the circle of Zinn.

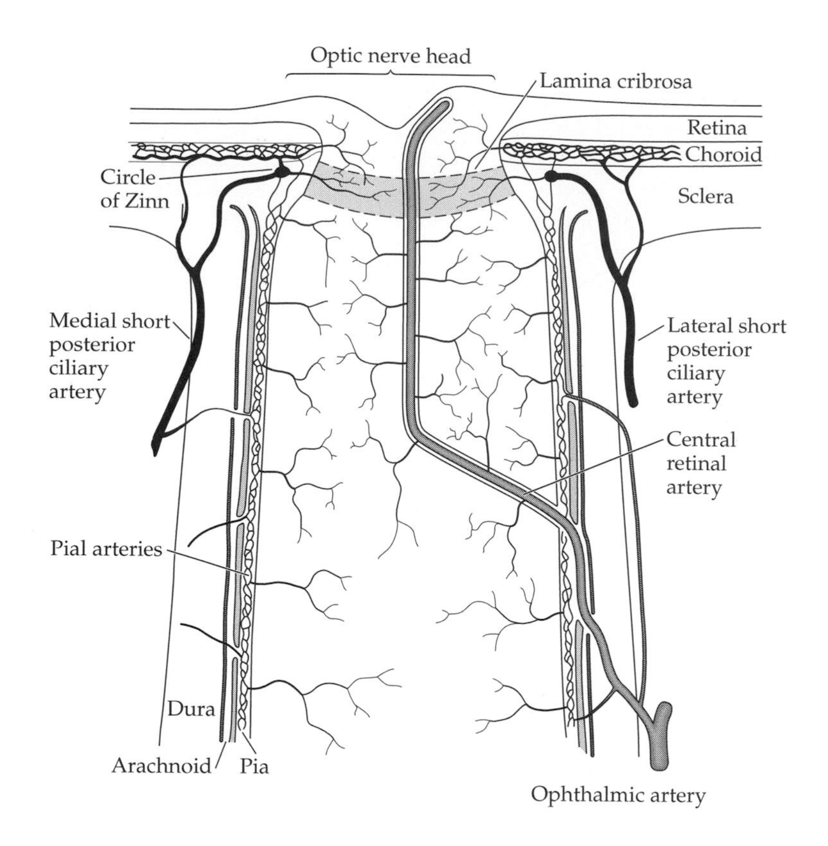

Figure 6.23
Blood Supply to the Optic Nerve and Nerve Head
The central core of the optic nerve is supplied and drained by the central retinal artery and vein, respectively. The surrounding portions of the nerve and nerve head are supplied by the short posterior ciliary arteries and their dependent branches. (After Hogan, Alvarado, and Weddell 1971.)

Branches from the pial plexus supply the outer part of the optic nerve (the core of the optic nerve is supplied by the central retinal artery).

Unlike the arteries supplying the choroid, the branches of the short posterior ciliary arteries that supply the optic nerve and nerve head are anastomotic at various levels of the branching hierarchy. The circle of Zinn is formed by anastomoses, and there are other anastomoses among the arterioles arising from the circle. Although the distribution of the arterioles to the capillary beds may be segmental—evidence on this point is unclear—the numerous shunts in the system should keep all parts of the nerve head supplied when one or a few arterial branches are blocked. (This arterial supply has been described as end-arterial and segmented in the past, and these features have been used to explain certain clinically observable defects, but the supply is *not* end-arterial.)

The superficial part of the optic nerve head adjacent to the vitreous chamber is supplied by small branches from the central retinal artery, but most of the nerve head and the optic nerve just behind the eye are supplied exclusively by the short posterior ciliary artery system through the circle of Zinn and branches from the pial plexus. More posteriorly, dual supply comes from the pial arteries and branches from the central retinal artery. The two arterial systems are anastomotic in this region; there are a few direct shunts between the pial plexus and the central retinal artery, and capillary beds within the nerve are often supplied by branches from both systems. A branch from the central retinal artery to the pial plexus as the artery enters the nerve has also been reported as common.

The short posterior ciliary arteries sometimes contribute to the supply of the inner retina

The short posterior ciliary arteries generally do not contribute to the blood supply of the inner retina. In some individuals, however, one or several **cilioretinal arteries** may come from the circle of Zinn or the choroidal circulation around the nerve head. The cilioretinal arteries run forward to the surface of the nerve head, where they mingle with branches of the central retinal artery on the retinal surface. Cilioretinal arteries usually run temporally from the nerve head, supplying the part of the retina between the fovea and the nerve head. This supply appears to supplement the normal supply from the central retinal artery, providing an alternative route for blood to reach this part of the retina and making it less vulnerable to blockage in the central retinal artery system. For individuals who have them (estimated at 25 to 30% of all eyes), the cilioretinal arteries are a bonus, a kind of vascular insurance policy.

The central retinal artery enters the eye through the optic nerve and ramifies to supply the inner retina

The **central retinal artery** arises from the ophthalmic artery just inside the orbit where the ophthalmic artery runs below the optic nerve. It enters the optic nerve by passing through the nerve's meningeal sheaths (dura mater, arachnoid, and pia mater) and then runs directly forward in the center of the nerve to the center of the optic nerve head within the eye (see Figure 6.23).

As the central retinal artery emerges within the eye at the optic nerve head, it divides into large superior and inferior branches, from which dependent branches arise that radiate across the retinal surface (Figure 6.24). Arterial branches run radially from the nasal side of the nerve head, but the temporal branches arc above and below the fovea, eventually radiating in toward the fovea from all directions. In general, the paths followed by the major arterial branches mimic the paths followed by ganglion cell axons converging on the optic nerve head (see Chapter 16). The branches of the central retinal vein tend to parallel the arterial branches.

The blood vessels running over the retinal surface send small branches into the retina to supply capillary beds at four levels in the thicker parts of the retina (Figure 6.25); toward the periphery, where the retina is thinner, there may be fewer, less well defined levels of capillaries. The retina has three cellular layers (see Chapter 13), the outermost of which is associated with the photoreceptors. The capillary beds line the borders of the middle and inner cell layers (the inner nuclear and ganglion cell layers, respectively). Thus the deepest of the capillary beds is near the axon terminals of the photoreceptors, but still rather distant from metabolic and photoresponsive portions of the receptors. These critical parts of the photoreceptors lie closer to the choriocapillaris. In short, the central retinal artery supplies the inner retina, meaning everything *except* the photoreceptors. And since the inner layers of the retina are absent at the center of the fovea (see Chapter 16), there are no capillaries in the foveal center (see Figure 6.25), a fact that emphasizes the critical dependence of the photoreceptors on the choroidal circulation.

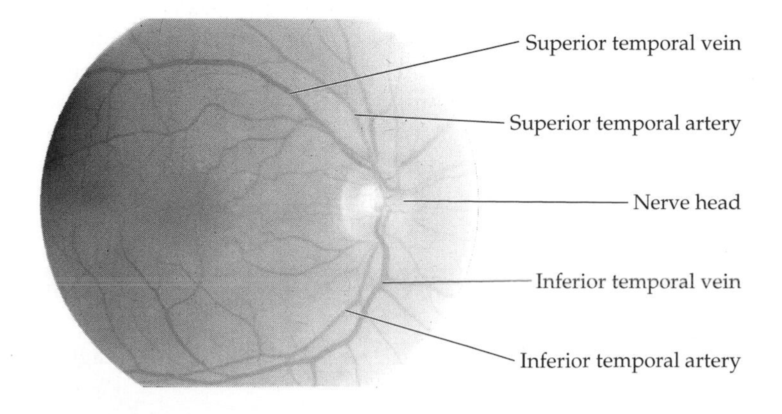

Figure 6.24
Branches of the Central Retinal Artery and Vein
This ophthalmoscopic view of the back of the eye shows the optic nerve head, from which branches of the central retinal artery and vein run over the surface of the retina. The fovea lies roughly at the center; it cannot be seen, but the dependent arterioles and venules radiate in toward it from all sides. Veins are typically larger in diameter than arteries.

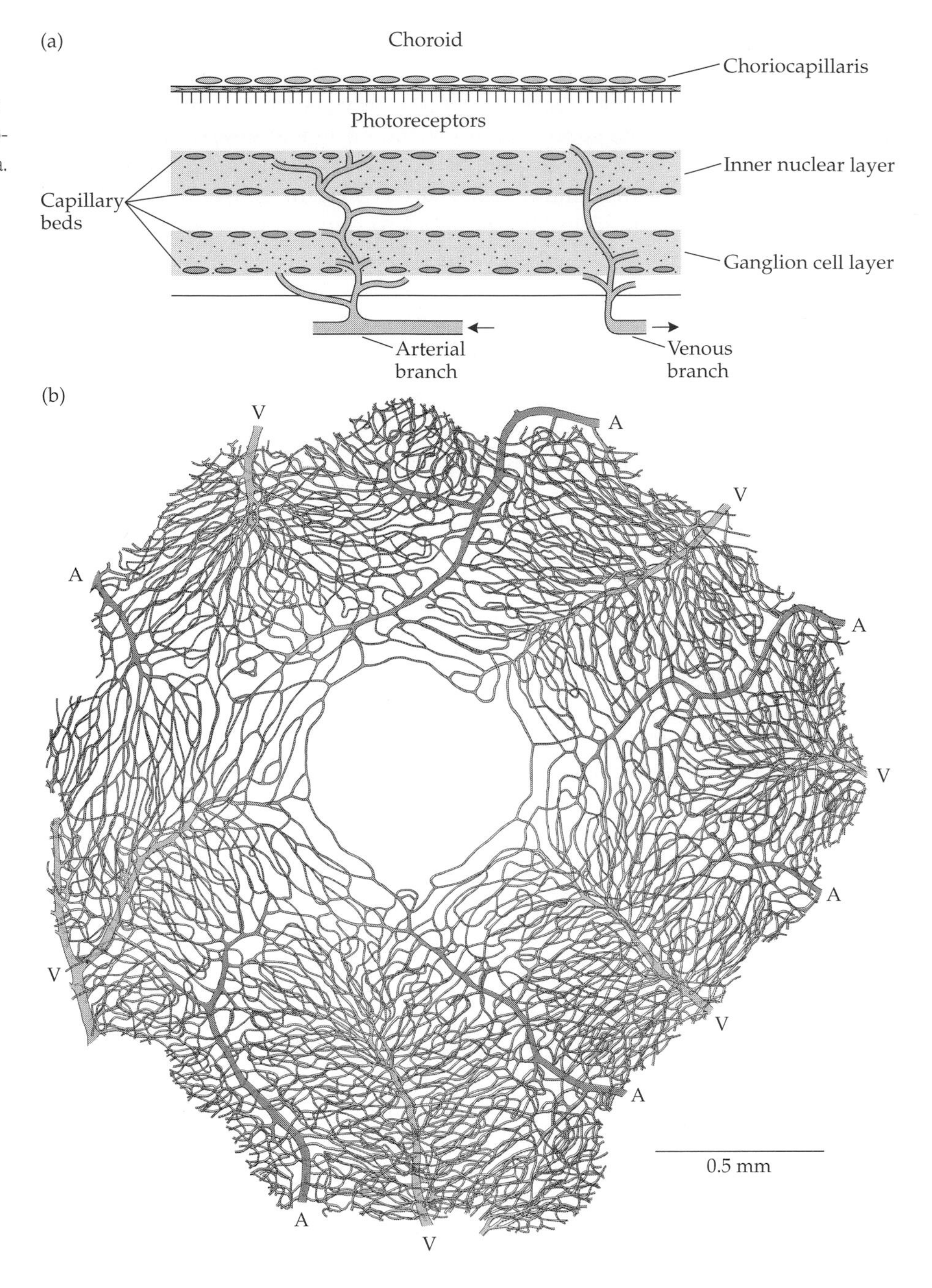

Figure 6.25
Capillary Beds in the Inner Retina
(*a*) Branches from the arteries and veins on the retinal surface run into the retina to supply and drain capillary beds within the retina. In the central retina, layers of capillaries are located on either side of the ganglion cell layer and of the inner nuclear layer. (*b*) Capillaries are absent in the central part of the fovea. A, artery; V, vein. (*b* courtesy of R. W. Rodieck, after Snodderly, Weinhaus, and Choi 1992.)

Although the retina has a dual blood supply, neither supply can make up for the loss of the other. Photoreceptors will survive occlusion of the central retinal artery, but the neurons of the inner retina will degenerate. Conversely, inner retinal neurons will survive the loss of the choroidal supply but photoreceptors will not. In either case, however, the perceptual result is blindness in the part of the visual field corresponding to the damaged retina.

There are no shunts anywhere in the branching hierarchy of the central retinal artery, and there is no overlap in the regions supplied by the terminal arterioles; it is a segmented, end-arterial system. The same pattern is mirrored in the drainage by the central retinal vein. As a result, the retinal circulation is vulnerable to both arterial and venous branch occlusions (see Chapter 16), since neither can be compensated for by alternative routes of supply or drainage. And, for the reasons mentioned earlier, regional defects in supply or drainage can trigger detrimental neovascularization.

The endothelium in the retinal vessels is of the barrier type, similar to blood vessels in the brain, and one can justifiably speak of a blood–retina barrier. In this respect, the inner retinal circulation is very different from the supply to the photoreceptors by the choriocapillaris, where the capillaries are highly permeable. For the outer retina, the blood–retina barrier is established by the retinal pigment epithelium (see Chapter 13). Another difference concerns the innervation of the smooth muscle in the arterioles. There is no innervation to the retinal vessels, and control of blood flow appears to be entirely local. The choroidal vessels have sympathetic (and possibly parasympathetic) innervation that is used for compensatory regulation from the central nervous system (see Chapter 5).

The central retinal vein exits the eye through the optic nerve

The retinal capillary beds are drained by small venules that merge to form larger veins running parallel to the branches of the central retinal artery (see Figures 6.24 and 6.25). The veins join at the optic nerve head as a single vessel—the **central retinal vein**—that runs back in the optic nerve in company with the artery (see Figure 6.23). Normally the central retinal vein emerges from the nerve along with the artery and then joins the superior ophthalmic vein; alternatively, it may run through the superior orbital fissure to connect directly with the cavernous sinus, but a shunt to the superior ophthalmic vein is normally retained.

Since venous pressure is normally much lower than arterial pressure, the central retinal vein and its branches are sensitive to pressure exerted by surrounding tissues and fluids; if the surrounding pressure exceeds venous pressure, the vein will collapse and normal venous return will be obstructed. Abnormally high pressure in the arteries may obstruct veins as the arteries cross them, and the branches of the central retinal vein are also exposed to the intraocular pressure. The normal range of intraocular pressures is no problem, but higher intraocular pressures can reduce venous return and, for values above 40 mm or so, can produce venous collapse at locations where the veins are particularly vulnerable (e.g., at the optic nerve head where the veins bend at right angles to run back into the nerve).

The central retinal vein is also vulnerable at the point where it leaves the optic nerve. The nerve is surrounded by meninges, and the vein is exposed to cerebrospinal fluid in the arachnoid space as it passes through. If cerebrospinal fluid pressure exceeds central retinal vein pressure, the vein will collapse, thereby impeding or totally preventing the exit of blood from the retina. The resultant venous congestion and edema make the optic nerve head bulge into the vitreous chamber; this is **papilledema**. Elevated cerebrospinal fluid pressure is only one possible cause of papilledema, however, and other abnormalities produce bulging of the nerve head without edema.

The vortex veins drain most of the uveal tract

There are usually four **vortex veins**, one for each posterior quadrant of the eye (Figure 6.26), but the number varies, ranging from three to six. The vortex veins are more or less the venous complement of the posterior ciliary artery system; they drain the iris, part of the ciliary body, and all of the choriocapillaris. The numerous dependent branches of the vortex veins are not named, and one may think of a vortex vein as simply a large river of blood that is being fed by a tributary system wherein the smallest venules draining the capillary beds join to form larger venules, which join to form even larger venules, and so on, until the final confluence just before the vortex vein exits the eye through the sclera. Normally, the two vortex veins on the upper side of the eye empty into the superior ophthalmic vein, while the two lower vortex veins go to the inferior ophthalmic vein.

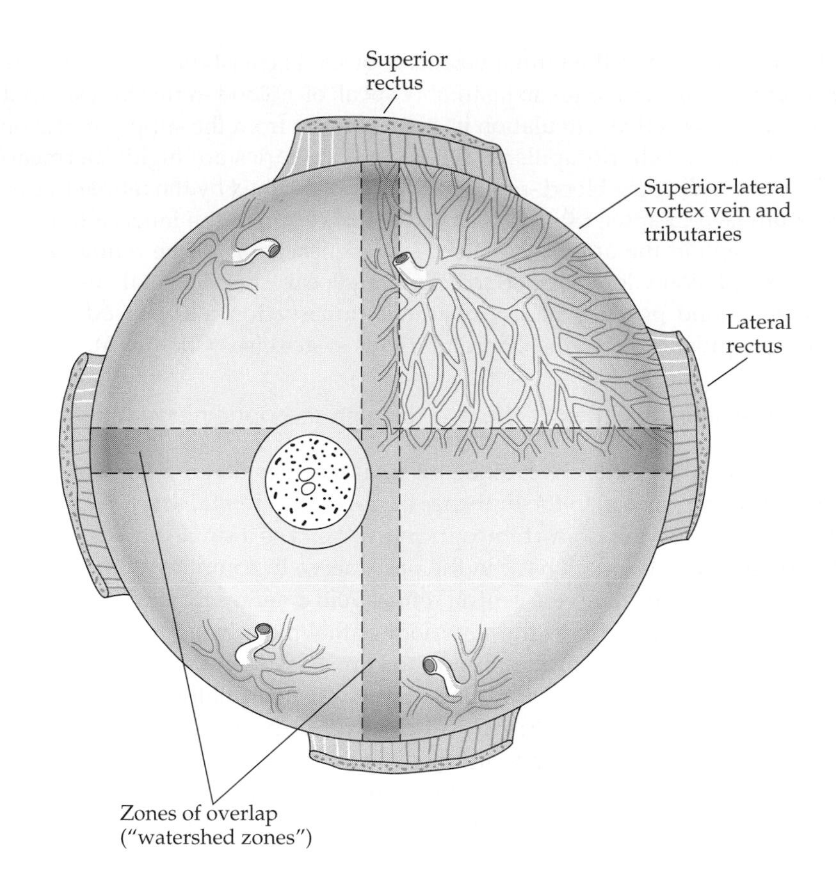

Figure 6.26
The Vortex Veins
In the typical case, in which there are four vortex veins, each one drains a quadrant of the choroid. What have been called watershed zones (shaded) divide the choroid vertically and horizontally to delimit the drainage quadrants. Branches from neighboring vortex veins overlap or form anastomoses within these zones. (After Hayreh 1990.)

In the iris, small veins run radially outward from the region of the sphincter muscle toward the root of the iris, collecting blood from small venules as they go; when they reach the ciliary body, the iris veins continue posteriorly through the ciliary body to the choroid (see Figure 6.17), where they merge with the choroidal veins draining into the vortex veins. The iris seems to have little or no drainage via the anterior ciliary venous system.

The venous drainage of the capillary beds in the ciliary processes also goes through the vortex veins. The capillary beds feed into venules running directly to the choroid, where they become tributaries of the vortex veins (see Figure 6.17). Drainage of blood from the muscular portion of the ciliary body probably depends on location, with the posterior part of the muscle draining into the vortex vein system and the anterior part, which is the site of the **ciliary plexus** (see Figure 6.14), draining into the anterior ciliary vein system.

Finally, all of the small venules that drain the choriocapillaris are part of the vortex vein system. Around the optic nerve, small veins drain the pial plexus and the outermost parts of the lamina cribrosa and optic nerve head. These veins usually merge with the veins in the choroid, thus becoming part of the vortex vein system, but they may merge to form small posterior ciliary veins that go to the ophthalmic veins directly. The central portions of the nerve head and the nerve proper are drained by small branches of the central retinal vein.

The vortex veins have segmented drainage fields, but they are heavily anastomotic

Each vortex vein and its tributaries has been described as draining a part of the choroid with little or no overlap between adjacent drainage fields. The borders between the drainage fields have been called watersheds and depicted as vertical and horizontal strips intersecting near the posterior pole (see Figure 6.26).

Thus, the temporal quadrant above the horizontal plane is drained by the superior-temporal vortex vein, while the inferior-temporal quadrant is drained by the inferior-temporal vortex vein. (Assuming the normal configuration of four vortex veins.)

If this description is correct, the watersheds should contain a line (in the ideal case) above which postcapillary venules run upward to a superior vortex vein, and below which they run downward to an inferior vortex vein. That is, the vortex vein system is segmented, and drainage fields for the different vortex veins do not overlap.

But as the labeling of the watersheds on Figure 6.26 implies, this interpretation is probably not appropriate. The tributary branches of individual vortex veins are heavily anastomotic at all levels of the branching hierarchy, and shunts are likely to occur within the watershed zones. These facts turn the original notion of a watershed upside down; rather than being regions that could be markedly affected by venous blockage, they are better thought of as regions having the *least* vulnerability because there are alternative drainage paths.

Fluorescein angiography (see Box 16.1) indicates that experimental occlusion of one vortex vein (in monkey) produces a deficit in drainage, as indicated by persistent fluorescence in the affected quadrant after it departed the unaffected parts of the choroidal circulation. But within a matter of hours, a subsequent test shows little or no deficit—that is, no persistent fluorescence. The restoration of the normal pattern of flow implies the existence of alternative drainage routes that can compensate for the initial disruption of choroidal drainage. The most likely alternative routes are shunts between the different vortex vein drainage fields.

The presence of shunts at all levels of the branching hierarchy means that each vortex vein operates as a single functional unit instead of being a collection of independent subunits. Shunts between the different vortex veins extend this idea to the entire choroid and other regions drained by the vortex veins; it is a very large functional unit whose interconnections make possible near-normal drainage in the face of blockages within the system.

Supply and Drainage of the Eyelids and Surrounding Tissues

The lacrimal gland is supplied by the lacrimal artery and drained by lacrimal veins

The **lacrimal artery** (see Figures 6.7 and 6.8) is one of the largest branches of the ophthalmic artery, usually arising from the ophthalmic artery before it passes below the optic nerve; it runs forward in the orbit along with the lacrimal nerve. The artery passes along the medial side of the lacrimal gland (i.e., between the gland and the eye) and supplies it through one or more small branches. These small arteries branch extensively with the lacrimal gland to form dense, widespread capillary beds in close association with the secretory lobules (see Chapter 7). After passing by the lacrimal gland, the lacrimal artery penetrates the orbital septum, whereupon it sends branches to the eyelids and continues to the skin of the face.

The **lacrimal veins** are small and variable in both number and location. They normally run above the eye as they pass from the lacrimal gland to the more medially located superior ophthalmic vein, but they may merge with one of the superior vortex veins instead.

The eyelids are supplied by branches of the lacrimal, ophthalmic, facial, and infraorbital arteries

Figure 6.27 shows the arterial supply to the eyelids. After the lacrimal artery passes by the lacrimal gland, it enters the upper eyelid and gives rise to the **lat-**

Vignette 6.2

Blood Vessels inside the Eye

THE MAJOR ARTERIES AND VEINS that supply and drain the tissues and organs of the body were discovered and described fairly accurately by Vesalius and his immediate successors (see Vignette 3.2), but the small vessels embedded within organs like the eye eluded them. These vessels are small, they are buried within the tissues and are not directly visible, and they deteriorate quickly in the absence of adequate preservation. These problems have not been fully solved, and important details about the arrangement of capillary beds are still being investigated.

Modern studies of vascular anatomy have their origin in techniques pioneered by a Dutch anatomist, Frederick Ruysch (1638–1731). Ruysch was an apothecary who became a physician and an anatomist, spending his entire career at the University of Leiden, where he had trained in medicine (he retired from his professorship, objecting all the way, at the age of 89!). His most important innovation was a method for tissue preservation that utilized the blood vessels to deliver the preserving fluid to the tissues. Since very few cells are far from a capillary, using the blood vessels to deliver a fixative is quite efficient, vastly superior to simply immersing a piece of tissue in fluid and hoping that diffusion will do the job. This perfusion technique is now standard practice.

Ruysch was unusually skillful. In a bizarre display of technical virtuosity, he made a collection of preserved animals, embryos, and human infants, all of which were said to look extremely lifelike. Peter the Great was so taken with the collection that he bought it, all 900 items, and shipped it back to Russia. Ruysch proceeded to do it all again, and after his death, the second collection of mummies ended up in Austria, where some of them can still be seen.

Preservation by perfusion works not only for tissues supplied by blood vessels but also for the blood vessels themselves; thus the vasculature within an organ can be studied by careful dissection, particularly when aided by the injection of a substance into the vessels to make them more visible; viscous ink, colored latex, and plastic have all been used. Good tissue preservation combined with meticulous dissections using low magnification allowed Ruysch to begin the description of the internal vasculature of the eye, particularly of the choroid.

Ruysch was also an ardent and apparently quite knowledgeable botanist—in later years he was supervisor of Leiden's botanical garden—and in one of those little coincidences of history, one of the next great contributors to understanding the vasculature of the eye was also a botanist, whose name and flower can be found in many summer gardens.

Johann Gottfried Zinn (1727–1759), for whom the zinnia was named, spent his brief adult life studying minutiae—stamens, pistils, and other flower parts, and small blood vessels embedded in the eye. At age 28, he published the first book devoted exclusively to the anatomy of the eye (*Descriptio Anatomica Oculi Humani Iconibus Illustrata*) and illustrated it with seven full-page plates, two of which are shown here as Figures 1 and 2. (All seven plates are reproduced in volume 4 of Hirschberg's *History of Ophthalmology*.) Zinn is remembered in anatomical terminology: The rectus muscles originate in the *common tendon of Zinn,* the arterial *circle of Zinn* lies in the sclera around the optic nerve, and the suspensory ligament of the lens was originally called the *zonule of Zinn.*

Zinn's illustrations are extremely good. The drawing at upper left in Figure 1 (his Fig. 2), for example, shows the long posterior ciliary arteries, their anastomoses to form the major arterial circle, and the branches running from the arterial circle into the iris; this is the first clear illustration of the major arterial circle. The drawing to the right (his Fig. 3) shows the minor arterial circle near the margin of the pupil. The two lower drawings are different views of the ophthalmic artery system in relation to the eye in situ. Unfortunately,

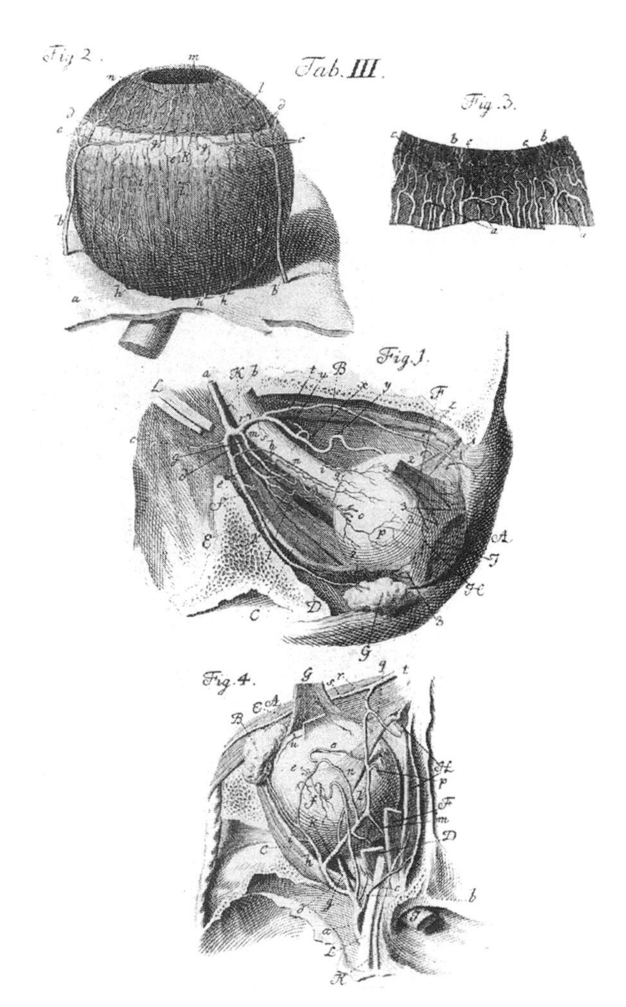

Figure 1
Arterial Supply to the Eye
The long ciliary arteries and their anastomoses to form the major arterial circle are shown at upper left (Fig. 2), the minor arterial circle at upper right (Fig. 3). The lower drawings (Fig. 1 and Fig. 4) illustrate the main branches of the ophthalmic artery within the orbital cavity. (From Zinn 1780.)

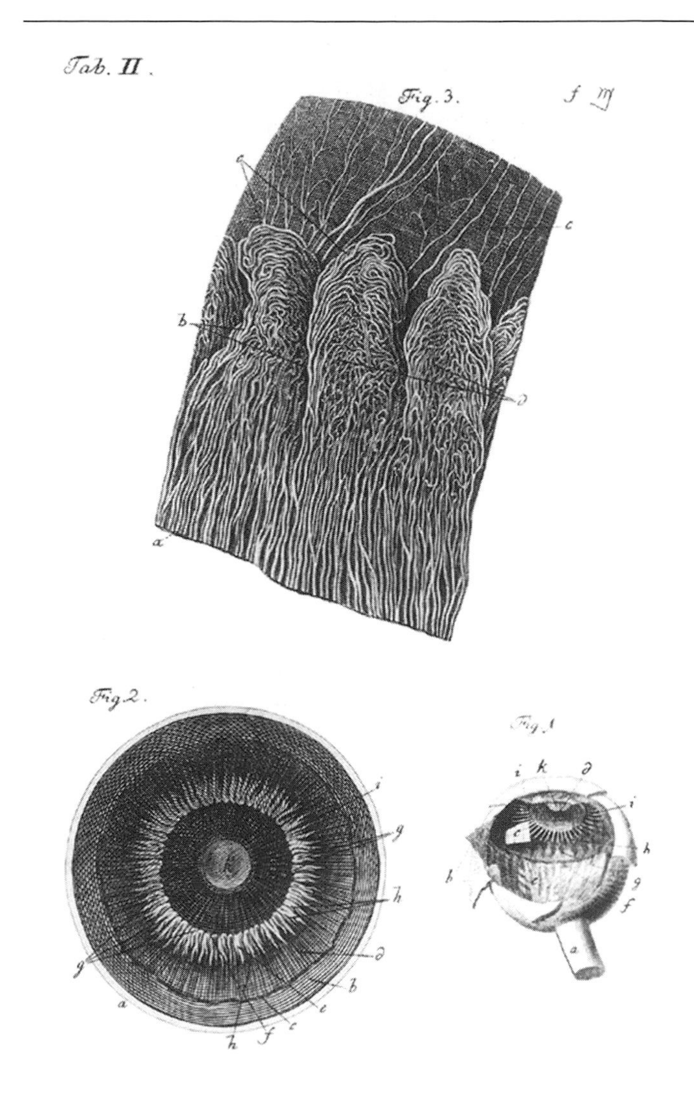

Figure 2
The Ciliary Body and Ciliary Processes
The view of the eye at lower right (Fig. 1) features a cutaway section through which the iris, lens, and some of the ciliary processes can be seen. The entire ring of ciliary processes, the corona ciliaris, is seen from behind at lower left (Fig. 2). The enlarged drawing at the top (Fig. 3) shows the ciliary processes with their epithelial covering removed, revealing the dense complexes of small blood vessels that form the processes. (From Zinn 1780.)

Zinn was a bit too enamored of the vascular system; he denied that the iris contained any musculature, explaining pupil constriction as an engorgement of the iris with blood and pupil dilation as a reduction of blood flow. Better histological techniques established the presence of the iris muscles some 50 years after Zinn's death.

In Figure 2, the two lower drawings illustrate the ring of ciliary processes from behind (Zinn's Fig. 2) and from outside (his Fig. 1). But Zinn reveals the most important feature of the ciliary processes in the larger drawing at the top—namely, the dense complexes of small blood vessels that give the ciliary processes their form and underlie their function. Versions of this illustration have appeared numerous times in subsequent publications by other investigators, with increasing refinement resulting from better injection techniques and increasing magnification for dissection, but the essentially vascular nature of the ciliary processes was established by Zinn. The contribution of more modern investigators was making clear the role of the ciliary processes in the formation of aqueous humor, thereby giving meaning to the masses of small vessels packed into the ciliary processes.

The vascular castings that form an essential part of contemporary descriptions of the ocular vasculature are the products of considerable technical sophistication and modern materials (see Box 6.1), but the basic ideas that underlie them are clearly recognizable in the methods developed by Ruysch and used by Zinn. These earlier admirers of flowers and vascular systems would undoubtedly appreciate the delicate, almost floral beauty of the eye's vasculature traced in plastic.

eral palpebral arteries that run into both the upper and lower eyelids. On the medial side, there are corresponding **medial palpebral arteries**, which arise either directly from the ophthalmic or from the dorsal nasal artery.* The lateral and medial palpebral arteries run toward one another in the lids, where their anastomoses form the **palpebral arcades**. The arcade vessels nearest the lid margins are the **marginal arcades**; those farther away from the lid margins are the **peripheral arcades**. These vessels lie deep in the tissue of the eyelids (between the tarsal plates and the orbicularis muscle; see Chapter 7), and numerous small branches run from the major arcade vessels to supply the orbicularis and other deep tissues. The peripheral arcade in the lower lid may not be complete or as well developed as its counterpart in the upper lid; it is not present in Figure 6.27.

In addition to the palpebral arteries, several additional arteries feed into the palpebral circulation, all into the inferior arcade. First is a branch from the **infraorbital artery**, which has emerged onto the face through the infraorbital

*Surgeons often substitute "infratrochlear" artery for "dorsal nasal" artery, which is the anatomist's traditional term. I have stayed with the anatomical terminology.

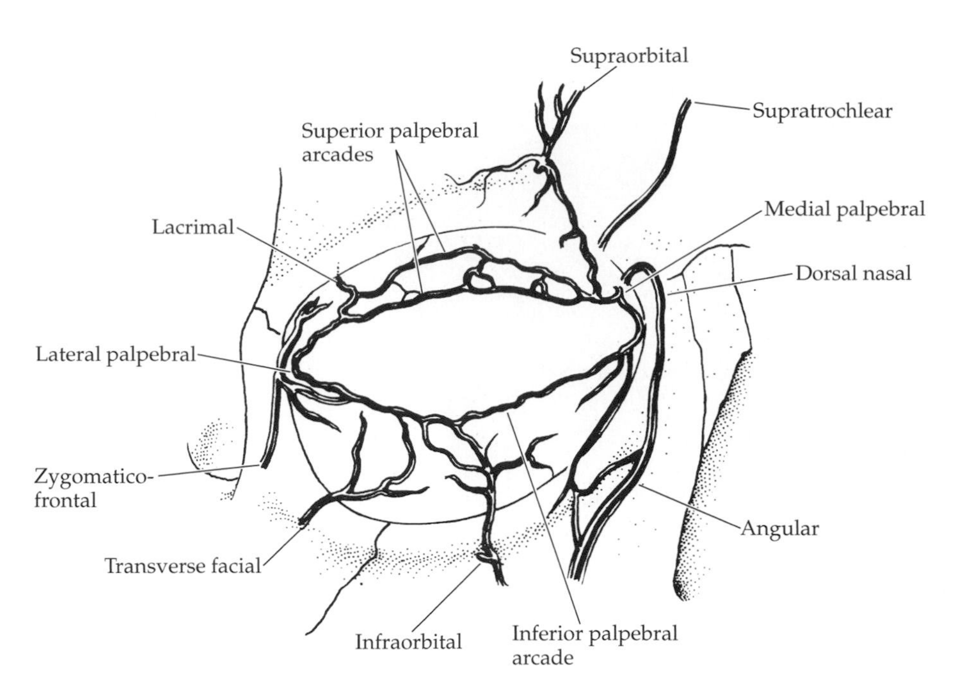

Figure 6.27
Supply of the Eyelids and Terminal Branches of the Ophthalmic Artery
The arcades of blood vessels in the eyelids are supplied through branches of the ophthalmic artery (lacrimal and medial palpebral arteries) and from arteries on the face that are part of the external carotid artery system (infraorbital, zygomatico-facial, and transverse facial arteries). The branches of the ophthalmic artery that run to the face—supraorbital, supratrochlear, dorsal nasal (infratrochlear), and lacrimal—supply regions comparable to the sensory nerves of the same name.

foramen, and additional branches from three of the large arteries of the face—namely, the **zygomatico-facial artery** and the **transverse facial artery** on the inferior-lateral aspect of the orbit and a branch from the **facial artery** on the medial side. The facial artery branch, the **angular artery**, normally ends as an anastomosis with the dorsal nasal artery. This means that the blood supply to the eyelids represents an anastomosis between the internal carotid (via the ophthalmic and its branches) and the external carotid (from which the infraorbital, zygomatico-facial, transverse facial, and facial arteries derive their blood). Obstructions in the ophthalmic artery system may be offset by blood reaching the eyelids through the arteries derived from the external carotid artery.

The veins of the eyelids do not form definite arcades, and they are so numerous and variable that they have not been named. They drain the lids by various routes into the superior and inferior ophthalmic veins, the infraorbital vein, and on the nasal side, the angular and facial veins. The large number of superficial arteries and veins in and around the eyelids and the loose structure of the lids provide considerable opportunity for blood to leak from the vessels and spread diffusely into the tissue following trauma. The resulting bruise is a "black eye."

The terminal branches of the ophthalmic artery leave the orbit to supply the skin and muscles of the face

The terminal branches of the ophthalmic usually arise from it just past the anterior ethmoidal arterial branch; Figure 6.27 shows part of their distribution on the face after they leave the orbit. The **supraorbital artery** runs above the eye to pass through the supraorbital notch (or foramen) along with the supraorbital vein and nerve. It branches extensively to supply the skin and muscles of the upper eyelid, the brow, and the forehead. The **supratrochlear artery** runs above the trochlea as it leaves the orbit to supply the skin and muscle along the midline of the brow and forehead.

Although the supraorbital and supratrochlear arteries normally do not supply any orbital structures, the final terminal branch of the ophthalmic, the **dorsal nasal artery** (or infratrochlear artery), does have branches within the orbit. As the dorsal nasal artery runs below the trochlea to exit the orbit, it sends small branches to the lacrimal sac and the region nearby, probably including part of the orbicularis muscle (specifically, to the lacrimal portion; see Chapter 7). Outside the orbit, its branches supply the skin and muscle along the side of the nose and around the medial canthus of the eyelids; it has an anastomosis with the angular artery (see Figure 6.27). The dorsal nasal artery does not have a corresponding vein of the same name, but the supraorbital and supratrochlear veins, discussed shortly, are found on the face and do not enter the orbit as discrete entities.

The infraorbital artery runs under the orbital floor

The **infraorbital artery** is a branch of the maxillary artery and therefore part of the *external* carotid circulation. It enters the orbit through the inferior orbital fissure along with the infraorbital nerve and vein and runs through the infraorbital groove and canal to the infraorbital foramen in the maxilla. It not only contributes to the inferior palpebral arcade, as already mentioned, but also supplies skin and muscle in the region of the face below the orbit.

Its companion vein, the **infraorbital vein**, drains part of the region supplied by the artery and follows the same path back into the orbit. The infraorbital vein drains into the venous pterygoid plexus below the orbit, but it may also have connections to the inferior ophthalmic vein.

The orbital veins are connected to the veins of the face, the pterygoid plexus, and the nose

Figure 6.9 gave a schematic representation of the major veins in and around the orbit. Although blood drains from the eye by the superior and inferior ophthalmic veins to the cavernous sinus, the orbital and external venous systems are extensively interconnected, and there are several possible routes for the drainage of venous blood from any particular region. (These peripheral veins do not have valves. Thus, there is no anatomical restriction on the direction of blood flow.)

The **supraorbital** and **supratrochlear veins** parallel the corresponding arteries on the forehead, but they do not follow the arteries into the orbit; instead, they merge at the junction between the superior ophthalmic vein and the **angular vein** at the upper medial corner of the orbit. The angular vein continues down alongside the nose to become the large **facial vein**. As the facial vein runs below the orbit, it forms anastomoses with both the inferior ophthalmic and the infraorbital veins. These anastomoses between the ophthalmic veins and the large veins of the face have two implications: First, it is possible for blood to leave the eye without going through the cavernous sinus (i.e., there is an alternative route if the normal one is blocked), and second, blood-borne infections from the face can be transmitted to the cavernous sinus with the possibility of a cavernous sinus thrombosis.

The other significant anastomoses are with the large veins in the nasal cavities; here, the shunts are the anterior and posterior ethmoidal veins. Connections between the inferior ophthalmic vein, the infraorbital vein, and the venous pterygoid plexus are also present, thus creating anastomoses between the orbital drainage and that of the maxillary sinuses. (In addition, the pterygoid plexus, which drains mainly into the external jugular via the maxillary vein, often has a direct shunt to the cavernous sinus, which ultimately drains into the internal jugular.) Again, these connections provide not only potential alternative routes for draining blood from the eye, but also routes for infections to invade the cavernous sinuses.

Development of the Ocular Blood Vessels

Primitive embryonic blood vessels appear very early in the eye's development

When the embryo is still small and the placenta is forming, there is little need for a vascular system; the cells of the embryo are quite active, but they can be supplied by diffusion from the surrounding amniotic fluid. But simple diffusion is inadequate after the embryo is a millimeter or so long, and the problem of supply is particularly important in the embryo, where the rates of cell proliferation and protein synthesis are high. In short, some sort of primitive vascular system is required early in development.

As a result, the cardiovascular system is the first functioning system to appear in the mammalian embryo, and the heart is the first functioning organ. Because of intrinsic pacemaker cells, the heart begins to beat and to pump blood during the third embryonic week, while it is little more than an expanded tube with blood vessels leading in or out of its two ends. The first blood vessels have very little of their adult structure; they are basically double-walled tubes, of which the inner is endothelium and the outer is mesoderm, which will become the smooth muscle and connective tissue of the mature vessels. Oxygen is transported by a special form of fetal hemoglobin that is replaced after birth.

The first ocular blood vessels appear about the time that the optic vesicle begins to invaginate to form the optic cup—during the fourth week of gestation. By this time, both a primitive internal carotid artery and a primitive internal jugular vein have formed, and they lie very close to the optic stalk (Figure 6.28). A short branch from the carotid artery ramifies in an anastomotic meshwork around the developing optic cup to become the first version of the choriocapillaris; a few small branches drain it into the primitive jugular vein. As the optic stalk grows longer and the optic cup moves away from the developing brain, the vessel connecting the internal carotid artery and the choriocapillaris is called the **primitive dorsal ophthalmic artery**. A **primitive ventral ophthalmic artery** feeds the primitive choriocapillaris; it is less prominent and eventually forms an anastomosis with the dorsal ophthalmic and becomes the medial posterior ciliary artery.

Several parts of the early ocular vasculature are transient and do not appear in the mature eye

As the eye and its surroundings grow, there is a great deal of relative change in position of the various structures. In particular, the dorsal ophthalmic artery shifts ventrally to lie below rather than above the optic stalk (see Figure 6.28). From this position, the dorsal ophthalmic sends a branch into the cleft—the choroidal fissure—along the bottom of the optic cup and optic stalk; this branch is now the **hyaloid artery**. It is accompanied by a **hyaloid vein**.

Figure 6.28
Early Development in the Ophthalmic Artery System
Major elements of the vascular system of the eye appear between weeks 4 and 7 of gestation. The choriocapillaris is first to appear, followed by the transient vasculature for the developing lens. Anastomoses and disappearance of the primitive ventral ophthalmic artery begin to produce branches resembling the mature system. (After Mann 1964.)

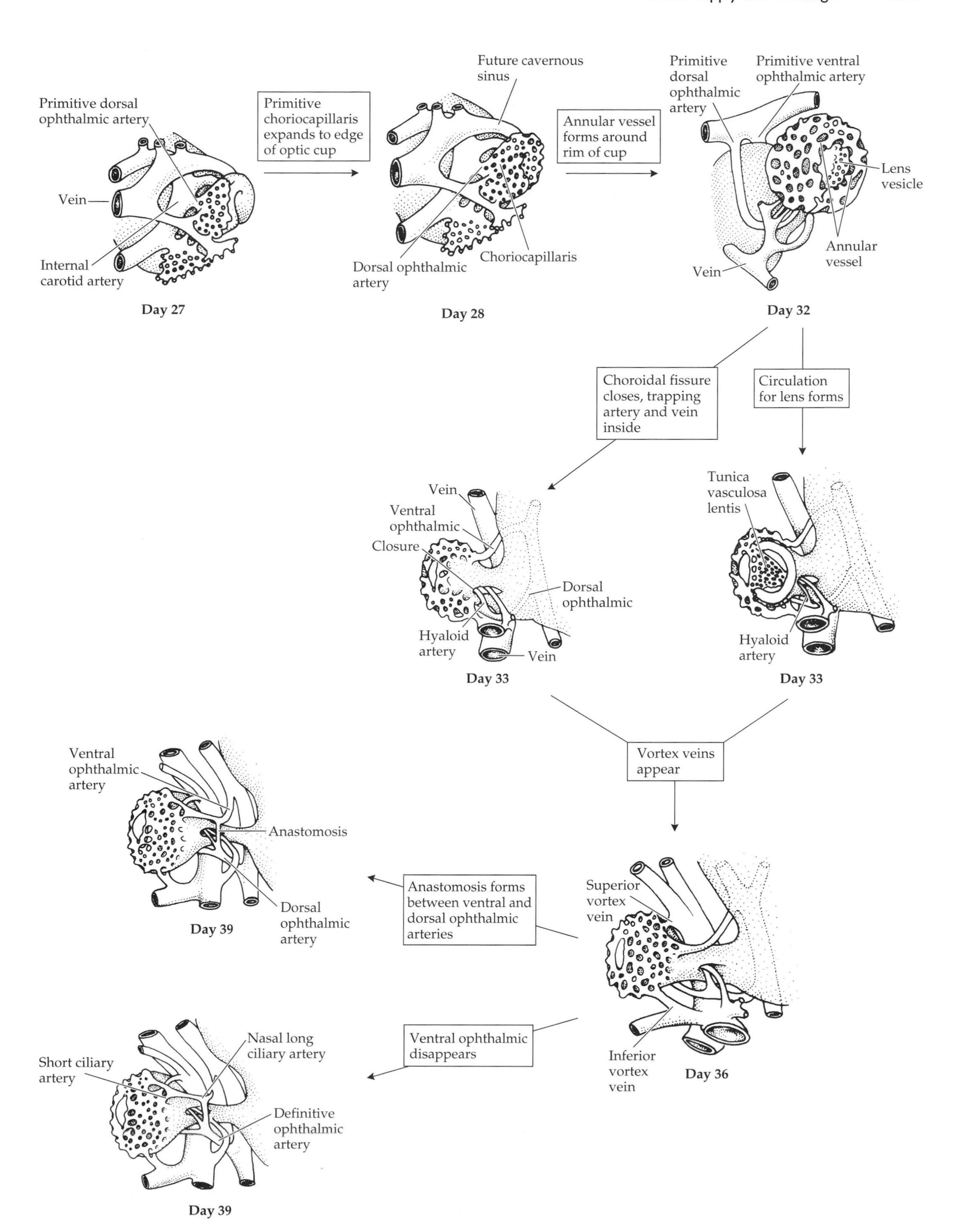
Primitive dorsal ophthalmic artery
Vein
Internal carotid artery
Day 27
Primitive choriocapillaris expands to edge of optic cup
Future cavernous sinus
Dorsal ophthalmic artery
Choriocapillaris
Day 28
Annular vessel forms around rim of cup
Primitive dorsal ophthalmic artery
Primitive ventral ophthalmic artery
Lens vesicle
Annular vessel
Vein
Day 32
Choroidal fissure closes, trapping artery and vein inside
Circulation for lens forms
Vein
Ventral ophthalmic
Closure
Dorsal ophthalmic
Hyaloid artery
Vein
Day 33
Tunica vasculosa lentis
Hyaloid artery
Day 33
Vortex veins appear
Ventral ophthalmic artery
Anastomosis
Dorsal ophthalmic artery
Day 39
Anastomosis forms between ventral and dorsal ophthalmic arteries
Superior vortex vein
Inferior vortex vein
Day 36
Ventral ophthalmic disappears
Short ciliary artery
Nasal long ciliary artery
Definitive ophthalmic artery
Day 39

These vessels form the supply and drainage, respectively, for an extensive capillary bed surrounding the developing lens (see Figure 6.28) called the **tunica vasculosa lentis**. Although the adult lens is avascular, at this stage of development the lens is growing very rapidly—more rapidly than it will ever grow again (see Chapter 12)—and the need for a blood supply is great. This need is filled by the tunica vasculosa, to which the primitive long posterior ciliary arteries also contribute, creating an anastomosis between the hyaloid artery and the posterior ciliary artery system. Shortly thereafter, an anastomosis forms between the dorsal and ventral ophthalmic arteries, and the part of the ventral ophthalmic connecting it to the internal carotid artery disappears. The remaining vessel will become the medial posterior ciliary artery.

All this is done by about the sixth week, and over the next several weeks, numerous small vessels cover the anterior surface of the lens to create a **pupillary membrane**. As Figure 6.29 shows, the vascular structure in this region has begun to take the form of a vascular circle around the equator of the lens—thus heralding the beginning of the major arterial circle—connected to extensive vascular networks on both the anterior and the posterior surfaces of the lens. A large part of this extensive vasculature is there purely to serve the developing eye; it is transient, in other words, and having done its job, it disappears.

All of the hyaloid artery (and vein) system begins to atrophy at about the end of the first trimester (Figure 6.30). The tunica vasculosa posterior to the lens equator breaks down as atrophy of the intravitreal portion of the hyaloid artery shuts off the blood flow and the entire system disappears from within the eye. This atrophy, however, is only within the vitreous chamber; at the retina, the hyaloid artery begins to sprout new branches along the retinal surface to help supply the now rapidly developing retina. (Development of the retinal vasculature is discussed in more detail in Chapter 16.) This sprouting marks the beginning of the central retinal artery system, and with these changes, the hyaloid artery system ceases to exist. Similar events occur in the hyaloid vein, which becomes the central retinal vein.

The system of blood vessels covering the anterior surface of the lens—the pupillary membrane—is also transient, at least in part. This system of vessels persists while the iris develops around it (see Chapter 10). Two epithelial layers from the anterior rim of the optic cup grow between the vessels and the lens to

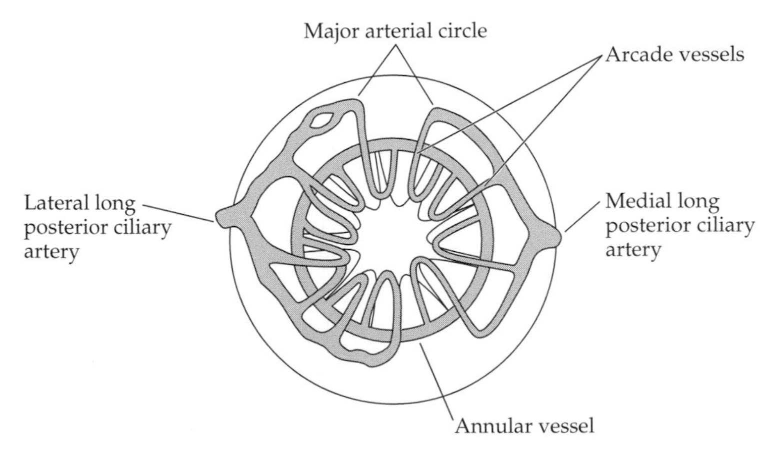

Figure 6.29
Formation of the Iris Supply and the Major Arterial Circle
The major arterial circle is formed by branches from the long posterior ciliary arteries. Branches from the major arterial circle run into the iris; their anastomoses will eventually form the minor arterial circle. The underlying annular vessel and its branches are transient; they soon atrophy and disappear. (After Mann 1964.)

form the two epithelial layers on the posterior side of the iris, while the iris stroma (fibroblasts, collagen fibers, and melanocytes) forms around and anterior to the vessels of the pupillary membrane. Late in gestation, the center portion of the pupillary membrane and iris stroma atrophy, thereby producing the pupil.

The anterior ciliary system forms later than the posterior ciliary system

The hyaloid artery system and the vessels associated with the primitive choriocapillaris are the first of the ocular vessels to appear, largely because they are responsible for the supply and drainage of the inner parts of the eye, which are the first to form. The vasculature for later-developing structures develops later. Thus, the anterior ciliary arterial and episcleral venous systems appear as the rectus muscles form and become more extensive as the sclera, limbus, and cornea differentiate.

This sequencing suggests that the intramuscular arterial circle and its contribution to the supply of the iris are relatively late additions and that the atrophy associated with formation of the pupil—that is, the disappearance of the pupillary membrane—is atrophy primarily within the posterior ciliary artery system from the major arterial circle. In any case, the result is a redistribution of arterial supply in the iris so that the intramuscular circle is the dominant supplier. This part of the story is largely conjecture, however; the intramuscular circle in the adult eye was identified fairly recently, and its development has not been studied (or, if it has been studied, it has not been distinguished from the major arterial circle).

Remnants of normally transient, embryological vasculature may persist in the mature eye

The large range of variation in the arterial and venous systems discussed earlier makes it difficult to talk about congenital anomalies in the ocular vasculature; as long as it works, almost anything can be considered to be normal or a normal variation. Several anatomical variations fall in a different category, however, because they are clearly the remains of structures formed in the embryonic eye that should have disappeared but did not.

One dramatic example of retained embryological structure is a **persistent hyaloid artery**. To be more precise, it is remnants of the hyaloid artery scattered along its original course between the optic nerve head and the posterior surface of the lens. This path, which contains the primary vitreous, is the **canal of Cloquet**; the canal is always present, but since the primary and later vitreous are normally transparent, it is visible only under special circumstances. Since most of the path followed by the persistent hyaloid artery runs away from the visual line of sight, it does not interfere with normal vision by the fovea; a problem arises only if the remnants near the posterior surface of the lens are particularly dense.

Another variant appears as a dense black spot more or less centered in the pupil. This black spot, called a **Mittendorf dot**, is a vestige of the hyaloid artery that has not atrophied, remaining stuck to the posterior lens capsule. Normally, a Mittendorf dot does not interfere with foveal vision, although under the right lighting conditions one may be aware of a dim, out-of-focus blob near the point of fixation that moves as the eye moves.

A third hyaloid artery remnant can appear on the optic nerve head; the normal cupping of the nerve head is obscured and there appears to be a projection of tissue into the vitreous. This is **Bergmeister's papilla**; it consists of an arterial remnant that has become invested with additional connective tissue and glial cells (astrocytes). Since this structure is superimposed on the normal blind spot created by the nerve head, it has no visual consequences.

(a) Week 5

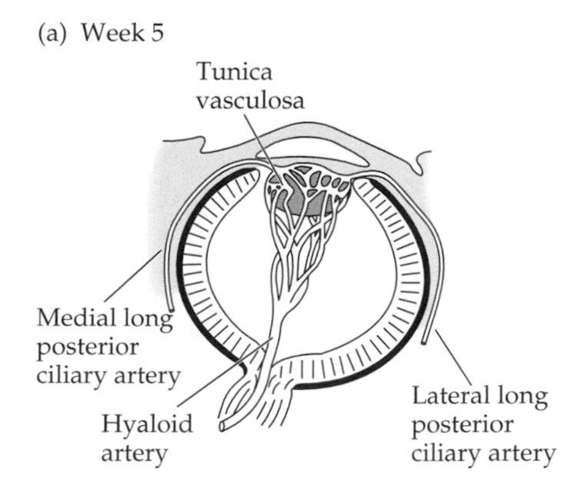

(b) Week 11

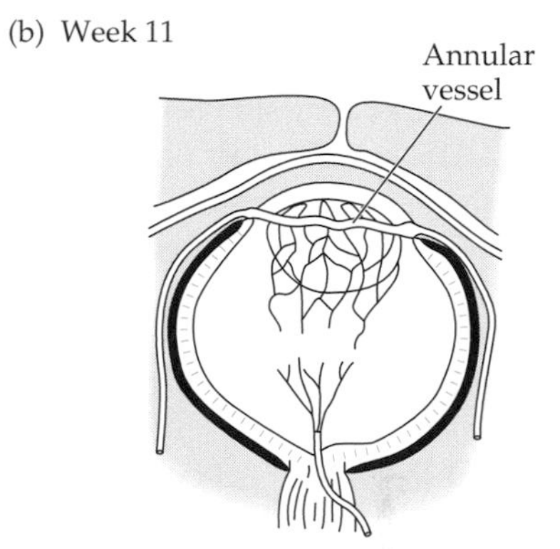

(c) Month 4

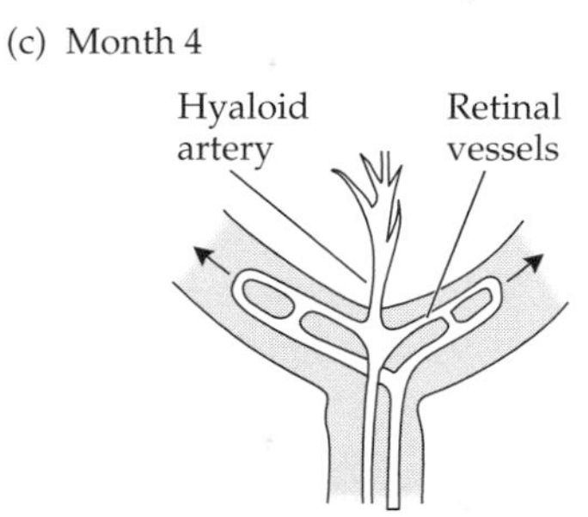

Figure 6.30
Zenith and Decline of the Hyaloid Artery System
(*a*) The hyaloid artery runs forward from the optic stalk to the developing lens, where its numerous branches form a dense arterial network, the tunica vasculosa, around the lens. The tunica vasculosa is also fed by the annular vessel, which is supplied by the long posterior ciliary arteries. (*b*) The hyaloid artery begins to atrophy by the eleventh week of gestation. (*c*) Shortly thereafter, sprouts from the hyaloid artery onto the surface of the retina mark the beginning of central retinal artery development. (After Mann 1964.)

The other transient vascular structure that sometimes fails to atrophy completely is the pupillary membrane; fragments of **persistent pupillary membrane** can often be seen crossing the pupil. The persistent membrane may range from a few strands of tissue, looking much like a bit of spiderweb crossing the aperture, to more extensive remnants that appreciably alter the shape of the pupil. Since structures in this plane are part of the aperture stop for the optical system, their effect, if any, is slight reduction of the amount of light reaching the retina.

References and Additional Reading

Distributing Blood to Tissues

Alberts B, Bray D, Lewis J, Raff M, Roberts K, and Watson JD. 1994. Renewal by simple duplication. Chapter 22, pp. 1147–1154, in *Molecular Biology of the Cell*, 3rd Ed. Garland Publishing, New York

Alm A. 1992. Ocular circulation. Chapter 6, pp. 198–227, in *Adler's Physiology of the Eye*, 9th Ed., Hart WM, ed. Mosby Year Book, St. Louis.

Bill A. 1984. Circulation in the eye. Chapter 22, pp. 1001–1034, in *Handbook of Physiology. Section 2: The Cardiovascular System*, Vol. 4, Part 2, Renkin EM and Michel CC, eds. American Physiological Society, Bethesda, MD.

Burger PC, Chandler DB, and Klintworth GK. 1983. Corneal neovascularization as studied by scanning electron microscopy of vascular casts. *Lab. Invest.* 48: 169–180.

Folkman J and Haudenschild C. 1980. Angiogenesis *in vitro*. *Nature* 288: 551–556.

Goss RJ. 1978. Vascular expansion. Chapter 6, pp. 120–137, in *The Physiology of Growth*. Academic Press, New York.

Levick JR. 1995. *An Introduction to Cardiovascular Physiology*, particularly Chapters 9 thorugh 12. Butterworth-Heinemann, Oxford.

Rhodin JAG. 1980. Architecture of the vessel wall. Chapter 1, pp. 1–31, in *Handbook of Physiology, Section 2: The Cardiovascular System, Vol. II: Vascular Smooth Muscle*, Bohr DF, Samylo AP, and Sparks HV, eds. American Physiological Society, Bethesda, MD.

Simionescu M and Simionescu N. 1984. Ultrastructure of the microvascular wall: Functional correlations. Chapter 3, pp. 41–101, in *Handbook of Physiology. Section 2: The Cardiovascular System*, Vol. 4, Part 1, Renkin EM and Michel CC, eds. American Physiological Society, Bethesda, MD.

The Ophthalmic Artery and Ophthalmic Veins

Bergen MP. 1990. Spatial aspects of the orbital vascular system. Chapter 33 in *Duane's Foundations of Clinical Ophthalmology*, Vol. 1, Tasman W and Jaeger EA, eds. Lippincott-Raven, Philadelphia.

Hayreh SS. 1962. The ophthalmic artery III. Branches. *Brit. J. Ophthalmol.* 46: 212–247.

Hayreh SS. 1963. Arteries of the orbit in the human being. *Brit. J. Surg.* 50: 938–953.

Hayreh SS and Dass R. 1962. The ophthalmic artery II. Intraorbital course. *Brit. J. Ophthalmol.* 46: 165–185.

Supply and Drainage of the Eye

Aiello LP. 1997. Vascular endothelial growth factor: 20th-century mechanisms, 21st-century therapies. *Invest. Ophthalmol. Vis. Sci.* 38: 1647–1652.

D'Amore PA. 1994. Mechanisms of retinal and choroidal neovascularization. *Invest. Ophthalmol. Vis. Sci.* 35: 3974–3979.

Fryczkowski AW, Sherman MD, and Walker J. 1991. Observations on the lobular organization of the human choriocapillaris. *Int. Ophthalmol.* 15: 109–120.

Funk R and Rohen JW. 1990a. *Basic Aspects of Glaucoma Research II: Functional Morphology of the Vasculature in the Anterior Eye Segment*. Schattauer, Stuttgart.

Funk R and Rohen JW. 1990b. Scanning electron microscopic study on the vasculature of the anterior eye segment, especially with respect to the ciliary processes. *Exp. Eye Res.* 51: 651–661.

Hayreh SS. 1975. Segmental nature of the choroidal vasculature. *Brit. J. Ophthalmol.* 59: 631–648.

Hayreh SS. 1990. *In vivo* choroidal circulation and its watershed zones. *Eye* 4: 273–289.

Hogan MJ, Alvarado JA, and Weddell JE. 1971.The limbus, Chapter 4, pp. 112–182; The choroid, Chapter 8, pp. 320–392; The optic nerve, Chapter 10, pp. 523–606, in *Histology of the Human Eye*. WB Saunders, Philadelphia.

Morrison JC, Fahrenbach WH, Bacon DR, Wilson DJ, and Van Buskirk EM. 1996. Microvasculature of the ocular anterior segment. *Microsc. Res. Tech.* 33: 480–489.

Morrison JC and Van Buskirk EM. 1983. Anterior collateral circulation in the primate eye. *Ophthalmology* 90: 707–715.

Morrison JC and Van Buskirk EM. 1984. Ciliary process microvasculature of the primate eye. *Am. J. Ophthalmol.* 97: 372–383.

Morrison JC and Van Buskirk EM. 1986. Microanatomy and modulation of the ciliary vasculature. *Trans. Ophthalmol. Soc. U.K.* 105: 131–139.

Nanjiani M. 1991. *Fluorescein Angiography: Technique, Interpretation, and Application.* Oxford University Press, Oxford.

Olver JM. 1990. Functional anatomy of the choroidal circulation: Methyl methacrylate casting of human choroid. *Eye* 4: 262–272.

Olver JM and Lee JP. 1989. The effects of strabismus surgery on anterior segment circulation. *Eye* 3: 318–326.

Olver JM and McCartney ACE. 1989. Orbital and ocular micro-vascular corrosion casting in man. *Eye* 3: 588–596.

Olver JM, Spalton DJ, and McCartney ACE. 1990. Microvascular study of the retrolaminar optic nerve in man: The possible significance in anterior ischaemic optic neuropathy. *Eye* 4: 7–24.

Olver JM, Spalton DJ, and McCartney ACE. 1994. Quantitative morphology of human retrolaminar optic nerve vasculature. *Invest. Ophthalmol. Vis. Sci.* 35: 3858–3866.

Raviola G. 1977. The structural basis of the blood-ocular barriers. *Exp. Eye Res.* 25 (Suppl.): 27–63.

Richard G. 1992. *Choroidal Circulation.* George Thieme, Stuttgart.

Snodderly DM, Weinhaus RS, and Choi JC. 1992. Neural-vascular relationships in central retina of macaque monkeys (*Macaca fascicularis*). *J. Neurosci.* 12: 1169–1193.

Yoneya S and Tso MOM. 1987. Angioarchitecture of the human choroid. *Arch. Ophthalmol.* 105: 681–687.

Supply and Drainage of the Eyelids and Surrounding Tissues

Tucker SM and Linberg JV. 1994. Vascular anatomy of the eyelids. *Ophthalmology* 101: 1118–1121.

Warwick R. 1976. The orbital vessels. Chapter 9, pp. 406–417, in *Eugene Wolff's Anatomy of the Eye and Orbit,* 7th Ed. WB Saunders, Philadelphia.

Zide BM and Jelks GW. 1985. Bones, vessels, and nerves. Chapter 1, pp. 1–12, in *Surgical Anatomy of the Orbit.* Raven, New York.

Development of the Ocular Blood Vessels

Cook CS, Ozanics V, and Jakobiec FA. 1991. Prenatal development of the eye and its adnexa. Chapter 2 in *Duane's Foundations of Clinical Ophthalmology,* Vol. 1, Tasman W and Jaeger EA, eds. Lippincott-Raven, Philadelphia.

Heimann K. 1972. The development of the choroid in man. *Ophthalmic Res.* 3: 257–273.

Mann I. 1964. *The Development of the Human Eye.* Grune and Stratton, New York.

Sellheyer K. 1990. Development of the choroid and related structures. *Eye* 4: 255–261.

Chapter 7

The Eyelids and the Lacrimal System

The fringed curtains of thine eye
advance,
And say, what seest thou yong'.

■ William Shakespeare, *The Tempest*

I don't know what it means
and I don't care because it's
Shakespeare and it's like having
jewels in my mouth when I say
the words.

■ Frank McCourt, *Angela's Ashes*

The lovely fragment of Shakespeare quoted here isn't difficult to understand. The fringed curtains of the eyes are the eyelids, and advancing them means squinting, in less elegant language. In a few well-chosen words, Prospero asks Miranda to close her eyelids down to slitlike apertures, which will increase the depth of field, reduce the effects of aberrations, and allow her to better see who is approaching in the distance. We do this all the time; although it is only a minor function of the eyelids, it is a useful one. (Habitual squinting, however, is often a sign of uncorrected myopia.)

The eyelids serve several functions in addition to the foregoing optical one: They protect the eye by closing reflexly to shield the cornea from objects hurtling toward it or by brushing away small foreign objects, they control the entry of light into the eye as does the shutter in a camera, they contribute important components to the tear film, and they maintain the smooth uniformity of the tear film by their wiping action during blinking. Since there is an important relationship between the eyelids and the lacrimal system, they will be considered together here.

The lacrimal system has three basic elements, all closely associated with the eyelids: the lacrimal gland and other glands that produce various components of the tears, the tears themselves, and the apparatus for draining the tears away from the eye. The tears create the optical surface for the cornea, some of their components help prevent infection of the eye by microorganisms, and when produced in quantity, they can flush small particles of dust and grit off the cornea.

Structure and Function of the Eyelids

Structural rigidity of the lids is provided by the tarsal plates

Controlled movements of the eyelids require some rigid structure to give the lids shape and to which muscles can be attached. Rigidity is provided by the **tarsal plates**, which are found in both the upper and the lower eyelids (Figure 7.1). They extend along the lid margins across the width of the lids, and their curved peripheral edges follow roughly the contour of the orbital margins. Both tarsal plates have about the same horizontal extent, around 29 mm, but the maximum vertical extent of the superior tarsal plate is greater than that of the inferior plate (10 mm versus 5 mm). The tarsal plates are curved to fit the contour of

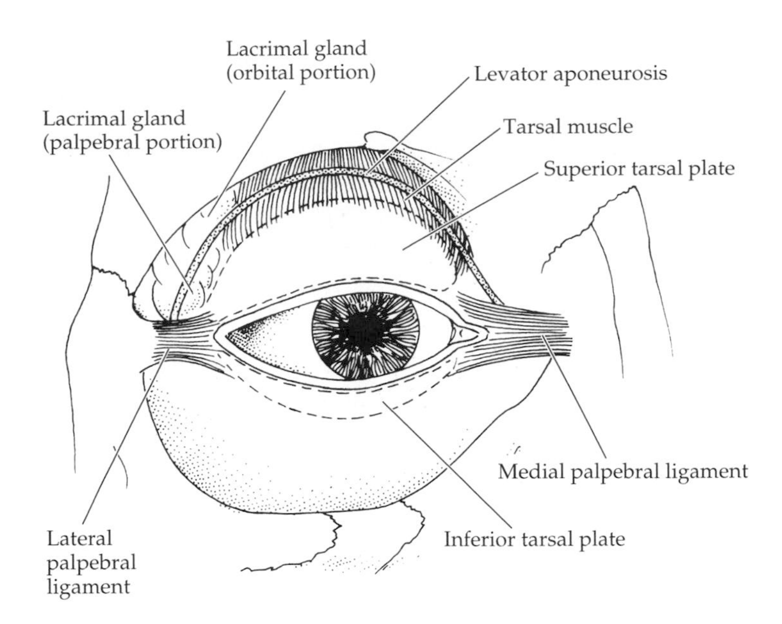

Figure 7.1
The Tarsal Plates
The superior and inferior tarsal plates are shown here as they are situated in the eyelids. The medial and palpebral ligaments, to which the orbicularis muscle attaches, lie in front of the plates.

the eye like the windshield wiper blades on an automobile; they are firm but flexible, and curved to fit closely against the glass.

The tarsal plates create divisions in the lids marked by several prominent folds in the skin of the lids (Figure 7.2). In both upper and lower lids, the part containing the tarsal plates is the **tarsal portion** of the lid; the rest, from the top of the superior tarsal plate to the orbital rim or from the bottom of the inferior tarsal plate to the orbital rim, is the **septal portion** (the orbital septum will be discussed later). Depending on how deeply the eyes are set into the orbits, these divisions are often marked externally by folds in the skin called the superior and inferior **palpebral furrows**. The superior palpebral furrow is usually the more pronounced, but there is much individual variation, and neither may be particularly obvious.

The tarsal plates are made of dense connective tissue in which glands are embedded

For all their structural importance, the tarsal plates are not well defined in histological sections, largely because they consist of dense connective tissue sur-

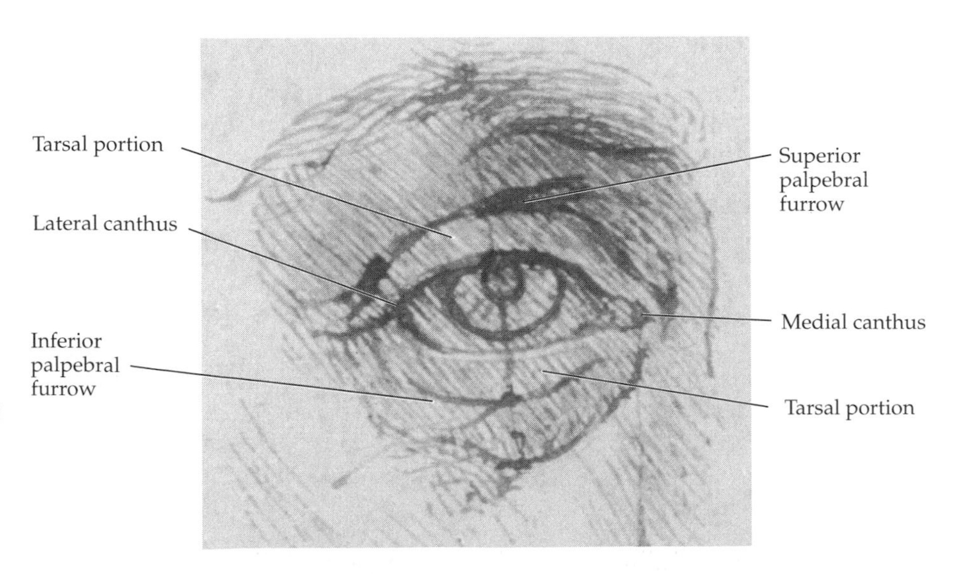

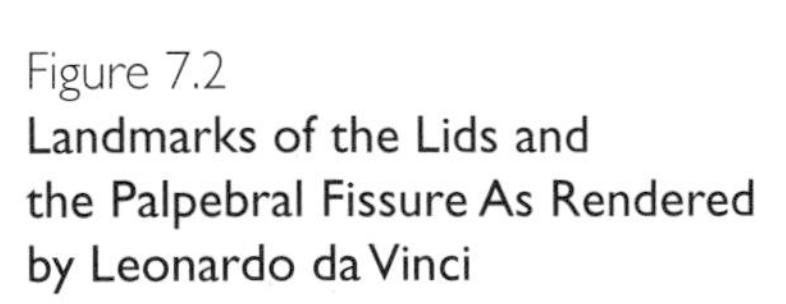

Figure 7.2
Landmarks of the Lids and the Palpebral Fissure As Rendered by Leonardo da Vinci
The palpebral fissure, the opening between the lid margins, forms two angles of intersection, the medial and lateral canthi. Creases in the eyelids, the superior and inferior palpebral furrows, occur near the edges of the tarsal plates, giving an external sign of the division of the lids into tarsal and septal portions. (Courtesy of the Turin Royal Library.)

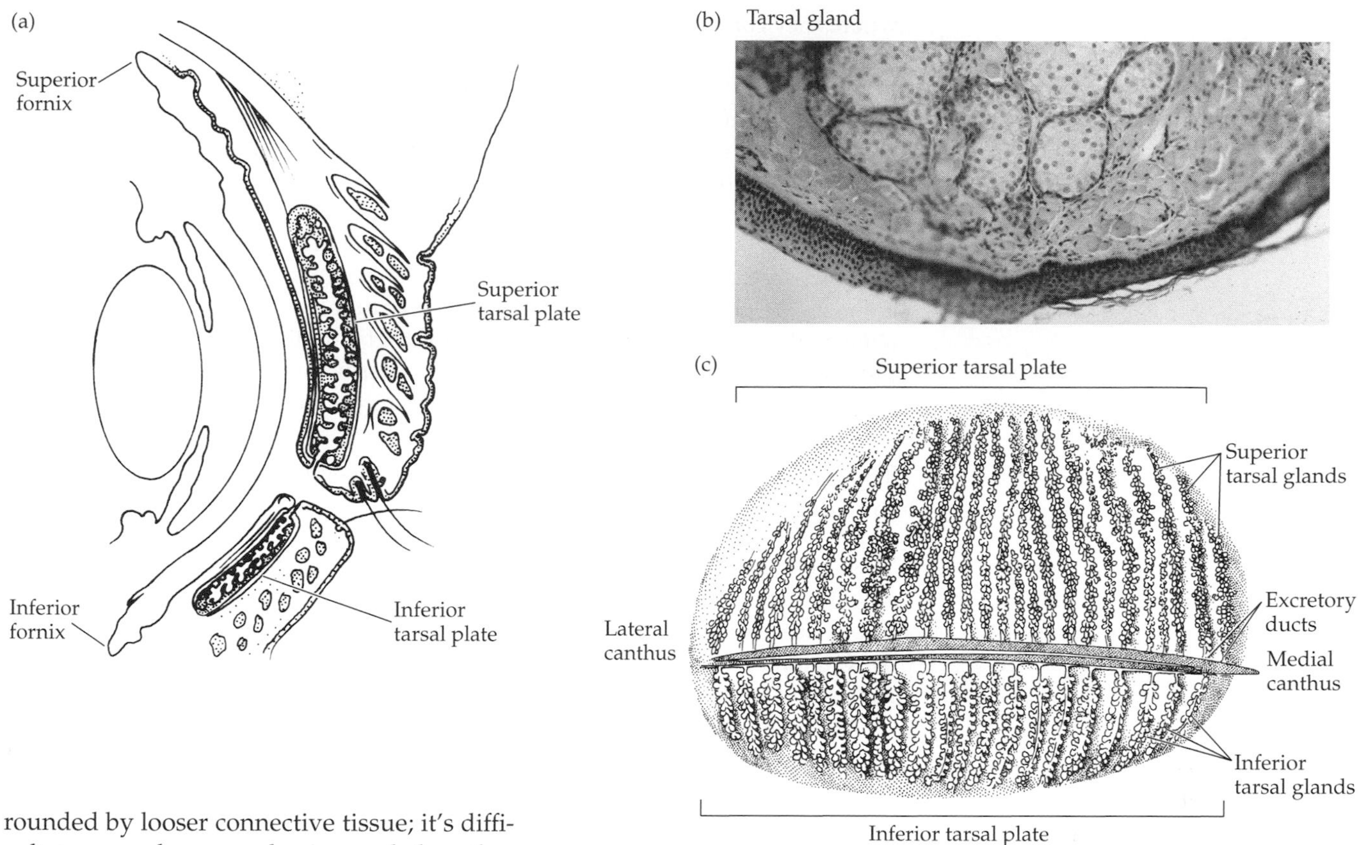

Figure 7.3
The Tarsal (Meibomian) Glands
(*a*) The tarsal glands are embedded in the tarsal plates. Their ducts empty onto the lid margins near the posterior side of the lids. (*b*) A section through one of the secretory lobules of the tarsal gland shows clusters of secretory cells separated and surrounded by connective tissue that forms the tarsal plate. (*c*) The tarsal glands extend from the lid margins to the farthest edges of the tarsal plates. This illustration shows 26 glands in the superior tarsal plate and 21 in the inferior plate, but the number varies among individuals. (*c* redrawn from Warwick 1976.)

rounded by looser connective tissue; it's difficult to see where one begins and the other ends. Therefore, the drawing of a vertical section through the upper eyelid in Figure 7.3*a,* showing the basic construction of the superior tarsal plate and its position within the eyelid, is an enhanced version. The dense, interwoven collagen forming the tarsal plate is not continuous; as the figure shows, openings within the tissue are filled with large secretory cells. These cell clusters are lobules of the **tarsal** (**meibomian**) **glands** that are embedded in the tarsal plates. A light micrograph of a secretory lobule (Figure 7.3*b*) shows the clusters of secretory cells that make up the lobules of the glands; the lumen into which they empty their secretion is not shown.

The tarsal glands are quite large; a single gland consists of a long collecting duct that receives numerous smaller ducts coming from the secretory lobules (Figure 7.3*c*). It has the appearance of a grapevine, with its central branch to which bunches of grapes (the secretory lobules) are attached. The largest of the tarsal glands in the upper tarsal plate are 8 or 9 mm long.

Numbers in the literature indicate up to 40 tarsal glands in the superior tarsal plate and up to 30 in the inferior plate. It is not clear how reliable these numbers are or how much variability there may be among individuals, but one of the few illustrations that purports to show all of the glands has 26 and 21 in the superior and inferior tarsal plates, respectively (see Figure 7.3*c*). Whatever numbers one uses, the inferior tarsal plate has about three-fourths the number of glands as the superior tarsal plate has. Their central collecting ducts open all along the lid margins at roughly 1 mm intervals; the duct openings are quite small, normally invisible without magnification.

Secretory cells in the tarsal glands produce a secretion that is oily because of its high lipid content. This oiliness prevents the tarsal secretion from mixing with water, which constitutes the bulk of the tears; the function of the tarsal gland secretion is based on this immiscibility. First, the oily secretion lines the lid margins between the openings of the ducts, forming a barrier to movement of the tears past the lid margins onto the skin. Second, the oil is deposited into the tears where they come in contact, and it forms a thin layer covering the entire surface of the tear film; this thin lipid layer prevents water in the tear film from evaporating as quickly as it would do otherwise.

An acute infection of a tarsal gland may cause the gland's duct to become plugged, producing noticeable swelling on the inner side of the eyelid; this swelling is called an **internal hordeolum**. The qualifier "internal" is used because the tarsal plate is closer to the back than to the front of the eyelid and the swelling bulges out on the inner surface of the lid, usually near the lid margin (there are external hordeola, but they are produced by infection of different glands, as we'll see later in the chapter). Granulomas of the tarsal glands are abnormal accumulations of cells within the glands that eventually produce a prominent bulge on either the inner or the outer lid surface. This bulge, called a **chalazion**, is typically farther from the lid margin than a hordeolum is.

The palpebral fissure is opened by muscles inserting onto or near the edges of the tarsal plates

The palpebral fissure is widened by elevation of the upper lid and, to a lesser extent, depression of the lower lid. As Figure 7.4*a* shows, two muscles elevate the upper lid—the **levator** (**palpebrae superioris**) and the **superior tarsal muscle** (**of Müller**)—and a single muscle depresses the lower lid—the **inferior tarsal muscle** (**of Müller**).

The levator originates at the apex of the orbit on the sphenoid bone, just above the origin of the superior rectus (see Figure 4.20) and runs directly above the superior rectus to the front of the orbit. The muscular portion of the levator ends within the orbit, just behind the orbital margin, where it expands into a broad flat tendon, the **levator aponeurosis** (Figure 7.4*b*). Fascial connections near the junction of muscle and tendon run laterally to the posterior edge of the lacrimal fossa and medially to the region of the trochlea, forming a connective tissue sling—the **superior transverse ligament** (or **Whitnall's ligament**)—that supports the levator as its tendon runs down into the upper lid. The levator aponeurosis extends across the entire width of the upper lid, inserting onto the anterior surface of the superior tarsal plate (or into the connective tissue surrounding the plate) and into the connective tissue of the orbicularis muscle lying in front of the tarsal plate. This arrangement means that contraction of the levator will exert an upward pull on the superior tarsal plate, thereby elevating the upper lid. The levator is innervated by the superior division of the oculomotor nerve (see Chapter 5).

The superior tarsal muscle arises near the origin of the levator aponeurosis, running behind the aponeurosis to the upper edge of the superior tarsal plate (see Figure 7.4*a*). Although the muscle does not have a rigid origin from bone, the normal tonus of the levator and the presence of the superior transverse ligament should provide a fairly firm origin from which to act; assuming the origin does not move much when the muscle contracts, it will pull upward on the tarsal plate and elevate the upper lid. Figure 7.4*c* shows the superior tarsal muscle from the front as if the tendon of the levator were transparent; in this illustration, the superior tarsal muscle extends almost to the medial and lateral ends of the tarsal plate, as does the levator aponeurosis.

The inferior tarsal plate and eyelid are moved downward by contraction of the inferior tarsal muscle, which runs between the bottom edge of the inferior tarsal plate and the connective tissue surrounding the inferior oblique and the inferior rectus near its insertion on the eye (see Figure 7.4*a*). In frontal view, the inferior tarsal muscle occupies about the same position in the lower lid as the superior tarsal muscle does in the upper lid (see Figure 7.4*c*). Again assuming that the ill-defined origin is stable, contraction of the inferior tarsal muscle will pull down on the tarsal plate. The inferior tarsal muscle is smaller than the superior tarsal mus-

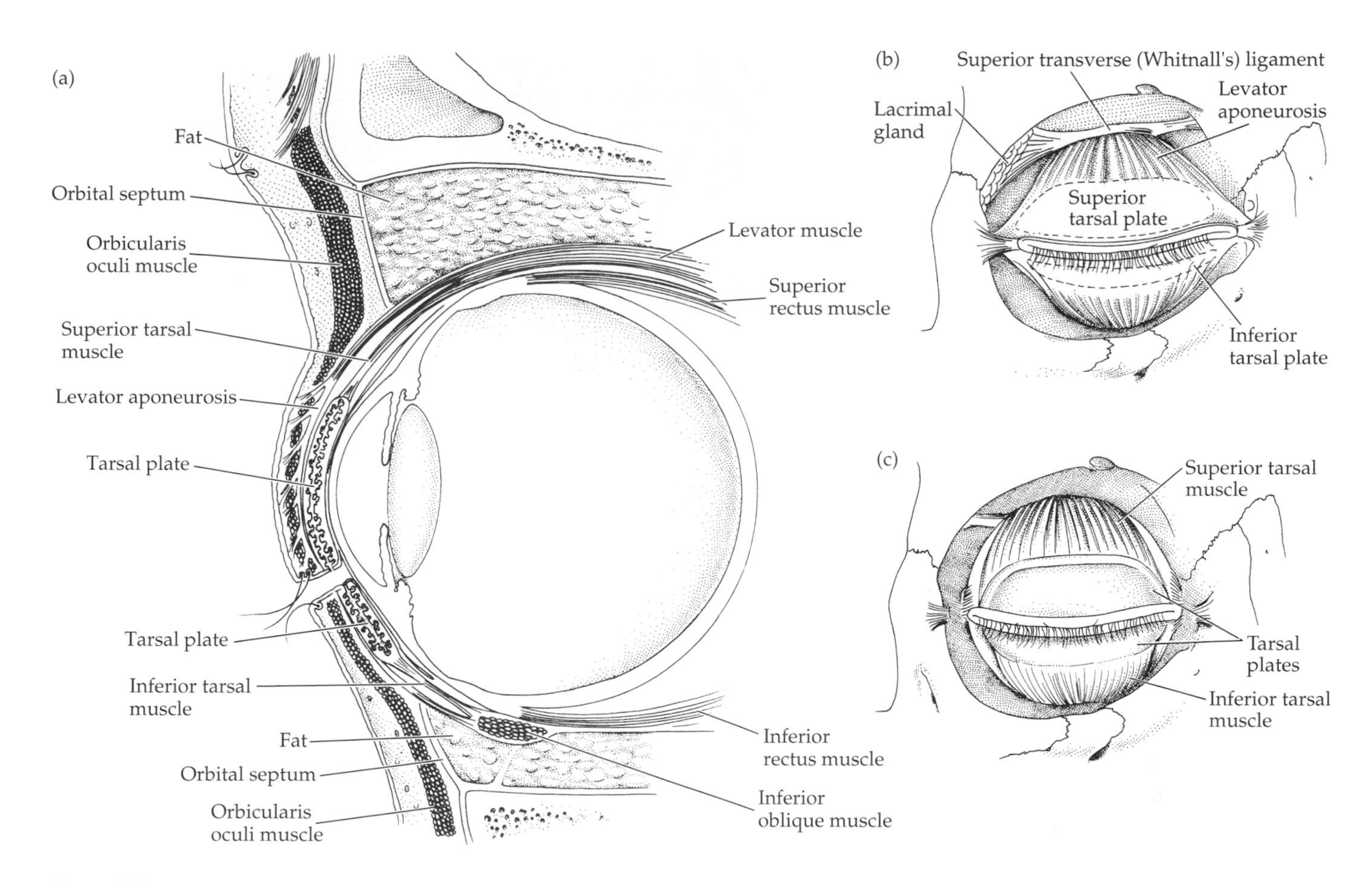

Figure 7.4
The Levator and the Tarsal Muscles of Müller
(*a*) In vertical section, the muscular portion of the levator is shown above the superior rectus. The tendon begins just behind the orbital rim and runs into the upper lid, attaching to connective tissue on the anterior surface of the tarsal plate and around the orbicularis muscle. The superior tarsal muscle runs from the connective tissue sheath around the levator down to the upper edge of the superior tarsal plate. The inferior tarsal muscle runs from the connective tissue sheath around the inferior rectus and oblique muscles to the lower edge of the inferior tarsal plate. The orbital septum fuses with the levator tendon above and with connective tissue on the anterior aspect of the inferior tarsal muscle below. (*b*) Viewed from the front, the levator tendon is a broad expansion (aponeurosis) that spreads out to the medial and lateral ends of the tarsal plate. The levator appears to be supported in the orbit by the superior transverse ligament, which runs from behind the lacrimal gland to the vicinity of the trochlea, fusing with the levator's connective tissue in passing. (*c*) When the levator is removed from the upper lid, as well as some of the connective tissue from the lower lid, the tarsal muscles can be seen, running from their insertions forward to fuse with the tarsal plates. The tarsal muscles are almost as broad as the tarsal plates.

cle, and because there is no other muscle to move the lower lid downward, it follows that the palpebral fissure is widened mostly by elevation of the upper lid.

Unlike the levator, which is striated muscle, the tarsal muscles are smooth muscle, innervated by the sympathetic system through poorly defined routes (see Chapter 5). The consequence of these differences is most obvious in the upper lid, where a loss of tonus in either the levator or the superior tarsal muscle causes the lid to droop—a condition called **ptosis** (Figure 7.5). The different types and routes of innervation for the upper eyelid muscles make ptosis an interesting problem in differential diagnosis; it may be associated with lesions in

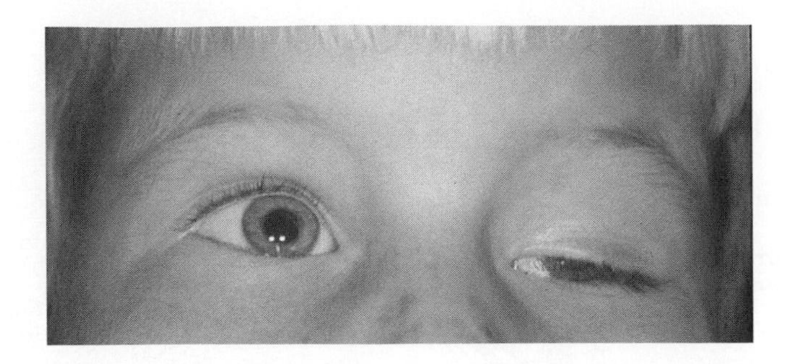

Figure 7.5
Ptosis
The left palpebral fissure is almost closed. This congenital ptosis is quite severe, much more pronounced than acquired ptosis normally is. (Courtesy of Robert Rutstein.)

the sympathetic system (as in Horner's syndrome; see Chapter 5), with lesions affecting the oculomotor nerve, or with the muscles themselves. In general, problems with the levator produce more ptosis because it normally has considerable tonus and is the more effective of the two muscles.

The palpebral fissure is closed by contraction of the orbicularis

The muscles for widening the palpebral aperture are muscles originating in the orbit and extending out into the lids, but the **orbicularis oculi** muscles used in lid closure are muscles of the face that extend into the lids from outside the orbits. The portions of the orbicularis within the lids lie anterior to the tarsal plates and to the other lid muscles.

Viewed from the front, the orbicularis for one eye covers the entire orbital opening, extending beyond the orbital margins onto the face (Figure 7.6*a*). Bundles of muscle fibers run in arcuate or elliptical courses, the ellipses becoming larger as the fiber bundles lie farther from the lid margins. Muscle fibers attach on the medial side of the orbit to the **medial palpebral ligament**, which in turn is attached to the frontal process of the maxilla, and at the lateral side of the orbit to the **lateral palpebral ligament**, which is attached to the zygomatic bone at the lateral orbital (Whitnall's) tubercle.* The interlaced ends of fiber bundles from the upper and lower lids at the lateral palpebral ligament form a horizontal line, the **lateral palpebral raphe**.

Individual muscle fibers do not run all the way between the medial and lateral palpebral ligaments. The fibers originating from one of the ligaments extend perhaps a third of the distance across the lid and insert into the connective tissue associated with the muscle. Fibers in the middle of the eyelids run toward the medial and lateral sides, but they never reach the ligaments (Figure 7.6*b*). The situation in the orbital part of the orbicularis is not known, but the muscle fibers there are probably relatively short as well. If so, substantial numbers of them will have both their origin and insertion in the muscle's connective tissue, and the elliptical fiber bundles will be made up of short segments created by overlapping muscle fibers.

The arrangement of muscle fiber bundles makes the orbicularis work like a sphincter, but one that cannot close down equally from all directions. Because of the attachments at the medial and lateral sides of the orbit, contraction results

*Strictly speaking, tendons attach muscle to bone and ligaments connect between bones, so the medial and lateral palpebral ligaments are actually tendons and are often referred to as the medial and lateral canthal tendons, respectively.

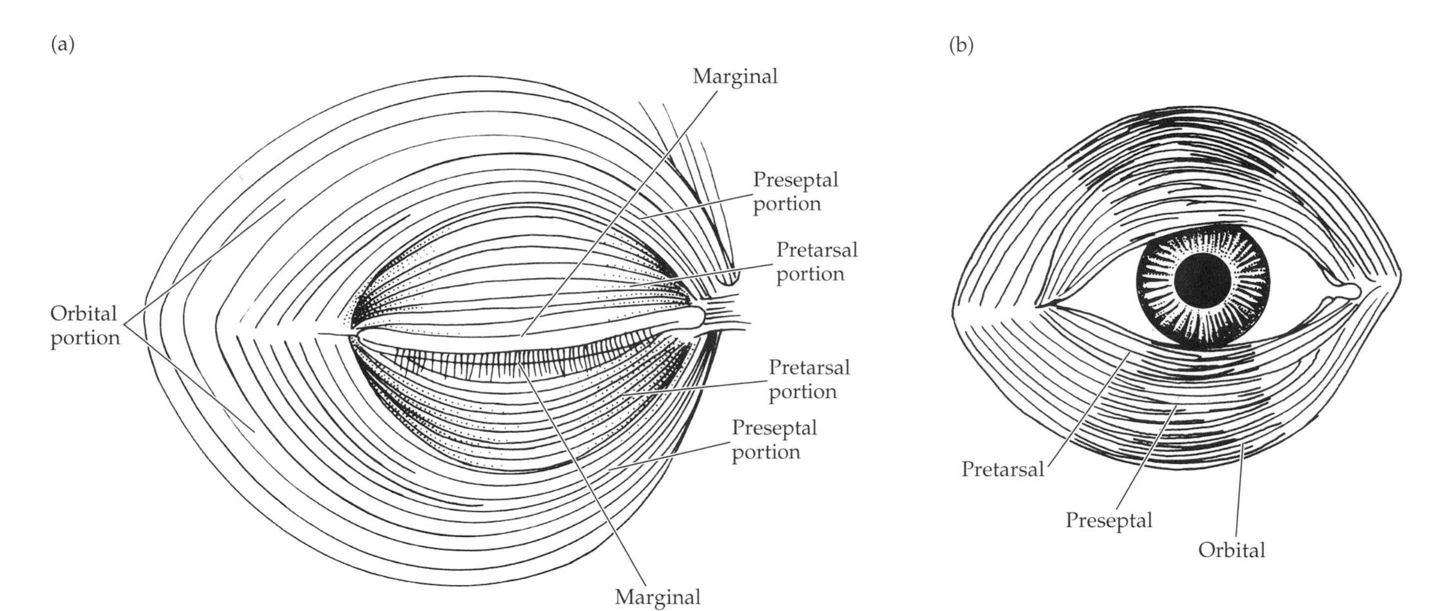

Figure 7.6
The Orbicularis Muscle and Its Insertions
(*a*) In front view, the bundles of fibers in the orbicularis are seen as running in arcuate courses within the eyelids and as large ovals extending past the orbital margin onto the brow and cheek regions of the face. Together, the pretarsal and preseptal portions of the muscle make up the palpebral portion—the part of the orbicularis within the eyelids. The orbital portion extends past the orbital margin onto the face and brow. (*b*) Orbicularis muscle fibers are not long enough to extend from the medial to the lateral sides. Many fibers have both ends attached to connective tissue within the muscle. (After Lander, Wirtschafter, and McLoon 1996.)

in movement of the upper lid down and the lower lid up (slightly), making the palpebral fissure more of a horizontal slit between the lid margins. Mechanically, this arrangement is both efficient and fast; from fully open to fully closed, the muscle fibers along the margin of the upper lid need to shorten by only about 25% of their resting length, and the contraction requires only 20 or 30 ms. Or to look at it another way, the area of the eye exposed when the palpebral fissure is fully open can be covered very quickly, which is appropriate for a protective reflex.

Figure 7.6 illustrates the several parts of the orbicularis. The dozen or so strands of the muscle adjacent to the lid margins constitute the **marginal** or **ciliary portion** (**muscle of Riolan**), the fibers lying in front of the tarsal plates constitute the **pretarsal portion**, and those from the tarsal plates out to the orbital margin constitute the **preseptal portion**; together, the pretarsal and preseptal portions make up the the **palpebral portion**. The rest of the muscle extending past the orbital rim onto the brow and cheek regions of the face is the **orbital portion**. (Another part is associated with the lacrimal sac and will be considered later.)

The part of the orbicularis extending past the orbital margin has attachments to the facial bones, mainly to the maxillary process of the frontal bone and to the frontal process of the maxilla. These attachments, along with the anchoring of orbicularis fibers to the medial palpebral ligament, mean that the sphincterlike action of the orbicularis when it contracts is more restricted on the medial side than it is on the lateral side.

The regional divisions of the orbicularis are not entirely arbitrary, and some functional distinctions also apply. Reflex and spontaneous blinking, for example, involve only the palpebral portion of the muscle, while the orbital portion is recruited for voluntary lid closure and is active in blepharospasm. The palpebral and orbital portions also differ in their contraction thresholds to applied electrical stimuli; the palpebral portion is the more sensitive. Tonus in the marginal portions is said to be responsible for holding the lid margins tightly against the eye, so much so that overaction of the marginal portion produces inturning of the lid margin (entropion). Thus one surgical approach to entropion is excision of the marginal portion of the orbicularis. This notion is reasonable, on the basis of the anatomy, but no evidence shows that the marginal portion has greater tonus than the rest of the pretarsal portion; thus there is no particular reason to make a functional distinction between them. If the marginal portion of the orbicularis has a special role in keeping the lid margins against the eye, it is probably a simple mechanical consequence of the fact that these muscle fibers are closest to the lid margins.

Blinking may be initiated as a reflex response or as a regular, spontaneous action

The orbicularis is innervated by branches of the facial nerve (see Chapter 5), and the impulses in the facial nerve axons are the efferent part of a reflex arc that begins with activation of nerve endings in the cornea and conjunctiva. Signals from the sensory endings are conveyed by the long ciliary nerves of the trigeminal system into the brainstem nucleus of the trigeminal, where connections are made with cells in the nucleus of the facial nerve.

The blink reflex is protective; the movement of the eyelids across the eye is an attempt to dislodge the object that activated the signals from the cornea or conjunctiva. Reflex blinks can be also elicited by visual stimuli, such as a small object moving rapidly toward the face or a very intense light flash, in which case the afferent part of the reflex arc is the optic nerve, or by stimuli to some of the other nerves innervating the skin around the orbits. Auditory stimuli, particularly if they are unexpected, can also elicit reflex blinking. Reflex blinks are generally not accompanied by eye movement larger than a few degrees, but voluntary lid closure is associated with significant rotations of the eyes upward and outward (**Bell's phenomenon**). The significance of Bell's phenomenon is uncertain.

Not all blinks are alike. Reflex blinks elicited by sensory stimuli are large excursions of the upper lid, usually covering the pupil and closing the palpebral fissure completely; they are very quick, lasting on the order of 150 to 200 ms. Spontaneous blinks, the sort that normally occur every few seconds, are different. The upper eyelid moves more slowly through a smaller amplitude; the blink lasts 250 to 300 ms, and the palpebral fissure does not close completely. These differences between reflex and spontaneous blinks are largely due to involvement of both pretarsal and preseptal portions of the orbicularis in reflex lid closure and greater involvement of the levator in elevating the upper lid after the reflex closure; thus the reflex blink is faster and larger in amplitude.

Although lid closure during a blink is basically an action of the orbicularis, the sequence of events, which is known by electrical recordings from the muscle during contraction (electromyography), begins with inhibition and relaxation of the levator, followed shortly thereafter by contraction of the orbicularis to move the upper lid down and the lower lid up (the lower lid moves most during voluntary lid closure, less in reflex blinks, and least in spontaneous blinks). By the time the lids reach their most closed position, the orbicularis has ceased to be active, and the reopening phase is essentially an action of the levator; any downward movement of the lower lid is mainly passive and due to gravity.

Most of us blink spontaneously every 3 to 7 s; 12 blinks per minute is about average. Although the eyelids do not completely close during spontaneous blinks, the upper lid usually drops low enough to cover most of the pupil. These interruptions of light entering the eye are very brief, however, and we rarely notice them, partly because visual sensitivity is reduced during blinks* and because the pupil is completely covered for only a small fraction of a blink, lasting 300 ms.

The frequency of spontaneous blinking varies. Infants blink infrequently (when awake); some animals seem almost never to blink (try a staring contest with a cat; the cat will always win if it doesn't get bored); and adults not only have individual variations in blink rate, but the rate for a given individual can change with emotional state, mental effort, illumination level, or atmospheric conditions, among other things. These variations suggest that spontaneous blinks are spontaneous only in the sense that the stimulus is not readily apparent.

Lid movements during spontaneous blinks move tears across the cornea

Lid closure during blinks has long been described as having both vertical and lateral-to-medial components, such that tears are not only swept vertically across the cornea during a blink, but also pushed medially toward the sites of drainage at the medial canthus. This action has been interpreted as a zipperlike closure of the palpebral aperture from lateral to medial, but that is not quite the case.

When the orbicularis contracts during a spontaneous blink, the lower lid shifts medially as the upper lid descends. The effect can be seen by observing fluorescent dyes or small carbon particles incorporated into the tear film. In the interval between blinks, the tear film and the tear meniscus along the lower lid margin are stationary, but during a blink, tears are wiped down across the cornea by the upper lid and moved medially along the margin of the lower lid in the opening phase of the blink (Figure 7.7).

The mechanical arrangement producing horizontal movement of the lower lid margin and vertical movement of the upper lid margin is not obvious. If the lower lid margin shifts medially because the attachment to the lateral palpebral ligament is elastic and not very restrictive, for example, a corresponding medial shift in the upper lid might also be expected, but it has not been reported. A simpler possibility is that fiber bundles in the lower lid run almost horizontally between the palpebral ligaments, so that their contraction is more nearly isometric. Fiber bundles in the upper lid, however, follow arcuate paths, allowing the lid margin to move vertically as the muscle fibers contract and shorten.

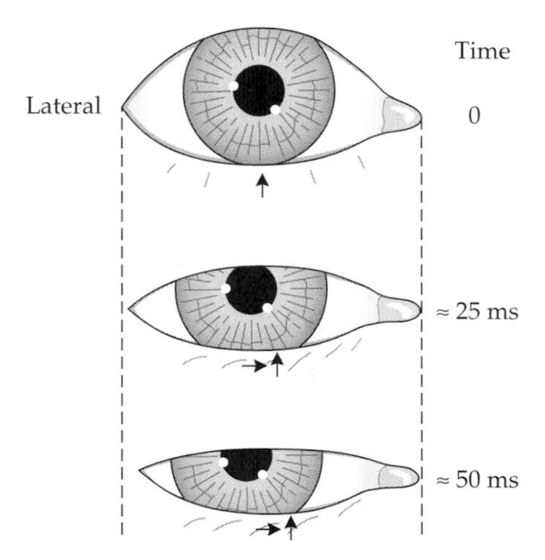

Figure 7.7
Movements of the Eyelids in a Spontaneous Blink
In this sequence, the palpebral fissure is maximally open at the beginning of a blink (time 0). At 25 ms later, the upper lid margin has dropped low enough to intersect the pupil; the lateral canthus has shifted medially, as has the whole lower lid margin (note the change in position and orientation of the lines below the lower lid margin). As the blink nears completion of its downward phase (50 ms), the pupil is more than half covered and the lateral canthus and lower lid margin have shifted more medially.

Overaction of the orbicularis may appear as blepharospasm or as entropion

The facial nerve is subject to irritative lesions, particularly where it exits from the stylomastoid foramen onto the face, and these lesions can produce spastic contraction of the facial muscles in various combinations. When the branches innervating the orbicularis are involved, the result is **blepharospasm**, which is a forcible, involuntary closure of the lids that is almost impossible to override voluntarily. There are other causes of blepharospasm, however; it may be produced as an exaggerated version of the blink reflex when there are painful stimuli to the anterior segment of the eye, or it may be idiopathic (cause unknown). In addition, the muscle spasms can be episodic or chronic.

Treatment for blepharospasm involves finding and eliminating the source of irritation, if this can be done, or reducing the effectiveness of neurotransmission

*This statement is worth a little thought. How can we show a reduction in sensitivity during blinks that interrupt any test stimulus a subject is asked to look at? The trick is bypassing the normal optical route by shining light into the eye through the roof of the mouth; the stimulus is diffuse, but it is unaffected by blinking and its intensity can be altered to produce very reliable threshold measurements at any time before, during, or after a blink. See Volkmann, Riggs, and Moore 1980 for details.

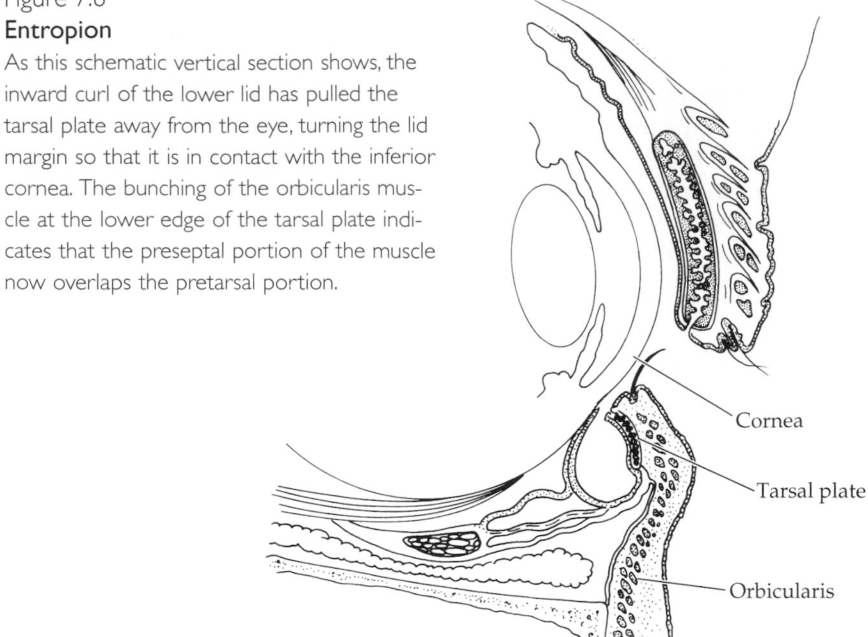

Figure 7.8
Entropion
As this schematic vertical section shows, the inward curl of the lower lid has pulled the tarsal plate away from the eye, turning the lid margin so that it is in contact with the inferior cornea. The bunching of the orbicularis muscle at the lower edge of the tarsal plate indicates that the preseptal portion of the muscle now overlaps the pretarsal portion.

at the neuromuscular junctions in the orbicularis. At present, one of the most effective ways to reduce the response of the orbicularis to the abnormal barrage of nerve impulses in essential blepharospasm is by injecting botulinum A toxin into the muscle. The toxin blocks the Ca^{2+}-mediated release of synaptic vesicles from the axon terminals, thereby reducing the amount of acetylcholine released from the terminals (see Chapter 4). Since the toxin binding to the receptor sites is long-lasting, an injection can provide several months of relief from blepharospasm; this symptomatic treatment does not address the underlying cause of the abnormal stimulation.

Sometimes part of the orbicularis exhibits chronic overaction, most commonly the preseptal portion of the muscle. This contraction, along with abnormally low antagonism in the opposing muscles, can cause the preseptal part of the muscle to override the pretarsal portion; as Figure 7.8 shows, the lid margin (the lower lid in this case) is rotated in toward the eye, producing a condition called **entropion**.

The inward rotation of the lid margins brings the eyelashes (cilia) into contact with the cornea and bulbar conjunctiva; the rubbing of the lashes across the cornea during blinks is painful and even if the corneal nerve endings adapt, continual rubbing can abrade the cornea, removing its normal protective surface and leaving the way open for inflammatory cell infiltration and infection. What begins as a simple change in position of the lid margins might eventually end up as a dense central corneal scar that requires a corneal graft to restore useful vision.

Entropion is not necessarily a result of spastic overaction in part of the orbicularis; trauma or severe infections of the lids can distort the lid margins with scar tissue formation during healing, producing a cicatricial entropion. Whatever the cause, correction of entropion is usually a surgical procedure.

Paresis of the orbicularis produces ectropion and epiphora

Underaction of the orbicularis has several effects. Lid closure is difficult or impossible, depending on the magnitude of the paresis (weakness), blinks are reduced in amplitude, and the lower lid everts—that is, it falls outward away from the eye (the upper lid is somewhat more elevated than normal, but gravity holds it against the eye). Chronic eversion of the lower lid is called **ectropion**. More commonly, however, ectropion is due to a gradual loosening or stretching of the connective tissue in the lower lid (a situation surgeons call excess "laxity") or of the palpebral ligaments; scar formation may also produce it.

Improper movement of tears across the eye, causing desiccation of the cornea and parts of the inferior conjunctiva, is an obvious consequence of a paralyzed orbicularis, but there is another, resulting from the ectropion. Since the lower lid has moved away from the eye, the narrow meniscus of tears at the junction of the lower lid margin and the globe is converted into a pool of tears in the cul-de-sac of the inferior fornix. In addition, the site on the medial lid margin where tears normally drain off (the inferior punctum, which will be discussed later) is everted away from the tear pool. Therefore, as tears accumulate, they have nowhere to go except over the lid margin onto the face, a phenomenon known as **epiphora**.

Epiphora is more than an aggravation. Although the epidermal cells of the skin contain keratin and are somewhat waterproof, the skin of the lids is very thin and does not tolerate continual wetting. In time, the outer layers of the skin break down and the tissue can ulcerate, producing still more tissue destruction. Epiphora requires surgery for relief; the various techniques combine removal of tissue from the lid with shortening of the medial or lateral palpepral ligaments so that the lower lid margin is brought back in contact with the eye.

Other glands in the lids are associated with the eyelashes

The eyelashes, or **cilia**, line both lid margins, just anterior to the ducts of the tarsal glands; there are more than 100 of them in the upper lid and about half as many in the lower lid. They curl away from the lid margins and appose as the lids close. Their function is probably protective, catching airborne dust and grit before it can reach the eye.

Like other hairs on the body, the eyelashes grow from follicles just below the skin surface (see Figure 7.9). A follicle is a deep pocket of cells specialized for synthesizing the proteins from which the lashes are assembled within the lumen of the follicle. The lashes are not permanent structures; they grow to their full length in a few months, last for a month or so longer if unmolested, and then are shed from their follicles. Unless the follicles are destroyed, they will begin to generate new lashes as soon as the old ones are shed.

The eyelash follicles have small glands associated with them (Figure 7.9). These **glands of Zeis** produce an oily secretion that lubricates the shafts of the lashes and may assist in the dust-trapping function of the lashes by making them somewhat sticky. The clusters of secretory cells that constitute Zeis's glands look very much like miniature lobules of the tarsal glands, but their collecting ducts are very short and are rarely seen in histological sections. Zeis's glands are one of those anatomical features that are easy to ignore as long as they are working properly, becoming of interest only when something goes wrong. But like the tarsal glands, Zeis's glands are subject to infection, with similar consequences—that is, soreness and swelling along the lid margin. Their swelling produces an obvious bulge on the outer surface of the lid, which is called an **external hordeolum**, or stye. Most external hordeola resolve spontaneously or in response to antibiotics and rarely require incision for drainage.

Figure 7.9 shows another small gland in the vicinity of the the lash follicles, but glands of this type are not a functional part of the eyelash–follicle system. These **glands of Moll** are enlarged, modified sweat glands, easily identified in histological sections by the ring of secretory cells around a large lumen (they are about the size of small arterioles and venules, but the cellular composition of their walls makes the distinction a simple one). The eyelid skin sweats, and Moll's glands contribute to it; their ducts empty onto the lid margins external to the cilia.

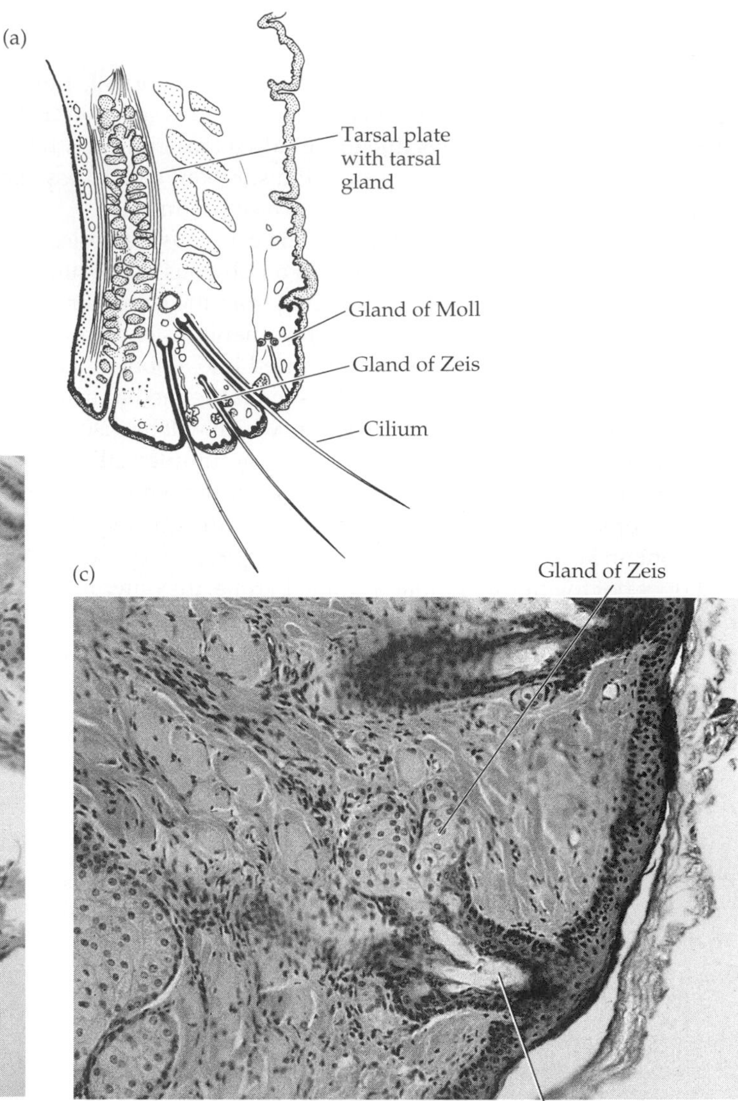

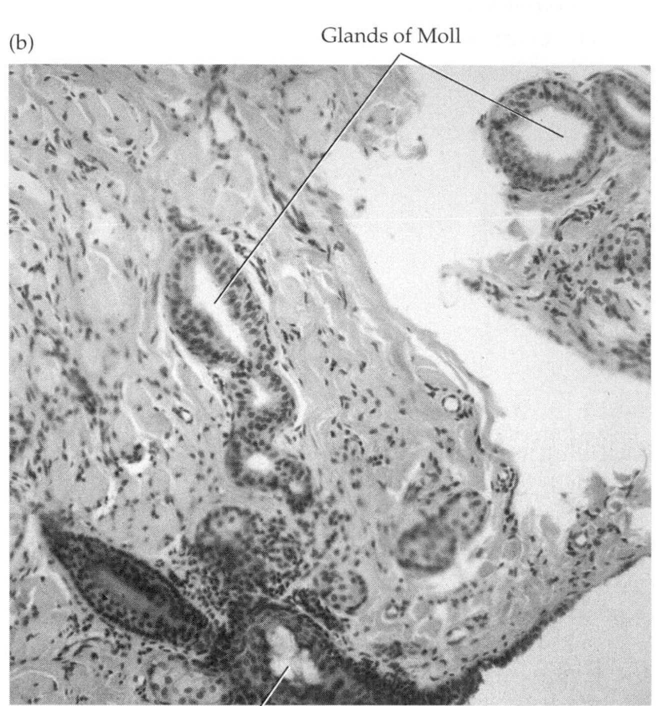

Figure 7.9
Glands of Zeis and Moll
The schematic section through the upper lid (*a*) shows the relationship of Zeis's and Moll's glands to a follicle of one of the cilia. Moll's gland (*b*) has large open lumina lined with a single layer of secretory cells. Zeis's gland (*c*) is attached to the follicle, and its secretion goes onto the cilium within the follicle.

The skin on the lids is continuous with the conjunctiva lining the posterior surface of the lids and covering the anterior surface of the sclera

More will be said later about the **conjunctiva** (see Chapter 9); here it enters into the discussion because it is associated with the eyelids as well as the eye. The conjunctiva is a tissue that has a layered surface epithelium overlying a connective tissue substrate or stroma, much like the layers of epidermis and dermis of the skin. In fact, the corresponding parts of the skin and the conjunctiva are continuous along the posterior margins of the eyelids, where there is an abrupt change as the epidermis loses its keratin to become the nonkeratinized epithelium of the conjunctiva and the thicker epidermis and dermis become thinner in their continuations as epithelium and stroma of the conjunctiva, respectively.

The **palpebral conjunctiva** lines the posterior surfaces of the eyelids; its epithelium is thin, only a few cell layers thick, as is its well-vascularized stroma. It is part of a continuous sheet of conjunctiva that covers the posterior lid surfaces and is then reflected back to cover the anterior sclera; the epithelium of this conjunctival sheet is continuous with the epithelium that covers the cornea (Figure 7.10). The reflection or folding back of the conjunctiva forms a deep pocket or cul-de-sac behind each of the lids; the region of the fold is the **fornix**, and the conjunctiva in the region is the **fornix conjunctiva**. Contact lens patients,

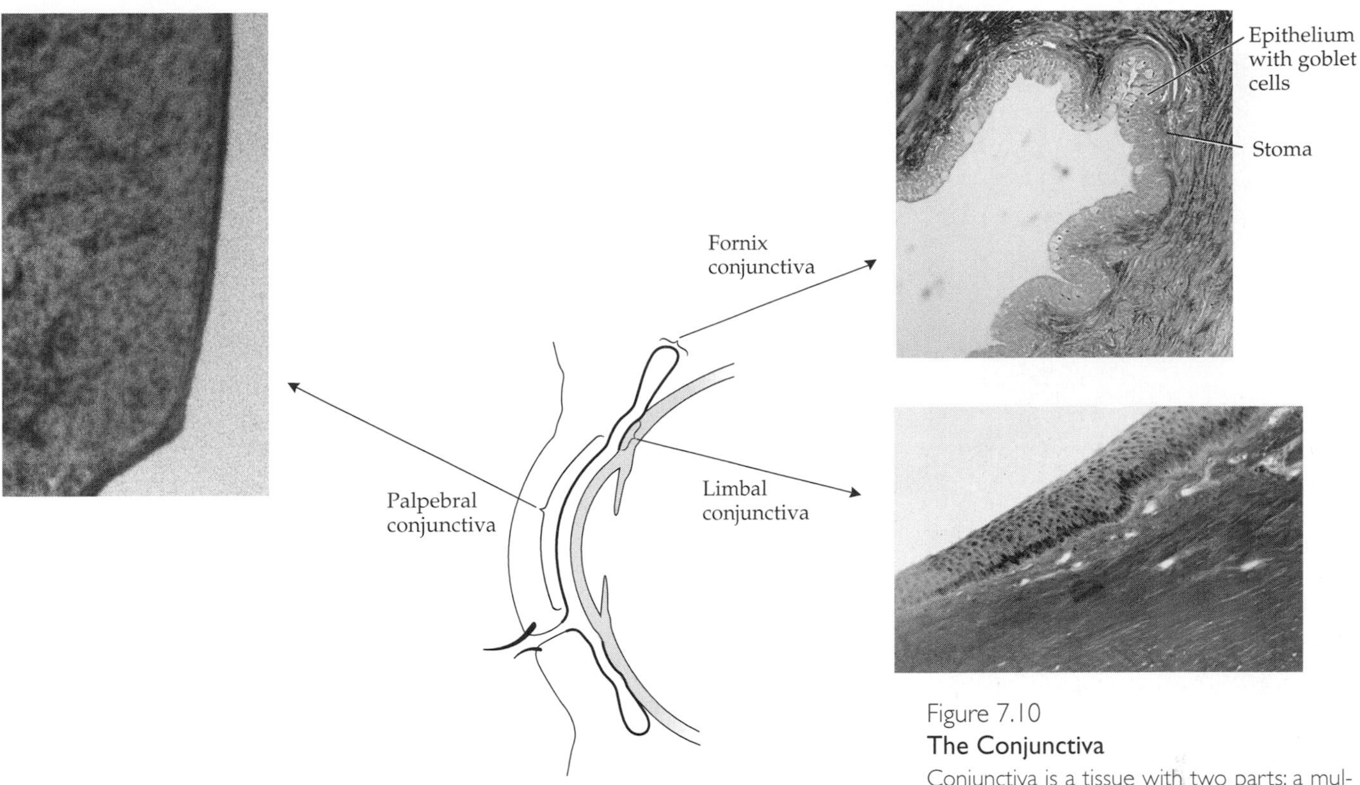

Figure 7.10
The Conjunctiva
Conjunctiva is a tissue with two parts: a multilayered outer epithelium and an inner connective tissue stroma. It lines the entire global side of the eyelids and is reflected back to cover the anterior part of the eye. Three different regions of the conjunctiva are shown here: palpebral, fornix, and limbal. Although all parts of the conjunctiva have epithelium and stroma, the thickness of the epithelium and the vascularity of the stroma vary from one region to another. Fornix conjunctiva and adjacent parts of the palpebral and bulbar conjunctiva also have mucus-secreting goblet cells in the epithelium.

particularly young ones, sometimes wonder if a contact lens can be displaced from the cornea and end up behind the eye; the answer is no, of course, although a lens lodged in the upper fornix can be positioned well back toward the equator. The continuity of the conjunctival sheet prevents escape from the cul-de-sac. From the fornix, the conjunctiva covers the anterior part of the eye as the **bulbar** and **limbal conjunctiva**.

Like the rest of the conjunctiva, the epithelial cells in the fornix are large cuboidal to columnar cells. Interspersed among the epithelial cells are unicellular, mucus-secreting **goblet cells**, so named because their shape in cross section is like a small vase or amphora (Figure 7.11*a*). These apocrine glands manufacture mucin, packaging it in secretory granules until the cell is full and then discharging the granules onto the conjunctival surface. The cycle of mucin production and secretion continues indefinitely (unlike the epithelial cells, goblet cells are not shed periodically). The fornix conjunctiva contains thousands of goblet cells, and their density is greater here than in the adjoining palpebral and bulbar conjunctiva. The distribution of goblet cells is far from uniform, however (Figure 7.11*b*); they are densest medially and inferiorly, and some are found in both palpebral and bulbar conjunctiva.

The mucus secreted by the goblet cells forms a thin layer of the tear film lying next to the corneal surface (see Figure 8.25); it also contributes to the accretion of hard, gritty material often found at the medial angle of the lids upon awakening in the morning. Stability of the precorneal film may depend on the presence of the mucoid layer and its adherence to microvilli on the corneal surface.

The orbital septum is a connective tissue sheet extending from the orbital rim to the tarsal plates

The **orbital septum** forms a barrier that prevents the herniation of orbital fat into the eyelids. As Figure 7.12*a* shows, the thin connective tissue sheet of the orbital septum fuses with the periorbita around the orbital rim, running from the rim

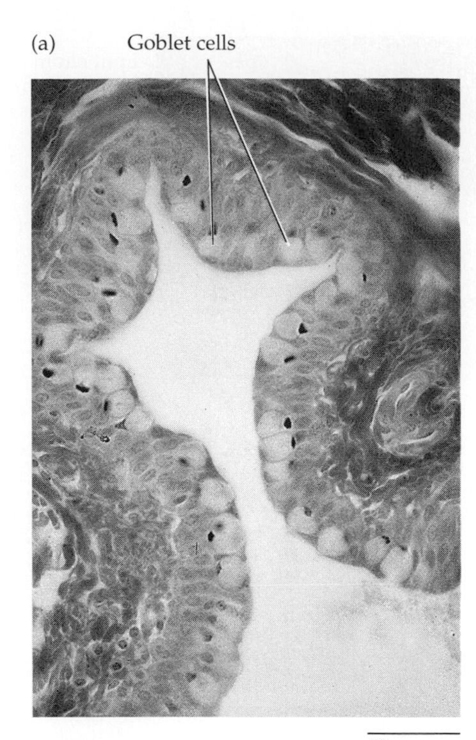

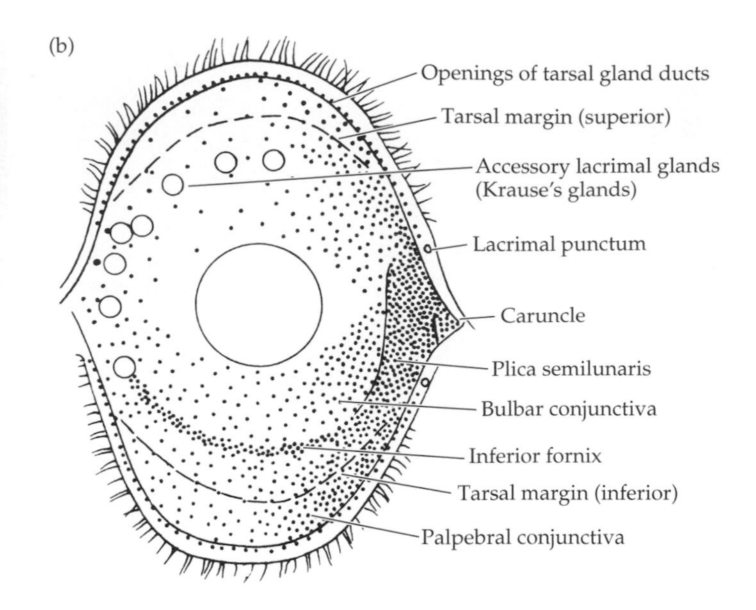

Figure 7.11
Goblet Cells
(*a*) Goblet cells are unicellular mucus-secreting glands distributed throughout the conjunctival epithelium, most densely in the fornix, which is shown here. (*b*) This drawing shows the lids completely everted, so the conjunctiva is spread out as a flat sheet of tissue. Each small dot represents several goblet cells. They are most numerous near the medial canthus and in a band extending across the inferior fornix. The large dots represent accessory lacrimal glands of Krause, which are located superiorly and laterally. The openings of the tarsal gland ducts are shown along the lid margins, as are the two puncta of the lacrimal drainage system.
(*b* after Kessing 1968.)

into both eyelids to fuse with the connective tissue around the tarsal plates. There are very few breaks in the continuity of the septum; branches of the medial and lateral palpebral arteries pass through it, but the most significant interruption accommodates the levator aponeurosis, which takes the place of the orbital septum a few millimeters above the tarsal plate and with which the orbital septum merges. With these exceptions, the septum is a continuous, flexible sheet joining bone at the orbital rim to the solid tarsal plates; unless it is perforated by trauma or by degeneration of the tissue, the orbital septum keeps the orbital fat in the orbit where it belongs.

There are three large pockets of fat below the eye and two above; all lie outside the extraocular muscles and their fascia sheaths (Figure 7.12*b*). The superior-lateral part of the orbit contains no fat pocket because of the presence of the lacrimal gland. Thus, a bulge in the lateral part of the upper lid that might indicate herniation of fat if it were elsewhere in the lids is instead a sign that the lacrimal gland has prolapsed into the lid.

The effectiveness of the septal barrier is partly a function of heredity, and it decreases with age or disease, allowing orbital fat to push forward into the lids; the result can be undesirable appearance and may lead to functional impairment (restriction of the visual field). The solution is fat removal and reconstruction of the septum. This bulging of the upper lid—called blepharochalasis—can be confused with, and may coexist with, folding and bulging of excess skin on the lid surface, a condition called **dermatochalasis**. Like other defects in the skin that appear with age, dermatochalasis is hereditary and may be exaggerated by the cumulative effect of sun exposure.

Most of the "bags under the eyes" about which people complain are relatively minor and largely due to loss of septal integrity and tissue elasticity, along with fluid accumulation in the loose tissue between the skin and the muscles of the lid and face. Cosmetic surgery can improve appearance markedly.

The shape and size of the palpebral fissure vary

The median length of the palpebral fissure in adults is about 29 mm from the lateral to the medial canthus, and the maximum width of the aperture is about 9.5 mm. As Figure 7.13 shows, the lower lid margin is normally tangential to the inferior margin of the cornea, but the upper corneal margin is overlapped slightly by the upper lid. Two structures are visible in the medial canthus—a crescent,

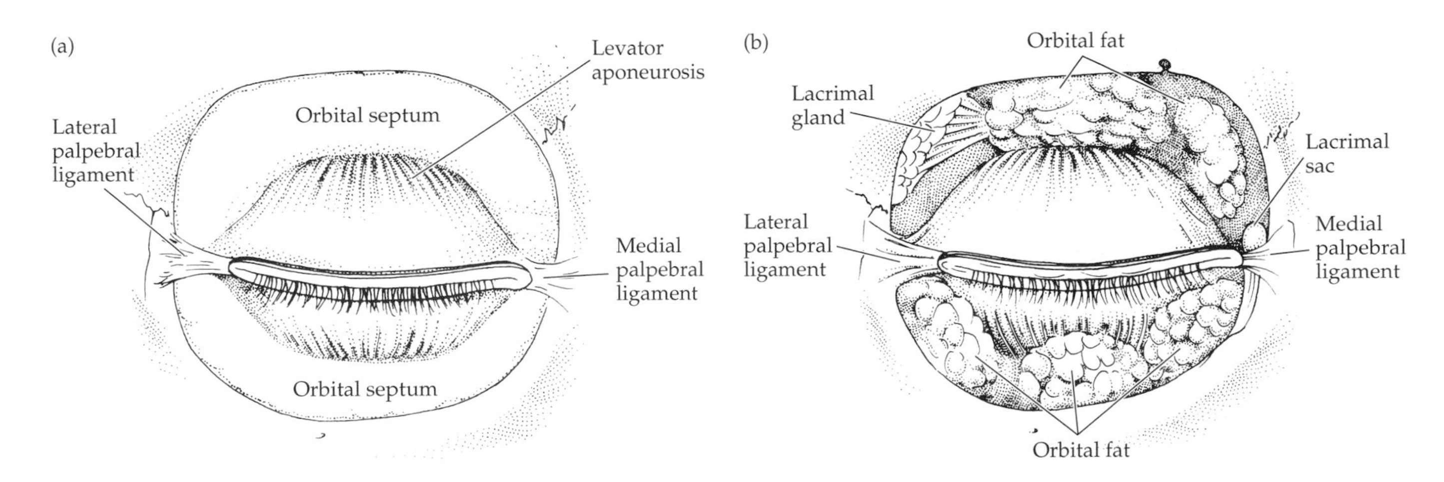

Figure 7.12
The Orbital Septum and Orbital Fat
(*a*) The orbital septum is a sheet of tissue extending from the periorbita at the rim of the orbit in toward the tarsal plates. It merges with the levator aponeurosis in the upper lid and with fascia overlying the inferior tarsal muscle in the lower lid. (*b*) Removal of the orbital septum reveals pockets of fat lying behind it within the orbit; there are typically three pockets below the eye and two above. There is no fat pocket in the superior lateral part of the orbit, where the lacrimal gland is located.

moon-shaped fold of conjunctiva—the **plica semilunaris**—and on top of it, a small tissue bulge called the **lacrimal caruncle**. Both the plica semilunaris and the caruncle are vestigial—remnants of the nictitating membrane or third eyelid in other species. Neither structure has any known function in humans, although the caruncle has some fine cilia and both structures contain small glands, including goblet cells.

Given the individual, racial, and sexual differences in human facial structure, it is not surprising to find significant variation in the size and shape of the palpebral fissure. For adults, the size variation can be summarized by adding standard deviations of ±2 mm to the 29 mm length and ±0.3 mm to the 9.5 mm width. The only importance of this variation is diagnostic: Narrow palpebral fissures that are normal for one individual may represent bilateral ptosis in another, or conversely, wide fissures could be taken as a sign of proptosis if they are not known to be habitual. Certain combinations of palpebral fissure size and shape are considered particularly aesthetic; my personal favorites belong to Sophia Loren. (This opinion is not based on direct observation, unfortunately.)

Most of the variation in shape of the palpebral fissure is due to the structure of the bridge of the nose and the skin on either side of it. Until fairly late in gestation, the fetus has a fold of skin, the **epicanthal fold**, or **epicanthus**, that covers the caruncle and plica semilunaris in the medial canthus. In Caucasians and Negroid races, the epicanthal fold has usually disappeared by the time of birth, unless birth is quite premature; it may persist in some cases, however. In some

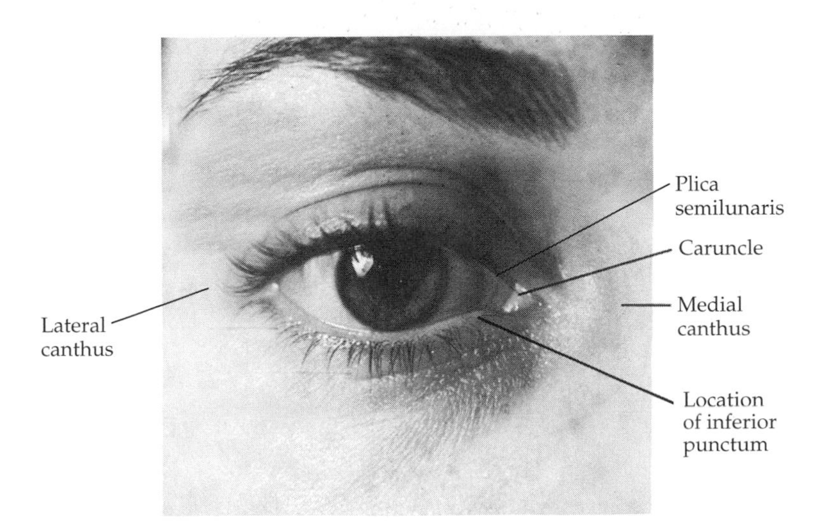

Figure 7.13
External Structure of the Medial Canthus
The medial canthus contains the plica semilunaris, an exaggerated fold in the conjunctiva, with the caruncle protruding from its surface. Small elevations on the lid margins mark the locations of the puncta for the lacrimal drainage system. (From Newell and Ernest 1978.)

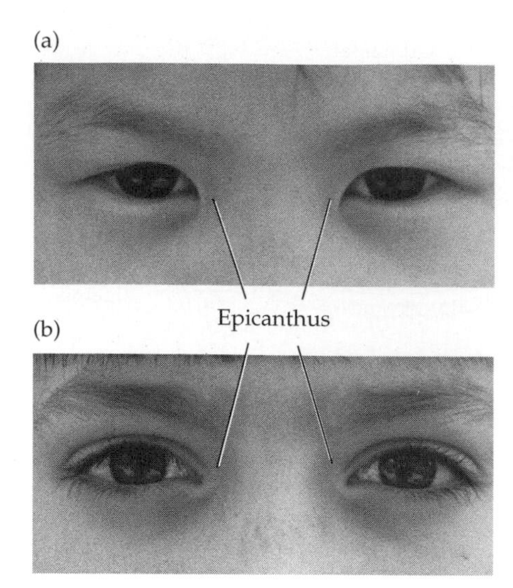

Figure 7.14
Variations in Lid Structure
(*a*) Asian children have a relatively flat bridge of the nose, an epicanthal fold of skin that covers the caruncle at the medial canthus, and no superior palpebral furrow. These features typically, though not always, persist into adulthood. Although Caucasian infants usually have a small epicanthal fold, it does not persist into childhood (*b*) (Photographs by Christopher Small.)

races, the bridge of the nose is less developed; it remains relatively flat as the face develops, and the epicanthal folds tend to be retained. This feature is particularly noticeable among Asians (Figure 7.14*a*).

The epicanthal fold often covers the true medial canthus, and the visible intersection of the upper and lower lids at the medial side is displaced laterally and down, giving the palpebral fissure the appearance of slanting downward from the lateral canthus to the medial angle of intersection; the palpebral fissure is also relatively narrow. In addition, the upper lid structure is somewhat different; the orbital septum fuses with the levator tendon much closer to the tarsal plate, thereby eliminating the superior palpebral furrow. These features are also seen among Native Americans, thus hinting at their link to Asia. The presence of the epicanthal fold and the absence of the superior palpebral furrow are variable among Asians, however, and many do not fit the stereotypical pattern.

Highly developed, retained epicanthal folds are associated with Down syndrome. The "oriental" shape of the palpebral fissures was the justification for the now discontinued terminology of "mongolism" for this syndrome.

The overall structure of the lids consists of well-defined planes or layers of tissue

Figure 7.15 is meant to make explicit a point alluded to throughout the foregoing discussion—namely, that the eyelids are constructed as layers of tissue in a well-defined sequence from the outer to the inner surface of the lids. In the tarsal portion of the upper lid, the layers include the very thin epidermis and dermis of the skin, the pretarsal portion of the orbicularis, connective tissue and the end of the levator aponeurosis, the tarsal plate, and the palpebral conjunctiva. The lower lid is the same, except there is no levator tendon.

The layering is somewhat different in the septal portion; for the upper lid, the sequence is skin, preseptal orbicularis, loose connective tissue, orbital septum, levator aponeurosis, tarsal muscle, and palpebral conjunctiva. Except for the levator tendon, the lower lid is the same. As indicated in Figure 7.15, most of the larger blood vessels are located in the connective tissue behind the orbicularis. The arteries are anastomoses between the medial and lateral palpebral arteries, forming what are called arcades (see Chapter 6). The peripheral arcade is in the septal portion; the marginal arcade lies in front of the tarsal plate.

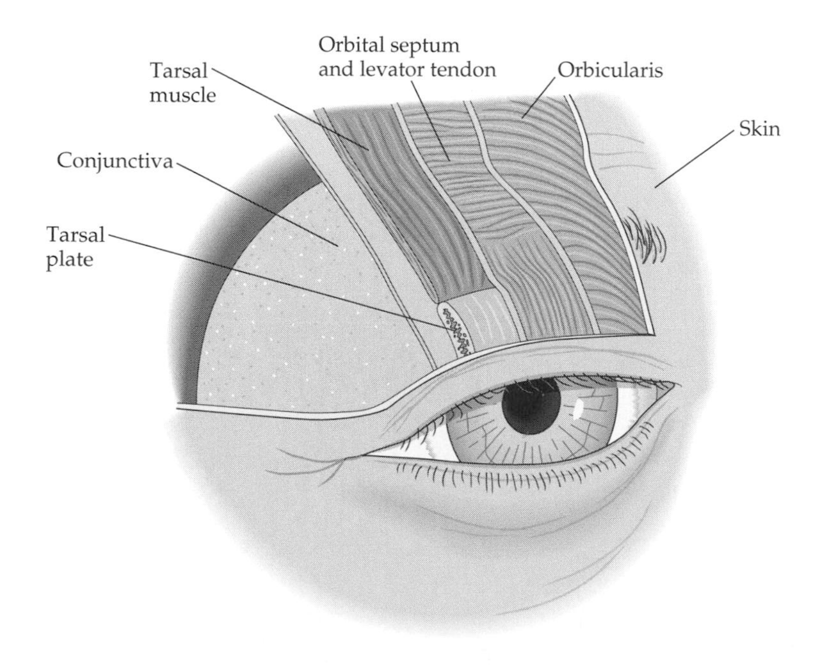

Figure 7.15
Three-Dimensional Structure of the Eyelids
In this drawing, portions of the upper lid have been cut away to show the layers within the lids. There are differences between the tarsal and septal portions of the lids (there is no septum in the tarsal portion and no tarsal plate in the septal portion) and between the upper and lower lids (the lower lid has no counterpart to the levator muscle).

One of the other differences between tarsal and septal portions of the lids is the amount of loose connective tissue in front of or behind the orbicularis; there is less connective tissue in the tarsal position of the lids, it is denser, and the layers are more tightly bound together. The septal portion has a much looser structure. The looseness of the septal portions of the lids contributes to the ease with which the tarsal portions can be moved and provides considerable space for fluid to accumulate from damaged blood vessels or into which fat can herniate if it is not restrained by the orbital septum.

Tear Supply and Drainage

Most of the tear fluid is supplied by the main lacrimal gland

Figure 7.12*b* gave us a glimpse of the **lacrimal gland** in the upper lateral portion of the orbit. It is a flattened, oval mass about 20 mm along its larger dimension, 12 to 15 mm along the smaller, and about 8 or 9 mm thick. The gland is partially divided by the levator aponeurosis into two connected lobes: a larger orbital lobe lying next to the frontal bone in the lacrimal fossa, and a smaller palpebral lobe. Because the palpebral lobe lies below the aponeurosis next to the eye, it can often be seen through the fornix conjunctiva when the upper lid is everted.

In its fine structure, the gland is essentially the same throughout, consisting of small clusters of secretory cells surrounding a lumen, with numerous secretory clusters (acini) assembled into small lobules (Figure 7.16*a*). The small openings in the centers of the acini and the granular appearance of the secretory cells, resulting from the secretory granules they contain, can be seen at higher magnification (Figure 7.16*b*). The smallest lumina merge to form larger ones, and all these tributaries eventually become a dozen or so large ducts that open into the upper lateral fornix.

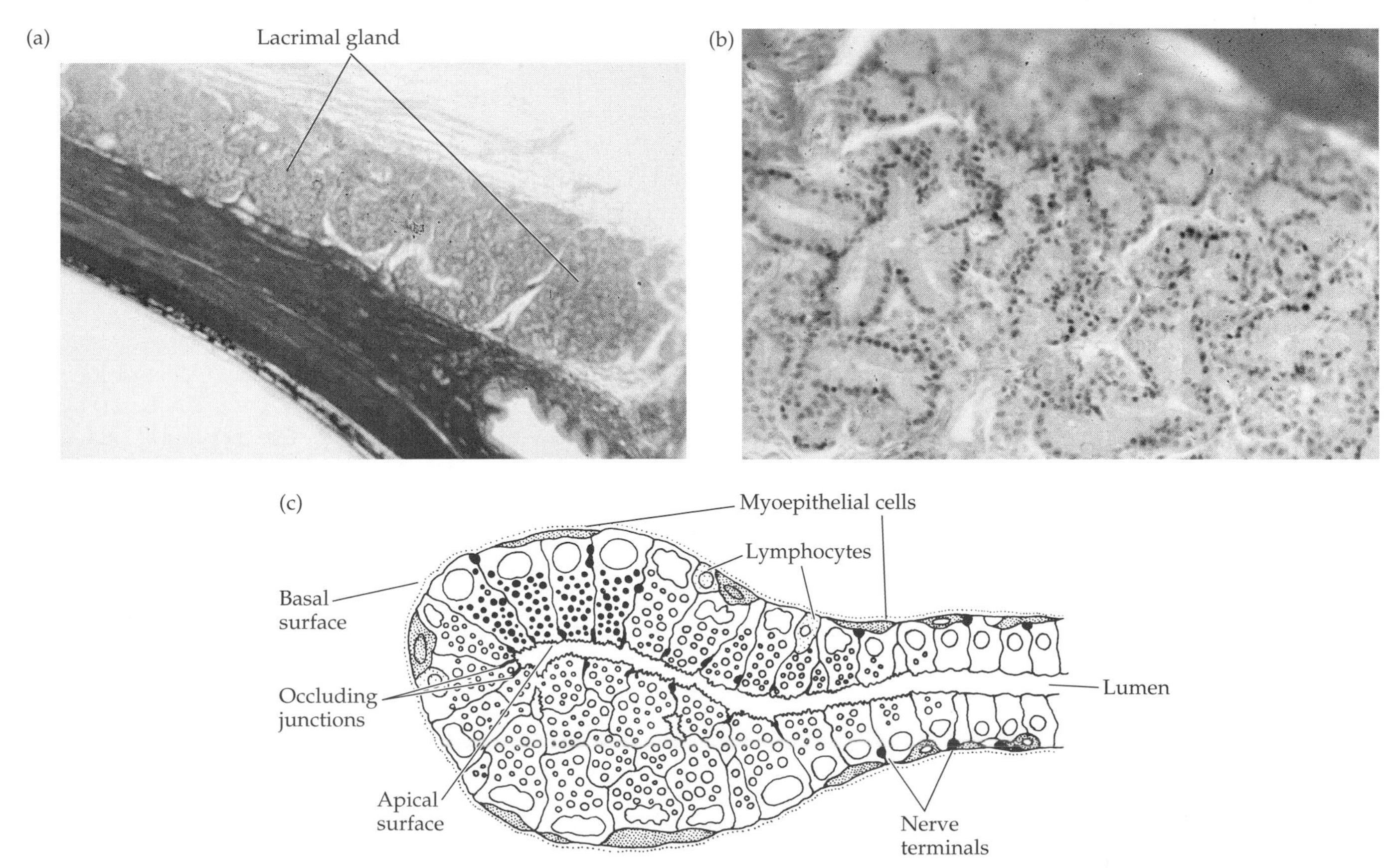

Figure 7.16
The Lacrimal Gland
(*a*) A portion of the lacrimal gland, showing lobules of secretory clusters (acini). (*b*) At high magnification, some of the small lumina at the centers of the acini can be seen. The cells have a granular appearance because of the secretory granules within them. (*c*) Several cell types are present in the secretory clusters, including myoepithelial cells and lymphocytes. The apical borders of the secretory cells have occluding junctions. Nerve terminals typically lie between cells near the basal surface. (*c* after on Ruskell 1975.)

Figure 7.16*c* shows more detail of a secretory cell cluster. Secretory cells surround a small lumen, into which their secretions pass; their apical surfaces are closest to the lumen, their basal surfaces are outermost. Occluding junctions near the apical surfaces bind the cells together and limit the movement of ions, small molecules, and water in the extracellular space. All of the secretory cells contain secretory granules, shown in Figure 7.16*c* as dots within the cells. (Some are darker than others, indicating that the contents of the secretory granules may differ among cells.) Lymphocytes and small myoepithelial cells lie next to the basal surface of the secretory cell cluster. The lymphocytes link the lacrimal gland and the immune system, but the role of the myoepithelial cells is unclear (myoepithelial cells contain contractile proteins and can operate like smooth muscle fibers; cells of this type form the dilator muscle in the iris; see Chapter 10). Small nerve terminals can be found between cells near the basal surface of the cell cluster.

The grapevine analogy used earlier for the structure of the tarsal glands works just as well here. The grapes are the acini (the small clusters of secretory cells), each bunch of grapes is a secretory lobule, and the twigs and branches are the collecting ducts merging toward the central stem. Imagine compressing the grapevine into a more compact mass; the result is the gross structure of the lacrimal gland.

At its resting level, tear fluid is produced at a rate of about 1 μl per minute, or about 1.5 ml each day (at this rate, the total production for a year is just over a half liter).* Most of this tear fluid is contributed by the main lacrimal gland; additional secretion of a similar type comes from the accessory lacrimal glands (Krause's glands and the glands of Wolfring). The accessory lacrimal glands look like miniature versions of the main lacrimal gland; there are eight to ten Krause's glands in the conjunctiva along the upper lateral fornix (see Figure 7.11*b*) and several times as many glands of Wolfring along the orbital margins of the tarsal plates (the reported numbers vary widely). Because the combined volume of the accessory lacrimal glands is calculated to be about one-tenth that of the main lacrimal gland, they are assumed to contribute about 10% of the aqueous component of tears.

Secretion by the lacrimal gland is regulated by autonomic inputs operating through a second-messenger system

The most obvious neural input to the lacrimal gland consists of parasympathetic fibers originating in the pterygopalatine ganglion (see Chapter 5), but this simple picture needs two qualifiers. First, the parasympathetic fibers show considerable immunoreactivity for vasoactive intestinal polypeptide (VIP) as well as for acetylcholine, and second, sympathetic fibers that arise from cells in the superior cervical ganglion enter the lacrimal gland.

Similarities between the distributions of VIP and acetylcholine suggest that the parasympathetic nerve endings contain both substances; this co-localization of a neuropeptide with a conventional neurotransmitter is a common observation. Blocking or removing the parasympathetic input to the lacrimal gland cells reduces secretory activity, probably in large part because of the reduction in acetylcholine release, but the role of VIP in lacrimal secretion is not clear.

Sympathetic inputs may influence secretion by the lacrimal gland, but the effects are indirect; the terminals identified as sympathetic by anatomical criteria are most closely associated with the vasculature of the gland; only parasympathetic terminals are found along the basal portions of the secretory cells. Yet there

*Resting level refers to the absence of obvious stimuli for tear production. It was once thought that this basal level of tears was produced solely by the accessory lacrimal glands, with the main lacrimal gland responding to reflex stimuli. But since the main and accessory lacrimal glands all appear to be innervated, they all are likely to contribute to tear production under all conditions. If there is a basal, unstimulated level of tear production, it probably occurs only when the epithelia of the cornea and conjunctiva have been anesthetized; anesthesia markedly reduces tear production.

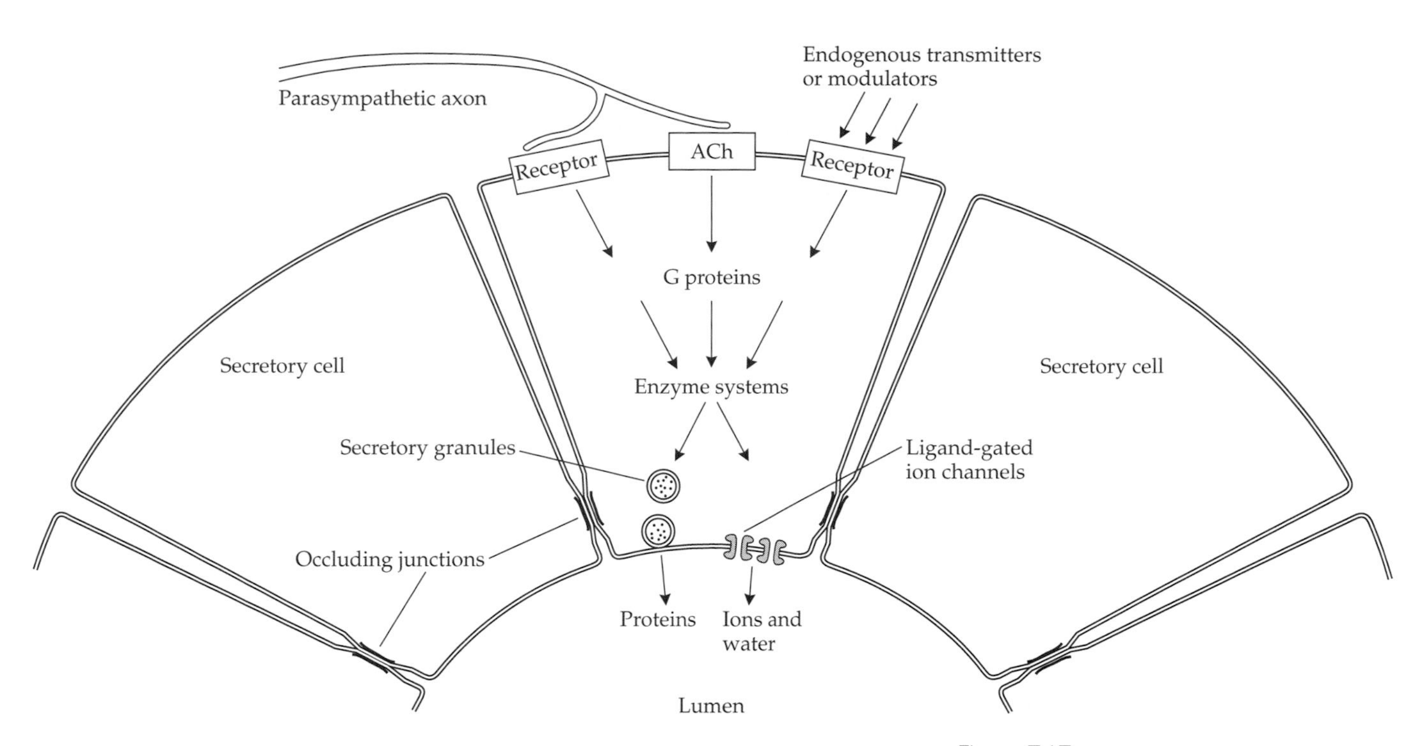

Figure 7.17
Stimulated Secretion by the Lacrimal Gland
The secretory cell membranes have G protein–linked receptors for acetylcholine (ACh) and other substances, probably VIP and norepinephrine at minimum. Binding of the neurotransmitter or modulator to the receptor activates G proteins that in turn activate one of several enzyme systems that can stimulate exocytosis of the proteins in the secretory granules or alter the status of ion channels in the membrane. (After Dartt 1989.)

are indications that sympathetic lesions will reduce output from the lacrimal gland; the question is whether this effect is on secretory activity or if it is secondary to changes in blood flow through the gland.

These possibilities (and uncertainties) are summarized in Figure 7.17, which is a schematic version of a secretory cell cluster, the inputs that may impinge upon it, and the major stages of intracellular events leading to secretion. One cell is shown as having several different types of receptors in its basal membrane; one is for acetylcholine, and the others are deliberately unspecified—provisional labels for VIP and norepinephrine might be appropriate, however.

The acetylcholine receptor is muscarinic; it is a metabotropic receptor that works through an intracellular **G protein**. (G proteins are discussed in more detail in connection with phototransduction; see Chapter 13.) For present purposes, consider the G protein to be a messenger, whose activation prompts it to seek out and activate intracellular enzyme systems involved in the production of secretory products from the cell. The other receptors are thought to do much the same thing, although they are activated by different ligands that bind to them (such as VIP), they use different G proteins, and they exert their effects through different enzyme pathways.

Proteins and ions are the main products of secretion by the lacrimal gland; water is produced as a consequence of the ion movement. Most of the proteins are packaged within the cells as small secretory granules; proteins are secreted when a secretory granule binds to the apical membrane and opens to release its contents into the lumen of the secretory cell cluster (exocytosis). The ion channels in the membrane appear to be ligand-gated channels in which the ion gates are opened and closed by binding or dissociation of an intracellular molecule (the ligand). The specific ligands involved in channel opening and closing are not known.

The composition of the lacrimal gland secretion varies with the secretion rate

Most of the proteins produced by the lacrimal gland cells are *regulated* proteins whose rate of output can be controlled through the sorts of pathways shown in Figure 7.17. A few, however, are not subject to external regulation; these are *constitutive* proteins, and their rates of production and secretion are intrinsic to the cell. One of the significant differences between regulated and constitutive proteins is their storage; regulated proteins are not secreted as they are formed but are stored in the secretory granules, or vesicles. Therefore, what is being regulated by external stimuli is not the rate of protein synthesis, but the rate of vesicular exocytosis.

The lacrimal secretion includes various proteins, the most prominent of which are secretory immunoglobulin A (sIgA), which is a constitutive protein, and three regulated proteins—lactoferrin, tear-specific prealbumin (TSP), and lysozyme. Lactoferrin is a bacteriostatic agent that acts by binding iron required by bacterial cells for metabolism. Lysozyme is bacteriolytic, not just rendering bacteria impotent, but destroying the bacteria that contain molecules targeted by lysozyme. Unfortunately, the spectrum of lysozyme is limited, and some important microorganisms are impervious to its action. Nonetheless, lactoferrin and lysozyme are important in defending the eye against infection; a decrease in their production may leave the eye vulnerable. Secretory IgA is also defensive, since it is an antibody that can act to neutralize a variety of potentially infectious agents. The function of the third regulated protein, tear-specific prealbumin, is unknown.

Stimulation may increase tear production by the lacrimal gland 100-fold or more above its resting level. The output of both water and regulated proteins appears to increase under these circumstances, but the relative concentration of regulated proteins drops somewhat (between 15 and 40%, depending on the specific protein) relative to the resting concentration. The concentration of secretory IgA drops precipitously, to 10% or less of resting concentration, indicating that stimulation of secretion has very little effect on this tear component. In short, reflex tears have more fluid and less dissolved protein than resting-level tears have, but much of the regulated protein secretion lags the fluid production by only small amounts.

Dry eye may result from a decreased amount of tears, abnormal tear composition, or both

Dry eye refers to a variety of disorders of the ocular tears that we experience as sensations of dryness of the eyes, discomfort, or presence of a foreign object. In most instances, the tear film loses its normal continuity and breaks up so rapidly that it cannot maintain its structure during the interval between spontaneous blinks.

Classification of dry eye is still evolving, but there are three large groups of disorders within the class of dry eye syndromes; these groups are related to deficiences in the major components of the tears. All of these tear abnormalities may have multiple causes; normal age-related decreases in secretion, trauma, innervational changes, hormonal effects, and so on, are all possibilities. Some forms of dry eye occur in conjunction with systemic pathology, most notably with rheumatoid arthritis, but many forms are idiopathic (cause unknown).

Perhaps the most common form of dry eye (**keratoconjunctivitis sicca**) is due to a decreased aqueous component in the tears. Secretion by the lacrimal gland declines with age, for unknown reasons, so most people who suffer from dry eye are middle-aged and older. Dry eye is more common in older women than in older men, suggesting a link between decreased tear production and hormone levels, but such a link has yet to be substantiated. In very severe cases, there are distinct changes in the lacrimal gland, including a loss of secretory tissue and an infiltration of lymphocytes. Severe dry eye of long standing may lead to deterioration of the corneal surface.

Abnormalities of the conjunctiva with decreased goblet cell activity or number produce a form of dry eye associated with decreased mucin content in the tears. The reduction in mucin affects the surface tension in the tear film, resulting in instability and rapid breakup of the tear film.

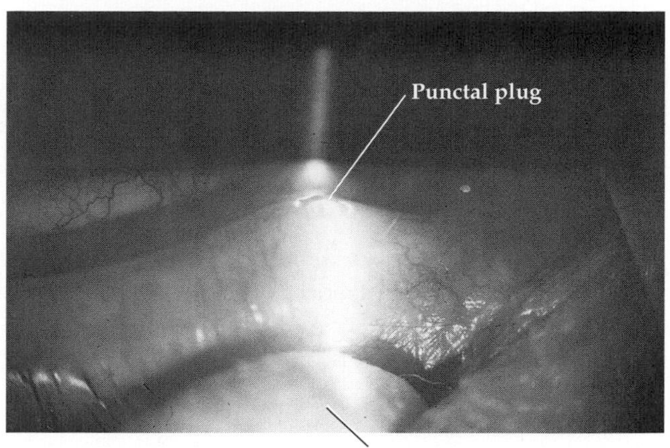

Figure 7.18
The Medial Canthus and Entrances to the Drainage System
(*a*) This photograph shows the inferior punctum when the eyelid has been everted by pressure from a cotton swab. (*b*) One of the possible solutions in cases of dry eye is blocking the puncta so that the reduced supply of tears is not drained away so quickly. Here, the inferior punctum has been blocked with a removable silicon plug. (Photographs courtesy of Leo Semes.)

Reduced secretion by the tarsal gland reduces the oily (lipid) component in the tears. This deficiency also leads to breakup of the tear film, because the watery component of the tears evaporates more rapidly in the absence or reduction of the oily outer surface layer.

In its mild form, dry eye is marginally symptomatic; a feeling of dryness and slight irritation of the eye, often increasing late in the day and exaggerated by wind and exposure, are the most common complaints. These mild annoyances may be exaggerated by contact lens wear; since the lenses require a fluid cushion to ride on and soft lenses require fluid to maintain their hydration, tear abnormalities affect them. These problems are usually solved with artificial tears. More severe cases may produce local dry spots on the cornea, epithelial erosion, and strands of epithelial cells and mucin attached to the cornea. More aggressive therapy for these cases requires blocking normal tear drainage.

Tears are drained off at the medial canthus and deposited in the nasal cavity

Two small elevations are positioned opposite one another on the upper and lower lids near the medial canthus, each with a small hole in the center (Figure 7.18*a*); the holes are the **puncta** through which the tears in the rivulets along the lid margins exit the palpebral fissure. The lower punctum can be seen when the lower lid is everted, as in Figure 7.18*a*. Blockage of tear drainage sites in patients suffering from dry eye means blocking the puncta; Figure 7.18*b* shows a silicon plug in the inferior punctum. (The puncta may also be closed permanently, by cauterization, or temporarily, with cyanoacrylate glue.)

The puncta are openings for small tubes of epithelium and connective tissue called the **canaliculi** (Figure 7.19). Initially, each canaliculus runs perpendicularly to the lid margin for about 2 mm, expanding as it goes. The expansion forms a small sac, the **ampulla**, after which the canaliculus turns medially and runs for another 8 mm or so. Since the canaliculi are behind the pretarsal portion of the orbicularis, their medial course takes them behind the medial palpebral ligament toward the nasolacrimal fossa. The canaliculi merge just before they reach the **lacrimal sac** lying in its fossa. An extension of the lacrimal sac, the **nasolacrimal duct**, runs down through the nasolacrimal groove and canal to the nasal cavity.

Anatomically, the drainage system could not be much simpler; there are two tubes—the canaliculi—leading the tears away from the lid margins, a sac collecting these contributions, and a duct depositing the collected tears into the nasal cavity. But there are a few important details. The canaliculi are very delicate structures, constructed as tubes of epithelium surrounded by a thin layer of connective tissue. They are easily damaged. Thus the initially vertical, then hor-

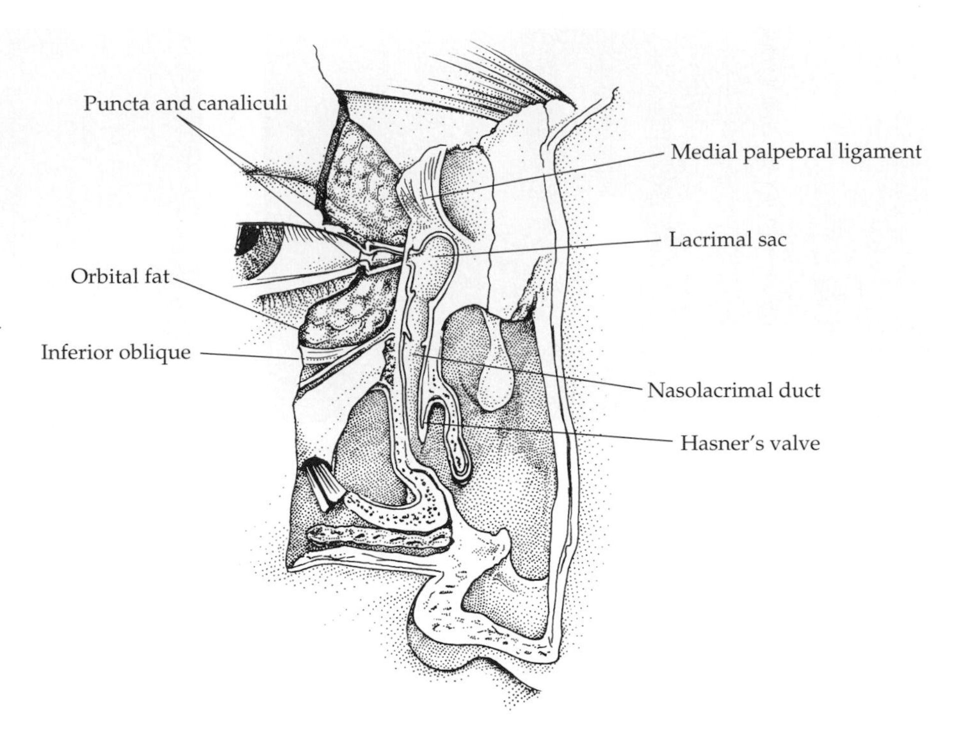

Figure 7.19
The Lacrimal Drainage System
The parts of the drainage system are shown in situ at the left and in isolation at the right. Tears enter the puncta on the upper and lower lid margins, flowing through the canaliculi to the lacrimal sac. The initial course of the canaliculi from the puncta is vertical, for about 2 mm, to the expanded regions (the ampullae), from which the canaliculi run horizontally to the sac. Tears in the lacrimal sac drain down through the nasolacrimal duct into the nasal cavity. A flap of tissue at the nasal end of the duct, the valve of Hasner, may prevent the movement of fluid and air from the nasal cavity to the drainage system.

izontal course of the canaliculi from the puncta to the lacrimal sac is significant when the canaliculi are being probed to find and clear obstructions; either the probe must be inserted perpendicular to the lid margin before being rotated to probe toward the nose, or the eyelid must be stretched laterally to straighten the kink in the canaliculus before probing horizontally. Failure to respect the anatomy can be traumatic.

The nasolacrimal duct has several flaps of tissue along its length, most prominently where the duct ends in the nasal cavity. These tissue folds have been called valves, implying that they exert some control on the direction of flow within the nasolacrimal duct, much as the valves in peripheral veins do. Although several of these tissue folds have names, only the tissue fold at the nasal end of the duct is consistently observed from one individual to the next; it is the **valve of Hasner** (see Figure 7.19).

Hasner's valve is like a hinged door; the door normally hangs down from its hinge, allowing free flow of tears from the nasolacrimal duct into the nasal cavity. But when pressure rises in the nasal cavity during sneezing or coughing, for example, the pressure differential between the nasal cavity and the nasolacrimal duct pushes the free end of Hasner's valve laterally so that it covers the end of the duct and prevents entry of the nasal contents into the duct. There is evidence for considerable variation, however; for example, the valve may be almost closed most of the time, opening only as sufficient tears accumulate in the duct to press the valve outward, or it may never close completely, permitting reflux from the nose into the duct, sac, and canaliculi.

Pressure gradients created by contraction of the orbicularis during blinks move tears through the canaliculi into the lacrimal sac

The lacrimal sac is in a tight place; it lies next to rigid bone of its fossa, it is just behind the strong medial palpebral ligament, and it is hemmed in on the lateral side by the **lacrimal portion** of the orbicularis, or **Horner's muscle** (Figure 7.20). In effect, the lacrimal part of the orbicularis is a bit of the pretarsal orbicularis that does not arise from the medial palpebral ligament; instead, it attaches to the pos-

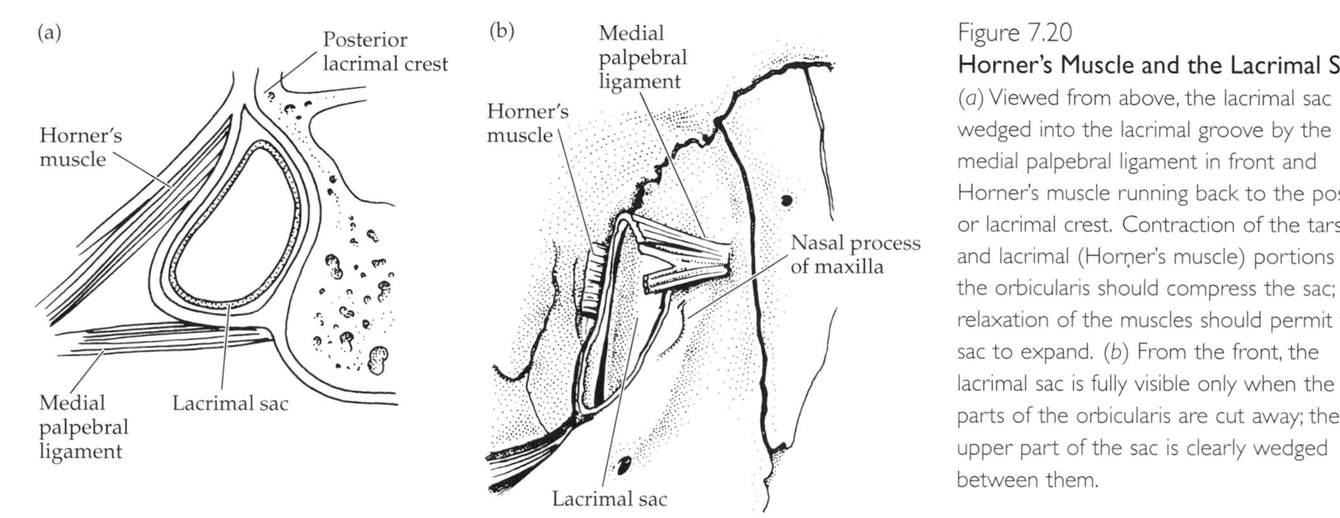

Figure 7.20
Horner's Muscle and the Lacrimal Sac
(*a*) Viewed from above, the lacrimal sac is wedged into the lacrimal groove by the medial palpebral ligament in front and Horner's muscle running back to the posterior lacrimal crest. Contraction of the tarsal and lacrimal (Horner's muscle) portions of the orbicularis should compress the sac; relaxation of the muscles should permit the sac to expand. (*b*) From the front, the lacrimal sac is fully visible only when the parts of the orbicularis are cut away; the upper part of the sac is clearly wedged between them.

terior lacrimal crest of the lacrimal bone, just behind the lacrimal sac (see Figure 7.20*a*). Assuming that this part of the orbicularis contracts along with the pretarsal part during blinks, the anatomical relationship between the muscle and the lacrimal sac suggests that Horner's muscle squeezes the lacrimal sac, producing pressure changes that affect the movement of tears through the drainage system.

There are at least two explanations for tear movement through the drainage system and various sorts of contradictory and inconclusive evidence related to the problem. Some evidence points to a squeezing or compression of the lacrimal sac during the closing phase of a blink. First, when very fine tubes connected to a manometer are inserted into the lacrimal sac, the recorded pressure in the sac increases transiently during the blink. Second, when radiopaque substances are added to the tears so that the drainage sytem can be visualized in situ, X-ray sequences show the middle of the lacrimal sac being compressed during a blink; it expands again after the blink. These observations are consistent with the anatomical relationships shown in Figure 7.20; as the lacrimal part of the orbicularis contracts, it should exert pressure on the lacrimal sac and compress it.

Compression of the lacrimal sac during the closing phase of a blink should send accumulated tears down into the nasolacrimal duct, and the subsequent decrease in pressure as the sac expands again should pull or suck tears into the sac from the canaliculi. But if increased pressure in the sac pushes fluid into the nasolacrimal duct during a blink, there is no obvious reason why fluid should not be pushed back into the canaliculi as well. And if this were the case, it is difficult to see why tears in the canaliculi would do much more than surge back and forth as the lacrimal sac contracted and expanded, with little net flow in any particular direction.

Some studies have concluded that the canaliculi are pumping tears into the lacrimal sac by undergoing a compressive sequence from the puncta to the sac, moving the tears along like squeezing toothpaste from a tube by flattening the tube upward from the bottom. Unfortunately, the evidence for this canalicular pumping is not convincing, except for an observation that the puncta are occluded during a blink, as other investigators have noted. The occlusion should prevent tears already in the canaliculi from being ejected back out though the puncta.

Figure 7.21 summarizes events that produce tear movement through the drainage system; this summary borrows from several studies and should be regarded as a hypothesis. Prior to the onset of a blink, the puncta, canaliculi, and lacrimal sac are maximally open; since there is no pressure gradient within the system, any tear movement into the puncta is by capillary action. As the blink begins, the puncta either come into contact so that their openings are blocked, as shown in Figure 7.21, or they are closed by the contraction of the orbicularis. In

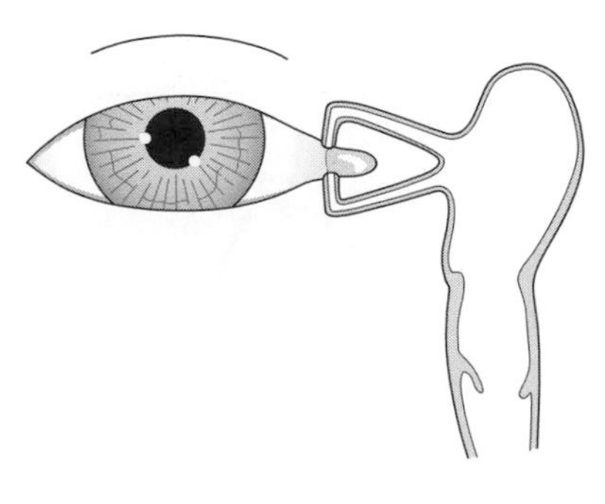

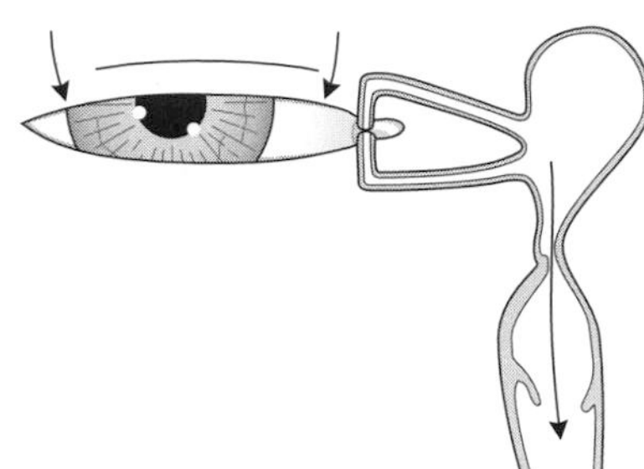

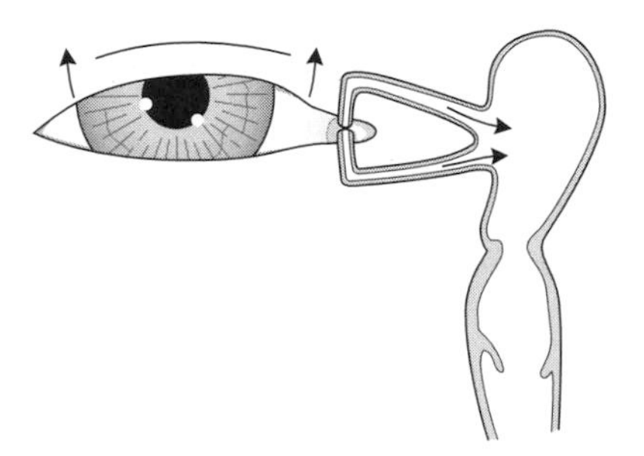

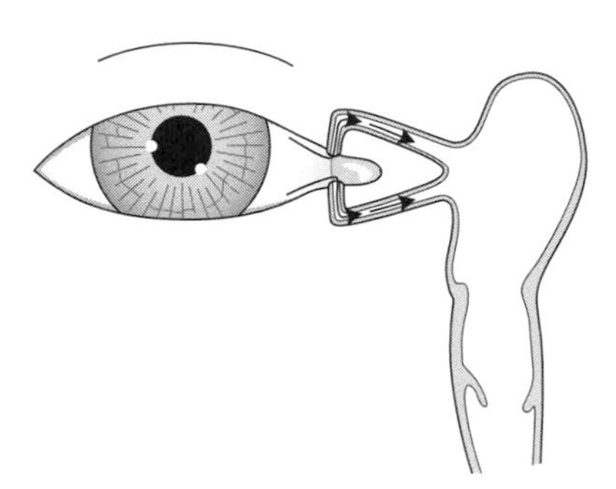

Figure 7.21
Drainage of Tears by Changes in the Drainage System during Blinks
Prior to a blink, the drainage system is fully open and no pressure gradient exists within the system. During the lid-closing phase of a blink, the puncta meet and close, and the canaliculi and lacrimal sac are compressed by contraction of the tarsal and lacrimal portions of the orbicularis. This compression produces positive pressure that moves tears into the nasolacrimal duct to the nasal cavity. As the lids reopen, the canaliculi and lacrimal sac expand before the puncta open, creating negative pressure within the system. This negative pressure sucks tears through the puncta into the canaliculi and lacrimal sac when the puncta separate as the lids continue to open. This sequence of events is not fully substantiated by direct evidence (see text).

any case, orbicularis contraction compresses both the canaliculi and the lacrimal sac. Insofar as the punctal occlusion is effective, the only possibility for fluid movement within the drainage system is toward the site of lowest pressure in the system, which is the nasolacrimal duct.

As the contraction of the orbicularis wanes, the lacrimal sac and the canaliculi expand again and the pressure within them decreases; if this expansion precedes reopening of the puncta, negative pressure develops in the canaliculi and lacrimal sac. Then as the puncta open, there is an influx of tears, which flow through the canaliculi toward the lacrimal sac.

Although this picture of lacrimal drainage is simple, it fits with the few pieces of solid evidence; the puncta close during a blink, the lacrimal sac is visibly compressed, pressure rises in the sac, and the main influx of tears occurs after the blink as the puncta reopen. As for other hypotheses involving pumping within the canaliculi, this is a case where Occam's razor is useful: "It is vain to do with more

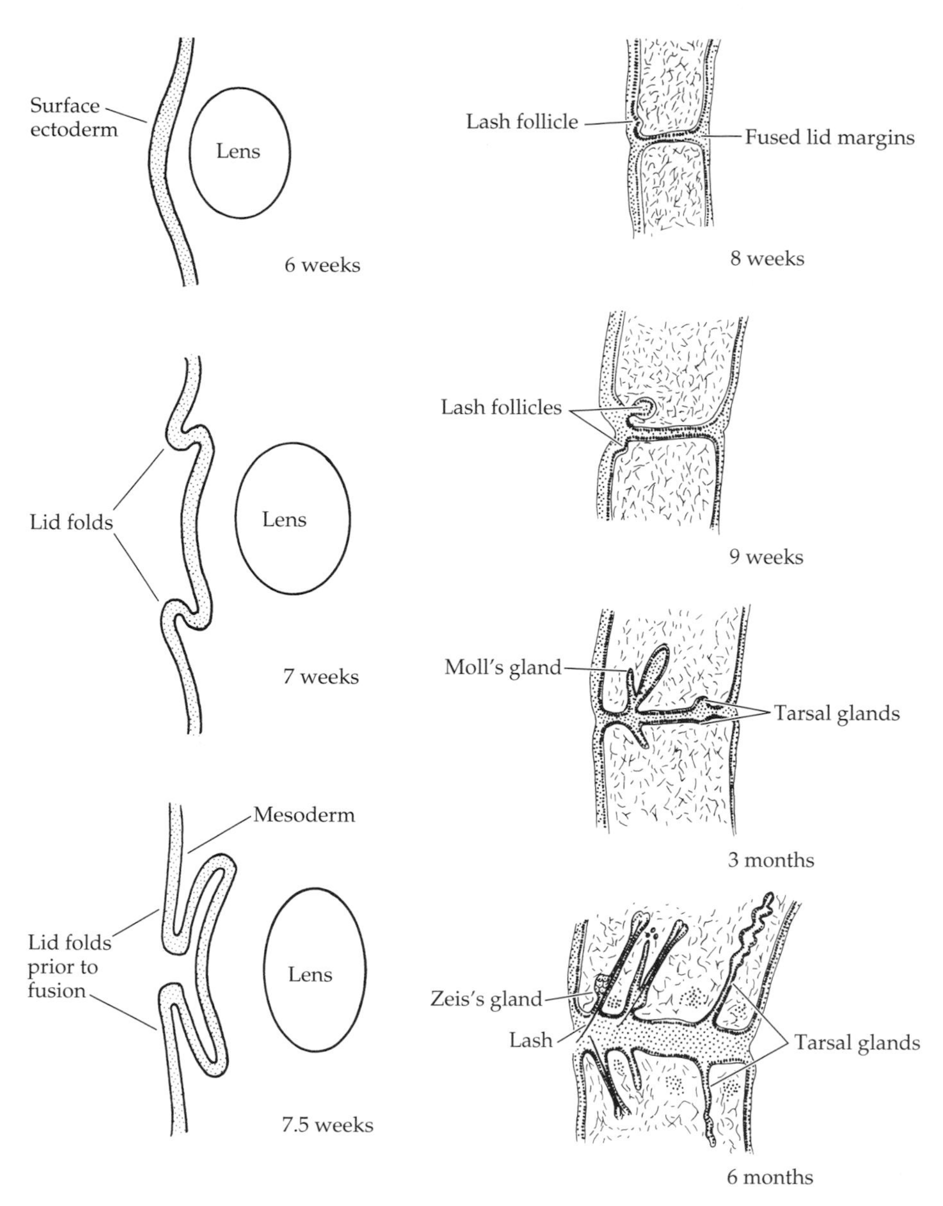

Figure 7.22
Development of the Eyelids
During the first 7 weeks of gestation, the lids appear as small folds in the surface ectoderm; the folds become more pronounced during the next week, and mesoderm moves into the interior of the folds. The upper and lower lid margins meet and fuse during the eighth week. The eyelash follicles and glands bud from the surface ectoderm, with the follicles appearing first, at about 9 weeks, followed by Moll's glands, the tarsal glands, and Zeis's glands. The eyelashes protrude from the follicles during the sixth month, and the adhesions between the lid margins begin to break down.

what can be done with less" or, as it is often paraphrased for young scientists, "When confronted by a multiplicity of hypotheses, choose the simplest."*

Formation of the Eyelids and the Lacrimal System

The eyelids first appear as folds in the surface ectoderm, which gives rise to the lid glands

Shortly after the lens vesicle separates from the surface ectoderm that will become the corneal epithelium, two horizontal folds develop in the ectoderm above and below the site of the future cornea (Figure 7.22); these folds will become the epidermis of the eyelid skin and the epithelium of the conjunctiva.

*William of Occam (c. 1285–?1349) was an English philosopher whose strongly empirical bent exerted considerable influence on subsequent Western thought. Another version of his famous maxim is, "Entities are not to be multiplied without necessity." A modern version is, "Keep it simple, stupid!"—which is good advice for a variety of circumstances. It is also difficult advice to heed.

As the folds become more pronounced, their interior fills with undifferentiated mesoderm, and they continue to grow until they meet, covering the developing eye that lies behind them. The developing lid margins meet and fuse together at about week 8. This fusion will remain intact for several months while the interior structure of the lids develops; the adhesions begin to break down during the sixth month and usually separate completely by the eighth month, thereby creating the palpebral fissure (in other species, lid fusion may persist for 2 to 6 weeks after birth).

Most of the structures within the lids appear after lid fusion; as Figure 7.22 shows, the eyelash follicles and the various glands appear first as small buds from the surface ectoderm during the second and third months. They differentiate and form their adultlike structures during the next several months. In terms of structure, formation of the tarsal glands is very similar to that of the lacrimal gland (see the next section), except that the side branches from the main duct are very short, and growth is mostly a matter of elongation of the original duct, sending out small side buds as it goes. Proliferation and differentiation of the ectodermal cells produce a type of secretory cell different from that in the main lacrimal gland.

As the tarsal glands form, fibroblasts differentiate from the mesoderm in the lid folds and begin to produce the collagen around the developing glands that will become the tarsal plates. Thus the glands and plates develop in close synchrony. The tarsal glands acquire their characteristic structure by the sixth month, and they are said to begin secretion at this time, but they do not reach their full size until late in gestation.

While the eyelash follicles and lid glands are developing from the ectoderm, the muscles and connective tissue differentiate from the surrounding mesoderm. Details about the sequence of events are scarce, but these events seem to proceed from anterior to posterior (roughly). Developmental foci for the orbicularis are present shortly after lid fusion, and the muscle is operational by the time the lid adhesions break down during the sixth month. The tarsal plate appears to begin to form somewhat later than the first appearance of the orbicularis, as fibroblasts derived from the mesoderm produce the collagen around the developing tarsal glands. The main exception in this timing sequence is the muscular portion of the levator, which appears in the orbit much earlier, shortly after the rectus muscles differentiate during week 6 (see Chapter 4). Its tendon does not form until development in the upper lid is several months along, however.

The lacrimal gland and the lacrimal drainage system derive from surface ectoderm

The main and accessory lacrimal glands first appear in the superior cul-de-sac as small buds from the ectoderm, which is in the process of becoming the conjunctival epithelium. The earliest stage of the lacrimal gland consists of eight or ten miniature versions of the mature gland, each having a short duct with cells clustered around its closed end. Further development extends the duct while sending out subsidiary branches like a growing shrub. Proliferation and differentiation of the cells around the growing duct system create numerous secretory cell clusters, each looking much like the original bud cluster from which the growth began. Since the growing duct systems are quite close to one another, they intertwine to become the compact mass of the mature gland. The accessory lacrimal glands presumably go through the same process, but less extensively and at different locations. Differentiation of the secretory cells continues into the postnatal period; the lacrimal gland does not begin secretion until several months after birth and requires several years to mature.

Investigators disagree about the development of the lacrimal drainage system; the main issue is whether the system forms in pieces that subsequently fuse, or forms as a single, continuous unit. Whatever the true situation may be, the drainage system begins as a solid rod of cells (or several short rods) that begin as indentations in the surface ectoderm and later separate from it. These columns of ectodermal cells extend horizontally from the medial canthus and then down toward the developing nasal cavity. A bifurcation in the cell column (or two clumps of cells) near the medial canthus marks the beginning of the canaliculi; Figure 7.23 (top) shows the alternative modes of formation.

By about 12 weeks, the system has two distinct horizontal sections, which are the future canaliculi, while the part bent down and in toward the nasal cavity will become the lacrimal sac and the nasolacrimal duct (see Figure 7.23). Degeneration within the solid core of cells begins at about this time, eventually forming an interior lumen within the system. This hollowing out of the drainage system begins in the region of the lacrimal sac, proceeding downward into the duct and laterally into the canaliculi. Normally, the lumen is completely open by the seventh month of gestation.

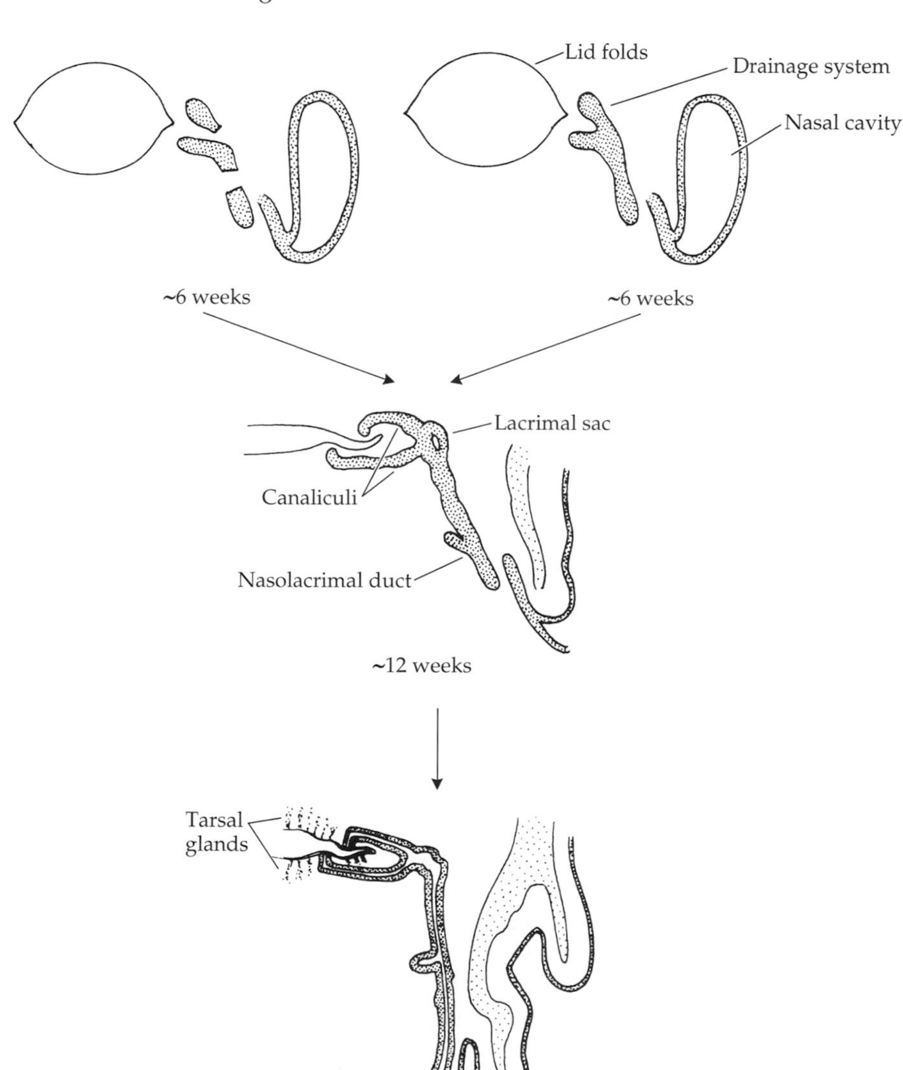

Figure 7.23
Formation of the Lacrimal Drainage System
Development of the lacrimal system begins around the sixth week of gestation as indentations of the surface ectoderm, either as several separate groups of cells (upper left) or as a single, continuous column of cells (upper right). By about 12 weeks, the cell columns are continuous, and the future canaliculi, lacrimal sac, and nasolacrimal duct are all recognizable; in addition, a lumen has begun to form in the lacrimal sac. By about 16 weeks, a lumen has formed throughout the system.

Most developmental anomalies in the eyelids and lacrimal system are problems in lid position or blockage of the drainage channels

Anomalies of eyelid structure are rarely isolated from anomalies of facial structure. Instances of congenital ectropion or entropion are usually related to malformations of the facial bones and the lateral and medial palpebral ligaments, for example, and colobomas of the lids, most commonly the upper, occur along with facial cleavages that produce other segmental defects, such as harelip. Persistent epicanthal folds, unusually large or small palpebral furrows, and unusually wide or narrow palpebral fissures are more common—all presumably results of uncoordinated growth of the bones and the lid tissues that attach to them. Most eyelid anomalies are therefore matters for cosmetic surgery, if anything; they rarely impair normal function.

Congenital defects in the lacrimal system are most common in the drainage channels, which may be blocked or lack some components, including the puncta and canaliculi. Blocked drainage channels represent a failure of the canaliculi, lacrimal sac, or nasolacrimal duct to complete the last phase of their development, which is hollowing out of the interior lumen; that is, the solid rods or cylinders of tissue fail to become tubes. The resulting lack of tear drainage may not be an immediate problem, since tear production in newborn infants is relatively small, but artificial drainage channels eventually need to be constructed. Similar surgery is required if part of the drainage system fails to form. Here, the fact that one part of the system can be normal and another part absent implies that the system has several developmental foci that normally fuse. If one of these ectodermal buds does not appear, it apparently cannot be replaced, and the drainage system thus remains incomplete.

Anomalous innervation can produce eyelid movements linked to contraction of muscles in the jaw

In some individuals, movements of the mandible when chewing are associated with rapid movements of one eyelid. This phenomenon, **Marcus Gunn's syndrome**, also called jaw-winking, is most noticeable in infants; it may persist into adulthood, but often with diminished intensity. The phenomenon is normally unilateral and accompanied by ptosis on the affected side.

Like some cases of Duane's retraction syndrome (see Chapter 4), Marcus Gunn's syndrome is most easily explained by an anomalous pattern of innervation affecting the levator and the pterygoid muscles of the mandible (the pterygoid muscles are implicated rather than the masseter muscle because lid movements are more easily elicited by side-to-side movement of the mandible, which is an action of the pterygoid muscles). Since the pterygoid muscles are innervated by axons from cells in the motor nucleus of the trigeminal nerve, the miswiring might involve replacement of oculomotor nerve axons going to the levator with axons from the trigeminal complex. The third and fifth nerve nuclei are not close to one another in the brainstem, however, and the intracranial paths followed by the oculomotor and mandibular nerves are divergent; misrouting of axons at these levels seems unlikely (though not impossible; evidence one way or the other is not available).

Some authors favor misrouting of supranuclear connections, either in the motor cortex, where the representations for the jaw muscles and the eyelids are near one another, or in the pathway between the cortex and the brainstem motor nuclei. Whatever the case, the result is reduced tonic activity in the levator, with ptosis, and anomalous signals to the levator muscle when the ipsilateral pterygoid muscles contract.

References and Additional Reading

Structure and Function of the Eyelids

Bedrossian EH. 1998. Embryology and anatomy of the eyelids. Chapter 5 in *Duane's Foundations of Clinical Ophthalmology*, Vol. 1. Tasman W and Jaeger EA, eds. JB Lippincott, Philadelphia.

Chung CW, Tigges M, and Stone RA. 1996. Peptidergic innervation of the primate meibomian gland. *Invest. Ophthalmol. Vis. Sci.* 37: 238–245.

Dutton JJ. 1996. Botulinum-A toxin in the treatment of craniocervical muscle spasms: Short- and long-term, local and systemic effects. *Surv. Ophthalmol.* 41: 51–65.

Evinger C, Manning KA, and Sibony RA. 1991. Eyelid movements: Mechanisms and normal data. *Invest. Ophthalmol. Vis. Sci.* 32: 387–400.

Gordon G. 1951. Observations upon the movements of the eyelids. *Brit. J. Ophthalmol.* 35: 339–351.

Hung G, Hsu F, and Stark L. 1977. Dynamics of the human eyeblink. *Am. J. Optom. Physiol. Optics* 54: 678–690.

Jester JV, Nicolaides N, and Smith RE. 1981. Meibomian gland studies: Histologic and ultrastructural investigations. *Invest. Ophthalmol. Vis. Sci.* 20: 537–547.

Kessing SV. 1968. Mucous gland system of the conjunctiva: A quantitative normal anatomical study. *Acta Ophthalmol. (Copenh.)* 95 (Suppl.): 1–133.

Lander T, Wirtschafter JD, and McLoon LK. 1996. Orbicularis oculi muscle fibers are relatively short and heterogeneous in length. *Invest. Ophthalmol. Vis. Sci.* 37: 1732–1739.

Newell FW and Ernest JT. 1978. *Ophthalmology, Principles and Concepts*, 4th Ed. CV Mosby, St. Louis.

Porter JD, Burns LA, and May PJ. 1989. Morphological substrate for eyelid movements: Innervation and structure of primate levator palpebrae superioris and orbicularis oculi muscles. *J. Comp. Neurol.* 287: 64–81.

Porter JD, Strebeck S, and Capra NF. 1991. Botulinum-induced changes in monkey eyelid muscle: Comparison with changes seen in extraocular muscle. *Arch. Ophthalmol.* 109: 396–404.

Tucker SM and Linberg JV. 1994. Vascular anatomy of the eyelids. *Ophthalmology* 101: 1118–1121.

Volkmann FC, Riggs LA, and Moore RK. 1980. Eyeblinks and visual suppression. *Science* 207: 900–902.

Warwick R. 1976. The ocular appendages. Chapter 3, pp. 181–237, in *Eugene Wolff's Anatomy of the Eye and Orbit.* WB Saunders, Philadelphia.

Whitnall SE. 1932. Eyelids. Part 2, pp. 109–252, in *The Anatomy of the Human Orbit and Accessory Organs of Vision*, 2nd Ed. Oxford University Press, London.

Wobig J. 1982. Eyelid anatomy. Chapter 7, pp. 78–87, in *Cosmetic Oculoplastic Surgery*, Putterman AM, ed. Grune & Stratton, New York.

Zide BM and Jelks GW. 1985. The eyelids. Chapter 3, pp. 21–32, in *Surgical Anatomy of the Orbit.* Raven, New York.

Tear Supply and the Lacrimal Glands

Bedrossian EH. 1996. The lacrimal system. Chapter 30 in *Duane's Foundations of Clinical Ophthalmology*, Vol. 1, Tasman W and Jaeger EA, eds. Lippincott-Raven, Philadelphia.

Dartt D. 1989. Signal transduction and control of lacrimal gland secretion: A review. *Curr. Eye Res.* 8: 619–636.

Dartt DA. 1992. Physiology of tear production. Chapter 3, pp. 65–99, in *The Dry Eye*, Lemp MA and Marquardt R, eds. Springer-Verlag, Berlin.

Dartt DA. 1994. Regulation of tear secretion. Pp. 1–9 in *Lacrimal Gland, Tear Film, and Dry Eye Syndromes: Basic Science and Clinical Relevance* (Advances in experimental medicine and biology, vol. 350), Sullivan DA, ed. Plenum, New York.

Dilly PN. 1994. Structure and function of the tear film. Pp. 239–247 in *Lacrimal Gland, Tear Film, and Dry Eye Syndromes: Basic Science and Clinical Relevance* (Advances in experimental medicine and biology, vol. 350), Sullivan DA, ed. Plenum, New York.

Doane MG. 1980. Interaction of eyelids and tears in corneal wetting and the dynamics of the normal human eyeblink. *Am. J. Ophthalmol.* 89: 507–516.

Egeberg J and Jensen OA. 1969. The ultrastructure of the acini of the human lacrimal gland. *Acta Ophthalmol. (Copenh.)* 47: 400–410.

François J and Neetens A. 1973. Tear flow in man. *Am. J. Ophthalmol.* 76: 351–358.

Fullard RW. 1994. Tear proteins arising from lacrimal tissue. Chapter 30, pp. 473–479, in *Principles and Practice of Ophthalmology: Basic Sciences*, Albert DM and Jakobiec FA, eds. WB Saunders, Philadelphia.

Hodges RR, Zoukhri D, Sergheraert C, Zeiske JD, and Dartt DA. 1997. Indentification of vasoactive intestinal polypeptide receptor subtypes in the lacrimal gland and their signal-transducing components. *Invest. Ophthalmol. Vis. Sci.* 38: 610–619.

Jordan A and Baum J. 1980. Basic tear flow: Does it exist? *Ophthalmology* 87: 920–930.

Lemp MA. 1992. Basic principles and classification of dry eye disorders. Chapter 4, pp. 101–131, in *The Dry Eye*, Lemp MA and Marquardt R, eds. Springer-Verlag, Berlin.

Lemp MA and Chacko B. 1991. Diagnosis and treatment of tear deficiencies. Chapter 14 in *Duane's Clinical Ophthalmology*, Vol. 4, Tasman W and Jaeger EA, eds. Lippincott, Philadelphia.

Lemp MA and Wolfley DE. 1992. The lacrimal apparatus. Chapter 2, pp. 18–28, in *Adler's Physiology of the Eye*, 9th Ed., Hart WM, ed. Mosby Year Book, St. Louis.

Rohen JW and Lütjen-Drecoll E. 1992. Functional morphology of the conjunctiva. Chapter 2, pp. 35–63, in *The Dry Eye*, Lemp MA and Margquardt R, eds. Springer-Verlag, Berlin.

Ruskell GL. 1975. Nerve terminals and epithelial cell variety in the human lacrimal gland. *Cell Tissue Res.* 158: 121–136.
Ruskell GL. 1985. Innervation of the conjunctiva. *Trans. Ophthalmol. Soc. U.K.* 104: 390–395.
Seifert P and Spitznas M. 1994. Demonstration of nerve fibers in human accessory lacrimal glands. *Graefes Arch. Clin. Exp. Ophthalmol.* 232: 107–114.
Sibony PA, Walcott B, McKeon C, and Jakobiec FA. 1988. Vasoactive intestinal polypeptide and the innervation of the lacrimal gland. *Acta Ophthalmol.* 106: 1085–1088.
Tiffany JM. 1995. Physiological functions of the Meibomian glands. *Prog. Ret. Eye Res.* 14: 47–174.

Lacrimal Drainage

Amanat LA, Hilditch TE, and Kwok CS. 1983. Lacrimal scintigraphy: III. Physiological aspects of lacrimal drainage. *Brit. J. Ophthalmol.* 67: 729–732.
Doane MG. 1981. Blinking and the mechanics of the lacrimal drainage system. *Ophthalmology* 88: 844–850.
Jones LT. 1961. An anatomical approach to problems of the eyelids and lacrimal apparatus. *Arch. Ophthalmol.* 60: 111–124.
Jones LT. 1973. Anatomy of the tear system. *Int. Ophthalmol. Clin.* 13 (1): 3–22.
Lemp MA and Weiler HH. 1983. How do tears exit? *Invest. Ophthalmol. Vis. Sci.* 24: 619–622.
Maurice DM. 1973. The dynamics and drainage of tears. *Int. Ophthalmol. Clin.* 13 (1): 103–116.
Murube J and Murube E. 1996. Treatment of dry eye by blocking the lacrimal canaliculi. *Surv. Ophthalmol.* 40: 463–480.
Zide BM and Jelks GW. 1985. Lacrimal apparatus. Chapter 4, pp. 33–39, in *Surgical Anatomy of the Orbit*. Raven, New York.

Formation of the Eyelids and the Lacrimal System

Anderson H, Ehlers N, and Mathiessen ME. 1965. Histochemistry and development of the human eyelids. *Acta Ophthalmol. (Copenh.)* 43: 642–668.
Mann I. 1950. The lids and lacrimal apparatus. Chapter VII, pp. 258–269, in *The Development of the Human Eye*. Grune & Stratton, New York.
Murube-del-Castillo J. 1983. Development of the lacrimal apparatus. Chapter 2, pp. 9–22, in *The Lacrimal System*, Milder B and Weil BA, eds. Appleton-Century-Crofts, Norwalk, CT.
Sevel D. 1988. A reappraisal of the development of the eyelids. *Eye* 2: 123–129.

Part Two

Components of the Eye

God is in the details.

■ Ludwig Mies van der Rohe

Chapter 8

The Cornea and the Sclera

The shape of the eye is spherical with a slight protuberance in front. . . . The anterior part of the eye is limited by a perfectly transparent membrane, called the cornea. The remainder of the exterior coating is an opaque white membrane, called the sclerotic coat. The cornea is set in the sclerotic coat, as a watch-glass is set in its frame.

■ William G. Peck, *Introductory Course of Natural Philosophy*

The cornea and sclera did not seem complicated when William Peck wrote his popular physics text in 1860, and in some respects they still don't. Anatomically, the cornea and sclera are two of the simpler tissues in the eye. They consist mostly of collagen, they contain only a few types of cells, and their structure is homogeneous—one bit of tissue is very much like any other. Both tissues are avascular, and their major structural elements are identical.

Yet the cornea and sclera seem very different. The cornea is transparent while the sclera is white and opaque, and they have different roles to play in the functioning of the eye: The cornea is the eye's primary refractive element; the sclera is the site of attachment of the eye to external structures (muscles and connective tissue) and has portals for the passage of blood vessels and nerves in and out of the eye. Together, the cornea and sclera are a roughly spherical, fibrous envelope whose important optical properties—curvature and axial length—are maintained by pressure from the intraocular fluids.

Most of the ensuing discussion will focus on the cornea because its special property of transparency is crucial to the operation of the eye as an optical system, because it is more vulnerable to external influences and age-related changes, and because it has become a prime target for modification and manipulation to correct refractive errors. Although the sclera constitutes about 85% of the external surface of the eye and its structural integrity is necessary for proper functioning of the eye, its structure will be considered mainly as it contrasts to that of the cornea; the sclera's largely mechanical function requires little explanation. The limbus, which is the region of transition between cornea and sclera, has some additional structures not found in either the cornea or the sclera; it will be considered separately, in Chapter 9.

Components and Organization of the Cornea and Sclera

The cornea, sclera, and limbus are made primarily of collagen fibrils

Collagen is a fibrous protein found throughout the body; it is a major component of all connective tissue, tendons, ligaments, bone, skin, and the outer coat of the eye, among other things. About 25% of all the protein in the body is collagen, making it the most common of all proteins.

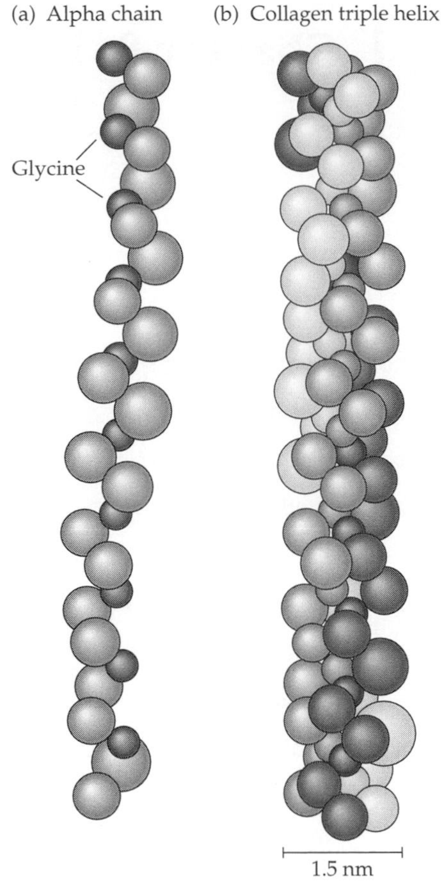

Figure 8.1
Structure of Collagen Molecules
(*a*) The alpha chains in collagen are helices formed from glycine and two other amino acids strung together like beads; different alpha helices have different sets of amino acids. (*b*) Collagen is made from three alpha chains of the same or different types that wind together to form a triple helix. (After Alberts et al. 1994.)

Collagen has a characteristic structure consisting of three chains of amino acids that spiral around one another to form a triple helix (Figure 8.1); each individual chain is an alpha helix. Collagen types differ in their component alpha helices, of which there are many possible varieties (although relatively few that will combine to form stable triple helices). In some collagen types, the three alpha helices are identical; in others, two of the chains are the same and one is different. At present, about 20 kinds of alpha chains have been distinguished, and they are used in more than 10 different types of collagen. Most of the collagen in the cornea and sclera is type I collagen, which constitutes around 90% of the total collagen in the body; it contains two different alpha helices.

The intertwined helices in a single collagen molecule form a thin strand about 1.5 nm in diameter and some 300 nm long, with nonhelical portions on the ends of the strand. The individual molecules form natural linkages, creating long assemblages of parallel molecules that are the collagen **fibrils**. The staggered arrangement of the parallel strings of molecules is highly regular and produces a characteristic banding or striation of collagen fibrils visible with electron microscopy (Figure 8.2). There seems to be no significant limitation on the length of collagen fibrils—several centimeters is not uncommon—and they can vary greatly in diameter, depending on collagen type and the tissue in which they are located. Corneal collagen fibrils are around 30 nm in diameter; those in the sclera may be several times as thick, up to about 150 nm. The strength of the linking between collagen molecules in the fibril is another variable that affects the overall strength of the tissue; linkages are particularly strong in the Achilles tendon, for example.

Collagen is embedded in a polysaccharide gel that forms the extracellular matrix

The collagen in the cornea and sclera is associated with polysaccharide molecules called glycosaminoglycans (GAGs). There are many GAGs, differing in the disaccharides that combine in the long polysaccharide chains; the dominant ones in the cornea and sclera are dermatan sulfate and keratan sulfate. Many of these long molecules will bind one of their ends to core proteins, thereby forming a complex structure called a **proteoglycan**; in effect, the long cablelike core protein acquires a dense fuzz of GAGs along its length. The core proteins in the corneal proteoglycans are not fully characterized, but there are at least two, one binding with keratan sulfate and the other with dermatan sulfate. In any case, the collagen fibrils in cornea and sclera are surrounded by and embedded in proteoglycans (Figure 8.3).

GAGs have some important structural properties. Although they are long chains, they cannot be arranged in compact strands like collagen molecules, nor can they be folded into compact globules; in other words, a GAG molecule occupies a fair bit of space. Also, these molecules have a strong net negative charge, which attracts positively charged ions, like Na^+, in great numbers. With the Na^+ comes water that is sucked into the molecular matrix, so tissues with large amounts of GAGs will take up considerable water if left to their own devices; the volume occupied by the glycosaminoglycans in the proteoglycans around the collagen fibrils depends to a large extent on the water content in the matrix. The combination of water and the distensible, space-filling nature of these molecules creates a gel around the collagen fibrils.

In the cornea, the spacing among collagen fibrils is a critical factor in corneal transparency, and the water content in the proteoglycan matrix needs to be maintained within fairly narrow limits to keep the spacing regular and constant. Because of all the GAGs in the extracellular matrix, however, the corneal stroma has a higher affinity for water than the amount optimal for transparency. As a result, many of the anatomical and physiological specializations in the cornea to

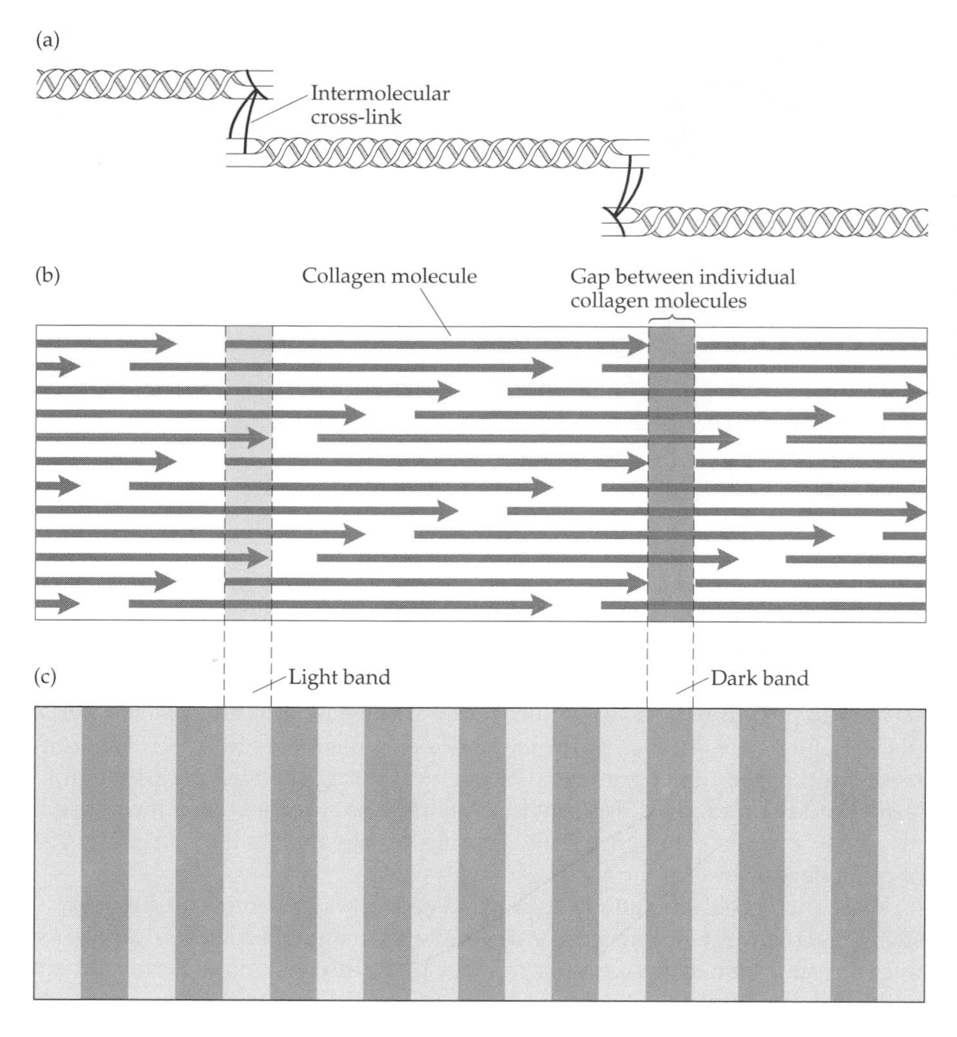

Figure 8.2
Structure of Collagen Fibrils
(*a*) Individual collagen molecules are cross-linked at their ends to form staggered arrays. (*b*) When assembled to form a collagen fibril, these arrays form a periodic repetitive pattern within the long fibril. (*c*) The gaps between molecules contrast with the regions of overlap to produce the characteristic banding of collagen fibrils seen in electron micrographs.

be considered shortly are directed toward keeping the stroma from imbibing as much water as its physical and chemical properties would demand.

The fibroblasts in the corneal and scleral stroma constitute a small fraction of the stroma's volume

The collagen and the proteoglycans in the cornea and sclera are produced and maintained by **fibroblasts**, a type of cell found in all connective tissues (although those in cartilage and bone are specialized and are called, respectively, chondroblasts and osteoblasts). The fibroblasts in the corneal stroma are often called **keratocytes** (in the older literature, "corneal corpuscles"), but whatever the name, they are fibroblasts.

An individual fibroblast has a large, flattened cell body from which numerous slender processes extend, giving the cell a spidery appearance (see Figures 8.4 and 8.7*a*). In culture, the fibroblasts normally grow until their processes touch and their coverage of the culture plate forms an irregular meshwork of cells and processes. Where processes from neighboring cells touch, the cells may have gap junctions, but these junctions are usually not numerous in adult fibroblast-containing tissues.

When connective tissue is forming during fetal development, the volume of fibroblasts relative to the extracellular matrix is high; the cells are generating both collagen fibrils and the matrix, so there are more cells than matrix. At maturity, however, the cellular content is low; the corneal stroma has only 2 to 3%

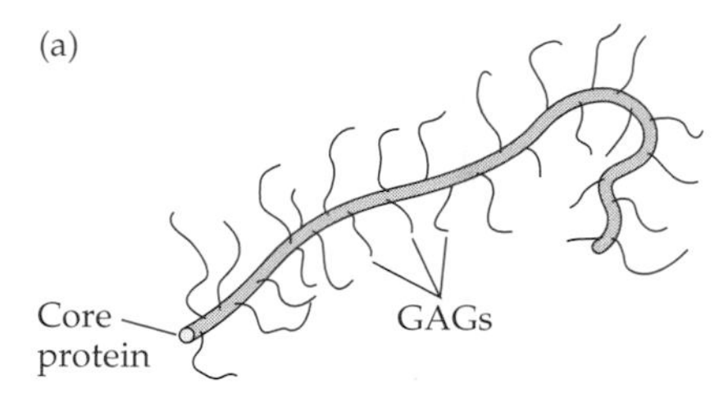

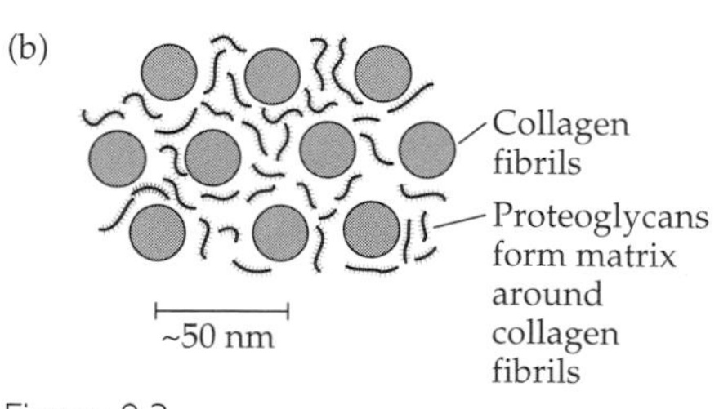

Figure 8.3
Proteoglycans and the Collagen Fibril Matrix
(*a*) A proteoglycan consists of a long core protein to which many space-filling glycosaminoglycans (GAGs) are attached.
(*b*) Proteoglycans and water form a gel-like matrix around the collagen fibrils.

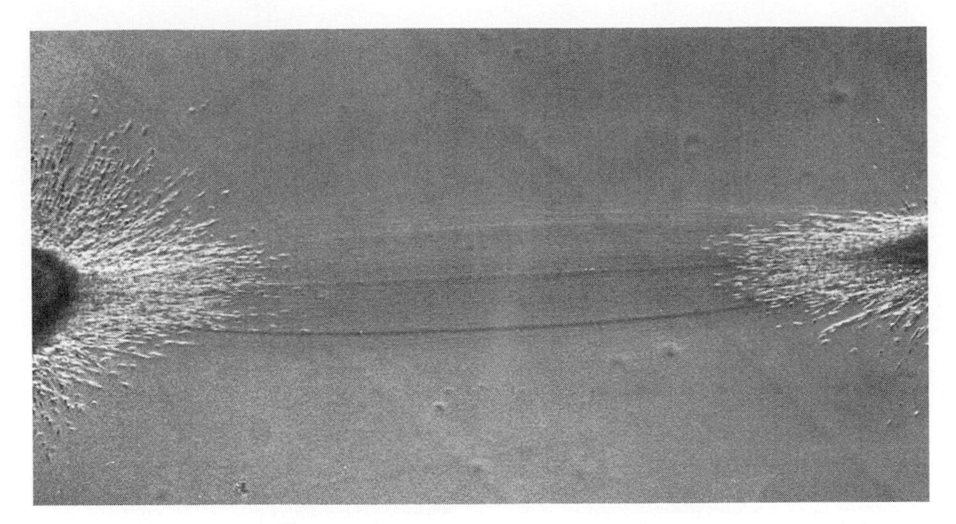

Figure 8.4
Fibroblasts Can Organize Collagen Fibrils
Two fibroblasts with their numerous radiating processes are shown in culture. They have attached themselves to collagen fibrils and pulled the fibrils into a nearly parallel bundle. (From Stopak and Harris 1982.)

fibroblasts by volume, and the fibroblast content of the sclera is probably even lower. Fibroblasts in the mature tissue maintain the protein content and organization of the collagen and the matrix and, if necessary, repair them after trauma. Fibroblasts are able to move around on appropriate substrates, as they can be seen to do in culture, and can influence the arrangement of the collagen fibrils (the fibroblasts are attached to the extracellular proteins) (Figure 8.4). This contact with collagen fibrils provides a mechanical linkage between nearby fibroblasts, in addition to any direct contacts they may have through gap junctions; it may be a way of coordinating their activities as they maintain the organization of the collagen fibrils in the tissue.

What the fibroblasts cannot do is regulate the water content of the matrix; such regulation requires other systems. Moreover, mature fibroblasts cannot produce the same kind of collagen, nor can they reproduce the original arrangement of the collagen fibrils.

Collagen fibrils in the cornea are highly organized; those in the sclera are not

Some of the critical similarities and differences between the corneal and scleral stroma are visible in the electron micrographs of Figure 8.5. Both tissues contain collagen, but in the cornea (Figure 8.5*a*), the collagen fibrils are very similar in diameter and very regularly spaced. Scleral collagen, in contrast, varies considerably in both diameter and spacing (Figure 8.5*b*); the diameter variation in the sclera may indicate that a portion of scleral collagen is not type I, or that we may be seeing the tapering ends of the long collagen fibrils. (Corneal collagen fibrils are said to run completely across the cornea, parallel to its surface. The basis for this statement is the fact that there is so little variation in fibril diameter from center to periphery of the cornea.)

Both parts of Figure 8.5 show collagen fibrils cut in cross section—hence the circular profiles—but in some regions this is clearly not the case. Fibrils have been cut either obliquely or nearly along the axes of the fibrils, revealing a basic laminar organization in the tissues; each lamina has fibrils running parallel to one another and, in the cornea, parallel to the corneal surface. The difference between adjacent laminae is the direction in which the fibrils run; in human cornea, fibrils in neighboring layers are roughly perpendicular to one another and are basically horizontal or vertical. The corneal **stroma** consists of about 200 of these laminae stacked on top of one another, much like the layers of plywood (Figure 8.6*a*); these stacked layers account for 90% of the thickness and volume of the cornea. The scleral stroma also has a laminar organization (Figure 8.6*b*),

(a) Corneal stroma

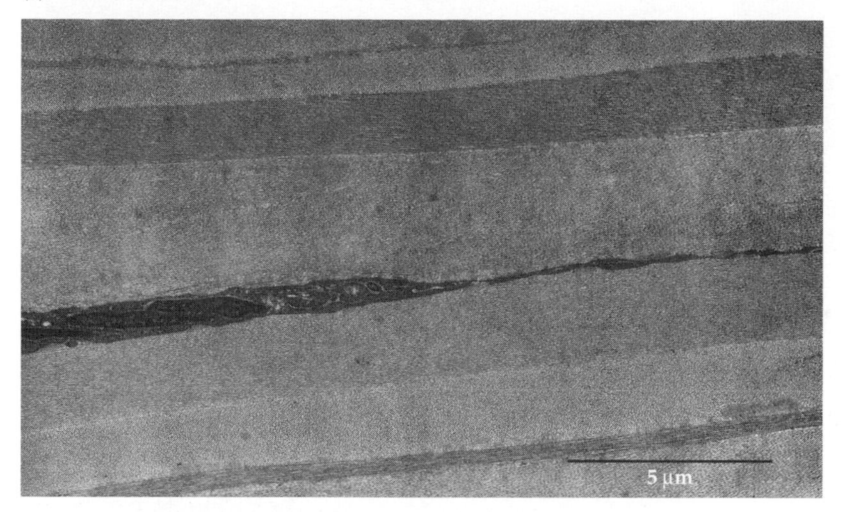

(b) Scleral stroma

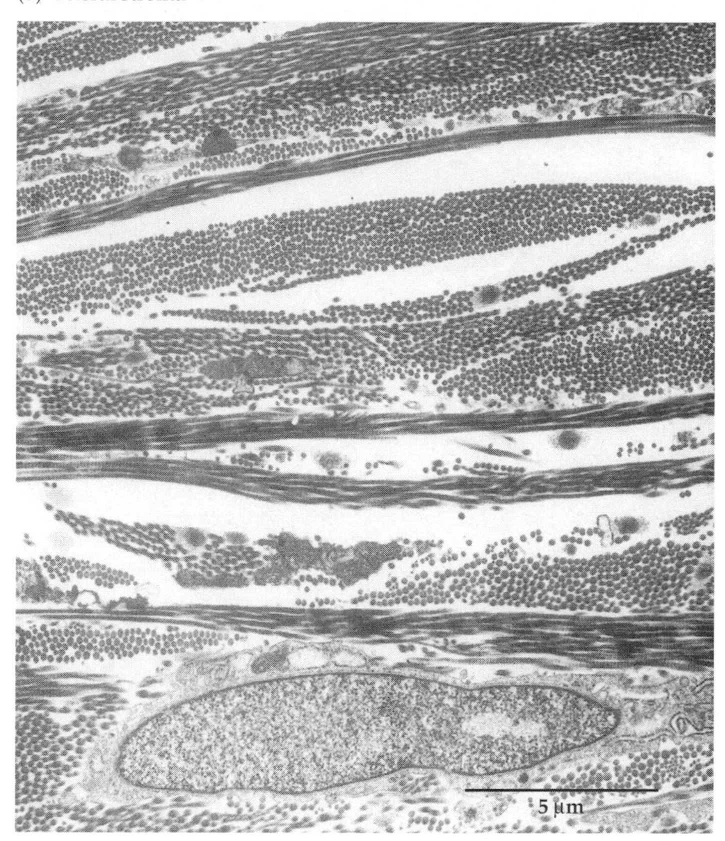

Figure 8.5
Layers of Collagen Fibrils Form the Corneal and Scleral Stroma
(*a*) The small, circular dots are collagen fibrils cut in cross section. They are all about the same diameter and are spaced at regular intervals. (*b*) In the scleral stroma, most of the collagen fibrils have a large diameter, but there is considerable variation. Fibril spacing also varies considerably. (From Komai and Ushiki 1991.)

but the laminae are not stacked as they are in the cornea. The scleral laminae interweave such that collagen fibrils within a lamina may have one end near the outer surface of the sclera and the other end near the inner surface.

This representation of stromal structure can be completed by adding fibroblasts, which form loose meshworks arranged in flat sheets running (usually) between laminae (Figure 8.7*a*). In the cornea, therefore, the fibroblast arrays are parallel to the surface, as are the laminae (Figure 8.7*b*); in the sclera, their arrangement is less planar.

The differing arrangements of the laminae in cornea and stroma give the two tissues different mechanical properties. In particular, the stacked, parallel laminae in the corneal stroma create natural planes of cleavage in the tissue, mean-

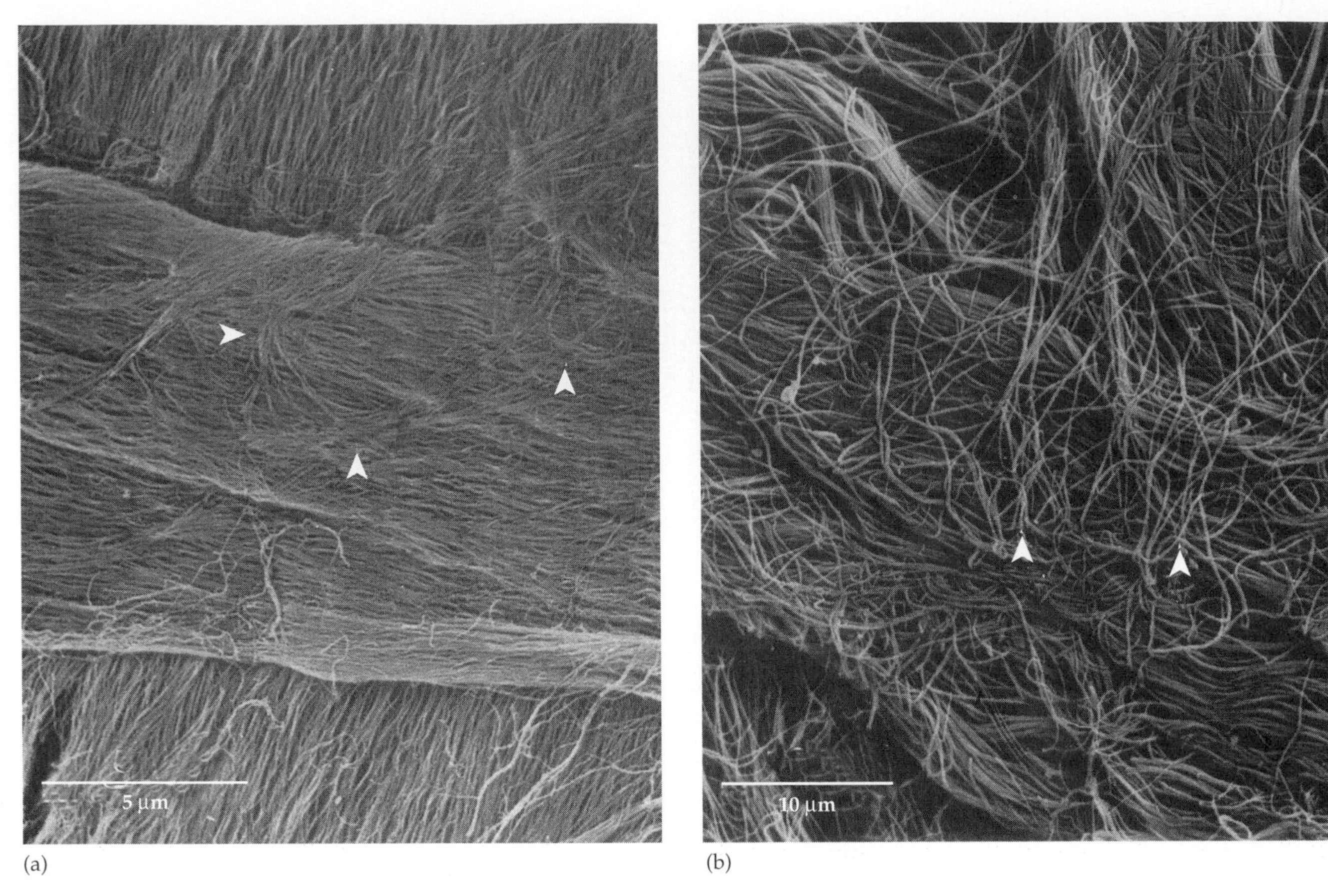

Figure 8.6
Arrangement of Collagen Fibrils in the Corneal and Scleral Stroma
(*a*) Two layers of collagen fibrils can be seen in this scanning electron micrograph of human corneal stroma. Fibrils in the two lamallae run at right angles to one another in the two layers, and small bundles (arrows) connect the two lamallae. (*b*) A similar view of the scleral stroma at lower magnification shows that the scleral collagen forms irregular bundles of fibrils rather than well-defined layers. Loose networks of collagen fibrils surround the larger bundles. (From Komai and Ushiki 1991.)

ing that it can be split along planes parallel to the corneal surface without the basic structural elements of the tissue being cut or damaged, and this is sometimes done in certain types of corneal grafting (discussed later in the chapter). The cleavage between laminae also appears frequently in histological sections as an artifact of tissue processing. Because of the three-dimensional interweaving of laminae in the sclera, however, separation occurs less easily and the sclera is more resistant to forces applied perpendicular to its thickness; if the sclera were constructed like the cornea, the force applied to the tissue by an extraocular muscle would tend to separate the laminae and pull the tissue apart.

The scleral stroma varies considerably from place to place; there are holes in the sclera where blood vessels and nerves pass through it, there is a major area of weakness at the lamina cribrosa where the optic nerve exits (see Chapter 16), the arrangement of fibrils and laminae is particularly complicated where the extraocular muscle tendons fuse with the sclera, and there is a thickened annular ridge—the scleral spur—at the limbus (see Chapter 9). In contrast, the corneal stroma is not only very regular in its structure, it is quite homogeneous; the structure of the stroma is the same throughout the cornea. For all its anatomical simplicity, however, the cornea is functionally the more complicated of the two tissues.

(a)

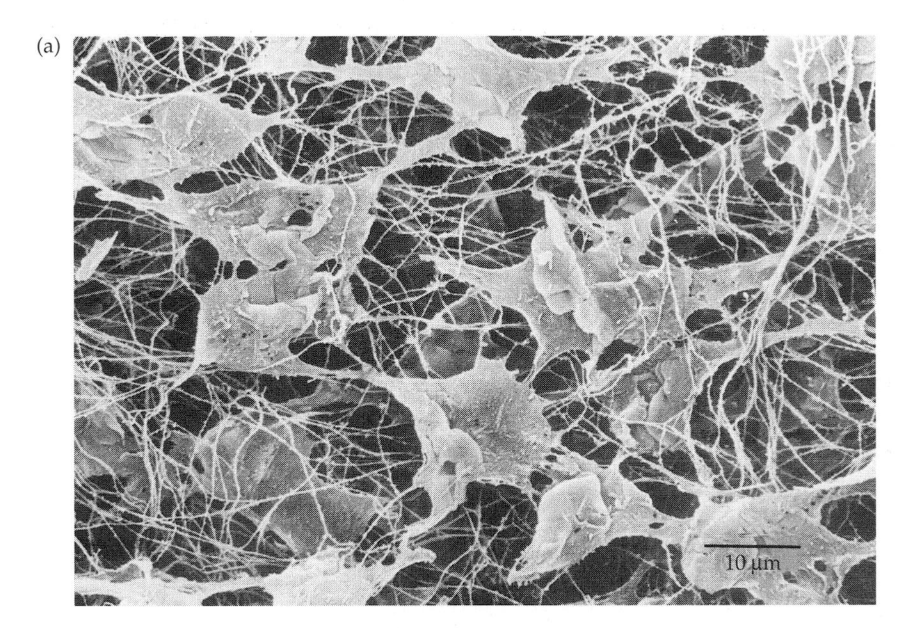

(b)

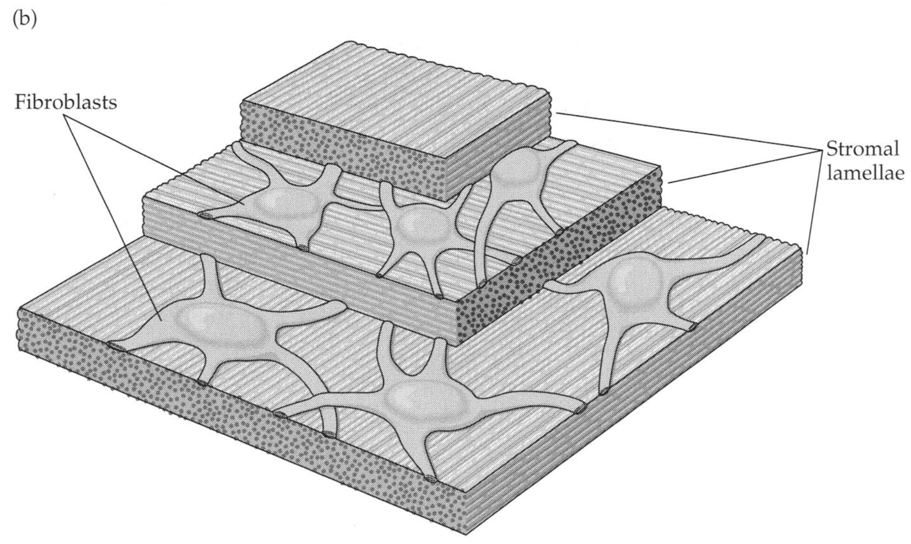

Figure 8.7
Fibroblasts and Their Relationship to the Collagen Layers
(*a*) Several layers of fibroblasts can be seen (in rat cornea) after most of the collagen fibrils have been removed. Processes radiate out from the cell bodies to contact neighboring cells. (*b*) A reconstruction of the corneal stroma shows the stacked layers of collagen with interposed layers of fibroblasts. (*a* from Nishida et al. 1988.)

The structure of the corneal stroma is altered in Bowman's layer

Named after its discoverer (see Vignette 8.2), Bowman's layer is the anteriormost part of the corneal stroma, lying next to the basal lamina of the epithelium; it is about 10 μm thick, recognizable in histological sections by its lack of fibroblasts and apparent lack of structure (Figure 8.8*a*). Originally, the layer was called Bowman's *membrane*, but because it is continuous with the stroma, though differently organized, it has come to be designated as a "layer" instead.

Collagen fibrils in Bowman's layer do not always run parallel to the corneal surface or to one another (Figure 8.8*b*); thus, the layer lacks the regular laminar structure of the stroma proper. It appears, in fact, that we are looking at the ends of collagen fibrils that have curved toward the surface and are interweaving in a feltlike arrangement. Such an arrangement would suggest that many of the anteriormost collagen fibrils in the stroma do not run all the way across the cornea; instead, they end in Bowman's layer, and their intertwining ends make up the layer.

There are problems with this interpretation. In particular, immunohistochemical results suggest that many of the collagen fibrils in Bowman's layer are not

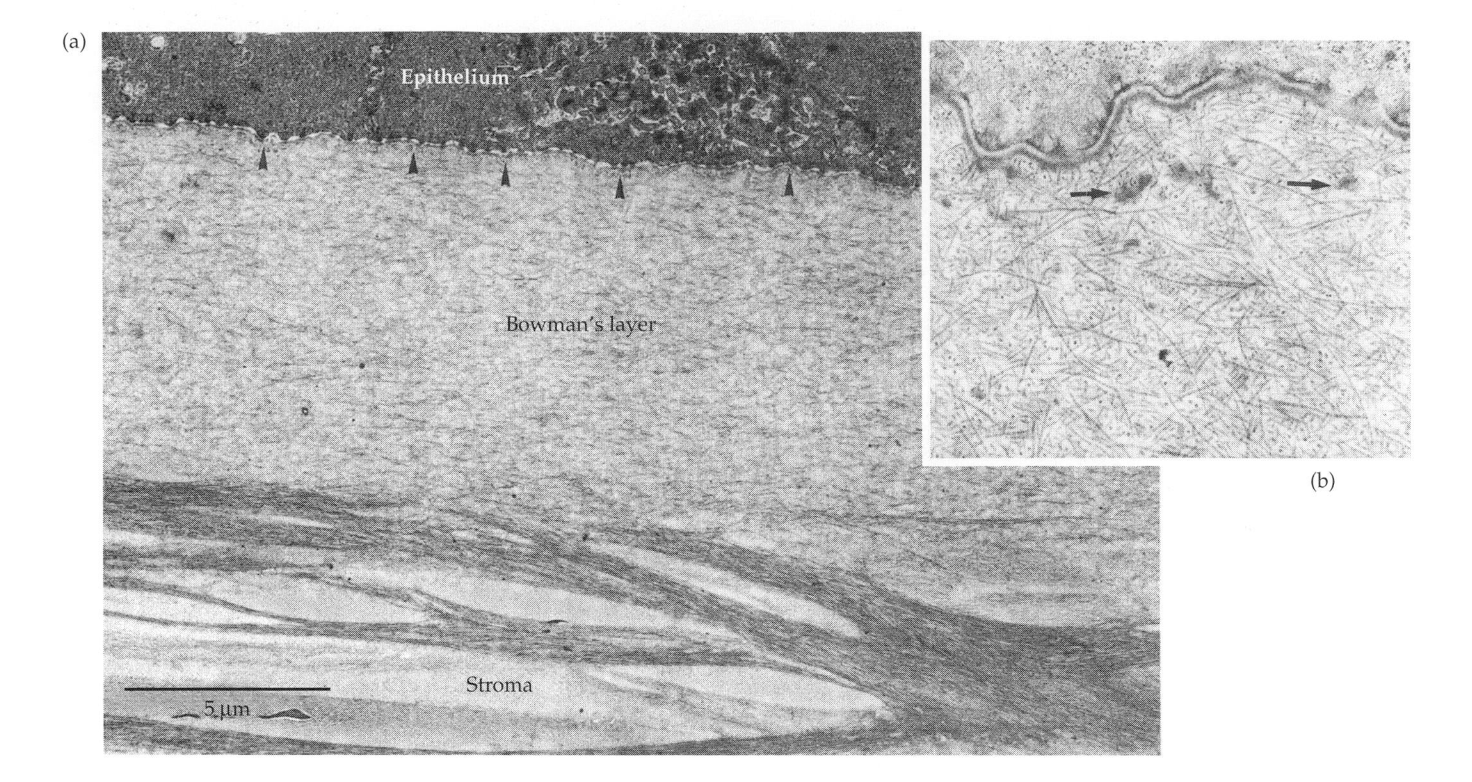

Figure 8.8
Bowman's Layer
(*a*) At low magnification, Bowman's layer is an acellular layer about 10 μm thick lying just beneath the epithelium. It is continuous with the underlying stroma. (*b*) At high magnification, Bowman's layer appears as a dense, irregular meshwork of interwoven collagen fibrils. Arrows indicate anchoring plaques. (*a* from Komai and Ushiki 1991; *b* from Gipson, Spurr-Michaud, and Tisdale 1987.)

the type I fibrils that are so dominant in the rest of the stroma. Immunohistochemistry also indicates that the stroma contains more than just type I collagen, but this story is still unfolding; immunolabeling of collagen types is subject to the same interpretive problems as other forms of immunohistochemistry (see Box 15.1), and because these are large complex molecules, the problems of false positives and negatives are especially difficult.

Figure 8.9 shows a schematic picture of the collagen fibril content and arrangement in Bowman's layer. Type I collagen fibrils are shown running in various directions through the layer after having separated from the anterior-most stromal lamella; whether these fibrils meet and terminate at the basement membrane of the overlying epithelium or continue across the cornea, perhaps reuniting with the stromal laminae, is not known. Other fibrils, thought to be type VII collagen, run almost perpendicularly to the basement membrane, often converging at electron-dense sites known as *anchoring plaques*. These type VII fibrils are continuous with anchoring fibrils from the hemidesmosomes that bind basal cells of the epithelium to the basement membrane. In short, Bowman's layer can be thought of as a layer in which the stroma is bound to the overlying basement membrane and epithelium; the binding interjects at least one other collagen type, along with a certain disarrangement of the type I collagen regularity that is seen in the rest of the stroma.

In general, the picture of collagen fibrils of one type linking stromal type I collagen to the epithelial basement membrane is seen in several species, although Bowman's layer is most obvious in primates. Why this region of the primate cornea should be free of fibroblasts when the same region in other species contains fibroblasts is not known.

Corneal transparency is a function of its regular structure

The most important feature of the cornea, and a striking difference between it and the sclera, is its transparency. This property, which the cornea shares with the lens, is easy to take for granted, but the question Why is the cornea transparent? is not

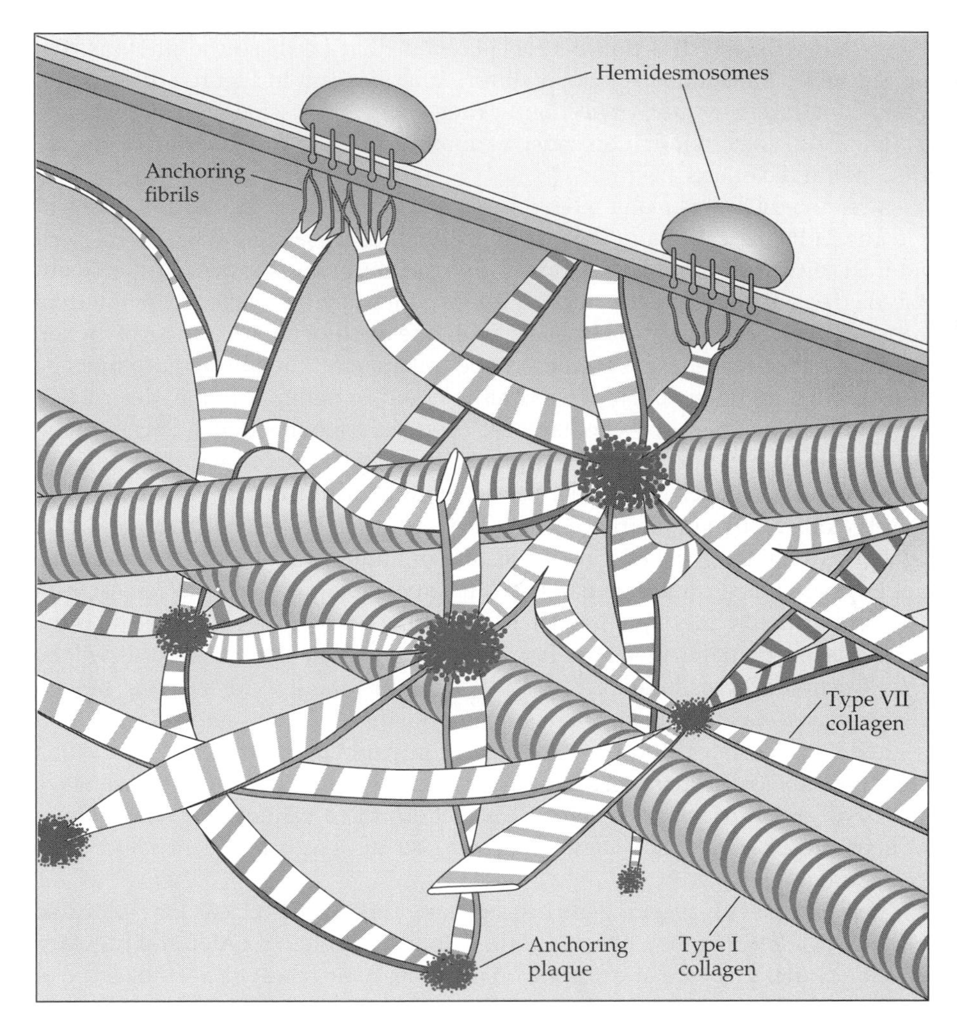

Figure 8.9
Anchoring Plaques and Fibrils in Bowman's Layer
In this schematic, Bowman's layer is shown as having two types of collagen: Large type I fibrils interweave in an irregular pattern, while smaller type VII fibrils meet at regions called anchoring plaques. Some of the type VII fibrils are continuous with anchoring fibrils that join the hemidesmosomes between the epithelial cells and their basement membrane. (After Gipson, Spurr-Michaud, and Tisdale 1987.)

trivial and the answer is not particularly simple. But it is important to have some idea about the normal conditions for transparency to better understand the conditions underlying transparency loss. Corneal transparency is a specific instance of the general problem of transparency—that is, why is any medium, including glass, transparent? The answer lies in the way in which photons of light interact with electrons in atoms and molecules, the stuff of which all matter is made.

There are several possible results when light interacts with molecules or atoms. Absorption is a transfer of energy from photons to electrons in which the incident photons seem to disappear; the transferred energy is dissipated by breaking chemical bonds (as in photopigments) or by being reradiated at another frequency (in the infrared as heat, for example). But other interactions between photons and electrons do not produce a dramatic loss of photons; the energy transferred to electrons is reradiated as photons at or near the same frequency (and energy level) as the incident photons. This sort of interaction occurs routinely, even in the media we call transparent; it is as if the atoms or molecules in the medium have become little oscillators, resonating under the impact of incident photons to become secondary sources of light.

Reradiated photons may disperse in any direction from the molecules in the medium, and the average behavior of many photons is indicated in one plane by an expanding circular wave front around a reradiating molecule (Figure 8.10*a*). In this optical microcosm, photons going back toward the original photon source are considered reflection, the light radiating obliquely or perpendicularly to the original path is scatter, and the light traveling in the same direction as the inci-

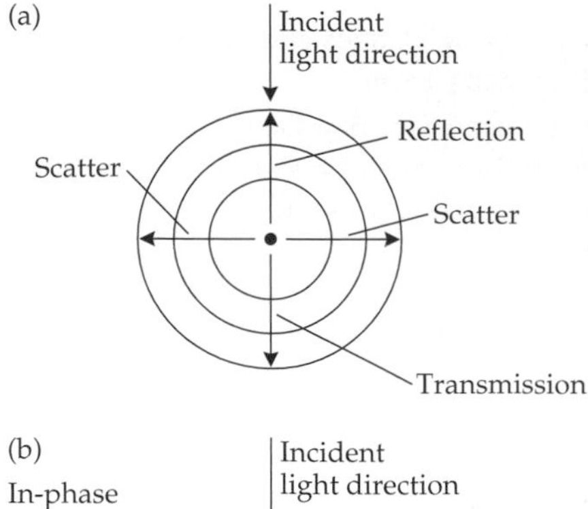

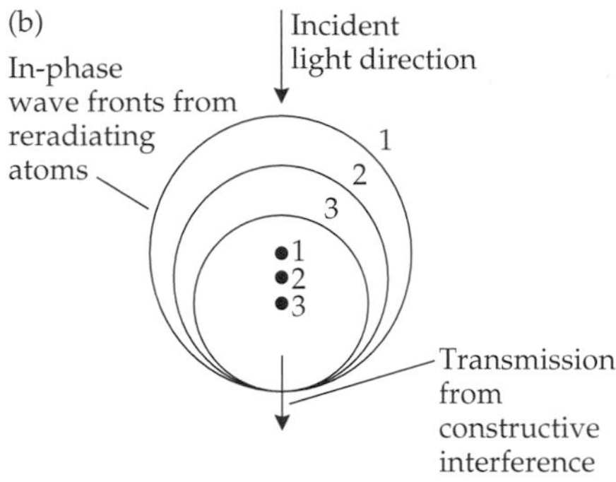

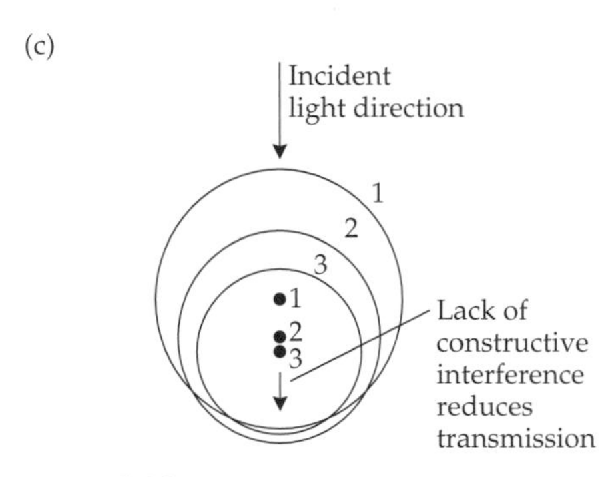

Figure 8.10
Transparency and Structural Regularity
(*a*) Incident light interacts with atoms, producing reradiation in all directions. Reradiation in the incident direction is transmission, in the opposite direction it is reflection, and the rest is scatter. (*b*) When atoms are regularly spaced, the in-phase wave fronts of reradiation coincide in the incident direction; this adding together (constructive interference) constitutes transmission and transparency. (*c*) When atoms are not regularly spaced, the in-phase wave fronts do not coincide; there is less constructive interference, less transmission, and less transparency. (After Higgins 1994.)

dent path is in the direction of transparency (that is, these are photons that appear to be transmitted by the medium). The diagram in Figure 8.10*a* implies that all media can produce reflection, scatter, and transmission of light; for transparency, however, reradiation must be maximal in the forward direction and minimal in the others.

A real medium has many more reradiating atoms or molecules than the single one in Figure 8.10*a*, and the energy from all of these secondary sources can interact, either by adding together or by canceling one another. Respectively, these effects are constructive and destructive interference; wave fronts interfere constructively when they are in phase with one another and destructively when they are out of phase. For a medium to be transparent, the interference must be constructive in the direction of transmission and partially or totally destructive in all other directions. But whether these conditions are met depends very much on the arrangement of the reradiating molecules in the medium.

Figure 8.10*b* makes the case, in a very simplified form, that transparency—that is, constructive interference in the forward direction—requires a great deal of regularity in the spacing of the molecules or atoms in the medium. The circles are in-phase wave fronts emanating from three equally spaced molecules; with this configuration, the only direction in which the wave fronts can coincide is the forward direction; the intensities from the three points will add, and this will be the direction of transmission. For other directions, the interference will be neither completely constructive nor destructive, so there will be some reflection and scatter, but their magnitudes will be small compared to the transmission. When the three molecules are spaced irregularly, as in Figure 8.10*c*, the in-phase wave fronts never coincide in the forward direction; there cannot be perfectly constructive interference, and the medium will be, at best, translucent rather than transparent.

By this line of argument, glass is transparent because its atoms are packed so closely together that they have an extremely high degree of order and regularity. Clouds in the sky are translucent, on the other hand, because the molecules of water in them are widely and randomly spaced. And the cornea, like glass, is transparent because its major optically interacting elements—the collagen fibrils—are of uniform size and spacing, as Figure 8.5*a* showed. The only substantive change from glass to cornea is a change of scale as we go from the dimensions by which atoms are separated to those by which collagen fibrils are separated; although this change is relatively large, the basic physical principles appear to remain the same. This explanation implies that anything making the stromal collagen fibrils less ordered will result in some loss of transparency—the magnitude of the loss presumably being a function of the degree of disorder.* The foregoing discussion summarizes a theory of corneal transparency proposed by Maurice (1957). Other analyses have shown that the collagen fibrils do not have to be spaced with perfect, crystalline regularity to account for the cornea's transparency, but the basic principle outlined here is still valid; namely, structural regularity and corneal transparency go hand in hand.

Collagen organization in the stroma and corneal transparency depend on intact epithelium and endothelium

An important structural difference between cornea and sclera is that the corneal stroma is sandwiched between two cellular layers: the **epithelium** on the anterior surface, and the **endothelium** on the posterior surface next to the anterior

*We do not know exactly the normal degree of order among the corneal collagen fibrils; the processing of the tissue for histology must affect the order to some degree. The wonderful transparency of the cornea may be an indication that the structure is even more ordered than it appears to be in histological sections.

(a) Epithelial damage

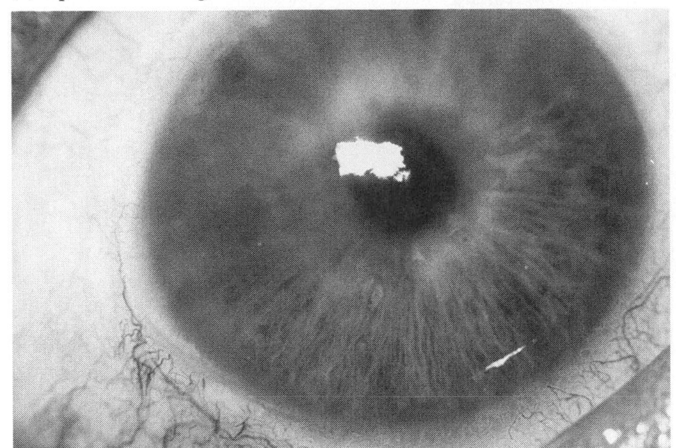

(b) Endothelial damage

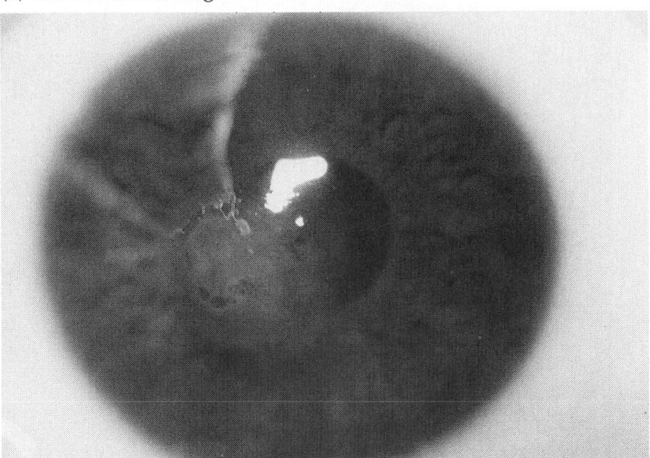

Figure 8.11
Corneal Haze and Reduced Transparency
(*a*) Edema, corneal thickening, and increased light scatter make this cornea appear hazy. (*b*) The haze in this cornea is due to degeneration and guttata in the endothelium (Fuch's endothelial dystrophy). (Courtesy of Roswell Pfister.)

chamber. Although the sclera is normally described as having an outer layer (the **episclera**) and an inner layer (the **lamina fusca**), they are relatively acellular and are little more than modifications of the scleral stroma. (In histological sections, the distinction between episclera and scleral stroma is made by the presence of small blood vessels in the episclera. The lamina fusca, the innermost layer of the sclera, appears different because it contains some melanocytes.)

The importance of the cornea's epithelium and endothelium can be easily demonstrated by seeing what happens to the cornea after experimental or accidental damage to these cell layers. Normally, we do *not* see the cornea except for the reflections from its surface. Figure 8.11*a*, however, shows a distinct hazy area in the cornea resulting from damage to the epithelium. Damage to the endothelium produces the same result, often more dramatically, as in Figure 8.11*b*.

The damage to the epithelium or endothelium and the resulting decreased transparency is accompanied by a measurable increase in corneal thickness. Both the change in corneal thickness and the increased scatter associated with transparency loss can be measured fairly accurately, but the essential message is the same as the qualitative assessment shown in Figure 8.11: Damage to either the epithelium or the endothelium produces corneal swelling because of increased water uptake, and a decrease of transparency is the inevitable result.

The corneal epithelium is a multilayered, renewable barrier to water movement into the cornea

The corneal epithelium consists of six to eight layers of cells stacked on top of one another to a thickness of around 50 μm (in central cornea, this is about 10% of the total corneal thickness) (Figure 8.12*a*). The cells differ in morphology; the bottom layer, which rests on a basement membrane next to Bowman's layer of the stroma, contains the columnar **basal cells**—the newest, most recently born cells in the epithelium. Just above the basal cells are several layers of **wing cells**, so called because of their irregular shape, with short projections (wings) extending from the cell. The wing cells were originally basal cells; as they moved or were pushed up out of the basal cell layer, the original columnar shape changed to the winglike form. And above the wing cells are several layers of flattened **squamous cells**; the squamous cells were once wing cells, and the anteriormost layer of squamous cells will soon be sloughed from the cornea into the tear film covering the anterior surface.

Virtually all the cells in the epithelium are born, transformed, and shed within about 10 days. The morphological transformation that corneal epithelial cells

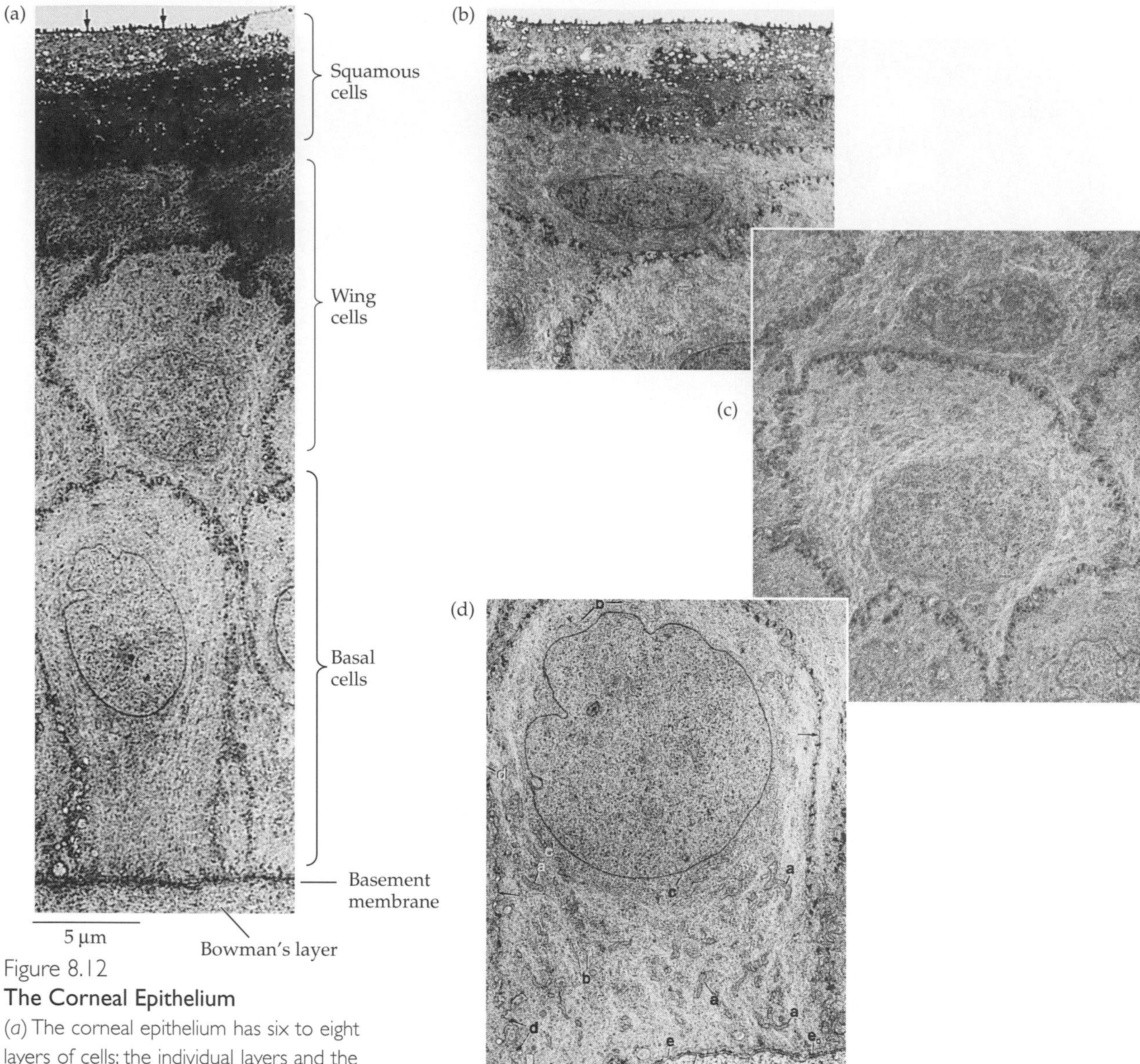

Figure 8.12
The Corneal Epithelium
(*a*) The corneal epithelium has six to eight layers of cells; the individual layers and the epithelium as a whole extend over the entire anterior surface of the cornea. A single layer of columnar basal cells rests on a basement membrane adjacent to Bowman's layer of the stroma. Two layers of wing cells are intermediate in shape between the basal cells and the layers of superficial, squamous cells. Arrows indicate small microvilli on the anteriormost squamous cells. (*b*) Squamous cells have long, overlapping borders with numerous anchoring junctions, seen here as small black dots along the cell borders. (*c*) Wing cells are irregular in shape—both cells shown here have downward protruding "wings"—with numerous electron-dense anchoring junctions. (*d*) The tall basal cell has anchoring junctions with the wing cell above, with its neighboring basal cells, and with the basement membrane. (From Hogan, Alvarado, and Weddell 1971.)

undergo as they change from columnar to cuboidal to squamous and slough off the surface is called **apoptosis**; it occurs in many kinds of cells throughout the body and represents a form of programmed cell death. The mechanisms underlying apoptosis are not known in any detail, but for the corneal epithelium, apoptosis begins when basal cells move or are pushed out of the basal layer into the layers of wing cells. Their sequence of morphological change and cell death is then inevitable.

The presence of dead and dying cells at the corneal surface is not a problem, because apoptosis is different from cell death due to trauma or infection. In these latter cases, cells do not shrivel and expire quietly; instead, they undergo **necrosis**, often swelling, bursting, and spilling out their intracellular contents. This release of intracellular materials from necrotic cells mobilizes inflammatory reactions, but it does not occur in response to apoptotic cell death. (Which is good, since apoptosis is a common phenomenon, occurring in most of the body's epithelia and endothelia and in many other types of cells. Cell death is a normal part of life, in other words.)

Differences in cell morphology in the epithelium are more apparent at higher magnification, and intercellular junctions can also be seen (Figure 8.12*b*–*d*). Squamous cells overlap extensively and adjacent cell membranes interdigitate, creating long and winding extracellular paths between cells (Figure 8.12*b*). The squamous cells are bound to each other and to the underlying wing cells by **desmosomes**, which at the magnification in Figure 8.12*b* are seen as small black dots all around the cell borders. The small projections from the most superficial cells are microvilli (small fingerlike projections) or microplicae (small folds) cut in cross section; they are shown more realistically in Figure 8.15. The small open spaces in the cytoplasm are vacuoles; they are characteristic of the squamous cells, and their number increases as the cells near their time of shedding from the epithelium.

Wing cells are roughly cuboidal, but they have projections that give the cells their name; the two shown in Figure 8.12*c* have downward-projecting wings. Like the squamous cells, the membranes of adjacent cells interdigitate, and numerous desmosomes bind them to their neighboring cells.

The columnar basal cells (Figure 8.12*d*) are bound to their basement membrane by **hemidesmosomes**, and full desmosomes bind cells to their neighbors. Adjacent cell membranes are very tightly apposed at additional cell junctions, but their details are not clear at the magnification of Figure 8.12*d*.

Cell junctions and contacts made by basal cells are illustrated in Figure 8.13. Desmosomes (Figure 8.13*a*) are associated with dense intracellular plaques from

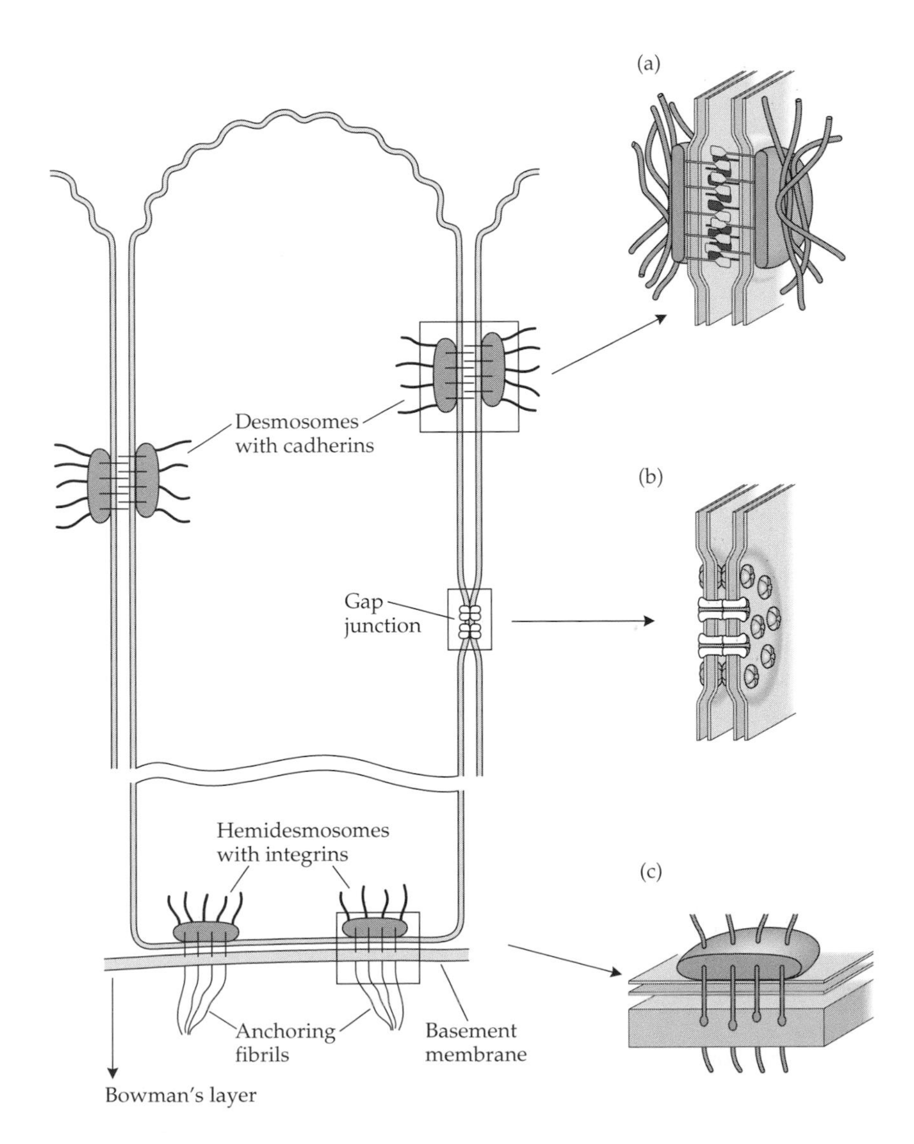

Figure 8.13
Basal Cell Junctions
(*a*) The schematic drawing on the left shows the desmosomes that bind adjacent cells together; the space between the two sides of the desmosome contains interleaving cadherin molecules. Desmosomes between other epithelial cells are similar. At right is a schematic model of the desmosome. (*b*) Basal cells also have communicating junctions—gap junctions—with one another, in which the adjacent cell membranes are very tightly apposed. (*c*) Hemidesmosomes bind the basal cells to the basement membrane with integrin molecules that are continuous with anchoring filaments in Bowman's layer. (After Alberts et al. 1994.)

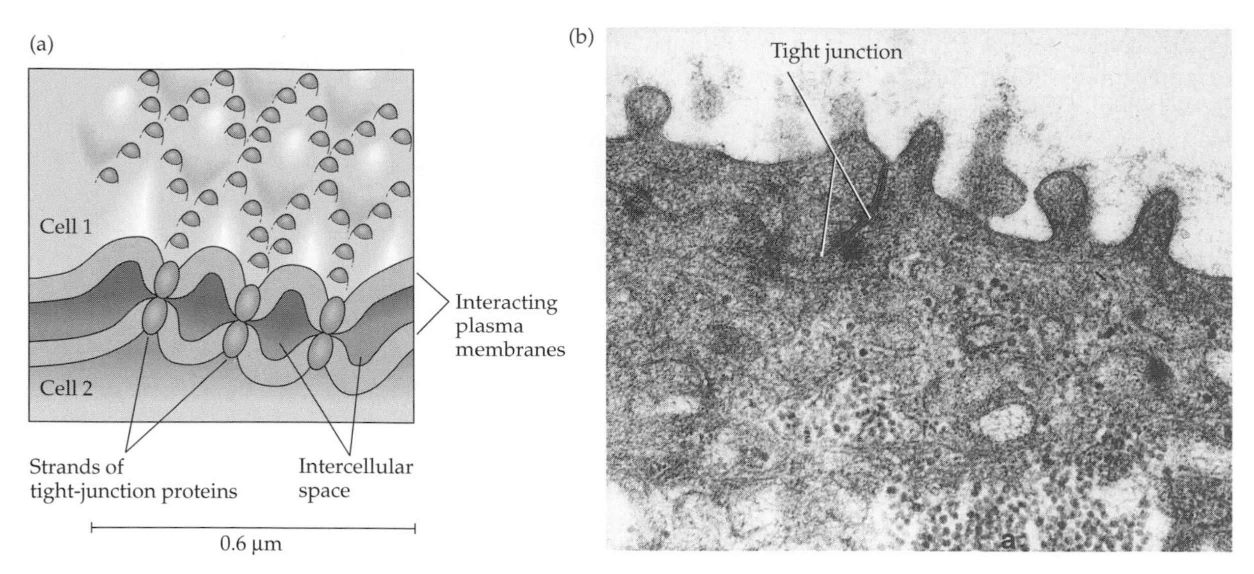

Figure 8.14
Occluding Junctions in the Squamous Cells
(*a*) An occluding (tight) junction binds adjacent cell membranes together along an interwoven belt made of special linking protein strands. (*b*) Tight junctions in the corneal epithelium are common between superficial squamous cells. (*b* from Hogan, Alvarado, and Weddell 1971.)

which specialized proteins (cadherins) extend into the extracellular space to link adjacent cells; within the cells, some of the intracellular filament proteins are linked to the plaques. Hemidesmosomes (Figure 8.13*c*), as the name implies, are like half a desmosome, with a dense plaque inside the cell and proteins extending from the plaque into the extracellular space. In this case, however, the binding is to a basement membrane, using another group of specialized proteins (integrins). Desmosomes and hemidesmosomes are cellular rivets; they hold all the cells in the epithelium together, bind the basal cells to the underlying basement membrane, and, through the intracellular connections of the desmosomes and hemidesmosomes with elements of the cytoskeleton, contribute to the rigidity and stability of both cells and tissue.

Desmosomes and hemidesmosomes are anchoring junctions, but the basal cells also have a form of communicating junction—namely, **gap junctions** (see Figure 8.13*b*; see also Figure 5.2). Gap junctions permit two-way exchanges of small molecules and ions between adjacent cells; they appear to be located only on basal cells in the epithelium, but their specific function is not known. These gap junctions were once thought to be small occluding junctions (specifically, *maculae occludens*), but more recent work has shown the presence of specific gap junction proteins.

An occluding junction is illustrated in Figure 8.14. This is a **tight junction**, specifically a *zonula occludens* (like adhering junctions, the tight or occluding junctions can exist either as a belt—zonula—or as a small spot—macula). The characterization as a zonula is significant, because it means the occluding junction encircles the cell; spot junctions have unoccluded spaces between them. These tight junctions are found primarily between the most superficial squamous cells, although they have been reported on deeper squamous cells and occasionally on wing cells. Their presence in the corneal epithelium is important because these are the only junctions that can completely close the extracellular space between adjacent cells and restrict the passage of small molecules, ions, and water. The tightness or permeability of the tight junctions is variable, however; they are more restrictive in some epithelia than in others. Those in the corneal epithelium compare favorably to the highly restrictive junctions in the urinary bladder, and it is reasonable to assume that their presence effectively blocks the movement of osmotically active molecules and water in the extracellular space.

The multiple cell layers in the epithelium (Figure 8.15), the various types of junctions, particularly tight junctions between squamous cells, and the interdigitating cell processes that create a long and tortuous extracellular space are all

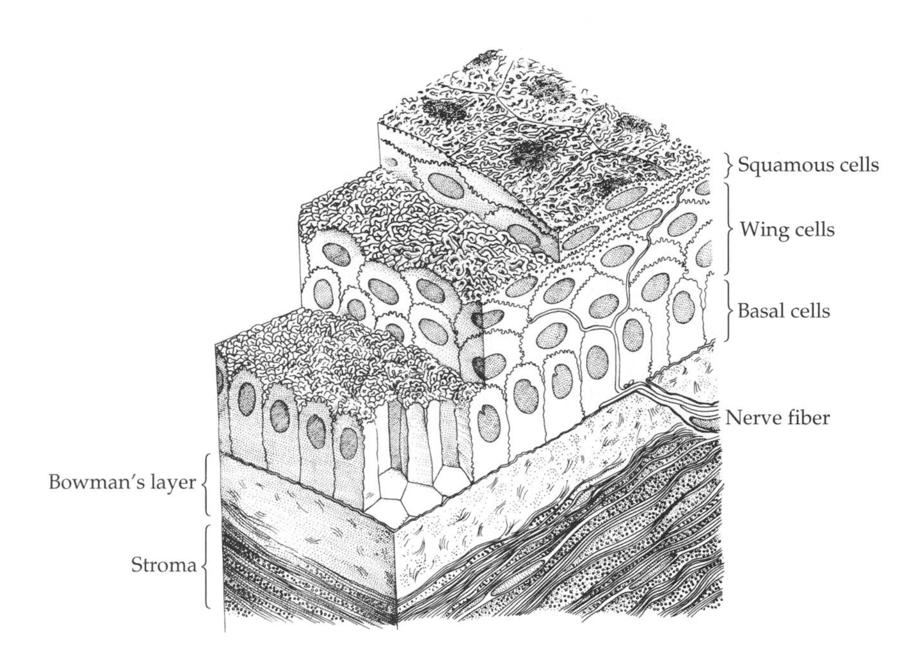

Figure 8.15
Structure of the Corneal Epithelium
The small piece of epithelium reconstructed here contains all the structural elements that, when extended, cover the corneal surface. The polygonal tiling of the squamous cells with their microvilli and microplicae is evident; this is the surface covered by the tear film. One branching corneal nerve fiber is shown reaching out to the layers of squamous cells. (After Hogan, Alvarado, and Weddell 1971.)

factors helping to make the corneal epithelium a barrier to the free movement of water from the tears to the stroma. And to make a comparison with the endothelium, which will be considered in the next section, the epithelium is a much more effective passive barrier; the endothelium is relatively leaky.

The corneal endothelium is a single layer of metabolically active cells

The single layer of cells that forms the corneal endothelium lines the entire posterior surface of the cornea and is in direct contact with the aqueous in the anterior chamber. The endothelium is separated from the stroma by Descemet's membrane, which is a tough, glassy layer about 10 µm thick. Figure 8.16*a* shows the structure of the endothelial surface, Figure 8.16*b* a section through the layer at high magnification. The endothelial cell layer is only about 5 µm thick, but the cells adjacent to one another are arranged so that the extracellular distance from the anterior chamber surface to Descemet's membrane is considerably longer; cells do not abut like bricks or paving stones—instead, their interfaces are tortuous and interdigitating (see Figure 8.16*b*). In addition, tight junctions are often found near the posterior surface of the cell layer; it seems, however, that these occluding junctions are spot junctions, rather than the zonular beltlike junctions formed by superficial epithelial cells. As a result, the endothelium resists the movement of molecules and water between the aqueous and the corneal stroma, but overall the endothelium is a less effective *passive* barrier than the epithelium.

The endothelial cells are very active metabolically, however, as is suggested by the presence of numerous mitochondria within the cells (see Figure 8.16*b*). Structurally, the mitochondria are unusual, having internal cross-members (cristae) running parallel, rather than perpendicular, to the long axis of the organelle. Much of the metabolic activity in the endothelium is associated with pumps that move ions of various species across the cell membranes; one consequence of the ion pumping is movement of water from the corneal stroma to the aqueous. (Ion pumping resulting in water movement also occurs in the epithelium, but it is less important in maintaining normal stromal water content than the endothelial pumping is.) The importance of this activity is readily seen when metabolic inhibitors are applied to the endothelium: The cornea takes on water, thickens, and decreases in transparency. The endothelium is a leaky membrane, but it compensates for the leakage by its ability to actively pump water out of the cornea.

(a) Endothelial surface

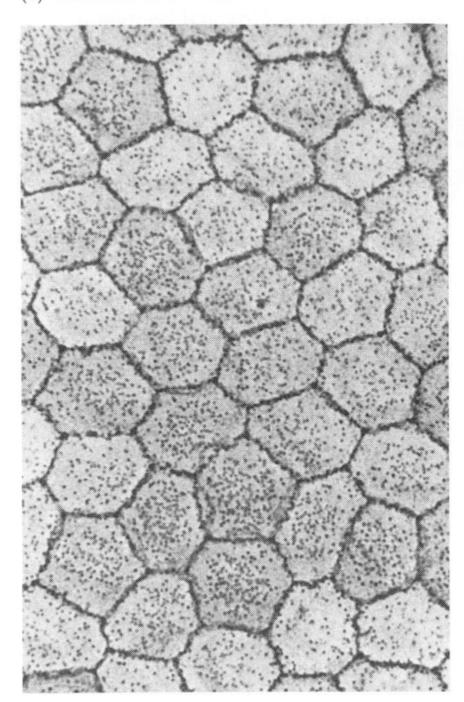

(b) Endothelial section

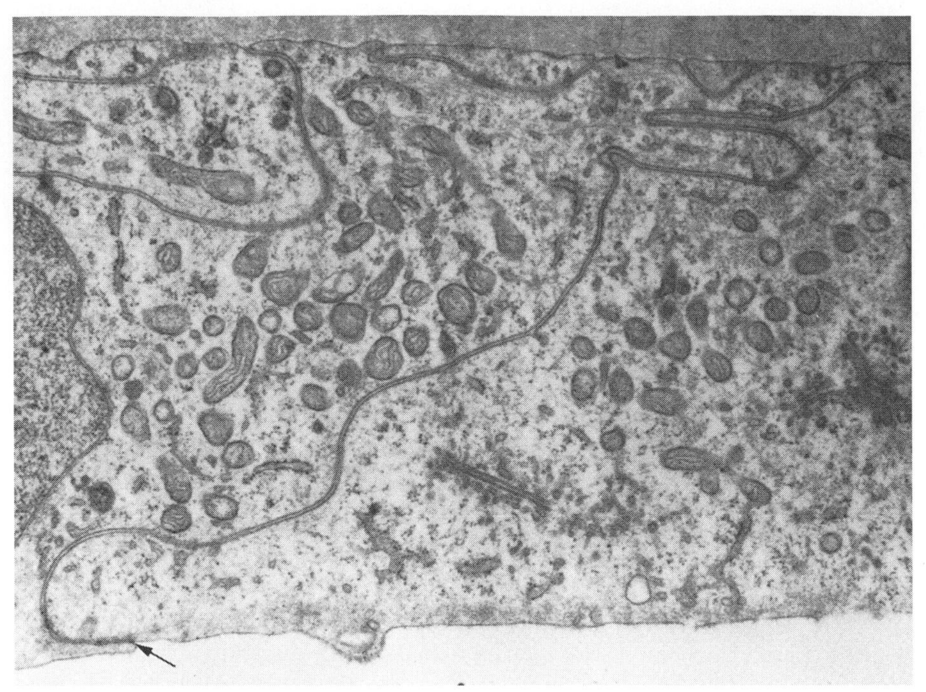

Figure 8.16
The Corneal Endothelium
(*a*) A surface view of the endothelium shows the tiling formed by the polygonal cells, the mean diameter of which is about 20 μm. (*b*) A section of the endothelium at high magnification shows the long, winding border between adjacent cells, an occluding junction (arrow), and numerous mitochondria in the cytoplasm. (From Hogan, Alvarado, and Weddell 1971.)

The endothelial cell tiling changes with time because cells that die cannot be replaced

Endothelial cells form an irregular tiling in which the individual cells are irregular polygons, usually with six but sometimes with other numbers of sides (see Figure 8.16*a*). Cell density decreases from around 3000 cells/mm^2 in the center of the cornea to around 1500 cells/mm^2 in the periphery. On average, the cells are about 20 μm across, but because cell size is inversely correlated with cell density, peripheral cells are roughly twice the size of central cells.

If the cells in the endothelium were a perfectly regular array with equal spacing between the centers of neighboring cells, they would all be identical hexagons, like the hexagonal cells in a beehive's honeycomb. But they vary both in size and in shape (not all of them are six-sided), which means that the spacing of the endothelial cells is not perfectly regular. When the irregularity increases, the variability in cell area (*polymegatheism*) and number of cell sides (*pleomorphism*) also increases. These measures of irregularity in the endothelial cell array have been shown to be affected by pathology, age, trauma, and prolonged contact lens wear—in all cases because cells lost from the endothelium for any reason cannot be replaced. The endothelium maintains its continuity by migration and expansion of the surviving cells.

The array of endothelial cells can be examined in vivo with a specular microscope, a high-power biomicroscope that looks at light reflected from the endothelial cells (Box 8.1). When young and old eyes are compared (Figure 8.17), the higher density and the greater regularity of the young endothelium are obvious. These differences are normal ones. Ongoing cell loss from the endothelium causes cell density to decline linearly with age such that the average density near the center of the cornea at 80 or 90 years of age is roughly half the value at birth. Assuming that the decline in cell density over time is the result of random cell death throughout the endothelium, the increased irregularity of the cell array with age is due to the expansion and migration of the remaining cells as they maintain a confluent cell layer (see Figure 8.19).

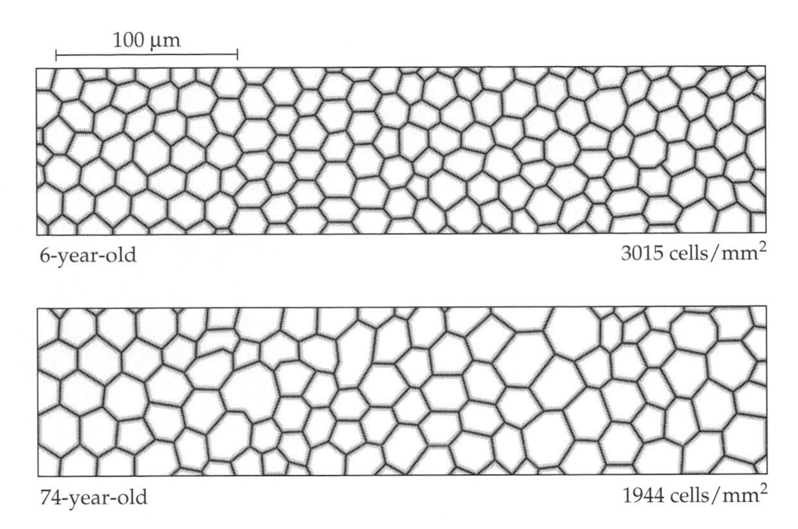

Figure 8.17
Age-Related Change in the Endothelium
Specular microscope views of young (6-year-old) and elderly (74-year-old) corneas show that cell density decreases with age, while cell variability (in both size and shape) increases. (After Bourne and Kaufman 1976.)

Increased variation in cell size and number of cell sides in the absence of significant overall differences in cell density is evident in the comparison of the endothelium of an individual who had worn hard contact lenses for 27 years with that of an age-matched control (Figure 8.18). These changes have been demonstrated in both hard and soft contact lens wearers, varying with duration of lens wear (the longer lenses have been worn, the greater the endothelial change) and lens type (the effect is less pronounced in soft contact lens wearers). It is not known how prolonged contact lens wear exerts its effect on the endothelium, but it appears to be associated with loss of local cell clusters. The endothelium becomes more irregular as cells expand and migrate into the regions of cell loss. More generally, trauma (including surgical incisions) and disease processes produce a similar form of irregularity in the endothelial cell array that exceeds the changes produced by normal aging.

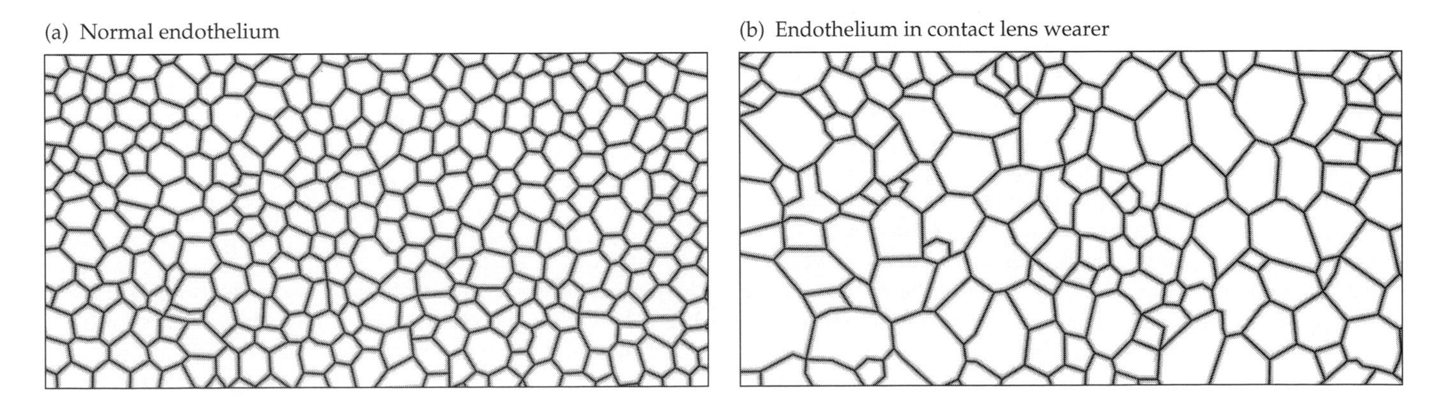

Figure 8.18
Endothelial Change after Long-Term Contact Lens Wear
(*a*) Specular microscopy of the endothelium shows some irregularity consistent with age (46 years), but it is significantly less than in the endothelium of an individual of the same age who wore contact lenses for more than 25 years (*b*). (After MacRae et al. 1986.)

[Box 8.1] Biomicroscopy of the Cornea

THE CORNEA IS normally transparent, and its structure is not meant to be visible. But like any other transparent medium, the cornea's transparency is not perfect; light reflects at interfaces with different refractive indices and scatters from structures within it. And since any abnormality or irregularity in the cornea reflects or scatters more light than the normal cornea, it follows that scattered and reflected light are the principal means by which we examine corneal structure in vivo.

Although large opacities and abnormalities in the cornea are visible with no optical aids, higher magnification is needed to see small defects. The clinical instrument used for this purpose is the biomicroscope, a low-power microscope that is usually adjustable for magnifications from about 5× to 40× and mounted horizontally to look directly at the eye. The biomicroscope is used with an illumination source whose angle of incidence can be varied relative to the viewing axis of the biomicroscope, and with a variable configuration of the beam of illuminating light. This combination of biomicroscope and variable light source is called a **slit-lamp**.

The first of two basic methods for illuminating the cornea is *direct* illumination, in which the light falls on the object one is trying to observe. There are various ways to employ direct illumination; the light beam can be wide and diffuse or small and focused, for example, and the different methods are useful for seeing different things. Figure 1*a* shows the configuration for seeing an "optical section" through the cornea; the light beam is very narrow—a slit—and focused on the corneal surface. As the narrow beam passes through the cornea, some light scatters at oblique angles, and it is this scattered light one sees with the biomicroscope. Figure 1*b* shows an optical section through a normal cornea, with thin bright lines defining the anterior and posterior surfaces. In Figure 1*c*, a small region of damaged epithelium scatters more light and thus appears brighter than the rest of the optical section.

Several figures in the main text show the endothelium viewed by specular reflection, and Figure 1*d* here shows the configuration for this procedure; to see the reflected light at its strongest, the angle of view is equal to the angle of incidence. The tiling of the endothelial cells can be seen by specular reflection in Figure 1*e*, even though the magnification here is fairly low. Higher magnification is needed to determine cell size or shape, as discussed in the text. The dark region in the middle of the reflecting endothelial array is abnormal, perhaps representing guttata (see Figure 8.22).

In the usual configuration, optical sections give an impression of relative thickness across the cornea, but they cannot be used to measure the absolute thickness. This can be done, however, by making two alternations. First, the beam of the illuminating slit must be perpendicular to the corneal surface, and second, the optical section must be split into upper and lower halves, one of which can be shifted relative to the other. The calibrated beam splitter in this configuration is a **pachometer**, a device for measuring thickness. Standard optical pachometers are accurate to about ±25 μm or about ±5% of normal central corneal thickness. Corneal surgeons must pay close attention to corneal thickness and its variation with location; they generally use ultrasonic pachometers, which are a form of A-scan ultrasonography (see Box 3.1) calibrated for the velocity of ultrasound through the cornea.

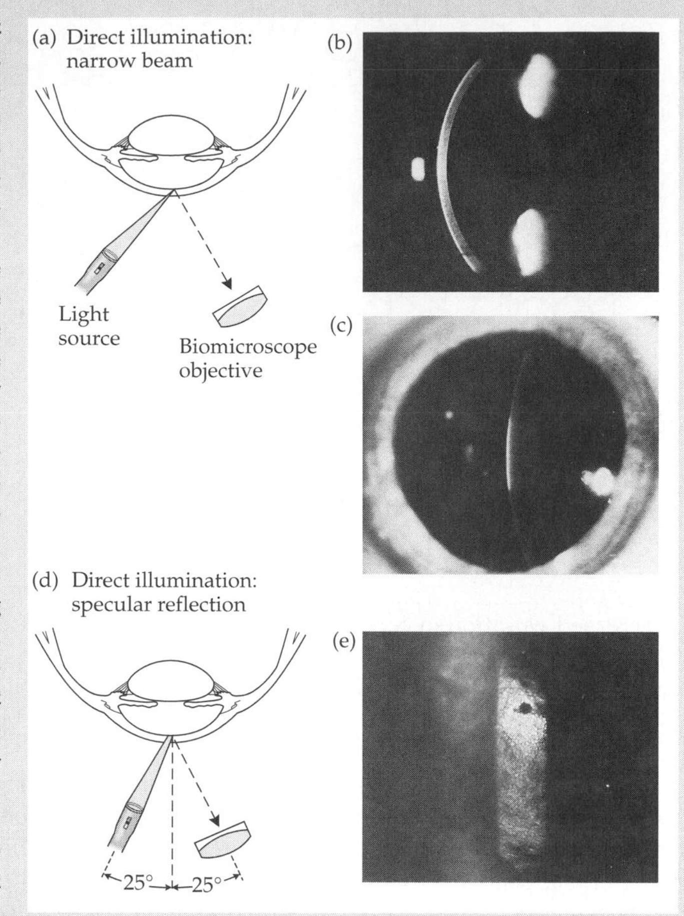

Figure 1
Corneal Structure with Direct Illumination
(*a*) An optical section through the cornea is created by use of a very narrow beam of light focused at the corneal surface and viewed from an oblique angle. (*b*) An optical section through a normal cornea. (*c*) An optical section through a cornea with an epithelial dystrophy; the damaged portion of the epithelium scatters more light than normal and appears much brighter than the rest of the optical section. (*d*) For viewing specular reflections, the illumination beam is a wide slit and the angles of light incidence and viewing are equal. In this case the specular reflection is from the endothelium (e). (Photographs from Martonyi, Bahn, and Meyer 1985.)

The second illumination method is *indirect* illumination, in which the object of interest is illuminated by light reflected from another structure; Figure 2 shows two of the several forms of indirect illumination. In the technique called *sclerotic scatter* (Figure 2*a*), the light beam is directed at the limbus. Light is scattered and reflected from this region, and some of this stray light is funneled across the cornea by internal reflections at the corneal surfaces. If

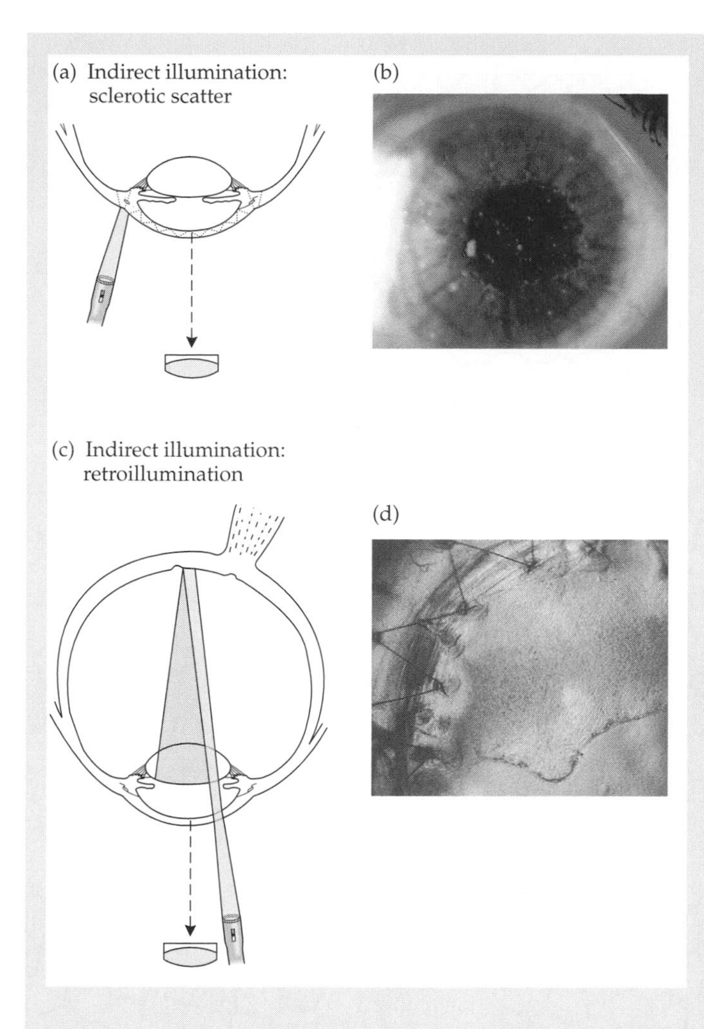

Figure 2
Corneal Structure with Indirect Illumination
In indirect illumination, the structure viewed is illuminated by light reflected from another structure. (*a*) In this example of sclerotic scatter, the light source is directed toward the limbus, but some of this light passes into the cornea by internal reflections. Small foreign bodies embedded in the cornea show up as brightly illuminated spots (*b*). (*c*) In retroillumination, a structure behind the cornea is illuminated, allowing irregularities to be seen as dark regions against a bright background. Here, the fundus is illuminated, and a region of graft rejection following keratoplasty is quite visible (*d*). (Photographs from Martonyi, Bahn, and Meyer 1985.)

this scattered light traveling within the cornea encounters small reflective particles, they will reflect some light in the direction of the biomicroscope objective, appearing as bright spots against a dark backgound. Small foreign bodies embedded in the cornea show up very clearly, as in Figure 2*b;* they would be very difficult to see with direct illumination.

In retroillumination (Figure 2*c*), the object of interest is not illuminated at all, but is seen as dark against a bright background. In Figure 2*d,* the retina and choroid (the fundus) are illuminated with a broad diffuse light beam, creating a uniformly illuminated background for structures lying anteriorly. The picture shows a region near the edge of a corneal graft; the mottled area within the graft bounded by a meandering dark line is an area of graft rejection. The iris can also be used as the background for retroillumination.

The foregoing examples are a small sample of the ways in which a slit-lamp can be used; its versatility makes it one of the most important tools for examining ocular anatomy in vivo. It is useful for examining not only the cornea but also other external structures—the eyelids, the conjunctiva, and the superficial vasculature. The structure of the iris can be examined with both direct illumination and retroillumination, as can the lens, which can also be optically sectioned much as the cornea can be. Special lenses can be added to focus on the fundus or to look into the anterior chamber angle (gonioscopy; see Box 9.1). In addition, the standard tonometer used to measure intraocular pressure is used in conjunction with the slit-lamp. Although the modern slit-lamp did not come into being all at once, the basic idea and earliest implementation can be credited to Allvar Gullstrand (1862–1930); his 1911 Nobel prize in physiology or medicine was not for this invention, but his contribution to our understanding of the anatomy of the living eye was enormous all the same.

The endothelium becomes less regular with time because cells lost from the array cannot be replaced and the process of filling the gaps, regardless of how they occur, always makes the array less regular, as Figure 8.19 illustrates. From an array of endothelial cells (Figure 8.19a), 5% are removed either at random or in a small cluster (Figure 8.19b). Because the endothelium cannot produce new cells to replace the deleted cells, the gaps must be filled by cell expansion and cell migration. Figure 8.19c shows how the endothelial cell arrays might appear after migration and expansion when the cell arrays have stabilized in their new configurations.* The new arrays have two features in common: Compared to the

*Figure 8.19 began as an endothelial cell array redrawn to be a Dirichlet tiling in which the cell sides are the perpendicular bisectors of lines drawn between the centers of neighboring cells (the centers and their connecting lines—the Delauney triangulation—are not shown). After deleting some cells, I moved the centers of cells nearest the gaps to reestablish a reasonably uniform coverage, and redrew the tiling. The final tilings are not unique solutions to the problem of filling gaps after cells have been deleted.

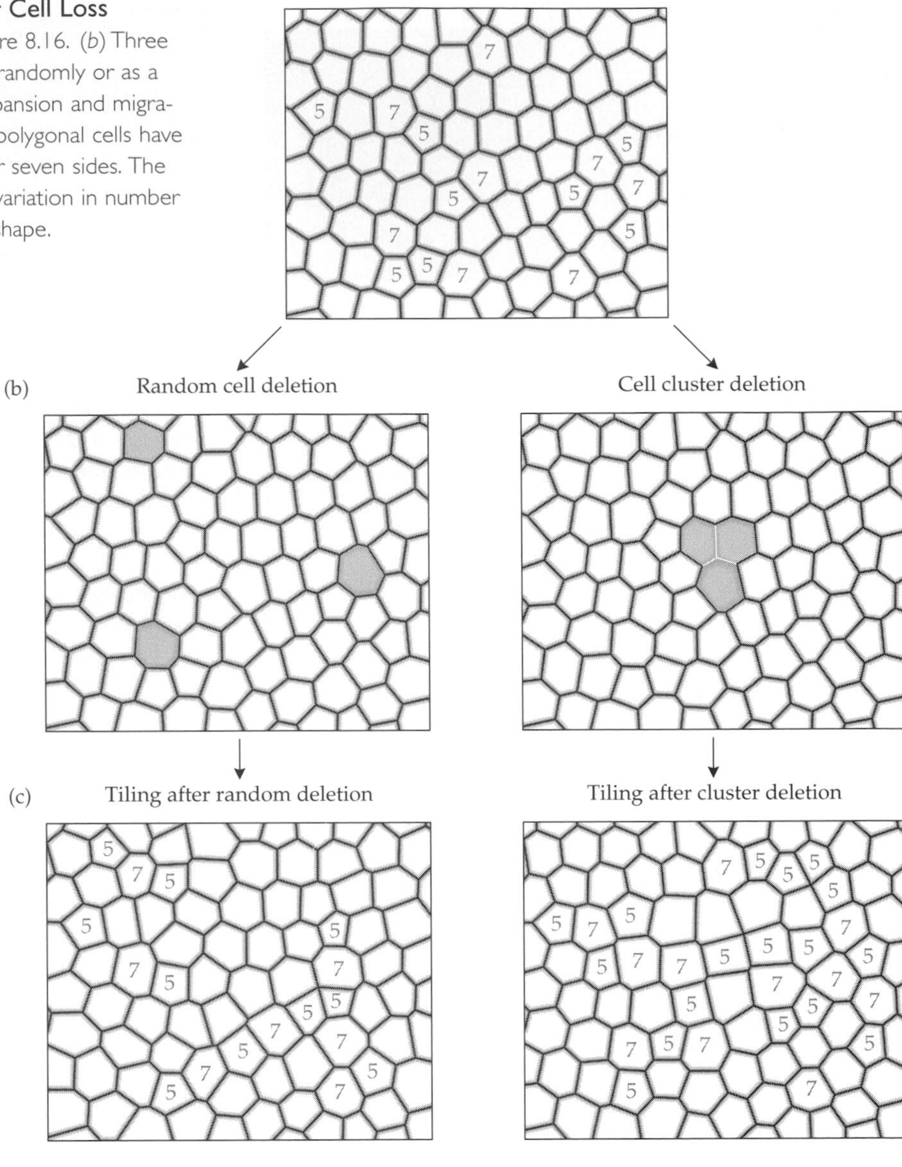

Figure 8.19
Changes in the Endothelial Cell Tiling after Cell Loss
(*a*) The endothelial cell array, redrawn from Figure 8.16. (*b*) Three cells are removed from the original tiling either randomly or as a small cluster. (*c*) The arrays change after cell expansion and migration to fill gaps left by the deleted cells. All the polygonal cells have six sides except those indicated as having five or seven sides. The rearrangement after cell deletion increases the variation in number of cell sides, in cell size, and in regularity of cell shape.

original array, more cells have five or seven sides, and the cells have more variation in size; that is, the new arrays exhibit more pleomorphism and polymegatheism. And to my eye, cell shape is more variable, in the sense that the polygons depart more from their regular configurations in which the lengths of the sides (whether the number of sides is five, six, or seven) are equal.

Since these changes in the endothelium can be measured in vivo, the results are potentially important to clinicians as a way to assess the cornea's health and normality. Unfortunately, as Figure 8.19 was meant to show, any cell loss from the endothelium *always* increases the variability in cell area and number of cell sides, regardless of the nature of the cell loss or its underlying cause. (And it will also increase other variables not measured by polymegatheism and pleomorphism.) Normal aging also has variable effects among individuals, so there seems to be no strong correlation between a particular type or degree of irregularity and a particular disease process or metabolic stress. The cornea's history is to some extent written in the irregularity of the endothelial cell array, but the record is not very specific.

Unless the endothelium is so badly compromised that it cannot maintain confluence, however, it can be markedly irregular and still function well enough to maintain corneal transparency. Cell density must drop below 500 cells/mm^2 before cell expansion and migration are unable to maintain continuity of the cell layer. The critical question concerns the ability of an irregular, disordered endothelial cell layer to compensate for additional stress or trauma. Are long-term contact lens wearers, to take one example, at greater risk of severe endothelial dysfunction as they age or require ocular surgery that penetrates the cornea? How large is the margin for error?

Descemet's membrane separates the endothelium from the stroma

Descemet's membrane has no obvious structure in light micrographs, appearing as a thin, clear band between the endothelium and the stroma (Figure 8.20*a*). It is considered to be a basement membrane for the endothelium, largely because it is secreted by the endothelium, but there are no hemidesmosomes or other anchoring junctions between Descemet's membrane and the endothelium. In young eyes, Descemet's membrane is around 3 or 4 μm thick—about the same thickness as the endothelium—but its thickness increases with age to around 10 to 15 μm.

Descemet's membrane is highly elastic, but the elasticity is a property of its collagen fibrils, not of elastin. At least five types of collagen have been reported

(a)

Descemet's membrane

Endothelium

50 μm

(b)

1 μm

Plane of section

(c)

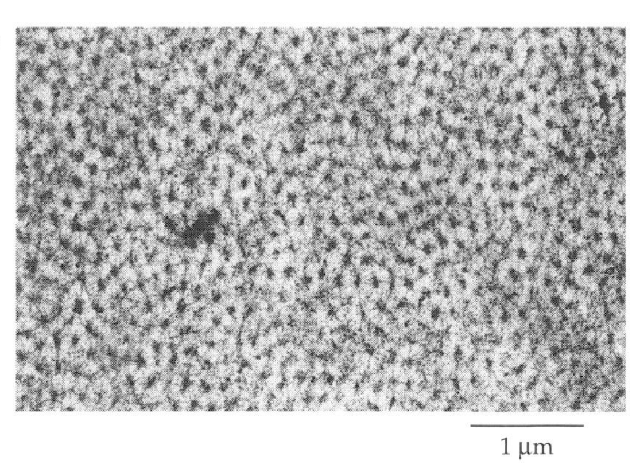

1 μm

(d)

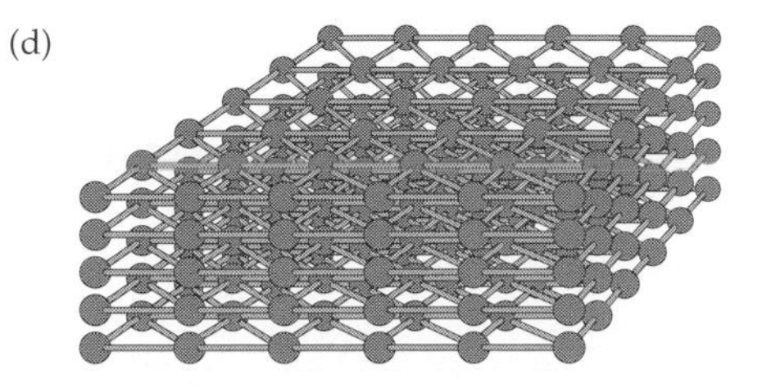

Figure 8.20
Descemet's Membrane
(*a*) At low magnification, Descemet's membrane is a structureless band lying between the posterior stroma and the endothelium. In this section, from a 50-year-old eye, the membrane is almost 10 μm thick. (*b*) At high magnification, some structural regularity becomes apparent as a pattern of small dots of high electron density. Note the absence of any anchoring junctions between the endothelium and Descemet's membrane. (*c*) A section cut parallel to the plane of the membrane shows the regularity of the internal structural very clearly. (*d*) This schematic view of the membrane shows the fine collagen fibrils arranged as a very regular lattice, with regularly spaced intersections (nodes). The spaces in the lattice would be filled by a proteoglycan matrix. (*a–c* from Hogan, Alvarado, and Weddell 1971.)

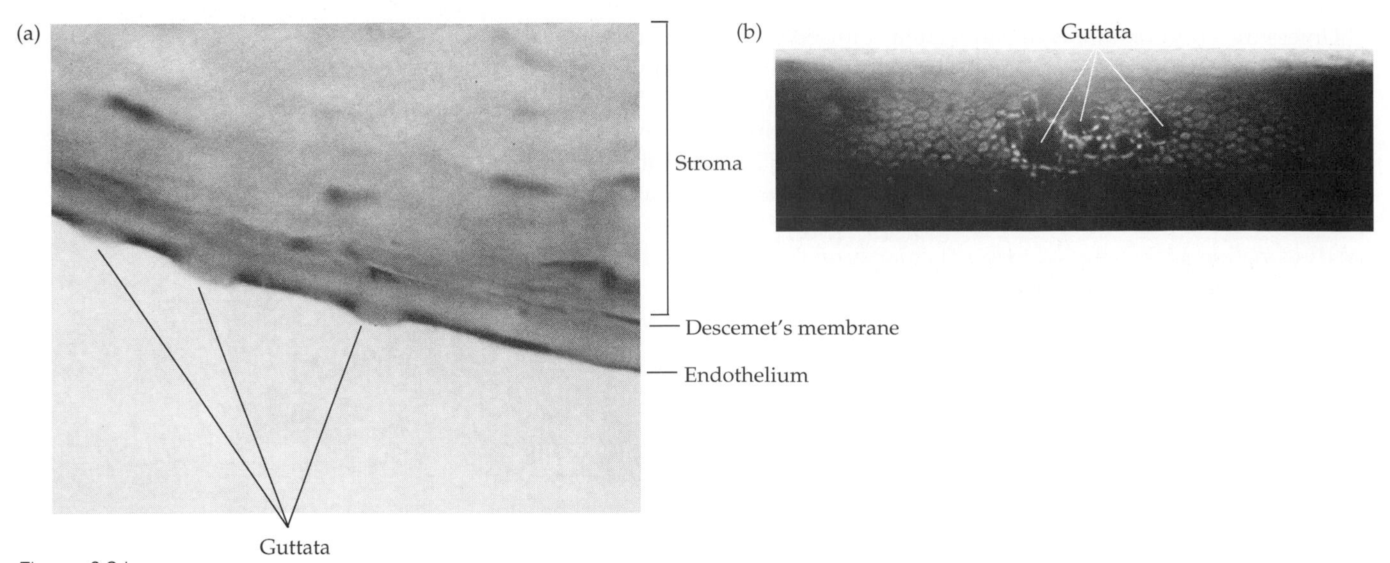

Figure 8.21
Guttata
(*a*) Guttata in a section through the cornea appear as small, localized bulges in Descemet's membrane; the endothelium is either thinned or absent over the surface of the bulges. (*b*) With specular microscopy, guttata appear as small, circular breaks or holes in the endothelial cell tiling. (From Arffa 1991.)

as present in Descemet's membrane, with distributions varying between the anterior banded zone of the membrane and the posterior unbanded portion (Figure 8.20*b*). Type VIII collagen is said to be dominant in the anterior banded zone, where it forms a regular, springy lattice (Figure 8.20*d*). The spaces in the collagen fibril lattice are probably filled with proteins, such as fibronectin, that attach Descemet's membrane to the stroma and to the endothelium.

Because Descemet's membrane is interposed between the endothelium and the stroma, any interchange between stroma and aqueous must cross both the membrane and the endothelium, but Descemet's membrane does not seem to affect this interchange in any significant way. Thus the function of the membrane is mostly structural, forming a tough, resistant barrier to perforation of the cornea. The membrane's resistance is demonstrated by severe ulceration of the corneal stroma anterior to Descemet's membrane; the pressure in the aqueous will force the membrane forward through the site of ulceration, much like a blister. The resulting **descemetocoele** may protrude beyond the anterior surface of the cornea without rupturing, demonstrating the toughness and elasticity of the membrane.

The increase in thickness of Descemet's membrane with age is normally quite uniform, but under some conditions the membrane exhibits localized thickenings that bulge posteriorly, accompanied by thinning of the overlying endothelium (Figure 8.21*a*). These local thickenings go by different names, depending on their location and etiology; where the thickenings are numerous and widespread, they are called **guttata**, and they are typically associated with endothelial pathology. In specular microscopy, guttata appear as dark holes in the endothelial cell mosaic (Figure 8.21*b*). Similar bulges in Descemet's membrane near the periphery of the cornea are called **Hassall–Henle bodies**. They are commonly observed in older eyes and in the corneas of longtime contact lens wearers.

Figure 8.22 summarizes the main features of the posterior stroma, Descemet's membrane, and the endothelium. The posterior stroma is slightly irregular, though not as irregular as Bowman's layer in the anterior stroma, and this may be a region in which there is some intermingling of collagen fibrils from the stroma and Descemet's membrane. The surface of Descemet's membrane next to the endothelium is smooth, with no indication of any anchoring structures for the endothelium. In Figure 8.22 we see for the first time a representation of the endothelial surface that is bathed in aqueous humor within the anterior chamber; the borders where cells meet are ruffled, with maculae occludens just below the ruffles, and microvilli are scattered sparsely over the epithelial cell surfaces. The function of the microvilli is unknown.

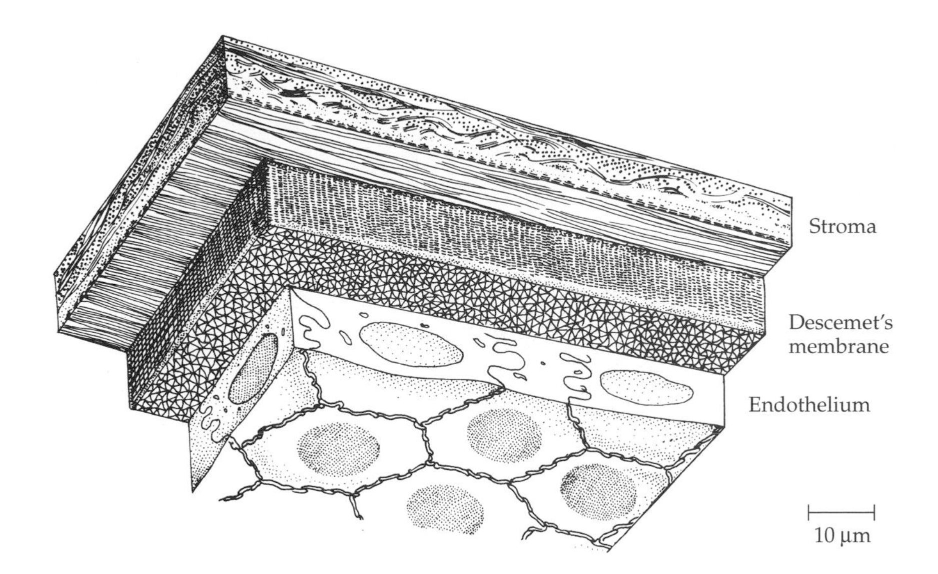

Figure 8.22
The Posterior Surface of the Cornea
This reconstruction shows the posterior stroma, Descemet's membrane, and the endothelium, the surface of which confronts aqueous humor in the anterior chamber. Some irregularity of the collagen layers can be seen in this part of the stroma, but Descemet's membrane is quite regularly organized; the fibrillar latticework is obvious on the surface view of the membrane. The endothelial surface has filigreed edges where the endothelial cells meet, and microvilli are scattered across the cell surfaces. (After Hogan, Alvarado, and Weddell 1971.)

Nerve endings in the cornea give rise to sensations of touch or pain, a blink reflex, and reflex lacrimation

As the part of the eye most exposed to trauma, the cornea has a well-developed sensory system. The cornea's sensitivity to mechanical pressure is extremely high, many times greater than the sensitivity of the skin generally, and perhaps 10 times higher than the sensitivity of the fingertips. Touching the cornea almost always produces a sensation of pain; the exception is stimuli just at threshold, which are perceived as touch. Sensations of heat or cold are debatable, but warm stimuli seem to produce a painful sensation rather than a feeling of warmth, while cool stimuli *may* be perceived initially as sensations of touch, becoming painful as the stimulus is made colder. Disagreeable as it is, the sense of pain is important; it provides a clear message that something is wrong and that something needs to be done about it.

Stimulation of the corneal nerve fibers can trigger reflex blinking or lacrimation, both of which are protective, operating to brush or wash small objects from the corneal surface. The afferent pathway for these reflexes is through the long ciliary nerves to the brainstem, where connections are made to the efferent parts of these reflex arcs—namely, motor innervation via the facial nerve (VII) to the orbicularis muscle of the eyelids for reflex blinking, and parasympathetic innervation of the lacrimal gland via the pterygopalatine ganglion for increased tear production (see Chapter 5). Neither reflex blinking nor reflex lacrimation helps with major trauma; their role is helping to maintain an intact optical surface in the face of a constant daily bombardment of dust and grit from various sources.

The epithelium contains a dense array of free terminals of nerve fibers from the long ciliary nerves

Although some corneal nerve fibers terminate in the stroma, the vast majority of the terminals are in the the epithelium, very close to the surface (see Figure 8.15). These terminals are not specialized in any obvious way. Terminal density is extremely high; an electron microscopic study reports an average of one terminal in each 20 μm square piece of cornea, which corresponds to one terminal per 400 μm^2, or about 2500 terminals/mm^2. Since the surface area of the cornea is around 130 mm^2, there must be nearly 325,000 nerve endings in the corneal epithelium. It is this dense innervation that makes the cornea so exquisitely sensitive to mechanical pressure on its surface.

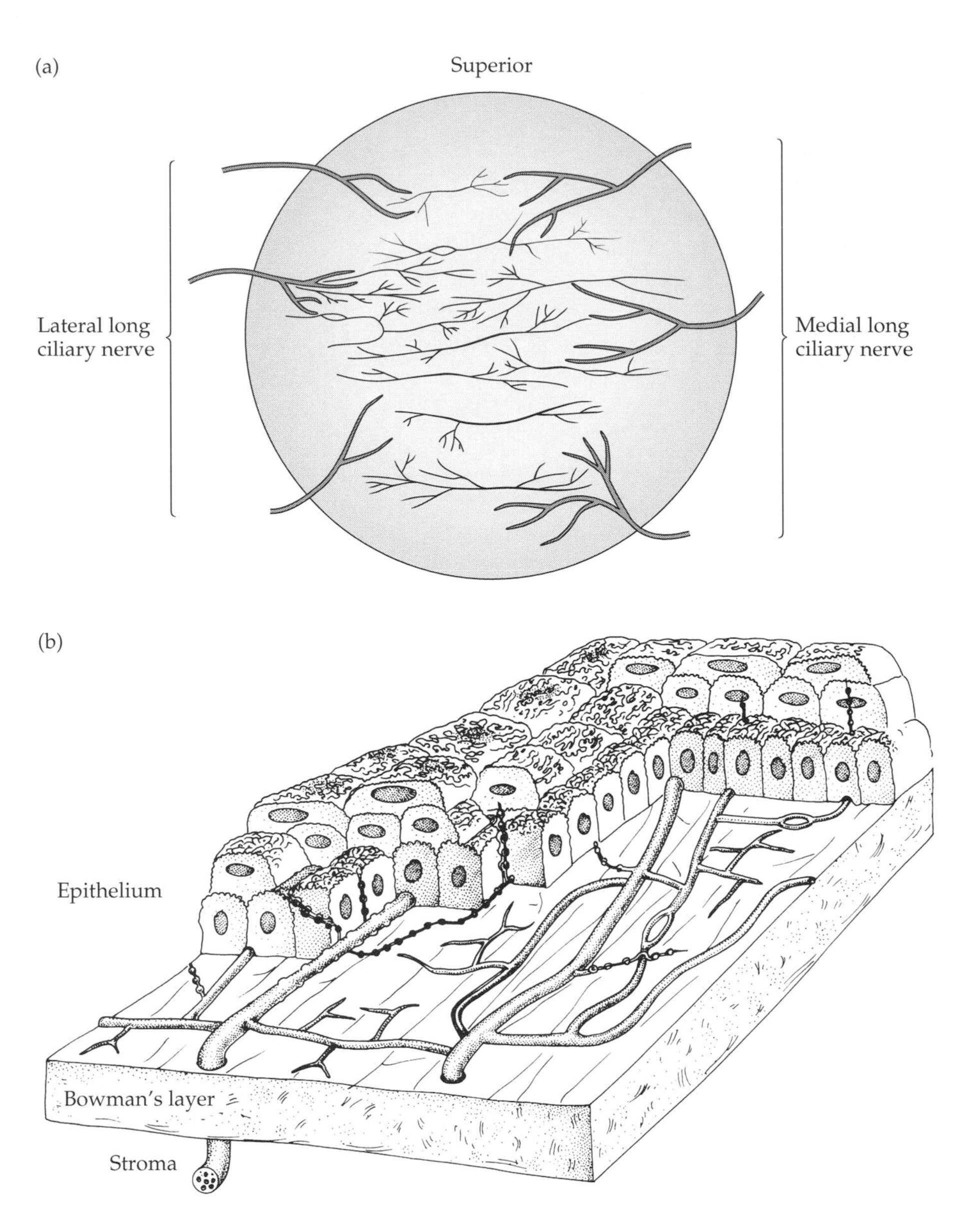

Figure 8.23
Corneal Nerves and Nerve Endings
(*a*) Large bundles of fibers (shaded) from the medial and lateral long ciliary nerves run radially through the corneal stroma. After turning upward to reach the epithelium, bundles of fibers (black) tend to run parallel to the horizontal axis, with subsequent branches running vertically. The arrangement is a roughly rectangular matrix of fiber bundles. (*b*) Fibers from the rectangular bundle array at the basal cell level run vertically through the epithelium to end among the squamous cells. (After Müller et al. 1997.)

The terminals in the epithelium belong to about 2000 nerve fibers from the long ciliary nerves that run in bundles through the stroma (Figure 8.23). Fibers leave the bundles singly or a few at a time and run up to the epithelium, where they run parallel to the corneal surface, branching extensively as they go. Small branches go up toward the surface, where they break up into a spray of terminal endings (see Figure 8.23). Each nerve fiber is estimated to have 100 to 200 terminal endings in the epithelium.

The area over which terminals from a single nerve fiber spread and the degree to which terminals from different fibers overlap are not known for human cornea, but in other species, the area of cornea innervated by a single fiber is highly variable, ranging from less than 1 mm^2 to perhaps a third of the total corneal area. Whatever the actual numbers are, the regions of the cornea innervated by different nerve fibers must overlap considerably so that a point stimulus to the corneal surface will activate multiple nerve fibers. This arrangement implies, first, that the innervation system is constructed so that even very small stimuli will be effective and, second, that the innervation of the cornea provides only a rough sense of spatial location; that is, it can give only an approximate location for a small object that has touched the cornea.

Corneal sensitivity can be measured quantitatively

Corneal sensitivity is measured with an **aesthesiometer**, which is essentially a thin, flexible, nylon filament of variable length. If the filament is very long, it bends readily when touched against a surface and therefore applies very little pressure; the shorter the filament, the less readily it bends, and the more pressure it applies. Accurate measurements of corneal sensitivity require very careful control of the filament's angle of incidence to the corneal surface (it should be perpendicular) and the maximum bending angle of the filament after it touches the surface. Studies using appropriate controls have shown that corneal sensitivity decreases from the center to the periphery and that sensitivity varies diurnally; the cornea is more sensitive in the morning than in the evening.

A loss of corneal sensitivity may not seem like a serious deficit, since only the blinking and lacrimation reflexes are lost or reduced. But such loss is not entirely trivial, since it may lead to drying of the cornea with a reduction of transparency or to the embedding of small objects in the corneal surface with subsequent ulceration. A reduced ability to recognize (to feel) corneal trauma or infection may lead to severe tissue damage before it is noticed.

The dense innervation of the cornea makes it subject to viral infection

Lesions that disrupt neural transmission are not the only disorders affecting sensory nerves. Another, generally familiar, disorder is infection of the nerves by members of the herpes family of viruses. In the most common, relatively minor form, the viral infection produces small, irritating blebs in the skin innervated by the infected nerve; a certain amount of resulting local tissue destruction normally heals without consequence. Viral infections of the trigeminal system, however, can involve the cornea; what might be small blebs in the skin can be significant ulcerations in the cornea that destroy its transparency and optical quality and provide a route for further secondary infection. These viral infections demand early recognition and aggressive treatment.

The Cornea as a Refractive Surface

The optical surface of the cornea is the precorneal film covering the surface of the epithelium

Squamous epithelial cells on the corneal surface have numerous microvilli (small projections) and microplicae (small folds) that are large relative to the wavelength of light; they make the surface irregular and therefore optically unsatisfactory. Covering this surface with a thin layer of tears, however, makes it optically smooth.

Tears contain secretions from several sources (see Chapter 7) that are either watery, oily, or mucoid; these components do not mix well and tend to form separate layers in the precorneal film (Figure 8.24). The mucoid layer lies next to the epithelial surface, the oily component is outermost, and the relatively thick watery layer lies between them. The microvilli and microplicae on the surface

Vignette 8.1

The Invisible Made Visible

I never once saw a cell through a microscope. This used to enrage my instructor. . . . "Try it just once again," he'd say, and I would put my eye to the microscope and see nothing at all, except now and then a nebulous milky substance—a phenomenon of maladjustment.

We tried it with every adustment of the microscope known to man. With only one of them did I see anything but blackness or the familiar lacteal opacity . . . a variegated constellation of flecks, dots, and spots. These I hastily drew. . . . "That's your eye!" he shouted. "You've fixed the lens so that it reflects! You've drawn your eye!"

James Thurber, *University Days*

WE WILL NEVER KNOW if Thurber was victimized by his high myopia, mechanical ineptitude, or an aversion to biology, but he seemed to realize that he was missing out on an important experience, a child's feeling of excitement and wonder on first seeing life at a microscopic scale (admittedly, this revelation usually occurs *outside of* a classroom). And this sense of awe on seeing things not seen before is one of the reasons that anatomists do what they do.

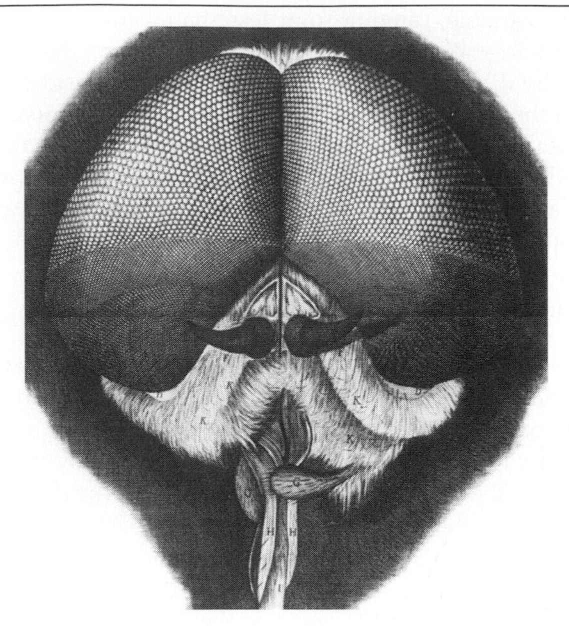

Figure 1
The Fly's Cornea
The multifaceted compound eyes of insects are impressive structures when seen at high magnification; this is the first illustration of its kind, showing the outer surface of a fly's eye. Each of the facets (ommatidia) has its own optical system and photoreceptors, as discussed in the Prologue. (From Hooke 1665.)

The existence of life forms that we cannot normally see was a sensation right from its first demonstration. When Robert Hooke (1635–1703) published his *Micrographia* in 1665, it generated enormous excitement among intellectuals in all fields; the government administrator and diarist Samuel Pepys wrote on January 21, 1665, the day after purchasing Hooke's book: "Before I went to bed I sat up till two o'clock reading Mr. Hooke's Microscopical Observations, the most ingenious book that I ever read in my life." Although Hooke did not invent the microscope and was not the first to publish on the subject, his book was the first to be devoted extensively to the normally unseen world, and his beautiful illustrations commanded attention in a way that no verbal descriptions of unseen wonders could do. His illustration of a fly's eye is shown in Figure 1.

Robert Hooke was not a specialist in biology or anatomy. He was, instead, a very precocious mathematician and physicist (he formulated Hooke's law of elasticity) and a superb designer and builder of mechanical devices. But with his interests in optics and instruments, it was natural that he would obtain one of the new microscopes and investigate its possibilities. Moreover, he was the Secretary and Curator of Experiments for the Royal Society, and part of his charge was the weekly demonstration of microscopical observations to his colleagues; many of these demonstrations are included in his *Micrographia*. This book, though perhaps the most widely known of his publications, was but one facet of many in his career.

On the other hand, the person whose name is usually associated with the development of microscopy devoted most of his adult life to the subject. Antoni van Leeuwenhoek (1632–1723) spent most of his life in Delft, his birthplace, and Amsterdam. In contrast to Hooke, he was largely self-educated, working first as a cloth merchant and later becoming a civil servant as a surveyor and as inspector of weights and measures. He spoke and wrote only Dutch, and is known to have left the Netherlands only twice in his long life.

Neither Leeuwenhoek nor Hooke invented the microscope. It is not possible to name the person who did, if in fact a single person were responsible, but the device is certainly a creation of Dutch instrument makers in the early 1600s. Most of the early instruments were compound microscopes that had both objective and eyepiece lenses mounted in cardboard tubes, and it was one of these instruments that Hooke used and illustrated (Figure 2*a*).

Leeuwenhoek took another approach to the magnification problem. As a draper, he was familiar with the low-power lenses used to inspect cloth, and he developed methods to grind and polish single lenses of higher and higher powers that became "simple" microscopes when mounted in a holder for support (Figure 2*b*). These small lenses were quite difficult to make, but a microscope with just one well-made lens had less aberration and better resolution than the combination of two lenses of lesser quality in the compound microscopes. This is the advantage that Leeuwenhoek exploited; he made more than 500 microscope lenses, he made them better than anyone else, and he didn't tell anyone how he did it.

Some of Leeuwenhoek's observations, most of them original, were published in his book *Anatomia* in 1685, but most of his observations are recorded in a long series of letters to the Royal Society, which translated and published them in its proceedings. Leeuwenhoek looked, seemingly, at everything on which he could bring his lenses to bear: plant cells, erythrocytes, spermatozoa, bacteria, protozoans, photoreceptors, lens fibers, ommatidia (facets) in insect eyes, sediments in wine, and so on. It is likely that he, and

(a)

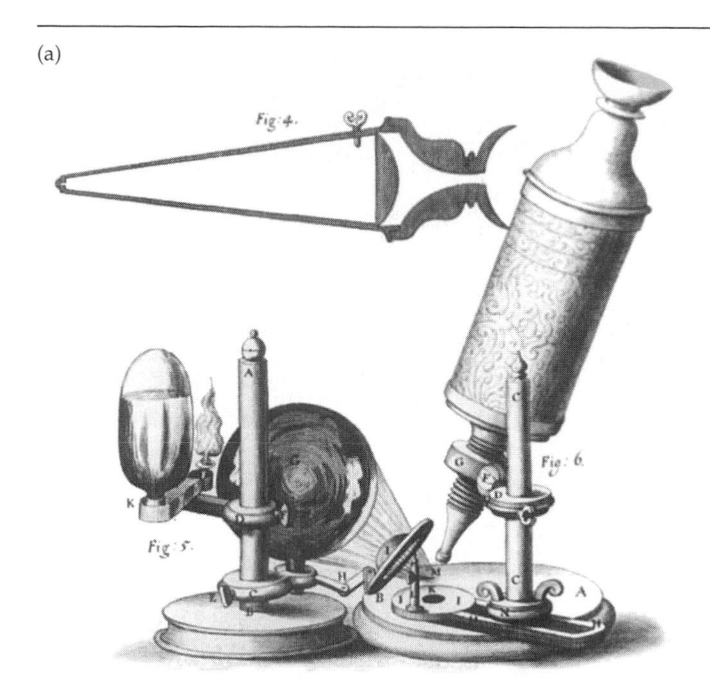

(b)

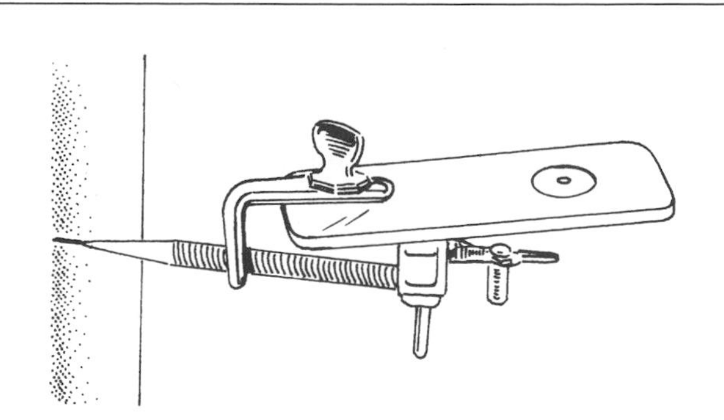

Figure 2
Two Early Microscopes
(*a*) Hooke's compound microscope in longitudinal section (upper left) and as it appeared in use. Illumination was provided by a bright flame, reflected and focused on the object of regard just below the microscope's objective. (*b*) Leeuwenhoek's microscopes were deceptively simple, consisting of a metal plate with a small hole in which a single, high-power lens was placed. In optical quality, this system was superior to the compound microscope used by Hooke. (*a* from Hooke 1665.)

Hooke before him, were like children with new toys—exhilarated, delighted, and unable to leave their toys alone—but the deliberate impersonality of scientific prose allows little of their emotion to show through. Nonetheless, the advent of microscopy was revolutionary, forcing scientists to think about life and nature in very different terms; recognizing microorganisms as agents of infection, for example, is possible only when the little buggers can be seen and one knows that such things exist.

Hooke and Leeuwenhoek were the agents of what Thomas Kuhn called a "paradigm shift" (*The Structure of Scientific Revolutions*, 1965), which is a new way of looking at things (literally, in this case) that moves our understanding to a level not appreciated before. Three centuries later, the light microscope is still a major tool for discovery and insight; it is hard to imagine doing serious anatomy without one, and this is a measure of its enormous importance. It is unfortunate that James Thurber was never able to join in the fun.*

*Thurber's vision became progressively worse later in his life, and he was nearly blind at the time of his death. Because of his poor vision, many of his wonderful, enigmatic cartoons depicting the war between men and women were drawn on enormous sheets of paper tacked to the walls of his office.

epithelial cells increase the surface area and probably help stabilize the mucoid layer of the precorneal film. The squeegeelike action of the lids during blinking (see Chapter 7) spreads the precorneal film evenly and uniformly across the cornea, thereby periodically renewing the optical surface.

The cornea's outline is not circular, its thickness is not uniform, and its radius of curvature is not constant

The cornea is the major refractive element of the eye, contributing about 40 D or more of the eye's 60 D of optical power (see Chapter 2). Its thickness varies from a minimum near the corneal apex (the geometric center of the cornea), where it is about 0.5 mm thick, to a maximum at the junction between cornea and limbus, where it is on the order of 0.67 to 0.75 mm thick (Figure 8.25). Since the anterior and posterior surfaces are not parallel and thickness increases peripherally, the posterior surface has a smaller radius of curvature than the anterior surface. Thus the corneal cross section is that of a convex–concave lens in which the power of the anterior surface is positive in sign and the power of the posterior surface is negative. The anterior surface dominates optically, however, because the change in refractive index between air and tissue is much greater than at the posterior tissue–aqueous interface.

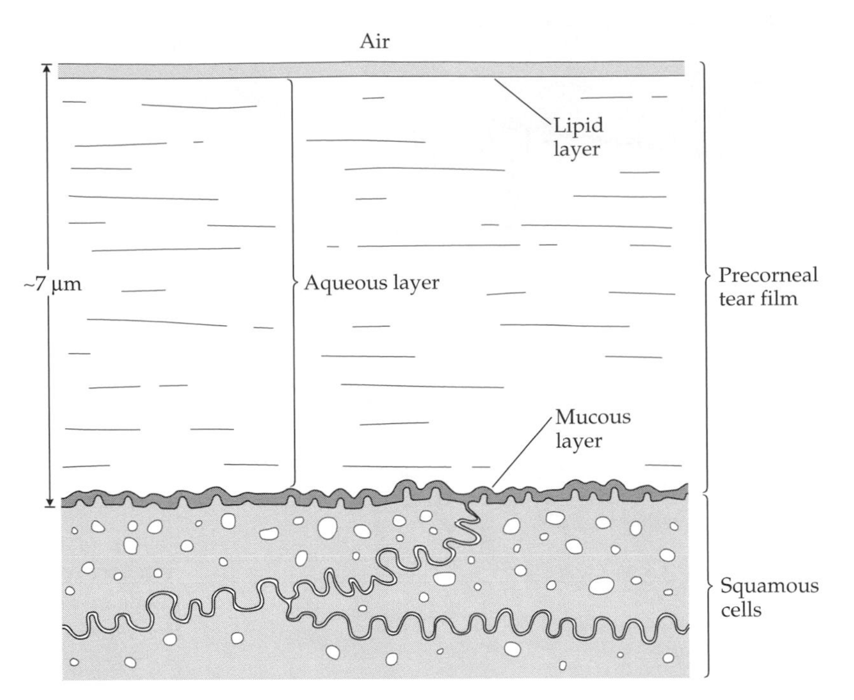

Figure 8.24
Corneal Epithelium and the Precorneal Tear Film
The precorneal tear film covering the corneal surface has three components: a thin mucous layer in which the microvilli from the squamous cells are embedded, a thin layer of lipids that forms the interface between the cornea and air, and the thick aqueous layer. Normally about 7 μm thick, the tear film is the optically smooth surface of the cornea. (After Holly and Lemp 1977.)

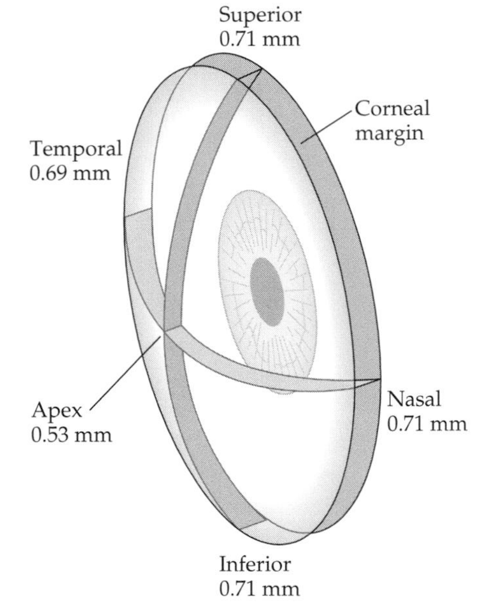

Figure 8.25
Variation in Corneal Thickness
Corneal thickness varies from center to periphery in all meridians; the vertical and horizontal meridians are shown here as if they were optical sections. The cornea is thinnest at the apex—around 0.5 mm—and thickest at the periphery—about 0.7 mm. Temporal cornea tends to be thinner than the nasal cornea, but the difference is very small. (After Waring 1992.)

Viewed from the front (Figure 8.26*a*), the outline of the cornea formed by the corneal–limbal junction is roughly elliptical; the longer horizontal axis of the ellipse is typically about 11.5 mm, the vertical axis around 10.6 mm (the standard deviations around these averages are about ±0.5 mm). The smaller vertical diameter is largely due to a superficial overlapping of the sclera onto the cornea; the outline of the posterior surface of the cornea is circular.

The cornea's shape is often described as toroidal, meaning that it is like a piece cut from the surface of a doughnut or bagel. But while a torus has constant horizontal and vertical radii of curvature, the cornea does not; corneal radii of curvature increase from center to periphery in all meridians, and the cornea flattens peripherally (Figure 8.26*b*). The curvature of the cornea is usually treated as being constant in a given meridian only within the **optical zone**, which is a circular region about 4 mm in diameter centered on the corneal apex. When examined critically, however, the cornea begins to flatten within 1 mm of the corneal apex; the curvature is always changing. When corneal curvature along one meridian varies symmetrically with respect to the corneal apex, it is best described as an ellipse.

For determining and correcting most refractive errors, the exact shape of an individual cornea is not terribly important. In correcting refractive errors with spectacle lenses, for example, the overall refractive error is being corrected, and the cornea's contribution to this error is not a factor in determining the appropriate correction. (If the corrected visual acuity is poorer than 20/20 [6/6], however, an abnormality in corneal shape needs to be considered.) Refractive corrections with hard contact lenses must take the average curvatures of the posterior surface of the contact lens and the anterior surface of the cornea into

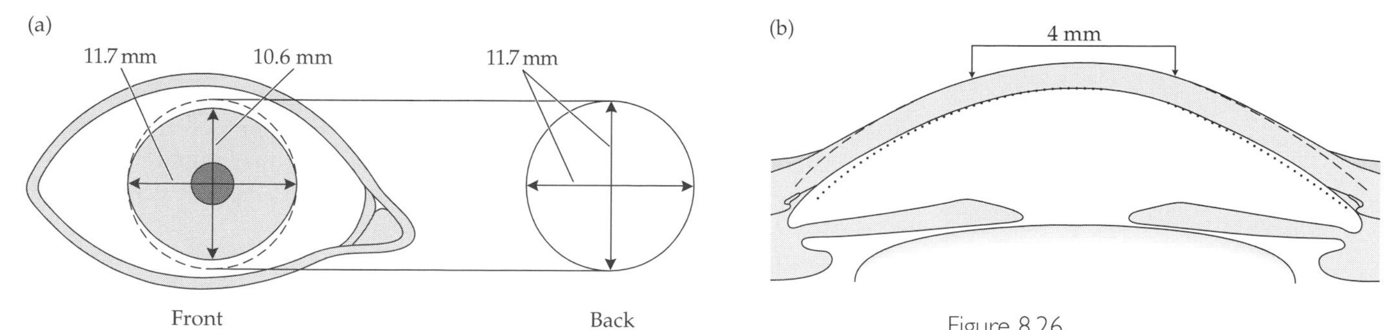

Figure 8.26
Approximation of Corneal Size and Shape
(*a*) Viewed from in front, the outline of the cornea is slightly elliptical, with a mean vertical diameter about 1 mm less than the mean horizontal diameter. From behind, the outline of the cornea is circular. (*b*) The flattening of the peripheral cornea can be seen easily when its cross section is compared to a circle (dashed line). The central 4 mm of the cornea—the optical zone—has a near constant radius curvature, shown here as 8 mm. Peripherally, the 8 mm radius circle deviates from the corneal surface, and the deviation is such that the corneal radius of curvature is larger; that is, the cornea is flatter. The posterior corneal surface has less peripheral flattening; here it is compared to a circle with a 7.5 mm radius (dotted line). (After Hogan, Alvarado, and Weddell 1971.)

account, but the exact shape of the cornea is not a factor; usually, the contact lens is a compromise between the steepest and flattest meridians of the cornea within the optical zone.

Surgical reshaping of the corneal surface to correct refractive errors has increased the importance of knowing the shape of the cornea. The surgical procedures differ in detail, but since their goal is to correct refractive error by changing the shape of the corneal surface, it follows that the nature of the surgical approach is aided by knowing the exact shape of the cornea that is to be modified.

The shape of the cornea is determined by comparison to a sphere

Modern devices for determining corneal shape are technically sophisticated, but most employ an old, simple principle. In its original form, Placido's disc is a flat circular plate with a series of concentric, equally spaced black and white circles (Figure 8.27*a*). A small peephole at the center of the circles allows one to look at or to photograph the reflection of the rings from any reflective surface. By knowing

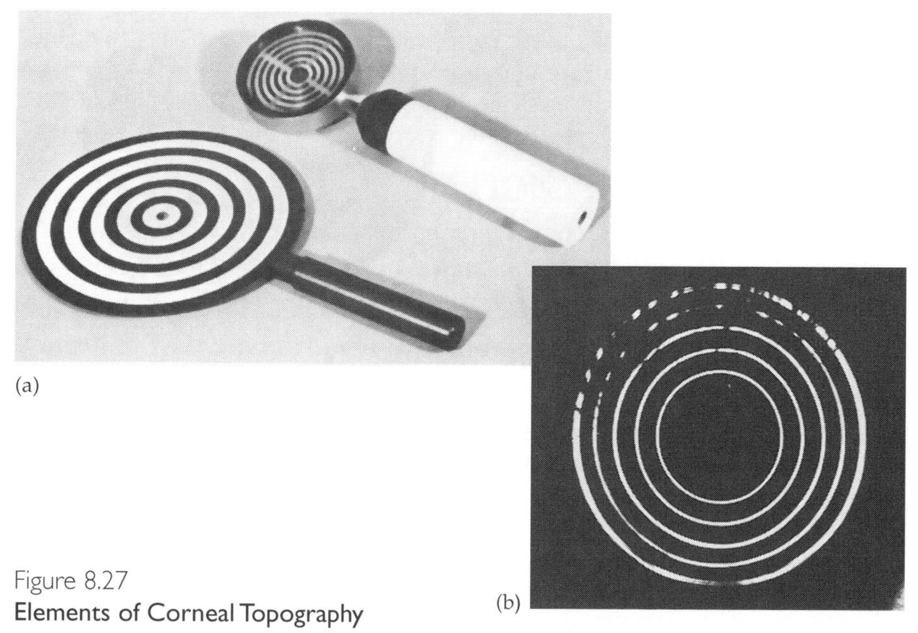

Figure 8.27
Elements of Corneal Topography
(*a*) At left is a version of the original Placido's disc, at right a modern internally illuminated version. The observer looks through a hole in the center of the disc to see the white rings reflected at the cornea. (*b*) Ring reflections from the human cornea. By direct inspection, these reflections appear to be circular and equally spaced; careful measurement, however, shows that the vertical diameter is slightly larger than the horizontal and that the center-to-center ring spacing changes from center to peripherally.

the spacing of the circles on Placido's disc, the distance between the disc and the reflecting surface, and the spacing of the reflected circles, we can calculate the radius of curvature for the reflecting surface. (It is essentially a comparison of image to object size after reflection. Given the ratio of image to object size—the magnification—one must calculate the radius of curvature that would produce it.)

When the cornea is the reflecting surface (Figure 8.27*b*), the reflected contours appear to be circular, but measurement will reveal that they are slightly elliptical. Measurements will also reveal that the separation between the contours changes from center to periphery. Qualitatively, these observations are consistent with what has already been said about the shape of the cornea—the radius of curvature varies with meridian and the surface flattens from center to periphery—but a detailed characterization will emerge only if the local radius of curvature (and power) from the distance between each pair of adjacent rings along a number of different meridians is calculated. Showing localized changes in the surface requires many calculations made at numerous points along many radii.

Digitizing the reflected images and using computers to do the calculations make it possible to represent the topography of any cornea in considerable detail. After the calculations are done, the surface can be represented in any of several ways. The most common, shown in Figure 8.28*a*, is contour mapping that uses shading to show regions whose dioptric power falls within certain limits. In this example, which is the cornea of my right eye, the lightest regions have the highest power (smallest radius), and the darkest regions have the least power, as the accompanying scale shows. (These maps are usually color-coded. Color adds no information to the display, but it can be an effective way to emphasize regions of special interest.)

If my cornea were perfectly spherical, the power would be constant everywhere and the map in Figure 8.28*a* would be a uniform shade of gray; this clearly is not the case, and the map indicates how my cornea departs from sphericity. The region of highest power (about 45 D) is an elongated region inferior and nasal to the pupillary axis. It is embedded in a larger region shaped like a distorted bowtie; the mean power here is around 44 D. In fact, all the power regions are elongated along the axis running between 110° and 290°. This elongation shows my corneal astigmatism; measured from 0°, the highest power is along the 110° meridian, and the low power meridian is roughly 90° less, around 20°. The minus cylinder axis of my refractive correction is about 20°.

Although the word "shape" was used earlier in this discussion, corneal maps seldom represent the physical shape of the cornea. Although they are contour maps, they are not like contour maps of Earth's surface showing the height or depth of the surface relative to a reference plane (such as sea level). Instead, these are optical surface mappings, giving information about the optical shape; the three-dimensional physical shape is implied—a small region of very high power can be interpreted as a local bulge, for example—but it is not given directly. (Some instruments will produce representations of the physical shape, however.) There are several advantages to specifying optical rather than physical shape, the first of which is that optical shape is easier to calculate. And in most cases, the optical configuration is more important. Contact lenses are characterized by their radii of curvature and powers, for example, and while refractive surgery produces a change in the physical shape of the cornea, the goal is a particular optical shape that eliminates the overall refractive error.

The cornea does not have a single, specifiable shape

The map for my left cornea (Figure 8.28*b*) differs from my right corneal map, as one would expect, but both show some astigmatism with a certain amount of bilateral symmetry (the highest-power meridians are rotated outward at the top

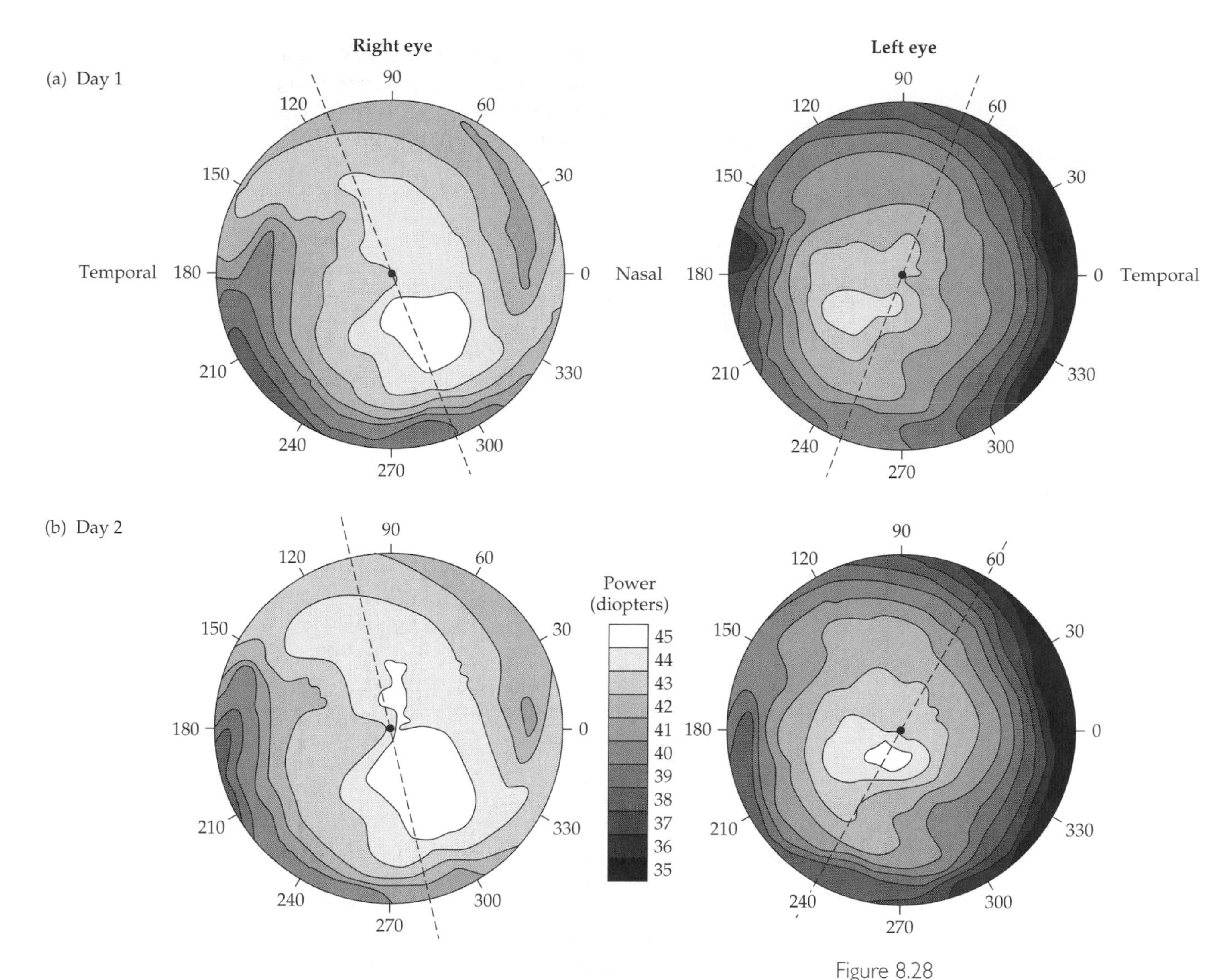

Figure 8.28
Maps of the Cornea's Optical Shape
In these optical contour maps, the contours are lines along which optical power is constant. The contours are at 1D intervals. The dashed line through the center of each map shows the meridian of highest average power. In the map for the right cornea (*a*), the vertical elongation of the contours indicates astigmatism; the highest power is along the 100°–290° meridian. The left cornea's map (*b*) has less astigmatism, but the meridians of highest power for the two corneas are bilaterally symmetric. Maps of the right (*c*) and left (*d*) cornea made a day later are similar, but not identical to the earlier maps (*a* and *b*, respectively). The axes of highest power have changed in the later maps and they are no longer symmetric.

away from the vertical, for example). And both conform to the general pattern of flattening from center to periphery. It may be more interesting to compare Figure 8.28*a* and *b* to Figure 8.28*c* and *d*, which are maps of my corneas made a day later. The maps are similar but not identical, raising a curious question: Which are the true maps of my corneas, or is there any such thing? Did my corneas change slightly from one day to the next, or is the difference due to the instrument?

There are no definitive answers to these questions, because the differences are probably a combination of biological and instrumental variation. Corneas do change during the day, and repeatability of the maps depends on focus and alignment of the measuring instruments. The problem is simply that corneas are like fingerprints or snowflakes; they are all similar, but each is unique. Thus the more closely we examine it, the more elusive the cornea's shape becomes; it is a will-o'-the-wisp.

Most of the small-scale, local variations from one cornea to the next are not known to have optical consequences that need to be considered in any form of refractive correction. Therefore, measures of the cornea's symmetry and approximations to the curvature variations are not only adequate, but useful. The simplest approximation is that provided by measurements with a keratometer, which normally gives the curvatures (or dioptric equivalents) for the meridians of maximum and minimum curvature within the optical zone. The cornea is

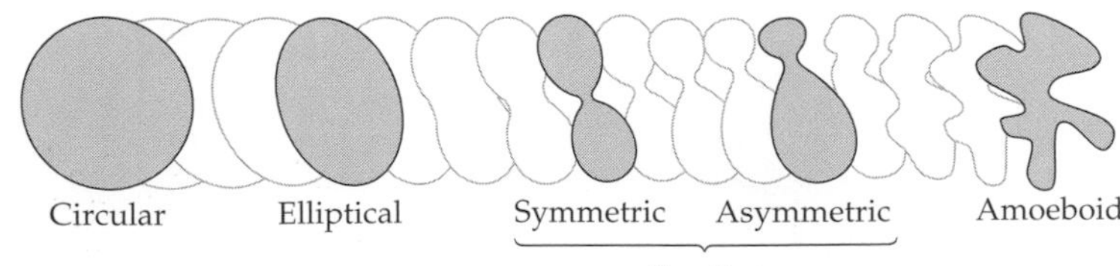

Figure 8.29
Central Patterns in Corneal Contour Maps
The regions of highest power in the corneal contour maps can often be classified by their similarity to certain geometric shapes. As shown here, they range from circular on the left, through ellipses of varying eccentricities, through bowtie- or dumbell-shaped figures, to unclassifiable, amoeboid patterns. The implied continuum in the drawing gives some impression of the variability in these patterns. The bowtie patterns are the most strongly associated with astigmatism. (After Bogan et al. 1990.)

assumed in this measurement to be symmetric around the visual axis; curvature changes occurring over distances less than 3 mm are ignored, as is the peripheral flattening of the cornea. This oversimplified view of the cornea's shape is adequate for most refractive corrections with spectacles or contact lenses, it is not helpful in assessing corneal irregularities or deformations like keratoconus (which will be discussed later in the chapter), however.

Corneal topographic maps contain much more detail about curvature variations over the corneal surface; these variations show up as different patterns of contours within the maps and different quantitative variations when the same regions of different maps are compared. In Figure 8.29, for example, maps are arranged so that the central regions of highest power undergo a kind of metamorphosis from a circular region on the left through elliptical and "bowtie" configurations to an irregular, amoeboid shape on the right. The changing patterns appear to reflect a change from very little astigmatism at the left to large amounts of regular astigmatism in the bowties near the right. My corneal maps in Figure 8.28*a* and *c* would be near the right as asymmetric bowties, for example. The meaning of the irregular patterns is not clear, however, and this is where most of the unusual and interesting variations will be found. Efforts to reduce the information in a corneal contour map to a set of meaningful descriptive statistics are under way.

The development of the instruments for plotting corneal topography has paralleled the increasing interest in corneal refractive surgery; the hope is that more complete and accurate characterization of an individual cornea will help determine the best way to alter it. Unfortunately, the surgical procedures alter the biomechanical properties of the cornea in not fully predictable ways; information about corneal topography is important, since that is the starting point for surgery, but it is a *guide* for action, not a rigorous prescription.

At present, corneal mapping is most useful for documenting change in an individual cornea. Corneal topography before and after refractive surgery, for example, may provide clues about the success or failure of a particular procedure, and it will certainly record what happened. Corneal maps can also reveal peripheral irregularities produced by surgery, trauma, or disease that should be considered when a patient is being fitted for contact lenses (standard keratometry will usually not detect such irregularities unless they are central, and they may not be obvious with a slit-lamp unless there is some opacification).

Contact lenses can affect corneal shape and structure directly or indirectly

Using contact lenses to correct refractive errors is essentially a matter of substituting one refractive surface for another; with a contact lens on the cornea, the anterior surface of the contact lens becomes the eye's primary optical component and the cornea's anterior surface is rendered ineffective. As normally used, neither hard nor soft contact lenses produce any significant, permanent change in corneal shape.

But hard contact lenses can mold the cornea to a new shape; in time, and with pressure from the eyelids, a lens whose posterior surface is flatter than the central corneal curvature can flatten the cornea, thereby reducing the refractive error in a myopic eye, for example. Deliberate reshaping of the cornea with contact lenses to correct refractive error is called *orthokeratology;* the hope is that a reshaped cornea will be stable, and thus no refractive correction will be necessary for significant periods of time.

The cornea's normal elasticity is not easily overcome, however. The reshaping produced by contact lenses persists for a relatively short time after the lens wear is discontinued. As a result, once a cornea has been reshaped by the desired amount, patients alternate between no refractive correction and wearing a "retainer" lens to keep the cornea in its new configuration. Corneal instability during the periods without a contact lens are reflected as fluctuations in visual acuity. (Orthokeratology has been advocated as a way to halt or slow the progression of myopia in childhood, but there is no convincing evidence that it works.*)

Unlike orthokeratology, the conventional use of contact lenses to correct refractive error does not produce significant changes in corneal shape, but it may create other problems. First, the contact lens is a foreign object resting on an exquisitely sensitive surface that has built-in mechanisms—reflex tear production and reflex blinking—meant to detect and remove foreign objects. These protective mechanisms are not always easy to circumvent, particularly with hard contact lenses that may rock back and forth on astigmatic corneas. The palpebral conjunctiva also has a rich supply of sensory nerve endings, and individuals whose eyelids are tightly apposed to the cornea may never adapt to the presence of contact lenses, hard or soft. Adaptation, the gradual desensitizing of the cornea's sensory nerve terminals, is necessary for wearing contact lenses comfortably. How this desensitization affects the protective reflexes is not clear, however.

Another problem is the degree to which contact lenses interfere with the oxygen supply to the cornea; both the epithelium and the endothelium have aerobic systems for metabolism that are necessary for their long-term normal operation. Most of the oxygen supply for the central cornea comes from oxygen dissolved in tears, and the concentration of oxygen is highest at the anterior surface (Figure 8.30); it drops rapidly across the epithelium because of oxygen consumption by the epithelial cells, and it continues to fall through the stroma. A slight rise in oxygen tension at the endothelial surface reflects the small but significant amount of oxygen dissolved in the aqueous humor. Closing the eyelids sharply reduces the oxygen available at the anterior surface, and a gas-impermeable contact lens can push the oxygen available to the epithelium below that at the endothelial surface (lowest curve in Figure 8.30).

Reducing the oxygen supply to the epithelium forces the tissue to use its anaerobic pathway for glycolysis, resulting in increased levels of lactate. Diffusion of lactate into the stroma causes increased water uptake into the stroma, swelling of the tissue with an increase in stromal thickness, some disarrangement of the collagen, and some decrease of transparency. All of this occurs normally when we are asleep and the lids are closed, but the return to aerobic conditions on awakening quickly reestablishes normal conditions; the cornea thins and any slight haze disappears. The trick with contact lenses is keeping the reduction of the cornea's oxygen supply to a minimum; otherwise, there can be persistent edema, transparency loss, and possibly some long-term structural and functional changes in the epithelium. Decreased density of desmosomes and decreased mitotic activity in the epithelium have been reported, and long-term contact lens wear has also been associated with changes in the posterior cornea; stria in Descemet's membrane, irregularity in the endothelial cell array, and peripheral guttata are some of the observations.

Interference with the oxygen supply has been virtually eliminated in recent years by the use of soft contact lenses; both their material and their thinness

*Recent advances in corneal topology are being incorporated into the practice of orthokeratology, prompting a change in the name for the procedure to "controlled kerato-reformation (CKR)." The biomechanics of the cornea are not likely to be affected, however.

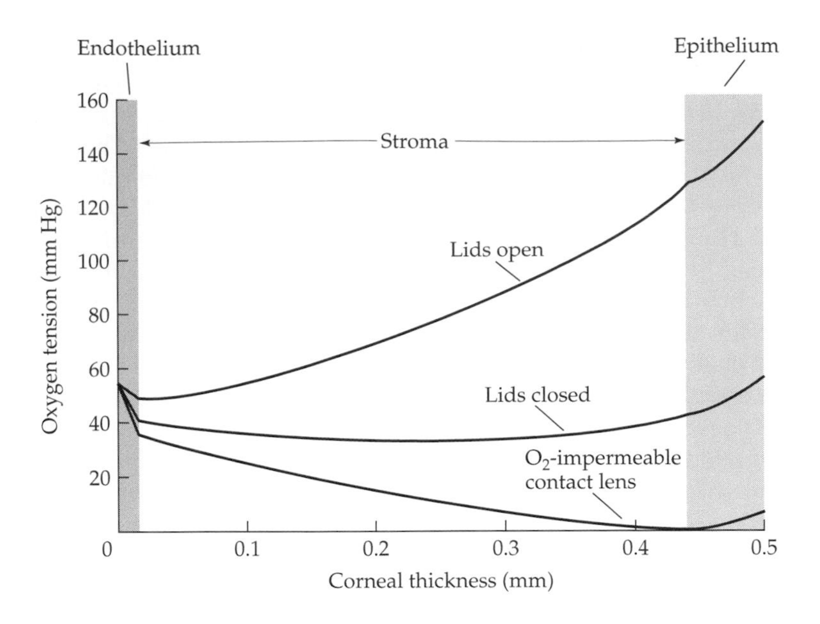

Figure 8.30
Variation in Oxygen Tension across the Thickness of the Cornea
The oxygen tension at the endothelial surface depends almost entirely on the oxygen dissolved in the aqueous humor, and it is constant at about 55 mm of mercury. Oxygen tension elsewhere in the cornea, however, depends on the availability of oxygen at the anterior surface. The three curves show the oxygen tension profiles calculated for three conditions: (1) normal, with the lids open; (2) with the lids closed, which makes oxygen tension throughout the stroma drop below the level at the endothelial surface; and (3) with a gas-impermeable contact lens (lids open), which reduces oxygen throughout the cornea below the endothelial level. (Replotted from Fatt, Freeman, and Lin 1974.)

make them highly permeable to oxygen. But while they have little effect on oxygen supply, they fit so closely to the corneal surface and cover so much of the surface that the normal precorneal tear layer is altered; some individuals cannot tolerate this change, particularly if their precorneal tear film tends to break up and rapidly lose its normal confluence. In addition, soft contact lenses of the extended-wear variety make a wonderful home for microorganisms, some of which are quite virulent and can destroy ocular tissue very quickly; soft lenses must be cleaned regularly and thoroughly (and carefully, since the lenses are easily torn).

Contact lens specialists must sort through a large array of variables involving lens materials, sizes, and curvatures, among others, to arrive at an appropriate correction for an individual eye; the procedure is highly empirical, and considerable trial and error may be required to find an optimal solution. In addition, practitioners now have a new set of problems, arising from the need to correct residual refractive errors in eyes that have had refractive surgery and have irregular corneal surfaces.

The shape and optical properties of the cornea can be permanently altered

Although corneal shape cannot be permanently altered by remolding, as in orthokeratology, it certainly can be changed by alteration of the structure; this is the goal of corneal refractive surgery. Refractive surgery methods are changing rapidly, so rapidly that the technique most widely used to alter the corneal shape when the first draft of this chapter was written (**radial keratotomy**, or **RK**) is on its way out as the final version nears completion. But since the effects of radial keratotomy on several million corneas will persist for decades to come, it still merits some discussion.

Radial keratotomy corrects refractive error by making small, precise incisions in the cornea, outside the central optical zone, to redistribute stress within the tissue and change corneal shape. Surgery for correction of a simple myopia with no significant astigmatism involves 4 to 16 radial incisions in a symmetric pattern, beginning 1.5 to 2 mm from the corneal apex and extending to within 0.5 mm of the limbus; the depth of the incisions is 90 to 95% of the corneal thickness at the edge of the optical zone (Figure 8.31*a*). The number of incisions is related to the magnitude of the refractive error; more incisions produce larger changes.

Because of the normal tension and elasticity of the tissue, the incisions initially gape slightly, and the redistribution of forces in the cornea flattens the optical zone while steepening the periphery (Figure 8.31*c*). Correction of astigmatism requires radially asymmetric incisions combined with tangential incisions (Figure 8.31*b*), or in a newer approach, arcuate incisions by themselves.

Depth, length, orientation, location, and the number of incisions are the primary variables underlying the shape change produced by radial keratotomy. Translating the required dioptric changes into a surgical plan in which the various incision parameters are specified is done with a model of the cornea's biomechanical properties—a model that is part theoretical and part empirical. Models can take several forms, varying in sophistication from rules of thumb to nomograms to computer programs, but the goal is to use the information about initial conditions—the cornea's original shape and the eye's refractive error—to produce recommendations for the optimal incision parameters. These recommendations are subject to modification based on the surgeon's experience and judgment—that is, on the surgeon's model of the cornea.

Whatever the pattern of incisions is, healing involves formation of epithelial cell plugs that are pushed out of the incisions by the formation of new collagen; if these processes of repair do not alter the gaping of the incisions, the inital flattening of the cornea will be permanent.

In a large clinical trial of radial keratotomy, myopia was reduced in all cases, and about 60% of the patients had no need for any additional refractive correction by spectacles or contact lenses (about half of these patients were accepting visual acuities of 20/25 [6/7.5] or poorer, however).* About 34% of operated eyes had refractive errors 1 year after surgery within ±0.50 D of emmetropia, and 60% were within ±1.00 D. Approximately 47% had uncorrected visual acuities after surgery of 20/25 or better. Success was dependent on the initial refractive error; in general, the lower the initial error, the better the surgical result. Only 26% of eyes with refractive errors larger than –4.50 D achieved uncorrected visual acuities of 20/25 or better after surgery.

(a)

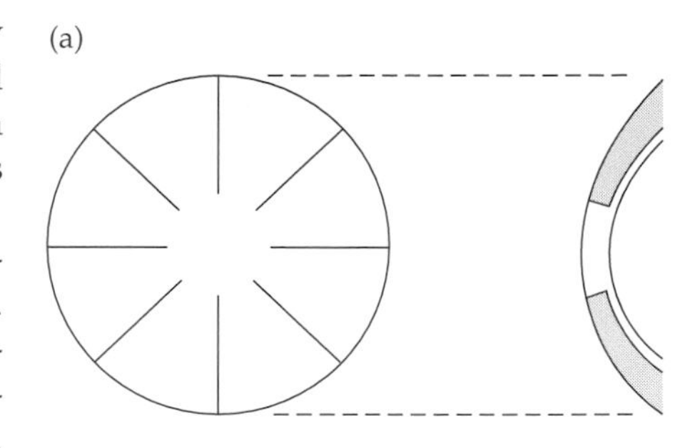

(b)

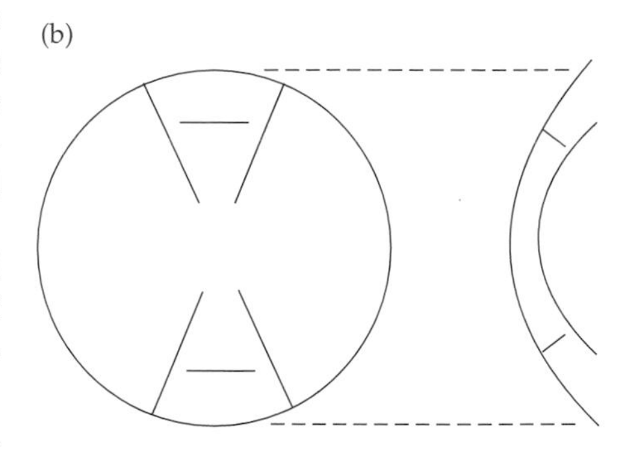

(c)

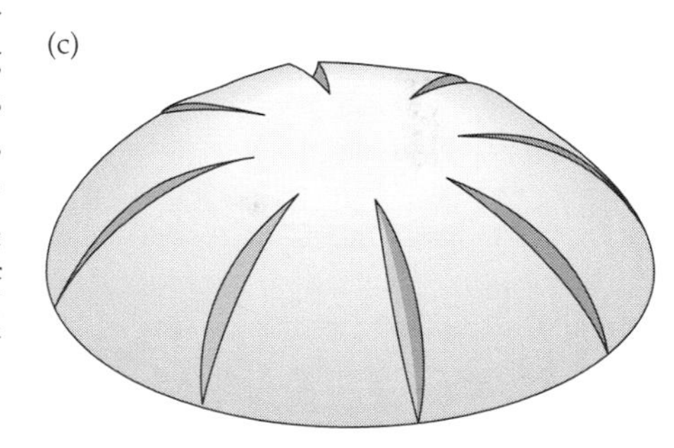

Figure 8.31
Principles of Radial Keratotomy
(*a*) This frontal view shows eight equally spaced radial incisions that have been made from the edge of the optical zone to the periphery to correct myopia. The vertical section shows the normal incision depth, which is more than 80% of the corneal thickness. (*b*) One strategy for correcting astigmatism uses tangential incisions placed perpendicular to the meridian of greatest power (the vertical meridian in this case). Incision depth is about 80%, as in *a*. (*c*) This representation of the corneal surface after surgical correction for myopia shows the gaping incisions (exaggerated), a relatively flat optical zone, and steepening of the peripheral corneal curvature. (After Waring 1992.)

Surgically reshaped corneas may change with time

Follow-up examinations of RK patients reveal that a significant percentage of corneas change postoperatively, as indicated by a progressive shift of refractive error in the *hyperopic* direction and detectable flattening of the corneas. Ten years after surgery, 43% of all eyes have undergone a hyperopic shift of 1.00 D or more, and 15% had a hyperopic shift greater than 2.00 D (about 3.5% show a *myopic* regression greater than 1.00 D). In most cases, the hyperopic shift was most rapid in the first few years after surgery, when the average refractive change was +0.21 D per year, with a smaller change thereafter (+0.06 D per year). In short, the average eye went from slightly myopic after surgery to about 1.00 D hyperopic 10 years later.

The hyperopic shift probably reflects a continual widening of the surgical incisions, even after they have healed by bridging with new tissue, producing continual, progressive peripheral steepening and central flattening of the cornea. (The incisions do not always heal completely.) Unfortunately, we do not know exactly what is happening, how long the changes will continue, or whether some eyes are more predisposed to change than others (the amount of hyperopic shift is not strongly associated with initial refractive error). Whether the change will continue or eventually stabilize is not known.

*All figures relating to the success of this technique are from the Prospective Evaluation of Radial Keratotomy (PERK) study. The results have been published in a series of papers during the past 15 years. Later studies with less restrictive protocols and certain technical refinements have had better success rates.

(a)

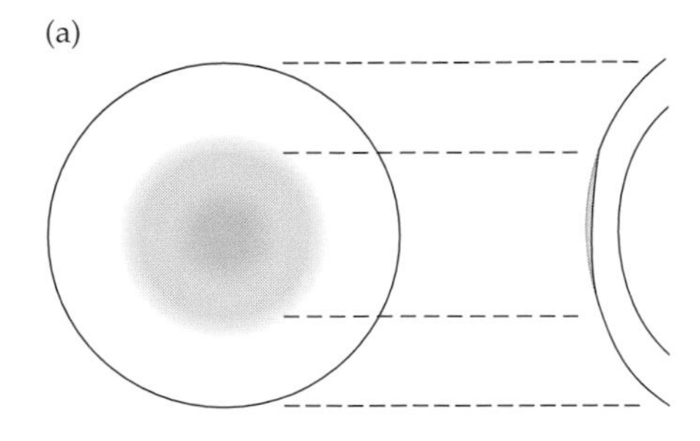

(b)

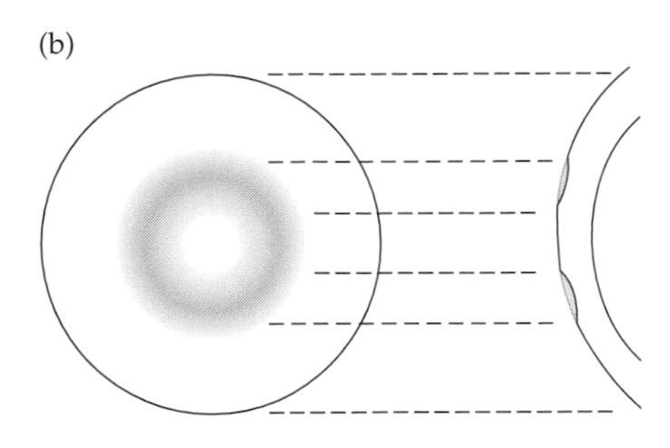

Figure 8.32
Elements of Photorefractive Keratectomy
(*a*) In PRK, myopia is corrected by the ablation of tissue across a central zone about 6 mm in diameter. As the sectional view shows, this approach makes the cornea flatter and thinner. (*b*) An annular ablation zone is an attempt to steepen the central curvature to correct for hyperopia. (After Waring 1992.)

Modest spherical changes in the hyperopic direction are of little consequence for pre-presbyopic patients; their accommodation will compensate for the hyperopic shift. They may require correction for near vision a few years earlier than normal as their accommodative amplitude declines, however, and with complete presbyopia, a distance correction may be necessary. Astigmatic changes have also been reported and may be large enough in some cases to require correction with spectacles or contact lenses.

Corneal shape can be changed by removing tissue

Radial keratotomy changes corneal shape by weakening the midperipheral cornea so that it bulges and the central cornea flattens. A newer method, called **photorefractive keratectomy** (**PRK**), uses a laser to reshape the surface by controlled ablation of corneal tissue within the optical zone. For a myopic eye, correction of the refractive error means that the curvature should be reduced (flattened), which is accomplished by the removal of more tissue centrally than peripherally (Figure 8.32*a*). Removing tissue peripherally and almost none centrally corrects hyperopia by steepening the curvature (Figure 8.32*b*). Correction of astigmatism requires the removal of more tissue in some meridians than in others. The amount of tissue ablated by the laser depends on the exposure time; the greater the exposure duration, the more tissue that is removed. After the eye is properly aligned and stabilized, tissue in a small region can be vaporized in a second or so; the procedure for one eye may require from 20 or 30 s to several minutes, depending on the amount of tissue to be removed.

RK and PRK differ both in technique and in principle; in particular, PRK focuses on the cornea's optical zone directly, removing both the epithelium and some of the stroma (the epithelium is removed by a combination of laser ablation and gentle scraping). Tissue repair after PRK surgery is largely epithelial replacement to cover the ablated region rather than formation of new collagen to bridge incisions, but a proliferation or migration of fibroblasts into the ablated region may sometimes produce scarring within the optical zone. The large disc of tissue removed by epithelial abrasion and laser ablation causes pain through damage to the dense network of nerve endings in the epithelium.

Photorefractive keratectomy has generated considerable interest because it seems less invasive than radial keratotomy, it is less dependent on the niceties of surgical technique, and it is less likely to seriously weaken the structure of the cornea. And the reports from clinical trials are quite positive; for small refractive errors, less than 4 D or so, almost 90% of the eyes do not need a postsurgical refractive correction. As with the earlier study, however, a substantial number of these patients are accepting visual acuities poorer than 20/20.

There are difficulties yet to be resolved. For example, vaporization of tissue by the laser is so rapid and cataclysmic that vortices of escaping gas produce local regions of calm—like the eye of a hurricane—in which the tissue is not completely ablated. The result is a small island or islands of surviving tissue within the ablation region; these are optically unacceptable, particularly near the corneal apex, and a second round of treatment may be required to remove them (they may disappear without treatment, however).

Another recurrent problem is postsurgical haze in the ablation region. A transient reduction in transparency is expected in view of the epithelial loss and the resultant stromal swelling, but there are reports of the haze persisting long after the epithelium has regenerated over the ablation zone; this is attributable in part to migration and accumulation of fibroblasts in the ablation zone. (The term "persistent haze" almost certainly refers to a thin layer of scar tissue.) A long-lasting decrease in transparency, with an attendant loss of best corrected visual acuity, is not an acceptable result.

The long-term effects of PRK are not known. There are no incision scars to change over time, as there are in RK, but the procedure makes the cornea 10 to 15% thinner, and this reduced thickness may predispose the tissue to long-term change. Recent reports show altered cell death among the fibroblasts; their loss from the anterior stroma over time seems to be a consequence of the epithelial abrasion. Follow-up studies of PRK patients a few years after surgery report a myopic regression in some patients; when the original refractive error is large, requiring extensive tissue ablation, the regression begins earlier and increases more rapidly. The underlying mechanism for the change is not known, but one possibility is increased curvature of the thinned cornea in response to intraocular pressure. An overabundance of new epithelium may also contribute to a postoperative steepening of the curvature.

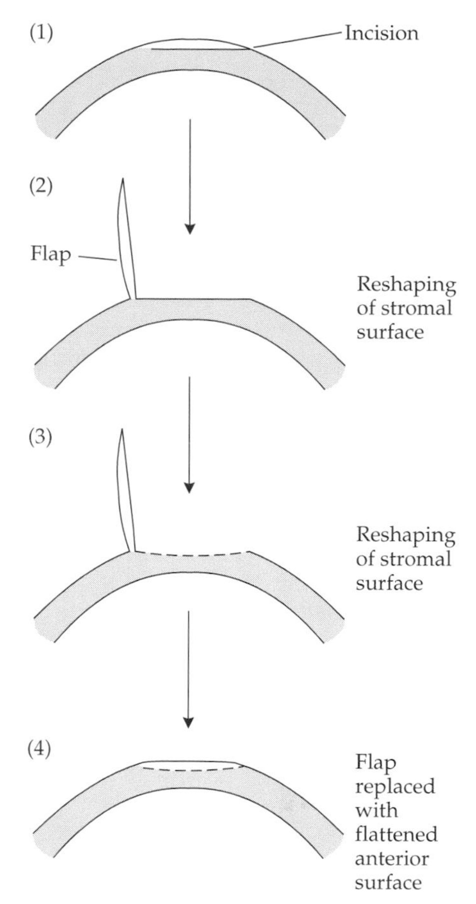

Figure 8.33
Reshaping the Interior Stroma
(1) An incision is made perpendicular to the pupillary axis, undercutting the optical zone of the cornea. (2) The undercut region is reflected back as a flap of tissue containing the optical zone. (3) The exposed stromal bed is reshaped by laser ablation, in this case making it concave. (4) The optical zone flap is placed back in position over the reshaped stromal bed, producing a reduction in curvature within the optical zone.

Stromal reshaping leaves the epithelium intact

One of the newest methods for corneal reshaping combines surgical removal and subsequent replacement of an anterior corneal cap on a stromal surface whose shape has been changed by laser ablation of the tissue; it is called **laser in situ keratomileusis** (**LASIK**; for reasons that will become obvious, the slang term is "flap and zap"). Figure 8.33 illustrates the basic procedure. An oscillating blade, a keratome, is set at a predetermined depth while the pressure in the eye is maintained at a high level by a vacuum ring. A slice made by the keratome undercuts the optical zone perpendicular to the pupillary axis, creating a "flap" of tissue with a nasal attachment, a kind of hinge, that allows the flap to be replaced precisely on the underlying stroma. When the flap is reflected back out of the way, the plane of the cut stromal surface can be reshaped by the same procedure that is used in PRK ("zap"); in Figure 8.33, the stromal surface has been made slightly concave. When the optical zone flap is put back in position on the reshaped stroma, it will be flatter than before, thus providing a correction for myopia. Had the stromal reshaping been in the direction of convexity, the final corneal curvature would be increased.

A possible advantage of LASIK is keeping the epithelium and Bowman's layer intact within the optical zone. Also, the method is said to be a better solution for the extremely high refractive errors that are difficult to correct with RK or PRK. But there is an opportunity for excess formation of new collagen along the interface between the reshaped stromal surface and the optical zone flap; a scar of this kind may produce unacceptable image degradation. It is much too early to know much about the long-term prospects for the LASIK procedure or any of the other evolving techniques for refractive surgery. Box 8.2 outlines some general reservations about refractive surgery.

Corneal grafts are used to repair optically damaged corneas

When a cornea has become opaque or highly irregular in shape as a result of trauma or disease, the only solution is a corneal transplant (**keratoplasty**), in which damaged tissue is replaced with normal tissue. Transplants of tissue from one person to another can be complicated by the immune system's recognition of the donor tissue as foreign, thereby inviting an attack. Successful organ transplants, in particular, require large doses of immunosuppressive drugs to circumvent the body's normal defenses. Because of its lack of blood vessels, the cornea occupies what has been called a "privileged site" with respect to the immune system, but in spite of this partial isolation, antigens from a donor cornea can still be recognized by the host tissue and may activate an inflammatory response. The inflammatory response may jeopardize the success of the transplant unless it is controlled.

Box 8.2 Some Reservations about Corneal Refractive Surgery

We have more need to distinguish what is desireable from what is feasible. . . . We shall have to learn to refrain from doing things merely because we know how to do them.

Sir Theodore Fortesque Fox, *The Purposes of Medicine**

WRITING IN A MORE innocent time, which was not so long ago, Theodore Fox told us how to recognize and avoid deep pitfalls. But his sage advice was not heeded, and we are busily engaged in doing things because we have the know-how, not because the things are desirable. Refractive surgery is one of those things, in my opinion.

The term "refractive surgery" refers to several different techniques used to alter corneal structure with the goal of eliminating or reducing a refractive error. The procedure has been controversial, but newer methods, better patient selection, and better short-term success rates in recent years have muted the debate. If a final visual acuity of 20/25 or better is the criterion for success, it is achieved in more than 80% of eyes after just one surgery, regardless of the surgical procedure used (RK, PRK, or LASIK; see text).

Refractive surgery is always elective, never necessary. There are no known vision-threatening conditions for which any form of refractive surgery is the best or the only solution; it neither cures nor prevents anything. But elective surgery is not necessarily a bad thing. Much reconstructive surgery, for example, is both elective and functionally or psychologically beneficial. A case has been made that refractive surgery is of this type—that is, elective surgery that makes the eye a better eye by reducing its refractive error and simultaneously improving the quality of life for people to whom refractive errors are a burden.

On the face of it, the notion that successful refractive surgery makes the eye better has some merit. If reduction of refractive error is the criterion for success, then virtually all eyes are improved by the procedure. A more stringent requirement that best *corrected* acuity after surgery be 20/25 or better still indicates substantial improvement; in only a small percentage of eyes is the best corrected acuity reduced by surgery (and even fewer suffer other complications). Finally, if a better eye is one that needs no refractive correction after surgery to achieve a visual acuity of 20/25 or better, more than 80% of eyes are improved.

But while successful refractive surgery may improve an eye by one or more of these criteria, refractive surgery isn't always successful, even in the best of hands. Some eyes require more than one operation for surgery to be declared successful, and some surgeries are never fully successful. A few eyes are made worse; that is, their best corrected acuity is reduced. Refractive surgery *may* produce a better eye, but it may not; the odds favor success, but there are no guarantees.

In any case, a Snellen acuity of 20/20 is not a very demanding criterion for spatial resolution. What we really need to know is how refractive surgery affects the optical quality of the eye; ideally, optical quality would be assessed by the modulation transfer function (see Chapter 2), but in practice, the contrast sensitivity function provides the best available information. In the few studies in which the contrast sensitivity function has been used as a measure of success, refractive surgery fails the test. Any form of refractive surgery reduces sensitivity to high spatial frequencies above 20 cycles/degree. Thus if the criterion involves not just the eye's focal length but the quality of the retinal image produced, refractive surgery never makes the eye better; in fact, it always reduces the eye's potential for best vision.

But is the reduction in image quality produced by refractive surgery significant in terms of the eye's performance? Is it important that the eye's best corrected acuity might be 20/20 instead of 20/15? Surely the practical benefits to the patient outweigh a little image degradation that is probably unnoticeable.

A friend who had refractive surgery (radial keratotomy) several years ago is pleased with the results. She has been freed from wearing spectacles or contact lenses and she is not aware that anything significant has been lost in the process; her vision is good enough and she is happy with it. I cannot argue with this as a short-term result, even though her eyes have been compromised to some extent. What I do not know is what will happen in the long-term. The optical quality of an eye inevitably declines with normal aging and with the maladies to which aging eyes are prone. Will the seemingly insignificant reduction in image quality introduced by refractive surgery in a young eye predispose it to greater-than-normal impairment later on? What will happen to these eyes in 40 or 50 years?

No one knows the answers to questions about the long-term consequences of refractive surgery, except to say that some eyes change with time. (Anecdotes about radial keratotomy go back almost 30 years, but data from controlled studies extend less than 15 years after surgery. Even less is known about the newer methods.) This means that refractive surgery is an ongoing experiment—or several experiments, since there are several types of surgery—in which significant questions cannot be answered for several decades. The questions concern the cornea's ability to stabilize after refractive surgery and to maintain stability in the face of normal aging changes. Refractive surgeons clearly believe that significant long-term problems are unlikely—I don't think they would be doing the surgery otherwise—but this is belief, not knowledge. Recent studies are reporting accelerated cell death and loss of fibroblasts from the anterior stroma as a result of refractive surgery, and the long-term implications are not clear.

If my friend who had radial keratotomy is representative of people whose surgeries have been successful, quality of life is the most important consideration in the equation—so much so that arguments for or against surgery on optical grounds are almost irrelevant. Whatever the underlying motivations are, the hope of being freed from cumbersome refractive corrections is paramount. People outside the vision care professions seem to be quite aware of the relationship between refractive surgery and quality of life, even if the association implies that refractive surgery is largely a high-tech, high-class form of cosmetic surgery. Needless to say, most refractive sur-

*This elegant and humane essay (*The Lancet* 2: 801–805, 1965) should be a subject of study and discussion for anyone in the health professions.

geons are not comfortable with this point of view (although a few of them take advantage of it in their advertising).

Since quality of life means different things to different people, decisions about the desirability of refractive surgery are highly individual. Unfortunately, the decisions have to be made on the basis of incomplete information; a truly informed decision is impossible. It is one thing to pit hope against uncertainty when doing nothing might be fatal, but it is altogether different when inaction is the prudent course, as it is when an unnecessary, irreversible alteration of one's corneas is being considered. Nobody loses this game by *not* having refractive surgery; the losers are those caught by the finite probability of surgical failure.

My concerns and reservations are academic, of course. Not only will refractive surgery be with us for the foreseeable future—there is a good deal of money at stake, to put it bluntly—but also the experiment will continue as new techniques are developed and put into practice. Those of us who are unenthusiastic about the procedure can only hope that its popularity will wane as people become more sophisticated about assessing risk and reward. I think it would have been better had refractive surgery not been pursued and developed; the situation, in a small way, reminds me of words Walt Kelly once put in a cartoonist's balloon coming from the mouth of Pogo the possum: *We have met the enemy, and he is us.*

Two basic procedures are used in corneal transplantation (Figure 8.34). Penetrating keratoplasty is a graft involving the full thickness of the cornea; a circular button of donor tissue, including both epithelium and endothelium, is substituted for a button removed from the recipient eye (see Figure 8.34*a*). Lamellar grafts replace perhaps the anterior two-thirds of the cornea, leaving the endothelium and Descemet's membrane of the recipient cornea intact (see Figure 8.34*b*). The lamellar graft involves removing a circular plug of tissue from the donor cornea and inserting it into a circular well of similar size in the host cornea.

Whether the graft is full-thickness or lamellar, a circular cutting device (a trephine) is used to remove both the donor and the recipient buttons (or plugs). The major variables are the size of the transplant, which is simply a function of the size of the trephine, and the thickness of the transplant; in general, the diameter of the transplant should be larger than the region of damaged recipient cornea, but not so large that it comes within 1.5 mm of the limbus and the peri-

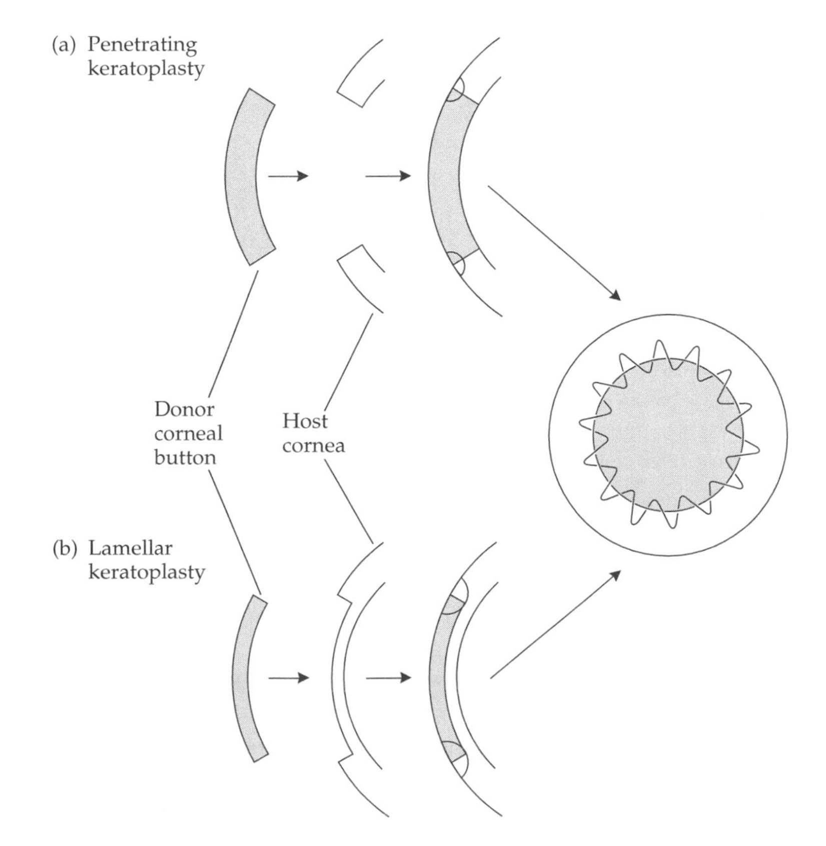

Figure 8.34
Principles of Keratoplasty
(*a*) Penetrating keratoplasty uses a full-thickness button of donor tissue sutured into a hole in the host cornea. (*b*) In lamellar keratoplasty, a button of tissue from a donor cornea is sutured into a corresponding well in the host cornea. In this illustration, the graft is about two-thirds of the corneal thickness. Both types of graft appear similar from the front; the running suture around the graft margin normally remains in place for years after the surgery.

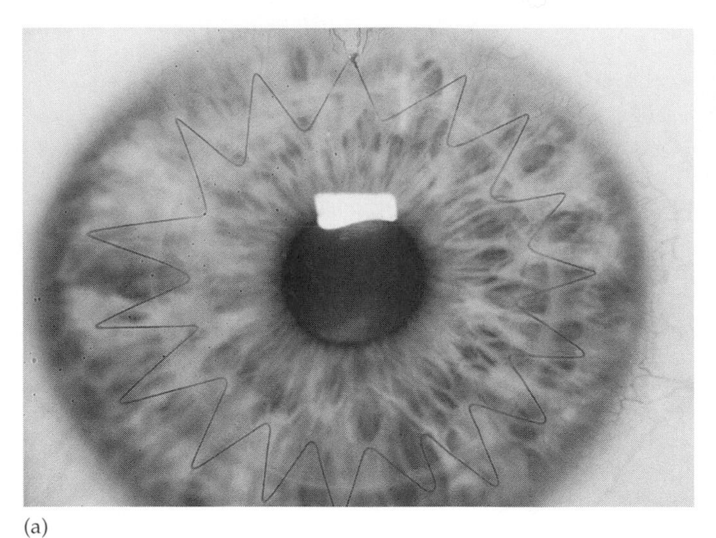
(a)

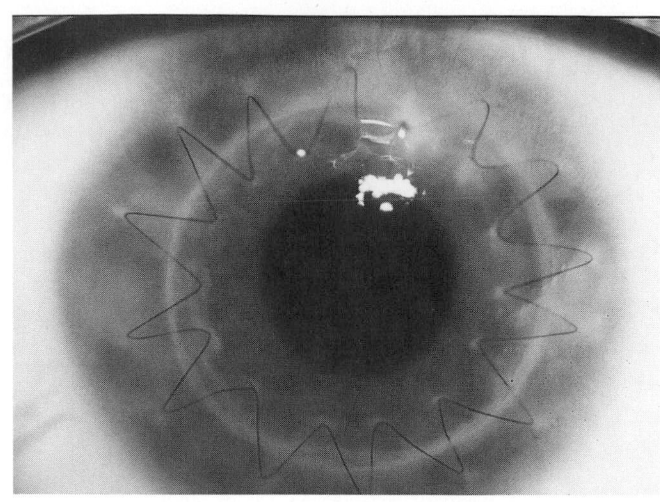
(b)

Figure 8.35
Corneal Grafts after Surgery
(*a*) In a successful graft, the cornea is clear and the margin of the graft is very difficult to see. (*b*) In a graft that is being rejected, the cornea is hazy from edema (stromal thickening), and inflammatory deposits are visible on the endothelium. (Courtesy of Roswell Pfister.)

corneal vasculature. The thickness of the transplant (full or lamellar) depends on whether the pathology or trauma to be corrected is relatively superficial or extends through the full thickness of the cornea. Once the donor tissue has been prepared and the corresponding button removed from the recipient eye, the donor tissue is placed into the host corneal bed and secured with interrupted sutures or a continuous suture around the margin of the graft; the sutures often remain in place for years after the surgery. Epithelium will quickly cover the junction on the anterior surface, and the strength of the wound will increase as new collagen forms at the junction between the donor and recipient stroma. This annular zone of new collagen will be the residual scar from the transplantation. In a full-thickness graft, endothelial cells migrate over the junction at the posterior surface.

As simple as this technique may seem in principle or may appear in illustration, numerous things can go wrong, and the surgeon's skill (or lack thereof) has considerable scope for expression. The amount of damage done to the endothelium of the donor button in being removed and handled, the precision with which the size and shape of the button match the hole in the recipient cornea, the uniformity of tension on the sutures used to secure the graft, and the nature of the underlying problem in the recipient cornea are just a few of the variables that can affect the outcome for a corneal transplant. The goal, of course, is survival of the donor tissue without rejection and with good optical clarity; in skilled hands, this goal is routinely met. A graft being rejected is edematous and cloudy in comparison to a successful graft (Figure 8.35).

Keratoplasty does not come under the heading of refractive surgery, since grafts are normally done to restore corneal transparency, not to change corneal shape. A significant exception is grafting to treat **keratoconus**, a condition in which the cornea has thinned and bulged, producing an irregular astigmatism and high myopia (Figure 8.36). The etiology of keratoconus is unknown, but it is usually progressive, and replacement of a large portion of the abnormal cornea with tissue from a normal donor eye can be a successful treatment strategy; large, full-thickness grafts are preferred. A keratoconic eye before and after graft-

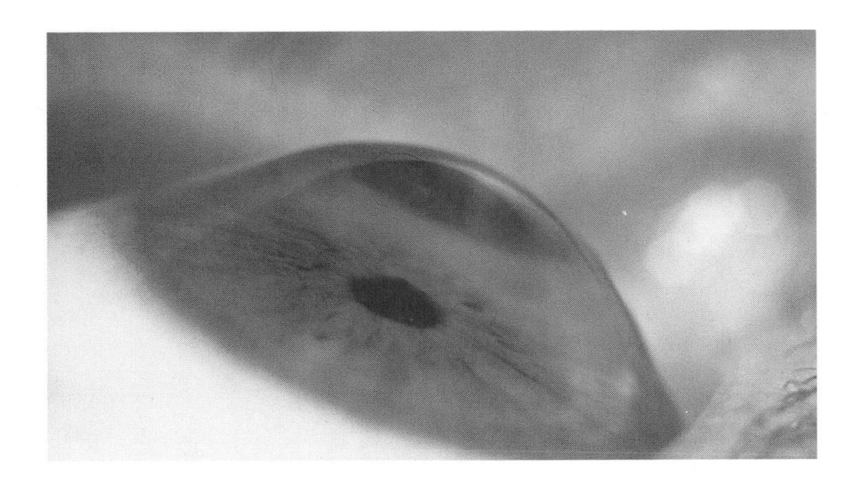

Figure 8.36
Keratoconus
In a photograph taken prior to surgery, the steep conical shape of the cornea characteristic of keratoconus is obvious. (Courtesy of Roswell Pfister.)

ing undergoes a remarkable change in shape produced by the graft; this change is accompanied by a significant reduction of the refractive error.

Corneal grafts always introduce changes in corneal shape. But unlike RK or PRK, corneal grafts are not designed to eliminate refractive error, and additional correction by spectacles or contact lenses usually is required. These corrections can be challenging, particularly if the graft has corneal irregularities or if an anisometropia exists; hard contact lenses can eliminate the effects of surface distortions and minimize refractive differences between the eyes.

Corneal Healing and Repair

The epithelium heals quickly and completely

The corneal epithelium normally replaces itself every 10 days or so; the cells now present in your epithelium were certainly not there 2 weeks ago. This replacement is the result of basal cell mitosis, displacement of basal cells into the wing and squamous cell layers, and sloughing of squamous cells into the tear film.* These processes are supplemented by a continual, slow migration of new basal cells from the periphery to the center of the cornea (Figure 8.37). The parents of the basal cells, the **stem cells**, are thought to reside as undifferentiated cells in the epithelium of the limbal conjunctiva (specifically, in the palisades of Vogt; see Chapter 9). After division and differentiation, their offspring migrate into the cornea, moving at a rate estimated to be about 100 μm per week. This migration suggests that the rates of sloughing and cell movement to the surface exceed the frequency of basal cell mitosis, thus requiring continual replenishment of the basal cell complement.

As long as the limbus is intact, damage to the corneal epithelium can be repaired quickly by accelerating its normal replacement activities. The release of intracellular components from damaged cells triggers a local response around the lesion site that temporarily shuts down mitosis and prompts a rapid lateral migration of cells into the gap; Figure 8.38 shows cells migrating into a lesion site. The migration is so fast—about 100 μm/hour—that an area several millimeters in diameter that is denuded of epithelium will be completely covered in less than 2 days. Mitosis resumes at an accelerated rate after the gap has been covered.

*One of the more direct ways to monitor the process in experimental animals is to expose the tissue to tritiated thymidine or one of its analogs. Thymidine is incorporated into dividing cells; cells that do not divide again remain strongly radioactive and can be followed as they move up toward the epithelial surface. The elapsed time from administration of the radiolabel to the disappearance of radioactive cells from the cornea is a measure of the epithelial turnover or replacement time.

Vignette 8.2

The Art of William Bowman

AT THE PEAK OF HIS CAREER, Sir William Bowman (1816–1892) was one of the leading ophthalmologists in Europe; the others were his friends Frans Cornelius Donders (1818–1889) and Albrecht von Graefe (1828–1870). Working in close communication with one another, they pioneered new methods for ophthalmic surgery, invented new instruments, and laid much of the groundwork for modern surgical methods. The lacrimal probes in use today were Bowman's invention.

Bowman's father was a banker by profession, but an ardent amateur naturalist and geologist by inclination and interest—interests he passed along to his son at an early age. When he was 16, William became apprenticed to a surgeon—the usual path to becoming a surgeon in those days—and began to study human anatomy in considerable detail. In addition to possessing a first-rate mind, Bowman had an extraordinary talent that set the course and guaranteed the success of his entire career; this gift was superb manual dexterity—an extraordinary linkage between eyes, brain, and hands.

For example, he was not a trained artist, but he could draw so well that in making woodblocks for illustrations in a publication, he drew directly on the block, a clear indication of his skill and confidence (most illustrators would work on paper, correcting their work before transferring the drawing to the woodblock). And Bowman's anatomical work included dissections of very small pieces of tissue, prepared by a new method adapted to his purposes. Today we would call it microdissection, but he did this work without any of the high-quality dissecting microscopes we now take for granted. Later, his skill made him one of the most gifted of the ophthalmic surgeons.

Bowman became an assistant surgeon at King's College in London in 1840, at the age of 24, and a year later he was made a Fellow of the Royal Society, one of the world's oldest and most prestigious scientific organizations.* What does it take to become a Fellow of the Royal Society at the age of 25? What it took for Bowman, and would certainly require today, was beginning to do original, innovative work by the age of 20.

In Bowman's case, his election was based on a thorough, accurate description of voluntary muscle fibers; he described their striation and the sarcolemma, among numerous other features, showing the generality of muscle fiber properties across many different muscles in a number of different species. Some aspects of the work were so good that they were unsurpassed until the advent of electron microscopy. Other research dealt with many different tissues, all at a microscopic level not previously explored (this, incidentally, was prior to the development of good methods for tissue sectioning and staining); Pacinian corpuscles and other specialized sensory structures, and the minute structure of tubules and blood vessels in the kidney, were among the subjects of his studies.

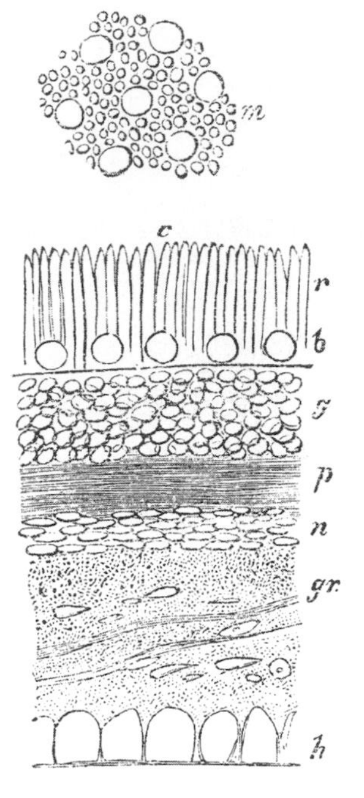

Figure 1
Vertical Section of the Human Retina
This section was from peripheral retina; it is about 140 mm thick. Most of the photoreceptors are rods (r), but cones are present; Bowman called them "bulbs," and their inner segments are indicated by the circles (b). Other layers of the retina are the outer nuclear layer (g), outer plexiform layer (p), and inner nuclear layer (n). The inner plexiform layer, ganglion cell layer, and nerve fiber layer are combined (gr); only a few ganglion cells are present. Bowman called the innermost layer containing bulbous spaces the "hyaloid" (h); if not artifactual, the spaces might be the terminal expansions of Müller's cells (see Chapter 16). The drawing above the retinal section labeled m is a tangential section through the photoreceptors at level b. The large circles are cone inner segments, and the others are rod inner segments. It is comparable to the photomicrographs in Figure 15.11 at 8.0 mm from the fovea. (From Bowman 1892.)

Then there was the eye. Bowman studied the retina (see Figure 1), discovered the radial fibers in the ciliary muscle (it was called Bowman's muscle for a time), and described the basic layered structure of the cornea as we now know it (Figure 2). Bowman's layer, once called Bowman's membrane, was his discovery; he called it the "anterior elastic lamina." He observed the stromal fibroblasts (others called them Bowman's corpuscles), reaffirmed the existence of Descemet's membrane, and recognized the presence of the epithelium and endothelium. Again, there was little left to be said about the structure of the cornea until it could be observed at a finer scale with electron microscopy.

In 1842, a year after becoming a Royal Society Fellow, he was awarded the Royal Society medal—another significant honor—and he got married. At this juncture, Bowman's career changed as he

*Election to the Royal Society is rarely taken lightly; I know no one who hasn't been excited about entering their name in the society's Charter Book, signed earlier by the likes of Captain James Cook, Benjamin Franklin, Hermann von Helmholtz, Isaac Newton, and Christopher Wren—not to mention William Bowman.)

Figure 2
Vertical Section of the Human Cornea
This drawing shows the anterior cornea to a depth of about 180 µm. Three layers of corneal epithelial cells lie on top of the "anterior elastic lamina" (*a*), the structure that we now call Bowman's layer. The horizontal cigar-shaped objects just below Bowman's layer are fibroblasts, and they are represented accurately as lying in rows between the corneal lamellae. The lines running obliquely through the corneal stroma from Bowman's membrane are probably parts of stromal layers that were displaced when the tissue was sectioned. (From Bowman 1892.)

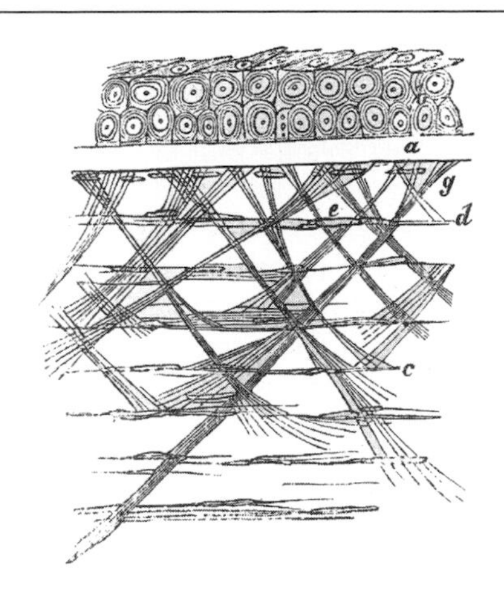

shifted from being a student of anatomy to being an ophthalmic surgeon and clinician (the need to support a family probably had something to do with this), and he followed these pursuits for the rest of his life. Bowman was apparently as brilliant a surgeon as he was an anatomist, and his earlier career surely aided his later one; surgeons need to know their anatomy very well, and Bowman was a master of the subject. In any case, he blazed a trail for later generations of surgeons to follow and improve upon.

In reading about William Bowman, I was struck by repeated references to his personality—in writing about people of great accomplishment, the accomplishments usually get most of the attention. But Bowman's personality comes through. He was a man who genuinely liked people and cared about them, who was devoted to his family and friends, and perhaps best of all, a man whose sense of humor kept his talents and accomplishments in perspective. His patients apparently loved him—not for his skill, but for his manner. It is said that he did not win all the honors he should have and was not elected to all the prestigious positions he merited, because he was too diffident, or too uninterested in these accolades, to campaign for them. But if one has had the security of being a member of the Royal Society and a winner of its medal since the age of 25, other honors might be unnecessary.

Basal cells are normally attached to their basement membrane by hemidesmosomes; if they are to migrate, these attachments must be broken to allow movement and re-formed after the cells have stopped migrating. When the basement membrane remains intact in a lesion site, the breakdown and subsequent reformation of hemidesmosomes appear to be normal features of basal cell migration. The presence of basement membrane is not necessary, however, because basal cells will migrate onto bare stroma, later secreting a basement membrane to which they can attach. In this case, complete healing of the lesion is protracted, even though the lesion may be covered with cells fairly rapidly. A new basement membrane begins to form within 6 weeks or so, but reestablishing normal attachments to the epithelium may take much longer. For this reason, destruction of the basement membrane in photorefractive keratectomy prolongs healing.

The epithelium is the part of the cornea with the greatest capacity for maintaining and rebuilding its structure, which, given its continual exposure to minor

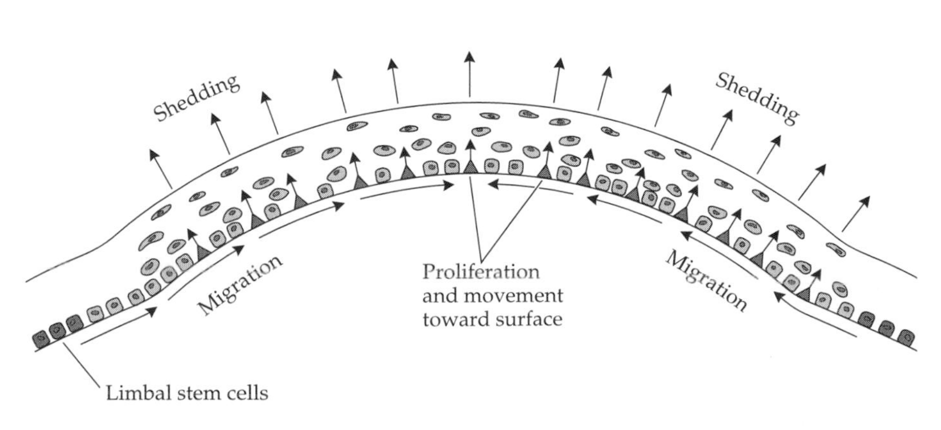

Figure 8.37
Renewal and Replacement of the Epithelium
Basal cells in the epithelium are in a proliferative phase of their life cycle, and cells generated in the basal cell layer move up into the epithelium as wing and later squamous cells that are eventually shed from the surface into the tear film. The stock of basal cells is replenished by stem cells residing in the limbus; the stem cells divide and differentiate into basal cells that migrate into the epithelium. (After Thoft and Friend 1983.)

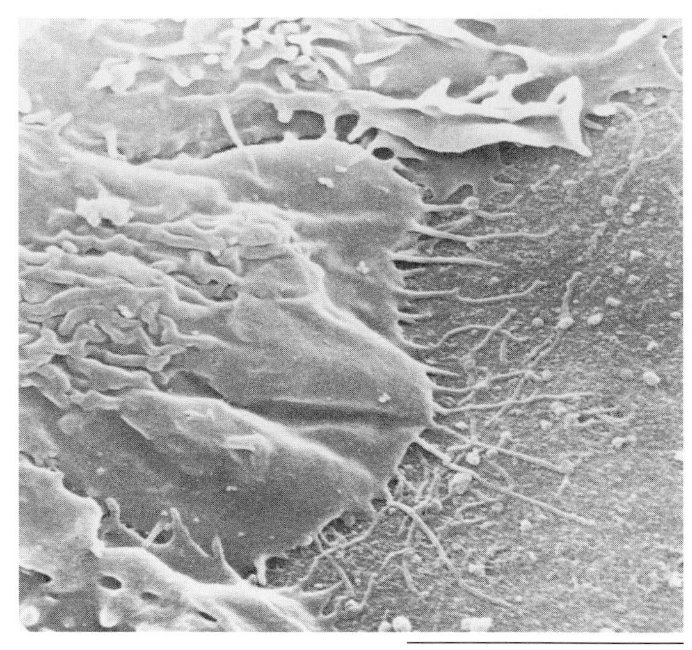

Figure 8.38
Epithelial Cell Response to Damage
In this scanning electron micrograph, the damaged part of the epithelium is to the right. Cells around the border of the damaged area can be seen flattening and sending out pseudopodia onto the basement membrane over which they are migrating into the region of damage. Migration rates are on the order of 100 μm/hour, so even large lesions can be covered quickly. (From Pfister 1975.)

trauma, is all to the good. There are very few conditions from which the epithelium cannot recover and heal. This recovery is contingent, however, on the viability of the stem cells in the limbal conjunctiva; if the stem cells are destroyed, a normal covering of epithelium cannot be regenerated or maintained.

Corneal healing may require limbal transplants

Direct evidence that stem cells for the corneal epithelium reside in the limbus is not available for the human eye, but the inability of the epithelium to repair itself when the limbal epithelium has also been damaged lends credibility to the notion. Extensive damage of this kind is often due to accidental splashing of alkali or other harsh chemicals onto the eye; repair may require grafting of new limbal tissue onto the eye as a source of new stem cells. Some abnormalities may also be related to an underlying problem with stem cells that can be treated by limbal tissue transplants. One such abnormality is *pterygium,* an abnormal growth of conjunctival tissue and blood vessels into the cornea that may be a consequence of altered stem cells that invade the cornea and subsequently signal vasoproliferation.

The strategy in these cases is to replace the damaged tissue with one or more pieces of healthy limbal tissue. Both autotransplantation (donor tissue is from one of the host's eyes) and homotransplantation (donor tissue comes from another individual) have been shown to work in providing a source of new stem cells. The transplanted pieces of tissue not only supply stem cells for regrowth of the corneal epithelium but also expand to reestablish the limbal and bulbar conjunctiva.

Repair of damage to the stroma produces translucent scar tissue

The stroma, including Bowman's layer, has nothing like the epithelium's capacity for repair. Although fibroblasts and the proteins they manufacture are probably not permanent, the turnover time is measured in years. Thus there is no rapid, ongoing maintenance process to be accelerated to heal the tissue. As a result, healing in the corneal stroma is slow, and any wounds to the cornea weaken it for long periods of time.

Because most of the stromal volume is collagen, repair means that new collagen must be synthesized to replace damaged collagen and fill gaps in the tissue. Damage to the stroma triggers a reorganization of the fibroblasts around the damaged region; they withdraw their pseudopodia and lose contact with one another, proliferate, and begin to synthesize new collagen. The original arrangement of the collagen fibrils and their lamellae is not replicated by the new collagen, however (see Figure 8.40); the resulting irregularity of the fibrillar arrangement reduces transparency. Also, replacement collagen is mostly type III, whereas the original collagen was type I; this difference also contributes to transparency loss.

The regions that have been repaired with irregular replacement collagen are scars, reminders of past trauma. A study of scars produced by penetrating corneal incisions done during cataract surgery indicates that the mechanical strength of the scars increases for about 4 years after surgery, at which time the strength of the scar tissue is similar to that of normal, undamaged tissue. (The results hint that scar strength may decline thereafter, but there are too few data to be sure.) These incisions were closed with sutures, as they are in keratoplasty, but they clearly take several years to heal completely. Unsutured incisions, like those in radial keratotomy, may take longer for complete healing, perhaps 6 years or more. It is not clear whether scars maintain their mechanical strength or weaken over longer periods of time.

The endothelium repairs itself by cell expansion and migration

Cells in the endothelium do not proliferate, but small cuts or lesions in the endothelium can be repaired quickly and easily. Undamaged endothelial cells around the lesion expand into the area and migrate until they make contact with one another, thereby restoring a continuous cell layer. If Descemet's membrane is damaged, as it would be in any penetrating incision, the endothelial cells produce new membrane materials after they have reestablished their coverage of the damaged region. After penetrating keratoplasty, the endothelium at the incision site is intact and the layer is continuous, but the cells are extremely large and have unusual shapes (they may become less irregular as the process of healing stabilizes) (Figure 8.39). However distorted the individual cells are, the repair mechanisms have worked and the endothelium has reestablished itself as a confluent layer; this is an exaggerated version of the endothelial changes that accompany aging and long-term contact lens wear.

When the lesion in the endothelium is small, repair by cell expansion and migration works well; cell density in the damaged region will be reduced (see Figure 8.39), but endothelial coverage is maintained. Problems arise when lesions are extensive, because the cells appear to be limited in how large they can become and how far they can migrate. A break in both the endothelium and Descemet's membrane requires that the gap be filled with stromal collagen. Although filling the gap in this way constitutes repair in the sense of mechanically bridging the wound, the structure is irregular. And since the endothelial covering is interrupted, the normal state of stromal dehydration cannot be maintained. Thus edema also contributes to a substantial loss of transparency.

Corneal graft incisions are repaired by the normal healing processes

One goal in penetrating keratoplasty is a good fit between the cut surfaces of the host and the donor tissue. Smooth transitions of the anterior and posterior surfaces from graft to host encourage rapid epithelial and endothelial coverage of the wound, thereby re-creating a relatively normal environment within which

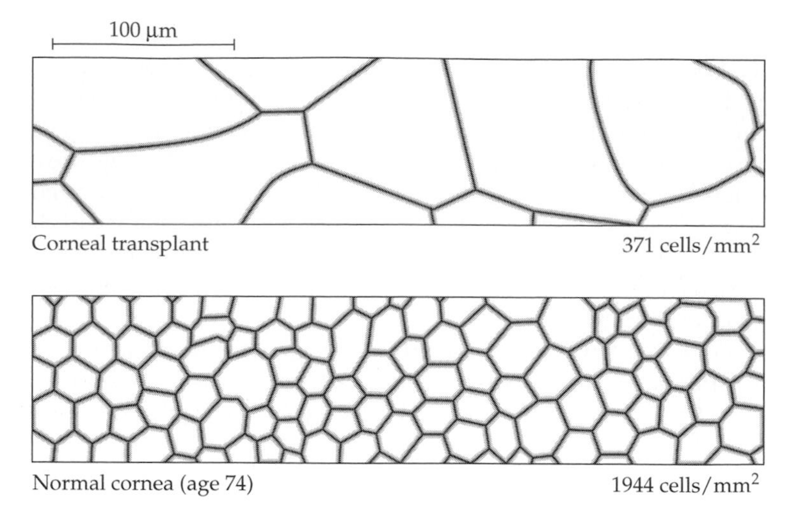

Figure 8.39
Repair in the Endothelium
A view of the endothelium after keratoplasty is compared with the normal cornea from an old eye. Since the endothelial cells cannot proliferate, they can reestablish their monolayer only by cell expansion and migration. In the region of the endothelium affected by the keratoplasty, the density of the endothelial cells is very low and they have become extremely large, about five times the normal diameter. (After Bourne and Kaufman 1976.)

the stromal repair can be done. A close fit between host and donor tissues also means that there are no major gaps that require longer periods of time to be bridged with new collagen. With a good match of graft and host tissue, the healing processes already described will proceed apace; that is, the incisions will attain the strength of normal corneal tissue in about 4 years (the long duration of stromal healing is the reason that graft sutures are left in place for so long). Healing of keratoplasty incisions may be affected by the inflammation associated with graft rejection, which disrupts normal repair, and some of the drugs used to counteract rejection also slow the healing process.

Healing of lamellar grafts is somewhat different, because there is a large surface where the base of the donor plug meets the bottom of the well in the host cornea; any binding together of these surfaces by the formation of new collagen will produce a layer of irregular collagen—a scar—extending across the entire graft area. Some scarring is inevitable, and the more irregular the surfaces of the host and donor tissues, the more scarring there will be. For this and other reasons, lamellar grafts are done less commonly than full-thickness grafts.

Radial keratotomy incisions are repaired by epithelial hyperplasia and collagen formation

Unlike graft incisions, RK incisions are meant to gape, and they are not closed with sutures. The initial covering of the incision is by the epithelium, which extends down along the walls of the incision, filling the gap with epithelium and creating an "epithelial plug" (Figure 8.40). After the plug has formed, the incisions are bridged with new collagen, beginning at the bottom of the cleft and progressing toward the surface. As the bridging continues, the epithelial plug shrinks or is pushed out of the wound until the stroma has been re-formed out to the normal epithelial layer. This process may take up to 5 years or so, at which time the incision is said to be completely healed.

In some cases, the epithelial plugs are very persistent, and they may never withdraw completely (some have persisted at least 12 years after surgery). The

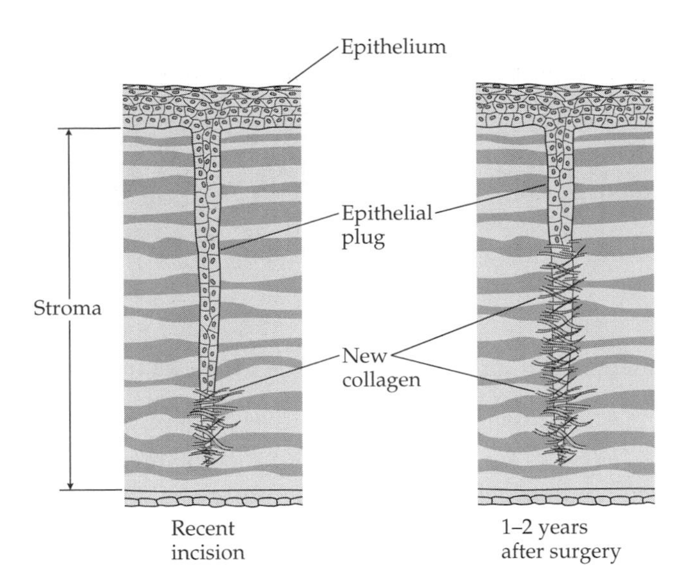
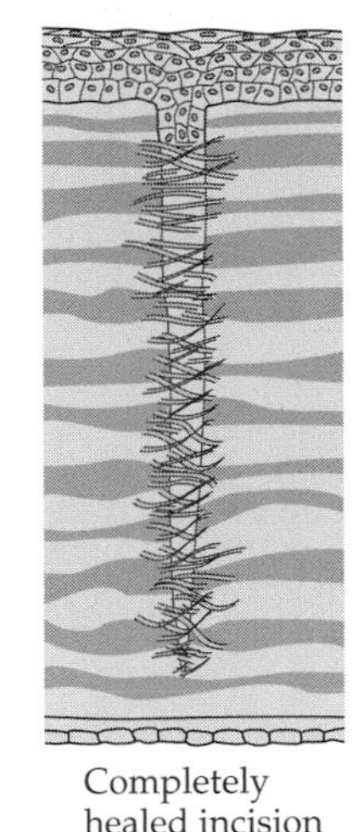

Figure 8.40
Repair of Radial Keratotomy Incisions
In the initial filling of the incision with epithelial cells, new collagen forms at the base of the incision to bridge across it. In time, as the collagen bridging proceeds toward the surface, the epithelial plug withdraws, eventually leaving a normal epithelium over the healed incision. (After Binder et al. 1987.)

persistence of an epithelial plug implies incomplete bridging of the incision by new collagen; this is one reason for concern about the long-term strength and stability of RK incisions. Another reason for concern is the fact that the incisions are relatively wide near the surface and collagen bridges across wide gaps may be inherently weaker than those between adjoining stromal surfaces.

Photorefractive keratectomy ablations are healed mostly by the epithelium

The first step in PRK, before ablation of the underlying stroma, is to remove the epithelium. Characteristically, the stromal surface after ablation is smooth (though not optically smooth), as if all the collagen fibrils were neatly clipped and shorn by the laser. This new surface is a suitable substrate for epithelial regrowth, and there is no obvious stimulus for the formation of new collagen by the fibroblasts. Why the fibroblasts respond so feebly is not known, but it may be due to the rapidity with which the surface is covered by epithelium (about 3 days) and the lack of necrotic tissue that elicits inflammation and scar formation. Fibroblasts do appear to migrate into the ablation zone, however.

The epithelium regrows by cell migration and accelerated mitosis, as described earlier. Although this regrowth is rapid, formation of a new basement membrane may take a year or so, and during this time the epithelium contains more than the normal number of cells and cell layers. This excess epithelial cell production and migration of fibroblasts into the region under the new epithelium may contribute to the postsurgical haze observed in some patients. Normally, the basement membrane of the epithelium is anchored to fibrils in Bowman's layer, but it is not clear that Bowman's layer ever re-forms after having been ablated.

Growth and Development of the Cornea

The epithelium and endothelium are the first parts of the cornea to appear

After the lens vesicle separates from the surface ectoderm, the layer of cells remaining on the surface is the precursor of the corneal epithelium, the epithelium of the conjunctiva (limbal, bulbar, fornix, and palpebral), and the epidermis of the skin. The adult differences among these epithelial tissues are presumably

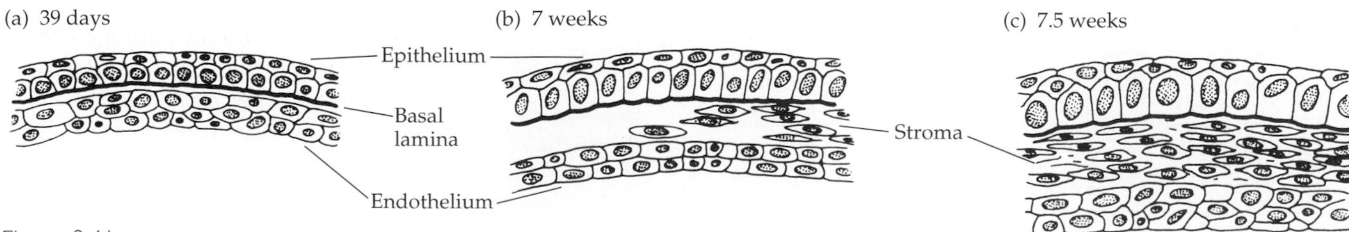

Figure 8.41
Early Stages of Corneal Development
(*a*) During the sixth week of gestation, neural crest cells migrate behind the epithelium to form a definitive endothelium. (*b, c*) Another wave of migration brings in more neural crest cells, which will become the stromal fibroblasts. (*d*) At 3 months of gestational age, Descemet's membrane has appeared and the posterior stroma is laminated. (e) Bowman's layer appears by 4½ months and lamination in the stroma has progressed farther anteriorly. (*f*) With the exception of lamination in the anteriormost part of the stroma, the cornea is complete by 7 months of gestation. (After Cook, Ozanics, and Jacobiec 1991.)

the results of regional differences in differentiation that occur (or are induced?) as their underlying connective tissues form. The part of the surface ectoderm just in front of the developing lens will become the corneal epithelium.

The transformation of surface ectoderm to corneal epithelium requires that a monolayer of cells becomes multilayered; this process begins at about 6 weeks, about the same time that undifferentiated cells from around the optic cup migrate in behind the epithelium (Figure 8.41). These migrating cells are derived from the neural crest and will become the *endo*thelium. The basement membrane for the epithelium appears by 7 weeks, but the six or so cell layers that characterize the adult epithelium are not acquired for another 3 or 4 months.

The endothelial cells form a complete layer, separated from the epithelium by a diffuse basal lamina, by week 7. They also form an interior lining for the future limbus, in particular for the trabecular meshwork. At this stage, the endothelium is several cells thick, and as development proceeds the cells gradually flatten and spread to create the adult monolayer of endothelial cells. Descemet's membrane, which is produced by the endothelium, appears shortly after the endothelial cells have formed a confluent layer; it becomes thicker throughout gestation and well into adulthood. Neither the endothelium nor the epithelium acquires its fully adult structure until the sixth or seventh month of gestation, at which time the previously translucent cornea becomes transparent.

The stroma is derived from neural crest cells associated with the mesoderm

After the endothelium has formed, another group of neural crest cells migrates from the mesoderm around the optic cup into the space between the developing epithelium and endothelium, thus occupying the region that will become the corneal stroma (see Figure 8.41*b* and *c*). These cells form the stroma by proliferating and differentiating as fibroblasts and then synthesizing collagen.

The stroma is exclusively cellular to begin with, consisting of several layers of fibroblasts. Collagen production and the formation of the laminae that characterize the adult cornea create most of the stromal volume and make the cellular fraction of the stroma much less significant. Stromal development proceeds from posterior to anterior (i.e., inside out); clear-cut laminae can be seen next to the endothelium at a time when the rest of the developing stroma has no obvious organization. The addition of collagen to the stroma is nearly complete at birth, when the cornea has acquired about 90% of its final diameter and all of its adult thickness.

The regular arrangement of the stromal collagen appears soon after collagen production begins

Although the fibroblasts in the future cornea, limbus, iris, and sclera all perform the same basic task—namely, producing collagen—something very different happens in the future cornea to produce the high degree of organization we see in the tissue. As Figure 8.41 suggests, the stroma exhibits a laminar organization

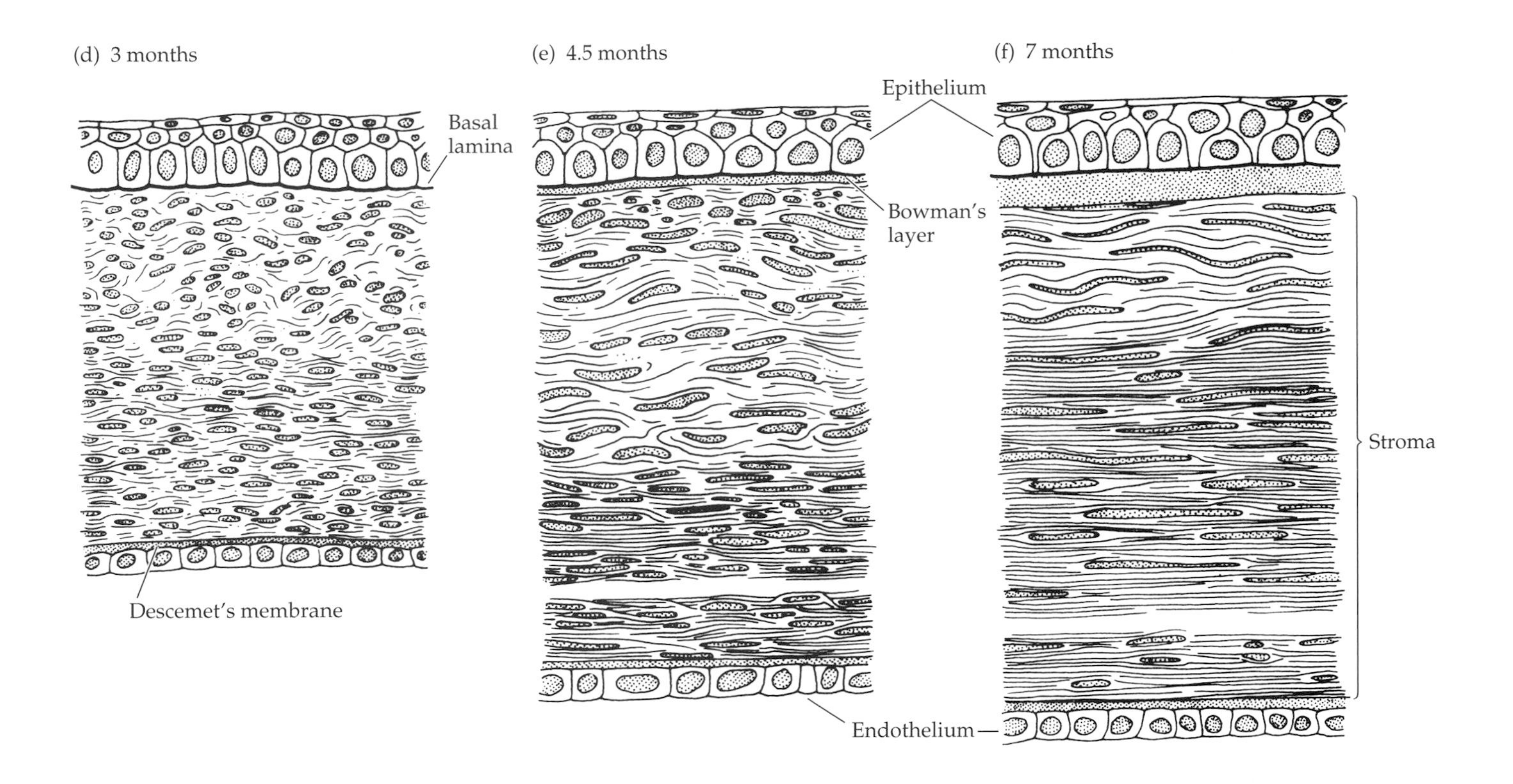

very early; collagen is being produced throughout the developing stroma, but it is being arranged into laminae first in the posterior part of the developing stroma. (Laminae will form in the scleral collagen as well, but they are not continuous, parallel sheets like those in the cornea.) The steady increase in stromal thickness throughout gestation is due to the addition of new laminae to the stack and the addition of collagen to each of the laminae.

Collagen production by a fibroblast is not a spatially random procedure with new fibrils being spewed out in all directions. Instead, the fibrils are highly oriented. According to one current model of collagen synthesis, procollagen, which is produced inside the cell, is secreted into a narrow extracellular channel, where the conversion to collagen and the assembly of collagen fibrils take place (Figure 8.42*a*). Fibrils from several parallel channels combine to produce a collagen fibril. This model suggests that forming a sheet of parallel collagen fibrils generated by many different fibroblasts requires that the cells be properly aligned with one another, such that their fibrils will be parallel (Figure 8.42*b*). To acquire and maintain a particular alignment, the fibroblasts probably need to communicate with one another; their gap junctions may have something to do with this interaction.

Based on this model of collagen formation, the first stage of stromal development would be organizing the fibroblasts closest to the endothelium into a communicating, interacting array of cells; they will produce collagen fibrils that are parallel in some direction. Then the set of fibroblasts just anterior to the first group becomes organized and begins to produce parallel collagen fibrils. Assuming that the successive arrays of fibroblasts are organized independently, the fibril directions in successive laminae should vary. By organizing a few hundred such fibroblast arrays in a posterior-to-anterior sequence, a structure very much like the corneal stroma should be produced.

If the laminae in the cornea are produced by sequential recruitment and activation of fibroblast arrays, the lack of such stacked laminae in the sclera implies that scleral fibroblasts go into production at about the same time, with no inner-to-outer sequence.

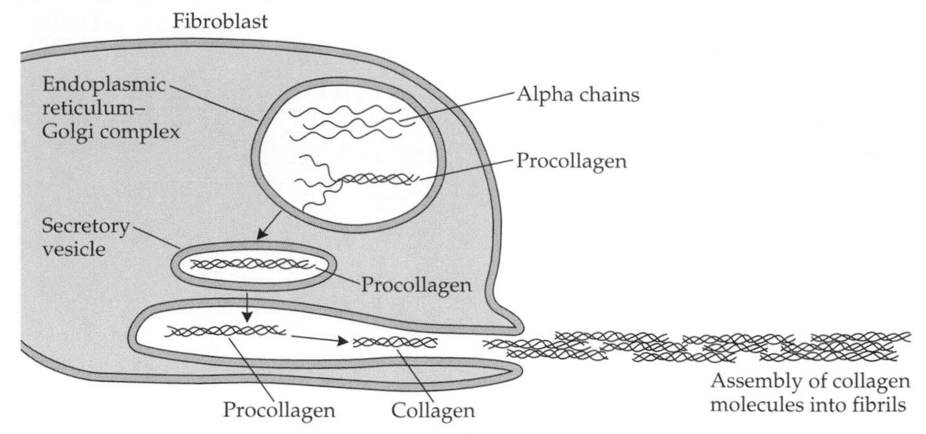

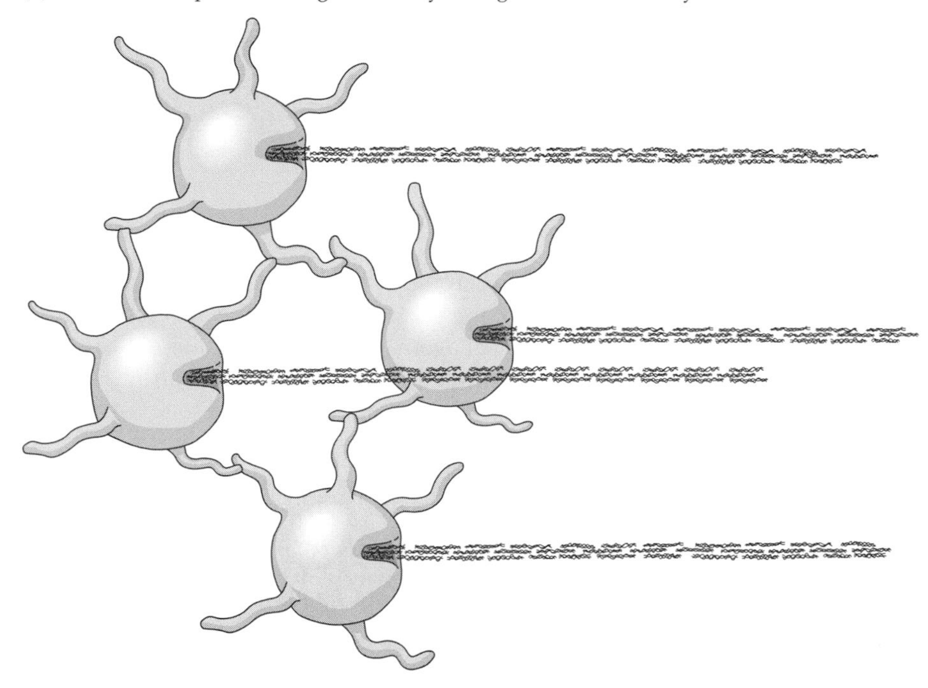

Figure 8.42
Collagen Production by Fibroblasts
(*a*) Collagen production begins with the intracellular assembly of alpha chains into molecules of procollagen. The procollagen is transported out of the cell by secretory vesicles, where it is converted to collagen and assembled into long collagen fibrils. (*b*) If there is a distinct orientation to collagen fibrils produced by a particular fibroblast, an array of fibroblasts with similar orientations can produce sheets of collagen in which the fibrils are parallel to one another. (This cartoon is conjecture.) (*a* after Alberts et al. 1994.)

Fibroblasts placed onto randomly arranged collagen fibrils in tissue culture can move around, pulling and tugging on the fibrils to change the fibril organization from random to more nearly parallel. To all appearances, the fibroblasts are shepherds and collagen fibrils are their flock. The extent to which fibroblasts in vivo can exercise this aligning, shepherding function is not clear, but active tuning of the fibril orientation by the fibroblasts may contribute to the final result.

Corneal growth continues for a few years postnatally

At birth, the cornea has already acquired its full adult thickness; it may even be a bit thicker than it will eventually become. The horizontal diameter is around 10 mm, about 90% of the adult value. Since the axial length of the eye and its other major diameters are just over two-thirds of their adult value, however, the cornea occupies about 25% of the eye's surface area at birth; in adults it accounts for only about 15% of the surface area of the eye. The infant's cornea, in other words, is more nearly complete than the rest of the eye; it will acquire its final size at about 3 years of age, but growth of the sclera will continue for another 10 or 12 years.

The combination of a relatively mature cornea with a relatively small eye accounts for the hyperopia of most infant eyes; the hyperopia normally decreases as the eye grows in axial length; and most eyes become emmetropic at 6 to 8 years of age. This is the age at which childhood myopia becomes so obvious, implying that the myopia develops because the growth of the sclera goes on too long, failing to match optically the already mature cornea (see the Epilogue).

Anomalous corneal development can produce misshapen or opaque corneas

The vast majority of full-term infants have horizontal corneal diameters very close to 10 mm, but the cornea may sometimes be appreciably larger or smaller, even when the eye size is normal. **Megalocornea** refers to the case in which the cornea is too large, while **microcornea** means it is too small; the abnormal corneas may be transparent, but there will be a very high refractive error. Megalocorneas may be very thin, as in keratoconus, requiring an extensive graft to restore some semblance of normality.

It is not entirely clear why these corneas failed to achieve a normal size. One possibility is exuberant early growth, producing megalocornea, or retarded growth, producing microcornea; the implication is that future corneal fibroblasts have differentiated in the proper region but then, for reasons unknown, have either run amok or become lazy. Another possibility is that the region of fibroblast differentiation is disproportionately large or small from the outset, with subsequent growth proceeding at a normal pace. In any event, the failure probably occurs early in corneal development.

There appear to be failures of fibroblast development as well. In **cryptophthalmos**, the eyelids are abnormal and the skin of the lids appears to continue across the eye, replacing the cornea with a skinlike tissue. This is an intractable situation; efforts to graft normal corneal tissue into the abnormal region have not been successful in giving vision to these eyes. A less extreme situation is one in which the corneal stroma is more like sclera than like normal cornea; it is called **sclerocornea**, and part or all of the cornea can be white and opaque. This would not be a case of the corneal fibroblasts failing to differentiate, but of their failing to organize properly; the etiology is unknown, however. Fortunately, corneal grafts can be successful in some cases of sclerocornea.

Another of the long list of congenital abnormalities is one in which fibroblasts all around the anterior chamber seem to have become confused about which tissue they belong to. The result is a variable and often extensive set of structural defects in the cornea, iris, and limbus known as **Peter's anomaly**. Parts of the cornea may be incompletely formed, large strands of the iris may run forward through the anterior chamber to fuse with corneal tissue, and so on. If the anomaly is not too severe, corneal grafts in combination with other surgical procedures can restore relatively normal structure and function.

The congenital defects listed here can occur in isolation, but since they appear to have their origins in early developmental failures, they are often associated with much more widespread defects involving the whole eye, the orbit, and the face.

References and Additional Reading

Components and Organization of the Cornea and the Sclera

Alberts B, Bray D, Lewis J, Raff M, Roberts K, and Watson JD. 1994. Cell junctions. Chapter 19, pp. 950–962, and The extracellular matrix of animals, Chapter 19, pp. 971–995, in *Molecular Biology of the Cell*, 3rd Ed. Garland, New York.

Arffa RC. 1991. *Grayson's Diseases of the Cornea*, 3rd Ed. Mosby Year Book, St. Louis.

Beebe DC and Masters BR. 1997. Cell lineage and the differentiation of corneal epithelial cells. *Invest. Ophthalmol. Vis. Sci.* 38: 1815–1825.

Beuerman RW and Pedroza L. 1996. Ultrastructure of the human cornea. *Microsc. Res. Tech.* 33: 320–335.

Borcherding MS, Blacik LJ, Sittag RA, Bizzelli JW, Breen M, and Weinstein HG. 1975. Proteoglycans and collagen fibre organization in human corneoscleral tissue. *Exp. Eye Res.* 21: 59–70.

Bourne WM and Kaufman HE. 1976. Specular microscopy of the human corneal endothelium *in vivo*. *Am. J. Ophthalmol.* 81: 319–323.

Bourne WM, Nelson LR, and Hodge DO. 1997. Central corneal endothelial cell changes over a ten-year period. *Invest. Ophthalmol. Vis. Sci.* 38: 779–782.

Daxer A and Frantzl P. 1997. Collagen fibril orientation in the human corneal stroma and its implications in keratoconus. *Invest. Ophthalmol. Vis. Sci.* 38: 121–129.

Foster CS and Maza MS de la. 1994. Structural considerations of the sclera. Chapter 1, pp. 1–32, in *The Sclera*. Springer, New York.

Gipson IK, Spurr-Michaud SJ, and Tisdale AS. 1987. Anchoring fibrils form a complex network in human and rabbit cornea. *Invest. Ophthalmol. Vis. Sci.* 28: 212–220.

Hogan MJ, Alvarado JA, and Weddell JP. 1971. The cornea. Chapter 3, pp. 55–111, in *Histology of the Human Eye*. WB Saunders, Philadelphia.

Johnson DH, Bourne WM, and Campbell RJ. 1982. The ultrastructure of Descemet's membrane. I. Changes with age in normal corneas. *Arch. Ophthalmol.* 100: 1942–1947.

Joyce NC. 1994. Cell biology of the corneal endothelium. Chapter 2, pp. 17–37, in *Principles and Practice of Ophthalmology: Basic Sciences*, Albert DA and Jakobiec FA, eds. WB Saunders, Philadelphia.

Komai Y and Ushiki T. 1991. The three-dimensional organization of collagen fibrils in the human cornea and sclera. *Invest. Ophthalmol. Vis. Sci.* 32: 2244–2258.

MacRae SM, Matsuda M, Shellans S, and Rich LF. 1986. The effects of hard and soft contact lenses on the corneal endothelium. *Am. J. Ophthalmol.* 102: 50–57.

Maurice DM. 1984. The cornea and sclera. Chapter 1, pp. 1–158, in *The Eye*, Vol. 1b, 3rd Ed., Davson H, ed. Academic Press, Orlando, FL.

Millodot M. 1984. A review of research on the sensitivity of the cornea. *Ophthalmic Physiol. Opt.* 4: 305–318.

Müller LJ, Pels L, and Vrensen GFJM. 1995. Novel aspects of the ultrastructural organization of human corneal keratocytes. *Invest. Ophthalmol. Vis. Sci.* 36: 2557–2567.

Müller LJ, Vrensen GFJM, Pels L, Cardoza BN, and Willekens B. 1997. Architecture of human corneal nerves. *Invest. Ophthalmol. Vis. Sci.* 38: 985–994.

Nishida T. 1997. Cornea. Chapter 1, pp. 3–27, in *Cornea*, Vol. 1, *Fundamentals of Cornea and External Disease*, Krachmer JH, Mannis MJ, and Holland EJ, eds. Mosby Year Books, St. Louis.

Nishida T, Yasumoto K, Otori T, and Desaki J. 1988. The network structure of corneal fibroblasts in the rat as revealed by scanning electron microscopy. *Invest. Ophthalmol. Vis. Sci.* 29: 1887–1890.

Pepose JS and Ubels JL. 1992. The cornea. Chapter 3, pp. 29–70, in *Adler's Physiology of the Eye*, 9th Ed., Hart WM, ed. Mosby Year Book, St. Louis, MO.

Pfister RR. 1973. The normal surface of corneal epithelium: A scanning electron microscope study. *Invest. Ophthalmol. Vis. Sci.* 12: 654–668.

Schimmelpfennig B. 1982. Nerve structures in human central corneal epithelium. *Graefes Arch. Clin. Exp. Ophthalmol.* 218: 14–20.

Stiemke MM, Watsky MA, Kangas TA, and Edelhauser HF. 1995. The establishment and maintenance of corneal transparency. *Prog. Ret. Eye Res.* 14: 109–140.

Stopak D and Harris AK. 1982. Connective tissue morphogenesis by fibroblast traction I. Tissue culture observations. *Dev. Biol.* 90: 383–398.

Tuft SJ and Coster DJ. 1990. The corneal endothelium. *Eye* 4: 389–424.

Waring GO, Bourne WM, Edelhauser HF, and Kenyon KR. 1982. The corneal endothelium: Normal and pathologic structure and function. *Ophthalmology* 89: 531–590.

Yee RW, Matsuda M, Schultz RO, and Edelhauser HF. 1985. Changes in the normal corneal endothelial cellular pattern as a function of age. *Curr. Eye Res.* 4: 671–678.

Yurchenco PD and Schittny JC. 1990. Molecular architecture of basement membranes. *FASEB J.* 4: 1577–1590.

The Cornea as a Refractive Surface

American Academy of Ophthalmology. 1996. Preliminary procedure assessment: Automated lamellar keratoplasty. *Ophthalmology* 103: 853–861.

Bogan SJ, Waring GO, Ibrahim O, Drews C, and Curtis L. 1990. Classification of normal corneal topography based on computer-assisted videokeratography. *Arch. Ophthalmol.* 108: 945–949.

Dingeldein SA and Klyce SD. 1989. The topography of normal corneas. *Arch. Ophthalmol.* 107: 512–518.

Fatt I, Freeman RD, and Lin O. 1974. Oxygen tension distribution in the cornea: A re-examination. *Exp. Eye Res.* 18: 357–365.

Hersh PS, Stulting RD, Steinert RF, Waring GO, Thompson, KP, O'Connell M, Doney K, Schein OD, and the Summit

Study Group. 1997. Results of phase III eximer laser photorefractive keratectomy for myopia. *Ophthalmology* 104: 1535–1553.

Higgins TV. 1994. How optical materials respond to light. *Laser Focus World* July: 91–98.

Holly FJ and Lemp MA. 1977. Tear physiology and dry eyes. *Surv. Ophthalmol.* 22: 69–87.

Maguire LJ. Keratometry, photokeratoscopy, and computer-assisted topographic analysis. Chapter 12, pp. 223–235, in *Cornea*, Vol. 1, *Fundamentals of Cornea and External Disease*. Krachmer JH, Mannis MJ, and Holland EJ, eds. Mosby Year Book, St. Louis.

Mandell RB. 1992. The enigma of the corneal contour. *CLAO J.* 18: 267–273.

Martonyi CL, Bahn CF, and Meyer RF. 1985. *Clinical Slit Lamp Biomicroscopy and Photo Slit Lamp Biomicroscopy*, 2nd Ed. Time One Ink, Ann Arbor, MI.

Maurice DM. 1957. The structure and transparency of the cornea. *J. Physiol.* 136: 263–286.

McDonnell PJ. 1995. Eximer laser corneal surgery: New strategies and old enemies. *Invest. Ophthalmol. Vis. Sci.* 36: 4–8.

Polse KA, Brand RJ, Schwalbe JS, Vastine DW, and Keener RJ. 1983. The Berkeley orthokeratology study, Part 2, Efficacy and duration. *Am. J. Optom. Physiol. Optics* 60: 187–198.

Rashid ER and Waring GO. 1989. Complications of radial and transverse keratotomy. *Surv. Ophthalmol.* 34: 73–106. (Also Chapter 23 in Waring, 1992.)

Saleh T, Waring GO, el-Maghraby A, Moadel K, and Grimm SB. 1995. Eximer laser in-situ keratomileusis (LASIK) under a corneal flap for myopia of 2 to 20D. *Trans. Am. Ophthalmol. Soc.* 93: 163–183.

Salz JJ, Maguen E, Nesburn AB, Warren C, Macy JI, Hofbauer JD, Papaioannou T, and Berlin M. 1993. A two-year experience with eximer laser photorefractive keratectomy for myopia. *Ophthalmology* 100: 873–882.

Seiler T and McDonnell PJ. 1995. Eximer laser photorefractive keratectomy. *Surv. Ophthalmol.* 40: 89–118.

Waring GO. 1992. *Refractive Keratotomy for Myopia and Astigmatism*. Mosby Year Book, St. Louis.

Waring GO, Lynn MJ, McDonnell PJ, and the PERK Study Group. 1994. Results of the prospective evaluation of radial keratotomy (PERK) study 10 years after surgery. *Arch. Ophthalmol.* 112: 1298–1308.

Winkle RK, Mader TH, Parmley VC, White LJ, and Polse KA. 1998. The etiology of refractive changes at high altitude after radial keratotomy: Hypoxia versus hypobaria. *Ophthalmology* 105: 282–286.

Corneal Healing and Repair

Alberts B, Bray D, Lewis J, Raff M, Roberts K, and Watson JD. 1994. Renewal by stem cells: Epidermis. Chapter 22, pp. 1154–1161, in *Molecular Biology of the Cell*, 3rd Ed. Garland, New York.

Assil KK and Quantock AJ. 1993. Wound healing in response to keratorefractive surgery. *Survey Ophthalmol.* 38: 289–302.

Binder PS, Nayak SK, Deg JK, Zavala EY, and Sugar J. 1987. An ultrastructural and histochemical study of long-term wound healing after radial keratotomy. *Am. J. Ophthalmol.* 103: 432–440.

Binder PS, Waring GO, Arrowsmith PN, and Wang C. 1988. Histopathology of traumatic corneal rupture after radial keratotomy. *Arch. Ophthalmol.* 106: 1584–1590.

Bryant MR, Szerenyi K, Schmotzer H, and McDonnell PJ. 1994. Corneal tensile strength in fully healed radial keratotomy wounds. *Invest. Ophthalmol. Vis. Sci.* 35: 3022–3031.

Cameron JD. 1997. Corneal reaction to injury. Chapter 8, pp. 163–182, in *Cornea*, Vol. 1, *Fundamentals of Cornea and External Disease*. Krachmer JH, Mannis MJ, and Holland EJ, eds. Mosby Year Book, St. Louis.

Collin HB, Anderson JA, Richard NR, and Binder PS. 1995. *In vitro* model for corneal wound healing; organ-cultured human cornea. *Curr. Eye Res.* 14: 331–339.

Crosson CE, Klyce SD, and Beuerman RW. 1986. Epithelial wound closure in the rabbit cornea: A biphasic process. *Invest. Ophthalmol. Vis. Sci.* 27: 464–473.

Davenger M and Evensen A. 1971. Role of the pericorneal papillary structure in renewal of corneal epithelium. *Nature* 229: 560–561.

Helena MC, Baerveldt F, Kim W-J, and Wilson SE. 1998. Keratocyte apoptosis after corneal surgery. *Invest. Ophthalmol. Vis. Sci.* 39: 276–283.

Helena MC, Meisler D, and Wilson SE. 1997. Epithelial growth within the lamellar interface after laser in situ keratomileusis (LASIK). *Cornea* 16: 300–305.

Honda H, Ogita Y, Higuchi S, and Kani K. 1982. Cell movements in a living mammalian tissue: Long-term observation of individual cells in wounded corneal endothelium of cats. *J. Morphol.* 174: 25–39.

Kruse FH. 1994. Stem cells and corneal epithelial regeneration. *Eye* 8: 170–183.

Matsuda M, Sawa M, Edelhauser HF, Bartels SP, Neufeld AH, and Kenyon KR. 1985. Cellular migration and morphology in corneal endothelial wound repair. *Invest. Ophthalmol. Vis. Sci.* 26: 443–449.

Miller SJ, Lavker RM, and Sun T-T. 1997. Keratinocyte stem cells in cornea, skin, and hair follicles. Chapter 11, pp. 331–362, in *Stem Cells*, Potten CS, ed. Academic Press, London.

Pfister RR. 1975. The healing of corneal epithelial abrasions in the rabbit: A scanning electron microscope study. *Invest. Ophthalmol. Vis. Sci.* 14: 648–661.

Pfister RR. 1994. Corneal stem cell disease: Concepts, categorization, and treatment by auto- and homotransplantation of limbal stem cells. *CLAO J.* 20: 64–72.

Rocha G and Schultz GS. 1996. Corneal wound healing in laser *in situ* keratomileusis. *Int. Ophthalmol. Clin.* 36: 9–20.

Sharma A and Coles WH. 1989. Kinetics of corneal epithelial maintenance and graft loss. *Invest. Ophthalmol. Vis. Sci.* 30: 1962–1971.

Simonsen AH, Andreassen TT, and Bendix K. 1982. The healing strength of corneal wounds in the human eye. *Exp. Eye Res.* 35: 287–292.

Thoft RA and Friend J. 1983. The X, Y, Z hypothesis of corneal epithelial maintenance. *Invest. Ophthalmol. Vis. Sci.* 24: 1442–1443.

Wagoner MD. 1997. Chemical injuries of the eye: Current concepts in pathophysiology and therapy. *Surv. Ophthalmol.* 41: 275–313.

Wilson SE and Kim W-J. 1998. Keratocyte apoptosis: Implications on corneal wound healing, tissue organization, and disease. *Invest. Ophthalmol. Vis. Sci.* 39: 220–226.

Growth and Development of the Cornea

Cook CS, Ozanics V, and Jacobiec FA. 1991. Prenatal development of the eye and its adnexa. Chapter 2 in *Duane's Foundations of Clinical Ophthalmology*, Vol. 1, Tasman W and Jaeger EA, eds. Lippincott-Raven, Philadelphia.

Hay ED. 1980. Development of the vertebrate cornea. *Int. Rev. Cytol.* 63: 263–322.

Murphy C, Alvarado J, and Juster R. 1984. Prenatal and postnatal growth of the human Descemet's membrane. *Invest. Ophthalmol. Vis. Sci.* 25: 1402–1415.

Tripathi BJ, Tripathi RC, and Wisdom JE. 1996. Embryology of the anterior segment of the human eye. Chapter 1, pp. 3–38, in *The Glaucomas*, Vol. 1, *Basic Sciences*, 2nd Ed., Ritch R, Shields MB, and Krupin T, eds. Mosby Year Book, St. Louis.

Waring GO and Rodrigues MM. 1993. Congenital and neonatal corneal abnormalities. Chapter 9 in *Duane's Foundations of Clinical Ophthalmology*, Vol. 1, Tasman W and Jaeger EA, eds. Lippincott-Raven, Philadelphia.

Chapter 9

The Limbus and the Anterior Chamber

I have been able to see in human eyes blood vessellike structures containing highly diluted blood, or a fluid clear as water, and leading from the limbus . . . towards the equator . . . finally emptying into the conjunctival or episcleral vessels.

■ K. W. Ascher, *Aqueous Veins*

Ascher's description of clear fluid flowing into the episcleral veins was the first convincing piece of evidence for aqueous flow in the human eye; prior to his observation, there were only hints that the fluid filling the eye's anterior chamber was anything but stagnant, having little if anything to do with the normal functioning of the eye. Small wonder that blindness produced by glaucoma was quite mysterious. But the fact that aqueous was flowing out of the eye implied that the fluid must be continually produced. Moreover, the relationship between aqueous production and aqueous drainage must have something to do with the elevated pressure known to exist inside the eye. Sixty years and many thousands of publications later, we are in a much better position to understand how such a tiny volume of fluid can be so important.

This chapter is one of three that will address anatomical issues related to **glaucoma** or, more accurately, the glaucomas; there are several different disorders within this broad category, sharing only the feature of potential visual field loss and blindness. This chapter focuses on the route and the structures involved in draining aqueous out of the eye and how one goes about altering the drainage route when it impedes normal outflow. Part of Chapter 11 will deal with the mechanism of aqueous production from blood flowing through the ciliary processes and how this production can be altered. Finally, Chapter 16 will discuss structural changes in the optic nerve head that are associated with glaucomatous field defects and blindness.

The Anterior Chamber and Aqueous Flow

The anterior chamber is the fluid-filled space between the cornea and the iris

Figure 9.1 shows the anterior chamber and its landmarks. Under normal conditions, the chamber is optically empty, filled with the transparent aqueous humor; the space is bounded posteriorly by the iris and lens and anteriorly by the cornea and the internal structures of the limbus.

If the posterior surface of the cornea were perfectly spherical and the iris were flat from side to side, the anterior chamber could be described as a segment of a

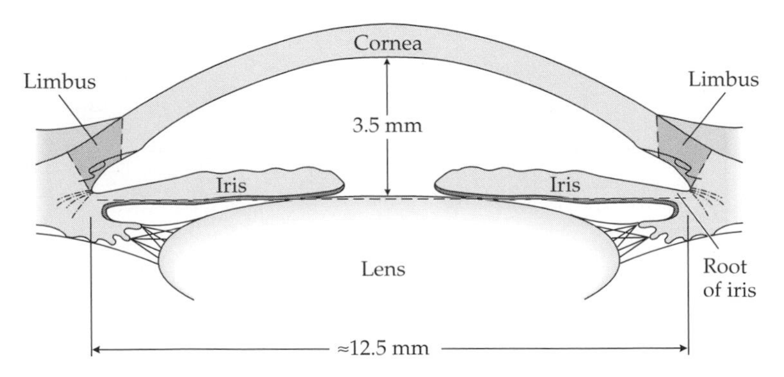

Figure 9.1
The Anterior Chamber
The anterior chamber is the space bounded by the posterior surfaces of the cornea and limbus and the anterior surfaces of the iris and lens. The chamber is about 3.5 mm in depth and about 12.5 mm in diameter, with a volume of around 260 μl.

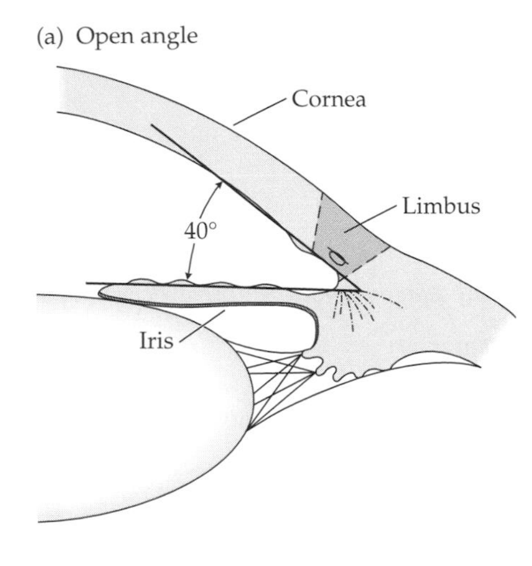

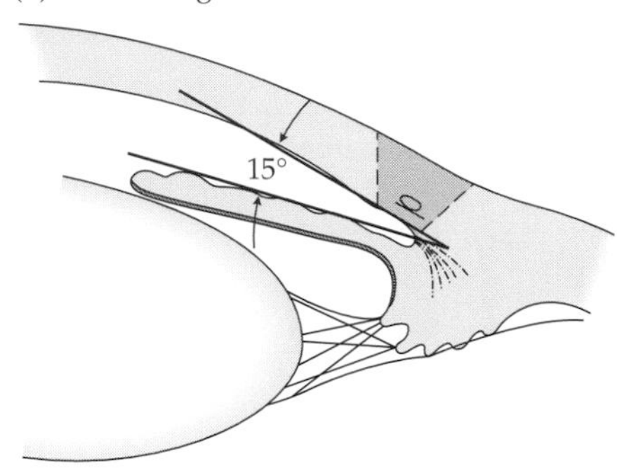

Figure 9.2
The Angle of the Anterior Chamber
(*a*) The chamber angle, defined by the intersection of the cornea and limbus with the iris, is shown here with lines drawn parallel to the corneal and iridial surfaces. The angle formed by these intersecting lines is about 40°, which is a normal open angle. (*b*) The lens is placed farther forward, the anterior chamber depth is reduced, and the chamber angle is only 15°. Although this angle is not completely closed, it is abnormally narrow and might close if the pupil were dilated and folds near the root of the iris were to project forward against the limbus.

sphere, but neither condition is quite accurate; in particular, the iris projects slightly forward from its periphery to the pupil. Thus, the **depth of the anterior chamber**, which is the distance along the pupillary axis from the posterior surface of the cornea to the anterior surface of the lens, is less than the distance from the posterior surface of the cornea to the plane of its peripheral joining to the ciliary body. Values for the average depth of the anterior chamber vary; the figure given here is 3.5 mm (from Weekers et al.) for an adult eye.

The depth of the anterior chamber varies among individuals who have no refractive error (the standard deviation is around ±0.35 mm), and additional variation is associated with age and refractive error. In general, the depth of the anterior chamber decreases with age (by about 0.1 mm per decade); it tends to be deeper in myopia and shallower in hyperopia.

In the plane of the **iris root**, where the iris joins the ciliary body, the maximum width of the chamber is around 12.5 mm—about 1 mm wider than the cornea; among other things, this dimension means that the point where the iris and ciliary body meet the limbus is difficult to view directly from in front, and special optical techniques are required (see Box 9.1). On the basis of the dimensions given here, the volume of fluid in the anterior chamber will be between 240 and 280 μl; thus 260 μl is a reasonable estimate. The volume of the anterior chamber accounts for 4% of the eye's total volume (about 6.5 ml).

The angle of the anterior chamber varies in magnitude

Since the iris rests on the anterior surface of the lens, the fact that the depth of the anterior chamber varies means that the orientation of the iris will also vary. A shallow anterior chamber implies that the lens is placed relatively far forward, for example, and this position will be associated with an iris whose pupillary margin is set farther forward in the anterior chamber than the root of the iris. At the other extreme, a very deep anterior chamber implies that the pupil will lie slightly behind the plane of the iris root. These variations affect what is called the **angle of the anterior chamber**, the angle formed by the intersection of the iris with the structures of the limbus (Figure 9.2).

Most of the aqueous leaves the eye through the limbus, and it is important for the fluid to have free, unimpeded access to the drainage structures, which is the case if the angle of the anterior chamber is relatively wide, on the order of 40° or so (Figure 9.2*a*). In clinical terms, this is an **open angle**. With shallow anterior chamber depths, however, the angle is more acute (Figure 9.2*b*), and it may be so narrow as to be classified as a **closed angle**. Clinicians do not have access to these sectional views of the angle, but they have several other ways to examine the angle and several systems for grading the degree to which the angle is open or closed, depending on how much of the limbal structure is hidden from their direction of view by the iris (see Box 9.1).

The main points here are, first, that the angle is variable in a normal population and, second, that glaucoma may exist with either open or closed angles. Thus the angle of the anterior chamber is used to make a distinction between two classes of glaucoma—namely, open-angle and closed-angle glaucomas. In general, glaucomas in these two classes are treated differently. The strategy in closed-angle glaucoma is to relieve or mitigate the flow-impeding effect of the closed angle; in open-angle glaucoma the strategy is to reduce the rate of aqueous production or increase the rate of aqueous outflow.

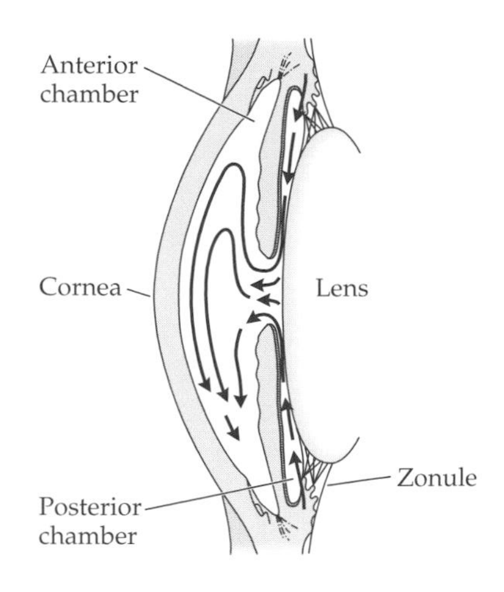

Figure 9.3
Aqueous Flow into and through the Anterior Chamber
The flow of aqueous is from the posterior chamber, through the pupil, into the anterior chamber. Aqueous flowing into the anterior chamber is somewhat warmer than aqueous already in the chamber that has been in contact with the relatively cool cornea. Convection currents carry newly arrived aqueous upward and then down along the posterior surface of the cornea.

Aqueous is formed by the ciliary processes and enters the anterior chamber through the pupil

Aqueous is derived from blood flowing through the ciliary processes (see Chapter 11) and secreted into the posterior chamber behind the iris (Figure 9.3). It flows between the posterior surface of the iris and the anterior surface of the lens, entering the anterior chamber through the pupil. Fluid entering the anterior chamber will be warmer than the fluid already there (the cornea is several degrees cooler than other ocular tissues because of its avascularity and evaporative cooling), and the incoming aqueous will rise as it flows outward toward the cornea. As the fluid column reaches the cornea and becomes cooler, it flows downward along the endothelium.

The aqueous normally lacks proteins or any suspended particulate matter, which accounts for its transparency and low degree of light scatter, but these conditions can change dramatically following infections in the uveal tract or loss of pigment from the iris. Debris released into the aqueous sometimes sticks to the corneal endothelium, thus revealing something of the pattern of flow. Figure 9.4*a* shows one example; the pattern is called **Krukenberg's spindle**, produced in this instance by pigment that was exfoliated (shed) from the iris. Since the aque-

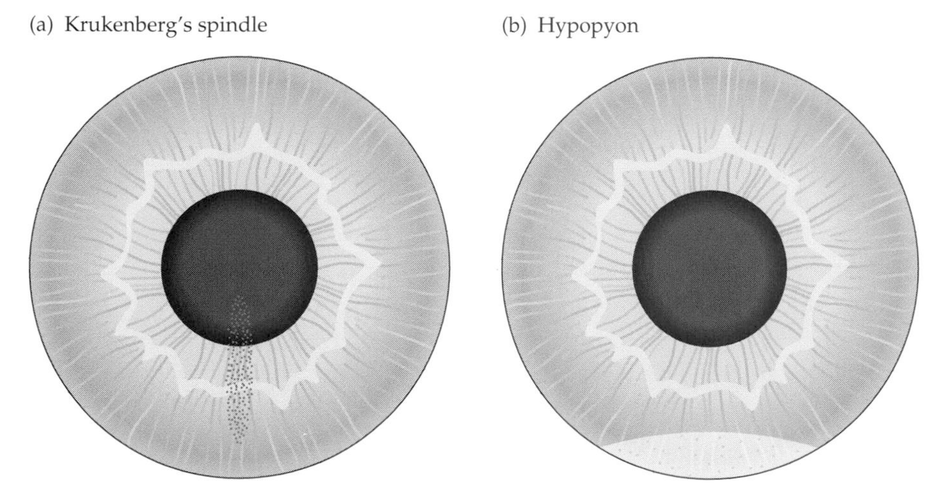

Figure 9.4
Particulate Debris in the Anterior Chamber
(*a*) Krukenberg's spindle is a vertical band of pigment deposited on the posterior surface of the cornea; a possible source of the pigment is the iris, from which it is carried to the cornea by the aqueous flow. (*b*) A layer of pus in the inferior anterior chamber angle is called a hypopyon; in this case, it is secondary to an infection in the iris and ciliary body. (Redrawn from Catania 1995.)

[Box 9.1] Through the Looking Glass: Gonioscopy

How can we tell if the angle of the anterior chamber is open, closed, or somewhere in between? Looking at the apex of the angle from in front is impossible because it is hidden behind the tissue of the limbus—that is, by the limbal stroma and conjunctiva. And using the biomicroscope at an oblique angle of view (Figure 1*a*) will not work; the optical surfaces of the cornea will defeat us, hiding the apex of the angle behind a veil of transparency.

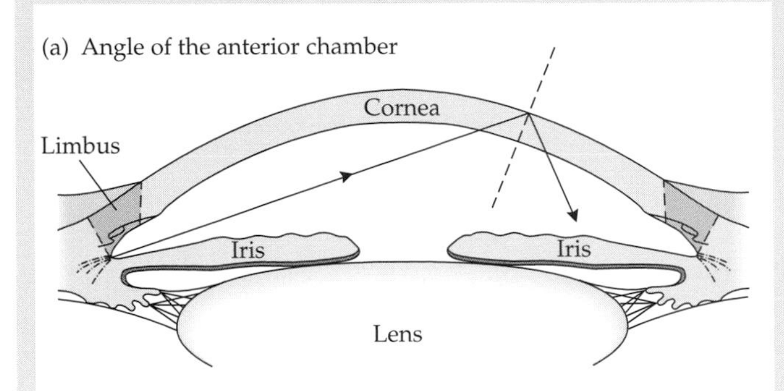

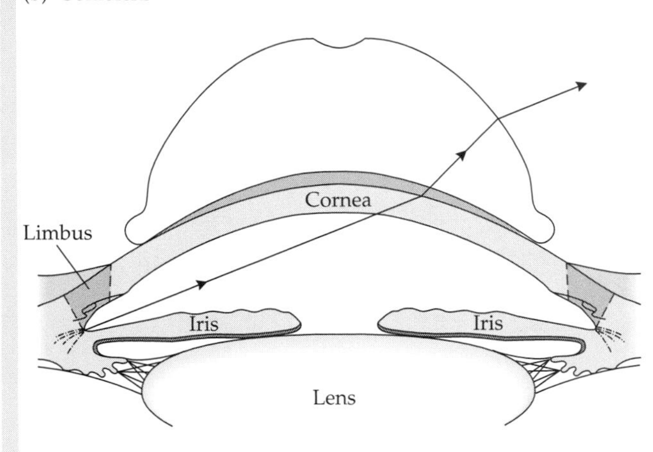

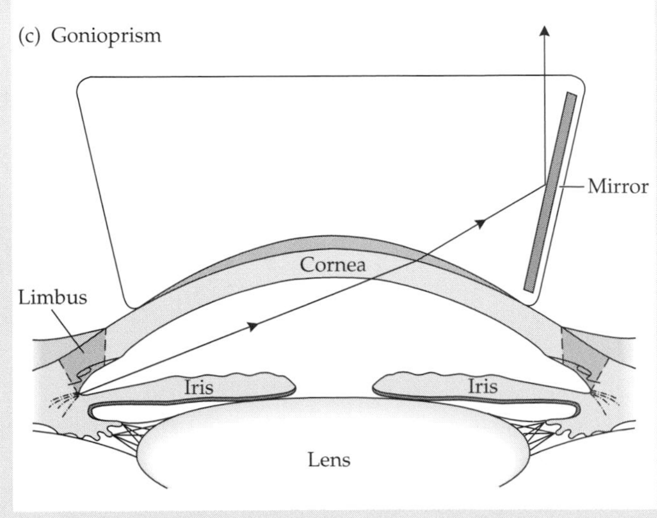

To see the apex of the angle, light reflected from structures in the apex must be able to pass through the cornea and enter the optical axis of the biomicroscope, but internal reflection at the interface between cornea and air prevents it. When light passes from a medium of higher to one of lower refractive index, it will be refracted only if the angle of incidence (i) is less than the critical angle (i_c) at which $\sin i_c = n/n'$, where n' is the higher index of refraction. For the cornea, the $i_c \approx 46°$. As Figure 1*a* shows, light coming from the apex of the angle will strike the region of the corneal pole such that $i \approx 60°$. Since this value exceeds that of the critical angle, the light will be totally reflected. For the part of the cornea closer to the apex of the chamber angle, the angle of incidence will be even higher and light there will be totally reflected as well.

The only possibility for seeing the apex of the angle is that the incidence angle will drop below the critical value somewhere along the corneal surface opposite the apex. Although the angle of incidence does decrease as we move to the opposite side, it doesn't decrease enough. By my calculation, $i = i_c$ about when we reach the limbus on the opposite side, but the limbal tissue will block the light. Thus without some special technique, we cannot look directly into the apex of the chamber angle.

Eliminating the cornea–air interface eliminates the internal reflection problem; this can be done with a glorified contact lens whose refractive index is equal to that of the cornea. Thus, $\sin i_c = n'/n' = 1$ and $i_c = 90°$. A critical angle of 90° is quite generous, and for practical purposes we can forget about the cornea's optical properties. These special contact lenses permit us to perform **gonioscopy**—that is, to examine the apex of the chamber angle. There are two forms of gonioscopy—direct and indirect—for which the different lenses are shown in Figure 1*b* and *c*.

The goniolens, used for direct viewing, allows us to do what we tried in the first place, that is, to look into the angle from an oblique angle of view. Although the goniolens also has a critical reflectance angle at its outer surface, its construction is such that light from the apex of the chamber angle is nearly perpendicular to the surface (i.e., the angle of incidence is very small) and light passing the surface is refracted rather than reflected. Therefore, we can

Figure 1
Principles of Gonioscopy
(*a*) Since light coming from the apex of the angle meets an interface at the corneal surface where the refractive index decreases, the light will be reflected unless the angle of incidence is less than the critical angle. There is no place across the corneal surface where this condition holds, and light from the apex always undergoes internal reflection. (*b*) A goniolens eliminates the cornea–air interface along with the usual internal reflection. The steep curvature of the anterior surface of the goniolens reduces the angle of incidence for light coming from the apex below the critical angle; light exits the lens without internal reflection. (*c*) Gonioprisms also eliminate the cornea–air interface. Light from the apex is reflected by an internal mirror such that its angle of incidence at the anterior surface of the prism is near zero. Again, there is no internal reflection at the interface.

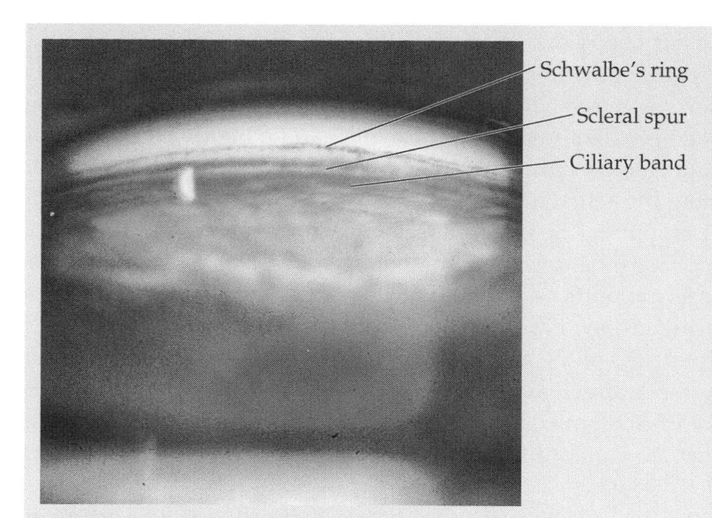

Figure 2
The Angle Viewed by Gonioscopy
Limbal structures are seen as parallel lines or streaks running around the chamber angle. Detailed structure is not visible, because of the low magnification, but the structures can be identified by their relative anterior–posterior location. Schwalbe's ring is the most anterior; the ciliary body band is the most posterior. The angle here is open. (Courtesy of Leo Semes.)

use a biomicroscope in conjunction with the goniolens to look directly into the chamber angle.

Indirect gonioscopy uses gonioprisms, which contain reflective surfaces arranged such that light coming from the chamber angle is redirected along the optic axis of the eye into the biomicroscope. Figure 1*c* shows a single reflecting surface, but gonioprisms may incorporate three or four mirrors so that one can view different parts of the chamber angle without having to rotate the gonioprism on the eye.

Figure 2 shows a gonioscopic view into the chamber angle. The angle is open, based on the structures of the limbus that are visible; they are Schwalbe's ring (or line) separating the cornea from the trabecular meshwork, the anterior part of the meshwork, the scleral spur viewed through the meshwork, and the ciliary band, which is the small part of the ciliary body between the root of the iris and the trabecular meshwork.

Several classification systems relate gonioscopic appearances to the anatomical width of the angle by assigning grades to the different appearances (unfortunately, high grades in one system may represent closed angles but in other systems open angles, so one needs to know what system is being used). The following grades characterize the Scheie classification:

Classification	Appearance
Wide open	All structures visible
Grade I	Ciliary band obscured
Grade II	Scleral spur obscured
Grade III	Posterior trabecular meshwork obscured
Grade IV	Only Schwalbe's ring visible

Grades I through III are narrow angles, Grade IV is closed; thus the picture in Figure 2*b* is a Grade IV closed angle.

Gonioscopy allows one to distinguish not only between primary open-angle and closed-angle glaucoma, but also among the various secondary forms of open-angle glaucoma. In Figure 2, for example, the angle is wide open, but there are heavy bands of pigment in the trabecular meshwork and in Schwalbe's ring. In this case, which is my right eye, the pigment deposits have accumulated with age and the intraocular pressure is normal. In some cases, however, the pigment accumulation may produce elevated intraocular pressure and a secondary glaucoma.

ous tends to flow directly outward from the pupil into the chamber, the pigment deposit is usually located centrally; the most likely explanation for the vertical elongation of the deposit is flow from superior to inferior along the endothelium. Material suspended in the aqueous may settle by gravity into the inferior part of the angle, where the accumulation is called a **hypopyon** (Figure 9.4*b*) if the debris (white blood cells and proteins) results from inflammation, or a **hyphema** if there is blood in the chamber as a result of trauma.

Since aqueous is continually being produced and flowing into the anterior chamber, it must be removed from the eye at about the same rate at which it is produced. If the rate of drainage (outflow) lags the rate of production, the pressure inside the eye—the **intraocular pressure** (**IOP**)—will rise, with potentially deleterious effects.

Aqueous drains from the eye at the angle of the anterior chamber

The human eye has two routes for aqueous drainage, both associated with the apex of the anterior chamber angle. The minor route is into the tissue of the ciliary body (Figure 9.5). Fluid percolates into spaces in the connective tissue surrounding the muscle fibers in the ciliary body and then moves into the potential space between the ciliary body and the sclera (the supraciliary space). The fluid is probably removed by veins in the ciliary muscle and anterior choroid that drain into the vortex veins (see Chapter 6) and by lymphatic vessels. Aqueous drainage by this route is called **uveoscleral flow**; it can be highly significant in some species, including other primates, but in humans it is relatively small. Although determining the actual amount of any fluid flows in the eye is accompanied by some uncertainties, the uveoscleral flow accounts for some 10 to 15% of aqueous drainage from the anterior chamber. It spite of its normally minor role, however, uveoscleral outflow is the target of one of the newest, most effective drugs for reducing intraocular pressure (discussed later in the chapter).

The remaining 85 to 90% of the aqueous drainage is through structures of the limbus, specifically the **trabecular meshwork** lining one side of the chamber angle and the **canal of Schlemm**, a large modified blood vessel encircling the chamber angle. From the canal of Schlemm, aqueous flows into the superficial episcleral veins.

Under normal conditions, 2.5 to 3.0 μl of aqueous leaves the anterior chamber each minute by both routes of outflow combined. To put this value in perspective, the entire volume of the anterior chamber would be emptied in just under 2 hours if the aqueous were not continually resupplied. The aqueous in the chamber is therefore renewed 12 to 13 times each day.

A possible third route of drainage is blood vessels in the iris. The anterior surface of the iris contains large openings, and the iris stroma is very loosely organized (see Chapter 10), allowing aqueous to come into direct contact with the iris vasculature. The use of various sorts of tracer molecules makes it possible to

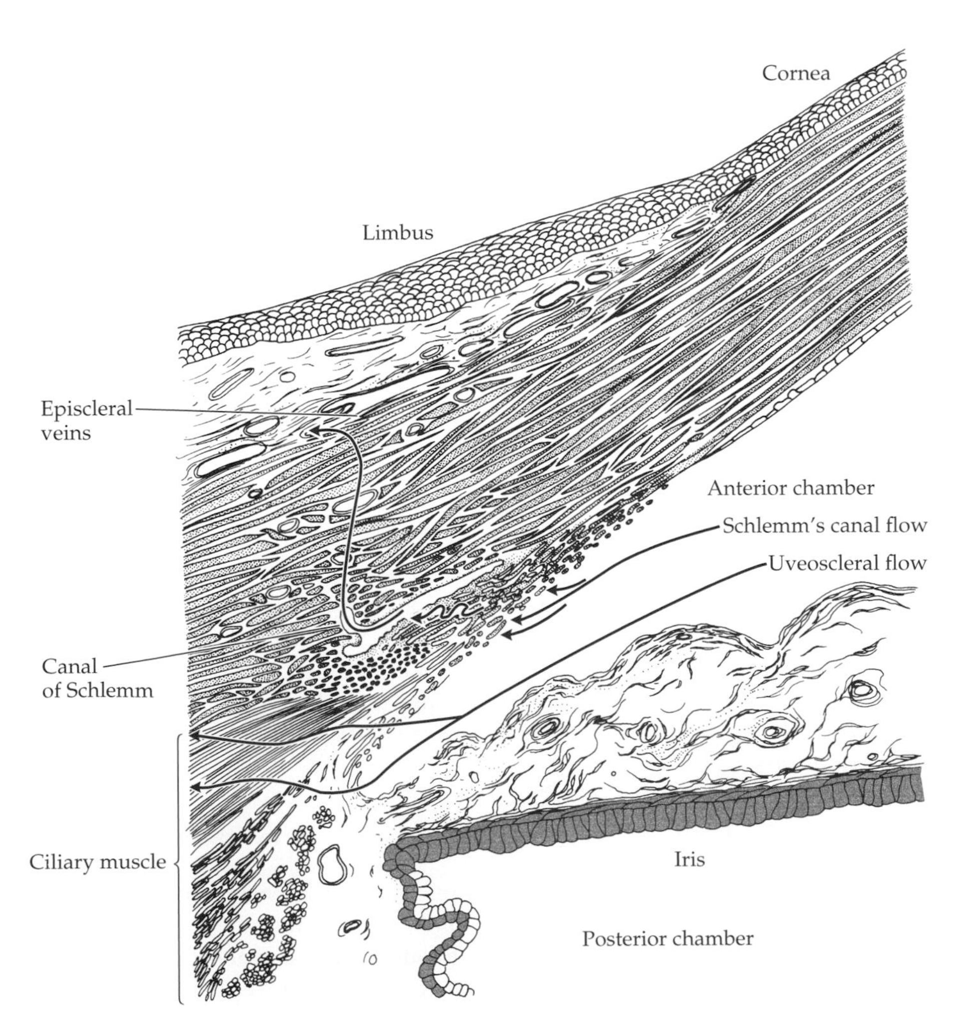

Figure 9.5
Routes of Aqueous Outflow
The solid lines show the major route for aqueous outflow, through the trabecular meshwork to the canal of Schlemm, and then to the episcleral veins by way of venous plexuses in the limbal stroma. The dashed lines indicate the secondary route, the uveoscleral flow, which passes through small spaces between ciliary muscle fibers to the supraciliary space between the ciliary body and the sclera. 10% to 15% of aqueous outflow is by the uveoscleral route.

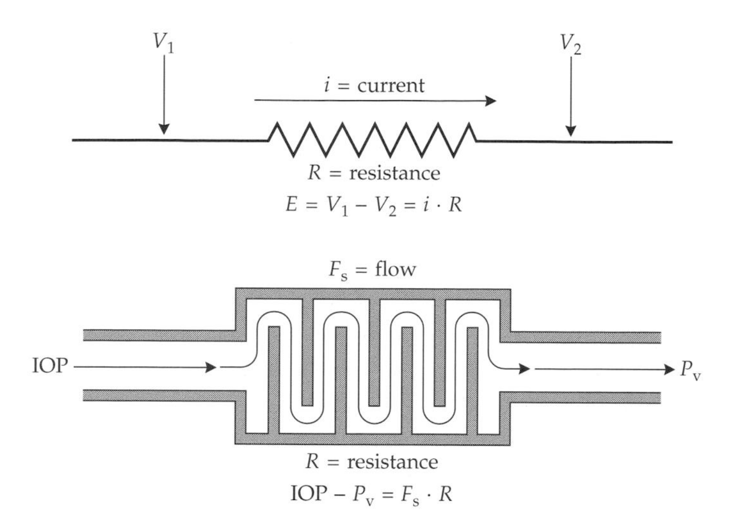

Figure 9.6
The Aqueous Flow Equation
In electrical circuits, the product of current flow (i) and resistance (R) is equal to the difference in voltage ($V_1 - V_2$) across the resistive element. The analogous case for aqueous outflow states that the product of aqueous outflow (F_s) and outflow resistance (R) is equal to the difference between the intraocular pressure (*IOP*) and episcleral venous pressure (P_v).

show that aqueous from the anterior chamber passes through the walls of the iris vessels and in species lacking the canal of Schlemm—rabbits, for example—there can be considerable aqueous outflow via the iris vessels. In humans, however, the flow of aqueous into the iris vasculature is accompanied by fluid flow back from the vessels to the aqueous, making the *net* fluid outflow by this route essentially zero.

Intraocular pressure depends on the rate of aqueous production and the resistance to aqueous outflow

The basic equation defining intraocular pressure is analogous to Ohm's law relating the flow of electric current to voltage (potential difference), that is, $E = iR$, where E is the potential difference ($V_1 - V_2$) between two points along a path of current flow (i), and R is a measure of the resistance to electrical current flow (Figure 9.6*a*). Here, we substitute the difference between pressure within the eye (IOP) and pressure in the episcleral veins (P_v) for V_1 and V_2, respectively, designate aqueous outflow via the canal of Schlemm as F_s (Figure 9.6*b*) and write the equation as follows:

$$\text{IOP} - P_v = F_s \cdot R \tag{9.1}$$

Rearranging this equation to isolate intraocular pressure as the dependent variable produces

$$\text{IOP} = P_v + F_s \cdot R \tag{9.2}$$

This equation is often rewritten to use the reciprocal of resistance, which is an expression of the ease or "facility" of outflow, that is, $R = 1/C$. In this formulation, C corresponds to the electrical parameter of conductance. Thus,

$$\text{IOP} = P_v + F_s/C \tag{9.3}$$

Simple as it is, this equation has some important implications. First, the outflow resistance is an anatomically dependent variable. It says something about the degree to which the trabecular meshwork, the canal of Schlemm, and the small veins carrying aqueous from the canal to the episcleral veins impede the fluid flow. That is, the anatomy is responsible for the outflow resistance (or facility). Much of the rest of this chapter will be concerned with the structural details of these tissues.

In addition, varying outflow resistance can have large effects on intraocular pressure. As the resistance increases (or facility decreases), the intraocular pressure rises. Conversely, if there is no resistance (facility is very high), which would be the case when there were a large direct connection between the anterior chamber and the episcleral veins, the intraocular pressure will drop to the level of the episcleral venous pressure (normally around 8 or 9 mm Hg).* Therefore, when the intraocular pressure becomes abnormally high, it can be reduced to normal levels by surgical increase of the outflow facility, by alteration of the structure of the limbal tissue, or pharmacologically, if part of the outflow path is both variable and pharmacologically accessible.

Equations 9.2 and 9.3 do not have a term relating to aqueous inflow, but it is easy to add. The outflow through the canal of Schlemm (F_s) must represent the total inflow to the anterior chamber (F_{in}) minus the outflow by the uveoscleral route (F_{us}); that is, $F_s = F_{in} - F_{us}$, which is simply a restatement of the fact that total inflow and total outflow will be equal ($F_{in} = F_s + F_{us}$). Substituting this relationship in Equations 9.2 and 9.3 yields

$$\mathrm{IOP} = P_v + (F_{in} - F_{us}) \cdot R \tag{9.4}$$

and

$$\mathrm{IOP} = P_v + (F_{in} - F_{us})/C \tag{9.5}$$

In this form, these equations point to another way to control intraocular pressure—by changing the rate of inflow, which depends on (and is almost identical to) the rate of aqueous production. Reducing aqueous production will reduce intraocular pressure by reducing aqueous inflow (F_{in}), while increasing aqueous production will increase pressure. The mechanism of aqueous production and strategies for altering the rate of production are discussed in Chapter 11.

These equations describe linear relationships between intraocular pressure and aqueous inflow, as modified by outflow resistance, and one of them (Equation 9.4) has been plotted in Figure 9.7 for several different values of outflow resistance. The lines relating intraocular pressure to inflow rate intersect on the ordinate at a pressure value equal to the episcleral venous pressure (P_v), set here at 8 mm Hg, and their differing slopes have values equal to outflow resistance (R). The uveoscleral flow is assumed to be constant at 0.3 μl/min. (The assumption of constant uveoscleral flow is an approximation, since it may vary somewhat with intraocular pressure. The change is small, however, and the assumption of constancy has little effect.)

Experimental studies in monkey eyes indicate that the relationships plotted in Figure 9.7 are valid over a wide range of intraocular pressures, but at extremely high values (over 50 mm Hg) or low values (less than episcleral venous pressure), the simple linearity between IOP and inflow rate will change. Over most of the range shown in Figure 9.7, however, the linear relationship is a good and useful approximation. The shaded circle on the graph indicates average or "normal" values for flow and intraocular pressure; note that the resistance need change by a factor of only 2.5 (from R = 3.0 to R = 7.5) to move the intraocular pressure well into the upper 5% of the normal pressure range (see Box 9.2).

Some studies have indicated the presence of a small *decrease* in the rate of aqueous production as intraocular pressure increases, indicating not only that

*Pressures are specified in millimeters of mercury (mm Hg), which is the force per unit area exerted by a mercury column *n* mm high. Flow rates are expressed as volume per unit time—specifically, μl / min. Resistance (R) has the units of $[(1/\mu l \cdot min^{-1}) \cdot mm\ Hg]$ and facility (C) has the unit of $(\mu l \cdot min^{-1} \cdot mm\ Hg^{-1})$. Therefore, IOP (mm Hg) = P_v (mm Hg) + F_s $(\mu l \cdot min^{-1}) \cdot R$ $[(1/\mu l \cdot min^{-1}) \cdot mm\ Hg]$. By cancellation, the quantity $F_s \cdot R$ (or F_s/C) has the unit of mm Hg.

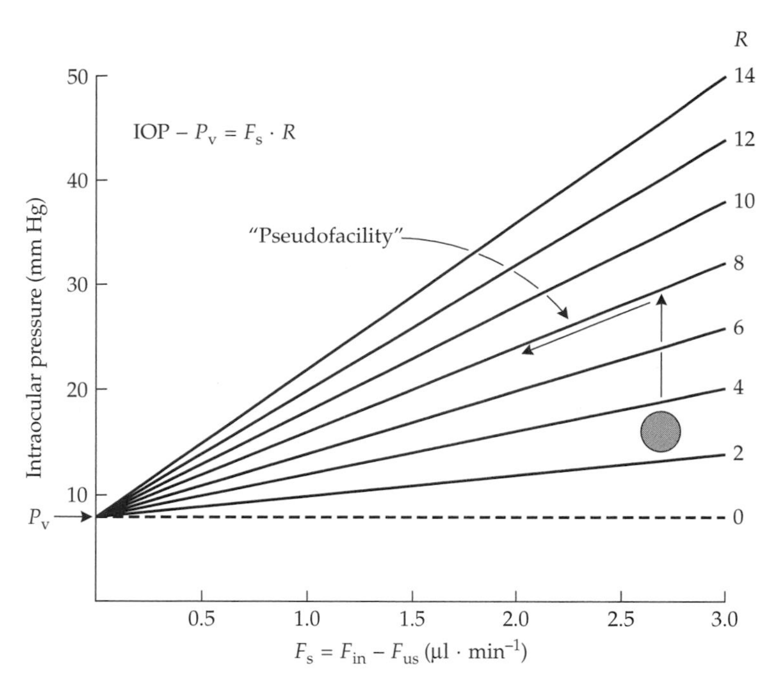

Figure 9.7
Intraocular Pressure as a Function of Outflow and Resistance
The relationship between intraocular pressure and aqueous outflow is represented by a family of curves whose slopes are different values of outflow resistance. When resistance is zero (dashed line), IOP is constant at the same value as the episcleral venous pressure. At the other extreme, high outflow resistance means that IOP changes dramatically with small changes in outflow. The normal values of outflow and IOP lie within the shaded circle. Increasing resistance 2.5 times would cause the IOP to rise, as the upward-pointing arrow shows; if the increased pressure affected inflow, this "pseudofacility" would produce a subsequent decline in pressure, as shown by the arrow sloping down and left.

inflow affects IOP but also that IOP affects inflow. In Equation 9.4, this addition to the relationship would reduce both F_{in} and the term $(F_{in} - F_{us}) \cdot R$. But since this interaction between IOP and inflow rate is not explicit in the equations, it could easily be taken for a reduction in outflow resistance or its equivalent, an increase in outflow facility. It is therefore called *pseudofacility.*

In terms of the relationship between IOP and aqueous inflow, pseudofacility is compensatory, making an increase in IOP less significant than it would be otherwise. For example, if an obstruction in the outflow path increased resistance (decreased facility), the IOP would rise. But in the presence of pseudofacility, the rise in IOP (upward-directed arrow in Figure 9.7) would produce a decline in aqueous production and inflow, and this decline should operate to reduce IOP again (arrow pointed down and left in Figure 9.7).

Although the term "pseudofacility" appears throughout the literature, most recent studies of the relationship between aqueous production and intraocular pressure show that these two variables have little reciprocal interaction unless the intraocular pressure becomes so high as to shut down the blood vessels in the ciliary processes. Such a shutdown, however, is more system collapse than compensatory interaction, and short of this point we are free to regard intraocular pressure as a dependent variable; that is, the rate of aqueous production affects IOP, but IOP does not affect the rate of aqueous production. Or to put it another way, pseudofacility is a pseudoproblem.

Finally, the foregoing equations represent steady-state conditions, in which variables such as inflow rate or episcleral venous pressure are average values

[Box 9.2] Estimating the Pressure Within: Tonometry

THE ONLY DIRECT, accurate method for measuring intraocular pressure is to insert a hollow tube into the anterior chamber and connect the tube to a manometer, a device for measuring pressure in millimeters of mercury (mm Hg). Because it damages the cornea, however, direct manometry is an experimental method that cannot be used with human subjects. Clinical measurements of intraocular pressure are necessarily indirect, and their validity relies on experimental comparison to direct manometric measurements in other species.

The oldest technique for estimating intraocular pressure is "palpation," a fancy word for pushing on the eye with one's finger to see if it feels hard—that is, if it resists the application of pressure. Although high intraocular pressure can be detected in this way, palpation is not sensitive and does not quantify the magnitude of the pressure. These problems have been resolved over many years of development and experimentation, so modern devices for measuring intraocular pressure without invading the eye—**tonometers**—are both sensitive and quantitative. There are several different types of tonometers, but they all measure pressure indirectly by determining the amount of force required to deform the eye by a known amount and using some underlying assumptions to relate this force to intraocular pressure.

Figure 1 shows the working principles of an **applanation tonometer**. The tonometer probe has a flat surface that is pressed against the cornea until a known area of the cornea is flattened, or applanated; the force required to flatten this area can be used to estimate the intraocular pressure. If the cornea were a thin, dry membrane, like a small balloon, the force applied to the eye by the tonometer (F_t) would equal the product of the area flattened (A) and the intraocular pressure (IOP). Thus, $F_t = \text{IOP} \cdot A$, and the intraocular pressure is simply the applied force divided by the area flattened: $\text{IOP} = F_t/A$.

But the cornea is neither thin nor dry. The thickness of the cornea means that it will resist deformation even in the absence of any intraocular pressure; therefore, flattening of the cornea must overcome two opposing forces: the intraocular pressure and the cornea's inherent resistance to deformation (F_c). The tear film covering the corneal surface produces an attractive force (F_s) through surface tension that assists the tonometer probe; the resultant force that flattens the cornea is therefore the sum of the force applied by the tonometer and the corneal surface tension. As summarized in Figure 1*b*, $F_t + F_s = \text{IOP} \cdot A + F_c$. Thus, $\text{IOP} = (F_t + F_s - F_c)/A$.

The forces due to corneal resistance and to fluid surface tension complicate matters, and they are handled somewhat differently by different tonometers. For the Goldmann applanation tonometer, which is the major clinical standard, the size of the applanation area has been selected so that the attractive force (F_s) and the corneal resistive force (F_c) are roughly equal in magnitude. In effect, these forces cancel one another so that the applanation force is directly proportional to the intraocular pressure.

Since the measurements made with a tonometer are indirect, there are several sources of possible error. One is the size of the applanation area, which must be constant to a high degree of accuracy; *Goldmann tonometers,* which are used with a biomicroscope, are designed so that the size of the area can be observed and made always the same. Equating the corneal resistive force and the fluid attractive force is more of a problem. The tonometers have been calibrated so that these forces cancel for normal corneas, but abnormal situations may alter this equation. Abnormally thin corneas, such as are found in keratoconus, may offer less resistance to deformation than do normal corneas, leading to spuriously high estimates of intraocular pressure. Conversely, corneas with higher-than-normal resistance to deformation will cause intraocular pressure readings to be lower than expected. Fortunately, these errors are unlikely to be large (i.e., more than a few millimeters of mercury), and by knowing when over- or underestimates are likely, clinicians can adjust their expectations accordingly.

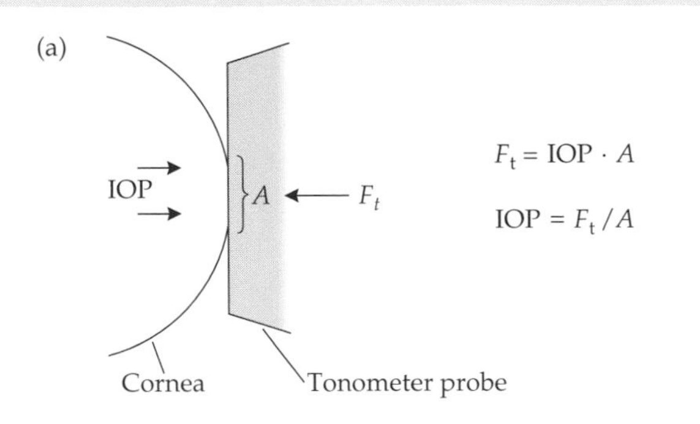

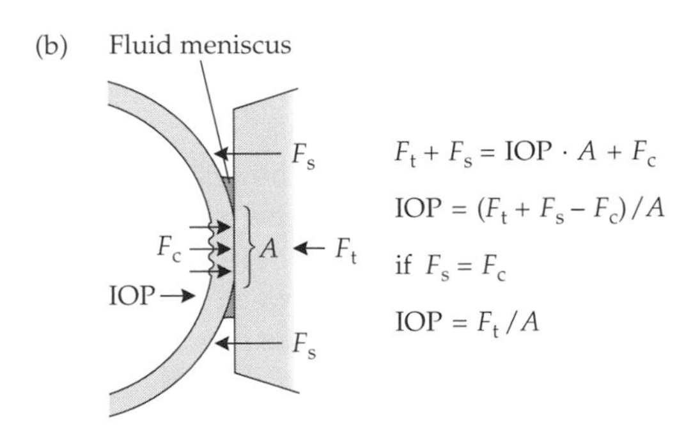

Figure 1
Principles of Applanation Tonometry
(*a*) If the cornea could be regarded as a very thin membrane, the force required to deform it would equal the intraocular pressure divided by the area of the region that was flattened. (*b*) Since the cornea has substantial thickness, it has some mechanical rigidity that resists the effect of the force applied to it by the tonometer probe; the resistive force (F_c) acts in the same direction as the IOP. Another force (F_s), which is due to the fluid meniscus where the probe contacts the cornea, is attractive and acts in the same direction as the applied force of the tonometer probe. These additional forces must be taken into account in the measurement of intraocular pressure.

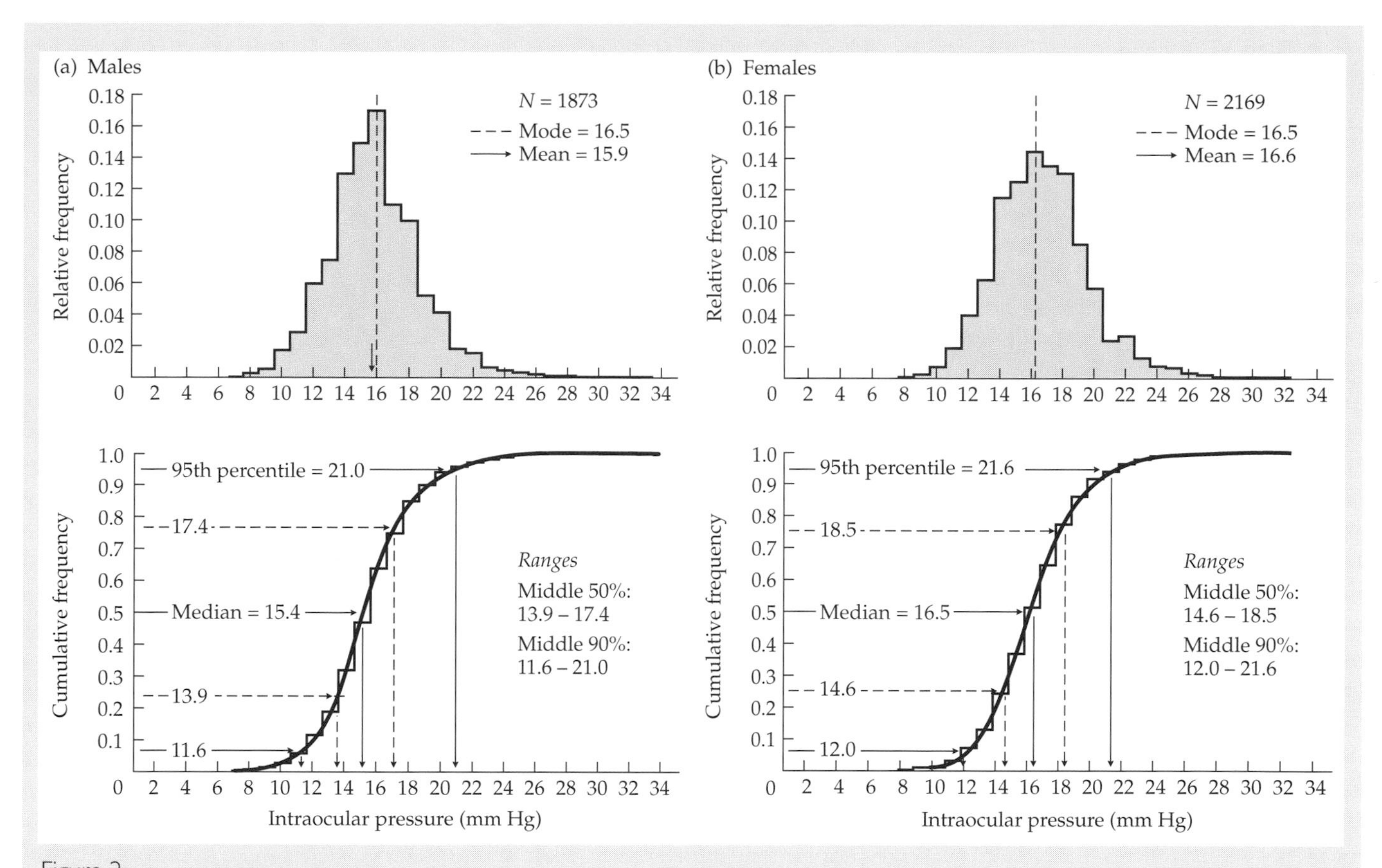

Figure 2
Intraocular Pressures in the Normal Population
(*a, b*) Relative frequency distributions of intraocular pressures in male and female eyes, respectively. The most common value for both male and female eyes was 16.5 mm Hg, but the shapes of the distributions differ; the distribution for males is less symmetric. The corresponding cumulative distribution functions show that the values for females are slightly higher, a difference that is statistically significant. The upper 5% of the population have IOP readings greater than 21 mm Hg and would be classified as ocular hypertensive. (Replotted from Hollows and Graham 1966.)

Although Goldmann tonometers are the most widely used in clinical practice, two other applanation tonometers are currently in use. *Noncontact tonometers* use a puff of air to flatten the cornea and a photoelectric sensing system to detect the flattening; unlike other tonometers, they do not require corneal anesthesia. *MacKay–Marg tonometers* measure the force required to displace a spring-mounted plunger by 10 μm as it touches the cornea. All the sensing is electronic, and these instruments provide a record on graph paper of the pressure measurement. Although MacKay–Marg tonometers were intended to be used without corneal anesthesia, they are more reliable when a topical anesthetic is used.

An older form of tonometer, the *Shiøtz tonometer,* works by indenting the cornea with a small plunger that can be loaded with weights of different magnitude. The Schiøtz tonometer is less accurate than the others, in part because it displaces a relatively large volume of aqueous in making its measurement, but it provides a continuous reading of pressure that the others do not. For certain purposes, such as monitoring short-term changes in intraocular pressure, there is no alternative to the Schiøtz tonometer.

What does an intraocular pressure reading on an individual eye tell us? For example, my optometrist recently measured the intraocular pressure in my eyes; she found values between 14 and 15 mm Hg, said "That's fine," and went on to other things. Why?

She was comparing the readings from my eyes to her expectation of "normal," an expectation based on information about the normal range of intraocular pressures in the population. Intraocular pressure distributions have been studied several times, and two sets of data are shown in Figure 2*a* (males) and 2*b* (females) for samples of about 2,000 eyes each. As the vertical line through the peaks of the distributions (the mode) shows, the most commonly recorded IOP for both men and women was around 16.5 mm Hg, with ranges from 7 to 33 mm Hg (men) and 8 to 32 mm Hg (women). None of the eyes in this study had glaucoma, so these distributions should represent a normal population of Caucasian adults (ages ranged from 40 to 75 years). The fact that the distributions are asymmetric creates some minor problems in using the usual parametric statistics to describe the distributions, but these issues can be circumvented by replotting the data as cumulative frequency distributions (Figure 2*c* and *d*).

In this form, the percentages of the population having different intraocular pressures are more obvious: The median pressure (cumulative frequency = 0.5 or 50%) is 15.4 mm Hg for males and 16.5 mm Hg for females, meaning that half of the population had IOP readings of this magnitude or smaller. It is also clear that most pressures fell into a limited range; the middle 50% of the population (between the 25th and 75th percentiles) had pressures within

3.5 to 4 mm Hg of one another, and the middle 90% (between the 5th and 95th percentiles) fell within 9.5 mm Hg of one another. To put it another way, 95% of the population had intraocular pressures less than 21.6 mm Hg. (The difference between males and females, incidentally, is genuine, though other studies show less difference.) Since my intraocular pressures were close to the median value for males, they were regarded as normal.

Had my intraocular pressures been 6 or 7 mm Hg higher, they would have been regarded differently. Eyes with pressures above 21 mm Hg are classified as **ocular hypertensive**; that is, they are in the high end of the normal range—they have a significantly higher risk of developing glaucoma than eyes with lower pressures, although many never do. And herein lies the problem: The normal range of intraocular pressure overlaps the range of pressures in glaucomatous eyes. One can have glaucoma with a "normal" intraocular pressure, or one can have chronically high intraocular pressure *without* having glaucoma.

For an individual eye, an intraocular pressure up to about 30 mm Hg has no predictive value. For the population, however, the incidence of glaucoma increases dramatically with elevated pressures; this is why clinicians pay so much attention to IOP measurements. An elevated intraocular pressure is an alerting signal; the *chance* that the eye will develop glaucoma is significantly higher than for an eye with a normal pressure. Responsible clinicians monitor the visual status of patients with high IOP readings as if glaucoma were a certainty; by so doing, they can detect change and begin therapy early in the course of the disease if it occurs.

over extended periods of time. Venous pressure can vary in the short term, however, affected by such things as posture; the measured IOP will be different depending on whether a patient is seated or lying down. Tight collars may also elevate IOP by impeding venous return and elevating the episcleral venous pressure (this is one of my numerous excuses for not wearing a necktie).

Another factor affecting IOP that is not included in the flow equations is the variation in arterial pressure with the cardiac cycle or respiration. The arterial pulse within the ocular tissues is reflected as small synchronous fluctuations in the intraocular pressure riding on somewhat longer fluctuations produced by variation in mean arterial pressure with respiration (Figure 9.8). In most clinical measures of IOP, however, these small variations are effectively averaged out. There are also long-term diurnal variations in mean intraocular pressure, probably reflecting an underlying circadian rhythm in mean arterial pressure, but other factors are likely to be involved. Like other diurnal rhythms, the time of day at which IOP is at a maximum can vary widely among individuals. This variation means that a single IOP reading in the high end of the normal range is difficult to interpret; it could be the low point in a cycle that is abnormally high on average or the high point in a normal IOP cycle.

The Anatomy of Aqueous Drainage

The scleral spur is an anchoring structure for parts of the limbus and the ciliary body

In a histological section, the scleral spur is an unimpressive projection—a spur—of scleral tissue extending inward and forward in the limbus (Figure 9.9*a*). It is just posterior to the canal of Schlemm and often appears as more of a bump than a spur. A better way to think about the scleral spur is shown in Figure 9.9*b*, where it appears in a three-dimensional drawing in relative isolation from other structures (with the iris and ciliary body removed). The scleral spur is really a thickened ridge of the sclera running around the apex of the anterior chamber angle.

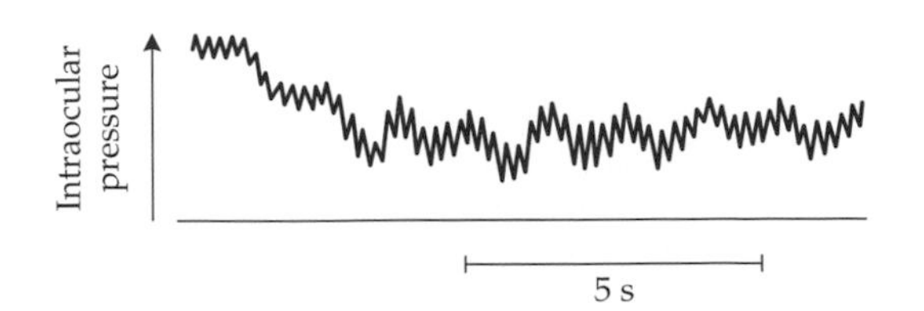

Figure 9.8
Short-Term Variations in IOP
This direct manometric pressure recording from a rabbit eye shows small, short-term variations in intraocular pressure with respiration and the arterial pulse. (After Fatt and Weissman 1992.)

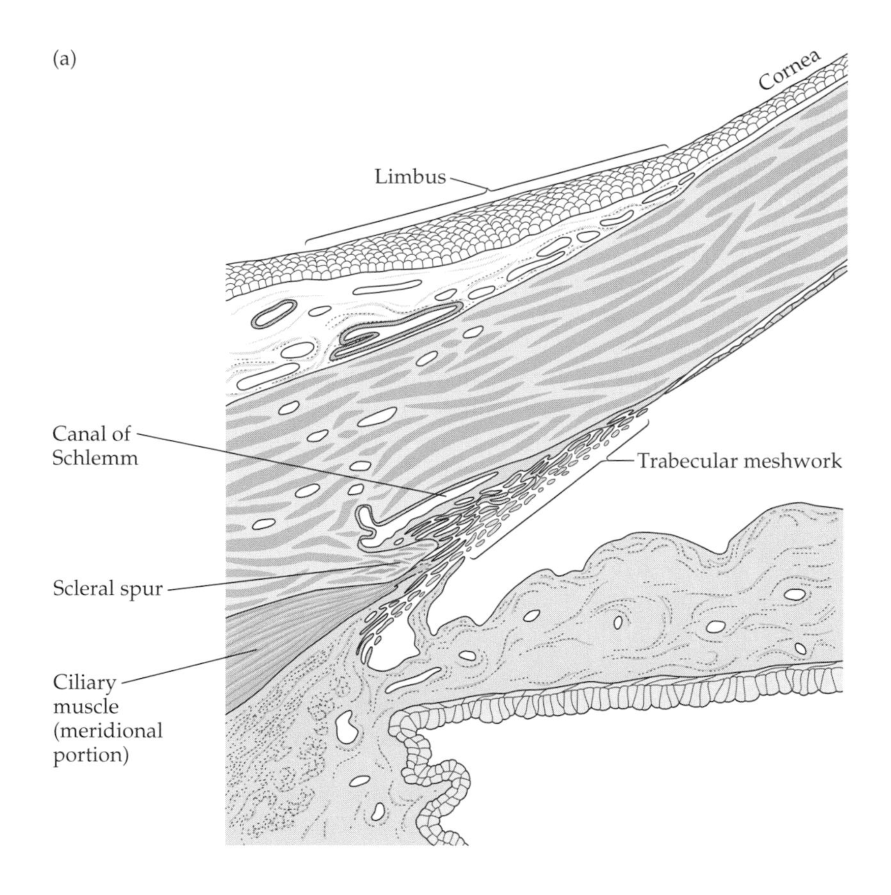

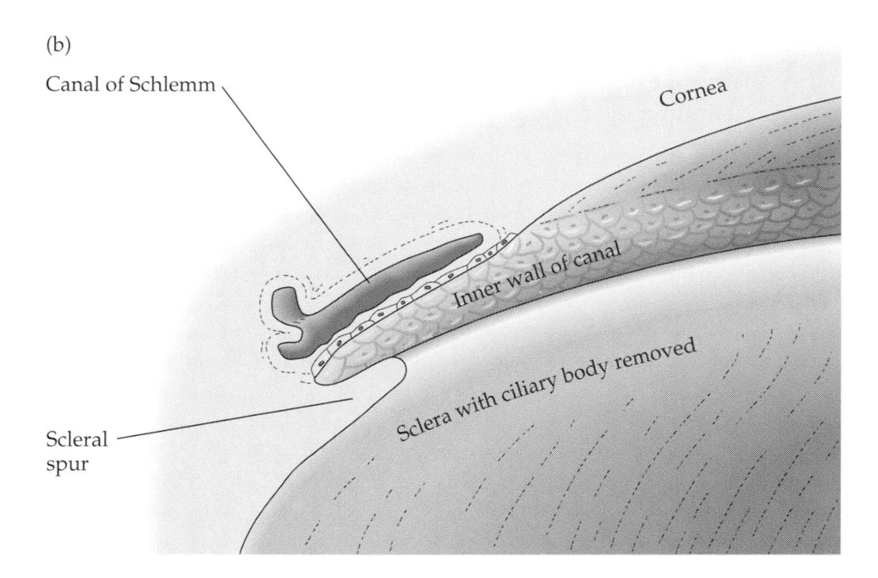

Figure 9.9
The Scleral Spur
(*a*) In a histological section, the scleral spur projects like a thumb inward and behind the canal of Schlemm. Strands of the trabecular meshwork attach to its anterior surface, while fibers of the meridional portion of the ciliary muscle fuse with its posterior surface. (*b*) In a schematic three-dimensional view with other structures removed, the scleral spur is seen as a thickened ridge of scleral tissue encircling the angle of the anterior chamber. (*a* after Hogan, Alvarado, and Weddell 1971.)

And since the spur is a thickened part of the sclera, it should also be a region that is more rigid than other parts of the sclera.

The function of the scleral spur is basically mechanical; it is a good place to attach other things. Most of the fibers of the ciliary muscle insert into (i.e., fuse with) the scleral spur on its posterior surface, as do some of the cords of the trabecular meshwork (see the next section) on its anterior surface. This point is relevant to the problem of aqueous flow. If the scleral spur were to be deformed as the ciliary muscle contracts, the deformation might affect the trabecular meshwork by changing tension on the trabecular cords arising from the spur or by altering the size of the lumen in the canal of Schlemm or both. In other words, contraction of the ciliary muscle could alter outflow resistance, insofar as the resistance is due to the trabecular meshwork.

One study of this problem compared the morphology of the scleral spur and trabecular meshwork in eyes that had been treated with pilocarpine prior to enucleation with that of a control group that had no pilocarpine treatment (pilocarpine is a parasympathomimetic agent that will produce contraction of the ciliary muscle). The force produced by contraction of the ciliary muscle bends the scleral spur posteriorly somewhat, and this distortion acts to open up the trabecular meshwork (see Figure 9.19). We will come back to the question of what this action does to aqueous outflow later; for the moment, the results indicate that the scleral spur cannot be considered a totally rigid structure.

Not all of the ciliary muscle fibers insert onto the scleral spur. Some muscle fibers or connective tissue extensions from the fibers terminate within the trabecular meshwork anterior to the scleral spur. Assuming that the muscle fibers or their extensions are firmly attached to the cords of the trabecular meshwork, this feature is another way in which contraction of the ciliary muscle could exert tension on the meshwork and alter its structure. Again, how this might affect outflow resistance remains to be considered.

The trabecular meshwork is made of interlaced cords of tissue extending from the apex of the angle to the margin of the cornea

As Figure 9.10 shows, the trabecular meshwork is well named; it is a mesh formed by strands or cords of tissue with open spaces between the strands. There are two parts or regions within the trabecular meshwork. The **uveal portion** lies closer to the anterior chamber, and its cords arise from the uveal tract—specifically, from the root of the iris and the ciliary body. In addition, some large strands from the anterior surface of the iris merge with the uveal portion of the meshwork; these are **iris processes** (see Figure 9.10). About 100 of these processes are scattered around the circumference of the chamber angle. With the excep-

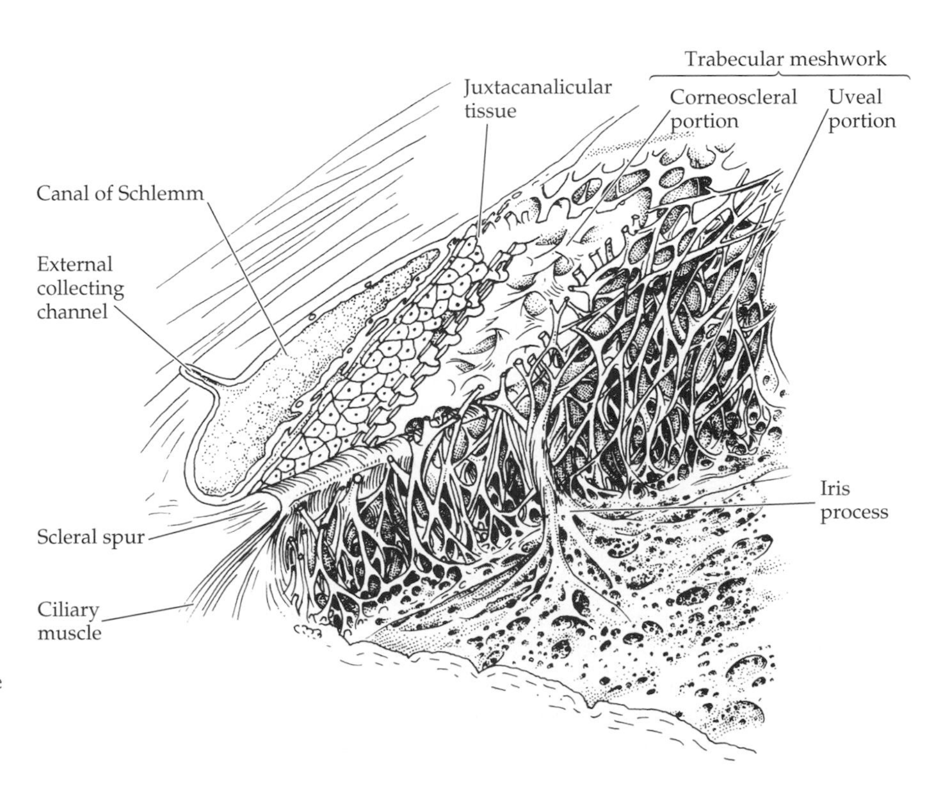

Figure 9.10
The Trabecular Meshwork
Slender cords forming the uveal portion of the trabecular meshwork lie closest to the chamber angle, where they are joined by an occasional extension of tissue from the iris (an iris process). Cords deeper in the meshwork—the corneoscleral portion—are broader and flatter, while the open spaces are fewer and smaller. The canal of Schlemm is separated from the trabecular meshwork by the juxtacanalicular tissue, which is a loose assemblage of endothelial cells, fibroblasts, and collagen fibers.

tion of the iris processes, the cords in the uveal portion are slender, having large gaps between the interlaced cords. Although fluid moving through this part of the meshwork must follow a somewhat circuitous route, and thereby encounter some resistance to flow, the resistance should be relatively slight; the spaces in the meshwork range up to 100 μm or so in diameter.

Fluid should encounter more resistance in the **corneoscleral portion** of the meshwork. These cords attach to the scleral spur and end near the termination of Descemet's membrane of the cornea (see Chapter 8); they are broad and flat relative to the uveal cords, and there is less open space between them. It may be more appropriate to think of the corneoscleral portion of the meshwork as several layers of tissue sheets with holes punched in the sheets (see Figure 9.10). The holes are not in register, so fluid flowing through a hole in one sheet must flow laterally before it encounters a hole in the next sheet through which it can pass. On average, the holes in this part of the meshwork are about half the size of the spaces in the uveal portion.

Schwalbe's ring separates the trabecular meshwork from the cornea

The peripheral edge of the cornea can be defined very precisely in a histological section such as that in Figure 9.11; where Descemet's membrane disappears, the cornea ends and the limbus begins. The limbal tissue next to the cornea, interposed between the cornea and the anteriormost strands of the trabecular meshwork (arrows), is **Schwalbe's ring**. The tissue of Schwalbe's ring is similar to the rest of the limbal stroma, consisting mainly of fibroblasts and collagen fibers and covered on the surface next to the anterior chamber by a lining of endothelial cells that are continuous with the corneal endothelium.

Schwalbe's ring has no particular functional significance except as an insertion region for strands of the trabecular meshwork. Its main importance is as a clinical landmark; it is the most anterior of the limbal structures visible with gonioscopy (see Box 9.1) and the last structure to disappear in more severe grades of angle closure. If Schwalbe's ring is not visible, the angle is truly closed.

The trabecular cords have a collagen core wrapped with endothelial cells

Although the size and shape of the trabecular cords change throughout the meshwork, becoming broader and more straplike in the corneoscleral part, they all have a similar construction. As Figure 9.12 shows, the central core is a bundle of collagen fibers running parallel to the axis of the cord; some of the collagen is coiled, however, making the cords elastic and capable of being stretched when force is applied to them. The collagen core is surrounded by a sheath of endothelial cells, whose basement membrane separates them from the core.

The endothelial cells form a continuous surface that covers all the cords in the meshwork, and this endothelial surface is in contact with aqueous flowing through the meshwork. The trabecular endothelium is also continuous with the corneal endothelial cells through the endothelium covering Schwalbe's ring (see Figure 9.11), and the two endothelia have some properties in common. Like their corneal counterparts, trabecular endothelial cells are broad and thin, they have anchoring junctions with their neighbors, and they can both change their shape and move around. It is not clear if their number declines with age, as happens with the corneal endothelium, but there is little evidence for mitotic activity.

One of the more striking differences between corneal and trabecular endothelium is the phagocytic activity of the trabecular cells. They are very good at ingesting melanin pigment granules and other particulate debris; this characteristic has lead to descriptions of the trabecular meshwork as a self-cleaning filter.

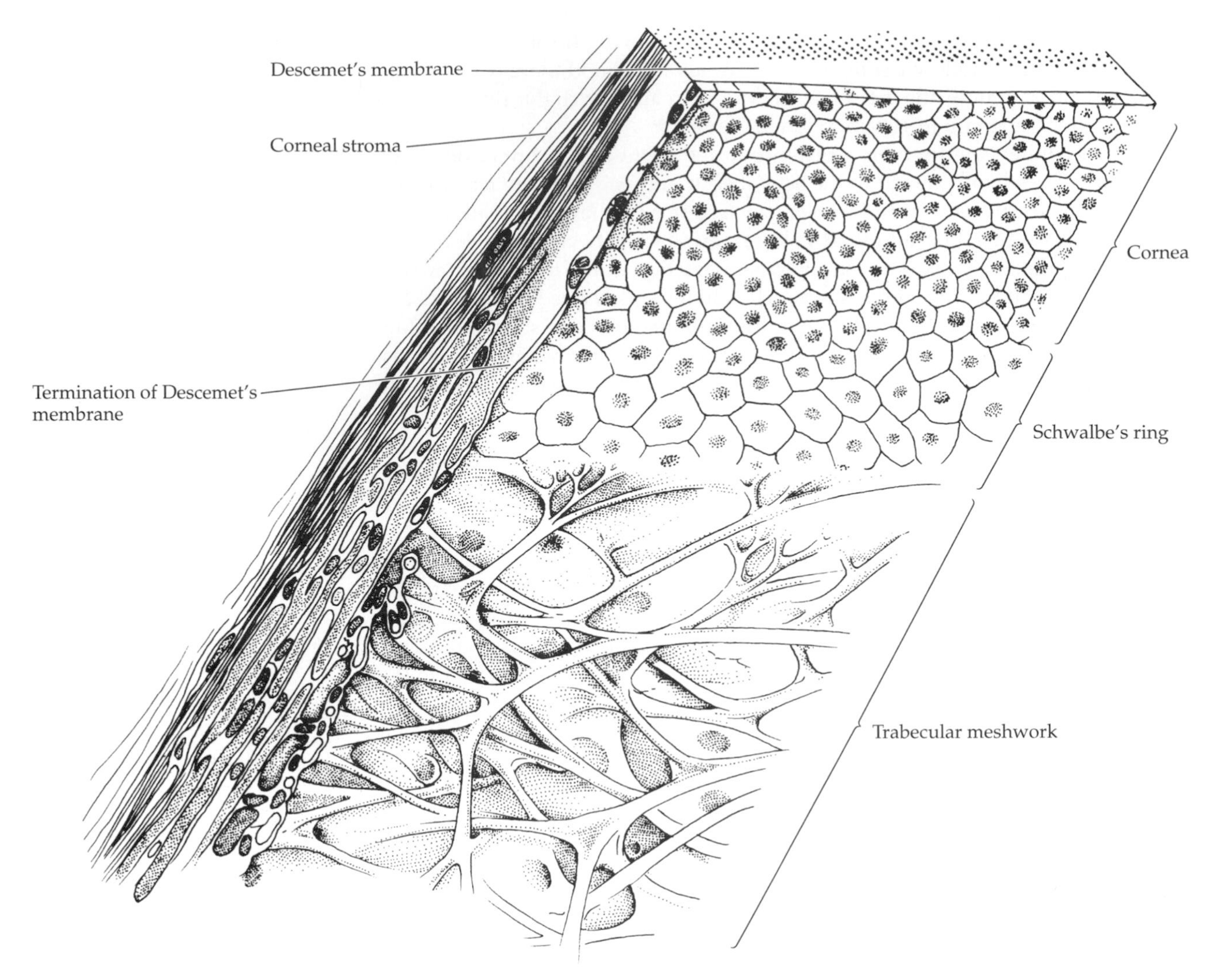

Figure 9.11
Transition from Cornea to Limbus: Schwalbe's Ring
The edge of the cornea in a histological section is defined by the termination of Descemet's membrane; beyond that, the tissue is part of the limbus. The transition zone between the edge of the cornea and the first cords of the trabecular meshwork is Schwalbe's ring. Although it is histologically undistinguished, it can be a prominent landmark in gonioscopy.

It is less obvious that the cells can digest and break down the material they ingest, however, and it has been suggested that restrictions to aqueous flow result in some instances because the endothelial cells have become bloated with ingested material they cannot dispose of. This retention of ingested material may also account for the increased pigmentation of the trabecular meshwork with age (see Figure 2 in Box 9.1).

The trabecular meshwork cells, which include some fibroblasts, as well as the endothelial cells, are responsible for synthesizing proteoglycans found in the meshwork. The presence of proteoglycans means that the spaces in the meshwork are not completely empty; part of the space will be occupied by a proteoglycan gel. How much space the proteoglycans occupy is not known—in the extreme case they completely fill the trabecular spaces—but our impression of resistance to fluid flow based on the size of the anatomical space is probably too conservative. The presence of the proteoglycans, even if they only wrap each cord in a sheath, will make the space for free fluid flow smaller and the resistance higher. Detailed studies of the extracellular matrix in the trabecular meshwork have begun, and the full story of these components and their functional role is unfolding. The presence of the proteoglycans raises the possibility of modifying outflow resistance with agents that alter the proteoglycan matrix.

The major source of outflow resistance is the juxtacanalicular tissue separating the canal of Schlemm from the trabecular spaces

More than 30 years of electron microscopy has failed to reveal any direct connections between the lumen of the canal of Schlemm and the fluid-filled spaces of the meshwork. In other words, the aqueous must pass a barrier of tissue to enter the canal, and since this is the only obstruction of its kind in the entire outflow system, it is likely to be the dominant site of resistance to outflow.

The relationship between the canal and the nearest part of the trabecular meshwork is shown at high magnification in Figure 9.13. There are several layers of cells between the trabecular spaces and the lumen of the canal, two of which are the trabecular endothelial cells and the endothelium of the canal; other cells come between these two endothelial layers. The name most often used for these cell layers is the **juxtacanalicular tissue**, a regrettable term meaning "the tissue lying next to the canal" (the terms "cribriform layer" and "endothelial meshwork" have also been used).

The juxtacanalicular tissue is a mélange of endothelial cells, fibroblasts, fine collagen fibrils, and a proteoglycan matrix. The cellular components often are connected by tight junctions, and the endothelial cells lining the canal have zonulae occludens.

There are, in other words, structures that will resist flow between cells; the tight junctions are not of the highly restrictive type, however, and fluid flow between cells is not impossible. So while this tissue does not form an impregnable barrier to aqueous movement, it certainly makes flow more difficult; the final barrier seems to be the endothelium that lines the canal itself.

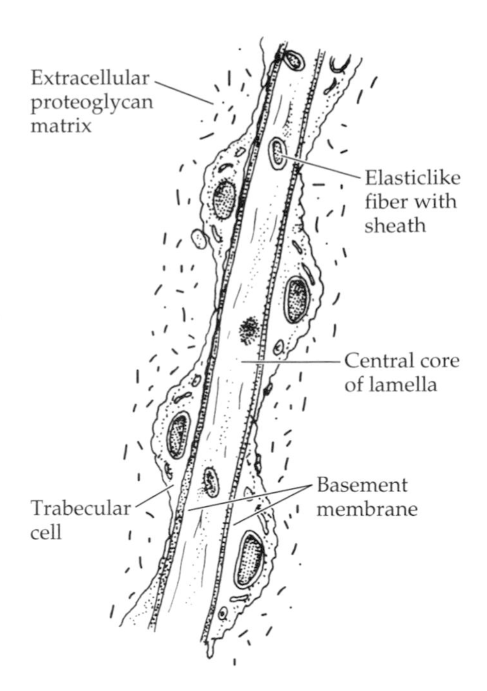

Figure 9.12
Structure of Trabecular Cords
The trabecular cords have a central core of collagen fibers with a few elastic fibers as well. The collagen core is wrapped by a sheath of trabecular endothelial cells and their basement membrane. A proteoglycan matrix of unknown thickness surrounds each of these cords.

The canal of Schlemm encircles the anterior chamber angle

The first complete picture of the canal of Schlemm is relatively recent, produced by the injection of neoprene rubber into the canal to make a casting of it. This feat has been repeated very few times; the original picture produced by Ashton in 1951 is redrawn here as Figure 9.14. When viewed from the front as it is in Figure 9.15, the complete circle formed by the canal is clear, as are small, short branches on the inner and outer edges of the circle. The canal does not always have a

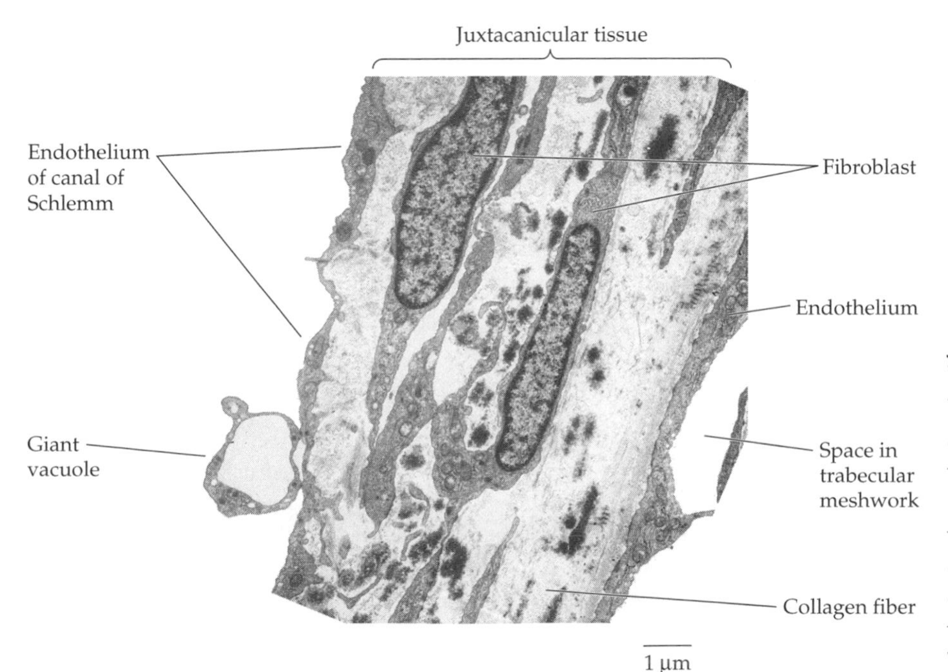

Figure 9.13
Juxtacanalicular Tissue and the Inner Wall of the Canal of Schlemm
The inner wall of the canal of Schlemm has an endothelial lining, as do the spaces in the trabecular meshwork. These two endothelial layers are separated by the juxtacanalicular tissue, which is mainly a loose arrangement of fibroblasts and collagen. A large portion of outflow resistance will be generated across this tissue. (From Hogan, Alvarado, and Weddell 1971.)

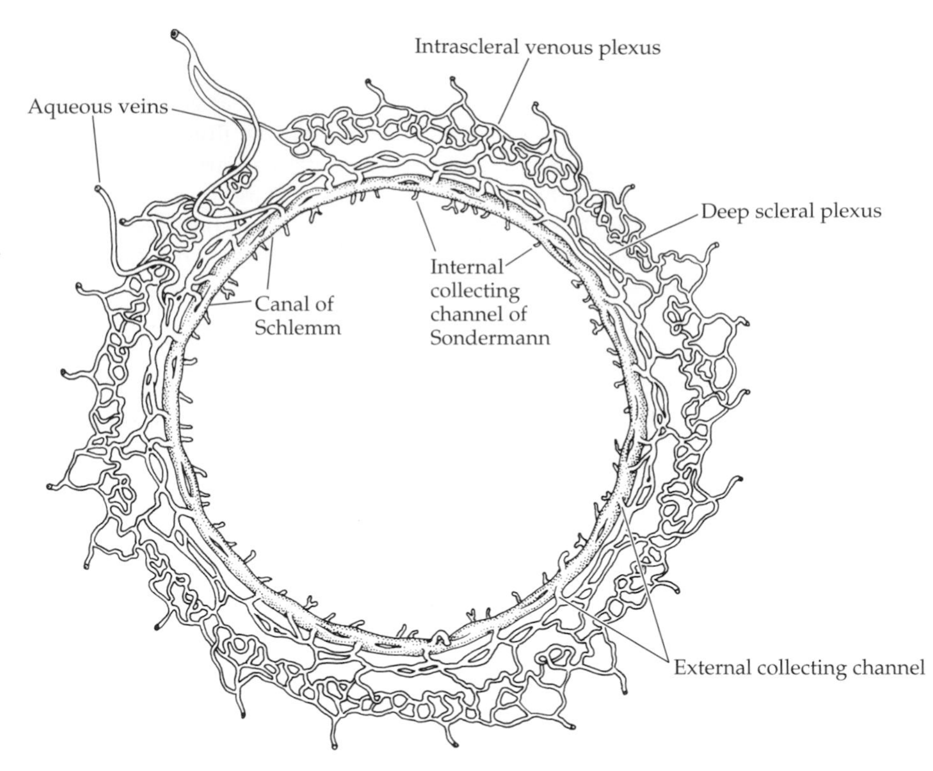

Figure 9.14
The Canal of Schlemm
The canal is a large annular vessel encircling the angle of the anterior chamber; its lumen sometimes divides into two parts. The inner side of the canal, which is closest to the anterior chamber, has small extensions, the internal collecting channels (of Sondermann); their ends are closed and are not continuous with spaces in the trabecular meshwork. Larger external collecting channels extend from the outer surface of the canal, through which aqueous is drained into small veins in the limbal stroma and ultimately into the episcleral veins. An aqueous vein is an external collecting channel that bypasses the limbal veins, carrying aqueous directly to the episcleral veins. The limbal and aqueous veins are found all around the circumference of the canal; only some of them are shown here. (After Hogan, Alvarado, and Weddell 1971, based on Ashton 1951.)

single lumen; occasionally it divides into two parts, diverging and then reuniting like a river flowing around an island in the stream. Although these divisions might reduce the ease with which fluid can move from one place to another in the canal (as some authors suggest), free, unimpeded flow in the canal is more likely the normal situation.

Aqueous enters the canal all along its inner circumference and probably also through the small inner branches (these are the **internal collecting channels** or the **canals of Sondermann**). Once in the canal, aqueous soon leaves by way of the **external collecting channels**, which are the small branches on the outer circumference. The two routes by which aqueous can go from the external collecting channels to the episcleral veins will be considered later; both routes offer little resistance to fluid flow.

The internal collecting channels are more of a problem; a few investigators suggest that they are artifacts generated by the experimental techniques used to study the anatomy of the canal. Most of the evidence, however, indicates that the negative result—failure to see the internal channels—is the real anomaly. Other questions relate to the function of the internal collecting channels: Does aqueous enter these small tributaries of the canal by the same mechanism by which it enters the canal proper, do they offer an easier access than through the wall of the canal, or are the "collecting" channels just small functionless diversions from the flow of aqueous into and around the canal?

The internal collecting channels provide no direct connections between the canal and the spaces of the trabecular meshwork; that is, their ends within the trabecular meshwork are closed and any fluid entering the connecting channels must pass through some tissue. Whether aqueous enters the channels by exactly the same process by which it enters the canal is less certain, but there is no compelling reason to regard aqueous inflow to the channels as qualitatively different from aqueous inflow to the canal itself.

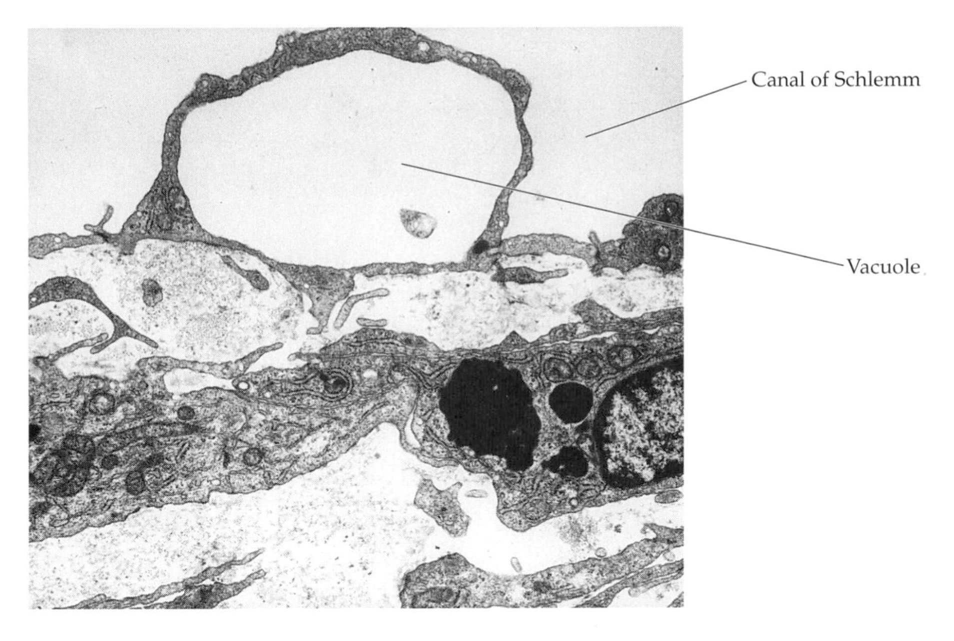

Figure 9.15
Giant Vacuoles in the Endothelium Lining the Canal of Schlemm
An electron microscopic section through the inner wall of the canal of Schlemm shows a large endothelial bulge into the lumen of the canal as the endothelium encloses a space, presumably a fluid-filled bubble. This blister-like structure is a giant vacuole. The fluid probably entered the vacuole through a pore on the side of the endothelium that is next to the juxtacanalicular tissue. (From Hogan, Alvarado, and Weddell 1971.)

Aqueous enters the canal of Schlemm by way of large vacuoles in the endothelial lining of the canal

Electron micrographs of the inner wall of the canal of Schlemm show large spaces surrounded by endothelium bulging into the lumen of the canal (Figure 9.15). These spaces have been interpreted from the outset as structures similar to blisters—accumulations of fluid that will be released when the blister breaks. This impression is strengthened by seeing them with scanning electron microscopy, which seems to show some of the blisters as if they have just burst, leaving small holes in the lining of the canal. Similar holes, or "pores," can also be seen when the inner endothelium of the canal is viewed from the trabecular meshwork side.

There are several reasons for thinking that these blisters, usually called **giant vacuoles**, have something to do with the entry of aqueous into the canal of Schlemm. They are found almost exclusively on the inner wall of the canal next to the trabecular meshwork, which is the side of the canal from which aqueous must enter. In addition, their number varies directly with intraocular pressure—the higher the pressure prior to removal of the tissue for histology, the greater the number of giant vacuoles. (Rearranging Equation 9.2 with flow as the independent variable produces the equation $F_s = (1/R) \cdot (\text{IOP} - P_v)$. Increasing IOP should increase outflow, and if outflow is related to the number of vacuoles, that number should also increase.) Finally, vacuole formation fits nicely with the conception of outflow being a passive, pressure-dependent phenomenon; the mechanism should be insensitive to cellular metabolic activity and undiscriminating in its removal of aqueous components. This is brute force removal, and the presence of vacuoles is consistent with it.

But there are questions about the nature of the vacuoles and the mechanism of their formation and rupture. For example, is a vacuole an intracellular bubble of aqueous that has accumulated within the cytoplasm of a single endothelial cell, or is it an extracellular fluid bubble that one or more cells have managed to surround? Do cell membranes have to rupture to release aqueous into the canal or is it a matter of membranes separating from one another to allow the fluid to escape from a compartment? Does the endothelium of the canal have a sponge-like structure such that the vacuoles—the spaces in the sponge—are fairly permanent features of the endothelial wall that increase and decrease in size as circumstances dictate?

The most secure answer is that vacuoles are not intracellular inclusions; at very high magnification, one can see the vacuole surrounded by cell membrane, meaning that it is not inside the cell as a cytoplasmic bubble. It looks very much as if the endothelial cell has surrounded a droplet of aqueous that has percolated through the juxtacanalicular tissue and will release it into the canal by reversing the process by which the droplet was surrounded.

A schematic version of the capture and release of aqueous by the endothelial cells (Figure 9.16) begins with the formation of a small invagination on the basal side of the endothelial cell. The invagination—presumably filled with aqueous—enlarges and deepens, extending toward the lumen of the canal. Eventually, the membrane lining the invagination will come in contact with the cell membrane lining the canal; at this juncture we must imagine that the membranes fuse, joining inner to outer membrane, then open at one point, creating a small pore through which the aqueous can flow into the canal. And since both the basal and the luminal surfaces have pores, aqueous can flow directly from the juxtacanalicular tissue spaces to the lumen of the canal. The flow should continue as long as the pores on both sides of the endothelium remain open.

After the contents of the vacuole enter the canal, the process continues with membrane fusion to close the basal pore, followed by a progressive elimination of the invagination emptying into the canal. This process restores the endothelial cell to its initial condition and sets the stage for another repetition of the cycle.

Although this scenario fits with anatomical facts, as it was meant to do, it is an interpretation, showing a dynamic process reconstructed from anatomical snapshots. We don't know exactly what is happening, since there is no way to see the events as they occur. For example, small pores and vacuoles may always be present, expanding as needed to allow fluid movement into the canal and then shrinking again. Thus, the cycle of pore appearance and disappearance shown in Figure 9.16 may be an exaggeration. (The number and size of the pores appears to increase with increasing intraocular pressure, but this observation does not rule out the possibility that some pores are always present.)

The constant presence of pores and channels in the endothelium leads to the notion of the tissue having a structure much like that of a compressed sponge, with a complicated arrangement of pores and vacuoles providing a series of circuitous routes for aqueous to move from the spaces in the juxtacanalicular tissue into the canal of Schlemm. As fluid pressure builds up on the trabecular side of the system, the vacuoles should inflate and the pores should expand, thus permitting greater fluid flow.

To make matters more difficult still, it has been suggested that the pores are artifacts of tissue processing, meaning they are irrelevant or, at the other extreme, that the pores are too numerous and too large to account for the normal level of outflow resistance; this also says that pores have little or nothing to do with outflow.

Although opinions differ among the experts, I have taken the anatomical observations at face value; that is, there are vacuoles, there are pores in one or

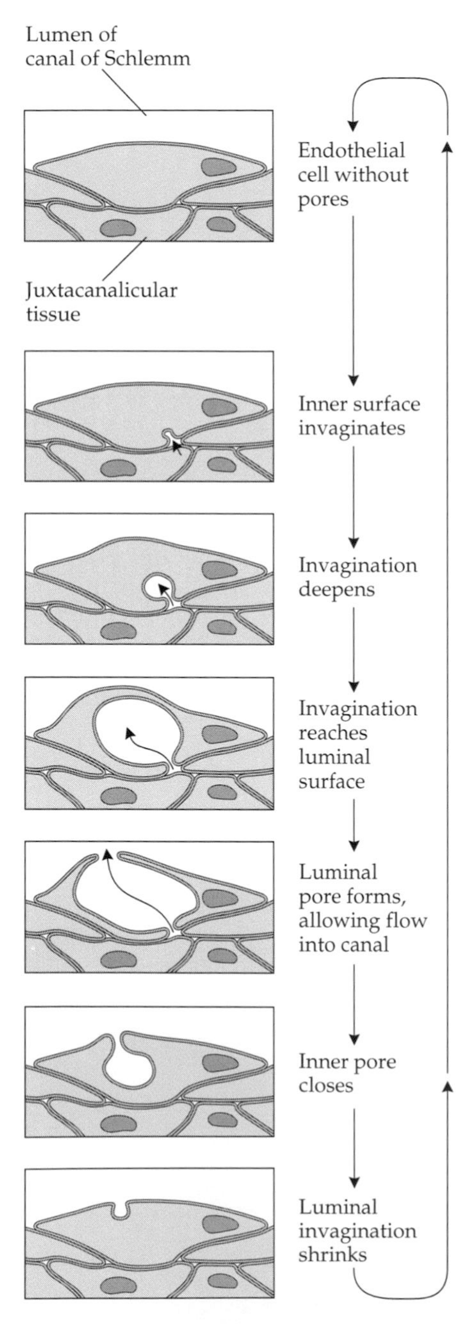

Figure 9.16
Movement of Aqueous across the Endothelium Canal of Schlemm
The sequence begins at the top; the lumen of the canal of Schlemm is uppermost, separated from the juxtacanalicular tissue by an endothelial cell. Aqueous begins to move across the endothelial barrier when the basal surface of the cell invaginates. As the invagination deepens and expands, it becomes a large, fluid-filled vacuole that is surrounded by the endothelial cell except at the basal pore. Eventually, a pore opens on the luminal surface, releasing aqueous from the vacuole into the lumen of the canal. Closure of the basal pore and shrinkage of the vacuole returns the cell to its original condition, ready for another repetition of the cycle. (After Tripathi 1971.)

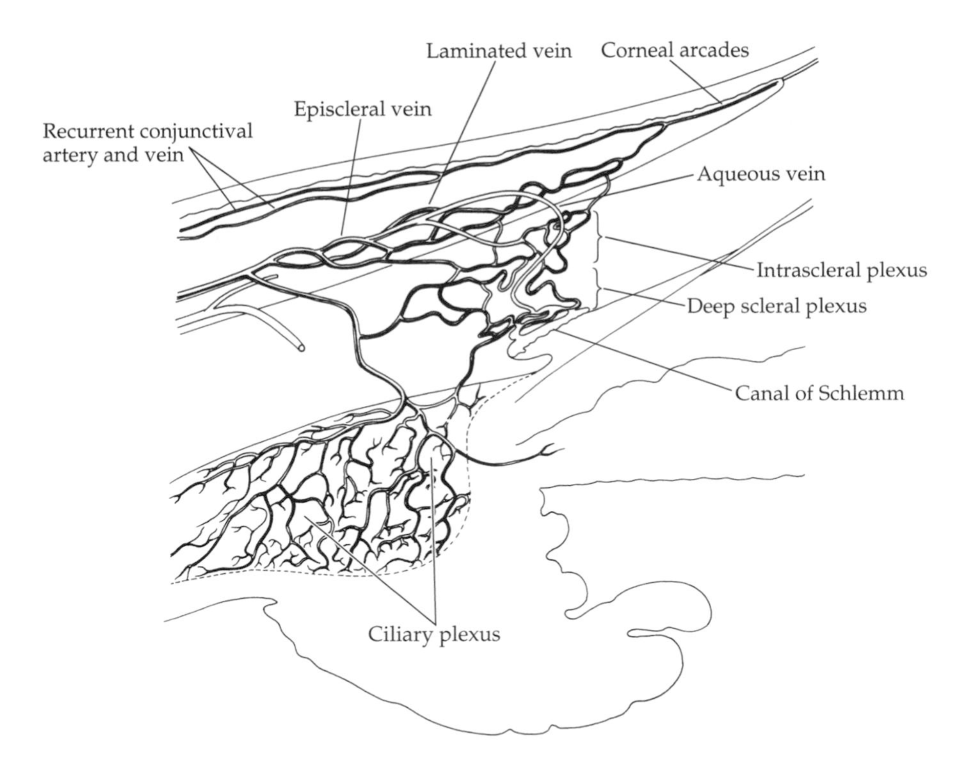

Figure 9.17
Drainage of the Canal of Schlemm and the Limbus
The main route of aqueous outflow is through the external collecting channels of the canal to small veins in the limbal stroma; these veins constitute the deep scleral plexus. From here, aqueous flows to a more superficial set of anastomotic veins, the intrascleral plexus, and finally to the episcleral veins. Both the deep scleral and the intrascleral plexuses contain blood as well as aqueous, since they are connected to the extensive ciliary plexus in the ciliary body. Aqueous veins either leave the canal of Schlemm directly or arise very close to it in the deep scleral plexus, carrying clear aqueous to the episcleral veins.

both surfaces of the endothelial cells, and the numbers and sizes of vacuoles and pores wax and wane with increasing and decreasing intraocular pressure. Although tissue preparation for electron microscopy can be harsh and the possibility of artifact is always present, the observations have been repeated by different investigators using somewhat different methods. Moreover, the pores, which have been particularly controversial, do not look like artifacts; the cell membranes are continuous around their edges, for example, making them look like genuine holes rather than places where the cell has pulled apart by shrinkage or mechanical stress during processing.

Since we cannot watch an endothelial cell go through the cyclic appearance and disappearance of pores that are depicted in Figure 9.16, we can only guess at the truth. The interpretation given here draws on the natural abilities of endothelial cells to move, to change their shape, and to engulf particles. One can imagine the cells trapping and expelling droplets of aqueous because these are simple elaborations of things we already know the cells can do.

Aqueous drains out of the canal into venous plexuses in the limbal stroma

The external collecting channels (see Figure 9.14) convey aqueous from the canal to the venous drainage system; they are not likely to offer any significant resistance to aqueous flow. A venous plexus, the **deep scleral plexus**, encircles the limbus just anterior to the canal of Schlemm (Figure 9.17); this plexus is the destination of the aqueous carried in the external collecting channels. From here, the aqueous moves toward the surface into the **intrascleral plexus** and finally to the episcleral veins. Both the deep and the intrascleral plexuses also carry blood, mostly from the anterior portion of the ciliary body and iris root, so aqueous mixes with blood very quickly after leaving the canal of Schlemm by this route. This path is the major route of aqueous drainage from the canal.

The **aqueous veins** observed by Ascher short-circuit the venous plexuses, arising directly from the canal or the deep scleral plexus near the canal and running directly to the episcleral veins (see Figure 9.17). Because the aqueous in

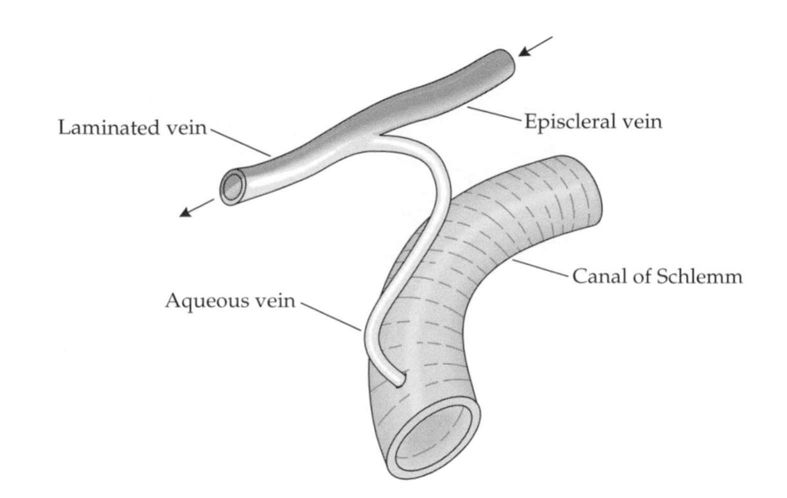

Figure 9.18
Formation of a Laminated Vein
The junction of an aqueous and an episcleral vein creates a laminated vein, one that appears to have two parallel stripes within it. The stripes are blood and aqueous that have not mixed.

these veins has not mixed with blood, they are transparent and difficult to see. When an aqueous vein joins an episcleral vein, however, the aqueous and blood do not mix immediately, creating what is called a **laminated vein**, one in which unmixed streams of aqueous and blood flow side by side (Figure 9.18).* Most eyes have a small, variable number of aqueous veins (and therefore laminated veins) that constitute a minor drainage route from the canal of Schlemm. Their existence is fortunate, for it allowed Ascher to deduce the existence of aqueous flow, but they are not known to have any special functional significance.

Pilocarpine reduces intraocular pressure, probably by an effect of ciliary muscle contraction on the structure of the trabecular meshwork

Pilocarpine is one of the oldest of the drugs used to treat glaucoma and one of the few that acts on the outflow system. An acetylcholine agonist, pilocarpine mimics the effect of the natural transmitter on smooth muscle, producing both ciliary muscle contraction and pupil constriction. Its therapeutic effect is a reduction in intraocular pressure.

Pilocarpine could lower intraocular pressure in three ways: It might reduce aqueous production, reduce outflow resistance secondary to pupil constriction, or reduce outflow resistance secondary to ciliary muscle contraction. Contraction of the iris sphincter can be ruled out because pilocarpine can reduce intraocular pressure in an eye that has had a complete iridectomy. In addition, of the few studies of pilocarpine and aqueous production, the most recent indicate that if pilocarpine has an effect, it is small and in the direction of *increased* production, which should elevate IOP. But the effect of pilocarpine on IOP disappears after disinsertion of the ciliary muscle (in which the ciliary muscle is surgically detached from the scleral spur all around the circumference of the limbus). Finally, several studies have concluded that contraction of the ciliary muscle, whether brought about by accommodation, by nerve stimulation, or by drugs, will reduce the resistance to aqueous outflow. This lowering of resistance is thought to be accomplished by distortion of the scleral spur (Figure 9.19) or by direct effects on the trabecular meshwork and juxtacanalicular tissue by the ciliary muscle fibers that insert there.

*Mark Twain mentioned a similar phenomenon in *Huckleberry Finn*. In Twain's day, the clear water of the Ohio River failed to mix with the muddy Mississippi for 100 miles downstream from the mouth of the Ohio. Clarity isn't everything, however. The "Child of Calamity," one of Twain's keelboatmen, opined that mud—like blood—is more nutritious: "Trees won't grow worth shucks in a Cincinnati graveyard. . . . It's all on account of the water the people drunk before they laid up. A Cincinnati corpse don't richen a soil any."

It is not entirely clear how ciliary muscle contraction alters the trabecular meshwork and juxtacanalicular tissue. Pulling on the trabecular cords from the direction of the scleral spur should change the *shape* of the trabecular spaces, but it is not obvious that it will change their *size*. (Imagine pulling on a rubber sheet with circular holes in it; if you pull in only one direction at a time, the circles will become ellipses, but their areas will not change significantly.) Therefore, this

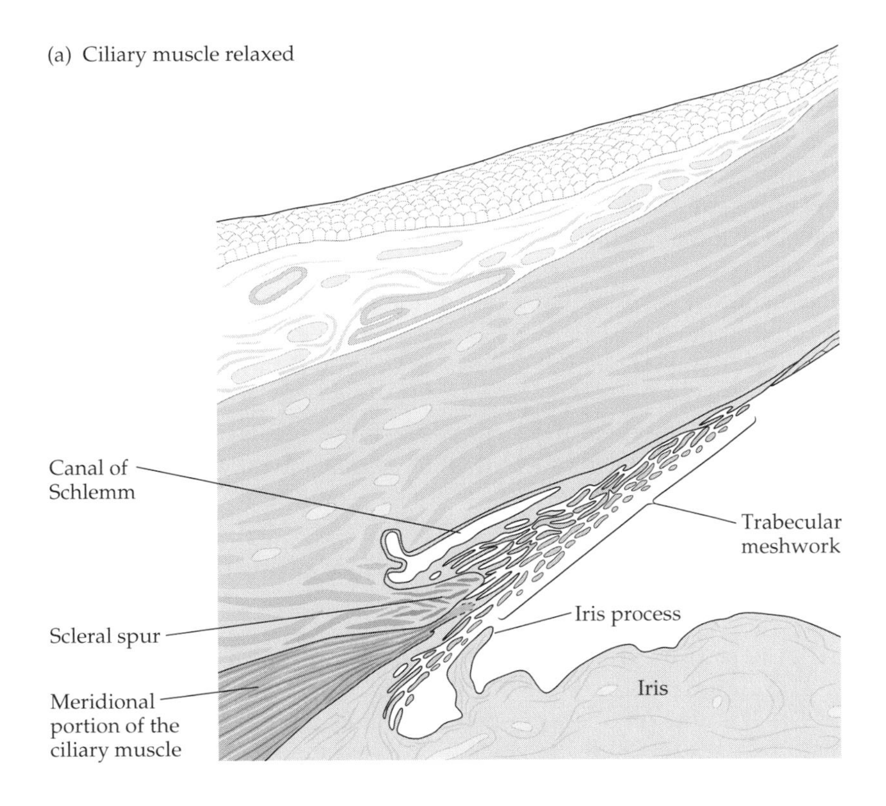

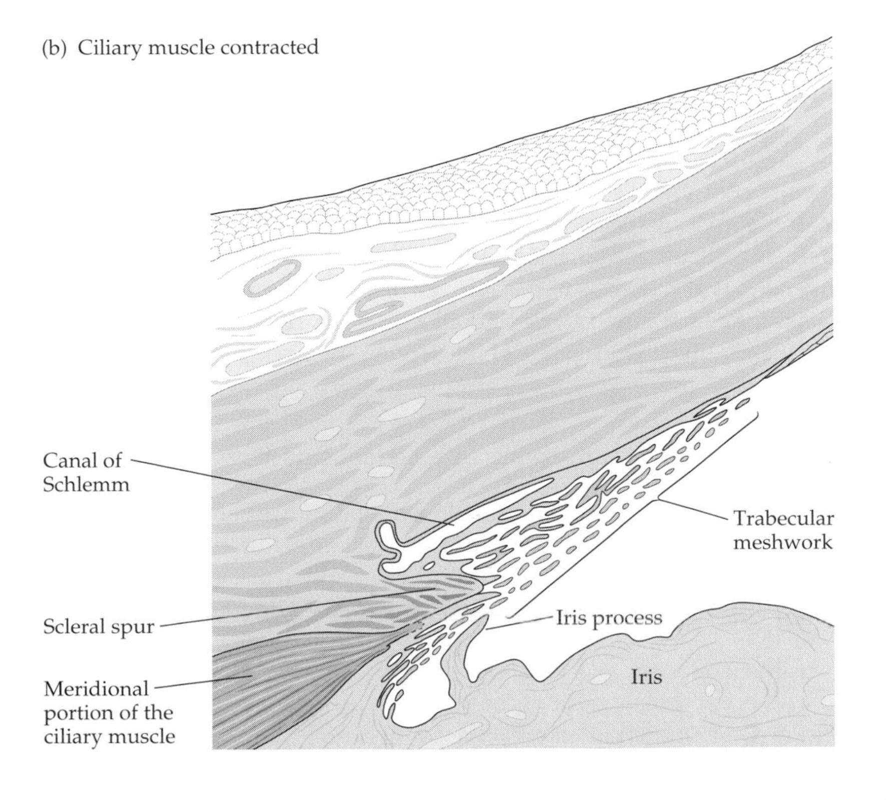

Figure 9.19
Ciliary Muscle Contraction and the Trabecular Meshwork
(*a*) When the ciliary muscle is relaxed, the scleral spur points toward the center of the anterior chamber and the sheets of the trabecular meshwork press against one another. (*b*) Strong contraction of the ciliary muscle bends the scleral spur back from its normal position. This bending may pull back and inward on the strands of the meshwork, creating more separation between the trabecular sheets and reducing outflow resistance.

kind of change in the trabecular meshwork geometry would not be expected to reduce outflow resistance significantly. A more likely possibility is increased separation between the layers of cords and sheets in the trabecular meshwork, as Figure 9.19 suggests; here, the bending of the scleral spur pulls back and *inward* on the meshwork, causing the trabecular sheets to separate. If we assume that something similar occurs in the juxtacanalicular tissue, where most of the outflow resistance resides, reduced outflow resistance is a reasonable consequence.

The reduction in outflow resistance and IOP with ciliary muscle contraction is useful when it occurs (it doesn't always work), but it should not be given much additional significance. The presence of ciliary muscle fibers or their extensions (sometimes called the ciliary muscle tendon) within the trabecular meshwork is fortuitous and therefore unlikely to be purposive. The number of ciliary muscle fibers entering the meshwork is a small proportion of the total, and they are probably strays whose number and organization vary from eye to eye. Their attachment to the trabecular meshwork instead of the scleral spur is unlikely to affect ciliary muscle action or to exert any consistent, significant effect on the resistive nature of the meshwork within the normal range of accommodation. The bending of the scleral spur will also be slight at normal accommodative levels and may vary among individuals depending on the prominence of the spur, which is itself variable.

In other words, the effect of ciliary muscle contraction on the trabecular meshwork may be important clinically, but under normal operating conditions, ciliary muscle contraction has little to do with aqueous outflow.

An effective way to reduce intraocular pressure seems to be to increase the uveoscleral outflow

The effort to reduce intraocular pressure has a new tool: The drug latanoprost (PhXA41) is a prostaglandin analog that can reduce intraocular pressure by about 25%. It is now routinely prescribed in combination with timolol (an agent that reduces aqueous production by the ciliary body; see Chapter 11).

Latanoprost is thought to reduce intraocular pressure by increasing uveoscleral outflow. This default conclusion is based on lack of evidence for any increased outflow through the trabecular meshwork and canal of Schlemm or for any decrease in aqueous production, leaving only the uveoscleral route as a possible explanation for the effect of latanoprost. Why a prostaglandin analog would alter the uveoscleral outflow in this way is not known.

Although latanoprost has been widely accepted as valuable in controlling intraocular pressure, it produces an anatomical change in some patients: a darkening of the iris at its periphery. There are at least two possible explanations for the darkening. One is migration of iris melanocytes (see Chapter 10) into the peripheral part of the iris. Another is a *loss* of cells and collagen from the periphery, making the iris less opaque and allowing one to see more of the darkness behind the iris. But the mechanism underlying iris darkening is not known, nor are its implications.

Surgery for glaucoma aims to increase aqueous outflow

Surgery for open-angle glaucoma is difficult, normally done as a last resort after other methods have failed to control progression of the disease. (Surgery for closed-angle glaucoma usually involves the iris and is discussed in Chapter 10.) Now that less invasive laser-based procedures for reducing outflow resistance are available, the laser surgeries are usually tried first.

In **trabeculoplasty**, a laser is used to burn a series of small (about 50 μm) holes in the trabecular meshwork; the holes do not penetrate to the canal of Schlemm. The burn spots may be placed all around the limbus, in which case there may be 80 to 100 of them, or half those numbers if only 180° of the limbus is included in the operation. The laser beam is directed to its target and viewed through a modified gonioscopy lens.

In essence, trabeculoplasty is an attempt to lower outflow resistance produced by the trabecular meshwork by punching holes through it. The procedure is usually successful in the short term; almost three-fourths of the patients have postsurgical reductions in IOP. Unfortunately, the effect is not long-lasting in most cases—within 5 years or less, the intraocular pressure regresses back to its preoperative, elevated level. The reason for the regression is not fully understood, but it is likely to be the trabecular meshwork's attempt to repair itself—to patch the holes—by endothelial cell migration and reconstruction of trabecular cords. If the meshwork's capacity for repair were greater, the regression of the IOP after surgery would probably be faster.

Where trabeculoplasty approaches the outflow problem through the trabecular meshwork, from the inner side, direct surgical procedures come from the outside. In the standard **trabeculectomy**, which has several variations, the surgeon first opens a small flap of sclera overlying the canal of Schlemm and then removes a small block of the trabecular meshwork anterior to the canal. When the scleral flap is sutured back in place, the incisions are not tightly closed; the scleral flap is deliberately made to be leaky. Aqueous can now flow through the large hole in the trabecular meshwork into the scleral flap, from which it can leak into the loose connective tissue of the overlying conjunctival stroma. The leaking aqueous creates a fluid-filled bulge or "filtering bleb" in the conjunctiva from which the fluid is taken up by the episcleral vessels.

The success of this and other surgical procedures is very difficult to specify; there are differing definitions of success, differing surgical procedures, and differing conditions for which the surgery is done. Moreover, this is a last-resort strategy that may be performed too late to lower intraocular pressure significantly. Even so, most patients have an immediate reduction of intraocular pressure into the normal range, and the reduction continues for at least 5 years in somewhat more than half of the cases. The point of the surgery, however, is not just to lower IOP but to halt any progressive visual field loss by doing so; significant, maintained reductions in intraocular pressure generally meet this objective.

The outer surface of the limbus is covered with episcleral tissue and a heavily vascularized conjunctiva

With the exception of the superficial blood vessels into which aqueous humor drains, all of the limbal structures discussed so far are buried within the limbus and cannot be seen by looking directly at the eye from in front. From this viewpoint, only the superficial layers are visible, and it is very difficult to say where the limbus ends and the sclera begins. For clinical purposes, the limbus is defined as an annulus of tissue, 1.5 mm wide, surrounding the cornea. Thus defined, the limbus contains all of the structures considered earlier, even if they cannot be seen directly.

Figure 9.20 shows the relationship of this clinically defined limbus to the anatomical limbus. The imaginary dividing line between the cornea and the limbus is defined by the terminations of Bowman's layer and Descemet's membrane; this line is very close to the transition from transparent cornea to white, opaque limbal tissue. The posterior limit of the limbus—the transition from limbus to sclera—is less clear-cut, but it is determined by a line running from the apex of the chamber angle out to the surface where there is a distinct thinning of the conjunctival epithelium. To put it another way, the epithelium of the limbal conjunctiva is somewhat different from the epithelium of the bulbar conjunctiva covering the sclera, and we can use that difference to define the posterior limit of the limbus. Viewed from in front, this transition is invisible even with a bio-

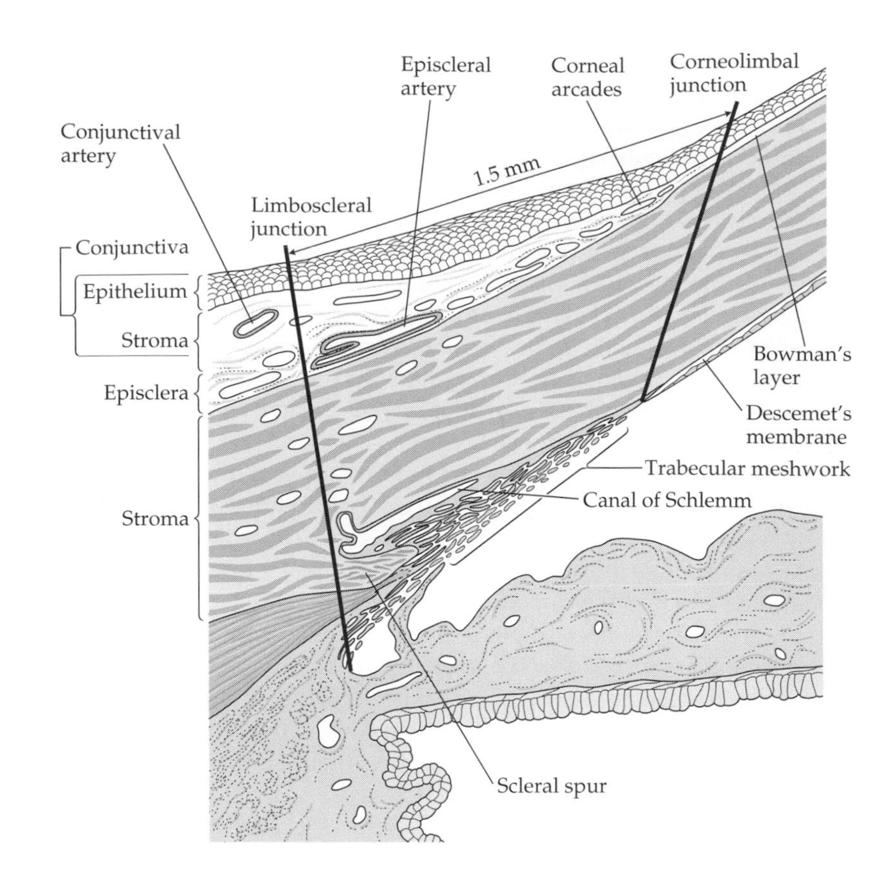

Figure 9.20
Limits and Structures of the Limbus
The boundary between cornea and limbus is marked by a line drawn through the terminations of Bowman's layer and Descemet's membrane, neither of which continues into the limbus. The boundary between limbus and sclera is less well defined, but it is shown here as a line drawn from the apex of the chamber angle to the region of transition between limbal and bulbar conjunctivae. From the outer surface inward, the layers of the limbus are the limbal conjunctiva, with its epithelium and stroma, the episclera, and the limbal stroma. The innermost structures are the canal of Schlemm, the scleral spur, and the trabecular meshwork.

microscope, but it lies about 1.5 mm from the margin of the cornea—hence the clinical definition of the limbus as a 1.5 mm wide annulus around the cornea.

The limbus contains all the aqueous drainage apparatus: scleral spur, trabecular meshwork, canal of Schlemm, and the venous plexuses lying within the limbal stroma. The limbal stroma is continuous with the corneal stroma on one side and with the scleral stroma on the other and represents a region of transition between the two; the limbal stroma is less regular than the corneal stroma but more regular than the scleral stroma. The superficial layers of the limbus are the **episclera**, which lies next to the limbal stroma and is an extension of the sclera's episclera into the limbus, and the limbal conjunctiva. The episclera consists of loose connective tissue, with the usual complement of collagen fibers and fibroblasts, and numerous blood vessels, which are branches of the episcleral arteries and veins.

The limbal conjunctiva, the outermost tissue of the limbus, has the same two parts found in other regions of the conjunctiva: an epithelium some eight to ten cells thick and a stroma that merges with the episclera. (It is important to make a clear distinction between the *limbal* stroma and the stroma of the limbal *conjunctiva;* they are quite different.) The epithelium of the limbal conjunctiva is continuous with the epithelium of the bulbar conjunctiva on one side and with the corneal epithelium on the other; it has more cell layers than other regions of the conjunctiva and more than the corneal epithelium. In addition, the limbal conjunctival epithelium lacks the goblet cells that are found in the bulbar and fornix conjunctival epithelia (see Figures 7.10 and 7.11).

The importance of the limbal conjunctival epithelium is primarily its relationship to the cornea; as discussed in Chapter 8, the basal cell layer of the limbal epithelium contains stem cells that differentiate and migrate into the corneal epithelium to replenish the cornea's complement of basal cells. Large-scale damage to the limbal conjunctiva may eliminate a source of new cells for the corneal epithelium, thereby jeopardizing the ability of the epithelium to renew and repair itself. Such severe damage may be offset and repaired by grafting of limbal conjunctiva from the other eye or from another person into the damaged region.

The stroma of the limbal conjunctiva contains numerous blood vessels supplied by or draining into the episcleral vessels. **Recurrent conjunctival arteries** carry blood outward from the cornea into the bulbar conjunctiva, which is drained by a parallel set of recurrent conjunctival veins. This arrangement creates two layers of vessels overlying the limbal stroma: the episcleral arteries and veins in the episclera, and the more superficial conjunctival vessels.

The other set of blood vessels in the limbal conjunctival stroma consists of those running in toward the margin of the cornea to form the **superficial marginal plexus**, from which the **corneal arcades** arise. The relationship of the corneal arcade vessels to the episcleral and conjunctival vessels is shown in Figure 9.17.

The vessels running to and from the corneal margin travel through radially oriented channels in the stroma of the limbal conjunctiva; the stromal channels are separated from one another by thickened ridges of epithelium. Collectively, these stromal channels, with their associated blood vessels and epithelium, are called the **palisades of Vogt**. (There are also small lymphatic vessels running parallel to the blood vessels in the palisades, but they have not been shown in Figure 9.20.) Although the palisades of Vogt are a distinctive anatomical feature of the limbal conjunctiva adjacent to the cornea, their functional importance resides solely in the corneal arcade vessels running within them; the corneal arcade vessels are the closest thing the cornea has to a blood supply, but they probably have little impact on the central cornea.

Development of the Limbus

The anterior chamber is defined by the iris growing between the developing cornea and lens

After the lens vesicle has separated from the surface ectoderm (see Chapter 1), it moves steadily back away from the surface, leaving a space between the lens and the future cornea. This space will enlarge as the eye grows and will become divided into the anterior and posterior chambers.

With the exception of the corneal epithelium, most of the structures around the anterior chamber are derived from neural crest cells that first migrate into the mesoderm around the optic cup and then move into the developing anterior segment of the eye. There are three major waves of neural crest cell migration (Figure 9.21). First, cells move in behind the surface ectoderm—the future corneal epithelium—and differentiate to form an endothelial lining for the space in front of the lens. The cells in this first migration become not only the corneal endothelium, but also the endothelial lining of the trabecular meshwork and Schwalbe's ring. Cells in the second migration move into the space just in front of the lens vesicle, forming a thin sheet of cells that will give rise to the iris stroma. This migration delimits the anterior and posterior chambers by drawing a line between them; the anterior chamber is the space in front of this cell layer, and the posterior chamber lies behind it. In the third and final phase, cells migrate into the cornea between the epithelium and endothelium; these cells will become the corneal fibroblasts.

Once the anterior chamber is defined by the developing cornea and iris, its growth is simply a matter of the space increasing in size as the eye grows. And since the eye continues to enlarge for 10 or 12 years after birth, the anterior chamber continues to enlarge as well. At birth, therefore, the structure and definition of the chamber angle are not quite complete; the scleral spur will enlarge and the iris root will shift posteriorly with respect to the spur.

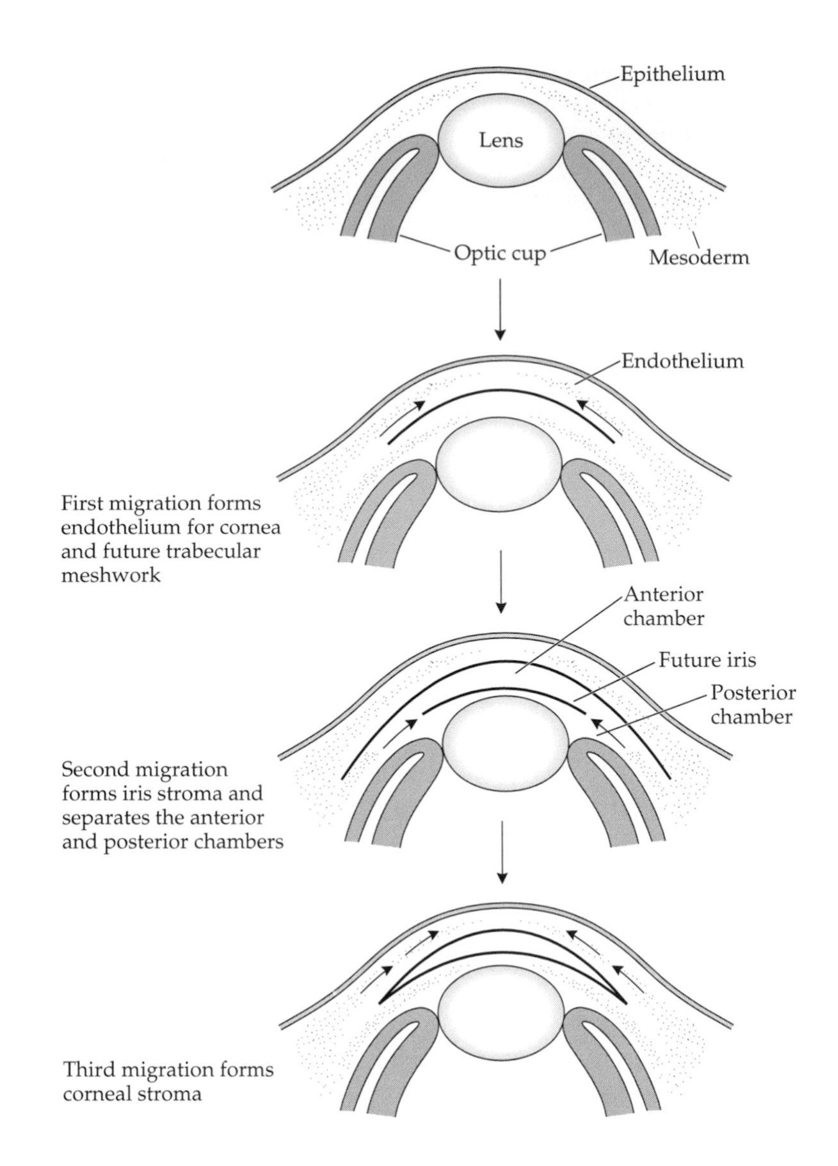

Figure 9.21
Formation of the Anterior Chamber
The structures that delimit the anterior chamber first appear as neural crest cells and migrate into the area from the surrounding mesoderm. The first migration forms the anterior boundary of the chamber by forming the corneal endothelium; the second migration forms the posterior boundary by forming the iris stroma. By putting the iris in place, the second migration also separates the anterior and posterior chambers. In a third phase of migration, the neural crest cells move into the cornea, providing the future fibroblasts for the corneal and limbal stroma. (After Tripathi, Tripathi, and Wisdom 1996.)

The angle of the anterior chamber opens during development as the root of the iris shifts posteriorly

At the end of the first trimester, the anterior chamber has become relatively large, but the apex of the angle has yet to appear; there is no sharp intersection between the iris and the future limbus, as if the space of the future angle were filled with undifferentiated tissue (Figure 9.22*a*). The future canal of Schlemm is present as a small blood vessel, however. As the tissue in this region differentiates—part of it contributing to the trabecular meshwork and part to the iris stroma—the scleral spur begins to enlarge, pushing inward behind the canal. These events have the effects, first, of beginning to define the structures of the outflow system and, second, of pushing the iris root farther back along the developing trabecular meshwork (Figure 9.22*b*).

As growth continues, more and more of the trabecular meshwork appears anterior to the iris root, as if the angle were progressively changing from closed to open (Figure 9.22*c*). Shortly before birth, the angle is open to the level of the scleral spur, but there is as yet no ciliary tissue between the iris root and the trabecular meshwork (Figure 9.22*d*); that is, the ciliary band seen gonioscopically in adult eyes with open angles has yet to appear. Otherwise, all the adult structures in the angle are present and functional at birth.

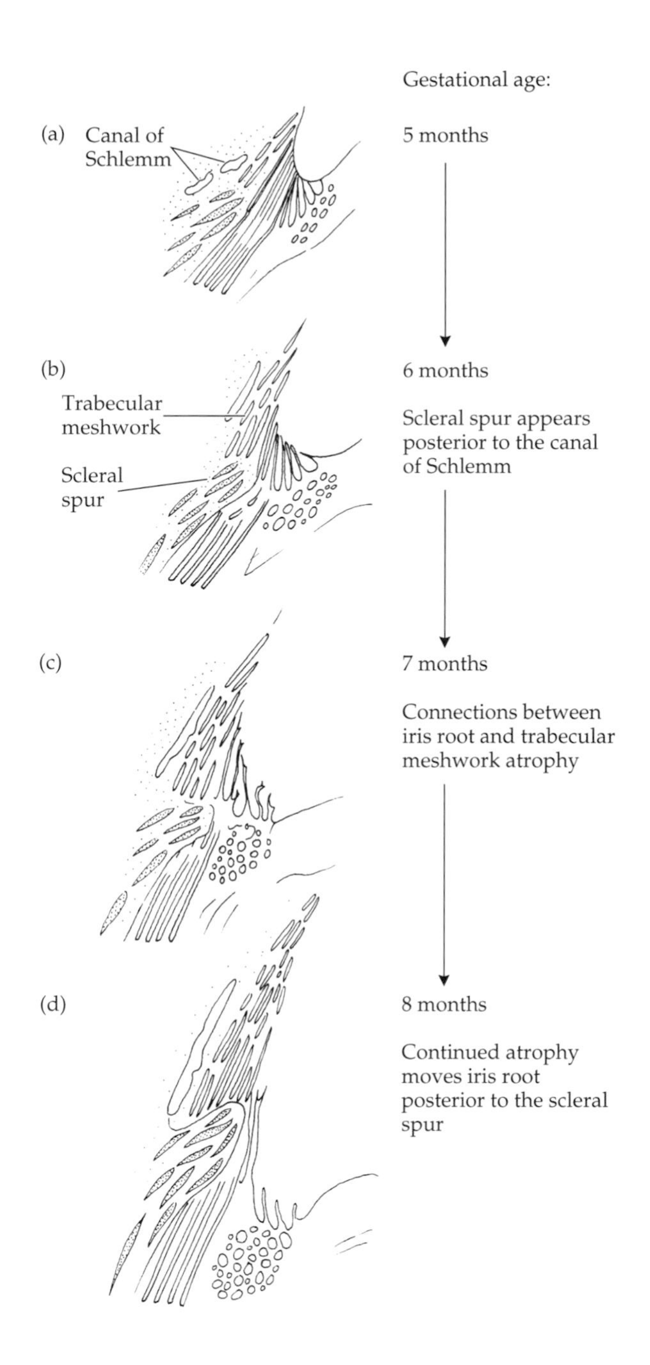

Figure 9.22
Formation of the Anterior Chamber Angle
(*a*) At the fifth month of gestation, the canal of Schlemm and the trabecular meshwork have begun to form, but there is no scleral spur. (*b*) One month later, the scleral spur has begun to protrude inward just posterior to the canal and trabecular meshwork. (*c*) By 7 months of gestational age, the connections between the iris and the trabecular meshwork are disappearing, either by atrophy or by reorganization of the tissue (see text). (*d*) At term, the scleral spur completely separates the ciliary muscle and the trabecular meshwork, and its position is now anterior to the root of the iris.

The changes at the apex of the angle depicted in Figure 9.22 are described there as "atrophy"—that is, disappearance of tissue joining the developing iris and trabecular meshwork; the apex of the angle is exposed and the angle opens as the tissue disappears. Moreover, the connections between iris and trabecular meshwork are regarded as a transient structure called the fetal tendon of the ciliary muscle, or the pectinate ligament. Although describing development of the chamber angle as atrophy of the pectinate ligament is useful for depicting the sequence of changes, this description may not accurately reflect the mechanisms involved. Specifically, the pectinate ligament may not be a true transient structure that is built up only to disappear, like the hyaloid artery, but rather an early stage in the development of the uveal part of the trabecular meshwork. If so, opening of the chamber angle is not a matter of atrophy but of continued differentiation and reorganization.

These alternatives would be solely of academic interest were it not for the existence of congenital glaucoma, which is a threat to vision that appears in infancy. Infants with congenital glaucoma typically have improperly formed chamber angles. Dealing with the malformation may depend critically on its underlying nature, however. A normal outflow system obscured by a persistent pectinate ligament may offer very different options for treatment from those offered by an outflow system whose development has been arrested before it can function normally.

The trabecular meshwork develops between the fourth and eighth months

Figure 9.22 shows most of the developmental sequence for the outflow system. The first structure to appear is the canal of Schlemm; the canal and the associated venous plexuses are present in rudimentary form at the end of the first trimester. The endothelium of the canal, like that of the blood vessels, is derived from the mesoderm and therefore differs from the trabecular endothelium, which is derived from neural crest cells. Initially, the canal of Schlemm functions as a blood vessel; it does not contain aqueous until the fifth or sixth month, when the cycle of aqueous production and outflow begins.

When the canal of Schlemm appears, the trabecular meshwork is still a clump of undifferentiated tissue at the apex of the developing anterior chamber angle. The cells at this stage are small and loosely organized, with spaces between them (Figure 9.23*a*). As they differentiate, they begin to produce collagen (Figure 9.23*a* and *b*) and to elongate as the collagen becomes the cores for the trabecular cords. This process is far from complete at the time aqueous outflow begins, but the characteristic spaces in the meshwork are obvious (Figure 9.23*b*). As more collagen is synthesized, the trabecular cords elongate, acquiring their endothelial cell wrappings, and the spaces between the interlaced cords in the trabecular sheets become longer and flatter (Figure 9.23*c*), taking on a relatively mature appearance by the eighth month.

Although the outflow system is fully functional at birth, the chamber angle and the outflow system continue to develop as the eye enlarges during the next 10 to 12 years. The postnatal changes are not dramatic, however, and consist mainly of adding tissue as the limbal circumference expands. But this expansion is considerably less than the growth of the whole eye; although the axial length of the eye at birth is about two-thirds of the adult value, the cornea and limbus are already close to their final size. So while growth continues in the limbus postnatally, the amount of tissue added is small and it is added slowly.

Most developmental anomalies in the limbus are associated with structural anomalies that affect other parts of the anterior chamber

Because large parts of the structure around the anterior chamber form as a result of the migration and differentiation of neural crest cells (see Figure 9.21), failures in neural crest cell development are often widespread, involving the cornea, limbus, and iris. The widespread, variable nature of the defects is reflected by their characterization as "syndromes," collections of improperly formed structures. Peter's anomaly, Axenfeld–Reiger syndrome, iridiocorneal endothelial (ICE) syndrome, anterior chamber cleavage syndrome, and Sturge–Weber syndrome are among the names given to these defects.

The distinctions among these anomalies are for the experts to make; the point here is that most of these anterior chamber anomalies are associated with congenital glaucoma. What differs is the proximate cause of the glaucoma; it may be incomplete development of the outflow system, including the canal of Schlemm, in one syndrome, or blockage of the angle by persistent endothelial covering or overdeveloped iris processes in others. Surgical rescue of eyes with develop-

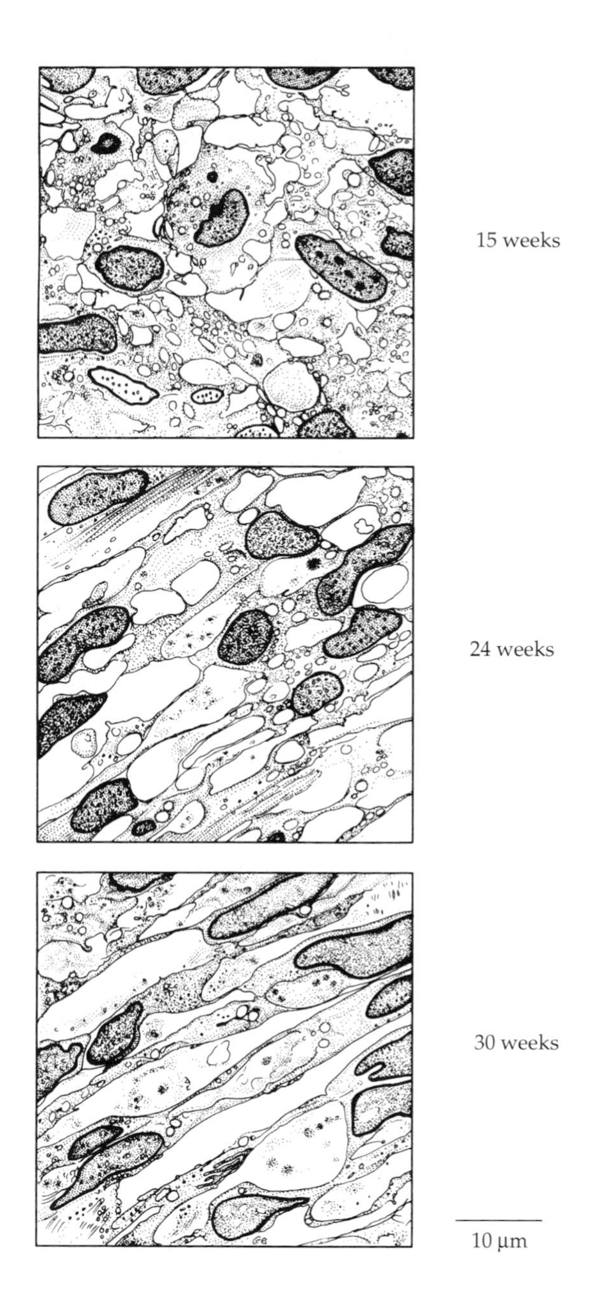

Figure 9.23
Development of the Trabecular Meshwork
(*a*) At 15 weeks of gestation, the trabecular meshwork cells have just begun to differentiate and to synthesize collagen. The cells are round and loosely organized. (*b*) Two months later, near the end of the second trimester, the cells have elongated, more collagen is present, and the intercellular spaces have enlarged. Aqueous flow begins about this time. (*c*) By the eighth month of gestation, trabecular cords are quite apparent and have acquired their typical structure: collagen cores wrapped by endothelial cells. The arrangement of cords into sheets or layers separated by spaces is now clear. (After Reme and d'Epinay 1981.)

mental anomalies of this type is extremely difficult; it may involve a combination of trabeculectomy, iridectomy, and keratoplasty, for example. Surgical intervention may not be feasible because the malformations are so extensive, but even when it is attempted, successes are rare.

References and Additional Reading

The Anterior Chamber and Aqueous Flow

Brubaker RF. 1991. Flow of aqueous humor in humans. *Invest. Ophthalmol. Vis. Sci.* 32: 3145–3166.

Catania LJ. 1995. *Primary Care of the Anterior Segment*, 2nd Ed. Appleton & Lange, Norwalk, CT.

Fatt I and Weissman BA. 1992. The intraocular pressure. Chapter 3, pp. 31–76, in *Physiology of the Eye: An Introduction to the Vegetative Functions*, 2nd Ed. Butterworth-Heinemann, Boston.

Hart WM. 1992. Intraocular pressure. Chapter 8, pp. 248–267, in *Adler's Physiology of the Eye*, 9th Ed., Hart WM, ed. Mosby Year Book, St. Louis.

Hollows FC and Graham PA. 1966. Intra-ocular pressure, glaucoma, and glaucoma suspects in a defined population. *Brit. J. Ophthalmol.* 50: 570–586.

Schottenstein EM. 1996. Intraocular pressure and tonometry. Chapter 20, pp. 407–445, in *The Glaucomas*, Vol. 1. *Basic Sciences*, 2nd Ed., Ritch R, Shields MB, and Krupin T, eds. Mosby Year Book, St. Louis.

Weekers R, Grieten J, and Lavergne G. 1961. Etude des dimensions de la chambre antérieure de l'œil humain. 1e partie: Considérations biometriques. *Ophthalmologica* 142: 650–662.

The Anatomy of Aqueous Drainage

Acott TS. 1994. Biochemistry of aqueous humor outflow. Chapter 1, Section IV, pp. 11.47–1.78, in *Glaucoma, Textbook of Ophthalmology*, Vol. 7, Kaufman PL and Mittag TW, eds. Mosby, St. Louis.

Alward WLM. 1994. *Color Atlas of Gonioscopy.* Wolfe (Mosby-Year Book Europe), London.

Ascher KW. 1942. Aqueous veins. *Am. J. Ophthalmol.* 25: 31–38.

Ashton N. 1951. Anatomical study of Schlemm's canal and aqueous veins by means of neoprene casts. Part I. Aqueous veins. *Brit. J. Ophthalmol.* 35: 291–303.

Bill A. 1970. Scanning electron microscopic studies of the canal of Schlemm. *Exp. Eye Res.* 10: 214–218.

Bill A and Mäepea O. 1994. Mechanisms and routes of aqueous humor drainage. Chapter 12, pp. 206–226, in *Principles and Practice of Ophthalmology: Basic Sciences,* Albert DM and Jakobiec FA, eds. WB Saunders, Philadelphia.

Epstein DL and Rohen JW. 1991. Morphology of the trabecular meshwork and inner-wall endothelium after cationized ferritin perfusion in the monkey eye. *Invest. Ophthalmol. Vis. Sci.* 32: 160–171.

Fine BS and Yanoff M. 1979. The anterior chamber angle. Chapter 11, pp. 251–270, in *Ocular Histology,* 2nd Ed. Harper & Row, Hagerstown, MD.

Fisch BM. 1993. *Gonioscopy and the Glaucomas.* Butterworth-Heinemann, Boston.

Goldberg MF and Bron AJ. 1982. Limbal palisades of Vogt. *Trans. Am. Ophthalmol. Soc.* 80: 155–171.

Gong H, Tripathi RC, and Tripathi BJ. 1996. Morphology of the aqueous outflow pathway. *Microsc. Res. Tech.* 33: 330–367.

Grierson I and Hogg P. 1995. The proliferative and migratory activities of trabecular meshwork cells. *Prog. Ret. Eye Res.* 15: 33–67.

Grierson I, Lee WR, and Abraham S. 1978. Effects of pilocarpine on the morphology of the human outflow apparatus. *Brit. J. Ophthalmol.* 62: 302–313.

Higgenbotham EJ. 1996. Will latanoprost be the "wonder" drug of the '90s for the treatment of glaucoma? *Arch. Ophthalmol.* 114: 998–999.

Hogan MJ, Alvarado JA, and Weddell JE. 1971. The limbus. Chapter 4, pp. 112–182, in *Histology of the Human Eye.* WB Saunders, Philadelphia.

Lerner LE, Polansky JR, Howes EL, and Stern R. 1997. Hyaluronan in the human trabecular meshwork. *Invest. Ophthalmol. Vis. Sci.* 38: 1222–1228.

Lütjen-Drecoll E and Rohen JW. 1992. Functional morphology of the trabecular meshwork. Chapter 10 in *Duane's Foundations of Clinical Ophthalmology,* Vol. 1, Tasman W and Jaeger EA, eds. Lippincott-Raven, Philadelphia.

Lütjen-Drecoll E and Rohen JW. 1996. Morphology of aqueous outflow pathways in normal and glaucomatous eyes. Chapter 5, pp. 89–123, in *The Glaucomas*, Vol. 1. *Basic Sciences*, 2nd Ed., Ritch R, Shields MB, and Krupin T, eds. Mosby Year Book, St. Louis.

Mishima HK, Masuda K, Kitazawa Y, Azuma I, and Araie M. 1996. A comparison of latanoprost and timolol in primary open-angle glaucoma and ocular hypertension. *Arch. Ophthalmol.* 114: 929–932.

Raviola G and Raviola E. 1981. Paracellular route of aqueous outflow in the trabecular meshwork and canal of Schlemm. *Invest. Ophthalmol. Vis. Sci.* 21: 52–72.

Rohen JW and Lütjen-Drecoll E. 1989. Morphology of aqueous outflow pathways in normal and glaucomatous eyes. Chapter 2, pp. 41–74, in *The Glaucomas,* Vol. 1, Ritch R, Shields MB, and Krupin T, eds. CV Mosby, St. Louis.

Rohen JW and Lütjen-Drecoll E. 1992. Functional morphology of the conjunctiva. Chapter 2, pp. 35–62, in *The Dry Eye,* Lemp MA and Marquardt R, eds. Springer-Verlag, Berlin.

Sit AJ, Coloma FM, Ethier CR, and Johnson M. 1997. Factors affecting the pores of the inner wall endothelium of Schlemm's canal. *Invest. Ophthalmol. Vis. Sci.* 38: 1517–1525.

Stamper RL and Tanaka GH. 1998. Intraocular pressure: Measurement, regulation, and flow relations. Chapter 7 in *Duane's Foundations of Clinical Ophthalmology*, Vol. 2., Tasman W and Jaeger EA, eds. JB Lippincott, Philadelphia.

Tripathi RC. 1971. Mechanism of the aqueous outflow across the trabecular wall of Schlemm's canal. *Exp. Eye Res.* 11: 116–121.

Tseng SCG. 1989. Concept and application of limbal stem cells. *Eye* 3: 141–157.

Van Buskirk EM. 1989. The anatomy of the limbus. *Eye* 3: 101–108.

Development of the Limbus

Reme C and d'Epinay SL. 1981. Periods of development of the normal human chamber angle. *Doc. Ophthalmol.* 51: 241–268.

Tripathi BJ and Tripathi RC. 1989. Neural crest origin of human trabecular meshwork and its implications for the pathogenesis of glaucoma. *Am. J. Ophthalmol.* 107: 583–590.

Tripathi BJ, Tripathi RC, and Wisdom JE. 1996. Embryology of the anterior segment of the human eye. Chapter 1, pp. 3–38, in *The Glaucomas*, Vol. 1. *Basic Sciences*, 2nd Ed. Ritch R, Shields MB, and Krupin T, eds. Mosby Year Book, St. Louis.

Chapter 10

The Iris and the Pupil

Although displaced from her heavenly haunts, Iris still reigns in one part of our universe. [When] the line passing from the sun through the pupil of the observing eye is extended out to sunlit mists, a rainbow appears to encircle . . . that axis of sight. But another far more modest ring of color encircles the same axis. It rests upon the aqueous organ of vision itself, between a black interior and a white exterior, at the threshold to an inner world, and "is the open door of it." That slight ring of color carries still the name of Iris—messenger of the gods.

■ Arthur Zajonc, *Catching the Light*

In Greek mythology, Iris was the original messenger from the Olympian gods to man. She was the daughter of Thaumas, god of wonder, and Electra, goddess of waters and oceans; we call this union of wonder and water a rainbow, and Iris is goddess of the rainbows along which she ran to deliver her messages. Her name is given to the colored ring of tissue around the pupil through which light—a message from the gods?—enters the interior of the eye.

The iris and its pupil admit light to the interior of the eye, regulate the amount of light admitted, and affect the quality of the retinal image. The color of the iris, the feature linking it to a rainbow, is a lovely epiphenomenon of light reflecting from melanin and blood.

Functions of the Iris and Pupil

The iris is an aperture stop for the optical system of the eye

All optical systems contain one or more *stops,* elements that have nothing to do with refraction and focus, but can affect image intensity, image spread, field of view, and depth of focus. A stop is an opaque sheet with a hole in it; light passes only through the hole because light falling on the opaque portion of the sheet is absorbed—stopped, in other words. The effect of a stop depends on its location in the optical system. In Figure 10.1*a,* the stop within the lens limits the size of the incident bundle of light, allowing less light through the system than would enter if the full diameter of the lens were used; it is an **aperture stop**. This stop has very little effect on the field of view of the lens, however, as shown by the fact that very oblique rays can pass through the center of the lens and reach the image plane. In Figure 10.1*b,* the stop is in front of the lens and rays coming from oblique positions cannot pass through the hole in the stop to reach the lens. In this case, the field of view is limited by a **field stop**. Cameras contain both aperture and field stops, although the aperture stop is the most obvious because it is what we regulate, along with shutter speed, to control the amount of light in the image. A camera usually has two field stops, one in the film plane and another in the lens system, and neither is variable.

In essence, the iris is an opaque sheet of tissue with a circular hole in it—the hole is the **pupil**—and its position in the optical pathway makes the iris operate as an aperture stop. The iris lies just in front of the lens (Figure 10.1*c*), a position that permits oblique rays to pass through the pupil to be imaged on the peripheral retina while limiting the size of the incident ray bundle. The field of view is restricted somewhat, but because this effect is small, the main function of the stop is to limit the amount of light passing through the system.

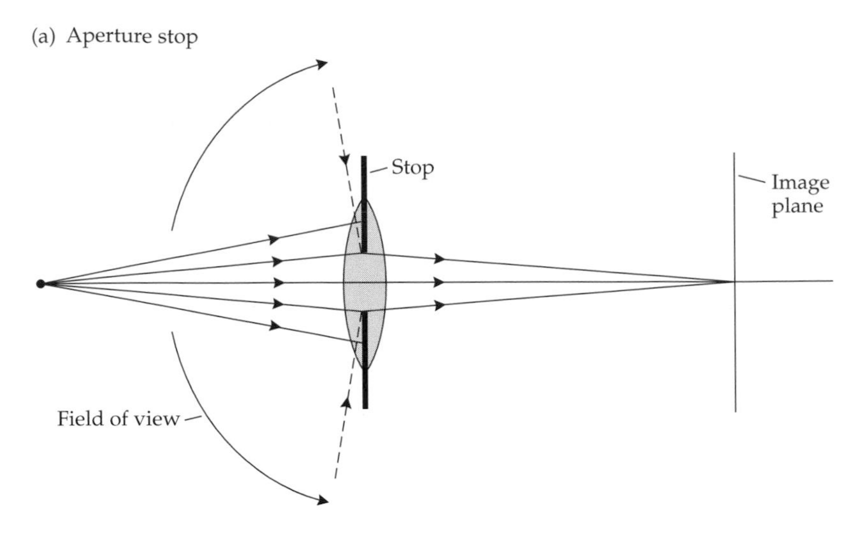

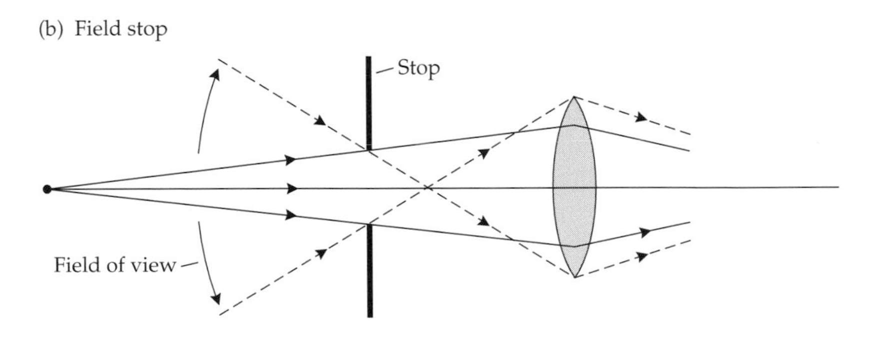

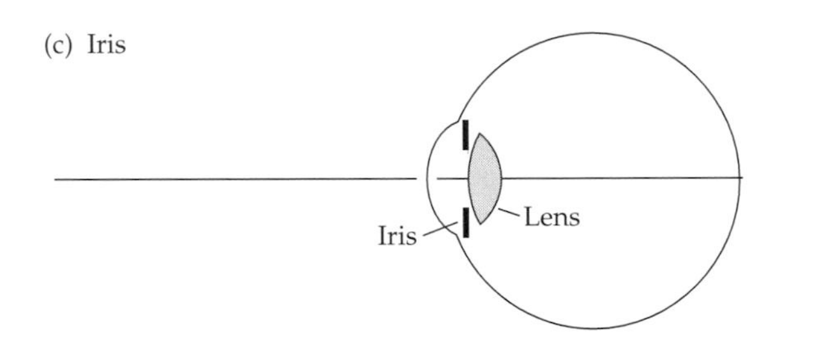

Figure 10.1
Aperture and Field Stops
(*a*) An aperture stop limits the amount of light passing through the lens (solid lines) from a light source on the optic axis. Light rays from very oblique sources (dashed lines) can still pass through the lens, which means that the field of view is very large. (*b*) A field stop has less effect on the amount of light that can reach the lens from a point source on the optic axis, but it restricts the field of view considerably. (*c*) A schematic section through the eye shows the iris lying next to the anterior surface of the lens, which indicates that it operates mainly as an aperture stop.

The entrance pupil is a magnified image of the real pupil

When we look at another person's eye, the pupil, as defined by the iris around it, is the object from which reflected light is refracted by the cornea. The reversed optical problem is illustrated schematically in Figure 10.2, where selected rays have been used to construct the position and size of the image when the real pupil is the object. Both corneal surfaces have been drawn, but the construction ignores the posterior surface. The constructed image of the real pupil is the **entrance pupil**. As shown, the entrance pupil is an erect, virtual image in the anterior chamber about 0.5 mm in front of the real pupil and—by calculation—about 1.13 times larger than the real pupil. (There is also an **exit pupil**, which is the image of the real pupil formed by the lens. The exit pupil is about 0.8 mm *behind* the real pupil and about 1.03 times larger than the real pupil.)

When we refer to "the pupil" or to "pupil size," we are necessarily talking about the entrance pupil, because that is what we can see and measure. Thus, statements about pupil size really mean "entrance pupil size"; the real pupil is a bit smaller.

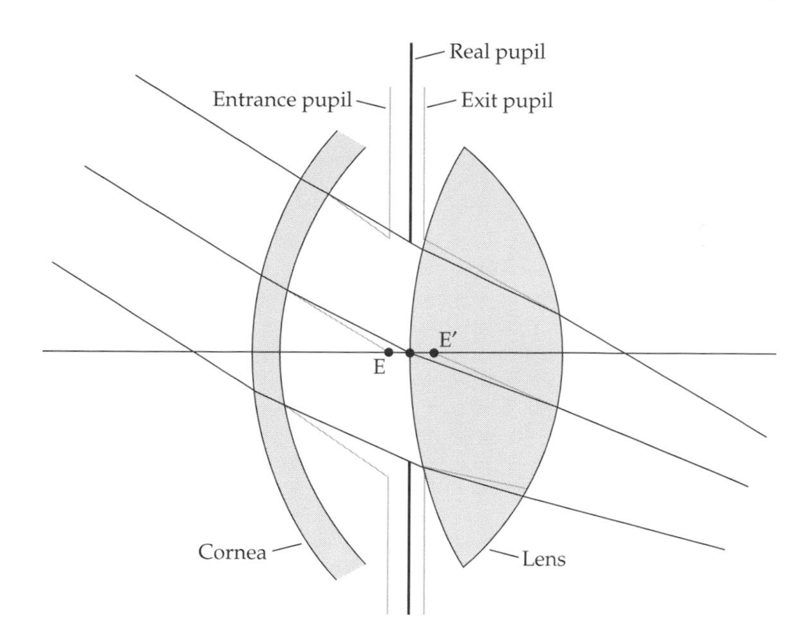

Figure 10.2
Entrance and Exit Pupils
Because of refraction by the cornea, when light rays passing through the center or grazing the edges of the real pupil are projected back into object space toward an observer, the rays will appear to arise from a slightly larger pupil that lies in a plane in front of the real pupil. This apparent pupil is the entrance pupil, whose center is at E. A similar construction taking refraction by the lens into account shows the smaller exit pupil, centered at E′ behind the real pupil. Unlike the entrance pupil, the exit pupil is not visible.

Variation of pupil size changes the amount of light entering the eye, the depth of focus, and the quality of the retinal image

In young adults, pupil diameter ranges from about 2 to 8 mm, and the most obvious effect of this variation in size is variation in the amount of light entering the eye. The amount of light passing through the pupil is proportional to its *area,* or to its diameter squared. Therefore, over the range of normal pupil size, the amount of light admitted to the eye will vary by a factor of 16 ($8^2/2^2$). As discussed in the next section, this variation helps keep the amount of light entering the eye within an optimal operating range by admitting more light when the ambient illumination is dim and admitting less light when the ambient illumination is bright.

Although optical systems have an optimal plane of focus for an object at a given distance, there is always some flexibility arising from the system's sensitivity to blur. In photographic imaging, for example, coarse-grained films are more tolerant than fine-grained films. For human vision, an object at a given distance from the eye can be moved slightly closer or slightly farther away without perceptible effects on its sharpness; this range of object movement within which acceptable focus can be achieved is the **depth of field** (Figure 10.3*a*). As the object moves, the image also moves, producing another range in image space within which focus is acceptable; this is the **depth of focus**. Depth of field and depth of focus are identical in dioptric terms; the term "depth of focus" will be used here because it refers specifically to the retinal image and perceived image clarity.

The magnitude of the depth of focus, which can be specified either as distance or as the dioptric equivalent of distance, varies inversely with pupil diameter: The smaller the pupil, the larger will be the depth of focus (Figure 10.3*b*). In the limiting case of a pinhole pupil, the depth of focus becomes very large, which is why pinhole apertures placed in front of the eye are an efficient way to screen for large refractive errors without measuring the refractive error. Making the depth of focus large reduces the effect of any refractive error, so if one has better acuity through a pinhole than without the pinhole, a refractive error is certainly present.

Experimental studies of depth of focus in the human eye can be summarized by an equation (from Campbell 1957) that relates depth of focus (E) to pupil diameter in millimeters (d_p) as follows:

$$E = \pm[(0.75/d_p) + 0.08] \quad (10.1)$$

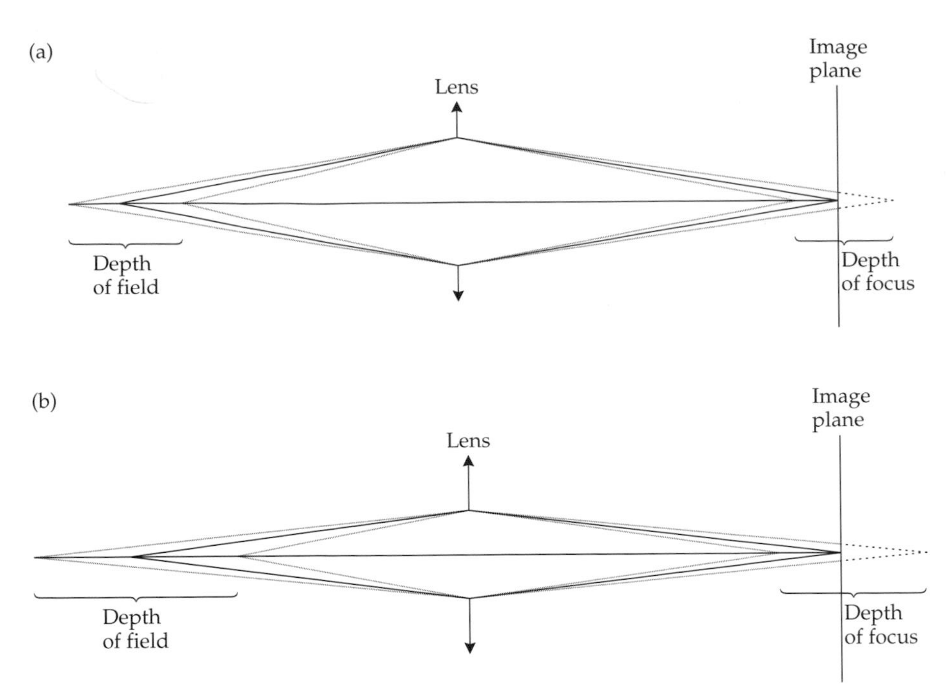

Figure 10.3
Depth of Field and Depth of Focus
(*a*) A target focused precisely in the image plane can be moved either closer to or farther from the lens and still appear to be sharply focused. The range of target movement for which the image is still acceptably sharp is the depth of field. The corresponding range of image displacements from optimum focus is the depth of focus. (*b*) When the aperture is made smaller, both depth of field and depth of focus increase. This effect is due purely to the geometry; the maximum permissible size of the blurred image is the same in both *a* and *b*.

For the normal range of pupil size, depth of focus will range from ±0.46 D or 0.92 D (pupil diameter of 2 mm) to ±0.17 D or 0.34 D (8 mm pupil). For the 0.87 mm pinholes recommended for clinical screening, the depth of focus is slightly over 2 D.* Pupil size affects image quality through the effects of diffraction and aberration, as discussed in Chapter 2. Increasing pupil size reduces image quality by increasing the effects of aberrations; decreasing pupil size reduces image quality by increasing the effects of diffraction. Although this seems like a "damned if you do, damned if you don't" situation, diffraction is less significant than aberration in the normal range of pupil size (see Chapter 2). Therefore, the smallest pupil size consistent with the ambient illumination level is the best solution.

Pupil size varies with illumination level, thereby helping the retina cope with large changes in illumination

Our eyes must work not only in the brilliance of the noonday sun but also when the world is illuminated only by starlight—a difference in light intensity of about 10 billion times (10 log units). But a photoreceptor sensitive enough to detect a few quanta of light from a distant star would be overwhelmed by the blaze of photons from the sun, and a photoreceptor operating comfortably in sunlight would be hopeless at detecting starlight. The eye deals with this problem in several ways, mainly by using different sets of photoreceptors for dim and bright light conditions and by internally adjusting the sensitivity of the photoreceptors and other retinal neurons (see Chapter 13). The pupil also contributes; its 16-fold variation in area can compensate for about 1.2 log units of the 10-log-unit operating range.

Figure 10.4*a* shows pupil diameter as a function of the ambient illumination level. The original data and the curve fit through them have been redrawn with straight-line segments to emphasize two different regions of the pupil diameter

*Although pinholes certainly increase the depth of focus and can be used clinically to offset the effects of refractive error, their use requires some care. First, the pinhole must be properly positioned on the visual axis. In addition, small pinholes reduce image contrast and increase image spread by diffraction, both of which reduce visual acuity. The optimum pinhole size, which has the largest depth of focus with the least image degradation, is between 0.75 and 1.00 mm. See Takahashi 1965.

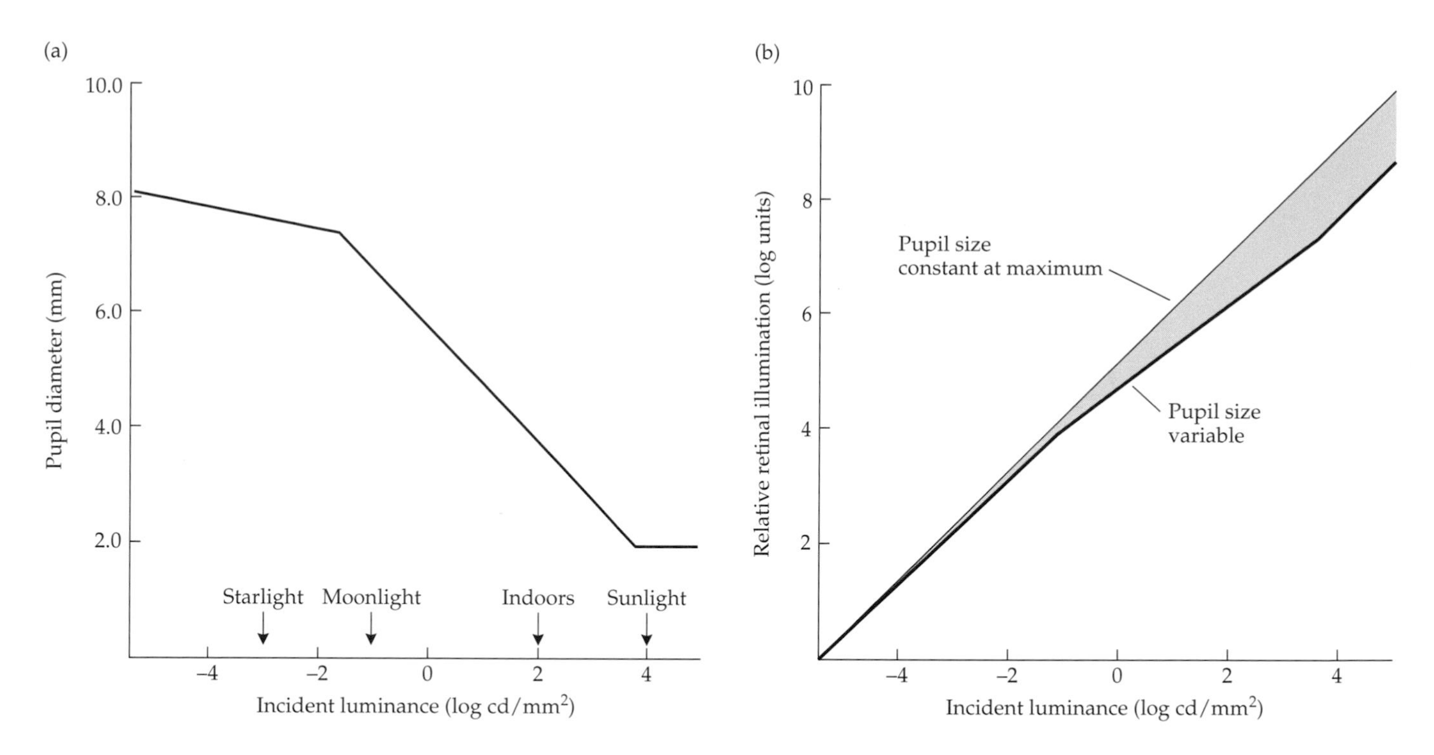

Figure 10.4
Pupil Size and Illumination
(*a*) At low luminance levels—less than bright moonlight—the pupil is relatively unaffected by the ambient light level, but pupil size decreases significantly as the light increases above the level of moonlight. The minimum pupil size, about 2 mm in diameter, is reached at a light level corresponding to bright sunlight. (*b*) If the pupil size never changed from its maximum, the amount of light reaching the retina would increase in direct proportion to the amount of light incident on the eye. Pupil constriction at high light levels means that retinal illumination increases less rapidly than ambient illumination increases. The reduction is shown by the shaded region between the two curves, which represent constant and variable pupil size, respectively. (*a* after Reeves 1920.)

change. As already mentioned, the pupil diameter ranges from 8 to 2 mm. With very dim illumination, the pupil is fully dilated to nearly 8 mm, and it gradually constricts as the illumination is increased, eventually contracting to about 2 mm under very bright light conditions. In the region between –5 and –1 log units, which would be considered night vision, the change in pupil diameter is relatively small: Diameter decreases by only 1 mm over a 4-log-unit (10,000-fold) increase in light intensity. In the daylight range, however, the change is more dramatic: Pupil diameter decreases by about 1 mm for each log unit of intensity change.

The effect of the change in pupil size with illumination level is shown in Figure 10.4*b*, where retinal illumination (the amount of light reaching the retina) has been plotted as a function of the ambient illumination level.* If pupil size were constant at its maximum diameter, the amount of light reaching the retina would be directly proportional to the amount of light striking the cornea. But because the pupil constricts with increasing illumination, the relationship changes and the retina is not exposed to the full magnitude of the ambient illumination range. Of the 5 log units of illumination change under daylight conditions, the retina sees slightly less than 4, thanks to the constricting pupil. Variation on the order of 100,000 times in ambient lighting conditions has been reduced to variation on the order of 10,000 times at the retina. (This is good, but any cat worthy of the name does much better; their slit pupils can vary in area by more than 2 log units, reducing a variation of 100,000 times in ambient illumination to a variation of 1,000 times at the retina.)

*Although the term "illumination" has a precise definition, I am using it here as a generic term for the amount of light. The quantity most commonly measured is *luminance*, which is the *luminous flux* (in lumens per unit solid angle, or candelas, abbreviated cd) emitted or reflected from a unit of surface area of a light source. Luminance is specified in cd/m^2. Retinal illuminance is the product of luminance and the pupil area *in mm^2*, specified in units called trolands (td). When corrected for transmission losses in the media of the eye, 1 td = 10^7 photons · sec^{-1} · mm^{-2} incident on the retina (approximately). The average luminance reaching your eye from this page should be about 80 cd/m^2; with a pupil 4 mm in diameter, the retinal illuminance would be about 1000 td. Photometry can be a messy subject, but for an accessible discussion, see the sections on photometric measures and quantum catch in Rodieck 1998.

Pupil size varies with accommodation and accommodative convergence

Simultaneous accommodation, convergence, and pupil constriction (miosis) is called the **near triad**. The linkage between accommodation and convergence is linear, characterized by the **AC/A ratio**, which is the amount of convergence elicited by each diopter of accommodation (Figure 10.5*a*). The AC/A ratio varies among individuals, but for any given person, it is robust and resistant to change by lenses, prisms, or any form of vision training. The relationship between pupil size and accommodation is also linear (Figure 10.5*b*); each diopter of accommodation elicits about 0.3 mm of pupil constriction (i.e., the **P/A ratio** is –0.3 mm/D). Note that pupil constriction and accommodative convergence do not cease when the accommodative response reaches its limit. This means that one should expect to see pupil constriction in normal presbyopes when they attempt to accommodate; the *absence* of pupil constriction is a sign of abnormality.

When the data in Figure 10.5 are replotted to show the relationship of pupil size to accommodative convergence, the relationship is linear throughout the entire range of convergence (Figure 10.5*c*); the slope of the line could be called the P/AC ratio, but doing so would imply a dependence that probably does not exist. The truly independent variable is the accommodative stimulus, which is a proxy for the amount of innervation. Because convergence and pupil constriction are both linear functions of the accommodative stimulus, they are also linearly related, but they are not dependent on one another.

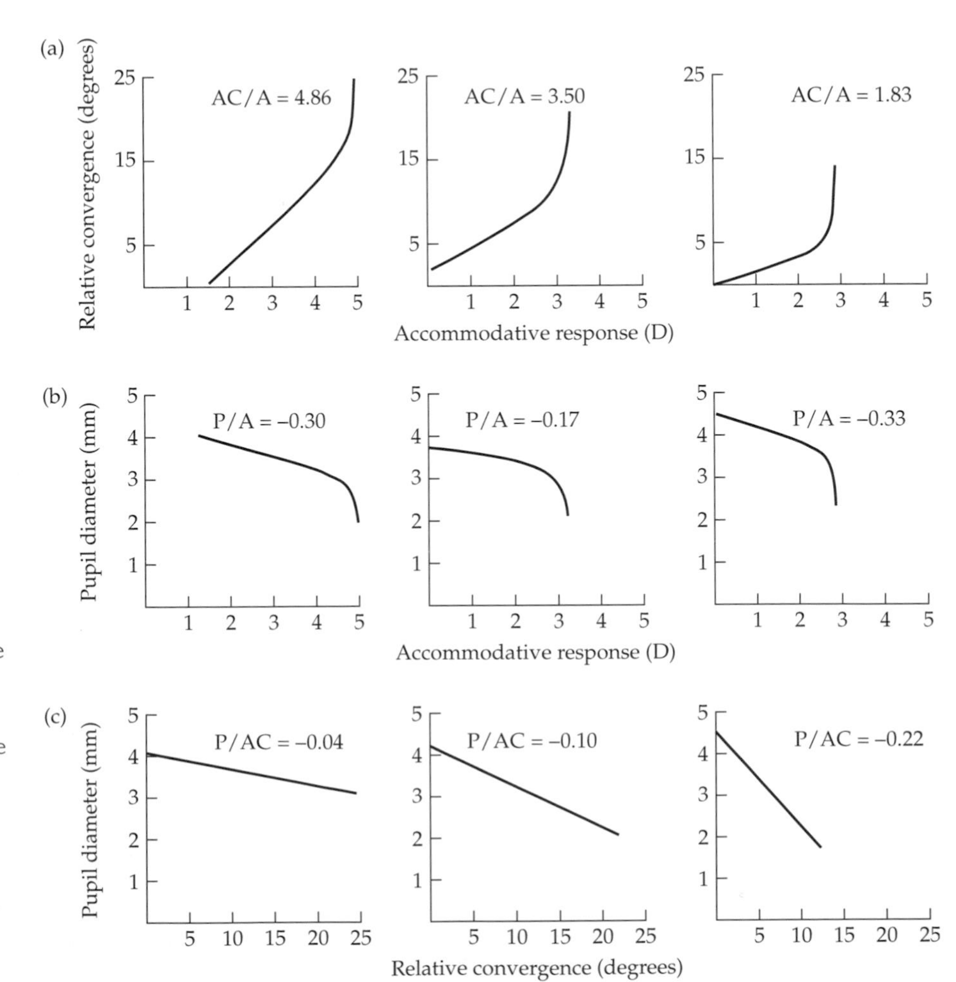

Figure 10.5
Pupil Size and the Near Response
(*a*) The convergence associated with the near response increases linearly over the range of accommodative responsiveness. The slope of the linear portion of the curves is the AC/A ratio. (*b*) Pupil size decreases linearly with accommodative response until the accommodative system can no longer respond. The slope of the linear portion of each curve is the P/A ratio. (*c*) The relationship between pupil size and convergence is quite linear over the entire range of convergence. (Replotted from Alpern, Mason, and Jardinico 1961.)

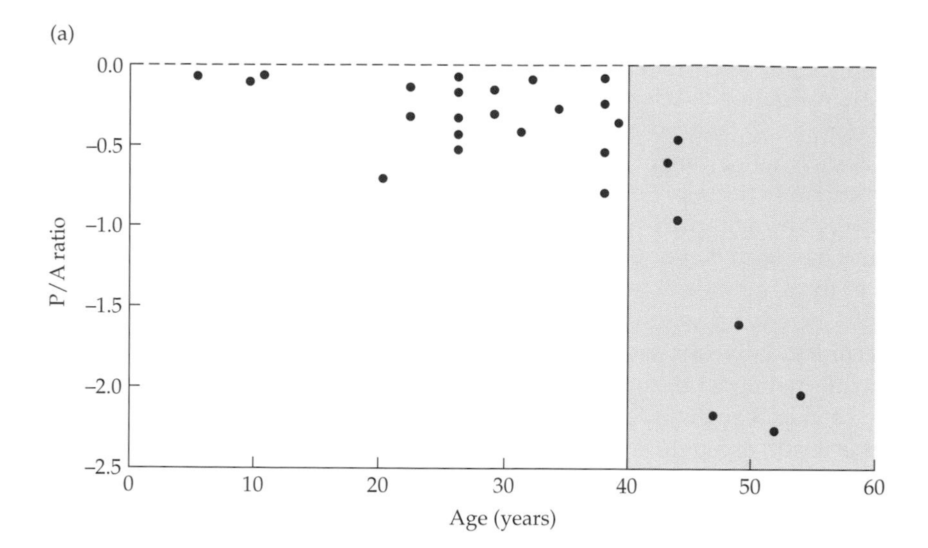

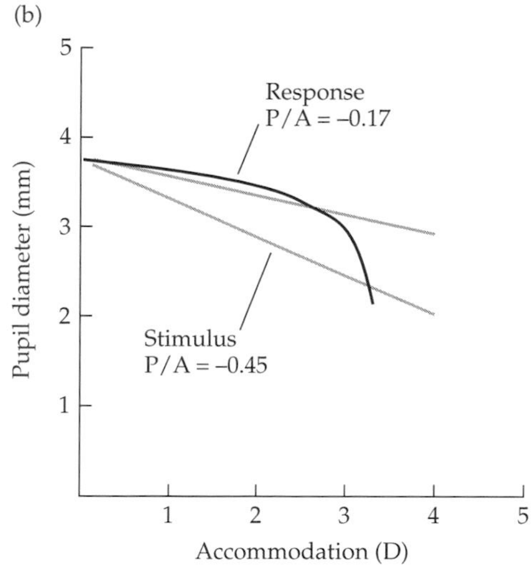

Figure 10.6
Pupillary Near Response and Age
(*a*) The data points show the ratio of change in pupil size per diopter of accommodation for a 4 D accommodative stimulus with subjects ranging in age from 5 to 55 years. The pupillary near response is small under the age of 10, but is noticeably larger in most individuals between the ages of 20 and 40. Data in the shaded region for ages 40 and above are unreliable because of a calculation artifact. (*b*) A true P/A ratio can be calculated only from the linear portion of the curve that relates pupil size to accommodative response. Assuming the response to have been 4 D in this case more than doubles the value of the P/A ratio. This is the calculation artifact for ages 40 and above in *a*. (*a* replotted from Schaeffel, Wilhelm, and Zrenner 1993.)

Under some circumstances, pupil size does *not* change during accommodation, conventional wisdom and the data in Figure 10.5 notwithstanding. But the circumstances in which accommodation occurs with no pupil response appear to be instances in which the *only* stimulus to accommodation is image blur produced by defocusing.

The stimulus to accommodation is complex and not fully understood (see Chapter 11), but blur by defocusing is only one of the possible stimuli. Under natural conditions, the full range of accommodative stimuli comes into play when we look from a distant object to a near object, and one or more of these stimulus variables elicits an associated pupillary constriction. But artificially restricting the stimulus to image blur eliminates pupil constriction even though accommodation still occurs; other variables, most probably changes in image size or position during accommodation are required to activate the pupil response.

Pupil constriction with accommodation produces a larger depth of focus, which demands less precision from the accommodative system. To be in good focus with a 5 mm pupil, for example, the system must be accurate to within ±0.23 D (from Equation 10.1), but with a 2.5 mm pupil, the criterion is ±0.38 D. Relaxing the criterion accuracy of the accommodative response by ±0.15 D may not seem like a significant gain, but each diopter of accommodation shifts the image plane relative to the retina by about 300 µm. Therefore from Equation 10.1, the focus must be accurate to within ±70 µm with a 5 mm pupil, but with a 2.5 mm pupil, the criterion is ±115 µm. Threading a needle with an eye 230 µm wide should be considerably easier than threading one with an eye 140 µm wide.

The pupillary near response is smaller in children than in adults

Although pupil constriction with accommodation (under normal viewing conditions) is quite obvious in adults, some evidence indicates that the association is nearly absent in children. The data in Figure 10.6*a*, for example, indicate that for children under 10 years of age, the P/A ratio is close to zero (about –0.1). Individual P/A ratios vary between –0.1 and –0.7 between the ages of 20 and 35 and then increase dramatically for the next 20 years. According to the authors of this study and others who have cited it, one should expect a large pupil constriction as part of the near triad in persons over age 40, less in young adults, and almost no constriction in young children.

But the P/A ratios were calculated on the basis of pupil *responses* to a 4.0 D accommodative *stimulus*, and as Figure 10.6*b* shows, the calculated P/A ratio will be exaggerated whenever the limit of the accommodative *response* is 4 D or less. In complete presbyopia, in which the accommodative response is near zero, the P/A ratio calculated in this way is very high. Thus, the large increase in the P/A ratio after age 40 (shaded region in Figure 10.6*a*) is a calculation artifact. More accurate P/A ratios cannot be derived from the original data, but the values would certainly decrease, probably to –0.5 or less.

Between the ages of 20 and 40, most of the P/A values are between –0.1 and –0.5, and the mean is about –0.35. If there is any trend from smaller to larger P/A ratios in the age range, it is extremely small and unlikely to be statistically significant. There is no reason to think that the linkage between pupil constriction and accommodative innervation changes at all after the age of 20.

For children under the age of 10, however, the P/A ratios are around –0.1 and clearly smaller than those for age 20 and older. I cannot find any reason to argue with this result, except to comment that the number of subjects was small and it is likely that the range of P/A values in this age group has been underestimated. Even so, the pupillary near response appears to be small in young children. If there is any age-related trend in the P/A ratio, it is occurring in the second decade of life, between the ages of 10 and 20, for which there are no data.

Assuming the smaller pupillary near response in children to be genuine, there is no obvious explanation for it. As a guess, however, the differences in the stimuli for accommodation and pupil constriction may be important. If the pupil response is driven by changes in retinal image position or size rather than by defocusing blur, inaccurate bifixation of the near targets by children might generate inappropriate signals for pupil constriction. Alternatively, the image displacement thresholds for pupil constriction may be relatively high in children. But guessing at the answer is not profitable at this stage; the experiments need to be repeated with a better experimental design.

The pupil is in constant motion, and it reacts quickly to changes in retinal illumination

Anything said thus far about pupil size refers to its average diameter over relatively long periods of time—several seconds at minimum. But when the pupil size is measured continuously, a fast, rhythmic fluctuation can be seen (Figure 10.7*a*); it can sometimes be observed directly in the pupil of another person who is looking at a distant object. The pupil is dilating and constricting about two times a second over a range of about 0.5 mm. This rhythmic oscillation, called **hippus**, becomes larger, more rapid, and more regular as the illumination level is increased. All this is normal, meaning that a statement about pupil size has an implicit "±0.2 mm" attached to it. The *absence* of hippus is an anomalous condition, implying a problem with the innervation of the iris or damage to the iris itself.

The recordings in Figure 10.7*a* also show the pupil's response to sudden increases in illumination; for the brightest light increase, the pupil responded after about 200 ms and constricted about 4.5 mm within 2.5 s. For dimmer stimuli, the latency is somewhat longer, and the constriction is smaller. The dependence on stimulus intensity is obvious in Figure 10.7*b*, which shows the pupil responses to brief light flashes (1 s duration) at relative intensities of 1, 4, and 8 log units. In each case, the light flash elicits a pupil constriction, followed by a slower dilation back toward the original pupil size. Increasing the flash intensity makes the pupil respond more quickly (i.e., decreases the latency), increases the response magnitude, and increases the time required for the pupil to return to its resting value. These changes are proportional to the logarithm of light intensity; roughly, each ten-fold (1 log unit) increase in intensity elicits about 0.37 mm of constriction in pupil diameter.

The rapid pupil response is important because large changes in ambient illumination can occur suddenly. Walking on a bright sunny day through patches of

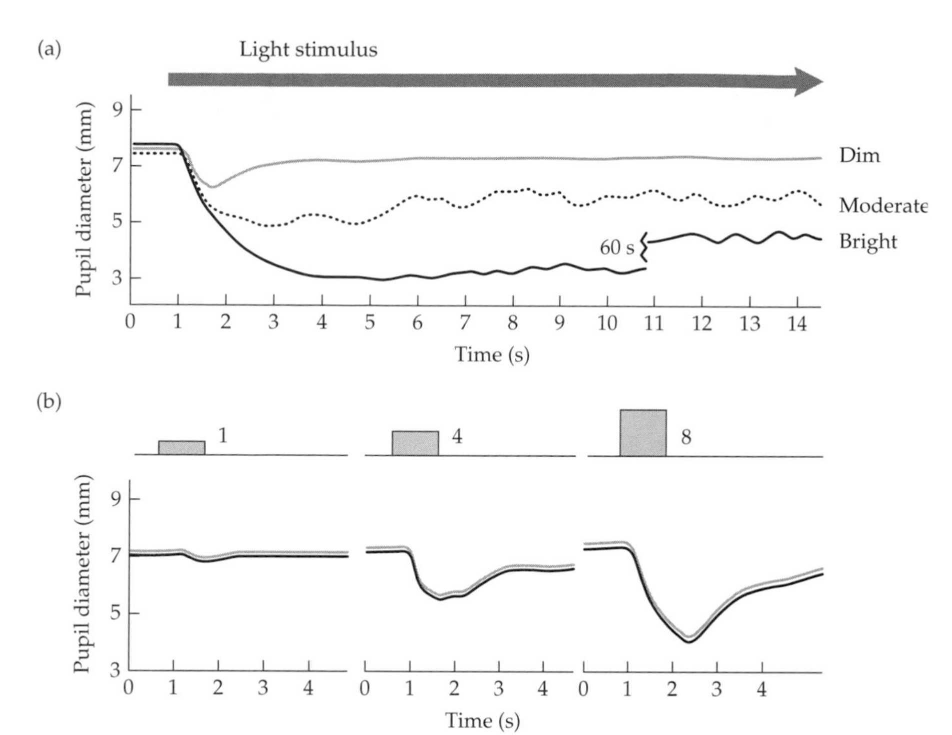

Figure 10.7
Pupillary Light Responses
(*a*) The three traces show continuous records of pupil diameter in response to prolonged light increase at three different intensities. The pupil constriction becomes faster and larger as brightness increases and shows a pronounced oscillation at moderate and bright intensities. This oscillation is hippus of the pupil. (*b*) Pupil responses to brief light flashes are faster, larger, and longer as stimulus intensity increases. Although the stimulus was delivered to the right eye only, whose response is shown by the solid lines, the response of the left pupil (dashed lines) was identical. The relative intensities of the light stimuli are indicated in log units. (After Lowenstein and Loewenfeld 1969.)

shade punctuated by brilliant sunlight demands continual, rapid adjustments of the eye's sensitivity. The transition from shade to sunlight elicits fast neural adjustments in the retina, assisted by rapid pupil constriction, that keep one from being dazzled by the glare. And this adjustment is not trivial. Going out into the sunlight with your pupils dilated is a very uncomfortable experience, because without assistance from pupil constriction, the retina cannot compensate adequately for the change in illumination.

Decreases in light intensity when walking from light to shade also produce adjustments of sensitivity, but they are slower. The pupil dilates less rapidly than it constricts (see Figure 10.7), and the retinal adjustments, which are proportional to the amount of bleached photopigment, are slower because photopigment is bleached by light more quickly than it can regenerate in darkness (see Chapter 13).

Decreased iris pigmentation in ocular albinism affects the optical function of the iris

Individuals with **albinism*** have little or no melanin pigment in the parts of their eyes where pigment is normally found. Their irises are typically pale, sometimes pinkish, because of the melanin deficiency in the iris. The iris is less opaque than normal, and light enters the eye through the tissue of the iris as well as through the pupil, creating a diffuse, irregular spread of light over the retina on which the retinal image is superimposed. Contrast is reduced and the retinal image is less sharply defined.

Visual acuity is typically reduced in albinos, and the reduced image contrast may be partially responsible, but some albino retinas do not have foveas (a condition known as foveal hypoplasia) and their cone densities may be abnormally low. When present, the reduced cone density probably contributes more to acuity loss than does the contrast reduction from light transmitted through the iris.

*Of the many types of albinism, the two main categories are *oculocutaneous* albinism, in which the reduction in pigmentation is in eyes, skin, and hair, and *ocular* albinism, in which the reduced pigmentation is confined to the eyes.

(The magnitude of the reduction in cone density is not known, however, nor is this reduction known to occur in *all* eyes affected by albinism.) Albinos are often photophobic, probably because of the inability of the iris to exclude bright light and the lack of retinal and choroidal pigment to absorb stray light within the eye. Although increasing illumination aids visual acuity in normals, bright images make things worse for albinos.

Other factors add to the visual deficits experienced by albinos. The deficiency in melanin synthesis also affects the routing of axons through the optic chiasm (see Chapter 5), producing anomalous wiring of the primary visual pathway. The wiring anomaly may be partially responsible for the nystagmus that so often accompanies albinism.

Structure of the Iris

The pupils in the two eyes are normally the same size and are decentered toward the nose

When the eyes are equally illuminated, the pupil diameters will be the same to within a fraction of a millimeter, as one can verify by measuring the pupils in Figure 10.8*a* and *b*. Not only are the pupil sizes nearly equal when dilated under dim light (Figure 10.8*a*) and when constricted under bright illumination (Figure 10.8*b*), but also they generally move in synchrony during reflex responses (see Figure 10.7*b*) and during their normal hippus.

Any detectable difference between pupil diameters is called **anisocoria**. If the anisocoria is less than 0.4 mm, as it is in Figure 10.8*a*, it is within the range of differences considered to be normal and it is called simple or physiological anisocoria. Differences greater than 0.4 mm under conditions of equal illumination are

(a) Dim illumination

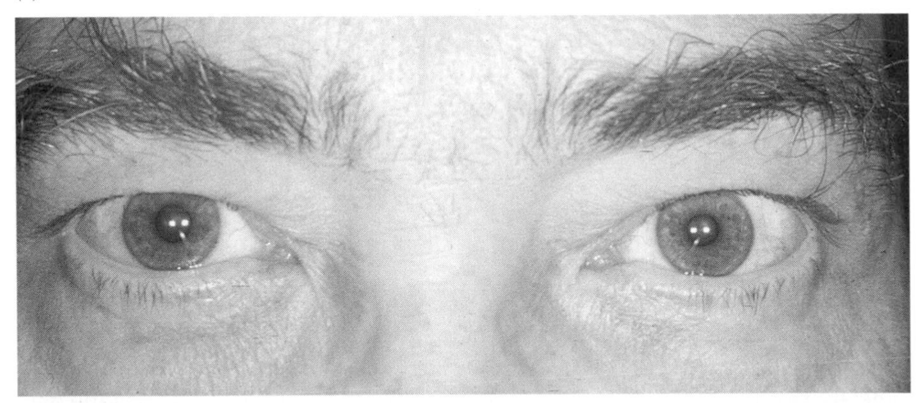

(b) Bright illumination

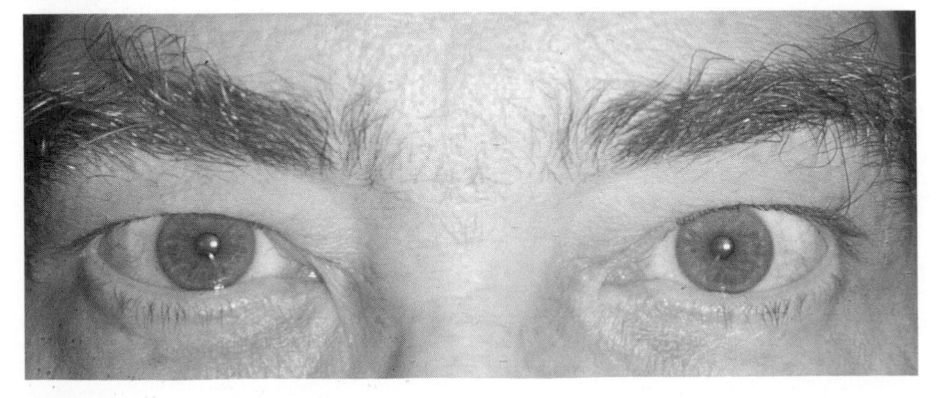

Figure 10.8
Normal Iris and Pupil
(*a*) This photograph of my eyes was taken after about 30 secs of adaptation to dim illumination; the pupils are somewhat dilated and are about 5 mm in diameter. The left eye's pupil is slightly smaller. (*b*) After adaptation to brighter illumination, the pupils are constricted to about 3 mm in diameter. The diameters are equal. The rim of iris visible on the lateral side of each pupil is slightly wider than the medial iris rim, indicating that the pupils are decentered slightly toward the nose.

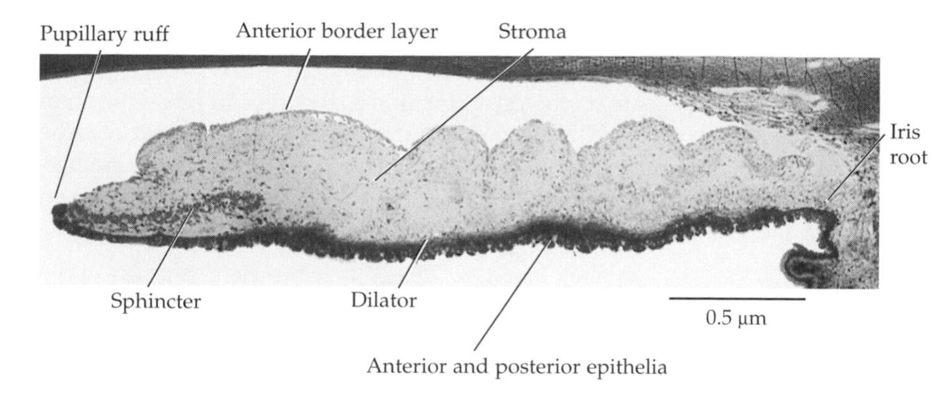

Figure 10.9
The Iris in Cross Section
The pupil in the eye from which this section was made was dilated; the distance between the margin of the pupil and the iris root is just over 2 mm, and the anterior surface has marked contraction folds. The main structural elements of the iris are indicated. (From Hogan, Alvarado, and Weddell 1971.)

abnormal; anisocoria of this magnitude indicates a problem somewhere in the reflex arc through which pupil size is regulated.

Normal pupils are circular, but their center generally does not coincide with the center of the circular iris. The decentration is small, but it can be seen in Figure 10.8 by comparing the width of the iris on the nasal side of the pupil to the width of the iris on the temporal side; the nasal iris is slightly narrower in both eyes, meaning that the pupils are decentered toward the nose, usually by about 0.5 mm. Any significant departure from circularity and decentration of more than 0.5 mm are abnormal; an abnormally decentered pupil is *ectopic*. Both irregular and ectopic pupils (shown in Figure 10.29) degrade retinal image quality because of off-axis imaging and increased aberration.

The iris is constructed in layers and regional differences in the iris are related to the different muscles within them

The structural requirements for an aperture stop are not demanding; the main one is pigment dense enough to absorb most of the incident light. Operating as a variable aperture stop complicates matters, however; there must be at least one muscle and innervation that links the musculature to neural control circuitry. Most of the interesting and meaningful structural features of the iris are related to the variable size of its aperture.

A cross section through one side of the iris reveals a sheet of tissue of variable thickness (Figure 10.9). It is thinnest where it joins the ciliary body, a junction called the **iris root**; damage to the iris by trauma appears most often as a tear at the iris root, where the iris rips away from the ciliary body. The iris is thickest about two-thirds of the way from the iris root to the pupil margin at a region called the **collarette**, which is the site of the minor arterial circle. (The relative locations of the collarette and minor arterial circle vary with pupil size; they are pushed closer to the pupil margin as the pupil dilates. The example in Figure 10.9 is markedly dilated.) The collarette marks a rough division of the iris into **pupillary** and **ciliary regions**; the pupillary region contains the sphincter muscle, and the ciliary region contains the dilator muscle. At the magnification in Figure 10.9, the sphincter is the only one of these two muscles that is clearly visible.

The iris is usually described as having five layers of tissue, although two of them are not well defined; they are indicated in Figure 10.9. The anterior surface of the iris is the **anterior border layer**; it varies in thickness and is not continuous, having large breaks or holes in it. Most of the iris consists of its **stroma**, which is a mixture of fibroblasts, melanocytes, collagen fibers, and blood vessels, but in terms of volume, a large fraction of the stroma is open space that is filled with aqueous in vivo. The **sphincter** and **dilator** muscles are considered together to be the muscular layer, but the muscles are not connected, except for a few muscle fiber strands, and the sphincter is considerably thicker than the dilator.

Finally, the posterior surface of the iris is lined with *two* layers of epithelial cells, the **anterior** and **posterior iris epithelia**. The terminology is sometimes confusing; both epithelial layers are on the posterior side of the iris, but one—the anterior epithelium—is in front of the other. The anterior iris epithelium is not to be confused with the anterior border layer. Although both epithelial layers are pigmented, the anterior epithelium has less pigment and is *myo*epithelium from which the dilator muscle arises.

The anterior border layer is an irregular layer of melanocytes and fibroblasts interrupted by large holes

When the iris is viewed from in front, we see the anterior border layer. It appears irregular, even at low magnification, with strands of tissue weaving back and forth in a generally radial pattern, interrupted by gaps and holes of varying size (Figure 10.10*a*). These holes are the **crypts of Fuchs**. They are significant only because they vary from one iris to another, giving each iris the individuality of a fingerprint (Figure 10.10*b*), and because aqueous can pass freely from the anterior chamber to the interior stroma of the iris.

The margin of the pupil is outlined with a ring of dark pigment known as the **pupillary ruff** (or pupillary frill). This pigment line is formed by the posterior epithelial layers as they wrap around the edge of the pupil (see also Figure 10.9). The pupillary ruff is more noticeable in light than in dark irises—a consequence of the better contrast—but it is normally present in all irises. Its only functional significance may be to provide a sharp, densely pigmented border for the pupil.

When the pupil is dilated (Figure 10.10*b*), prominent circular lines appear in the surface. These are **contraction folds**, which are deep valleys between ridges of iris tissue. Contraction folds are a natural consequence of pupil dilation. Although the iris stroma has a loose, open construction, there is an irreducible volume of tissue that cannot be compressed completely, so it must fold, forming regions of increased thickness. Normally, contraction folds are of no consequence, but if the angle of the anterior chamber is not fully open (see Chapter 9), folds near the apex of the angle may block the trabecular meshwork. The result can be elevated intraocular pressure because of the reduced aqueous outflow.

(a)

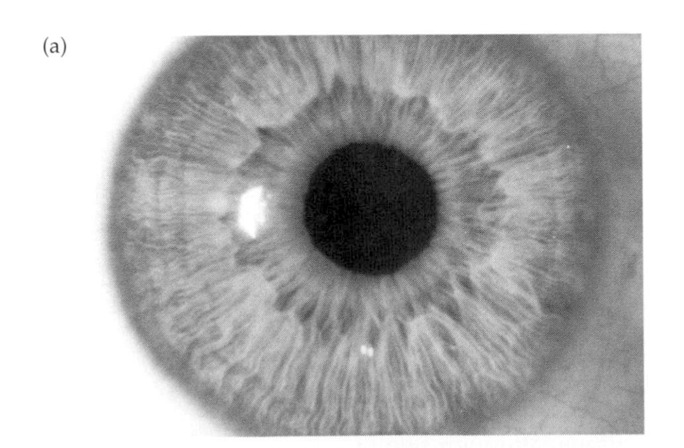

(b)

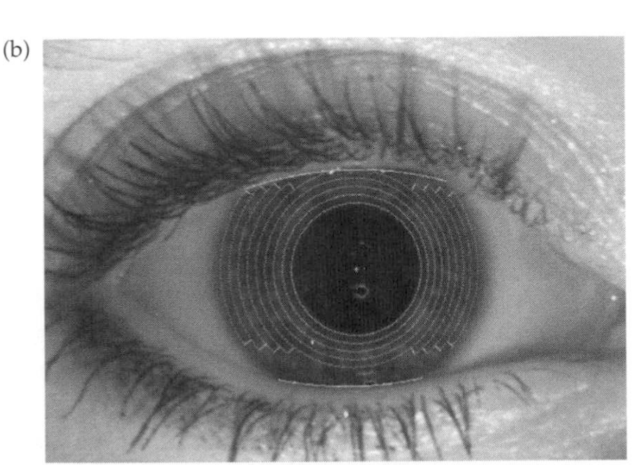

Figure 10.10
Surface Features of the Iris
(*a*) In this constricted pupil, irregular strands of tissue can be seen following meandering, radial paths from the periphery toward the pupil. The large dark gaps in the tissue are the crypts of Fuchs. (*b*) The uniqueness of iris patterns makes them ideal for biometric identification of individuals. Here, a digitally captured image of an eye is shown with zones of the iris mapped for computer analysis, which can distinguish individual patterns quickly and accurately. This technique is now being employed for building security, bank transactions, and other areas where identification is crucial. (Courtesy of IriScan, Inc.)

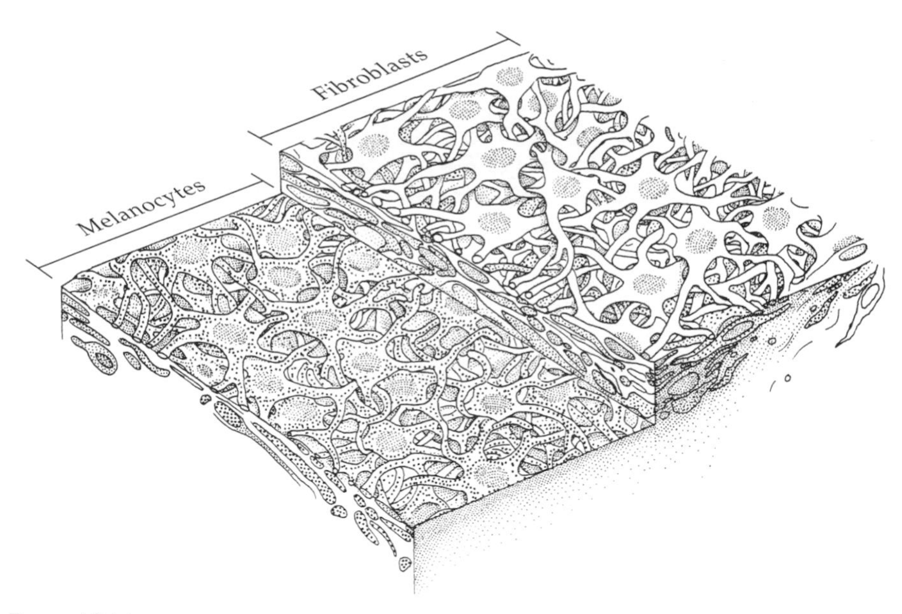

Figure 10.11
The Anterior Border Layer of the Iris
A three-dimensional reconstruction of the anterior border layer shows that most of the surface cells are fibroblasts, which look rather like fried eggs with long extensions of the white part. The processes interweave extensively, and the cells are connected by gap junctions (not illustrated). The underlying melanocytes appear similar, except for the pigment granules they contain. Some collagen fibers from the stroma weave between the cells. (After Hogan, Alverado, and Weddell, 1971.)

Clinicians use this effect of pupil dilation as a "provocative test" for subacute angle closure glaucoma; an increase in intraocular pressure of more than 8 mm of mercury after the pupil has been dilated (either by dark adaptation or by a pupil-dilating drug) is an indication of angle closure. A corollary to this effect is the need to use pupil-dilating drugs (**mydriatics**) that can be readily counteracted if the pupil dilation elevates intraocular pressure abnormally.

The anterior border layer (Figure 10.11) consists of interwoven strands of tissue interrupted by the crypts of Fuchs, which constitute breaks or holes in the layer. Some of the very fine strands are collagen fibers. The cells whose processes interweave to form the layer are melanocytes and fibroblasts. The fibroblasts tend to concentrate at the surface, overlying the melanocytes, making a rough subdivision between the fibroblasts that define the surface bounding the anterior chamber and melanocytes that fill in the layer below them. Although not shown here, fibroblasts communicate with their neighboring fibroblasts with gap junctions, as do the melanocytes with their compatriots.

Near the iris root, extensions from the anterior border layer run across the apex of the anterior chamber angle to merge with the trabecular meshwork. These **iris processes** are composed mainly of fibroblasts from the iris and endothelial cells from the trabecular meshwork.

The anterior border layer has two functions: It allows aqueous to penetrate the stroma of the iris, and by virtue of its melanocytes, it is the first light-absorbing layer in the iris, thereby helping to make the iris an effective aperture stop.

The iris stroma has the same cellular components as the anterior border layer, but loosely arranged

The iris stroma contains much empty space (Figure 10.12), which in vivo is filled with aqueous. Two types of cells are scattered here and there: The melanocytes

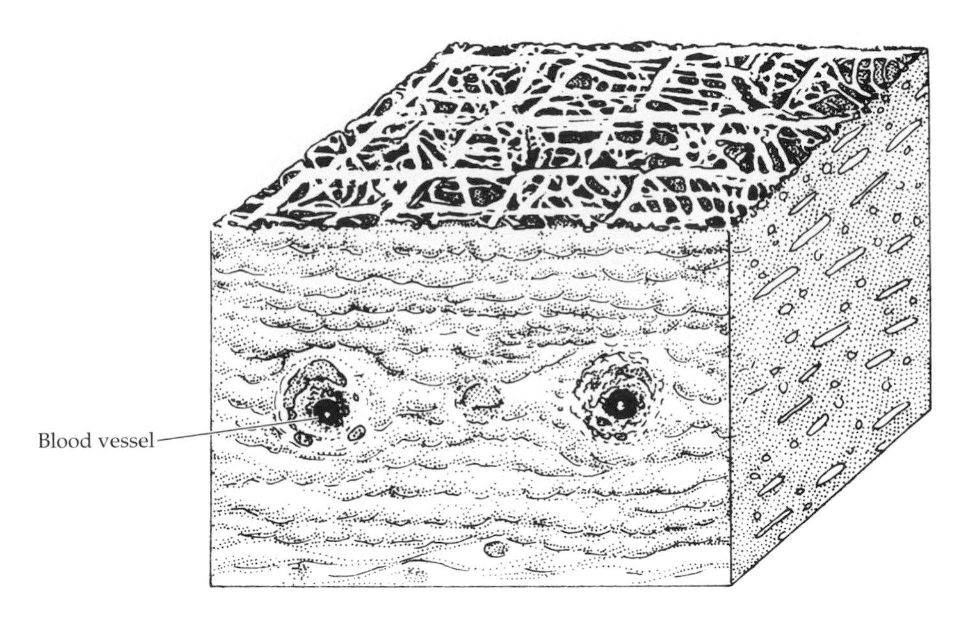

Figure 10.12
The Iris Stroma
Collagen in the iris stroma runs in bundles that tend to be oriented radially or tangentially to the pupil margin. Thus the spaces in the tissue are often rectangular. A rough layering is also apparent. Scattered fibroblasts and melanocytes are less oriented than the collagen. (After van de Zypen 1978.)

are revealed by their pigment granules—the rest are fibroblasts. Collagen fibers weave in and out. The only other notable components of the stroma are blood vessels and fine nerve fibers (which are not shown in Figure 10.12).

The three-dimensional structure of the iris stroma is like a coarse sponge whose tissue consists of collagen fibers and the interconnected processes of the melanocytes and fibroblasts. Since the collagen fibers tend to form bundles that run mainly either radially or tangentially, the holes in the sponge tend to be rectangular. Blood vessels and nerve fibers run through aqueous-filled spaces.

Trivial as this may sound, the stroma's structure—or lack of it—is consistent with the way the iris operates. The continual change in pupil size stretches the tissue as the pupil constricts and compresses the tissue as it dilates. Mechanically, these changes would be difficult if the iris stroma were a dense, compact tissue, because it would resist both expansion and compression. But the open structure of the stroma makes it workable. If you imagine holding a sponge under water, alternately squeezing and releasing it, you will have a good idea of how the stroma operates in its normal aqueous environment.

Figure 10.13 shows a unique cellular component of the stroma: a **clump cell**. This clump cell and others like it are most commonly found in the vicinity of the sphincter muscle. At least two types of clump cells can be distinguished by their microscopic appearance; the more common type, shown in Figure 10.13, is stuffed full of pigment and looks like a bloated melanocyte. Because their number and size appear to increase with age, clump cells are assumed to be similar to macrophages whose job is to clean up stray bits of melanin released from melanocytes or pigmented epithelial cells. The clump cells do not seem to break down the melanin they ingest, however, and they become more swollen as they age. Little is known about the life cycles of stromal melanocytes and pigment epithelial cells, but assuming there is some epithelial cell death and replacement, cleaning up the pigment released when the epithelial cells die is a useful service; melanin released into the aqueous can cause problems.

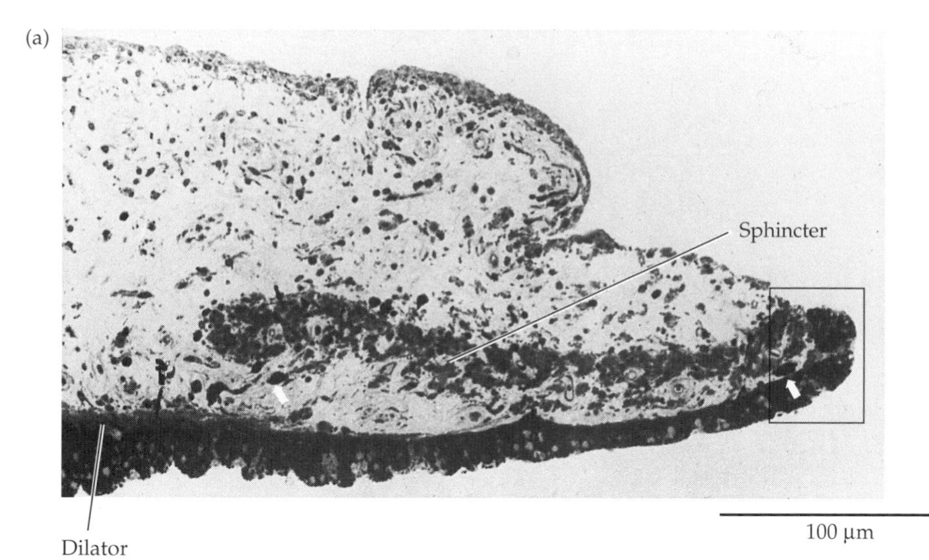

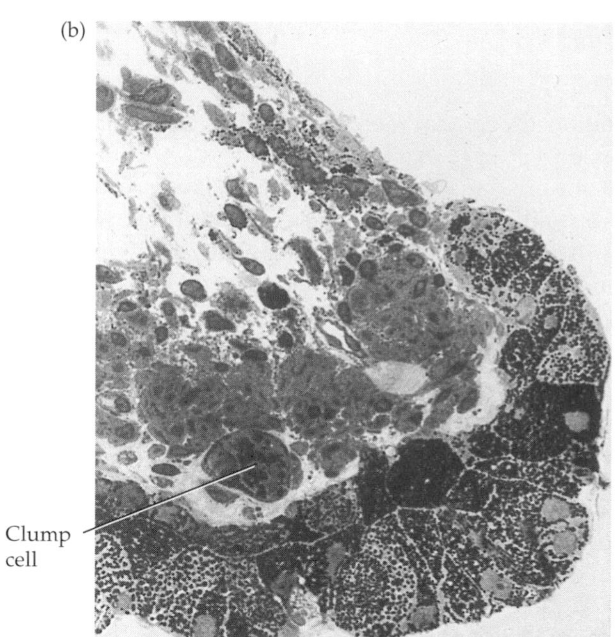

Figure 10.13
Iris Clump Cells
(*a*) Clump cells are commonly found in the vicinity of the sphincter muscle (arrows). They can be recognized at low magnification because they are larger and darker than melanocytes. (*b*) The clumps of melanin granules that give the cells their name show clearly at high magnification; note that the melanin granule density is considerably greater than in the surrounding melanocytes. (From Hogan, Alvarado, and Weddell, 1971.)

Small blood vessels run radially through the stroma, anastomosing to form the minor arterial circle and supply the iris muscles

Figure 10.14 shows the pattern of iris blood vessels. No distinction is possible here between arteries and veins, but all the vessels follow a meandering radial path from the iris root toward the pupil. Meandering of the vessels is probably related to variation in pupil size; the curves will straighten out when the pupil constricts and become more exaggerated when the pupil dilates. In other words, there is some slack in the system, minimizing mechanical restrictions as the tissue stretches and compresses. The arteries are branches from the major and intramuscular arterial circles in the ciliary body, and the veins drain through the ciliary processes into the vortex veins (see Chapter 6).

The vessels branch extensively at the collarette, and anastomoses among the branches form the incomplete **minor arterial circle**. Subsidiary branches from the minor arterial circle form radial arcades running toward the pupil margin and supplying the sphincter muscle. Other branches from the minor circle are recurrent, running along the dilator muscle toward the iris root. The veins par-

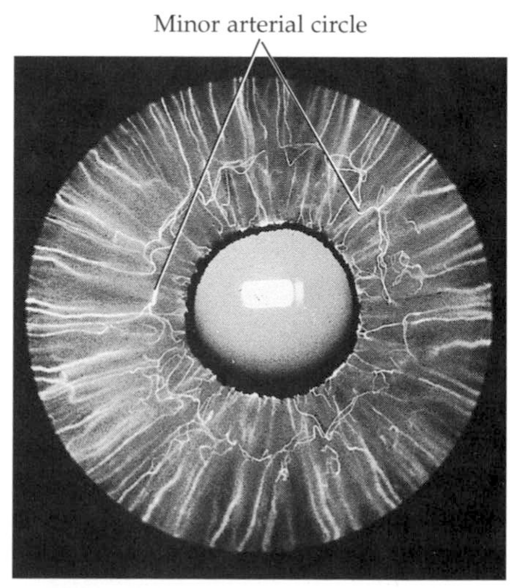

Figure 10.14
Iris Vessels and the Minor Arterial Circle
The iris vessels can be seen in vivo with fluorescein angiography. All vessels run radially between the iris root and the pupil margin, meandering as they go. Branching and anastomoses in the vicinity of the sphincter form the incomplete minor arterial circle. (Courtesy of Rosario Brancato.)

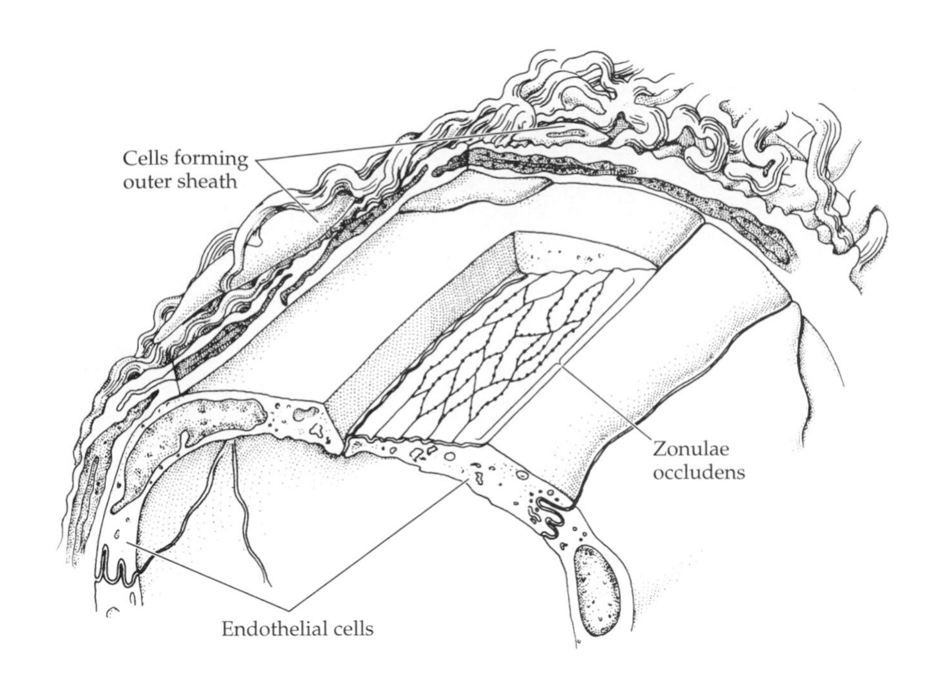

Figure 10.15
Structure of the Iris Vessels
The blood vessels in the iris are specialized to create a blood–aqueous barrier. The endothelial cells are connected by long complex junctions that are very tightly closed with extensive occluding (tight) junctions (zonulae occludens). Outside the endothelial wall, other cells (pericytes) and collagen fibers form a sheath that encloses the blood vessel. The sheaths may help keep the vessels from kinking as they are thrown into folds and compressed as the pupil dilates. (After Freddo 1996.)

allel the arterial distribution, except that there is no venous circle corresponding to the minor arterial circle.

The endothelium of the iris vessels, whatever their size, is constructed to restrict the movement of fluid and large molecules through the capillary walls. As Figure 10.15 shows, the endothelial cells overlap extensively and are connected by numerous anchoring and occluding junctions; molecules like horseradish peroxidase will not leak out of the iris vessels after being injected into them. This restrictive construction of the vessels is one of the reasons for thinking that they have little to do with aqueous outflow (see Chapter 9), even though they are immersed in aqueous.

Although the iris vessels have no smooth muscle in their walls, the larger vessels have a connective tissue sheath around the endothelium and an incomplete sheath of pericytes around the connective tissue (see Figure 10.15). This "tube within a tube" construction may keep the vessels from kinking. Kinks in the blood vessels would cut off blood flow just as kinks in a garden hose cut off the water supply. And since kinks form in garden hoses when they are moved around, not when they are stationary, it seems reasonable that the constantly moving iris vessels would have some sort of protection against kinking.

The sphincter and dilator occupy different parts of the iris and have antagonistic actions

Figure 10.16 is a drawing of the iris with the anterior border layer and stroma removed to show the iris muscles. The sphincter occupies the pupillary portion of the iris, by definition, and its muscle fibers run along circular arcs parallel to the pupil margin. The length of the muscle fibers is not known with any accuracy, but it is unlikely that an individual muscle fiber extends around more than one-fourth of the muscle's circumference. Thus, the muscle fibers interlace and are bound together by their connective tissue sheaths.

This circular arrangement of muscle fibers makes the action of the sphincter muscle obvious and gives it its name. When the fibers contract, they shorten and become straighter, making the circle of muscle smaller; it is a sphincter, whose action is **miosis**, or pupil constriction. Note, however, that relaxation of the sphincter from a given state of contraction will produce **mydriasis**, or pupil dilation, as the muscle circle expands. This response is important but passive; the active response—the muscle's action—is pupil constriction.

Fibers of the dilator muscle define the ciliary portion of the iris; they run radially along lines between the center of the pupil and the iris root. The dilator fibers are extensions or processes of the anterior epithelial cells, as we will see, and they are relatively short, probably not more than a few hundred micrometers in length. As in the sphincter, connective tissue binds the dilator muscle fibers together as a cohesive muscle.

Given their arrangement, contraction of dilator fibers generates forces pulling radially outward, thereby enlarging the pupil. That much is obvious, but it may be less clear how contraction of such short muscle fibers can produce several millimeters of change in the pupil diameter. The key point is that the only dilator fibers with origins fixed in position are those closest to the iris root. If they contract 200 µm, all other fibers move outward by 200 µm. Contraction of the next nonoverlapping set of fibers adds another 200 µm of outward movement, the next set 200 µm more, and so on, adding up to a large movement when all the muscle fibers are activated. The actual contraction does not occur in an outer-to-inner sequence, as this example implies, but the near simultaneous small contractions of individual fibers have the same additive effect.

Except for the extremes of the pupil diameter range, both the sphincter and the dilator always have some tonus, meaning that pupil size reflects a balance between the constrictive force exerted by the sphincter and the expansive force exerted by the dilator. It also means that there are two ways to constrict the pupil and two ways to dilate it. Constriction can be achieved by contraction of the sphincter *or* by relaxation of the dilator. Similarly, dilation can be the result of active contraction of the dilator *or* of relaxation of the sphincter.

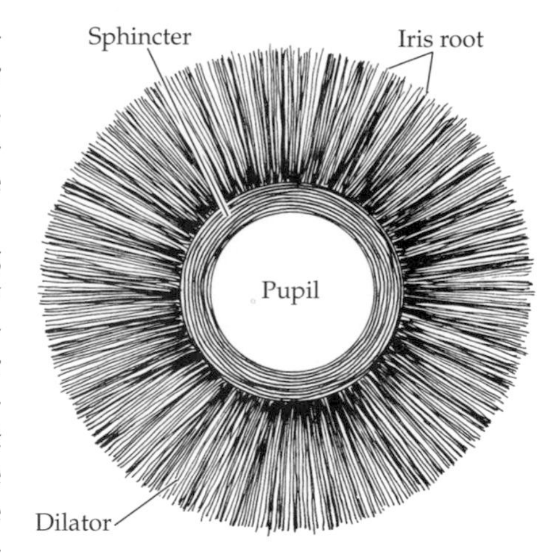

Figure 10.16
Orientation of Sphincter and Dilator Fibers
The sphincter lies closer to the pupil than the dilator does, and its muscle fibers run in circular arcs parallel to the pupil margin. The dilator occupies a much larger area of the iris, and its fibers run radially between the iris root and the outer edge of the sphincter muscle. A few strands of the dilator merge with the sphincter, but the two muscles overlap very little.

The sphincter is activated by the parasympathetic system, the dilator by the sympathetic system

As discussed in Chapter 5, the efferent pathway for innervation of the sphincter is via neurons in the Edinger–Westphal nucleus, whose axons terminate on cells in the ciliary ganglion. The ciliary ganglion cell axons travel into the eye in the short ciliary nerves and then to the sphincter. Any stimulus that activates the sphincter uses this two-neuron parasympathetic pathway to exert its effect (see Figure 10.22).

The corresponding sympathetic pathway to the dilator is slightly more complicated. The cells that directly innervate the dilator have their cell bodies in the superior cervical ganglion in the neck. From here, the axons travel up to the sympathetic plexus around the internal carotid artery, enter the orbit as the sympathetic root of the ciliary ganglion, and enter the eye in the short ciliary nerves. (There is *no* synapse in the ciliary ganglion.) The efferent pathway to the dilator is longer than that to the sphincter, so there are more places at which lesions can affect the sympathetic innervation. Cells in the superior cervical ganglion receive inputs from cells whose axons arise in the brainstem and travel down the spinal cord to synapse on other neurons at the first thoracic vertebra (T1), whose axons run out to the sympathetic trunk and up to the ganglion (see Figure 10.24).

Figure 10.17 illustrates neurotransmission at the muscles; acetylcholine is the transmitter to the sphincter, and norepinephrine is the transmitter to the dilator. Unlike the fast ligand-gated receptors at nicotinic synapses earlier in the parasympathetic pathway, the acetylcholine receptors on the sphincter fibers are muscarinic; the neurotransmitter does not open ion channels directly, but operates indirectly through a G protein and a second-messenger link between the acetylcholine receptor and the ion channel. Activation of the muscle is therefore somewhat slower than the fast activation of skeletal or extraocular muscles, and the pharmacologic agents that mimic or block neurotransmission also differ from those that act on conventional neuromuscular junctions.

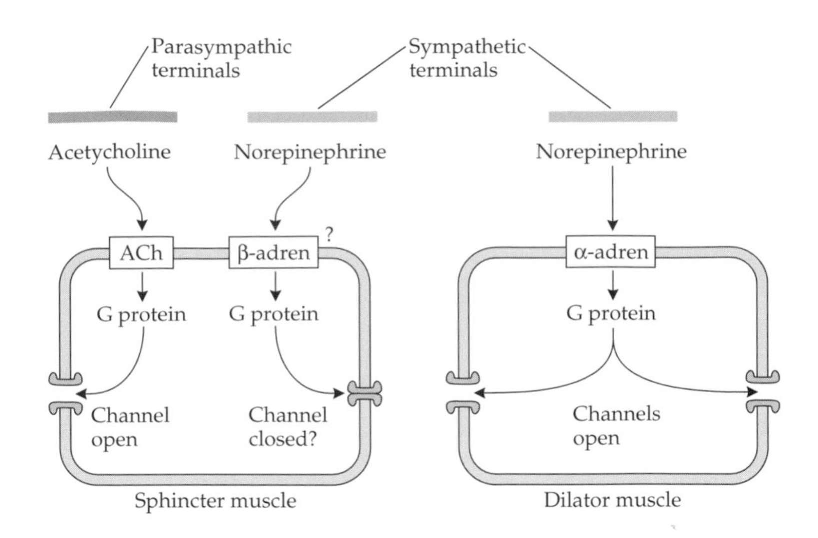

Figure 10.17
Elements of Neurotransmission at the Iris Muscles
Sphincter muscle fibers are activated by acetylcholine acting on metabotropic (G protein–linked) receptors (ACh). Ion channels are opened indirectly through the intermediate G protein and second-messenger pathway. The sphincter fibers also have some beta-adrenergic receptors (β-adren) on which norepinephrine can act to close ion channels; the question marks reflect the fact that this possible pathway for inactivating the sphincter fibers is not well documented. Dilator muscle fibers are activated by norepinephrine released from sympathetic nerve terminals. The norepinephrine binds to alpha-adrenergic receptors (α-adren) to open membrane channels through a G protein and second-messenger pathway.

A second receptor on the sphincter muscle is the beta-adrenergic receptor; it binds norepinephrine—the *sympathetic* neurotransmitter—and works through a G protein and second messenger to close ion channels, thereby reducing—that is, inhibiting—muscle fiber activation. If sympathetic axons going to the dilator also made contact with the sphincter at these beta-adrenergic sites, these inputs would constitute a form of reciprocal innervation; activation of the dilator would coincide with relaxation of the sphincter. Although the presence of beta-adrenergic receptors on the the sphincter has been reported several times, the proportion of these receptors relative to the acetylcholine receptors is not known with any precision, and very little evidence suggests that they are functionally important. Drugs that block the beta-adrenergic receptor sites might be expected to reduce any tonic inhibition of the sphincter and produce pupil constriction, but either this result does not occur or the effect is too small to be noticeable. Thus modulation of the sphincter's state of contraction appears to be largely a matter of modulating the synaptic drive from the parasympathetic system at the acetylcholine receptors.

Dilator fibers have only alpha-adrenergic receptors. The neurotransmitter is norepinephrine released from the terminals of the sympathetic axons. Muscle fiber activation in this system is also accomplished by a G protein and a second messenger. Because no receptors are known to inhibit the dilator, its tonus is modulated apparently solely by modulation of the rate of norepinephrine release from the sympathetic terminals. (But one quick qualifaction—the receptors will respond to endogenous epinephrine or norepinephrine, so dilator tonus may not always be tightly linked to synaptic events.)

The anterior pigmented epithelium is a myoepithelium, forming both the epithelial layer and the dilator muscle

In a light microscope section, such as the one shown in Figure 10.13*a,* the dilator muscle is a thin gray line separating the stroma from the first of two pigmented epithelial layers. The close contact between the dilator and the anterior epithelium is a clue to the true nature of the dilator; it certainly differs from the sphincter, which is separated from the epithelium by intervening stroma.

Electron microscopy reveals the myoepithelial character of the anterior epithelium (Figure 10.18). The cytoplasm around the epithelial cell nucleus contains the usual organelles and melanin pigment granules, but anterior to the nuclear region, the cytoplasm changes in appearance. In particular, long filaments or fibers can be seen running obliquely away from the region containing the cell nucleus. These filaments are made up of contractile proteins, and they lie within a long process extending from the nuclear region of the cell; the

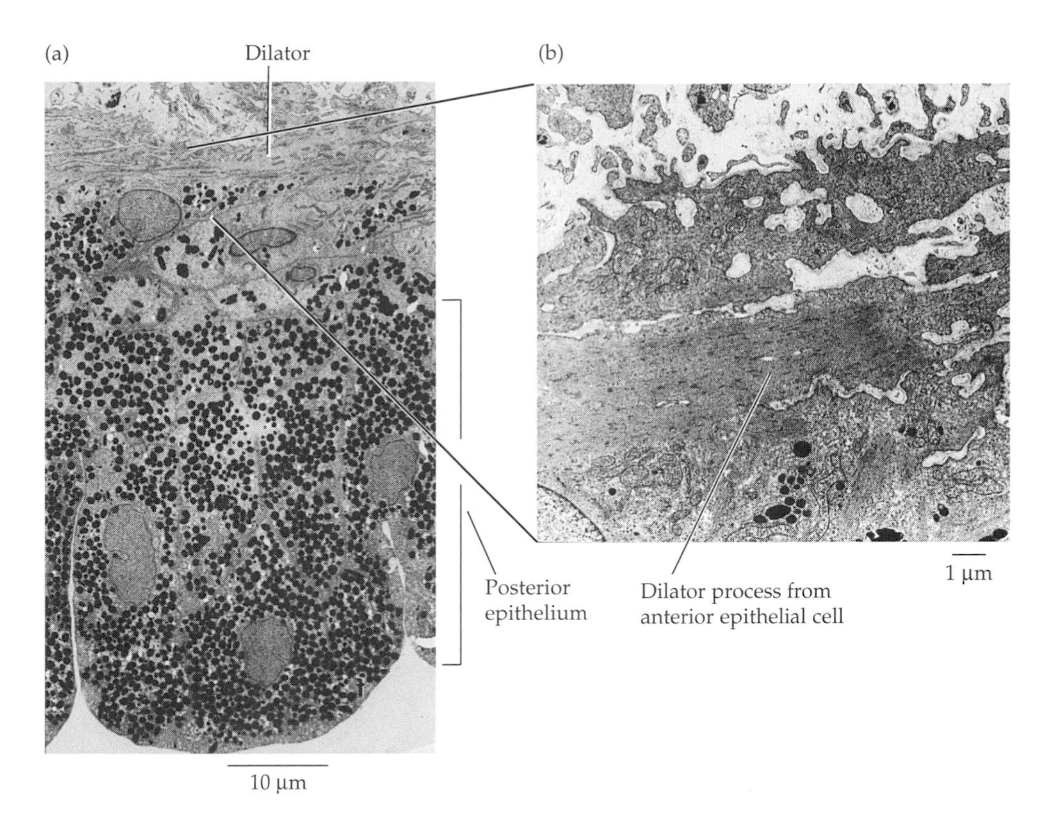

Figure 10.18
Anterior Epithelium and the Dilator
(*a*) A section through the posterior surface of the iris reveals three layers. The posterior epithelial cells are elongated and heavily pigmented. They are overlaid by flatter, less heavily pigmented cells that constitute the anterior epithelium. A layer of unpigmented, elongated processes above the anterior epithelial cell bodies is the dilator. (b) A view of the junction between dilator and anterior epithelium at high magnification shows that they are continuous—that is, all parts of the same cells; no cell membrane separates the dilator processes from the pigment-containing cytoplasm below. The dilator is therefore part of the anterior epithelium, the cells of which are thus myoepithelial. (From Hogan, Alvarado, and Weddell 1971.)

process is a muscle fiber whose mechanism of operation and contractile properties are the same as those of other smooth muscle fibers, including those of the sphincter. In short, one set of cells makes both the anterior epithelium and the dilator muscle, but because the cell bodies (the epithelial portion) and the cell processes (the dilator) look different in histological sections and have different functions, the different names are justifiable and worth retaining.

Like the fibers in the sphincter muscle, the contractile processes in the dilator have numerous gap junctions between adjacent processes (Figure 10.19). The resulting electrical coupling should make the muscle fibers contract in concert rather than in isolation, thereby ensuring that pupil dilation is radially symmetric; contraction by just one segment of the dilator would be pointless. Nothing is known about the number of muscle fibers innervated by a single motor axon, but it is probably large and there is likely to be considerable overlap, leading to multiple (polyneuronal) innervation of individual muscle fibers. This innervational pattern would also encourage simultaneous contraction of all muscle fibers.

In comparison to the posterior layer of epithelium, the anterior epithelial cells are more cuboidal and less densely pigmented (see Figure 10.19). The epithelial portions of the cells are connected to one another and to posterior

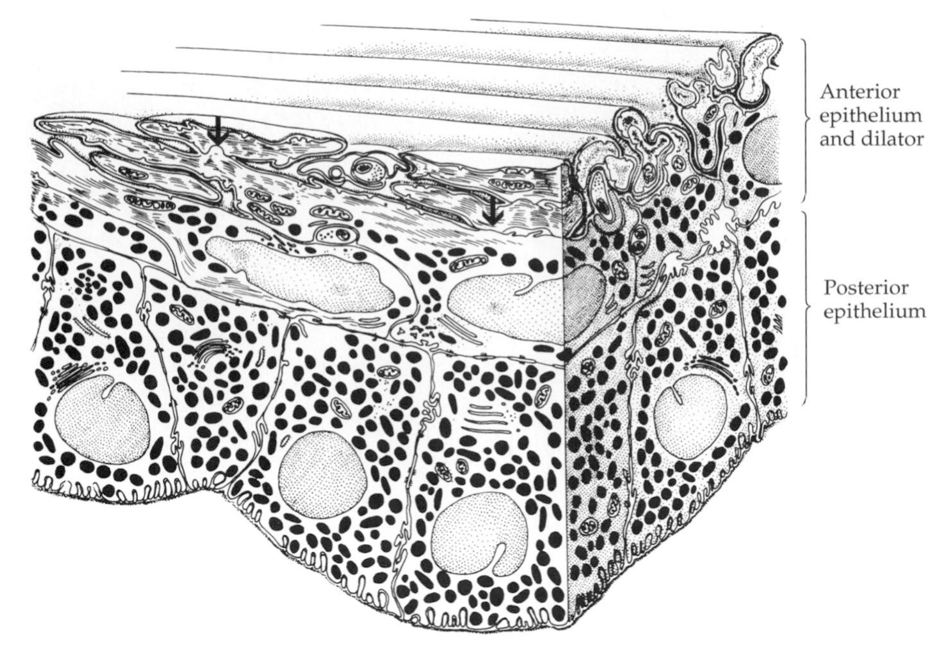

Figure 10.19
The Iris Epithelium
This reconstruction shows that the posterior epithelial layer cells are columnar and have a high density of melanosomes in the cytoplasm. The cells are bound to one another and to the anterior epithelial layer by desmosomes and other anchoring junctions. The anterior epithelial cells are flatter, have lower melanin density, and have long extensions that form the strands of the dilator. The epithelial portions of these cells are connected by desmosomes and anchoring junctions; the dilator processes make gap junctions (arrows) with one another. (After Hogan, Alavarado, and Weddell 1971.)

epithelial cells by gap junctions, as well as by desmosomes and other anchoring junctions. In the pupillary portion of the iris—the part containing the sphincter muscle—most of the anterior epithelial cells lack muscular processes (except for a few strands of tissue, there is no dilator muscle here) and therefore cannot be called myoepithelial cells like their counterparts in the ciliary portion of the iris. Despite this distinct anatomical difference, it would be surprising if the anterior epithelial cells in the two portions of the iris were genetically different. The cells in the pupillary portion of the iris undoubtedly once had, and may still retain, the ability to form muscular processes, but that ability presumably has been suppressed by the presence of the sphincter muscle, which may exert a negative or inhibitory induction during development.

As mentioned earlier, nothing is known about the life cycle of the iris epithelial cells, but they are probably not permanent. The presence of clump cells in the iris stroma, with their large complements of ingested melanin, suggests that anterior epithelial cells do die and degenerate over time, requiring a mechanism for cleaning up their complement of melanin pigment.

The posterior epithelial cells contact the anterior surface of the lens

The posterior epithelial cells are columnar and contain a higher density of melanin granules than the anterior epithelial cells. They have gap junctions and anchoring junctions with their neighbors and a junctional complex near the apical side consisting of anchoring and tight junctions arranged in belts around the cells. This junctional complex appears to be a diffusional barrier that prevents the movement of molecules from the stroma into the posterior chamber. The pigment density appears to be greater than in any of the melanocytes in the stroma or anterior border layer, making the posterior epithelium a major light-absorbing layer in the iris. Light absorption is its only known function.

Perhaps the most important feature of the posterior epithelium is its location. The posterior epithelium forms the anterior boundary of the posterior chamber, it is in direct contact with aqueous flowing through the posterior chamber, and it rests on the anterior surface of the lens, separated from the lens only by the thin layer of aqueous flowing between them. One consequence of these relationships is the fate of melanin pigment released from posterior epithelial cells if the cells die naturally or are damaged by trauma or infection. Pigment released from the posterior epithelial cells goes into the posterior chamber, where it is caught up in the normal aqueous flow through the pupil into the anterior chamber and eventually into the trabecular meshwork (see Chapter 9). Pigmentation in the trabecular meshwork appears to increase with age, suggesting a modest, ongoing loss of cells and pigment from the posterior epithelium (see, for example, Figure 2 in Box 9). The trabecular endothelial cells ingest the pigment, and under normal circumstances it causes no problem.

Higher rates of pigment release can occur because of cellular damage or "exfoliation syndromes" in which cell loss from the posterior epithelium is greatly exaggerated. Whatever the proximate cause of abnormally high pigment release is, the result can be excessive accumulation of pigment granules in the trabecular meshwork. The pigment granules block the passage of fluid through the meshwork either by plugging the pores in the meshwork directly or by unusual enlargement the trabecular endothelial cells that ingest the pigment. The result is an increased resistance to aqueous outflow and a **pigmentary dispersion glaucoma**.

Although the iris normally slides freely over the surface of the lens, it can become attached to the anterior lens capsule; adhesions between iris and lens are called **posterior synechiae**. (Adhesions between the iris and cornea—**anterior synechiae**—can occur if a damaged iris floats forward to come in contact with the corneal endothelium.) A posterior synechia may come to be noticed because it prevents symmetrical changes in pupil size. If the region of attachment is small, it can be broken by pharmacological dilation or constriction of the pupil, using the force of the contracting iris muscles to break the adhesion. This approach, however, risks tearing the lens capsule if the adhesion is too strong, and damage to the capsule can set the stage for cataract formation.

If the iris adhesion to the lens surface is extensive—in the extreme case adhering all around the margin of the pupil—the normal flow of aqueous through the pupil will be blocked. Aqueous produced by the ciliary epithelium will accumulate in the posterior chamber, causing the iris to bulge forward into the anterior chamber (Figure 10.20). This condition is called **iris bombé** (from the French verb *bomber,* meaning "to bulge" or "to arch"). The abnormal bulging of the iris may narrow the anterior chamber angle considerably, thereby blocking aqueous outflow, and the obvious result will be a serious elevation of intraocular pressure. If the synechiae cannot be broken to allow normal aqueous flow, the solution is usually a partial iridectomy or the creation of drainage holes near the iris root to allow aqueous flow between the posterior and anterior chambers (see the next section).

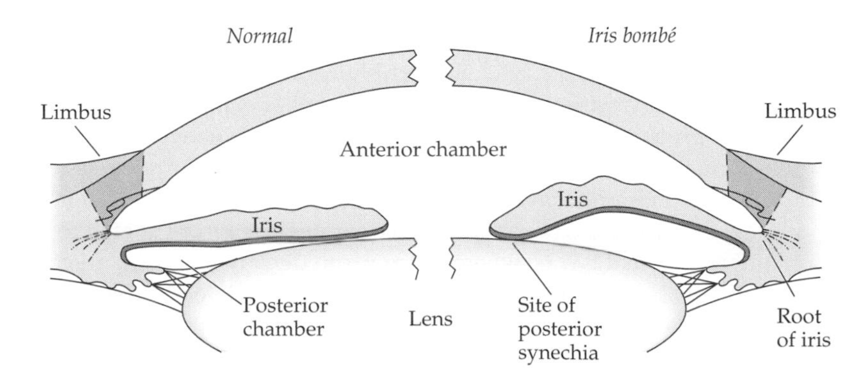

Figure 10.20
Iris Bombé
The normal relationship between the lens and iris is shown on the left. At right, the iris has attached to the lens—forming a posterior synechia—preventing aqueous flow into the anterior chamber. The fluid buildup in the posterior chamber pushes the iris forward so that it bulges into the anterior chamber. This condition is known as iris bombé.

(a) Partial iridectomy

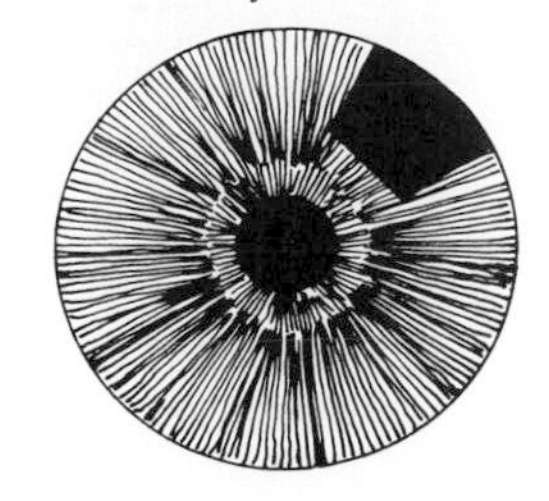

(b) Iridoplasty

(c) Treatment of iris bombé with iridoplasty

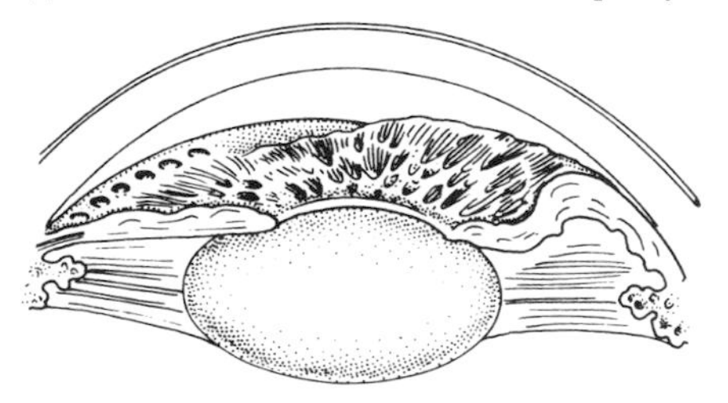

Figure 10.21
Iris Surgery for Closed-Angle Glaucoma
(*a*) In this partial iridectomy, a piece of the iris has been cut away to allow access to the anterior chamber angle. (*b*) In laser iridoplasty, a series of holes are punched through the iris near the iris root. (*c*) Iridoplasty eliminates iris bombé by allowing aqueous to flow directly from the posterior chamber to the filtration angle.

Surgery for closed-angle glaucoma often involves the iris rather than the limbus

Prior to the use of lasers in surgery, the strategy in closed-angle glaucoma was to remove one or more pieces of the iris out to the iris root (Figure 10.21*a*). Removal of tissue around part of the angle's circumference allows aqueous from both anterior and posterior chambers to flow into the trabecular meshwork. In general, the success rate for this type of surgery has been quite high, but this approach has the drawback of having to penetrate the eye in order to remove the iris tissue. Ocular penetration always carries a finite risk of complications.

High-power laser beams focused on the iris can accomplish the same task without opening the eye. The laser is used to burn a series of holes through the iris around the circumference of the iris root (Figure 10.21*b*). If the angle closure is secondary, as in iris bombé, the holes allow aqueous to move directly from the posterior to the anterior chamber, reducing the fluid accumulation in the posterior chamber so that the iris can drop back into a normal position and thus eliminate the angle closure (Figure 10.21*c*). In primary or anatomical angle closure, the relative positions of the iris and trabecular meshwork are *not* altered by this technique, but many holes in the iris will give direct access to a larger surface area of the trabecular meshwork. In other words, the angle is still closed, but the laser surgery has created greater access to the apex of the angle.

As in the laser trabeculoplasty used to create drainage channels in the trabecular meshwork (see Chapter 9), the surgical holes in the iris tend to fill in over time as the tissue attempts to heal the wound. In the iris, however, the holes are larger and the process of filling in is slower than in the trabecular meshwork. Regression to the preoperative condition is less common, and additional surgery is required less frequently than with trabeculoplasty.

Some Clinically Significant Anomalies of the Iris and Pupil

Changes in iris color after maturity are potentially pathological

Normal iris color, which is usually the same in both eyes, ranges from light blue through shades of gray and green to dark brown. The detailed anatomical differences between a blue iris and a brown one are not known, although the question is commonly asked, and people are still trying to figure it out. In general, however, color differences in the iris are a matter of melanin pigment density and distribution.

All irises have two sorts of light-absorbing pigments with different absorption spectra; one is melanin in the melanocytes and pigmented epithelial cells, and the other is hemoglobin in the blood vessels. Of these, the density and distribution of hemoglobin should vary less because the pattern and distribution of blood vessels in the iris is generally similar from one person to the next. Therefore, variation in iris color should be related to the density of melanin and to the location of the melanin relative to the blood vessels.

The distribution and density of melanin pigment throughout the body are quite variable among individuals and races. In general, dark skin pigmentation is associated with dark iris color, while lighter blue and gray iris colors are more common among persons with light skin pigmentation. The intermediate shades, such as the gray-green combination called hazel, are less amenable to generalization; these irises presumably have less melanin than brown irises, but whether the reduction is uniform throughout the iris or more pronounced in one layer or another is not known. (Variation in color appears *not* to be related to pigmentation in the epithelial layers, where the melanin density is always much the same.) Because much of the melanin in the body accumulates in tissues after birth, Caucasian infants usually have very light irises that acquire darker tones with age; some, of course, remain light in color.

The structure and color pattern of the iris are highly individual, and at some level of detail it is unlikely that any two irises are identical; they are like ocular fingerprints. Small differences in color or different spots of color ("freckles") between an individual's irises are to be expected. Significant difference in color between the two irises in an individual's eyes is called **heterochromia**. Heterochromia should be regarded as abnormal and sometimes pathological. Congenital heterochromia, which results from a developmental difference between the two eyes, is normally stable and of no clinical significance. Acquired heterochromia, however, often indicates that something is amiss; it may be associated with unilateral lesions in the sympathetic pathway to the iris, with melanomas, or with drug-induced changes in melanin synthesis or degradation, among other causes. In other words, unilateral changes in iris color indicate an underlying instability in the vasculature (affecting hemoglobin content) or the iridial melanocytes.

A drug-induced change in iris color has recently been reported that will be a clinical issue for some time to come. The drug latanoprost has proved to be very effective in reducing intraocular pressure in patients with primary open-angle glaucoma or ocular hypertension (see Chapter 9), but in some patients the color in the periphery of one or both irises darkens. The reason for the darkening is not known; thinning of the iris or migration of melanocytes into the region are just two of the possible explanations. The point is that iris color does not normally change over short periods of time; when it does, the iris is sending a signal, in color, that its circumstances have altered and all may not be well.

Differences between the two eyes in pupil size or pupillary responses to light are commonly associated with neurological problems

Pupil size can be thought of as an equilibrium condition, a balance between the sympathetic and parasympathetic innervations to the dilator and sphincter muscles. And since the innervational pathways to the two eyes are not completely independent, the signals going to the iris muscles in the two eyes are strongly correlated. Therefore, under conditions of equal illumination of the eyes, the pupils should be about the same size. Pronounced differences in pupil size (anisocoria) imply abnormal innervational differences.

Although anisocoria is an important indicator because it says something is amiss, it is ambiguous, because it says nothing about the source of the abnormality. For example, is one pupil abnormally large or is the other pupil abnormally small? Is one of the autonomic innervational systems relatively ineffective or is the other system overactive? These questions cannot be answered by observating the pupils under steady-state conditions; the innervational systems must be probed by seeing how the pupils react to varying illumination conditions.

The pupillary light reflex is the pupil constriction that occurs when light is shone in the eye. This is the **direct reflex**: The pupil constricts in the eye that is illuminated. The simultaneous pupil constriction of the unilluminated eye is the **consensual reflex**. The existence of both direct and consensual reflexes is a consequence of the reflex pathway shown in Figure 10.22. The reflex wiring has some specific points of uncertainty (see Chapter 5), but the basic points are clear. Signals from retinal ganglion cells go to cells in the ipsilateral olivary pretectal nucleus. From the pretectum, however, signals go to both Edinger–Westphal nuclei and thus to both iris sphincter muscles. Therefore, light shone in either eye will normally cause both pupils to constrict.

Clinical evaluation of pupillary light reflexes is most useful for detecting deficits in the afferent pathways—that is, in the optic nerves and tracts. The procedure, which is called the swinging-light test or Marcus Gunn's test, is a comparison of the pupil responses as the eyes are alternately illuminated; the ambient illumination is normally dim so that the test begins with dilated

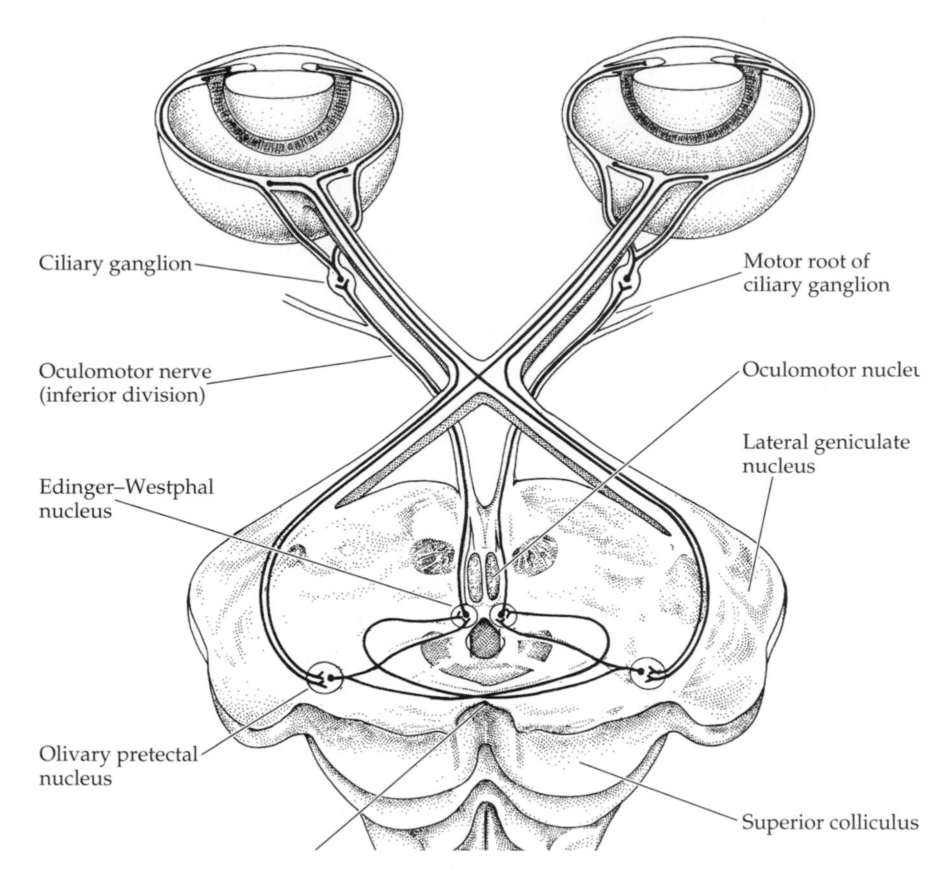

Figure 10.22
Direct and Consensual Pupillary Light Reflex Pathways
The afferent part of the pupillary light reflex pathways consists of the retinal ganglion cell axons terminating in the olivary pretectal nuclei. From the pretectum, cells project to *both* Edinger–Westphal nuclei, which lie next to the nuclei of cranial nerve III. The efferent pathways are from the Edinger–Westphal nuclei through the third nerve (inferior division) to the ciliary ganglion. Ciliary ganglion cells project to the iris sphincter via the short ciliary nerves and the choroid. Because of this arrangement, light shone in either eye will produce pupil constriction in both eyes. (After Kourouyan and Horton 1997.)

pupils. Equal, dim illumination of the two eyes produces equal pupil sizes (Figure 10.23, top). Illumination of the patient's right eye causes both pupils to constrict by equal amounts. In this case, however, shifting the light to the patient's left eye produces a small *dilation* of both pupils; they constrict again when the light is swung back to the right eye.

There is nothing wrong here in terms of an imbalance between direct and consensual reflexes; the pupils are always equal in size, whatever the illumination conditions. The problem is that illumination of the patient's left eye has less effect, both directly and consensually, than illumination of the right eye. This relative ineffectiveness indicates that the afferent signals from the left eye are at fault; something in the retina, the optic nerve, or the optic tract has reduced the signal strength. If the left eye were totally blind, illuminating the eye would have no effect and both pupils would dilate to their original pretest size. The difference between the two eyes in the pupillary response is a *positive* Marcus Gunn sign. (It has nothing whatever to do with Marcus Gunn's syndrome—jaw-winking—discussed in Chapter 7.)

A simple way to make this test yield quantitative values involves placing a neutral density filter in front of the *good* eye—the right eye in the foregoing example—and repeating the test. The filter density for which the difference in effect of illumination to the right versus the left eye disappears is a measure of the relative inefficiency of the affected eye.* If the balancing filter in front of the good eye had a value of 1 log unit, the affected eye would be judged ten times less efficient than the good eye.

One of the common sources of afferent pathway defects is **optic neuritis**, an infection of the optic nerve. The deficit may be anything from half as effective

*Optical density (D) is the logarithm to the base 10 of the ratio of light incident (I_{inc}) to light transmitted (I_{trans}); that is, $D = \log_{10} (I_{inc}/I_{trans})$. Therefore, a 1.0 density filter transmits 1/10 of the incident light, a 2.0 density filter transmits 1/100 of the incident light, and so on. $\text{Log}_{10} 2 \approx 0.3$, which means that increasing filter densities in steps of 0.3 log units will halve the transmitted light at each step.

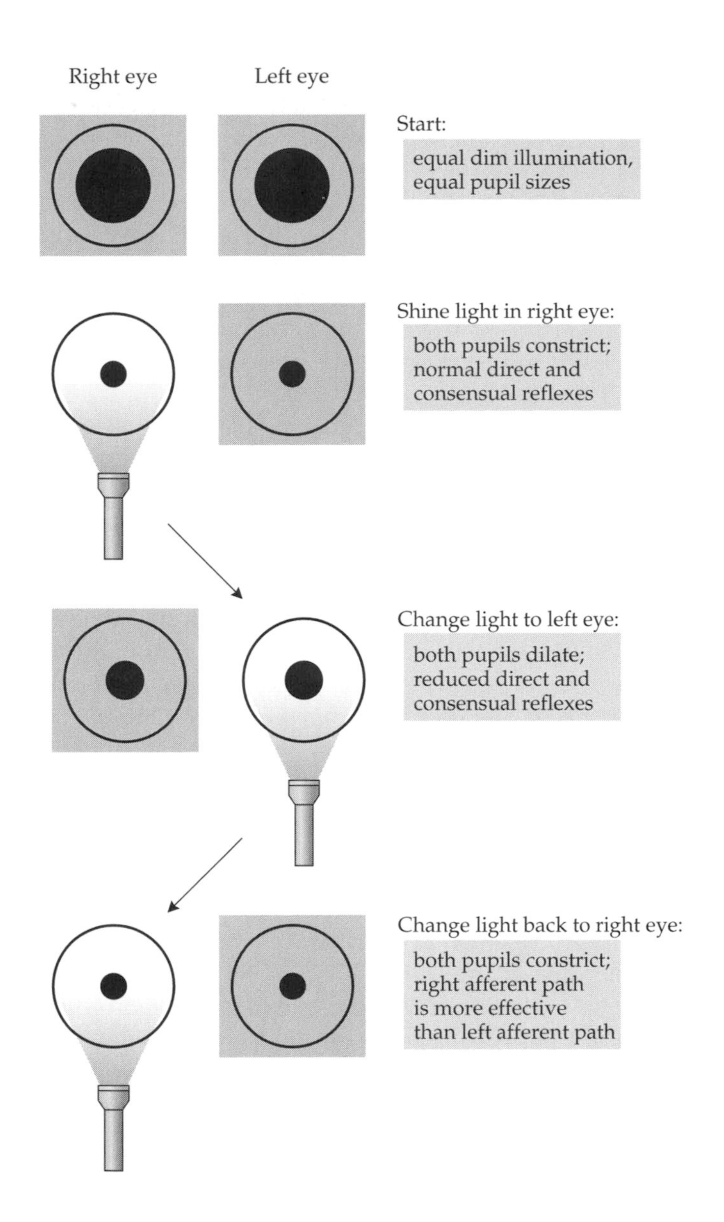

Figure 10.23
Demonstration of an Afferent Pathway Defect: Marcus Gunn's Test
As illumination is alternated between the right and left eyes in this example, illumination of the left eye is less effective than illumination of the right in producing pupil constriction; this is evidence for a *left* afferent pathway defect and is a *positive* Marcus Gunn sign. In the normal case, the effects of illuminating the right and left eyes would be identical.

(0.3 density units) to 1000 times less effective (3.0 density units) depending on the severity of the damage to the optic nerve. Although optic neuritis can be detected with other methods, including the ophthalmoscopic appearance of the optic nerve head, the simple swinging-light test can provide an early sign, because a small reduction in effectivity may be present before other diagnostic signs—like nerve head pallor—begin to appear. Various retinal pathologies (macular degeneration, central serous retinopathy, posterior retinal detachment, among others) are evident as afferent pathway defects, as are lesions of the optic tracts.

Anisocoria and unresponsive pupils are often associated with defects in the efferent part of the innervational pathways

Efferent pathway defects are more difficult to deal with, despite the apparent simplicity of the reflex pathway in Figure 10.22. The differences between direct and consensual reflexes must be considered, and the responses of the pupil to other stimuli—like accommodation—may also be important. To make matters worse, a lesion in the efferent parasympathetic pathway will make the pupil in the affected eye dilated and relatively unresponsive to light because of the reduced innervation to the sphincter. Although the dilated, unresponsive pupil is already a sign of a problem, the lack of response makes further testing difficult.

Reduced parasympathetic innervation can result from lesions at any point in the efferent pathway—the midbrain at or above the Edinger–Westphal nuclei, along the third cranial nerve to the point where parasympathetic fibers leave it within the orbit, or at the ciliary ganglion and short ciliary nerves. Damage to the iris itself or posterior synechiae may also produce a dilated, unresponsive pupil. Other observations, such as paretic extraocular muscles or the presence or absence of pupillary constriction when miotic drugs are applied, are used to localize the lesion site and suggest other tests to show the presence of tumors or aneurysms, for example.

Most of the literature on unresponsive, dilated pupils concerns a phenomenon whose underlying defect is unknown: **Adie's syndrome**. The first sign in these cases is an anisocoria of abrupt onset in which the dilated pupil has reduced light reflexes, whether elicited directly through the affected eye or consensually through the other eye. Pupil constriction with accommodation is supposedly unaffected, however, and there are no signs of third-nerve involvement. Taken together, these observations suggest a lesion site in the midbrain just prior to the Edinger–Westphal nucleus on the affected side; only here is the pupillary reflex pathway separate from the accommodative and oculomotor pathways.

Adie's syndrome is idiopathic, however; no midbrain lesions have been reported to explain the phenomenon. Some authors have suggested that the problem is within the orbit, perhaps due to a viral infection of the ciliary ganglion. But this suggestion only raises another question with no answer: A viral infection of the ciliary ganglion would have to be highly specific, affecting only the cells projecting to the pupil sphincter while sparing those targeting the ciliary muscle, and there is no obvious basis for such a specific infection.

Part of the problem here is contradictory statements in the literature—sometimes in the same article—concerning accommodative paresis in Adie's syndrome. The near response may be affected to some extent, but the effect may not persist. If so, the pupillary deficit is *not* independent of the accommodative pathway. Uncertainties of this sort prevent the linking of Adie's syndrome with any specific causative factor or even with a specific lesion site. At present, this form of tonic pupil seems to be a kind of diagnostic wastebasket for the cases of tonic pupil that have no clear explanation.

Matters are much simpler when a tonic pupil is associated with an obviously normal pupillary near reflex. In this situation—an **Argyll Robertson pupil**—the lesion site is in the midbrain, where it affects the pupillary pathway prior to the Edinger–Westphal nuclei without interfering with the accommodative inputs to the nuclei. Several decades ago, Argyll Robertson pupils were considered to be diagnostic for neurosyphilis, which produces midbrain lesions in the appropriate place. This association between Argyll Robertson pupil and neurosyphilis is still accurate, but other neuropathies can produce similar lesions, including those related to diabetes and alcoholism. Because advanced syphilis is now much less common than it once was, clinicians may be checking for Argyll Robertson pupil less frequently, but the association of this pupil deficit and other neuropathies suggests that the pupillary near reflex should always be evaluated.

The forms of tonic pupil mentioned thus far are associated with dilated pupils and a problem with the sphincter muscle or its parasympathetic innervation. Pupil size and mobility are also affected by deficits in sympathetic innervation to the dilator, and this innervation loss is reflected as pupil constriction. The classic example of a sympathetic deficit is **Horner's syndrome**, discussed in Chapters 5 and 7; the full-blown syndrome is a combination of miosis (pupil constriction), ptosis (drooping of the upper eyelid), and anhidrosis (lack of sweating by the facial skin). The postganglionic sympathetic pathways from the superior cervical ganglion to different target structures begin to diverge as the fibers run up the

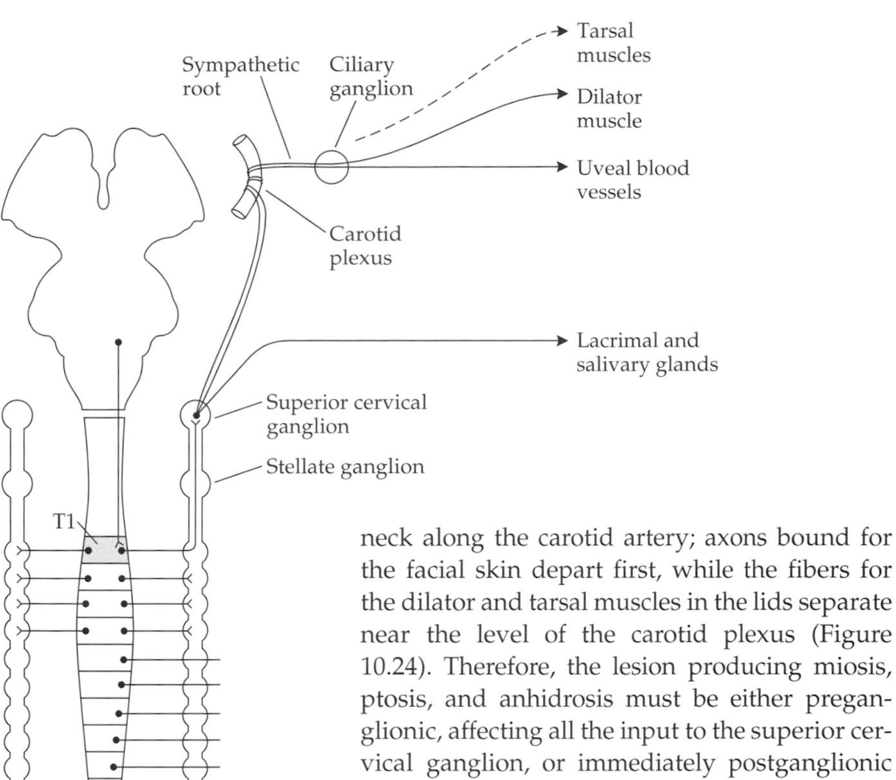

Figure 10.24
Sympathetic Pathways to the Eye and Its Surroundings
The afferent portion of the sympathetic pathway is a two-neuron link from the brainstem through the spinal cord and out to the superior cervical ganglion. Axons from the cells in the ganglion ascend along the carotid artery; some branch off to innervate sweat glands in the facial skin while the others continue up to the carotid plexus. Axons leave the plexus and enter the orbit as the sympathetic root of the ciliary ganglion, through which they pass into the short ciliary nerves and then into the choroid. A lesion at the superior cervical ganglion or anywhere along the afferent pathway will produce the symptoms associated with Horner's syndrome: ptosis, miosis, and anhidrosis.

neck along the carotid artery; axons bound for the facial skin depart first, while the fibers for the dilator and tarsal muscles in the lids separate near the level of the carotid plexus (Figure 10.24). Therefore, the lesion producing miosis, ptosis, and anhidrosis must be either preganglionic, affecting all the input to the superior cervical ganglion, or immediately postganglionic before the fibers depart to the facial skin. Unfortunately, these requirements do little to pinpoint a lesion site; it could be any place from the brainstem through the cervical part of the spinal cord, which is the route of the central pathway, between the spinal cord and the superior cervical ganglion (the preganglionic pathway), or in the upper part of the neck just above the superior cervical ganglia. More to the point, there are a variety of possible causes of a lesion—some benign and some not; localizing and identifying the lesion in Horner's syndrome is not simple, but it must be done.

Lesions in the sympathetic pathways closer to the eye do not produce classic Horner's syndrome, because the only possible effects are miosis and ptosis, which can occur together or separately, depending on the lesion site. Miosis and ptosis might be produced by a tumor in the vicinity of the cavernous sinus, for example, while isolated miosis is likely to be a lesion within the orbit or a defect in the dilator muscle itself.

Clinically useful drugs affecting pupil size fall into four functional groups

Because pupil size represents a balance, or equilibrium, between tonus of the opposing sphincter and dilator muscles, the pupil can be dilated either by activating the dilator or by inhibiting the sphincter, and it can be constricted either by activating the sphincter or by inhibiting the dilator. Thus the pharmacologic agents affecting pupil size are those that mimic or block the actions of the neurotransmitters to the sphincter and dilator, as Figure 10.25 summarizes.

Consider first the upper right-hand quadrant, in which dilation of the pupil is a result of relaxation of the sphincter. The easiest way to reduce the effectivity of acetylcholine at the sphincter is with agents that compete with acetylcholine

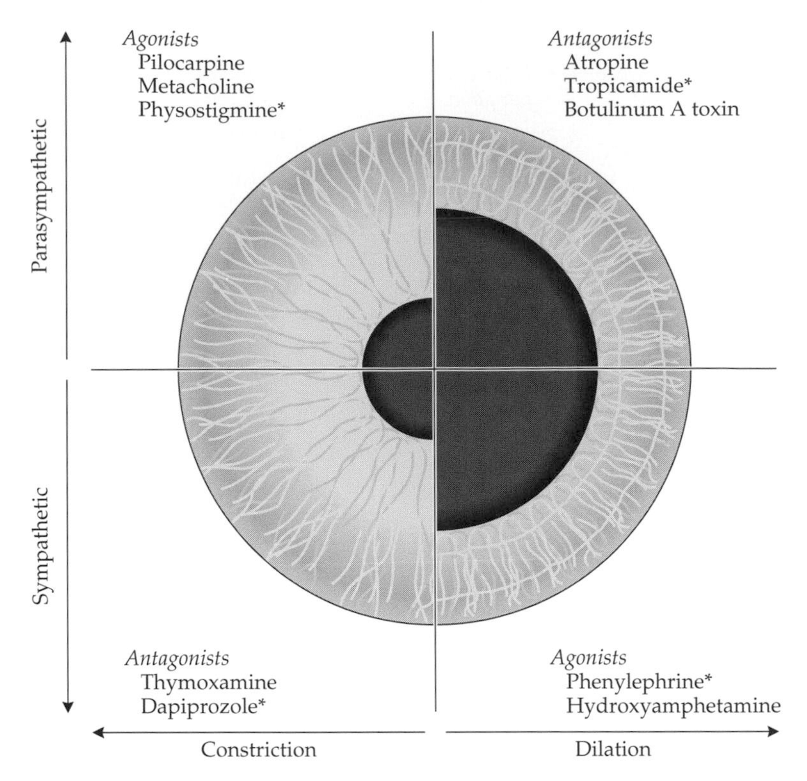

Figure 10.25
Some Drugs That Affect Pupil Size
The schematic iris and pupil in this illustration have been quartered by a vertical division between drugs that dilate or constrict the pupil and a horizontal division between drugs that affect the parasympathetic system or the sympathetic system. Asterisks denote the agents most often used clinically. The pupil can be dilated either by parasympathetic antagonists (e.g., tropicamide) or by sympathetic agonists (e.g., phenylephrine). It can be constricted by either parasympathetic agonists (e.g., physostigmine) or sympathetic antagonists (e.g., dapiprozole).

at the receptor sites on the muscle fibers; when a competitor occupies the receptor sites, naturally released acetylcholine cannot bind to produce muscle contraction. The oldest of these parasympatholytic agents are atropine and scopolamine, both of which also produce accommodative paresis (cycloplegia); the pupil dilation can last for several days. The ideal drug in this category would be one that had little effect on accommodation but caused a strong but short-lived pupil dilation. Of the other agents listed in Figure 10.25, none meets these criteria, but the best approximation is tropicamide, which is used routinely in clinical practice. Tropicamide appears to be readily diffusible, which makes it fast-acting, and to be loosely bound to receptor sites, which makes its action persist for only a matter of hours. It often has little effect on accommodation at doses that produce mydriasis, but it can be an effective cycloplegic if desired.

Neurotoxins, such as botulinum A toxin, reduce synaptic transmission by preventing acetylcholine release from the nerve terminals (see Chapter 4). In theory, botulinum A toxin could be used as a mydriatic, but its extreme persistence, among other contraindications, makes it inappropriate for this purpose.

The other route to pupil dilation is direct activation of the dilator muscle, which is accomplished by agents that either mimic the action of norepinephrine or promote its release from the sympathetic nerve terminals. Phenylephrine, which is better known under one of its trade names—NeoSynephrine—is the most commonly used agent in the category of sympathomimetic drugs; it is structurally similar to norepinephrine and supplements the action of the endogenous neurotransmitter release. In keeping with its sympathomimetic action, phenylephrine produces noticeable vasoconstriction of the conjunctival blood vessels and some widening of the palpebral fissure due to induced contraction of the superior and inferior tarsal muscles in the eyelids. Unlike the parasympatholytic drugs, phenylephrine has little effect on accommodation, and what effect there is may be attributed to vasoconstriction in the blood vessels of the ciliary body.

Hydroxyamphetamine (trade name Paredrine) is another effective mydriatic, but it acts on the nerve terminals to increase the normal transmitter release. The end result is more transmitter or transmitter analogue to bind to postsynaptic receptor sites and produce more muscle contraction. But because the drug acts on transmitter *release,* the pupil's response to hydroxyamphetamine can be used to test for the presence of a postganglionic lesion in the sympathetic pathway. If such a lesion has occurred, the nerve fibers will degenerate; with the nerve terminals gone, hydroxyamphetamine can have no effect and the pupil will not dilate. With phenylephrine, however, which acts directly on the muscle fibers, the pupil will dilate whether or not the nerve terminals are present. Hydroxyamphetamine has no detectable effect on accommodation.

Mydriatics are routinely used in clinical practice to obtain a good field of view for ophthalmoscopy, but there is little reason to constrict pupils in the course of a routine eye examination. Thus, miotics are less commonly used. Their main uses are to test the sphincter's ability to contract—a normal sphincter should contract in response to a parasympathomimetic drug—or to counteract the effect of a mydriatic that was administered earlier. The latter use is dignified by the term "mydriolytic," but the drugs are essentially miotics. On a long-term basis, miotics may be used in the management of glaucoma, but their effects in reducing intraocular pressure have nothing to do with their effect on pupil size (see Chapter 9).

The most effective miotics are those acting as agonists or potentiators of the cholinergic input to the sphincter muscle. Pilocarpine is the prototypical acetylcholine agonist at the sphincter; it exerts its effect by binding to and activating muscarinic acetylcholine receptors on the muscle fibers. Not surprisingly, pilocarpine also acts on the ciliary muscle, which also contracts. Most of the clinical use of pilocarpine centers on its accommodative rather than its miotic effects; the induced contraction of the ciliary muscle appears to enhance aqueous outflow and reduce intraocular pressure (see Chapter 9). For other purposes, however, the accommodative spasm that may result from pilocarpine administration limits its value. Metacholine is another of the parasympathomimetc drugs whose action is similar to that of pilocarpine.

When acetylcholine is released into the synaptic cleft at the neuromuscular junction, its action is normally short-lived because cholinesterases break the acetylcholine down into constituent parts for reuptake into the nerve terminals. Thus, another effective stategy for activation of the sphincter muscle is to reduce the amount of cholinesterase so that the normal transmitter release can be prolonged. Physostigmine and neostigmine are two of the commonly used drugs in this category of *anti*cholinesterases. Although both these agents and others with similar actions are effective miotics, they are generally used for other purposes, most notably for symptomatic relief of ptosis in myasthenia gravis (see Chapter 4).

Miosis will result not only from contraction of the sphincter but also from relaxation of the dilator, which can be accomplished by administering alpha-adrenergic blockers. These drugs compete with norepinephrine for receptor sites and prevent the neurotransmitter from having its normal effect. Two commonly used sympatholytic drugs are thymoxamine and dapiprozole, both of which block alpha-adrenergic receptors. They are usually employed to counteract the effect of a mydriatic drug; that is, they are used as mydriolytic agents.

Mydriasis produced by an acetylcholine receptor blocker, such as tropicamide, cannot be readily counteracted with any of the parasympathomimetic agents, because the receptor sites on the sphincter muscle are blocked to both endogenous acetylcholine and its analogues or potentiators. Thus, any reduction in pupil size must be accomplished by relaxation of the normal tonus of the dilator, which is what alpha-adrenergic blockers do. Both thymoxamine and dapiprozole reduce tropicamide mydriasis to near the predrug state in about an hour.

Mydriasis produced by a sympathomimetic agent, such as phenylephrine, can be counteracted with a parasympathomimetic drug, such as pilocarpine; the sphincter is by far the stronger of the two muscles, and in a tug-of-war between them, it will always win. The problem, however, is that the simultaneous use of sympathomimetic and parasympathomimetic drugs makes the anterior chamber shallower, increasing the risk of angle closure. Therefore, the usual way to counteract phenylephrine-induced mydriasis is to administer one of the alpha-adrenergic blockers. This technique works, presumably, because the receptor blockers bind more readily or more strongly to the receptor sites than does phenylephrine. Whatever the case, thymoxamine and dapiprozole can eliminate phenylephrine mydriasis in about an hour; without the alpha-adrenergic blocker, the mydriasis would persist for 12 hours or more.

Development of the Iris

The iris stroma forms first by migration of undifferentiated neural crest cells

As discussed and illustrated earlier in reference to formation of the anterior chamber (see Figure 9.23), there are three early waves of cell migration into the region between the developing lens and the surface epithelium (Figure 10.26). The migrating cells come from the mesoderm around the optic cup, but they are neural crest cells, not mesodermal cells. Cells in the first wave of migration will become the corneal and trabecular endothelium, and the third wave produces the corneal stroma. The second migration, which is the one of interest here, produces a complete, thin layer of cells in front of the lens and the rim of the optic cup (see Figure 10.26). The tunica vasculosa lentis—the complex of transient blood vessels supplying the developing lens—lies between the new tissue layer and the lens.

The cells in this new layer will differentiate and proliferate, becoming fibroblasts and melanocytes for the stroma and the anterior border layer. They have nothing to do with forming either the sphincter or the dilator, and the stromal blood vessels are of mesodermal origin, produced as branches from the arterial circles in the ciliary body. In effect, the blood vessels penetrate the developing stroma from the developing ciliary body.

Note that the newly formed iris lacks a pupil. The original stroma is a complete layer from which the pupil will be created by tissue atrophy and cell migration.

The epithelial layers and the iris muscles develop from the rim of the optic cup and are therefore of neuroectodermal origin

After the first layers of iris stroma have formed, the two layers of cells at the rim of the optic cup proliferate and grow inward between the lens and the iris stroma (Figure 10.27*a*). The cells in the layers become smaller with time, and by the time their inward growth ceases (around month 6 of gestation), they have begun to form anchoring junctions and take on the appearance of conventional epithelium.

Both iris muscles are formed from the anterior epithelial layer, but the sphincter appears considerably earlier than the dilator. The first sign of the sphincter is a bulge or buckle in the epithelium near the margin of the future pupil (Figure 10.27*b*). As the bulge increases in size, it begins to pinch off from the epithelium, eventually separating completely to create a distinct cell cluster. It is not clear if this process occurs continuously all around the future pupil margin or if there

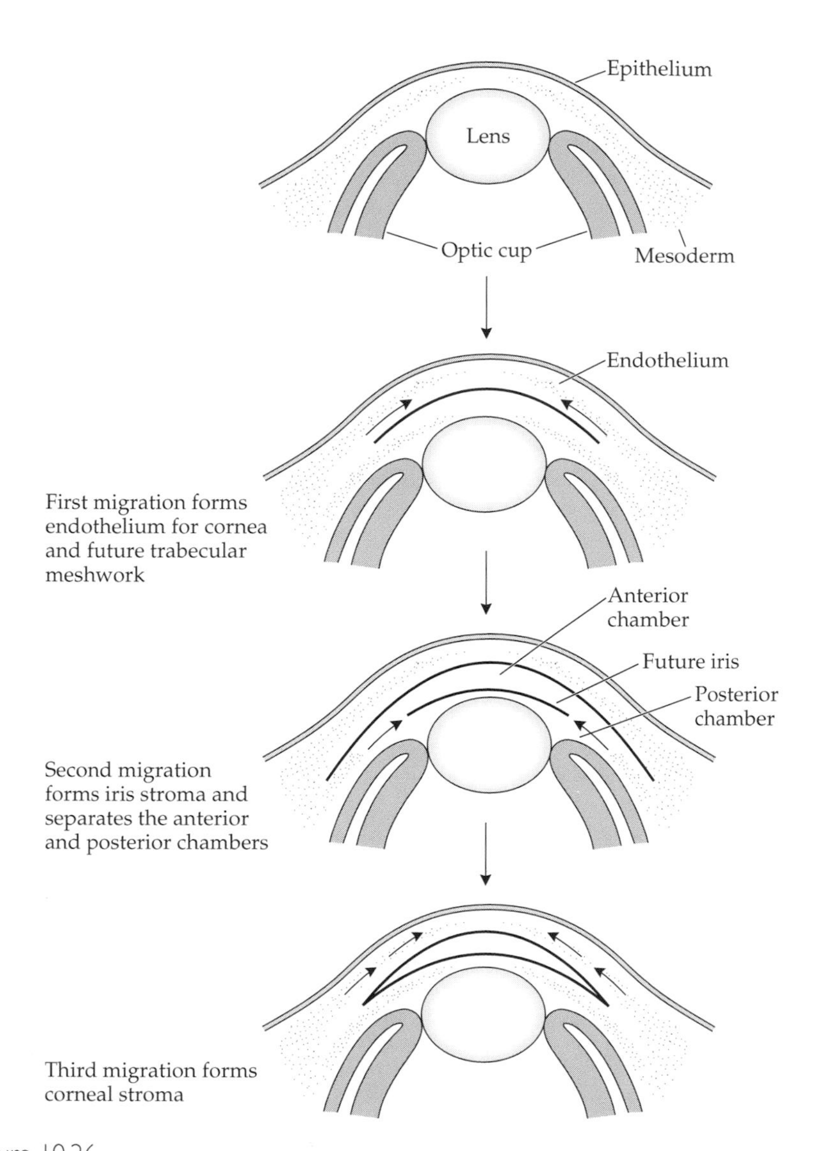

Figure 10.26
The Iris Appears in the Second of the Three Waves of Cell Migration
The anterior chamber and its surrounding structures are generated by waves of neural crest cell migration from the surrounding mesoderm. Cells in the first and third waves migrate into the future cornea. The second migration, however, creates a layer of tissue just in front of the lens. This tissue layer is the precursor of the iris stroma.

are multiple disconnected developmental loci from which the fibers of the sphincter form; whatever the case, cells have budded from the epithelium into the location of the future sphincter. Once this cell budding is done, formation of the muscle is a matter of proliferation and differentiation of the epithelial cells to become muscle fibers. As the fibers elongate and interlace, a continuous annular muscle is created.

Because the dilator fibers are sprouts from the anterior epithelial cells, formation of the dilator is simpler; the epithelial cells need only to grow out the small processes that will become the muscle fibers and synthesize contractile proteins within the processes (Figure 10.27*d*). In the mature iris, the territories occupied by the sphincter and dilator have almost no overlap; the dilator runs in to the outer rim of the sphincter and stops there. This territoriality appears to be a developmental phenomenon in which cells in the anterior epithelium that lie next to the developing sphincter do not differentiate as myoepithelium; that is,

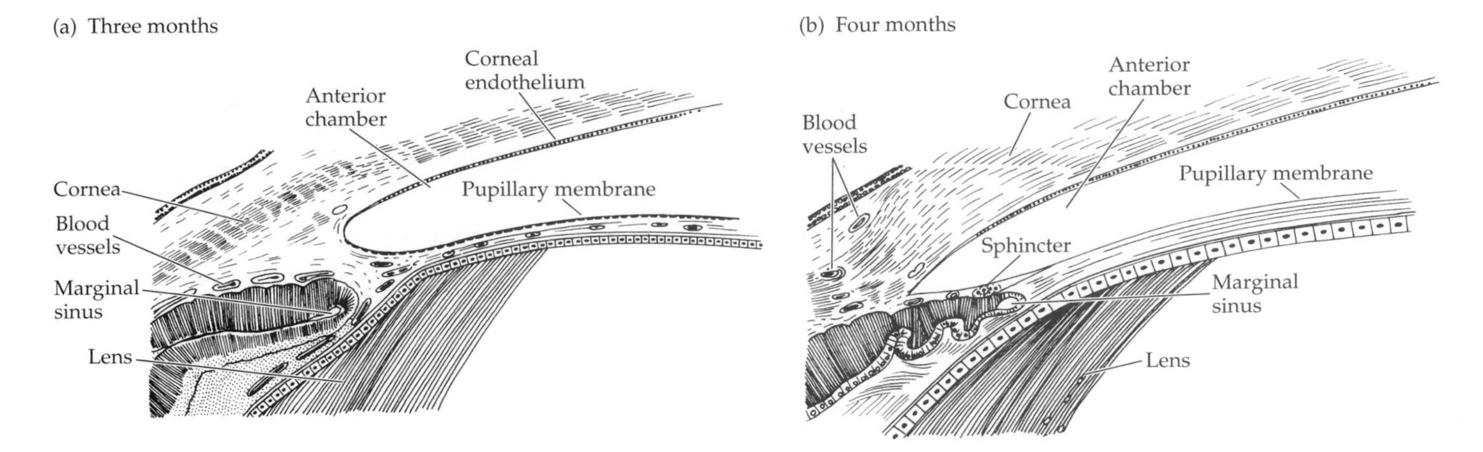

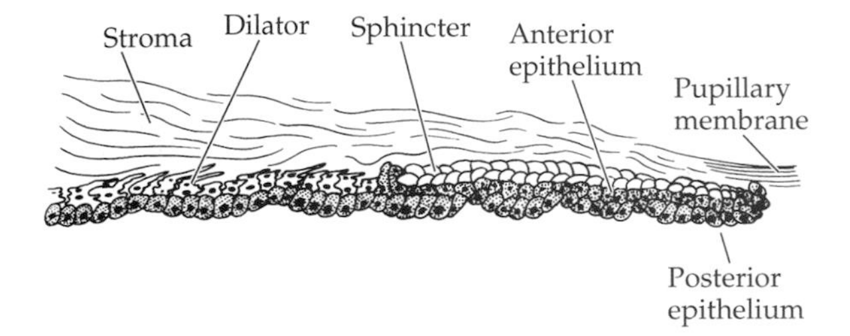

Figure 10.27
Development of Epithelium and Iris Muscles
(*a*) After the migrating neural crest cells have formed the primitive iris stroma, cells at the rim of the optic cup begin to proliferate and push inward along the posterior surface of the pupillary membrane. (*b*) The sphincter appears at this time, about 4 months into gestation, as a small cluster of cells that has budded off from the edge of the advancing optic cup rim. The sphincter cells move into the iris along with the advancing epithelial layers. (*c*) The epithelium is largely complete by the sixth month of gestation, at which time the anterior epithelial cells differentiate further and begin to form the processes of the dilator.

any expression of their ability to elongate and synthesize contractile proteins is inhibited in some way, probably by a signal from the developing sphincter muscle cells.

The pupil is the last feature of the iris to appear

Viewed from in front, the iris is nearly complete by the end of the fifth month of gestation (Figure 10.28*a*), with recognizable muscle and epithelial layers, blood vessels, and so on, but it still lacks a pupil. Although the inward growth of the epithelial layers has not progressed all the way to the center of the iris, thereby creating an epithelium-free zone, there is no hole in the overlying stroma. The epithelium-free zone marks the site of the future pupil, but the pupil must be created by removal of the tissue overlying the epithelium-free region.

During the sixth month of gestation, the arcades of blood vessels that have penetrated to the center of the iris (Figure 10.28*b*) begin to change. Circulation in the most central branches slows, then ceases, depriving the central tissue of any supply. The closed-down vessels atrophy back to the region of the sphincter, and the tissue around the atrophied vessels also begins to change. Both cell degeneration and migration clear tissue out of the central area to leave a hole in the iris; this hole is the pupil.

The early iris lacking a pupil is sometimes called the pupillary membrane, and one of the most common developmental anomalies of the iris is a **persistent pupillary membrane**. The postnatal remnants of the pupillary membrane may be just a few cobweblike strands of tissue or a more extensive aggregation of tissue that makes the pupil appear misshapen.

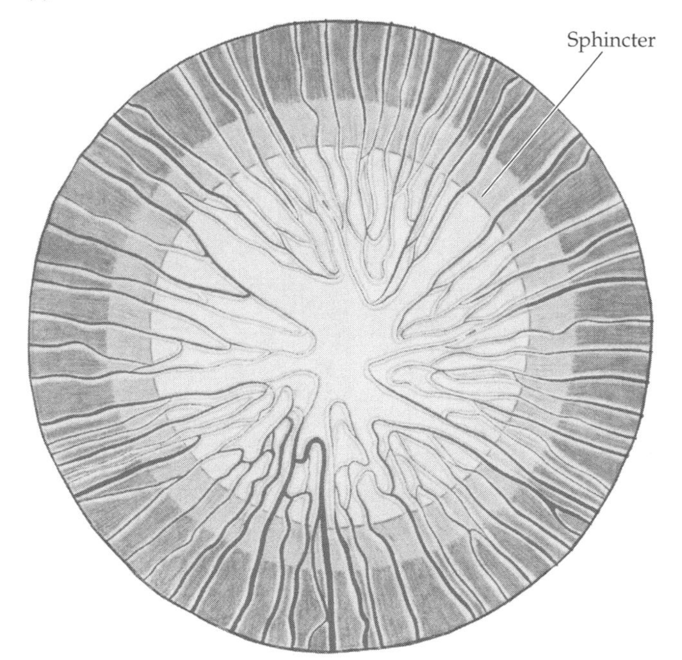

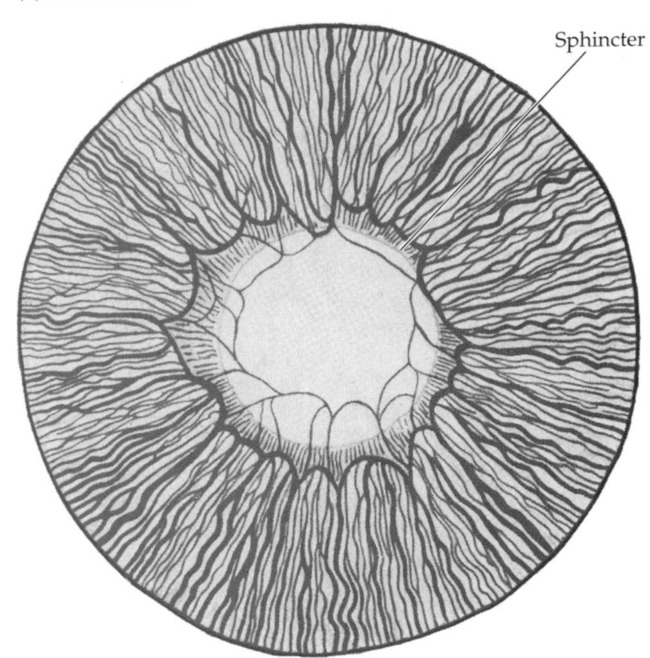

Figure 10.28
Formation of the Pupil
(*a*) The iris at 5 months is a complete, unbroken layer of tissue with a well-developed vascular system; only the blood vessels are shown here, as seen from in front. The epithelium, with the overlying sphincter at its margin, has advanced about halfway into the iris. (*b*) At 7 months, the epithelial advance is complete, and the blood vessels and associated tissue are withdrawing from the epithelium-free zone in the center of the iris. The tissue withdraws by atrophy and cell migration; once the withdrawal is complete, the pupil will be free of any obstructing tissue. (After Mann 1950.)

Most postnatal development of the iris is an addition of melanin pigment

Like the cornea, the iris is very near its adult size at birth; there is very little tissue to be added and it is done during the first few postnatal years. The most noticeable postnatal change is in iris color. In Caucasian infants, the iris is usually blue, becoming darker in the next several years to produce the range of iris colors seen among individuals: light blue to very dark brown.

The anterior epithelium acquires its pigment fairly early in gestation, while still part of the optic cup; the posterior epithelium, which is derived from the inner, unpigmented layer of the optic cup, acquires its pigment gradually during the fifth and sixth months. This leaves the postnatal color change in the iris to be explained by gradual accumulation of melanin in the stroma and anterior border layer. The final color of the iris—blue, gray, green, brown—appears to depend on the density and location of the added pigment, but the relationship of a specific color to particular densities and locations is not known.

In non-Caucasian races, infants have more melanin in their skin at birth and generally have brown irises. The brown color usually darkens with age, however, indicating that melanin continues to accumulate in these iris stromas also.

Segmental defects and holes in the iris result from unsynchronized or failed growth of the optic cup rim

If the rim of the optic cup fails to proliferate and push in over the anterior surface of the lens, the iris does not form, except for a bit of tissue at the rim of the optic cup; the result is **aniridia** (total absence of the iris). This rare genetic defect is associated with the homeotic *Pax-6* gene. A more common occurrence is a more or less normal iris with abnormal sectors or holes. The sector defects are **colobomas** (see Chapter 1). Typical colobomas (Figure 10.29*a*) are located in the

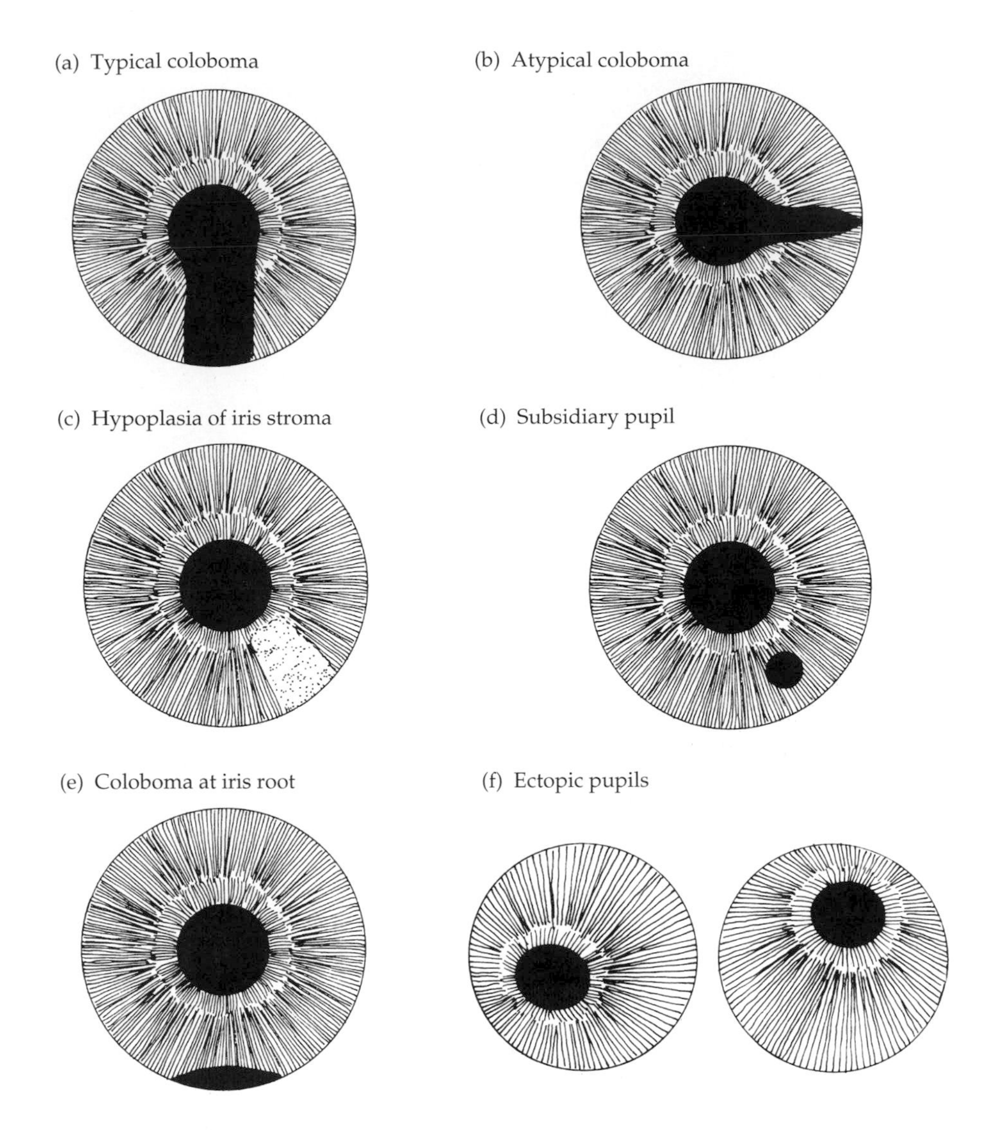

Figure 10.29
Anomalies of the Pupil and Iris

lower part of the iris and are associated with improper closure of the choroidal fissure at the rim of the optic cup. Atypical colobomas (Figure 10.29*b*) in other parts of the iris also occur, and while they are thought also to represent failures in development of the optic cup rim (and failed induction of overlying tissue formation), they cannot be associated with the choroidal fissure and the reason for their occurrence is not known.

Sector defects may appear not as complete absence of tissue in a particular region, but as a region of thinned, abnormally formed stroma (Figure 10.29*c*); defects of this sort are **hypoplasias**, meaning a reduced amount of cell formation. These defects could be inductive failures of less severity than colobomas, but another possibility is an overabundance of the normal tissue atrophy that forms the pupil.

The normal holes of the anterior border layer (the crypts of Fuchs) do not extend completely through the iris. Complete holes may occur as developmental anomalies, however, appearing as a small subsidiary pupil (Figure 10.29*d*) or as a gap at the iris root where the tissue seems to have broken away from the ciliary body (Figure 10.29*e*). How these discontinuities form is not known. They may represent local anomalies in the primitive epithelium over which tissue fails to form because of the lack of inductive signals, or, as some have suggested, the holes may be places where the tissue atrophy of the pupillary membrane has spilled over into surrounding tissue.

Any of these departures from a normal solid iris with a central circular pupil are optically inferior; image quality is inevitably degraded by light entering the eye outside of the normal pupil. In addition, the normal regulation of the amount of light entering the eye is compromised; even if the pupil in an abnormal iris retains some mobility, the aperture can never be closed to a normal minimum, because of the presence of a hole or coloboma.

An ectopic pupil is improperly centered in an otherwise normal iris

Normal pupils are never centered perfectly within the iris, but the decentration is rarely by more than 0.5 mm or so. An **ectopic pupil** (Figure 10.29*f*), however, is markedly out of position. Ectopic pupils are usually bilateral, and the displacement is usually symmetric, although the direction and amount of decentration can be quite different in the two eyes. Some pupil mobility may be present, but it is normally much reduced.

The cause of ectopic pupils is not entirely clear, but the anomaly is often associated with abnormal positioning of the lens (ectopia lentis). This association with a lens anomaly suggests that the ectopic pupils are not simply the by-product of radially asymmetric growth of the optic cup margin—a situation that should affect only the iris. Instead, the defect is attributed to the abnormal persistence of vessels in the tunica vasculosa lentis surrounding the developing lens. Either by mechanical asymmetry or by a growth-inhibiting effect of the persistent blood vessels, the iris grows less readily and the zonule develops less completely on one side. The decentration of pupils and lenses usually produces considerable myopia and poor image quality that are very difficult to correct.

A persistent pupillary membrane may be the result of either insufficient tissue atrophy or tissue hyperplasia

A persistent pupillary membrane may consist of anything from a few strands of tissue running across the pupil to an almost complete blockage of the aperture. Both extremes can be explained by insufficient tissue atrophy as the pupil is forming; a slight failure of atrophy would leave a few tissue strands, while almost no atrophy would leave the pupil filled with tissue. The latter case, however, is probably accompanied by abnormal tissue growth. A persistent pupillary membrane must be extensive for it to cause any serious optical problems, and the extra membrane can usually be removed surgically if necessary. An abnormally thick iris is of concern because of possible narrowing of the anterior chamber angle.

Cysts are fluid or semisolid material enclosed in a sac or capsule. Congenital iris cysts apparently begin as small evaginations from the epithelial layers as the iris is forming. Why such abnormal bulges appear in the first place is not known—a local defect in the overlying primitive stroma?—but once begun, the normal restrictions to epithelial cell growth no longer apply; cells proliferate and the bulges enlarge. The evaginations may detach from the epithelium or remain in contact with it, but they protrude forward into the stroma; if they are large enough, they extend beyond the anterior border layer and are directly visible. Congenital iris cysts may remain small, or they may exhibit an anomalous growth pattern and enlarge for many years. They are usually benign and rarely become large enough to interfere with light entering the pupil or to block the anterior chamber angle.

References and Additional Reading

Functions of the Iris and Pupil

Alpern M, Mason GL, and Jardinico RE. 1961. Vergence and accommodation. V. Pupil size changes associated with changes in accommodative vergence. *Am. J. Ophthalmol.* 52: 762–767.

Campbell FW. 1957. The depth of field of the human eye. *Optica Acta* 4: 157–164.

Fulton AB, Albert DM, and Craft JL. 1978. Human albinism: Light and electron microscopy study. *Arch. Ophthalmol.* 96: 305–310.

Lowenstein O and Loewenfeld IE. 1969. The pupil. Chapter 9, pp. 255–337, in *The Eye,* Vol. 3, *Muscular Mechanisms,* 2nd Ed., Davson H, ed. Academic Press, New York.

Reeves P. 1920. The response of the average pupil to various intensities of light. *J. Opt. Soc. Am.* 4: 35–43.

Rodieck RW. 1998. Plotting light intensity, pp. 151–158, and Photometry, pp. 453–469, in *The First Steps in Seeing.* Sinauer, Sunderland, MA.

Schaeffel F, Wilhelm H, and Zrenner E. 1993. Inter-individual variability in the dynamics of natural accommodation in humans: Relation to age and refractive errors. *J. Physiol.* 461: 301–320.

Spring KH and Stiles WS. 1948. Variation of pupil size with change in the angle at which the light stimulus strikes the retina. *Brit. J. Ophthalmol.* 32: 340–346.

Stakenburg M. 1991. Accommodation without pupillary constriction. *Vision Res.* 31: 267–273.

Takahashi E. 1965. The use and interpretation of the pinhole test. *Optom. Weekly* 56: 83–86.

Structure of the Iris

Brancato R, Bandello F, and Lattanzio R. 1997. Iris fluorescein angiography in clinical practice. *Surv. Ophthalmol.* 42: 41–70.

Fine BS and Yanoff M. 1979. The iris. Chapter 10, pp. 197–214, in *Ocular Histology: A Text and Atlas,* 2nd Ed. Harper & Row, Hagerstown, MD.

Freddo TM. 1996. Ultrastructure of the iris. *Microsc. Res. Tech.* 33: 369–389.

Hogan MJ, Alvarado JA, and Weddell JE. 1971. Iris and anterior chamber. Chapter 6, pp. 202–259, in *Histology of the Human Eye,* WB Saunders, Philadelphia.

Hutchinson AK, Rodrigues MM, and Grossniklaus HE. 1995. Iris. Chapter 11 in *Duane's Foundations of Clinical Ophthalmology,* Vol. 1, Tasman W and Jaeger EA, eds. Lippincott-Raven, Philadelphia.

Loewenfeld IE. 1993. *The Pupil: Anatomy, Physiology, and Clinical Implications,* Vol. 1. Wayne State University Press, Detroit, MI. (There is virtually nothing of importance about the iris and pupil that is not discussed somewhere in this book's 1590 pages.)

van der Zypen E. 1978. The arrangement of the connective tissue in the stroma iridis of man and monkey. *Exp. Eye Res.* 27: 349–357.

Wobmann PR and Fine BS. 1972. The clump cells of Koganei: A light and electron microscopic study. *Am. J. Ophthalmol.* 73: 90–101.

Some Clinically Significant Anomalies of the Iris and Pupil

Gamlin PDR, Zhang H, and Clarke RJ. 1995. Luminance neurons in the pretectal olivary nucleus mediate the pupillary light reflex in the rhesus monkey. *Exp. Brain Res.* 106: 177–180.

Heller PH, Perry F, Jewett DL, and Levine JD. 1990. Autonomic components of the human pupillary light reflex. *Invest. Ophthalmol. Vis. Sci.* 31: 156–162.

Imesch PD, Wallow IHL, and Albert DM. 1997. The color of the human eye: A review of the morphologic correlates and of some conditions that affect iridial pigmentation. *Surv. Ophthalmol.* 41 (Suppl. 2): S117–S123.

Jaanus SD and Carter JH. 1995. Cycloplegics. Chapter 9, pp. 167–182, in *Clinical Ocular Pharmacology,* 3rd Ed., Bartlett JD and Jaanus SD, eds. Butterworth-Heinemann, Boston. (Although the title refers to cycloplegic agents, most of them are also mydriatics of the parasympatholytic variety.)

Jaanus SD and Nyman N. 1995. Mydriatics and mydriolytics. Chapter 8, pp. 151–166, in *Clinical Ocular Pharmacology,* 3rd Ed., Bartlett JD and Jaanus SD, eds. Butterworth-Heinemann, Boston.

Kardon R and Weinstein J. 1997. The iris and the pupil. Chapter 9 in *Duane's Foundations of Clinical Ophthalmology,* Vol. 2, Tasman W and Jaeger EA, eds. Lippincott-Raven, Philadelphia.

Kourouyan HD and Horton JC. 1997. Transneuronal retinal input to the primate Edinger–Westphal nucleus. *J. Comp. Neurol.* 381: 68–80.

Mays LE and Gamlin PDR. 1995. Neuronal circuitry controlling the near response. *Curr. Opin. Neurobiol.* 5: 763–768.

Thompson HS. 1992. The pupil. Chapter 12, pp. 412–441, in *Adler's Physiology of the Eye,* 9th Ed., Hart WM, ed. Mosby Year Book, St. Louis.

Development of the Iris

Cook CS, Ozanics V, and Jakobiec FA. 1991. Prenatal development of the eye and its adnexa. Chapter 2 in *Duane's Foundations of Clinical Ophthalmology,* Vol. 1, Tasman W and Jaeger EA, eds. Lippincott-Raven, Philadelphia.

Mann I. 1950. *The Development of the Human Eye.* Grune & Stratton, New York.

Chapter 11

The Ciliary Body and the Choroid

On contraction, the ciliary muscle could pull the posterior end of the zonule forwards nearer the lens and reduce the tension of the zonule.... If the pull of the zonule is relaxed in accommodating for near vision, the equatorial diameter of the lens will diminish, and the lens will get thicker in the middle, both surfaces becoming more curved.... It would seem that the changes in form of the lens could be explained on this basis.

■ Hermann von Helmholtz, *Helmholtz's Treatise on Physiological Optics*

Three-fourths of the uveal tract consists of the ciliary body and choroid. The choroid is the larger of the two in terms of surface area, lining all of the posterior segment and extending well forward of the equator into the anterior segment, where it meets the ciliary body. Although the ciliary body and choroid differ in structure and function, they have some elements in common. Some structures, like the pigmented epithelium, are continuous from one region to the other, and others, like Bruch's membrane, the ciliary muscle fibers, and some of the vasculature, overlap the dividing line between choroid and ciliary body. In addition, both choroid and ciliary body are about blood, blood flow, and what is done with the blood; the choroid's main job is supplying the needs of the retinal photoreceptors, and one of the ciliary body's jobs is using the blood to make aqueous.

The ciliary body has another job. A large part of it is muscle that is mechanically linked to the lens. This linkage, and the sphincterlike action of the ciliary muscle, are responsible for regulating the curvature and optical power of the lens by a mechanism outlined by Helmholtz almost 150 years ago.

These three functional issues—aqueous formation, accommodation, and photoreceptor blood supply—will be the main topics of the discussion that follows. Because much of the essential structure of the choroid is vascular and is also discussed in Chapter 6, it will receive less attention here than the ciliary body does.

Anatomical Divisions of the Ciliary Body

The ciliary processes characterize the pars plicata

Figure 11.1 shows the ciliary body from behind—that is, looking from the vitreous chamber out toward the pupil. The circular, translucent region in the center of the picture is the lens. It is surrounded by radial lines beginning a millimeter or so from the perimeter of the lens and extending outward another 4 to 5 mm; these radial lines are the **ciliary processes**. The term "processes" is a bit misleading; they are actually linear folds or ridges on the inner surface of the ciliary body. The number of ciliary processes is usually given as 70 to 80; I counted 68 of them in Figure 11.1, but the actual number is unlikely to be of much importance.

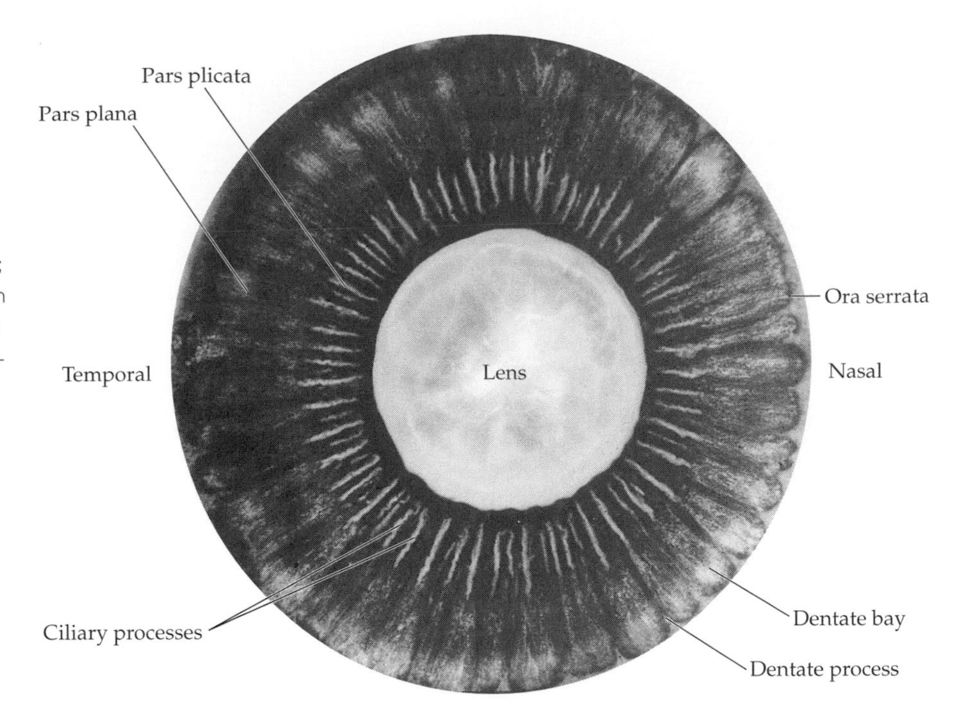

Figure 11.1
The Ciliary Body Viewed from Inside the Eye
This picture was made by hemisecting the eye at the equator so that the inner surface of the anterior segment could be seen. The lens is in the center, surrounded by a series of radial lines, which are the ciliary processes; by definition, the part of the ciliary body with processes is the pars plicata. The pars plicata is surrounded by the pars plana, which terminates in a scalloped edge, the ora serrata, where the retina begins. (From Hogan, Alvarado, and Weddell 1971.)

The region of the ciliary body occupied by the ciliary processes is the **pars plicata**, the folded part of the ciliary body (the Latin *plica* means "fold"). The pars plicata is an annulus of tissue encircling the eye just outside the lens (Figure 11.1). It contains more than the ciliary processes, however. Other tissue, specifically the bulk of the ciliary muscle, is in the pars plicata as well, although it cannot be seen in Figure 11.1. To put it another way, the term "pars plicata" refers to the entire anterior third (roughly) of the ciliary body; it is defined by the presence of the ciliary folds (processes), but there is more to it than that.

The collective term for all of the ciliary processes is **corona ciliaris**, the ciliary crown, so called because its radiating folds look like the rays of a crown (see Figure 11.1). The effect of all the folds on the inner surface of the pars plicata is a larger surface area than the tissue would have if there were no folds. Because the aqueous humor is secreted here, the folds increase the surface area across which materials can diffuse.

The remaining two-thirds of the ciliary body, which is not included in the pars plicata, makes up the **pars plana**, so called because it lacks any major folds and is therefore smooth (the Latin *plana* means "plane," "flat," or "smooth"). By definition, the pars plana has no ciliary processes and consists mostly of muscle fibers and an inner lining of epithelium. The outer limit of the pars plana, and therefore of the ciliary body, is marked by a serrated or scalloped line encircling the region (see Figure 11.1). This line is the **ora serrata**, which is the anteriormost limit of the retina. The ciliary body is in front of the ora serrata; the retina and choroid are behind it.

In looking at the ora serrata and the pars plana adjacent to it, someone must have been reminded of the relationship between our teeth and gums, where the gums extend up between the teeth but are scalloped down along the sides. Thus the extensions of the ora serrata are called **dentate processes**, and the portions of the pars plana bounded by neighboring dentate processes are called **dentate bays** (Figure 11.2; see also Figure 11.1). Each dentate process is associated with a small ridge extending toward the pars plicata, leading to the space between major ciliary process. These small ridges are the sites of attachment for fibers of the zonule, which connects the ciliary body to the lens.

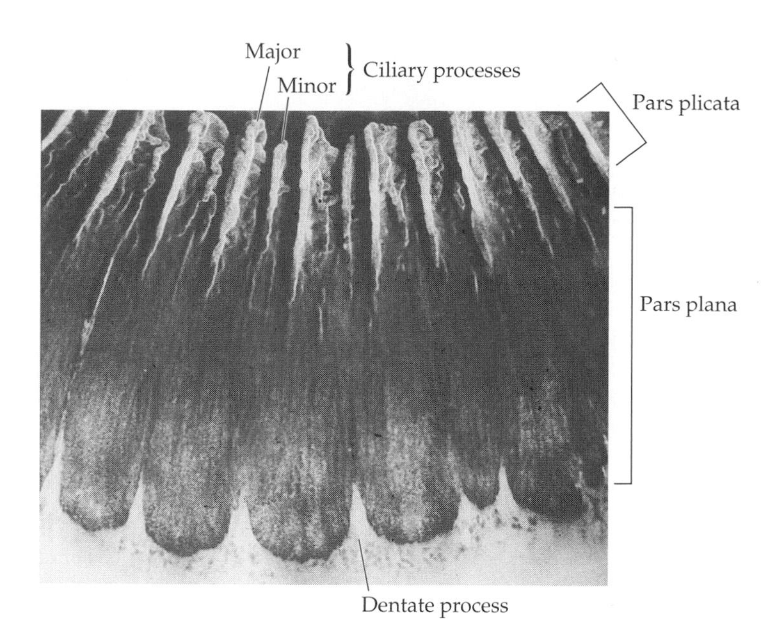

Figure 11.2
Inner Surface of the Ciliary Body
The scalloped ora serrata has small projections, the dentate processes, that extend anteriorly toward the ciliary processes, creating corresponding dentate bays in the pars plana. The ciliary processes are seen to be folds or ridges of tissue, alternating between large folds (major processes) and smaller ones (minor processes). (From Hogan, Alvarado, and Weddell 1971.)

The ciliary muscle extends through both pars plicata and pars plana

Another way to look at these divisions of the ciliary body is in a histological section cut along the eye's horizontal plane (Figure 11.3). The large mass of tissue enclosed by a rectangle is the pars plicata; the irregular plumes on its inner surface are one of the ciliary processes (the ciliary processes are not perfectly straight, and they weave in and out of the plane of section, giving the false impression of an incomplete structure). The inner surface of the ciliary body, including the ciliary processes, is covered with two layers of epithelium—one with melanin pigment, the other without pigment. The ciliary processes are full of small blood vessels, but most of the tissue in Figure 11.3 is muscle—smooth, unstriated muscle extending from the anterior choroid to the scleral spur and trabecular meshwork.

Because the ciliary muscle extends throughout the ciliary body, we need to make another division in addition to the pars plicata and the pars plana—specifically, a division between ciliary muscle and ciliary processes. This is the major functional division, separating the ciliary body's role in accommodation—done by the muscle—from its role in aqueous formation, which is done in the ciliary processes.

The Ciliary Processes and Aqueous Formation

The ciliary processes are mostly filled with blood vessels

Some of the most revealing illustrations of the ciliary processes are also some of the oldest, notable for their detailed rendering of the blood vessels within the processes (see, for example, Figure 2 in Vignette 6.2). Ciliary processes are mainly masses of small blood vessels, a description that speaks better than anything else to their function: They are blood-filled folds in the ciliary epithelium where the blood can be brought into close contact with the secretory epithelium. Figure 11.4 presents a contemporary view of a ciliary process, made by vascular casting (a technique described in Box 6.1). The photograph also shows the main source of the blood coming to the small vessels in the ciliary processes—that is, branches from the major arterial circle. These branches run into each of the ciliary processes, where the small arterioles break up into dense capillary beds.

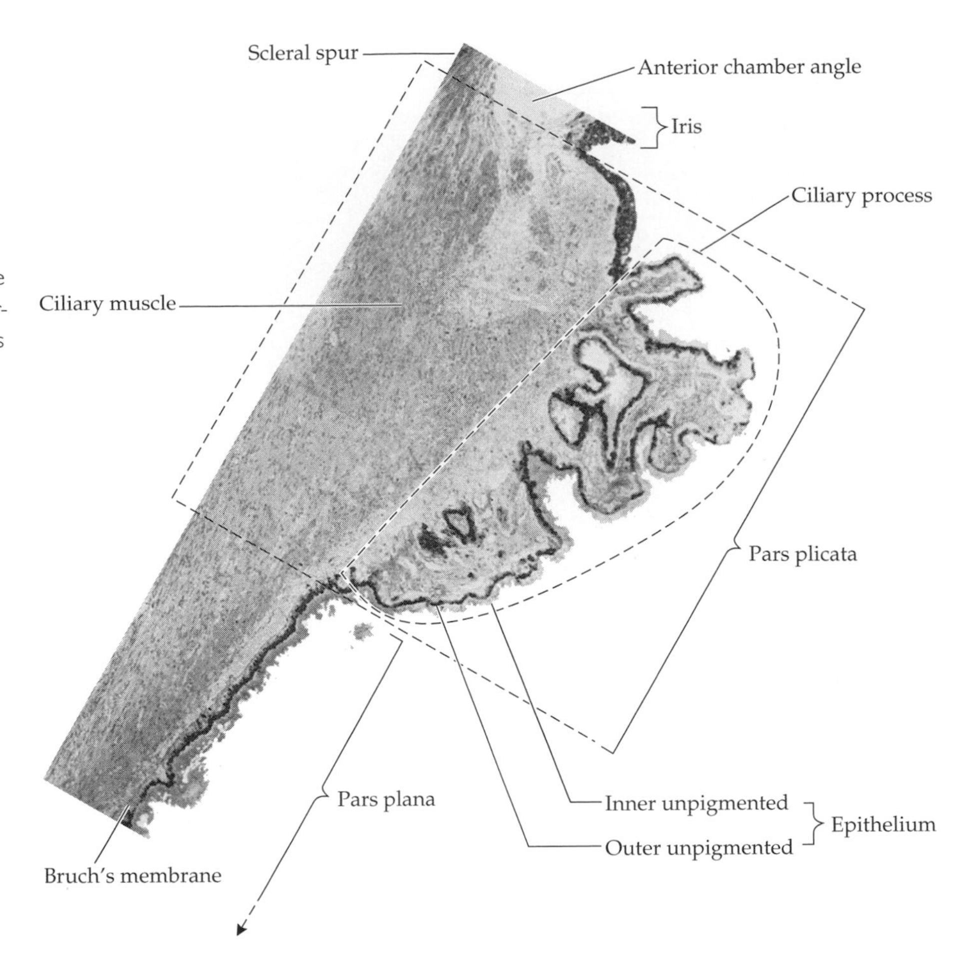

Figure 11.3
The Ciliary Body in Longitudinal Section
The ciliary muscle extends from the scleral spur to the choroid (not shown), although some of its fibers are confined to the pars plicata. The ciliary process and the entire inner surface (at right) of the ciliary body are lined with two layers of epithelium, the innermost of which lacks melanin pigment. Bruch's membrane (discussed later in the chapter) extends forward from the choroid into the pars plana.

Figure 11.5 shows a simplified version of the supply and drainage in the ciliary processes. Short arterioles run from the major arterial circle to the ciliary processes, where they form dense capillary networks; there appear to be several separate capillary beds in each ciliary process, each supplied by a small arteriole. Capillaries merge in the posterior part of each ciliary process, creating several

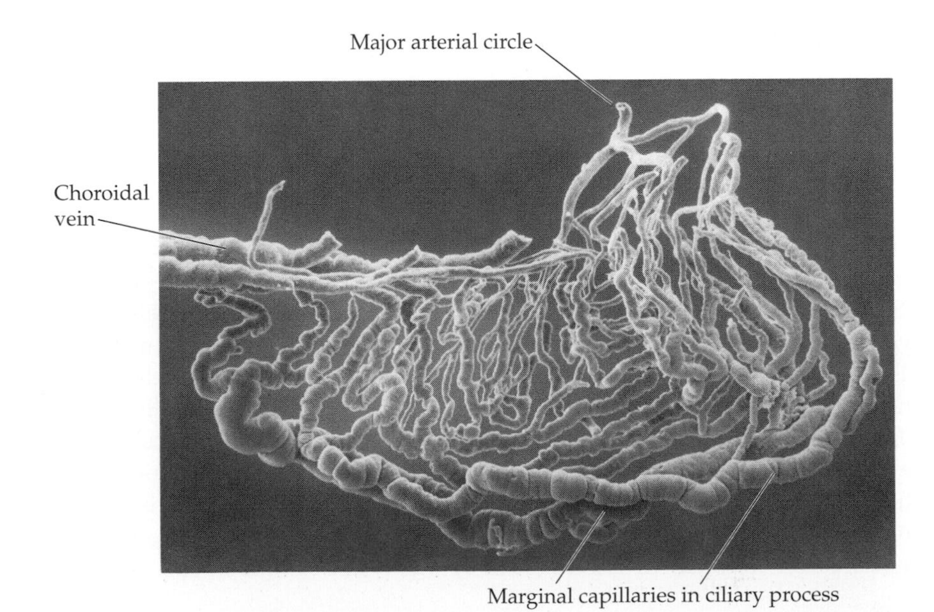

Figure 11.4
Blood Vessels in the Ciliary Processes
A vascular casting shows the profusion of ciliary process vessels in considerable detail. The small arterioles feeding the large capillaries arise from the major arterial circle. The capillaries run back through the process, merging as venules near the anterior edge of the pars plana. (From Morrison and Freddo 1996.)

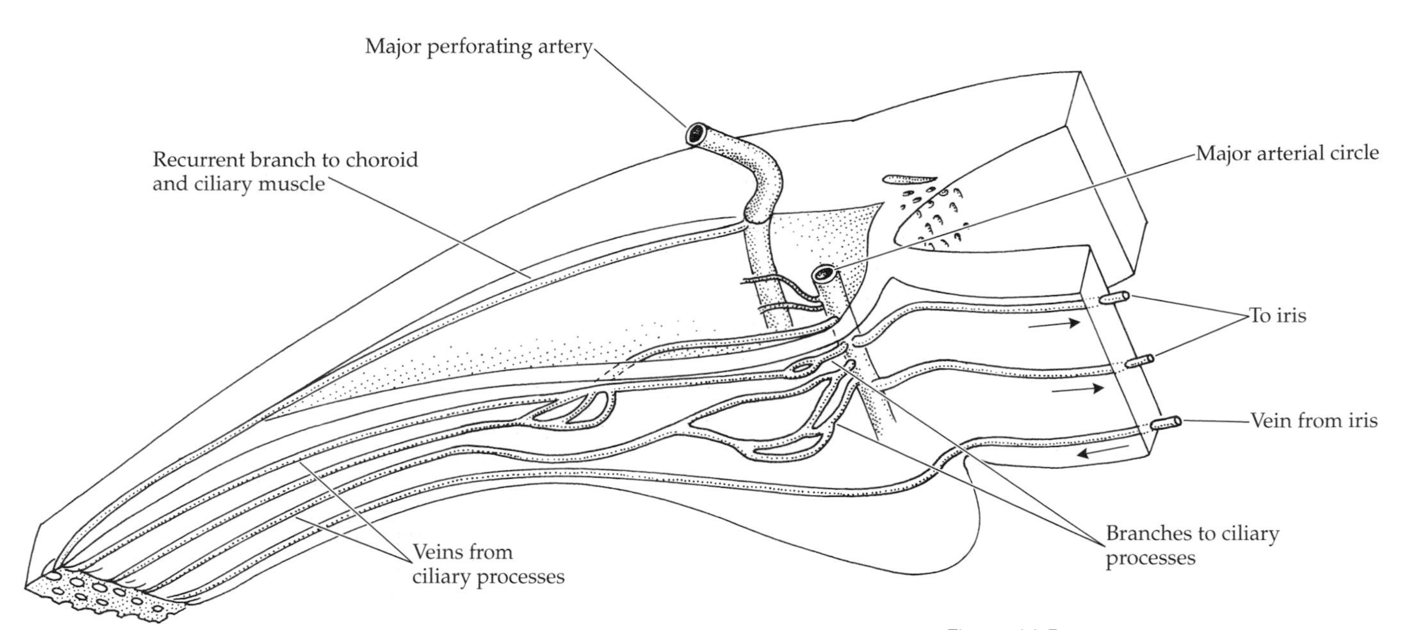

Figure 11.5
Blood Supply and Drainage of the Ciliary Processes
The major arterial circle sends branches into the ciliary processes, where they divide into large capillaries. The capillaries run at several levels of depth in the processes, but the large marginal capillaries run back along the ridges of the processes. The capillaries merge as small veins running back through the pars plana into the drainage fields of the vortex veins. (After Funk and Rohen 1990 and Morrison and Freddo 1996.)

small venules that drain each process. After running back to the choroid, the veins merge into the drainage field for one of the vortex veins (see Chapter 6).

In the older literature, the major arterial circle was considered to be the only arterial circle in the ciliary body, formed by anastomoses between branches from the anterior ciliary and long posterior ciliary arteries. This arrangement would give the ciliary processes a redundant arterial supply. As implied in Figure 11.5 and discussed in detail in Chapter 6, however, the anterior ciliary artery system forms another arterial circle—the intramuscular circle—and there is no consensus on the magnitude of its contribution to the major arterial circle. In short, the vessels in the ciliary processes are supplied by branches from the major arterial circle, but whether the blood is primarily from the long posterior ciliary arteries or from the anterior ciliary arteries as well is not clear.

The capillaries in the ciliary processes are highly permeable

Several of the major capillary systems in the eye are constructed to form blood–tissue barriers by having thick endothelial walls, overlapping cells, and numerous occluding junctions between the endothelial cells. These features are notable in the retinal capillaries, where there is a blood–retina barrier, and in the iris, where the capillaries are bathed in aqueous but have little interchange with the surrounding fluid. Capillaries in the ciliary processes are quite different; their structure makes them leaky.

Figure 11.6 shows a capillary in a ciliary process. All the small circles are closed fenestrations in the capillary endothelium—places where the endothelium is extremely thin, with nothing but cell membrane between the blood and the surrounding tissue. Not only is the endothelium liberally dotted with fenestrations, but the endothelial cells are not tightly fastened to one another; the cells do not overlap, and they have no prominent anchoring or occluding junctions.

These features in the endothelium are typically found only in the capillaries of the ciliary processes and the choroid. All other capillaries in the eye have either a tight restrictive endothelium (retina and iris) or an endothelium that lacks specializations for either high or low permeability. Both fluid and large molecules can easily move out of the highly permeable capillaries in the ciliary processes, where they will begin the transformation to aqueous humor.

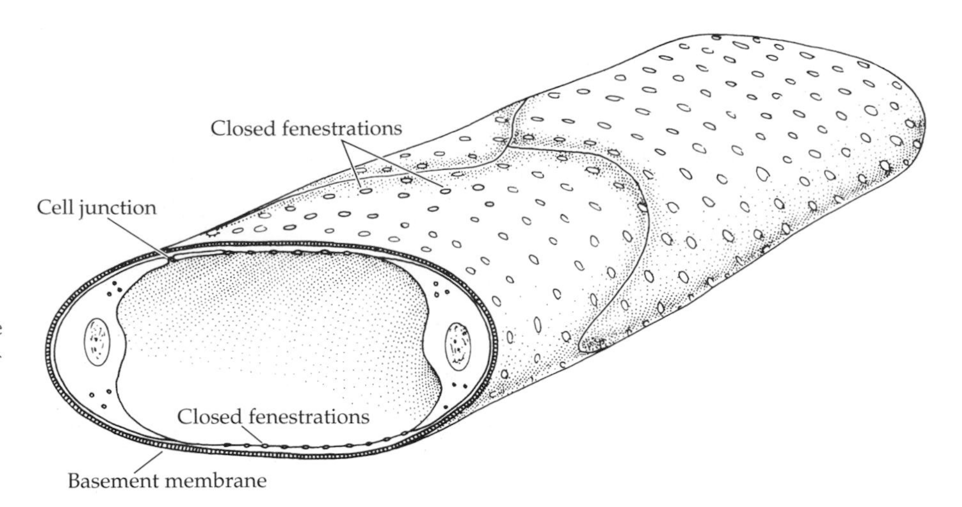

Figure 11.6
Capillary Structure in the Ciliary Process
This reconstruction of a ciliary process capillary shows several endothelial cells joining to form the tubular capillary. Adjoining cells are loosely attached with a few anchoring junctions. Fenestrations are located on both inner and outer surfaces, and the capillary is enclosed by the basement membranes of the endothelial cells. This structure is leaky. (After Alm 1992.)

Two layers of epithelium lie between the capillaries and the posterior chamber

The relationship between the blood in the capillaries of the ciliary processes and the aqueous-filled posterior chamber is readily apparent in Figure 11.7. At low magnification (Figure 11.7*a*), a section through a ciliary process shows blood vessels of various sizes embedded in a stroma consisting of loose connective tissue. The complex of vessels and stroma is contained in an epithelial cell lining consisting of a pigmented epithelial cell layer next to the vessels and stroma and a layer of unpigmented epithelium lining the space of the posterior chamber. Most of the blood vessels lie next to the pigmented epithelium.

At higher magnification (Figure 11.7*b*), two blood vessels are visible next to the epithelium. The larger of the two (at right) is a capillary cut in cross section with red blood cells inside; most capillaries are so small that red blood cells pass through them in single file, but this is clearly not true here. Capillaries in the ciliary processes are large. Figure 11.7*b* makes another important point: The distance between blood in the capillaries and aqueous outside the ciliary process is short, about 50 μm. Blood and aqueous are separated by only three layers of cells: the capillary endothelium, the pigmented epithelium, and the unpigmented epithelium. Whatever is being done to make aqueous from blood occurs over a very short distance within these three cell layers.

Aqueous formation involves metabolically driven transport systems

Because the walls of the capillaries are so permeable, fluid moving out of the capillaries into the stroma of the ciliary processes is very similar to blood plasma. But plasma and aqueous are different; compared to plasma, aqueous has a minuscule protein content, about half the amount of dissolved oxygen, a very high amount of ascorbate, and a significant amount of hydrogen peroxide. Concentrations of most ions and organic substances, such as glucose, are similar in the two fluids.

The low protein content of the aqueous is responsible for its high optical clarity. Viewed with a slit-lamp, the anterior chamber appears empty because the aqueous scatters less light than either the cornea or the lens. Haze in the aqueous from increased light scatter is called aqueous flare by clinicians; it is a sure sign of inflammation and tissue breakdown in one or more of the tissues bordering the anterior or posterior chambers. The large, randomly distributed molecules that are released by tissue breakdown and increased capillary permeability increase light scattering.

(a)

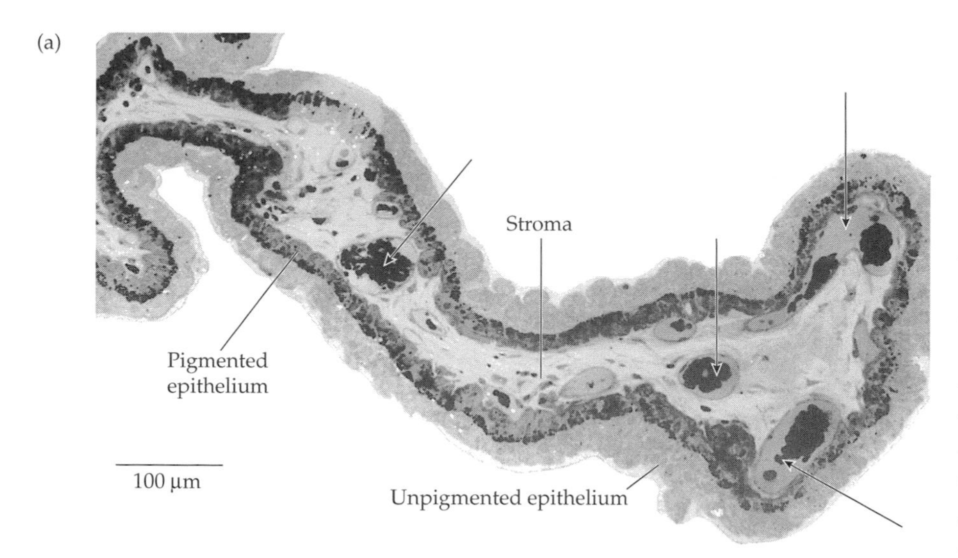

(b)

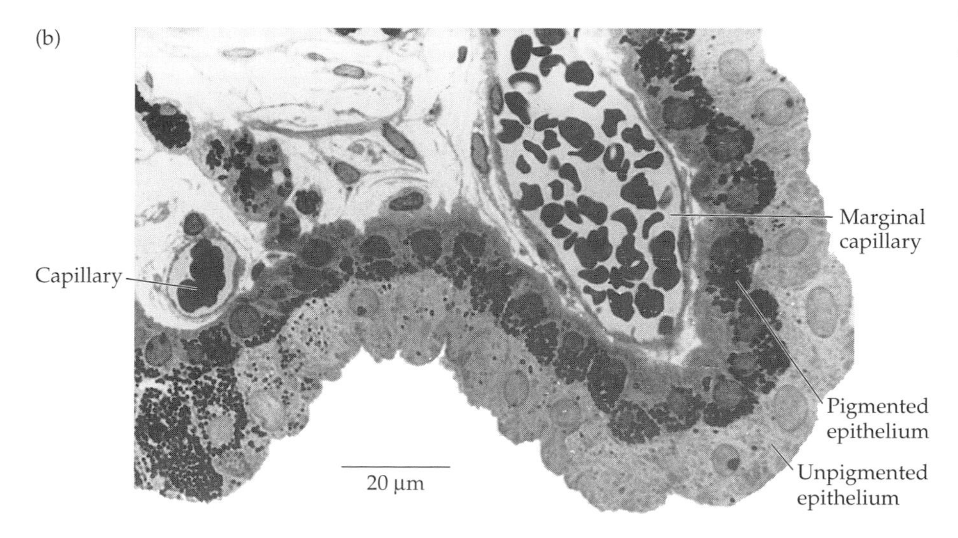

Figure 11.7
Capillaries and Epithelium in the Ciliary Processes
(*a*) The ciliary processes are completely lined by two layers of epithelium. The pigmented epithelium lies next to the blood vessels (arrows) and the stroma, the loose connective tissue that surrounds the vessels. Much of the open space in the stroma is probably filled with fluid leaking from the capillaries. (*b*) At higher magnification, the close relationship between the capillaries, seen here with red blood cells still in them, and the epithelial layers is clear. Blood is separated from aqueous by a distance of about 50 µm and by three layers of tissue: the capillary endothelium, pigmented epithelium, and unpigmented epithelium. (From Hogan, Alvarado, and Weddell 1971.)

Oxygen, ions, and small organic molecules in the aqueous are important to the tissues bathed by the aqueous—the iris, lens, and cornea in particular. These tissues not only draw on the aqueous for oxygen, glucose, and other things, but also dump their metabolic by-products into the aqueous. The relatively high lactate content of the aqueous is a consequence of this dumping. The high ascorbate concentration in the aqueous is not understood, however, largely because its function in the eye is not clear. High ascorbate levels in the aqueous are characteristic of diurnal mammals; nocturnal species have ascorbate levels similar to those of plasma. This association has led to a hypothesis that ascorbate is protective, acting as an intraocular filter by absorbing potentially damaging ultraviolet radiation.

The differences between aqueous and plasma must be related to the mechanism that underlies aqueous formation. At least three mechanisms need to be considered: diffusion, ultrafiltration, and active, metabolically driven transport. *Diffusion* is the movement of molecules along a gradient from higher to lower concentration of the molecular species. Small, uncharged molecules move easily across cell membranes by diffusion, and much of the movement of O_2, CO_2, and water occurs by diffusion (for water, the process is called osmosis). Because the aqueous is mostly water, the mechanism that underlies water movement is particularly important; most of the water moves across the ciliary epithelium by osmosis, as it happens, but the osmotic driving force is dependent on other systems.

Ultrafiltration is the movement of fluid and solutes along a hydrostatic pressure gradient from higher to lower pressure. At first glance, ultrafiltration seems to be involved in aqueous formation because the pressure in the capillaries, estimated to be somewhat less than 30 mm Hg, is higher than the average intraocular pressure of 15 mm Hg. The hydrostatic force pushing fluid from plasma to aqueous is countered by oncotic pressure of the tissue, however, which is produced by the high protein concentration in the plasma; this concentration gradient acts in the direction of bringing water from the aqueous to the plasma, in opposition to the hydrostatic pressure gradient (Figure 11.8). Ultrafiltration is a factor in aqueous production only to the extent that the hydrostatic pressure difference exceeds the opposing oncotic pressure, and the net pressure for ultrafiltration is small, probably negligible.

Active transport is a way of moving ions and molecules across cell membranes against opposing concentration or electrical gradients. If diffusion represents movement downhill, active transport is movement uphill. But uphill movement requires energy to drive the ionic pumps; when metabolic poisons are used to deprive the pumps of energy, aqueous production drops to about one-third of normal—a result showing active transport to be a major factor in aqueous production.

The ciliary epithelium is anatomically specialized as a blood–aqueous barrier

The two layers of ciliary epithelium have several unusual features (Figure 11.9). Because the pigmented layer derives from the outer layer of the optic cup while the unpigmented layer derives from the invaginated layer of the cup, the pigmented and unpigmented cells touch at their apical surfaces, as do the two pigmented epithelial layers in the iris. Their basement membranes are therefore on opposite sides of the epithelium; that is, the basement membrane of the pigmented layer is next to the stroma of the ciliary processes, and the basement membrane of the unpigmented layer lines the posterior chamber in direct contact with the aqueous. Cells in both layers have complicated folds next to their basement membranes, a feature often seen in secretory epithelia. These basal folds increase the cell surface area considerably.

Like most epithelia, the ciliary epithelia have desmosomes joining neighboring cells within each layer and some desmosomal binding between the two lay-

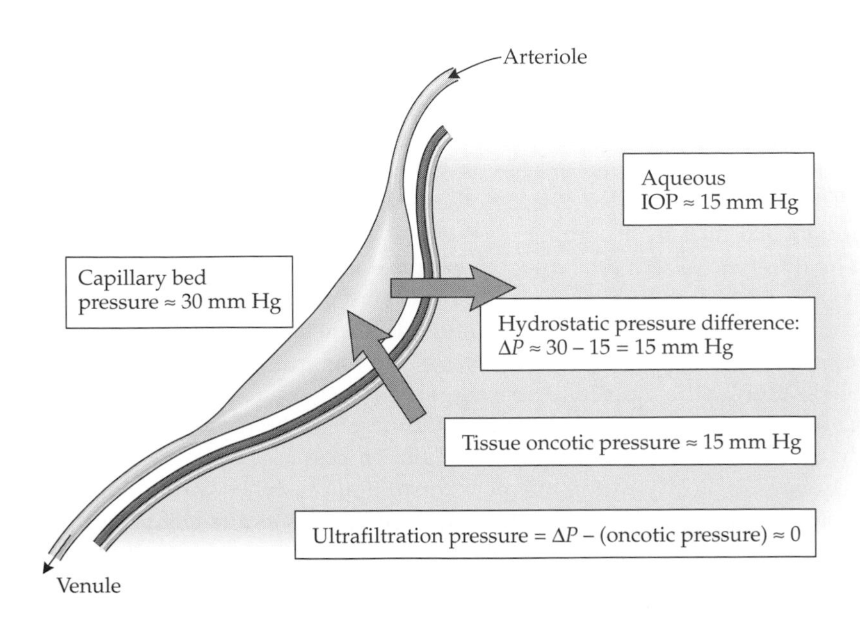

Figure 11.8
Pressures That Affect Bulk Fluid Flow from Blood to Aqueous
The capillaries in the ciliary processes are shown with an internal pressure of about 30 mm Hg. Since the intraocular pressure (IOP) in the aqueous is about 15 mm Hg, there is a hydrostatic pressure difference of 15 mm Hg that could produce flow *from* blood *to* aqueous. This hydrostatic pressure is opposed, however, by oncotic pressure in the tissue, estimated to be about 15 mm Hg, which could produce flow *to* blood *from* aqueous. The net pressure—the ultrafiltration pressure—is therefore near zero and unlikely to be a factor in aqueous formation under normal circumstances.

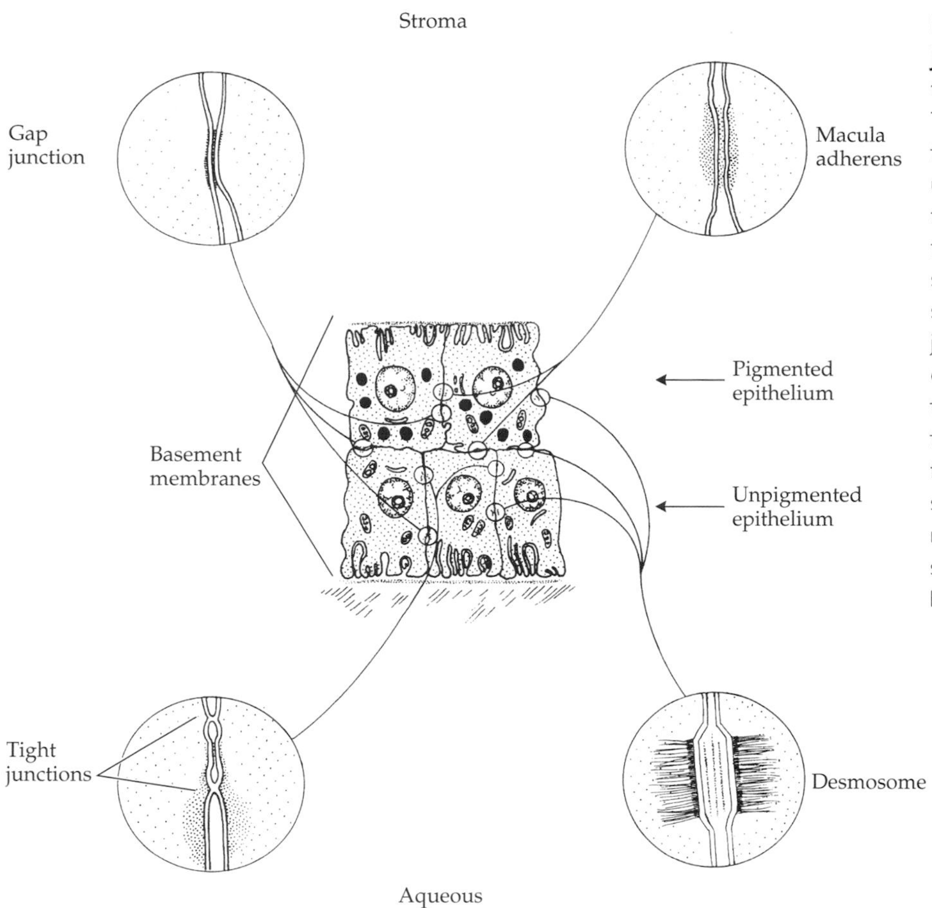

Figure 11.9
Junctions among Cells in the Ciliary Epithelium
The epithelial layers of the pars plicata join at their apical surfaces; the basal surface of the unpigmented epithelium faces the aqueous-filled posterior chamber, while the basal surface of the pigmented epithelium faces the stroma of the ciliary processes. Both basal surfaces have deep infoldings. Anchoring junctions, desmosomes, and gap junctions connect the two cell layers, as well as cells in the same layer. Tight junctions are confined to the unpigmented epithelium, where they form belts around the cells near their apical surfaces. The tight junctions constitute a barrier that prevents the movement of fluid and small molecules into the aqueous by passing between cells. (After Raviola 1977.)

ers. Other junctions are more interesting, however. Communicating gap junctions are common between layers, for example, prompting some speculation that the two anatomical layers should be thought of as a single functional unit. Whatever the case, these epithelial cells have unusual opportunities for electrical and molecular communication. One type of junction is confined to the unpigmented epithelial cells; these are tight occluding junctions of the zonular variety that close down the extracellular space near the apical side of the unpigmented epithelial cell layer (see Figure 11.9).

The pigmented and unpigmented epithelia of the ciliary processes are constructed so that fluid and solutes cannot move from the stroma to the posterior chamber by going around the cells; they must move through the cells. Movement through the cells is made easier by the large basal surface area for diffusion and the numerous gap junctions, while movement around the cells is blocked by the belts of tight junctions around the apices of the unpigmented epithelial cells. The remaining question concerns the nature of the movement through the epithelial layers and the development of an osmotic gradient for water flow.

Ions are transported around the band of tight junctions to produce an osmotic gradient in the basal folds of the unpigmented epithelium

The basal folds of the pigmented epithelium face the stroma and provide a large area for diffusion of water and some small molecules into the cells. Further diffusion to the unpigmented epithelium is also easy, made easier still by the gap junctions between cells in the two layers. The movement of ions and polar molecules, such as glucose, is a more difficult proposition, normally requiring transmembrane channels for ions or specific carrier proteins for other molecules. These mechanisms are difficult to isolate and study; most of what we know comes from studies of pigmented ciliary epithelial cells grown as cultures. Several systems for transporting Na^+, K^+, HCO_3^-, and Cl^- in various combinations and directions,* as well as an ascorbate transport system, have been found.

This story is incomplete, but the pigmented epithelium seems to have the requisite active transport mechanisms to move ions and other molecules across its membrane. Whether similar mechanisms move these ions and molecules to the unpigmented epithelium or the movement is through the gap junctions is not known. Whatever the details, the essential ingredients reach the unpigmented epithelium, where they are made into aqueous.

The final stage of aqueous production is illustrated in Figure 11.10, which shows a basal cleft between two unpigmented epithelial cells; the cleft opens to the posterior chamber at the base and is tightly closed near the apices of the cells by a tight junction. Several pumps, or active transport systems, are located in the membranes just below the occluding junction toward the basal side of the cell. One of the transport systems exchanges Na^+ and K^+, another probably produces a net outflux of HCO_3^-, and another may involve Cl^-. These metabolically driven systems move ions into the basal cleft, which is continuous with the posterior chamber. As a result, the ion concentration is relatively high in the cleft, creating an osmotic gradient that causes water to diffuse from cells to clefts and from the clefts out into the posterior chamber. The water, with its ions and dissolved molecules, is now aqueous.

Aqueous production varies during the day and declines with age

The rate of aqueous production cannot be measured directly in humans. Instead, it is assumed to be proportional to the flow of aqueous through the anterior chamber—a flow that can be measured. To measure aqueous flow, the corneal stroma is loaded with fluorescein by topical application, and the dye slowly diffuses from the stroma into the anterior chamber. Aqueous flow is calculated from measurements of the amount of aqueous fluorescence at different times. The decrease in aqueous fluorescence over a given interval of time is a measure of the rate at which fluorescein is being removed by the flow of aqueous.

As determined by this method, aqueous flow should be a measure of aqueous production by the ciliary epithelium, but some of the aqueous secreted into the posterior chamber may diffuse into the iris, vitreous, or lens before entering the anterior chamber. To the extent that this diffusion occurs, measurements of aqueous flow will underestimate aqueous production by a small amount. A more serious problem with using aqueous flow as a surrogate for aqueous production is the dependence of aqueous flow on the outflow resistance (see Chapter 9). Changes in aqueous flow represent changes in aqueous production only when outflow resistance remains constant.

Aqueous flow varies significantly during the day or, more accurately, varies between waking and sleeping hours. When we are awake, aqueous flows at twice the rate (about 3.1 μl/minute) that occurs while we sleep (1.4 to 1.6 μl/minute). Subjects who are not permitted to sleep do not show as much reduc-

*Transport does not always involve movement of a single molecule in a given direction—*uniport*. Two different molecules may be transported simultaneously in the same direction (*symport*) or in opposite directions (*antiport*). These systems may occur together, making for considerable complexity. If one molecule is involved in both symport and antiport, for example, it may be quite critical for the movement of other molecules, but its net concentration across the cell membrane may not change, making its role difficult to detect. Part of the problem in understanding the details of uptake and secretion by the ciliary epithelium is lack of information about the types of transport systems utilized by the cell membranes.

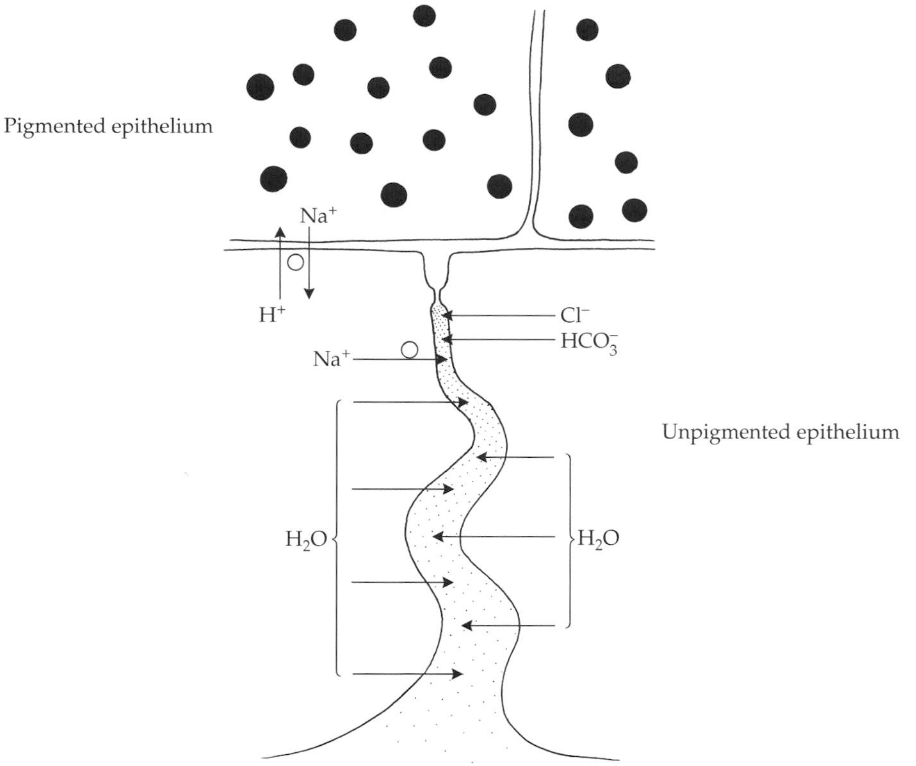

Figure 11.10
Ion and Water Flow in Aqueous Formation
Several ion pumps, not all fully characterized, affect the movement of Na^+ and H^+ between the epithelial layers and the movement of Cl^-, Na^+, and HCO_3^- out of the unpigmented cells into the deep basal clefts between adjoining cells. The ionic concentration in the basal cleft produces an osmotic pressure that drives water into the cleft. (After Cole 1977 and Diamond and Bossert 1967.)

tion in aqueous flow during their normal sleep hours, suggesting that the phenomenon is not just a matter of time of day but is partly dependent on the sleep state. The mechanism for this variation in aqueous flow is not known, however. The answer probably lies in levels of endogenous hormones that vary between waking and sleeping, but some of the obvious ones, such as melatonin, are not known to affect aqueous production.

When daytime rates of aqueous flow are measured in subjects of different ages, the results show a small but steady decline with age (Figure 11.11). Flow is highest in the decade between 10 and 20 years and lowest in the last decade, from 80 to 90 years, decreasing by about 3% each decade (the line was fitted to the data by eye, ignoring the first data point, at age 5 years). If aqueous flow is proportional to aqueous production, the data suggest that aqueous production declines by about 25% over the course of a long life. This is not a drastic change; to turn the statement around, aqueous flow in an elderly eye is still three-fourths of its highest value in youth. It is interesting that glaucoma is generally a disease of older eyes, since the age-related decline in aqueous production should act to *reduce* intraocular pressure.

The gradual decline in aqueous production might be related to long-term changes in the ciliary epithelium, perhaps something as simple as the same kind of gradual cell loss that occurs in the corneal endothelium. Estimates place the number of unpigmented ciliary epithelial cells around 4 million. Losing 1 million of them over 80 years or so means that 100 cells must be lost each day, without replacement. This loss does not seem unreasonable, but we know nothing about the epithelial cell death rate and the capacity of the epithelium for replacement. Declining aqueous production might be attributable to declining levels of hormones or neuromodulators that normally regulate the production of aqueous. Unfortunately, regulation of aqueous production is another aspect of the ciliary epithelium that is not well understood.

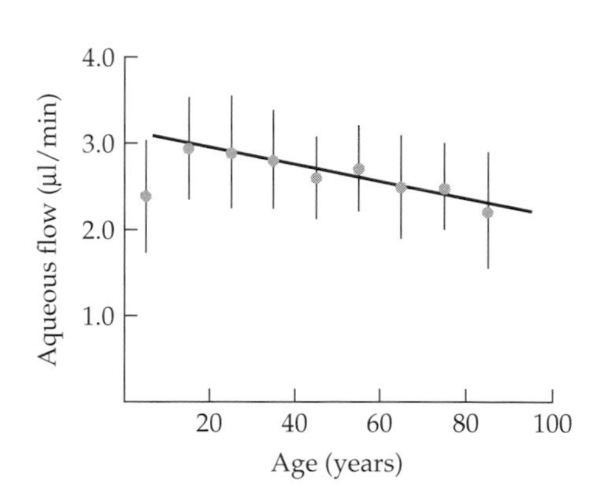

Figure 11.11
Aqueous Flow as a Function of Age
After the age of 15 or so, when the eye is nearly full grown, aqueous flows at a rate of about 3 µl/minute. Flow declines with age by about 25%, to around 2.25 µl/minute at age 80. If aqueous flow is a reliable proxy for aqueous production, the data also indicate a 25% decline in production from youth to old age. Vertical lines with each data point show ±1 standard deviation. (Replotted from Brubaker et al. 1981.)

The major classes of drugs used to reduce aqueous production interact either with adrenergic membrane receptors or with the intracellular formation of bicarbonate ions

Given the importance of active transport systems in the production of aqueous, an obvious way to reduce aqueous production is to reduce the rate of transport. The rate of transport can be reduced by several methods, one of which is to reduce energy to the pumps by using metabolic poisons—for example, the drug ouabain. Generally, however, agents like ouabain are too toxic or too widespread in action to be used clinically. Drugs with more specific effects are required.

Carbonic anhydrase inhibitors, of which acetazolamide is the prototype, offer the requisite specificity. Bicarbonate ions are formed within cells by the reaction $H_2O + CO_2 \rightarrow H^+ + HCO_3^-$, which is catalyzed by the enzyme carbonic anhydrase. Inhibiting the action of the enzyme reduces the amount of bicarbonate available for transport and reduces aqueous production. As used clinically, acetazolamide and related drugs can reduce aqueous production by about one-third.

Most of the agents used to reduce intraocular pressure interact with adrenergic membrane receptors on the ciliary epithelial cells, one of which is illustrated schematically in Figure 11.12. The cell is shown as having both alpha- and beta-adrenergic receptors, as well as others for neuropeptide Y, vasoactive intestinal polypeptide (VIP), and acetylcholine. When a receptor site binds the appropriate neurotransmitter or neuromodulator, the conversion of ATP to cAMP is altered through an intermediate G protein. Some of the G protein linkages are inhibitory, acting to reduce the amount of cAMP and to reduce aqueous production; they are associated with alpha-adrenergic receptors, neuropeptide Y receptors, and acetylcholine receptors. Two stimulatory G proteins are associated with beta-adrenergic receptors and VIP receptors. The possible strategies for reducing

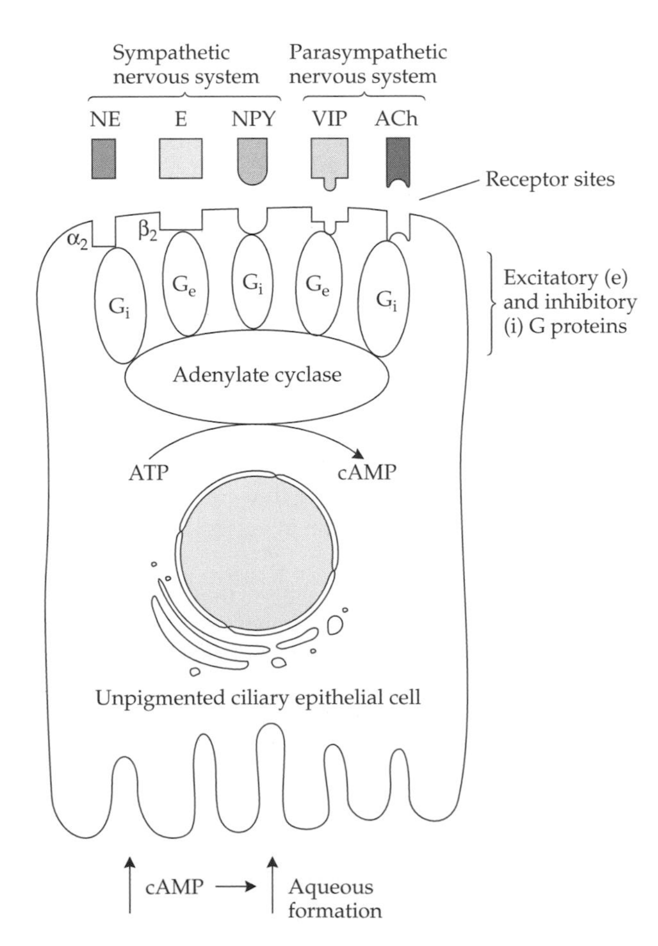

Figure 11.12
Epithelial Membrane Receptors and Effects on cAMP
Increased conversion of ATP to cAMP is thought to lead to an increase in aqueous production. This conversion can be affected positively or negatively by hormones or neuromodulators acting at membrane receptor sites that are linked by either excitatory or inhibitory G proteins to the enzyme adenylate cyclase. The drugs most commonly used at present are beta receptor blockers or alpha receptor agonists. ACh, acetylcholine; E, epinephrine; NE, norepinephrine; NPY, neuropeptide Y; VIP, vasoactive intestinal polypeptide. (After Nilsson and Bill 1994.)

aqueous production are to activate the receptors associated with inhibitory G proteins and to block the receptors associated with stimulatory G proteins.

At present, two alpha receptor agonists are in use for reducing intraocular pressure by reducing aqueous production: apoclonidine and brimonidine. Several nonselective beta receptor blockers are available, of which the most commonly used is timolol; it reduces aqueous production by preventing the normal stimulation of cAMP production by endogenous norepinephrine.

Although the ciliary epithelial cells have no nerve endings in contact with them, sympathetic axon terminals are in the vicinity, associated with the blood vessels (a situation similar to that in the lacrimal gland). Norepinephrine could come from this source, or it could be present as endogenous epinephrine produced by the adrenal gland (adrenergic receptors of all types are sensitive to both epinephrine and norepinephrine). Whatever the source, modulating the amount of the catecholamine in the vicinity of the ciliary epithelium should alter aqueous production by acting on the adrenergic membrane receptors.

Unfortunately, experiments designed to test this hypothesis have produced mixed and contradictory results. There is no easy way out of this impasse except to say, as many authors do, that we don't know how aqueous production is regulated. But given the ability of adrenergic agonists or beta blockers to reduce aqueous production, it is likely that ciliary epithelial cells have adrenergic receptors for a reason, and the reason is regulation of secretion.

The pars plana is covered by epithelial layers that are continuous with the epithelial layers of the pars plicata

Figure 11.13 shows the relationships of the ciliary body to the posterior chamber, lens, zonule, and vitreous chamber. The posterior chamber is bounded anteriorly by the iris, and posteriorly by the lens and zonule, and its outer circumference is lined with the ciliary processes. Because the vitreous extends forward to the posterior side of the zonule, it is closely related to the pars plana. In other words, the secretory epithelium of the pars plicata is associated with the posterior chamber, while anything secreted by the pars plana epithelium necessarily enters the vitreous.

Figure 11.14 compares the epithelial layers of the pars plana and the pars plicata. The inner unpigmented epithelium adjacent to the aqueous in the pars

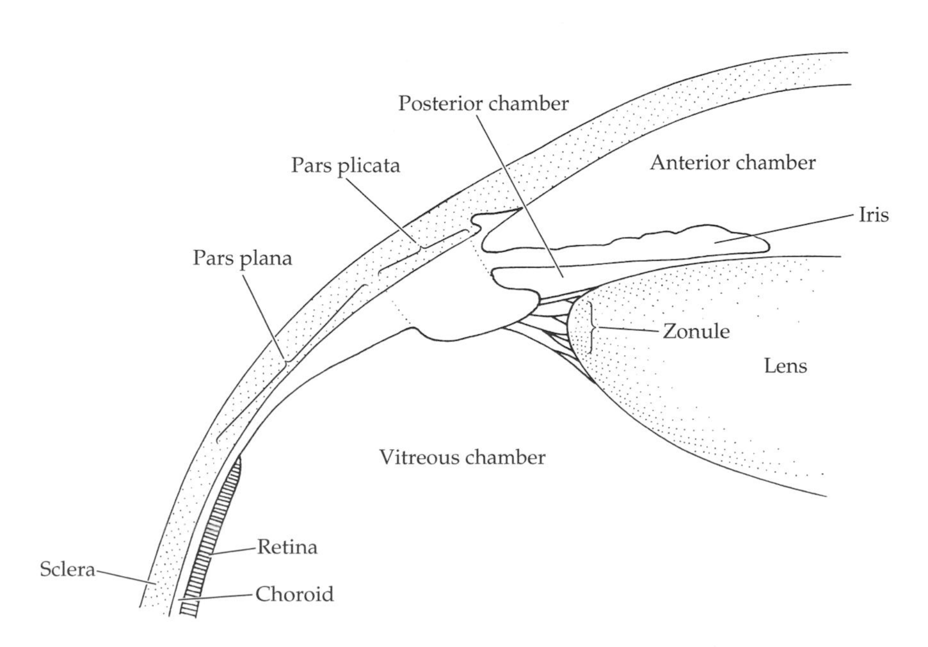

Figure 11.13
Relationships of the Ciliary Body to the Posterior Chamber and Vitreous
The posterior chamber is bounded by the iris, lens, zonule, and pars plicata. Thus, fluid moving through the epithelium of the ciliary processes enters the posterior chamber. The entire surface of the pars plana, however, faces the vitreous chamber, and all secretions from the pars plana enter the vitreous.

plicata is continuous with unpigmented epithelium next to the vitreous in the pars plana. The unpigmented epithelial cells in the pars plana, however, are columnar and considerably larger, and they form an irregular surface confronting the vitreous. The pigmented epithelia of the two regions are more alike, but pigmented cells in the pars plana are also more columnar.

Blood vessels in the stroma lie near the pigmented epithelium in both cases, but in the pars plana the epithelium is separated from the stroma by **Bruch's membrane**. Because this membrane is a choroidal structure that extends forward into the ciliary body, its structural details will be discussed later, when we look at the choroid. Its significance here is its location; as in the choroid, Bruch's membrane lies between epithelium and capillary beds, implying that interchanges between blood and epithelium in the pars plana are somewhat different from those in the ciliary processes.

Cells in the epithelial layers of the pars plana have less basal infolding than those of the ciliary processes, and the various intercellular junctions are also less numerous. Desmosomes join cells within and between layers, and there are local adhering junctions (maculae adherens) and gap junctions between the pigmented and unpigmented epithelial cells. The occluding (tight) junctions that are so important in the unpigmented epithelium in the pars plicata are also found in the unpigmented epithelium of the pars plana, but whether they are as tight and impermeable is not clear.

The blood vessels lying next to Bruch's membrane (see Figure 11.14) are quite similar to those in the ciliary processes; the capillaries are unusually large and have closed fenestrations, although the density of fenestrations is lower, perhaps about one-third the density in the capillaries of the ciliary processes. Because of these fenestrations, these capillaries are presumably quite leaky, raising the question of what, if anything, the pars plana epithelium does with the fluid that leaks into the stroma.

The vitreous is discussed in Chapter 12, but we need to mention a few of its features here. First, the vitreous has some gel-like qualities, but there is considerable flow of water through it; in spite of the much larger volume of the vitreous chamber, the substantial water component of the vitreous is replenished almost as quickly as the aqueous in the anterior chamber. In addition, some substances, such as ascorbate, are present in the vitreous at much higher concentration than in blood plasma, pointing to the possibility of active transport of the sort involved in aqueous formation. Much of the water entering the vitreous seems to come by way of the pars plana, and active transport may also play a part.

A large part of water movement into the vitreous across the epithelial layers in the pars plana is by diffusion—osmosis—as it is across the epithelium of the ciliary processes, but ultrafiltration plays a more significant role here. The oncotic pressure that opposes ultrafiltration in the ciliary processes (see Figure 11.8) is lower across the pars plana epithelium because there is less difference in protein concentration between blood and vitreous than between blood and aqueous. Thus the difference in hydrostatic pressure between the blood vessels and the vitreous will not be canceled out. Moreover, the large volume of water movement would demand a very high rate of secretion if it were done solely by a mechanism like that used to produce aqueous. What is less clear is the role of active transport in creating osmotic gradients for water movement or for concentrating certain substances, such as the ascorbate, in the vitreous; these issues have received very little attention in comparison to the pars plicata epithelium.

Uncertainties and unknowns notwithstanding, the pars plana epithelium seems to be in much the same business as the epithelium in the pars plicata. It has the same anatomical features, though less developed, and occupies the same critical position between leaky blood vessels and another fluid whose character is different from blood. What the pars plicata epithelium does for the aqueous, the pars plana epithelium does for the vitreous, though perhaps less actively.

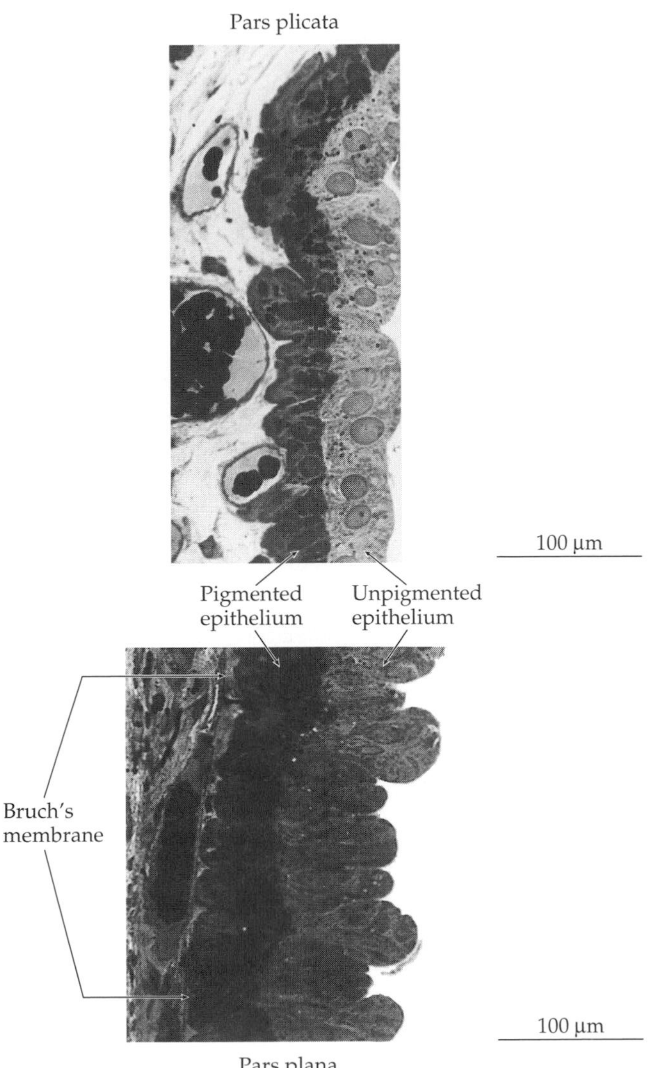

Figure 11.14
Epithelial Layers in the Pars Plicata and Pars Plana
An obvious difference between epithelial layers in the two parts of the ciliary body is the larger size and more columnar shape of the unpigmented epithelial cells in the pars plana. The cells are also less regular in size. In addition, the epithelial layers in the pars plana are separated from the stromal capillaries by Bruch's membrane, which has no equivalent in the pars plicata. (From Hogan, Alvarado, and Weddell 1971.)

The Ciliary Muscle and Accommodation

The ciliary muscle has three parts with a complex geometry

The three parts of the ciliary muscle, identified in the mid-nineteenth century, are indicated on the longitudinal section through the ciliary body in Figure 11.15. The outermost part of the muscle, which lies next to the sclera, is the **meridional portion** (originally called Brücke's muscle). It is easily recognized in longitudinal histological sections because the muscle fibers run parallel to the plane of section and the long streaks of muscle fibers are obvious.

The part of the muscle lying closest to the ciliary processes is the **circular portion** (originally called Müller's muscle). In this portion the muscle fibers are perpendicular to the plane of section, and the cut ends of the fibers are small circles. As Figure 11.15 shows, there is often enough separation between the circular portion and the rest of the ciliary muscle that it could easily be thought of as a separate muscle.

Finally, the **radial portion** (originally called Bowman's muscle) consists of everything else that cannot be attributed to the meridional or circular portions; the division between meridional and radial portions is particularly difficult to define. Because the muscle fibers here are neither parallel nor perpendicular to the section plane, their cross sections are elongated ellipses.

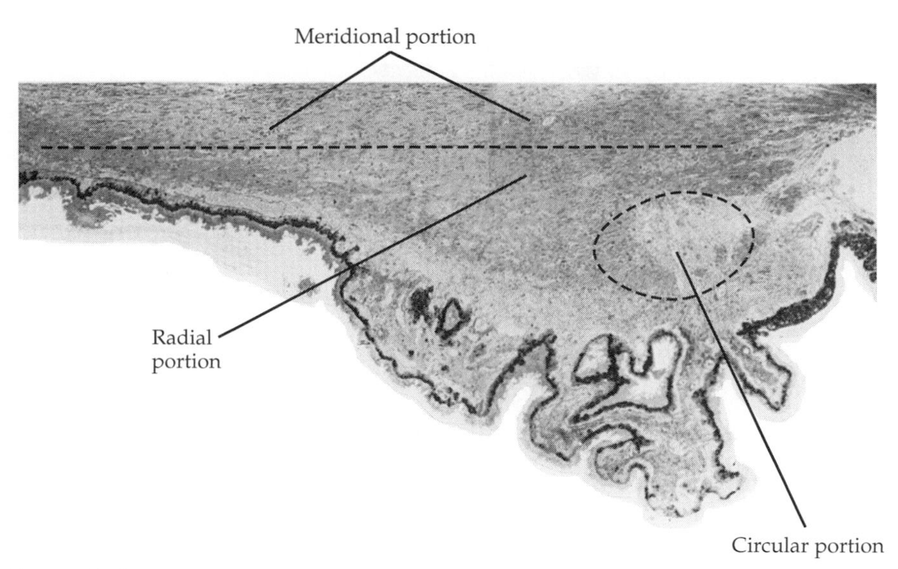

Figure 11.15
The Ciliary Muscle in the Pars Plicata
The ciliary muscle is traditionally divided into three parts. The outermost, lying next to the sclera, is the meridional portion. The innermost fibers nearest the ciliary processes constitute the circular portion, and the radial portion lies between the meridional and circular portions. The divisions are not sharp; the dashed lines in the photograph show approximate boundaries.

Although it is clear that the ciliary muscle has different regions in which the muscle fibers run in different directions, it is less clear how we should interpret and think about the muscle's structure. Is this one muscle with three subdivisions, or is it three interconnected muscles? In either case, where are the origins and insertions of the muscle fibers? Does the muscle have one action or several? What is it, or what are they?

None of the studies on the architecture of the ciliary muscle are definitive; that is, all the available data and observations leave room for interpretation and legitimate differences of opinion. But two studies (Calasans 1953 and Rohen 1964) agree on several major points: Ciliary muscle fibers are grouped in small bundles ensheathed by connective tissue, different bundles intersect and cross as they run through the muscle, and the angles of intersection vary in different parts of the muscle—from very acute angles of intersection in the meridional portion of the muscle to very wide, almost 180°, angles in the circular portion.

Figure 11.16 shows schematically the structure of the ciliary muscle as I interpret the evidence. All fiber bundles, and therefore all muscle fibers, attach at one end to the scleral spur or the adjacent connective tissue in the ciliary body at the root of the iris. In the outermost part of the muscle, the meridional portion, bundles with nearby attachments to the scleral spur run back toward the choroid, intersect near the level of the ora serrata, and end with small fingerlike expansions (called **muscle stars**) in the anterior choroid. This arrangement produces diamond-shaped spaces between the interlaced bundles of muscle fibers.

Fiber bundles in the radial portion of the muscle follow the same pattern, but the intersecting muscle fiber bundles cross at nearly right angles, and they do not run all the way to the choroid. They end in the connective tissue of the anterior pars plana just external to the epithelial layers. At the scleral spur, the bundles in the radial portion attach more toward the tip of the scleral spur than do those of the meridional portion.

In the innermost circular portion of the muscle, the fiber bundles intersect at a very obtuse angle, nearly 180°. They are almost perpendicular to the fibers in the meridional portion and therefore run tangentially around the ciliary body.

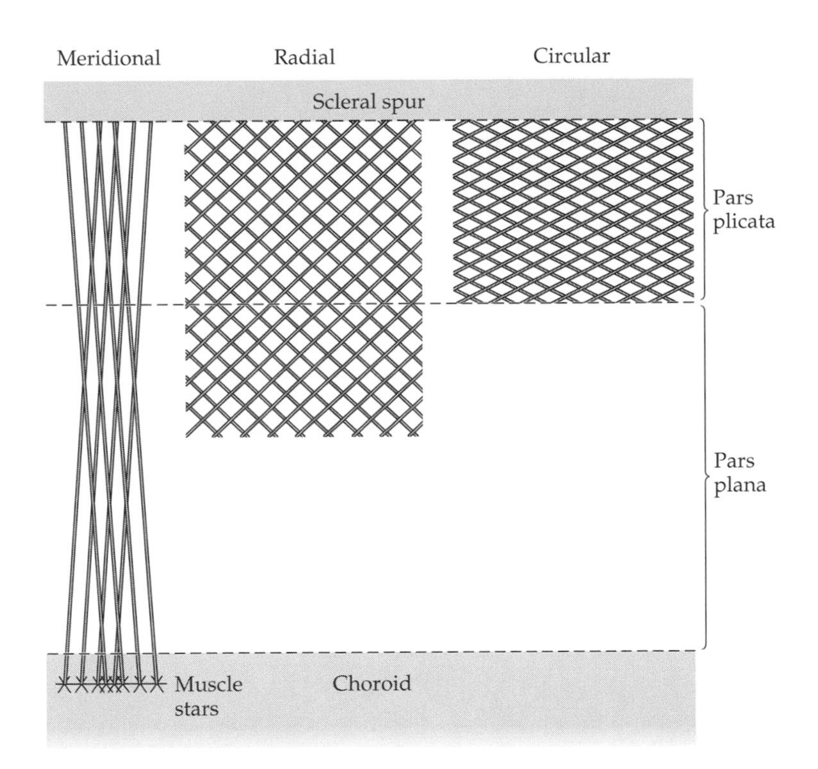

Figure 11.16
Ciliary Muscle Fiber Organization
In this schematic, the ciliary muscle fiber bundles are viewed from outside the muscle. The meridional fibers run from the scleral spur to the choroid, where they end in expanded muscle stars. Fiber bundles intersect at very acute angles. Radial muscle fiber bundles intersect at right angles and run between the scleral spur and the anterior portion of the pars plana. The deep circular fiber bundles have the largest angles of intersection, which makes them run nearly tangential to the inner surface of the ciliary body.

One end of the fiber bundles is in the connective tissue near the scleral spur; the other is in connective tissue near the junction of the pars plicata and pars plana.

In short, the main difference between different parts of the ciliary muscle is the angle of intersection made by the interlaced fiber bundles and the distance they extend posteriorly from the scleral spur. These features probably change gradually from the outside to the inside of the muscle, which would mean there are no sharp divisions between one part and the next, but this is not known with any certainty.

Figure 11.17 summarizes the organization of the ciliary muscle in a more realistic way. The view is from the outside with the sclera removed, and two successive cutaways show the deeper parts of the muscle. The muscle stars belonging to fibers in the meridional portion (not shown) lie in the outermost part of the choroid (the epichoroid or suprachoroidea), fusing there with connective tissue and perhaps to some extent with the overlying sclera. Fibers in the deeper parts of the muscle have the equivalent of muscle stars that fuse with Bruch's membrane throughout the pars plana; the most superficial of the radial muscle fibers arise most posteriorly in the pars plana; the deepest fibers in the circular portion arise in the most anterior region of the pars plana.

Contraction of the ciliary muscle produces movement inward toward the lens so that the muscle behaves like a sphincter

The ciliary muscle fibers are innervated by parasympathetic neurons in the ciliary ganglion that receive their input from axons of cells in the Edinger–Westphal nucleus.* Sympathetic fibers entering the muscle appear to be targeting the blood vessels; if they innervate the ciliary muscle fibers, it is at beta-adrenergic receptors that are not involved in muscle contraction.

*As noted in Chapter 5, some portion of the innervation to the ciliary muscle may not arise from the cells in the ciliary ganglion. Some of the small ganglia scattered throughout the choroid may be part of the ciliary muscle's innervation. This issue has been controversial for 20 years or so and has yet to be fully resolved, but any route of innervation outside the ciliary ganglion probably is a minor one.

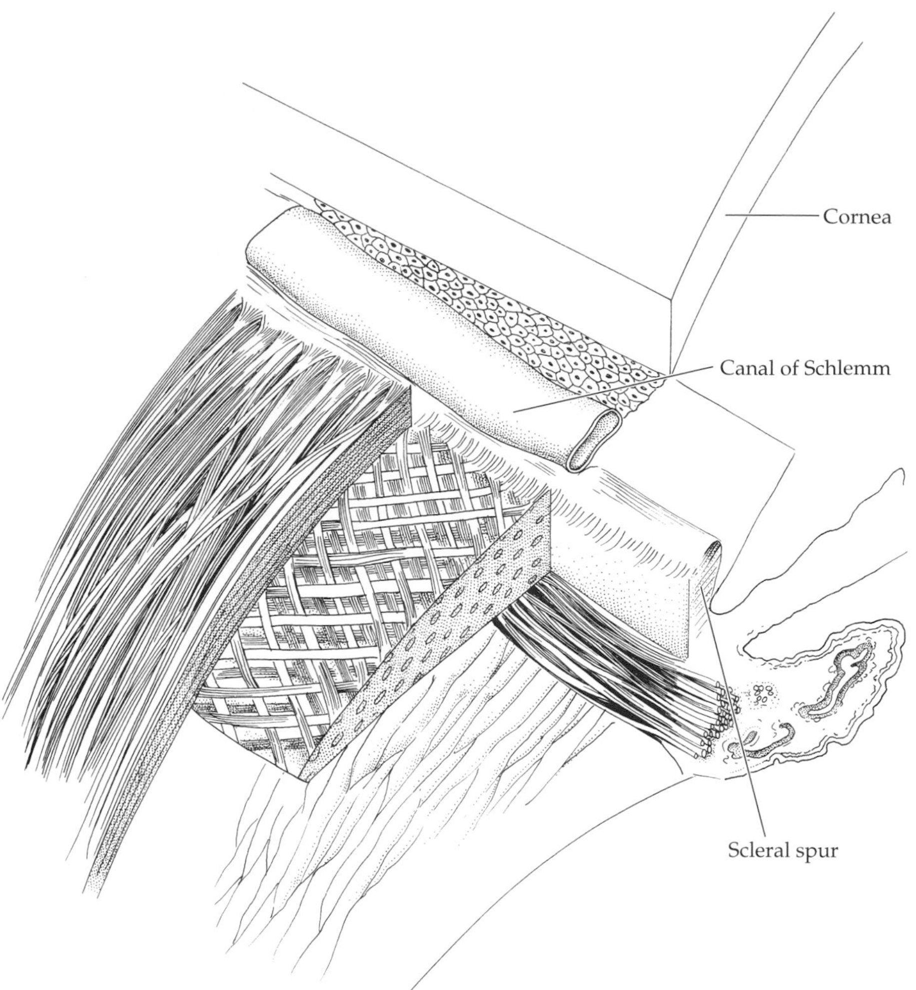

Figure 11.17
Three-Dimensional Structure of the Ciliary Muscle
The intersecting bundles of ciliary muscle fibers are shown here with portions of the muscle cut away to make visible the different parts of the ciliary muscle. The meridional portion is farthest left and the circular portion is at the right. Intersecting pairs of fiber bundles diverge more and more widely from the meridional to the circular portion of the muscle, and the innermost circular fibers form a sphincter. The circular fibers are confined to the pars plicata, the radial fibers extend a bit into the pars plana, and the meridional fibers extend into the anterior choroid. (After Hogan, Alvarado, and Weddell 1971.)

Given the complexity of the ciliary muscle's structure, the functional consequences of muscle contraction are not obvious, particularly for the meridional portion of the muscle, where the fibers run between relatively immovable points and contraction is largely isometric. But it is easy to see the overall consequence of ciliary muscle contraction. When the iris is surgically removed (a procedure called iridectomy), the tips of the ciliary processes ringing the equator of the lens can be seen from in front. When the ciliary muscle is relaxed (Figure 11.18*a*), there is a distinct space between the lens and the ciliary processes. When the muscle contracts (Figure 11.18*b*), the ring of ciliary processes moves inward closer to the lens, indicating that the ciliary muscle behaves as a sphincter.

For practical purposes, this description of the ciliary muscle as a sphincter is quite adequate, but as we saw earlier, not even the fibers in the circular portion of the muscle have a truly circular, sphincterlike arrangement. How do we get sphincterlike action from the geometry of the muscle fibers depicted earlier?

The result of contraction in the circular portion of the muscle is easiest to visualize from in front (Figure 11.19*a*). None of the interlaced muscle fiber bundles are shown as extending long distances around the ciliary body (their length is not known), but they have been drawn long enough so that their paths are curved; this is taken to be the relaxed condition. When the fibers contract, they shorten, and they will become straighter as a result; the straightening of the muscle fibers as they contract shifts the center of each fiber inward toward the center, thereby closing the sphincter.

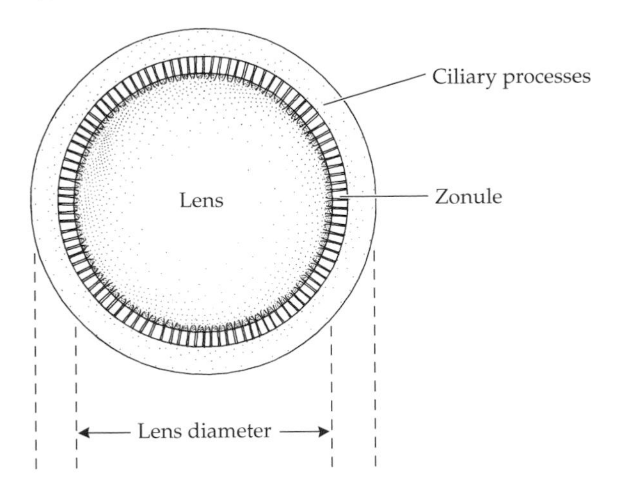

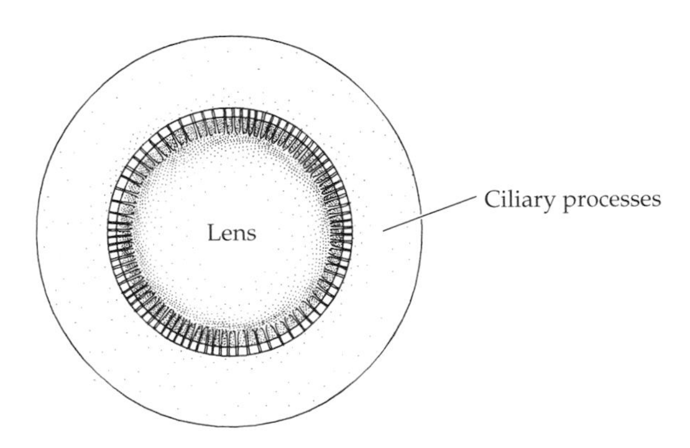

Figure 11.18
Movement of the Ciliary Body When the Ciliary Muscle Contracts
The lens, zonule, and ciliary processes are shown here from the front with the iris removed. (*a*) When the ciliary muscle is relaxed for zero accommodation, the lens diameter and diameter of the ring of ciliary processes are at their maxima. (*b*) With maximal accommodation, contraction of the ciliary muscle has a sphincterlike effect, reducing the diameter of the ciliary process ring. This action reduces tension in the zonule, allowing the lens to change shape, a change reflected here as a reduction in lens diameter. (After Koretz and Handelman 1988.)

Contraction of the meridional and radial parts of the muscle should also produce inward movement, but this effect is better visualized in a longitudinal section (Figure 11.19*b*). Here, a few of the bundles in the meridional portion are shown in the relaxed state; their paths as they run forward from the choroid to the scleral spur are gently curved. When the muscle fibers contract and shorten, their paths become straighter, thereby pushing the inner portion of the ciliary body inward toward the lens. Since this straightening occurs throughout the meridional muscle around the circumference of the ciliary body, the overall effect is to aid the sphincterlike action of the circular muscle. The radial portion of the muscle should have effects on contraction that combine the basic elements of meridional and circular muscle fiber contraction, again with the net result of aiding a sphincterlike closing of the ciliary body around the lens.

There are at least two other results of muscle fiber contraction in the meridional part of the muscle. One is a slight forward movement of the anterior choroid and retina, a shift that can be demonstrated with the appropriate psychophysical experiments. The other is some posterior bending of the scleral spur, with increased opening of the trabecular meshwork and the lumen of the canal of Schlemm (see Chapter 9).

The zonule provides a mechanical linkage between ciliary muscle and lens

Zonule (of Zinn, or suspensory ligament of the lens) is a collective term for the many fine strands running between the ciliary body and the lens, holding the lens

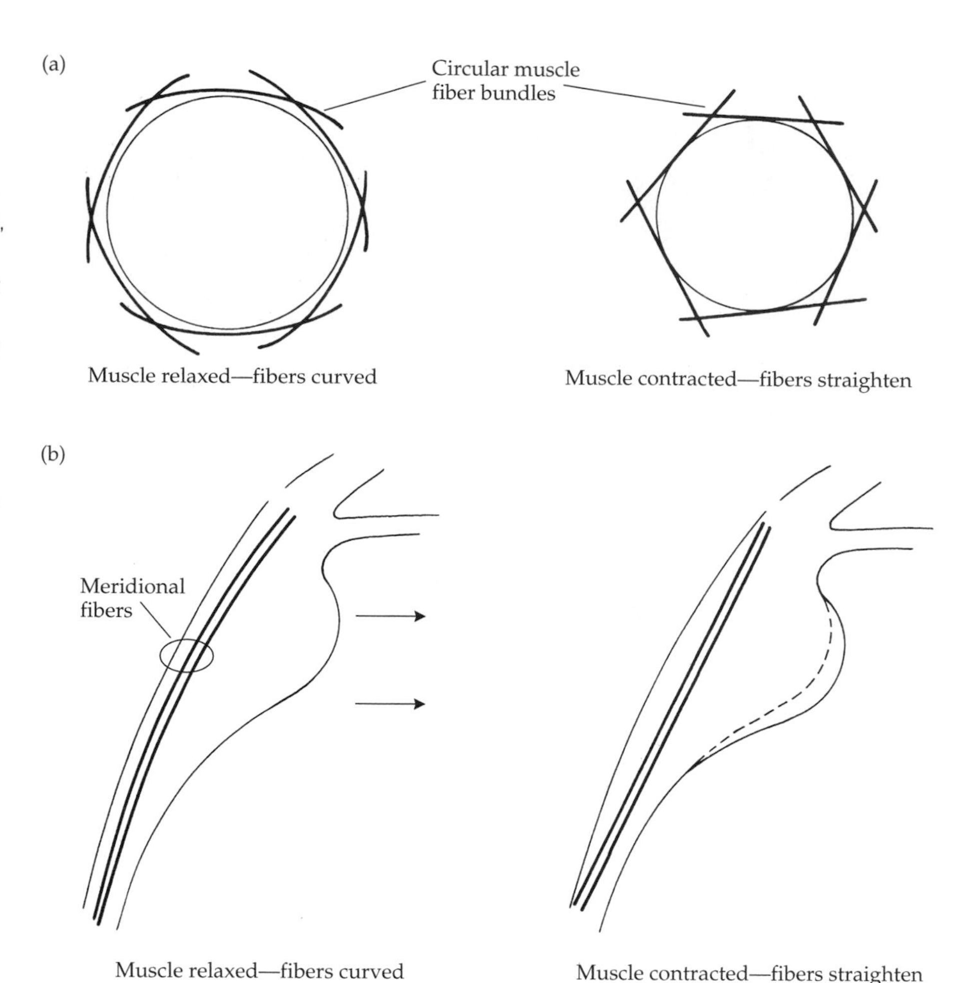

Figure 11.19
Effects of Ciliary Muscle Fiber Contraction
(*a*) A schematic version of the ciliary muscle viewed from in front shows bundles of circular muscle fibers. When the muscle is relaxed, the fiber bundles curve around the circumference of the aperture formed by the ciliary processes. When the muscle contracts, the fiber bundles shorten and become straighter, thereby pushing inward toward the lens. (*b*) Meridional fiber bundles viewed in longitudinal section also have gently curved paths as they run from the choroid to the scleral spur. They shorten and straighten when they contract, again pushing inward toward the lens. The effect is not as pronounced, however, as that of the circular fibers.

in place like guy ropes supporting a tent. Zonular fibers are attached to the ciliary body by fusion with the basement membrane of the unpigmented epithelium (Figure 11.20). The attachments are concentrated along the ridges running up through the pars plana from the dentate processes and in the valleys between ciliary processes. Zonular fibers rarely attach to the ciliary processes themselves. At the lens, the fibers fuse with the lens capsule (see Chapter 12).

The zonular fibers are apparently not made up of collagen as had long been assumed, but are more closely related to elastic fibers—that is, those containing elastin as the dominant protein. As a result, they are elastic, capable of being stretched somewhat and developing tension within as a result.

The geometry of the zonule has been described in various ways; Figure 11.21 shows the features agreed upon by most investigators. Most of the zonular fibers originate in the pars plana and run forward between the unpigmented epithelium and the anterior hyaloid membrane of the vitreous (see Chapter 12). Along the way, the fibers can have secondary attachments to either the epithelium or the hyaloid membrane. At the level of the ciliary processes, most zonular fibers pass through the valleys between processes and then go to the anterior surface of the lens, to the posterior surface, or to the lens equator. The direct equatorial route has the fewest followers. Not all of the fiber bundles pass through the valleys between the ciliary processes; some run along ridges of the ciliary processes, and others pass along the sides of the ridges. In either case, there can be small secondary attachments to the basement membrane of the unpigmented epithelium.

(a) The ciliary body and the zonule

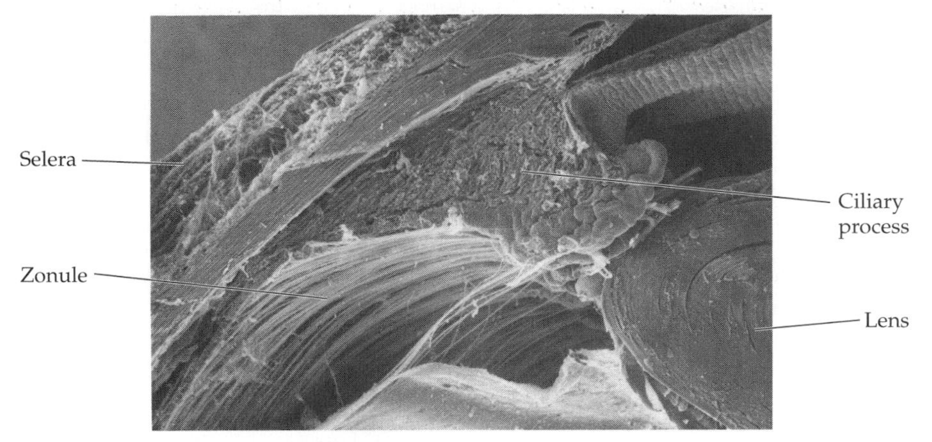

(b) Zonular fibers and pars plana epithelium

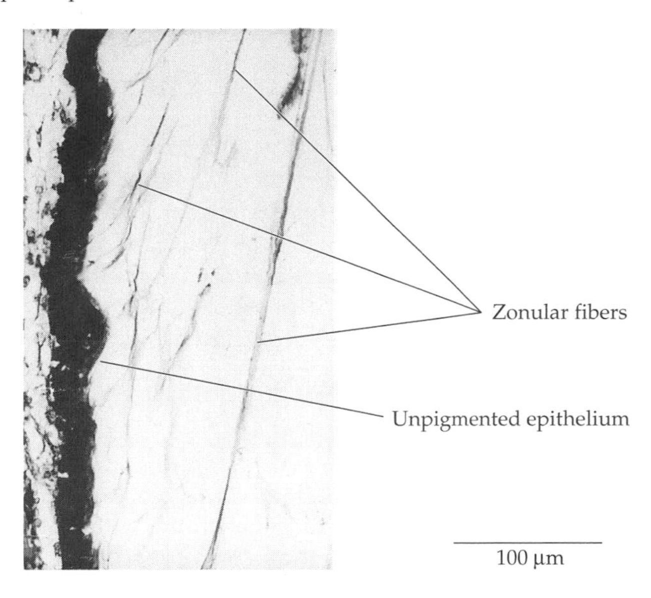

Figure 11.20
Attachment of the Zonule to the Ciliary Epithelium
(*a*) This scanning electron micrograph shows strands of the zonule coursing over the surface of the pars plana to run between the ciliary processes toward the lens (the lens is not in its normal position, however). (*b*) This low-power electron micrograph shows strands of the zonule in the region of the pars plana. The strands turn inward near the epithelium appearing to merge with it. (*a* from Morrison and Freddo 1996; *b* from Hogan, Alvarado, and Weddell 1971.)

The main points of disagreement concern the existence of fibers originating from the pars plicata and the strength of fiber attachments to the pars plicata. Figure 11.21 is deliberately vague on these points, although a site of attachment to the pars plicata epithelium is indicated. The question is whether the fibers attaching here are primary fibers that join the bundles running forward from the pars plana or secondary "tension fibers" that form pulleys or slings through which the primary fibers run, much like the relationship between the trochlea and the superior oblique tendon.

The important point is that the bundles of zonular fibers, whether they originate in the pars plana or in the pars plicata, are anchored to the valley regions between the ciliary processes. Like the trochlea of the superior oblique, the anchor point becomes the functional origin. Being able to think about the zonule in these terms simplifies matters considerably; regardless of the actual paths followed by the bundles of zonular fibers as they run from the ciliary body to the lens, tension developed in the zonule will be reflected as force directed radially outward from the equator of the lens toward the anchoring points on the pars plicata. These anchoring points are very close to the circular portion of the ciliary muscle, separated from the muscle only by the epithelial layers and some connective tissue. Thus, changes in the state of contraction in the muscle are communicated directly to the zonule and from the zonule to the lens.

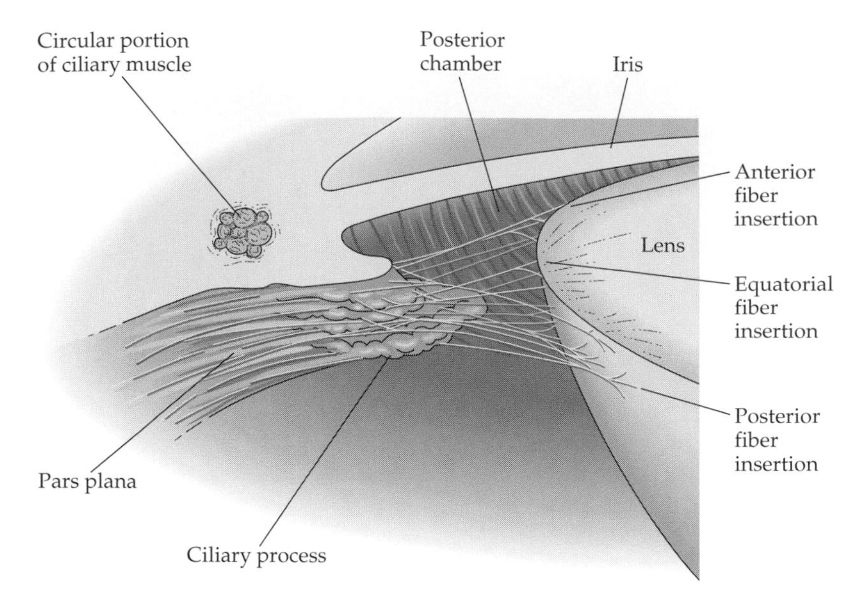

Figure 11.21
Organization of the Zonule
The zonule forms three groups of fibers as it attaches to the lens: An anterior group inserts anterior to the lens equator, a posterior group inserts posterior to the equator, and a smaller group inserts directly at the equator. It is not clear that these three insertion groups have different origins on the ciliary body, however. Most of the zonular fibers arise from the pars plana, running forward between the ciliary processes to take one of the three paths to the lens. These fibers appear to be attached in the valleys between ciliary processes by small fibrillar slings; other fibers may arise directly from this part of the pars plicata. The arrow indicates a site of attachment to the pars plicata. (Based on Streeten 1995.)

Accommodation is a result of ciliary muscle contraction

Figure 11.18*a* shows the zonule fibers running radially from their attachments on the anterior lens surface to the valleys between the ciliary processes. The space between the lens and the ciliary processes is exaggerated slightly, but this is essentially the situation that exists when the ciliary muscle is relaxed and the optical system of the eye is in focus for distant objects. Since we can think of the ciliary muscle as a sphincter, we are seeing the sphincter when it is relaxed and its aperture is fully open. In this state, the bundles of zonular fibers are under tension and are pulling outward on the lens.

Contraction of the ciliary muscle, either by an accommodative stimulus from a near object or by parasympathomimetic drugs, causes the ciliary sphincter to close in around the lens (see Figure 11.18*b*). This closure reduces the tension in the zonular bundles and therefore reduces the outward pull around the equator of the lens; at the limit of ciliary muscle contraction no tension is applied to the zonule, and the bundles of fibers are thrown into folds.

As Helmholtz showed so long ago (Figure 11.22), in the unaccommodated condition with the ciliary muscle relaxed and the zonule pulling outward on the lens, the lens has its maximum diameter, minimum thickness, and flattest radii of curvature. As the ciliary muscle contracts, the outward pull by the zonule around its equator is reduced; the equatorial diameter of the lens becomes smaller, its thickness increases, and its curvatures, particularly on the anterior surface, increase. The overall result is increased optical power.

At this level of description, the accommodative system is simple and unlikely to be challenged; the description has been around since Helmholtz made all the measurements on lens curvatures and thickness during accommodation (see

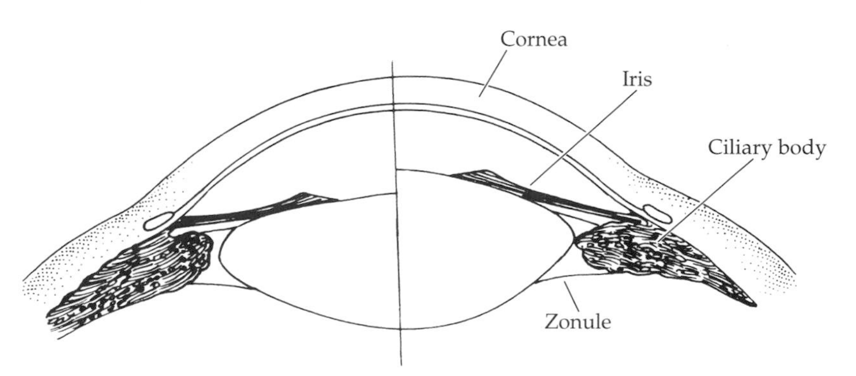

Figure 11.22
Contraction of the Ciliary Muscle and Change in Lens Shape
The left half of the drawing shows the unaccommodated state; the ciliary muscle is relaxed, there is tension in the zonule, and the lens is in its most flattened configuration. On the right, the ciliary muscle has contracted. The lens is thicker and more steeply curved, and the anterior surface is closer to the cornea. The change in lens shape depicted here corresponds to about 15 D of accommodation. (After Helmholtz 1924.)

Vignette 11.1). These measurements have been verified many times. But Helmholtz did not pursue the matter of how the lens changes its shape; he noted that it changed shape, characterized the change, assumed the change was due to inherent elasticity of the lens, and left it there. The arguing begins with the question, What is the mechanism underlying the change in lens shape?

I will save the details for Chapter 12, but the basic problem concerns the nature of the forces involved when the lens increases its curvature. The lens is plastic,* and the force applied to it when the ciliary muscle is relaxed flattens the lens. When the ciliary muscle contracts, however, the external deforming force is reduced and another force must be responsible for the subsequent change in lens shape. The question to be pursued in Chapter 12 is whether that force is an elastic restoring force inherent to the lens substance, an elastic force applied by the lens capsule, a hydrostatic pressure applied by the vitreous that is secondary to ciliary muscle contraction, or something else. For the moment, however, it is important to know only *how* the lens responds to contraction and relaxation of the ciliary muscle.

The primary stimulus to accommodation is retinal image blur

The efferent part of the accommodative reflex pathway consists of the neurons in the Edinger–Westphal nucleus that send signals to the ciliary muscle via the relay in the ciliary ganglion. The afferent part of the pathway is the optic nerve; accommodation is a visually guided reflex that normally requires a visual stimulus. (It is possible to accommodate voluntarily in the absence of a stimulus, however, and most persons can be taught to do it.) When we are looking at something in the distance, the retinal image of the target is in focus, whereas the images of closer objects are blurred to some extent. When we decide to look at a nearby object, the accommodative system generates signals to elicit ciliary muscle contraction and an increase in lens power that is sufficient to minimize the image blur for the object of interest. Accommodation is normally accompanied by pupil constriction, which increases the depth of focus, and convergence, which maintains foveal bifixation; this combination of accommodation, miosis, and convergence is called the **near triad**.

This mechanism seems straightforward, but image spread—blur—can arise from several sources: defocusing, spherical aberrration, or chromatic aberration, among others. Which components of blur is the system responding to? This is a

*Plastic substances, such as modeling clay, can be deformed by an external force, but when the external force is removed they remain in the deformed shape. Elastic materials, such as rubber bands, have a restoring force that acts as a kind of memory. A rubber band can be deformed by an external force, but when the force is removed, the rubber band returns to its original shape. The lens is certainly plastic, and all or part of it is elastic as well.

Vignette 11.1

The Source

Physiological Optics is the science of the visual perceptions by the sense of sight. The objects around us are made visible through the agency of light proceeding thence and falling on our eyes. This light, reaching the retina, which is a sensitive portion of the nervous system, stimulates certain sensations therein. These are conveyed to the brain by the optic nerve, the result being that the mind becomes conscious of the perception of certain objects disposed in space.
Accordingly, the theory of the visual perceptions may be divided into three parts: 1. The theory of the path of light within the eye.... 2. The theory of the sensations of the nervous mechanisms of vision.... 3. The theory of the interpretation of the visual sensations.

Hermann von Helmholtz, *Helmholtz's Treatise on Physiological Optics*

AS CONCEIVED AND PRACTICED by Hermann Ludwig Ferdinand von Helmholtz (1821–1894), physiological optics is a wide-ranging endeavor whose goal is understanding vision. Nowadays, we would say it is multidisciplinary and call it "vision science." In spite of Helmholtz's broad definition, the term "physiological optics" has increasingly come to mean only the first of his three theories cited above—that is, the optics of the eye.

Helmholtz was a superb scientist, and there is still gold to be mined from reexamining the ideas and observations left to us in his monumental textbook. One key idea concerns the mechanism of accommodation. For all the recent fussing over details, Helmholtz made several critical observations that have never been refuted and must be incorporated in any detailed model of the accommodative system: The ciliary muscle contracts in a sphincterlike fashion, producing a decrease of tension in the zonule that permits the lens to change its shape; the anterior surface curvature increases, while the equator decreases in diameter and increases in thickness. Helmholtz also ruled out some alternatives, including anterior–posterior movement of the lens, which occurs in some species, a change in corneal curvature, or a change in the axial length of the eye. Characteristically, he did not comment on issues about which he lacked information; there is no mention of lens elasticity or plasticity, and no speculation about the underlying cause of presbyopia.

Our current understanding of the basis for color discrimination, with its three cones and photopigments (see Chapter 13), derives directly from Helmholtz, who resurrected an elegant idea proposed a half century earlier by Thomas Young (to whom Helmholtz gave full credit, although the idea had generally been ignored). Helmholtz also repeated many of Clerk Maxwell's observations on color mixtures and proposed—again correctly—that dichromats lack one of the cone pigments. (The rods were a problem, however; the retinal anatomy of the time could not show any connections between rods and ganglion cells, and Helmholtz was not convinced they were visual cells.) As another example of his profound insight, a throwaway sentence buried in a long discussion of something else pointed out the fundamental difference between ordinary visual acuity and hyperacuities, such as thresholds for aligning vertical line segments; one is a resolution task, while the other, hyperacuity, involves detection of spatial *location*. Almost everyone missed this point for more than a century, until it was given new life by Gerald Westheimer.

In addition to his monumental work on vision, Helmholtz is rightly regarded as one of the nineteenth century's preeminent physicists, credited for his contributions to a variety of problems in acoustics, electricity and magnetism, optics, and one of the most fundamental laws of physics, that of the conservation of energy (energy cannot be created or destroyed; it can only be changed from one form to another). Helmholtz was not a physicist by training, however. Although he wanted to study physics at university, his family could not afford the tuition. But government stipends were available for medical study, requiring 8 years of service after training, and this was the course Helmholtz pursued.

As a medical student in Berlin, Helmholtz came under the influence of the physiologist and anatomist Johannes Müller (author of the "law of specific nervous energies") and his first major piece of research for his doctoral thesis focused on neuroanatomy; he used the insect nervous system to show that nerve fibers are extensions of nerve cell bodies—very basic stuff. Helmholtz became an army surgeon in 1843 at age 22 and began a series of experimental studies on nerve and muscle that he pursued in addition to all his clinical duties. The experimental work gained attention, as well as early release from medical service in 1848; he became professor of physiology at Königsberg and married his first wife, Olga, in 1849. The daguerreotype reproduced as Figure 1 shows him at this time—at age 27, a physician of some experience and a rising star in science.

Soon after beginning his work on vision, Helmholtz discovered something during a few weeks of fiddling with lenses and mirrors

Figure 1
Helmholtz as a Young Man
The original daguerreotype was made in 1848, when Helmholtz was 27 years old and had just finished his stint as an army surgeon. (From Koenigsberger 1906.)

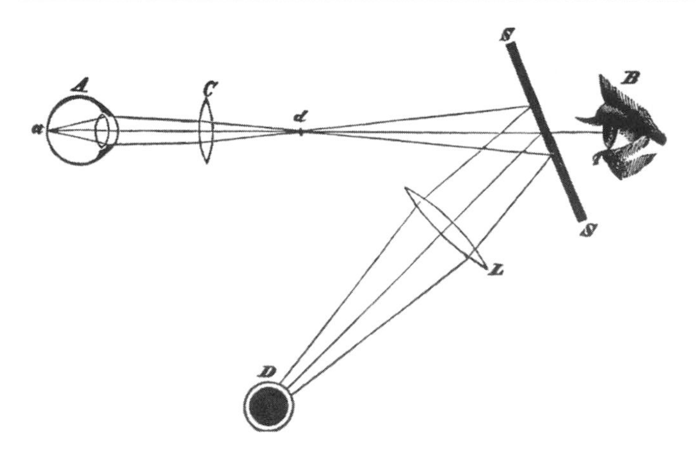

Figure 2
The Ophthalmoscope
This version of the optics for direct ophthalmoscopy shows the subject's eye at A and the observer's eye at B. Light from the illumination source (D) enters the subject's eye after reflection from a plane mirror (S). Light reflected from the subject's fundus is viewed by the observer through an aperture in the mirror. The subsidiary lenses, C and L, are used to produce a fourfold magnification of the subject's fundus. "a" is the focal point on the retina and "d" is an intermediate focal point. (From Helmholtz 1924.)

that made him famous—the technique for direct ophthalmoscopy.* His friend Ernst Brücke had recently shown that some of the light entering the human eye is reflected back from the fundus; if an observer could figure out a way to image this reflected light in his own eye, he would be able to see the other person's fundus. The trick, shown in Figure 2, is to use a mirror with a small aperture in it placed at a 45° angle to the subject's line of sight. Light from the illumination source is reflected off the mirror into the eye, and the observer can see the light reflected from the eye through the aperture. If the subject is emmetropic, the reflected rays exit the eye in parallel and can be imaged directly on the observer's retina (assuming emmetropia and no accommodation). This "splendid toy," as Helmholtz called it in private, was enthusiastically adopted by ophthalmologists and was a source of fascination to the public as well. Helmholtz enjoyed the attention, but he was also embarrassed by it, which says something significant about the man; the ophthalmoscope was important, but it was not a major intellectual achievement, and none knew that better than he.

There are many things to like about Helmholtz. No stuffy, self-important man could have made a habit of climbing alpine peaks and viewing the world upside down, bending over and looking back between his legs so that he could appreciate and describe the effect on one's perception. I also appreciate a problem he encountered while teaching anatomy. A complaint to the education ministry alleged that the anatomy was "inadequate" because Helmholtz insisted on discussing physiology as well as well as anatomy (the complaint was dismissed as "stupid," as was another about Helmholtz using cosine functions in a physiological optics course). Bringing different disciplines to bear on a problem is the hallmark of Helmholtz's style, and his ability to do it well was a major factor in his success. Even more stunning, however, is the full range of his interests and abilities. He was an excellent mathematician, though mostly self-taught, he was widely read in philosophy and literature (in his native German, as well as in French and English, both of which he spoke), he read Greek and Latin and dabbled in Hebrew and Arabic, he was a lover of music and a pianist, a lifelong student of art (he often lectured on vision and the visual arts), and an enthusiastic, observant, open-minded traveler.

And Helmholtz was also certainly not a narrowly focused, fact-grubbing experimentalist, as indicated by the following comment:

> *Art and science are essentially distinct in their external aspects and techniques; but I am none the less convinced of the profound internal relationships between them. Art, too, strives to acquaint us with reality, with psychological truths, though it expresses them in the wholly different form of sensual manifestation, and not in that of concepts. Eventually, however, the complete phenomenon connotes the conceptual idea and the two are ultimately united in the whole.*

Truths about the nature of our world can be approached from different directions—you heard it from Helmholtz.

For all his success and fame, Helmholtz did not always have an easy time of it. He was plagued by migraine and other ailments throughout his life, by his own estimate spending one day in seven incapacitated. His wife Olga died from a prolonged illness after they had been married less than 10 years, leaving him a clinically depressed young widower with two small children. Soon after, he was fortunate to meet and fall in love with a young woman, Anna von Mohl, who became his second wife and the mother of two more children. Having been brought up and educated primarily in Paris, Anna was an accomplished linguist, and the Helmholtz residence in Berlin became a gathering place for intellectuals from diverse fields and countries. But there was recurring grief at the deaths of the older children at early ages; only the daughter Ellen, who married into the Siemens family of industrialists, lived a normal span of years.

As he became older, Helmholtz suffered from unpredictable fainting spells; he fainted while descending a staircase on board a ship returning from the United States to Europe and was injured badly enough to have persistent vertigo and diplopia for some months afterward. And about 8 months later, he experienced the first of several strokes that lead to his death in September 1894. Until the day of the first stroke, he had been working on revisions for a new edition of the *Handbuch der Physiologischen Optik*.†

*The principle had been devised independently a few years earlier by the mathematician Charles Babbage (1792–1871), better known for his calculating machines, which were forerunners of modern computers. Helmholtz did not know of Babbage's work at the time, and he gets credit for making a practical instrument. Similarly, Helmholtz worked out the principle for indirect ophthalmoscopy, but Gullstrand is credited for inventing the clinical technique.

†The English translation is titled *Helmholtz's Treatise on Physiological Optics*, edited by J.P.C. Southall and published in three volumes during 1924 and 1925 by the Optical Society of America. It was translated from the third German edition, which is an amalgam of the earlier editions, though mainly the first, with commentaries and updates by Gullstrand, von Kries, and Nagel. It was reprinted by Dover Publications in 1962.

difficult question that has yielded a variety of answers over the years. It seems, however, that the accommodative system uses any information available to it, including defocus blur, image spread from aberrations, changes in image size, and judgments of apparent distance. Although defocus blur may be the primary stimulus, other cues can dominate, depending on the circumstances.

Accommodation is not a particularly simple reflex, and its afferent pathway involves higher levels of visual information processing at the cortical level—so much so that a recent review of the subject summarized it in this way: "The whole cognitive apparatus of the person is involved." How the cognitive apparatus signals its decisions to the Edinger–Westphal nuclei is not known in detail.

Accommodative amplitude decreases progressively with age

Benjamin Franklin noted that "in this world nothing can be said to be certain, except death and taxes." Actually, nothing is certain but death, taxes, and presbyopia, and Franklin, who is credited with inventing bifocals, was undoubtedly aware of it. Anyone whose refractive error is properly corrected has the experience, around age 45, of being unable to extend their arms far enough to hold a book or newspaper and see the text clearly. No one escapes. Myopes can cheat and read without their refractive correction, but everyone experiences the loss of accommodative amplitude that we call **presbyopia**.

Accommodative amplitude can be measured in several ways, but for practical purposes, it is the shortest dioptric distance from the eye at which fine detail, like small print, can be seen clearly (the reciprocal of the actual distance in meters is the dioptric equivalent). Measured in this way, accommodative amplitude is the maximum accommodative stimulus to which the eye can respond (the test is done unilaterally, but the amplitudes in the two eyes are normally similar). The maximum stimulus, however, is not the amplitude of the accommodative response, largely because of the eye's depth of focus; the response will generally be smaller than the stimulus, by 1 D or so. Accommodative response is rarely measured clinically, and the simpler measurement of *stimulus* amplitude is used as a proxy for the response amplitude. The difference between stimulus amplitude and response amplitude can be important, however, and one should remember that the difference exists.

The onset of presbyopia becomes apparent when the accommodative amplitude becomes less than one's normal reading or working distance. This distance varies from person to person, but it is usually on the order of 40 to 50 cm. Thus, when the amplitude drops below 2.0 to 2.5 D, the reading distance must be increased, and sooner or later, one's arms cannot be extended far enough. Although we tend to think of presbyopia as something that occurs suddenly, certifying one's arrival at middle age, only our awareness of the declining amplitude is sudden. In fact, accommodative amplitude declines progressively from a very young age.

Figure 11.23*a* shows accommodative *stimulus* amplitude as a function of age. The amplitude is highest for the youngest group included in the study, averaging about 14 D at age 8 years. From here, the trend is progressively downward, reaching a minimum value of 1 D between the ages of 50 and 55 years. The decline is fairly linear; about 1 D is lost every 5 years to age 30, after which the rate of decline accelerates, dropping to just under 4 D by age 45. At this point the loss becomes noticeable.

The decline of accommodation during the critical 2 decades from ages 40 to 60 is better seen in Figure 11.23*b*, which shows both stimulus and response measures of accommodative amplitude. By age 45, the maximum accommodative response is less than 1 D, and it is zero 10 years later, at an age when the stimulus amplitude is around 1.5 D (1 D in the previous data; the criteria for the stimulus amplitudes were different in the two studies). After 50 years of age, the

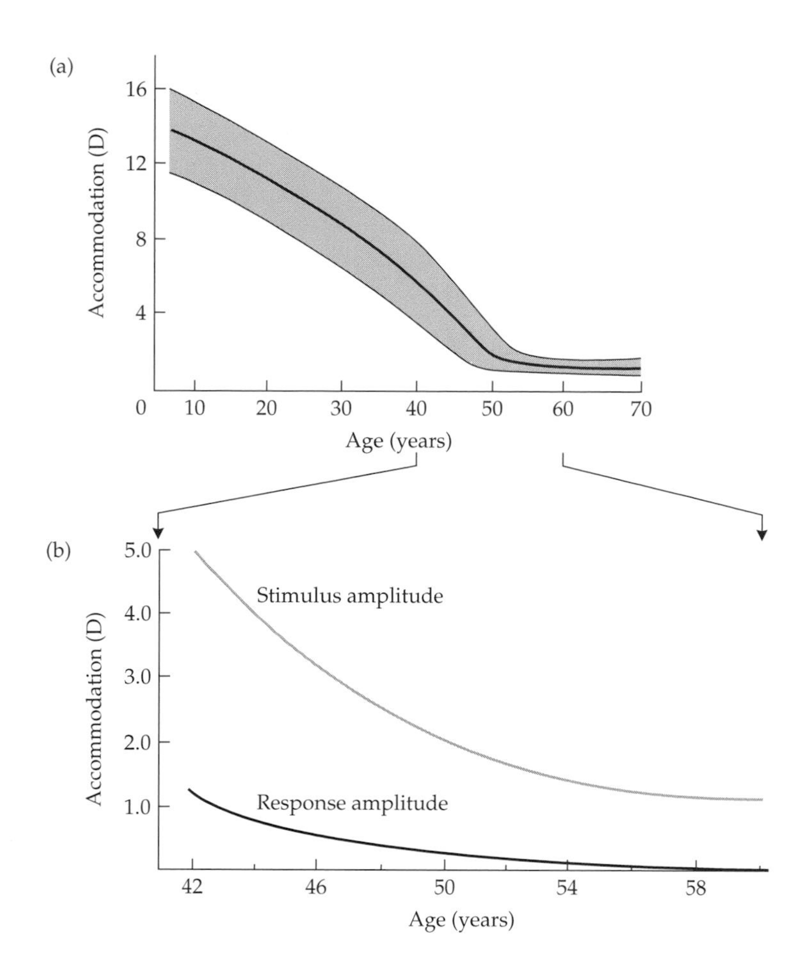

Figure 11.23
Accommodative Amplitude as a Function of Age
(*a*) The amplitude of accommodation declines steadily to about age 55, remaining stationary thereafter. The curves here show mean amplitude and the range in which 95% of the observations in this study lay. (*b*) Expansion of the range from ages 40 to 60, including a comparison of the stimulus and response amplitudes of accommodation. The accommodative response is less than 1 D by age 45 and is effectively zero by age 55. (*a* replotted from Duane 1912; *b* replotted from Hamasaki, Ong, and Marg 1956.)

accommodative response is essentially zero, but the depth of focus permits a range of clear vision from 1 m or so out to infinity. Although this is an extensive range, it does not include the critical working distances at an arm's length or less.

Presbyopia has no cure; we cannot halt or slow the loss of accommodation. The best we can do is compensate for the loss by providing more positive power for near vision, either as reading glasses if one is emmetropic, or as an addition (a near "add") to an existing distance correction. Until fairly recently, added corrections took the form of bifocals or trifocals with one or two regions of fixed added power in the lower part of the spectacle lens that one looks through when reading. Newer progressive addition lenses have a vertical strip several millimeters wide extending down from the center of the lens in which the added power changes gradually to its maximum. This graded addition provides regions at which intermediate distances will be in good focus, a definite advantage over conventional bifocals and trifocals. (The disadvantage is a restricted clear field of view at close working distances.) Near corrections can also be incorporated in contact lenses, but this form of correction has been difficult to implement successfully.

Presbyopia is not a consequence of reduced innervation to the ciliary muscle

Although we have no direct way to monitor the innervational signal going to the ciliary muscle, the pupil constriction and convergence that normally accompany accommodation can be taken as indirect measures of the accommodative signal. Within the available range of accommodation, the increase in convergence and decrease in pupil size are linearly related to the accommodative response (Figure

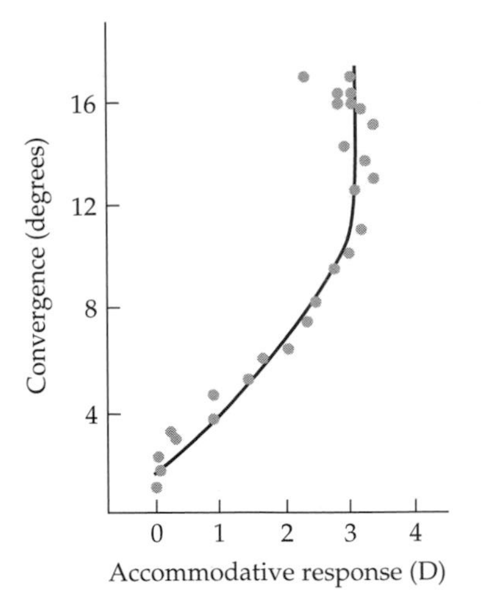

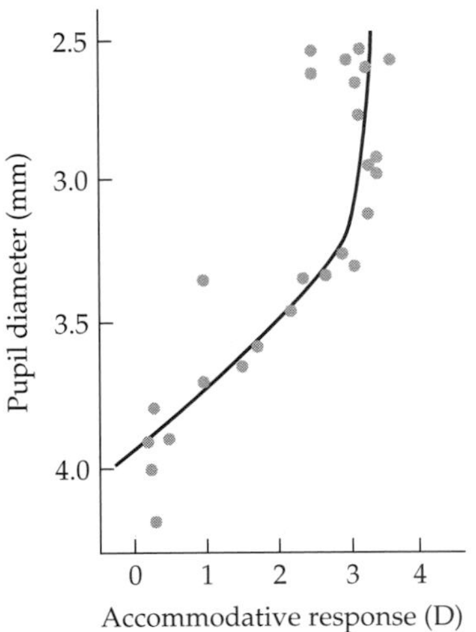

Figure 11.24
Convergence and Pupil Diameter as Functions of Accommodative Response
Accommodation, convergence, and pupil constriction are linked as the near triad. Convergence increases linearly and pupil diameter decreases linearly as accommodative response increases. These data are from an individual whose accommodative amplitude was limited at about 3 D. Although the accommodative response could not increase past this value, increased accommodative *stimuli* elicited more convergence and pupil constriction, as the upward spikes in the data show. (Replotted from Alpern, Mason, and Jardinico 1961.)

11.24). The slopes of the lines give the AC/A ratio (accommodative convergence to accommodation) and the P/A ratio (pupil size to accommodation) (see Chapter 10); here the AC/A ratio is about 3°/1 D and the P/A ratio is about –0.3 mm/1 D. The convergence response after the limit of accommodation has been reached, at 3.0 D of accommodation in this case, is interesting. Even though the subject could not increase his accommodative response beyond this limit, convergence continued to increase, producing a spike in the curve.* In a similar way, pupil size continues to decrease after accommodation has halted, producing a sharp deflection.

Because the signals for the convergence and miosis that normally accompany accommodation persist after the accommodative response has reached its limit, we can assume that the accommodative signal persists as well; therefore, accommodation is limited by something other than a limitation of its neural command. The situation is similar in presbyopia, in which the accommodative response is negligible; accommodative stimuli will still elicit convergence and miosis, and the same conclusion applies. Absence of accommodation is not due to the absence of a neural signal.

The lack of a direct measure of the neural signal to the ciliary muscle means one can argue with the foregoing conclusion. For example, accommodative neurons, pupillary neurons, and medial rectus neurons diverge at some point, and perhaps only the efferent pathway to the ciliary muscle is affected. But the progressive decrease in accommodative amplitude for the first 50 years of life implies the existence of a progressive failure in this particular neural pathway, an effect without precedent. Thus while alternative explanations of this type are possible, they are complicated and not very plausible. The existence of a continued signal to accommodation after the system can no longer respond is by far the simplest interpretation.

Aging of the ciliary muscle is unlikely to be a significant factor in presbyopia

Assuming that presbyopia is not due to a lack of neural commands for accommodation, we must look to the ciliary muscle, the zonule, and the lens. The loss of accommodation could be a failure of the muscle to contract, a failure of the lens to change shape when the muscle *does* contract, a failure of the zonule to transmit imposed force to the lens, or some combination of these factors. But the evidence suggests to me that the ciliary muscle is a minor player in presbyopia.

Anatomical evidence for altered function of the ciliary muscle in older eyes consists of age-related increases in connective tissue within the ciliary muscle and in cellular inclusions in the muscle fibers that are characteristic of degeneration. The problem, however, is knowing what the changes mean. The increased amount of connective tissue and the atrophy of muscle cells, for example, are most evident in older eyes—eyes that lost their ability to accommodate years earlier. Connective tissue changes and muscle atrophy are progressive, but it is not clear that the changes are sufficiently extensive or severe to account for a loss of ciliary muscle contraction by age 50, if ever. In addition, other muscles show similar changes with age; are the changes in the ciliary muscle greater than those occurring in other smooth muscles that do not fail to contract in old age? In short, the age-related structural changes in the ciliary muscle are not necessarily associated with a loss of muscle function and with presbyopia.

*This set of data is a small memorial to Mathew Alpern (1920–1996), from whose eyes the results were obtained; I learned many years ago to call the continuing convergence after the limit of accommodative response the "Alpern spike." If my experience is representative, Mat never failed to encourage people in the next generation of vision scientists, and it is pleasant to remember his help while remembering his work.

The ciliary body, or at least the ring of ciliary processes around the lens, can be seen directly when the iris has been surgically removed (see Figure 11.18). In a study of accommodation in rhesus monkey, a species that also becomes presbyopic, electrodes implanted in the Edinger–Westphal nucleus of iridectomized monkeys were used to elicit accommodation by electrical stimulation of the Edinger–Westphal neurons. In young monkeys, stimulating the accommodative system in this way produced an inward contraction of the ring of ciliary processes (Figure 11.25*a* and *b*) and an increase in lens thickness and curvature. In older monkeys, the lens and ciliary changes were less pronounced, and in at least one case—a 24-year-old animal—there was very little change at all; the ciliary ring contracted very slightly, and the lens exhibited a correspondingly small change in shape (Figure 11.25*c* and *d*).

These results seem important, particularly since the oldest eyes showed so little response when the efferent accommodative pathway was activated, but the experimental design is flawed. The problem is knowing whether the stimulus was as effective in old monkeys as in young monkeys. Placing electrodes accurately into small nuclei deep in the brain is difficult in the best of circumstances, and investigators require some indication that the electrode is in the correct location. The usual signs are the kind of electrical activity recorded by the electrode or the consequences of electrical stimulation; in this case, the signs would be accommodation and pupil constriction when current passed through the electrode. But the monkeys in these experiments had iridectomies, so pupil constriction was not available as a guide to electrode placement. In young monkeys, accommodation resulting from electrical stimulation was a sign of proper electrode positioning, but what about in the older ones? Does the weakness of the accommodative response from electrical stimulation mean that the accommodative system was weaker or that the electrode was not in the correct location? Unfortunately, there is no way to know, and the experimental results cannot be interpreted unequivocally.

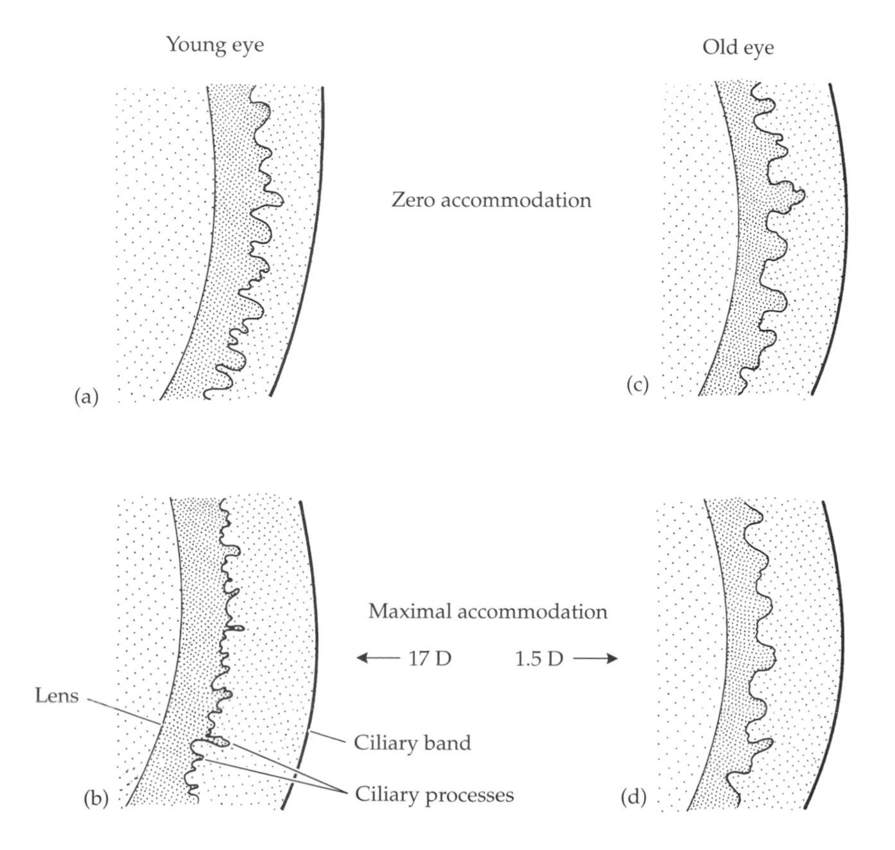

Figure 11.25
Stimulated Accommodation in Young and Old Monkey Eyes
After total iridectomy, the lens and the ring of ciliary processes can be seen in rhesus monkey eyes. In both animals, stimulating electrodes had been implanted in the vicinity of the Edinger–Westphal nucleus so that current could be passed through the electrodes to elicit accommodation. Stimulation in the young eye produced a strong accommodative response and a sizable inward movement of the ciliary processes with respect to the ciliary band (compare *a* and *b*). The effect was less obvious in the older eye; both the accommodative response and the ciliary process movement were quite small (compare *c* and *d*). (After Neider et al. 1990.)

In young rhesus monkey eyes, the shape of the ciliary muscle when completely relaxed (accomplished by use of an acetylcholine blocker) and its shape when fully contracted (as induced by a acetylcholine agonist) are very different (Figure 11.26*a* and *b*). The difference was considerably less in the eyes of older monkeys (Figure 11.26*c* and *d*), suggesting that the ciliary muscle in older eyes is unable to contract to the extent that it can in younger eyes. But the relaxed and stimulated ciliary muscle conditions were effected of necessity on different eyes. Since the results do not compare relaxed and contracted conditions of the *same* ciliary body, it is difficult to know what the comparison means.

One possibility for assessing ciliary muscle action is to record its electrical activity, but this approach has proved to be unworkable for technical reasons. Instead, in an indirect method called impedance cyclography, small surface electrodes are placed on the lateral and medial sides of the eye just over the ciliary body. Alternating current passed between members of one electrode pair on opposite sides of the eye is recorded by another pair; the difference between the amplitude of the current that passed through the eye and the amplitude of the recorded current is proportional to the alternating-current resistance (the impedance) of the tissue. Impedance decreases as the ciliary muscle contracts, and this change has been used as a measure of the contraction.

In a young adult, the accommodative response increases with the stimulus up to a limit of about 5 D, after which the response is constant (Figure 11.27*a*). The tissue impedance decreases linearly as the accommodative response increases (Figure 11.27*b*), spiking sharply upward when accommodation reaches its limit (decreases are plotted as upward deflections). Insofar as the impedance change reflects ciliary muscle contraction, these data indicate that contraction continues *after* the accommodative response has reached its maximum. Therefore, accommodation cannot be limited by a failure of ciliary muscle contraction. Other data show that the situation is similar in presbyopes; although the accommodative response is negligible, there is a clear decrease in the measured impedance as the accommodative *stimulus* increases. This evidence indicates that the ciliary muscle contracts even though the lens fails to respond.

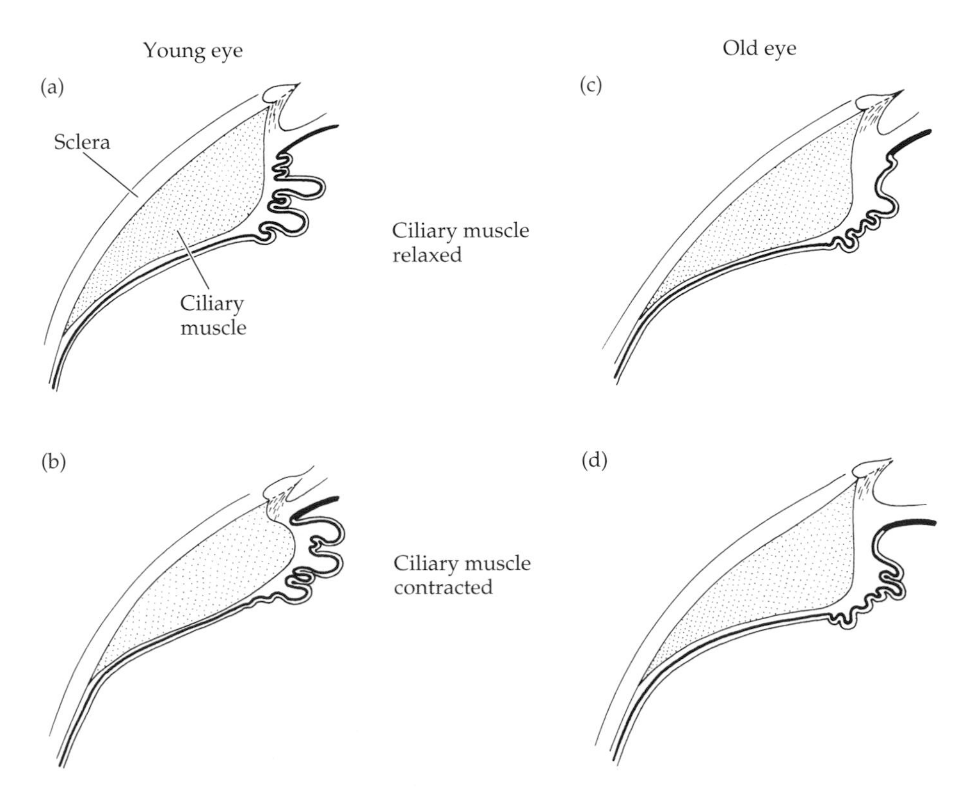

Figure 11.26
Drug-Induced Ciliary Muscle Contraction in Young and Old Monkey Eyes
Ciliary muscle relaxation was produced by atropine in the left eye and contraction was produced by pilocarpine in the right eye before the eyes were removed for histological processing. In young animals, there is an obvious difference between the two conditions: The contracted ciliary muscle has pulled the ciliary body anteriorly and inward (compare *a* and *b*). The difference between relaxed and contracted states is smaller in old monkeys (compare *c* and *d*). The old eyes also have connective tissue not observed in the young eyes. (After Lütjen-Drecoll, Tamm, and Kaufman 1988.)

Unfortunately, the most significant factor in the impedance measurements is not the change in muscle volume, but the change in blood volume within the ciliary muscle. The question now concerns the relationship between blood flow and muscle contraction. According to one line of argument, blood flow decreases directly as a function of muscle contraction—a contention that leaves the conclusion intact. If contraction and blood flow vary independently, however, the conclusion is questionable; the impedance decrease after accommodation has halted might reflect decreased blood flow rather than increased muscle contraction.

But other evidence suggests that the impedance cyclography data give the correct impression about muscle contraction. Paralysis of the ciliary muscle with a cycloplegic drug eliminates the impedance changes shown in Figure 11.27*b*, for example. Another line of argument involves the intraocular pressure–lowering effect of pilocarpine, which is thought to be produced by ciliary muscle contraction and bending of the scleral spur (see Chapter 9). Pilocarpine works well in presbyopic eyes, and *if* the rationale for pilocarpine's action is correct, the ciliary muscle must still be capable of contraction.

At present, no unequivocal evidence suggests that the ciliary muscle loses its ability to contract with age. Somewhat better evidence, however, shows that the ciliary muscle continues to contract after accommodation has ceased. Although this will continue to be a subject for study and debate, the ciliary muscle appears not to be responsible for the loss of accommodation that we experience as presbyopia.

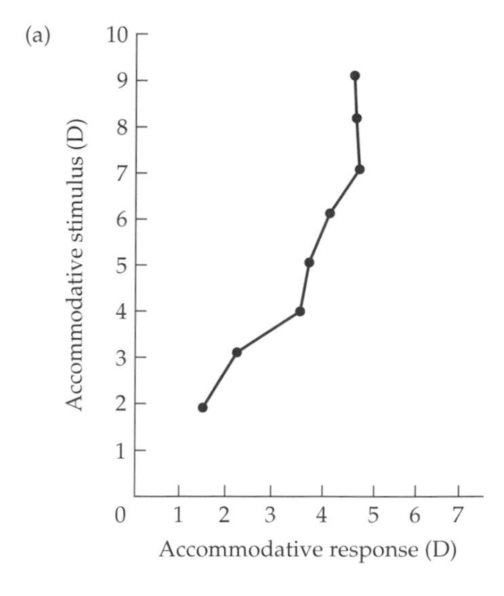

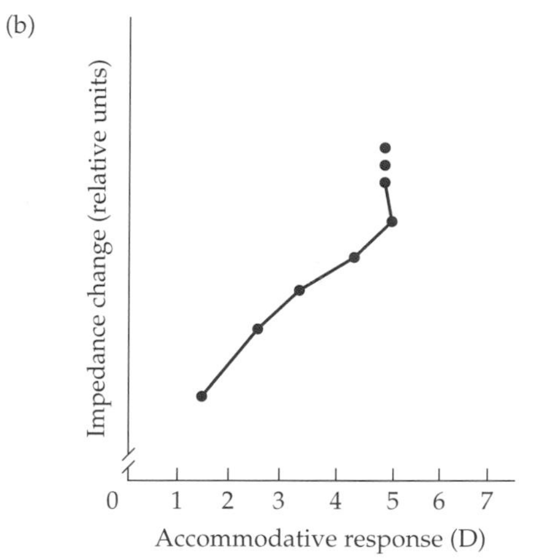

Figure 11.27
Evidence for Ciliary Muscle Contraction beyond the Accommodative Response Limit
(*a*) The accommodative response in a 29-year-old subject is limited at about 5 D and maintains this level as the accommodative stimulus increases. (*b*) Changes in impedance in the subject's eye increase as the accommodative response increases, and they continue to increase after the accommodative response has hit its limit. Insofar as the impedance changes are related to ciliary muscle contraction, the result indicates that ciliary muscle contraction does *not* limit the accommodative response. (Replotted from Saladin and Stark 1975.)

The Choroid

The choroidal stroma consists of loose connective tissue and dense melanin pigment

Most of the choroid is filled with blood vessels; the rest is the stroma, the tissue that surrounds the blood vessels. Part of the stroma is connective tissue—that is, loose strands of collagen and fibroblasts—but the more obvious component is dense melanin pigment found throughout the thickness of the choroid (Figure 11.28).

Some amount of pigment located behind photoreceptors is universal, even occurring in association with the sensitive "eyespots" in unicellular organisms, in which it gives the eye some directional sense (see the Prologue). In image-forming eyes, pigment absorbs stray light that would otherwise degrade the quality of the retinal image.

Some stray light comes from reflection at refractive surfaces or from scatter by small inhomogeneities in the optical media, but the most important source of stray light is the retinal image itself. Because the retina is laid out on the inner surface of a hemisphere, any point on the retina has an unobstructed view of all other points. Light in a bright retinal image that is not absorbed by the photopigments passes through the retina and is reflected back into the eye by the sclera; the retinal image becomes a secondary light source sending light to the rest of the retina. The result is veiling glare, reduced contrast, and reduced acuity. The choroidal pigment reduces this effect by absorbing light that has passed through the retina.

If external light sources are very bright, their retinal images inevitably act as secondary light sources within the eye; light reflected back out through the pupil causes the "red-eye" phenomenon in photographs taken with direct flash, for example. This reflection is not altogether bad; if all light entering the eye were absorbed and none were reflected back, there would be no such thing as ophthalmoscopy.

Some species take advantage of the retinal image light that is not absorbed by having a reflecting surface in the choroid to ensure that the light is reflected back. This choroidal reflector—the **tapetum**—is responsible for the yellowish or greenish glow in the pupils of cats or dogs when they face a bright light source. Tapeta are very common and are an adaptation for nocturnality. Under dim light conditions, good image quality with high spatial resolution is not important; the goal is to detect light—any light—with high sensitivity. Therefore, if any light escapes the photoreceptors, it will be reflected back by the tapetum so that the photoreceptors have another chance to capture it. But while a tapetum is an advantage in dim illumination, it could be a serious handicap under bright light conditions. The solution, in some species, is covering or uncovering the tapetum, usually by pigment migration; in these species the tapetum is covered for daytime vision and uncovered at night.

Blood vessels that supply and drain the capillary bed supplying retinal photoreceptors make up the main part of the choroid

Most choroidal blood vessels are branches of the short posterior ciliary arteries and tributaries of the vortex veins (see Chapter 6). Because the capillary bed is on the inner surface of the choroid and the largest vessels enter (or exit) on the outer surface, the successive branching makes the vessel diameters smaller as one goes from outer to inner layers (see Figure 11.28). The capillary bed—the **choriocapillaris**—lies next to the retinal pigment epithelium and is the source of supply for the photoreceptors.

The branching hierarchy has led to a designation of layers in the choroid. The outermost part of the stroma next to the sclera is the **suprachoroidea**; it has no particular functional role that differs from the rest of the stroma. Melanocytes

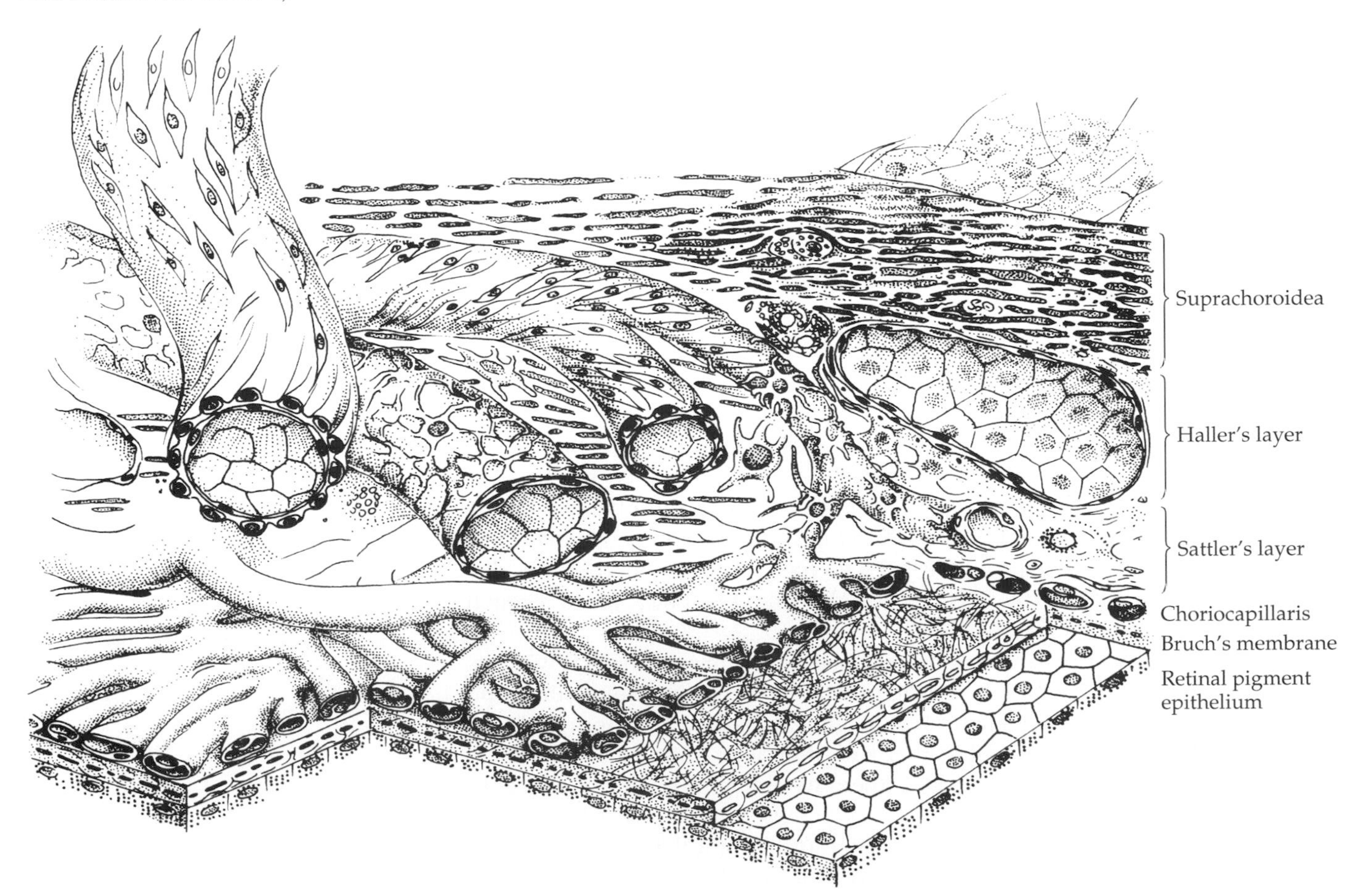

Figure 11.28
Structure of the Choroid
The outermost layer next to the sclera—the suprachoroidea—is connective tissue, a mixture of collagen, fibroblasts, and melanocytes. Most of the choroid consists of blood vessels, the diameter of which decreases from outer to inner as the vessels branch. These arteries and veins supply and drain the choriocapillaris, which is a single layer of large interconnected capillaries. The choriocapillaris is separated from the retinal pigment epithelium by Bruch's membrane. Nerve fibers running through the choroid, some to the arterioles, are not shown. (After Hogan, Alvarado, and Weddell 1971.)

from the suprachoroidea tend to infiltrate the inner part of the sclera, which is called the **lamina fusca**. There is an outer layer with larger blood vessels (**Haller's layer**) and an inner layer with smaller vessels (**Sattler's layer**). These layers are indicated in Figure 11.28, but neither the names nor the designations as layers are of much importance. These distinctions simply reflect the fact that the vessels in the branching hierarchy become smaller as one approaches the capillary bed, a point illustrated schematically in Figure 11.29.

We can take one of the short posterior ciliary arteries as an example; the artery will have several levels of branching before it reaches the capillary bed as multiple small arterioles (Figure 11.29*a*). In the choroid, however, the distance between a main artery and the capillary bed is short, meaning that the entire branching hierarchy is compressed (Figure 11.29*b*). Because of this compression, the choroid shows a layering of vessels by size, with the larger branches outermost and the smallest vessels near the inner surface.

Anastomoses are rare in the short posterior ciliary artery system, and the supply to the choriocapillaris is of the end-arterial type; the short posterior ciliary arteries are not interconnected to provide alternative routes of supply in the event that one fails. In addition, the regions of the capillary bed supplied by the separate short posterior ciliary arteries have little if any overlap; the supply is segmental. The venous drainage by the vortex veins is almost the opposite, however; anastomoses are common at all levels, and the drainage fields overlap. So while arterial occlusions can deprive local regions of the capillary bed of blood, venous occlusions should have relatively little effect on drainage of the choriocapillaris.

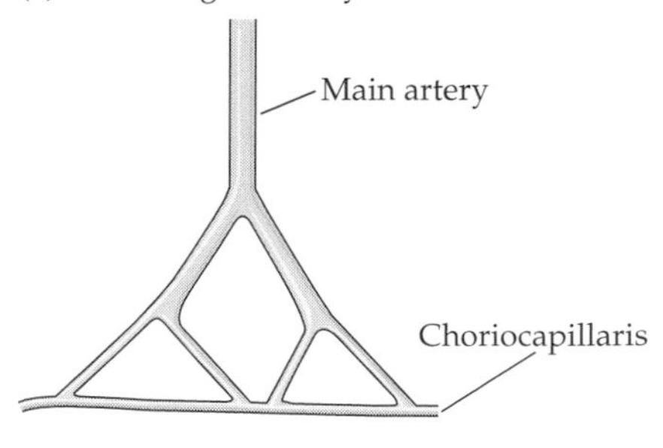

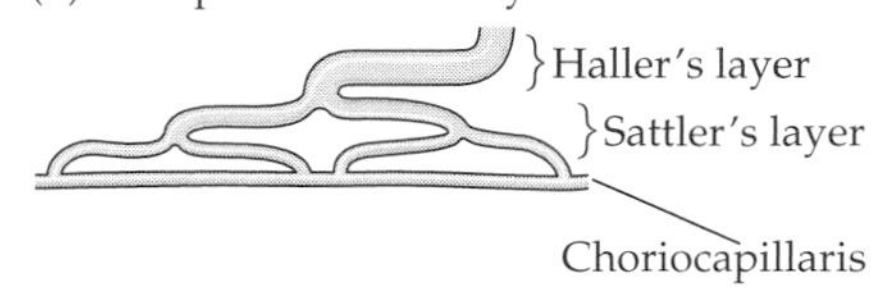

Figure 11.29
Elements of Choroidal Structure and Organization
The choroid is a branching hierarchy of blood vessels embedded in connective tissue and pigment. A four-tiered hierarchy (*a*) can be made to resemble the choroid by compression (*b*). The outer-to-inner sequence of large to small vessels creates what are called Haller's and Sattler's layers, respectively. If the hierarchy of veins draining the choriocapillaris were added to this picture, the impression that the choroid consists mostly of blood vessels would be heightened.

The choriocapillaris is heavily anastomotic but has local functional units

Figure 11.30*a* is a view of the choroidal vasculature from outside; the choriocapillaris, which looks like a sheet of lace at this magnification, is overlaid by arterioles and venules that supply and drain the capillaries. At higher magnification, with the overlying vessels removed, the choriocapillaris has the appearance of an interconnected meshwork (Figure 11.30*b*). Any two separate points in the sheet of capillaries are connected by numerous paths along which blood could flow from one point to the other. At first glance, this extensive interconnection seems to imply that the lack of anastomoses among the arteries supplying the choriocapillaris is unimportant; if the normal route of arterial supply to a particular location is blocked, numerous other routes through the capillary bed could carry the blood to its destination. But the relative numbers and locations of the arterioles and venules that supply and drain the choriocapillaris should affect the way in which blood flows through the capillaries.

Figure 11.31 shows two extremes in the number and distribution of arterioles and venules. When venules outnumber arterioles (Figure 11.31*a*), by about five to one in this case, the pattern of blood flow develops local subunits that are relatively independent of one another. Because blood flows from higher to lower pressure, its movement in the capillary bed is from the arterioles to the venules. But with several venules surrounding each arteriole, the venules create a kind of low-pressure drainage ditch: Blood flows into the ditch, but not past it. If one of the arterioles is blocked, however, there will be no pressure gradient within the region normally supplied by the obstructed arteriole and therefore no impetus for blood flow through the deprived region. At best, blood will move sluggishly into the region from surrounding areas that have intact arterioles. More probably, the blood will be stagnant. So while the choriocapillaris is heavily interconnected and anastomotic, under the conditions shown here, the regions supplied by one arteriole are more or less on their own; they do not have alternative sources of supply.

(a)

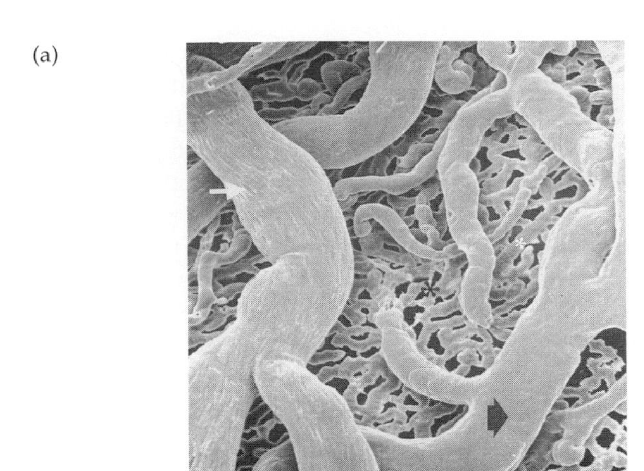

(b)

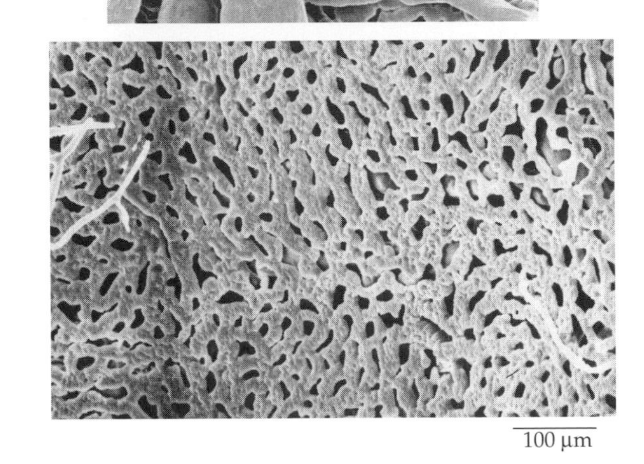

Figure 11.30
The Choriocapillaris
(*a*) This vascular cast is viewed from the scleral side, looking through the larger arterial (e.g., white arrow) and venous (e.g., dark arrow) branches to the choriocapillaris beneath. Large vessels overlie smaller ones, and the choriocapillaris (asterisk) shows as a delicate network of small vessels that are fed or drained by the larger ones. (*b*) The choriocapillaris near the posterior pole is a dense network of anastomosing capillaries; potentially, there should be many routes by which blood could flow between any two points in the capillary bed. This view is from the retinal side; the overlying vessels are branches of the central retinal artery. (From Olver 1990.)

At the other extreme, arterioles outnumber venules (Figure 11.31*b*). This is a situation of redundant supply; when one arteriole is blocked, blood can flow from neighboring arterioles to resupply the deprived area because there are no intervening venules to drain the blood away. This arrangement is more vulnerable on the drainage side of the capillary bed. Blocking a single venule will not stop blood flow through the drainage field belonging to the obstructed venule, but the pattern and direction of flow will change to follow the shift in the pressure gradient produced by the obstruction. The danger here is edema; when blood enters the capillary bed at a normal rate but with less-than-normal drainage capacity, the capillaries can expand, become abnormally leaky, and release fluid into the adjacent tissue.

The choriocapillaris varies in capillary density and in the ratio of arterioles to venules

The choriocapillaris changes from its center—the region behind the fovea—to the periphery, where the choroid meets the ciliary body. The most obvious change is in capillary density (Figure 11.32) or, to put it more directly, in the amount of blood per unit area. Capillary density is highest behind the fovea, declining gradually and then more sharply out to the periphery. This variation corresponds roughly with the change in photoreceptor density (both rods and cones) in the retina; photoreceptor density is highest in and around the fovea and declines steadily toward the periphery (see Chapter 15).

The relative numbers of arterioles that supply the choriocapillaris and venules that drain it also appear to vary from center to periphery, although this change is not well documented. The region behind the fovea has the largest ratio of arteri-

Figure 11.31
Blood Flow Patterns in the Choriocapillaris
Applying the simple rule that blood flows from high to low pressure, the flow in a portion of the choriocapillaris where venules outnumber arterioles would be as depicted in *a*. Blood flows away from the arterioles and toward the venules. If the arteriole were blocked, flow would be very much diminished or absent in the capillaries around the blocked arteriole. If the arterioles outnumbered the venules, however (*b*), blockage of a single arteriole would have little effect on the pattern of flow.

(a) Venules outnumber arterioles

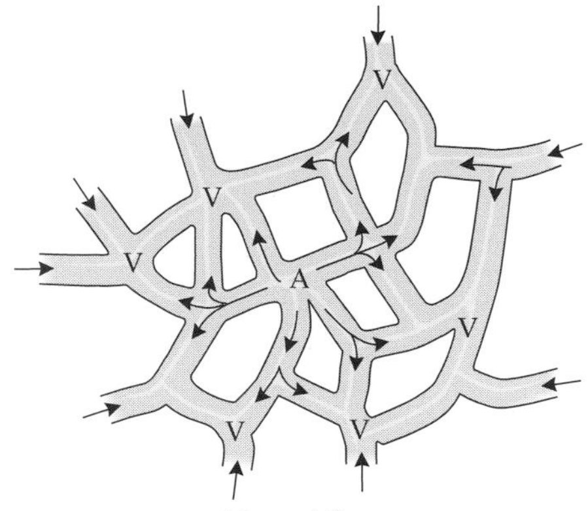

Normal flow

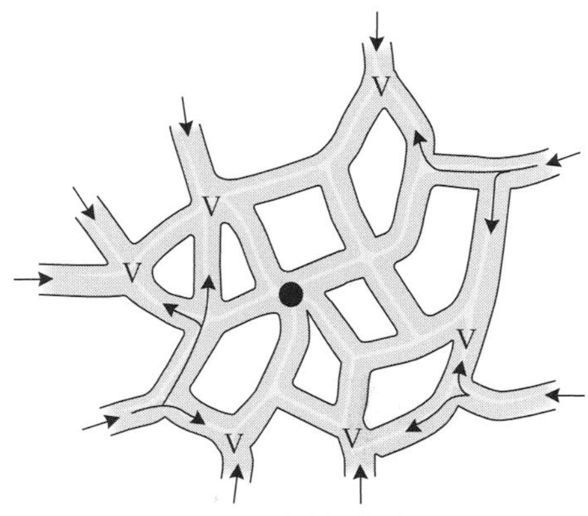

Arteriole blocked

(b) Arterioles outnumber venules

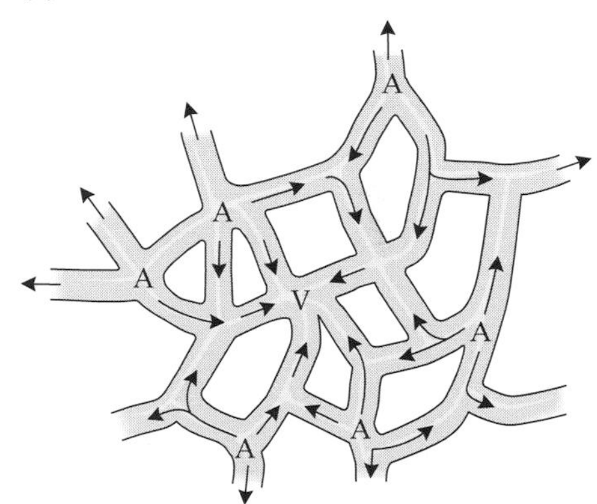

Normal flow

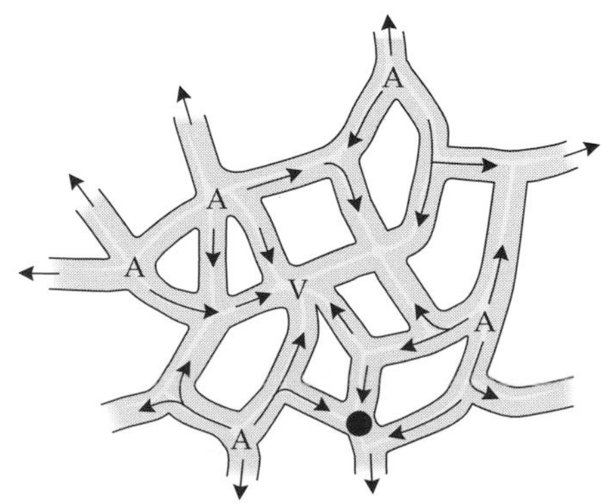

Arteriole blocked

oles to venules, giving the all-important foveal cones a redundant blood supply. The cost, however, seems to be a higher risk of edema in the foveal region; foveal edema is common and often idiopathic (an example is central serous retinopathy). The choriocapillaris 20 or 30° from the center has a dramatically different ratio—venules outnumber arterioles—and functional subunits can be demonstrated with fluorescein angiography. As a result, this part of the choriocapillaris may be more susceptible to local blockages in the arterioles. In the periphery of the choroid, the numbers of arterioles and venules are about the same, and both are sparsely distributed, in keeping with the low capillary density.

Capillaries in the choriocapillaris are specialized for ease of fluid movement across the capillary endothelium

Like the capillaries in the ciliary processes, those in the choriocapillaris are large, they have endothelial cells that are not tightly bound to one another, and their endothelium has closed fenestrations. The only notable difference is the location of the fenestrations: Capillaries in the ciliary processes have fenestrations on all sides (see Figure 11.6), but in the choriocapillaris the fenestrations are mainly on the inner side, facing the retina.

From this information, together with the earlier discussion about capillaries in the ciliary body, it is clear that the majority of the capillary beds supplied by the posterior ciliary arteries, both long and short, are specialized for ease of fluid movement; the only exception is capillary beds in the iris supplied by branches from the major arterial circle.

Bruch's membrane lies between capillaries and pigmented epithelium in both the choroid and the pars plana of the ciliary body

Figure 11.33 shows Bruch's membrane, lying between the capillaries of the choriocapillaris and the pigment epithelial cells of the retina (in Figure 11.14 it was shown within the pars plana). The membrane is very thin—around 2 μm in young eyes—and it has no obvious structure at low magnification. At high magnification, however, layers are visible. The inner and outer surfaces of Bruch's membrane are basement membranes—the inner one for the retinal pigment epithelial cells and the outer for the choriocapillaris. Neither of these basement membranes is attached by hemidesmosomes to the adjacent cell layers. The outer basement membrane is discontinuous: It is present next to the endothelium of the capillaries but absent in the spaces between capillaries.

The dark-appearing central layer consists of coarse, interlaced elastic fibers. This elastic lamina is sandwiched between two layers of collagen fibers, each layer about 1 μm thick in young eyes. The collagen fibers are loosely arranged, with considerable matrix around them, and fibers from one collagenous layer may cross over to the other by passing through the central elastic fiber layer. Collagen fibers in the outer collagen layer are less likely to run parallel to the surface of the membrane, and where there are breaks in the outer basement mem-

(a) Behind the fovea

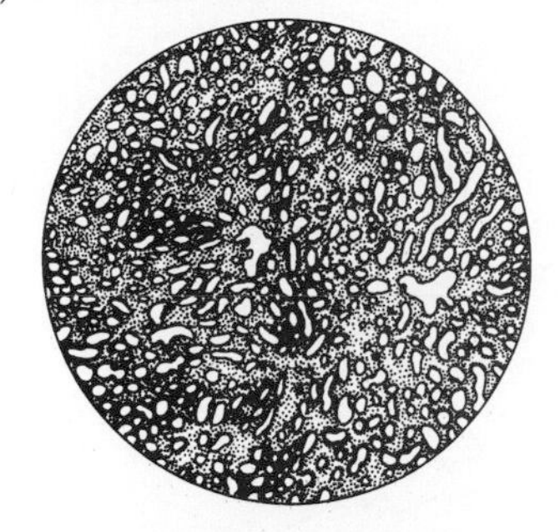

(b) At the equator

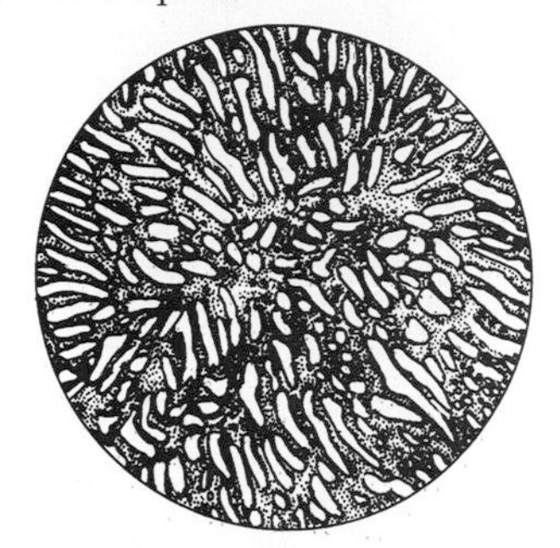

(c) At the periphery

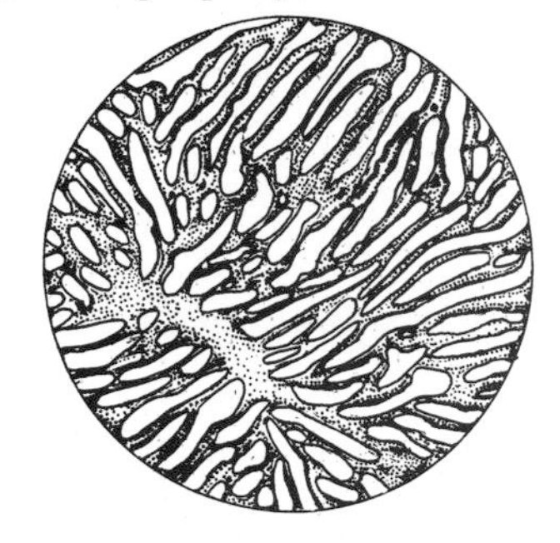

Figure 11.32
Choriocapillaris Density as a Function of Location
In these drawings of the choriocapillaris, the capillaries are gray and spaces between them are white. In the part of the choriocapillaris behind the fovea (*a*), the capillary density is quite high, as indicated by the preponderance of capillary space over open white space. Capillary density is lower at the equator (*b*), and lower still at the periphery (*c*) of the choroid. (After Hogan, Alvarado, and Weddell 1971.)

brane between capillaries, the membrane collagen runs up between the capillaries to merge with stromal collagen fibers. This arrangement may help bind Bruch's membrane to the choriocapillaris.

Anything that passes between the choroidal blood and the retina, in either direction, must cross Bruch's membrane. (In the pars plana of the ciliary body, the same can be said about exchanges between blood and vitreous.) But while the location is clearly critical, the effect of Bruch's membrane on the normal trafficking between cells and blood is not known; the primary blood–retina barrier is created by the pigmented epithelial cells (see Chapter 13). And while one can speculate about a role for Bruch's membrane in terms of mechanical support and stability of the retina, it is a matter of speculation. The importance of Bruch's membrane becomes apparent only when it fails to do whatever it normally does.

Like Descemet's membrane in the cornea, Bruch's membrane gradually thickens with age, reaching an average thickness of 6 μm or so late in life. The collagenous layers, particularly the inner one, become thicker by accumulating some sort of granular or filamentous material, while the elastic layer in the center becomes less densely organized; it may develop cracks and holes where the layer is discontinuous. Large breaks in Bruch's membrane are commonly associated with edema, which may lead to accumulation of fluid in the potential space between the retinal pigment epithelium and the photoreceptors and to retinal detachment. The stagnant fluid may be invaded by fibroblasts and other cells, accompanied by necrosis of nearby tissues, including the retina. The association of a discontinuous membrane with edema suggests that under normal conditions Bruch's membrane plays a role of limiting fluid movement inward toward the retina.

Although the age-related thickening of Bruch's membrane is small, from 2 to 6 μm, local increases in thickness can be much larger. Accumulations of material in the inner collagenous layer of Bruch's membrane form inwardly bulging regions called **drusen** (Figure 11.34). There are various types of drusen, probably with different etiologies, but they all have the effect of pushing the photoreceptors away from their blood supply in the choriocapillaris. If the separation becomes too large as the drusen increase in size, the overlying pigment epithelium and photoreceptors may degenerate. Accumulations of large drusen that become confluent may therefore produce sizable scotomas in the visual field.

Development of the Ciliary Body and Choroid

The ciliary epithelium arises from the optic cup, the ciliary muscle from neural crest cells

As discussed in Chapter 10, the epithelial layers of both iris and ciliary body are derived from the rim of the optic cup. The migration of mesenchymal tissue in front of the lens along with the anterior portion of the tunica vasculosa lentis divides the anterior segment into the spaces that will be the anterior and posterior chambers (Figure 11.35). Later, the two cell layers at the margin of the optic cup differentiate to become epithelial cells and then push inward over the anterior surface of the lens and the tunica vasculosa lentis. The result is an epithelial substrate for the undifferentiated cells in the future iris stroma and ciliary body. Because the epithelial layers are created as the optic cup margin grows over the lens, it follows that the ciliary epithelium is in place first, followed by the iris epithelium. In accordance with the inside-to-outside gradient of development (see Chapter 1), the ciliary epithelium appears after the lens and the tunica vasculosa have formed but before elaboration of the ciliary muscle.

The ciliary muscle does not differentiate until about week 16, although some important events have already occurred. Like the cellular components of the corneal, limbal, scleral, and iridial stromas, the cells that will become ciliary

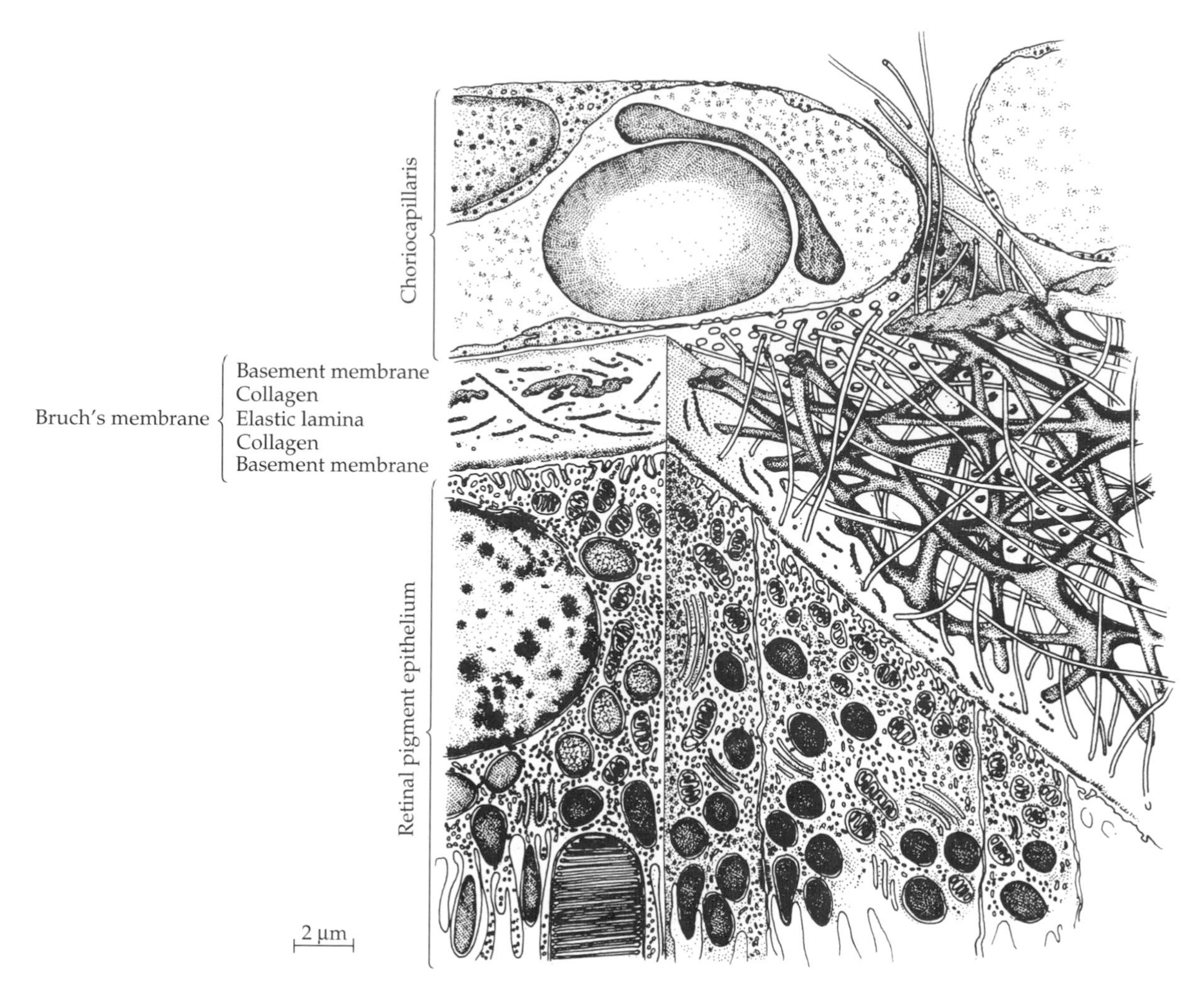

Figure 11.33
Bruch's Membrane, the Choriocapillaris, and the Retinal Pigment Epithelium
Parts of two large capillaries in the choriocapillaris lie above Bruch's membrane, part of which has been reduced to its major fibrous components—namely, the elastic fibers in the middle layer and the collagen fibers on either side of the elastic lamina. Some collagen fibers run up into the space between the capillaries, mingling there with stromal collagen fibers. The inner basement membrane is next to the basal surfaces of the retinal pigment epithelial cells, which have numerous infoldings on their basal surface but have no anchoring junctions to the basement membrane. One photoreceptor outer segment is shown in its pigment epithelial cell socket. All interchanges between the photoreceptor and the capillaries pass through Bruch's membrane and the pigment epithelium. (After Hogan, Alvarado, and Weddell 1971.)

muscle are migrants—neural crest cells that have moved into the undifferentiated mesenchyme around the optic cup (see Figure 11.35). Once the neural crest cells are in place in the ocular mesenchyme, the cells await the appropriate signals for determination and differentiation.

The cellular components of the uveal tract and the outer coat of the eye are of neural crest origin (excluding the epithelial layers and the vascular endothelium), but it is not known how a wandering neural crest cell decides to become a ciliary muscle cell instead of a corneal fibroblast. Neural crest cells are unlikely to have decided to become one thing or another before arriving on the scene. The answer, presumably, is that, just as one cannot know in advance who one might fall in love with, the fate of the neural crest cells probably depends on timing, proximity, and chemistry. In other words, it is a matter of induction, as well as where a cell happens to be when the appropriate signals are emitted.

Formation of the ciliary epithelium may be induced by the lens

Experimental studies in bird and rat embryos indicate that the presence of the lens is necessary for the development and elaboration of the ciliary epithelium (and for the iris epithelium as well). If the lens vesicle is removed from the developing eye before the extension of the optic cup margin that forms the epithelial layer, the extension does not occur. Instead, the margin of the optic cup appears to become retina.

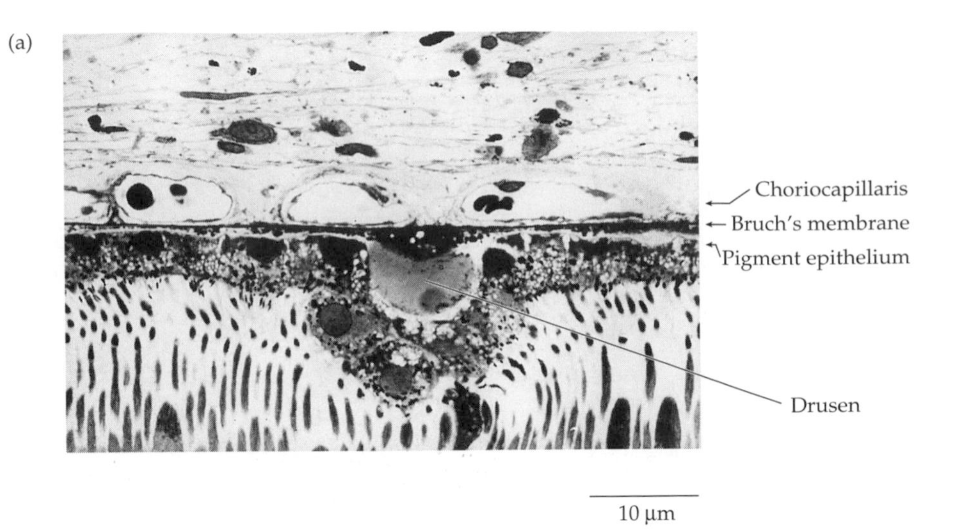

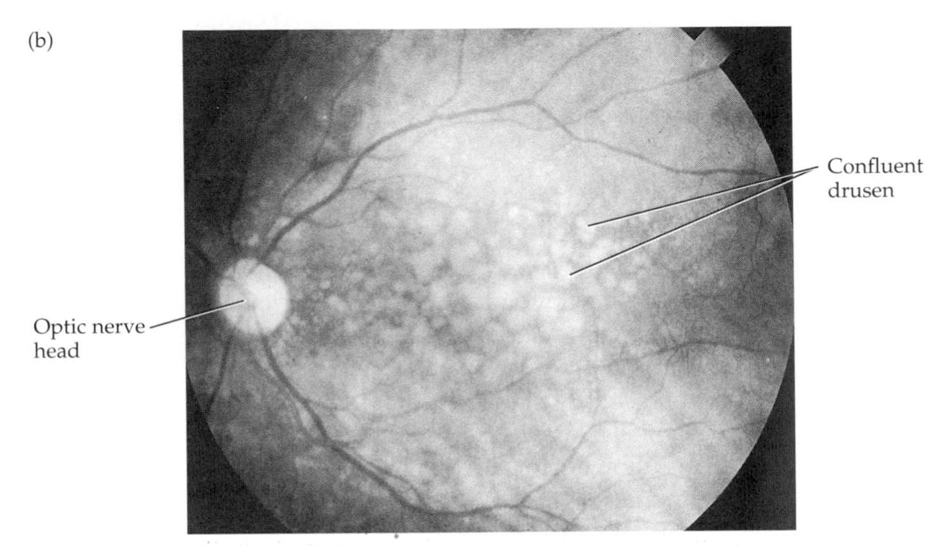

Figure 11.34
Drusen
(*a*) A small drusen appears as a local thickening of Bruch's membrane, overlaid with deformed and degenerating pigment epithelium. (*b*) Fundus photograph of confluent "soft" drusen throughout the region around the fovea. (*a* from Hogan, Alvarado, and Weddell 1971; *b* from Alexander 1994.)

Transplanting an extra lens vesicle to an abnormal location in the side of the optic cup will result in an extra anterior segment, complete with cornea, anterior chamber, and at least part of the iris. Presumably, the presence of iris epithelium implies that ciliary epithelium has been produced as well, but that detail has not been examined critically.

These incomplete results suggest, with the usual qualifier about species differences, that the lens is necessary for formation of the ciliary epithelium. More specifically, because the early lens is mostly epithelium around a small core of early lens fibers, the lens epithelium is thought to be the source of the inducing signals. As in most cases of apparent induction, the nature of the inducing signal is unknown.

Formation of the ciliary muscle may be induced by the ciliary epithelium

It would make sense for the ciliary epithelium to trigger differentiation of the ciliary muscle cells; this source for the signal would guarantee that the muscle formed in the correct location. Unfortunately, the evidence on this point is skimpy and circumstantial. It boils down to two facts. First, ciliary epithelium precedes ciliary muscle in the normal developmental sequence; the epithelium is therefore present in time to provide inducing signals. Second, if the ciliary

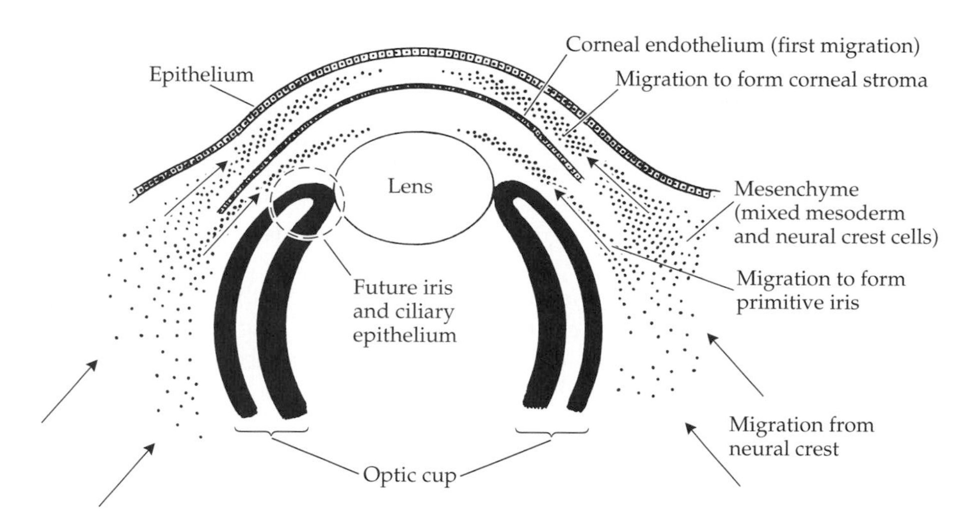

Figure 11.35
Origins of the Ciliary Body
The eye cup is surrounded by mesenchyme that consists of undifferentiated mesoderm and cells from the neural crest. The neural crest cells migrate in front of the lens in several waves, the second of which forms the primitive iris and divides the anterior segment into anterior and posterior chambers. After the primitive iris stroma has formed, the two layers at the rim of the optic cup push forward just in front of the lens to form, first, the epithelial layers of the ciliary body and, second, the iris epithelium. The ciliary muscle forms from the neural crest cells lying next to the primitive ciliary epithelium. (After Tripathi and Tripathi 1989.)

epithelium does not form—because the lens vesicle was removed, for example—the ciliary muscle also fails to form. Both facts suggest, but fall short of proving, that the epithelium induces formation of the muscle. The idea should be regarded as a hypothesis awaiting test.

The ciliary muscle begins to form during the fourth month and continues to develop until term

Few details of ciliary muscle development are known. Developmental foci appear in the mesenchyme outside the ciliary epithelium during the fourth month, marking the appearance of the muscle as differentiated muscle cells. If there is any gradient in differentiation of different parts of the muscle, it is not known, but in terms of *organization* within different portions of the muscle, the process seems to go from outside to inside. Meridional fibers are said to be detectable, stretching from the choroid and inserting into the scleral spur, early in month 7. Circular fibers, however, are not apparent for another month; one is free to suppose that the radial fibers form between these two extremes during this time.

Given the difficulty in distinguishing the parts of the muscle in adult eyes, it is hard to know exactly what the foregoing statements mean. Most probably, muscle fibers destined to be a part of the meridional or circular portions of the muscle are present before month 7, but the muscle has yet to organize toward its adult configuration. In any case, the organization is likely to begin before it can be observed in histological sections.

If the ciliary muscle is at all like the other ocular muscles, its development is likely to be incomplete at birth. It certainly increases in size for some years as the eye grows, but its internal organization, perhaps including some of its innervation, probably continues to develop for some time (whether months or years is unknown) into the postnatal period.

The muscles associated with the eye originate from different germinal tissues

With this consideration of ciliary muscle development, all of the eye's muscles have been discussed and a summary comment is in order. Namely, differences among these muscles should not be surprising, in light of their different origins. The extraocular muscles are derived from mesoderm (see Chapter 4), the iris muscles from neural ectoderm of the optic cup (see Chapter 10), and the ciliary muscle from the neural crest.

I offer this comment for two reasons. First, it differs from the traditional story, particularly with respect to the ciliary muscle, which was long regarded as being of mesodermal origin. Second, it may offer some guidance for comparisons between muscles. Unless other muscles in the body that are now thought to be of mesodermal origin turn out to be neural crest derivatives (e.g., the muscles of the middle ear), the ciliary muscle may be incomparable, unique among muscles.

The ciliary processes form in synchrony with the vascular system in the ciliary body

Aqueous production begins shortly before the ciliary muscle differentiates in the fourth month. It is not clear that this early aqueous is the same as adult aqueous or that it is produced in quite the same way, but its presence implies the existence of a primitive ciliary circulation and a differentiated epithelium.

Vasculogenesis in the mesenchyme around the optic cup begins early in gestation, forming the basis of the future choroid first and of the ciliary circulation not long after. The tissue folds we see as adult ciliary processes begin to appear as soon as the basic capillary system is in place. As the radial rows of blood vessels become more elaborate, dense, and space-consuming, they bulge inward, the epithelium is thrown into folds, and the ciliary processes appear. The processes appear during the fifth month and grow steadily more prominent throughout gestation and into the postnatal period.

The zonule is produced by the ciliary epithelium

The zonule has been referred to as the tertiary vitreous because it appears during the third month of gestation, at the same time as the anterior vitreous begins to take on a membranous character. This terminology implies that the zonule is formed as part of the vitreous or by cells in the vitreous, but this seems not to be the case.

Zonular fibers are not made up of collagen, like those in the vitreous or the lens capsule, but are more like elastic fibers. For this reason, the zonule contains unusually high concentrations of specific amino acids. Introducing radioactive forms of these amino acids into the developing eye makes it possible to see what

Figure 11.36 ▶
Development of the Choroidal Circulation
For ages 2 to 6 months, the figure shows two views of the choroid: from outside the eye (*a*) and in a histological section (*b*). The choriocapillaris forms first, and at 2 months it covers the entire posterior segment out to the site of the future ora serrata. The first branches of the vortex veins and the posterior ciliary have just begun to connect into the capillary network. During the third month, the arteries and veins elaborate, the sclera begins to form outside the choroid, and the vessels that will become the circle of Zinn appear around the optic nerve. Branching of the arteries and veins continues to increase during the fourth month; the major arterial circle appears in the ciliary body, and the layers of the choroid become apparent. By the sixth month, the choroid is complete, and the characteristic high destiny of capillaries at the posterior pole is evident. (After Heimann 1972.)

structures have used them in molecular construction. The amino acids incorporated into zonular fibers turn out to be taken up by ciliary epithelial cells and later appear in the part of the zonule closest to the ciliary body, suggesting that the ciliary epithelium is in the business of synthesizing the zonule. If this result

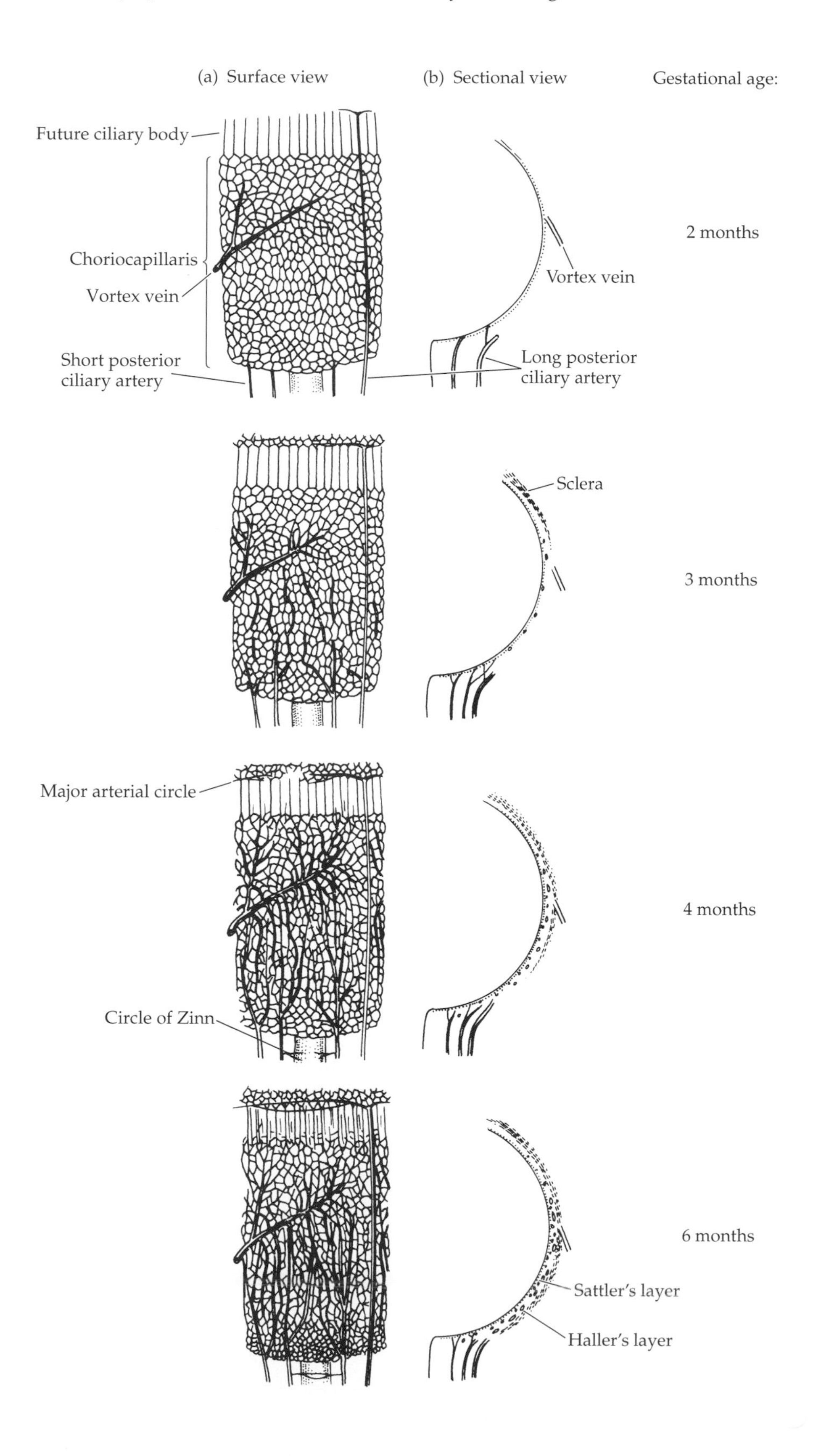

is correct (the experiment has been done only a few times in nonprimate eyes), it implies a neat division of labor among cells around the posterior chamber. Lens epithelial cells produce the lens capsule, vitreal cells produce vitreal collagen and hyaluronic acid, and ciliary epithelial cells produce the special molecules incorporated in the zonular fibers.

Synthesis of new zonule probably continues into the postnatal period, perhaps throughout life—in company with lens enlargement—but this is unknown territory.

The choroidal vasculature has two developmental gradients: center to periphery and inside to outside

Angiogenesis in the choroid is probably induced by the retina, specifically by the developing retinal pigment epithelium. Shortly after the pigment epithelium begins to differentiate at week 3, primitive capillaries appear in the mesenchyme outside the eye cup at the posterior pole. Initially, these new vessels are unconnected segments, recognizable as capillaries only by the open lumina in the segments and the membrane fenestrations that are so characteristic of the adult choriocapillaris. As the pigmentation in the retinal epithelium spreads from the posterior pole toward the rim of the optic cup, the capillary formation follows along. This process establishes the center-to-periphery gradient, in which the capillaries that first formed at the posterior pole elaborate their structure as new capillaries form at the expanding edge of the capillary bed. At 2 months, the primitive choriocapillaris extends to the equator of the eye (Figure 11.36).

A capillary bed is not much use without arteries to feed it and veins to drain it; these major vessels begin to appear during the third month. Branches from the future short posterior ciliary arteries appear first at the posterior pole and subsequently at more peripheral locations in accordance with the center-to-periphery gradient. The main point is that the choriocapillaris forms first in the inner choroid next to the retina, and the outer parts of the vascular system are then added for supply and drainage. The choroid at 4 and 6 months is shown in Figure 11.36 with the short and long posterior ciliary arteries in place.

Note that the capillary beds in the ciliary body are not part of the choroid's developmental gradient. Ciliary capillaries appear not long after the first choroidal capillaries at the posterior pole, and the ciliary circulation is well advanced before the peripheral choriocapillaris ever appears. The developing long posterior ciliary arteries seem to skip over the undeveloped part of the choroid to go directly to their target in the ciliary body.

The choroidal stroma is organized around the arteries and veins growing in to the choriocapillaris, meaning that the vascular development leads the stromal development. Formation of the choroidal stroma involves differentiation of melanocytes and fibroblasts, as well as collagen production.

Because the choroid must expand as the eye increases in size, both angiogenesis and production of new stroma should continue for years after birth. Once the eye has stopped growing, however, breakdown and replacement of vascular endothelium will continue throughout life.

References and Additional Reading

The Ciliary Processes and Aqueous Formation

Alberts B, Bray D, Lewis J, Raff M, Roberst K, and Watson JD. 1994. Principles of membrane transport. Chapter 11, pp. 508–522, *Molecular Biology of the Cell,* 3rd Ed. Garland, New York.

Alm A. 1992. Ocular circulation. Chapter 6, pp. 198–227, in *Adler's Physiology of the Eye,* 9th Ed., Hart WM, ed. Mosby Year Book, St. Louis.

Brubaker RF. 1991. Flow of aqueous humor in humans. *Invest. Ophthalmol. Vis. Sci.* 32: 3145–3166.

Brubaker RF, Nagataki S, Townsend DJ, Burns RB, Higgins RG, and Wentworth W. 1981. The effect of age on aqueous humor formation in man. *Ophthalmology* 88: 283–287.

Caprioli J. 1992. The ciliary epithelia and aqueous humor. Chapter 7, pp. 228–247, in *Adler's Physiology of the Eye*, 9th Ed., Hart WM, ed. Mosby Year Book, St. Louis.

Cole DF. 1977. Secretion of the aqueous humor. *Exp. Eye Res.* 25 (Suppl.): 161–176.

Diamond JR and Bossert WH. 1967. Standing gradient osmotic flow: A mechanism for coupling water and solute transport in epithelia. *J. Gen. Physiol.* 50: 2061–2083.

Friedenwald JS. 1949. The formation of the intraocular fluid. *Am. J. Ophthalmol.* 32 (6, Part 2): 9–27.

Funk R and Rohen JW. 1990. Scanning electron microscopic study on the vasculature of the human anterior eye segment, especially with respect to the ciliary processes. *Exp. Eye Res.* 51: 651–661.

Hara K, Lütjen-Drecoll E, Prestele H, and Rohen JW. 1977. Structural differences between regions of the ciliary body in primates. *Invest. Ophthalmol.* 16: 912–924.

Hogan MJ, Alvarado JA, and Weddell JE. 1971. Ciliary body and posterior chamber. Chapter 7, pp. 260–319, in *Histology of the Human Eye.* WB Saunders, Philadelphia.

Krupin T and Civan MM. 1996. Physiologic basis of aqueous humor formation. Chapter 12, pp. 251–280, in *The Glaucomas*,Vol. 1. *Basic Sciences*, 2nd Ed., Ritch R, Shields MB, and Krupin T, eds. Mosby Year Book, St. Louis.

Maren TH. 1994. Biochemistry of aqueous humor inflow. Chapter 1, Section II, pp. 1.35–1.46, in *Textbook of Ophthalmology*, Vol. 7, *Glaucoma*, Kaufman PL and Mittag TW, eds. Mosby Year Book Europe, London.

Mclaughlin CW, Peart D, Purves RD, Carré DA, Macknight ADC, and Civan MM. 1998. Effects of HCO_3^- on cell composition of rabbit ciliary epithelium: A new model for aqueous humor secretion. *Invest. Ophthalmol. Vis. Sci.* 39: 1631–1641.

Morrison JC and Freddo TF. 1996. Anatomy, microcirculation, and ultrastructure of the ciliary body. Chapter 6, pp. 125–138, in *The Glaucomas*, Vol. 1. *Basic Sciences*, 2nd Ed., Ritch R, Shields MB, and Krupin T, eds. Mosby Year Book, St. Louis.

Morrison JC and Van Buskirk M. 1986. Microanatomy and modulation of the ciliary vasculature. *Trans. Ophthalmol. Soc. U.K.* 105: 131–139.

Nilsson SFE and Bill A. 1994. Physiology and neurophysiology of aqueous humor inflow and outflow. Chapter 1, Section II, pp. 1.17–1.34, in *Textbook of Ophthalmology*, Vol. 7, *Glaucoma*, Kaufman PL and Mittag TW, eds. Mosby Year Book Europe, London.

Raviola G. 1977. The structural basis of the blood-ocular barriers. *Exp. Eye Res.* 25 (Suppl.): 27–63.

Raviola G and Raviola E. 1978. Intercellular junctions in the ciliary epithelium. *Invest. Ophthalmol. Vis. Sci.* 17: 958–980.

Reiss GR, Werness PG, Zollman PE, and Brubaker RF. 1986. Ascorbic acid levels in the aqueous humor of nocturnal and diurnal mammals. *Arch. Ophthalmol.* 104: 753–755.

Tamm ER and Lütjen-Drecoll E. 1996. Ciliary body. *Microsc. Res. Tech.* 33: 390–439.

The Ciliary Muscle and Accommodation

Alpern M, Mason GL, and Jardinico RE. 1961. Vergence and accommodation. V. Pupil size changes associated with changes in accommodative vergence. *Am. J. Ophthalmol.* 52: 762–767.

Calasans OM. 1953. Arquitetura do músculo ciliar no homen. *Ann. Fac. Med. Univ. Sao Paulo* 27: 3–98. (This reference is included for completeness; I have not been able to obtain a copy. Some key illustrations are reproduced in Rohen 1964.)

Duane A. 1912. Normal values of the accommodation at all ages. *Trans. Sect. Ophthalmol. Am. Med. Assn.* 381–391.

Fisher RF. 1977. The force of contraction of the human ciliary muscle during accommodation. *J. Physiol.* 270: 51–74.

Hamasaki D, Ong J, and Marg E. 1956. The amplitude of accommodation in presbyopia. *Am. J. Optom.* 33: 3–14.

Helmholtz H. 1924. Mechanism of accommodation. Chapter 12, pp. 143–172, in *Helmholtz's Treatise on Physiological Optics*, Vol. 1, Southall JPC, ed. Optical Society of America, Rochester, NY.

Hofstetter HW. 1944. A comparison of Duane's and Donders' tables of the amplitude of accommodation. *Am. J. Optom.* 21: 345–362.

Koretz JF and Handelman GH. 1988. How the human eye focuses. *Sci. Am.* 259 (1): 92–99.

Kruger PB and Pola J. 1986. Stimuli for accommodation: Blur, chromatic aberration and size. *Vision Res.* 26: 957–971.

Lütjen-Drecoll E, Tamm E, and Kaufman PL. 1988. Age-related loss of morphologic responses to pilocarpine in rhesus monkey ciliary muscle. *Arch. Ophthalmol.* 106: 1591–1598.

McLin LN and Schor CM. 1988. Voluntary effort as a stimulus to accommodation and vergence. *Invest. Ophthalmol. Vis. Sci.* 29: 1739–1746.

Moses RA. 1987. Accommodation. Chapter 11, pp. 291–310, in *Adler's Physiology of the Eye*, 8th Ed., Moses RA and Hart WM, eds. CV Mosby, St. Louis.

Neider MW, Crawford K, Kaufman PL, and Bito LZ. 1990. In vivo videography of the rhesus monkey accommodative apparatus: Age-related loss of ciliary muscle response to central stimulation. *Arch. Ophthalmol.* 108: 69–74.

Rohen JW. 1964. Ciliarkörper (Corpus ciliare). Chapter V, pp. 189–239, in *Handbuch der mikroskopischen Anatomie des Menschen*, Vol. 3, Part 4, *Das Auge und seine Hilfsorgane*, Mollendorf WV and Bargmann W, eds. Springer-Verlag, New York.

Saladin JJ and Stark L. 1975. Presbyopia: New evidence from impedance cyclography supporting the Hess-Gullstrand theory. *Vision Res.* 15: 537–541.

Streeten BW. 1995. The ciliary body. Chapter 13 in *Duane's Foundations of Clinical Ophthalmology*, Vol. 1, Tasman W and Jaeger EA, eds. Lippincott-Raven, Philadelphia.

Swegmark G. 1969. Studies with impedance cyclography on human ocular accommodation at different ages. *Acta Ophthalmol. (Copenh.)* 47: 1186–1206.

Tamm S, Tamm E, and Rohen JW. 1992. Age-related changes of the human ciliary muscle: A quantitative morphometric study. *Mech. Ageing Dev.* 62: 209–221.

The Choroid

Alexander LJ. 1994. *Primary Care of the Posterior Segment*, 2nd Ed. Appleton & Lange, Norwalk, CT.

Bernstein MH and Hollenberg MJ. 1965. Fine structure of the choriocapillaris and retinal capillaries. *Invest. Ophthalmol.* 4: 1016–1025.

Feeney-Burns L and Ellersieck MR. 1985. Age-related changes in the ultrastructure of Bruch's membrane. *Am. J. Ophthalmol.* 100: 686–697.

Fryczkowski AW, Sherman MD, and Walker J. 1991. Observations on the lobular organization of the human choriocapillaris. *Int. Ophthalmol.* 15: 109–120.

Hayreh SS. 1990. *In vivo* choroidal circulation and its watershed zones. *Eye* 4: 273–289.

Hogan MJ, Alvarado JA, and Weddell JE. 1971. Choroid. Chapter 8, pp. 320–392, in *Histology of the Human Eye.* WB Saunders, Philadelphia.

Olver JM. 1990. Functional anatomy of the choroidal circulation: Methyl methacrylate casting of the human choroid. *Eye* 4: 262–272.

Torczynski E. 1982. Choroid and suprachoroid. Chapter 22 in *Duane's Foundations of Clinical Ophthalmology,* Vol. 1, Tasman W and Jaeger EA, eds. Lippincott-Raven, Philadelphia.

Yoneya S and Tso MOM. 1987. Angioarchitecture of the human choroid. *Arch. Ophthalmol.* 105: 681–687.

Development of the Ciliary Body and Choroid

Beebe DC. 1986. Development of the ciliary body: A brief review. *Trans. Ophthalmol. Soc. UK* 105: 123–130.

Heimann K. 1972. The development of the choroid in man. *Ophthalmic Res.* 3: 257–273.

Mann I. 1950. *The Development of the Human Eye.* Grune & Stratton, New York.

Sellheyer K. 1990. Development of the choroid and related structures. *Eye* 4: 255–261.

Tripathi BJ and Tripathi RC. 1989. Embryology of the anterior segment of the human eye. Chapter 1, pp. 3–40, in *The Glaucomas,* Vol. 1, Ritch R, Shields MB, and Krupin T, eds. CV Mosby, St. Louis.

Chapter 12

The Lens and the Vitreous

The native structure of the human lens is truly unique. No other part of the body even faintly resembles this compacted, disc-shaped mass of protein-filled cells of a single lineage, all of which (except for a single layer, one cell thick) are dying or dead. Not only the structure but also the habitat of the lens is unique. Surrounded by a capsule, suspended in a watery fluid, permanently deprived of contact with any other cell (living or dead), the lens is isolated to an unprecedented degree. . . . its peculiar characteristics have a marked effect on the way in which it gradually decays with the passage of time.

■ Richard W. Young, *Age-Related Cataract*

The lens contributes about one-third of the eye's total dioptric power of 60 D or so when the eye is in its unaccommodated state. In doing so, however, the lens is at a disadvantage. Unlike the cornea, where the refractive surface is an interface between air and tissue, the lens resides in a fluid environment whose refractive index is appreciably higher than that of air. Thus the lens must have a refractive index that is higher still if it is to make a significant optical contribution. That it does so attests to some of its unique features, which we will consider in this chapter.

In keeping with its optical function, the lens must be transparent in the visible region of the spectrum; this requirement also places special demands on the underlying structure, particularly in a tissue that is cellular and many layers thick. In addition, the lens is the only variable refractive element in the eye, changing shape and optical power in response to forces imposed on it by the ciliary muscle.

These properties of the lens give rise to the clinical issues of transparency loss, the opacification we call **cataract**, and the loss of accommodation (**presbyopia**). Strictly speaking, neither cataract formation nor presbyopia has much to do with the gross or cellular structure of the lens; cataracts and lenticular changes in presbyopia are primarily *intra*cellular events, and if structural change is involved, it is structure at the molecular level. For this reason, and because the primary lens molecules are so unusual, the ensuing discussion will emphasize the internal organization of lens cells, as well as the way in which these cells are put together.

The vitreous is included in this chapter in part because it is in physical contact with the lens and may influence it during accommodation and in part because the lens and vitreous share the property of transparency; the vitreous is also subject to transparency loss. In other words, the vitreous is an important part of the environment of the lens. Not so long ago, there wasn't much to say about the vitreous itself; it appeared to have more in common with a bowl of Jell-O than with anything one might call structured. But the vitreous is neither amorphous nor homogeneous; it has cells, structural elements, physical attachments to the retina, lens, and ciliary body, and regional differences in composition. Some of these features can have important clinical implications.

Structure of the Lens

Some unusual proteins, the crystallins, are the dominant structural elements in the lens

Ninety percent of the lens is protein—a higher percentage than in any other tissue in the body. The lens proteins can be divided into two large classes. One group includes insoluble proteins, which are mainly membrane and cytoskeletal proteins; because the lens is cellular, a substantial fraction of lens protein is of this insoluble kind. But the dominant group consists of soluble proteins—that is, hydrophilic proteins. Among the soluble proteins, those of primary interest are the **crystallins**, which are largely responsible for the special properties of the lens.

Crystallins in the human lens fall into several groups; the main ones are the α-, β-, and γ-crystallins. They differ in molecular weight, N-terminal structure, and three-dimensional structure, among other things, and subgroups have been described within these basic categories. For our present purposes, however, the properties the crystallins have in common are more important than their differences, significant as these may be; certain of these shared properties are crucial to our understanding of the normal and abnormal lens.

The amino acid chains in the crystallins are very long, as indicated by their high molecular weight, and it is important to have some idea about the three-dimensional structure of these large molecules. Are they long, linear strands, like collagen, or small, compact globules, like G proteins? Or something else? The answer is not entirely clear for the the larger α- and β-crystallins, but they are thought to be similar to the γ-crystallins, the structure of which is illustrated in Figure 12.1*d*. Figure 12.1*a–c* review protein structure.

Proteins are important molecules because of their seemingly infinite variety of shapes. Their amino acid chains are flexible, able to twist or fold in numerous ways. Specific proteins have specific patterns of twisting or folding, however, determined largely by the sequence of amino acids in the chain and the electrical charge, or lack thereof, inherent to the different amino acids. Because of the sequence of charged or uncharged regions along the chain, certain kinds of folds or twists are preferred as some regions along the folded chain attract each other, while other patterns are avoided. Predicting the patterns from the amino acid sequence is rarely possible, however, so understanding the folds and twists that make up the three-dimensional configuration (the tertiary structure*) is an experimental problem.

When proteins are crystallized so that their basic three-dimensional structure is repeated at regular intervals throughout the crystal, X rays passing through the crystal are absorbed and scattered in a repetitive fashion that appears as an array of exposed dots on a photographic film. By working back through the mathematical description of absorption and scatter, we can sometimes use these diffraction patterns to deduce the underlying molecular structure. Watson and Crick's double-helical model of DNA, for example, was based in part on X-ray diffraction patterns made by Rosalind Franklin from crystalline DNA. But a double helix is relatively simple. Many proteins do not crystallize easily, and many structural shapes within the amino acid sequence have X-ray diffraction patterns that either are unknown or are indistinguishable from one another. In

*The amino acid sequence in a protein is its *primary* structure. Certain relatively simple patterns of twists (alpha helices) or folds (beta sheets) constitute its *secondary* structure. Sequences of helices and sheets fold into compact three-dimensional shapes called *domains*, which are the modules of a protein's *tertiary* structure. The result is a *monomer*, which is the structure formed by a particular sequence of amino acids (a polypeptide). Two monomers can combine to form a *dimer*, three to form a *trimer*, and many to form a *polymer*, all of which are *quaternary* structures, because the level of complexity is greater than that of a tertiary structure. Figure 12.1*a–c* review these concepts.

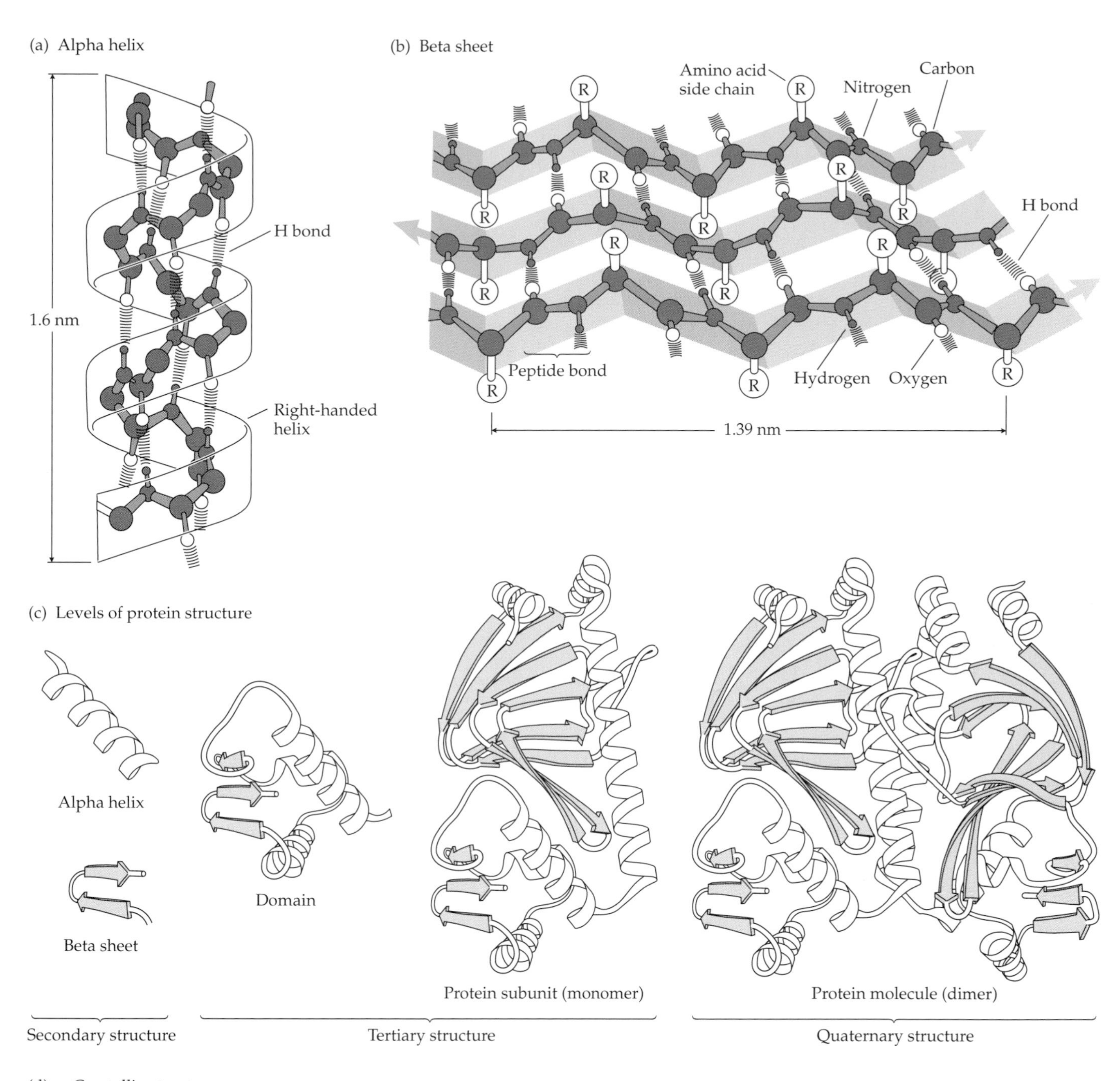

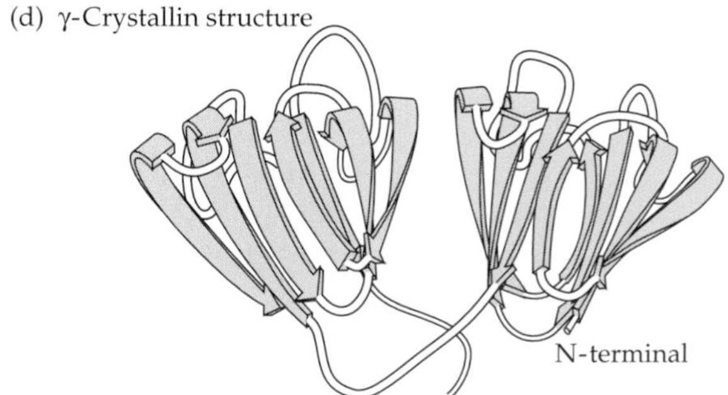

Figure 12.1

Elements of Protein and γ-Crystallin Structure

Large proteins are combinations of simpler elements. (*a*) In an alpha helix, the amino acid chain forms a spiral shape maintained by hydrogen bonds. (*b*) Beta sheets are regions where the amino acid chain has been folded into parallel strips that are bonded together. (*c*) Helices and sheets are secondary structures that are joined in various sequences to form small domains, and different domains are joined to form a protein subunit or monomer; domains and monomers are tertiary structures. Quaternary structures are linkages of monomers to form dimers or complexes of higher order. (*d*) The tertiary structure of γ-crystallin has two domains that contain beta sheets, and the whole molecule is compact and globular. The other crystallins are larger and tend to combine as dimers. (*a–c* after Alberts et al. 1994; *d* after Wistow 1990.)

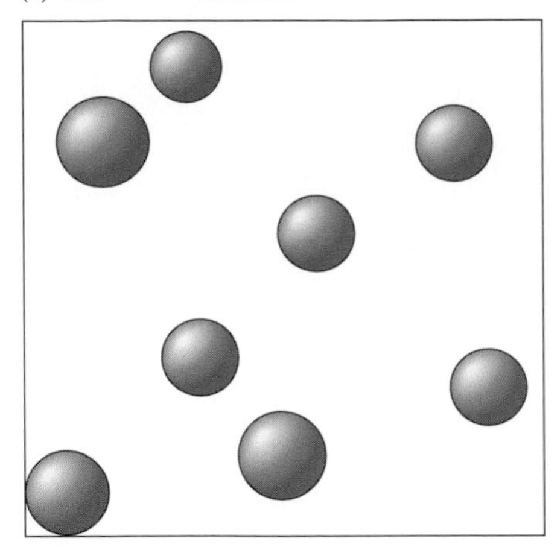

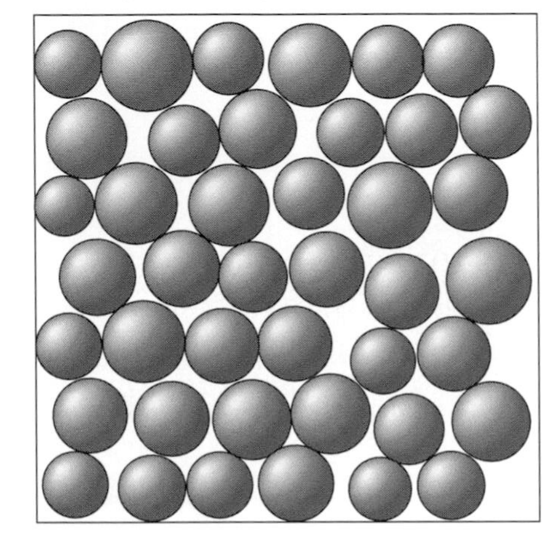

Figure 12.2
Packing Density Affects Regularity of Spacing
(*a*) A small number of marbles in the bottom of a box will have a random arrangement in which the spacing between neighbors varies considerably. (*b*) A large number of marbles packed tightly necessarily make the array more regular, even if the marbles vary in size. Crystallin molecules behave in the same way.

general, larger proteins have more complex geometry, and it is more difficult to work out their tertiary structures.

Simple coils in the polypeptide are called alpha helices (see Figure 12.1*a*); their presence is often detectable in the X-ray diffraction pattern of a protein. The turns of the helix are joined with hydrogen bonds, making the helix a stable element of the overall structure. Another stable pattern is the beta sheet (see Figure 12.1*b*), in which the basic amino acid chain is folded so that parts running back and forth parallel to one another can be linked by hydogen bonds. By convention, beta sheets are represented in molecular models by broad, flat arrows with the arrowhead pointing in the direction of the C terminus. Sequences of alpha helices and beta sheets are called *motifs* that fold into compact forms; motifs connected by *loops* in the amino acid chains produce globular structures called *domains*.

γ-Crystallin isolated from bovine lenses has two domains, created from interconnected beta sheets (see Figure 12.1*d*). The domains are very compact, and the whole molecule can be thought of as a football-shaped globule. The twists in the beta sheets have the effect of keeping all the nonpolar amino acids, which are hydrophobic, on the inside of the domains, leaving the polar, hydrophilic amino acids exposed on the exterior of the globule; the molecule is therefore compact and water soluble. The structures of the α- and β-crystallins are not known to this level of detail, but both are thought to have compact domains and neither is thought to have alpha helices as structural elements.

Dense, uniform packing of the crystallins within lens cells is responsible for lens transparency

Structural regularity at the atomic or molecular levels is critical for transparency of any medium. In glass, it is regularity of atomic spacing, and in the cornea, regularity of collagen fiber spacing (see Chapter 8). In the lens, the crucial factor is the regularity with which the crystallins are packed into lens cells.

The key word is "packing." Think of a lens cell as a sausage; the cell membrane is the sausage casing filled with material of some sort. For the lens cells, the casing is filled mostly with crystallin globules. If there are enough of these molecules in each cell, they will be packed together at high density; because of the dense packing, crystallins must have a fairly regular arrangement, even if they are somewhat different in size (Figure 12.2). Thus the crystallin molecules determine lens transparency; both their solubility and their compact, globular structure permit their uniform distribution in the lens cell cytoplasm. It follows that irregularity in the crystallin arrangement should produce transparency loss. Much evidence supports this conclusion, and irregular clumping and insolubility of the crystallins may be the proximate cause of cataract formation.

Crystallins are highly stable molecules, making them some of the oldest proteins in the body, but they can be changed by light absorption and altered chemical environments

Cells in the developing lens differentiate early, with crystallins appearing in the embryonic lens around week 6. Because lens cells are permanent (the firstborn cells are still present in the adult lens), crystallins are thought to be similarly long-lived. In view of this permanence, stability of the lens proteins becomes important; as long as the crystallins retain their original configuration and properties, we can assume that they will play a constant functional role and the lens will retain its youthful transparency and flexibility.

But there is evidence that lens proteins change as they age. In Figure 12.3*a*, which shows the fractions of soluble and insoluble proteins from lenses of different ages, it is clear that the insoluble fraction rises inexorably throughout life (these were normal lenses with no evidence of opacity). The increase in insolu-

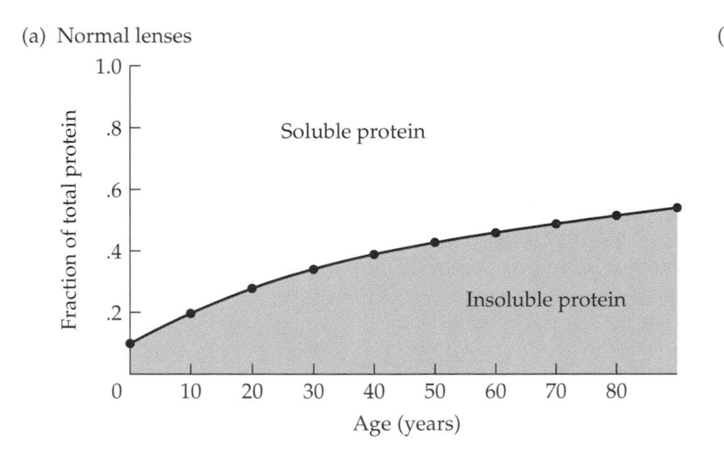

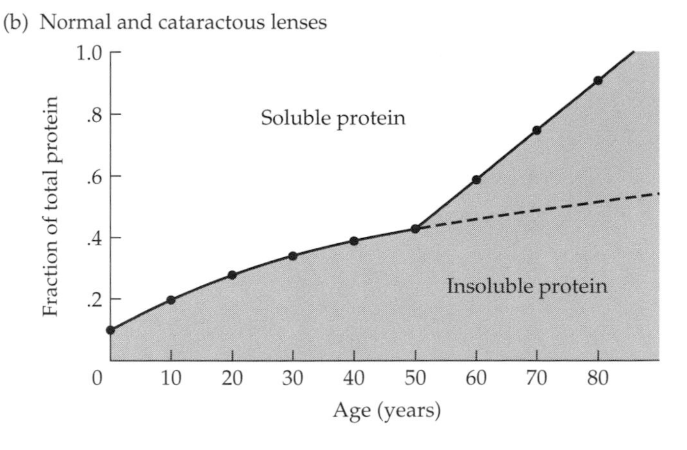

Figure 12.3
Aggregation of Lens Proteins with Age and Cataract
(*a*) In normal lenses, the insoluble fraction of lens protein increases gradually with age. The relationship is nonlinear, and the rate of increase of the insoluble fraction slows over time. (*b*) When lenses with cataracts are included in the sample, the opacities are reflected after age 50 as a more rapid increase of the fraction of insoluble protein in comparison to normal lenses. (After Spector 1984.)

ble protein cannot be due solely to the continual addition of insoluble membrane and cytoskeletal proteins as cells are added to the lens; in any new lens cell, the volume of crystallins, which normally make up the soluble fraction, is considerably larger than the insoluble fraction. Soluble protein must change to an insoluble form as the lens ages.

When cataractous lenses are included in the sample (Figure 12.3*b*), the normal increase in the fraction of insoluble protein is exaggerated after the age of 50, the age at which cataracts usually begin to appear. This association between opacification and the amount of insoluble lens protein hints that opacification is an acceleration of a normal, ongoing process of protein change.

Crystallins can combine as dimers or higher-order aggregates that have high molecular weights (greater than 10^6 daltons, compared to less than 10^4 daltons for the monomeric forms). Their correspondingly large sizes make the intracellular protein packing less regular and therefore reduce transparency, as discussed earlier. And as the proteins aggregate, they become insoluble. Thus the increase of insoluble protein content with age appears to reflect increasing aggregation of the crystallins. There are now questions about mechanism—why do crystallins in normal lenses aggregate?—and the relation of the mechanism to the abnormal case—what might increase the rate of aggregation to the point of opacity?

Changes in protein structure and their aggregation can be brought about in various ways, one of which is ultraviolet absorption. Photons in the ultraviolet portion of the spectrum are highly energetic, and when this energy is transferred to the electrons in a molecule's atoms (this is what is meant by "absorption"), the energy must be dissipated in some way. Energy may be dissipated by breaking chemical bonds or by reradiating photons at another frequency. The breaking of chemical bonds can lead directly to a change in molecular structure that can increase aggregation of the lens proteins. Even reradiated energy, which is often in the infrared region as heat or in the visible part of the spectrum as fluorescence, can break bonds or accelerate chemical reactions that change molecular structure.

Wavelengths of less than 300 nm or so (the ultraviolet component called UV-A) are absorbed strongly by the cornea, and little of this very energetic short-wavelength radiation penetrates the eye. (Incidentally, when speaking of the cornea or the lens as being transparent, the implicit assumption is transparency to wavelengths in the *visible* part of the spectrum, from about 400 to 700 nm. The tissues can absorb strongly at other wavelengths and therefore be opaque to ultraviolet or infrared radiation.) The lens and ascorbic acid in the aqueous absorb ultraviolet in the range from 300 to 400 nm (UV-B).

Ultraviolet radiation is absorbed in the young lens by the lens proteins, but pigments gradually accumulate in the lens cells, eventually producing a pale yellowish color in the lens. These pigments are based on tryptophan, and they absorb ultraviolet radiation strongly. (In some animals, such as arboreal squirrels, the yellow lens color is intense; it is probably an adaptation that suits the strongly diurnal habit of squirrels.) Up to a point, these pigmentary changes may be protective, like the increased melanin concentration in the skin in response to ultraviolet exposure, but in some cases the lens color strengthens and deepens, becoming brownish and opaque; this is a **brunescent cataract**. Some of the pigments formed in the lens fluoresce, meaning that they first absorb ultraviolet radiation and then reemit light at longer wavelengths; this reemitted light can also be absorbed by the lens proteins, leading to structural changes, aggregation, and insolubility.

Another way in which protein structures can be altered is by oxidation, a process that can be accelerated by ultraviolet absorption (photooxidation). Although the lens is bathed in a source of free oxygen radicals from hydrogen peroxide in the aqueous, it has a built-in protective mechanism in the form of a high glutathione concentration. Glutathione is a tripeptide that works as an antioxidant by combining with oxygen so that other molecules, such as the proteins, will not be oxidized. For unknown reasons, however, the level of glutathione in the normal lens declines with age, leaving the lens more vulnerable to oxidation. Moreover, the glutathione concentration drops sharply in cataractous lenses, pointing to opacification as a consequence of a failed protective mechanism.

The changes in lens proteins with age are accompanied—of necessity—by exposure to ultraviolet radiation, whose cumulative direct and indirect effects can alter protein structure. The proteins tend to aggregate and become insoluble, and if the aggregation is sufficiently large, it will be manifest as a reduction in transparency. We should therefore expect the ocular lens to become less transparent and to scatter more light as it ages, and this is almost always the case; that is, reduced transparency and increased light scatter with age are normal consequences of living in a lighted environment. The remaining and more controversial question, to be taken up later, concerns the factors that could exaggerate this normal process to produce opacities.

α-Crystallins may play a special role in maintaining native crystallin structure over time

Protein denaturing (unfolding) and other changes are not problems in many tissues, either because the cells are continually being recycled or because they can recycle their proteins. In the lens, however, there are no mechanisms for replacement of damaged crystallins, so a crystallin synthesized prenatally must last for the lifetime of the individual, however long that may be. But there may be a way to undo some damage; that repair may be part of the job of the α-crystallins.

Many proteins fold into their native states spontaneously, and if they become denatured they may be able to refold when the denaturing agent is no longer present. Other proteins need assistance in folding, however, and the crystallins are of this type. Folding assistance is provided by special proteins called *molecular chaperones,* which belong to the group known as heat shock proteins because they protect other proteins from thermal denaturing. α-Crystallin appears to be a molecular chaperone for the β- and γ-crystallins; it helps them fold properly as they are synthesized and helps them refold if they are denatured. The effect and its importance are demonstrated by heating of a solution of β-crystallin; it becomes cloudy because of the denaturing and aggregation of the β-crystallin. A solution with β- and α-crystallin remains transparent when heated, however.

If something similar occurs in vivo, the stability and longevity of the crystallins is explained, as is the long-term transparency of the lens. The various factors that can degrade the crystallins—UV absorption, oxidation, and so on—may

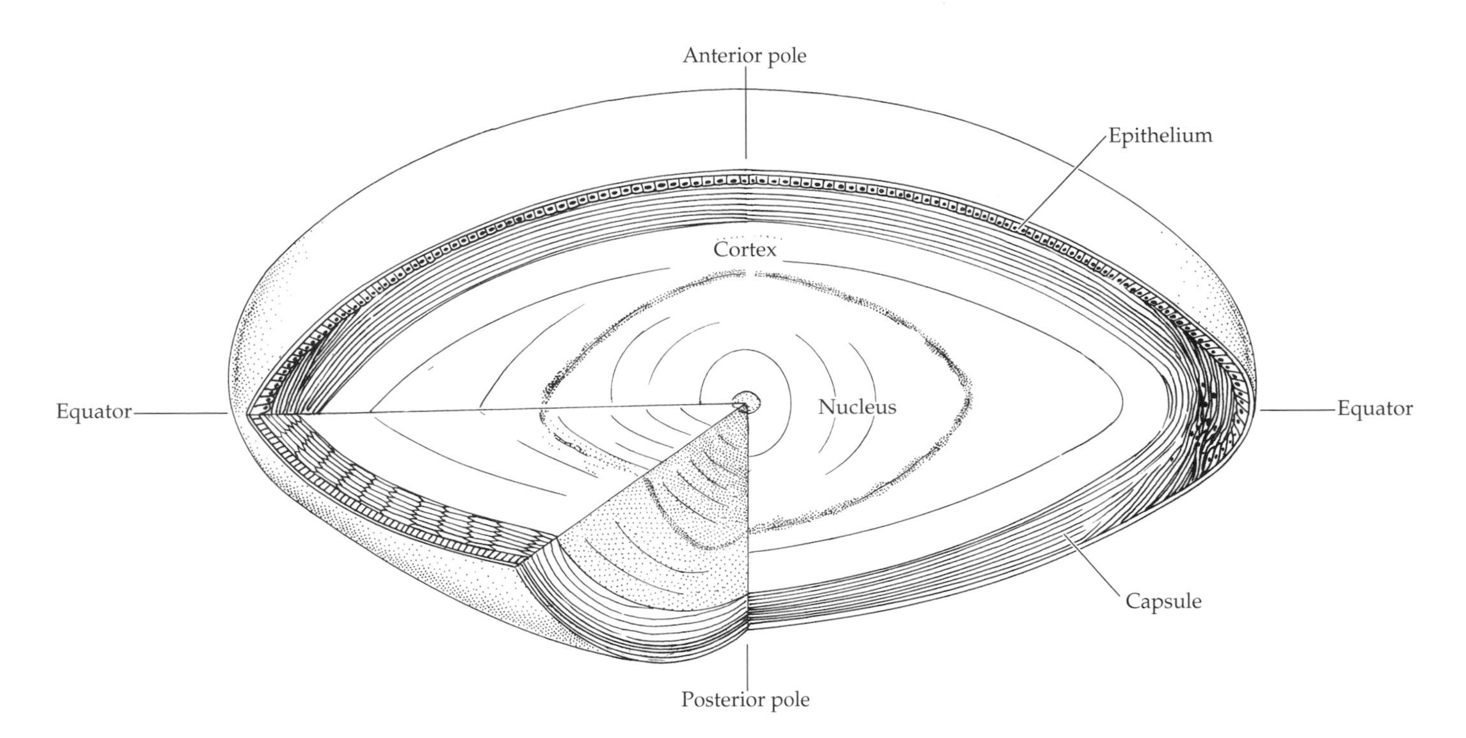

Figure 12.4
Structure of the Adult Lens
The lens is constructed from a series of concentric shells or layers. The shells have been drawn here thicker than they really are. The inner part of the lens is the nucleus, which can be subdivided into the innermost portion formed during the first 3 months of gestation (embryonal nucleus), a portion added until birth (fetal nucleus), and a portion added postnatally (adult nucleus) (see also Figure 12.18). The outer part of the lens, the cortex, forms after the age of 10. The anterior surface has a single layer of epithelial cells, and the lens is enclosed in a collagenous capsule. The lens poles are the points at the centers of the anterior and posterior surfaces, and the equator encircles the lens at its maximum diameter, which is roughly midway between the poles.

be routinely offset by the chaperone function of α-crystallin. By preventing denaturing and aggregation, the α-crystallin also ensures that the lens remains transparent. It follows that the normal decrease in transparency with age may be related to a gradual loss of the chaperoning ability of α-crystallin, perhaps because it also aggregates with time and has no chaperone of its own to protect it.

This story is fairly new and therefore incomplete; the in vivo properties of α-crystallin are still under investigation. It is clear, however, that α-crystallin is not just a structural protein with a strictly passive role in lenticular function. Moreover, its presence in other tissues indicates that its influence and importance extends well outside the eye. But exactly what α-crystallin does in the lens under normal physiological conditions, how its functions are affected over time, and what this may have to do with transparency loss, particularly with cataract, have yet to be determined.

The lens is formed of long, thin lens fibers arranged in concentric shells to form a flattened spheroid

In longitudinal section, the lens has an onionlike structure, with layer upon layer of cells (or fibers) built up around a small central core. As Figure 12.4 suggests, the cells are quite long, wrapping around in their layers from anterior to posterior. As we will see, this arrangement is a direct consequence of lens growth, which proceeds from inside out, adding a new shell on top of (outside) the existing structure.

The anterior–posterior dimension of the lens is smaller than either its vertical or its horizontal diameters, which means that its essentially spherical form is flattened. It can be described as having anterior and posterior surfaces, separated by the **equator**, which is the circumference of a transverse section through the lens at its maximum diameter. The lens **poles**, anterior and posterior, are the geometric centers of the anterior and posterior lens surfaces; a line through the poles is the optic axis of the lens.

The central portion of this flattened sphere is the **nucleus**, which is the oldest part of the lens; the outer, younger part is the **cortex**. No anatomical landmark separates nucleus from cortex, and any line drawn to demarcate them is somewhat arbitrary, but differences in light scatter between the regions can be seen with a slit-lamp, and there are differences in details of cellular structure. In the absence of an anatomical division, however, think of the cortex as the outer third (roughly) of the adult lens and the rest as nucleus. Other subdivisions of the nucleus will be considered later.

The lens substance consisting of all the concentric shells of lens fibers has an epithelial layer on the anterior surface (the **anterior epithelium** of the lens; the posterior epithelium disappeared early in the formation of the lens), all of which is enclosed by the **lens capsule**. The strands of the zonule attach to the lens by merging with the capsule.

Lens fibers in each shell meet anteriorly and posteriorly along irregular lines

Figure 12.5 shows two lens shells, each with most of the lens fibers omitted for clarity. The smaller of the two shells is from the lens nucleus (Figure 12.5*a*); the view is of the anterior surface of the shell, with the posterior surface visible

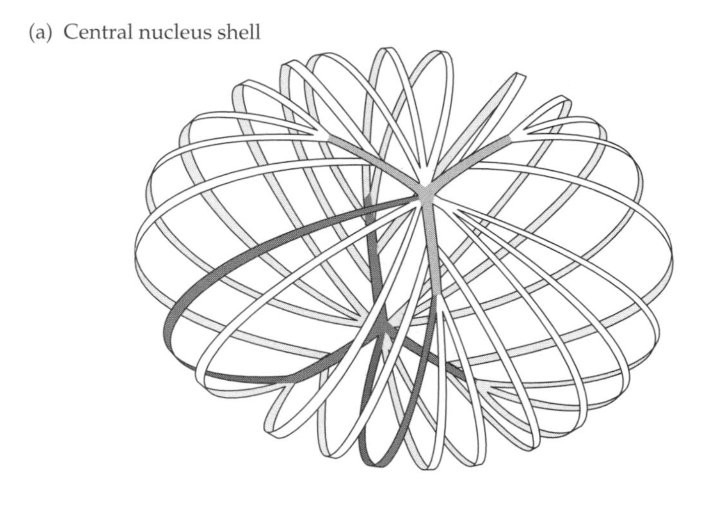

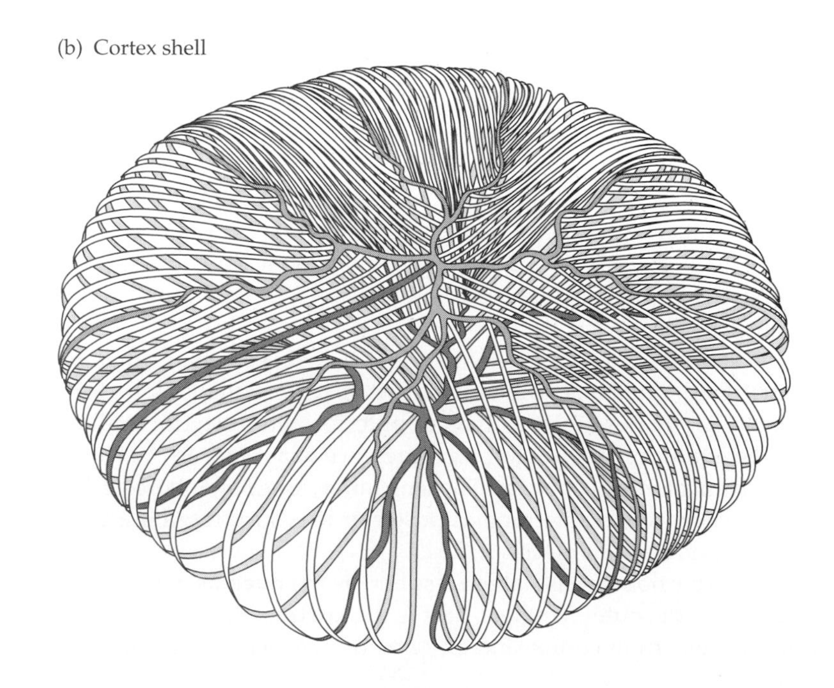

Figure 12.5
Lens Fibers and Lens Shells
(*a*) This lens shell, from the central nucleus, was formed in utero. Most of the lens fibers have been omitted for clarity. No fiber stretches from pole to pole. The fibers join along linear sutures, which in this case take the shape of the letter Y. The Y suture is upright on the anterior surface of the shell. (*b*) A shell from the lens cortex is considerably larger, has more lens fibers, and irregular, branching suture lines. As in the shell in *a*, no lens fiber reaches from pole to pole. (After Hogan, Alvarado, and Weddell 1971.)

through and behind it. If all lens fibers were shown, there would be no gaps between them and the shell would be a continuous surface. The fibers that make up the shell wrap around the lens from anterior to posterior, but no fiber extends completely from anterior pole to posterior pole, or vice versa. Note, for example, that the darkly shaded fiber originating at the anterior pole ends on the posterior surface well short of the posterior pole. Similarly, the shaded fiber from the posterior pole stops well short of the anterior pole. Lens fibers do not reach from pole to pole; moreover, they do not meet at a single point on either the anterior or the posterior surface of the shell. Instead, fibers meet along lines called **lens sutures**.

In this simple shell from the nucleus, the sutures are straight lines radiating out from the lens poles. A fiber with one end near a pole on one surface will have its other end well out along the suture line on the other surface, while fibers originating partway along a suture will end partway along a suture on the other surface. Here, there are three suture lines on each surface, intersecting at the poles to form a *Y* on the anterior surface and an inverted *Y* on the posterior surface. These Y sutures in the nucleus are often visible with a slit-lamp, and the orientation of the *Y* indicates position relative to the center of the lens: *Y* upright is anterior, *Y* inverted is posterior.

The other shell (Figure 12.5*b*) is larger, because it is in the lens cortex, and more complicated, because it has more lens fibers. Lens fibers are not only much longer in this shell, but they are also so numerous that simple straight-line sutures will not accommodate all of them. To provide more length of suture along which fibers can join, the sutures branch, forming a more complex pattern. The original Y sutures are still present, but they are buried within the lens. A tracing of the suture pattern from the interior to the exterior of the lens shows a gradual transition, a kind of metamorphosis, from the simple *Y* of the nucleus to the more complicated branching sutures in the cortex (Figure 12.6*a*).

The sutures are not just lines along which lens fibers meet, but lines along which they interlock (Figure 12.6*b*). It is as if the ends of the lens fibers were hands, with fingers interlaced to hold them together. The result of this arrangement is probably better mechanical stability of the tissue. If the lens were not continually changing its shape in response to the changing force imposed on it by the zonule, this cellular interlocking might not be necessary; there would be no strong forces to pull the lens fibers apart, and ordinary anchoring junctions between cells might be sufficient. But given the forces imposed, something more may be necessary to keep fibers from separating, and the interlocking at sutures is part of the story (there is more, which will be addressed shortly).

Because of the irregular structure of the sutures, they scatter more incident light than does the rest of the lens. This scatter, which also represents a slight reduction in transparency, is the reason that the sutures are visible under appropriate illumination conditions. The sutures may also be responsible for the radiating flare around bright light sources that becomes more obvious with age.

Lens shells are bound together with miniature locks and keys, a kind of biological Velcro

When lens fibers are cut perpendicular to their long axes (see Figure 12.6*b*), the cross sections are like flattened hexagons, stacked in columns that fit nicely together. But if there is a need to keep fibers in a shell together at their ends, along the suture lines, there is also a need to keep adjacent fibers from separating along their length and to keep overlapping shells from pulling apart. Figure 12.7 illustrates the mechanisms for keeping lens cells together.

In the outer cortex, just under the epithelium, neighboring cells in a shell interdigitate along their adjoining edges and have some anchoring junctions with the overlying epithelium. On occasion, cells in different shells have small

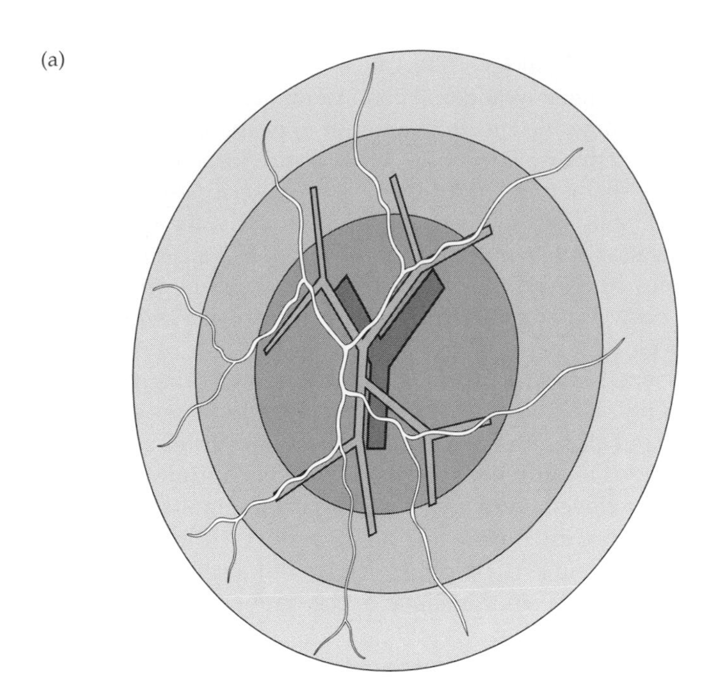

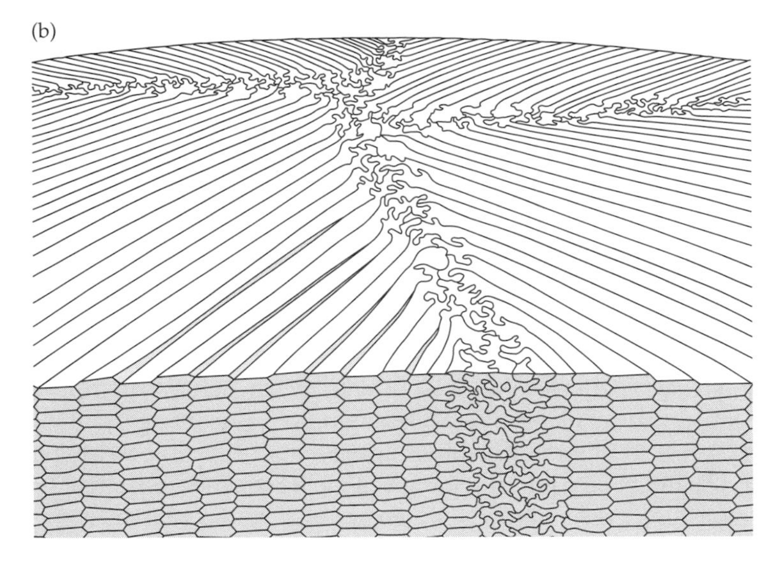

Figure 12.6
Lens Sutures
(*a*) The structure of the lens sutures becomes more complex as the lens grows and new shells with more lens fibers are added. The simple Y suture is found only in the fetal nucleus, the part of the lens formed prior to birth. Shells in the adult nucleus and cortex have many more fibers, so there must be more linear extent of suture along which the fibers can join. (*b*) The regular organization of the lens fibers is disrupted at the sutures where the ends of lens fibers interdigitate. Because of this irregularity, sutures scatter more light than the rest of the lens does. (After Hogan, Alvarado, and Weddell 1971.)

fingerlike projections that interdigitate with their neighbors above and below (there are also numerous small ridges, or microplicae, that fit into a complementary set on adjacent cells). The projections are like small bent fingers, hooks, or keys that fit into invaginations in the neighboring cell.

These fingers and invaginations, or keys and locks, create a biological Velcro, where small flexible elements interlock to bind two surfaces together. (Velcro actually has small hooks on one surface and small protruding loops on the other.) As the cutaway portion of Figure 12.7 shows, the inner and outer surfaces of the lens fibers are liberally covered with these small keys and recipient locks, all of which bind the lens shells together.

The density and distribution of the locks and keys change through the cortex and into the nucleus. The outermost shells have relatively few locks and keys holding shells together; here, most of the attachments are to neighboring cells in the same shell. Deeper in the cortex, attachments between shells become more prominent, and attachments within shells less prominent. The locks and keys continue into the nucleus, but they are less well defined, in large part because the nuclear fibers are generally less regular in structure.

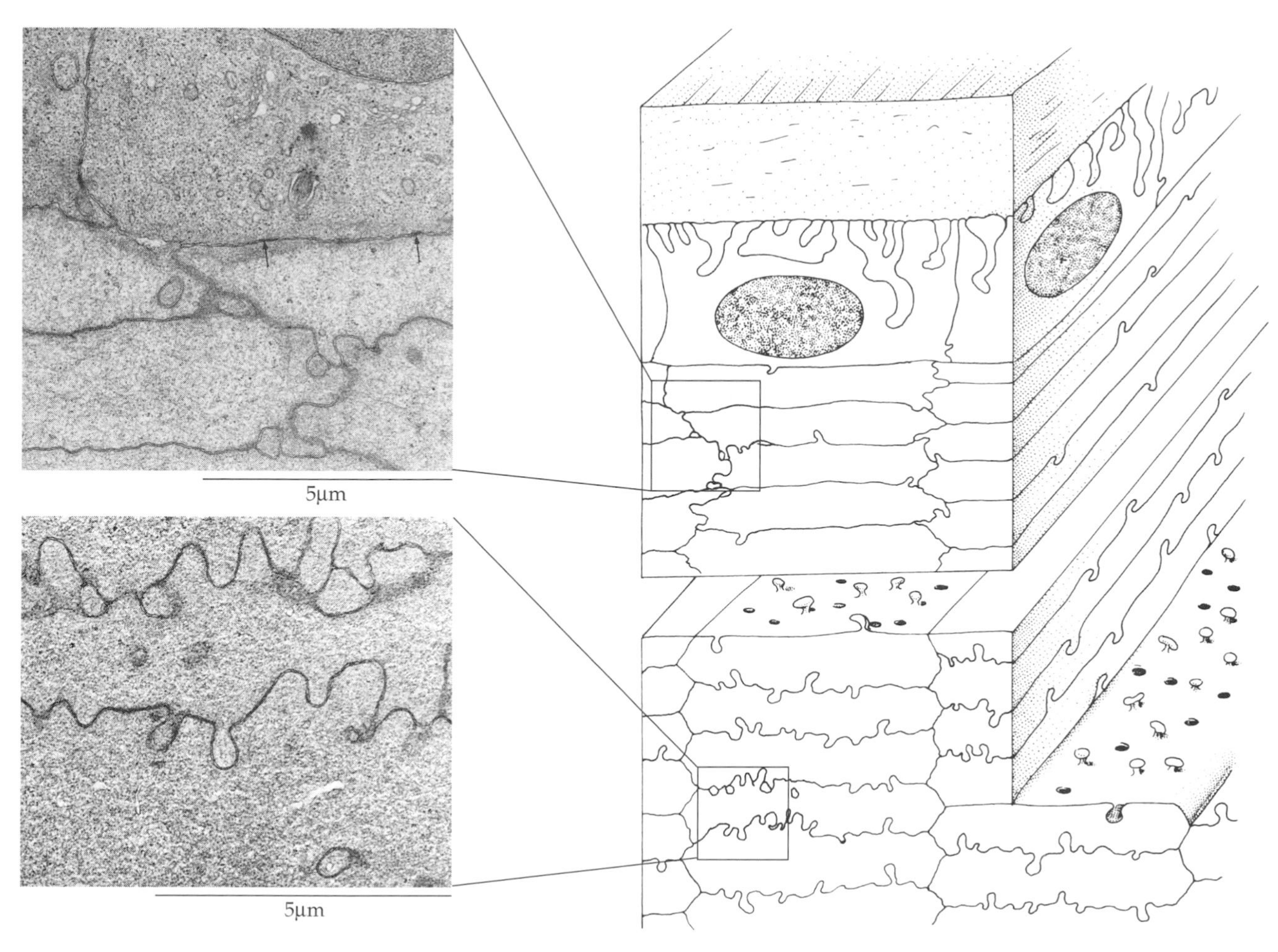

Figure 12.7
Interlocking of Lens Fibers
The outermost layers of lens fibers interdigitate mostly along their edges, where neighboring cells in a shell abut. The layer of lens fibers next to the epithelium has small occluding junctions (arrows) with the epithelium. Lens fibers deeper in the cortex are less regular in cross section as the fibers acquire interdigitating processes along their broad surfaces, linking them to the fibers just above and below in the same column of cells. The side-to-side interdigitations are most prominent in the outer cortex, becoming much reduced deeper in the lens. Interdigitations within a cell column, however, become more numerous in the deeper layers. (Electron micrographs from Hogan, Alvarado, and Weddell 1971.)

Since the lens is enclosed by a capsule, it wouldn't fall apart if these interdigitations between cells and at the lens sutures did not exist. These specializations may be responsible for maintaining internal structural relationships, however. It is easy to imagine neighboring lens fibers slipping over or under one another and becoming misaligned as the lens changes shape, for example. But since interlocking of the lens fibers is prominent in some species, such as fish, that either have no accommodation or accommodate by moving the lens rather than by reshaping it, mechanical stability is probably not the only consideration. Keeping cells together in a regular arrangement without gaps and spaces is probably a factor in transparency, though it is perhaps less important than intracellular regularity, and because of considerable communciation throughout the lens by gap junctions between lens fibers, precise relationships between fibers may be important. Whatever the underlying reason for the interlocking of lens fibers, it means that a freshly removed lens is a very cohesive tissue; even with the capsule removed, it must be peeled apart.

The anterior epithelium is the source of new cells for the lens

The lens adds new shells to the existing structure by recruiting new cells from the lens epithelium, which can be thought of as a reservoir of lens fibers-in-waiting. When needed, the epithelial cells differentiate and become lens fibers that drop out of the cell cycle—meaning they can no longer proliferate—and go about their business of being lens fibers.

Cells in the epithelium vary in morphology from one location in the epithelium to another (Figure 12.8). All epithelial cells have infoldings on their basal sur-

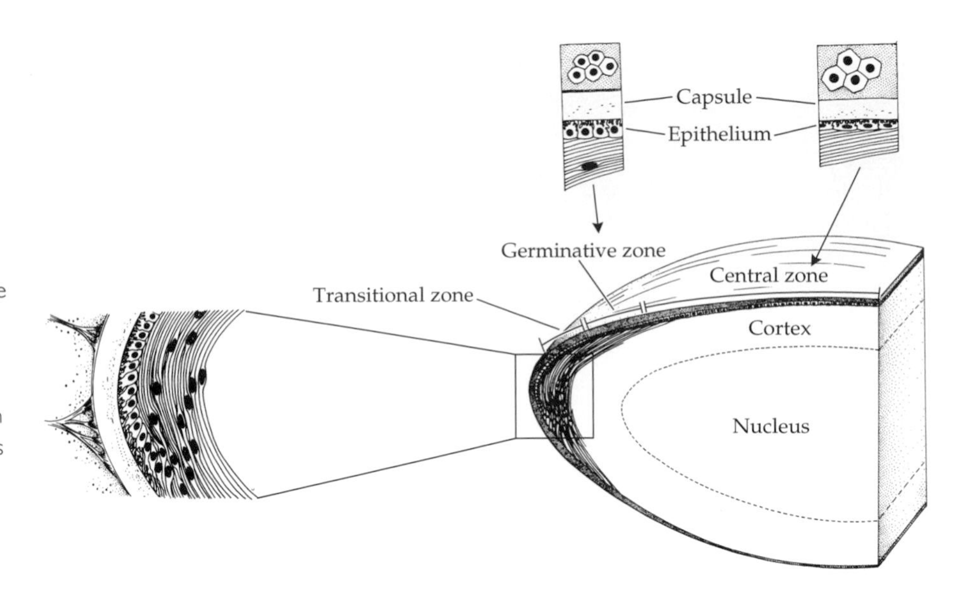

Figure 12.8
Morphology and Functional Zones in the Lens Epithelium
Epithelial cells vary in morphology from the anterior pole to the equator. In the central zone surrounding the anterior pole, they are flattened and hexagonal in outline. Closer to the equator in the germinative or proliferative zone, the cells are more columnar, smaller in area, and higher in density. New cells for the lens are being generated here. In the transitional zone and at the equator, the cells begin to elongate and rotate so that their long axes are parallel to the cortical surface. Crystallin synthesis is first detected in these cells.

faces, next to the lens capsule, but cells near the anterior pole are relatively flat, while those closer to the equator are columnar, much like the basal cells in the corneal epithelium. The transition in shape is gradual, and cells midway between the pole and the equator are cuboidal.

This morphological variation is associated with functional variation. In the central region around the anterior pole, the flattened epithelial cells are quiescent, showing little or no sign of mitotic activity. In the region of cuboidal cells, however, cells are proliferating, and closer to the equator, where the cells have become columnar, there is evidence for differentiation; these cells are beginning to synthesize crystallins. At the equator itself, the cells are beginning to elongate and rotate so that their long axes are parallel to the surface of the lens. These changes can be summarized by dividing the epithelium into zones (see Figure 12.8): A central zone of inactive cells is surrounded by an annulus of epithelium in which the cells are proliferating and differentiating—the germinative, or proliferative, zone—which in turn is surrounded by an annulus of cells at the equator that are elongating—the transitional, or marginal, zone.

These zones in the epithelium are not hard-edged in the sense of having a clear dividing line separating one from the other. Some cell division occurs within the inactive central zone, for example, but it occurs infrequently, increasing very gradually and then more rapidly away from the anterior pole; the rapid rise in proliferation is the transition to the proliferative zone. Proliferation declines toward the equator, but it probably does not fall to zero until well into the equatorial transitional zone.

Although the central zone of the epithelium, which comprises almost half of the epithelial cells, is usually nonproliferative, the cells have not lost the ability to proliferate; all that is missing is the appropriate stimulus. If the epithelium is damaged, proliferation in the central zone is part of the repair mechanism. Under normal circumstances, however, most of the cells in the central zone are old cells, formed in embryogenesis and busy with their biochemical business ever since. Yet if called on, even in an aging eye, they can exhibit a youthful capacity for reproduction, a fact that has caught the attention of gerontologists, who would like to know how these cells can seemingly be impervious to senescence. They are stem cells for the lens epithelium and therefore stem cells for the lens itself. It is as if Adam and Eve were still around, countless generations after they first appeared, still ready, willing, and able to produce more offspring if needed to keep the species going.

Cells in the germinative zone, where proliferation is at its peak, do not become noticeably smaller and do not pile up in more than one layer, which means that cells must be continually migrating toward the equator into the transitional zone. The epithelial cells have differentiated at this stage and have begun to produce some of the crystallins that will characterize them as mature cells. When they reach the transitional zone, they begin to do the other thing that characterizes lens cells: They start producing new cell membrane so that they can elongate.

Elongating epithelial cells at the equator become long lens fibers that form new shells in the lens

As the epithelial cells at the equator elongate, they also rotate so that the long axis of the cell is parallel to the lens surface. In this orientation, it is possible for one of the growing tips of the cell to extend forward next to the overlying epithelium while the other tip extends back, next to the capsule. These new lens fibers continue to extend in both directions until they meet other growing fiber tips; when other fibers are encountered, interdigitations form along the lines of contact, creating the lens sutures.

This process of cell extension and formation of lens shells is illustrated in Figure 12.9. Cells all around the equator elongate in approximate synchrony, forming a complete layer of new fibers in a belt that extends anteriorly and posteriorly from the equator. As the fibers continue to elongate, the belt widens to include more of the anterior and posterior surfaces. Some fibers stop growing at the ends of the sutures; the rest continue to fill in the parts of the shell around the lens poles.

Proliferation and differentiation in the epithelium are continuous processes, as is lens fiber elongation and shell formation, but new lens fibers are produced faster than shells can be completed. As a result, a new shell begins to form at the equator before its immediate predecessor has covered the lens from pole to pole. Thus there may be several incomplete shells in the adult lens cortex.

The size of the lens and the number of lens fibers increase throughout life

The addition of shells to the lens begins in the embryo and continues throughout life, as evidenced by the presence of mitotic figures in the germinative zone

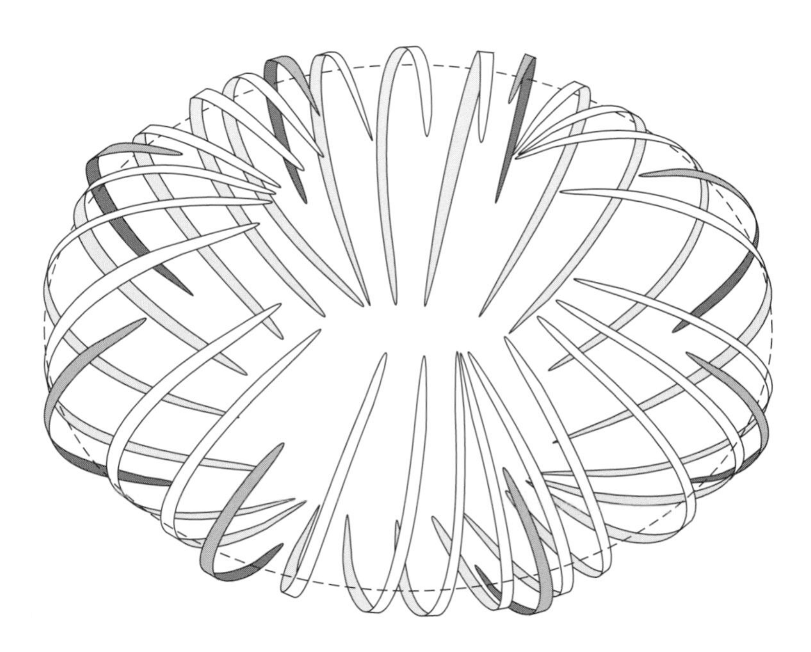

Figure 12.9
Lens Fibers Forming New Shells
New lens fibers form a solid belt around the equator, as implied here (some have been omitted for clarity). As their growing tips extend forward and back over the existing lens surface toward the poles, the belt widens; it is shown here just before fibers meet to form sutures. Cells for the next shell have already begun to elongate at the equator, forming a second, narrower belt in which only a few cells are illustrated.

of the lens epithelium and a continual increase in equatorial diameter and thickness, which is the distance between the anterior and posterior lens poles. (Because lens fibers do not increase in size with age, the increased size of the lens must be due to increased cell number.) From age 5 to age 80, the equatorial diameter of the lens increases by almost 2 mm at a constant rate of about 20 μm per year (Figure 12.10*a*); thickness increases at about the same rate during this time (Figure 12.10*b*). Although data for the first 10 years of life are sketchy, there appears to be a rapid growth phase for the first 5 postnatal years, during which the diameter increases by about 2 mm, thereby matching in 5 years all the subsequent increase in diameter.

Lens fibers in the cortex, where they are being added to the lens, are about 2 μm thick and 12.5 μm wide. The number of fibers in a shell is the number of fibers that can fit side by side around the equator; that is, the number of fibers in a shell equals the circumference in micrometers divided by 12.5 μm (Figure 12.11*a*). And because circumference is a function of diameter, the number of fibers in a shell will also be a function of lens diameter. Thus, the relationship between diameter and age also shows how the number of fibers in new lens shells increases with age. For example, at age 20, when the lens diameter is 8.67 mm and the circumference is 27.24 mm, the next shell added will contain 2176 fibers (27.24/0.0125). At age 80, the diameter is 9.64 mm, the circumference is 30.28 mm, and the number of fibers in the outermost shell is 2422 (30.28/0.0125).

As the lens enlarges, the epithelium and its proliferative zone expand to maintain coverage of the anterior surface. Assuming that epithelial cell size is roughly constant, the increase in epithelial cell number will be proportional to the increase in surface *area,* which is proportional not to lens diameter (d) but to lens diameter squared (d^2). The lens diameter increases by a factor of 1.1 over 60 years (9.64 mm/8.67 mm); the epithelial surface area increases by a factor of 1.24 ($[9.64]^2/[8.67]^2$). Therefore, the number of cells available for differentiation into lens fibers increases more rapidly than the number of new fibers required; supply exceeds demand, in other words. Because of the excess number of potential lens fibers, the number of mitotic cells per unit area of the epithelium can decline with age, as it is known to do, and still meet the requirement for new lens cells.

Each new lens shell has one more fiber than the previous shell and about five new shells are added each year after the age of five

Each new lens shell has more lens fibers than any earlier shell, so the number of epithelial cells that differentiate to become lens fibers must increase throughout

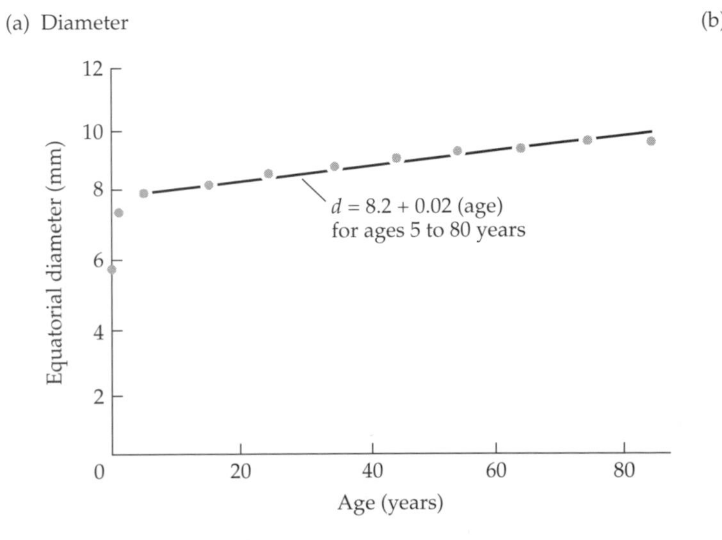

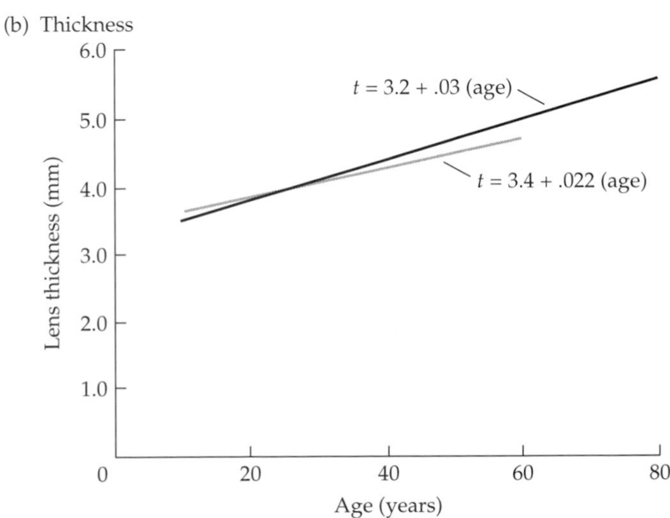

Figure 12.10
Changes in Equatorial Lens Diameter and Lens Thickness with Age
(*a*) Lens diameter (d) is slightly less than 6 mm at birth. It increases rapidly for about 5 years and then increases at a steady rate of 20 μm/year. (*b*) Lens thickness (t) also increases steadily after age 10, adding between 20 and 30 μm per year. (Diameter data replotted from François 1963; thickness data from Brown 1976 and Rafferty 1985.)

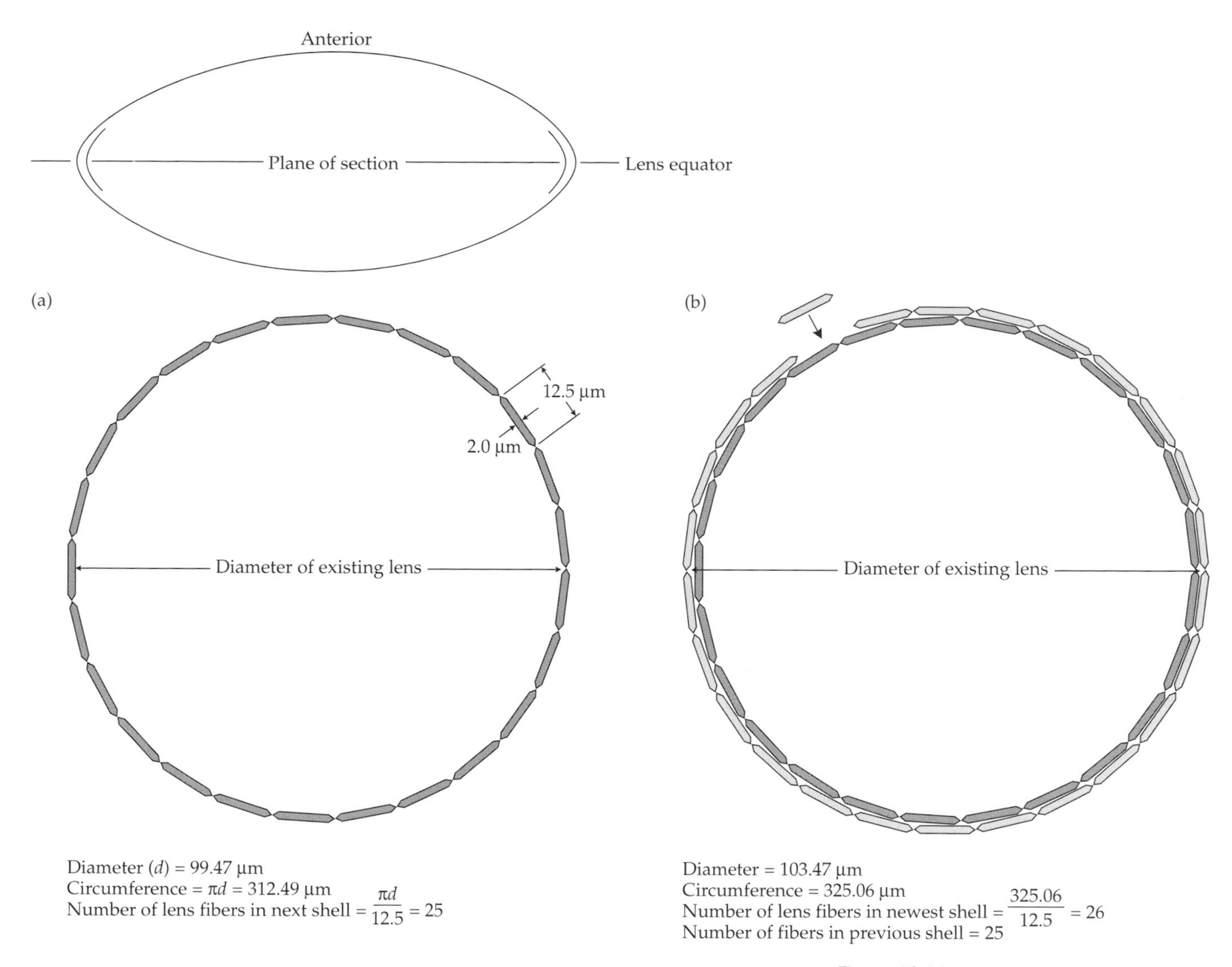

Figure 12.11
The Number of Fibers in a Lens Shell Is a Function of Fiber Size and Equatorial Circumference
(*a*) New lens fibers are about 12.5 μm wide and 2.0 μm thick; they are drawn here 1000 times larger and all dimensions and calculations are in millimeters. The number of fibers in a new lens shell is therefore the circumference of the existing lens divided by 12.5. (*b*) The newest shell always has one more lens fiber than the previous shell.

life; in that sense, lens fiber production also increases throughout life. This increase is not large, however. Each new shell adds 4 μm to the existing lens diameter and about 12.5 μm to the circumference, regardless of the current lens diameter. (A lens with diameter d will have a circumference equal to πd. Increasing the diameter by an increment Δd increases the circumference to $\pi(d + \Delta d)$. The difference in the circumferences is $\pi(d + \Delta d) - \pi d = \pi\Delta d$. If $\Delta d = 4$ μm, $\pi\Delta d = 12.6$ μm, regardless of the value of d.) Thus, each successive shell has only one more fiber than the previous shell (Figure 12.11*b*).

Since lens diameter increases by about 20 μm each year and each new shell adds 4 μm to the diameter, about five new shells are added yearly. About 300 shells will form between the ages of 20 and 80, and a shell added at age 80 has about 300 more fibers than a shell added at age 20.

An aged lens has about 2500 shells and 3.6 million lens fibers

Estimates of the total number of fibers in the adult lens range from 2100 cells (probably a typographical error) to 6.5 million cells, a number that is probably inflated by measurement errors. Assuming that truth lies between the extremes, the following calculation provides a more reasonable estimate of the number of lens fibers.

If each new lens shell has one fiber more than the preceding shell, the numbers of fibers in the shells form an arithmetic progression of the form n_0, $n_0 + \Delta n$, $n_0 + 2\Delta n$, $n_0 + 3\Delta n$, and so on, where n_0 is the number of fibers in the first shell and Δn is the number of fibers added in each new shell; in this case, $\Delta n = 1$. For a series like this with N terms (N is the total number of shells), the number of fibers in the last shell (n_L) is given by the equation

$$n_L = n_0 + (N - 1)\Delta n \tag{12.1}$$

and the sum of the terms in the series, which will be the total number of lens fibers (n_T), is given by the equation

$$n_T = (N/2)(n_0 + n_L) \tag{12.2}$$

These relationships are shown graphically in Figure 12.12.

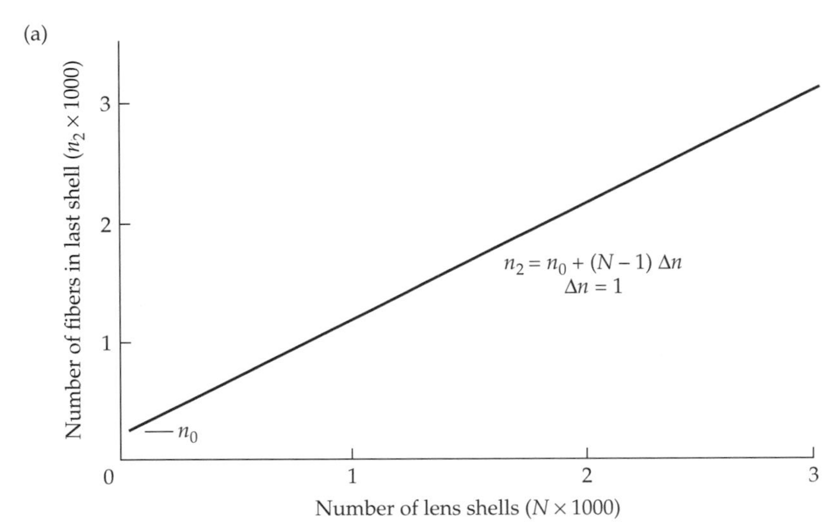

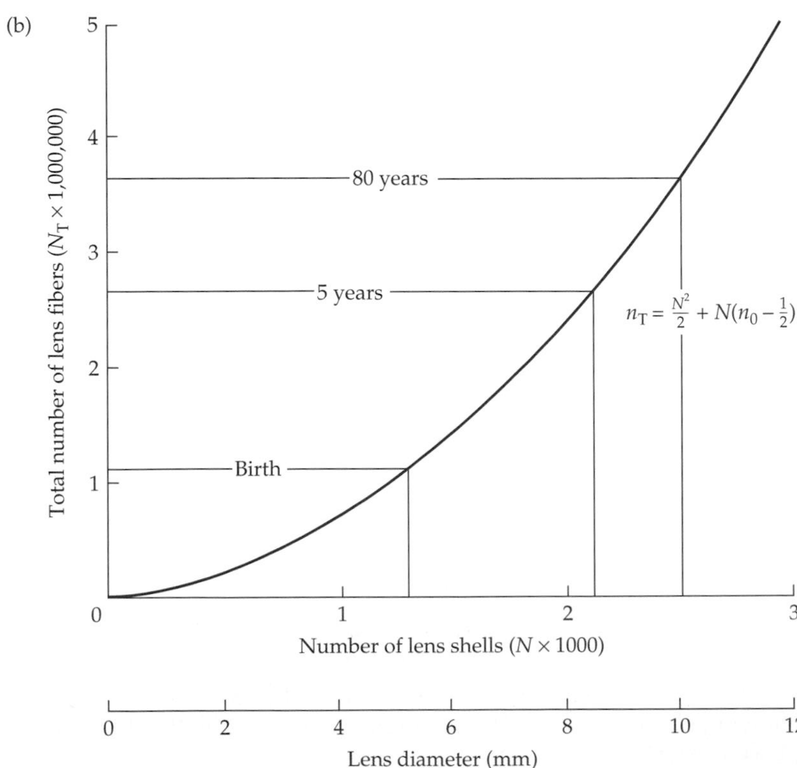

Figure 12.12
Number of Lens Fibers as a Function of the Number of Lens Shells
(*a*) The number of fibers in the newest shell increases linearly with the number of lens shells. (*b*) The total number of lens fibers is proportional to the number of lens shells squared. The equation for n_T is the result of substituting Equation 12.1 for n_L into Equation 12.2.

To apply these equations, we need some reasonable values for n_0 and N. The first lens shell to form will be very small and cannot have many fibers, so a guess that $n_0 = 200$ is probably not far off the mark. The total number of shells is more difficult, but lens diameter and shell thickness provide a reasonable estimate; if each shell adds 4 μm to the diameter and an 80-year-old lens is 9.64 mm in diameter, $N = 9.64/0.004 = 2410$. Then if one fiber is added for each new shell ($\Delta n = 1$), the number of fibers in the last shell (n_L) is 2609, and the total number of lens fibers (n_T) is 3,384,845 (plus a few hundred primary lens fibers in the embryonic nucleus).

These numbers are reasonable lower limits on the numbers of lens shells and fibers; they depend most critically on the value selected for N (the total number of lens shells) and the amount that each new shell adds to the lens diameter (4 μm). If the oldest lens fibers in the nucleus shrink in volume, as some have suggested, the value of 2410 for N will be too low because some of the 4 μm increases in shell thickness will not be fully reflected in the final lens diameter. The data that purport to show nuclear compression in the lens are not particularly convincing (and some data do not show compression), but the effect can be taken into account by assuming that there are somewhat more shells than the calculated value; for simplicity, take the number of shells in the mature (80-year-old) lens to be 2500. This value yields 2699 as the number of fibers in the last shell and 3,623,750 as the total number of lens fibers.

Knowing the precise number of fibers in the lens may not be particularly important, but estimating the relative numbers of lens fibers at different ages gives substance to any discussion of lens growth rates (see Figure 12.12*b*). For example, if 5 shells are added each year during youth and adulthood, the 10-year-old lens had 2150 shells, 350 fewer than the 80-year-old lens. Its outermost shell had 2349 fibers, and the lens as a whole had 2,740,145 fibers. If the growth rate can be extrapolated back to 5 years of age—it is a bit questionable—the 5-year-old lens had 2125 shells and 2,681,750 fibers. Over 75 years, the lens added 375 shells and almost a million lens fibers.

At birth, the lens is about 6 mm in diameter. On the basis of this value, the number of fibers in the most recent shell (n_L) should be about 1508 ($= \pi d/0.0125$), which—assuming 200 as the starting number of shells and using Equation 12.1—predicts that the number of lens shells (N) will be 1309. Then according to Equation 12.2, the total number of lens fibers (n_T) will be 1,117,886. Thus, in the first 5 postnatal years, the lens adds about twice as many shells and 1.5 times as many fibers as in the subsequent 75 years of life. And, of course, the million plus fibers in the lens at birth were all generated in the previous 8 months in utero. If we take the 80-year-old lens as a target, 31% of the lens fibers are present at birth, and almost 75% of them are present by 5 years of age.

The lens capsule encloses the lens shells and epithelium

The cellular parts of the lens are enclosed by the acellular **lens capsule** (Figure 12.13). Although the capsule contains collagen fibrils, it has very little obvious structure because the fibrils do not have a regular arrangement; they seem to run in all directions parallel to the surface of the capsule. The fibrils are primarily type IV collagen embedded in a glycosaminoglycan matrix, producing a flexible, meshlike arrangement that makes the capsule elastic even though it contains none of the elastin fibrils that characterize elastic tissues.

The capsule forms on the lens vesicle as a basement membrane of the epithelium before the posterior epithelial cells elongate to fill the lumen of the lens vesicle, thus ensuring that the membrane encloses the lens completely. It also means that the posterior capsule remains thin, because only the remaining epithelial cells on the anterior surface can contribute to building up the capsule's thickness. Subsequent growth of the capsule is from the inside, with new capsu-

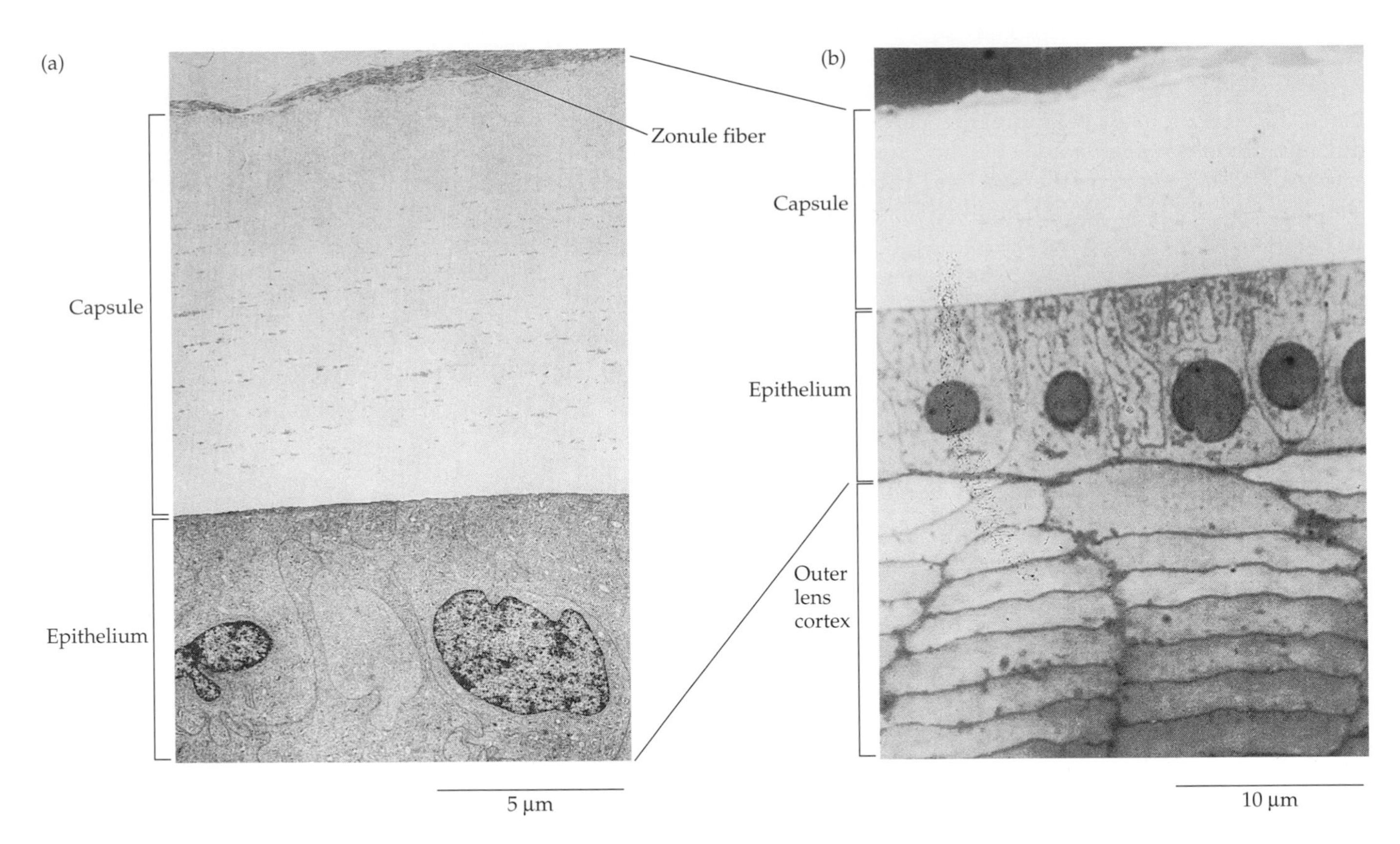

Figure 12.13
The Lens Capsule
(*a*) Strands of the zonule fuse with the outer surface of the capsule; anchoring junctions on the inner surface bind it to the epithelium. (*b*) The lens capsule consists of fine collagen fibers in a matrix, shown here overlying the epithelium that makes the capsule. (From Hogan, Alvarado, and Weddell 1971.)

lar material deposited next to the basement membrane of the epithelium, as shown by radiographic labeling with tritiated amino acids; with time, the label moves from the epithelium to the inner surface of the capsule and eventually to the outer surface.

When the capsule first appears as a basement membrane, the lens is very small. Material is added only to the anterior capsule as the lens grows, but the part of the capsule just anterior to the equator when the lens is small will end up posterior to the equator when the lens is larger; in effect, all of the posterior capsule in the adult lens was once adjacent to anterior epithelial cells and has therefore had some opportunity to be built up over its original basement membrane thickness. But the part of the capsule closest to the posterior pole has had the least opportunity for elaboration, and it is therefore the thinnest part of the capsule, around 2 to 3 μm thick.

This notion of the capsule gradually slipping around from anterior to posterior as the lens grows implies that the capsule increases in thickness from the posterior pole to the equator and onto the anterior surface. In general, this is the case, but the situation is complicated by the insertion of the zonular fibers anterior and posterior to the equator; merging of the zonular fibers with the capsule increases the thickness in these regions. Moreover, capsule thickness varies with age as more material is deposited in the anterior capsule and as the zonular insertions shift relative to the equator as the lens grows.

These variations in thickness are summarized in Figure 12.14, which exaggerates the dimension scale for capsule thickness to show variation more clearly. At 22 years of age, the maximum capsule thickness, about 19 μm, is located just anterior to the equator; from the equator, thickness falls off to about 8 μm at the anterior pole and about 3 μm at the posterior pole. The pattern changes in older eyes; capsule thickness increases all across the anterior surface, and the point of maximum thickness shifts anteriorly from the equator. The two modes in the curve for age 55 years indicate the locations of zonular fiber insertions on

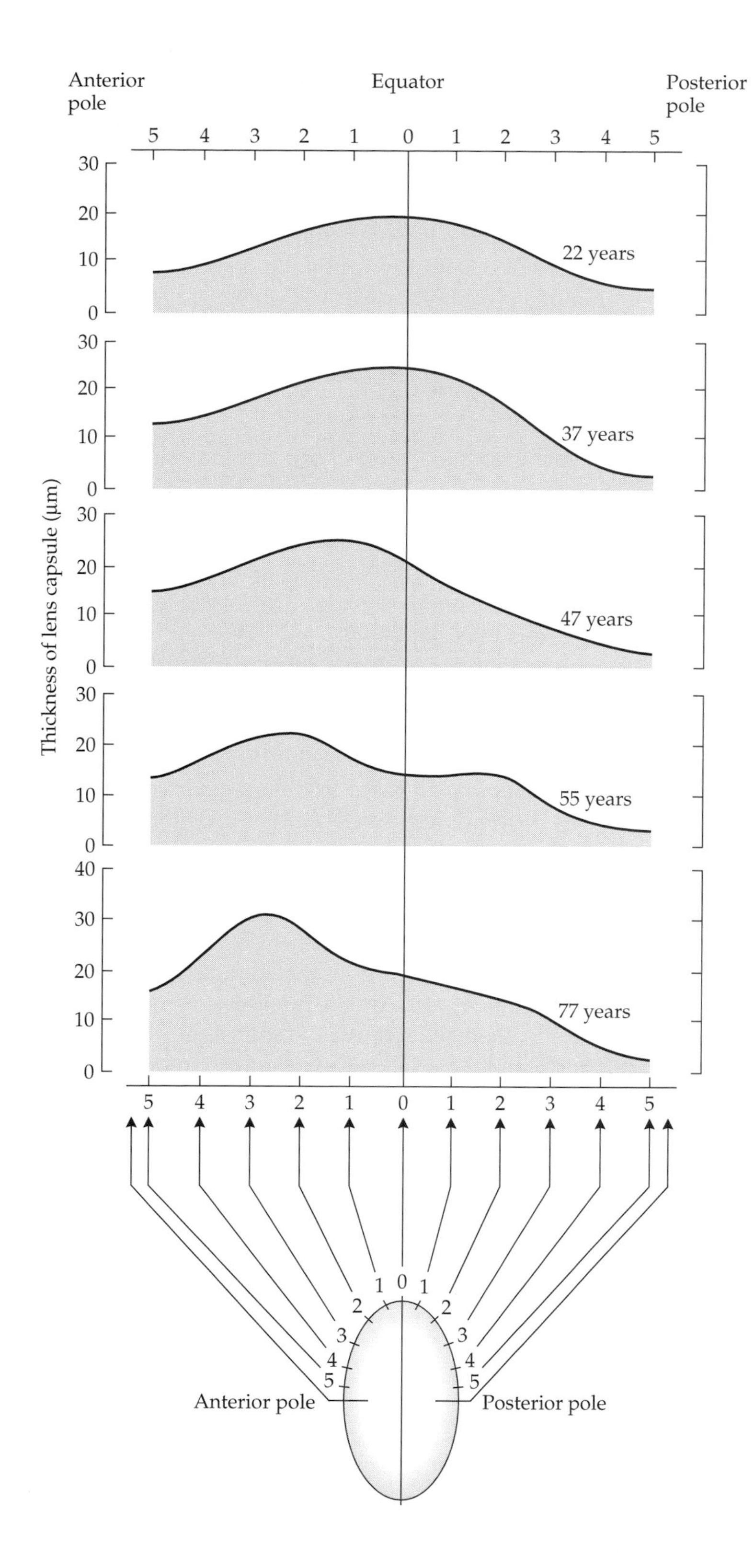

Figure 12.14
Variation in Capsule Thickness with Location and Age
The capsule never has a uniform thickness. In young eyes it is thickest at the equator, but the region of maximum thickness gradually shifts anteriorly; eventually the thickest region is about halfway between the equator and the anterior pole. (After Fisher and Pettet 1972.)

the anterior and posterior surfaces. There is almost no change in the capsule thickness at the posterior pole and little change elsewhere on the posterior surface except near the equator.

Variation in capsule thickness has been invoked to explain how the lens changes its shape during accommodation. Fincham, in particular, argued that the annular region of maximum thickness on the anterior surface would force a more pronounced change in curvature near the anterior pole than would be produced if the capsule thickness were uniform. Curvature should change most where the capsule is thinnest.

As it turns out, the change in anterior surface curvature during accommodation is not uniform, but the effect is subtle and not of the sort envisioned by Fincham and others. Moreover, the region of maximum capsule thickness does not shift onto the anterior surface until accommodation is nearly gone. In young eyes with large accommodative amplitudes, the thickest part of the capsule is at the equator. The pronounced regional variations in capsular thickness in older lenses cannot have much to do with the way in which lenses in young eyes change shape as they accommodate.

The locations at which the zonule inserts onto the lens change with age

The detailed structure of the zonule is not fully understood, but it is not made of collagen fibrils as generally stated in the older literature. The primary component is a type of microtubule commonly found in elastic tissues, where the microtubules are surrounded by elastin fibrils and a polysaccharide matrix. The microtubules in the zonule are assembled into long strands that are collected as bundles running between the basement membrane of the ciliary body's unpigmented epithelium and the lens capsule. Fiber bundles run over the surface of the capsule, tapering down to fine points as individual fibers penetrate the capsule, where attachment proteins bind them to the capsular collagen. Although these attachments of single strands and small bundles may not seem impressive, the strength of the zonular attachment is in the numbers. A view of the zonule all around the lens (Figure 12.15) makes it clear that many thousands of fibers are involved, giving a better feeling for the potential strength of this linkage between lens and ciliary body.

At early stages of lens and zonule growth, the zonular fibers are distributed fairly evenly within their zone of attachment from anterior to posterior of the equator. As time passes, the fibers redistribute themselves into three overlapping groups: the densest attachment anterior to the equator, another posterior to the

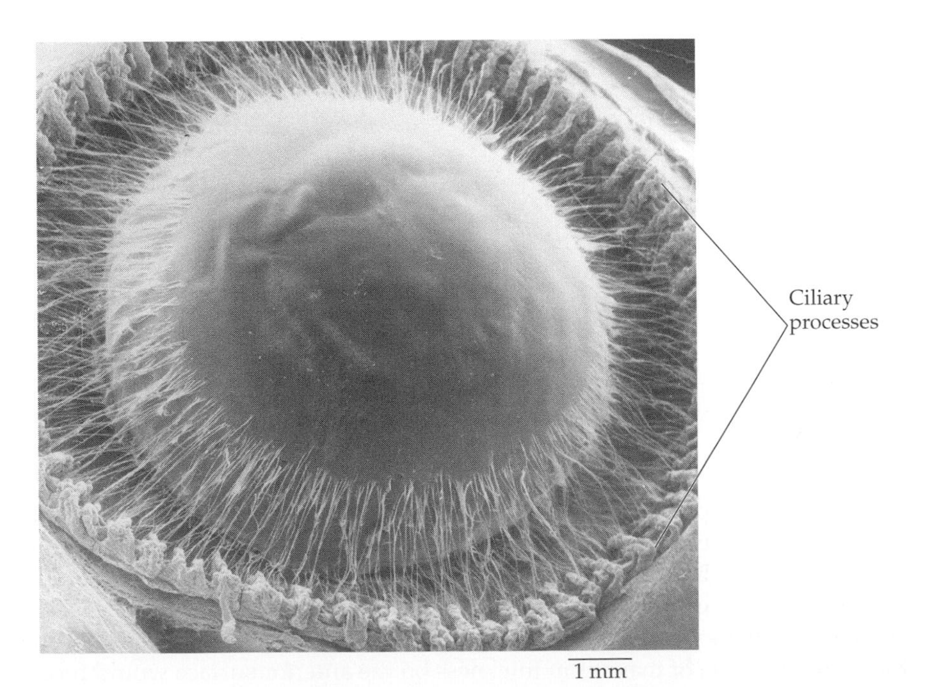

Figure 12.15
The Zonule
This view of the lens from the front shows many strands of the zonule running from the valleys between the ciliary processes to their attachments on the lens. Other strands, only some of which are visible, attach to the equator and to the posterior capsule. The zonular fibers support the lens as guy wires support a tent. (From Streeten 1992.)

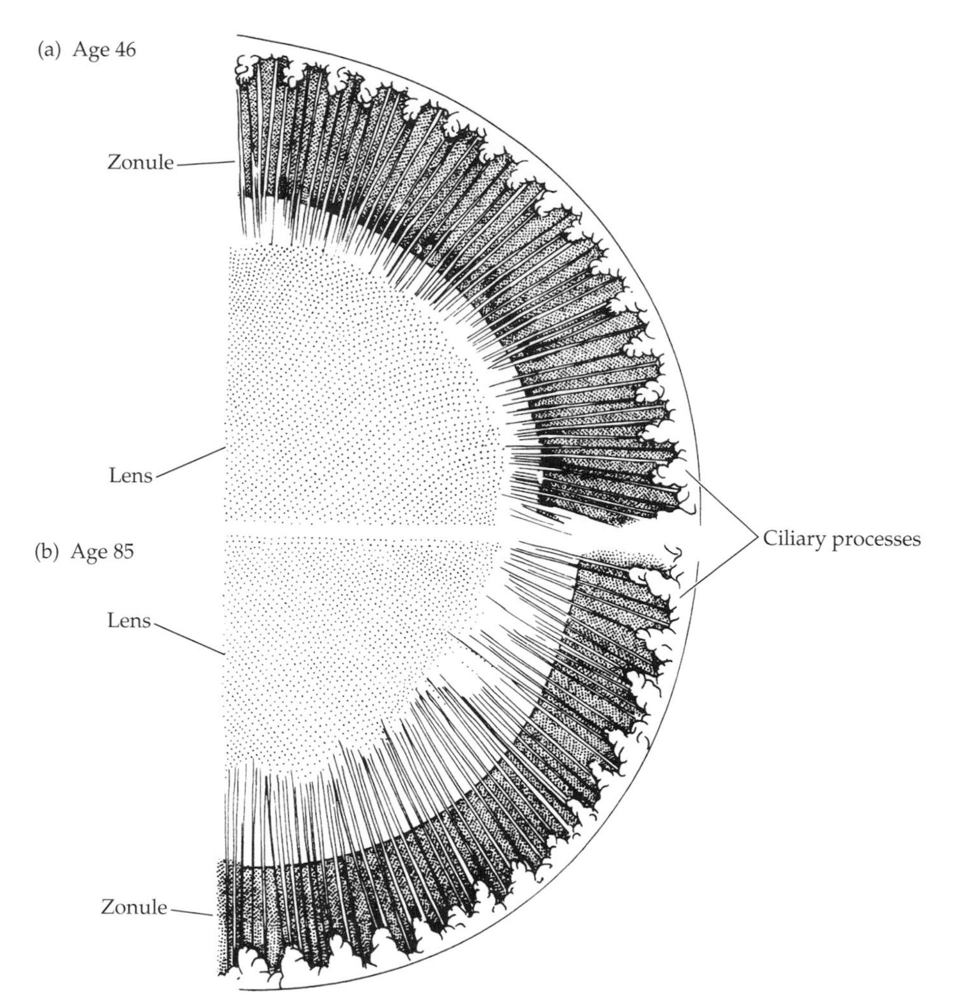

Figure 12.16
Shift of the Zonule Insertion with Age
Comparison of a 46-year-old eye (*a*) with an 85-year-old eye (*b*) shows the lens closer to the ciliary body and the limit of the zonule insertion farther from the lens equator in the older eye. The limit of the zonule insertion is the same distance from the ciliary body in both eyes, however. (After Farnsworth and Shyne 1979.)

equator, and a third group attached at the equator itself. This rearrangement of the zonule appears to continue throughout life. Figure 12.16*a*, for example, shows frontal views of the lens, zonule, and ciliary processes in a 46-year-old eye. In comparison to the lens in Figure 12.16*b*, which is from an 85-year-old eye, it is smaller and its zonular fibers insert closer to the equator. The distance from the zonule insertion to the ciliary body is the same in both eyes, however; in other words, the distance from the zonule insertion to the anterior lens pole is constant, regardless of the age and size of the lens. As the lens grows, the zonule insertion shifts anteriorly *relative to the equator,* but its position relative to the anterior pole does not change. (Although it is not shown here, the posterior zonule attachment does the same thing. It shifts posteriorly relative to the equator, but lies at a constant distance from the posterior pole.) Insofar as the thickest part of the lens capsule is where the zonule inserts, this description is consistent with the capsule thickness data shown in Figure 12.14.

The shift in the zonule's insertions relative to the equator has been implicated as a factor in presbyopia; the underlying idea seems to be that changing the zonule's position on the lens alters the efficiency of the mechanical linkage between the lens and the ciliary body, producing a smaller reduction in zonular tension when the ciliary body contracts in older eyes and therefore less change in lens shape. But the obvious shift in zonule position relative to the equator in Figure 12.16 occurs too late; most of the accommodative amplitude is already gone by age 45. The difficult task is using smaller changes in zonule position to explain the progressive loss of accommodation during the preceding 40 years; it is not obvious (to me, at any rate) how that might work.

Vignette 12.1

Putting the Lens in Its Proper Place

UNTIL RELATIVELY RECENTLY, certainly within my lifetime, an opacity of the lens was a serious problem. The cataractous lens could be removed, but with a significant risk of complications and the subsequent need for a very unwieldy spectacle correction. Nowadays, surgeons can remove a cataractous lens and replace it with an intraocular lens in 15 or 20 minutes, with little risk and the realistic expectation of excellent visual acuity. Getting to this point, however, has been a very long journey.

Operations for cataracts are quite ancient, dating back the better part of 3000 years. The method, still used by Georg Bartisch in the sixteenth century (see Vignette 4.2), was called *couching*. The surgeon inserted a needle through the temporal side of the eye near the equator to impale the opacity, break it free, and move it down to the inferior part of the eye away from the optic axis. On the basis of the anatomy as it was then understood, practitioners thought they were removing an opaque veil that had formed in front of the lens. In other words, the opacity was not known to involve the lens at all.

It is impossible to know what rate of success this operation had, but it probably went wrong more often than not, for at least two reasons. First, the risk of infection from an unsterile needle inserted into the eye was very high (Bartisch understood this well enough to insist on clean silver needles). Second, any serious rupture of the lens capsule and exposure of the lens proteins was almost sure to invoke a massive immunological reaction; lens proteins are normally so isolated from the body's immune system that they are targeted for destruction as foreign proteins. Neither infections nor immune reactions bode well for the future health of the eye, and a great many eyes that underwent couching probably ended up in worse shape than before the surgery.

The man who first understood where the lens was located in the eye was the teacher of William Harvey (see Vignette 6.1) and the discoverer of the valves in peripheral veins—Girolamo Fabrici, or Hieronymus Fabricius ab Aquapendente (1533–1619). Aquapendente is a small town north of Rome, not far from Orvieto; Fabricius spent his youth there, and took part of his name from the village, but left while in his teens to study Greek, Latin, and philosophy at the University of Padua. He subsequently studied medicine and surgery, receiving his anatomical training from Fallopius (see Vignette 4.1), and eventually taking his teacher's place as the professor of anatomy.

Figure 1 is a plate from *De Visione Voce Auditu,* which Fabricius published in 1600 to document his dissections on the eye, the larynx and pharynx (*voce*), and the ear. Most of these drawings are of cow or sheep eyes, but the first five in the top row are of the human eye (numbers 38, 29, 39, 30, and 31). Judging from the illustrations, Fabricius generally used the same peel-away dissection procedure employed by Vesalius and Eustachius, but somehow he obtained a different result. The lens, shown isolated in drawing 39, has a position in situ on the anterior surface of the spherical vitreous body, as indicated in drawings 38 and 31. The lens position is shown more clearly for the bovine eye in drawings 35 and 36, where streaks of pigment from the ciliary processes lie on the vitreous body around the lens.

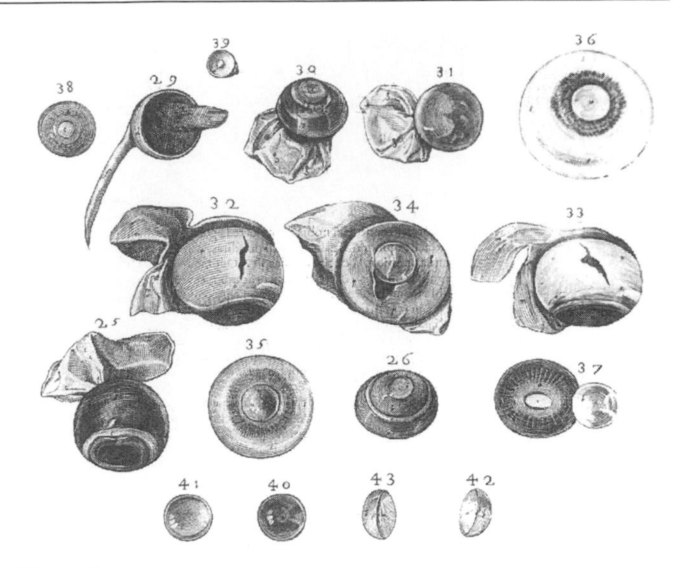

Figure 1
The Lens and the Vitreous
These drawings show the lens and vitreous body after the outer coats of the eye have been removed. Numbers 38, 29, 39, 30, and 31 in the top row are human eyes; the others are cow or sheep eyes. The message with respect to the lens is the same in all cases; the lens lies on the anterior surface of the vitreous body and it is encircled by the ciliary body. The lens is not in the center of the eye as had long been thought. (From Fabricius 1600.)

Fabricius's conception of the location of the lens is clear; it is on the anterior surface of the vitreous body, where it forms a boundary for the anterior chamber. It is emphatically *not* at the center of the eyeball, as it had been depicted up to this time.

This discovery is now about 400 years old and so much a part of our common knowledge about the eye that it seems almost trivial. In terms of how it altered ideas about how the eye works, however, its importance was monumental. Knowing where the lens was located allowed Johannes Kepler (1571–1630) to calculate—for the first time—that the eye formed an inverted image in the plane of the retina. The eye was mapping the external world, in miniature, onto the retina, and a recent, unsubstantiated proposal that the retina had something to do with vision was suddenly an almost inescapable conclusion. Moreover, a cataract could no longer be thought of as an opaque veil between the lens and the iris; the space between the lens and the iris had just disappeared. The opacity had to be in the lens itself. Removing a cataract now meant removing the lens, and surgeons were able to start intelligently down the path that has lead to modern surgical procedures.

As the latest in the sequence of great anatomists at Padua that had begun with Vesalius (see Vignette 3.2), Fabricius had the longest and most contentious academic career. He had nasty quarrels with the German student contingent after publicly reprimanding them for uncouth behavior and bad Latin, ongoing disputes with various administrators and functionaries at the university, and numerous legal wrangles with members of his family, including his illegitimate son. The phrase "difficult personality" comes to mind when reading about his life. On the other hand, he was a first-rate anatomist, a gifted teacher, and a highly regarded physician; Galileo was one of his patients. Fabricius published several books on anatomy, particularly on embryology, and planned a massive illustrated anatomical text that would have been his version of the famous *Fabrica* of Vesalius. This effort was unfinished at his death, but more than 100 color plates are housed in the library of San Marco in Venice.

Fabricius designed the dissecting theater in Padua (see Figure 1 in Vignette 6.1), and the building was completed under his supervision in 1595. Recently restored, the dissecting theater is named in his honor; it is a reminder of a time when the best measures of man were made in Padua.

The Lens as an Optical Element

The refractive index of the ocular lens varies from one part of the lens to another

Although the cornea is the eye's major refractive element, the refracting surfaces have so little separation (0.5 mm) that it can be treated optically as a simple thin lens, whose principal planes and nodal points coincide at a single location. The lens is more complicated for at least two reasons. First, it is not thin; in the adult lens, the surfaces are separated by almost 4 mm at the poles, and this thickness contributes to overall lens power. Second, the lens does not have a single, uniform refractive index; there seems to be a continuous change of index—a gradient—from a lower index (1.37) at the outer cortical surface to a higher index (about 1.5) in the nucleus. The lens is a gradient index system.*

Two consequences of a refractive index gradient are shown in Figure 12.17, which traces ray paths through a model lens for two conditions: a uniform index and a gradient index from center to periphery of the lens. The principal difference between the two cases is the location at which the refraction occurs. For the uniform index lens (Figure 12.17*a*), refraction occurs only at the surfaces where the index changes; elsewhere, the ray paths are straight lines, including the ray paths within the lens. With the gradient index lens (Figure 12.17*b*), refraction occurs not only at the surfaces, but also within the lens as rays encounter the continually changing index. The resulting ray paths are curved rather than straight. As a result, a gradient index lens with the same curvatures as a uniform index lens has a higher dioptric power. This is an advantage for the eye's lens, allowing it to offset some of the reduction in power it suffers by being immersed in fluid.

Spherical lens surfaces produce considerable spherical aberration when the lens is used at a wide aperture. The eye has several built-in corrections for spherical aberration, including asphericity of both the cornea and the lens surfaces (discussed shortly), but the index gradient in the lens also contributes to the correction. A gradient index lens with spherical surfaces behaves as if the surfaces were aspheric (see Figure 12.17*b*).

A difference in refractive index between the nucleus and the cortex of the human lens has been recognized for a long time. In Gullstrand's exact schematic

*Gradient index optics is not a new subject. James Clerk Maxwell gave theoretical consideration to the so-called fish-eye lens in the 1850s, Helmholtz discussed index gradients, and Gullstrand wrote extensively about the subject in his sections of *Helmholtz's Treatise on Physiological Optics*. Serious interest is relatively recent, however, inspired by the advantages of gradient indices in optical fibers used for data transmission. Gradient index fibers are a reality, but gradient index lenses are difficult to manufacture and their analysis involves mathematical complexities well beyond those of uniform index lenses; nature's lenses, however, particularly those in aquatic eyes, typically have gradient indices (see the Prologue). My first encounter with gradient index optics when I was a graduate student was a humbling, near disastrous experience.

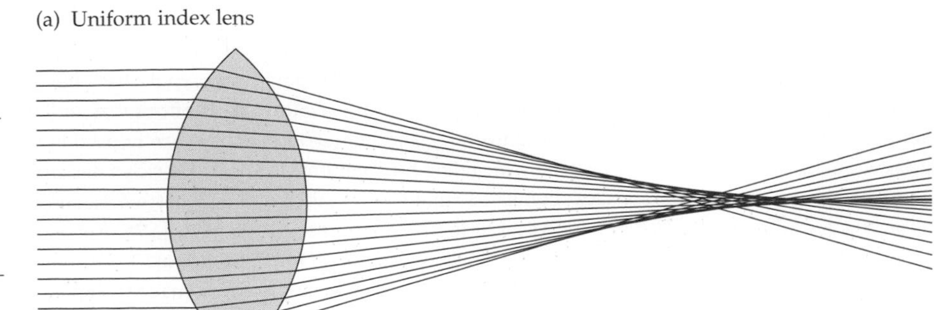

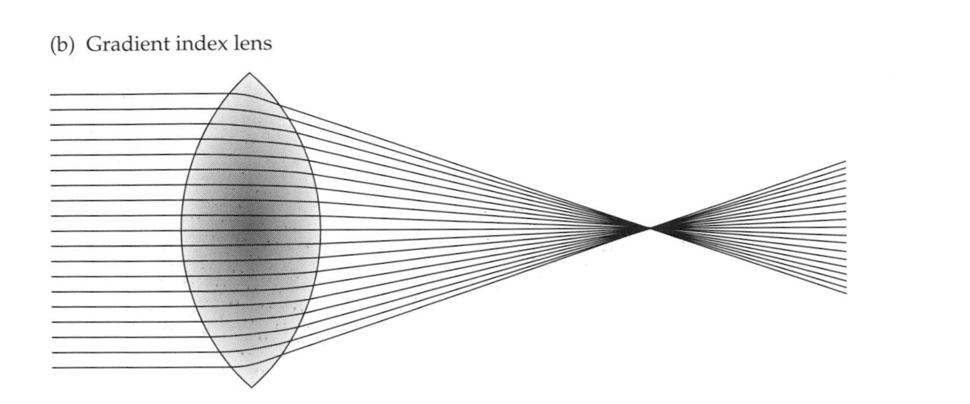

Figure 12.17
Uniform and Gradient Index Lenses
Refraction occurs only at the surfaces of a uniform index lens (*a*), but throughout a gradient index lens (*b*) because of the continually changing refractive index. Compared to a uniform index lens with the same curvatures and mean index, the gradient index lens has higher power and little or no spherical aberration.

eye, for example, the lens was modeled as having two refractive indices—one for the cortex and another, higher index for the nucleus (see Figure 2.13). This model better approximates the real lens than does a model lens having a uniform refractive index, but it still has less power than a model lens with an index gradient.

The details of the refractive index gradient in the human lens are not known. One possibility is that it is like fish lenses, having a central core—the nucleus—with a uniform index surrounded by a cortex in which the index increases from the lens surface though the cortex. Insofar as refractive index is a function of protein concentration, the model fits nicely; the protein concentration is constant within the nucleus and declines from the nucleus to the lens surfaces, and the refractive index should do the same. Studies in vitro indicate that increasing the concentration of crystallins produces a higher refractive index, and there is no reason to think that the effect would be different in vivo. If the lens proteins became progressively more concentrated in older lens fibers, the increasing index from cortex to nucleus would be a direct consequence of the way in which the lens grew. Thus we have another example of these unique proteins giving this tissue some special properties.

Lens transparency is related not only to protein regularity but also to water content, which is maintained by ion pumping in the epithelium

Although the lens is immersed in water (even the vitreous is mainly water), its water content is low, only 60% or so. Like the cornea, the lens loses transparency if it takes on more water than normal, and maintaining the normal water content requires active metabolically driven ion pumps. Almost by default, the pumping must be done by the epithelium; with the exception of the newest and most superficial lens fibers, metabolic activity in the rest of the lens is at low ebb, ceasing altogether in the center of the nucleus. The modest needs of cells deep in the lens appear to be met by intercellular diffusion of ions and small molecules through the system of gap junctions between lens fibers. (As implied earlier, the specialized structures that bind lens fibers together may maintain the alignment of this extensive set of gap junctions.)

Several ion species and amino acids are actively transported across the epithelium, but the dominant ion flux is mediated by Na^+–K^+ pumps, which transport Na^+ inward and K^+ outward. The osmotic gradient created by differences in ion concentrations between the lens and the aqueous produces enough water outflow to maintain the normal state of hydration.

The lens proteins scatter more light as their concentration decreases, and their concentration decreases as water content rises. Therefore, transparency requires that the water content be maintained at a relatively low level, which it is—normally. But there are normal age-related losses in transparency and more severe losses associated with cataract. If protein concentration based on water content were the only factor involved, water content would have to increase gradually with age and more precipitously in cataract, presumably because the epithelium was no longer pumping as effectively as it once did. That is, diminished epithelial function may contribute to normal and exaggerated losses of transparency with age.

Although the role played by the epithelium in maintaining lens hydration and transparency is not in doubt, its contribution to transparency loss is not understood. Some studies report large changes in lens Na^+ and K^+ concentrations with cataracts, indicating a breakdown in epithelial pumping, while others show no change, for example. At present, evidence for declining epithelial function with age is inconclusive, and since transparency loss can be explained *without* invoking epithelial changes, explanations for cataract formation generally do not include the epithelium in any significant way.

The lens contains several different optical zones

In addition to the divisions of the lens into cortex and nucleus, the nucleus can be further subdivided (Figure 12.18). The smallest part, the **embryonal nucleus**, represents the lens after the first 6 weeks of gestation and contains nothing but primary lens fibers; it lies within the **fetal nucleus**, which represents the lens at birth. The rest of the lens, which includes the adult nucleus and the cortex, is the result of postnatal lens growth. The adult nucleus probably corresponds to the first decade of life, with all subsequent growth creating the lens cortex.

Diagrams like the one in Figure 12.18*b* are easy to draw, since the size of the lens at various stages of development is known, but it is difficult to locate corresponding divisions in vivo. Scatter from a slit of light passing through the lens often reveals zones with different light-scattering properties, but there are more optical zones than there are developmental subdivisions. As Figure 12.18*a* shows, however, scatter from the outer part of the lens differs from the inner portion, and for clinical purposes this difference provides a working distinction between cortex and nucleus.

The bright bands of scattered light are difficult to interpret. There are more of them in older lenses than in younger lenses (Figure 12.19), suggesting that changes are occurring in the lens that affect its light-scattering properties. The bands of scattered light indicate discontinuities of some kind; abrupt changes in refractive index or a physical separation between lens shells are two possibilities, but there is little evidence for either. Another suggestion relates the bands of light scatter to changes in the suture pattern; since the sutures are sources of scatter, there may be more scattering, or at least a change in scattering, as the sutures become more complex in older lenses. The underlying implication, however, is that the suture pattern changes are large and abrupt, and it is not clear that this is the case.

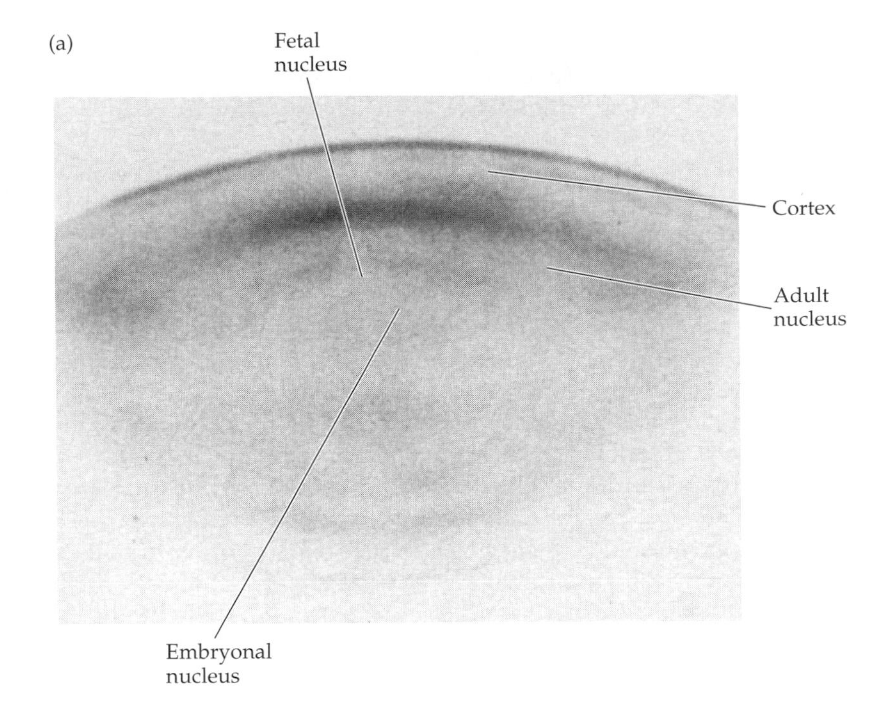

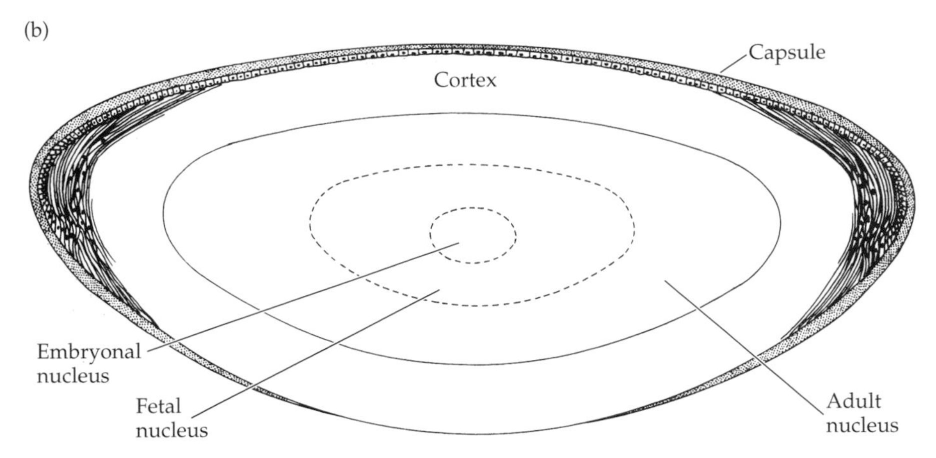

Figure 12.18
Optical and Developmental Zones in the Lens
(*a*) An optical section through the lens viewed with a slit-lamp shows several bands of scattered light. Some of them can be associated with the lens regions defined by their time of growth (*b*), but there are often zones of scatter within lens regions. (Photograph from Koretz and Handelman 1988.)

Although the reason for the bands of light scatter is not known, they are interpreted as isoindical lines—that is, surfaces along which the refractive index is the same. The main isoindical lines correspond to the anterior and posterior lens surfaces, but those within the lens are treated in the same way: lines along which the refractive index is constant and, by implication, lines across which the refractive index changes abruptly. These assumptions are important, since much analysis of the lens depends on their correctness; if the assumptions are wrong, so are some of the interpretations based on them.

The lens surfaces are parabolic and therefore flatten gradually from the poles to the equator

Helmholtz (see Vignette 11.1) depicted the lens surfaces 150 years ago as aspheric, but the nature of the asphericity was uncertain because of limitations in the methods for measuring lens surface curvatures (he assumed they were parabolic). Sections through the surfaces can be viewed directly, however, with a modification of slit-lamp observation in which light scatter from a slit of light passing through the lens is viewed from an oblique angle and corrected for the obliqui-

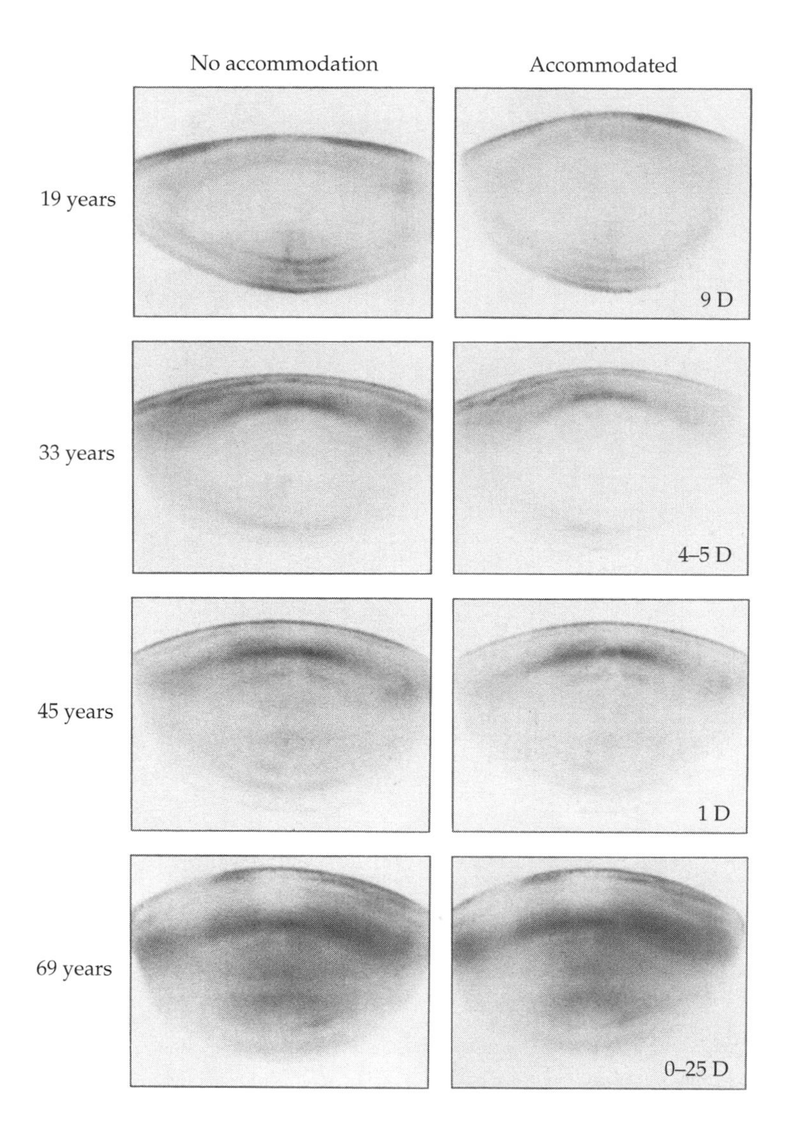

Figure 12.19
Changes in Lens Surface and Optical Zone Curvatures with Age and Accommodation Level
Optical sections through the lens are shown for four ages and for unaccommodated and maximally accommodated states. Lens surface and internal curvatures increase with age, as does lens thickness. For a given age, the surface and internal curvatures increase with accommodation. (From Koretz and Handelman 1988.)

ty of view (a method known as Scheimpflug imaging*). All surfaces that scatter light can be visualized, photographed, and measured with considerable precision. Helmholtz was on the right track, as he often was.

Figure 12.19 shows lenses at different ages and different levels of accommodation; the lenses at younger ages (19 and 33 years) have particularly obvious bands of light scattering within the lens cortex. By enlarging the lens photographs and digitizing points along the curves into an x,y coordinate system, descriptive equations for the curves can be determined. For the lens surfaces and all the interior bands of light scatter, the curves are best fit by the equation $y = a + cx^2$, which describes a family of parabolas. The origin of the coordinate system—the point where $x,y = 0,0$—is the anterior pole, the y axis is the line through the anterior and posterior poles, and the x axis is parallel to the equatorial plane (Figure 12.20).

*Spatial distortions occur in photographs when the film plane is not parallel to the object plane; tilting a camera upward to photograph a tall building, for example, makes the parallel sides of the building converge in the photograph. Large-format view cameras have tilting backs and lens holders that permit the optical axis to be directed upward while keeping the film plane vertical, thus eliminating this kind of distortion. In Scheimpflug imaging, the camera back can be rotated around a *vertical* axis so that it is parallel to the slit of light that illuminates the lens, thereby eliminating distortion created by the oblique viewing angle. The images still require correction for refraction by the cornea, however.

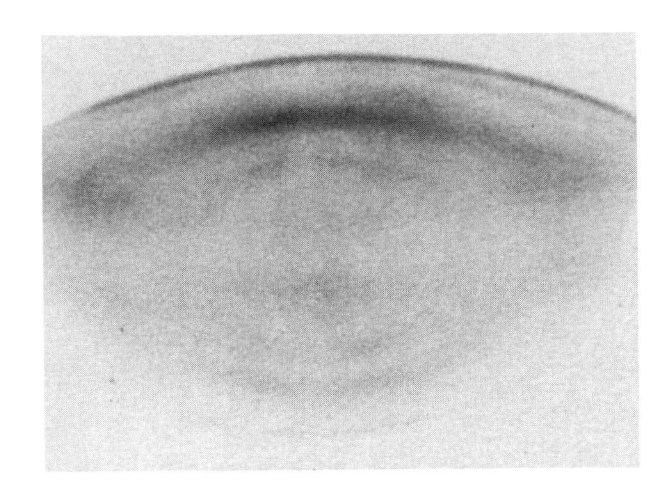

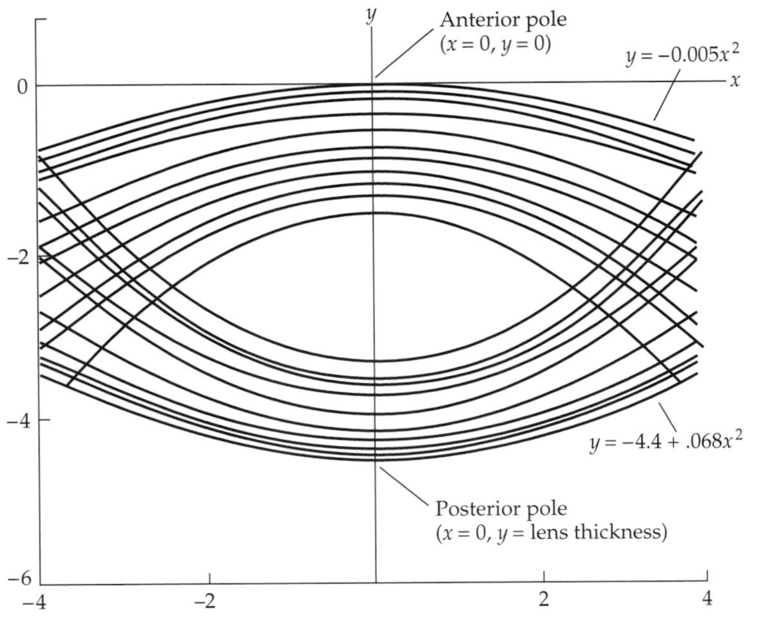

Figure 12.20
Lens Surfaces Can Be Described by Parabolas
Using an x,y coordinate system in which the y axis is the anterior–posterior axis through the lens and the point $x,y = 0,0$ is the anterior pole, the lens surfaces and the internal optical zones can be described by an equation of the form $y = a + cx^2$. The photograph and the constructed parabolas are for a 45-year-old lens. (Photograph from Koretz, Handelman, and Brown 1984.)

The best-fit parabolas for the anterior and posterior surfaces and some of the interior zones of a 45-year-old lens are plotted in Figure 12.20. The interiormost curves are clearly not circular, but circles with the same polar radii of curvature as the front and back surface parabolas are nearly indistinguishable from the parabolas over most of the central part of the surface. (A large increase in magnification is necessary to distinguish between circles and parabolas.) Toward the equator, however, the parabolas are flatter than circles that are based on the polar radii of curvature.

Being able to characterize the sections cut through the lens surfaces as parabolas is a useful simplification; revolving the parabolas around the lens axis (y axis) forms a paraboloid whose surface can be fully described by just two parameters. In the equation $y = a + cx^2$, a is a position term, stating where a surface intersects the lens axis relative to the anterior pole; for the anterior surface, a is zero, and for the posterior surface, a is in the vicinity of 4 mm—that is, the lens thickness. The curvatures of the parabolas are governed by the parameter c, with larger values generating steeper, more highly curved parabolas.*

*c is a "shape parameter" that can be used in the standard equation for local curvature: $\kappa = y''/(1 + y'^2)^{3/2}$. In this case, $\kappa = 2c/[1 + (2cx)^2]^{3/2}$ because $y' = 2cx$ and $y'' = 2c$. The *radius* of curvature, ρ, equals $1/\kappa$. At the poles, where $x = 0$, $\kappa = 2c$ and $\rho = 1/2c$.

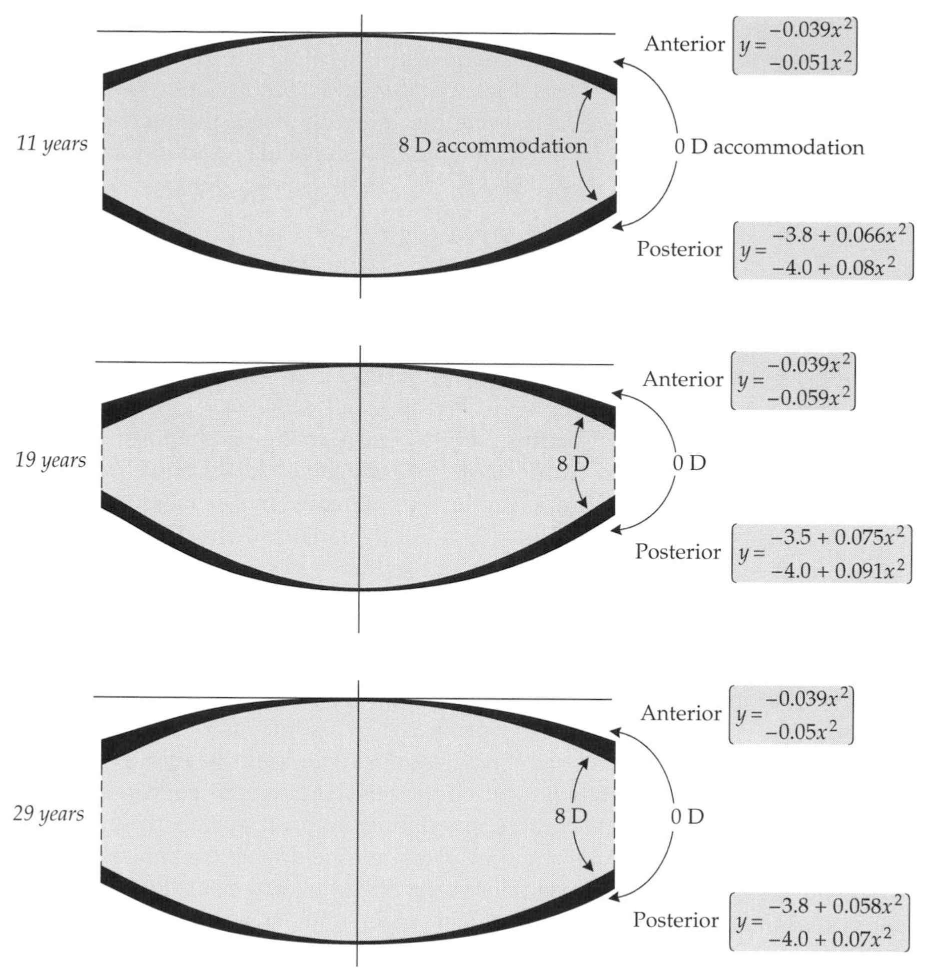

Figure 12.21
Anterior and Posterior Lens Surfaces Increase Their Curvatures with Accommodation
The parabolas that describe the anterior and posterior lens surfaces are shown for the unaccommodated state and for an accommodative stimulus of 8 D for three different ages. The shaded regions show the difference between accommodated and unaccommodated curvatures. Small thickness changes during accommodation are not shown. Both the anterior and the posterior surfaces become more curved with accommodation. The anterior surface is less curved to begin with, and it changes about twice as much as the posterior surface does. (After Koretz, Handelman, and Brown 1984.)

Since the surface equation, $y = a + cx^2$, applies to lenses of all ages and all levels of accommodation that have been examined, the descriptive parameters a and c can be used to compare lens shapes as they vary with age and accommodation.

Both anterior and posterior lens surfaces become more curved with accommodation, but the anterior surface change is larger

Figure 12.21 shows the surfaces from young eyes at two accommodative levels (0 and 8 D). The surfaces become more curved with accommodation, but the anterior curvature always increases more than the posterior curvature. The lens also becomes thicker during accommodation, but that change is not shown here; the anterior pole shifts anteriorly about 0.2 mm during 8 D of accommodation.

Because of the increased curvatures of the lens surfaces, the dioptric power at both surfaces and the overall power of the lens increase, but the curvature changes at the lens surfaces are not enough to account for a refractive change of 8 D. In fact, they account for just over 4 D of the 8 D accommodative stimulus (the actual accommodative response was undoubtedly somewhat less than 8 D, but the surface changes still fall short by several diopters). The remainder must be due to curvature changes at the interior optical zones, but without knowing the refractive indices within the lens, the contributions at individual interior zones cannot be calculated. Even so, the conclusion is unavoidable: The change in lens power during accommodation is due to curvature changes throughout the lens, as the interior curvature changes suggest (see Figure 12.19). Although the data are recent, Gullstrand came to this conclusion about a century ago and incorporated it into his "exact" schematic eye.

The effect of the increased lens thickness during accommodation is less obvious because two opposing effects come into play. By itself, the increased lens thickness acts to *decrease* lens power slightly. This conclusion follows from the equation for lens power:

$$F = F_1 + F_2 - (t/n)F_1F_2 \qquad (12.3)$$

where F_1 and F_2 are the anterior and posterior surface powers, respectively, t is lens thickness, and n is the average refractive index of the lens. As thickness increases, the negative term in the equation becomes slightly larger, thereby decreasing total lens power.

This small effect is offset because the thickness change is manifest as a forward movement of the lens. The posterior pole is against the vitreous, and it cannot shift posteriorly. Therefore, when the lens increases in thickness, the anterior pole moves forward and the depth of the anterior chamber decreases. Reducing the distance between cornea and lens *increases* the effective power of the cornea and lens combination. This, too, is a consequence of Equation 12.3 when F_1 and F_2 are the cornea and lens powers, respectively, and t is the anterior chamber depth. (Making t smaller reduces the negative term in the equation.) This increase in the effective cornea and lens power will be slightly larger than the decrease in effective lens power produced by its increased thickness. Thus the net effect of the changing lens thickness during accommodation is positive, adding to the dioptric increase produced by the changing lens curvatures.

Some older explanations of accommodation invoked surface changes that were not uniform—in particular, an exaggerated bulging of the central portion of the anterior surface and a posterior surface change only near the equator. As the curves in Figure 12.21 demonstrate, however, both lens surfaces change systematically and continuously. Changes confined to a local region on one surface or another either do not occur or are too small to be detected.

The lens thickens with age and its curvatures increase, but unaccommodated lens power does not increase with age

Since the lens grows by adding shells and new shells are added throughout life, all measures of lens size also increase. The increases in equatorial diameter and lens thickness were shown in Figure 12.10; both dimensions increase linearly by about 20 μm per year after age 10.

After the eyeball stops growing at around age 15, the distance between the cornea and the posterior lens pole doesn't change significantly. As a result, the continuing increase in lens thickness produces a steady decrease in anterior chamber depth. As already noted, the decreasing anterior chamber depth increases the effective power of the cornea and lens combination, and the eye's refractive power should therefore increase slightly with age.

If the growth of the physical dimensions of the lens were proportional—that is, if each new shell were uniform in thickness—the increase in lens size would automatically produce increased radii of curvature and reduced optical power. It would be like expanding a small circle to become a larger one; the small circle has a smaller radius and greater curvature than the larger circle. Thus, the increasing thickness of the lens should be accompanied by a *decrease* in optical power for two reasons: Increasing thickness decreases power (as we have seen already), and uniform expansion increases the radii of curvature (decreases curvature).

Measurements by Donders in the 1860s indicated that lens curvatures decrease with age—that is, the radii increase and the lens flattens—and his data were used for many years to explain age-related hyperopia. But more recent

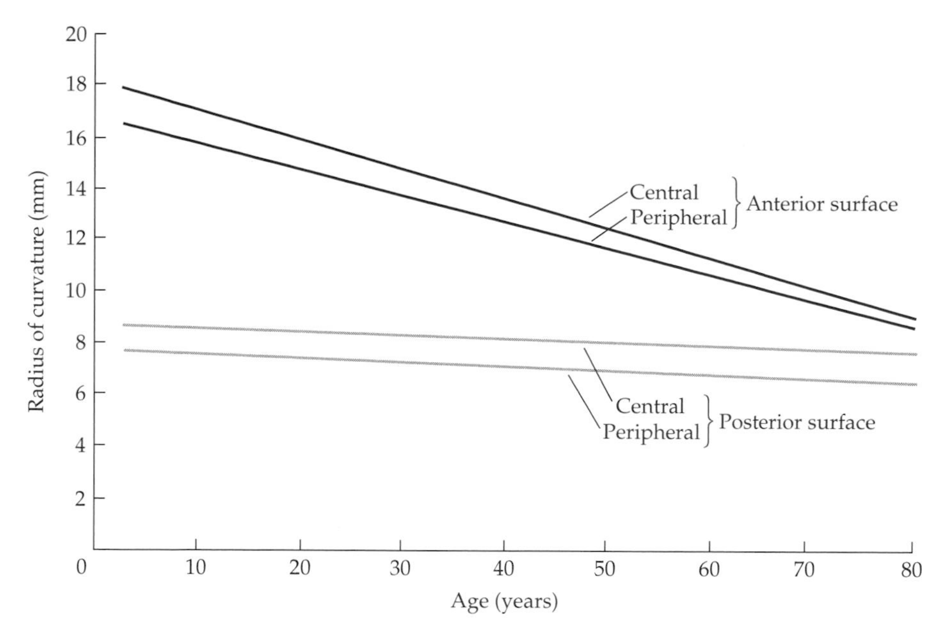

Figure 12.22
Curvature of the Anterior Surface Increases with Age
These regression lines are fitted to data (not shown) relating anterior and posterior lens surface radii of curvature to age. Both the central and the peripheral anterior-surface curvatures increase with age (the radii of curvature *decrease*). The posterior surface does not change significantly. (After Brown 1974.)

studies give a different picture. For example, the regression lines in Figure 12.22 show radii of curvature at the anterior and posterior poles and at locations 3 mm peripheral to the poles (both anterior and posterior) as functions of age. Accommodation was relaxed in all cases by administration of a cycloplegic drug. Contrary to Donders's measurements, all radii appear to *decrease* with age, meaning that the surface curvatures and dioptric powers are *increasing*.

The regression lines for the posterior radii are only marginally different from zero, meaning that the posterior radii do not change systematically with age. The anterior radii change markedly, however, and probably at the same rate (the slopes of the regression lines are statistically indistinguishable). The change in the central radius of curvature is quite large, about 1.2 mm per decade, decreasing 8.5 mm from 17.5 mm at age 10, to 9 mm at age 80. In dioptric terms, the surface power changes from about 4.4 D at age 10 to 8.7 D at age 80. (If the dioptric powers seem small for such highly curved surfaces, remember that the difference in refractive index between aqueous and lens is also small.)

Add to this the effect of decreasing anterior chamber depth with age and we should expect increasing *myopia* with age, amounting to perhaps 5 D from the second to the eighth decade of life. The questions—the so-called lens paradox—are these: How have these eyes remained emmetropic? Why is the increased lens power *not* reflected as age-related myopia?

It is worth noting that the lens paradox exists only because of one experimental study and a reanalysis of the original data; it would be nice to have independent confirmation. But assuming the data are correct, the increasing lens curvatures might be offset by an age-related *decrease* in the refractive index of the lens, corneal flattening, decreased axial length of the eye, or refractive-index *increases* in the aqueous or vitreous. None of these are known to occur, however. The most likely possibility is an age-related decrease in the refractive index of the lens, which might occur as the result of an increase in lens water content with a subsequent reduction in lens protein concentration (refractive index is proportional to protein concentration). Some data indicate that water content of the lens does increase with age, but whether this increase is enough to produce the necessary reduction of protein concentration is not known.

So there it stands: The lens is said to increase in thickness and curvature with age, and if it does, it should increase in optical power. But it doesn't seem to, and we aren't sure why.

The increased lens surface curvatures in accommodation are primarily a consequence of tissue elasticity

The sequence of changes during accommodation have been understood and agreed upon for many years. Contraction of the ciliary muscle from its relaxed state reduces tension exerted by the zonule on the lens, and the lens consequently becomes thicker and more curved. A problem arises, however, when we consider *how* the lens changes shape. What are the forces involved, and how are they generated?

The dominant hypothesis for many years, which is largely attributable to Fincham, designates the lens capsule as the main source of the forces that produce the change in lens shape. In this conception, the capsule is highly elastic, meaning that after it is deformed by tension from the zonule, it attempts to return to its resting state when the deforming force is removed, just as a stretched rubber band does when it is released. The lens nucleus and cortex are plastic, capable of being molded to different configurations by external forces, but retaining any new configuration after the deforming force is removed. In everyday terms, the lens is like a rubber balloon—the capsule—filled with a sticky, viscous fluid—the nucleus and cortex. The external deforming force applied by the relaxed ciliary muscle through the zonule pulls on the lens around the equator and flattens it. When this force is removed by contraction of the ciliary muscle, the elastic recoil of the capsule remolds the lens to its resting shape, which is more highly curved. Where the capsule is thinnest, as at the anterior pole, the change in shape is most pronounced.

This story is quite sensible, which accounts for its long tenure as an explanation of shape change of the lens during accommodation, but it is difficult to test rigorously and it does not fit with some of the anatomy. An obvious shortcoming, discussed earlier, has to do with capsular thickness; there are significant differences in anterior capsule thickness, but they are significant only in presbyopic eyes. Young, actively accommodating eyes show little variation in anterior capsule thickness. Moreover, the surface curvatures of the lens are always parabolic, regardless of accommodative level; there is no prominent bulging at the anterior pole or anywhere else.

Part of the story is correct; the capsule *is* elastic, and its modulus of elasticity has been measured. But recent biomechanical models of the lens can account for changes in lens shape only when the force applied by the capsule to the lens substance is applied uniformly—that is, applied equally to all parts of the lens surface, both anterior and posterior. In technical jargon, the lens capsule is a "force distributor."

What this means in terms of the capsule's function is illustrated in Figure 12.23. In the unaccommodated state (Figure 12.23*a*), the outwardly directed force applied by the zonule around the lens equator is transmitted by the capsule to the lens substance as an *inward* compressive force that molds the lens to a relatively flattened configuration. To exert a compressive force, the capsule must be near its limit of stretch; if this were not the case, the force applied by the zonule would be dissipated in stretching the capsule, rather than being redirected inward to compress the lens.

Removing the outward force applied by the zonule, as in maximal accommodation (Figure 12.23*b*), removes the compressive force transmitted through the capsule, allowing the elastic recoil of the stretched capsule to reshape the lens. In effect, the force is redirected so that it is inward around the equator and outward across the lens surfaces. Thus the lens in accommodation becomes smaller in equatorial diameter, thicker along its anterior–posterior axis, and its surfaces become more curved.

Another modification in the story concerns the elasticity of the lens nucleus and cortex, which were originally conceived as plastic and passive. This conception is not quite correct; the lens substance is also elastic, though less so than the

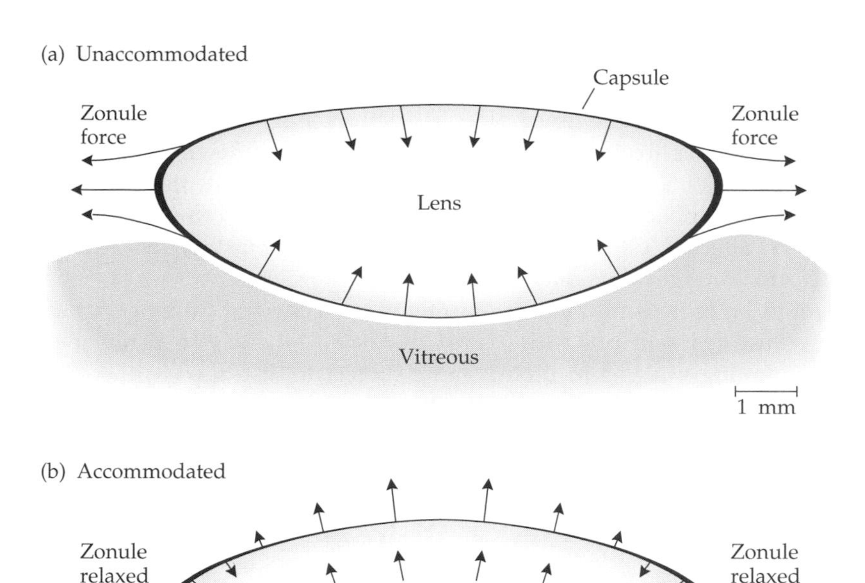

Figure 12.23
Forces That Affect Lens Shape
(*a*) In its unaccommodated state, the outward pull of the zonule on the stretched lens capsule produces an inward force from the capsule on both lens surfaces; this force flattens the lens. (*b*) During accommodation, the zonule is relaxed and the inherent elasticity of both capsule and lens substance are reflected as outwardly directed forces, except at the equator, where the force is directed inward. The lens surface curvatures increase, anterior–posterior thickness increases, and the equatorial diameter decreases. The small arrows within the lens indicate the contribution of internal elasticity. (Lens surface curves derived from Koretz, Handelman, and Brown 1984.)

capsule, and this internal elasticity will also contribute to the increased lens thickness and curvature when the compressive force is removed (see Figure 12.23*b*).

Finally, we need to address the location of the lens in vivo; its posterior surface is in contact with the surface of the vitreous body, where it sits in a bowl-shaped depression—the patellar fossa—on the anterior vitreal surface. The vitreous is mostly water and therefore incompressible, and any force applied to it will be resisted—that is, met with an equal, opposing force. For this reason, the posterior pole of the lens cannot move backward as lens thickness increases during accommodation, and all the thickness change is reflected as forward movement of the anterior pole and decreased anterior chamber depth.

One proposed accommodative mechanism assigns a more active role to the vitreous, such that force applied to the vitreous by the contracting ciliary muscle is transmitted by the vitreous to the lens. Roughly speaking, the vitreous might squeeze the posterior lens to generate the shape change in accommodation. This "hydraulic suspension" theory of accommodation appears to be an interesting idea in search of some evidence that will test the model. A few observations are consistent with the idea, but any contribution by the vitreous is probably quite small.

The presbyopic lens is aging, fat, and unresponsive

Shape changes undergone by young lenses during accommodation are markedly reduced in older lenses. In Figure 12.24, for example, the surface curves for a 45-year-old lens are shown for the unaccommodated condition and during attempted maximal accommodation. This 45-year-old lens ("aging") is almost 0.5 mm thicker ("fat") than younger lenses, and on attempted accommodation, its anterior surface shows very little change ("unresponsive"). Moreover, none of the internal optical zones in the anterior part of the lens exhibit any change. It is

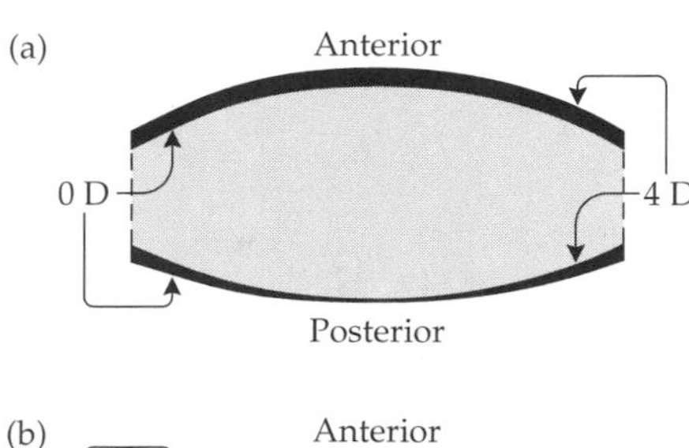

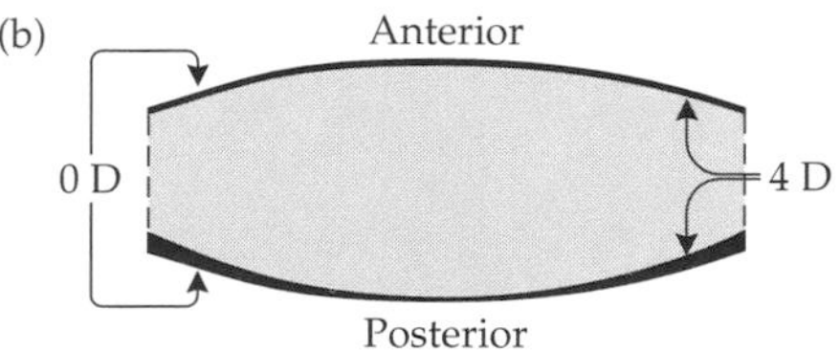

Figure 12.24
Accommodation at the Onset of Presbyopia
The changes in anterior and posterior surface curvatures for a 45-year-old lens with a 4 D accommodative stimulus are shown with the change in lens thickness included (*a*) and without it (*b*). The lens shows a slight increase in both thickness and posterior surface curvature, but these changes are smaller than in younger eyes. The anterior surface does not change appreciably. (After Koretz, Handelman, and Brown 1984.)

as if the anterior lens has become locked in its unaccommodated configuration.

The same cannot be said of the posterior surface, however. Its curvature increases slightly with attempted accommodation, although the change is smaller than in younger, freely accommodating lenses. This difference between anterior and posterior lens surfaces is curious, and therefore interesting; if the lens has become less pliable with age, it is not obvious that its front and back halves should be affected differently. (Differential loss of plasticity is more often assumed to occur between nucleus and cortex.)

I would be more comfortable with this story if there were more data. Conclusions based on only one lens—as in this case—are always questionable, and the Scheimpflug imaging method is more subject to distortion at the posterior lens surface. But these results are the only indication that forces imposed on the lens differ between the anterior and posterior surfaces, thereby hinting that external pressure from the vitreous *might be* involved in posterior surface changes during accommodation. If the lens is becoming less pliable, internal elastic forces may eventually be insufficient to change lens shape on either surface, but some change in the posterior surface may persist in presbyopia because of the undiminished force transmitted through the vitreous to the lens from the contracting ciliary muscle. If so, we have an indication that the vitreous does play a minor role in accommodation and another piece of evidence that ciliary muscle contraction persists in presbyopia despite the failure of the lens to respond.

Before turning to a final consideration of the failure of the lens to respond as it ages, one more piece of information needs to be added; it has to do with the lens becoming fat. Not surprisingly, the progressive growth of the lens increases not only its physical dimensions but also its weight, as Figure 12.25 shows.

During the period when accommodation is declining—say from age 10 to age 50—the lens increases in bulk and mass, gaining weight steadily from about 160 mg at age 10 to 220 mg at age 50—nearly a 40% weight gain. Increased size is accompanied by increased weight, just as one would expect, but it is not clear that internal elastic forces increase proportionally. An older, heavier lens, simply

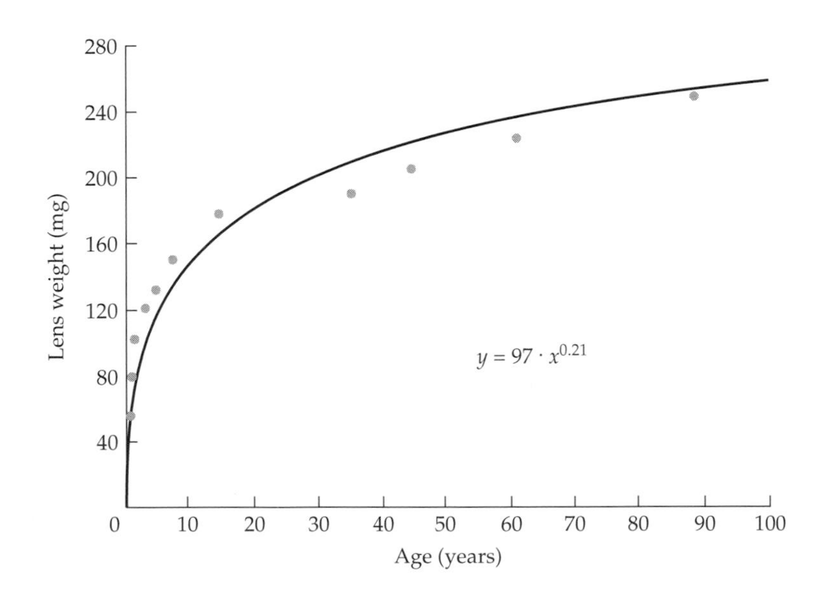

Figure 12.25
Lens Weight as a Function of Age
The weight of the lens increases exponentially with age. The curve is an exponential function where weight = 97 · $\text{age}^{0.21}$; the equation is appropriate for 1 year and older. (In logarithmic coordinates, the increase in weight is a straight line whose equation is log weight [mg] = 1.99 + 0.21 log age [years].) (After Harding and Crabbe 1984.)

by virtue of increased bulk and mass, should be more difficult to reshape than a younger, lighter lens. I don't know of a simpler explanation for accommodative loss with age.

Presbyopia is largely, if not solely, associated with age-related changes in the lens

A point made in Chapter 11 is worth repeating; namely, presbyopia is not something that happens suddenly. By age 45, most people realize that their near point of accommodation is farther away than the normal reading or working distance, but only the realization is sudden. Accommodative amplitude has been declining steadily for the past 40 years or so (see Figure 11.23), and any explanation of presbyopia must account for this long-term decline. As discussed in Chapter 11, the evidence for age-related ciliary muscle failure is outweighed by evidence that the ciliary muscle can contract after accommodative amplitude has been reduced to zero. And because there is no reason to suspect a progressive decline in the innervational commands to the accommodative system, we are left with the lens and the zonule as the possible sources of accommodative decline.

Although the geometry of the zonule or the geometry of its insertion on the lens may change with age and contribute to presbyopia, the argument for the zonule as the cause of presbyopia is weak. The zonule's insertion does shift relative to the lens equator (see Figure 12.16), but the changes are relatively late and are not evident until accommodation is almost gone. The geometry of the zonule itself—that is, the origins and insertions of the bundles of zonular fibers and the angle at which they approach the lens—is not helpful for at least two reasons. First, the zonule changes as the anterior segment of the eye grows—the posterior attachments to the ciliary body, for example, shift posteriorly relative to the lens equator—but these are early changes, complete by age 10 or so, and they cannot account for the subsequent accommodative decline. Second, it is not clear that the zonule has an ordered structural arrangement that is consistent in detail from eye to eye (see Chapter 11). Any hypothesis about accommodation or presbyopia that includes zonular "geometry" as a key assumption is probably not realistic.

Which brings us to the lens. Of the different structures that could contribute to presbyopia, only the lens grows and changes continuously throughout the 50 or so years of accommodative decline. From birth onward, the lens increases in diameter and thickness (see Figure 12.10) and in volume and mass (see Figure 12.25), it adds many new cells (see Figure 12.12), it changes in curvature (see Figure 12.22), and it probably changes in refractive index. There is a strong—indeed, unique—association between lenticular change and accommodative decline.

To be sure, a strong association does not necessarily mean that two variables are causally related. Pupil size decreases as accommodative convergence increases (see Chapter 10), for example, but the association is due to the mutual linking of pupil size and convergence to accommodation, not to a linking between pupil size and convergence. People with higher levels of education are less likely to develop cataracts than those with modest education (see the next section), but the causal variable is occupational exposure to sunlight. The association between age-related lens changes and accommodative loss is certainly strong, but it could be as spurious as the association between cataracts and lack of education.

But that said as a cautionary note, the association between lenticular change and accommodative loss is difficult to dismiss. For one thing, there are no other respectable candidates for the role of causal variable. For another, it is easy to suggest a plausible link between lens change and accommodative loss. The suggested link, which is not a new idea, is simply that as the lens increases in bulk

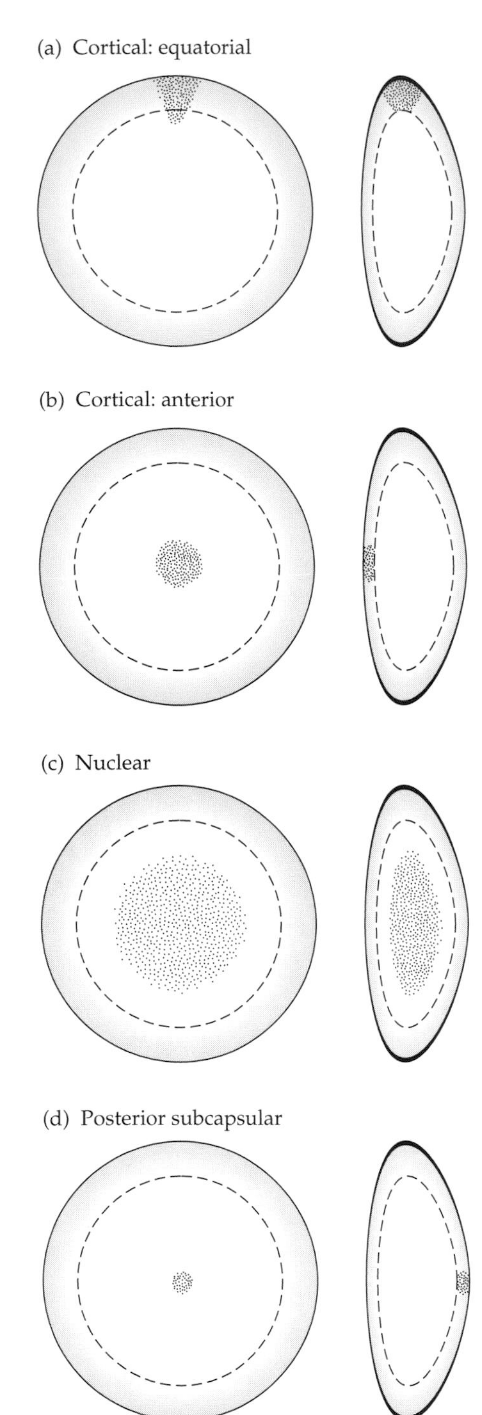

Figure 12.26
Catacts Classified by Location

and mass, the inherent elastic forces are rendered less effective in changing lens curvature. An ant may be able to move a small rubber tree plant, but as the plant grows, a time comes when the ant can't (as the old song says). In addition, there is some evidence that elasticity decreases with age, a change that would also decrease the capacity to alter curvature.

Unfortunately, telling details are scarce. Elasticity and plasticity are probably no more uniform within the lens than refractive index is, and the mechanical properties are at least as elusive as the optical one. Biomechanical models of the lens are helpful in putting constraints on the mechanical properties, but they also await the test of experimental data. Of the various hypotheses, the simplest is that the lens is the guilty party in the matter of presbyopia (Occam's razor again; see Chapter 7). It is guilt by association, however, based on circumstantial evidence; this issue has yet to be settled conclusively.

Cataracts, most of which are age-related, take different forms and can affect any part of the lens

Cataracts—opacities in the lens—occur at all ages from birth through adulthood. Congenital and juvenile-onset cataracts are not common, however; the vast majority of cataracts appear after the age of 50. The incidence of cataract is so strongly associated with advancing age that almost everyone over the age of 70 has some degree of transparency loss in the lens. Worldwide, cataracts account for almost half the cases of blindness, and they know no boundaries; all people—regardless of race, sex, national origin, or socioeconomic status—are potentially at risk.

Cataracts differ in optical density, shape, color, size, and location, creating difficulties in how we think about them because it is not always clear which of these differences are important in terms of the cataract's etiology. This discussion uses the simplest subdivision that seems meaningful—a coarse subdivision by location: Cataracts are cortical, nuclear, or posterior subcapsular.

Most age-related cataracts are **cortical cataracts**, in which the initial opacity is confined to the outer lens shells. A cortical cataract can be peripheral in the lens, which makes it an equatorial cataract (Figure 12.26*a*) or located more centrally near either the anterior (Figure 12.26*b*) or the posterior pole. The second most prevalent group consists of **nuclear cataracts**, in which the opacity first appears in the center of the lens (Figure 12.26*c*). A third, less common group consists of **posterior subcapsular cataracts** (Figure 12.26*d*), which lie in the outermost layers of the cortex at the posterior pole. The prevalence of cortical and nuclear cataracts in the population varies with age. Cortical cataracts are more common up to about age 65; nuclear cataracts dominate thereafter.

Cataracts absorb and scatter more light than do normal regions of the lens, thereby producing more light spread and less contrast in the retinal image. The result is reduced visual acuity, and acuity loss is related to the density and extent of the opacity. Small equatorial cataracts are the only sort to leave acuity unaffected; because they are peripheral, light entering through the normal, undilated pupil does not encounter the cataract. Any other cataract does affect acuity, however; nuclear and posterior subcapsular cataracts are the major offenders. In time, any cataract may expand to involve the entire lens, reducing vision to just the perception of light. Cataract surgery is discussed in Box 12.1.

Because cataracts are so common and affect so many people, efforts to understand how they form and to delineate the underlying causal factors have generated an enormous, often confusing body of literature. For example, a recent review of epidemiological studies listed six major categories of risk factors for cataracts, repeated here in the order given: social and personal factors, ultraviolet radiation, diabetes, diarrhea, antioxidants, and drugs. The social and personal factors included education, gender, smoking, alcohol, and hypertension. (A risk factor, incidentally, is not necessarily something that *causes* cataract, although a risk factor may turn out to be a causal factor. Risk factors increase the likelihood of cataract by enhancing the effects of causal factors.)

Epidemiological studies are difficult, as demonstrated by the silly conclusion that incidence of cataract is inversely related to educational level. The commonsense explanation is that people with little education more often end up in outdoor jobs; that is, the real issue is occupational exposure to sunlight—not education. Statisticians know this, of course, and they use methods designed to factor out such extraneous variables. But the fact that education level keeps popping up as a factor in cataractogenesis is probably a sign that the statistical models are inappropriate for the data sets.

Problems of this sort are unavoidable. Trying to remove the influence of the dominant variable is difficult, perhaps impossible, and it may be difficult to measure the dominant variable in any meaningful way. In this case, the dominant variable is likely to be sunlight, something to which we are all exposed in varying degrees and something for which the degree of exposure is difficult to specify.

All the major and minor risk factors notwithstanding, the argument has been made that most cataracts in adults are due to the cumulative effects of exposure to sunlight, specifically to the ultraviolet component of solar radiation.* In this view, cataracts are exaggerations of normal transparency loss, where the exaggeration has come about by greater-than-normal exposure to sunlight or by other factors giving normal exposure a greater-than-normal impact.

As mentioned earlier, absorption of high-energy photons can not only break chemical bonds directly but also break them indirectly by increasing normal rates of oxidation or by producing heat through reradiation. All of these effects alter protein structure, causing the native monomeric or dimeric forms to link together as large, insoluble aggregates. These insoluble aggregates reduce the normal regularity of protein packing in the lens cells and thus reduce transparency. Nuclear and cortical cataracts appear to form by this process of protein alteration from UV absorption; the main difference between them being the longer time period required for the effects to accumulate in the nucleus.

Posterior subcapsular cataracts differ from the others in their mechanism of formation. Nuclear and cortical cataracts are alterations occurring within normal lens fibers, but posterior subcapsular cataracts contain abnormal cells. Interference with lens cell production at the equator produces lens fibers that do not elongate properly, and they migrate to the posterior pole, where their irregular structure creates an opacity. But while this process differs from the opacification in nuclear and cortical cataracts, the causal agent appears to be the same. Here, however, absorption of UV light interferes with epithelial cell differentiation and protein synthesis rather than with the protein structure in mature cells. (One piece of related evidence is the unusually high incidence of posterior subcapsular cataracts among survivors of the atomic bomb blasts at Hiroshima and Nagasaki. Atomic radiation interferes with cellular DNA more readily than ultraviolet radiation does.)

The lens normally changes with age, not only in gross features of size and weight but also at the level of molecular structure. Much of this normal senescence may be related to ultraviolet radiation exposure, and it is not unreasonable to think of cataractogenesis as an acceleration or exaggeration of normal lens aging. Or to put it another way, we will all have some degree of lens transparency loss if we live long enough; those of us who have had the highest expo-

*Much of this discussion is derived from *Age-Related Cataract* by R. W. Young (1991). The author, who is a basic scientist, draws a clean line through a messy clinical problem, but the book is rarely cited; the specialists may feel that he doesn't understand the nuances and complexities of their subject. I found it the best thing I read on the subject, but I'm not really a clinician either. Among other things, Young pioneered the use of radiolabeled amino acids to study cellular regeneration and renewal processes in the eye, including the lens, and was the first to demonstrate shedding and renewal of the photoreceptor outer segment (see Chapter 13).

[Box 12.1] Cataract Surgery

CATARACTOGENESIS OPERATES AT the molecular level, and cataractous lenses are removed because of a failure of biochemistry, not because of a failure of anatomy. Other than the increased light scatter and absorption that characterize these opacities, cataracts in their early stages do not affect lens morphology. Only when cataracts become mature do structural changes appear; lens fibers develop vacuoles, lens shells separate, the epithelium breaks down, and the lens substance may eventually liquefy. Where there is good access to medical care, however, cataractous lenses are rarely allowed to proceed this far.

There is no cure for cataract. No therapeutic agent in use or development has been shown to halt or reverse cataractogenesis. And given the nature of the molecular changes in cataract formation, the prospects for finding a therapeutic agent are exceedingly poor. Returning the aggregated lens proteins to their original form by undoing cross-linking produced by UV exposure or other causes might by possible in theory, but doing so without adverse effects to other proteins in the lens, the eye, and elsewhere seems unlikely.

The ancient method of "couching" a cataract was a lens displacement technique; the lens was not removed from the eye, but after being broken free of the zonule, the lens was moved down and back into the vitreous chamber, away from the optic axis. Lens extraction was not performed routinely until the nineteenth century, and the procedure was an *intracapsular* extraction wherein the lens was removed from the eye with the capsule intact. Keeping the capsule intact was an important element in successful surgery. Lens proteins that escape from the capsule have been so long isolated from the body's immune system that they may be the target of severe immunological responses, generating inflammation throughout the eye and possibly leading to blindness.

Nevertheless, modern cataract surgery is almost always *extracapsular.* While the technique leaves the posterior capsule in place, it also ensures that all other parts of the lens are removed completely. The procedures have become highly refined in the last 20 years or so, but surgical methods are still evolving, with numerous themes and variations employed by different surgeons; only the basic elements will be considered here.

Figure 1 schematically illustrates an extracapsular lens extraction. An incision is made either through the peripheral cornea or through the posterior limbus, behind the canal of Schlemm but in front of the iris root. A viscoelastic material is often injected into the anterior chamber to keep the chamber inflated and provide working room between the lens and cornea. A cutting tool inserted through the incision is used to open the anterior capsule and expose the interior of the lens. Most of the lens removal is done with a small probe that has two functions: It fragments the lens tissue with ultrasound and removes the fragments through a suction line. In most procedures, first the anterior cortex and the nucleus are carved out by fragmentation and aspiration, and then the remaining cortex is removed. The latter step is a more delicate procedure, since the goal is to leave the posterior capsule intact.

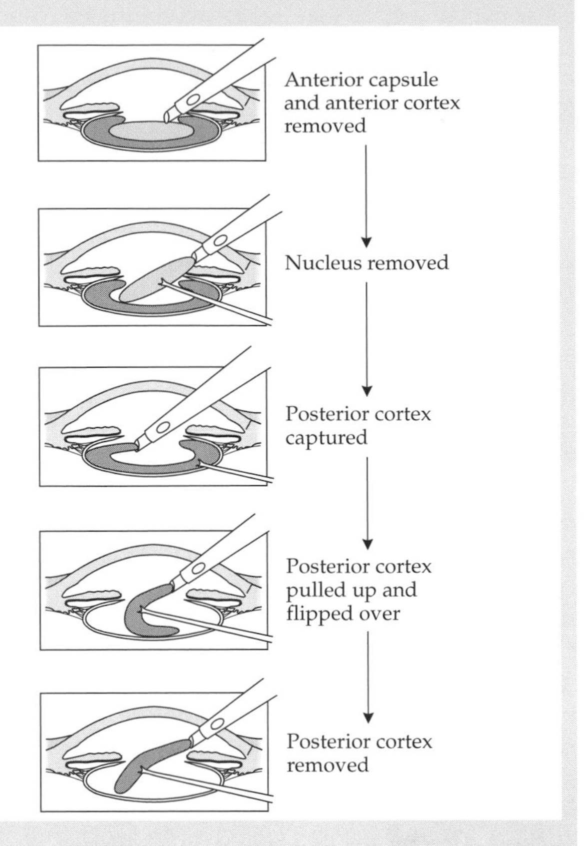

Figure 1
Lens Extraction
Lens extraction is done in several distinct stages. After the anterior capsule and anterior cortex are removed, the nucleus is separated from the posterior cortex and removed. The posterior cortex is removed by the lifting of one edge; when flipped over, it peels away from the posterior capsule, which remains in place. There are numerous variations on this basic procedure. (After Jaffe and Horwitz 1992.)

An eye without a lens is **aphakic** and severely hyperopic. Correction of the hyperopia with spectacles is not a good solution, however, because the very high power lens in the spectacle plane magnifies the retinal image considerably and introduces aberration and distortion. The unequal image sizes in the two eyes (aniseikonia) produced by spectacle correction of monocular aphakia may render normal binocular vision impossible. A contact lens works better; it is much closer to the anterior nodal point of the eye, so it magnifies the image less, and it is less annoying than a heavy, cumbersome spectacle correction. This is a good solution to an aphakic correction if the patient can adapt to contact lens wear. Unfortunately, contact lenses require reasonable manual dexterity for insertion and removal, as well as careful cleaning and handling, and they may exaggerate symptoms in patients with dry eye (see

Chapter 7). Some elderly cataract patients may find these requirements difficult to meet.

Because spectacle and contact lens corrections of aphakia are so often unsatisfactory, most extracapsular lens extractions are combined with insertion of a plastic or silicone intraocular lens. Intraocular lenses are made in several different designs—for insertion in the anterior chamber or, more routinely, for insertion in the posterior chamber in roughly the same position as the natural lens. Figure 2*a* shows one of the posterior chamber lenses. The lens is about 6 mm in diameter and has two springy arms that hold the lens in position (Figure 2*b*). The lens arms press outward against the ciliary body just anterior to the ciliary processes; this pressure and support from the intact posterior capsule keep the intraocular lens stable and centered with respect to the pupil and the visual axis.

In a recent variation, intraocular lenses have been made of flexible silicone. They can be rolled up as a small tube, thereby requiring only a very small incision for their insertion, after which the lens opens up inside the eye to its original shape. (A rigid lens requires an incision slightly larger than the lens diameter.) Although there have been some questions about the stability and optical quality of the flexible lenses, the small incisions make for less traumatic surgery and they do not require sutures for closure (cyanoacrylate glue—"crazy glue"—is often used instead of sutures).

Penetrating the eye for any reason risks complications of various kinds—bleeding into the anterior chamber or vitreous, inflammatory reactions, abnormal tissue adhesions, and so on—and cataract surgery is not immune to these sequelae. Nor is the technique as simple as it has been outlined here. But current success rates, without complications, are very high, and the surgery—usually done on an outpatient basis—is fast; in 20 minutes or so, an eye that is functionally blind from cataract can be restored to normality.

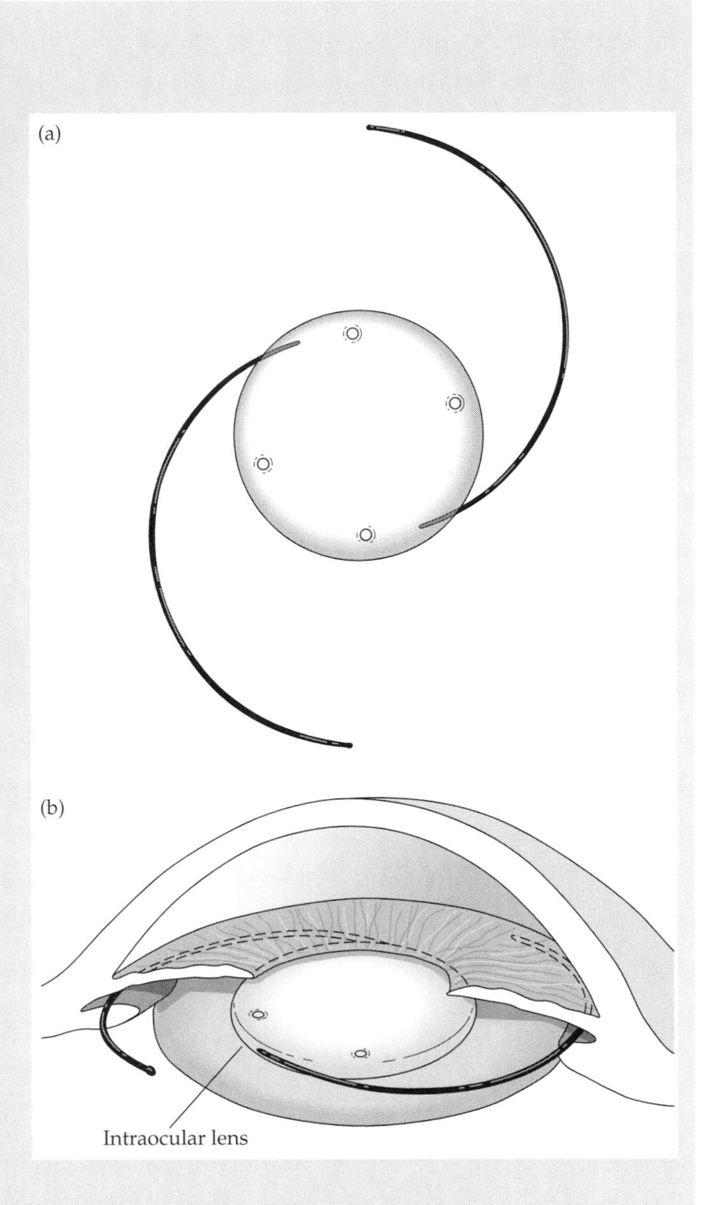

Figure 2
An Intraocular Lens
(*a*) There are different designs of intraocular lenses, but most have some form of springy arms attached to the lens to hold it firmly in place. The small holes in the lens allow it to be hooked and maneuvered. For a posterior chamber lens, the arms press against the ciliary body just behind the root of the iris (*b*). (After Jaffe and Horwitz 1992.)

sures to ultraviolet radiation will have more and earlier transparency loss, as will those who have been exposed to other accelerants of biochemical aging.

This is where other risk factors come into play. Trauma increases the risk of cataract formation, as do some metallic salts (e.g., cobalt and selenium), some ocular drugs (particularly cholinesterase inhibitors), and some systemic drugs (corticosteroids and chlorpromazine). Smokers may be more subject to nuclear cataracts, and heavy alcohol consumption may increase the risk for all types of cataract. Abnormal blood sugar levels increase the risk of cataract, and diabetics have been said to be at higher risk, but the evidence for increased prevalence of cataracts among controlled diabetics is contradictory.

Because of the numerous risk factors that may bear on the problem, cataract is often described as "a multifactorial disease," thereby implying that it's all very complicated and mysterious, and perhaps it is. But this description may obscure rather than illuminate the dominant factor, which is ultraviolet exposure, its contribution to normal senescence, and its cumulative impact on lens biochemistry.

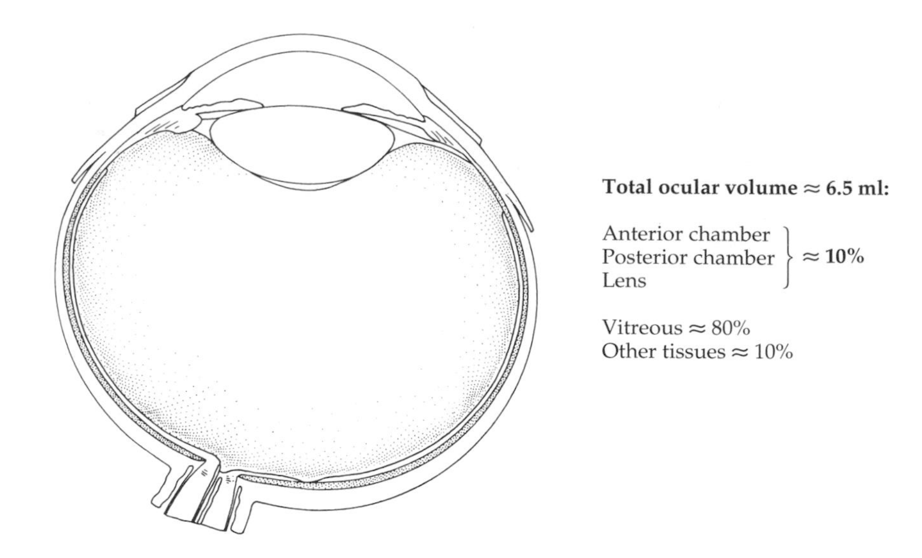

Figure 12.27
The Vitreous
The vitreous is shaded in this cross section through the eye. It occupies about 80% of the eye's total volume and is therefore the largest component of the eye.

Although some loss of lens transparency is an inevitable fact of life, transparency loss to the point of cataract formation may be preventable. Like the production of basal and squamous cell carcinomas in the skin, the effect of sunlight in cataractogenesis is cumulative; preventative measures must begin early and be practiced routinely. UV-blocking sunglasses are an obvious part of a solution.* As a global public health issue, however, the problem is making sunglasses acceptable and accessible to populations at high risk.

The Vitreous

The vitreous is the largest component of the eye

Eighty percent of the interior volume of the eye is vitreous (Figure 12.27), and the longest part of the optical path from the cornea to the retina is through the vitreous. The vitreous not only is in contact with the retina, ciliary body, and lens, but it also has structural and functional interactions with all of them.

None of these facts would be of the slightest interest if the vitreous were homogeneous, featureless, and inert. As long as it remained transparent and provided mechanical support for surrounding structures, it would be easy to ignore. But the vitreous has structure, it is not always transparent, and it is not by any means an inert glob of gelatin. As I write this, a large, translucent floater in the vitreous of my left eye is drifting back and forth across my central visual field. The floater is a result of a posterior vitreal detachment and I mention it here as evidence that the vitreous has some structure worth talking about.

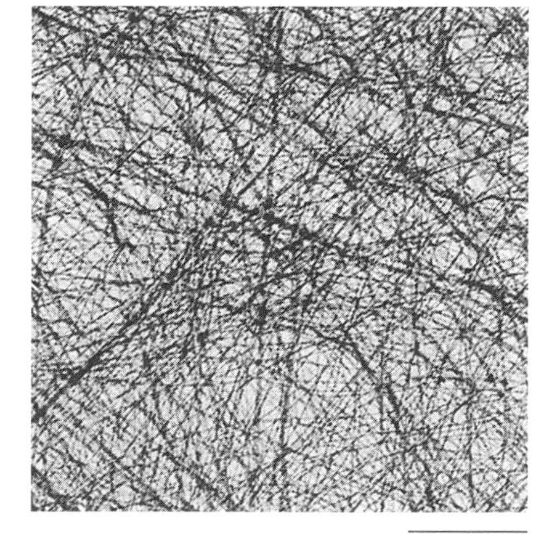

Figure 12.28
Vitreous Collagen
An electron micrograph of the vitreous shows that it has a fibrillar structure. The fibers are collagen. (From Balazs 1994.)

The primary structural components of the vitreous are collagen and hyaluronic acid

As Figure 12.28 shows, the vitreous has fibrils in it; higher magnification would show the characteristic banding of collagen. Immunohistochemical labeling and measures of amino acid content indicate that most of the vitreal collagen is type II, a close relative of the collagen in cartilage. Both type IV and type IX collagen may also be present as very minor components.

*The vision clinic associated with the institution in which I work requires that all spectacles prescribed, sunglasses or not, have UV-blocking coatings. Even if UV absorption by the lens is not the only factor in cataract formation, reducing the amount of ultraviolet exposure by this simple expedient makes very good sense.

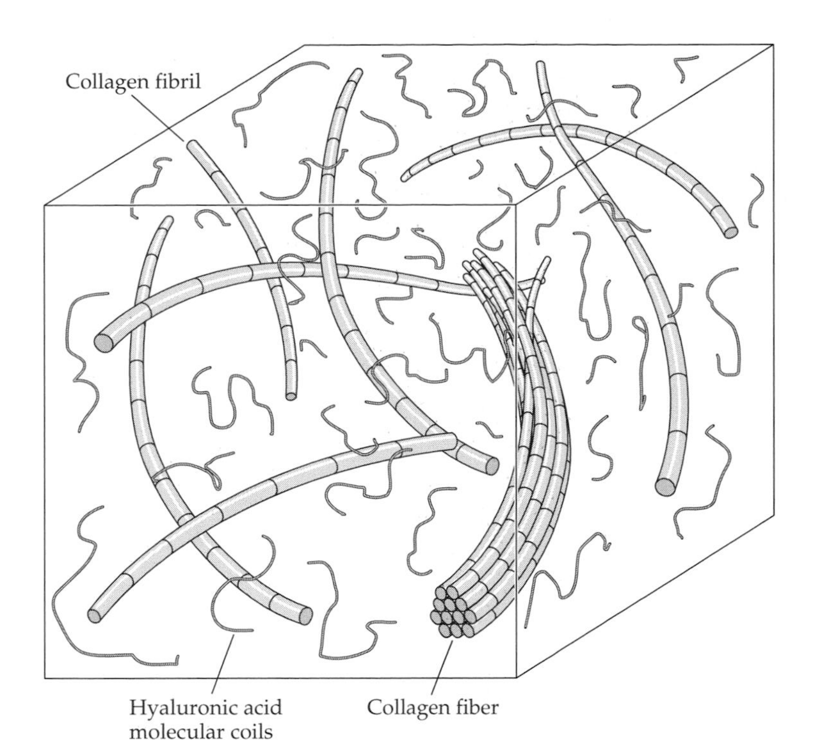

Figure 12.29
Vitreous Collagen and Its Matrix
The vitreous collagen fibrils run in different directions, sometimes aggregating as larger fibers or fiber bundles. The proteoglycan matrix contains the glycosaminoglycan hyaluronic acid; most of the vitreous volume is the matrix and its bound water. (After Sebag and Balazs 1989.)

As in the cornea and other collagen-based tissues, vitreal collagen is associated with a proteoglycan matrix—that is, core proteins to which glycosaminoglycans are linked (see Figure 8.3). In the vitreous, the core proteins are chondroitin and heparan sulfate, which are present in other tissues; the glycosaminoglycan is **hyaluronic acid**.

In terms of vitreal structure, the important features of hyaluronic acid are a tremendous affinity for water and a very large size for the hydrated molecule. Thus hyaluronic acid and water are likely to constitute a large part of the total vitreous volume. The three-dimensional structure of hyaluronic acid is thought to be a long helix or coil whose tightness of coiling varies with water content and some ionic concentrations. Since the volume of a fully hydrated hyaluronic acid molecule is several thousand times larger than the unhydrated form, any changes in vitreous volume are likely to be attributable to the hyaluronic acid, whose volume is very much larger than the volume of the collagen with which it is associated.

The elemental structure of the vitreous consists of long collagen fibers running in different directions, each fiber wrapped with proteoglycan matrix. As Figure 12.29 illustrates, the hyaluronic acid is dehydrated, so there appears to be considerable space between fibers in the collagen meshwork. Normally, the open space is filled with water and the expanded hyaluronic acid molecules. But the fact that there is so much empty space in the structure emphasizes an important point: The vitreous has structure, but the main component of the vitreous, some 99% by weight, is water.

Thus the structure of the vitreous is water bound to a delicate, widely spaced meshwork of collagen and the space-filling hyaluronic acid in the proteoglycans. The collagen and hyaluronic acid give the vitreous its viscous, gel-like character and distinguish it from the equally watery aqueous. The possibility of changes in the gel over time suggest that the fundamental character of the vitreous changes, a point to be considered later, and differences in the density of the collagen meshwork may give the vitreous structure at a coarser scale, which will be considered next.

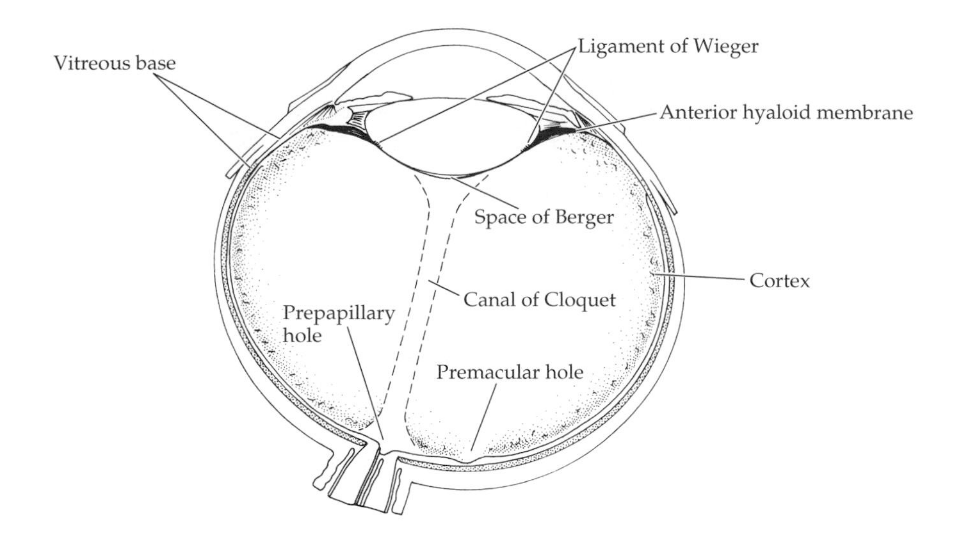

Figure 12.30
The Vitreous Cortex and Other Structural Elements
The outermost part of the vitreous is its cortex; collagen fiber density is highest here, particularly anteriorly, where the collagen forms the anterior hyaloid membrane. Collagen fibers are attached to the retina, to the pars plana at the vitreous base, and to the lens at the ligament of Wieger. The lens rests in a depression in the anterior surface of the vitreous, the patellar fossa, with a potential space (of Berger) between the lens and the vitreous. The canal of Cloquet once contained the embryonic hyaloid artery and is the location of the primary vitreous in adult eyes. The vitreous cortex is interrupted over the fovea (the premacular hole) and the nerve head (the prepapillary hole).

The external layer of the vitreous—the vitreous cortex—attaches the vitreous to surrounding structures

The structure at the outer vitreal surface is different from the inner vitreous. In particular, the density of collagen near the surface is considerably higher, although it is not highly organized; this outer layer of the vitreous is the **vitreous cortex**. The thickness of the cortex varies; it is around 200 μm thick on average, but is generally thicker on the anterior vitreal surface next to the ciliary body and lens than on the posterior surface next to the retina. The collagen at the outer surface of the cortex merges with the basement membrane of the ciliary epithelium (see Chapter 11) and the inner limiting membrane of the retina (see Chapter 13). Thus, the vitreous cortex is attached to the structures just outside the vitreous.

The external structures and the locations of the most prominent vitreal attachments are shown in Figure 12.30. The vitreal collagen is most dense in the anterior cortex. Its fusion with the basement membrane of the pars plana epithelium and the inner limiting membrane of the anterior retina is particularly strong, which has led to this region being called the **vitreous base**. The attachment is so firm that it is difficult to separate the vitreous from the pars plana by dissection. Dense collagen in the cortex extends forward and inward from the vitreous base to form a distinct layer—the **anterior hyaloid membrane**—that is commonly seen in light microscopic sections; it constitutes the anterior surface of the vitreous.

The posterior surface of the lens contacts the vitreous within a shallow, bowl-shaped depression called the **patellar fossa**. In young eyes, but less so in older ones, the anterior hyaloid membrane within the fossa attaches to the lens capsule in an annular region that is roughly coincident with the insertion of the posterior zonular fibers—this attachment is **Wieger's ligament** (also known as Egger's line, or the hyaloideocapsular ligament). Wieger's ligament is not known to be functionally significant, but in company with the denser collagen in this region and the cupping of the posterior lens by the vitreous, it lends some credibility to the idea that vitreal pressure on the lens plays a minor role in accommodation.

In histological sections, there is often a small separation between the posterior lens capsule and the surface of the patellar fossa. This is the **retrolental space of Berger** (or of Erggelet), but it is not clear if this space exists in vivo or if it is an artifact of tissue shrinkage during histological processing. Characterizing it as a "potential space" that opens whenever there is tissue shrinkage, for whatever reason, is probably more realistic. Similarly, there is said to be a space between the anterior hyaloid membrane around the patellar fossa and the most posterior strands of the zonule. If it exists in vivo, this **canal of Petit** would be an annular slit running around the lens near the equator. Again, it is difficult to be sure that the canal of Petit is normally present; it may appear only after some tissue shrinkage. Its presence as a potential space, however, indicates that the vitreous cortex does *not* attach to the zonule to any significant degree.

The cortex has ringlike attachments to the retina around both the fovea and the optic nerve head, but it is absent over these structures, creating what are called the *premacular* and *prepapillary holes.* Although the vitreous cortex has some attachment to the inner limiting membrane over the entire retina, except for the fovea and the optic nerve head, the attachments at the margins of the premacular and prepapillary holes are the strongest.

Inhomogeneity of the vitreous structure produces internal subdivisions in the vitreous

The most obvious internal region of the vitreous, which can sometimes be seen ophthalmoscopically, is the **canal of Cloquet** (see Figure 12.30), which runs through the center of the vitreous from the base of the patellar fossa to the optic nerve head. (The expansion of the canal at the nerve head is called the area martegiani.) The canal of Cloquet marks the site of the embryonic hyaloid artery and the first vitreous to form in the eye; the vitreous within the canal is therefore the *primary* vitreous, and the remainder, outside the canal, is the *secondary* vitreous. In adult eyes, primary vitreous lacks collagen fibers and is separated from secondary vitreous by fine remnants of the hyaloid artery wall.

Because of its transparency, internal subdivisions and structures of the vitreous are difficult to visualize, but it can be done, as with lens or cornea, by taking advantage of the small amount of light scatter from internal structures. The technique uses illumination of the isolated vitreous by a slit of light under conditions suitable for dark-field microscopy. After the sclera, choroid, and retina are dissected away, the vitreous—still attached to the anterior segment of the eye—is suspended in a physiological solution that maintains normal hydration. A beam of light is shone through the vitreous parallel to the anterior–posterior axis, and any light scatter from structural elements can be seen when viewed from the side (i.e., perpendicular to the incident slit of light) against a black background (Figure 12.31*a*).

When illuminated in this way, bright strands can be seen coursing through the vitreous from front to back. The strands are too large to be individual collagen fibers; they are probably fiber bundles. In general, there are more of these bundles in the outer part of the vitreous nearer the cortex than in the central region near the canal of Cloquet. When examined at high magnification, the bundles appear to run roughly parallel to the vitreal surface from the vitreous base back to the region around the fovea. The low collagen density over the fovea produces the premacular hole in the fiber bundle pattern (Figure 12.31*b*), as does the lack of collagen in front of the nerve head. The obvious ring around the premacular hole is the site of attachment to the retina.

The vitreous changes with age

The parts of the vitreous that lack collagen fibers and the associated hyaluronic acid matrix are more liquid and less gel-like in character than the collagen-containing portions. These liquid and gel components are separable by filtration, allowing one to assess the relative amounts of liquid and gel components in the vitreous.

At birth and for some time thereafter, the vitreous has no liquid component, but the liquid portion is measurable by age 2 and increases steadily throughout life (Figure 12.32*a*). The gel volume shows a rapid early increase, 3 or 4 decades

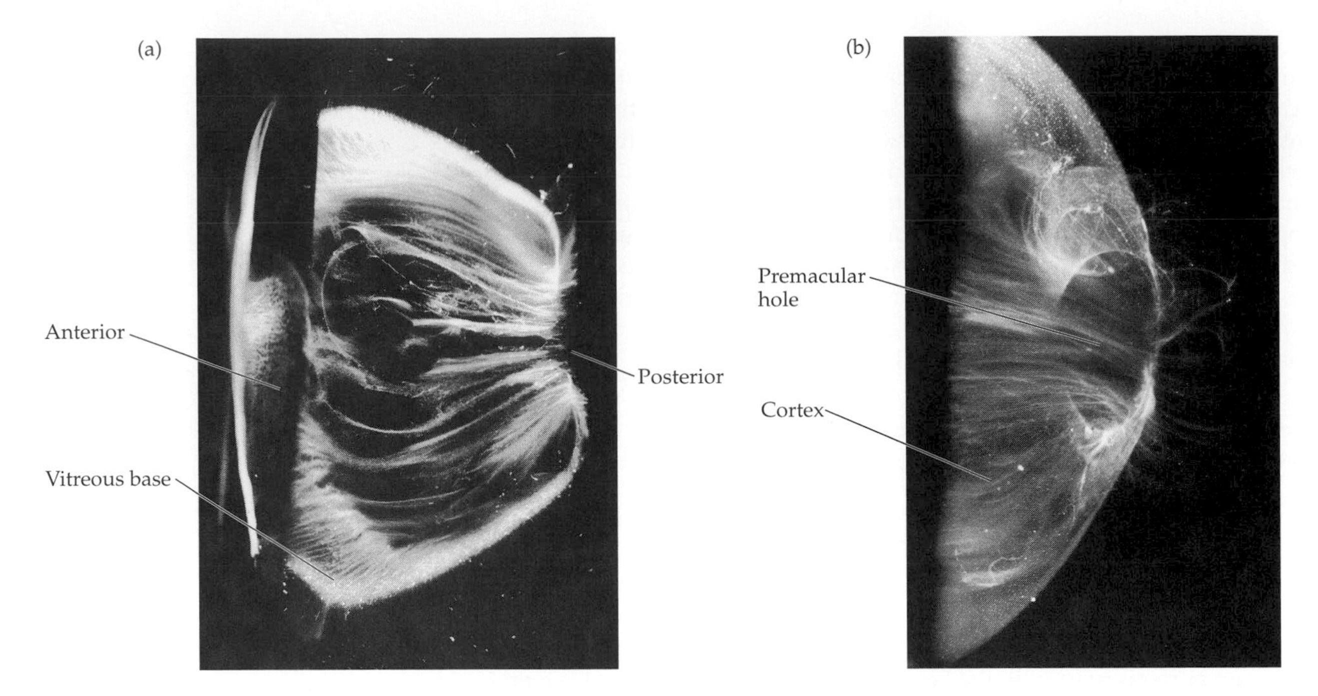

Figure 12.31
Fiber Bundles in the Vitreous Cortex
(*a*) When the isolated vitreous is illuminated by a slit of light against a dark background, bundles of collagen can be seen running through the vitreous cortex from the vitreous base back toward the posterior pole. (*b*) At higher magnification, the premacular hole and the ring of dense collagen around it show clearly. (*a* from J. Sebag and E. A. Balazs. 1985. *Trans. Ophthalmol. Soc. U.K.* 104: 123; *b* from J. Sebag and E. A. Balazs. 1984. *Surv. Ophthalmol.* 28 (Suppl.): 493.)

of stability, and a pronounced decline late in life. These data have been used as evidence that vitreous liquefaction is a phenomenon of the aging eye because of the decline in the gel portion in later years, but it overlooks the steadily increasing volume of liquid vitreous and the overall increase in total vitreous volume. Replotting the data to show the *relative* volumes of gel and liquid vitreous gives a different picture (Figure 12.32*b*).

At 2 or 3 years of age, when a liquid component can first be measured, about 95% of the vitreous is a collagen gel and only 5% or so is liquid. But as the data show, the gel portion decreases and the liquid portion increases steadily throughout life; the gel fraction declines to 80% of the total vitreous by age 25, to 60% by age 65, and to less than 50% by age 90. The regression line through the data points has the equation $y = 0.93 - 0.005x$, where y is the gel fraction and x is age. Beginning with 90% gel at age 5, the gel fraction decreases (or the liquid fraction increases) by 5% of the total every 10 years.

This continual, linear increase in the liquid fraction of the vitreous suggests that there is a gradual change in the vitreous collagen. At birth, collagen fibers are everywhere, so there is little or no liquid component, but over time, collagen must either be removed from the vitreous or redistributed so that it is less uniform, aggregating the collagen in some regions and leaving other regions free of collagen.

What evidence there is favors a redistribution of collagen. Measures of total collagen content show little change after the eye has completed growth, which argues against collagen breakdown in selected parts of the vitreous. In electron micrographs, collagen is found throughout the vitreous at all ages, but as individual fibers in young eyes and as fiber *bundles* in older eyes. The prominent strands of collagen seen in dark-field microscopy are a feature of older eyes, not of young ones.

Figure 12.33 shows a schematic two-dimensional version of what may be happening. In Figure 12.33*a*, the small dots can be thought of as randomly distributed collagen fibers cut in cross section. Proteoglycans and their associated water in the immediate vicinity of the collagen fibers will be bound to the fibers, as indicated by the shaded annuli around each of the collagen dots. This part of the array, where fibers and proteoglycans are close together, filling the plane, is the gel portion. The unfilled white space is where unbound proteoglycans and

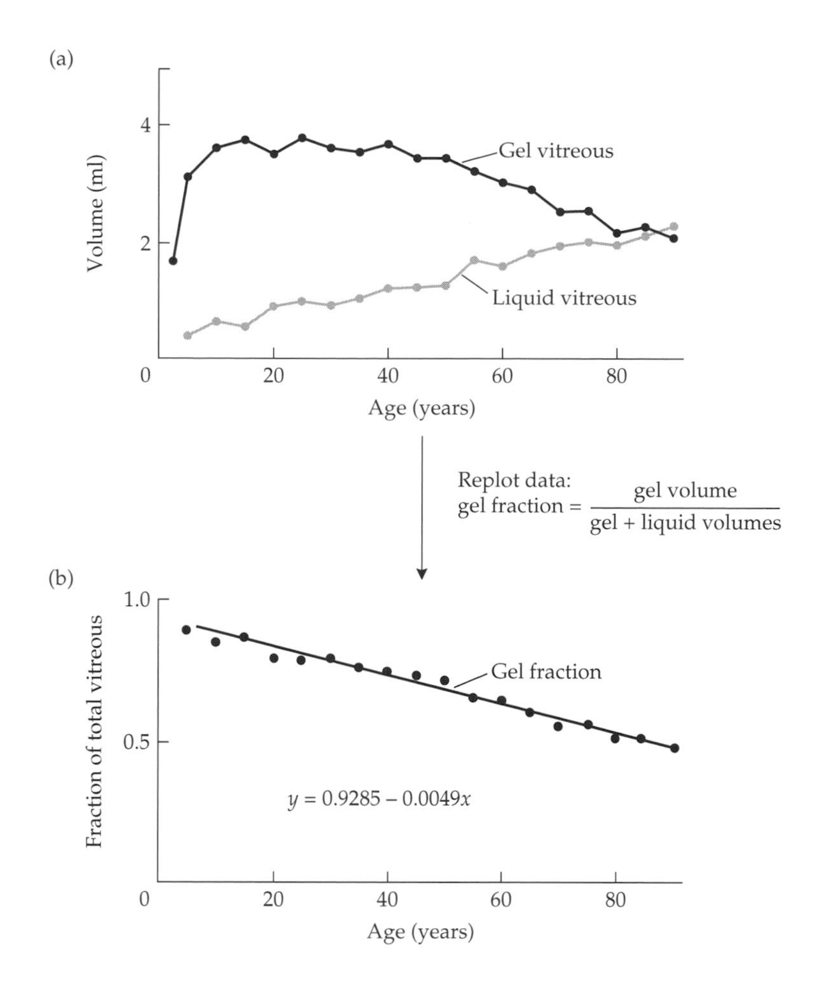

Figure 12.32
Liquid and Gel Fractions of the Vitreous Change with Age
In terms of absolute volume, the gel component declines as the liquid component increases (*a*). Although this result has been cited as evidence for vitreous liquefaction as an age-related change, replotting the data to show gel and liquid components as fractions of the total vitreous volume (*b*) changes the interpretation. Vitreous liquefaction begins early in life and continues throughout. (Replotted from Balazs and Denlinger 1982.)

water could reside; it represents the liquid component. In this random arrangement, the amount of unfilled white space is small, as is the liquid component in the young vitreous.

Figure 12.33*b* contains the same number of collagen fibers as in Figure 12.33*a*, but some have been put together in clusters (bundles cut in cross section), with each cluster surrounded by a proteoglycan shell. The obvious result is an increase in the open area where unbound proteoglycans and water can reside; this increases the liquid fraction at the expense of the gel fraction, as occurs in older eyes.

The obvious question about this model of vitreous liquefaction concerns the mechanism for the aggregation of collagen fibers into bundles: Why and how does it happen? The honest answer is that we don't know. Various plausible sug-

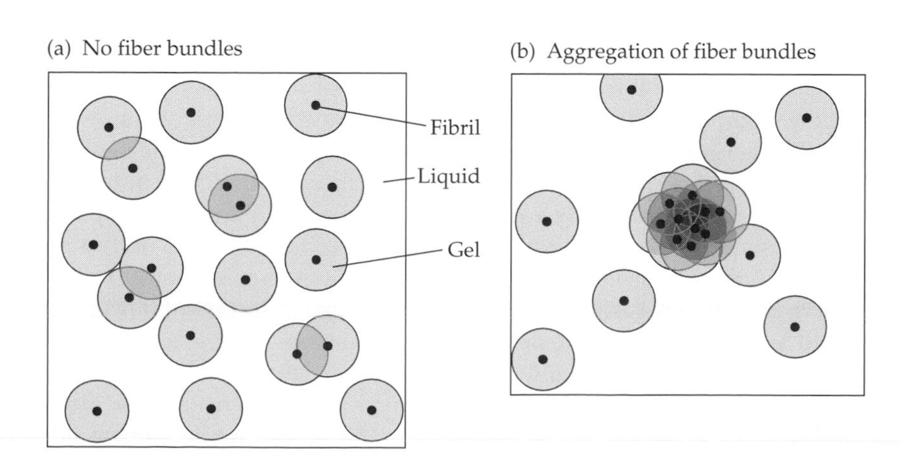

Figure 12.33
Aggregation of Vitreal Collagen Reduces the Volume of Vitreous Gel
(*a*) When multiple collagen fibers and their associated matrix (shown here in cross section) do not form bundles, the gel is a substantial fraction of the total area. (*b*) Bundling is portrayed as coalescence of the matrix to reduce the proportion of gel and increase the proportion of the liquid component.

gestions involve structural changes in hyaluronic acid, collagen, or both that increase the tendency for cross-linking between collagen fibers. An intriguing possibility for the underlying causal agent is the same as for age-related cataract in the lens—namely, cumulative absorption of ultraviolet radiation and its direct and indirect effects on molecular structure. This explanation would fit nicely with the progressive, lifelong process of vitreous liquefaction and provide another solid reason for limiting exposure to sunlight. Unfortunately, there are no epidemiological data on this issue, and since specifying the degree of vitreous liquefaction is not a simple task, such data are unlikely to be forthcoming anytime soon.

Shrinkage of the vitreous gel may break attachments to the retina

(a) Superior
Liquid
Incomplete vitreal detachment
Gel

(b)

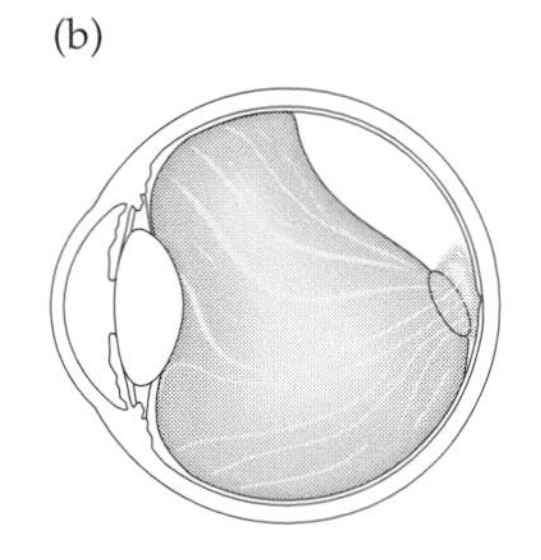

(c)

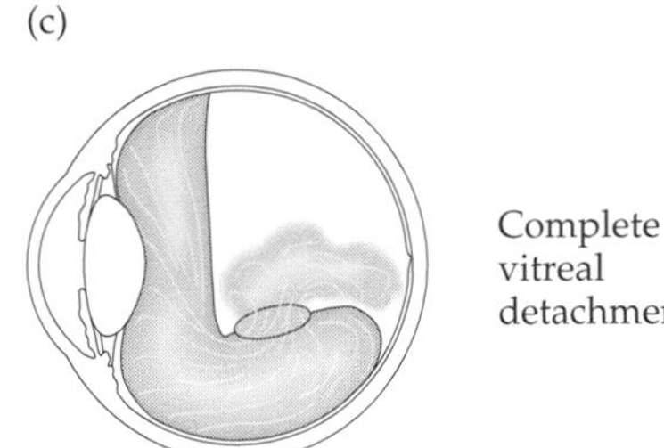

Figure 12.34
Posterior Vitreal Detachment
(*a*) Vitreal detachments usually begin with a separation of the gel vitreous from the superior retina. The space is filled by liquid vitreous. (*b*) A more extensive detachment may break the perimacular attachments, allowing gel vitreous to extrude through the premacular hole. (The same thing could be occurring at the prepapillary hole, which is not shown in this plane of section.) (*c*) In a complete detachment, the gel vitreous has detached from much of the inferior retina, and the original macular region has prolapsed forward and down. (After Eisner 1990.)

Posterior vitreal detachment (or PVD in the list of clinical acronyms) is a separation of the posterior vitreous cortex from the inner limiting membrane of the retina. (The qualifier "posterior" implies the existence of anterior vitreal detachment. There is such a thing, but because the vitreous is so tightly attached at the vitreous base, anterior vitreal detachment is rare and invariably due to trauma.) Posterior vitreal detachment is common in older eyes, reportedly affecting about 65% of people over the age of 65.

Various degrees of uncomplicated posterior vitreal detachment are illustrated schematically in Figure 12.34. When the detachment is complete (Figure 12.34*c*), the strongest attachments to the retina around the foveal area and the optic nerve head have broken away and much of the vitreal cortex in the superior part of the eye has collapsed downward. Liquid vitreous has moved into the space between vitreous cortex and retina, presumably by passing through the premacular and prepapillary holes in the vitreous cortex.

In terms of the volume of vitreous within the vitreous chamber, nothing has changed; the overall effect is simply a rearrangement of the liquid and gel components, and by itself a posterior vitreal detachment is not a traumatic event. Many people have a vitreal detachment of which they are unaware until it is observed in the course of a routine eye examination. Others realize that something has happened because large, obtrusive floaters suddenly appear in the visual field; the floaters, as in my case, are the attachment rings around the premacular or prepapillary holes. Although the vitreous may continue to detach from the superior retina up to the ciliary body and the vitreous base, in the absence of large floaters, there is no visual impairment.

But posterior vitreal detachment is not always uncomplicated; complications arise when the vitreous does not separate from the retina cleanly and completely. If some vitreal attachments do not break away as the rest of the vitreous is detaching, enough stress may be exerted to tear the retina, particularly if the retina has already been weakened by degeneration. Once the retina has torn, liquid vitreous can flow behind the photoreceptors and separate them from their normal contact with the retinal pigment epithelium (see Chapter 13); this is a retinal detachment, which is a serious problem. The risk of sebsequent retinal detachment makes posterior vitreal detachment a matter of concern.

Another complication arises when vitreal attachments do not break free from retinal blood vessels. Athough the small, vulnerable capillaries lie within the retina, the arterioles and venules are on the surface, sandwiched between the retina and the vitreous cortex (see Chapter 16). Venules in particular are prone to leak if put under stress; thus, if attachments on the blood vessels do not break away readily, the pull exerted by the detaching vitreous may result in hemorrhage into the vitreous. Add hemorrhage to a retinal detachment, and the surgeon will have a proper mess to deal with; blood can obscure the operating field, surgery becomes less precise, and the success rate decreases.

The cause of posterior vitreal detachment is not known, but the ongoing redistribution and change in vitreal collagen probably play a role. Contraction of the collagen as it is redistributed into bundles, for example, could exert increasing stress on the vitreal attachments, eventually causing the weakest of them to break. Because the posterior attachments are weaker than the anterior ones, such as the vitreous base, the posterior vitreous cortex detaches. Alternatively, the attachments themselves may progressively weaken. If such weakening occurred around the premacular vitreal hole, as some have suggested, liquid vitreous could move from the hole into the new space between the retina and the vitreal cortex, thereby separating the two structures even more.

Altered activity of cells normally present in the vitreous or the introduction of cells from outside the vitreous may produce abnormal collagen production and scar formation

Although nothing has been said about it thus far, the vitreous has a small cellular component. Most of these cells, the **hyalocytes**, reside in the vitreous cortex next to the retina; the anterior vitreous cortex is relatively acellular. The hyalocytes look much like fibroblasts, though less spidery, and they are so sparsely distributed that neighboring cells are not in contact with one another. As their name implies, hyalocytes probably produce hyaluronic acid for the vitreous; they contain the requisite synthetic enzymes and can produce hyaluronic acid in vitro. They certainly function as macrophages, ingesting and eliminating debris from the vitreous, and they may also produce some collagen. About 10% of the cells in the vitreous are fibroblasts, located mainly in or near the vitreous base. They are presumed to be involved in the synthesis and maintenance of vitreal collagen, but they have not been caught in the act.

Under normal circumstances, the populations of hyalocytes and fibroblasts are unobtrusive: They are few in number, they do not lie on the light path to the fovea, and they are easy to forget. They can be provoked into unusual activity, however, and thereby present a serious threat to vision.

Proliferative vitreoretinopathy (PVR) is very much like scar tissue that forms and spreads within the vitreous cortex, except that the wound the scar is meant to heal is absent. For some reason, native vitreous cells, perhaps inspired and aided by cells from outside the vitreous, give up their normal dispersed mode of living, collecting in groups to generate abnormal amounts of collagen and matrix (both collagen and matrix may differ from the normal types). Leakage from retinal blood vessels may add to the problem, with clotting of the released blood plasma in the tissue. The result is a disorganized mass of collagen, phagocytic and collagen-producing cells, matrix, and clotted blood, all sitting in the vitreous cortex, where none of this belongs. There are invariably abnormal attachments to the retina.

Figure 12.35 illustrates some of the dense, opaque tissue at several places on the retina in proliferative vitreoretinopathy. Abnormal tissue like this over the fovea has a devastating effect on vision.

Interference with the light path to the retina is bad enough, but the effect of abnormal tissue attachments to the retina is worse. The abnormal tissue contracts and produces local folds and buckled regions in the retina that constitute small regions of retinal detachment. If the retinopathy is widespread, there can be numerous regions of detachment, some of them large. (A similar situation arises in proliferative diabetic retinopathy, but the abnormal tissue growth is initiated differently.)

The underlying cause of proliferative vitreoretinopathy is unknown, but it is one of the possible consequences of retinal detachment. In this case, the initiator may be foreign cells—retinal pigment epithelial cells or retinal glial cells, for example—that enter the vitreous through a retinal tear or are introduced accidentally when the eye is penetrated during surgery. Once in the vitreous, these invaders may trigger abnormal tissue production by the native vitreous cells.

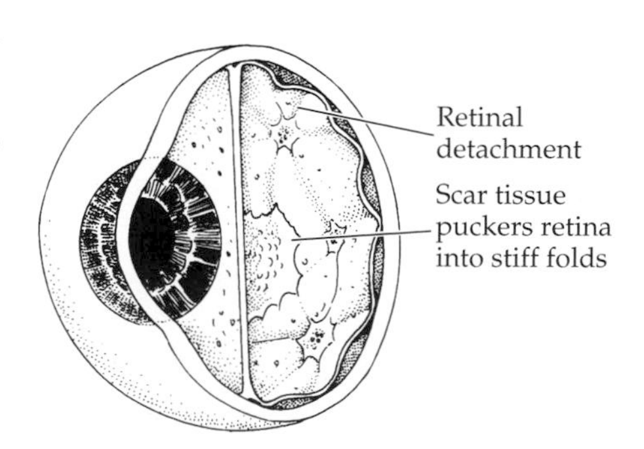

Figure 12.35
Proliferative Vitreoretinopathy
Scar tissue that forms in PVR puckers and detaches the retina in addition to blocking the light path to the retina.

At present, the only treatment for proliferative vitreoretinopathy is surgical removal of the scar tissue from its retinal attachments, which is a difficult and delicate business. The chance of restoring normal vision to the affected area is almost zero, but some visual function is better than none, particularly if the fovea is involved. Because surgery comes too late to prevent visual impairment, the search is on for ways to halt the progress of abnormal tissue formation and to break down whatever abnormalities have formed. One strategy currently under investigation uses monoclonal antibodies (see Box 15.1) that recognize and bind to the abnormal vitreal cells. The antibodies are linked to a UV-absorbing pigment; exposing the labeled tissue mass to pulses from a UV-emitting laser photically destroys the abnormal cells to which the antibodies are bound, effectively removing the abnormal tissue. The hope is that this technique will work in human patients.

Vitrectomy removes abnormal portions of the vitreous

Until the mid-1970s, conditions like proliferative vitreoretinopathy could not be treated with any genuine hope of success. Among other things, surgeons could not justify the risk involved in making incisions into the vitreous chamber large enough to admit their surgical instruments. The breakthrough came with the development of multipurpose intraocular probes, hollow needles through which tiny cutting tools, suction lines, miniature forceps, and other devices could be inserted and manipulated. The puncture wound made by such a probe inserted into the eye is small and easily sealed. Combine these delicate tools with high-magnification operating microscopes (and some very steady hands at the controls), and one has the elements of **vitrectomy**.

Figure 12.36 shows schematically, but to scale, vitrectomy instruments used to remove a piece of scar tissue from the vitreous cortex. One instrument has small cutting blades at the end of a rigid shaft rotating at high speed within the intraocular probe. It chops up vitreal collagen strands and removes the pieces by suction. Another instrument, not shown here, is a miniature scissors, the tiny blades of which project perpendicularly from the tip of the probe; it can clip attachments

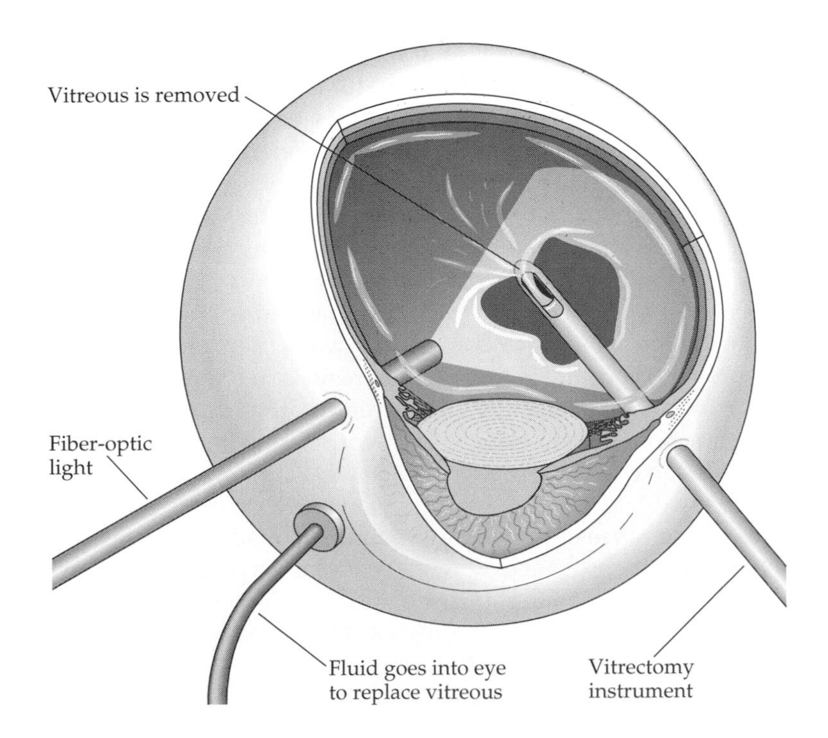

Figure 12.36
Vitrectomy
Vitreous surgery requires a special instrument that combines a miniature cutting tool with a suction line to remove vitreous and tissue that has been cut away. A small fiber-optic probe provides bright illumination within the eye, and another small needle is used to pump fluid into the eye to replace the vitreous that has been removed.

between the scar and the retina. Two other probes are shown in the eye in Figure 12.36. One contains a fiber-optic bundle for illumination; the other is a conduit for fluid used to replace the removed vitreous.

Although it is possible to separate abnormal vitreous masses from the retina and to remove the debris from the eye, vitrectomy is intricate, difficult surgery; it cannot be rushed, and some surgeries may last 4 to 8 hours. In addition, vitrectomy often requires removal of the ocular lens, cataractous or not, to provide maneuvering room and a better view of the operating field. Thus, vitrectomy patients usually have an intraocular lens implant in addition to the other procedures.

Vitrectomy has become an important technique for saving eyes from inevitable blindness, and its high risk of complications is far outweighed by the potential benefit. But vitrectomy cannot restore what has already been lost; for this reason, efforts to better understand vitreoretinopathies will continue in the hope that their progression can be halted or reversed without surgery.

Development of the Lens and Vitreous

The lens forms from a single cell line

Most of the elements of lens formation and growth were discussed earlier in this chapter and in Chapter 1. Figure 12.37 is a reprise of these earlier discussions. The sequence of lens formation begins with an induced thickening of the surface ectoderm overlying the developing optic vesicle. This **lens placode** buckles inward and invaginates in synchrony with invagination of the optic cup and becomes the **lens vesicle**. At about 5 weeks, the lens vesicle pinches off from the surface ectoderm, becoming a hollow sphere with an exterior lining of cells. These few cells, all derived from the surface ectoderm, are the source of the adult lens—lens fibers, epithelium, capsule, the lot.

The cells on the posterior side of the hollow sphere, the side closest to the optic cup, elongate into the center of the sphere and fill the empty space; this is their sole contribution to the lens, with the possible exception of secreting a bit of lens capsule. The interior cells lose their nuclei and will be buried within the center of the lens as the lens grows around them. Thus, all subsequent lens growth comes from the remaining cells on the anterior surface, cells now identifiable as the anterior lens epithelium.

Next, the lens cells begin to secrete the capsule and start the sequence of proliferation, differentiation, and elongation that adds new cells at the lens equator. The first new cell brood forms a shell around the solid interior core of cells, and successive generations form additional shells, each overlying its predecessor. As discussed earlier, this process continues in the same way throughout life. At birth, the lens has a complete capsule into which zonular fibers have inserted, an anterior epithelium, about 1.1 million lens fibers, and 1300 lens shells. The lens is 5.5 to 6.0 mm in equatorial diameter and 3.2 mm in sagittal thickness.

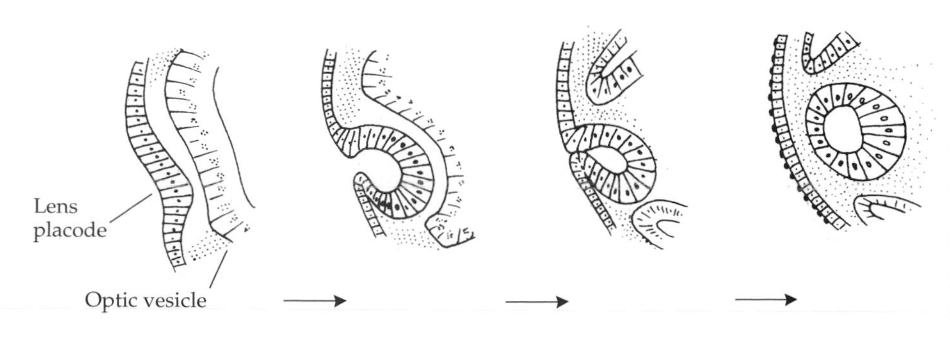

Figure 12.37
Formation of the Lens
The lens begins as a thickening of the surface ectoderm overlying the optic vesicle. This lens placode buckles and invaginates to become the lens vesicle. After the vesicle pinches off from the surface ectoderm, the cells on the posterior surface nearest the optic cup elongate to fill the hollow interior of the vesicle. All subsequent lens growth is by the addition of shells or layers of lens fibers over this mass of elongated posterior epithelial cells.

Most failures of lens development are manifest as congenital cataracts

Congenital cataracts—those evident at birth or shortly thereafter—can take a variety of forms and range from small, punctate opacities to opacification of the entire lens. Like cataracts appearing later in life, they may be nuclear, cortical, polar, equatorial, or positioned along the lens sutures.

Some rare forms of congenital cataract are associated with gross abnormalities of lens morphology, but most are like adult cataracts in having normal structure at the cellular level. For these, the opacity is presumably a failure at the scale of molecular structure, as it is in age-related cataracts. But whether the failure is improper synthesis of lens proteins or later changes induced by an abnormal lens environment in the embryo is not known. Congenital cataracts consisting of small, granular opacities along the lens sutures are more likely to be in a different category in terms of their etiology; here improper formation of the sutures, with otherwise normal lens fibers, appears to be the fault.

One sort of congenital cataract has an obvious origin; the cataract is associated with an indentation or notch on the inferior margin of the lens. This is a typical coloboma of the lens, usually accompanied by an absence of zonular fibers in the interior quadrant and a ciliary coloboma. All these defects are a consequence of incomplete closure of the embryonic choroidal fissue (see Chapter 1). Because of the incomplete zonule, a colobomatous lens will be improperly positioned; this can also happen when the coloboma affects the zonule without involving the lens. When the lens is not attached to the ciliary body along its inferior margin, it is pulled upward by the intact parts of the zonule (or it is not held down in position by the inferior zonule). In Figure 12.38, the lens is displaced horizontally, so the displacement is not due to a typical coloboma. In this case, the zonule did not form properly on the lateral side as part of a congenital connective tissue disorder. Whatever the cause, this condition is called **ectopia lentis** or a **subluxated lens**.

It is interesting that such a thing as lens coloboma exists. After all, the lens is derived from surface ectoderm and never has direct contact with any part of the developing optic cup—unlike the iris, ciliary body, zonule, and choroid, in which colobomas resulting from incomplete closure of the choroidal fissure are to be expected. But the lens is not immune to the power of induction by the optic cup. The elongation of posterior lens cells to fill the hollow primordial lens is

Figure 12.38
Ectopia Lentis
The lens is displaced medially and its edge bisects the pupil vertically. In this case, the displacement is associated with a congenital disorder of connective tissue called Marfan's syndrome. (Courtesy of John F. Amos.)

probably triggered by signals from the optic cup, and later stages of lens growth—whether proliferation, differentiation, or both—appear to depend on induction for their successful completion.

But here is a question. Ocular growth generally may involve inductive signals from the retina. Other parts of the eye stop growing at some time, presumably when signals to grow are turned off; they add new cells only to replace those that have completed their life cycles. Why does the lens keep growing?*

The primary vitreous forms around the embryonic hyaloid artery

At the peak of its development, the hyaloid artery runs through the developing vitreous chamber to supply a dense network of blood vessels around the lens, the **tunica vasculosa lentis**. But well before this stage, even as the hyaloid artery is forming and the choroidal fissure closes, the first elements of the vitreous appear, most noticeably fibroblasts accumulating inside the vitreous chamber, just in front of the future optic nerve head (at about week 5).

As the hyaloid artery extends toward the lens, it throws off numerous small capillarylike branches in the vitreous chamber, a system called the **vasa hyaloidea propria**. Cells and fluid accumulate around this vascular network around the end of week 7. Collagen has just made its appearance in this primitive vitreous.

The secondary vitreous, initially acellular, forms outside the vasa hyaloidea propria

The origin of the secondary vitreous, which will constitute most of the adult vitreous, is a bit mysterious, but part of it may be produced by the developing retina. There are very few cells in any case, and the first part of the secondary vitreous is a matrix of type II collagen. The collagen appears first at the posterior part of the vitreous chamber near the nerve head, expanding forward and laterally around the primary vitreous. As this expansion continues, the primary vitreous with its vasa hyaloidea propria is compressed centrally toward the position it will assume as the canal of Cloquet.

From this time onward, the hyaloid artery system and its primary vitreous wane in importance, while the surrounding secondary vitreous continues to expand. The small vessels in the vasa hyaloidea propria break down first, followed by the tunica vasculosa lentis and eventually by the hyaloid artery itself, in which blood flow stops around the end of the seventh month (by this time, branches from the hyaloid artery at the nerve head have become the elements of the central retinal artery system; see Chapter 16).

While all this deterioration in the hyaloid artery system is in progress, proteoglycan production begins in the secondary vitreous, cells thought to be precursors of hyalocytes migrate in from somewhere (the primary vitreous?), and substantial hyaluronic acid production begins (around the fifth month). This process continues to birth and beyond, with the secondary vitreous enlarging as long as the eye itself continues to grow—that is, for 12 years or so.

Most developmental anomalies in the vitreous represent incomplete regression of the hyaloid artery system

In the normal course of hyaloid artery regression, the tissue in the artery wall atrophies, and any remnants are reduced to invisible fragments. But there can be

*It's something to think about. Does the lens continue to grow because its cells are permanent while most other cells in the eye are short-lived and must be continually replaced? If so, what about the retina, whose neurons are permanent? Mammalian retinas do not add new cells indefinitely. I don't have an answer.

a failure to atrophy, leaving the intact, translucent artery in place, looping back through the vitreous from the lens to the optic nerve head. This **persistent hyaloid artery** is a dramatic departure from normality.

More commonly, just a few pieces of the hyaloid artery persist. One is a small fragment adhering to the posterior surface of the lens; with an ophthalmoscope, it appears as a small black spot known as a **Mittendorf dot**. At the other end of the canal of Cloquet, an arterial remnant on the surface of the optic nerve head is often sheathed with nerve head glial cells, creating a slender projection of tissue into the vitreous from the center of the nerve head; this structure is called **Bergmeister's papilla**.

These anomalies are innocuous. Bergmeister's papilla lies on the nerve head and is therefore in the center of the normal blind spot. A Mittendorf dot, though central in the visual field, is small and poorly focused on the retina; it is rarely noticeable. Persistence of the entire hyaloid artery does produce a sizable obstruction and shadow in the temporal visual field of the affected eye, but it will not compromise vision unless the artery is quite opaque and enlarged at the posterior surface of the lens.

A condition called **persistent hyperplastic primary vitreous** (PHPV for short) is an anomaly of anterior vitreous development in which persistence of the anterior portion of the hyaloid artery is combined with abnormal tissue formation and attachment to the posterior lens capsule. The cells that have proliferated abnormally are thought to be retinal glial cells (astrocytes) acquired at an early stage of development when the vitreous chamber was small and the retina and lens less widely separated. The resulting opacity requires surgical correction.

References and Additional Reading

Structure of the Lens

Alberts B, Bray D, Lewis J, Raff M, Roberts K, and Watson JD. 1994. Protein structure, Chapter 3, pp. 111–128, and The birth, assembly, and death of proteins, Chapter 5, pp. 212–221, in *Molecular Biology of the Cell*, 3rd Ed. Garland, New York.

Benedek GB. 1997. Cataract as a protein condensation disease. *Invest. Ophthalmol. Vis. Sci.* 38: 1911–1921.

Borkman RF and McLaughlin J. 1995. The molecular chaperone function of α-crystallin is impaired by UV photolysis. *Photochem. Photobiol.* 62: 1046–1051.

Costello MJ, Oliver TN, and Cobo LM. 1992. Cellular architecture in age-related human nuclear cataracts. *Invest. Ophthalmol. Vis. Sci.* 33: 3209–3227.

Farnsworth PN and Shyne SE. 1979. Anterior zonular shifts with age. *Exp. Eye Res.* 28: 291–297.

Fisher RF and Pettet BE. 1972. The postnatal growth of the capsule of the human crystalline lens. *J. Anat.* 112: 207–214.

Harding CV, Reddan JR, Unakar NJ, and Bagchi M. 1971. The control of cell division in the ocular lens. *Int. Rev. Cytol.* 31: 215–300.

Hogan MJ, Alvarado JA, and Weddell JE. 1971. Lens. Chapter 12, pp. 638–677, in *Histology of the Human Eye*. WB Saunders, Philadelphia.

Horwitz J. 1993. The function of alpha-crystallin. *Invest. Ophthalmol. Vis. Sci.* 34: 10–22.

Kuszak JR. 1995. The ultrastructure of the epithelial and fiber cells in the crystalline lens. *Int. Rev. Cytol.* 163: 305–350.

Kuszak JR, Peterson KL, and Brown HG. 1996. Electron microscopic observations of the crystalline lens. *Microsc. Res. Tech.* 33: 441–479.

Kuwabara T. 1975. The maturation of a lens cell: A morphologic study. *Exp. Eye Res.* 20: 427–443.

Rafferty NS. 1985. Lens morphology. Chapter 1, pp. 1–60, in *The Ocular Lens: Structure, Function, and Pathology*, Maisel H, ed. Marcel Dekker, New York.

Spector A. 1984. The search for a solution to senile cataracts. *Invest. Ophthalmol. Vis. Sci.* 25: 130–146.

Streeten BW. 1992. Anatomy of the zonular apparatus. Chapter 14 in *Duane's Foundations of Clinical Ophthalmology*, Vol. 1, Tasman W and Jaeger EA, eds. Lippincott-Raven, Philadelphia.

Taylor VL, Al-Ghoul KJ, Lane CW, Davis VA, Kuszak JR, and Costello MJ. 1996. Morphology of the normal human lens. *Invest. Ophthalmol. Vis. Sci.* 37: 1396–1410.

Weekers R, Delmarcelle Y, and Luyckx J. 1975. Biometrics of the crystalline lens. Chapter 5, part B, pp. 134–147, in *Cataract and Abnormalities of the Lens*, Bellows JG, ed. Grune & Stratton, New York.

Wistow G. 1990. Evolution of a protein superfamily: Relationships between vertebrate lens crystallins and mircroorganism dormancy domains. *J. Mol. Evol.* 30: 140–145.

The Lens as an Optical Element

Brown N. 1974. The change in lens curvature with age. *Exp. Eye Res.* 19: 175–183.

Brown N. 1976. Dating the onset of cataract. *Trans. Ophthalmol. Soc. U.K.* 96: 18–23.

Brown N. 1979. Photographic investigation of the human lens and cataract. *Surv. Ophthalmol.* 23: 307–314.

Burd HJ, Judge SJ, and Flavell MJ. 1999. Mechanics of accommodation of the human eye. *Vision Res.* 39: 1591–1595.

Cook CA, Koretz JF, Pfahnl A, Hyun J, and Kaufman PL. 1994. Aging of the human crystalline lens and anterior segment. *Vision Res.* 34: 2945–2954.

Delaye M and Tardieu A. 1983. Short range order of crystallin proteins accounts for eye lens transparency. *Nature* 302: 415–417.

Dolin PJ. 1994. Ultraviolet radiation and cataract: A review of the epidemiological literature. *Brit. J. Ophthalmol.* 78: 478–482.

Fagerholm PP, Philipson BT, and Lindström B. 1981. Normal human lens—The distribution of protein. *Exp. Eye Res.* 33: 615–620.

Fernald RD and Wright SE. 1983. Maintenance of optical quality during crystalline lens growth. *Nature* 301: 618–620.

Fincham EF. 1937. The mechanism of accommodation. *Brit. J. Ophthalmol.* 8 (Suppl.): 5–80.

Glasser A and Campbell MCW. 1998. Presbyopia and the optical changes in the human crystalline lens with age. *Vision Res.* 38: 209–229.

Harding JJ and Crabbe MJC. 1984. The lens: Development, proteins, metabolism and cataract. Chatper 3, pp. 207–492, in *The Eye*, Vol. 1B, *Vegetative Physiology and Biochemistry*, Davson H, ed. Academic Press, Orlando, FL.

Hemenger RP, Garner LF, and Ooi CS. 1995. Change with age of the refractive index gradient of the human ocular lens. *Invest. Ophthalmol. Vis. Sci.* 36: 703–707.

Jaffe NS and Horwitz J. 1992. The technique of cataract surgery. Chapter 11, pp. 11.1–11.90, in *Lens and Cataract*, Vol. 3, *Textbook of Ophthalmology*, Podos SM and Yanoff M, eds. Gower Medical, New York. (Chapter 1 contains a good, concise summary of lens anatomy and embryology.)

Koretz JF. 1995. Accommodation and presbyopia. Chapter 16, pp. 270–282, in *Principles and Practice of Ophthalmology: Basic Sciences*, Albert DM and Jakobiec FA, eds. WB Saunders, Philadelphia.

Koretz JF, Cook CA, and Kaufman PL. 1997. Accommodation and presbyopia in the human eye: Changes in the anterior segment and crystalline lens with focus. *Invest. Ophthalmol. Vis. Sci.* 38: 569–578.

Koretz JF, Cook CA, Kuszak JR, and Kaufman PL. 1994. The zones of discontinuity in the human lens: Development and distribution with age. *Vision Res.* 34: 2955–2964.

Koretz JF and Handelman GH. 1988. How the human eye focuses. *Sci. Am.* 259 (1): 92–99.

Koretz JF, Handelman GH, and Brown NP. 1984. Analysis of human crystalline lens curvature as a function of accommodative state and age. *Vision Res.* 10: 1141–1151.

McCarty CA and Taylor HR. 1996. Recent developments in vision research: Light damage in cataract. *Invest. Ophthalmol. Vis. Sci.* 37: 1720–1723.

Moore DT. 1980. Gradient-index optics: A review. *Appl. Optics* 19: 1035–1038.

Paterson CA and Delamere NA. 1992. The lens. Chapter 10, pp. 348–390, in *Adler's Physiology of the Eye*, 9th Ed., Hart WM, ed. Mosby Year Book, St. Louis.

Pierscionek BK and Chan DYC. 1989. Refractive index gradient of human lenses. *Optom. Vision Sci.* 66: 822–829.

Siebinga I, Vrensen GFJM, de Mul FFM, and Greve J. 1991. Age-related changes in local water and protein content of human eye lenses measured by Raman microspectroscopy. *Exp. Eye Res.* 53: 233–239.

Sivak, JG. 1980. Accommodation in vertebrates: A contemporary survey. Pp. 281–330 in *Current Topics in Eye Research*, Vol. 3, Zadunaisky JA and Davson H, eds. Academic Press, New York.

Streeten BW and Eshaghian J. 1978. Human posterior subcapsular cataract. *Arch. Ophthalmol.* 96: 1653–1658.

Tardieu A and Vérétout F. 1991. Biophysical analysis of eye lens transparency. Chapter 6, pp. 49–55, in *Presbyopia Research*, Obrecht G and Stark LW, eds. Plenum, New York.

van Alphen GWHM and Graebel WP. 1991. Elasticity of tissues involved in accommodation. *Vision Res.* 31: 1417–1438.

West SK and Valmadrid CT. 1995. Epidemiology of risk factors for age-related cataract. *Surv. Ophthalmol.* 39: 323–334.

Wyatt HJ. 1988. Some aspects of the mechanism of accommodation. *Vision Res.* 28: 75–86.

Wyatt HJ and Fisher RF. 1995. A simple view of the age-related changes in the shape of the lens of the human eye. *Eye* 9: 772–775.

Young RW. 1991. *Age-Related Cataract.* Oxford University Press, Oxford.

Young RW. 1992. Sunlight and age-related eye disease. *J. Natl. Med. Assoc.* 84: 353–358.

The Vitreous

Balazs EA. 1994. Functional anatomy of the vitreus. Chapter 17 in *Duane's Foundations of Ophthalmology*, Vol. 1, Tasman W and Jaeger EA, eds. Lippincott-Raven, Philadelphia. (Note: The author employs "vitreus" as a noun and "vitreous" as the corresponding adjective.)

Balazs EA and Denlinger JL. 1982. Aging changes in the vitreus. Pp. 45–57 in *Aging and Visual Function*, Sekuler R, Kline D, and Dismukes K, eds. Alan R. Liss, New York.

Eisner G. 1990. Clinical anatomy of the vitreous. Chapter 16 in *Duane's Foundations of Ophthalmology*, Vol. 1, Tasman W and Jaeger EA, eds. Lippincott-Raven, Philadelphia.

Foulds WS. 1987. Is your vitreous really necessary? The role of the vitreous in the eye with particular reference to retinal attachment, detachment and the mode of action of vitreous substitutes. *Eye* 1: 641–664.

Hogan MJ, Alvarado JA, and Weddell JE. 1971. Vitreous. Chapter 11, pp. 607–637, in *Histology of the Human Eye.* WB Saunders, Philadelphia.

Sebag J. 1992. The vitreous. Chapter 9, pp. 268–347, in *Adler's Physiology of the Eye,* 9th Ed., Hart WM, ed. Mosby Year Book, St. Louis. (This chapter is an updated version of Sebag J. 1989. *The Vitreous—Structure, Function, and Pathobiology.* Springer-Verlag, New York. The book includes more material on pathobiology.)

Sebag J and Balazs EA. 1989. Morphology and ultrastructure of human vitreous fibers. *Invest. Ophthalmol. Vis. Sci.* 30: 1867–1871.

Smiddy WE and Chong LP. 1995. Vitrectomy. Chapter 9, pp. 113–148, in *Color Atlas of Ophthalmic Surgery: Retinal Surgery and Ocular Trauma.* JB Lippincott, Philadelphia.

Development of the Lens and Vitreous

François J. 1963. *Congenital Cataracts.* Royal VanGorcum, Assen, Netherlands.

Grainger RM, Henry JJ, Saha MS, and Servetnick M. 1992. Recent progress on the mechanism of embryonic lens formation. *Eye* 6:117–122.

Kleiman NJ and Wargul BV. 1994. Lens. Chapter 15 in *Duane's Foundations of Clinical Ophthalmology,* Vol. 1, Tasman W and Jaeger EA, eds. Lippincott-Raven, Philadelphia.

Mann I. 1950. *The Development of the Human Eye.* Grune & Stratton, New York.

Sang DN. 1987. Embryology of the vitreous. Congenital and developmental anomalies. Chapter 2, pp. 11–35, in *The Vitreous and Vitreoretinal Interface,* Schepens CL and Neetus A, eds. Springer-Verlag, New York.

Chapter 13

Retina I: Photoreceptors and Functional Organization

It is light above all, according to the direction from which it comes and along which our eyes follow it, it is light that shifts and fixes the undulations of the sea.

■ Marcel Proust, *Within a Budding Grove*

The eyeball . . . consists primarily and essentially of a sheet of nervous matter visually endowed*—that is, capable of being so affected by light, that, when duly connected with the sensorium, what we call* sight or perception of light, *is the result.*

■ William Bowman, *Lectures on the Eye*

The Retina's Role in Vision

The retina detects light and tells the brain about aspects of light that are related to objects in the world

By virtue of its photoreceptors, the retina detects light in the retinal image and transforms the light energy into a communicable form of chemical energy; this process is **phototransduction**, and all retinas do it—even the simplest. The photoreceptors are laid out in a large array (about 100 million of them in the human eye), much like the molecules of silver halides or dyes in photographic films, and as a group they transform the pattern of light and shade in the retinal image into a corresponding pattern of gradations in neural activity. Also, from the different photopigments they contain, the photoreceptors begin the process of giving color to a world that has none (the sensation of color, like vision itself, is a product of the brain).

The signals from 100 million photoreceptors constitute a massive amount of information to be processed if the retinal image is to be reconstructed to form a mental image of the world; numbers in the gigabyte range are not unreasonable. Processing this information by taking into account each photoreceptor's signal would be a slow process—very much slower than the important, dynamic changes in the retinal image as we look around or as objects move. What the retina is doing, therefore, is editing the information, selecting salient features of the retinal image and packaging them in a form that can be quickly and economically conveyed to the brain by the million or so retinal ganglion cell axons in the optic nerve.

But what is a "salient feature" and how are salient features extracted from the rest? Salience, as far as the retina is concerned, appears to be related to changes in light intensity—in space, time, or both. An abrupt spatial change in intensity, for example, is an edge or border, a temporal change is a sudden brightening or dimming in one part of the image, and a spatiotemporal change is movement of part of the image. To put it another way, the most informative aspects of the retinal image are the variations in light intensity. By signaling variations and ignoring those parts of the retinal image where light intensity is constant over space and time, the retina provides the brain with important information about the retinal image and yet does not overload it with repetitive, redundant facts.

The question of how salient features are recognized, extracted, and signaled is a question about how the retina works. In other words, how are photoreceptor signals about the *amount* of light striking them transformed to signals about *changes* in light over space and time? The general answer is that the retina doesn't make decisions about which aspects of the retinal image are important and which are not; recognition is not part of its job. Instead, its circuitry automatically gives preference to spatiotemporal changes in the retinal image by comparing signals from different photoreceptors (the spatial dimension), and signals occurring within certain periods of time (the temporal dimension). The flow of signals through the circuitry depends on these comparisons revealing differences—differences between the amounts of light falling on different photoreceptors or differences between the timing of signals arriving at a particular point in the circuit.

Objects are defined visually by light and by variations in light reflected from their surfaces

In thinking about what the retina is doing, it is worth considering the problem confronted by the visual system as a whole. The retinal image is the input to the system—it is what the retina sees—and the image is nothing more than variations in light intensity mapped onto a two-dimensional surface (the retina is hemispherical, but that is topologically equivalent to a plane surface; hence it is two-dimensional). These variations of intensity in the image are produced by light reflected from three-dimensional objects that have a vast array of shapes, sizes, textures, reflectances, and conditions of illumination.

As noted in the quote from Marcel Proust at the beginning of this chapter, light defines surfaces, their physical boundaries, their shapes, and their textures.* Since the reflectance of an object usually differs from that of its surroundings, we expect a variation in light intensity in the retinal image that defines the border or contour of the object and to some extent defines its shape, but there will also be variation in the amount of light reflected from the surface of the object. Some parts of the object's surface may differ in reflectance because they have different textures; the reflectance of a cat's damp nose, for example, is different from the reflectance of its fur, and this difference produces a fur-to-nose variation of light intensity in the retinal image of a cat. Or surface reflectances may be the same, but the change in shape of the surface—that is, its three-dimensional structure—means that different parts of the surface have different orientations with respect to the prevailing light source and therefore reflect different amounts of light into the eye, thereby creating another variation in light intensity in the retinal image.

A simple example of inferring three-dimensional shape from variations in light intensity on a plane surface is shown in Figure 13.1, beginning on the left with a line drawing of an irregular geometric shape. Using the information on relative intensity in the middle panel to shade the different parts of the figure in

*Marcel Proust (1871–1922) was an extraordinarily keen observer of his world who made numerous insightful comments on vision and perception in the course of his storytelling. His writing is notable for its rich visual imagery.

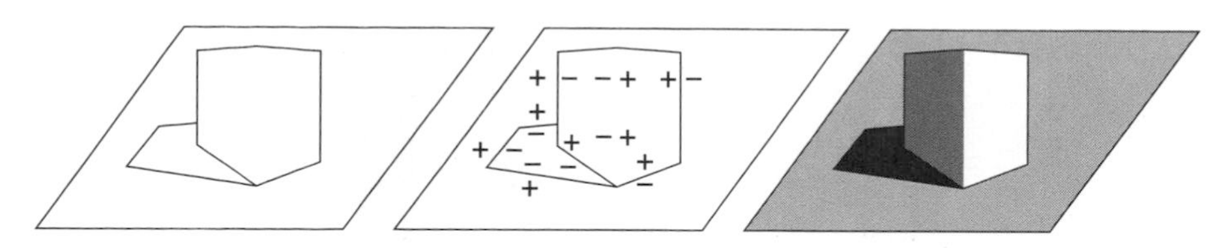

Figure 13.1
Intensity Variations Can Define Three-Dimensional Shape
With appropriate shading, flat two-dimensional shapes appear to be solid three-dimensional objects, in this case, a rectangular block standing upright on a horizontal surface. Relative brightness is indicated by + (brighter) and – (dimmer).

terms of relative brightness yields the figure on the right, which can easily be interpreted as a rectangular block and the shadow cast by the block on its supporting surface. There are some unanswered questions here (for example, does our interpretation require prior experience and knowledge of blocks and shadows?), but these questions do not affect the notion that changes in light intensity across a two-dimensional surface can provide clues to three-dimensional shape.

Another example of the power of light-intensity variations to convey information about shapes of surfaces is shown in Figure 13.2. The upper pattern of light and shade is normally seen as a series of bulges rising above the surface plane, the pattern below as a series of dimples in the surface. These perceptions are the result of our visual system's ability to infer surface contour from intensity variations and the system's use of a perceptual assumption that light normally comes from above.*

It is one thing to indicate the kind of information the visual system has at its disposal to generate its representation of the world and quite another to say how this representation is created or how the representation is related to neural activity; at this stage, we do not know how it is done. (But there are plenty of interesting ideas afloat and considerable activity in the field of higher visual processes.) For our present purposes, however, we have a very important and powerful idea to work with—namely, the notion that the light reflected from the surfaces of three-dimensional objects produces variations in light intensity within the retinal image that, if detected and signaled, should provide a rich flow of information from which the structure of real-world objects can be inferred and represented. The primary task of the human retina is detecting, selecting, and signaling; inference and representation are left to the brain. (Retinas in some species do much more in the way of inference about the nature of objects, in some cases to the extent that biologically important decisions are made in the retina with little or no help from the central nervous system.)

The retina makes sketches of the retinal image from which the brain can paint pictures

An intuitive feel for the operation of the retina might begin with Figure 13.3; the upper photograph can be thought of as the retinal image and the cartoonlike version below as the information transmitted by the retina to the brain. Although the photograph has been reduced to nothing but lines, there is still a great deal of detail present and the leaves and links in the fence are recognizable. To borrow a word from the late David Marr, the retina's editing process is generating a "sketch" of the world as depicted in the retinal image.† Although Marr used the word "sketch" in several specific ways, its vernacular meaning has some

*This is one of what V. S. Ramachandran calls the nervous system's "bag of tricks," a set of assumptions about the world that are built into the organization of the brain. "Light comes from above" is not a surprising assumption, because that is the natural condition under which visual systems evolved.

†David Marr wrote his book *Vision* in the late stages of an illness from which he died at age 35. *Vision* is a provocative summary of his earlier work and thinking on the subject, one that deliberately challenges more traditional ideas about what vision is and how best to understand it. In borrowing his word "sketch," I have used it more loosely and less richly than he did.

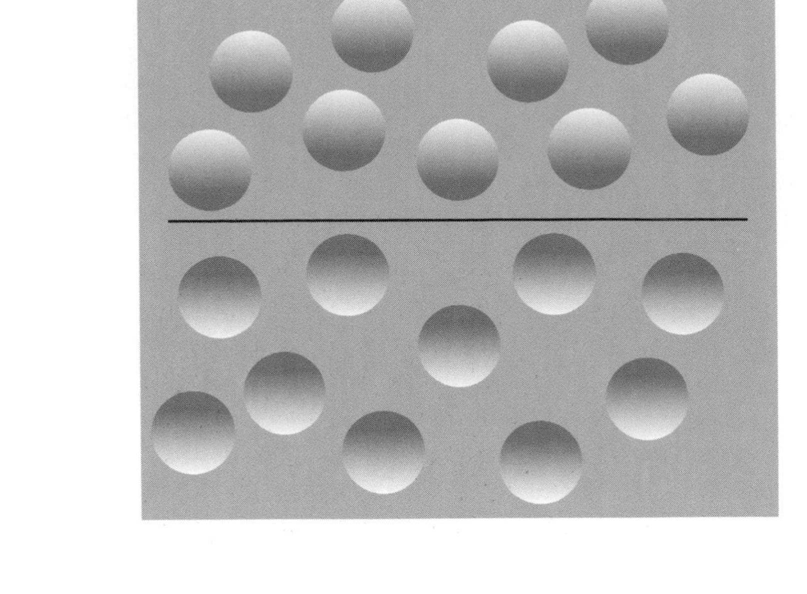

Figure 13.2
Shading Changes Apparent Shape
The apparent bulges (top) and dimples (bottom) reverse their appearance when the page is turned upside down. (After Kleffner and Ramachandran 1992.)

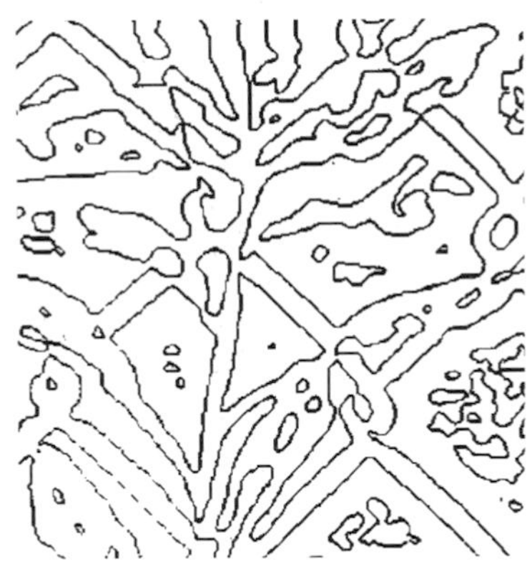

Figure 13.3
Filtering an Image Can Show Only the Locations of Intensity Variations in the Image
The filtered version of the original picture (top) is a sketch (bottom) in which the lines indicate intensity changes in the original. The retina provides the information to make a sketch like this. The sketch was generated by transforming the original picture with filters designed to identify regions where variations in light intensity went above or below the average light level. (From Marr and Hildreth 1980.)

interesting and potentially relevant implications. A good artist can make a sketch very rapidly, for example, capturing much that is essential in a complex scene or object with a few pencil strokes, and several sketches of the same object at different levels of detail can emphasize different critical elements of the object—its bulk and form in a coarse sketch, its textures in a fine sketch, and so on.

The retina also makes several sketches of the same scene, but unlike an artist drawing one sketch after the other, different kinds of sketches are transmitted to the brain simultaneously as parallel streams of information. This simultaneous transmission makes the processing rapid. Compared to serially extracting different sets of information from the retinal image, the time required to generate several sketches is no greater than the time to produce a single sketch. (The primate retina does not include lines or line segments in its sketches, however; we will come back to this issue in Chapter 14.)

An idea of the information that might be contained in these sketches is given in Figure 13.4. The original photograph (Figure 13.4*a*) is of a sculpture by Henry Moore—a complex, abstract object with many variations in light reflected from its surface. The three sketches in Figure 13.4*b, c,* and *d* were made by filters that detect light-intensity variations at progressively coarser scales. Thus, the sketch in Figure 13.4*b* contains information about local surface features and small background details (such as the trees), while the coarsest level (Figure 13.4*d*) offers mostly information about the form and physical edges of the object. It is not obvious how the information in these sketches will be used by the brain to interpret and represent the object—this *is* the problem if we are to understand vision—but different sketches will have to be compared in several ways to interpret their wealth of information about the object and its structure. The point is that the sketches contain information that is directly related to the physical properties of real-world objects.

Before going on to consider how retinal neurons produce these sketches, some reservations about the sketch idea deserve mention. First, we do not know how many sketches there are; psychophysical studies and theoretical analyses suggest that the human visual system uses four or five for rendering the spatial information in the retinal image. At the moment, however, the anatomical and physiological data from primate retina do not support the existence of more than three information pathways operating at different spatial scales (although this discrepancy may merely indicate that the right sort of data are not yet available). Second is the problem of color vision and how wavelength information is incor-

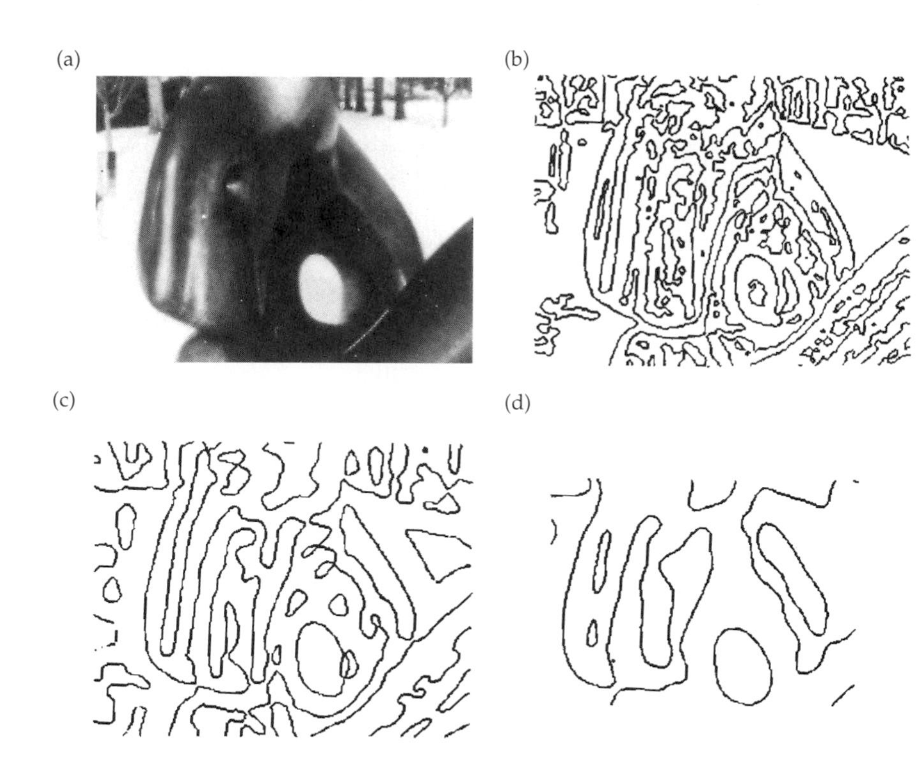

Figure 13.4
Filters of Different Sizes Provide Sketches with Different Levels of Detail
Filtering of the original picture (*a*) at three scales from fine to coarse (*b–d*) produced these sketches with different amounts of detail. The retina also generates several sketches at different scales. (From Marr and Hildreth 1980.)

porated in the sketches; it seems that some sketches contain wavelength information while others do not.

Also, my use of the word "sketch," and of Figure 13.4 as an illustration of sketches, emphasizes spatial information that is static and can be rendered with pencil on paper—that omits temporal change and is therefore too restrictive. For example, the rabbit retina contains pathways dealing with direction and velocity of image motion, and the frog retina, to take an extreme case, has a pathway signaling the presence and location of moving, buglike objects. It is unlikely that the human retina contains bug detectors, but it may well have movement- or velocity-sensitive pathways whose sketches are not purely spatial. The sketches, in other words, may be more abstract than those illustrated here.

Functional Organization of the Retina

Photoreceptors catch photons and produce chemical signals to report photon capture

The human retina is a thin sheet of tissue containing photopigments, the sine qua non of vision. Photopigments differ from pigments such as hemoglobin in that they are photolabile, meaning not only that they absorb energy from incident photons but also that they are structurally altered by the energy transfer; this structural alteration is the primary visual event from which all the rest of vision arises. The transfer of energy from photons to photopigments is the *only* interaction between light and the eye that is significant for vision. Without this change, there is no vision.

The photopigments are contained within photoreceptors. In the human retina, as in all other vertebrates, the cylindrical photoreceptors stand side by side in a layer along the posterior border of the retina, separated from the choroid by a single layer of cells, the pigmented epithelium of the retina. This arrangement means that light must pass through the rest of the retinal tissue to reach the photoreceptors, but the thinness of the retina, and certain specializations (the fovea), mean that this passage is of little consequence.

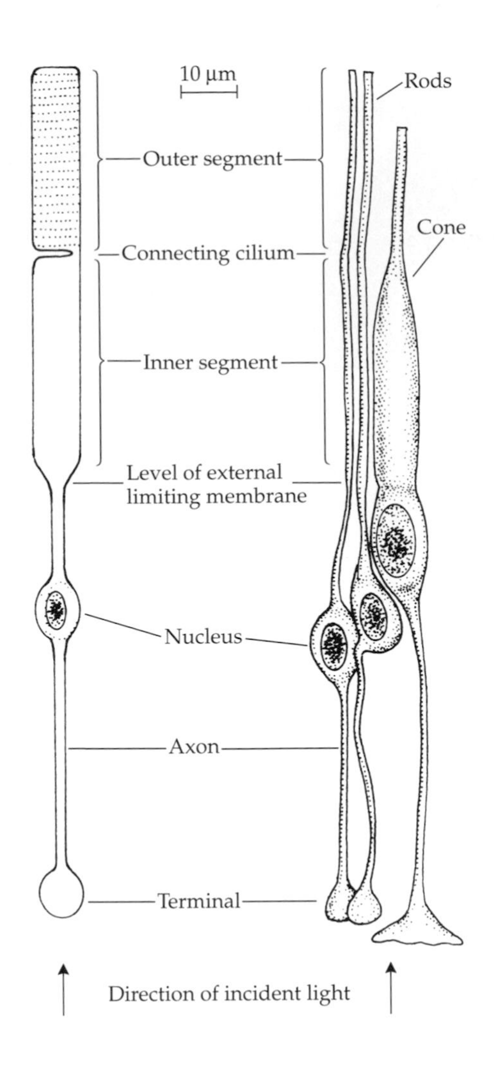

Figure 13.5
Parts of Photoreceptors
The diameter of the schematic photoreceptor at left has been exaggerated. The primate rods and cone at right are drawn with the correct proportions. (Rods and cone after Polyak 1941.)

The major elements of a photoreceptor are shown in Figure 13.5. The photopigment lies within the **outer segment** adjacent to the pigment epithelium. The next compartment, the **inner segment**, is filled with mitochondria; it is the photoreceptor's metabolic center. Outer and inner segments are joined by a narrow **connecting cilium**. The inner segment is continuous with the **nucleus** and an axon ending as the **synaptic terminal**. The terminal contains synaptic vesicles with the neurotransmitter glutamate; the release of glutamate when the terminal's membrane depolarizes is the photoreceptor's mode of communication with other retinal neurons.

The inner segment of the photoreceptor is the site of metabolically driven ion pumps that sustain constant electric-current flow around the photoreceptor; part of the current flows into the outer segment and part to the inner segment and terminal. The outer segment membrane contains ion channels whose status, open or closed, is affected by the absorption of light in the photopigment in the outer segment. The absorption of light by the photopigment sets off the cascade of phototransduction events that *close* the ion channels, as will be considered in more detail later. The closure of these channels makes the potential across the photoreceptor membrane more negative—that is, hyperpolarizing—and the rate of neurotransmitter release *decreases*. In short, absorption of light decreases the rate of transmitter release, and odd as it may sound, vertebrate photoreceptors are more depolarized and release transmitter at their maximum rate in the dark.

Photoreceptors in the human retina utilize four different photopigments: The **rods** or rod photoreceptors contain a well-characterized pigment called **rhodopsin**; the other three pigments are contained in the **cones**—one pigment in one cone photoreceptor. The photopigments differ in the wavelength of light they absorb maximally, their kinetics, and some other features, but their operation in the production of signals in the various photoreceptors is fundamentally the same.

Photoreceptor signals are conveyed to the brain by bipolar and ganglion cells

We can regard the detection of light as the essential event in vision, and its communication to the brain as the essential element of visual information; the signal says that light has been detected. This information is conveyed by what may be called vertical, or direct, pathways through the retina (Figure 13.6). The basic elements of the vertical pathways are shown in Figure 13.6*a*. The photoreceptor makes synaptic contact with the second-order vertical neuron, a **bipolar cell**, which has a "bipolar" form: The cell body lies between the dendritic end, which receives inputs from the photoreceptor, and the axon terminal end, which is presynaptic to a third-order vertical neuron, a **ganglion cell**. The dendrites of the ganglion cell integrate bipolar cell inputs, and the initial segment of its axon converts these signals to action potentials that travel down the axon. (The optic nerve is the bundle of all retinal ganglion cell axons going to the brain.)

This scheme is too simple to be true, but one feature is accurate and very important: It suggests that action potentials in a single optic-nerve axon can be related to the absorption of light in a single photoreceptor, independent of all other photoreceptors. This is a fair description of the elemental wiring of foveal cones, and it means, as we will consider later, that neighboring cones can detect different amounts of light and faithfully transmit these differences to the brain. This is the basis for our ability to see fine spatial detail.

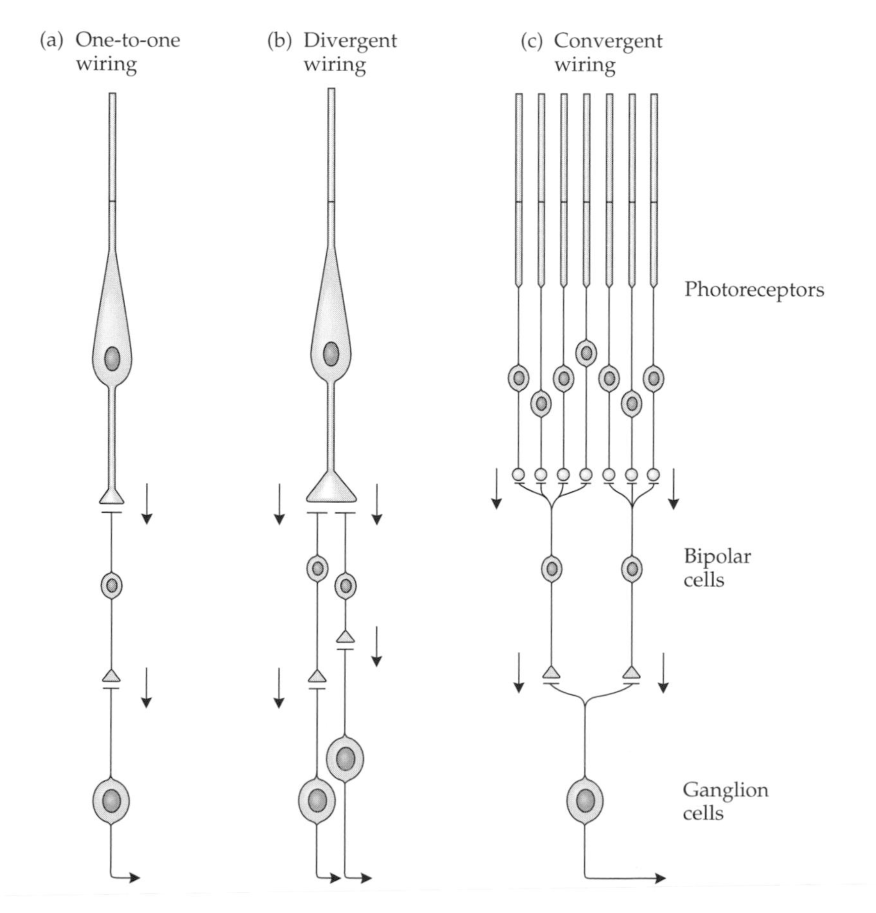

Figure 13.6
Wiring Patterns in the Vertical Pathways through the Retina
The simplest vertical pathway is one-to-one (*a*), in which each ganglion cell receives signals originating in a single cone. From the cone's perspective, however, the wiring is divergent (*b*) since each cone sends signals to more than one bipolar cell and thus to more than one ganglion cell. Convergent wiring (*c*) is characteristic of the peripheral retina, where each ganglion cell collects signals originating in many photoreceptors. Arrows indicate the direction of information flow.

Cones normally contact several bipolar cells, which makes the wiring, from the cone's perspective, *divergent*. As a result, the cone signal travels down several vertical pathways. As drawn in Figure 13.6*b*, however, each ganglion cell is still reporting on signals originating in a single cone.

In *convergent* wiring (many-to-one), there are a number of photoreceptors, a smaller number of bipolar cells, and a single ganglion cell (Figure 13.6*c*); this type of wiring is characteristic of the retina outside the fovea. Since inputs converge on a single ganglion cell, that cell reports on the aggregate activities of many photoreceptors. This system is not good for resolving fine detail in the retinal image, because the signals generated by the ganglion cell do not, and cannot, indicate which of the photoreceptors absorbed light. Convergent wiring, however, is very good for detecting and reporting the presence of small amounts of light; the more photoreceptors the ganglion cell samples, the higher the probability that the requisite number of photoreceptors will absorb a photon and that their combined signals will reach the ganglion cell's threshold for signaling. The extrafoveal retina utilizing convergent wiring in the vertical pathways contains both rods and cones, and many ganglion cells in this part of the retina receive signals from both of them.

Lateral pathways connect neighboring parts of the retina

If the vertical pathways illustrated in Figure 13.6 were the only pathways in the retina, the output from the ganglion cells to the brain would be a fairly direct reflection of the amount of light absorbed by the photoreceptors to which the ganglion cells were connected by the intervening bipolar cells. Light falling on nearby photoreceptors would be of little consequence. There are lateral pathways at two levels of the retina, however, and their presence means that signals supplied directly to vertical pathways from one or more photoreceptors are influenced indirectly by signals from neighboring photoreceptors.

The first of these lateral pathways is made up of **horizontal cells** (Figure 13.7), which are neurons in the outer retina that receive inputs from photoreceptors and make connections between neighboring photoreceptors There are at least two types of horizontal cells in the human retina, with different patterns of connectivity, but for the moment the important feature is the general nature of the horizontal cells' influence: The signals going to the bipolar cells in the vertical pathways are a result of direct photoreceptor input, as well as input from surrounding photoreceptors as mediated by the horizontal cells.

At first glance, this kind of lateral connection might seem detrimental; it implies that the signals reaching a ganglion cell will *always* be a consequence of activity in more than one photoreceptor, a form of convergent wiring that would interfere with the resolution of fine spatial detail in the retinal image. The effect of the lateral pathway, however, depends on the sign or polarity of the lateral input. If the effect were positive—that is, of the same sign as the direct photoreceptor input—it would be another form of convergent wiring and it would coarsen the grain for resolution. But if the effect were negative—that is, opposite in sign to the direct photoreceptor input—the result would be different; as will be discussed in Chapter 14, negative lateral inputs transform the vertical pathway from one that signals the amount of light to one that signals illumination *contrast* (difference in illumination) between photoreceptors and their neighbors. And as mentioned earlier, information in the retinal image is associated with change. A difference in illumination between regions is a spatial change, and these lateral pathways appear to enhance the signaling of spatial changes in illumination. This enhancement should help, rather than hinder, spatial resolution.

Lateral pathways in the inner retina are formed by **amacrine cells**, whose processes establish connections between vertical pathways where the bipolar cells and ganglion cells interact. Amacrine cells may receive inputs from bipolar

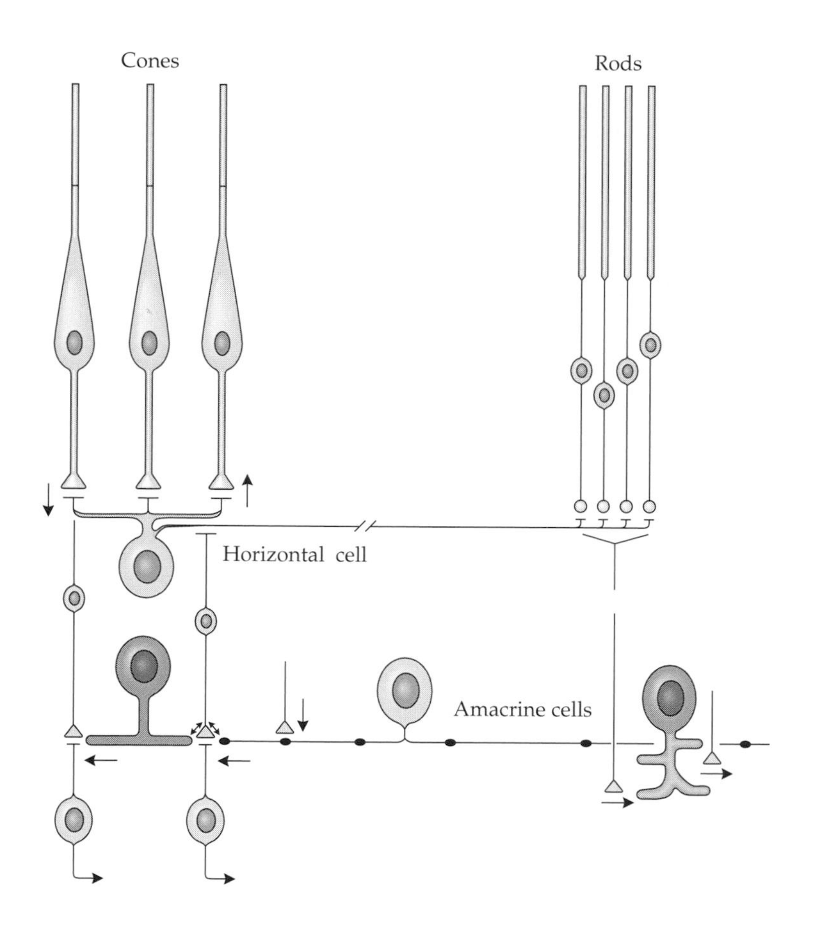

Figure 13.7
Lateral Pathways Are Formed by Horizontal Cells and Amacrine Cells
Horizontal cells make connections among photoreceptors in the outer retina. Amacrine cells of different sizes make connections among vertical pathway neurons in the inner retina. Arrows indicate possible directions of information flow.

cells and other amacrine cells; their outputs may be to bipolar cells, other amacrine cells, and ganglion cells. Three amacrine cells are shown in Figure 13.7; two small amacrine cells connect nearby vertical pathways and a large one makes connections over a large distance.

Since there are only a few types of horizontal cells, only a few different lateral pathways are possible in the outer retina. Amacrine cells, however, come in numerous distinguishable types; they differ in the extent of their lateral spread, in the precise level within the inner retina at which their processes branch, in the neurotransmitters or neuromodulators they contain, and presumably in their effects on information flow. The possible number of lateral pathways in the inner retina is accordingly large.

At present, we know so little about amacrine cells that any discussion must be very general. It is clear that amacrine cells make connections among neighboring vertical pathways, and it is likely that some of these connections have the sort of contrast or edge-enhancing function attributed to the horizontal cell pathways. Amacrine cell inputs also seem to make some of the ganglion cells remarkably sensitive to temporal changes, so that they signal the instant at which light intensity increases or decreases. Photoreceptors, horizontal cells, and bipolar cells have sustained responses to light, meaning that the response persists at a fairly constant level throughout the duration of the light stimulus. The situation changes in the inner retina, however, where some amacrine and ganglion cells have "transient" responses, consisting of a brief burst of activity when light intensity increases (ON), decreases (OFF), or both (ON–OFF). Possible mechanisms for this conversion from sustained to transient responses will be considered later; for the moment, the key point is that lateral amacrine cell pathways are intimately involved.

Recurrent pathways may assist in adjusting the sensitivity of the retina

The elements of the vertical pathways (photoreceptors, bipolar cells, and ganglion cells) and of the lateral pathways (horizontal and amacrine cells) have been recognized and considered for many years by scientists thinking about retinal operations. More recently, other cells, called **interplexiform cells**, have been shown to be a functionally significant component of some retinas. Most of the interplexiform cells studied thus far are generally similar to some of the large, laterally spreading amacrine cells. As Figure 13.8 shows, however, they have an important distinguishing feature: a process, sometimes several, arising from the cell body or nearby branches, and extending up to the outer retina in the vicinity of the photoreceptor terminals. In a few species, electron microscopy indicates that the interplexiform cells receive inputs from bipolar cells and amacrine cells in the inner retina and they have outputs to horizontal cells and occasionally bipolar cells in the outer retina. In short, the flow of information carried by these cells is opposite the direction of the bipolar cell pathways from outer to inner retina.

Several different types of interplexiform cells using different neurotransmitters have been identified in nonmammalian retinas. It is not clear if these cell types perform the same operation, but their general pattern of connectivity from inner to outer retina raises the possibility that one or both types are involved in the adjustment of retinal sensitivity. If signals in bipolar cells, for example, were becoming too strong as they reached the inner retina, a sample of this activity by

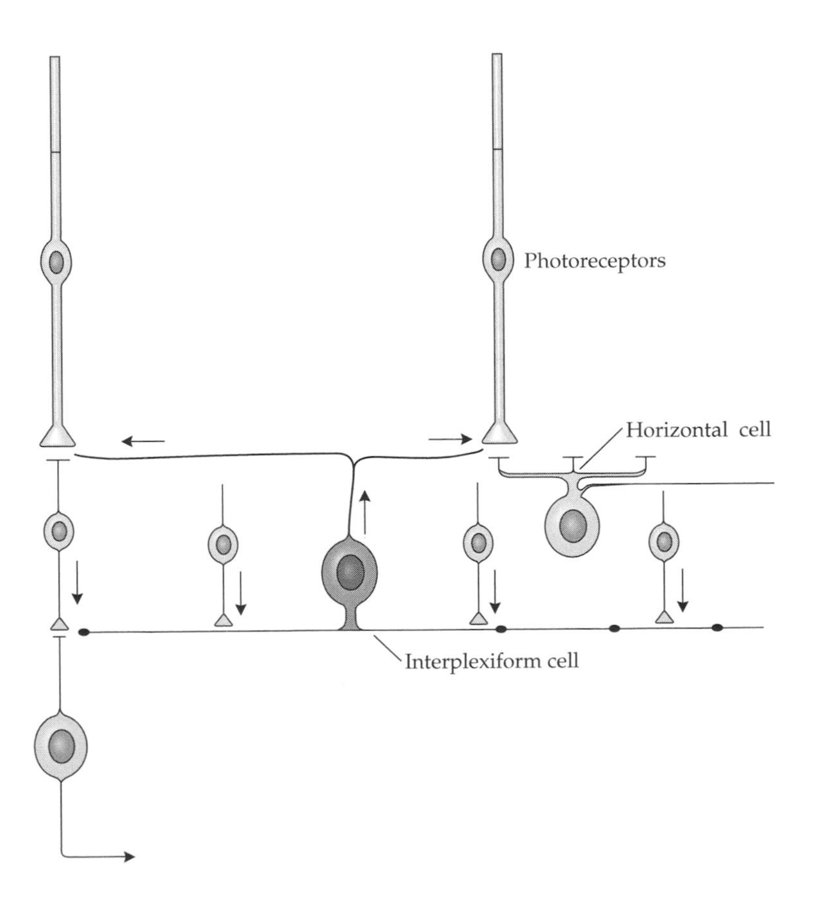

Figure 13.8
Interplexiform Cells May Constitute Recurrent Pathways
Arrows show the presumed path of signal flow. The interplexiform cell receives inputs from bipolar cells, and its outputs are at the level of the photoreceptor terminals and horizontal cells.

the interplexiform cells could be conveyed to the outer retina and used to turn down the amount of signal amplification, thereby reducing the bipolar signals in the inner retina back to an acceptable range. Conversely, the recurrent pathway could be used to increase the amplification if the bipolar signals became too small. These changes could be generated by changes in convergence within the vertical pathways; more convergence would increase sensitivity, and less convergence would decrease it.

Evidence in a few species of fish shows that some interplexiform cells perform a sensitivity adjustment operation. It is less certain what the interplexiform cells of mammals do, and as far as the human retina is concerned, we know only that interplexiform cells are present. They may have a role to play in sensitivity adjustment, but until we know more about their number and distribution, we can only speculate. If it turns out that interplexiform cells in the human retina are rare and scattered, it will be difficult to make a case for any significant functional role. But if their processes in the outer retina spread out so that there are no large gaps between processes from neighboring cells, it is more likely that the cells are there for a reason, not just by developmental accident.

I suspect that interplexiform cells do play a role in regulating some aspect of information flow through the human retina. At this point, however, all we can do is note their presence, consider a possible function, and move on to other things. Stay tuned.

The retina has anatomical and functional layers

Much has been made of the retina's beautiful layering ever since it became known—and with good reason. As we go from the outer to the inner retina, we can talk about a hierarchy of anatomical structure, in which new elements are added at each stage, and a hierarchy of functional operations, in which information flow is modified in stages. The basic structure is shown in Figure 13.9.* Traditionally, the retina has been shown with ten layers, and they are indicated (at left of the micrograph) with the names they usually carry. Other important distinctions can be made, however; they have been added (at right of the micrograph) to the traditional labeling, and their relation to the elements of the vertical and lateral pathways is shown in the schematic.

The outermost layer of the retina is the **pigment epithelium of the retina** (**RPE**); it lies between the outer segments of the photoreceptors and Bruch's membrane of the choroid, and consists of a single continuous layer of roughly cuboidal cells that contain melanin pigment. The pigment epithelium is not directly involved in any of the neural events of vision, but it is critical for the normal functioning and well-being of the photoreceptors, a topic to which we will return later in the chapter.

Photoreceptors extend through four of the traditional layers: Their outer and inner segments lie in the **photoreceptor layer**, their nuclei form the **outer nuclear layer**, and their axons and terminal endings make up the **outer plexiform layer**. The **external limiting membrane** along the border between the inner segments and the outer nuclear layer is a series of tight junctions between photoreceptors and **Müller's cells**, which are the main glial cells in the retina (see Chapter 16). Since the photoreceptor layer contains only part of the photoreceptors, I will rarely use the term, referring instead to the outer or inner segments. Similarly, the outer plexiform layer has two different components, the photoreceptor axons and their terminals. The terminals form a well-defined layer called the **outer synaptic layer**. The axonal portion of the outer plexiform layer in the

*Anatomists have illustrated the retina with its outer layers uppermost for the past century and a half. In the clinical literature, however, it is often shown with the photoreceptors down. I don't know why. I follow anatomical tradition here.

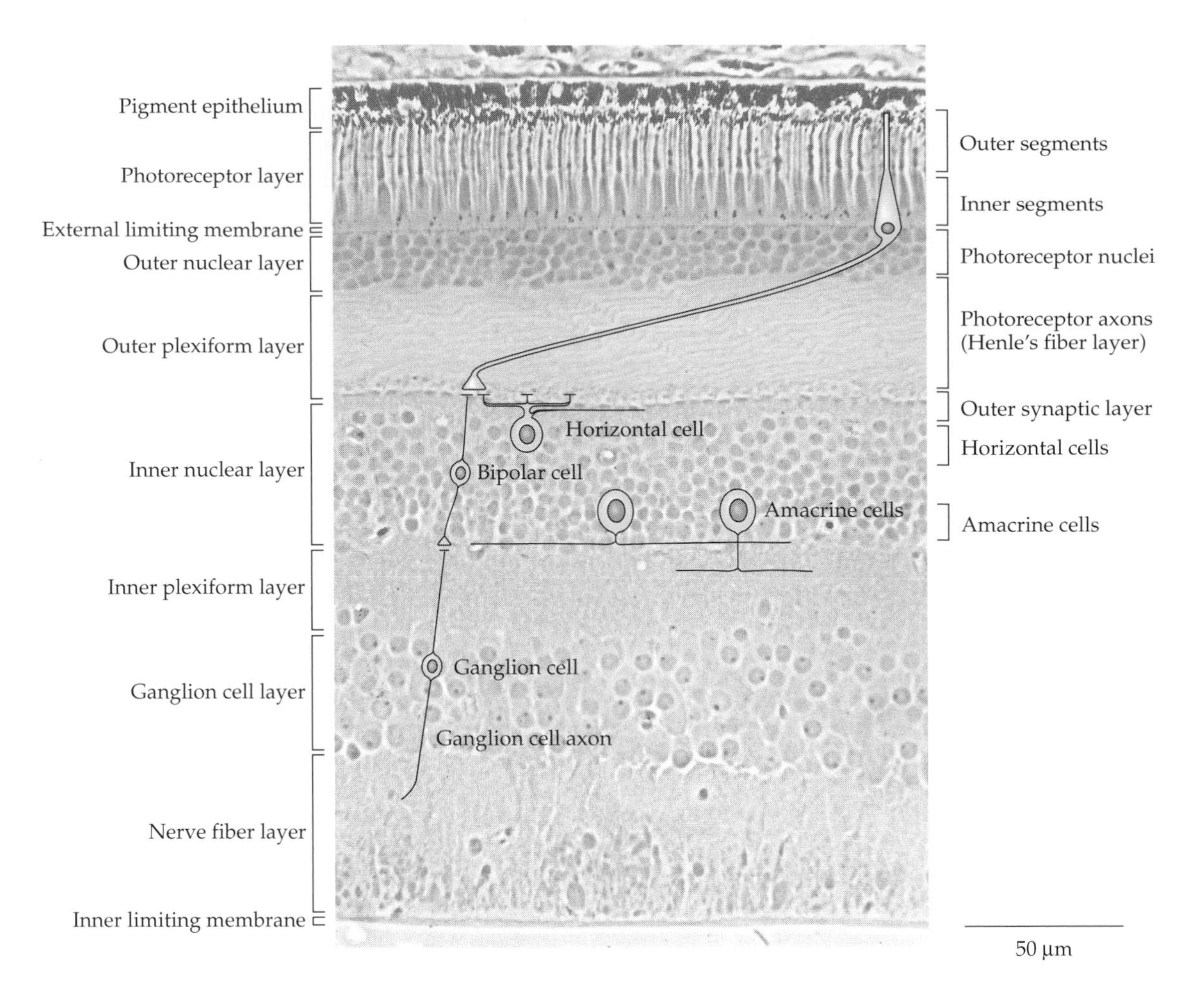

Figure 13.9
Layers of the Human Retina
(Background) Photomicrograph of a section through the retina near the fovea, which is out of the picture to the right; most retinal layers are thicker here than in more peripheral parts of the retina. The schematic (foreground) shows the locations of the major classes of neurons. (Photomicrograph from Boycott and Dowling 1969.)

central retina is **Henle's fiber layer**; otherwise, the reference will be to photoreceptor axons.

Functionally, these outer layers can be thought of in terms of the photoreceptors: The pigment epithelium is involved in the transport of materials to and from the photoreceptors, the photoreceptor layer is the site of light absorption and phototransduction, the outer nuclear layer contains the cells' genetic instructional sets, and the outer synaptic layer is the site of photoreceptor communication with other neurons. The function of the external limiting membrane is less clear, but it may create a resistance to the flow of ionic currents around photoreceptors that will affect the way voltage changes in the outer segments are reflected at the photoreceptor terminals.

The **inner nuclear layer** contains the nuclei of all the cells that are intermediate between the photoreceptors and the ganglion cell output from the retina—namely, the bipolar cells in the vertical pathways, all of the horizontal cells and most of the amacrine cells in the lateral pathways (some displaced amacrine cells have their cell bodies in the ganglion cell layer), and the interplexiform cells of the recurrent pathways. The nuclei of Müller's cells are also located here. The cell bodies of the horizontal cells lie mainly along the border of the inner nuclear layer and the outer synaptic layer, while amacrine and interplexiform cells have their somata along the border with the inner plexiform layer. Bipolar cell bodies occupy the central part of the inner nuclear layer. Functionally, the inner nuclear layer is the site of the cells' genetic and synthetic machinery, and while there is a flow of information in the form of ionic current through the inner nuclear layer, the important interactions among cells occur elsewhere, in the plexiform layers.

The dendrites of the bipolar cells extend up to the outer synaptic layer to receive inputs from the photoreceptors. Also, all of the lateral connections by the horizontal cells occur at the level of the photoreceptor terminals and bipolar dendrites, as do any connections made by the recurrent processes from the interplexiform cells. Thus, the outer synaptic layer is the place where photoreceptor signals are passed along to bipolar cells and modified by the lateral connections of horizontal cells and—perhaps—the recurrent pathways from interplexiform cells.

The bipolar cells have their axon terminals in the inner plexiform layer, where they make contact with ganglion cell dendrites and the processes of amacrine and interplexiform cells. Thus, the inner plexiform layer is the site of the final set of interactions among vertical and lateral pathways that produce the signals leaving the retina via the ganglion cell axons.

The inner plexiform layer has been known for many years to be laminated, that is, to have sublayers within it. These sublayers are very obvious in some animals, and Ramón y Cajal (see Vignette 14.2) noted five of them in fish and seven or more in birds and reptiles. Most mammalian retinas do not show clear histological divisions to mark sublayers, but these layers are present nonetheless. In fact, one of the major functional subdivisions was suggested in the divergent pathway in Figure 13.6, which showed the bipolar cells terminating in different parts of the inner plexiform layer, one ending in the outer (distal) part of the inner plexiform layer and one ending closer to the ganglion cells (proximal). This point will be taken up in more detail later, but signals about increasing (ON) and decreasing (OFF) light intensity go to different parts of the inner plexiform layer; OFF signals go to the distal part, ON signals to the proximal part.

Different lateral amacrine cell pathways are also associated with specific sublayers in the inner plexiform layer. This story will also be taken up again later, in Chapter 14, but it can be summarized by the statement that the amacrine cell pathways are highly ordered; the inner plexiform layer can be thought of as a series of functional sheets, stacked one on top of another. The interactions among bipolar cell terminals, amacrine cell processes, and ganglion cell dendrites within these sheets or sublayers represent the final stage of the editing of information flowing through the vertical pathways on its way to the brain.

The **ganglion cell layer** contains the cell bodies of ganglion cells and displaced amacrine cells. The presence of the amacrine cell bodies here makes the name of the layer somewhat misleading, particularly since a sizable fraction of the cell bodies, up to 40% in some species, are not ganglion cells. Functionally, however, the displaced amacrine cells operate like other amacrine cells, and all their interactions take place in the inner plexiform layer. The ganglion cell dendrites are in the inner plexiform layer, where they receive bipolar cell and amacrine cell inputs, while their axons are in the next layer inward, the **nerve fiber layer**.

At this point, the retina's job of editing information is complete, thanks to the interactions in the inner plexiform layer, and the remaining task is transmission. In terms of transmission, the ganglion cells are the most conventional of the retina's neurons; their axons leave the cell bodies at axon hillocks, where action potentials arise that are propagated along the axons to the target nuclei in the brain. Although there is evidence in some species for electrical coupling between neighboring ganglion cells, there are no comparable observations in primate retinas and it is generally assumed that the activity in a ganglion cell has no direct effect on its neighbors. (Neural activity in neighboring ganglion cells can be highly correlated, however, because they may share elements of their input pathways.)

The innermost retinal layer, the **inner limiting membrane**, is formed by terminal expansions of Müller's cells, which extend from the inner surface of the retina out to the photoreceptor layer. As discussed in Chapter 12, vitreal collagen fuses with the inner limiting membrane, an arrangement that may detach the retina when the collagenous portion of the vitreous shrinks in volume.

Vignette 13.1

"Everything in the Vertebrate Eye Means Something"

LIKE MANY GRADUATE STUDENTS, I worked part-time as a teaching assistant to supplement a modest fellowship stipend. My first job, organizing and overseeing laboratories for a course on the anatomy of the eye, was the seed from which this book slowly grew. After being told to look for microscope slides used by the professor who had taught the course until his death the previous year, I soon found them—well over a thousand slides—strewn on shelves, dumped in boxes, and covered with dust. As I cleaned, examined, and sorted the collection, I was surprised by retinas with cones shaped like miniature Chianti bottles, bipolar cell nuclei like giant cobblestones paving the inner nuclear layer, and Latin names of species I did not recognize. I had inherited the legacy of Gordon Lynn Walls (1905–1962), one of the great students of comparative ocular anatomy.

Walls originally trained in engineering at Tufts University in Massachusetts, but he was not good at mathematics, so he changed fields and became a zoologist. His publications on the eye began in 1928, on the eye of the deer mouse, a product of his work for a master's degree at Harvard; he received his doctorate of science from the University of Michigan in 1931. His subsequent career took him to the University of Iowa, to Wayne State University in Detroit, to Bausch & Lomb during World War II, and finally to Berkeley. His best-known work is his now classic book, *The Vertebrate Eye and Its Adaptive Radiation* (Cranbrook Institute of Science, 1942, reprinted by Hafner Publishing Company, 1963). Many of the slides I so carefully cleaned as a teaching assistant were the basis for illustrations in his book, and the parts he drew are easily recognizable.

I never met Gordon Walls, but there are still many people around who knew him. Some people knew him better than they would have liked; however brilliant his work and his writing, he could be a difficult personality—egotistic, opinionated, arrogant, and caustic. Yet the affectionate tone of his obituary (in *Vision Research*, volume 3, pages 1–7, 1963), reveals other, more appealing aspects of his character, however rarely they surfaced. And he was a gifted, inspiring teacher, one of those rare individuals who could elicit enthusiasm for the subject in his students (in spite of his being a dictator in the classroom). Anyone who schedules laboratory experiments on dark adaptation only at night in the dark of the moon, in a darkened building with only red safelights is out to fascinate, not just to inform.

Walls was not the first to study the comparative anatomy of the eye, but he was one of the first to consider ocular structure in a broad biological context, trying to relate the themes and variations among vertebrate eyes to the visual environments of different species (hence the word "adaptive" in the title of his book) and to their evolutionary history.* He was enthralled with the variations he saw or read about in the work of others, and he enthusiastically explored such issues as pupil sizes and shapes, lens pigmentation in squirrels, the various forms of reflective elements in the choroid (tapeta), photomechanical changes in photoreceptors, shapes and locations of foveas, and so on.

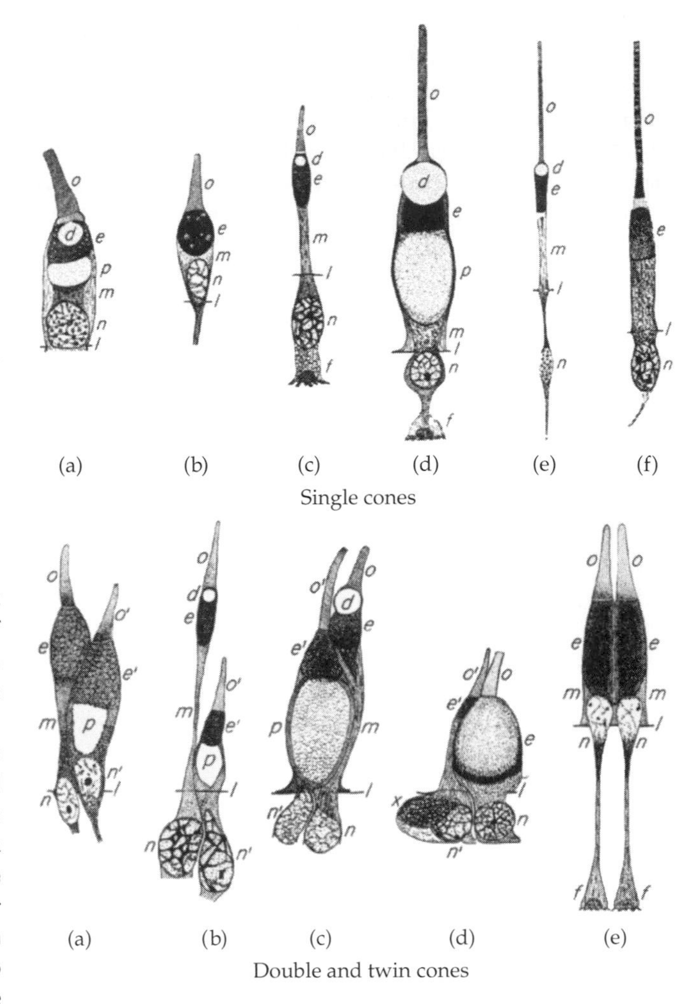

Figure 1
Varieties of Cones
The single cones are from sturgeon (a), goldfish (b), frog (c), snapping turtle (d), marsh hawk (e), and human (f). The double cones are from bowfin (a), frog (b), painted turtle (c), and European grass snake (d). The twin cones are from the bluegill (e). Parts of the cones are labeled as follows: *d*, oil droplet; *e*, ellipsoid (part of inner segment); *f*, footpiece (cone terminal); *l*, external limiting membrane; *m*, myoid (part of inner segment); *n*, nucleus; *o*, outer segment.

*In 1943, just a year after Walls published his book, another on the same subject was published in Paris. *Les yeux et la vision des vertébrés* by André Rochon-Duvigneaud is a wonderful book that has languished in relative obscurity, quite undeserved; as far as I know, it has not been translated into English. The book's cheap paper cover and the author's brief comment in the preface about *circonstances difficiles* are understated reminders of life in occupied Paris during World War II.

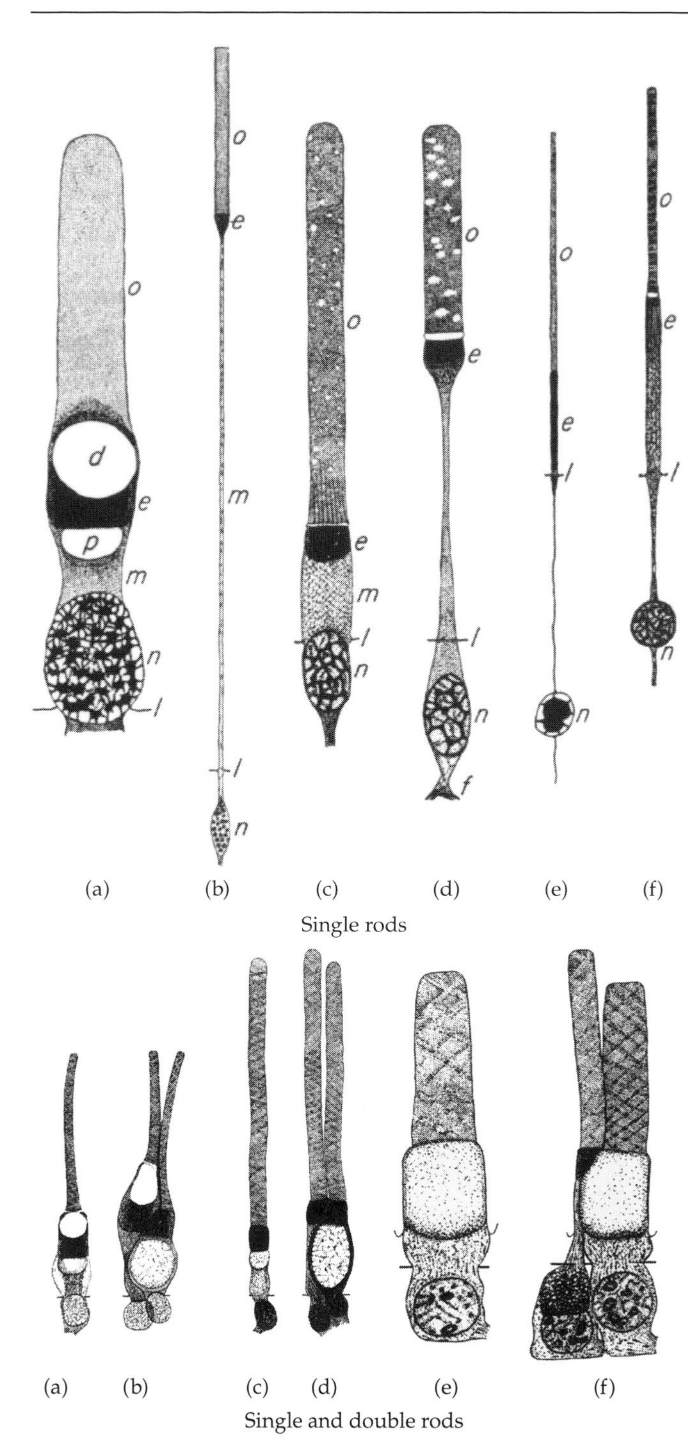

Figure 2
Varieties of Rods
The single rods are from lungfish (a), goldfish (b), frog "red" rod (c), frog "green" rod (d), flying squirrel (e), and human (f). The pairs of single and double rods are from the night lizard (a and b), banded gecko (c and d), and the spotted night snake (e and f). Labels are the same as in Figure 1.

Photoreceptors were Walls's real passion, however, and he described their comparative anatomy in loving detail; Figures 1 and 2 contain some of his drawings. One of his main contributions was describing the wonderful variety of photoreceptors—double cones, twin cones, double rods, rodlike cones, conelike rods—as well as telling us where they are found and why, in his opinion, they are there. (The terms "cones," "rods," and "photoreceptors" do not appear in the index of Walls's book; he called them by their older name, "visual cells.") Walls concluded that the first vertebrate photoreceptors were cones, and that rods evolved later—a view that he expressed vigorously as a "transmutation" hypothesis; he was basically correct, as more recent and better evidence has shown.

Although Walls left us with some wonderful images of the eye's structure, his real artistry was his command of language; he could paint pictures with words. Whether or not he was correct in what he said (he wasn't always), he is a delight to read, and that is one of the reasons his work still commands attention. He began a discussion of reflecting tapeta in vertebrate eyes thus: "Perhaps you will ponder for a moment the antiquity of vanity. Even if you are not so philosophical, you will at least dwell on the antiquity of mirrors." And an account of his efforts to trace obscure references in the literature led to describing a library card catalog as having "its brazen tongue in its oaken cheek." (In the same article, Walls refers to reference librarians as "nature's noblewomen"; a bit sexist, perhaps, but clearly a compliment.)

His best-known line is the title of this essay: "Everything in the vertebrate eye means something." That sentiment, which has an Aristotelian ring to it, is another reason people find inspiration in Walls's writings. He was interested in vision in its broadest terms, and there are very few aspects of vision that did not come under his purview. Nocturnality, diurnality, color vision, binocularity, acuity, sensitivity, eye movements, and their variations in the vertebrate phylum were all fair game for a man who must have been one of the world's great compulsive explainers.

Walls's most important message was not original, but his expression of it is peerless. He told us that eyes and visual systems take many forms—variations that are inherently interesting and rich in biological significance and implication (see the Prologue). Mimicry, camouflage, navigation and orientation, courtship displays, prey catching and predator avoidance, habitat selection, and a host of subtler social interactions that go on in the world involve vision—invariably a vision that differs from our own. (We use our vision in the human versions of these activities, of course.) Understanding human vision is important for societal and cultural reasons, but it is only one version of a ubiquitous and varied biological phenomenon. And thinking about vision in other animals is a useful reminder that our eyes and visual system evolved in the same natural environment that most of the animal kingdom still inhabits, conditions very different from the artificial environments that we have recently created for ourselves.

Walls's personal vanity notwithstanding, his work can be taken as a reminder that our species did not invent the mirror in which we admire ourselves: "The inventor of the mirror was no hawk-nosed youth dreaming in the shadow of a half-built pyramid. Indeed, no man at all, but an armored shark gliding over the bottom ooze of the warm Devonian sea."*

*"It's done with mirrors." *Chicago Naturalist* 1: 103–109, 1938.

Catching Photons: Photoreceptors and Their Environment

Each photoreceptor contains one of four photopigments, each of which differs in its spectral absorption

The human retina has four photopigments, one rod photopigment and three cone photopigments, all of similar construction. Each photopigment has two elements: a large protein portion, which is called an **opsin**, and a small attached molecule called ***retinal***, which is an aldehyde of vitamin A.* The *retinal* is the portion of the photopigment first affected by light absorption and it is considered to be the active part, or **chromophore**, of the opsin–*retinal* complex. In the human photopigments, only opsin differs from one pigment to another; the same *retinal* is used for all.

The differences among photopigment opsins have an important consequence—namely, differences among the photopigments in the wavelengths of light that they preferentially absorb. Thus a photopigment's **absorbance spectrum**, which is the amount of light absorbed as a function of wavelength, is its signature, distinguishing it from other photopigments.

In principle, an absorbance spectrum is easy to determine (Figure 13.10). It is a matter of having a solution of photopigment, passing light of a given wavelength through the solution, and measuring the amount of light incident on the solution (I_{inc}) and the light exiting the solution (the light transmitted, I_{trans}); the difference, $I_{inc} - I_{trans}$, is the amount of light absorbed (I_{abs}). The **optical density** (D) and the **absorbance** (A) are both described by the logarithm of the ratio of incident to transmitted light, that is, $\log (I_{inc}/I_{trans})$.

But as indicated in Figure 13.10, the amount of light absorbed depends not only on the pigment's intrinsic spectral absorption properties, but also on the concentration of the pigment in solution and the length of the path that the light beam must traverse through the pigment solution. These features are expressed formally in terms of absorbance as

$$A = e_{\lambda} \cdot c \cdot l$$

where c is concentration, l is path length, and e_{λ} is the spectral **extinction coefficient**. The quantity of interest is e_{λ} because it is a property of the photopigment molecule, but it may be difficult to isolate if concentration or path length is not known. The strategy is to use the ratio $A_{\lambda}/A_{\lambda_{max}}$. In this ratio, which is the *relative* absorbance, concentration and path length cancel out. The relative absorbance is unity at the peak absorption wavelength and less than 1 at other wavelengths. Thus a plot of relative absorbance as a function of wavelength provides a unique signature for the photopigment molecule, showing how well the molecule absorbs light at different wavelengths.

Absorbance spectra for the human photopigments are shown in Figure 13.11. Each pigment absorbs maximally at a different wavelength (its λ_{max}) that can be used to characterize the pigment. The rod photopigment, rhodopsin, has its peak absorption near a wavelength of 496 nm, which would be seen as blue-green light. The three cone pigments have absorption maxima in the blue part of the spectrum (419 nm), green (531 nm), and what is usually called red when referring to the photopigment but is really yellow (558 nm).

These absorbance spectra were obtained from individual photoreceptors, not pigments in solution.† The method used, called microspectrophotometry, is dif-

*The old gambit that mothers used to get children to eat carrots—"carrots are good for your eyes"—contains an element of truth. Carrots are rich in vitamin A, and vitamin A deficiencies do affect vision, since the *retinal* in the photopigments is derived from vitamin A stored in the liver. Normal diets have plenty of vitamin A from sources other than carrots.

†William A. H. Rushton (1901–1980) used his method of "retinal densitometry" to provide believable spectra for the human cone pigments in vivo, and he took the opportunity to name

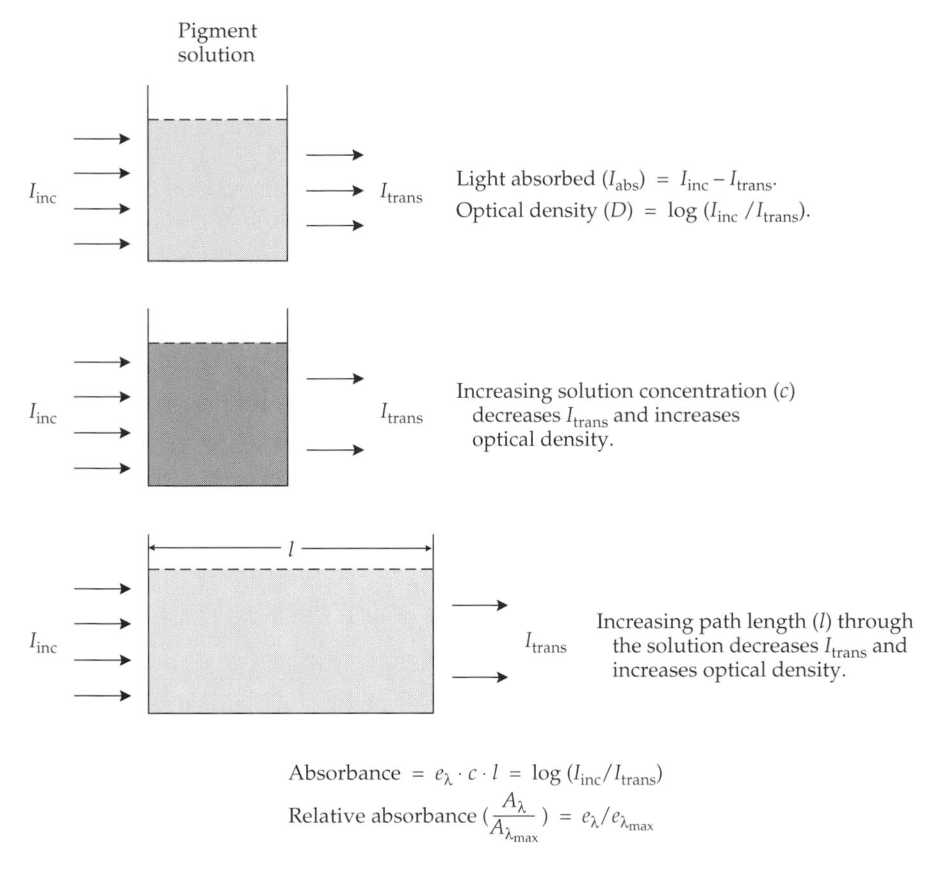

Figure 13.10
Measuring Absorbance
The amount of light absorbed by a pigment in solution depends not only on the characteristics of the pigment, but also on its concentration and the distance that light travels through the solution. When neither concentration nor path length is known, the relative absorbance (or relative density) is used to factor out both concentration and path length. See text for additional explanation.

ficult because the amount of photopigment in a photoreceptor is limited. The light passing through the photoreceptor bleaches some of the pigment, which means that the concentration of the pigment changes during the course of measurements at different wavelengths. Moreover, bleached photopigment accumulates that may have absorbance characteristics different from those of the parent photopigment. Thus the raw data must be corrected to produce the curves in Figure 13.11. But once this is done, absorbance measurements from different photoreceptors always coincide with one of the four curves, which is evidence that each photoreceptor contains one, and only one, photopigment. Any mixture of photopigments would necessarily produce different absorbance spectra.

them: erythrolabe, chlorolabe, and cyanolabe (red-, green-, and blue-seizing). It is unfortunate that his names have not stuck, because the common alternatives (L, M, S; or long-, medium-, and short-wavelength-sensitive) are mundane and no more precise. Rushton's magnificent lecture *cum* performance entitled "Three Cambridge Students of Light and Colour" (Isaac Newton, Thomas Young, and James Clerk Maxwell) implied that a fourth name—his—should be added to this grand tradition.

I once questioned some of Rushton's early data on the blue-cone pigment, which prompted—in his inimitable style—a vigorous and thorough exposure of my ignorance about color vision. I was right to be skeptical, as it turned out, but I have avoided the subject ever since.

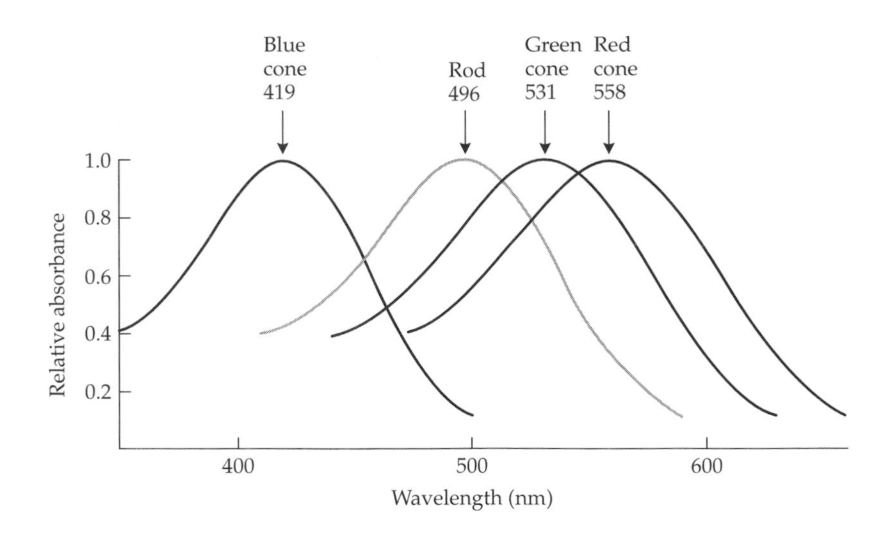

Figure 13.11
Absorbance Spectra for Human Visual Pigments
These curves are based on measurements from individual photoreceptors. Since each photoreceptor contains only one photopigment, these absorbance spectra are signatures for the photoreceptors. There is one kind of rod photoreceptor and three kinds of cone photoreceptor (red, green, and blue). (Replotted from Dartnall, Bowmaker, and Mollon 1983.)

A photoreceptor is characterized by the single photopigment it contains. All rods contain rhodopsin, so there is just one type of rod photoreceptor. Cones may have any one of the three other pigments, so there are three types of cone photoreceptors. The cones are familiarly referred to as red, green, or blue cones, on the basis of the part of the spectrum that their photopigment absorbs most strongly. (A noncommittal, less colorful designation is long-, medium-, and short-wavelength-sensitive cones, or L, M, and S cones.) Since normal trichromatic color vision utilizes all three cone photopigments, it has long been suspected that individuals with profound color deficiencies (dichromats) lack one of the photopigments. This is true, and since each cone contains only one pigment, the basis of the defect is the lack of the gene that normally allows a subset of cones to synthesize one of the pigments. (The total number of cones is not affected. What should have been red cones, for example, end up being green cones instead.)

Although the terminology "red cone" or "green cone" makes them sound very different, their absorbance spectra are not far apart, and the spectral regions in which they absorb overlap considerably. But the small difference is significant. Red cones absorb long-wavelength light to which the green cones are almost totally insensitive, and green cones absorb at shorter wavelengths that are not absorbed by the red cones. With the exception of the wavelength at which the two absorbance spectra intersect, no wavelength will have equal absorption by both photopigments. In other words, a single wavelength will have different effects, and this difference makes wavelength discrimination possible.

Color vision requires more than one photopigment

Even if we know that a cone contains a particular photopigment, the photoreceptor's signal says nothing about the wavelength of absorbed photons. Photoreceptors signal the amount of light they absorb—how many photons they catch—and as long as the wavelength is within their absorbance spectrum, only the number of photons caught is relevant. To put it another way, wavelength discrimination would not be possible with just one pigment.

The problem is illustrated in Figure 13.12. With a single photopigment, there are many possible pairs of wavelengths that produce the same absorption and the same signal from the photoreceptor. In fact, any two wavelengths can be confused because it is impossible to know if the given magnitude of a photoreceptor signal resulted from absorption of a large fraction of photons from a weak light source to which the pigment was quite sensitive or absorption of a small fraction

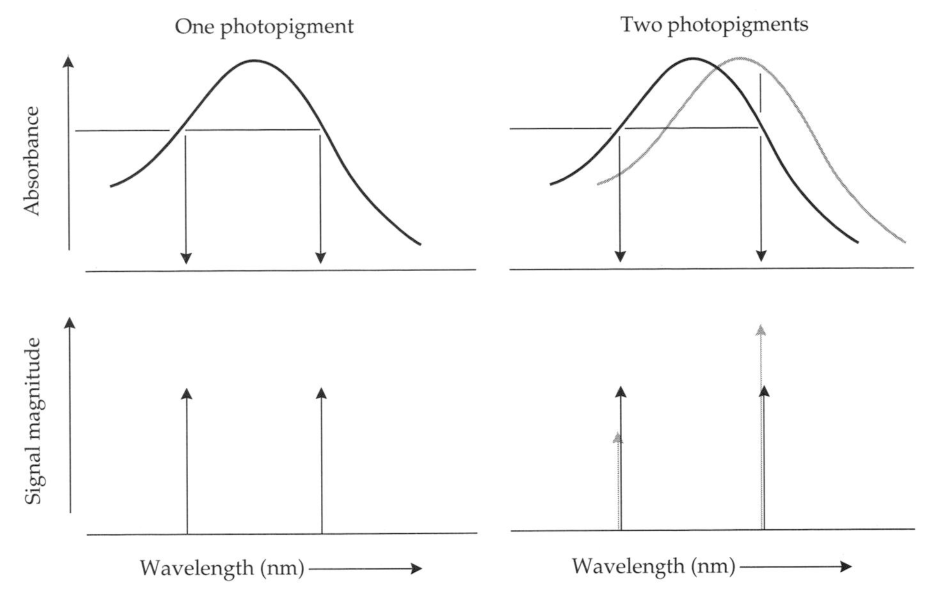

Figure 13.12
Wavelength Discrimination Requires More Than One Photopigment
Photoreceptor signals depend on the number of photons absorbed, not on their wavelength. Thus signals based on a single photopigment are ambiguous; lights of different wavelength can produce the same signal strength. When two photopigments are present, it is impossible to find two wavelengths that produce identical signals from both photoreceptors.

from a strong light source to which the pigment was relatively insensitive. Thus, with just one pigment, it is impossible to say if a change in photoreceptor signal strength is the result of a change in the wavelength of light, a change in the intensity of light with no change in wavelength, or a change in both.

With more than one pigment, however, wavelength discrimination is possible. The two wavelengths that are absorbed equally by a single photopigment are absorbed differently by a second one. If the signals from the photoreceptors containing these two pigments can be compared, there is a basis for saying that the two lights have different wavelengths. Also, the trade-off between wavelength and intensity that can fool a single photopigment does not work when two pigments are present.

In general, wavelength discrimination is possible with two different photopigments and is even better with more than two. Most mammals get by with blue and green cone systems, most primates also have the third (red) system, and some birds and insects have a fourth cone system, in the ultraviolet region of the spectrum. Our system of three cone pigments allows us to discriminate wavelengths of light near the middle of the spectrum that differ by only 2 or 3 nm.

At light levels where the cones are operating, the rods are very insensitive, and the behavioral spectral sensitivity curve for the eye is the curve designated "cones" in Figure 13.13. This is the photopic (light-adapted) spectral sensitivity function for the human visual system, produced by the combination of information from all three cone systems. The curve peaks near 555 nm, meaning that the visual system is most sensitive to yellow-green light under light-adapted, cone-dominated conditions.

When the eye is adapted to dark, however, the spectral sensitivity function shifts along the wavelength axis so that its peak is now in the blue-green area near 500 nm. This scotopic (dark-adapted) spectral sensitivity function is almost identical to the absorbance spectrum for rhodopsin (see Figure 13.11), which is one of the pieces of evidence that the curve is determined solely by the rods. And since the rods are detecting dim lights under dark-adapted conditions, there can be no wavelength discrimination, because only one photopigment is in use. Rods and their rhodopsin have nothing to do with our color vision.

The change from maximum sensitivity in the yellow-green under light-adapted conditions to maximum sensitivity in the blue-green under dark-adapted conditions is the **Purkinje shift**, named for Jan Purkinje (1787– 1869), who also described the entoptic phenomenon called the Purkinje tree (see

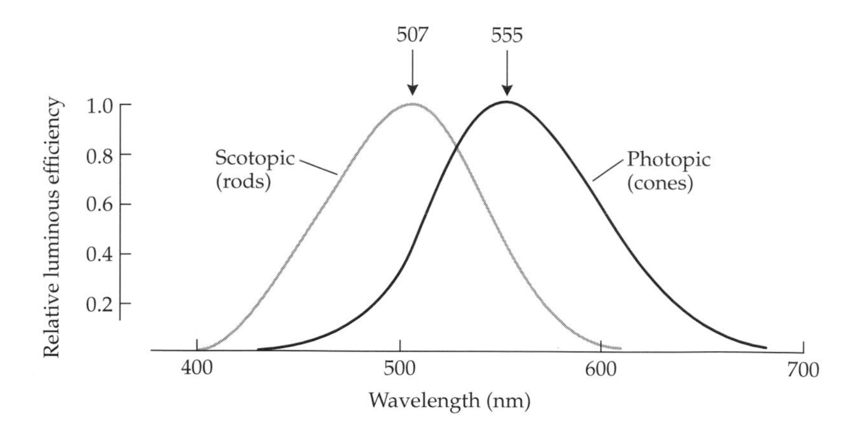

Figure 13.13
Luminous Efficiency Functions for the Standard Human Observer
Relative luminous efficiency is closely related to relative sensitivity. The standard observer is most sensitive to light of 507 nm wavelength under scotopic (dark-adapted) conditions and to light with 555 nm wavelength under photopic (light-adapted) conditons. (Plotted from data in Wyszecki and Stiles 1967.)

Chapter 16). The Purkinje shift and subsequent observations by such anatomists as Max Schultze and Heinrich Müller (see Vignette 14.1) on the dominance of rods in nocturnal species and cones in diurnal species led to what was called the duplicity theory of vision and the description of mixed rod-and-cone retinas like our own as duplex retinas.

The photopigments are stacked in layers within the outer segments of the photoreceptor

The outer segments of photoreceptors have a distinctive structure (Figure 13.14). They contain a series of **lamellae**, or **discs**, each of which is a flattened membranous sac, rather like a tiny deflated red blood cell. Discs in the outer segments of rods are not continuous with the outer segment membrane (except at the base of the outer segment, where they form; see Figure 13.33*a*); they are held in place by cytoskeletal attachment proteins (peripherins) around the rims of the discs. The outer segments of cones are different. The lamellae are continuous with the outer segment membrane, and the cone discs are actually deep folds in the membrane.

Although the rod and cone outer segments illustrated here differ in shape, with the cone outer segment tapering from base to tip, this is not always the case. The outer segments of foveal cones are as cylindrical as rods. The difference in the disc structure between rods and cones is always present, however.

The photopigment molecules are embedded in the disc membranes, as shown for rhodopsin in Figure 13.15; the situation is similar for the cone pigments. Each of the two paired disc membranes is a bilayer of phospholipids into which the photopigment molecules are inserted; seven helical portions of the opsin traverse the bilipid membrane, enclosing the *retinal* chromophore within a kind of cage. Both sides of the discs in the outer segments of rods are studded with photopigment molecules. Since there are up to 1000 discs in rod outer segments and each is covered with photopigment molecules, the total number of molecules in a rod will be very large; the number is finite, however, and very bright light can bleach a substantial fraction of the resident photopigment molecules.*

*How large *is* the number? Each side of an outer segment disc on a rod has an area of about 3 μm^2, giving a total disc surface area of around 6 μm^2. Estimates of the density of rhodopsin molecules are on the order of 25,000/μm^2, which means that each disc would have around 150,000 rhodopsin molecules. If there are 1000 discs in the outer segment of the rod, the total complement of rhodopsin molecules per rod will be around 150 million. By comparison, bright ambient sunlight produces a retinal illumination of almost a million photons per rod per second. If rhodopsin did not regenerate, it would be bleached away in about 2.5 minutes.

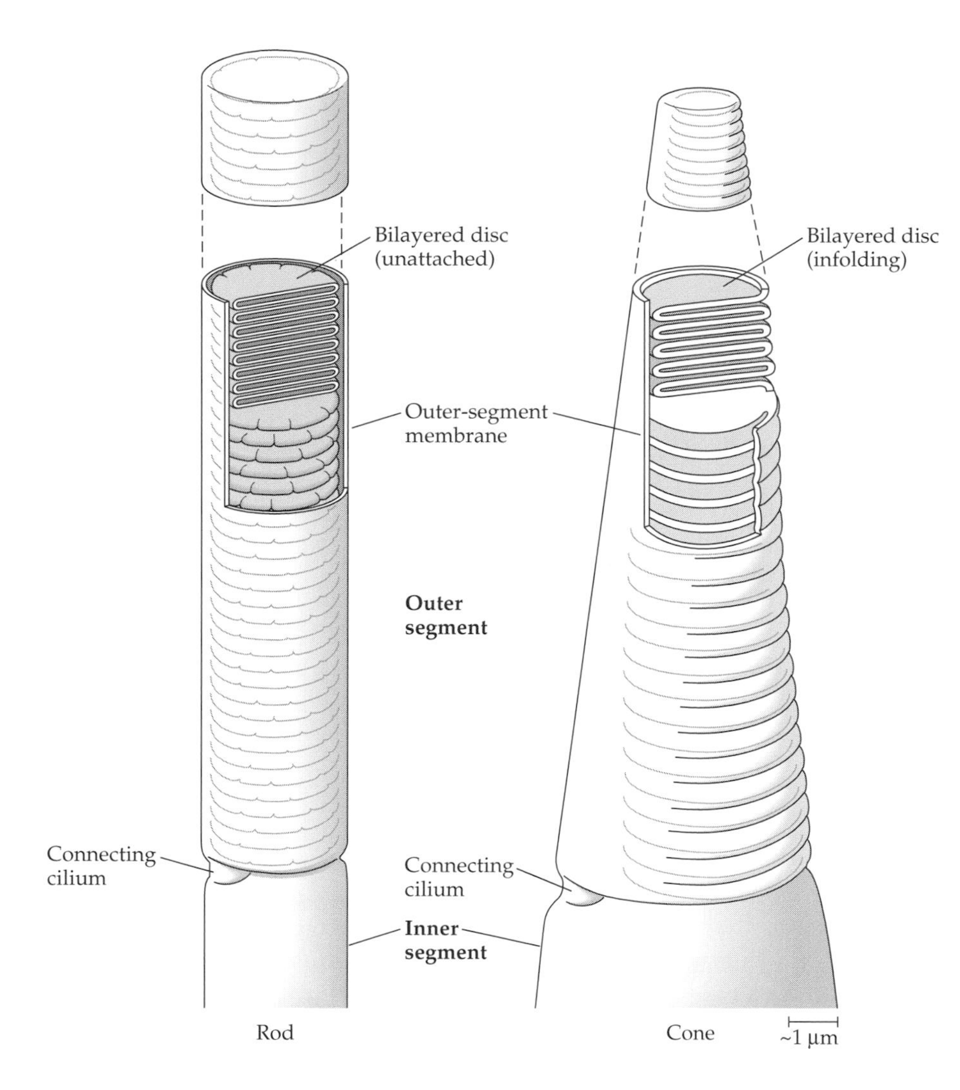

Figure 13.14
Structure of the Outer Segments of Rods and Cones
The outer segments of both rods and cones are filled with bilayered discs in which the photopigment molecules are embedded. Cone discs are deep folds of the membrane that remain continuous with the outer segment membrane. Rod discs are not continuous with the outer segment membrane, except at the base of the outer segment, where the discs form. Neither the membranes nor the discs are drawn to scale; they are thinner in real life than they appear here.

Light absorption produces a structural change in the photopigments

In its dark state, the *retinal* in a photopigment has a particular configuration, one of several **stereoisomeric** forms it can assume, in which it clings to a particular site on the opsin. As Figure 13.15 shows, the long opsin molecule loops in and out of the disc membrane, and the *retinal* is covalently bound to a lysine residue on transmembrane domain 7 within the disc membrane; one can think of the opsin as forming a snug, form-fitting pocket in which the *retinal* resides. Absorption of light, which is a transfer of energy from photons to the photopigment, raises the energy level of the *retinal* sufficiently that one of its double bonds breaks and the *retinal* changes its shape to another stable stereoisomer (Figure 13.16). Technically, the change is from the 11-cis isomer to the all-trans isomer of *retinal*. This change, which is the only direct effect of light on the eye and is the initiating step for vision, exposes a charged site on the opsin, which prompts a restructuring of the protein itself into its activated form. The overall change of a rhodopsin molecule from its inactivated form before absorbing a photon to its activated form is designated by the transaction $R \rightarrow R^*$.

The activated rhodopsin (R^*) persists for a brief period of time, during which it can interact with other molecules on the disc membrane. Additional changes in configuration end with the separation of the all-trans *retinal* from the opsin and its conversion to all-trans retin*ol*. The photopigment is then said to be **bleached**.

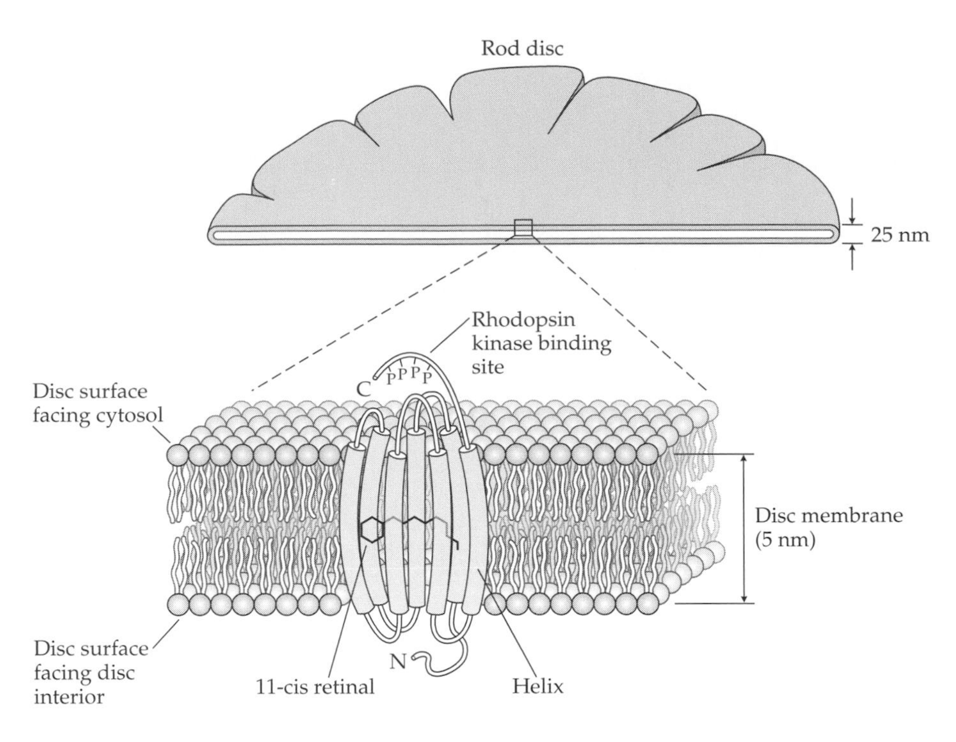

Figure 13.15
Rhodopsin in the Disc Membrane
An enlarged view of a small piece of disc membrane shows the photopigment opsin with seven helices running through the layers of phospholipids that constitute the membrane. The helices form a kind of cage for the photosensitive *retinal.* The C terminus of the opsin and the other loops of amino acids on the outer surface of the disc membrane are the sites at which opsin interacts with other molecules in the phototransduction cycle. (After Stryer 1986.)

Bleached photopigment cannot be affected by light; thus the more photopigment that is bleached, the less sensitive the eye will become. Restoration of sensitivity requires that the 11-cis *retinal* be reconstructed and reattached to the opsin; this is an enzyme-driven process called **regeneration**, which we will consider in more detail shortly. One of the differences among photopigments is their regeneration rate: Cone pigments regenerate faster than rhodopsin. Under conditions of steady illumination, an equilibrium is established such that the rates of bleaching and regeneration are equal and the amount of bleached photopigment will remain constant. In reaching this equilibrium, however, the rods, with their slower regeneration rate, will have had more pigment bleached than the cones before the equilibrium is established; therefore, for any given level of steady illumination, the rod pigment will always be more bleached than the cone pigments.

Structural change in the photopigment activates an intracellular second-messenger system using cGMP as the messenger

Phototransduction begins with the transfer of energy from photons to photopigments and ends with a change in glutamate release by the photoreceptor terminal to signal the photon absorption. The initial part of the process is a cascade of intracellular events resulting from activation of a photopigment molecule, as summarized in Figure 13.17.

Photoreceptor disc membranes contain a **G protein**, a multimeric (multipart) assembly of subunits that bind opsin on one side and the nucleotides **GDP** (guanosine diphosphate) or **GTP** (guanosine triphosphate) on another. The G protein in rods is called transducin and designated G_T; in its normal inactivated state, G_T binds a molecule of GDP to one of its subunits. When activated rhodopsin (R*) encounters a molecule of G_T, GDP is exchanged for GTP, and the GTP-carrying α subunit of G_T separates from the rest of the molecule; it is now activated and designated G*, which is the subunit with GTP. Thus, $R^* + G_T \rightarrow G^*$.

Like activated rhodopsin, the activated G protein has a brief lifetime in which it can interact with other molecules. The molecule it chooses is the enzyme

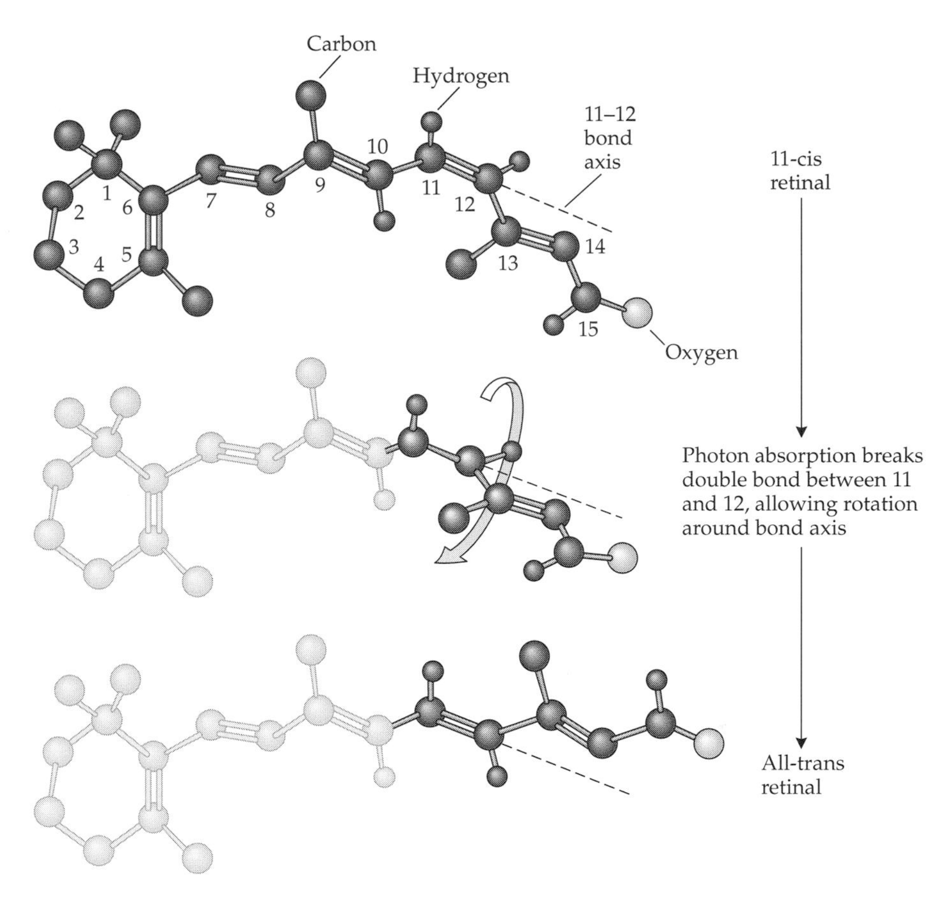

Figure 13.16
Photon Absorption Changes the Form of *Retinal*
Retinal is an aldehyde of vitamin A that has several stable configurations. Before photon absorption, it is bound to opsin in its 11-cis configuration. Photon absorption breaks the double bond between carbons 11 and 12, which permits rotation to the all-trans configuration. This is the initiating step in vision. (After Stryer 1987.)

cGMP phosphodiesterase (**PDE**). PDE is a multimer with two sites at which it can interact with G*. The interaction activates PDE: G* + PDE → PDE*. In this case, activation exposes a site on the PDE molecule that allows it to exert its catalytic effect on its target molecule, which is **cGMP** (3′,5′-cyclic guanosine monophosphate).

To this point, all events have taken place on the disc membrane. Rhodopsin, G_T, and PDE all reside there, and their activated forms act there. But the ultimate effect of this chain of events is to alter the status of ion channels in the outer segment membrane, which is some distance away. What is required is a molecule residing in the cytoplasm that can act as a go-between from the discs to the outer segment membrane channels. The messenger is cGMP.

When a molecule of cGMP diffuses near PDE* on a disc, the exposed catalytic site on PDE* will effect a change from cGMP to 5′-GMP (5′-guanosine monophosphate). The important event here is not the production of 5′-GMP; this molecule has no direct role in phototransduction. What matters is eliminating a molecule of cGMP from the intracellular pool of cGMP. The status of the ion channels in the outer segment membrane depends on the concentration of cGMP within the cell; eliminating cGMP molecules from the pool reduces the intracellular cGMP concentration. Lowering the cGMP concentration closes membrane channels, thereby altering the flow of current into the photoreceptor.

A G protein–mediated second-messenger system is not unique to photoreceptors. Many G proteins can interact with various sorts of receptors (the photopigments are the receptors in the rods and cones) and then activate other intracellular messengers, such as cAMP, cGMP, or Ca^{2+}. G protein–mediated secretory systems were considered, in less detail, for the lacrimal gland (see Chapter 7) and for aqueous production by ciliary epithelial cells (see Chapter 11). The neurotransmitters that normally interact directly with ion channels in cell

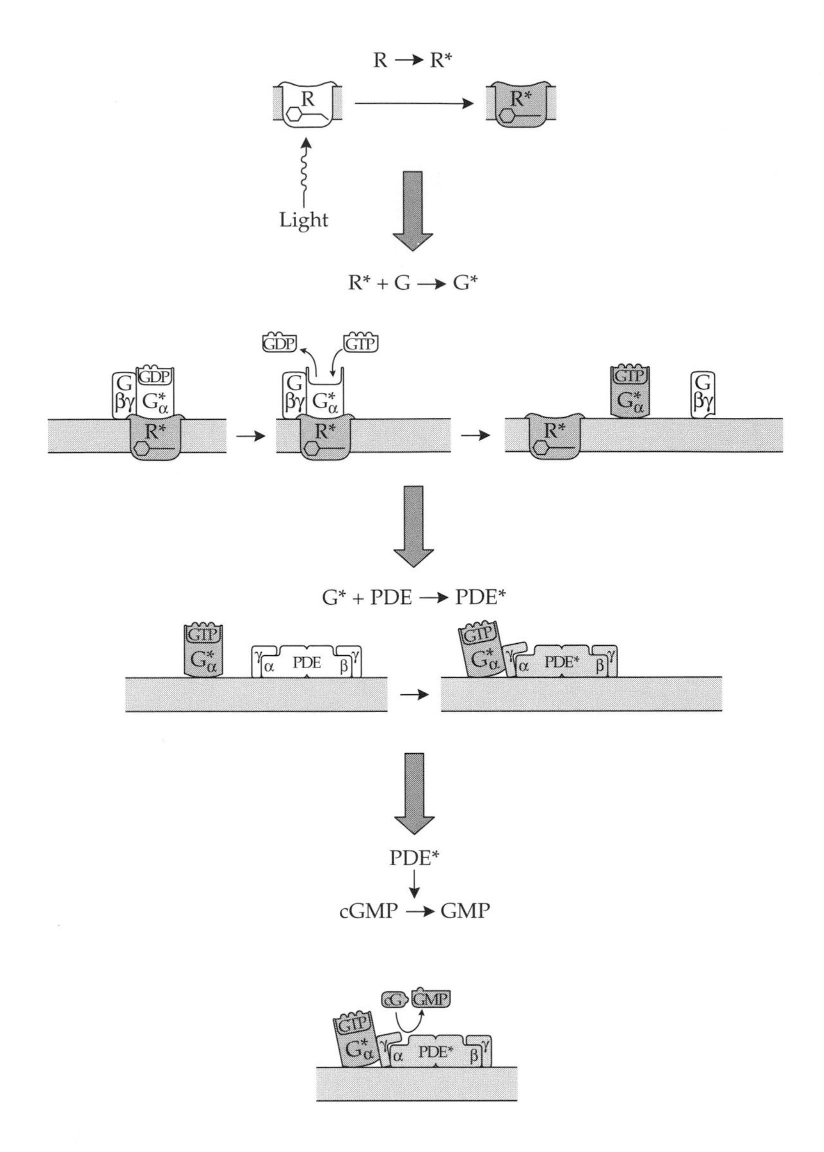

Figure 13.17
The Phototransduction Cascade
Events in the photoreceptor disc membrane initiated by photon absorption are shown here as interactions between molecules, which are represented by icons. There are four main steps in the cascade: (1) Rhodopsin is activated by photon absorption, (2) activated rhodopsin activates a G protein (G → G*), (3) activated G protein activates cGMP phosphodiesterase (PDE), and (4) activated cGMP phosphodiesterase converts cGMP to GMP. The reduction in cGMP concentration will cause outer segment membrane channels to close. (After Rodieck 1998.)

membranes can also interact with other receptors that trigger second messengers via G proteins. Acetylcholine, for example, acts directly at the nicotinic synapses and indirectly, via a G protein–linked receptor, at muscarinic synapses (see Chapter 5); a G protein–linked glutamate receptor at transmission between cones and bipolar cells will be considered later in the chapter. Also, most blood-circulated hormones and the neuromodulators work through G protein linkages, as do transduction mechanisms in the olfactory and gustatory systems.

A decrease in cGMP concentration closes cation channels, decreases the photocurrent, and hyperpolarizes the photoreceptor

Most of the ionic current flowing into the outer segment consists of Na^+ ions, as well as some Ca^{2+} ions, that pass through channels in the membrane. These channels, one of which is illustrated schematically in Figure 13.18, are not voltage sensitive. Instead, they are ligand-gated channels whose status—open or closed—is controlled by the binding of the ligand, which in this case is cGMP. The channels are open when they bind cGMP molecules and closed when they do not. The availability of cGMP for binding depends on the concentration; as the concentration increases, the number of molecules available for binding increases. Thus the

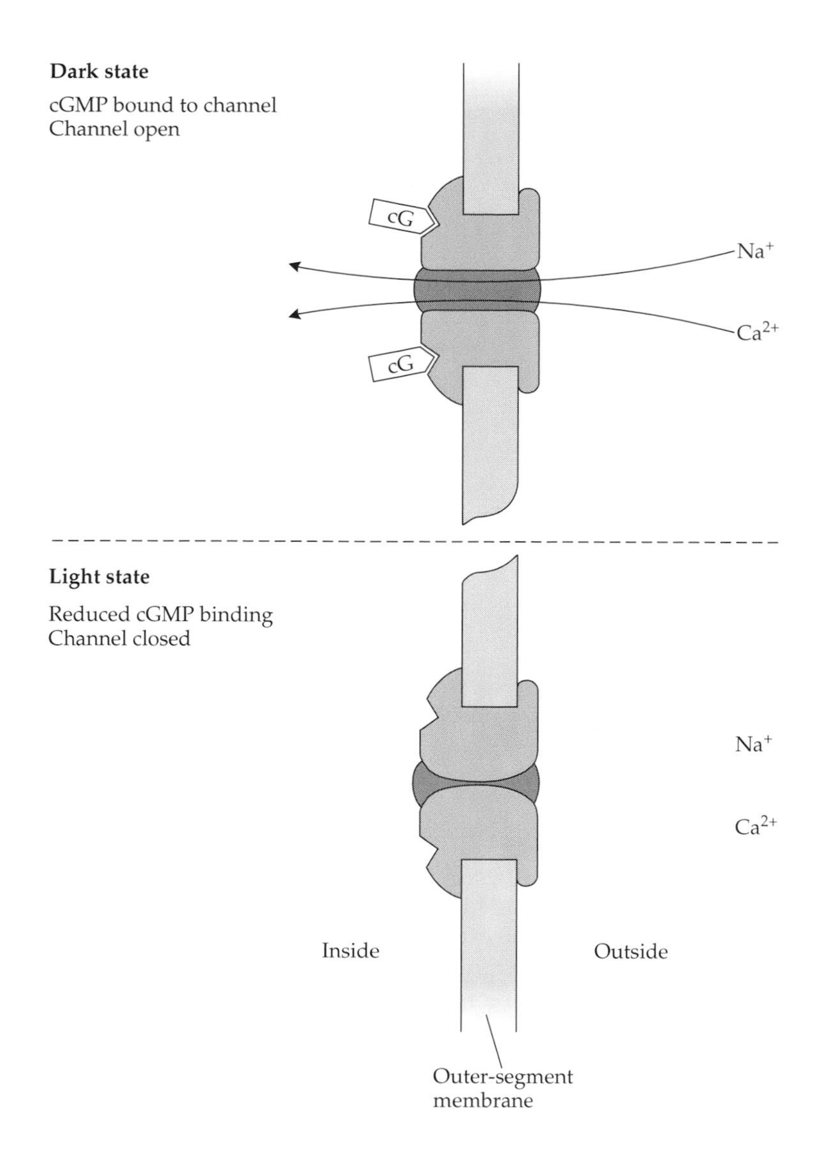

Figure 13.18
cGMP-Gated Cation Channels in the outer segment Membrane
When the cGMP concentration is high, which is the dark condition for the photoreceptor, cGMP is available to bind to membrane channels. Channels spend more time in the open state when cGMP is bound, thereby permitting Na^+ and Ca^{2+} to flow into the cell. The phototransduction cascade initiated by photon absorption reduces the cGMP concentration, which closes the cGMP-gated channels.

number of open channels will increase as the concentration of cGMP rises, and channels will close as the cGMP concentration falls.

Since the absorption of light *reduces* the amount of cGMP, channels will close and the number of Na^+ ions flowing into the cell will decrease. But to see the effect of this decrease in Na^+ flow on the cell's membrane potential, we need to consider other ion channels and pumps in the photoreceptor membrane (Figure 13.19).

When the photoreceptor is in the dark, both Na^+ and K^+ are flowing in and out of the photoreceptor. Na^+ enters the outer segment membrane through open cGMP-gated channels and through a cation exchanger (Na^+ is swapped for K^+ and Ca^{2+}). It is pumped back out of the cell by metabolically driven Na^+–K^+ pumps in the inner segment membrane. K^+ enters the inner segment through the Na^+–K^+ pump and flows out through selective K^+ channels in the inner segment membrane and the cation exchanger in the outer segment membrane. The result of these ionic currents is a cell membrane potential of about –40 mV (inside negative); this potential is a compromise between the tendency of the Na^+ inflow to make the inside more positive (depolarized) and the tendency of the K^+ outflow to make it more negative (hyperpolarized).

Photon absorption upsets this equilibrium. As Na^+ channels in the outer segment membrane close following light absorption, the inward Na^+ current decreases and the K^+ current becomes more dominant; light absorption has no effect on the cation exchanger or on the Na^+–K^+ pump. Positive charge is now

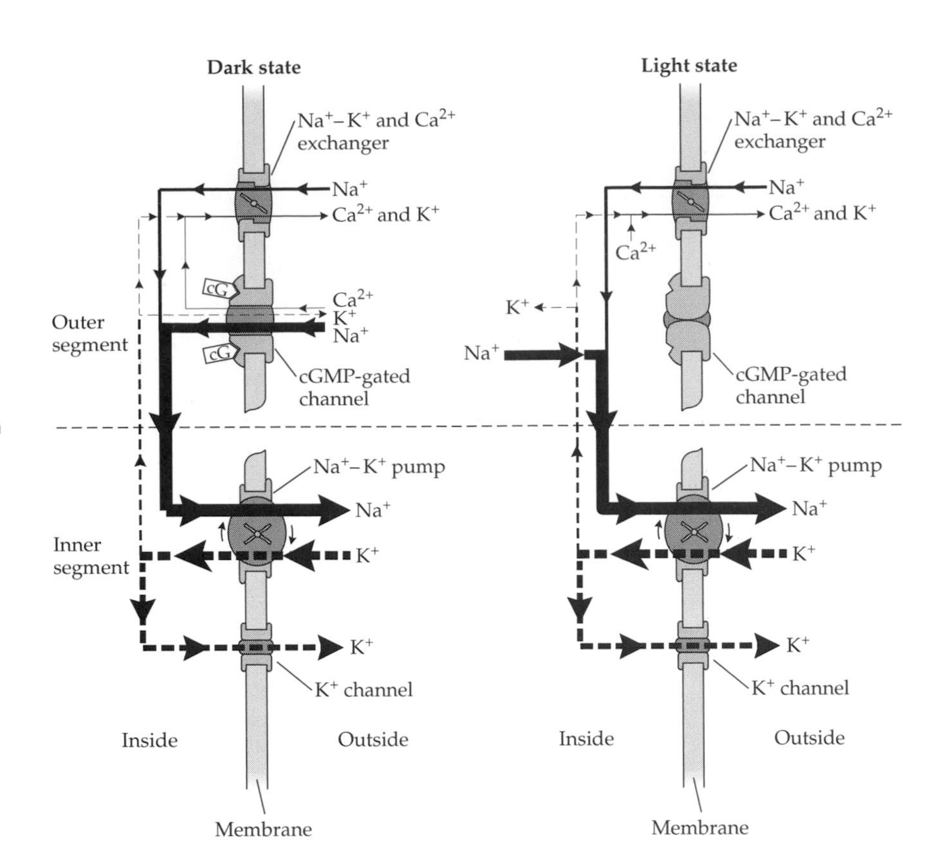

Figure 13.19
The Photocurrent Is Reduced by Light Absorption
Most of the cuurent flowing into the outer segment in the dark is a Na^+ current through the cGMP-gated channels. Na^+ outflow across the inner segment membrane is linked to K^+ inflow by membrane pumps. Channel closure following light absorption reduces the inward Na^+ current with little effect on other sources of inflow and outflow. This reduction in positive charge within the outer segment makes the transmembrane potential more negative; that is, the receptor hyperpolarizes.

moving out of the cell faster than it is flowing in; the interior of the cell thus becomes more negative relative to the exterior. This negative swing of the potential across the cell membrane is hyperpolarization.

Although only the current flowing into the outer segment through the cGMP-gated channels is altered by photon absorption, its effects are felt thoughout the cell. The decreased Na^+ current into the outer segment reduces the amount of resident positive charge, and the outer segment hyperpolarizes. But for the photoreceptor to alter its rate of glutamate release from its terminal, which is rather distant, the change in the outer segment membrane potential must also produce a change in the membrane potential at the terminal. This change occurs because ions can move with considerable freedom through the interior of the cell, and they will do so whenever there is an imbalance of electric charge. In this case, the increased negativity in the outer segment tends to draw positively charged ions from other parts of the cell until the electric potential is the same everywhere. Some of these positive ions will be drawn from the terminal of the photoreceptor, and their departure makes the terminal less positive, hence more negative. This hyperpolarization of the terminal reduces the rate of glutamate release.

Absorption of one photon can produce a detectable rod signal

A superb psychophysical study done more than 50 years ago by Hecht, Shlaer, and Pirenne (1942) concluded that the absorption of one photon is sufficient to excite a rod, but it took another 40 years to see the rod signal directly and to confirm that absorption of a single photon produces a detectable signal. Some responses by rods to single photons are shown in Figure 13.20. The small upward deflections are changes in current flowing across the outer segment membrane of a monkey rod; the smallest of them are produced by the absorption of single photons. (By convention, the upward deflections in the current traces represent *decreased* current flow into the cell. The method for recording photoreceptor current is illustrated in Box 14.1.)

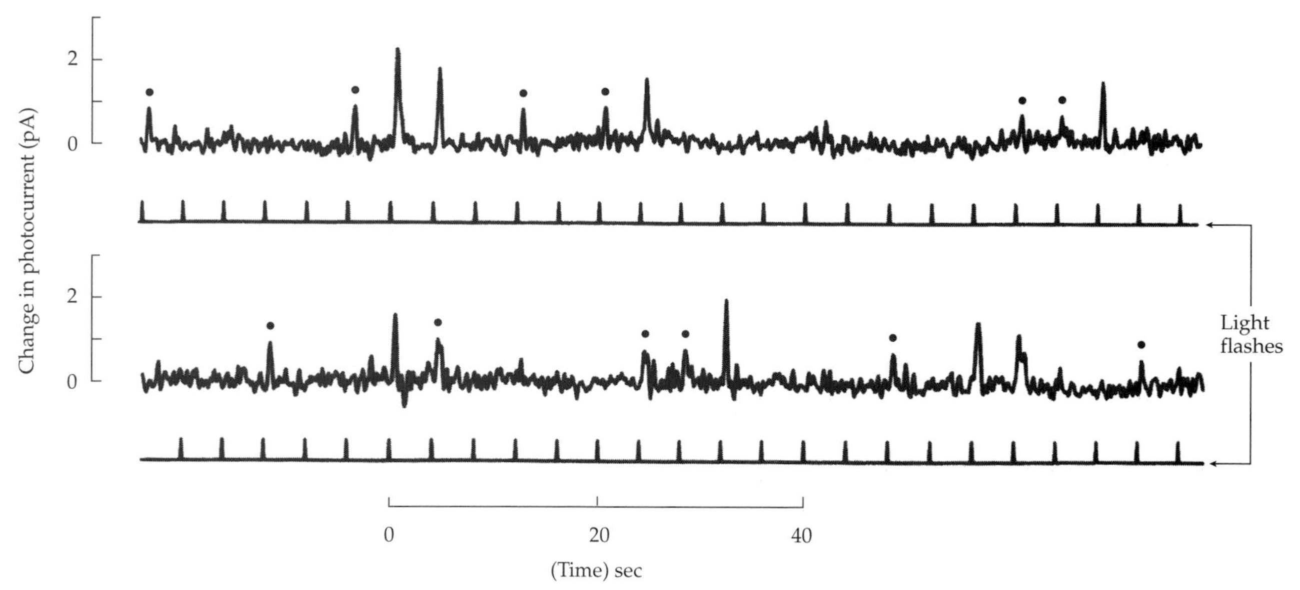

Figure 13.20
Rod Responses to Single Photon Absorptions
Very dim light flashes produce small *decreases* in rod photocurrent (shown as upward deflections of the current traces), the smallest of which are about the same magnitude as and correspond to the absorption of a single photon (identified by dots). Larger responses indicate absorption of more than one photon; absence of response indicates no absorption. (After Baylor, Nunn, and Schnapf 1984.)

The rod's ability to signal the absorption of a single photon raises a question: How can the breaking of a single bond in one *retinal* molecule produce a detectable change in a rod's signal? Neither the amount of energy transferred nor the change in one molecule out of 150 million is significant by itself. These small changes must be amplified; Figure 13.21 outlines how.

Absorption of one photon activates only one rhodopsin molecule. But photopigment molecules are not locked in position in the disc membranes; they can move rapidly in the plane of the membrane, like molecular sailboats buffeted by shifting breezes, exposing their interaction sites to multiple G proteins in a very short time. Having activated one G protein molecule, the activated rhodopsin sails on to other G proteins; during its activated lifetime (about 50 ms), rhodopsin will activate about 800 G protein molecules; that is, 1 R* → 800 G*.

The step in which G* activates PDE does not involve amplification. The relationship is one-to-one: 1 G* → 1 PDE*, and thus 1 R* → 800 PDE*. One activated PDE molecule, however, can catalyze the breakdown of more than one molecule of cGMP during its activated lifetime. Roughly 6 cGMP molecules are converted to GMP by each activated PDE. Thus, 1 PDE* → 6 cGMP → GMP, and 1R* → 4800 cGMP → GMP.

In terms of numbers of molecules, the original event has been amplified 4800-fold. Activation of one rhodopsin molecule has taken 4800 cGMP molecules out of circulation so that they are no longer available to keep the cGMP-gated Na^+ channels open. This may not sound like much, but in terms of the photoreceptor's pool of unbound cGMP, 4800 molecules constitute nearly 1% of the free cGMP molecules in the cell. Of more importance, this fall in the cGMP concentration causes a significant fraction of the open Na^+ channels to close; 175 Na^+ channels represent about 2% of the 10,000 or so channels that are open at any given moment. Since the flow of current into the outer segment is proportional to the number of open channels, closing 2% of the channels reduces the inward

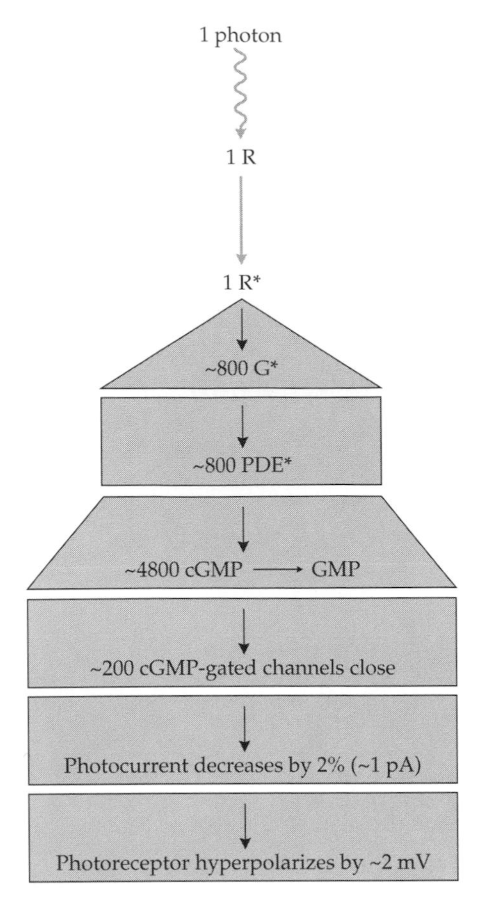

Figure 13.21
Amplification in the Phototransduction Cascade
Absorption of a single photon, an event affecting only one molecule, is amplified to produce a conversion of 4800 cGMP molecules to GMP, which reduces the cGMP concentration by about 2%. Enough cGMP-gated channels close to cause a change in photocurrent of about 1 pA (2%). The main amplification stage is the activation of 800 G protein molecules by one activated rhodopsin molecule. (Data from Rodieck 1998.)

current by 2%, which is on the order of 1 pA (see Figure 13.20). The corresponding voltage change across the cell membrane is about 1 mV hyperpolarization.

Amplification within the cone phototransduction cascade is similar to that in rods, and the evidence suggests that cones, when fully adapted to dark, can also signal the absorption of single photons. The change in photocurrent to single absorptions is smaller in cones than in rods, however. Although we can perceive the near simultaneous absorption of single photons by five to ten rods, cones need to be absorbing multiple photons before their signals reach threshold. Part of the difference is inherent to the photoreceptors and part to the neural wiring; the rod system is highly convergent, which allows rod signals to be pooled and summed, but the cone pathways have relatively little convergence.

Photocurrent in the outer segment decreases in proportion to the number of absorbed photons

The changes in current flow into the outer segments of a rod and a red cone from a monkey retina are shown in Figure 13.22. (The different timescale makes the smallest responses look very different from those in Figure 13.20.) In the dark, the rod has a steady inward current flow of –34 pA, which decreases transiently (becomes less negative) in response to a brief flash of light. In the series of responses shown here, the smallest response was to the absorption of just 3 photons, and the inward current decreased by about 2 pA. Increasing the number of photons per light flash increases the size of the decrease in current; for the brightest flash (860 photons) the decrease in current was a full 34 pA, effectively taking the flow of current to zero. Note that the effect of the bright stimuli lasts well beyond the duration of the light flash; if a second light flash were presented to the rod, it would have very little additional response, no matter how bright the second flash. This prolonged response represents adaptation, and for some time after a bright light flash, the rod will be less sensitive than it was before the stimulus.

If the rod always counted photon absorptions as individual events—that is, if the response were directly proportional to the number of photons absorbed—the maximum response in this series would be almost 300 times larger than the smallest response, which is clearly not the case. The actual relationship is shown when the response magnitude is plotted against the logarithm of the number of photons absorbed; over most of the range, the response magnitude (decrease in inward photocurrent) is proportional to the log number of photons absorbed. When the photon numbers are very small, the response does not increase rapidly, and in this region, the rod is counting photons. At the other extreme, where the curve flattens out for large numbers of photons, increases in flash intensity produce no increase in response. Here, the photocurrent has dropped to zero because all the membrane channels are closed and the rod is said to be *saturated*.

Recordings of photocurrents from cones show a similar dependence of response magnitude on flash intensity, with two significant differences. First, the numbers of photons per light flash are significantly larger than those of the rods, ranging from about 190 photons for the smallest cone response to 36,000 photons for the largest response. Cones signal flash intensities that are well out of the rod range. Second, the cone photocurrents lack the sustained character of the rod responses. Even when the cone photocurrent is driven to zero, it quickly returns to its original value and then swings the other way, indicating an *outward* current flow. This overshoot quickly returns to the original steady photocurrent level that existed before the light flash.

The plot of response versus flash intensity in the cone is similar in shape to that of the rod, except it is shifted to the right, reflecting the higher flash intensities needed to elicit the responses. Since the rod and the cone were fully adapted to the dark, the relative shift of the cone curve shows that rods are most sensi-

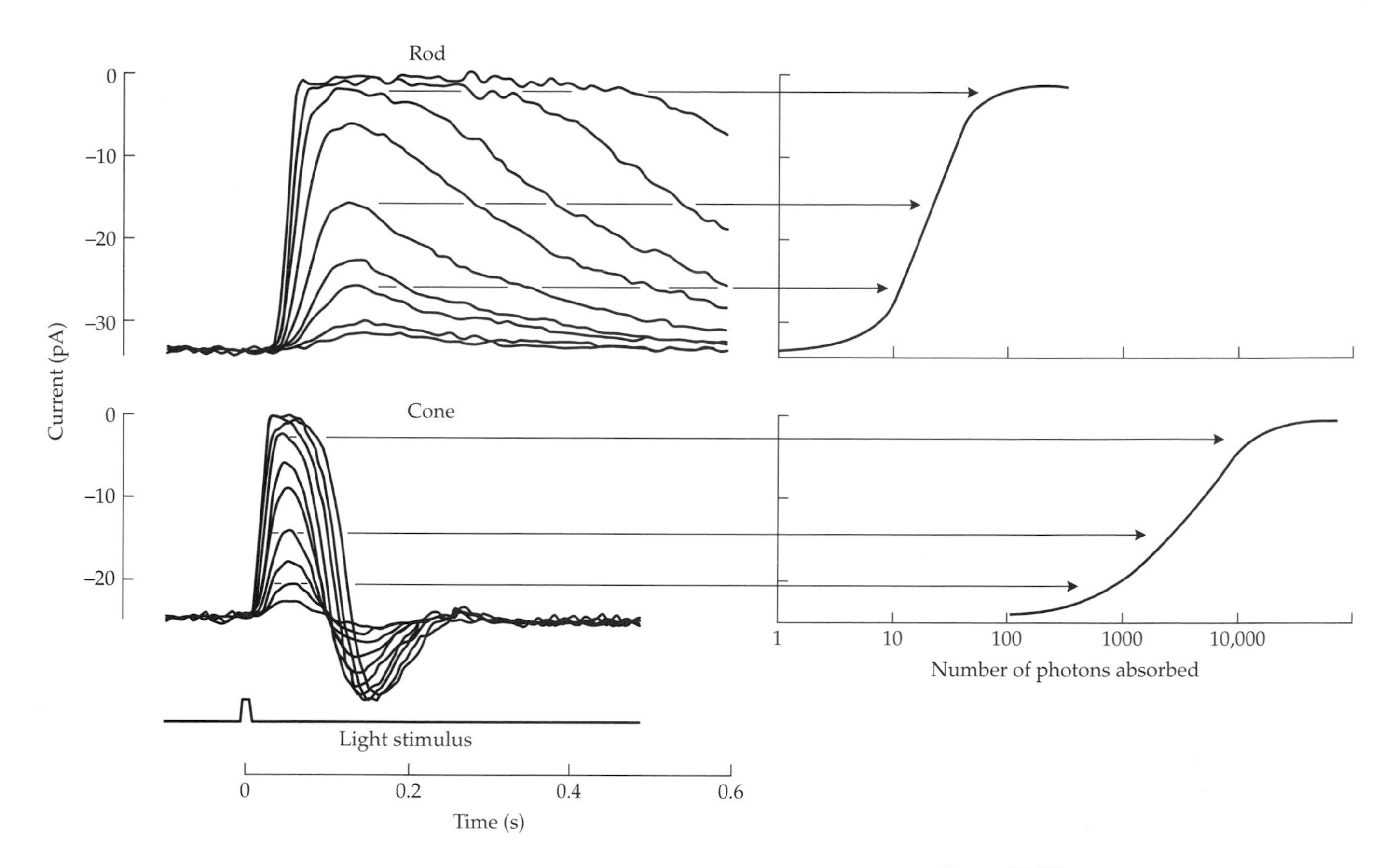

Figure 13.22
Responses of Rods and Cones to Light Flashes of Different Intensity
The magnitude of change in the photocurrent in both rods and cones increases with increasing intensity of brief light flashes (left). Responses by rods are considerably more prolonged than those of cones. The increase in response is directly proportional to the logarithm of the number of photons absorbed over a range of 1 log unit (right). Rods respond to much lower light levels than cones; the rod response is at its maximum before the cone response begins to increase appreciably. (After Baylor 1987.)

tive at low flash intensities, whereas cones operate at high intensities for which rods are saturated and insensitive. The initial rise of the cone response occurs as the rod response reaches its maximum; thus, the task of detecting light flashes is handed over from rods to cones with no break in flash detection.

The relatively transient nature of the cone response keeps the cone from saturating at high flash intensities. Although the cone response may be at its maximum (zero current) for a short time, it is restored so quickly that the cone will be sensitive to a second light flash in a way that the rod cannot be. The mechanism underlying this difference between rods and cones is not fully understood, but part of the explanation involves differences in the kinetics of the phototransduction cascade in rods and cones; events proceed more quickly in the cones.

Responses from red, green, and blue cones are similar in shape and timing, although there appear to be differences in response magnitude. The variation of their responses to light flashes that have equal numbers of photons but different wavelengths—their spectral sensitivity curves—fit very nicely with the absorbance spectra of the cone photopigments shown in Figure 13.11. And to reinforce a point made earlier, stimuli of two different wavelengths produce identical responses from a cone when equal numbers of photons are absorbed. Thus the signal from a cone, like that from a rod, relates to the number of photons captured and says nothing about their wavelength.

Photopigments activated by photon absorption are inactivated, broken down, and then regenerated

The photoreceptor responses in Figure 13.22 persist beyond the duration of the stimulus, largely because most events in the phototransduction cascade have lifetimes of 50 ms or more. These lifetimes are the durations of the activated forms of rhodopsin, G_T, and PDE. None of these activated states can be allowed

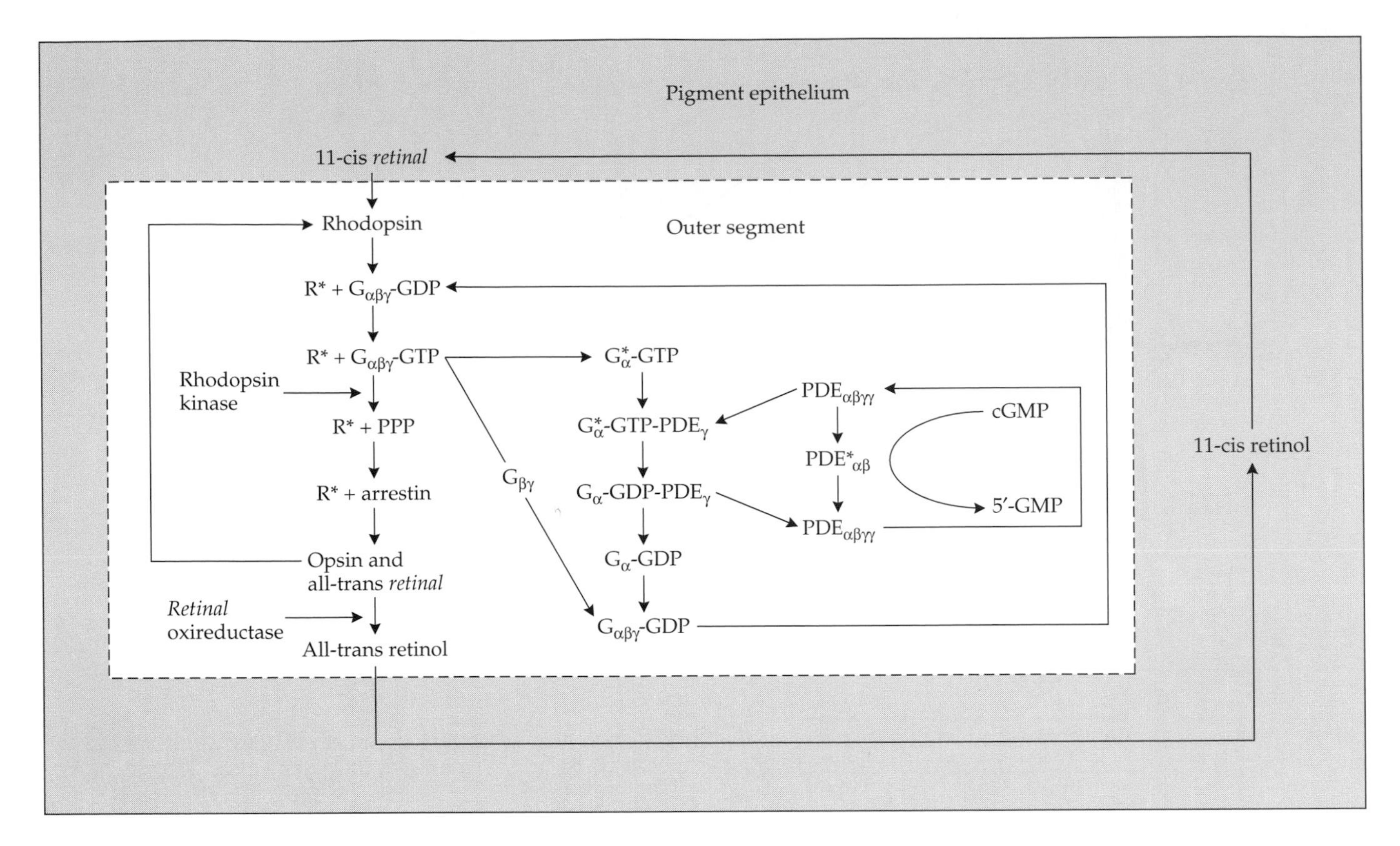

Figure 13.23
Rhodopsin, Transducin, and Phosphodiesterase Cycles
All the major players in phototransduction go from inactivated to activated forms and back again. The rhodopsin cycle is the longest, and part of the process of regenerating inactivated rhodopsin occurs in the pigment epithelium (shaded region). See text for discussion. G, G protein; PDE, cGMP phosphodiesterase; R, rhodopsin.

to persist indefinitely; the activation must be terminated and the different molecules must be restored to their unactivated conditions so that they can take part in the cascade triggered by new photon absorptions. Figure 13.23 illustrates some of the steps in inactivation and restoration of the native states of rhodopsin, transducin, and PDE.

Activated rhodopsin (R*) is phosphorylated by an enzyme, rhodopsin kinase. The addition of phosphates to the C terminus of the opsin (see Figure 13.15) produces enough structural change in the opsin that another protein, **arrestin**, can bind to the opsin at the site where the activated rhodopsin normally interacts with the G protein. Blocking the site prevents further interaction with G_T and brings the rhodopsin's ability to activate G_T to an end. Moreover, arrestin promotes the separation of all-trans *retinal* from the opsin; this is the final stage of photopigment bleaching. When all-trans *retinal* dissociates from the opsin, it diffuses into the cytosol of the photoreceptor outer segment, leaving the opsin behind in the disc membrane. Another enzyme (all-trans retinal oxidoreductase) changes the all-trans *retinal* to all-trans retin*ol*.

Regeneration of the photopigment requires converting the all-trans *retinal* to 11-cis *retinal* that can recombine with the opsin. All but the recombination takes place in the pigment epithelium, not in the photoreceptor; the details will be considered shortly (see Figure 13.29).

Transducin (G_T) goes through a similar cycle. The activated form, G*, is part of the original molecule (the α subunit) with GTP attached. After binding to and activating PDE, GDP is substituted for GTP, and the now inactive fraction of the G protein separates from PDE and recombines with the β and γ subunits to form the original G_T. The specific triggers for the changes in the G_T cycle are probably enzymes, but they have not been fully characterized.

The PDE cycle is very simple, insofar as it is known. PDE is activated when G* attaches and exposes one of the catalytic sites for cGMP; it returns directly to the inactivated state when the G* separates from it.

Both the G_T and the PDE cycles are relatively fast, in the sense that the transition from the activated form back to the original inactivated configuration is brief compared to the lifetimes of the activated molecules. This is not the case for regenerating rhodopsin, however. Regeneration takes considerably more time than anything else, and this time course dictates the fraction of a photoreceptor's pigment bleached by a steady source of illumination. Since the regeneration rate is constant, as the rate of photon absorption and bleaching exceeds the regeneration rate, the fraction of photopigment bleached at equilibrium will rise. Thus the brighter the ambient illumination, the greater the fraction of pigment bleached. But since the regeneration rate is significant, the fraction of bleached photopigment will always be less than 1 (that is, it is impossible to bleach all of it).

Photoreceptor sensitivity is modulated by intracellular Ca^{2+}

The transduction scheme outlined in the preceding discussion is of recent origin; at one time the favorite candidate for the second messenger in the photoreceptor light response was Ca^{2+}, not cGMP. The interest in Ca^{2+} was not misplaced, however, because it does have an important role to play, even though it is not in the direct phototransduction cascade.

The cGMP-gated channels in the outer segment membrane pass both Na^+ and Ca^{2+} when they are open, so the closing of the channels will reduce both the Na^+ and the Ca^{2+} influxes. The decrease in internal Ca^{2+} concentration has at least two effects (Figure 13.24). One is *increased* activity of a cGMP-synthesizing enzyme, which results in a gradual increase in cGMP. The second effect is an *increase* in the affinity of the cGMP-gated channels for cGMP, which makes the existing level of cGMP more effective. Both of these effects act in the same direction: Increased cGMP and increased affinity of the channels lead to the reopening of channels that had been closed following light absorption.

The reopening of cGMP-gated channels in the face of constant, ongoing light absorption means that the photoreceptor is adapting, readjusting its sensitivity to suit the current state of ambient illumination. The effect can be seen in a comparison of photoreceptor responses to a light stimulus when the photoreceptor is adapting (the Ca^{2+} system is operating normally) and when it is not adapting (the Ca^{2+} system is made inoperative), as in Figure 13.24.

Under normal circumstances, the photoreceptor signal rises quickly to a maximum and then declines to a steady level that is maintained for the duration of the stimulus. (The stimulus duration is quite long in this example—around 25 s—and the stimulus does not produce saturation of the rod response.) In the absence of the Ca^{2+} system, however, the initial rise to a maximum is not followed by a decline; the signal remains at the maximum. Without Ca^{2+}, channel affinity does not increase, and there is no increase in cGMP production to reopen any of the channels closed by light absorption.

The steady level of photocurrent produced by a prolonged light stimulus—that is, by steady background illumination—decreases (becomes less negative) as the background illumination increases. In the dark, the photocurrent is at its most negative level, and the absorption of relatively few photons produces a significant change in the photocurrent. In the presence of background light that makes the resting photocurrent less negative, however, relatively more photons will be needed to produce any further change in the photocurrent. This effect is illustrated for a rod in Figure 13.25.

As background illumination increases, the resting or steady-state photocurrent becomes progressively smaller (moves closer to zero). For any specific level of background illumination, a key question concerns the amount of additional light required to elicit a response of a given size—say 5% (0.05) of the maximum possible response (the maximum response has been normalized to unity). Two

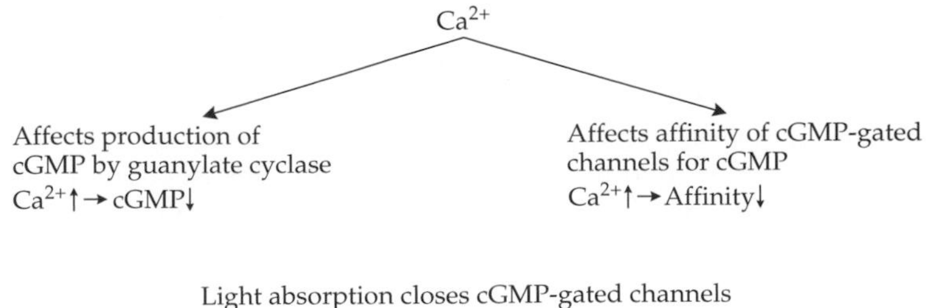

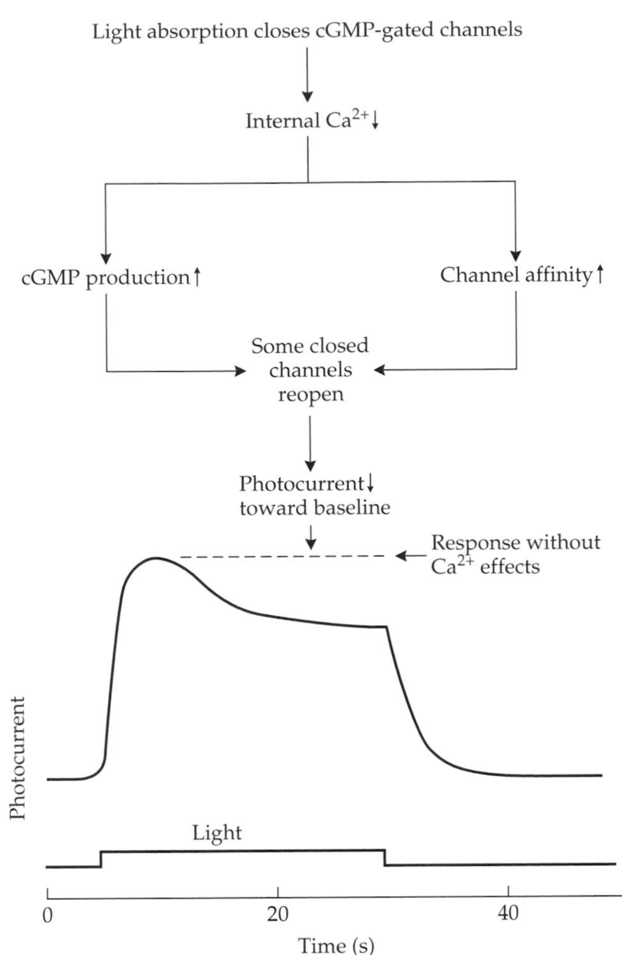

Figure 13.24
Calcium Levels Modulate Phototransduction
Because Ca^{2+} affects the level of cGMP and affects the affinity of cGMP-gated channels for cGMP, the reduction of internal Ca^{2+} following light absorption tends to offset some of the effects of the phototransduction cascade, making some closed channels reopen. As a result, some of the initial reduction of the photocurrent is offset, an effect that does not occur in the absence of internal Ca^{2+}. (Photocurrent data from Nakatani and Yau 1988.)

locations on the curve, one at low background and another at high background, have been isolated and expanded with background intensity on a linear scale rather than on the logarithmic scale of the main curve. At low illumination, a 5% response change requires a change from 1 to 2.6 photons/μm^2/s; a light flash of 1 photon will do the job, in other words. The slope of the curve here—0.03 (0.05/1.6 photons)—is a measure of sensitivity. At the higher background intensity, where the steady photocurrent level has shifted to a less negative value, a similar calculation gives a value of 55 photons required for a 5% (0.05) increase in the response. Here, the sensitivity is 0.0009 (0.05/55 photons). Thus the increase in background intensity has *decreased* the photoreceptor's sensitivity by a factor of 33 (0.03/0.0009).

Since the slope of the response-versus-intensity curve is a measure of sensitivity, plotting its derivative (the instantaneous slope) gives the change in sensitivity directly; this has been done in Figure 13.25*b*, and it shows what we expect: Sensitivity declines as background intensity increases. This desensitization is the photoreceptor component of light adaptation.

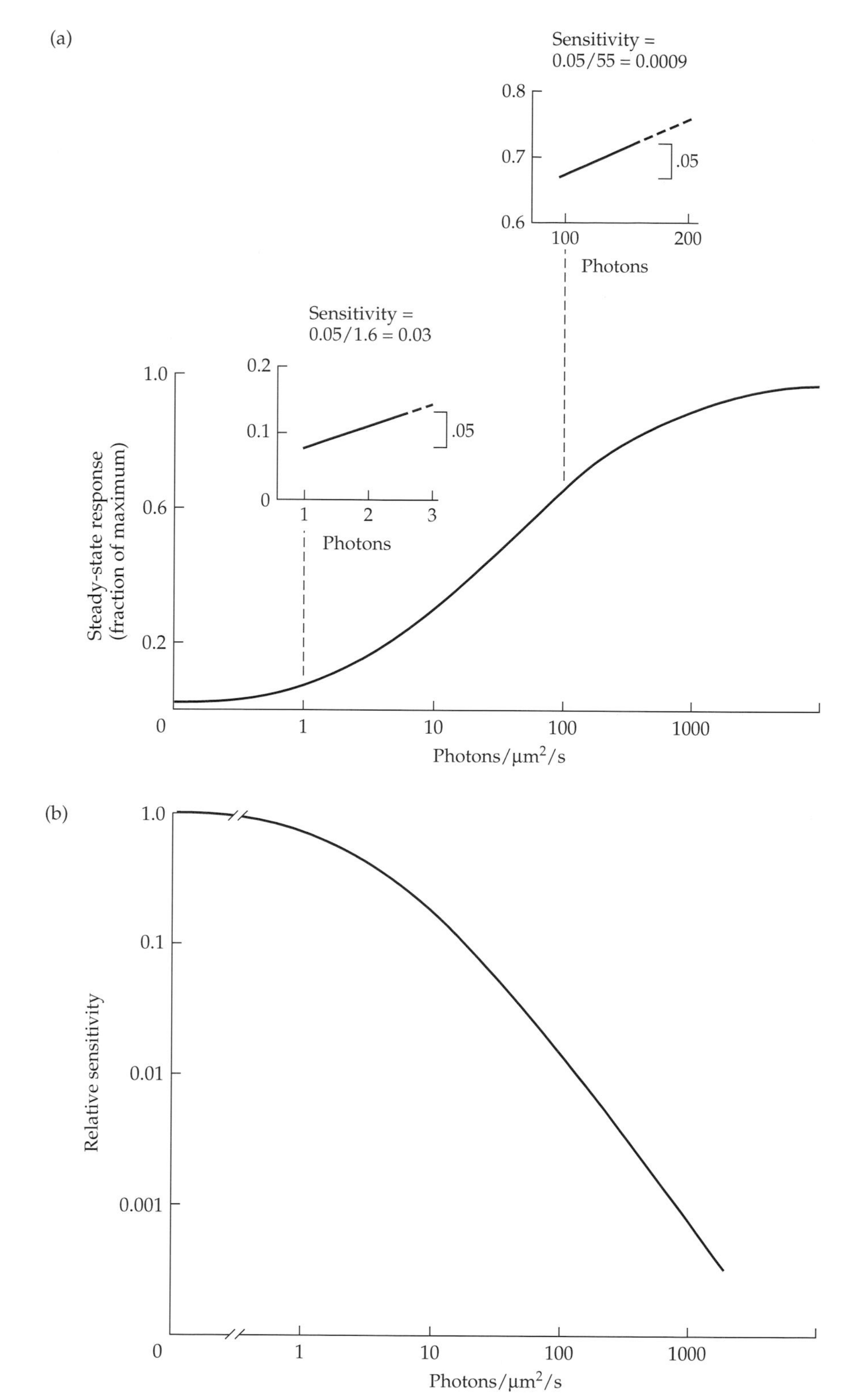

Figure 13.25
Light-Induced Changes in Rod Sensitivity
(*a*) As the steady-state level of illumination increases, the rod's steady-state response (reduction of photocurrent) moves closer to its maximum, as shown in this response plot of a salamander rod. The increasing illumination level produces a *decrease* in the rod's sensitivity, which is the slope of the curve on a linear intensity axis (insets). About 35 times more photons are required for a 5% change in response when the illumination level is 100 photons than when it is 1 photon.
(*b*) Relative sensitivity declines about tenfold for each tenfold increase in light intensity. (Replotted from Koutalos and Yau 1996.)

After a rod has been desensitized by light and its photocurrent is near zero, the photocurrent will gradually recover in darkness. This gradual change back to the maximum photocurrent level in the dark is accompanied by an *increase* in sensitivity, which is the photoreceptor component of dark adaptation.

As far as we know, the mechanisms underlying adaptation are similar in rods and cones, but cones desensitize less to light and they recover sensitivity in the dark more quickly. The curve of response versus background intensity for a cone would change less rapidly than the rod curve in Figure 13.25*a* does, and the response would not approach zero at the highest background intensities. These differences are due not to differences in transduction processes but to differences in kinetics—cone pigments may be activated for shorter times, for example—and differences in the magnitude of the effects elicited by Ca^{2+} in cones.

Changes in photoreceptor sensitivity account for less than half of the retina's sensitivity increase in the dark and sensitivity decrease in the light

When the eye is fully adapted to dark and at its most sensitive, it can detect light levels so low that a single rod will absorb a photon only once every hour or so. There are a lot of rods, however, and another way of saying the same thing is that the absorption of five to ten photons by different rods within a second will be detected. Any way one cares to state it, the dark-adapted eye is extremely sensitive. But bright reflected sunlight is more than a billion times more intense, and it bombards photoreceptors with almost a million photons each second. No photoreceptor could operate over this enormous range with constant sensitivity.

The visual system copes with this problem by decreasing its sensitivity as the level of illumination rises. The change is measured by the minimum amount of light that can be detected under different illumination conditions. When an observer is adapted to light and then put in total darkness, the visual threshold declines in a few minutes to a plateau, where it remains for several minutes longer before continuing its decline to a final plateau, where threshold is at its absolute minimum after 30 minutes or so in darkness. The initial threshold decline is attributable to cones and the later phase to rods. From the fully dark-adapted condition, thresholds remain at the plateau as background illumination is increased from total darkness and then begin to rise as illumination continues to increase. Like the dark-adaptation curve, there are two phases of threshold increase during light adaptation: Thresholds are determined by rods at low light levels, and cones take over at higher levels.

Sensitivity is the reciprocal of threshold, so sensitivity rises in two stages as we adapt to the dark and declines in two stages as we adapt to the light. The change in sensitivity (or threshold) is mathematically related to the amount of bleached photopigment; in general, the more photopigment bleached, the lower the sensitivity. But this is not the same as saying that adaptation is something done exclusively by the photoreceptors, even though they contain the photopigments. The extent of the photoreceptors' contribution can be seen in a comparison of the sensitivity curves for individual rods and cones with the sensitivity curve for the visual system as a whole; the curve shows decreasing sensitivity during light adaptation (Figure 13.26).

Increasing background intensity reduces the psychophysically determined sensitivity at low light levels, where rod sensitivity is unaffected. Rod sensitivity does not begin to decline until overall sensitivity has already decreased 1000-fold, and only then are the rod and overall sensitivity curves declining in parallel. When rod sensitivity drops below cone sensitivity, the cones enter the picture. The psychophysical sensitivity change at higher background levels closely parallels the cone sensitivity function.

The difference between the photoreceptor sensitivity curves and the overall sensitivity curve must be due to events occurring after the photoreceptor signals have been generated in the outer segments (the photoreceptor sensitivities were based on recordings of outer segment photocurrent). Of the 6 log units of sensitivity change shown here, about half the amount is accounted for by sensitivity changes in the rods and cones. The rest, which is called network adaptation, must be done by neural interactions in the retina. (One eye can be light- or dark-

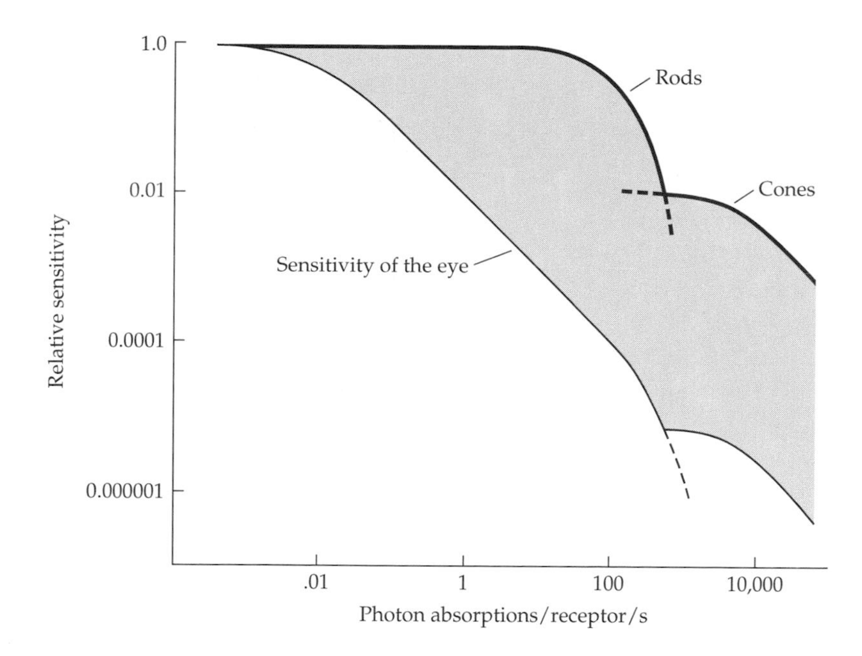

Figure 13.26
Comparison of Photoreceptor and Psychophysical Sensitivity Changes
The lower curve shows the eye's sensitivity, based on psychophysical threshold measurements, declining in several stages as the background illumination increases; the sensitivity in bright light is more than a million times less than in the dark. The upper curves show comparable rod and cone sensitivities derived from recordings of outer segment photocurrent. The difference between the photoreceptor sensitivity curves and the psychophysical curve (shaded region) is network adaptation. (Photoreceptor sensitivity curves from Baylor, Nunn, and Schnapf 1984 [rods] and Schnapf et al. 1990 [cones]; psychophysical data based on Aguilar and Stiles 1954 and Stiles 1949.)

adapted without affecting sensitivity in the other eye, which indicates that adaptation occurs before the site of interaction between signals from the two eyes. That points to the retina, and studies of retinal ganglion cells show that their changes in sensitivity parallel the psychophysical changes.)

Most of the photoreceptor sensitivity changes can be explained by the effects of Ca^{2+} on the phototransduction cascade, but the subsequent network adaptation is more difficult to pin down. Some possible mechanisms will be considered later, in Chapter 14, particularly in discussing interactions between photoreceptors and horizontal cells, but a definitive explanation will *not* be forthcoming.

The tips of photoreceptor outer segments are surrounded by pigment epithelial cell processes

The retinal pigment epithelium is a single layer of somewhat flattened cells lying between the photoreceptor outer segments and Bruch's membrane of the choroid. This location is a critical one, placing the pigment epithelium between the photoreceptors and their blood supply from the choriocapillaris (Figure 13.27), and it should come as no surprise that the relationship between photoreceptors and pigment epithelium is both intimate and important.

Like most epithelial layers, the cells of the retinal pigment epithelium (the RPE in clinical parlance) are bound together with desmosomes and other anchoring junctions. Of greater importance, high-resistance occluding junctions near their apical side constitute a significant blood–retina barrier. The pigment epithelium has a basement membrane, which is the innermost layer of Bruch's membrane, but the cells have few if any hemidesmosomes to anchor them to their basement membrane. Instead, the basal surfaces of the pigment epithelial cells have complex infoldings of a sort often associated with cells involved in fluid transport (see Figure 13.27). (Recall the inner epithelium on the ciliary processes, for example. See Figure 11.10.) The pigment epithelium seems to adhere fairly tightly to Bruch's membrane, but there is no obvious structural basis for this binding.

On the photoreceptor or apical side of the pigment epithelium, the cells have long processes extending well into the photoreceptor layer, surrounding the tip of each photoreceptor in a sheath of pigment epithelium (Figure 13.28). This

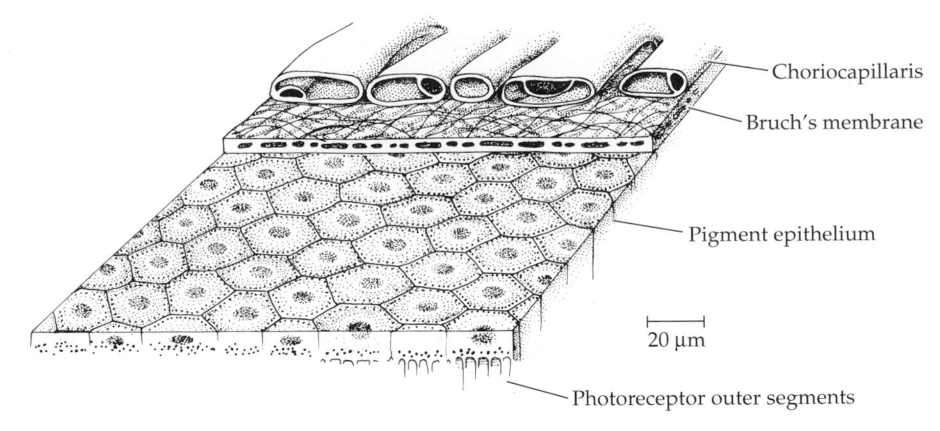

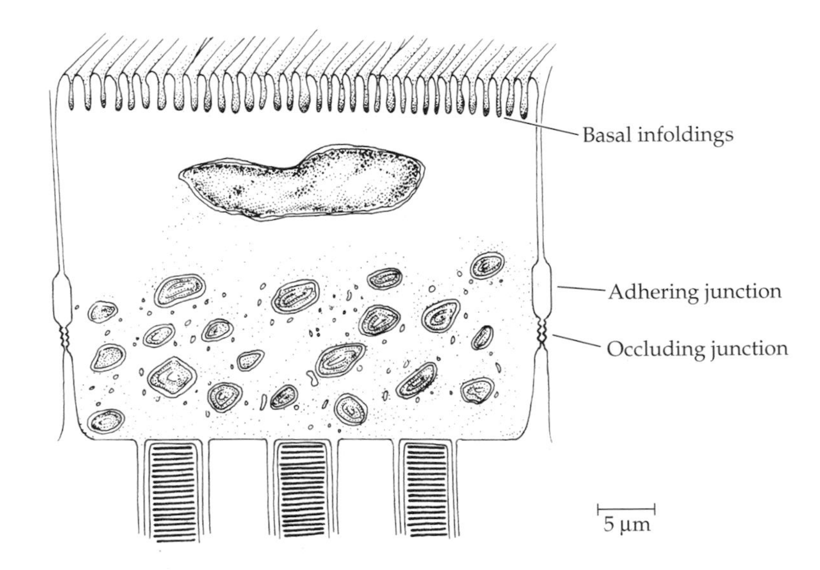

Figure 13.27
Structure of the Pigment Epithelium
The pigment epithelial cells are roughly hexagonal, packed together in a single layer between Bruch's membrane and the photoreceptor outer segments. At higher magnification, one can see infoldings of the cell membrane on its basal surface and processes extending from the apical surface along the outer segments of the photoreceptors. Adjacent cells have adhering junctions and tight occluding junctions near the apical side. The dark inclusions in the cytoplasm are pigment-containing melanosomes.

arrangement creates a kind of socket into which the outer segment of the photoreceptor fits like a plug. Most of the melanin pigment is in the apical part of the pigment epithelial cells, and some of it lies within processes extending down between the photoreceptors. Pigment in this location should absorb stray light not absorbed by photopigments and help optically isolate individual photoreceptors. In some species, pigment migrates in the epithelial cell processes as the adaptational state of the retina changes, moving farther down along the outer segments in light adaptation and withdrawing outward during dark adaptation. If this occurs at all in the human retina, it is very slight.

The pigment epithelium and the interphotoreceptor matrix are necessary for photopigment regeneration

Photoreceptor outer segments and the surrounding epithelial cell processes are always separated by a narrow space, with no sign of occluding or anchoring junctions between the cell membranes. This extracellular space is filled with a gel, the **interphotoreceptor matrix**, which is a mixture of proteins, glycoproteins, and glycosaminoglycans. Some components of the matrix are produced by photoreceptor inner segments, and the composition of the matrix around cones is somewhat different from the rod matrix. The interphotoreceptor matrix may help photoreceptors and pigment epithelium adhere by virtue of its viscosity, but it certainly does more than passively fill space; the matrix contains several components that are essential for the normal operation of the photoreceptors.

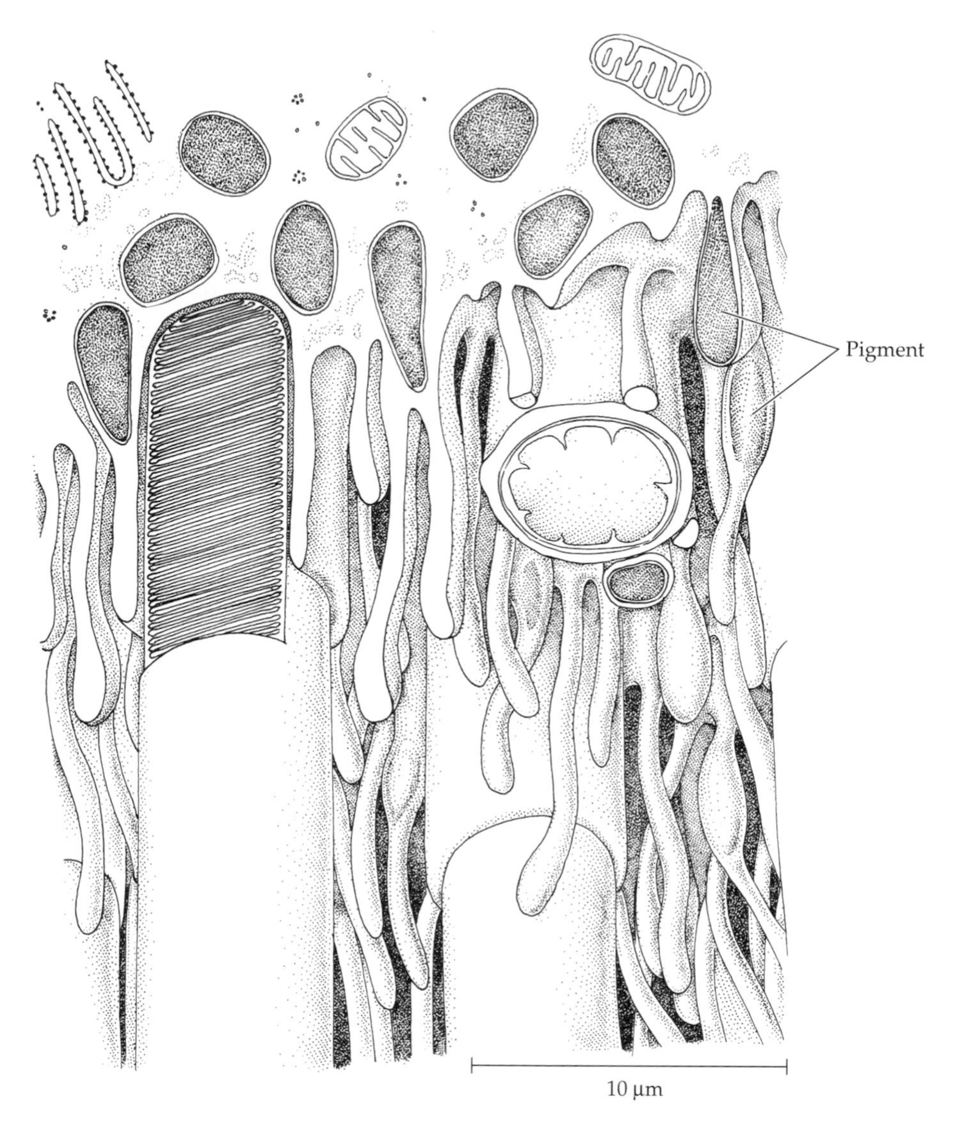

Figure 13.28
Pigment Epithelial Cells Surround the Outer Segments of the Photoreceptors
The processes on the apical surface of the pigment epithelium form complete sheaths around the outer segments of the photoreceptors, forming sockets into which the outer segments plug. Some processes do not ensheath receptors but contain melanin pigment.

One of the major proteins in the matrix, called **interstitial retinoid-binding protein**, or **IRBP**, is necessary for the regeneration of bleached photopigment. As mentioned earlier, fully bleached photopigment consists of two pieces, opsin and all-trans retinol. To regenerate the photopigment, the retinol must be converted by appropriate enzymes back to 11-cis *retinal* and reattached to the opsin. The enzymes are not resident in the photoreceptors, however; they are in the pigment epithelium. Therefore, all-trans retinol produced by bleaching must be transported to the pigment epithelium, converted to 11-cis *retinal,* and transported back to the photoreceptor; IRBP carries retinol to the pigment epithelium and returns *retinal* to the photoreceptor, as summarized in Figure 13.29.

Neither *retinal* nor retinol can survive long outside the photoreceptors; they are readily broken down by enzymes they encounter. Wherever they go, they must be bound to other molecules that protect them from breakdown; these other molecules are binding proteins, one of which is the IRBP in the interphotoreceptor matrix.

When all-trans retinol produced by photopigment bleaching leaves the outer segment, it is picked up by IRBP and carried to the adjacent epithelial cell membrane, where it is transferred to the interior of the cell. (The mechanism of transport across the cell membranes is not known.) Inside the epithelial cell, the retinol is picked up and carried around by cellular retinoid-binding protein (CRBP). Several enzymes convert the retinol from the all-trans to the 11-cis form while it is attached to CRBP. Another enzyme (11-cis retinol oxidoreductase) converts the retinol to *retinal* and it is transferred to another binding protein, cellular *retinal*-binding protein (CRalBP).

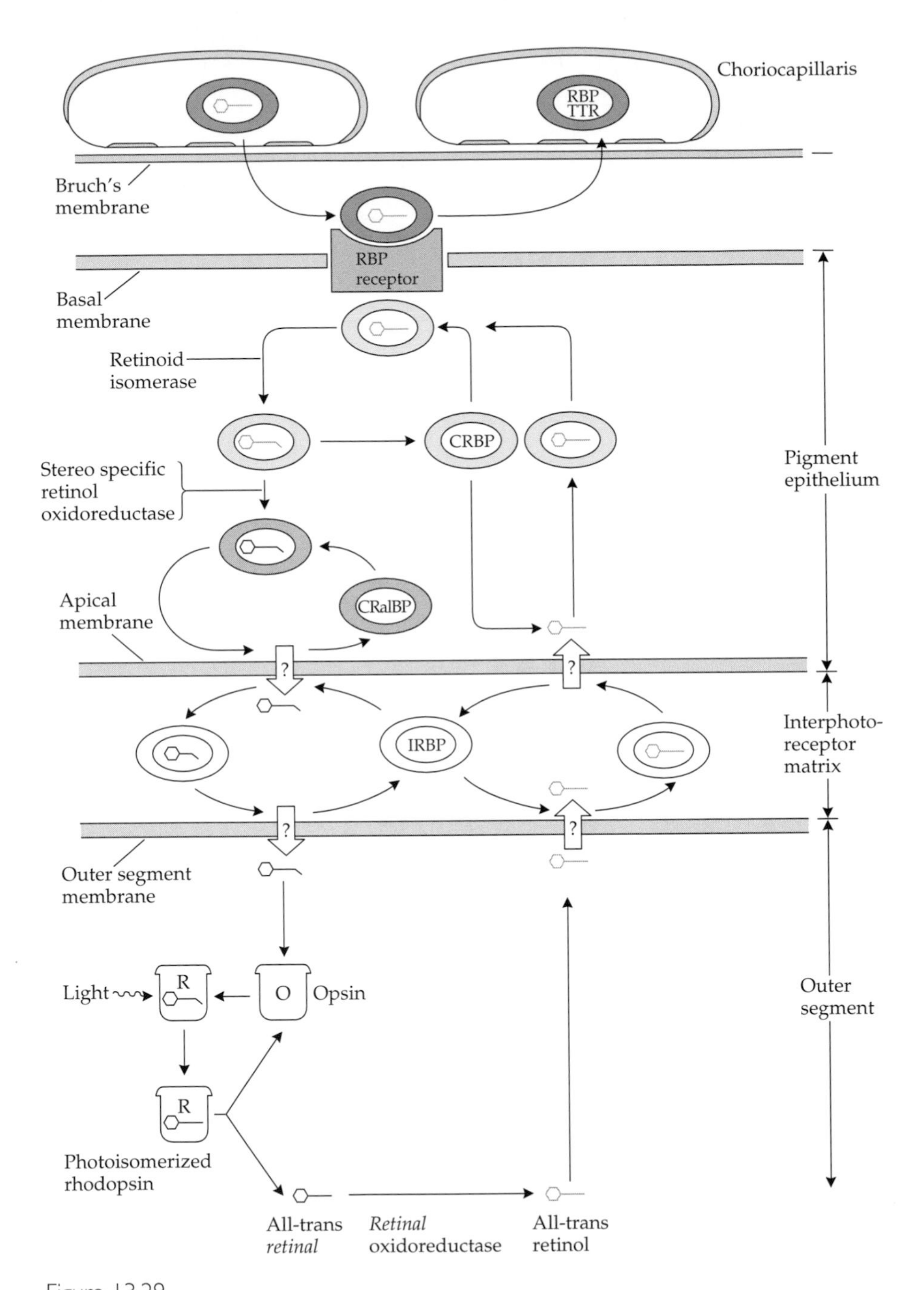

Figure 13.29
The Visual Cycle
All-trans retinol is supplied by the blood, and it is a product of photopigment bleaching. It must be converted to 11-cis *retinal* in the pigment epithelium and taken back to the outer segment of the photoreceptor, where it can combine with opsin to form an inactivated photopigment molecule. At all stages, the retinoids must be carried by binding proteins. Several enzymatic reactions in the pigment epithelium are only implied here, and the symbol "?" indicates where transport mechanisms are unknown. (After Bok 1990 and Rodieck 1998.)

Now that the conversion has been made to 11-cis *retinal,* which is the form required to regenerate the photopigment, all that is left is transport back to the photoreceptor. After IRBP carries the 11-cis *retinal* back through the interphotoreceptor matrix, the *retinal* enters the photoreceptor and recombines with opsin. Regeneration is complete.

Figure 13.29 also shows how new retinol enters the photopigment cycle. It is stored in the liver and carried in the blood by a complex carrier called RBP-TTR (retinol-binding protein and transthyretin). This aggregate of retinol plus carrier can diffuse out of the choriocapillaris and cross Bruch's membrane to the basal surface of the pigment epithelium. Here, the aggregate docks with receptors that remove the retinol from its carrier and transport it across the membrane to another carrier (CRBP). The RBP-TTR complex returns to the blood.

Although Figure 13.29 is schematic and the anatomical elements are not accurately scaled, the importance of maintaining normal anatomical relationships among photoreceptors, pigment epithelium, and choroid should be obvious. All the molecular interactions may be working as they should, but the cycle of photopigment regeneration and resupply fails when photoreceptors are separated from their matrix: Retinoids cannot be transported to or from the photoreceptors, and bleached photopigment cannot be regenerated. This is one of several reasons to worry about retinal detachment.

Both rods and cones undergo a continual cycle of breakdown and renewal

Until about 30 years ago, people assumed photoreceptors to be permanent structures, much like other neurons that have dropped out of the cell cycle and have little capacity for renewal or regeneration. This conception changed dramatically following an observation on the fate of radiolabeled amino acids in rods. Photoreceptors take up extracellular amino acids and use them to construct proteins. By tagging some of the extracellular amino acids with a radioactive label (usually tritium), one can see what happens to them (Figure 13.30). Initially, the label diffuses throughout the inner segment and terminal of the rod, but within

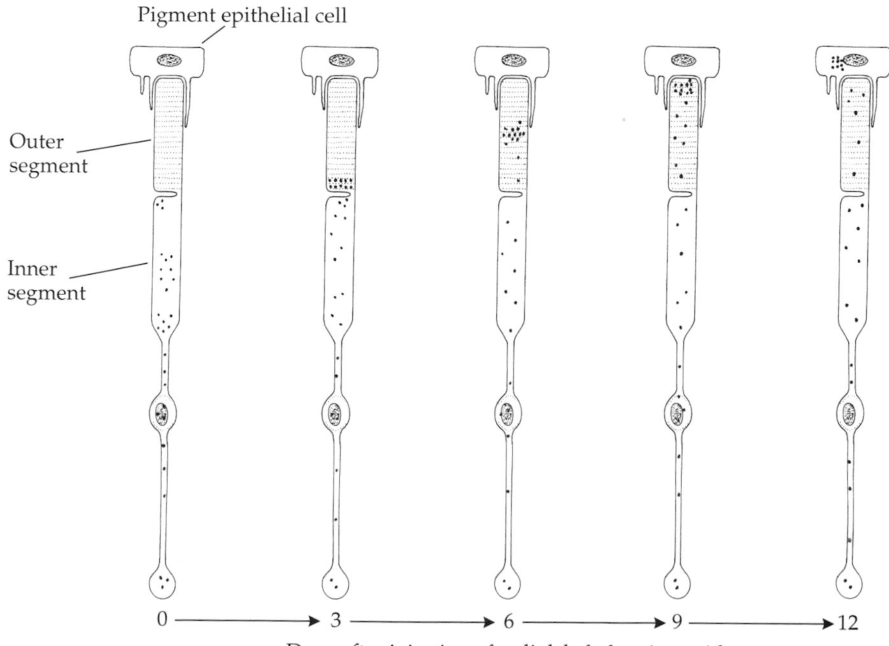

Figure 13.30
Outer Segment Discs Are Formed and Shed in about Two Weeks
The small dots in this schematic rod represent the locations of radiolabeled amino acids at various times after their incorporation into the cell. A distinct band of label forms at the base of the outer segment within a few days. It moves up through the outer segment and into the pigment epithelium in 12 days, indicating that new discs formed at the base of the outer segment on day 1 are shed from the tip of the rod in less than 2 weeks. (After Young 1978.)

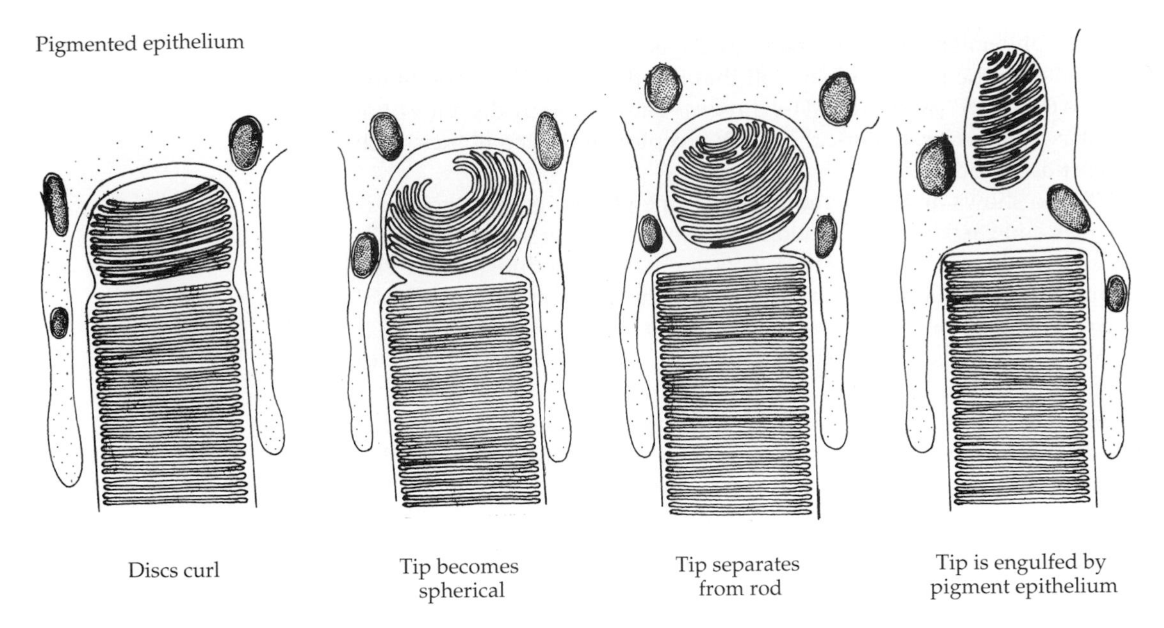

Figure 13.31
Disc Shedding in the Outer Segment
A slight pinching in the outer segment membrane as 10 to 20 discs begin to curl upward at their edges marks the beginning of disc shedding. As the discs continue to curl into a compact spherule enclosed by outer segment membrane, they separate from the rest of the outer segment, after which the detached photoreceptor tip is surrounded and engulfed by a pigment epithelial cell. As an inclusion in the pigment epithelium, the shed outer segment tip is now a phagosome. (After Young 1971.)

a few days it coalesces as a band of label at the base of the outer segment. Over the next week or so, the band of label moves up the outer segment to its tip, then leaves the rod for the adjacent pigment epithelial cell.

This sequence of events reflects the continual formation of new outer segment discs at the base of the outer segment and a corresponding removal of "old" discs at the tip of the outer segment. The release or removal of the outer segment tips is called **shedding**; discs shed by the rods are phagocytosed by the pigment epithelium, which digests them and disposes of the breakdown products to the choriocapillaris.

When the same experiment is tried with cones, the radioactive label never forms a compact band; instead, it spreads throughout the outer segment, leaving the impression that the outer segments of cones do not shed and rebuild their disc membranes. Recall, however, the difference between the rod and cone discs; those in rods are detached from the outer membrane, while the cone discs are continuous folds of the membrane. Although the amino acids taken up by the cones may be incorporated initially in newly formed discs, the continuity between the discs and outer segment membranes allows the radioactive molecules to diffuse throughout the outer segment membrane. But there is another way to see the result of outer segment shedding, for both cones and rods.

As Figure 13.31 shows, shedding begins as 10 to 20 discs at the tip of the outer segment curl up and the outer segment membrane pinches in between them and the underlying discs. The small ball of discs separates from the outer segment, whose membrane seals across the end of the outer segment, and the pigment epithelial cell engulfs the detached discs. Since the phagocytosed outer segment tip retains some of its lamellar structure for a time, these **phagosomes** are easily recognizable in the pigment epithelial cells. Thus the first evidence for cone shedding was an observation of phagosomes in pigment epithelial cells at the center of the fovea. (The center of the fovea has no rods.) Unless phagosomes can migrate between pigment epithelial cells, the conclusion is straightforward: Cones undergo the same processes of disc formation and shedding as rods.

The local time as I write this section is early afternoon, which means that my rods and cones are probably not shedding their outer segment tips. Shedding is not continuous; it happens in bursts at a particular time of day. Experimental animals maintained on a cycle of 12 hours of light and 12 hours of dark shed outer segment tips of rods late in the dark cycle or early in the light cycle and shed cone tips late in the light cycle. The shedding of outer segment tips is one of the body's **circadian rhythms** ("circadian" means "about a day"), wherein a particular function waxes and wanes with a roughly 24-hour period. Therefore, to see evidence of outer segment tip shedding in cones, one must look at the right time of day; this requirement may be one reason it was not observed earlier. Also, each rod may shed its outer segment tip every day, while cones shed less frequently, perhaps every other day; if so, phagosomes from outer segments of cones would be less numerous and less obvious.

We do not have quite enough evidence to know if shedding is initiated by the photoreceptors or by the pigment epithelium. That is, do photoreceptors discard discs at their own initiative or does the pigment epithelium eat them when it chooses? Some investigators have reported seams appearing around the photoreceptors before any other change, suggesting that the photoreceptors initiate the sequence, but this initial step could be a response to a signal of some sort from the pigment epithelium. The possibility that abnormal photoreceptor degeneration is a kind of normal shedding gone awry means that initiation and control of shedding will be studied in more detail.

The inner segments of photoreceptors assemble the proteins to construct the outer segment membranes

The inner segments of both rods and cones are connected to their outer segments by a narrow, cylindrical column called the **connecting cilium**. As Figure 13.32 shows, the connecting cilium contains an assembly of microtubules arising from basal bodies of the sort found in nonmotile cilia, thus hinting that photoreceptors derive from ciliated cells. The connecting cilium is considerably narrower than the inner and outer segments, but it is not clear if this narrow neck impedes or facilitates the transport of materials from the inner to the outer segment.

The inner segments of photoreceptors are packed with mitochondria; the unusually high mitochondrial density is their most distinctive feature. The density of mitochondria is not constant, however. It is highest in the region shown in Figure 13.32, which is near the outer segment; this region stains more darkly in histological sections and is called the ellipsoid, in keeping with the shape of the darkly stained region within the inner segment.

The mitochondria are less numerous closer to the base of the outer segment, where there are more microtubules and microfilaments and where the endoplasmic reticulum and Golgi apparatus are found (they are shown schematically in Figure 13.33). This part of the inner segment is called the myoid. In other species, the myoid contains contractile proteins running parallel to the photoreceptor axis; contraction and relaxation of this miniature muscle moves the photoreceptor away from or toward the pigment epithelium. Generally, cones move away from the epithelium when the eye is adapted to light and move back toward the epithelium during adaptation to dark. In primate photoreceptors, the myoid has little contractile protein, and there is little evidence of photoreceptor movement.

The inner segments also contain the Golgi apparatus and endoplasmic reticulum that are involved in protein synthesis, and they are the site of protein assembly for the construction of new outer segment disc membrane. The elements of this construction are illustrated in Figure 13.33*a*; step 1 shows the discs as outfoldings of the outer membrane at the top of the connecting cilium. Steps 2 through 4 show growth of the membrane between adjacent folds progressing around the circumference of the folds, eventually meeting and fusing to separate a disc from the outer membrane (in rods). This process is the same in cones, except that it does not go to completion and the cone discs retain continuity with the outer membrane.

All the different proteins used in the outer segment, including the photopigment opsin, are synthesized in the inner segment. As Figure 13.33*b* shows, an opsin precursor is manufactured in the rough endoplasmic reticulum and trans-

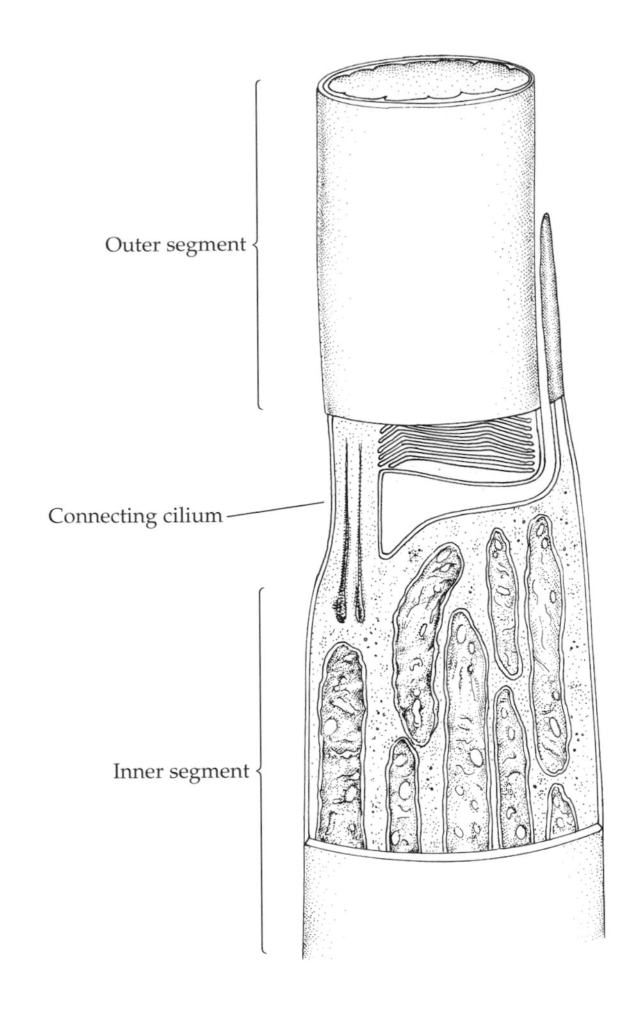

Figure 13.32
The Junction between Inner and Outer Segments
This cutaway drawing represents a rod. The inner and outer segments are connected by a thin stalk, the connecting cilium, which has two basal bodies and nine pairs of tubules (two are shown here), which are characteristic of nonmotile cilia. This part of the inner segment (the ellipsoid) is filled with long mitochondria, microfilaments, and microtubules. Other cytoplasmic structures, such as the endoplasmic reticulum and Golgi apparatus, are farther down in the inner segment, closer to its base. The cutaway portion of the outer segment is the region in which new outer segment discs form.

ported to the Golgi apparatus, where it is trimmed to its final form. Transport vesicles carry the opsin to the inner segment membrane adjacent to the outer segment and insert opsin into the membrane by fusing with it. By diffusing along the membrane of the connecting cilium, the opsin becomes a part of the membrane outfoldings that will be made into the discs.

In addition to continually synthesizing proteins for the outer segment disc membranes, the inner segment synthesizes other molecules that are transported to the terminal end of the photoreceptor. All this synthesis requires energy, as do the Na^+–K^+ pumps in the inner segment membrane (see Figure 13.19), which accounts for the high density of mitochondria in the inner segments. And all these energy-dependent operations require a steady supply of oxygen and other essential elements, which makes the photoreceptors highly dependent on their blood supply and quite vulnerable to any interruption in the supply.

Other than shape and size, there are no significant anatomical distinctions between the inner segments of rods and those of cones. Outside the fovea, the inner segments of cones are larger than those of rods, and they taper as they run up to join the outer segments, giving them a distinctly conical shape. Foveal cones, however, have cylindrical inner segments that are very difficult to distinguish from the inner segments of rods.

The inner segments form tight junctions with Müller's cells; these junctions are the external limiting membrane

Müller's cells will be described in more detail later (see Chapter 16). Here it is sufficient to know that they are large glial cells that extend from the inner limit-

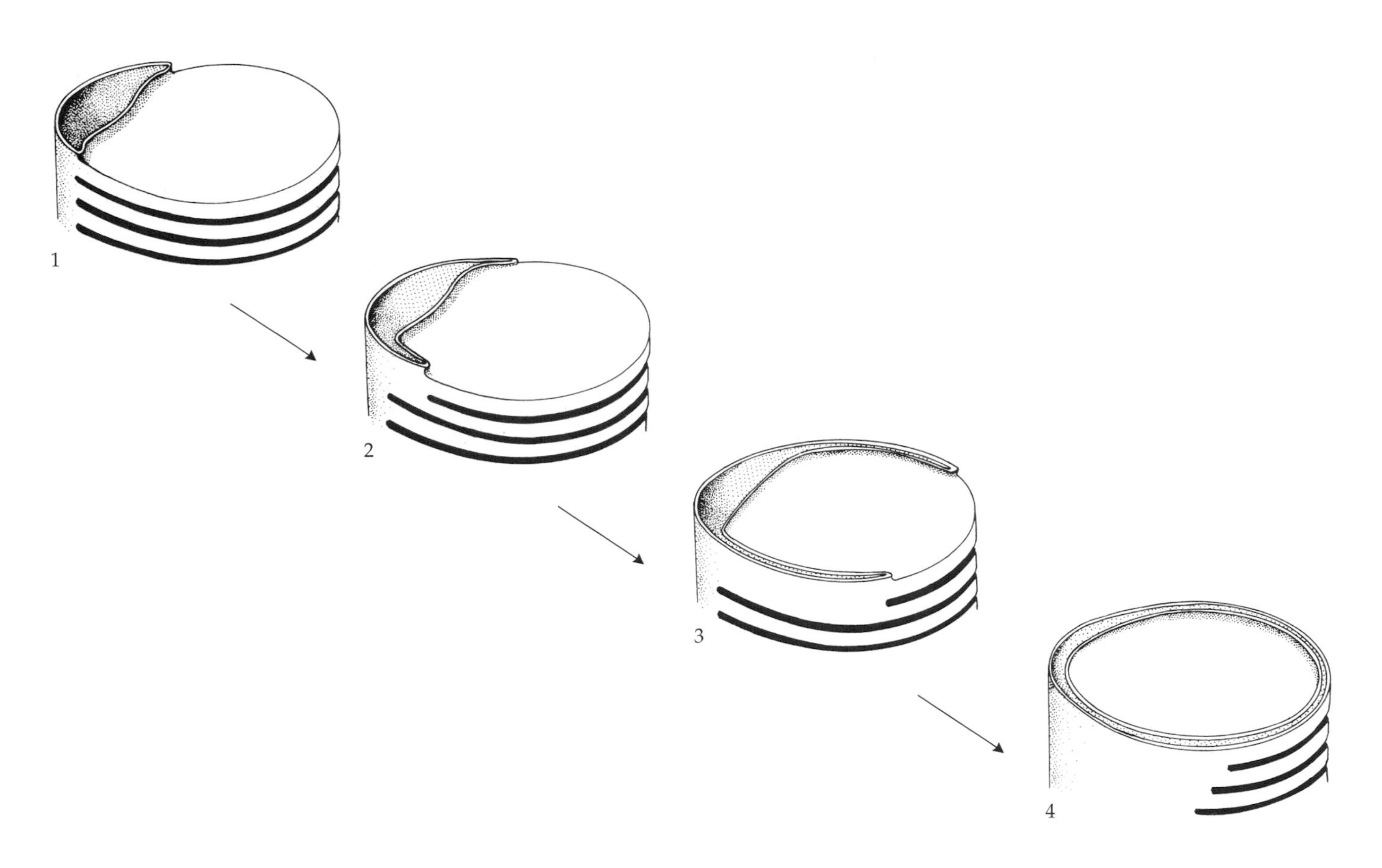

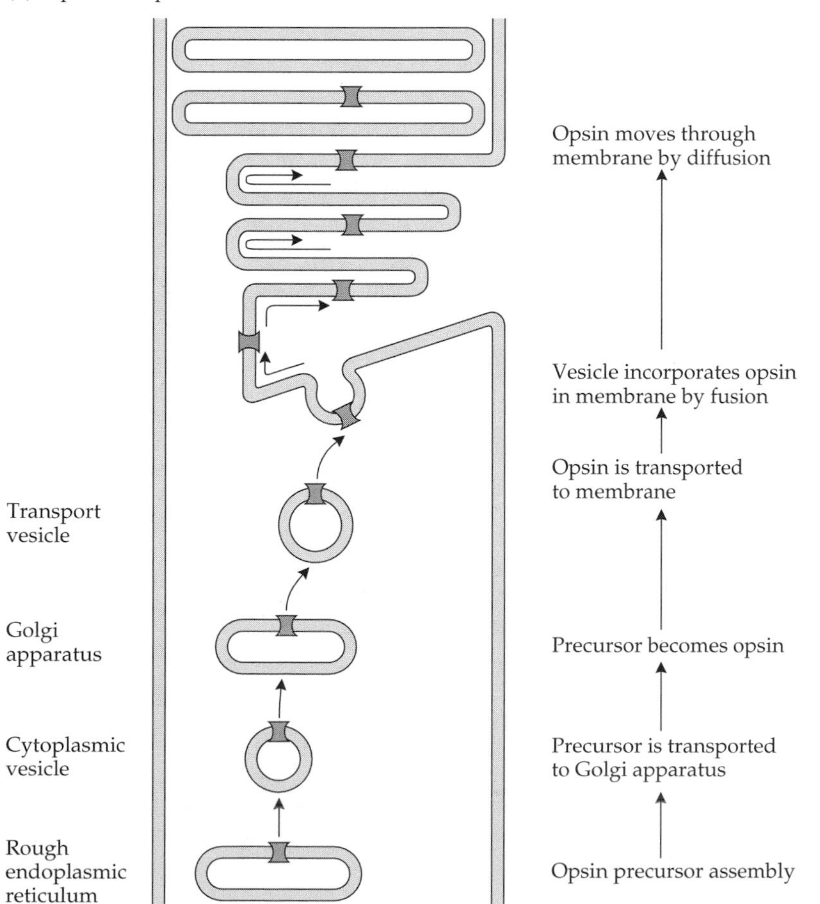

Figure 13.33
Disc Formation and Opsin Incorporation in Disc Membranes
(*a*) In both rods and cones, new discs form at the base of the outer segment as outfoldings of the membrane at the top of the connecting cilium. The membrane between two outfoldings grows along the circumference of the outfolding, eventually meeting and fusing to separate the disc from the outer segment membrane (in rods). Fusion does not occur in cones, and the discs never separate completely from the outer membrane. (*b*) Opsin is synthesized in the inner segment and inserted into the membrane by vesicular fusion. As the opsin diffuses in the membrane, it is incorporated in the outfoldings and then into the discs. (*a* after Steinberg, Fisher, and Anderson 1980; *b* after Bok 1985.)

ing membrane of the retina out to the photoreceptor layer. Almost all neurons in the retina are surrounded by extensions and processes from Müller's cells, and the photoreceptors are no exception. The relationship of Müller's cells to the photoreceptors is shown in Figure 13.34.

The inner segments of neighboring photoreceptors—whether rods or cones—do not touch one another; another cellular element always intervenes. The photoreceptors and the interposed cellular processes show adjacent spots of high electron density at points where the membranes are closely apposed; these are local anchoring junctions (maculae adherens). These junctions are between photoreceptors and Müller's cells, between adjacent Müller's cells, and occasionally between photoreceptors. Thus, the external limiting membrane is not a true membrane, but an array of adhering junctions lying in a plane along the base of the inner segments of the photoreceptor. The function of the adhering junctions is presumably structural, like epithelial anchoring junctions, and there is no reason to think that they play any role in basic photoreceptor operations. As mentioned earlier, however, the resistance to ionic-current flow by the external limiting membrane may affect the way current flows around the photoreceptors.

Photoreceptor terminals contain specialized synaptic structures where contacts are made with postsynaptic neurons

This discussion is about to skip from the level of the inner segments to the photoreceptor terminals, but this leap is not meant to imply that the nuclei and the

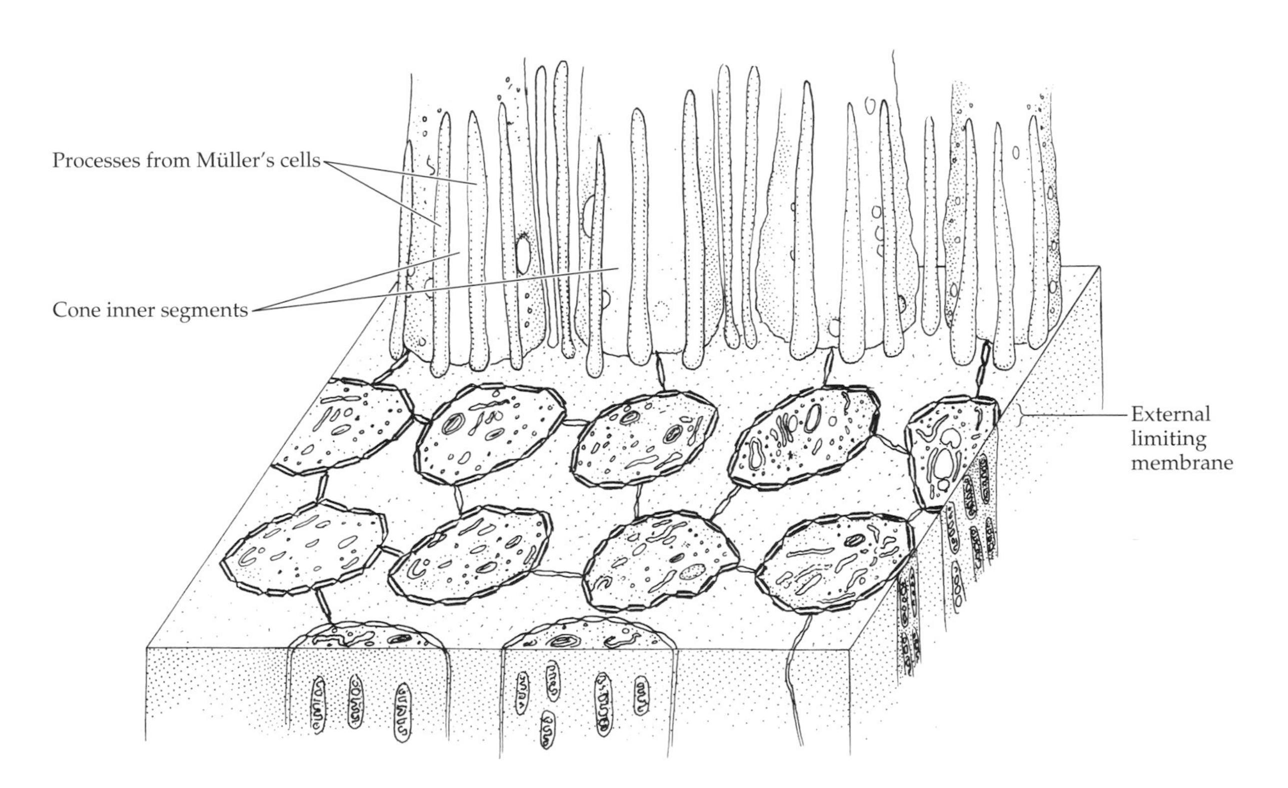

Figure 13.34
The External Limiting Membrane
The bases of the inner segments of cones are separated by Müller's cells that also send small processes up between the inner segments. As indicated in the cutaway portions of this drawing, Müller's cells and the inner segments of cones share numerous anchoring junctions, and these junctions are confined to a plane about 1 μm thick. The plane containing the junctions is the external limiting membrane, which is clearly not a membrane in the usual sense of the word.

sometimes long processes running to the terminals are unimportant. They are, of course, necessary elements; the nucleus is the repository of all the instructions for the protein synthesis taking place in the inner segments, and the axonal process is the route of cytoplasmic transport of materials to and from the terminals. The only significant anatomical difference between the nuclei of rods and those of cones is their placement relative to the inner segment: Outside the fovea, cone nuclei are usually just next to the inner segment, lining the inner edge of the external limiting membrane, while rod nuclei are usually displaced farther into the outer nuclear layer, where they pile up to form much of the layer's thickness. In foveal cones the nuclei are at some distance from the inner segments, partway along the axonal processes connecting the inner segments to the cone terminals.

Rod and cone terminals differ markedly in structure (Figure 13.35). Rod terminals are relatively small and roughly spherical, around 2 μm in diameter; they are called **spherules**. Cone terminals, called **pedicles** because of their footlike appearance, are flatter, broader, and less regular in shape, often having small radiating processes called **telodendria**. Cones contact one another indiscriminately, and they contact rod spherules at gap junctions on the telodendria, thus establishing possible interactions among neighboring photoreceptors. Electrical communication between cones might make them dependent on one another and create a pooling of cone signals that is detrimental to spatial resolution, which requires independent cone signals. But the role played by the gap junctions is not clear. It may be that conductance across these junctions can be varied; the gap junctions might be closed down at high luminance levels, maintaining cone independence, and opened at low light levels to permit some pooling of cone signals when high spatial resolution is no longer important.

Differences in size and shape make it easy to distinguish between rod spherules and cone pedicles, but there is very little difference among pedicles belonging to the three types of cones. In regions where the cone density is high, cone pedicles form a layer along the border of the outer plexiform and inner nuclear layers—that is, just next to the horizontal cells—while the rod spherules line up in several rows just outside the pedicles, a bit farther away from the inner nuclear layer.

Both cone pedicles and rod spherules have specialized synaptic regions, formed around deep pits (**invaginations**) in the base of the terminal (see Figure 13.35), and each invagination is associated with a specialized structure inside the terminal called a **synaptic ribbon**. In cross section, a ribbon appears as a very dark streak or bar next to a thickened, electron-dense strip of terminal membrane at the base of the invagination, but they are actually elongated and may be up to 1 μm long. The membrane specialization next to the ribbons runs along with them and forms a synaptic ridge at the base of the invagination. Ribbons are surrounded by small synaptic vesicles, some of which are attached to the ribbons by very fine filaments; the ribbons are thought to capture and collect the vesicles, positioning them for release of their contents into the invagination.

The invaginations are filled by processes from neurons that are postsynaptic to the photoreceptors; typically, there are three postsynaptic processes—two lateral, deeply penetrating processes from horizontal cells (shaded in Figure 13.35) and a central, less intrusive bipolar cell dendrite. The entire arrangement at the invagination, with its three postsynaptic processes, has led to its being called a **triad** synapse, or synaptic triad. (There may be more than three postsynaptic elements, but the term "triad" is still used.) In rod spherules, the central bipolar process is always associated with a single type of bipolar cell, but the situation at the cone pedicles is less certain, because there are numerous bipolar cell types in cones.

Cones make another type of contact outside the invaginations, along the bases of the pedicles (see Figure 13.35). These basal or flat contacts are not conventional synapses, since they are not associated with sizable accumulations of vesicles on the presynaptic side of the junctions; the dense, thickened membranes at

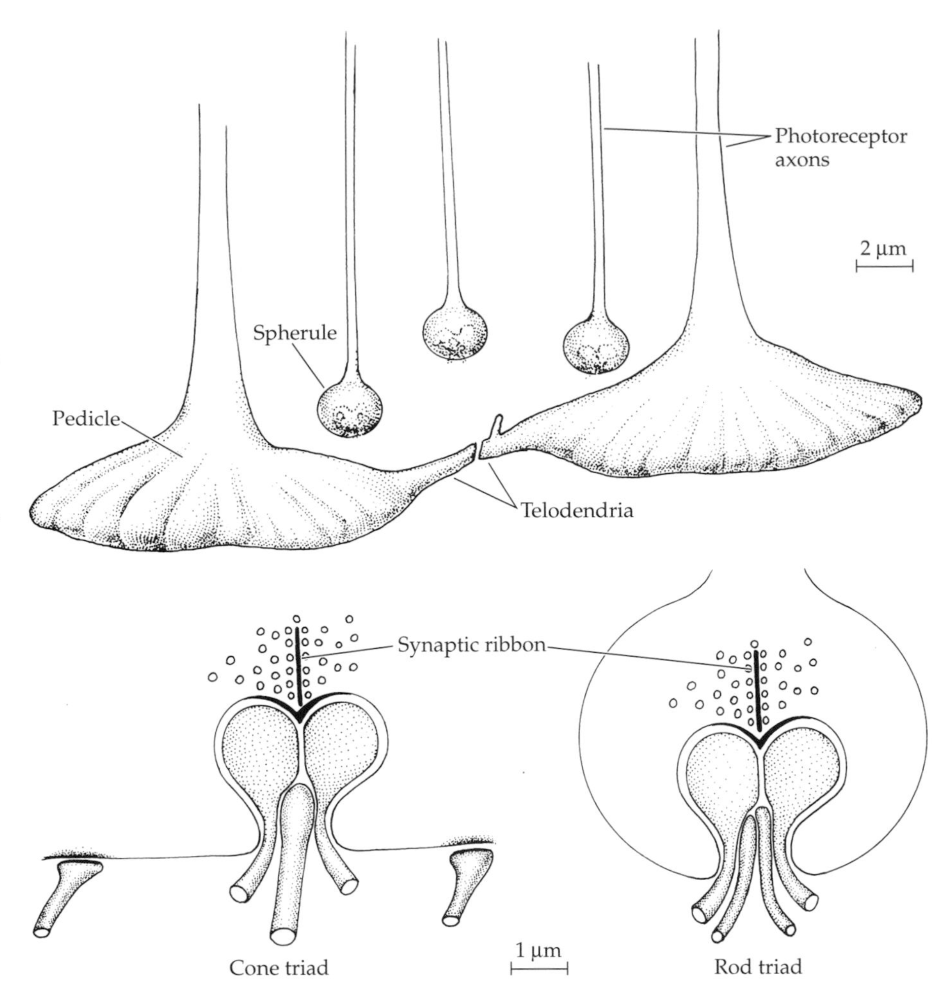

Figure 13.35
Photoreceptor Terminals and Triad Synapses
Rod spherules are small, about 2 µm in diameter; cone pedicles are 10 to 15 µm in diameter. Small processes, the telodendria, often extend from the pedicles to contact other pedicles or spherules at gap junctions. Both spherules and pedicles contain small invaginations in their bases, into which postsynaptic processes insert. The terminals contain dense synaptic ribbons surrounded by synaptic vesicles. Lateral processes in the invaginations (shaded) are from horizontal cells; the central process (two processes in rod triads) are from bipolar cells. Cone pedicles also make contact with bipolar cell processes along their bases at various distances from the invaginations.

these contacts are probably specialized anchoring junctions. Nonetheless, the flat contacts are the sites of neurotransmitter interaction between the cone pedicles and dendrites from bipolar cells. Rod spherules do not have basal contacts with their bipolar cells.

Although many of the flat contacts in cones are well away from the invaginations, some are very close. These nearby contacts are with the same class of bipolar cells whose dendrites penetrate into the invaginations (ON bipolar cells). The more distant flat contacts are with OFF bipolar cells. These distinctions will be considered in Chapter 14.

Because of their small size, rod spherules have very few invaginations, usually just one and rarely more than three. The cone pedicles are larger, however, and even the foveal cone pedicles have a half dozen or so invaginations; the pedicles become larger outside the fovea, and 40 or more invaginations can be expected (Figure 13.36). The large number of cone triads is related to signal divergence in the cone system: There are at least eight bipolar cell pathways in cones, and each cone probably contributes to all of them.

Photoreceptors signal light absorption by decreasing the rate of glutamate release from their terminals

Photoreceptors communicate with other neurons by releasing glutamate, which is stored in vesicles in the terminals. In spite of the specialized structures associated with the invaginations, like the ribbons and synaptic ridges, and the uncon-

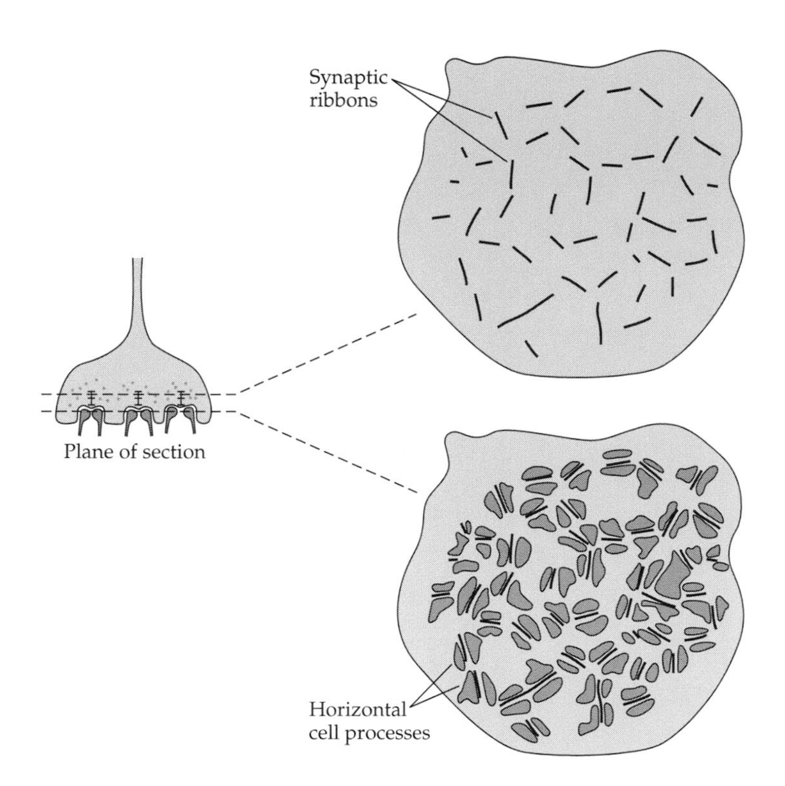

Figure 13.36
A Cone Pedicle Sectioned in the Plane of the Synaptic Ribbons
This drawing of a peripheral cone pedicle has 46 synaptic ribbons and 96 lateral processes (shaded) from horizontal cells. A foveal cone will have only a few ribbons in its pedicle, and rod spherules invariably have a single ribbon. (After Chun et al. 1996 and Rodieck 1998.)

ventional nature of the flat contacts on cone pedicles, the mechanism for glutamate release is like that of acetylcholine release at neuromuscular junctions (see Figure 4.18).

Photoreceptor terminal membranes contain voltage-gated Ca^{2+} channels that open when the terminal *depolarizes*—that is, when the inside of the terminal becomes more positive (Figure 13.37). The Ca^{2+} influx through the open channels sets off a cascade of interactions among specialized presynaptic proteins. Details of this cascade are not fully understood, but the effect is clear: Synaptic vesicles move, fuse with the terminal membrane at the synaptic cleft, and then open to release their contents. After release, the empty vesicles are reincorporated in the terminal (endocytosis) and eventually refilled with glutamate.

Glutamate release increases when the terminal depolarizes, but the change elicited by light absorption throughout the photoreceptor is *hyper*polarization. Thus, the photoreceptors signal light absorption by decreasing, not increasing, the rate of glutamate release. It follows that glutamate release will be highest in the dark and at its lowest under very bright light conditions. If glutamate is the photoreceptor's mode of communication, it may seem odd that the least amount of glutamate is being released when the photoreceptor is absorbing the most light. This is an extreme condition, however; a better way to think about photoreceptor signaling is to consider an intermediate state of moderate illumination.

With moderate illumination, glutamate will be released at a rate between maximum and minimum, and the changing illumination of the photoreceptor will modulate this rate up and down; the release of glutamate drops as illumination increases and rises as illumination decreases. So the change in glutamate release signals both increasing and decreasing illumination, as it should; it just happens that the glutamate signal is of opposite sign to the level of photon absorption.

In rod spherules, all the glutamate release is within the invagination near the invaginating horizontal and bipolar cell processes, and that is probably true for cones as well. But if this is the case, how does glutamate act on the bipolar cell dendrites that make the flat contacts along the base of the cone pedicles? Is glutamate also released from the base of the pedicle outside the invaginations? And if it is not, how does glutamate exert its effects at these locations?

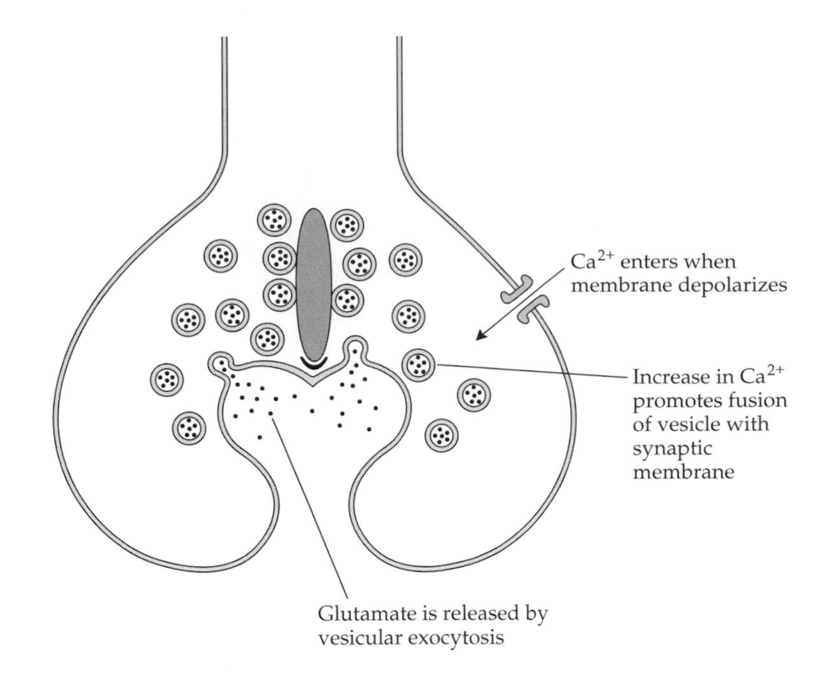

Figure 13.37
Glutamate Release at a Rod Spherule
Glutamate release depends on Ca^{2+} channels in the terminal membrane. Ca^{2+} flows in as the membrane depolarizes, and the increase in intracellular Ca^{2+} promotes fusion of the vesicle with the synaptic membrane near the ribbon. Glutamate in the vesicles is released by exocytosis; that is, the vesicle opens up into the synaptic cleft. The mechanism is the same at cone triads.

Glutamate is probably released only within the invaginations. But as is usually the case with neurotransmitters, release is followed by a cleanup operation to keep the transmitter's action from being overly prolonged. In some cases, enzymes break down the neurotransmitter, as acetylcholinesterase does at the neuromuscular junction. In others, uptake systems transport the neurotransmitter back into the cell from which it was released or into other surrounding cells. If we think of the release sites as the source of transmitter and the uptake sites as a drain, neurotransmitter flows from source to drain.

The uptake sites for glutamate appear to be distributed along the base of the cone pedicles (and on the ever-present Müller's cells), suggesting a flow of glutamate along the base of the pedicles from the invaginations to the uptake sites. As this glutamate flow encounters bipolar cell dendrites that have glutamate receptors, it will interact with them just as it does with bipolar cell dendrites within the invaginations. This arrangement no longer bears much resemblance to focal conventional synapses or neuromuscular junctions, but it allows a single cone to communicate simultaneously with a significant number of postsynaptic cells.

Glutamate release from a photoreceptor is subject to modification by activity in other photoreceptors

The consequence of all the topics discussed in this chapter is that the rate of glutamate release from a photoreceptor terminal is inversely proportional to the amount of light the photoreceptor catches. Release declines as more photons are absorbed and increases as fewer photons are absorbed; it's that simple. But the implication that the signal conveyed by a photoreceptor to postsynaptic neurons is totally independent of events outside the photoreceptor is not necessarily correct. If other neurons can affect the membrane polarity of a photoreceptor terminal, they will affect glutamate release and thereby modify the photoreceptor's signal to mean something other than just the number of photons absorbed by the photoreceptor. And this happens. Interactions with photoreceptor terminals by other neurons constitute an initial stage of editing signals from photoreceptors. Just as no man is an island, no photoreceptor acts entirely on its own.

References and Additional Reading

The Retina's Role in Vision

Barlow HB. 1972. Single units and sensation: A neuron doctrine for perceptual psychology? *Perception* 1: 371–394.

Barlow HB. 1986. Why can't the eye see better? Chapter 1, pp. 3–17, in *Visual Neuroscience*, Pettigrew JD, Sanderson KJ, and Levick WR, eds. Cambridge University Press, Cambridge.

Kleffner DA and Ramachandran VS. 1992. On the perception of shape from shading. *Percept. Pyschophysiol.* 52: 18–36.

Marr D. 1982. *Vision*. WH Freeman, San Francisco.

Marr D and Hildreth E. 1980. Theory of edge detection. *Proc. R. Soc. Lond. B* 207: 187–217.

Wilson HR and Bergen JR. 1979. A four mechanism model for spatial vision. *Vision Res.* 19: 19–32.

Functional Organization of the Retina

Boycott BB and Dowling JE. 1969. Organization of the primate retina: Light microscopy. *Philos. Trans. R. Soc. Lond. B* 255: 109–176.

Dacheux RF and Raviola E. 1994. Functional anatomy of the neural retina. Chapter 18, pp. 285–309, in *Principles and Practice of Opthalmology, Basic Sciences*, Albert DM and Jakobiec FA, eds. WB Saunders, Philadelphia.

Dowling JE. 1987. *The Retina: An Approachable Part of the Brain.* Harvard University Press, Cambridge, MA.

Dowling JE and Boycott BB. 1966. Organization of the primate retina: Electron microscopy. *Proc. R. Soc. Lond. B* 166: 80–111.

Hogan MJ, Alvarado JA, and Weddell JE. 1971. The retina. Chapter 9, pp. 393–522, in *Histology of the Human Eye*. WB Saunders, Philadelphia.

Kolb H. 1994. The architecture of functional neural circuits in the vertebrate retina. *Invest. Ophthalmol. Vis. Sci.* 35: 2385–2404.

Marc RE. 1995. Interplexiform cell connectivity in the outer retina. Chapter 15, pp. 369–393, in *Neurobiology and Clinical Aspects of the Outer Retina*, Djamgoz MBA, Archer SN, and Vallerga S, eds. Chapman and Hall, London.

Polyak SL. 1941. *The Retina*. University of Chicago Press, Chicago.

Rodieck RW. 1998. *The First Steps in Seeing*. Sinauer Associates, Sunderland, MA.

Ramón y Cajal S. 1893. La rétine des vertébrés. *La Cellule* 9: 17–257. English translations in Rodieck RW. 1973. *The Vertebrate Retina*, pp. 777–904. WH Freeman, San Francisco, and Thorpe SA and Glickstein M. 1972. *The Structure of the Retina.* Charles C Thomas, Springfield,IL.

Wässle H and Boycott BB. 1991. Functional architecture of the mammalian retina. *Physiol. Rev.* 71: 447–480.

Catching Photons: Photoreceptors and Their Environment

Aguilar M and Stiles WS. 1954. Saturation of the rod mechanism of the retina at high levels of stimulation. *Optica Acta* 1: 59–65.

Barlow HB. 1972. Dark and light adaptation: Psychophysics. Chapter 1, pp. 1–28, in *Handbook of Sensory Physiology*, Vol. VII/4, *Visual Psychophysics*, Jameson D and Hurvich LM, eds. Springer-Verlag, Berlin.

Baylor DA. 1987. Photoreceptor signals and vision. *Invest. Ophthalmol. Vis. Sci.* 28: 34–49.

Baylor DA. 1996. How photons start vision. *Proc. Natl. Acad. Sci. USA* 93: 560–565.

Baylor DA, Nunn BJ, and Schnapf JL. 1984. The photocurrent, noise and spectral sensitivity of rods of the monkey *Macaca fascicularis*. *J. Physiol.* 357: 575–607.

Besharse JC and Defoe DM. 1998. Role of the retinal pigment epithelium in photoreceptor membrane turnover. Chapter 8, pp. 152–172, in *The Retinal Pigment Epithelium*, Marmor MF and Wolfensberger TJ, eds. Oxford University Press, New York.

Bok D. 1985. Retinal photoreceptor-pigment epithelium interactions. *Invest. Ophthalmol. Vis. Sci.* 26: 1659–1694.

Bok D. 1990. Processing and transport of retinoids by the retinal pigment epithelium. *Eye* 4: 326–332.

Chader GJ. 1989. Interphotoreceptor retinoid-binding protein (IRBP): A model protein for molecular biological and clinically relevant studies. *Invest. Ophthalmol. Vis. Sci.* 30: 7–22.

Chader GJ, Pepperberg DR, Crouch R, and Wiggert B. 1998. Retinoids and the retinal pigment epithelium. Chapter 7, pp. 135–151, in *The Retinal Pigment Epithelium*, Marmor MF and Wolfensberger TJ, eds. Oxford University Press, New York.

Chun M-H, Grünert U., Martin PR, and Wässle H. 1996. The synaptic complex of cones in the fovea and in the periphery of the macaque monkey retina. *Vision Res.* 36: 3383–3395.

Dartnall HJA, Bowmaker JK, and Mollon JD. 1983. Human visual pigments: Microspectrophotometric results from the eyes of seven persons. *Proc. R. Soc. Lond. B* 220: 115–130.

Hecht S, Shlaer S, and Pirenne M. 1942. Energy, quanta and vision. *J. Gen. Physiol.* 25: 819–840.

Jacobs GH. 1998. Photopigments and seeing—Lessons from natural experiments. *Invest. Ophthalmol. Vis. Sci.* 39: 2205–2216.

Koutalos Y and Yau K-W. 1996. Regulation of sensitivity in vertebrate rod photoreceptors by calcium. *Trends Neurosci.* 19: 73–81.

Lolley RN and Lee RH. 1990. Cyclic GMP and photoreceptor function. *FASEB J.* 4: 3001–3008.

Molday RS. 1998. Photoreceptor membrane proteins, phototransduction, and retinal degenerative diseases. *Invest. Ophthalmol. Vis. Sci.* 39: 2493–2513.

Nakatani K and Yau K-W. 1988. Calcium and light adaptation in retinal rods and cones. *Nature* 334:69–71.

Nathans J. 1987. Molecular biology of visual pigments. *Annu. Rev. Neurosci.* 10: 163–194.

Nathans J, Thomas D, and Hogness DS. 1986. Molecular genetics of human color vision: The genes encoding blue, green, and red pigments. *Science* 232: 193–202.

Panda-Jonas S, Jonas JB, and Jakobcyzk-Zmija M. 1996. Retinal pigment epithelial cell count, distribution, and correlations in normal human eyes. *Am. J. Ophthalmol.* 121: 181–189.

Pugh EN and Lamb TD. 1990. Cyclic GMP and calcium: The internal messengers of excitation and adaptation in vertebrate photoreceptors. *Vision Res.* 30: 1923–1948.

Rao-Mirotznik R, Harkins AB, Buchsbaum G, and Sterling P. 1995. Mammalian rod terminal: Architecture of a binary synapse. *Neuron* 14: 561–569.

Schnapf JL, Kraft TW, and Baylor DA. 1987. Spectral sensitivity of human cone photoreceptors. *Nature* 325: 439–441.

Schnapf JL, Nunn BJ, Meister M, and Baylor DA. 1990. Visual transduction in cones of the monkey *Macaca fascicularis*. *J. Physiol.* 427: 681–713.

Steinberg RH, Fisher SK, and Anderson DH. 1980. Disc morphogenesis in vertebrate photoreceptors. *J. Comp. Neurol.* 190: 501–518.

Steinberg RH, Wood I, and Hogan MJ. 1977. Pigment epithelial ensheathment and phagocytosis of extrafoveal cones in human retina. *Philos. Trans. R. Soc. Lond. B* 277: 459–471.

Stiles WS. 1949. Increment thresholds and the mechanisms of colour vision. *Doc. Ophthalmol.* 3: 18–163.

Stockman S, Macleod DI, and Johnson NE. 1993. Spectral sensitivites of the human cones. *J. Opt. Soc. Am. (A)* 10: 2491–2521.

Stryer L. 1986. Cyclic GMP cascade of vision. *Annu. Rev. Neurosci.* 9: 87–119.

Stryer L. 1987. The molecules of visual excitation. *Sci. Am.* 257 (1): 42–50.

Tsukamoto Y, Masarachia P, Schein SJ, and Sterling P. 1992. Gap junctions between the pedicles of macaque foveal cones. *Vision Res.* 32: 1809–1815.

Wyszecki G and Stiles WS. 1967. *Color Science*. Wiley, New York.

Yau K-W. 1994. Phototransduction mechanism in retinal rods and cones. *Invest. Ophthalmol. Vis. Sci.* 35: 9–32.

Young RW. 1971. Shedding of discs from rod outer segments in the rhesus monkey. *J. Ultrastruc. Res.* 34: 190–203.

Young RW. 1978. Visual cells, daily rhythms, and vision research. *Vision Res.* 18: 573–578.

Zinn KM and Benjamin-Henkind JV. 1979. Anatomy of the human pigment epithelium. Chapter 1, pp. 3–31, in *The Retinal Pigment Epithelium*, Zinn KM and Marmor MF, eds. Harvard University Press, Cambridge, MA.

Chapter 14

Retina II: Editing Photoreceptor Signals

Peeping through my keyhole, I see within the range of only about thirty percent of the light that comes from the sun; the rest is infrared and some little ultraviolet, perfectly apparent to many animals, but invisible to me. A nightmare network of ganglia, charged and firing without my knowledge, cuts and splices what I do see, editing it for my brain.

■ Annie Dillard, *Pilgrim at Tinker Creek*

The Editing Process

Retinal information processing can be described by a small set of operations that are applied throughout the retina. Events in photoreceptors from photon absorption to glutamate release lead inward from the pigment epithelium to the retina's outer synaptic layer. At this juncture, the photoreceptor signals are subject to modification and editing by the rest of the retina's neurons—and there are a lot of different neurons in the rest of the retina. The discussion so far has dealt with only four types of photoreceptors and the pigment epithelial cells; more than 50 cell types remain.

Fifty cell types create a dense patch of neuronal shrubbery; before plunging into this thicket, some idea of the destination may make the details more significant and meaningful. The following statements summarize much of the discussion to come; they begin with photoreceptors and trace the flow of signals through the retina. In isolation, these elements of neural operations and organization are fairly simple—simple enough that most of them can be illustrated with a graphic icon; their combined application is what gives the retina its complexity and sophistication.

- A photoreceptor's signal is proportional to the number of photons caught by the photoreceptor.

Response

Log photons

- Lateral interactions mediated by horizontal cells emphasize differences in signals between neighboring photoreceptors.

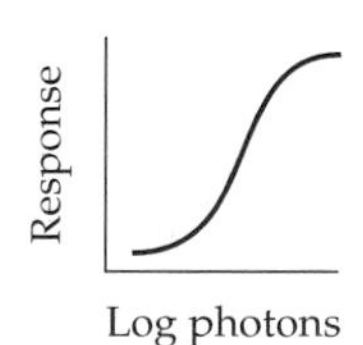

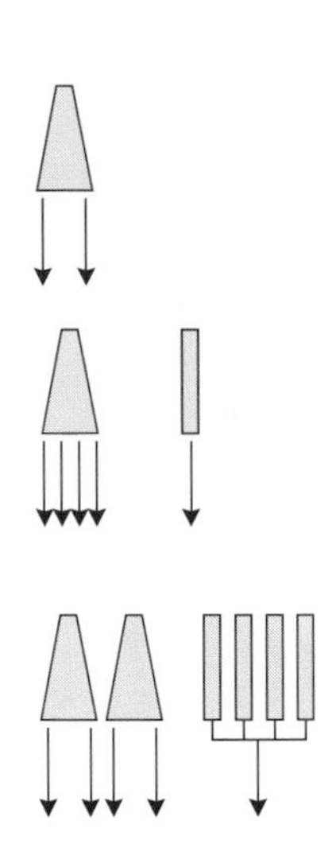

- Photoreceptor signals are divided into multiple, parallel streams of information conveyed by the bipolar cells.
- Cone signals are conveyed through more parallel pathways than rod signals are.
- Cone pathways are concerned with local events occurring in just one or a few cones; signals from many rods are pooled in the rod pathways.

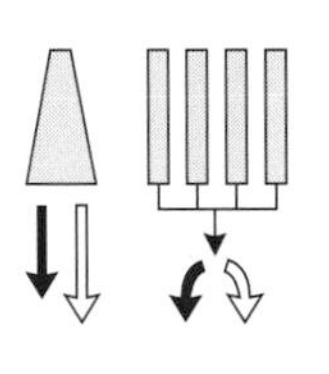

- Signals about increasing and decreasing light intensity are conveyed by separate ON and OFF pathways; these pathways are separated at the outer synaptic layer for cones and at the inner plexiform layer for rods.

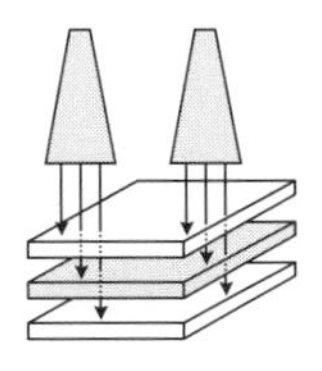

- The parallel streams of information conveyed by the bipolar cells are reorganized at the inner plexiform layer into sheets or sublayers stacked on top of one another, like overlaid maps in an atlas; the information in each sheet is continuous across the retina, without gaps or overlap.

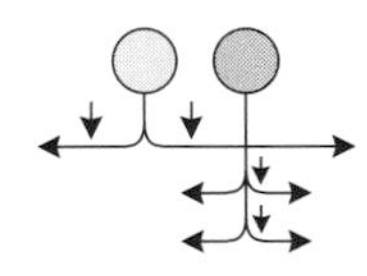

- Information in the different inner plexiform sublayers is exchanged and compared within sublayers (laterally) or between sublayers (vertically and laterally) by amacrine cells.

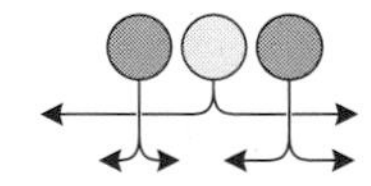

- Interactions mediated by amacrine cells can be made over short (local) to long distances; local interactions make specific modifications of information flow without loss of spatial resolution, midrange interactions probably modify the temporal features of signal flow, and long-range interactions may adjust the sensitivity with which information is transmitted.

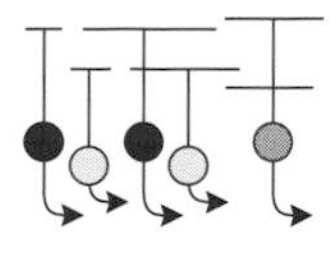

- Ganglion cells sample different sheets of information that have undergone different modifications by the amacrine cell connections; the size of the sampled regions varies among ganglion cells. Because of these sampling differences, ganglion cells differ in the information they convey to the brain.

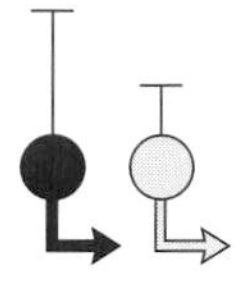

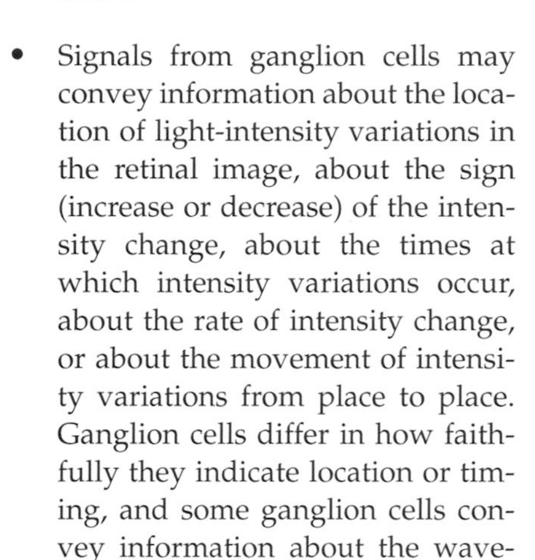

- Signals from ganglion cells may convey information about the location of light-intensity variations in the retinal image, about the sign (increase or decrease) of the intensity change, about the times at which intensity variations occur, about the rate of intensity change, or about the movement of intensity variations from place to place. Ganglion cells differ in how faithfully they indicate location or timing, and some ganglion cells convey information about the wavelength composition of the light.

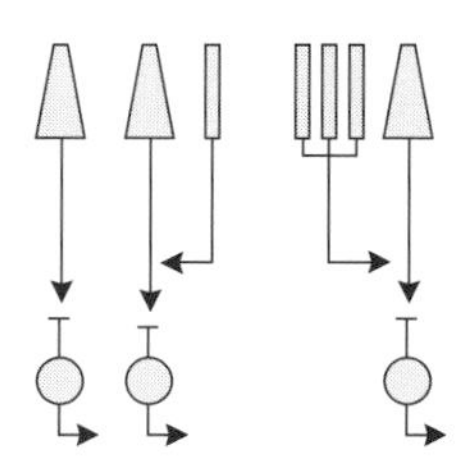

- Differences in the streams of information flow from ganglion cells in different parts of the retina are due to differences in the relative proportions of the different types of photoreceptors and the grain at which they are sampled. Different regions of the retina differ primarily in degree, not in kind.

Photoreceptor signals are about amounts of light; ganglion cell signals to the brain are about how the light is distributed through space and time. Photoreceptor signals are made more specific by dividing them into different pathways that can be edited separately and recombined in different ways. Editing involves comparisons between signals in the vertical pathways.

Interactions among Photoreceptors, Horizontal Cells, and Bipolar Cells

Horizontal cells integrate photoreceptor signals

Two horizontal cells are shown in Figure 14.1, as if viewed through the photoreceptor layer, to see the cells spread out below. Each of them has large processes radiating out from the cell body and numerous branchlets ending in fine, knoblike terminals; we saw these knobs earlier (see Figure 13.35) as they inserted as lateral processes into invaginations in cone pedicles. Both cells have a long axon-like process extending several hundred microns or more from the cell body.

The two horizontal cells are not identical, however. One of them—the **HI** horizontal cell—has a long axon (not shown here in its entirety) that ends in a clus-

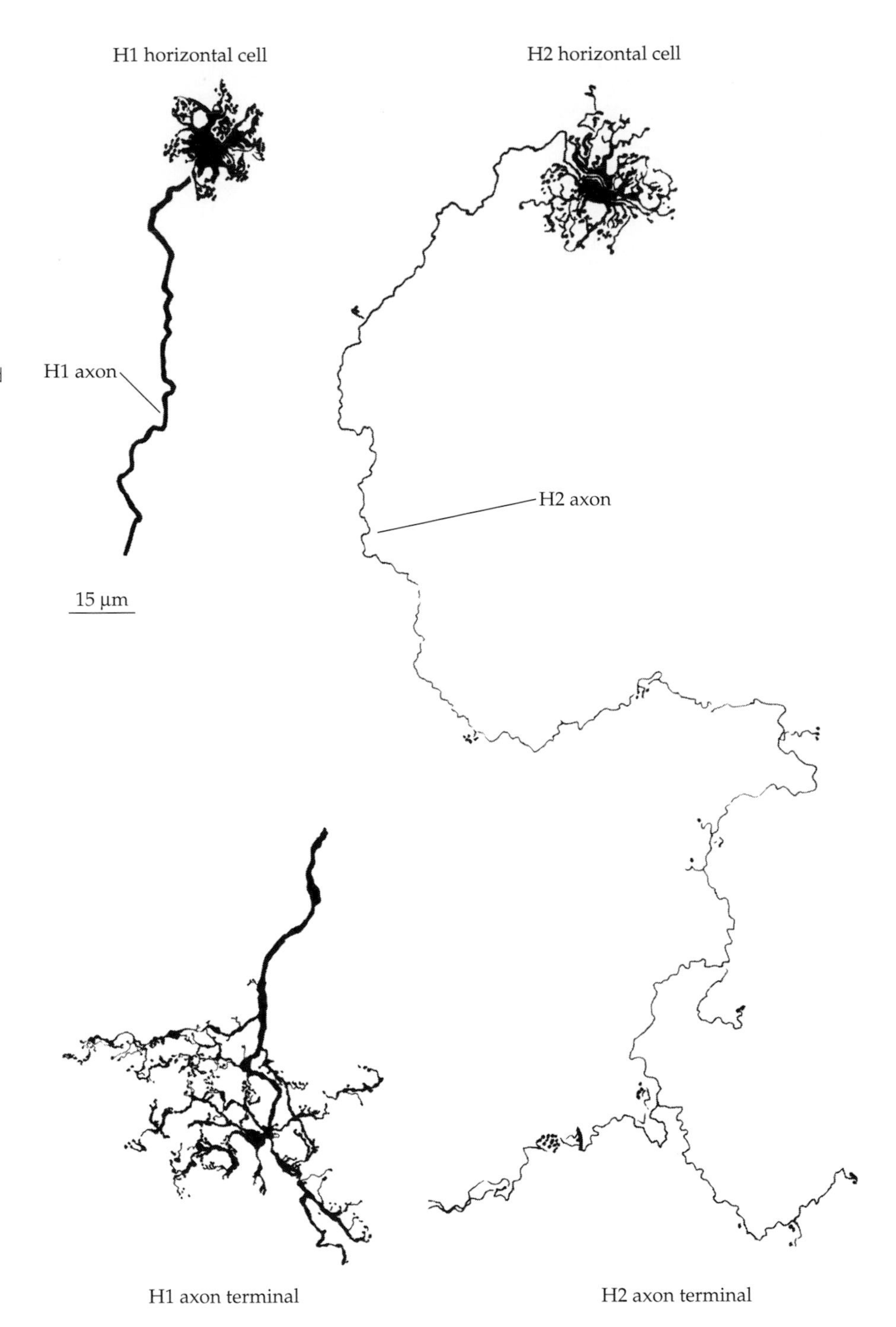

Figure 14.1
Two Types of Horizontal Cells
These horizontal cells were stained by the Golgi method in macaque monkey retina; they are viewed from above as if seen through the overlying photoreceptors. Both cells have several stout processes arising from their cell bodies (top) that end in small clusters of fine branches at which they contact photoreceptors. The two types, H1 and H2, are distinguished by the density and extent of branching around the cell body and the nature of the axonal process. The H1 cell body and axon terminal shown here belong to different cells. (From Kolb, Mariani, and Gallego 1980.)

ter of branching terminals similar to those around the cell body. But the terminals around the H1 cell body contact cones, and the terminals of its axon contact rods, where they are the lateral processes in the rod triads. Although interactions between distant clusters of rods and cones might be possible with this arrangement, the two ends of the H1 cell do not communicate electrically. Horizontal cells do not generate action potentials, and electrical changes at one end of the cell decay rapidly along the long, thin axon—in other words, the axon isn't really an axon. It serves mainly as a conduit for materials synthesized in the cell body to be conveyed to the terminal. Functionally, an H1 horizontal cell is two cells in one: The processes around the cell body collect signals from cones, while the terminal arborization collects signals from rods.

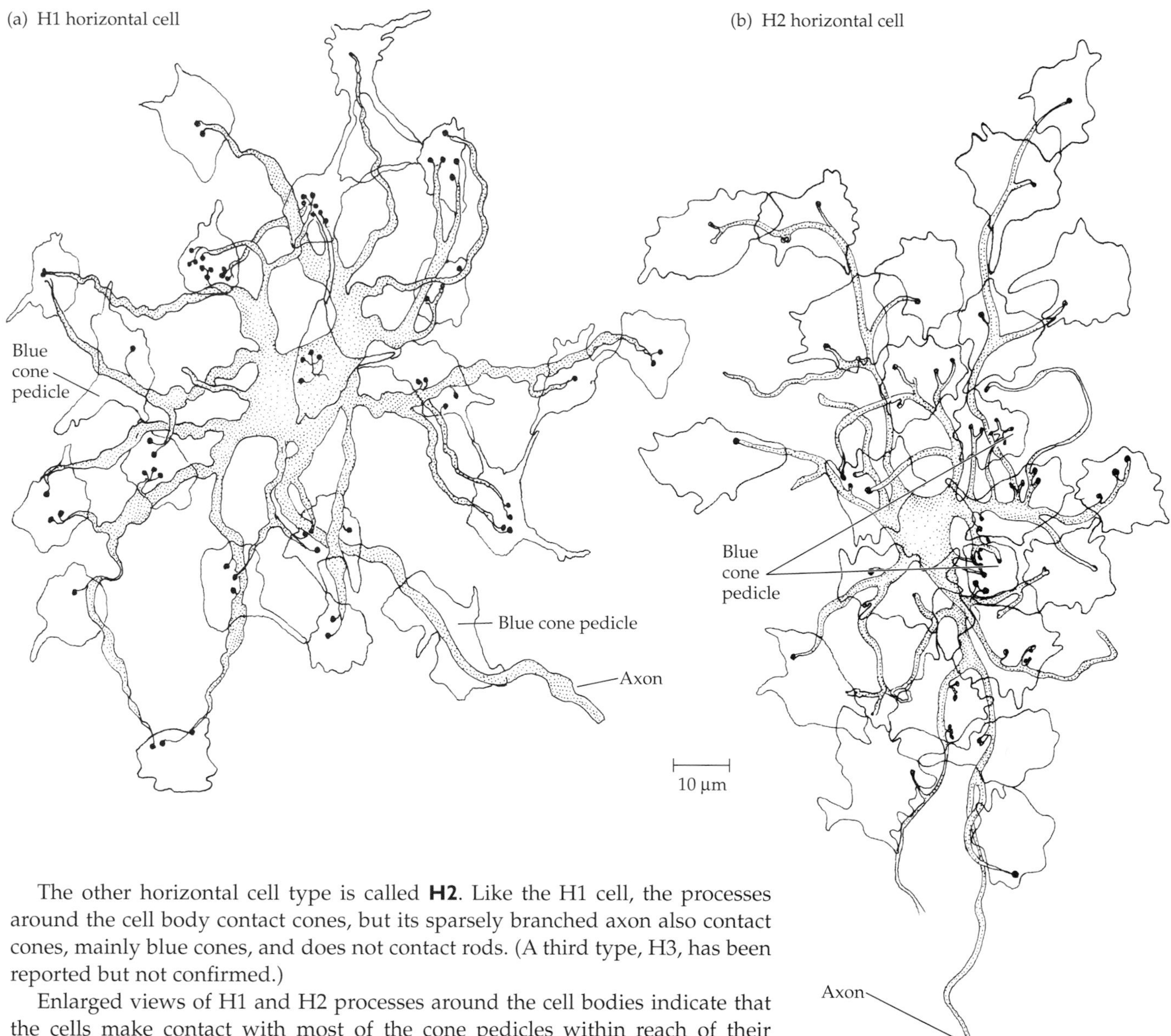

The other horizontal cell type is called **H2**. Like the H1 cell, the processes around the cell body contact cones, but its sparsely branched axon also contact cones, mainly blue cones, and does not contact rods. (A third type, H3, has been reported but not confirmed.)

Enlarged views of H1 and H2 processes around the cell bodies indicate that the cells make contact with most of the cone pedicles within reach of their processes (Figure 14.2). The cone inputs to the H1 cell are primarily, perhaps exclusively, from red and green cones—it has only a single contact with one of the two blue cone pedicles within reach of its dendrites. The dominant input to the H2 cell is from blue cones, whose pedicles are contacted by five or more terminal endings. There are some red and green cone inputs as well, but the pedicles are usually contacted by just one or two terminals.

In general, cone inputs to the horizontal cells are not specific by cone type. H1 horizontal cells receive inputs primarily from red and green cones. H2 horizontal cells receive inputs from all three cone types, but they *prefer* blue cones and must in some sense seek them out. The inputs from the different cone types have similar effects on the horizontal cells; when a cone hyperpolarizes in response to light, it elicits hyperpolarization in the postsynaptic horizontal cell regardless of the cone type.

As a result, horizontal cells are not particularly sensitive to stimulus wavelength (Figure 14.3). When the stimulus wavelength is controlled to selectively activate one or another of the cone types, H1 horizontal cells respond to inputs from both red and green cones, but not to inputs from blue cones. H2 horizontal cells respond to red, green, or blue cone inputs, but they respond *best* to blue. In

Figure 14.2
Horizontal Cells and Their Relationships to Cone Pedicles
The cell bodies and surrounding processes of H1 and H2 horizontal cells are shown in relationship to the overlying cone pedicles. The H1 cell has only one contact with one of two blue cones in its field (*a*); the H2 cell has many contacts with two blue cone pedicles (*b*). All of the cone pedicles shown in outline had numerous synaptic ribbons, but these cells usually contributed to just a few of them, rarely to more than five. The other triads will be contacted by processes from other horizontal cells of both types. (After Ahnelt and Kolb 1994.)

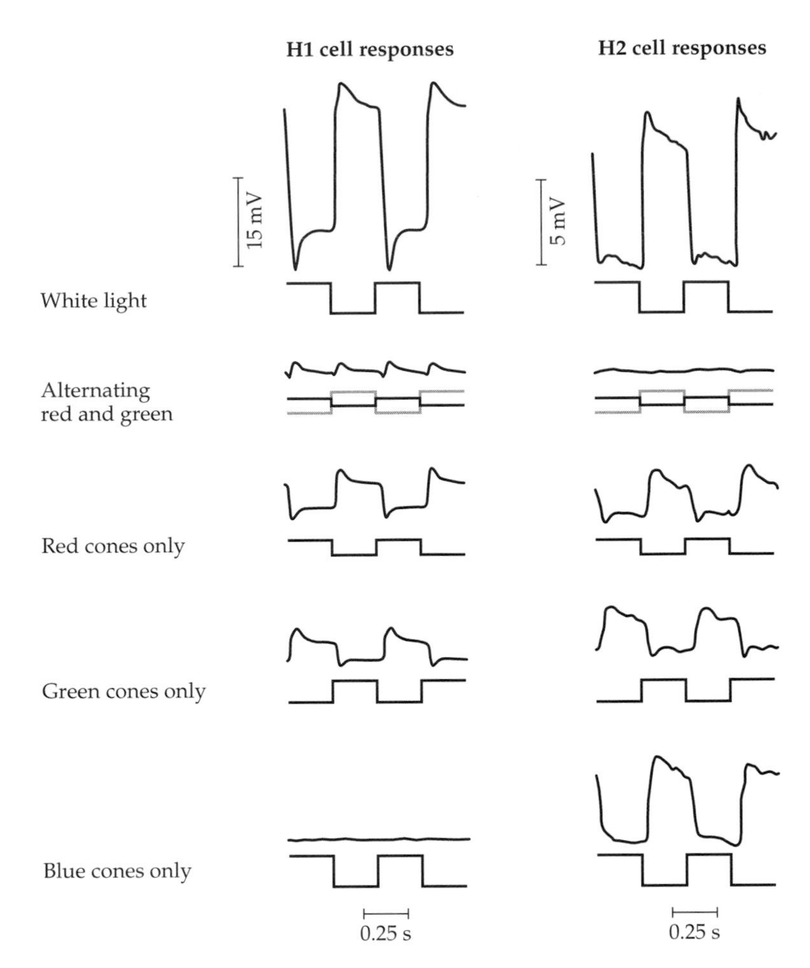

Figure 14.3
Responses of Horizontal Cells to Stimuli of Different Wavelength
The columns for H1 and H2 horizontal cells show voltage responses (hyperpolarizing downward) to different stimuli. H1 cells show little evidence for input from blue cones, but blue cones dominate in the H2 cell responses. (After Dacey et al. 1996.)

all cases, however, the horizontal cell response is hyperpolarization. A stimulus alternating between red and green produces responses at both the red and green phases; that is, the horizontal cells cannot distinguish between the two colors.

This lack of cone specificity has implications for color vision. As noted in Chapter 13, wavelength discrimination requires a comparison of signals from at least two cone types. But the comparison is most effective when the signals have opposite signs; green cone input might produce depolarization, for example, and red cone input would produce hyperpolarization. The horizontal cell signal would then convey the information "green and not red."* Since cone signals are being added to one another as they impinge on the horizontal cells, however, magnitude and polarity of the horizontal cells' responses cannot provide information about which of the cone types elicited the response. A horizontal cell will respond over a wider range of the wavelength spectrum than either a red or green cone, but like the individual cones, its response is related to the amount of light, not to its wavelength.

*This differencing operation in color vision is called "color opponency" and it is well established in the psychophysics of color vision as a basic mechanism for wavelength discrimination. The key comparisons are between red and green and between blue and yellow (red plus green). In terms of retinal neurons, color opponency requires red and green signals or blue and yellow signals of opposite polarities.

Horizontal cells receive inputs from photoreceptors and send signals of opposite sign back to the photoreceptor terminals, using GABA as the neurotransmitter

Triads in rod spherules and cone pedicles show specializations for transmission from photoreceptors to horizontal cells and bipolar cells, but there are no other signs of synapses. In particular, there is no anatomical sign that horizontal cells communicate with either photoreceptors or bipolar cells. Yet horizontal cells do not just receive inputs from photoreceptors; they also make inputs *back* to the photoreceptors. Moreover, the horizontal-to-photoreceptor feedback is opposite in sign to the photoreceptor signals impinging on the horizontal cells.

The influences of horizontal cells on photoreceptors are very difficult to study in mammalian retinas; the small size of photoreceptors makes recordings with micropipette electrodes extremely difficult, and the photocurrent recordings from outer segments do not show what is happening at the photoreceptor terminals. Therefore, most evidence comes from larger photoreceptors in nonmammalian species, and we must assume that something similar goes on in primates. But with this qualifier, Figure 14.4 shows that a cone's response decreases in size as the light stimulus is made larger to include surrounding photoreceptors (the amount of light falling on the test cone is always the same). The difference in response to small and large stimuli persists over a 1000-fold change in stimulus intensity.

This effect might have nothing to do with horizontal cells, because the gap junctions between the telodendria on cone terminals (see Figure 13.35) could allow cones to interact with one another, but such interaction can be ruled out. Injecting current into a horizontal cell to mimic the effect of light stimuli mediated through

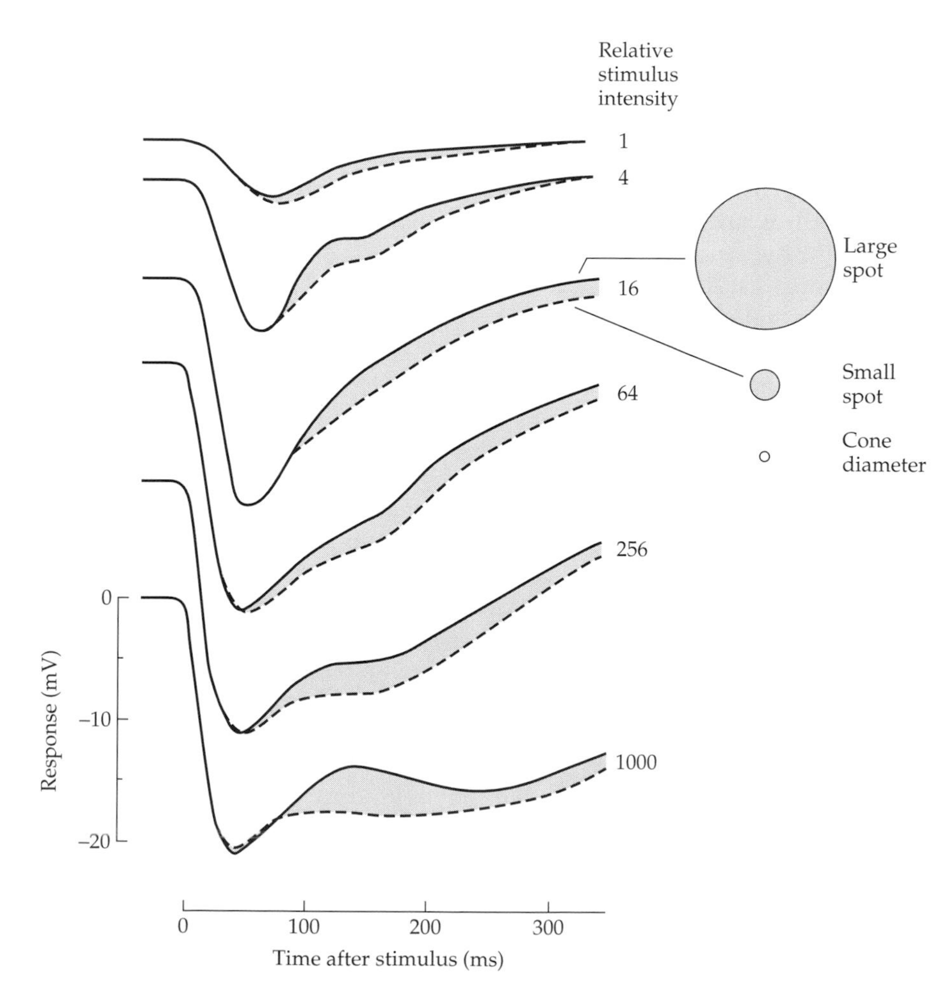

Figure 14.4
Effects of Horizontal Cell Feedback on Cone Responses
These voltage (not current) recordings are from a turtle cone; the hyperpolarizing direction is downward. Each pair of records shows the difference (shaded) between responses to small and large stimuli centered on the cone. Response pairs at different stimulus intensities have been displaced downward by arbitrary amounts. (After Baylor, Fuortes, and O'Bryan 1971.)

the cones has the same effect as a large light stimulus: The photoreceptor's response is smaller when the horizontal cell is activated directly. Since neighboring cones will not be active under these circumstances (there is no light falling on them), any interactions they might have with the test cone cannot be producing the observed change in the test cone's response. Horizontal cells must be responsible.

We do not know for certain that primate horizontal cells promote similar interactions among cones, nor do we know exactly how they might do it. But primate horizontal cells in and around the fovea contain **GABA** (gamma-aminobutyric acid) and its synthetic enzyme, glutamic acid decarboxylase; the same neurochemical signature—the presence of GABA and glutamic acid decarboxylase—is seen in horizontal cells of carnivores and most nonmammalian vertebrates. Thus, most if not all horizontal cells contain GABA, but there is no sign of synaptic vesicles in horizontal cells or of any distinct synapses between horizontal cells and photoreceptors. (There is evidence of synapses between horizontal cells and rods, however.) Is neurotransmission possible without synapses?

The flat bipolar contacts on cone pedicles discussed in Chapter 13 are an example of unconventional neurotransmission, and this is another. Vertebrate horizontal cells appear to use special GABA transporter molecules for neurotransmission. These GABA transporters operate in both directions, bringing GABA into the cell (ligand import) and taking it out (ligand export); the export-to-import ratio changes with membrane voltage, increasing as the cell depolarizes. Thus when horizontal cells depolarize, relative GABA export increases. In this way, horizontal cells could use GABA for neurotransmission and provide negative feedback onto photoreceptors, all without having conventional synaptic vesicles.

If so, the interactions mediated by horizontal cells may work as follows: When a cone hyperpolarizes in response to light, the glutamate release from its pedicle declines, and the postsynaptic horizontal cell hyperpolarizes. Hyperpolarization of the horizontal cell reduces its export of GABA at the invaginations in the cone pedicles. Since GABA is an inhibitory (sign-inverting) neurotransmitter, reducing its concentration at the cone pedicles will allow the pedicles' membrane potential to move in the *depolarizing* direction, which is a *reduction* of the normal light-evoked response. Alternatively, if the photoreceptors have depolarized (the dark condition), horizontal cell feedback will push them in the hyperpolarizing direction.

Assuming that this sort of feedback from horizontal cell to photoreceptor is operating in the primate retina, we need to consider what it is doing. What purpose is served by having photoreceptors inhibit one another through horizontal cells?

Horizontal cell connections emphasize differences in illumination between different photoreceptors

Several possible effects of horizontal cell feedback are illustrated schematically in Figure 14.5, which shows a horizontal cell receiving inputs from a small number of cones and having connections back to them. If the cones were equally illuminated, as they are in Figure 14.5*a*, and if the horizontal cell had no effect, all the cones would have the same level of response—say, 4 units of hyperpolarization. But because of the opposing horizontal cell feedback, the full 4 units of activity cannot be sustained, and the cone signals may be hyperpolarized only 2 units at the terminals (the effect is local and does not involve the outer segment).

At first glance, this effect may seem counterproductive. The cones have gone to considerable trouble to generate a response; why reduce it in this way? It seems that the cones will be *less* sensitive than they would be in the absence of horizontal cell feedback, because the full magnitude of the cone response to light will no longer be reflected in the change of glutamate release. But this is precisely the sort of sensitivity reduction we need to account for the network adapta-

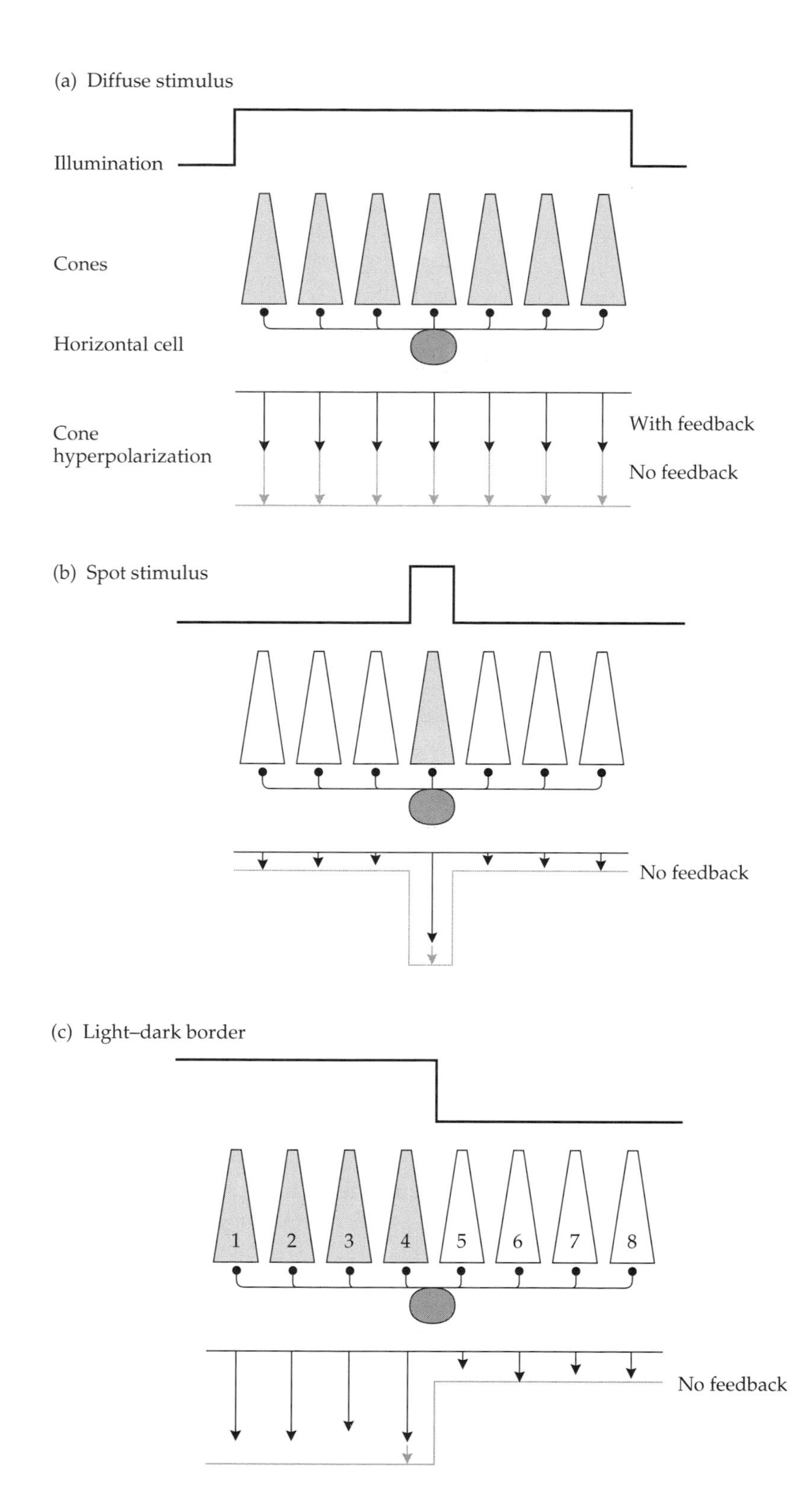

Figure 14.5
Possible Effects of Lateral Interactions Mediated by Horizontal Cells: Adaptation and Contrast Enhancement
(*a*) When neighboring cones are equally illuminated, they all receive equal amounts of negative feedback from the horizontal cell and their levels of hyperpolarization are less than they would be in the absence of feedback. (*b*) Illuminating just one of the cones means that the feedback will be quite small, affecting mainly the unilluminated cones, and the illuminated cone will respond almost as if no feedback were present. (*c*) When the stimulus is a light–dark border, horizontal cell feedback exaggerates the transition from light to dark in the cone responses, an effect known as contrast enhancement. See text for discussion.

tion discussed in Chapter 13 (see Figure 13.26). We saw that the changes in photoreceptor sensitivity mediated by Ca^{2+} in the outer segment do not account for all the sensitivity adjustment occurring in the retina. Now, however, we can consider horizontal cell feedback as a possible factor in network adaptation.

Network adaptation operates within a region commonly called the adaptation "pool," within which light falling on different photoreceptors can interact. The pools have finite size; adaptation in one region of the retina does not affect sensitivity in more distant regions. Estimates of the size of the *rod* adaptation pool in human parafoveal retina range from 0.1 to 0.5° of visual angle, or from 30 to 150 μm in diameter. The size of the H1 terminal arborizations fall within this range, which may mean that they set the size of the rod adaptation pools. If the horizontal cell processes contacting cones set the cone adaptation pool size, it would be smaller. The amount of horizontal cell feedback that contributes to network adaptation is not known, however, and there is likely to be more sensitivity adjustment in the inner plexiform layer.

Now consider Figure 14.5*b,* where only the central cone of the group is illuminated at the same intensity that would produce 4 units of hyperpolarization in the absence of horizontal cell effects. The surrounding unilluminated cones are inactive, and they have little effect on the horizontal cell. Thus the only opposing feedback to the illuminated cone will be feedback from its own signal; this will be modest, and the cone's activity level will remain near 4 units of hyperpolarization.

Since the central cone signals most strongly when it alone is illuminated, the effect of the horizontal cell feedback is to make it sensitive to contrast—that is, to differences in illumination. The amount of light falling on the central photoreceptor is the same as it was in the previous example, but the photoreceptor's signal is larger when there is a *difference* between the amount of light falling on it and the amount falling on its neighbors. This is a first step from signaling the amount of light to signaling *variation* in the amount of light in the retinal image.

The stimulus in Figure 14.5*c* is an edge, an abrupt transition from light to dark, in which cones to the left of the transition are strongly and equally illuminated while those to the right are equally in the dark. Cone 4, lying next to the edge on the illuminated side, will be most affected through the horizontal cell feedback by its immediate neighbors (3 and 5), of which one is illuminated (3) and one is not (5). Their net effects on cone 4 will cancel, as will the smaller influences of the next nearest pair (2 and 6). Cone 4 will signal strongly, unaffected by horizontal cell feedback. Cone 5 is in the dark, but its pairs of neighbors (4 and 6, 3 and 7) will have effects that offset one another because of their differing conditions of illumination. Cone 5 will have a minimal signal.

Cones 3 and 6 are in different situations; their immediate neighbors have the same illumination, bright for cone 3 and its neighbors, dark for cone 6 and its neighbors. As a result, horizontal cell feedback should reduce the signal in cone 3, because of strong feedback, and elevate it in cone 6 because the feedback is weak. The effects of more distant cone pairs should cancel.

We could go on to look at the other cones, but the overall result shown in the cone signal mapping is already clear. The strongest and weakest cone signals are from the cones on either side of the light-to-dark transition, and the effect is to exaggerate the difference between light and dark. This effect is *contrast enhancement,** and it is a direct consequence of horizontal cell–mediated interactions among neighboring photoreceptors. (Although it wasn't mentioned, contrast enhancement occurs in Figure 14.5*b* as well. Do you see why?)

In these contrived examples, all cones contribute equally to the horizontal cell, and the feedback onto the cones is proportional to the average of all the cone inputs. A more realistic case would have the horizontal cell feedback weighted by distance between cones, such that interactions among nearby cones were strong while interactions with more distant cones became progessively weaker with increasing distance. Weighting of horizontal cell feedback to make interactions among cones inversely proportional to their separation could be related to electrical conduction properties of the horizontal cell membrane; electric signals

*A detailed analysis of contrast enhancement was done years ago on eccentric cells in the compound eye of the horseshoe crab, *Limulus* (see the Prologue), by H. K. Hartline, Floyd Ratliff, and others. This and related work is described in Ratliff's book *Mach Bands*. Vertebrate retinas are very different from the *Limulus* retina, but the operating principles appear to be similar.

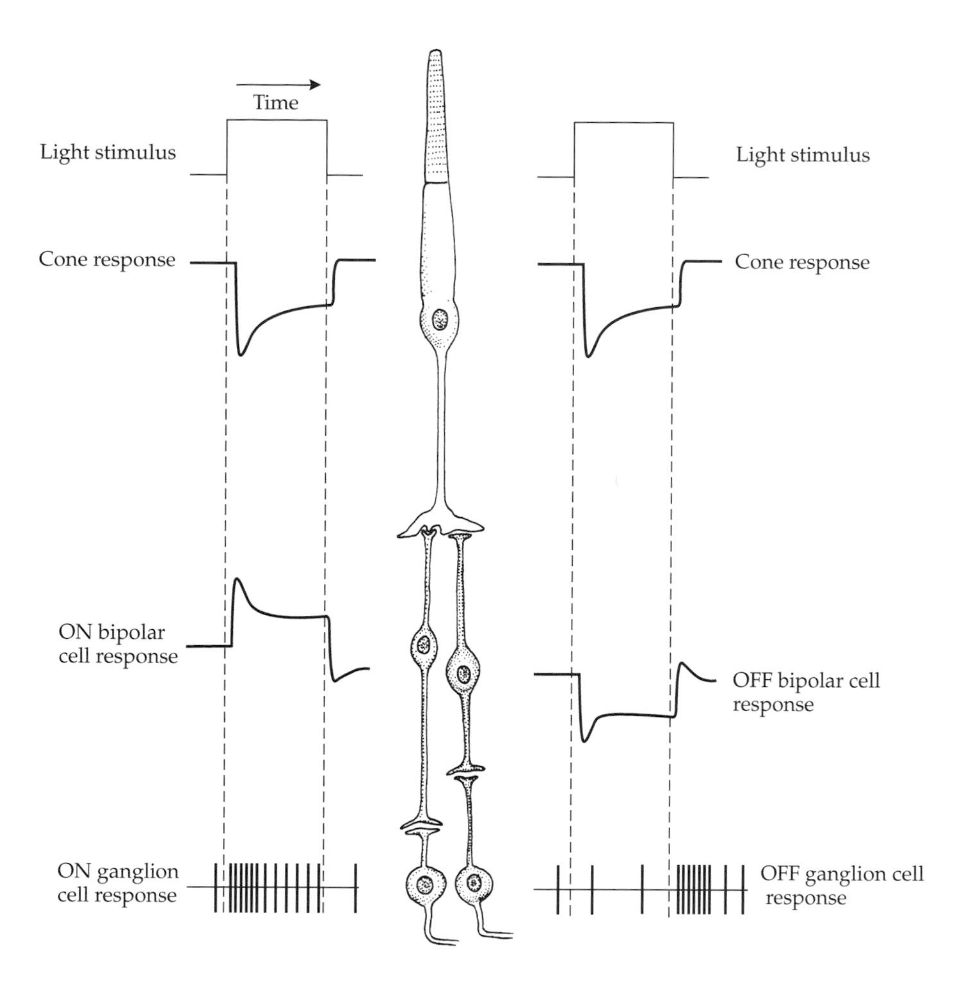

Figure 14.6
Cone Signals Are Divided into ON and OFF Pathways
Cones always send signals to more than one bipolar cell; two are shown here. One bipolar cell (shaded) makes flat contacts on the base of the cone pedicle, and its response to light is hyperpolarization, mimicking the cone response; it is called an OFF bipolar cell because it will depolarize when light intensity *decreases*. The other bipolar cell depolarizes to light; it is an ON bipolar cell. Ganglion cells contacted by these bipolar cells also give responses at light OFF or light ON. The light responses for the bipolar cells are schematic and based on results in nonprimate retinas.

decay with distance as they travel along horizontal cell processes. Cones widely separated in space will also be widely separated in distance along the horizontal cell processes, and an input to the horizontal cell by one cone will be much reduced in magnitude when it reaches the distant cone. (Horizontal cells are connected to one another through gap junctions, so the distances may exceed the spread of a single horizontal cell.)

If adaptation and contrast enhancement are due to horizontal cell feedback onto the cone terminals, the effects will be incorporated into the signals going from cones to bipolar cells. Thus, all cone bipolar cells should exhibit changes in sensitivity, as well as changes in response magnitude as the stimulus size varies. And if the terminal portions of H1 horizontal cells mediate similar interactions among rods, rod bipolar cells will also reflect the influence of horizontal cell feedback.

Finally, there are some caveats. Most of the foregoing analysis represents possible actions for horizontal cells in the primate retina that have yet to be confirmed or disproved. The story is plausible, by analogy with other species, but it is subject to modification as we learn more. And none of the analysis gives any reason for the existence of two or more types of horizontal cells.

Different glutamate receptors on cone bipolar cells cause increases and decreases in light intensity to be reported by ON and OFF bipolar cells, respectively

Absorption of light by a cone photoreceptor leads to a decrease in the rate of glutamate release from the cone terminal, and this change is monitored by two kinds of bipolar cells with opposing sensitivities to glutamate (Figure 14.6). One class reports increases in extracellular glutamate levels (the dark condition) by depolarizing and reports decreases in glutamate (light) by hyperpolarizing. Since the bipolar cell increases its own rate of neurotransmitter release when it depolarizes, it is called an **OFF bipolar cell**; when it depolarizes, it reports darkening. The synapses between cones and OFF bipolar cells are called **sign-conserving** because the two cells have the same response polarity: Both depolarize to darkness and hyperpolarize to light.

A second group of bipolar cells report increases in glutamate released from the cones (dark) by hyperpolarizing and decreases in glutamate (light) by depolarizing. They are called **ON bipolar cells** because they signal increasing illumination (ON) when they depolarize. The synapses between cones and ON bipolar cells are called **sign-inverting** because the two cells have opposite reponse polarities: When the cone hyperpolarizes, the ON bipolar cell depolarizes, and vice versa.

The situation depicted here is characteristic of cones, wherever they are in the retina; the presence of ON and OFF bipolar cells creates two distinct photoreceptor → bipolar cell → ganglion cell pathways. Therefore, the two sets of ganglion cells, like the bipolars that feed them, can be thought of as ON ganglion cells and OFF ganglion cells. As a result, cones have at least two private lines to the brain, one for increases and the other for decreases in light intensity; note, however, that the original idea of a single ganglion cell reflecting the activity of a single cone still applies.

It was thought for some time that the dendrites of the ON bipolar cells were the central processes within the invaginations of the cone triads, while those of the OFF bipolar cells made the flat or basal contacts on the cone pedicles. This belief led to the notion that the property of sign inversion was associated with the invagination and the triad synapse. It is now known, however, that ON bipolar cells may also make contacts along the base of the cone pedicle; although these contacts tend to be near the invaginations, they are clearly not within them, so sign inversion cannot be attributed to the synaptic triads.

Splitting the cone signal into ON and OFF channels through sign-inverting and sign-conserving synapses is done with a single neurotransmitter, glutamate, acting on different glutamate receptors. OFF bipolar cells have glutamate receptors linked directly to cation channels that are opened when glutamate binds to the receptor. These ionotropic, or ligand-gated, channels are responsible for the sign-conserving nature of the cone to OFF bipolar cell synapses. ON bipolar cells have metabotropic glutamate receptors whose effects on ion channels in the cell membrane are mediated through a G protein. In this case, the binding of glutamate to the receptor *closes* ion channels. As more glutamate is released when the cone depolarizes, channels in the bipolar cell membrane close and the membrane hyperpolarizes; the effect is sign inversion.

Signals from both red and green cones go to midget bipolar cells, which are specific for cone type, and to diffuse bipolar cells, which are not cone specific

The ON and OFF cone pathways each contain several types of bipolar cells. **Midget bipolar cells** receive inputs from either red or green cones. Thus in terms of their cone input and the polarity of their response, there are four types of midget bipolar cells: red-ON, red-OFF, green-ON, and green-OFF. The term "midget" refers to the small spread of their dendritic and axonal arbors (Figure 14.7*a*). In the central retina, a midget bipolar cell receives direct input from just one cone, so its dendritic arbor is no larger than the cone pedicle. The bipolar cell's terminal, which contacts just one ganglion cell, is correspondingly small. This one-to-one relationship between midget bipolar cells and cones on the dendritic end, and between midget bipolar cells and ganglion cells on the axonal

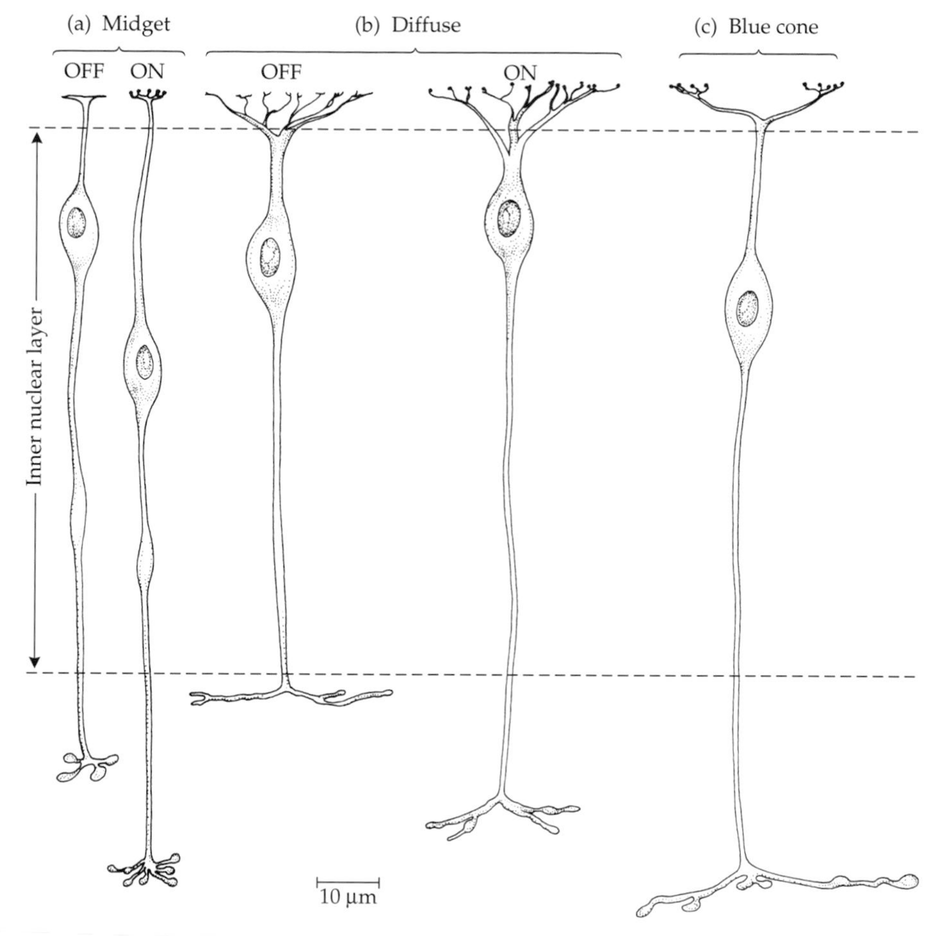

Figure 14.7
Cone Bipolar Cells
There are three major groups of cone bipolar cells: midget, diffuse, and blue cone bipolar cells. (*a*) Midget bipolar cells in the central retina (shown here) receive input from a single cone, either red or green, and contact a single ganglion cell at their terminal. ON and OFF midget bipolar cells terminate at different levels of the inner plexiform layer. (*b*) Diffuse bipolar cells always receive inputs from several cones, both red and green. There are six types of diffuse bipolar cells—three ON and three OFF—only two of which are shown here. (*c*) Blue cone bipolar cells receive inputs from only blue cones, and their broad, expanded terminal arbor lies next to the ganglion cell layer. Only this one type of blue cone bipolar cell has been definitely established.

end, is a distinctive property of midget bipolar cells in the central 15° or so of the retina around the fovea. Outside this central region, dendritic arbors of midget bipolar cells become larger because they are receiving inputs from several cones, but the cones are always of the same type, either red or green. The dendrites of these peripheral midget bipolar cells must spread more widely to contact the sparsely distributed cones.

Diffuse bipolar cells always receive inputs from *both* red and green cones, and they may have some blue cone input as well. In any case, diffuse bipolar cells should be less selective for stimulus wavelength than the midget bipolar cells and should respond well over a wider range of the spectrum. Because the diffuse bipolar cells always contact multiple cones, even in the fovea, their dendritic arbors are more extensive than those of the midget bipolar cells, as are their terminal in the inner plexiform layer; hence the term "diffuse" (Figure 14.7*b*).

Polyak observed some bipolar cells in his Golgi-stained retinas whose dendritic and axonal arbors were sparsely branching over larger areas than those of diffuse bipolar cells at the same retinal location; they have been seen in more recent studies and are now called "giant" bipolar cells. Since the giant bipolar cells have not been observed routinely, they have not been studied extensively, but they are thought to receive only input from cones and to constitute two cell types, one ON and one OFF in terms of their axonal arborization levels. The only firm conclusion, however, is that giant bipolar cells are not common; they may be simply a developmental variation that occurs here and there in the retina. If so, they need not be considered in thinking about how the retina works.

Diffuse bipolar cells are certainly a separate category, and on the basis of the inputs they receive, there should be just two types—ON and OFF—but there are actually six, three ON and three OFF. This sixfold distinction is made not on any known differences in response properties or connectivity in the outer synaptic layer, but on the levels at which the cells terminate in the inner plexiform layer. The implications of this apparent duplication of cone signals in the diffuse bipolar cell pathways will be considered later. For the moment, however, the key point is that the signal from each red or green cone is divided eight times; each cone provides inputs to two midget bipolar cells and six diffuse bipolar cells.

Blue cones have their own bipolar cells

Unlike red and green cones, blue cones do not divide their signals into midget and diffuse pathways, with the possible exception of a minor contribution to the diffuse bipolar cells. And it is not yet clear that they divide their signals into ON and OFF pathways. What we do know is that certain bipolar cells receive inputs from only blue cones and terminate in the inner, ON part of the inner plexiform layer (Figure 14.7*c*). (By way of confirmation, recordings from some ganglion cells show a strong blue-ON response component, as shown later in Figure 14.23). These **blue cone bipolar cells** always receive inputs from several cones at minimum, so they cannot be regarded as the blue version of the red and green midget bipolar cells, but since they receive only blue cone inputs, they are unlike the diffuse bipolar cells that have mixed cone inputs. They are a separate cell type.

In a way, blue cone bipolar cells are more like rod bipolar cells (described in the next section) than like the other cone bipolar cells. Each bipolar cell receives multiple inputs from a single photoreceptor type; they appear to be ON, or depolarizing, cells—there is no known OFF companion bipolar cell (although possible sightings have been reported); and the blue cone bipolar cells and rod bipolar cells terminate closer to the ganglion cells than any other bipolar cells do. These similarities may not be significant, but it will be interesting to see if blue cone signals are divided into ON and OFF pathways in the inner plexiform layer as is done for the rod signals. (A blue-OFF pathway is assumed to exist, and there is some physiological evidence for it.) Perhaps blue cone bipolar cells and rod bipolar cells are similar because of common ancestry.

The existence of a separate bipolar cell pathway for blue cones does suggest an answer to a question raised earlier about the existence of two types of horizontal cells. Since H1 horizontal cells deal with red and green cones almost exclusively, the separate blue cone pathway presumably requires a similar set of lateral interactions in which blue cones are included. That would seem to be the role of the H2 horizontal cells.

Rods have sign-inverting synapses to rod bipolar cells, which do not contact ganglion cells but send signals to the cone pathways through an amacrine cell

Rods have their own bipolar cells (Figure 14.8). Polyak called them "mop" bipolar cells, referring to the expanded, shaggy dendritic arbor. Individual dendrites extend to different levels in the outer synaptic layer to contact several rows of rod spherules lying outside the layer of cone pedicles; hence the moplike appearance. But where cones have a number of bipolar cells, there is only one class of rod bipolar cell and the rod-to-bipolar synapse is always sign-inverting; when light intensity increases, the rods hyperpolarize, their rate of glutamate release declines, and the rod bipolar cells depolarize. Sign inversion is mediated by a G protein–linked glutamate receptor (but not the one used by ON cone bipolar cells). In the terminology used for the cone bipolar cells, *all* rod bipolar cells are ON bipolar cells.

Another difference between the vertical pathways for rods and cones is that the rod bipolar cells do not make synaptic contacts with ganglion cells. Instead,

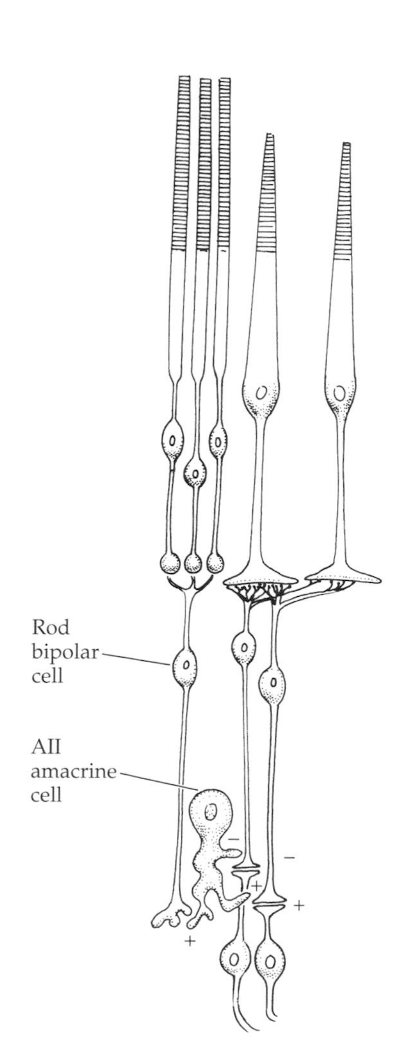

Figure 14.8
Rod Bipolar Cells and AII Amacrine Cells
Rod signals go to a single type of bipolar cell. These rod bipolar cells do not contact ganglion cells directly. Instead, rod bipolar cell signals go to AII amacrine cells that make sign-conserving (+) contacts through gap junctions with the terminals of ON cone bipolar cells and sign-inverting (–) glycine synapses onto the terminals of OFF cone bipolar cells. To leave the retina, rod signals must therefore ride piggyback on the cone pathways, as the cartoon emphasizes. (After Strettoi, Raviola, and Dacheux 1992; cartoon from the same source.)

the output from rod bipolar cells goes to small **AII amacrine cells*** whose output is onto the terminal end of *cone* bipolar cells. Thus there are extra elements in the rod pathway such that the flow of information is rod → rod bipolar cell → AII amacrine cell → cone bipolar cell → ganglion cell. If the terminals of midget or blue cone bipolar cells receive inputs from AII amacrine cells, there is no evidence for it; the contacts are assumed to be onto the terminals of diffuse bipolar cells.

The signal from the AII amacrine cell goes to both ON and OFF cone bipolar cells. It is conveyed to the ON cone pathway by gap junctions between the rod amacrine cells and the terminals of the ON bipolar cells. This arrangement means that a depolarization in the rod bipolar cell will depolarize the AII amacrine cell (the synapse from bipolar cell to amacrine cell is sign-conserving). The AII amacrine cell, in turn, will depolarize the ON bipolar cell terminal through the gap junctions, which are also sign-conserving. Thus the ON polarity of the rod bipolar cells is matched to the polarity of the ON cone bipolar cells.

*The different amacrine cells distinguished by morphology are designated by "A," for amacrine cell, followed by a numeral from 1 up to 20-something, depending on the species. The numbers have no special meaning. AII amacrine cells are a separate class (there is an A2 group), which makes the terminology inconsistent. Some amacrine cells have acquired other names, such as the "starburst" amacrine cells, as they have become better known. AII amacrine cells are often called "rod" amacrine cells because of their key position in the rod pathway, but they also receive inputs from cone pathways, so the neutral terminology, "AII amacrine cells," is used here.

The AII amacrine cells contact the OFF cone pathway by a chemical synapse onto the OFF bipolar cells; the inhibitory neurotransmitter **glycine** is used at these synapses. As a result, depolarization of the AII amacrine cell will hyperpolarize the OFF bipolar cell terminal. Thus ON is inverted to OFF in keeping with the interaction onto the OFF cone pathway.

Because of this arrangement, rod signals are divided like cone signals to report both increases (ON) and decreases (OFF) in illumination. Unlike the cones, however, the rod signals are divided in the inner plexiform layer through the actions of the AII amacrine cells. With the rod signals riding piggyback on the cone pathways at the level of the cone bipolar cell terminals (see Figure 14.8), there can be no purely rod-driven ganglion cells; whether a ganglion cell reflects rod or cone activity will depend on which photoreceptor system is the more sensitive and which system is preempting the bipolar cell terminals going to the ganglion cells.

Interactions among Bipolar Cells, Amacrine Cells, and Ganglion Cells

Bipolar cell terminals in the inner plexiform layer release glutamate at synapses to amacrine or ganglion cells and receive inputs from amacrine cells

Bipolar cells are in the business of relaying signals generated by photoreceptors and modified by horizontal cells to the next site of interaction in the inner plexiform layer. The bipolar cell terminals differ depending on the level at which the cells terminate in the inner plexiform layer (see the next section) and in the extent to which their terminal branches spread laterally, but they have at least two features in common: They all contain synaptic ribbons associated with their synapses, and they all use glutamate as their neurotransmitter.*

The characteristic synapse made by bipolar terminals contains a synaptic ribbon, shown in Figure 14.9 in cross section, lying next to a small synaptic ridge, along which the cell membrane is very densely stained. Synaptic vesicles throughout the terminal contain glutamate. Because two postsynaptic processes lie next to the synaptic ridge, this configuration is called a **dyad**. The variable from one dyad to the next is the nature of the postsynaptic processes. At cone bipolar cell dyads, the postsynaptic elements may be amacrine cell processes, ganglion cell dendrites, or one of each. Since rod bipolar cells do not contact ganglion cells, the postsynaptic processes always belong to amacrine cells.

There are also synapses onto the bipolar cells at or near their terminals; one is shown in Figure 14.9. These inputs to the bipolar cells will always be from amacrine cells (ganglion cells in mammalian retinas do not make synapses onto other retinal neurons). The synapse shown in Figure 14.9 is part of a **reciprocal synapse**; the bipolar cell has a synapse onto the amacrine cell, which has a synapse back onto the bipolar cell. This arrangement is like the relationship between photoreceptors and horizontal cells, even though horizontal cells do not make conventional synapses, and it may play a similar role in contrast enhancement. And like the horizontal cell feedback to photoreceptors, the amacrine cell feedback to bipolar cells is probably mediated by GABA except for possible reciprocal interactions between OFF cone bipolar cell terminals and the glycinergic AII amacrine cells.

*In primates, some cone bipolar cells contain glycine, but probably because of leakage of glycine molecules across the gap junctions from the AII (glycine-containing) amacrine cells to ON cone bipolar cell terminals; in this case, the glycine has nothing to do with neurotransmission. The weight of evidence indicates that *all* bipolar cells use glutamate as an excitatory neurotransmitter.

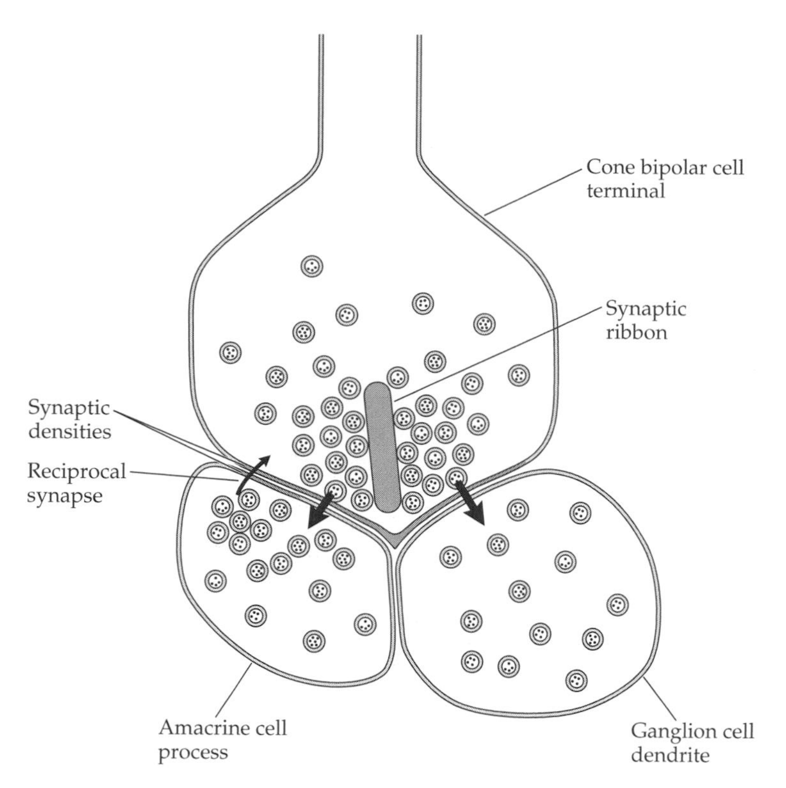

Figure 14.9
Synaptic Dyads in Bipolar Cell Terminals
Bipolar cell terminals have ribbons at the synapses they make to other neurons. Because there are always two postsynaptic processes, the synapses are called dyads. The two postsynaptic processes are always from amacrine cells at rod bipolar cell dyads, but at cone bipolar cell dyads they may be amacrine cell processes, ganglion cell dendrites, or one of each. A nearby synapse from the postsynaptic amacrine cell back to the bipolar cell constitutes a reciprocal synapse.

The details of interactions between bipolar cell terminals and specific types of amacrine cells are known for only a few cases—not enough to tell us if reciprocity is the exception or the rule for amacrine cells.

Bipolar cells terminate at different levels within the inner plexiform layer, thereby creating functional sublayers

A variety of experimental methods in a large number of species show the following pattern to be a general one: Rod bipolar cells terminate closest to the ganglion cells, ON cone bipolar cells terminate closer to the middle of the inner plexiform layer, and OFF cone bipolar cells terminate close to the border with the inner nuclear layer. In other words, all signals from ON bipolar cells go to the proximal half (roughly) of the inner plexiform layer, while signals from OFF bipolar cells go to the distal half (roughly). Because of this arrangement, any ganglion cells that are to receive direct OFF bipolar cell input must have their dendrites in the distal sublayers, while those receiving direct ON bipolar cell input must have dendrites in the proximal sublayers.

Figure 14.10 shows the bipolar cell termination levels for primate retina. Diffuse cone bipolar cells have six termination levels within the inner plexiform layer, ranging from near the border with the inner nuclear layer to the border of the ganglion cell layer, and these termination levels are used to recognize and name the different types of diffuse cone bipolar cells (Figure 14.10*a*). Since there

Vignette 14.1

The Retina Comes to Light

GREEK ANATOMISTS RECOGNIZED THE RETINA as one of the eye's tunics, and Herophilus named it *amphiblestroïodés*, which means "netlike or net-shaped tunic." Translated from Greek to Arabic and from Arabic to Latin, the term becomes *retina,* derived from *rete*, meaning a fine meshwork, like a fishnet. Thus the delicacy of the retinal tissue was recognized very early. As for the retina's function, the first clear idea we have comes from the Greek physician Galen (see Vignette 2.1), to whom the prominent retinal blood vessels suggested a primarily nutritive function.

Galen could not conceive of the retina as a light-sensitive tissue, because he subscribed to a version of Plato's emanation theory of light, in which a kind of visual spirit is generated within the brain, transmitted to the eye, and exits from the lens through the pupil to interact with the environment. If there is anything here that can be called sensitive, it is the lens. But since Galen realized that the retina surrounds the vitreous as an extension of the optic nerve, he considered a possible role for the retina in protecting the visual spirit from distortion as it passed from the optic nerve to the vitreous. Still, the main function of the retina was believed to be providing nutrition for the all-important lens.

As long as emanation theories of vision held sway, the retina could be no more than a bit player in the visual process. This situation began to change in the eleventh and twelfth centuries as Arab scholars criticized the emanation theory and resurrected Aristotle's idea that light comes from objects to the eye (see Vignette 2.1). The philosopher Averroës (Ibn Rushd), for example, thought that the retina was crucial in terms of *transmission* of the visual sensation to the brain and that it must be sensitive to whatever impressions it received from the vitreous. Although the lens was still regarded as the primary light-sensitive structure, the retina came to be considered more important in the visual process than it had been.

About 400 years after Averroës wrote on the subject, an anatomist from Basel, Felix Platter (1536–1614), suggested that the retina is the primary sensitive tissue. Platter lacked evidence to support his claim, and he simply assumed that the eye works much like a *camera obscura* (pinhole camera). This assumption implied that an image corresponding to the eye's field of view would be extensive, covering a large part the eye's interior surface. Since the retina is the only structure whose extent corresponds to this large image, it is the obvious candidate for receiving light. In this view, the lens is simply another optical element through which light passes on its way to the retina. (Leonardo da Vinci apparently had come to this conclusion almost a century earlier, but he thought the optic nerve head was the light-sensitive tissue.)

Theoretical and experimental studies of image formation by Johannes Kepler, Christoph Scheiner (see Vignette 2.2), and others early in the seventeenth century provided the critical knowledge that the retinal image was extensive and distributed over the posterior surface of the eye as a point-to-point mapping of external space. And an anatomical misconception of long standing was also eliminated: It was determined that the optic nerve head is not on the visual axis, as it had always been depicted, but is located several millimeters nasally—well away from the site of most acute vision. Thus, with a large image focused in the plane of the retina and with the optic nerve head out of the picture, only the retina could conceivably be responsive to light in the image.

Although the retina and its workings were now seen to be of paramount importance in the visual process, knowledge of the retina's structure had hardly changed in 2000 years. The retina was known only as a delicate, rather fibrous tissue with large blood vessels, attached to the optic nerve. Scholars could deduce some of its properties, like its point-by-point sampling of the retinal image (Descartes), but the most important structural features were concealed from seventeenth-century eyes. The invention of the microscope in the eighteenth century changed everything.

One of the first important microscopic observations was made on rabbit retina, a species I have promoted—more or less seriously—as containing answers to all the important questions about how the retina works. In rabbits, myelination of the ganglion cell axons extends past the nerve head, forming broad bands of opaque, whitish axons running horizontally across the retina; the bands of myelinated fibers are visible to the naked eye and appear to have an underlying fibrous structure. This much was known to the anatomist Felice Fontana (1730–1805), although he drew the fiber bands as if they were vertical; he realized that the myelinated fiber bands in the rabbit retina might offer solid evidence for the existence of nerve fibers hinted at by Antoni van Leeuwenhoek and others. And when he looked with his microscope, Fontana found them, thereby establishing a route for communication between retina and brain. (It's not clear whether Fontana saw individual axons or small axon bundles, but his interpretation was certainly correct.)

Fontana's observation was repeated by Gottfried Reinhold Treviranus (1776–1837), who also saw numerous thin, closely packed cylindrical structures that he believed to be the visual cells—the ones that might somehow interact with light. (Again, van Leeuwenhoek may have already seen them.) The problem was knowing where they were. The vitreal and choroidal surfaces appear much alike when the retina is removed from the eye, and the few clues to its original orientation require prior knowledge of its structure. So in dealing with the problem for the first time, Treviranus assumed that visual cells should be in a position to confront incoming light directly, with no other tissue in the light path. Thus he assigned them a position next to the vitreous, with the rest of the retina behind them.

This conception of photoreceptor location had a very short lifetime. In just a few years, observations on retinas in situ confirmed the presence of the little cylindrical structures, but placed them in a "bacillary layer" lying next to the choroid, not next to the vitreous. But this correction again shifted attention back to the vitreal surface of the retina because there were no observable connections between the bacillary layer and the nerve fibers. If the

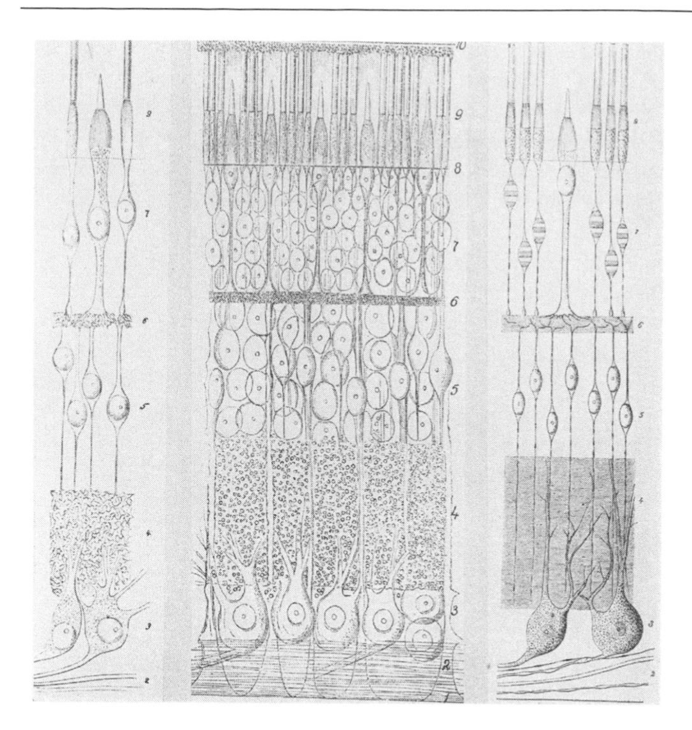

Figure 1
Layers of the Retina
These drawings were made from sections of human retina stained with nonselective dyes; cell nuclei stain darkly, but cytoplasmic staining is light or absent. The numbered layers were designated as follows: (1) inner limiting membrane, (2) optic nerve fibers, (3) ganglion cells, (4) inner granular layer, (5) inner nuclear layer, (6) outer granular layer, (7) outer nuclear layer, (8) outer limiting membrane, (9) rods and cones, (10) pigment epithelium. The radial fibers of Müller are shown running from the outer granular layer to the inner limiting membrane. (From Schultze 1873.)

nerve fibers are communicating with the brain, and are not connected to anything else in the retina, the nerve fiber layer must contain the photosensitive elements. Besides, incoming light encounters the nerve fiber layer first. The little cylinders observed by Treviranus must be doing something else.

Heinrich Müller (1820–1864) vindicated Treviranus in the 1850s. By using fixatives that hardened retinal tissue enough to section it perpendicular to its surface, Müller not only described (and named) the layers of the retina, but he also saw fibers running from the bacillary layer to the vitreal surface. The presence of radial fibers meant that the cylinders in the bacillary layer had a way to transmit some sort of signal to the nerve fibers and from there to the brain.

Better still, Müller knew that the blood vessels on the vitreal surface of the retina were responsible for the Purkinje tree, which is the visible shadow cast by the blood vessels when the eye is flooded with bright, diffuse illumination. The shadow moves as the light source moves, and Müller reasoned that this movement could occur only if the photosensitive layer lay some distance behind the blood vessels; since the nerve fibers lie next to the blood vessels, the nerve fibers could not be photosensitive, because there could be no movement of the blood vessel shadows at this level. Thus, the little cylinders in the bacillary layer must be the visual cells.

Müller's radial fibers were what we now call Müller's cells, which are the large glial cells of the retina. They do not have the function Müller ascribed to them, but there are radial elements connecting photoreceptors to ganglion cells—namely, the bipolar cells; they are shown in Figure 1 in drawings of the human retina made by Max Schultze (1825–1874) about 1870. All ten layers are present, there are two kinds of visual cells with different shapes, and there is a well-defined set of bipolar cells interposed between the visual cells and the ganglion cells. This picture offers a clear possibility of a route by which the effects of light on photoreceptors can be transmitted to the brain.

But why two kinds of visual cells? Schultze was a comparative anatomist and physiologist, and in looking at retinas from different species, he was struck by differences in their proportions of rods and cones; retinas in strongly diurnal animals had mostly cones, while retinas of strongly nocturnal species contained mostly rods. This observation led to the conclusion that cones are for daylight vision and, by implication, for color vision and high spatial resolution, while rods are for highly sensitive vision at low light levels; this, in essence, is the "duplicity" or "duplexity" theory of vision.

Schultze's drawings are clear in terms of defining the retina as a layered structure made up of many small elements of different kinds, but they are less clear about the nature of these elements. Are they individual cells, with definite boundaries and isolated cytoplasm, or are they small parts of a large, continuous meshwork? One can't tell from Schultze's drawings; although the staining methods used by Schultze and his contemporaries could distinguish between parts of cells, like nucleus and cytoplasm, they could not distinguish between the cytoplasm of one cell and its neighbor (assuming they are individuals).

The idea of the retina as a net or meshwork is quite old—embodied in its name in fact—so thinking about it as a network is a natural thing to do. And taking the histology at face value reinforces the notion of the retina, and the nervous system generally, as a "reticulum"—another variation of "net"—in which there is complete cytoplasmic continuity. Communication throughout the reticulum by fluid flow (or by the newly discovered electric currents) is therefore easily explained, with variations in the rate and magnitude of flow attributable to the size of branches within the reticulum. The alternative notion that the retina consists of many small, individual elements (neurons) is a more difficult hypothesis; individual elements cannot be visualized, and there is no obvious way for individual elements to communicate as they must.

The drawing of the retina in Figure 2 was published by Ferruccio Tartuferi in 1887, about 15 years after Schultze published his. It was the result of a new method for staining neural tissue in which gold or silver compounds formed insoluble precipitates within the tissue; the inventor of the method, Camillo Golgi, was Tartuferi's mentor. One of the great advantages of the Golgi method is its selectivity; only parts of the tissue are stained, allowing them to be visualized in isolation from the rest of the tissue in which they are embedded. But this advantage comes with a price: It is impossible to control what is stained from one retina to the next. The Golgi method has often been called capricious.

(continued)

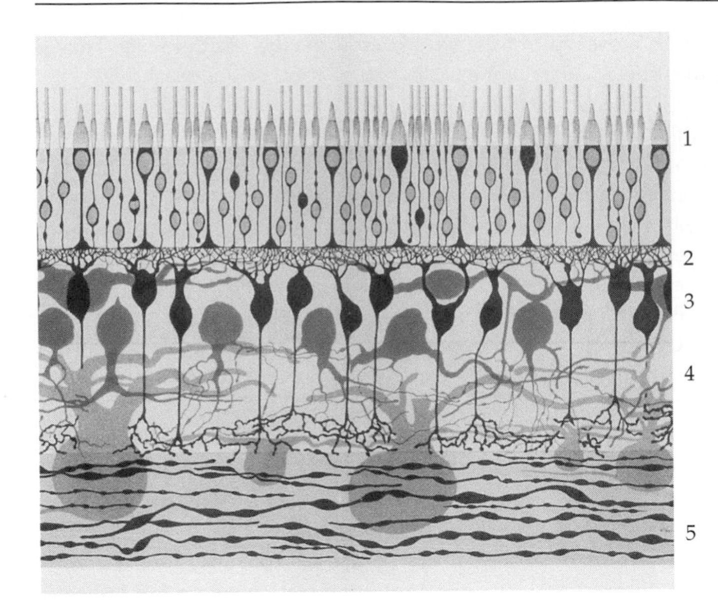

Figure 2
The First View of the Retina Using Golgi's Method
The numbers refer to rods and cones (1), outer reticular layer (2), inner nuclear layer (3), inner reticular layer (4), and optic nerve fibers (5). The term "reticular" makes the original interpretation explicit; the drawing shows pieces of the reticulum, not isolated cells. (From Tartuferi 1887.)

In Tartuferi's drawing, we seem to be seeing individual elements of the retina, essentially in the form we have come to know them. Rods and cones, horizontal cells, bipolar cells, amacrine cells, and ganglion cells are all represented in the illustration. But that is an interpretation in hindsight; while the drawing can be viewed as a representation of individual neurons, it is also possible that it shows selected fragments of the reticulum. If this drawing were the only evidence available, one interpretation would be as good as the other. Tartuferi chose the latter course, interpreting the results in terms of a reticulum and describing small cytoplasmic bridges between neighboring elements that preserved the continuous nature of the reticulum.

The notion of a reticulum is reinforced by Figure 3, which shows ganglion cells in the human retina stained by Alexander Dogiel (1891) using another new method, employing the dye methylene blue. (The original illustration, like many of Dogiel's drawings, is an exquisite color engraving that puts most modern illustrations to shame.) The methylene blue technique is aesthetically pleasing because the beautiful stained cells gradually emerge from a uniform, colorless background (it is necessary to watch so that the reaction can be stopped at the appropriate point); when the method works properly, it selects a portion of the cells in a given layer, but not all of them. As a general rule, however, it is rarely possible to separate cells from a group like that in Figure 3 and confidently designate them as individuals. Thus another staining method leaves the same impression; it seems that the network imagined by the Greeks when they named the retina is an anatomical fact to be reckoned with in thinking about how the retina works.

But that was before Ramón y Cajal (see Vignette 14.2).

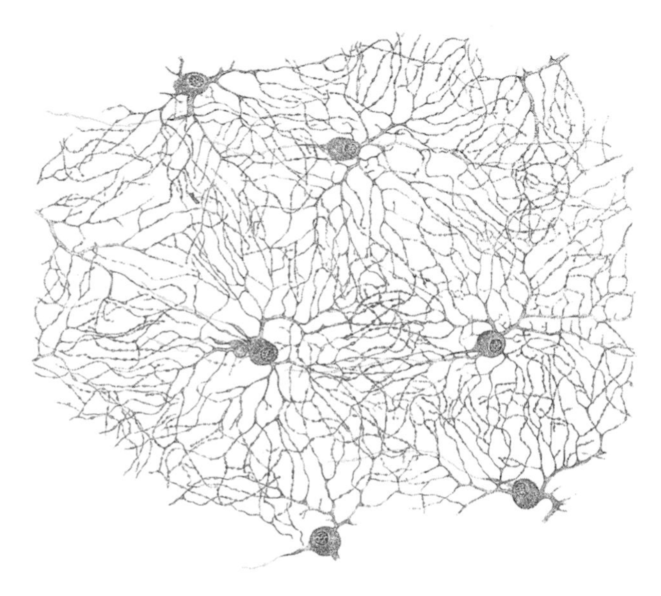

Figure 3
Ganglion Cells Stained with Methylene Blue
Stained ganglion cells in human retina are shown as they appear when the retina has been spread out flat and the cells are viewed from above. With pictures like this, it is little wonder that the retina was thought to be a reticulum. (From Dogiel 1891.)

are six termination levels, diffuse bipolar cells are designated DB1 through DB6, where level 1 is closest to the inner nuclear layer (distal) and level 6 is closest to the ganglion cell layer (proximal). Levels 1 through 3 are the OFF sublayers, levels 4 through 6 the ON sublayers.

This arrangement means that the diffuse bipolar cells represent six parallel streams of information from the cones running side by side through the inner nuclear layer. The information streams become separated from one another in the inner plexiform layer. For example, a DB1 cell and a DB6 cell may carry signals from the same cones, but their terminals and the outputs from those terminals are on opposite sides of the inner plexiform layer. In fact, the diffuse bipolar cell pathways from a small set of cones form a small column of terminals extending through the inner plexiform layer. Thus the entire array of cone signals has been

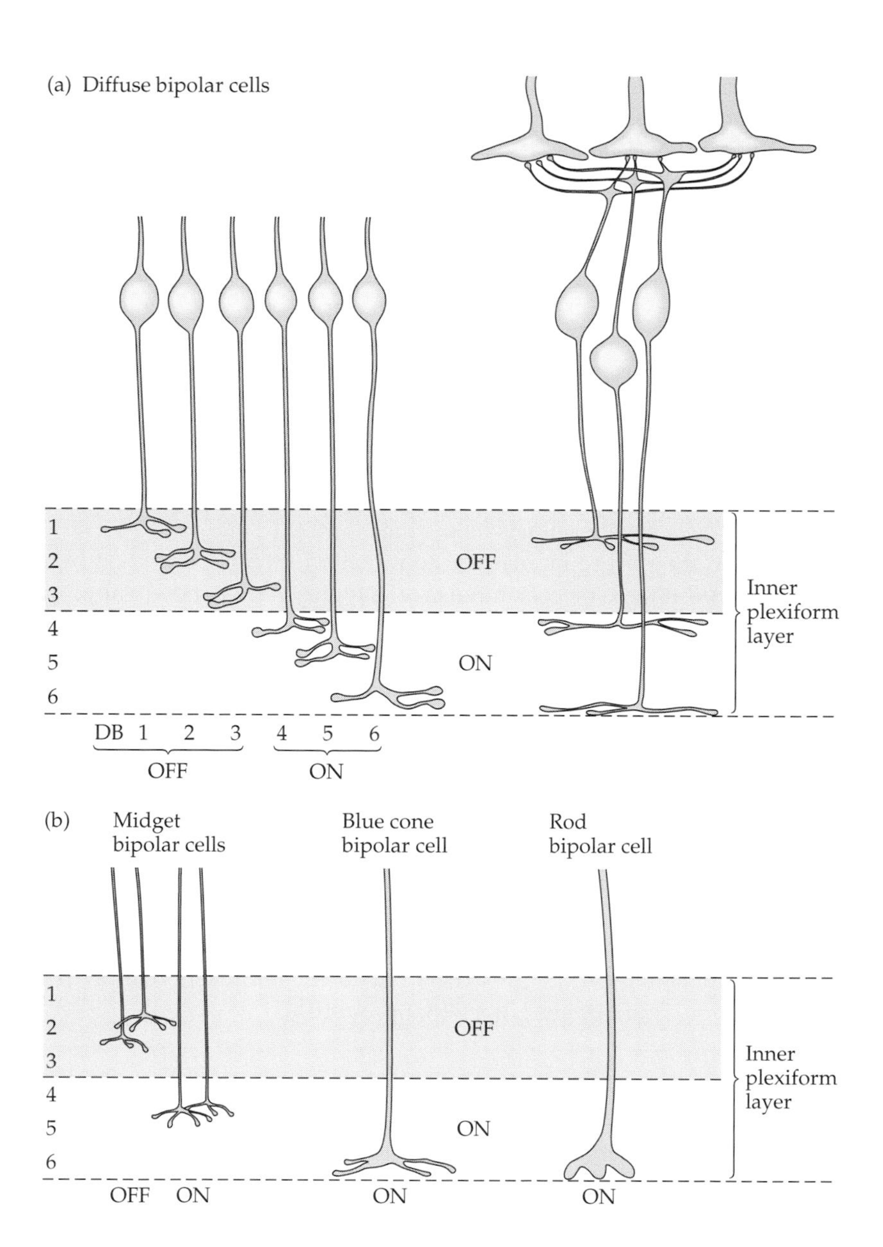

Figure 14.10
Termination Levels of Bipolar Cells in the Inner Plexiform Layer
(*a*) Diffuse bipolar cells have six distinct levels of termination in the inner plexiform layer, and these levels can be used to define sublayers. A small group of cones is therefore represented by a column of diffuse bipolar cell terminals extending through the inner plexiform layer. (*b*) Midget bipolar cells terminate mainly in levels 2 and 5, blue cone bipolar cells in level 6, and rod bipolar cells in level 6. (After Boycott and Wässle 1991.)

mapped six times in the inner plexiform layer by the diffuse bipolar cells, and each of the diffuse bipolar cell pathways occupies its own level, or stratum, of the inner plexiform layer. The half dozen parallel streams of information sampled from the cones are stacked on top of one another as sublayers.

What this sublayering implies depends on the signals conveyed by the diffuse bipolar cells. Assuming that they differ in the details of their inputs from cones and horizontal cells, they may differ in their responses to visual stimuli and convey different sorts of information. This line of reasoning implies that the inner plexiform layer contains, at minimum, six functionally different sublayers. At the moment, however, there is little evidence to suggest differences among diffuse bipolar cell signals; this may change as we learn more about them, but the better working hypothesis is that the diffuse bipolar cell pathways are not significantly different. Thus the inputs from the diffuse bipolar cells to the inner plexiform layer represent six identical mappings—three ON and three OFF—of the retinal image seen by the cones.

Until now, we have emphasized parallel pathways for different sorts of information—blue signals, red-ON signals, red-OFF signals, rod signals, and so on. Why are several identical streams of ON and OFF signals carried by diffuse bipolar cells? Wouldn't one of each be sufficient? The answer may be that functional divergence in these cone pathways begins in the inner plexiform layer instead of the outer synaptic layer, just as it does in the rod pathway. Each set of diffuse bipolar cells can contact different sets of postsynaptic neurons by virtue of their different levels of termination, thus initiating different sequences of events. This hypothesis retains the basic idea of functionally distinct sublayers in the inner plexiform layer, but it places the reason for their distinctiveness on the amacrine and ganglion cell processes in the different sublayers, not on different streams of diffuse bipolar cell information coming into the sublayers.

Midget bipolar cell terminals are most often located at levels 2 (OFF) and 5 (ON), with some spillover into levels 3 and 4 (Figure 14.10*b*). There seems to be no difference in termination levels according to the input cone type, however; red-OFF and green-OFF midget bipolar cells terminate at level 2, while red-ON and green-ON midget bipolar cells terminate at level 5. Blue cone bipolar cells and rod bipolar cells have terminals in level 6.

In addition to the diffuse bipolar cell termination levels, the midget bipolar cells (four types) and the blue cone bipolar cells represent five more mappings of the cone array, while the rod bipolar cells create a single mapping of the rod array.

Separating information streams from different vertical pathways by sublayers in the inner plexiform layer has implications for cells that are postsynaptic to the bipolar cells' pathways. Sublayer 1, for example, is a source of OFF cone signals that are not specific for cone type, while sublayer 2 contains both cone-specific and nonspecific OFF signals, as well as rod signals from AII amacrine cells. Thus amacrine cells or ganglion cells receiving inputs exclusively in layer 1 have possibilities for inputs that are different from those of amacrine or ganglion cells that receive their inputs exclusively in layer 2; the possible interactions and modifications of information flow may also differ. Thus, the sublayer or sublayers in which amacrine and ganglion cells spread their dendrites are an important anatomical distinction; the different levels of dendritic ramification imply different sources of input and different functional properties.

Amacrine cells vary in the extent over which they promote lateral interactions among vertical pathways and in the levels of the inner plexiform layer in which they operate

Among the many things we don't know about amacrine cells is how many types there are. Some 25 to 30 different types can be distinguished in the human retina by various anatomical criteria, but we do not know if all the morphological variations are meaningful in terms of differences in what the cells do. Because of this uncertainty, discussions about amacrine cells are necessarily discussions about possibilities—that is, the kinds of things amacrine cells *might* do in light of their known anatomical variability.

We do know that amacrine cells differ greatly in size. At any given location in the retina, some amacrine cells have processes extending less than 100 μm, while at the other extreme some cells may have processes spreading 1000 μm or more. For the sake of discussion, we can think of amacrine cells as coming in three sizes: small, medium, and large (Figure 14.11). Small-field amacrine cells spread their processes less than 100 μm or so, large-field cells have processes spreading more than about 500 μm, and medium-field cells are the ones in between. Although these size ranges are arbitrary and need adjustment for different parts of the retina, they work reasonably well in the sense that most amacrine cell types consistently fit into a particular size bin.

Size gives a conservative estimate of the distance over which amacrine cells can mediate interactions among vertical pathways, and these distances offer clues about what the amacrine cells can or cannot do. Interactions mediated by small-field amacrine cells, for example, must be local—that is, among vertical

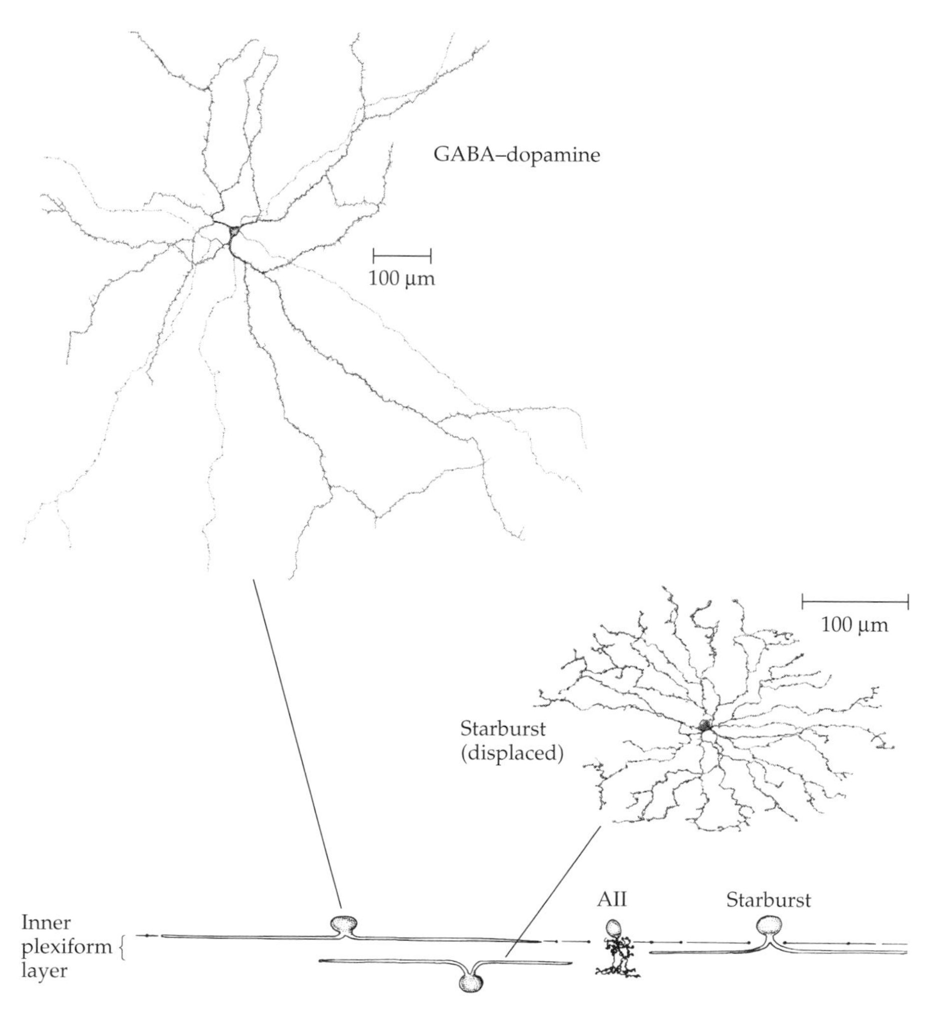

Figure 14.11
Varieties of Amacrine Cells
Amacrine cells vary considerably in morphology, and the human retina has between 25 and 30 different amacrine cell types. The major vairables are lateral extent (small-, medium, and large-field), ramification level in the inner plexiform layer, and ramification pattern (unistratified, diffuse). Four types are shown here. The large-field GABA–dopamine amacrine cell has both axons and dendrites lying in layer 1 of the inner plexiform layer. Medium-field starburst amacrine cells come in both ordinary and displaced forms with processes ramifying in layers 2 and 5, respectively (only the displaced starburst cell is shown in the flatmount view). Small-field amacrine cells are represented here by an AII amacrine cell shown in the sectioned view. (After Dacey 1990, Rodieck 1989, and Strettoi, Dacheux, and Raviola 1992.)

pathways lying close to one another. Interactions mediated by large-field amacrine cells are much more extensive, and for the largest amacrine cells may involve almost 1% of the retina's total area; vertical pathways a millimeter or so apart may interact through the auspices of the large-field amacrine cells. Finally, medium-field amacrine cells mediate interactions over distances that are neither local nor extensive.

Whatever small-field amacrine cells do, the interactions they promote will be among nearby neighbors; even if the small-field amacrine cells were part of a convergent wiring scheme, their small size means that a relatively fine spatial grain would be preserved. (And since the neurotransmitter would almost certainly be an inhibitory one, as discussed in the next section, excitatory convergence is unlikely.) At their largest, the small-field amacrine cells are comparable in size to the horizontal cell arborizations in the outer plexiform layer, and they are good candidates for another stage of contrast enhancement. In fact, they may underlie the most significant contrast enhancement in the retina.

Very extensive interactions promoted by large-field amacrine cells cannot preserve spatial grain (unless small parts of the cells can function independently of one another); they are likely to be involved in integrative processes in which the weighting of signals onto the amacrine cell spreads throughout the amacrine cell's large field of influence. Operations like adjusting sensitivity as a function of average retinal illumination may be one of the possible roles for these large cells.

The implications of intermediate-range interactions mediated by medium-field amacrine cells are less obvious, but on the basis of what some of them seem to be doing in other retinas, they may promote sensitivity to temporal or to spatiotemporal change—that is, sensitivity to the rate of change in light intensity, to retinal image movement, or to movement velocity.

Distinct roles for amacrine cell–mediated interactions over short, medium, and long distances suggest why there are so many types of amacrine cells. Since there are at least six definable sublayers in the inner plexiform layer, each representing an information stream from diffuse bipolar cells, each sublayer may require its own set of local, intermediate, and long-range interactions. If so, this requirement puts the number of amacrine cell types at 18—three in each of six sublayers. (This much duplication may not exist, but it is probably not far off the mark.) Many of the medium- and large-field amacrine cells are clearly stratified, with their processes confined to a single plane within the inner plexiform layer. A few of the better known examples are shown in Figure 14.11.

Some amacrine cells have processes that are not confined to a single sublayer (see Figure 14.11). Sometimes the processes lie in two sublayers, which makes them bistratified, but in primate retina, the more common pattern is "diffuse" branching in several sublayers, with no clear preference for one or the other. Diffuse branching is particularly noticeable among the small-field amacrine cells, of which the AII amacrine cells are prime examples. Their processes are found in sublayers 2, 3, 5, and 6.

Diffuse or multistratified branching patterns have a functional implication not shared by unistratified branching; namely, they raise the possibility of interactions *between* sublayers as well as *within* sublayers. Amacrine cells with diffuse or multistratified branching can mediate interactions between different kinds of vertical pathways. Again, the AII amacrine cells illustrate the point. They receive inputs from rod bipolar cells in sublayer 6 and relay it, with opposite signs, to cone bipolar cell terminals in OFF sublayers 2 and 3 and ON sublayers 4 and 5. And because the AII amacrine cells are among the smallest of the small-field amacrine cells, their interactions are so local that they can be considered part of the vertical rod pathways.

If these general notions about the morphology and functional roles of amacrine cells are at all realistic, the plethora of amacrine cell types is easily explained. Interactions over different spatial ranges within sublayers and interactions between different sublayers quickly add up to the 25 to 30 amacrine cell types that have been described.

Amacrine cells exert their effects mainly at glycine and GABA synapses, while several other neurotransmitters or neuromodulators play subsidiary roles

Amacrine cells are notorious for the variety of neuroactive substances they contain and might use for communication; a recent review of neurochemistry of the inner retina listed 14 possibilities. When this neurochemical diversity is added to the large number of amacrine cell types, making sense of all the interactions that might occur seems impossible, or at least very difficult. Fortunately, there are some important simplifications.

When the primate retina is assayed for specific amino acids, all retinal neurons are found to contain at least one of three classic neurotransmitters: **glutamate**, **GABA**, and **glycine**. Retinal neurons invariably contain more than one amino acid, and the number and relative amount of the amino acids vary from one cell group to another. The particular amino acids that a cell contains and their relative amounts constitute a signature that is something like a zip code. It identifies cells as belonging to a particular, distinguishable group. Cells may contain other potential neurotransmitters or neuromodulators in addition to their amino acid complement, but one of the three G's always dominates; their distribution is shown in Figure 14.12.

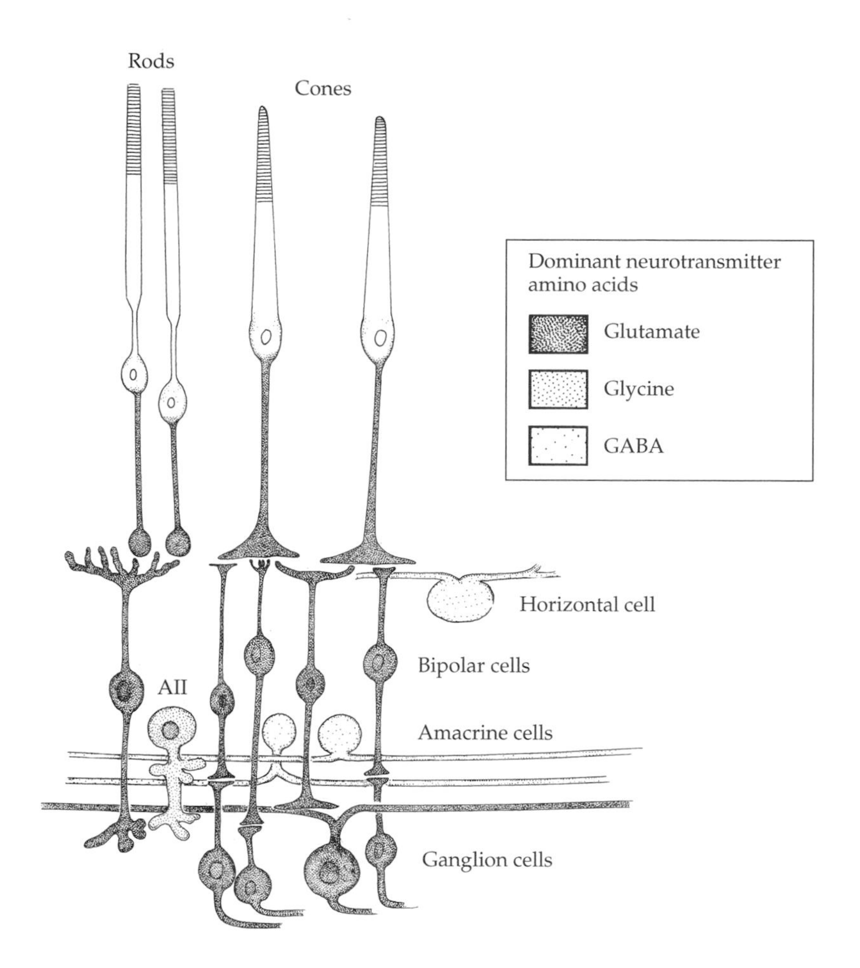

Figure 14.12
Distribution of Glutamate, GABA, and Glycine in Retinal Neurons
The dominant neurotransmitters in the retina are the amino acids glutamate, glycine, and GABA. Glutamate is the neurotransmitter in the vertical pathway neurons: photoreceptors (both rods and cones), bipolar cells of all types, and ganglion cells. Glycine is utilized by AII amacrine cells and other unidentified small-field amacrine cells. GABA is used by all horizontal cells in an unconventional way and by most amacrine cells. The GABA-rich amacrine cells may co-localize other neurotransmitters (glycine, acetylcholine) or neuromodulators (dopamine, substance P).

As noted earlier, glutamate is used by all photoreceptors and bipolar cells for neurotransmission, and the only other neurons in the retina that have high glutamate levels are ganglion cells. Thus glutamate is strictly associated with communication in the vertical pathways; the lateral pathways must use other neurotransmitters, and we have already seen that horizontal cells are dominated by their GABA signature. Now for the amacrine cells.

In terms of their amino acid signatures, amacrine cells belong to one of three groups: (1) those having GABA as the dominant element in the signature, but lacking glycine; (2) those having glycine as the dominant element, but lacking GABA; and (3) those having GABA and glycine in nearly equal amounts. The GABA-rich cells can be divided into four subgroups and the glycine-rich cells into three subgroups. (This subgrouping is rather like the increased specificity gained by going from five-digit to nine-digit zip codes.) Thus there are at least 8 subgroups of amacrine cells—4 GABA-rich, 3 glycine-rich, and 1 GABA–glycine. But how does this grouping simplify anything?

First, the dominance of GABA and glycine in the amacrine cells means that most synaptic interactions mediated by amacrine cells are at sign-inverting (inhibitory) synapses. It says, in general, that the amacrine cell lateral pathways conduct their operations in much the same way that the horizontal cell lateral pathways in the outer retina do.

Second, most if not all of the glycine-rich amacrine cells are small-field cells, and a substantial portion of these are AII amacrine cells. We can justifiably think of glycine in terms of local interactions and—much of the time—in association with the AII amacrine cells.

Finally, about 90% of the amacrine cell synapses in the inner plexiform layer are GABA synapses. The percentage of GABA synapses onto bipolar cell terminals is even higher, suggesting that all of the inputs from amacrine cells to bipolar cells are inhibitory through the GABA system, with the exception of the glycine synapses from AII and other small-field amacrine cells. Most of the conventional synapses from amacrine to amacrine cell or from amacrine to ganglion cell must also be GABA synapses, and most of the morphological types of amacrine cells must fall into the GABA-rich group. In short, most of the interactions in the inner plexiform layers are done with GABA.

But now for a few qualifications. Acetylcholine is used as a neurotransmitter by two well-characterized sets of amacrine cells (the starburst amacrine cells), and other amacrine cells contain neuromodulators. Dopamine is the major neuromodulator, but several neuropeptides are also present. Where do these substances fit into the GABA story?

The answer is that GABA-rich amacrine cells may contain other neurotransmitters or neuromodulators, just as they contain other amino acids; this phenomenon is known as **co-localization**. The differences between subgroups of GABA-rich cells likely are related to the other substances they co-localize. Thus what have been called "cholinergic amacrine cells," "dopaminergic amacrine cells," or "substance P–containing amacrine cells" are better thought of as GABA amacrine cells that also contain acetylcholine, dopamine, or substance P. These other substances, as well as GABA, may be used for neurotransmission or neuromodulation. In other words, an amacrine cell may have at least two sorts of interactions with other neurons.

Assessing the importance of acetylcholine, dopamine, or neuropeptides in the grand scheme of things is difficult. Although these substances are confined to specific and sometimes small populations of unistratified amacrine cells (in primate retina), their effects are likely to be significant at their sites of action and therefore of functional importance. But in terms of the common interactions in the inner plexiform layer, none of these other substances has been shown to have the widespread impact of GABA, which is clearly the dominant medium of exchange for amacrine cells. Thus we can think of most amacrine cell interactions in terms of GABA and its effects, but the presence of other neuroactive substances co-localized with GABA requires some mental reservations—it isn't totally GABA.

The effects of neurotransmission depend on postsynaptic receptors

The issue of multiple neurotransmitter receptor types came up earlier in our discussion of the autonomic innervation of the smooth muscle and glands of the eye (see Chapters 7, 10, and 11) and again in our consideration earlier in this chapter of ON and OFF bipolar cell responses to glutamate released from the cones. The major distinction is between *ionotropic* receptors associated with ligand-gated ion channels and *metabotropic* receptors that affect channels through an intermediate G protein. Most neurotransmitters act at at least one receptor in each category. As we saw earlier for glutamate released from the cone terminals, the existence of different receptors that produce different effects from the same neurotransmitter can make it difficult to characterize a transmitter as strictly excitatory or inhibitory; it depends on the receptors.

There are at least six types of glutamate receptors, four of which are ionotropic and mediate excitation by their action on Na^+ channels. At present, most amacrine cells and ganglion cells that receive inputs from bipolar cells appear to have one or another of the ionotropic glutamate receptors, and the synapses are sign-conserving. Like the rod and ON cone bipolar cells, however, some amacrine cells may have metabotropic glutamate receptors. This story is still developing, but the dominant action of glutamate in the inner plexiform layer is likely to be excitation (sign conservation).

Matters are simpler for glycine, because there is only one type of glycine receptor in the retina. It is an ionotropic Cl^- channel receptor, and the action of glycine is therefore always sign-inverting (inhibitory).

Although there are three types of GABA receptors ($GABA_A$, $GABA_B$, and $GABA_C$), their general effect is always hyperpolarization of the postsynaptic cell; that is, they are all associated with sign-inverting synapses, even though their mechanisms of operation differ ($GABA_B$ is metabotropic, the others are ionotropic). Ganglion cells appear to be dominated by $GABA_A$ receptors, but amacrine cells and bipolar cell terminals may have any of the three types. The distributions of these receptor types are currently being investigated, and a definitive statement, particularly for primate retina, will have to wait for more data.

GABA receptors are all associated with sign-inverting synapses, so we may always think of GABA as an inhibitory transmitter, but the different receptors work at different speeds, have different time courses of action, and have different sensitivities to GABA. Inhibition appears to be a more finely tuned process than we thought, and inhibitory amacrine cell interactions appear to be able to modulate signals in subtle ways that we are just beginning to recognize.

Amacrine cell connections centering on the AII amacrine cells illustrate difficulties in understanding amacrine cell operations

We do not fully understand the connections made by any particular type of amacrine cell, and we know even less about the larger circuits in which they participate. The most complete story centers on the AII amacrine cells that were illustrated earlier (see Figure 14.8) as part of the rod signal pathway. Other connections not shown earlier make the AII amacrine cells part of *cone*-driven circuits that may modify the flow of rod signals through the AII amacrine cells. These additional connections illustrate the difficulties we face in trying to understand what amacrine cells do and how they do it.

One AII amacrine cell is illustrated in detail in Figure 14.13*a*. (Everybody likes AII amacrine cells. They're distinctive, easy to isolate, and look a bit like the little 'droid, R2D2, in the film *Star Wars*.) The connections shown earlier in Figure 14.8 were rod bipolar cell inputs on its terminals at level 6 in the inner plexiform layer, gap junctions with ON cone bipolar cell terminals at level 5, and synapses to OFF cone bipolar cell terminals at level 2. But there are more. Some amacrine cells have inputs to the AII amacrine cell at level 6 in close association with the rod bipolar cell inputs, and others make contact at the neck of the AII amacrine cell near its cell body. And some of the cone bipolar cells to which the AII amacrine cell has inputs also have inputs back onto the AII amacrine cell (that is, there are some reciprocal synapses). Finally, the AII amacrine cell is connected to other AII amacrine cells by gap junctions.

One of the other amacrine cells providing inputs to the AII amacrine cells is a GABA–dopamine amacrine cell (Figure 14.13*b*). It is one of the large-field amacrine cells, with processes that may extend more than a millimeter from the cell body. Actually, there are two sets of processes. Some are relatively thick and do not extend more than several hundred micrometers from the cell body; they are essentially dendrites, and diffuse bipolar cells provide inputs to them. The other processes are more like axons; they are extremely long and thin, with small varicosities (swellings) where contacts are made with other cells, including the AII amacrine cells. Most processes, whether dendrites or axons, run at level 1 in the inner plexiform layer, so their bipolar cell inputs are probably from DB1 (OFF) cells and their outputs are those near the cell bodies of the AII amacrine cells. Their synapses onto the AII amacrine cells are probably GABA-ergic, but the dopamine in these amacrine cells may also act on the AII amacrine cells at dopamine receptors not associated with synapses.

The other amacrine cell to be considered is also from the large-field group; it is called an A17 amacrine cell (Figure 14.13*c*). The body of the A17 amacrine cell lies in the inner nuclear layer, and long processes trail down and out though the inner plexiform layer to level 6, where most of the field of processes lies. Viewed from the retinal surface, the numerous processes form a dense, radial pattern,

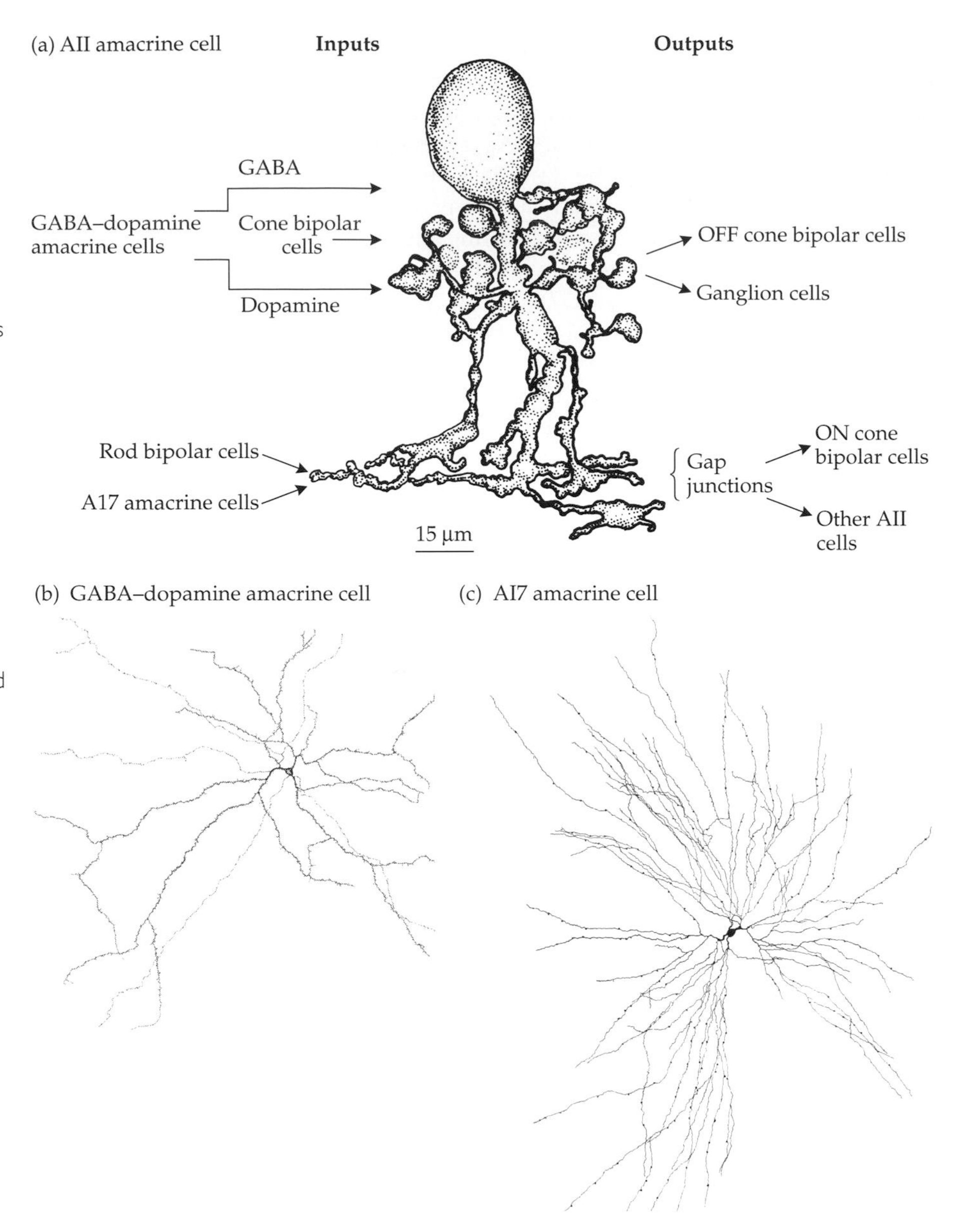

Figure 14.13
Three Types of Amacrine Cells
(*a*) AII amacrine cells are a small-field cell type that occupy a critical site of interaction between rod and cone pathways. They receive inputs, shown on the left side of the cell, from rod and cone bipolar cells, GABA–dopamine amacrine cells, and A17 amacrine cells. Outputs, shown on the right, are mainly to ON and OFF cone bipolar cells (probably diffuse bipolar cells) and to other AII amacrine cells. (*b*) The large-field GABA–dopamine amacrine cells have very extensive axonal processes that contact the AII amacrine cells in the OFF portion of the inner plexiform layer. (c) Large-field A17 amacrine cells that convey signals from rod bipolar cells contact the AII amacrine cells in the ON region. This example is from cat retina. (*a* after Strettoi, Raviola, and Dacheux 1992; *b* from Dacey 1990; *c* from Nelson and Kolb 1985.)

like spokes in a wheel, around the cell body. A17 amacrine cells contain GABA, which probably mediates their synaptic interactions, and a neuropeptide. The A17 amacrine cells make inputs to AII amacrine cells at level 6, as well as reciprocal synapses with the rod bipolar cell terminals.

The schematic in Figure 14.14 shows the relationships among these amacrine cells, the rod bipolar cells, and the diffuse bipolar cells; it is the basic rod pathway from Figure 14.8 with the GABA–dopamine and A17 amacrine cells added. Although the circuit does not seem particularly complicated, it is a good example of how quickly small uncertainties can add up to perplexity.

Consider the GABA–dopamine amacrine cell; it is clearly in a position to mediate between cones and rods as they jockey for control of the cone pathway through which both their signals must flow. Presumably, an increasing level of cone signaling could be transmitted to the AII amacrine cell and effectively switch it off, or turn down its gain, so that the rod pathways would have less access to the cone bipolar cell terminals on which they ride piggyback. But attractive as this idea may be, it is difficult to know how the adjustments of AII amacrine cell transmission are made. For example, is it done with GABA or with

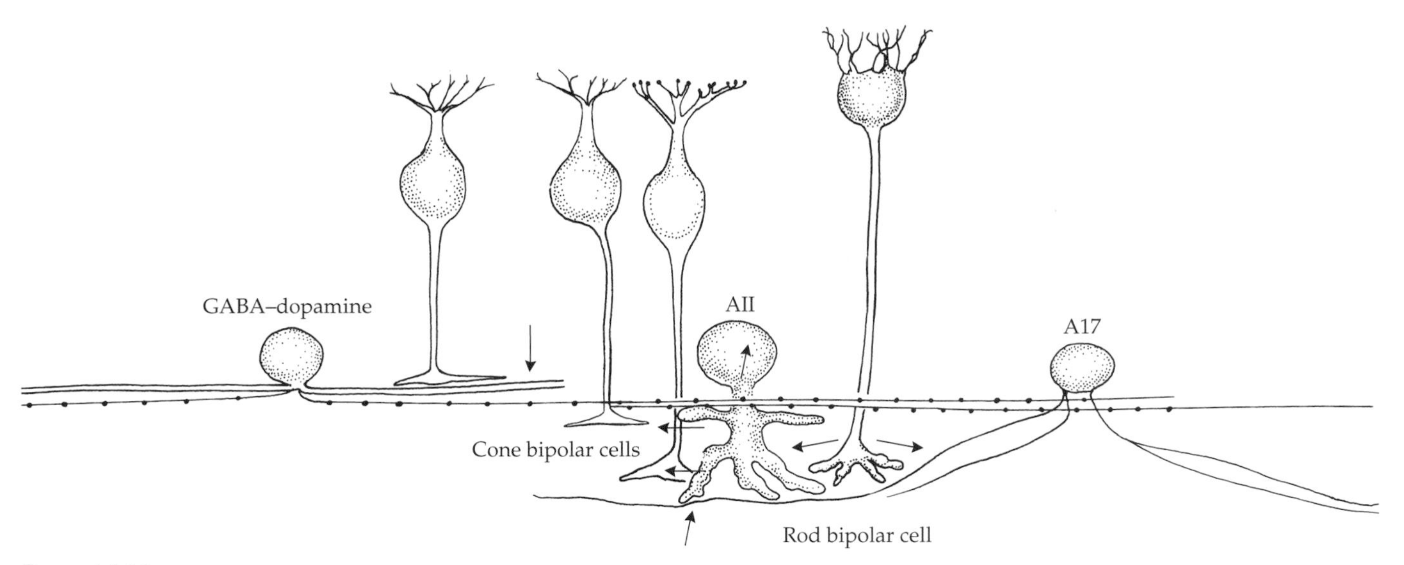

Figure 14.14
Amacrine Cell Circuits and the AII Amacrine Cells
AII amacrine cells receive inputs from rod pathways through both rod bipolar cells and A17 amacrine cells. Their cone pathway inputs are mediated by the GABA–dopamine amacrine cells. Integration of these diverse influences, which are both local and distant, by the AII amacrine cells should be reflected in their outputs to the cone bipolar cells.

dopamine? These two transmitters will not have the same effects on the AII amacrine cell. And if the AII amacrine cell is to be switched off or turned down as cones become more active at higher light levels, how do we explain the OFF bipolar cell input to the GABA–dopamine amacrine cell? It should be *less* active (less depolarized) as illumination increases.

So we're stuck—the seemingly innocuous addition of one amacrine cell type to the rod pathway has backfired. Something is missing, wrong, or both.

We can do a little better with the A17 amacrine cell that was added to the rod pathway. Most of the inputs to the A17 amacrine cells are from rod bipolar cells, which suggests that they are integrating rod bipolar cell signals in much the same way as horizontal cells integrate photoreceptor signals. Moreover, the A17 amacrine cells make synapses back onto the rod bipolar cell terminals; that is, most of the interactions are at reciprocal synapses, and the A17 amacrine cell synapses are GABA (inhibitory) synapses. This situation also looks much like the relationship between photoreceptors and horizontal cells. By analogy, A17 amacrine cells may be involved in adjusting sensitivity at the level of the rod bipolar cell terminals. Stronger rod bipolar cell signals over a wide area should produce more inhibitory feedback from the A17 amacrine cells to keep the bipolar cell terminal potentials within their operating range. If so, this adjustment of sensitivity affects a much larger area than adjustments mediated by horizontal cells in the outer synaptic layer do. Note also that the adjustment would occur *before* rod signals are divided into ON and OFF pathways by the AII amacrine cells.

The synapses from the A17 amacrine cell to the AII amacrine cell should inhibit the AII amacrine cell as the A17 amacrine cell depolarizes and GABA release is elevated. Since the rod bipolar cells feeding the A17 amacrine cell depolarize to light, increasing illumination and stronger rod bipolar cell signaling should push the AII amacrine cells in the hyperpolarizing direction. Thus the rod pathway may have its own mechanism for altering transmission through the AII amacrine cells as illumination goes from lower to higher levels. And as we will consider shortly, the feedforward inhibition from A17 to AII amacrine cells may alter the time course of the rod pathway signal, making it more transient.

Much of the discussion in this section is speculation. Even the circuit diagram under discussion is an assemblage from cat and rabbit retinas that may or may not be applicable in detail to primate retina. (The different cell types are present in primate retina, however.) The message concerns the difficulties we face in trying to understand how the retina works; amacrine cells are a major obstacle. Although they are certainly in the business of altering signal flow through the vertical pathways by integrating and comparing activity in different vertical pathways, the details we know usually raise more questions than they answer. In terms of telling a complete and coherent story about how the retina works, this impasse is frustrating and disappointing, but it is reality for the time being.

There is another way to get an idea about what amacrine cells do, however, and that is to see what ganglion cells do. Differences between ganglion cell responses and signals in the bipolar cell pathways must be consequences of amacrine cell actions. This approach offers a way to see *what* amacrine cells do, even if we cannot explain in detail *how* they do it.

Ganglion Cell Signals to the Brain: Dots for the Retinal Sketches

Most ganglion cells are midget or parasol cells

Midget ganglion cells receive inputs from midget bipolar cells. In and around the fovea, only one midget bipolar cell contacts each midget ganglion cell, so it is not surprising that the dendritic arbors of the midget ganglion cells are quite small, about the size of the midget bipolar cell terminal (Figure 14.15*a*). The midget ganglion cell class contains two anatomical groups, one with dendrites ramifying in the OFF (distal) portion of the inner plexiform layer and another with dendrites in the inner ON portion. As shown here, the ramification levels are 2 (OFF) and 5 (ON). Thus the separation of ON and OFF pathways created in the outer synaptic layer is preserved; OFF midget ganglion cells receive inputs from OFF midget bipolar cells, and a similar relationship holds for ON midget bipolar cells and ON midget ganglion cells.

Parasol ganglion cells receive their name from their extensive unistratified branching, which looks as if the cell were holding an umbrella over its head (Figure 14.15*b*). The dendrites spread more widely than those of the midget ganglion cells, and most of their bipolar cell input is probably from diffuse bipolar cells. But like the midget ganglion cells, parasol cells come in ON and OFF varieties based on the different levels at which their dendrites ramify in the inner plexiform layer. The parasol cells ramify closer to the middle of the inner plexiform layer than the midget ganglion cells do; the dendrites are in levels 3 (OFF) and 4 (ON).

Both midget and parasol ganglion cells are found at all retinal locations, from fovea to periphery. At any given location, midget ganglion cells are smaller and more numerous than the parasol cells, points that will be addressed in more detail later. Together, these two types account for more than 80% of all ganglion cells.

Although midget and parasol ganglion cells constitute the majority of ganglion cells in the human retina, another 20 types of ganglion cells have been described and two of them have been included in Figure 14.15. One is bistratified, meaning that its dendrites ramify at two levels of the inner plexiform layer. Typically one set of branches is in the OFF portion and another is in the ON portion of the inner plexiform layer. Bistratified ganglion cells differ mainly in size, such that we can talk about small-field bistratified (Figure 14.15) and large-field bistratified ganglion cells at any given location in the retina. The bistratified ganglion cells are very much minority populations, accounting for 10% or less of all ganglion cells.

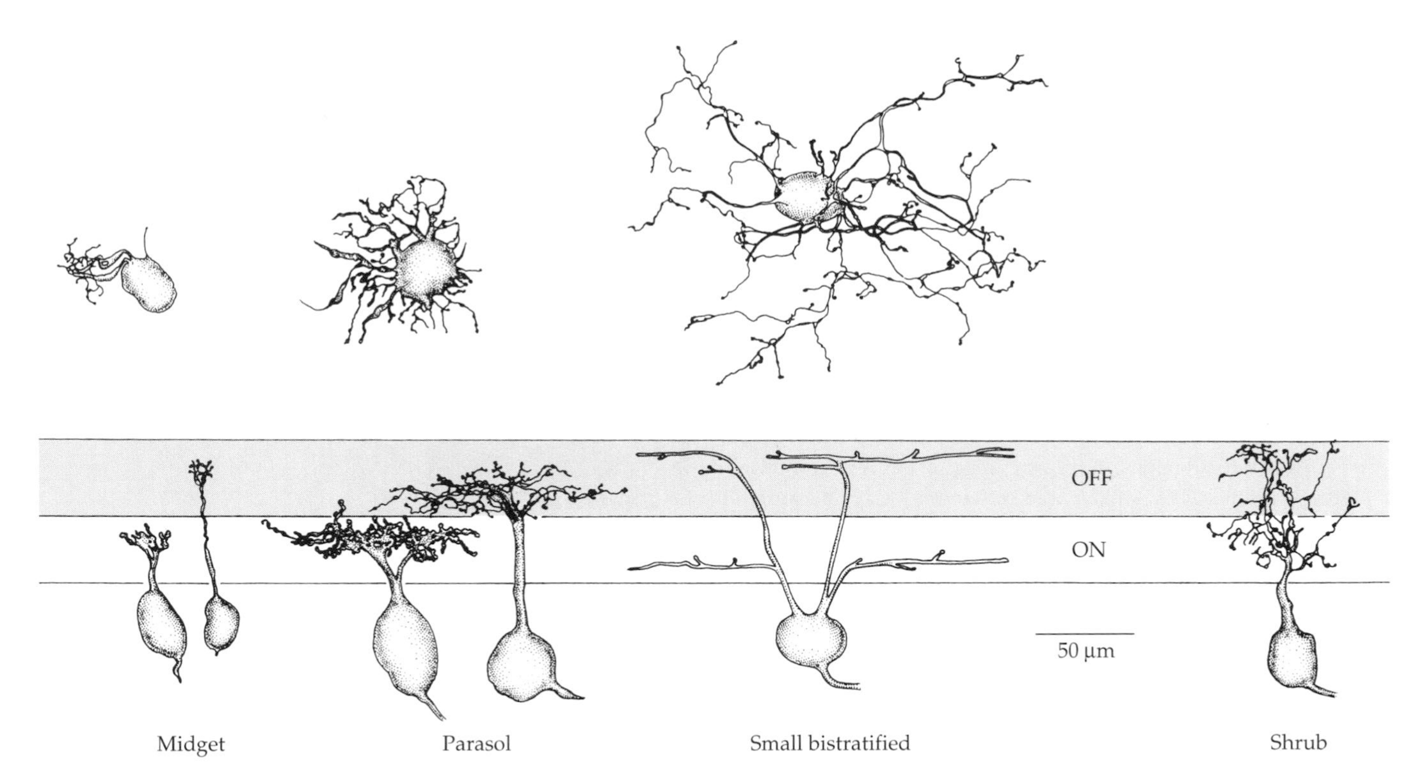

Figure 14.15
Midget, Parasol, Small Bistratified, and Shrub Ganglion Cells
Ganglion cells are shown as they appear in the flat-mounted retina (above) and in section (below). Identical cells are not shown in both views. Midget ganglion cells have very small dendritic arbors that contact only one midget bipolar cell in the central retina; their dendrites may ramify in either the ON or OFF regions of the inner plexiform layer. Parasol cells also come in ON and OFF varieties; their unistratified dendrites spread 3 or 4 times wider than those of the midget ganglion cells. Their dendrites contact diffuse bipolar cells. Small bistratified ganglion cells have dendrites ramifying in both the ON and OFF regions, while "shrub" ganglion cells have dendrites ramifying diffusely throughout the inner plexiform layer. Flatmount views of shrub ganglion cells are not available. (After Dacey 1993, Polyak 1941, and Rodieck, Binmoeller, and Dineen 1985.)

As mentioned earlier in our discussion of amacrine cells, bistratification implies that the bistratified neuron, a ganglion cell in this case, can sample different sets of bipolar cell and amacrine cell signals, where a unistratified cell is more restricted. These bistratified ganglion cells should be receiving both ON and OFF inputs and may be expected to signal both increases and decreases of illumination. An obvious question is why such an arrangement exists. What is the point of separating ON and OFF signals, only to recombine them on the bistratified ganglion cells?

Part of the answer has to do with the idea of signaling change in light intensity falling on the retina. Where ON cells and OFF cells signal the direction of change—increase or decrease—bistratified cells could signal change, pure and simple. The signal could be saying "something just happened here," and that might be a useful starting point for the brain in its process of evaluating *what* happened. Also, the inputs to the different levels of dendritic branching might vary in their wavelength specificity, and this variation might be involved in coding information about color. This seems to be the case for the small bistratified ganglion cells.

Other ganglion cells have diffuse dendritic branching patterns and were originally called "shrub" ganglion cells because of their bushy appearance (Figure 14.15); their dendrites are in a position to sample a variety of bipolar cell and amacrine cell inputs. The main point to take from our look at these cells, which will not be mentioned again, is that the subsequent discussion about ganglion cells and what they do in terms of supplying information to the brain is incomplete. Some portion of the retinal output, more than 10% or so, is unknown territory. But just because the portion is small does not necessarily mean that it is unimportant.

The oddest of the ganglion cells is the biplexiform cell (see Mariani 1982), which has a dendritic arbor in the ON portion of the inner plexiform layer and another arbor in the outer plexiform layer, where it receives direct inputs from rods. In apparent violation of the rule that rod bipolar cells do not contact ganglion cells, the inputs to the biplexiform cells in the inner plexiform layer are said to be from rod bipolar cells. Like giant bipolar cells, the question here is what the biplexiform cell morphology means; is it a genuine cell type that is always present as a certain fraction of the ganglion cell population, or is it a kind

Vignette 14.2

The Shoemaker's Apprentice

Figure 1
Santiago Ramón y Cajal
Cajal made this self-portrait in 1890, when he was about 38 years old and in the midst of his work on the retina. His right hand conceals the squeeze bulb he used to trigger the camera shutter. (From Ramón y Cajal 1937.)

SOME YEARS AGO, I HAD MY PICTURE TAKEN in Alicante, Spain, as I stood below a street sign along the "Avenida Dr. Ramón y Cajal." Every neuroanatomist pays homage to Ramón y Cajal in one way or another, often by referring to work he did a century ago as if it were done yesterday. Having my picture taken by the street named for Ramón y Cajal was an act of respect, and this vignette is another. We are indebted to Ramón y Cajal because his thinking is part of every neuroscientist's daily life; he showed us how the nervous system is constructed and in so doing developed fundamental ideas about how it works. And he didn't particularly want to be a scientist.

Santiago Ramón y Cajal (1852–1934) was born in Petilla de Aragon, a village in northeastern Spain in the foothills of the Pyrenees. His father, Justo Ramón, was a physician whose aspirations to better things kept the family on the move from one situation to another until they eventually settled in Ayerbe, which is near Zaragoza, the principal city of Aragon.

By his own admission, Ramón y Cajal was a tough, wild boy with a bent for fighting, petty vandalism, and generally raising hell; he was expert with his fists, clubs, and slung stones. But art was his passion, which he pursued zealously to the exclusion of almost everything else (including going to school), often disappearing into the hills for days at a time to work without interruption. He overcame his parents' refusal to support his efforts by making or stealing the necessary paints, brushes, or paper, and any freshly whitewashed wall in the village was soon covered with his drawings—more vandalism. Being put in jail for playing hooky (at his father's request) had little effect.

Justo Ramón had ambitions for his eldest son, but with no feeling for art and a disdain for the bohemian, often impoverished lifestyle that accompanied it, he refused to allow the boy any artistic pursuits. To instill some discipline and sense in his son, Don Justo sent Santiago to a school where severe beatings and short rations were daily fare, and legally bound him as an apprentice so that there would be no extra time for fooling with art; he was apprenticed first to a barber and later to several shoemakers, where his manual dexterity and artistry earned him the privilege of making fancy boots for wealthy ladies. (In spite of their later reconciliation, Cajal's preference for his mother's surname rather than his father's probably reflects the early tensions between father and son.)

The father's well-meaning experiment was not a success. The more he was punished, the more Cajal rebelled. He continued to fill the margins of his textbook pages with drawings, no matter how often or how severely he was beaten for doing it. But the inadequate diet at the school threatened his health and forced a compromise; a decent school and the opportunity to take a formal art course elicited a promise from Santiago to apply himself to his schoolwork. Art blossomed again; Cajal's teacher was so impressed by his ability that he pleaded with the father to permit his son to pursue an artistic career. Cajal's first art course was also his last, however, in part because of his father's continued opposition, but also because he discovered that good teachers can make many subjects interesting; he agreed to study medicine.

After passing his initial qualifying examinations in medicine at the age of 21, Cajal spent several years as an army physician, much of the time in Cuba, where he contracted malaria. He returned to Spain in poor health, but found a position as demonstrator of anatomical dissection at the University of Zaragoza. Two years later, after a bout of tuberculosis, he married,* fathered the first of eight children, and moved to his first major academic position at the University of Valencia.

Cajal concentrated on histology for several reasons: It was cheap (he covered the cost of his research by tutoring physicians in anatomy), he could work in the kitchen at home in the company of his family, and it satisfied some of his artistic feelings; he found great beauty in cells and tissues, and he could draw and photograph them to his heart's content. (He was an avid photographer and made his own photographic emulsions. The picture of Cajal in his laboratory—Figure 1—is a self-portrait.) The course of his life's work was set when he saw slides of nervous tissue stained by Golgi's method; individual cells were apparently stained in their entirety by a silver precipitate, appearing dense black against a clear background. Cajal improved the method, although he always carefully attributed it to Golgi (who had almost abandoned it out of frustration with its capriciousness) and began a series of studies that eventually grew to several hundred papers and books on neuroanatomy.

After a few years in Valencia, Cajal moved to Barcelona, which he left reluctantly after 5 years for Madrid, where better facilities were available for his work. (Cajal loved the Catalonian people and

*Neither Ramón y Cajal nor his wife had money or social standing, so their marriage was a love match rather than an arranged marriage of convenience. In his autobiography, however, Cajal includes a picture of his wife without telling us her name, and she and the children seem almost incidental to his recounting of his life. Cajal's views on marriage and family were very traditional, and writing about his life *en famille* would have been an unthinkable invasion of privacy. But he lapses occasionally, as in quoting a friend's remark: "Half of Cajal is his wife."

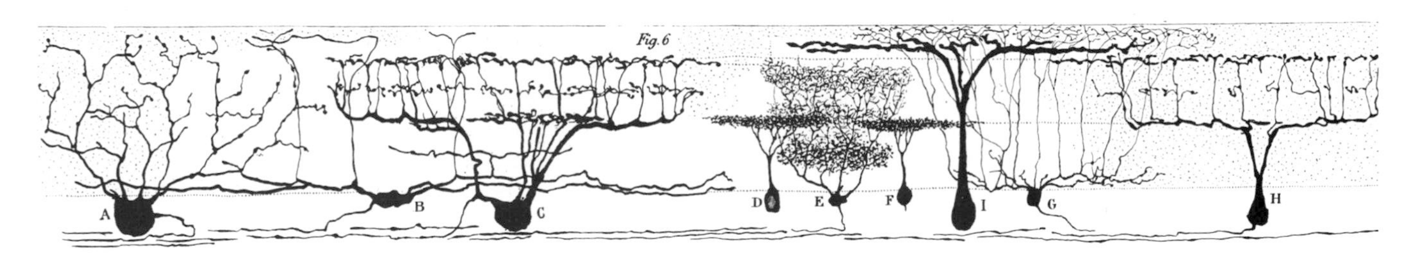

(a) Ganglion cells of the lizard retina

the city of Barcelona—hence his reluctance to leave; life along Los Ramblas was as enjoyable then as it is now.)

When Cajal began to study the nervous system, opinion and evidence favored the reticulum hypothesis in which the nervous system is conceived as a great interconnected web of continuous cytoplasm. Even the earliest results from Golgi's method were interpreted in this light (see Vignette 14.1). But if the reticulum exists, repeated applications of Golgi's method should produce different pictures as some fragments of the reticulum are stained at one time and other fragments stained at the next application. If cells are individual elements with no cytoplasmic connections, however, they should look the same each time they are stained. Discrediting the reticulum meant burying it under massive evidence that particular morphologies (shapes) are stable and robust, unvarying with time and circumstances. Ramón y Cajal provided the evidence. In 1906, he shared the Nobel prize with Golgi, who attempted to take all the credit for himself in his acceptance speech.

Cajal enjoyed studying the retina. His drawings of cells stained in dog and lizard retinas are shown in Figure 2; these drawings summarize repeated observations from many retinas. Retinal neurons can be grouped as classes (photoreceptors, bipolar cells, and so on) and as types within a class (ganglion cells have different dendritic morphologies, for example). The cell classes and many of the cell types are found in retinas of different species. These stable, repeatable observations are consistent with the existence of individual neurons, but they fit the reticulum hypothesis only with great difficulty.

Cajal's advocacy of the neuron doctrine was embraced enthusiastically by some (Charles Sherrington, for example—see Vignette 5.1), but general acceptance was initially slow in coming. Part of the problem was Cajal's linguistic isolation. Although he could read French, German, and a little English, and he spoke some French, he wrote only in Spanish. Most scientists did not know the language, and consequently Cajal's work and its importance did not become widely known until it had been translated into French. But monographs like *La rétine des vertébrés* (1892), from which Figure 2 was taken, and his monumental book *Histologie du système nerveux de l'homme et des vertébrés* (1909–1911) made Cajal the preeminent figure in neuroanatomy.* There were still dissenters, of course, and one of Cajal's last papers before his death was yet another defense of the neuron's individuality. (Several decades later, electron microscopy proved unequivocally that he was right.)

*Both books are now available in English translation. Unfortunately, the English version of the neurohistology book does great disservice to Cajal's original illustrations.

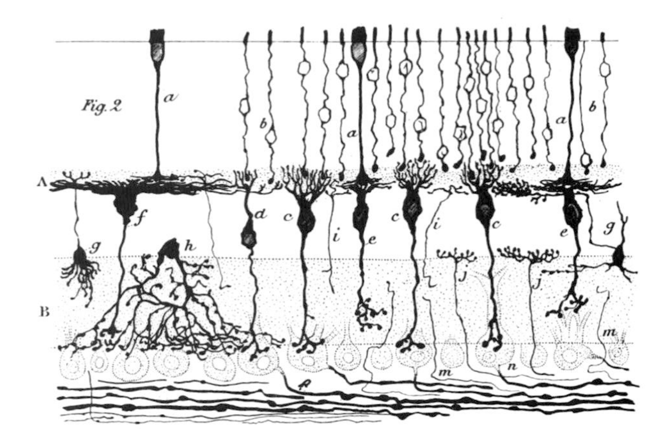

(b) Neurons of the dog's retina

Figure 2
Retinal Neurons
(*a*) Cajal's drawings of ganglion cells in the retina of the green lizard *Lacerta viridis*. Bird and lizard retinas have very thick inner plexiform layers with obvious sublayering. The ganglion cells shown here exhibit several of the possibilities for dendritic branching in the inner plexiform layer—for example, branching in a single sublayer (D, F, and I), branching at two or three well-defined sublayers (B, C, G, and H), or more diffuse arborization (A and E). Cajal was quite aware that these different branching patterns suggested different sources of input and therefore different visual response properties. (*b*) This composite drawing of the dog retina shows cones (a) and rods (b), cone bipolar cells (e), rod bipolar cells (c and d), and "giant" bipolar cells (f). There are one type of amacrine cell (h) and two interplexiform cells (g). The two processes labeled "i" may be from interplexiform cells, but those labeled "j" are apparently centrifugal axons from the brain. Canine retinas have been little studied since Cajal's time, a situation that has only recently changed. (From Ramón y Cajal 1892.)

In the process of showing that neurons are the anatomical elements of the nervous system, Cajal made numerous original observations that are now taken for granted. The list is much too long to repeat, which fails to do justice to his originality, but these are some examples: growth cones on developing axons; the major cell classes in the cerebellum; decussation patterns in the chiasm; and in the retina, rod and cone bipolar cells, interplexiform cells ("cells rarely stained"), and many types of amacrine cells ("spongioblasts") and ganglion cells. All of these observations were repeated in a wide variety of species.

(continued)

As demonstrated by one of his biographers, it is easy to romanticize the melding of art and science in Cajal's work. But there was a relationship. Whatever else he became, Cajal was an artist by inclination and experience; he was extremely observant, he was sensitive to details and nuances, and he trusted his vision to reveal the truth, perhaps because he knew how the eye can be fooled (*trompe l'oeil*). Thus he believed from the start that the neural elements picked out and isolated by Golgi's stains were exactly what they appeared to be—individual nerve cells, discrete and elemental.

Paul Klee noted that art is not simply a faithful rendering of visible objects; its task is to reveal things that might not otherwise be seen (or as stated more bluntly by Picasso, "Art is a lie that makes us realize the truth"). Cajal's drawings fit this conception of art. Never in his long career did he see retinal sections identical to those in Figure 2, for example. He saw bits and pieces at different times and in different places, assembling them in his mind to represent the retina as it *must* be, not as it appeared in a single microscopic section. The practice is now commonplace. But Cajal's assemblages of neurons are not just instructive; the grace and harmony of their composition reveal the artist at work.

In 1924, Cajal received a visit from Stephen Polyak, a young physician and scientist from Croatia. For the next 2 weeks, Cajal and his assistant showed Polyak slides and imparted their methods; Polyak "left Madrid like a Mohammedan after visiting Mecca." The result, which Cajal did not live to see, was Polyak's application of the master's techniques to human and other primate retinas, species not studied in detail by Cajal himself. Polyak's book *The Retina*, published in 1941, is one of the standards against which more recent work is judged and the source of several illustrations in this book.

Life is short, but Art is long.* Ramón y Cajal's art—his revelation of hidden truths—is enduring.

*"Life is short, and Art is long; the occasion fleeting, experience fallacious, and judgement difficult" (Hippocrates of Kos, circa 400 B.C.). Translated by Francis Adams in *The Genuine Works of Hippocrates*. Other translations substitute "Science" for "Art" and "experiment" for "experience."

of accident of retinal development that occurs once in a while? The answer is not known and cannot be known in the absence of a specific label that isolates these cells so that their numbers and distribution can be studied. At present, biplexiform cells are one of the retina's enigmatic curiosities.*

The small region of the world seen by a ganglion cell is its receptive field

Action potentials recorded from a single retinal ganglion cell reveal that visual stimuli influence the cell's behavior only within a very small region of the eye's visual field. Following Charles Sherrington's original use of the term (see Vignette 5.1), the **receptive field** of a visual neuron is that region of space in which stimuli will affect the cell's activity; the influence may be positive (increase in membrane depolarization or increase in action potential frequency) or negative. There will, of course, be a small region of the retina that corresponds in location and angular size to the receptive field in visual space. In general, the size of the receptive field indicates the number of photoreceptors to which the neuron is directly or indirectly coupled (Figure 14.16).

All retinal neurons, including photoreceptors, have receptive fields. Receptive fields of single photoreceptors are tiny, covering only a few minutes of arc in visual angle; their size is a combination of the photoreceptor's field of view plus the extent of horizontal cell feedback to the photoreceptor. Neurons farther along in the vertical pathways have receptive fields whose size reflects the amount of convergence onto them; diffuse bipolar cells have larger receptive fields than midget bipolar cells, for example. And ganglion cell receptive fields in the primate retina range in size from hardly larger than that of one cone, because of the one-to-one wiring, to several degrees in diameter in the peripheral retina, where there is considerable convergence onto the ganglion cells.

A now traditional way to characterize a ganglion cell's receptive field is with a receptive field map. A map is generated by recording the cell's responses to small spots of light flashed ON and OFF at different locations in the visual field;

*Preliminary results from work in progress by Boycott, Dacey, and Wässle suggest that biplexiform cells are really H2 horizontal cells whose cell bodies have been displaced into the ganglion cell layer, probably as a developmental mistake. (H. Wässle, personal communication.)

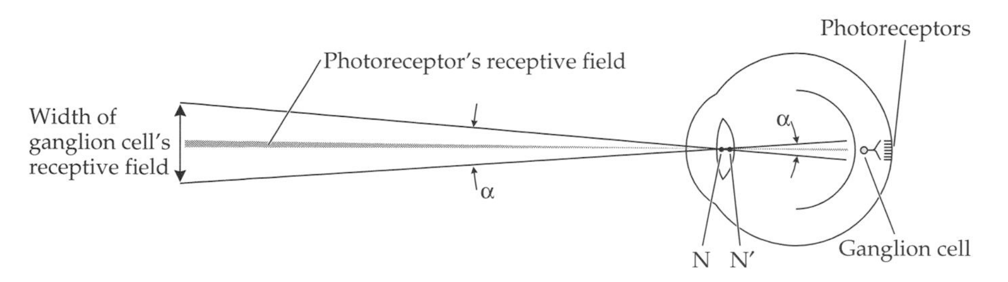

Figure 14.16
A Receptive Field
The receptive field of a visual neuron is the region of space within which stimuli will affect the neuron's behavior. The angular size of the receptive field at the anterior nodal point of the eye (N) corresponds to an angle subtended on the retina at the posterior nodal point (N′). The retinal region includes all the photoreceptors whose responses impinge on the neuron—a ganglion cell in this case—through the intervening bipolar and amacrine cells. Small receptive fields contain few photoreceptors.

the map is a record of what the cell did when the spot was flashed at a given location. By convention, "+" indicates that the cell's activity increased when the spot was turned ON, "–" indicates an increase in activity when the spot was turned OFF, and "0" indicates no detectable change in activity (there can also be combinations, such as "±," or indications of decreased activity, such as "⊕" for a decrease at light ON).

The maps in Figure 14.17 are from ganglion cells in cat retina. They are "concentric" receptive fields, so called because there is a central region in which the cell responded to the light spot turned ON or OFF at different locations and a surrounding annular region (the surround) in which the response was of the opposite polarity. The two possibilities, as shown here, are ON center with an OFF surround and OFF center with an ON surround. These maps make the important point that receptive fields need not be spatially homogeneous; the effect on the cell's behavior can depend on the location and geometry of the stimulus. (But receptive field geometry does not always predict important features of a cell's responses. Sensitivity to movement velocity or movement direction is not revealed by the map of responses to flashed stimuli.) More generally, the size and location of a cell's receptive field should be known even if the field is not mapped; failure to respect this principle has produced some egregious errors.

Dependence on stimulus geometry is shown by a comparison of the response to a spot of light in the center of the receptive field with the response to an annulus of light in the surround. Both are effective stimuli, but the polarity of the response has shifted completely between center and surround stimulation.

The center and surround regions of the concentric receptive fields in Figure 14.17 are mutually antagonistic. As the lowest panels in Figure 14.17*b* show, a stimulus large enough to cover the entire receptive field evoked center and surround responses, but these reponses are smaller than those elicited by center or surround stimuli alone. This is true for both the ON center and the OFF center cells. Thus, the effect of a light in the center of the ON receptive field is ON excitation *and* OFF inhibition; in the surround, the effect is ON inhibition *and* OFF excitation. (These are reversed in the OFF center receptive field.) The result of this organization is a dependence of the response magnitude and timing (ON, OFF, or both) on the stimulus geometry; the maximum ON response, for example, will be elicited by a spot of light that fills the center of the receptive field without encroaching on the surround.

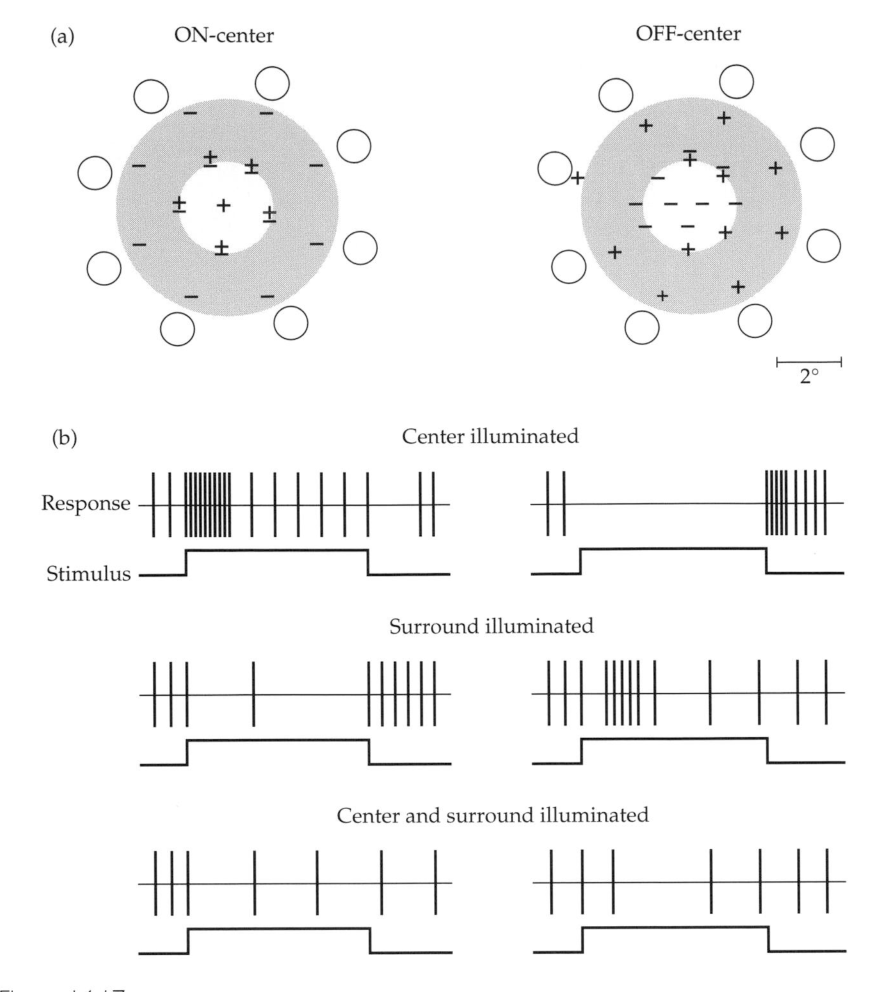

Figure 14.17
Concentrically Organized Receptive Fields and Responses to Light
(*a*) These two receptive field maps are from cat retinal ganglion cells; the dimension scale is in degrees of visual angle. (*b*) These records show how the cells responded to light falling in the receptive field center, surround (shaded), or both. Simultaneous illumination of center and surround produces a relatively small response because of the center–surround antagonism. The "responses" are not genuine action potential recordings.

The concentric organization of excitation and inhibition makes ganglion cells sensitive to contrast rather than to average light intensity

The decreasing responses as stimuli increase in size are similar to the situation discussed earlier for photoreceptors, and the interpretation is the same: Ganglion cells with concentric receptive field maps respond best when there is a difference in light intensity between center and surround. They are signaling *contrast*. Signals from a cell with an ON center receptive field will encode positive contrast (light increasing in the center), and signals from OFF center cells encode negative contrast (light decreasing in the center).

Responses from the receptive field centers are due to direct inputs from ON or OFF bipolar cells, but the origin of the antagonistic or inhibitory surrounds is less clear. A crucial issue is whether surrounds are generated by horizontal cell interactions that become a feature of the bipolar cell signals or whether they are a result of amacrine cell interactions in the inner plexiform layer. Opinion has swung between these possibilities for some time because there is no definitive evidence to eliminate one or the other. The weight of circumstantial evidence and the relative ease with which the features of the ganglion cell responses can be explained favor amacrine cells, however.

Antagonistic inputs for an ON-center cell might come from neighboring ON bipolar cells whose signals are inverted through amacrine cells (Figure 14.18). A small stimulus activating only the central bipolar cell would not affect neighboring bipolar cells, and the result would be an ON response from the ganglion cell. A larger stimulus would produce a smaller ON signal to the ganglion cell because of amacrine cell–mediated inhibition from surrounding bipolar cells at light ON. If only the surrounding bipolar cells were active, the effect would be inhibition at light ON, followed by a transient OFF response produced by a transient component in the bipolar cells' responses. Note that this scheme has all the interactions taking place at the same level within the inner plexiform layer—the level where the ganglion cell dendrites and the ON bipolar terminals are in contact.

A comparable arrangement works for OFF center ganglion cells. The center is formed by direct OFF bipolar cell input that is opposed through inputs from surrounding OFF bipolar cells after sign inversion mediated by amacrine cells.

Although one can imagine other workable arrangements using signals from ON and OFF bipolar cells in combination, the available evidence, though limited, favors the simple circuits in Figure 14.18. Specifically, it is possible to block the activity of ON bipolar cells with 2-amino-4-phosphonobutyric acid (APB), an agent that binds to the metabotropic glutamate receptors of ON bipolar cells. ON center ganglion cells are unresponsive to any stimuli affecting either the center or the surround of the receptive field in the presence of APB, but OFF center ganglion cells respond normally. These results indicate that all signals impinging on the ON center ganglion cells, directly or indirectly, must be from ON bipolar cells and all inputs to the OFF center ganglion cells must be mediated by OFF bipolar cell pathways.*

Other details could be added to these wiring diagrams. In particular, the strength, time course, and spatial extent of the amacrine cell interactions are sure to be important variables. The effects of amacrine cell inputs to the bipolar cell terminals will differ from amacrine cell inputs to the ganglion cell dendrites, which are used here. Even so, the circuits are plausible, and they are a useful reference for discussing what ganglion cells do as a consequence of the inputs they receive.

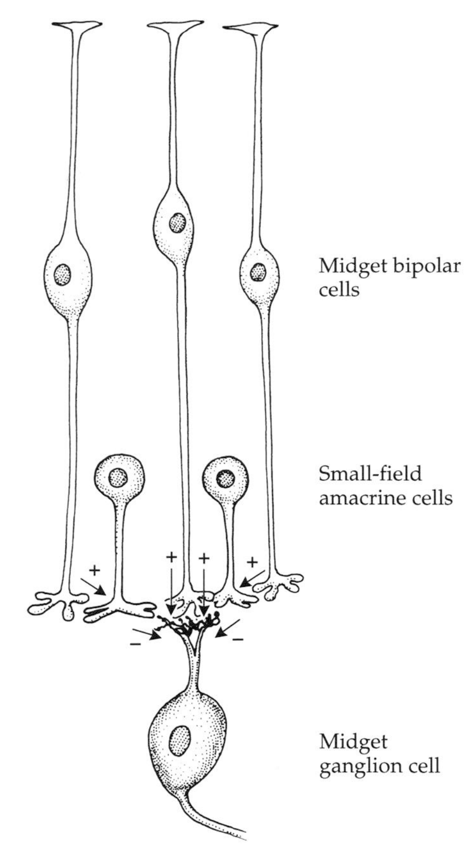

Figure 14.18
A Hypothetical Circuit for Concentric Receptive Field Organization
The sign, ON or OFF, of a ganglion cell's receptive field center is determined by its direct bipolar cell input. The opposing surround is shown here as generated by amacrine cell inputs that have inverted neighboring bipolar cell signals of the same type as the center bipolar cell. In this scheme, horizontal cell interactions in the outer plexiform layer make only a minor contribution to the ganglion cell's receptive field surround.

Ganglion cell receptive fields can be thought of as filters that modify the retinal image

The overall effect of the inputs to the ganglion cells can be summarized by a description of the direct and indirect bipolar cell inputs according to their relative strengths at various points within the ganglion cell's receptive field. As Figure 14.19*a* shows, direct bipolar cell input is modeled as a positive function that peaks in the center of the receptive field and declines smoothly to zero on either side of the peak. Indirect bipolar cell input is represented as a negative-going function whose peak is relatively small but is much wider, forming the surround component of the receptive field. Both curves are Gaussian functions—the same functions used to describe normal probability distributions—and when they are combined, the result is a **difference of Gaussians**, in which the central

*The same results have been used to argue that ganglion cell surrounds are formed by horizontal cell interactions that produce inhibitory surrounds for the bipolar cells (see Schiller 1992). In this case, no amacrine cell connections are needed, because both center and surround in an ON center ganglion cell would be the result of direct ON bipolar cell input. But invoking horizontal cells to generate ganglion cell surround inhibition makes it difficult to explain surround excitation (at light OFF in an ON center receptive field, for example) unless primate bipolar cells have some special response properties that we don't know about.

[Box 14.1] Studying Individual Neurons

NEURONS ARE THE functional units of the retina and of the nervous system generally, and while they may be connected in large networks, their structure and operation as individual elements are important to an understanding of the retina or the brain. The anatomy of single neurons is accessible through various staining methods, such as Golgi stains or immunohistochemical labeling (see Box 15.1), and microelectrodes can be used to record the electrical activity of single cells, but these methods are limited because they cannot directly link a cell's morphology to its physiological properties. Thus much contemporary study of retinal neurons combines anatomical and physiological methods.

Here is one of the simpler methods. To begin, the investigator opens an enucleated eye around the equator, removes the vitreous from the eyecup, and separates the retina and choroid from the sclera after immersing the eyecup in an appropriate physiological solution. This dissection is generally done under dim red illumination to avoid bleaching the photopigments. After the retina has been isolated, it is placed on a microscope stage, after being flattened by having some radial incisions cut into it or by being everted over a small dome. When details like proper physiological bathing solutions and oxygenation of the solution are attended to, isolated retinas can remain viable for 8 hours or so.

Mammalian retinal neurons are small targets, and the chances of blindly sticking an electrode into one are also small. For improved visualization of the cells, a fluorescent dye added to the bathing solution makes cell bodies glow under ultraviolet illumination. A neuron's responses to visual stimuli can be monitored if changes in the membrane potential (voltage) are recorded with a fine glass pipette inserted into the cell. By observation through the microscope, a micropipette can by inserted into a cell body with delicate micromanipulators. The micropipette is filled with a salt solution that conducts electric current, to permit the electrical recordings, and a dye added to the electrolyte can be injected into the cell through the pipette tip to reveal the morphology of the cell after histological processing. The figures in this chapter showing ganglion cell morphology in the human retina are results of this technique, as are the cells in Figure 1 here.

The dyes most commonly injected into cells are either fluorescent dyes, such as Procion Yellow or Lucifer Yellow, or those that can be made to form a dark reaction product, such as horseradish peroxidase or Neurobiotin. The dye molecules must be small enough to pass easily from the very small aperture in the tip of the micropipette, must diffuse or be transported throughout the cell, but must not be able to pass from one cell to the next (sometimes, however, it is desirable to have a dye that will cross the gap junctions between cells). An interesting variant of this technique is double labeling, in which an intracellular dye is combined with an immunohistochemical label; it is possible, for example, to demonstrate that neurons having a particular morphology contain a particular molecular species, such as a neurotransmitter or a neuromodulator.

The same sort of micropipettes used for dye injection can be used to make intracellular recordings of a cell's electrical responses to stimuli, but ganglion cells, which generate large action potentials, are easily studied with extracellular electrodes; that's why we know a great deal more about ganglion cell responses than about responses of other retinal neurons. The problem with most mammalian retinal neurons is their lack of action potentials and their small size; intracellular recordings are required, but even micropipette diameters of 1 μm are large relative to some of the cell bodies. Consequently, most of our information about response properties of horizontal, bipolar, and amacrine cells comes from nonmammalian retinas, whose neurons are typically large.

(a)

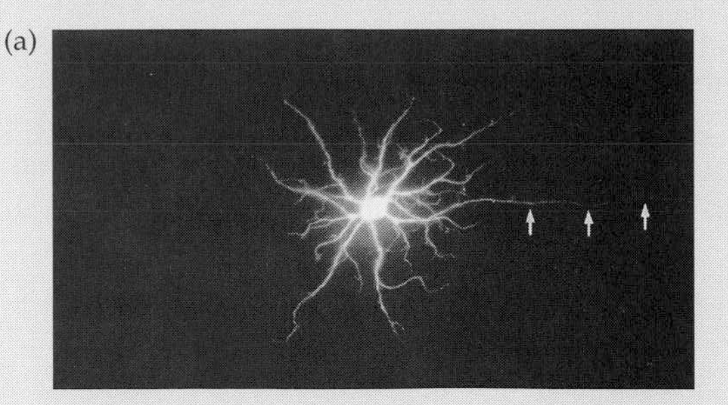

(b)

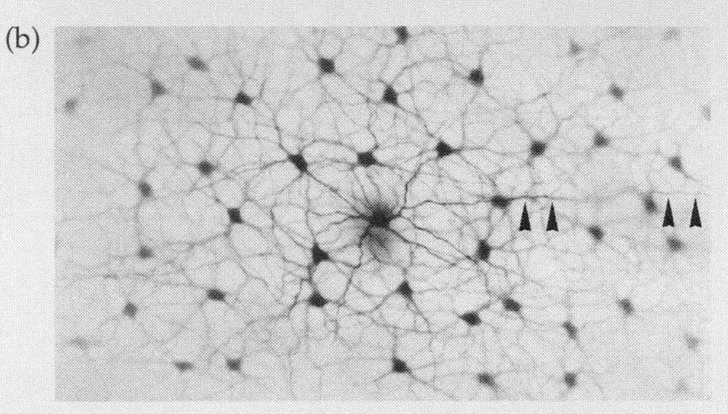

(c)

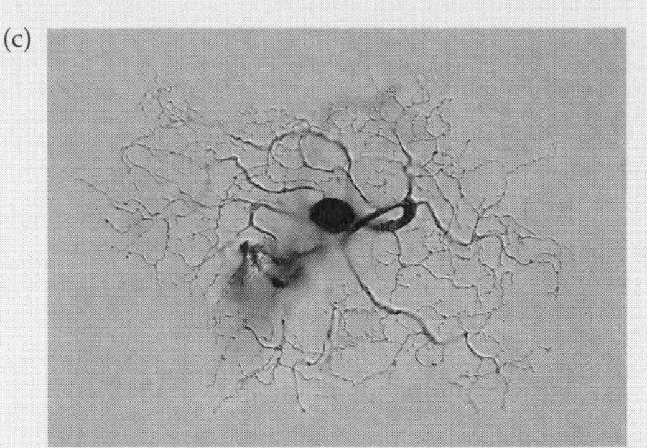

Figure 1
Neurons Stained by Intracellular Dye Injection
(*a*) Injection of the fluorescent dye Procion Yellow into a rabbit horizontal cell shows the portion of the cell around the cell body with the axon (arrows) running to the axon terminal ramification (not shown). (*b*) Injection of a single horizontal cell with Neurobiotin, which will pass through gap junctions, labels a cluster of horizontal cells. (*c*) The complex dendritic arbor of a midget ganglion cell in the human peripheral retina is revealed by an intracellular injection of horseradish peroxidase. (*a* and *b* from Vaney 1991; *c* from Dacey 1993.)

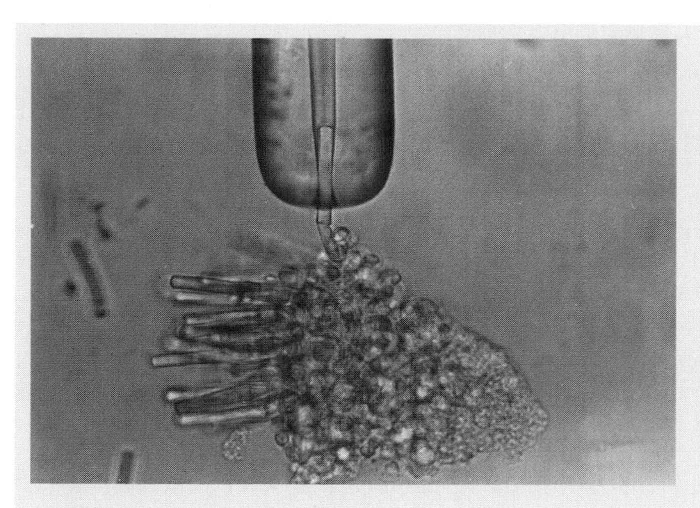

Figure 2
Recording from Individual Photoreceptors with a Suction Electrode
A small piece of toad retina is shown here with numerous photoreceptors jutting out. The outer segment of one rod has been drawn by gentle suction into the orifice of a pipette that surrounds the outer segment; a stimulating light can be shown directly on the rod outer segment. The rod is about 2.5 μm in diameter. (Courtesy of Denis Baylor.)

The limitations of intracellular recording are particularly vexing at the photoreceptor level, where the cell responses are the initial signals in the chain of visual events and are therefore of the utmost importance. The trick here is not to record voltage by inserting a small electrode into the cell but to record current flow with a large electrode *around* the cell. Figure 2 shows photoreceptors in the toad retina with their outer segments protruding (the pigment epithelium was removed and the retina folded; the photoreceptors are along the edge of the fold). One outer segment has been coaxed by gentle suction into the opening at the tip of a fluid-filled glass pipette. Because the pipette surrounds the photoreceptor rather than impaling it, the pipette can be relatively large and it does no damage to the photoreceptor. Better yet, when the pipette is coupled to the appropriate electronic circuitry, the recorded signal corresponds to almost all of the current flowing into the outer segment. This feature makes the technique extremely sensitive; the responses to single photon absorptions shown in Figure 13.20 were recorded with this technique, and it has produced most of the information we have about responses of primate photoreceptors.

Figure 3 illustrates a significant refinement of the suction electrode method. Here the pipette diameter is extremely small, but instead of being used to impale the cell, the pipette is placed in contact with the cell membrane and gentle suction is applied to pull the membrane into tight contact with the pipette tip. Pulling the pipette away from the cell removes the small patch of membrane, and one can record current flow across this tiny bit of isolated membrane. This technique is patch-clamp recording; "patch" refers to the small piece of membrane, and "clamp" denotes the electronics that keeps the potential across the membrane constant (since voltage = current × resistance, keeping the voltage constant means that any change in current flow must be associated with a change in membrane resistance).

Patch-clamp recording is often used to study the characteristics of membrane channels, like the cGMP-gated channels in photoreceptors, since the opening and closing of channels is associated with changes in membrane resistance. The method is so sensitive that resistance changes associated with the opening or closing of a single channel can be recorded. And as indicated in Figure 3, there are several patch-clamp configurations. The flow of current across the patch can be monitored with the patch still part of the cell membrane, with the patch isolated to expose either the inner or outer surface, or with the patch removed to record directly from the interior of the cell. This last configuration, called a whole-cell patch-clamp, has become a useful alternative to intracellular recording with a pipette inserted into a cell; it may make the small neurons of the primate retina accessible in a way they have not been before.

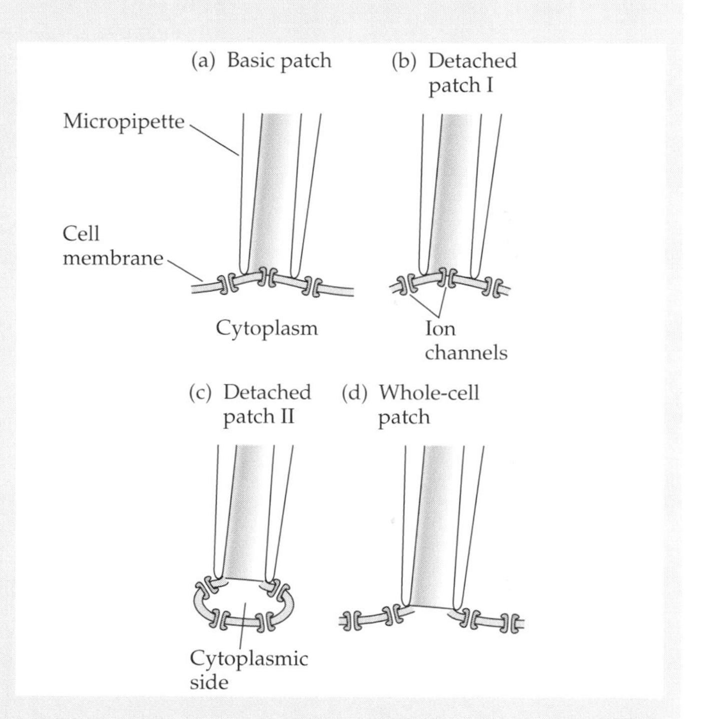

Figure 3
Patch Clamping
When a very fine glass pipette is pressed against a cell's membrane and gentle suction is applied through the pipette, a very tight seal forms, such that the pipette records only the current flowing through the "patch" of membrane to which it is applied. This can be done with the cell membrane intact (*a*), with the patch detached so that the inner surface of the membrane is exposed (*b*), or with the patch detached so that the outer face of the patch is exposed (*c*). In (*d*), the cell membrane is intact except for the small patch at the tip of the pipette; this configuration is equivalent to intracellular recording, but it damages the cell much less. The electronic systems coupled to the fluid inside the pipette hold the voltage across the membrane constant (the "clamp") so that current flow due to changes in membrane resistance by the opening or closing of ion channels can be recorded.

peak is flanked by troughs or depressions. Revolving this curve around its vertical axis produces a three-dimensional shape somewhat like a sombrero (see Figure 14.19*b*) that can be thought of as the ganglion cell's sensitivity function; the sensitivity function shows how well the ganglion cell would respond to small spots of light flashed at various points in its receptive field when both excitatory and inhibitory effects are taken into account.

The sensitivity functions for ON and OFF ganglion cells will be the same as long as the vertical axis is interpreted as sensitivity to increasing light intensity for ON cells and as sensitivity to decreasing light intensity for OFF cells.

Ganglion cells having this kind of sensitivity function would signal strongly the presence of small spots of light (or dark) coinciding with the peak of the sensitivity function and would signal less strongly any larger spots that fell into the depression around the central peak. But the more interesting issue is what the ganglion cells will do when confronted with natural stimuli—that is, with subtler variations of light intensity in the retinal image; except for fireflies, the natural world has relatively few instances of small, isolated light spots flashing on and off.

Because these sensitivity functions can be expressed as equations, it is possible to predict how ganglion cells will respond when various patterns of light and dark fall within their receptive fields. (Assuming that certain linearity conditions are met, the mathematical process is convolution of the sensitivity function with an equation describing the light stimulus. In some respects, the difference-of-Gaussians function can be used like the optical point spread function [see Chapter 2] to predict the system's response to any stimulus.) The sensitivity function can be thought of as a kind of filter through which the ganglion cell sees the world, and a picture viewed through sensitivity function filters should show us how ganglion cell signals relate to the original image. Several examples are shown in Figure 14.20.

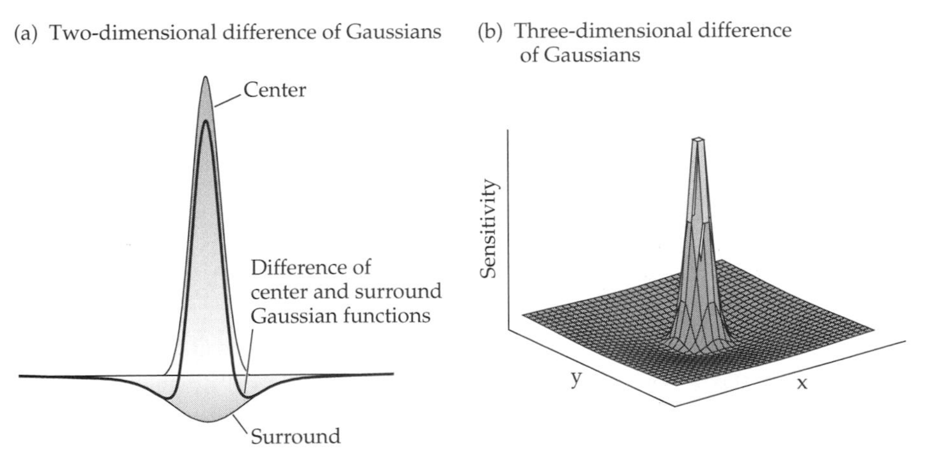

Figure 14.19
Sensitivity Functions for Concentrically Organized Receptive Fields: The Difference-of-Gaussians Model
(*a*) The sensitivity profiles of the center and surround mechanisms are conceived as Gaussian functions with different amplitudes and widths. Center–surround antagonism is indicated by inversion of the surround's sensitivity function. Adding the *inverted* surround function to the center function is the same as taking the difference; hence the resultant difference-of-Gaussians curve that describes the steady-state properties of the receptive field. (*b*) In three dimensions, the central peak is surrounded by a trough; the function has been described as a "Mexican hat." (*a* from Rodieck 1998; *b* from Wandell 1995.)

The pictures in Figure 14.20*a* are of a vine growing on a chain-link fence, a plant in front of a stuccoed wall, and the back of a leaf. Their filtered images (Figure 14.20*b*) are all somewhat blurred, which means that some of the fine details in the original pictures were smaller than the diameter of the filters. (There were more than 100,000 filters in the array, which is a lot, but not so many that each filter is a tiny point. To be realistic representations of ganglion cell receptive fields, the filters must have a finite width.)

It is interesting that the filtered images are recognizable. But more interesting still are features in the filtered images that were not present in the original pictures. The backgrounds are gray, for example, regardless of whether the original backgrounds were bright or dark; the filters have effectively factored out the average illumination level to better emphasize change relative to the average. And the changes in illumination have been exaggerated, a feature that is particularly obvious in the image of the chain-link fence; the dark fence wires are flanked by strips of white not present in the original. This is a striking example of contrast enhancement, which also works well to emphasize the veins and texture of the leaf in the bottom photograph.

To make a statement about what ganglion cells are doing, we need only to imagine that the shades of gray in the filtered images are levels of action potential frequency sent down the ganglion cell axons. Thus ganglion cells are in the business of signaling contrast—the variations of light intensity relative to the average intensity—and in the process, they are stretching the truth, exaggerating the true contrast by making it look larger than it really is. Some of this contrast enhancement begins with horizontal cell feedback in the outer plexiform layer, but is probably augmented considerably by interactions mediated through amacrine cells.

Figure 14.20
Emphasizing Variations in Image Intensity
The original images (*a*) were transformed by filters (*b*) with characteristics similar to the sensitivity functions of ganglion cell receptive fields (difference of Gaussians). (From Marr and Hildreth 1980.)

Sensitivity functions of ganglion cell receptive fields differ in size and in the strength of their inhibitory components

Dendritic fields of ganglion cells at any particular location on the retina differ in size, so the areas over which they receive direct bipolar cell inputs also differ in size. Translated to receptive fields, this statement is equivalent to saying that sizes of receptive field centers differ among ganglion cells. Receptive field center size is related to the sharpness of the peak in the receptive field's sensitivity function: Small receptive field centers are associated with narrow, sharply peaked sensitivity functions, while larger centers have sensitivity functions that are broader and flatter. Thus the size of the receptive field center in Figure 14.20 is related to the degree of blurring; larger centers are associated with more blur.

Sizes of dendritic fields for midget and parasol ganglion cells are plotted in Figure 14.21. Where they first appear along the edge of the fovea, dendritic fields of midget ganglion cells are about 10 μm or less in diameter (about 2′ of visual angle), while the parasol cells are three to four times larger. Dendritic field sizes for both types of cells increase with increasing distance (eccentricity) from the fovea, and the difference between them decreases somewhat. In peripheral retina, midget ganglion cells are about 100 μm in diameter (20′ of visual angle) and parasol cells are about 300 μm in diameter (1° of visual angle).

Since the diameter of a ganglion cell's dendritic field is closely related to the size of its receptive field center—that is, to the width of the peak in the sensitivity function—every location in the retina is being sampled at two different scales. Midget ganglion cells will have narrow, sharply peaked sensitivity functions, while those of nearby parasol cells will be considerably broader. The midget ganglion cells will provide the most detailed sampling of the retinal image, to which parasol cells will add information about intensity variations occurring over larger distances. Other types of ganglion cells probably differ in

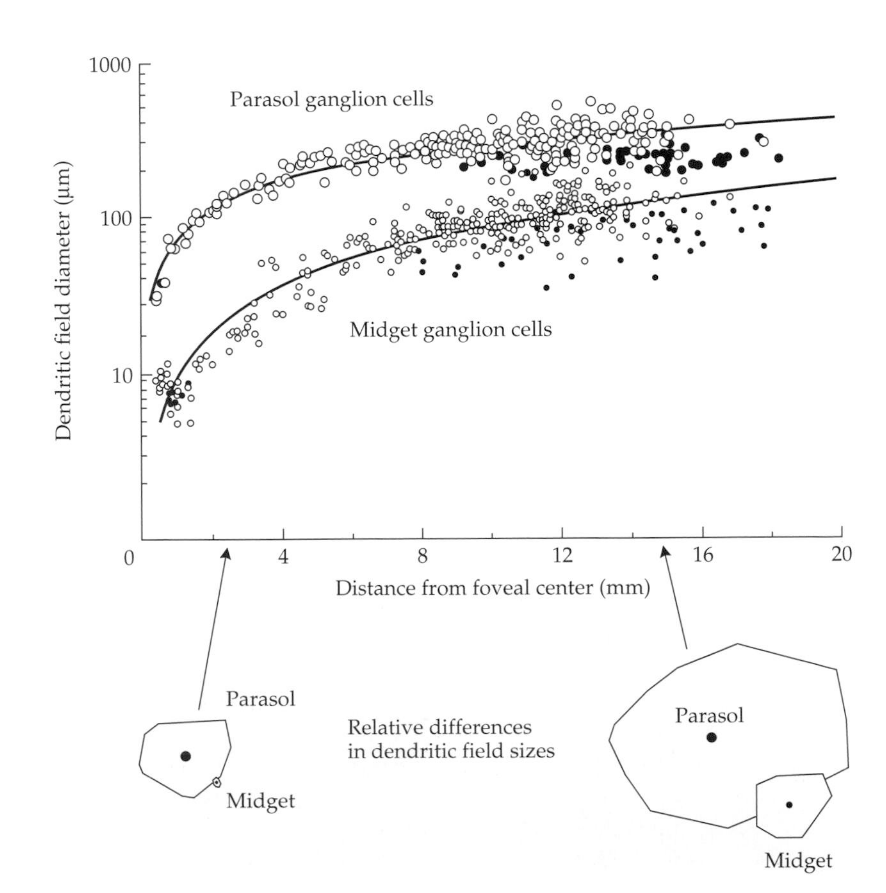

Figure 14.21
Sizes of Midget and Parasol Ganglion Cell Dendritic Fields as Functions of Retinal Eccentricity
These data are from human retinal ganglion cells that were injected with horseradish peroxidase. The logarithmic scale for dendritic field diameter compresses the size range, but relative sizes at two locations are shown below the graph. (Replotted from Dacey and Petersen 1992.)

size from both the midget and the parasol ganglion cells, thereby providing more sampling at different scales. (Small bistratified cells are somewhat larger than parasol cells, but data on other cell types are not available.)

Dendritic fields of the OFF midget ganglion cells are consistently smaller, about three-fourths the diameter of the ON midget cell dendritic fields. This difference implies that the most detailed sampling is done by the OFF midget ganglion cells.

Since the receptive field center corresponds to the dendritic field, the total receptive field is larger than the dendritic field because of the antagonistic surround. So another variable to be considered is the size of the antagonistic surround relative to the size of the receptive field center. Harking back to the neural circuit diagram in Figure 14.18, we would expect the size of the surround to be associated with the distance over which the amacrine cells are bringing in signals from surrounding bipolar cells. The depth and extent of the trough around the peak of the sensitivity function will depend on the strength and extent of the amacrine cell interactions. Contrast enhancement should be most pronounced when there is strong surround antagonism and a deep trough in the sensitivity function.

We know very little about the relationships between centers and surrounds in primate ganglion cells; the dendritic and receptive fields are so small that it is difficult to map subdivisions within them. But stimuli of different sizes produce different responses of different magnitude, as do spot or annular stimuli; these differences are consistent with the kind of center–surround organization described in other species. Even so, estimating the strength or extent of the antagonistic surrounds is difficult, and lack of data demands some guesswork about the nature of the antagonistic surrounds.

The sensitivity functions for midget ganglion cells are probably like those in Figure 14.19. Strong surround antagonism extends over a region about twice the diameter of the receptive field center, producing a sharply peaked sensitivity function with a deep trough around the central peak. In keeping with the increasing size of the midget ganglion cell dendritic fields in peripheral retina, the function will become broader and less sharply peaked in peripheral retina.

Parasol cells probably have weaker but more extensive surround antagonism. The sensitivity functions should therefore be broad, because of the large dendritic field sizes, with less obvious troughs around the central zone. The implication is that contrast enhancement is a less prominent operation in parasol cells than it is in midget ganglion cells. If the filtering shown in Figure 14.20 were done by parasol cell sensitivity functions, the filtered images would have more blur and less exaggeration of intensity variations.

Ganglion cell signals differ in their reports on stimulus duration and on the rate of intensity change

Photoreceptors, horizontal cells, and bipolar cells respond to stimuli with sustained hyperpolarizations or depolarizations. The responses persist at fairly constant levels until the stimulus changes again, and the change of illumination is marked by small transient potentials before or after the sustained phase. Amacrine cells may also have sustained responses, but some are more transient, changing their level of activity only when the stimulus changes. (Some of these transient amacrine cells have action potentials as well.) The difference between sustained and transient responses is most striking at the ganglion cell level, where some cells fire action potentials at an increased rate for many seconds (but not minutes, because of adaptation) and others fire only brief, high-frequency bursts of action potentials when the stimulus changes (Figure 14.22). These extremes in ganglion cell signaling have been observed in most species, and they appear to be present in primate ganglion cells as well.

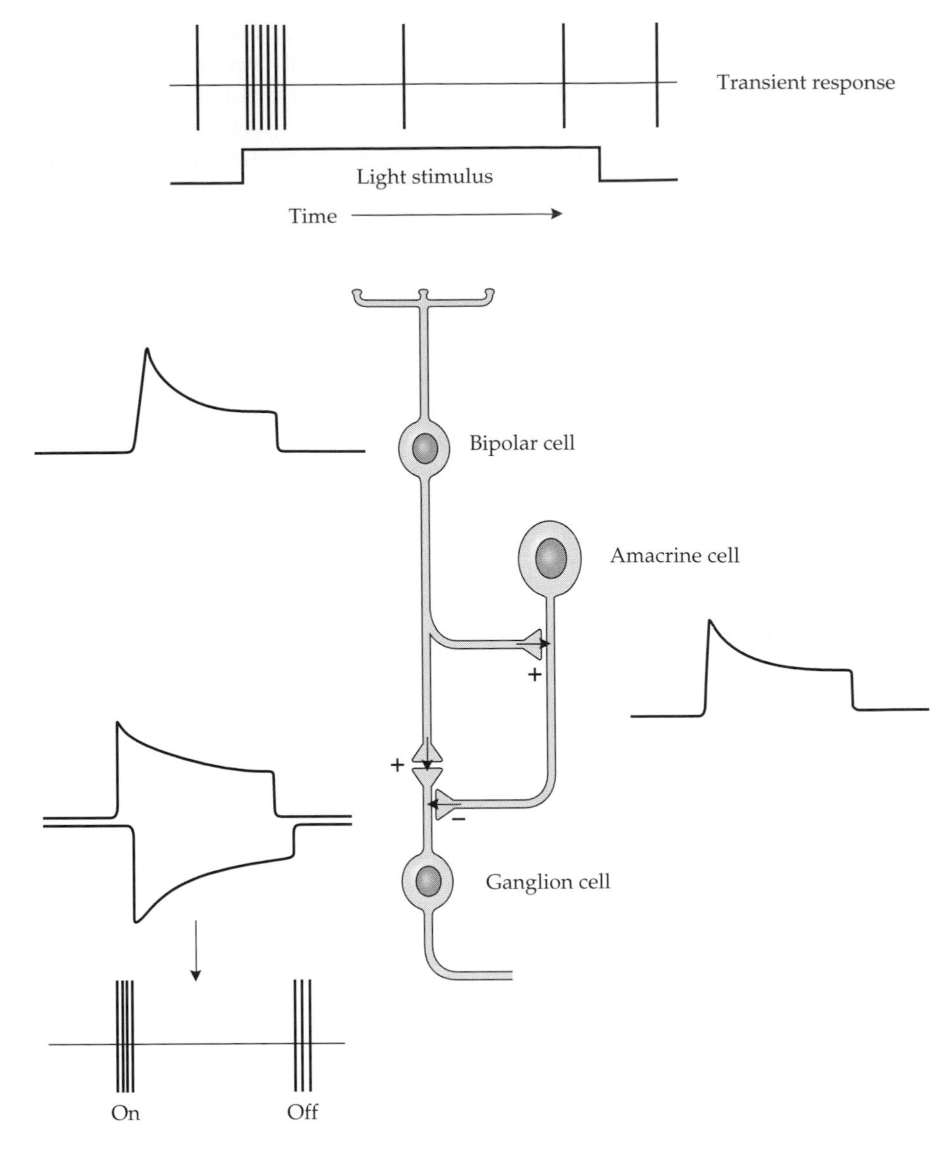

Figure 14.22
Sustained and Transient Responses from Ganglion Cells
Sustained responses persist for the duration of the stimulus; transient responses are brief bursts of activity within a few milliseconds of the light-intensity change (ON, OFF, or both). One circuit that might generate transient responses involves feedforward inhibition mediated by amacrine cells. The ganglion cell responds only during the slight delay between bipolar cell excitation and the amacrine cell inhibition (ON) and perhaps as a rebound from the amacrine cell inhibition (OFF). Other mechanisms are also possible.

Since the vertical pathway neurons have sustained responses down to the level of the bipolar cell terminals, transient responses from ganglion cells must be due to amacrine cell interactions (unless one assumes that some ganglion cell membranes have special properties). This conversion of sustained responses to transient responses is one of the better indicators we have of the kinds of things amacrine cells do.

Exactly how sustained bipolar cell signals are transformed into transient ganglion cell signals is not known, but a simple possibility is illustrated by the circuit in Figure 14.22. Here, an ON bipolar cell's signal is divided between a ganglion cell and an amacrine cell, using sign-conserving synapses in both cases. The amacrine cell, which also has a sustained response, makes a sign-inverting synapse onto the ganglion cell. The delay between the arrival of the direct bipolar cell signal to the ganglion cell and the amacrine cell's inhibitory effect produces a transient excitation of the ganglion cell that is then clipped off by the amacrine cell inhibition. A similar arrangement would work with OFF bipolar cells and amacrine cells to produce a transient OFF signal from a ganglion cell. (It could also be used to generate transient signals in amacrine cells.)

Other possible circuits can be devised, but this example makes the point that amacrine cell interactions are crucial, whatever the details of the circuit may be.

As a result, transiently responding ganglion cells can be expected to receive a substantial portion of their input from amacrine cells. In primate retina, parasol cells have transient response properties, and about 80% of the synapses on their dendrites are from amacrine cells. Midget ganglion cells receive 20% or less of their input from amacrine cells, and their responses reflect the more sustained character of their direct bipolar cell inputs. Thus we have another difference between midget and parasol ganglion cells: They differ not only in size but also in the way they signal intensity change.

It is interesting that this rough dichotomy between sustained and transient cells is consistently observed in vertebrate retinas, but what does it mean in terms of natural visual stimuli? One way to think about cells with sustained responses is that they are signaling the magnitude of an intensity change; the larger the change (brightening for ON cells, dimming for OFF cells), the greater the response. Thus as the retinal image moves back and forth, its variations in intensity will be reflected in varying action potential frequency in the midget ganglion cells, with small frequency variations for small intensity changes and larger variations for larger intensity changes.

Parasol cells with transient responses do something different; they report a stimulus change without much commentary on the size of the change. As the retinal image moves around, little bursts of activity from the parasol cells signal each new variation in light intensity. But these cells are also sensitive to the *rate* at which light intensity changes; abrupt changes at sharp edges or borders will elicit vigorous activity even if the magnitude of the change is relatively small, while a larger but more gradual change in intensity will produce a smaller signal. (In more formal terms, the parasol cell signals are proportional to the time derivative of their contrast sensitivity function.)

Both midget and parasol cells convey information about contrast and its rate of change, but the major component for midget ganglion cells is the magnitude of contrast, while the rate of contrast change dominates the parasol cells' signals.

Midget ganglion cells have wavelength information embedded in their signals, but only small bistratified cells are known to convey specific wavelength information

As discussed earlier, signals from a single cone type are not sufficient for wavelength discrimination. Signals from at least two different cones must be compared, and the ideal comparison is a differencing operation—color opponency—that allows a cell to signal "red and not green," for example. The issue here is the extent to which the differencing of signals from different cone pathways is done in the retina.

Diffuse bipolar cells contact cones indiscriminately, and they cannot be carrying any wavelength-specific information. And since parasol ganglion cells receive their bipolar cell inputs from diffuse bipolar cells, parasol ganglion cells are out of the picture as far as color vision is concerned. Parasol cells are interested in changes in light intensity and largely indifferent to the spectral composition of the light.

At the moment, the small bistratified ganglion cells provide the cleanest example of color opponency (Figure 14.23). These cells respond to increasing blue light (blue-ON) and to decreasing yellow light (yellow-OFF; yellow is the additive combination of red cone and green cone signals). When blue and yellow intensity changes are pitted against one another—that is, when the stimulus alternates between blue and yellow *with no change in mean intensity*—the cells still respond vigorously, indicating that they can distinguish between blue and yellow even when the different wavelengths produced equal photon absorptions. (Changing wavelength without changing the level of photon absorption is a condition called isoluminance.)

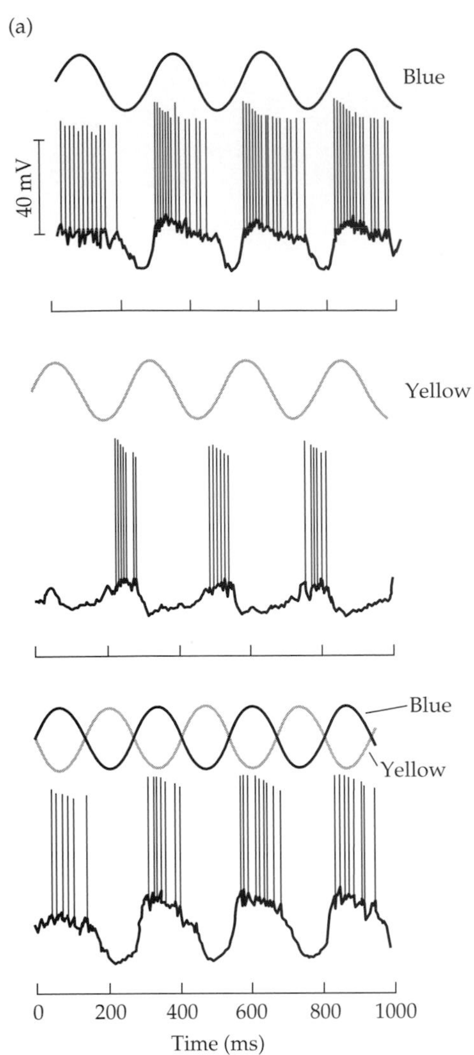

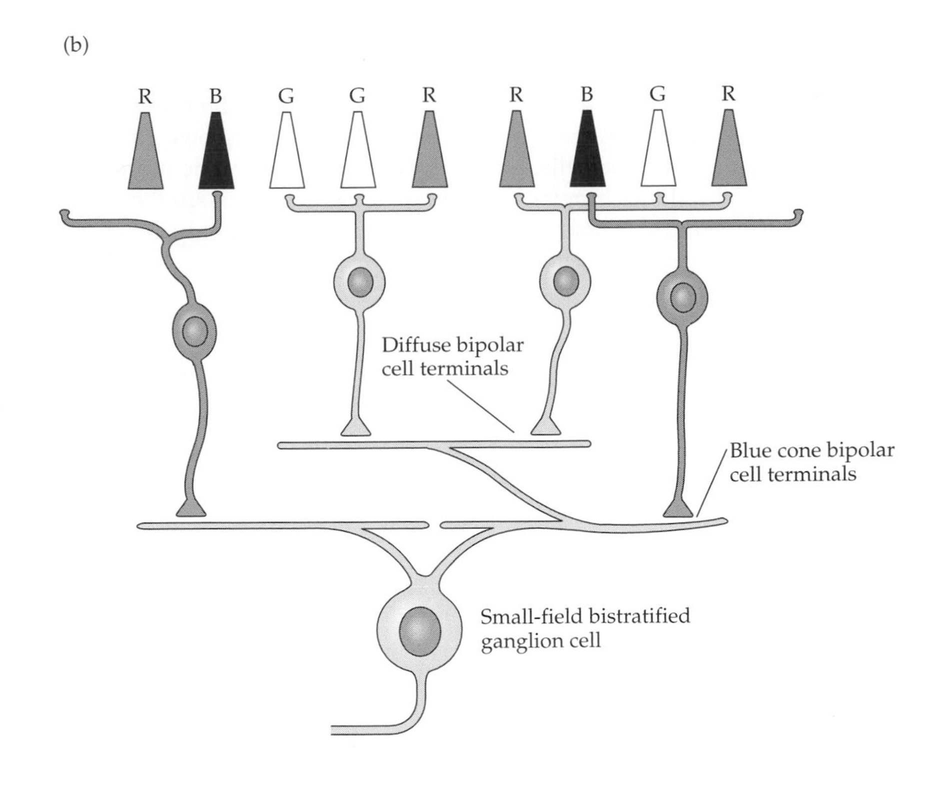

Figure 14.23
Small Bistratified Ganglion Cells Distinguish between Blue and Yellow
(*a*) These action potentials were recorded from a small bistratified cell in response to (1) a blue stimulus alternating from bright to dim (top), (2) an alternating yellow stimulus (middle), and (3) an alternation between blue and yellow (bottom). The cell responds to increasing blue (ON) and decreasing yellow (OFF), and it can distinguish between blue and yellow even when the net photon absorption is constant. (*b*) This drawing shows the probable input circuitry, with diffuse bipolar cell inputs in the OFF region and blue cone bipolar inputs in the ON region. B, blue; G, green; R, red. (*a* after Dacey and Lee 1994; *b* after Dacey 1994.)

The probable circuitry underlying these wavelength-specific responses is quite simple. The blue-ON component is due to input from blue cone bipolar cells to the ganglion cell dendrites in the proximal (ON) part of the inner plexiform layer, while the opposing yellow-OFF inputs are at the distal dendrites. As shown here, the yellow-OFF inputs are from diffuse bipolar cells that receive inputs from both red and green cones. (There are probably some amacrine cell inputs as well, but nothing is known about them.)

Midget ganglion cells in peripheral retina lack color opponency. A peripheral midget ganglion cell responds quite well to intensity modulation of white light and to wavelength modulation that is selective for either red or green cones (Figure 14.24). Except for the more sustained character of the responses, the midget ganglion cell is similar in this respect to parasol cells. And because both red cone and green cone inputs elicit positive responses from the cell, there is no opponency. Although midget bipolar cells in the peripheral retina receive inputs from a single cone type and thus maintain their cone specificity, the midget bipolar cells seem to converge indiscriminately on the peripheral midget ganglion cells. As a result, the red cone and green cone inputs to the ganglion cells are mixed and additive, and the midget ganglion cell cannot discriminate between red cone and green cone signals.

But midget ganglion cells have generally been assumed to be the source of wavelength discrimination between red and green, so the issue of red–green opponency now focuses on the central midget ganglion cells. We know they receive input from a single midget bipolar cell and therefore have a strong excitatory drive from just one cone type; the vertical pathways are clearly specific for cone type. But wavelength discrimination is not possible with a single cone type; another input is required, preferably an opposing one. Where does it come from?

A color opponency signal cannot be mediated through horizontal cells, as we considered earlier. Although there are differences between H1 and H2 cells in their relative amounts of blue cone input, neither of these cell types responds to different wavelengths with signals of different polarity; they cannot, for exam-

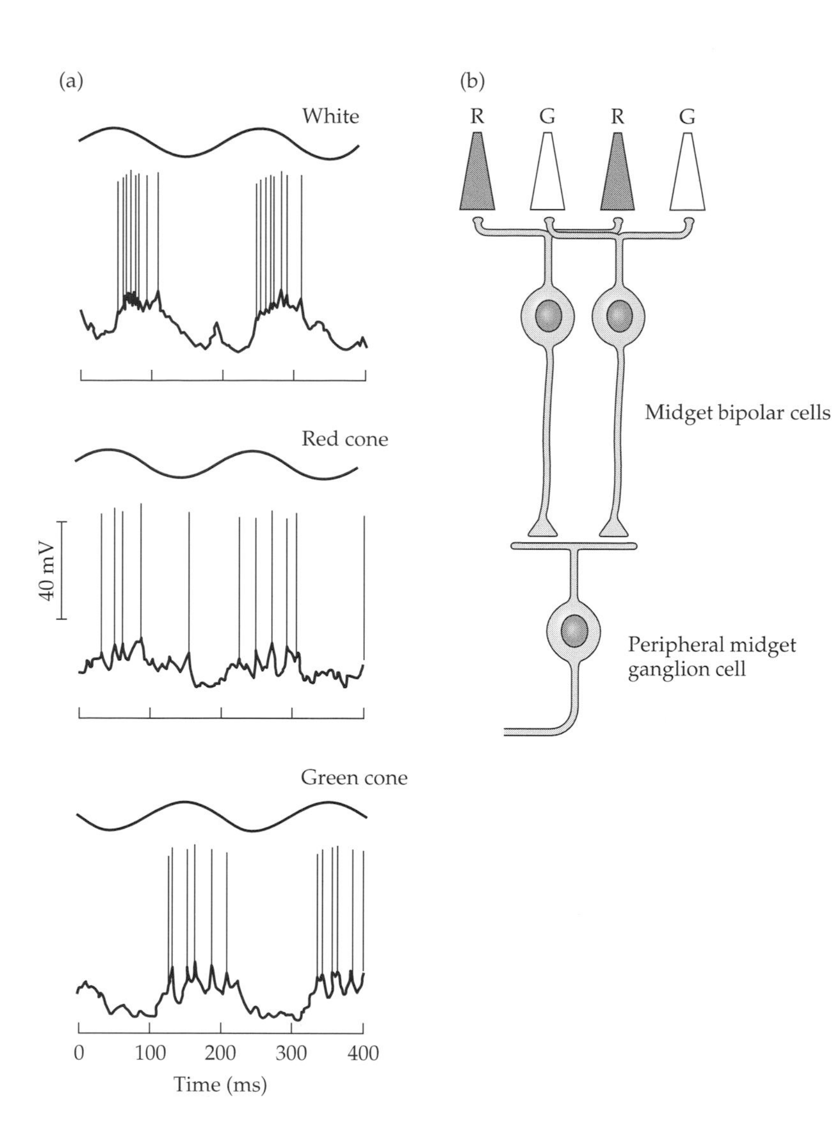

Figure 14.24
Responses of Peripheral Midget Ganglion Cells Add Red and Green
(*a*) These action potentials were recorded from a peripheral midget ganglion cell. The stimuli were (1) varying intensity of white light (top), (2) varying intensity of a red cone isolating stimulus (middle), and (3) varying intensity of a green cone isolating stimulus (bottom). The cell responded to increasing white light (ON), to increasing red cone input (ON), and to increasing green cone input (ON). (*b*) The responses in *a* could be a result of this circuitry, in which the midget ganglion cell receives inputs from both red-ON (R) and green-ON (G) midget bipolar cells. (*a* after Dacey 1996; *b* after Dacey 1994.)

ple, add a green-OFF component to a red-ON pathway. Another possibility for wavelength-opposing signals is lateral connections through amacrine cells. A green-ON bipolar cell signal could be inverted through an amacrine cell, for example, and fed into the red-ON pathway as an opposing green-OFF signal. The problem is finding an amacrine cell to do the job; small-field amacrine cells that contact midget ganglion cells receive inputs from both red and green midget bipolar cells (and a little input from diffuse bipolar cells as well). These amacrine cells would provide an opposing yellow (red + green) signal, as summarized in Figure 14.25.

Because the amacrine cells contacting the midget ganglion cells appear to be carrying additive red and green signals, their contribution to the midget ganglion cell's receptive field will be red + green (yellow) antagonism. Moreover, the combination of red and green inputs to the surround will be the same for all midget ganglion cells. Whatever the center's cone type or polarity (red or green, ON or OFF), the surround will be yellow, with a sign opposite that of the center.

But depending on the relative strengths of the center and surround inputs, this arrangement might still produce some opponency. A green ON center cell, for example, would have an opposing red plus green signal—that is, *minus* red and green. If the subtractive green signal were weaker than the positive green signal from the direct midget bipolar cell input, the net result would be a positive green signal, somewhat reduced, and a negative red. That would be color

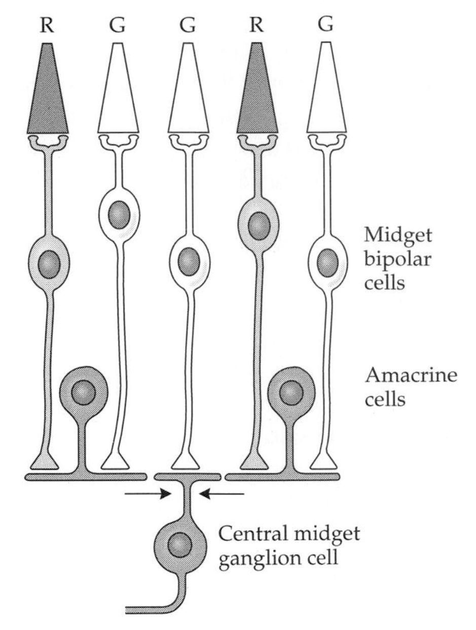

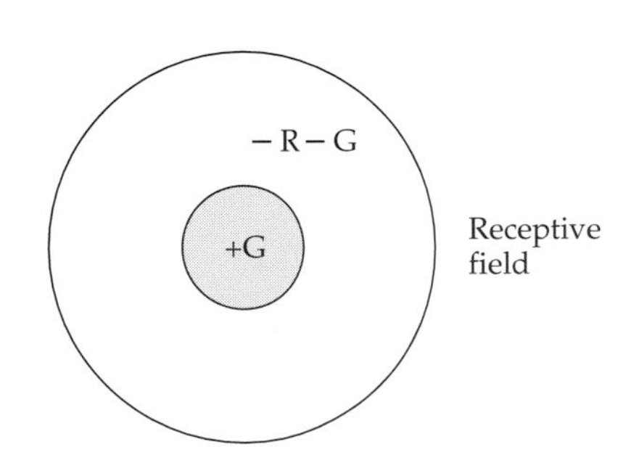

Figure 14.25
A Possible Circuit for Central Midget Ganglion Cells
This midget ganglion cell receives direct input from a green-ON midget bipolar cell, and the ganglion cell would have a green-ON center. The flanking amacrine cells receive inputs from both red-ON and green-ON bipolar cells, so the sign-inverting amacrine cell inputs would be – red and – green. The amacrine cells would provide the same signals to nearby red-ON center ganglion cells. A similar arrangement would be expected in the OFF portion of the inner plexiform layer. (After Calkins and Sterling 1996.)

opponency, although of less than optimal form. Subsequent interactions later in the visual pathway could sharpen the wavelength differencing and would probably be necessary to account for our ability to discriminate wavelength.

Finally, red–green opponent signals may be conveyed by one of the other ganglion cell types that we don't know much about. The small bistratified ganglion cells with their blue-ON/yellow-OFF characteristics were a surprise, and the retina may have more surprises in store.

The strongest conclusions are, first, that the parasol ganglion cells have nothing to do with color vision and, second, that the small bistratified cells have a clear role in discrimination between blues and yellows (but is there a parallel blue-OFF/yellow-ON system?). The midget ganglion cells in central retina may exhibit some color opponency, but we have no direct evidence for it, which is why this section began with the statement that midget ganglion cells have wavelength information *embedded* in their signals. The full potential of their cone specificity is probably elaborated in the central nervous system.

Axons from midget and parasol ganglion cells go to different layers in the lateral geniculate nucleus

The human lateral geniculate nuclei are layered structures, with the layers numbered 1 through 6 from bottom to top (see Chapter 5). Layers 2, 3, and 5 receive inputs from the ipsilateral eye, while layers 1, 4, and 6 receive contralateral inputs. Layers 1 and 2 at the bottom of the nucleus contain relatively large cells and are called the magnocellular layers. The other layers, which make up the bulk of the nucleus, contain smaller cells and are thus called parvocellular layers (Figure 14.26).

Experimental lesions in the monkey lateral geniculate nucleus, which is very similar to that of humans, have different effects depending on the layers included in the lesion. Color vision is unaffected by lesions confined to the large-cell layers (1 and 2), but it is profoundly affected by lesions in the small-cell layers. High-frequency spatial resolution follows the same pattern: Lesions in the large-cell layers don't affect it, but lesions in the small-cell layers do. Recordings from lateral geniculate neurons indicate that parasol cells provide inputs to cells in layers 1 and 2, while midget ganglion cells are the primary source of input to cells in the parvocellular layers.

The clear conclusion from this work is that the information used for wavelength discrimination and detailed spatial vision is confined to the small-cell layers, which have been shown to receive input from midget ganglion cells. Parasol ganglion cells send their axons to the large-cell layers, thus fitting nicely with our understanding of them: They do not carry color information, and their spatial resolution is relatively poor.* Also, the large-cell layers have relatively few cells, in keeping with the relatively low number of parasol cells.

Since lesions in the small-cell layers affect color vision and the midget ganglion cells make up most of the input to the small-cell layers, there is good reason to associate red–green discimination with the midget ganglion cell system, as we did earlier. But the small-field bistratified (blue–yellow) cells also send their axons to the parvocellular layers (see Figure 14.26), and other, as yet undiscovered cells may also carry wavelength-specific information. In other words, the association of the small-cell layers of the lateral geniculate nucleus with color

*Terminology can be a problem here. The association between midget ganglion cells and the parvocellular layers has led to the name "P cells" for the former, while parasol cells are "M cells" because they project to the magnocellular layers. But other ganglion cell types must project to either the magno- or parvocellular layers and this terminology cannot refer to specific types of retinal ganglion cells; it should be avoided. This point is important only because some of the primary literature requires the following translations: P cell = midget ganglion cell, and M cell = parasol ganglion cell.

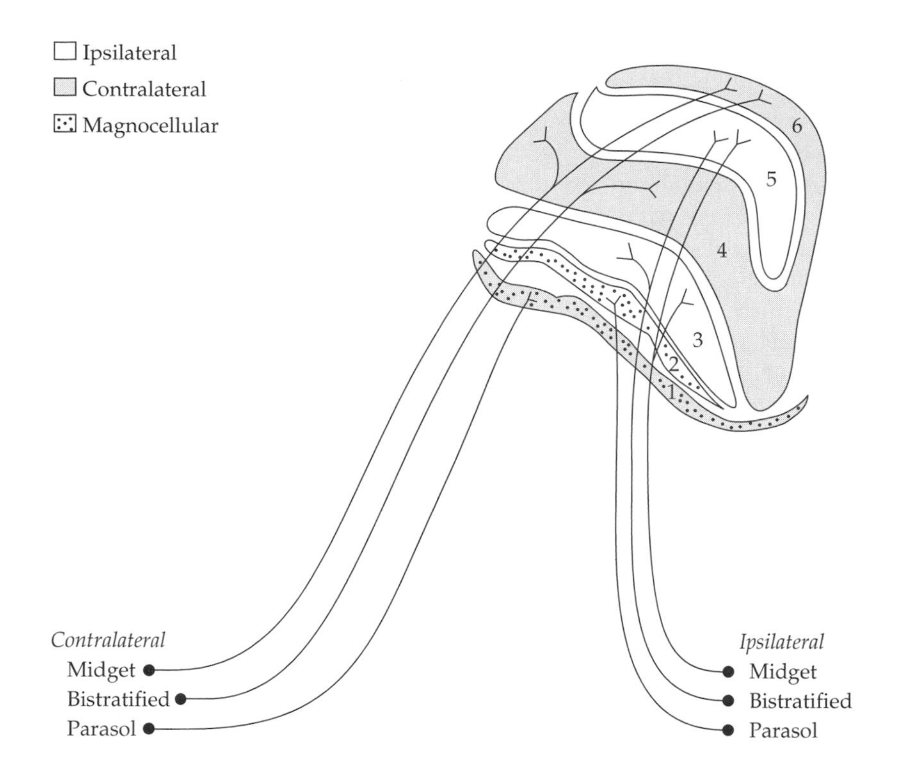

Figure 14.26
Projections of Ganglion Cell Axons to the Lateral Geniculate Nucleus
In the portion of the lateral geniculate nucleus containing six complete layers, layer 6 is uppermost and layer 1 is at the bottom. Layers 2, 3, and 5 receive ipsilateral inputs, and layers 1 and 2 are the layers with large cells (magnocellular layers). Parasol ganglion cells project to the magnocellular layers, while both midget and small bistratified ganglion cells project to the remaining parvocellular layers.

discrimination is sound, but it does not mean that the majority input from midget ganglion cells is the only color-encoding pathway.

The projections of other ganglion cell types about which little is known may be to other targets, such as the superior colliculus or pretectal nuclei (see Chapter 5), or they may contribute to the projections to the lateral geniculate nucleus shown here.

Ganglion cell responses are the elements of retinal sketches

Most stimuli used to map ganglion cell receptive fields and to characterize their response properties are unnatural; the simple geometries (spots or annuli), sharp discontinuities between light and dark (edges), abrupt onset and offset, and sinusoidal grating patterns that are so useful experimentally are rarely encountered in the real world. Thus we need to consider more abstract images consisting of diffuse blobs and subtle gradations of light and shade. What, for example, would be the response of retinal ganglion cells to an image like that of the Henry Moore sculpture in Figure 14.27*a*?

There is no answer to this rhetorical question because the experiment has not been done. But if such an image were presented to the retina and kept constantly in motion to simulate the normal ocular tremor, there are clearly numerous intensity variations in the image to which both midget and parasol ganglion cells would respond. In general, midget ganglion cells should prefer more discrete, localized intensity variations, like those corresponding to changes in surface reflectance or contour, while parasol cells should signal the more extended intensity variations, like those at the physical boundaries of the object.

If we were to record the signals from *all* midget ganglion cells at a particular instant, we could create a map of the activity, placing a dot on the map wherever there was activity in a midget cell. This snapshot of midget ganglion cell activity might look something like Figure 14.27*b*; if the dots were connected by lines, it would be identical (by design) to the fine-grained sketch shown in Figure 13.4*b*. Drawing in connecting lines is not legitimate, however, because individual midget ganglion cells provide no information about the orientation of the intensi-

(a) The object

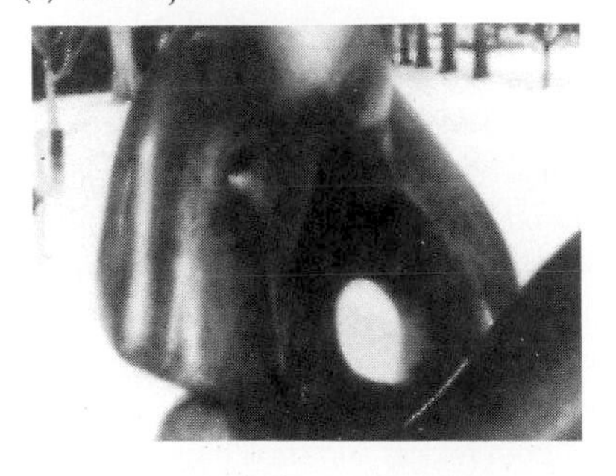

(b) A midget ganglion cell sketch

(c) A parasol ganglion cell sketch

Figure 14.27
Instantaneous Retinal Sketches as Dots of Activity from Midget and Parasol Ganglion Cells
(*a*) This is the same sculpture by Henry Moore that was shown in Figure 13.4. (*b*) Local changes in light intensity in the scene will be detected by midget ganglion cells, whose pattern of activity at a given instant might be like this fine-grained pattern of dots. (*c*) Parasol cells will report changes in intensity at a coarser scale; there will be fewer dots of activity and they will be farther apart. (*a* from Marr and Hildreth 1980.)

ty gradients to which they respond; they respond equally well to all orientations. "Drawing in the lines" requires comparisons among different cells to find active neighbors that should be connected; this is not done at the retinal level, at least not in primates, so the retinal sketches are patterns of dots, not continuous lines (in primates, oriented line detectors first appear in the primary visual cortex).

A similar map of parasol cell activity might look like Figure 14.27*c;* it is meant to mimic the coarse-grained sketch in Figure 13.4*d*. Although this sketch is still a pattern of dots—parasol cells cannot signal the orientation of intensity variations any more than midget ganglion cells can—the dots are larger and farther apart, reflecting the larger size of the parasol cells' receptive fields, and the pattern differs from the midget ganglion cells' activity map, reflecting the parasol cells' preference for less localized intensity changes.

These dot patterns representing the ganglion cells' activity at a given instant will change in the next instant because of image motion occurring for any or all of several reasons: eye movement, observer movement, or change in ambient illumination (if the object were something other than a massive piece of sculpture, object movement would be another possibility). Thus, it is not unreasonable to think of the retina's output as several simultaneous animated cartoons drawn with dots of neural activity; the brain's job is deciphering these continually changing patterns to interpret and represent the continually changing relationships among objects, light, and observer.

Thinking of the retina as a generator of dot patterns may seem both oversimplified and simpleminded, but what I am calling "dots" carry a great deal of freight. In the first place, retinal activity dots specify location. The topographic ordering of retinal ganglion cell projections to the brain (see Chapter 5) means that the spatial relationships within the retinal image are preserved in the ganglion cell signals, so the idea of a mapping, with different ganglion cells corresponding to different spatial locations on the map, is reasonable; the dots show the different locations of the active cells.

A retinal dot corresponds to an event in the visual world; dots are derived from the retinal image. But the set of all possible dot patterns is as large as the set of all possible retinal images; thus the occurrence of a *specific* dot of activity is a relatively rare event, and because it is rare, it is laden with information. (In a technical sense, information content is inversely related to the probability of occurrence. Rare events are more informative and interesting than common ones.) The information concerns the occurrence, size, and rate of change of an intensity gradient at a particular location in the visual world. Although an individual dot neither specifies the direction or orientation of the intensity gradient nor indicates whether the gradient is local or part of an extended contour, all these features can be extracted from the retinal dots by a comparison of dots or dot patterns.

Comparisons within a given dot pattern will indicate the presence and orientation of extended contours, for example; this is "drawing in the lines." Comparisons between ON and OFF dot patterns can establish the directions of gradients. Comparisons of dot patterns at different scales, such as those provided by the midget and parasol ganglion cells, may distinguish between surface contours and object borders. And because of image motion, the dot patterns referring to a particular object will be highly correlated from moment to moment, even though the activity in any particular cell is changing as the image moves. Thus comparisons of the dot *patterns* over time will also be useful.

All of the ganglion cell types discussed here will generate separate dot patterns. The major ones correspond to the ON and OFF midget ganglion cells and the ON and OFF parasol ganglion cells, but there are more. Midget ganglion cells can be subdivided by wavelength sensitivity into red- and green-center cells, for example, and the small bistratified ganglion cells, the blue-ON cells, are another source of dots. And there are other ganglion cells about which little is known. Their presence is a case of "the more the merrier": Adding more cell groups and more dot patterns to the major ones will not alter the basic conception of retinal function; it only enriches the possibilities for specific information outflows from the retina.

All vertebrate retinas studied thus far have ganglion cells that can be put into categories having some of the important features of the midget and parasol cell groups in primates. But in other species, ganglion cells with more complex response properties often dominate. Color opponency, for example, may be much more obvious at the ganglion cell level, and its elaboration may begin with horizontal cell interactions. And in other retinas, such as rabbit retina, ganglion cells signaling direction and velocity of motion in the retinal image constitute almost a third of the ganglion cell population, while cells with concentric receptive field maps and sustained responses to light, like midget ganglion cells, are relatively uncommon.

These differences between the retinas of primates and of other vertebrates do not mean that other species extract information from the retinal image to which we have no access. The differences mean only that information about wavelength or the direction of image motion begins to be extracted at an earlier stage of the visual pathway in some species. The same information is embedded in signals from the midget and parasol cell systems in primate retina, and it can be extracted later by the appropriate comparisons between responses from neighboring ganglion cells.

In a broader sense, the differences among species in the specific information outflows from the retina probably reflect the kinds of visual information that are of immediate importance to a species. Rabbits, for example, are objects of prey; early detection of motion and a quick, stereotyped response to it are critical to their survival. For most primates, who are hunters, gatherers, and manipulators of their environment, speed of response is less important than a wealth of spatially detailed information about the world around them. Thus the primate retina expends its neural resources on detecting and signaling a lot of spatially detailed general information from which the brain can pick and choose at its leisure; the primate retina expends relatively little of its neuronal resources on selecting and transmitting highly specific information.

References and Additional Reading

Interactions among Photoreceptors, Horizontal Cells, and Bipolar Cells

Ahnelt P and Kolb H. 1994. Horizontal cells and cone photoreceptors in human retina: A Golgi-electron microscopic study of spectral connectivity. *J. Comp. Neurol.* 343: 406–427.

Baylor DA, Fuortes MGF, and O'Bryan PM. 1971. Receptive fields of cones in the retina of the turtle. *J. Physiol.* 214: 265–294.

Boycott BB and Wässle H. 1991. Morphological classification of bipolar cells of the primate retina. *Eur. J. Neurosci.* 3: 1069–1088.

Calkins DJ, Tsukamoto Y, and Sterling P. 1996. Foveal cones form basal as well as invaginating junctions with diffuse ON bipolar cells. *Vision Res.* 21: 3373–3381.

Chun M-H, Grünert U, Martin PR, and Wässle H. 1996. The synaptic complex of cones in the fovea and in the periphery of the macaque monkey retina. *Vision Res.* 36: 3383–3395.

Dacey DM, Lee BB, Stafford DK, Pokorny J, and Smith VC. 1996. Horizontal cells of the primate retina: Cone specificity without spectral opponency. *Science* 271: 656–659.

Dacheux RF and Raviola E. 1990. Physiology of HI horizontal cells in the primate retina. *Proc. R. Soc. Lond. B* 239: 213–230.

Djamgoz MBA. 1995. Diversity of GABA receptors in the vertebrate outer retina. *Trends Neurosci.* 18: 118–120.

Hopkins JM and Boycott BB. 1997. The cone synapses of cone bipolar cells of primate retina. *J. Neurocytol.* 26: 313–325.

Kolb H, Mariani A, and Gallego A. 1980. A second type of horizontal cell in the monkey retina. *J. Comp. Neurol.* 189: 31–44.

Kouyama N and Marshak DW. 1992. Bipolar cells specific for blue cones in the macaque retina. *J. Neurosci.* 12: 1233–1252.

Linberg KA and Fisher SK. 1988. Ultrastuctural evidence that horizontal cell axon terminals are presynaptic in the human retina. *J. Comp. Neurol.* 268: 281–297.

Mariani AP. 1983. Giant bistratified bipolar cells in monkey retina. *Anat. Rec.* 206: 215–220.

Röhrenbeck J, Wässle H, and Boycott BB. 1989. Horizontal cells in the monkey retina: Immunocytochemical staining with antibodies against calcium binding proteins. *Eur. J. Neurosci.* 1: 407–420.

Rushton WAH. 1965. Visual adaptation: The Ferrier Lecture, 1962. *Proc. R. Soc. Lond. B* 162: 20–46.

Shiells R and Falk G. 1995. Signal transduction in retinal bipolar cells. *Prog. Ret. Eye Res.* 14: 223–247.

Smith RG. 1995. Simulation of an anatomically defined local circuit: The cone-horizontal cell network in cat retina. *Vis. Neurosci.* 12: 545–561.

Sterling P, Smith RG, Rao R, and Vardi N. 1995. Functional architecture of mammalian outer retina and bipolar cells. Chapter 13, pp. 325–348, in *Neurobiology and Clinical Aspects of the Outer Retina*, Djamgoz MBA, Archer SN, and Vallerga S, eds. Chapman and Hall, London.

Wässle H, Boycott BB, and Röhrenbeck J. 1989. Horizontal cells in the monkey retina: Cone connections and dendritic network. *Eur. J. Neurosci.* 1: 421–435.

Wilson HR. 1997. A neural model of foveal light adaptation and afterimage formation. *Vis. Neurosci.* 14: 403–423.

Interactions among Bipolar Cells, Amacrine Cells, and Ganglion Cells

Calkins DJ, Schein SJ, Tsukamoto Y, and Sterling P. 1994. M and L cones in macaque fovea connect to midget ganglion cells by different numbers of excitatory synapses. *Nature* 371: 70–72.

Calkins DJ and Sterling P. 1996. Absence of spectrally specific lateral inputs to midget ganglion cells in primate retina. *Nature* 381: 613–615.

Dacey DM. 1990. The dopaminergic amacrine cell. *J. Comp. Neurol.* 301: 461–489.

Grünert U and Martin PR. 1991. Rod bipolar cells in the macaque monkey retina: Immunoreactivity and connectivity. *J. Neurosci.* 11: 2742–2758.

Grünert U and Wässle H. 1990. GABA-like immunoreactivity in the macaque monkey retina: A light and electron microscopic study. *J. Comp. Neurol.* 297: 509–524.

He S and Masland RH. 1997. Retinal direction selectivity after targeted laser ablation of starburst amacrine cells. *Nature* 389: 378–382.

Kalloniatis M, Marc RE, and Murry RF. 1996. Amino acid signatures in the primate retina. *J. Neurosci.* 16: 6807–6829.

Kolb H. 1997. Amacrine cells of the mammalian retina: Neurocircuitry and functional roles. *Eye* 11: 904–923.

Kolb H, Linberg KA, and Fisher SK. 1992. Neurons of the human retina: A Golgi study. *J. Comp. Neurol.* 318: 147–187.

Koontz MA and Hendrickson AE. 1990. Distribution of GABA-immunoreactive amacrine cell synapses in the inner plexiform layer of macaque monkey retina. *Visual Neurosci.* 5: 17–28.

Marc RE, Murry RF, and Basinger SF. 1995. Pattern recognition of amino acid signatures in retinal neurons. *J. Neurosci.* 15: 5106–5129.

Masland RH. 1996. Processing and encoding of visual information in the retina. *Curr. Opin. Neurobiol.* 6: 467–474.

Masland RH, Rizzo JF, and Sandell JH. 1993. Developmental variation in the structure of the retina. *J. Neurosci.* 13: 5194–5202.

Nelson R and Kolb H. 1985. A17: A broad-field amacrine cell in the rod system of the cat retina. *J. Neurophysiol.* 54: 592–614.

Peterson BB and Dacey DM. 1999. Morphology of wide-field, monostratified ganglion cells of the human retina. *Visual Neurosci.* 16: 107–120.

Purves D, Augustine GJ, Fitzpatrick D, Katz LC, LaMantia A-S, and McNamara JO. 1997. Neurotransmitters. Chapter 6, pp. 99–119. Neurotransmitter receptors and their effects. Chapter 7, pp. 121–144, in *Neuroscience.* Sinauer Associates, Sunderland, MA.

Rodieck RW. 1989. Starburst amacrine cells of the primate retina. *J. Comp. Neurol.* 285: 18–37.

Schiller PH. 1992. The ON and OFF channels of the visual system. *Trends Neurosci.* 15: 86–92.

Stafford DK and Dacey DM. 1997. Physiology of the A1 amacrine: A spiking, axon-bearing interneuron of the macaque monkey retina. *Vis. Neurosci.* 14: 507–552.

Sterling P. 1998. Retina. Chapter 6, pp. 205–253, in *The Synaptic Organization of the Brain*, 4th Ed. Shepherd G, ed. Oxford University Press, New York.

Strettoi E and Masland RH. 1996. The number of unidentified amacrine cells in the mammalian retina. *Proc. Natl. Acad. Sci. USA* 93: 14906–14911.

Strettoi E, Raviola E, and Dacheux RF. 1992. Synaptic connections of the narrow-field bistratified rod amacrine cell (AII) in the rabbit retina. *J. Comp. Neurol.* 325: 152–168.

Vaney DI. 1997. Neuronal coupling in rod-signal pathways of the retina. *Invest. Ophthalmol. Vis. Sci.* 38: 267–273.

Wässle H, Grünert U, Chun M-H, and Boycott BB. 1995. The rod pathway of the macaque monkey retina: Identification of AII-amacrine cells with antibodies against calretinin. *J. Comp. Neurol.* 361: 537–551.

Zhang J, Jung C-S, and Slaughter MM. 1997. Serial inhibitory synapses in retina. *Visual Neurosci.* 14: 553–563.

Ganglion Cell Signals to the Brain: Dots for the Retinal Sketches

Calkins DJ, Tsukamoto Y, and Sterling P. 1998. Microcircuitry and mosaic of a blue–yellow ganglion cell in the primate retina. *J. Neurosci.* 18: 3373–3385.

Dacey DM. 1993. Morphology of a small-field bistratified ganglion cell type in the macaque and human retina. *Visual Neurosci.* 10: 1081–1098.

Dacey DM. 1994. Physiology, morphology and spatial densities of identified ganglion cell types in primate retina. Pp. 12–34 in *Higher-Order Processing in the Visual System* (Ciba Foundation Symposium 184). Bock GR and Goode JA, eds. Wiley, Chichester, England.

Dacey DM. 1996. Circuitry for color coding in the primate retina. *Proc. Natl. Acad. Sci. USA* 93: 582–588.

Dacey DM and Lee BB. 1994. The "blue-on" opponent pathway in primate retina originates from a distinct bistratified ganglion cell type. *Nature* 367: 731–735.

Dacey DM and Petersen MR. 1992. Dendritic field size and morphology of midget and parasol ganglion cells in the human retina. *Proc. Natl. Acad. Sci. USA* 89: 9666–9670.

De Monasterio FM and Gouras P. 1975. Functional properties of ganglion cells of the rhesus monkey retina. *J. Physiol.* 251: 167–195.

Jacoby R, Stafford D, Kouyama N, and Marshak D. 1996. Synaptic inputs to ON parasol ganglion cells in the primate retina. *J. Neurosci.* 16: 8041–8056.

Kier CK, Buchsbaum G, and Sterling P. 1995. How retinal microcircuits scale for ganglion cells of different size. *J. Neurosci.* 15: 7673–7683.

Kolb H and Dekorver L. 1991. Midget ganglion cells of the parafovea of the human retina: A study by electron microscopy and serial section reconstructions. *J. Comp. Neurol.* 303: 617–636.

Mariani AP. 1982. Biplexiform cells: Ganglion cells of the primate retina that contact photoreceptors. *Science* 216: 1134–1136.

Marr D and Hildreth E. 1980. Theory of edge detection. *Proc. Roy. Soc. Lond. B* 207: 187–217.

Masland RH. 1996. Unscrambling color vision. *Science* 271: 616–617.

Merigan WH and Maunsell JHR. 1993. How parallel are the primate visual pathways? *Annu. Rev. Neurosci.* 16: 369–402.

Polyak SL. 1941. *The Retina*. University of Chicago Press, Chicago.

Rodieck RW. 1991. Which cells code for color? Pp. 83–93 in *From Pigments to Perception*, Valberg A and Lee BB, eds. Plenum Press, New York.

Rodieck RW. 1998. Cell types. Chapter 11, pp. 226–265. Seeing. Chapter 14, pp. 328–360, in *The First Steps in Seeing*. Sinauer Associates, Sunderland MA.

Rodieck RW, Binmoeller KF, and Dineen J. 1985. Parasol and midget ganglion cells of the human retina. *J. Comp. Neurol.* 233: 115–132.

Vaney DA. 1991. Many diverse types of retinal neurons show tracer coupling when injected with biocytin or Neurobiotin. *Neurosci. Lett.* 125: 187–190.

Wandell BA. 1995. The retinal representation. Chapter 5, pp. 111–152, in *Foundations of Vision*, Sinauer, Sunderland, MA.

Chapter 15

Retina III: Regional Variation and Spatial Organization

What does it mean to see? The plain man's answer (and Aristotle's, too) would be, to know what is where by looking. In other words, vision is the process of discovering from images what is present in the world, and where it is.

■ David Marr, *Vision*

The retinal image of a familiar environment is probably never repeated exactly; variations in lighting, movement of objects from time to time, and differences in our precise direction of view all argue for the uniqueness of the retinal image at any given moment. The retina must therefore be prepared for anything because it cannot know in advance what is where.

The retina's strategy for dealing with this uncertainty is a kind of tool kit with a standard set of tools meant to cope with any situation. It is like an artist's tool kit, which would include at least several drawing pencils in varying degrees of hardness. A very hard pencil (5H) would be used for preliminary sketches or for adding fine detail to an elaborate sketch. A very soft pencil (5B), with its dark, broad line, adds emphasis and heightens contrast in deep shadows. With an array of pencils (about 15 in the standard series), an artist can handle all aspects of a sketch, from delicate, highlighted textures to dark, almost featureless shade. The retina's tool kit is the different types of ganglion cells whose signals—the dots—convey different shades of meaning about the retinal image.

If signals from the ganglion cells are to convey the same sorts of information at the same level of detail throughout the retina, each part of the retina needs to be able to do the same things—that is, to have the same tools. But each part's having its own tools represents a massive duplication of tool kits. If the retinal image were to be sketched in the same level of detail everywhere, there would have to be a lot of tool kits and a lot of artists working under very crowded conditions. Excessive duplication and overcrowding can be prevented only by limiting the amount of detail in the retinal sketches.

Making Retinal Sketches out of Dots: Limits and Strategies

The detail in a sketch is limited by dot size and spacing, and cones set the dot size in the central retina

Light intensity in the retinal image varies continuously from point to point, and the job of the photoreceptor array is reporting these intensity variations as faithfully as possible. But light varies continuously in the image, while photoreceptors are *discrete* elements having finite size and separation. If light intensity varies over distances smaller than the size or spacing of the photoreceptors, these small-scale variations cannot be reported by the photoreceptors and potentially meaningful information will not be transmitted to the brain.

The problem is illustrated in Figure 15.1, where a sinusoidal variation in light intensity falls on a line of cones; for illustrative purposes, the cone signals are shown as proportional to the average amount of light falling on them. If the sinusoid has a low frequency relative to the center-to-center spacing of the cones, as in Figure 15.1*a*, the cones sample the intensity variation periodically and their signals reproduce the essential features of the sinusoid. Increasing the frequency of the sinusoid so that its peaks and troughs fall on adjacent cones is also acceptable; their signals will be an alternating sequence of strong and weak, accurately reflecting the alternating sequence of maximum and minimum light intensities (Figure 15.1*b*). In this case, the photoreceptors are sampling at twice the frequency of the sinusoid. But if the frequency of the sinusoid is higher than twice the cone sampling frequency, maxima and minima fall within or between cones and the signals do not correspond to the intensity variation (Figure 15.1*c*); the cones are undersampling. In the last example, the signals produced by the cones in response to the high-frequency sinusoid are identical to those produced in response to the low-frequency sinusoid in Figure 15.1*a*. This anomalous result of undersampling is **aliasing**; we will come back to it later.

The limiting case, the **Nyquist frequency**, occurs when the frequency of the sinusoid is half that of the photoreceptor spacing; it is roughly equivalent to the cutoff frequency for transmission through optical systems (see Chapter 2) in the sense that it represents the highest frequency that can be transmitted by the system. Sinusoidal frequencies higher than the Nyquist frequency will be undersampled by the photoreceptor array and will not be faithfully represented in the photoreceptor signals. Thus the Nyquist frequency represents the limit to spatial resolution imposed by the finite size and spacing of the photoreceptors.

For the linear array of cones shown in Figure 15.1*b*, the Nyquist frequency has a spatial period (wavelength) that is twice the distance between cone centers. Since frequency is the reciprocal of the period, it follows that the Nyquist frequency (f_N) is simply

$$f_N = 1/2d \text{ or } f_N = 0.5/d$$

where d is the cone separation in degrees of visual angle (1 μm of linear distance equals 1/300°). If the center-to-center cone separation were 1 μm, the Nyquist frequency would be 150 cycles/degree. This formula for the Nyquist frequency is good only for cones in a square array, where each cone center is at the corner of a small square. As we will see later, the array is actually triangular (see Figure 15.4), which requires a slight modification in the formula:

$$f_N = (2/\sqrt{3})(1/2d) \text{ or } f_N = 0.577/d$$

With these formulas in hand, we can use the known center-to-center distance between cones to calculate their Nyquist frequency. In the center of the fovea, the average cone spacing is 2.5 μm, which corresponds to 0.00833° (2.5/300). The Nyquist frequency, f_N, is therefore 69.3 cycles/degree (0.577/0.00833), using the second of the two formulas.

This value for the Nyquist frequency of the central foveal cone array is comparable to the optical cutoff frequency for the eye's optical system with a 4 mm pupil diameter (around 60 cycles/degree; see Chapter 2). Cone spacing and the optical system are therefore fairly well matched. The cones are not more closely spaced than they need to be to convey the highest spatial frequencies trans-

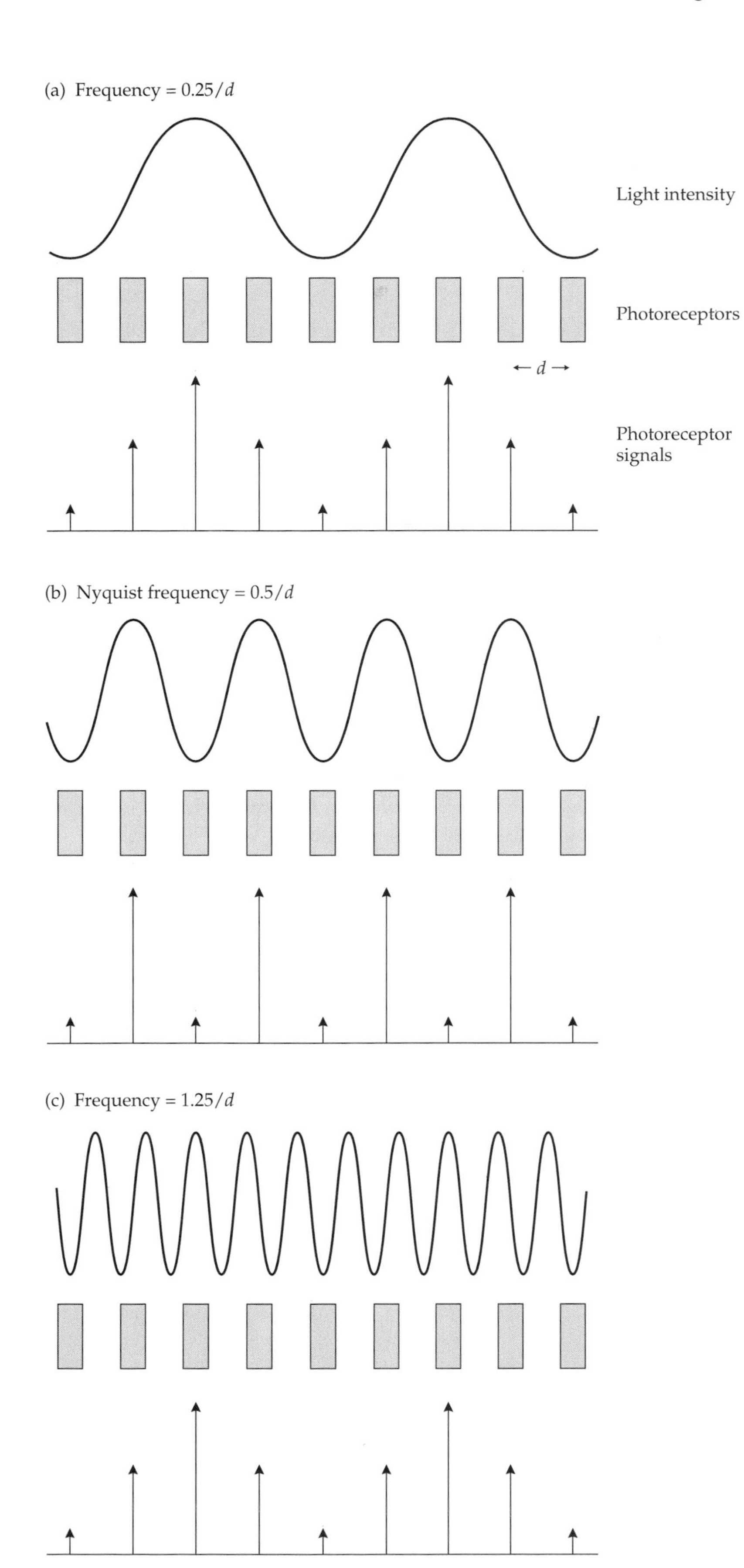

Figure 15.1
Discrete Sampling of Continuously Varying Light Intensity
Sinusoidally varying light intensity falling on a one-dimensional array of cones produces signals that are proportional to the average amount of light falling on each cone; the photoreceptors have a constant center-to-center spacing (represented by *d*). (*a*) When the frequency of the sinusoid is low, the cone signals reproduce the sinusoid. (*b*) When the sinusoid has twice the wavelength of the cone spacing, cones alternate between maximum and minimum signals. This condition is called the Nyquist frequency. (*c*) A high-frequency sinusoid cannot be reproduced faithfully by the cone signals; here, the signals are identical to the low-frequency sinusoid in part *a*, a condition called aliasing.

mitted by the optics; thus the cones are not oversampling, and they are not so widely spaced that spatial frequencies in the retinal image will not be conveyed faithfully—the cones are not undersampling either.

The equivalence of the foveal-cone Nyquist frequency, the optical cutoff frequency, and the maximum spatial resolution for the visual system as a whole is consistent with the discussion of cone wiring in previous chapters. Since each foveal cone has direct lines to the brain through the midget bipolar and ganglion cells such that each midget ganglion cell is reporting on the activity in a single cone, cone spacing is the limiting factor in resolution. If there were convergence, the same spatial resolution for the system as a whole would require tighter cone packing (and a higher Nyquist frequency) because resolution would be determined by the spacing of cone clusters corresponding to single ganglion cells, not by individual cone spacing.

But as long as the one-to-one wiring holds, the limiting factor in spatial resolution will be the Nyquist frequency set by the cone spacing (assuming that the optical system can transmit comparably high spatial frequencies). Thinner, more closely spaced cones will have a higher Nyquist frequency that should permit better spatial resolution, and this is part of the reason that raptorial birds have better spatial resolution than we do.

Where there is convergence in the wiring, however, resolution will be more closely related to center-to-center spacing of ganglion cell dendritic fields than to the cone spacing, as we will consider later. But in terms of calculating Nyquist frequencies, the concept still applies. If the centers of ganglion cell dendritic fields are 20 μm apart on average, the Nyquist frequency will be on the order of 7.5 cycles/degree, assuming a rectangular array. Thus we would expect the parasol ganglion cells receiving inputs from foveal cones to have a relatively low Nyquist frequency because there is convergence in the pathway from cones to diffuse bipolar cells to parasol cells.

The entire retinal image cannot be sketched in great detail

If the entire retina were constructed with a cone array like that in the fovea, the number of cones would be enormous. Foveal cone density approaches 200,000 cones/mm^2 and the total retinal area is about 1000 mm^2, so the retina would need 200 million cones and at least as many ganglion cells. The optic nerve would be larger than the eye, which shows as simply as anything the impossibility of having a closely spaced cone array everywhere in the retina.

And there are other reasons for not pursuing this kind of strategy. For one thing, the optical system does not work as well peripherally as it does centrally. Aberrations increase with off-axis imaging, and the contrast in the high-frequency components of the image will be very much reduced in comparison to the on-axis part of the image falling on the fovea. Spacing peripheral cones as the cones in the fovea are spaced would be oversampling to no effect. Most of the peripheral cones would be redundant, unable to contribute anything to the resolution of fine detail, because the detail is not present in the retinal image.

Also, a cone array of uniformly high density would confront the brain with more information than it could deal with; there isn't enough room to accommodate all the extra neurons that would be required. The area of the brain's cortex devoted to information arriving from the fovea would have to be expanded if all of the retinal input were to be dealt with in the same way, and the required expansion would exceed the total area of the cortex by a hundred times or more.

These sorts of limitations require another approach. Most visual systems deal with the problem by sketching only a small part of the retinal image in great detail. Rather than trying to see the entire visual world in detail all at once, eye or body movements are used to sample in detail over time. The part of the retina specializing for detailed sketching can be used to examine different parts of the total image in succession, thereby building up a detailed representation as parts of the image are examined in turn. Thus there is an important relationship between eye movements and retinal organization. The smaller the part of the retina with a high Nyquist frequency, the more need there will be to move the region around to examine other parts of the world's image on the retina.

Most retinas are organized around points or lines

All diurnal animals have cones in their retinas, and all face the same general limitations to creating detailed retinal sketches. Where they differ is in the amount of detail required and the location of important detail in their visual world. They have evolved two basic solutions to deploying their cone resources across the retina.

Figure 15.2 shows schematic cone distributions for two extreme cases in which cone distributions are plotted in terms of density—that is, number of cones per square millimeter of retinal area. In the first example, cone density has a single maximum, a central peak from which density declines symmetrically in all directions out to the edge of the retina. This central peak of cone density defines the **area centralis** (Figure 15.2*a*), and it is the point of maximum spatial resolution. Because this point provides the greatest detail in the retinal sketch, it is the center of attention as far as the visual system is concerned; the neural pathways representing these central cones occupy the largest part of the visual cortex, and eye movements are made with this point as the center of reference.

An area centralis is found in many species, particularly those whose lifestyle demands resolution of fine detail. Carnivores, primates, birds, and lizards are some of the main examples. The human retina is organized around a point of maximum cone density, as we will see shortly, but the most striking examples are found among the birds, some of which have two points of high cone density in each retina.

Another common pattern for distributing cones is a ridge of maximum cone density extending horizontally across the retina. This ridge is called a **visual streak** (Figure 15.2*b*); it is common among the ungulates, like sheep or deer, fish, turtles, rabbits, squirrels, and also birds, particularly seabirds, in which it may be combined with an area centralis. A visual streak in reduced form is often combined with an area centralis.

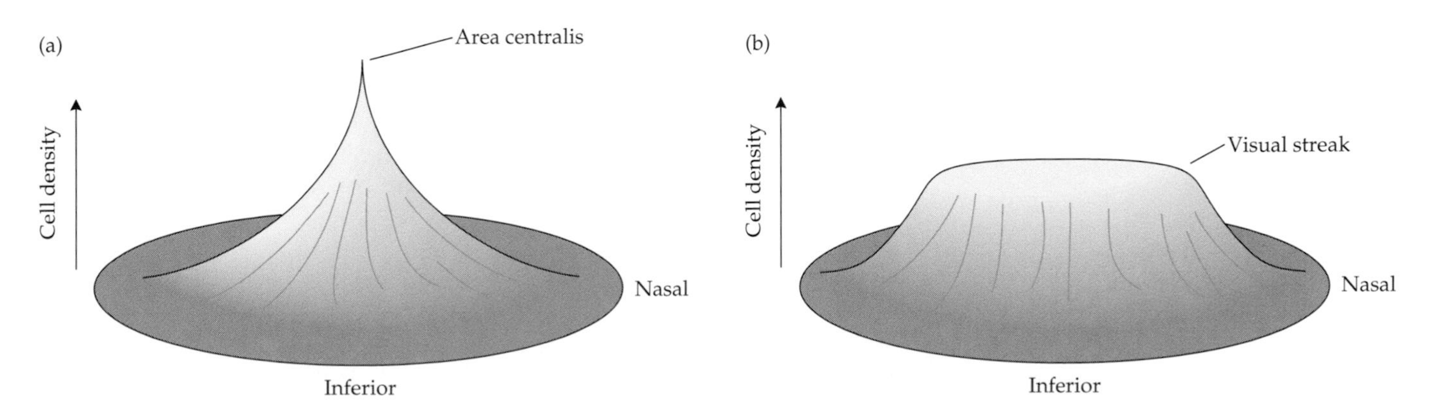

Figure 15.2
Area Centralis and Visual Streak
The retina is depicted here as a flat circular disc from which photoreceptor density in cells/mm^2 rises as a vertical dimension. (*a*) Retinas with an area centralis have a point of maximum cell density around which the retina is organized. (*b*) A visual streak is a line of high cell density that may extend across most of the retina's horizontal extent. The area centralis and visual streak shown here are idealized versions of cat and rabbit retina, respectively.

The presence of a visual streak probably says something about the visual importance of the horizon and events likely to occur along the horizon—a fox sneaking up on a rabbit, for example. A visual streak has the disadvantage of spreading the cones out at lower density than if they were concentrated in one small area, but it obviates the more elaborate patterns of eye or body movements and their attendant control systems demanded by an area centralis. A rabbit doesn't need to redirect its eyes to look at different objects along the horizon; it can see them all equally well (or poorly, as the case may be). Although the most obvious eye movements in primates are rapid saccades (see Chapter 4), rabbits rarely move their eyes rapidly and they do not require the neural circuitry to do so.

Thus the way in which a retina distributes its cones to sketch part of the retinal image in detail has repercussions in other parts of the visual system.

Retinal sketches should be continuous, with no unnecessary blank spots

Although the size and spacing of dots in the retinal sketch will vary from one part of the retina to another, the goal is to be able to sketch everywhere, to be able to place dots anywhere on the page. In other words, the retinal sketch should not have any blank areas, and the retinal image should be sampled without interruption over its entire extent. This is done by arranging vertical pathway neurons, or parts of the neurons, so that they constitute **tilings**.

The patterns in Figure 15.3 are mosaics or arrays of elements with specific properties that make them tilings. These four examples look different because of differences in the shape and number of the elements—the tiles—from which the tilings are constructed, but all of them *cover the plane surface with no overlap and with no gaps between tiles*. As these examples illustrate, a tiling may be constructed with a simple regular polygon used over and over again, there may be two or more regular tiles, the tiles may be irregular in shape, there may or may not be repeating patterns within the tiling, or the individual tiles may all be different and the overall pattern may be random, or nearly so.

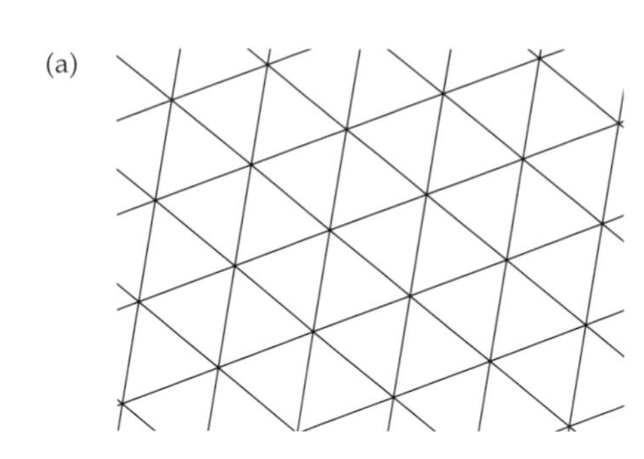

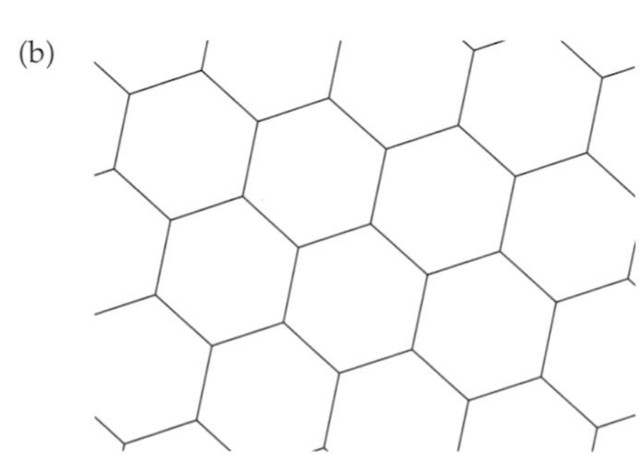

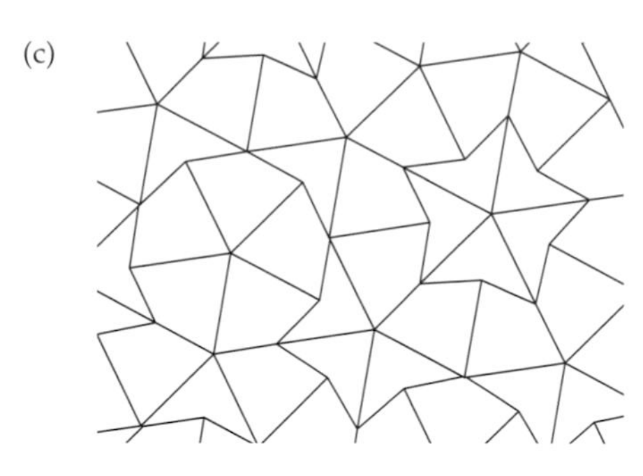

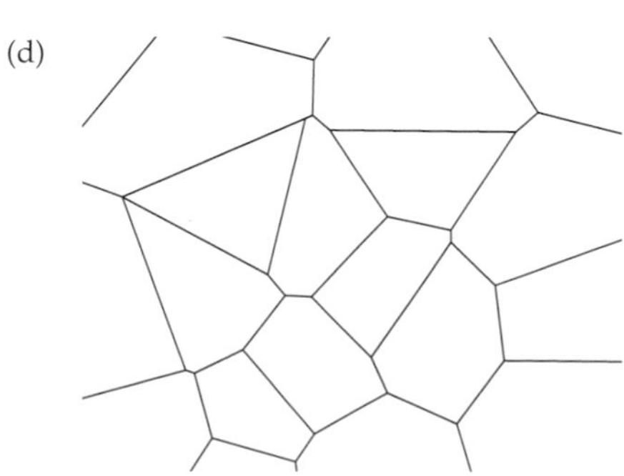

Figure 15.3
Examples of Tilings
(*a*) A regular tiling by equilateral triangles. (*b*) A regular tiling by hexagons. (*c*) A two-element Penrose tiling. (*d*) An irregular, aperiodic tiling by Dirichlet domains.

We have already seen examples of tilings elsewhere in the eye. Basal cells in the corneal epithelium form a nearly regular hexagonal tiling, and corneal endothelial cells tile as irregular polygons. In fact, all continuous layers of epithelium or endothelium are tilings; the retinal pigment epithelium and the epithelial layers in the ciliary body, iris, and lens all tile their surfaces. (The lens shells are examples of tiling a spherical surface.) There is nothing mysterious about the ubiquity of epithelial cell tilings; if the layers had gaps, their function would be compromised—thus anchoring junctions keep cells in contact and maintain the integrity of the tiling. But tilings formed by retinal neurons are not so simple. Most of them are irregular, their function is less obvious, and their presence is more difficult to demonstrate.

The photoreceptor tiling is the easiest to see and to appreciate. At the level of the inner segments, foveal cones have roughly hexagonal cross sections packed tightly and neatly together with a bit of Müller's cell cytoplasm as grout between the tiles (Figure 15.4). Points at the center of each inner segment can be connected to form nearly equilateral triangles, which is why the array is described as triangular. This tiling varies across the retina because of the changing density and size of the tiles—the rods and cones—but except at the optic nerve head, it is complete. Incoming photons will always encounter an inner segment to guide them to the photopigment. And looking back to the source of the photons, this amounts to a seamless representation of the visual world, one without gaps or multiple representations.

Having established this unbroken sampling of visual space with the photoreceptor tiling, the retina still has the problem of maintaining it. Photoreceptor

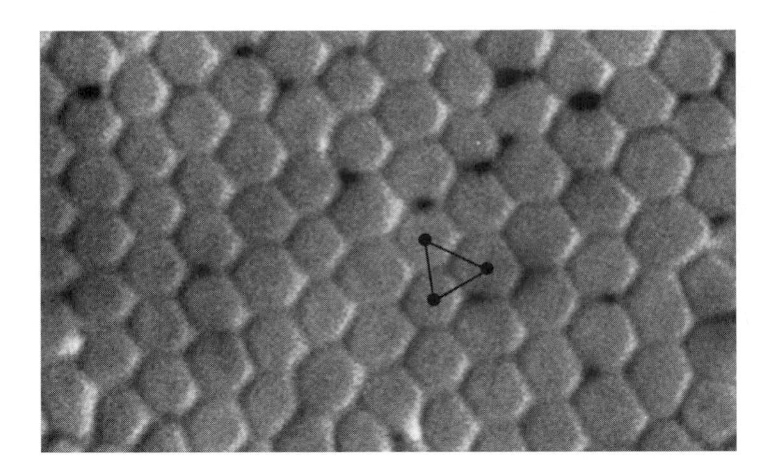

Figure 15.4
Cone Tiling in the Fovea
This optical section was taken through the cone inner segments near the level of the external limiting membrane. The underlying pattern of organization is triangular, as shown by the lines connecting the centers of three neighboring cones. (From Curcio et al. 1990.)

cross sections are solid, space-filling objects with well-defined edges, like pieces of ceramic tile, and it is easy to think of them as elements in a tiling. Information transmitted by photoreceptors, however, is being passed from pedicles and spherules to dendritic fields; pedicles and spherules may be space-filling structures, but the irregular, branched dendritic arbors of diffuse bipolar cells bear little resemblance to a piece of tile. How can something like a dendritic arbor or a branched axon terminal, which have no definite boundaries or edges, act as a tile in a tiling?

If we assume that spaces between branches in a dendritic arbor are small compared to the terminals from which the arbor is sampling, we can think of the arbor as continuous and simply draw lines connecting the tips of the dendrites, thereby enclosing the arbor within an irregular polygon. But when this is done for ganglion cells (Figure 15.5), the irregular dendritic fields overlap and are therefore not a tiling; they are a **covering**.

Coverings differ from tilings in that their elements overlap; my desktop, for example, is covered with papers (and the coverage is continuous), but the papers overlap; they cover, but they don't tile. The amount of paper on my desk can be specified by the **coverage factor**, which is the number of pieces of paper overlying each point on the desktop. To calculate the average coverage factor, I need to count the pieces of paper, measure the desktop to find its area, and measure a piece of paper (assuming they are all the same size) to find its area. The paper

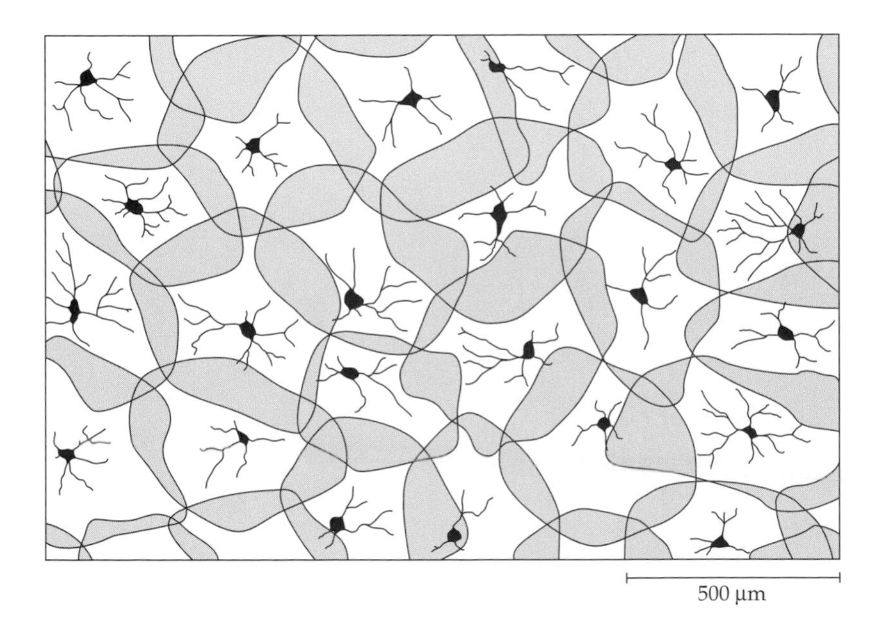

Figure 15.5
Overlapping Ganglion Cell Dendritic Fields
Cell bodies of one type of ganglion cell from cat retina are shown with some of their larger dendrites and the irregular boundaries of their dendritic fields. Regions where dendritic fields overlap are shaded. The coverage factor for these cells is 1.3 (see text for explanation). (After Wässle, Peichl, and Boycott 1981.)

density (number of pieces of paper divided by desk area) multiplied by the average area of a piece of paper will be the coverage factor. That is,

$$\text{Coverage factor} = (N_{\text{paper}}/A_{\text{desk}}) \times (A_{\text{paper}})$$

It works the same way for ganglion cell dendritic fields. Mean cell density multiplied by the mean dendritic area gives the coverage factor.

Coverage factors distinguish between tilings and coverings; by definition, the coverage factor of a tiling is 1.0. Anything else is not a tiling. Values less than 1 indicate that the covering is incomplete, with gaps in it; values greater than 1 indicate overlap. For the dendritic fields in Figure 15.5, the overlap factor is 1.3, and thus they cannot represent a tiling.

But 1.3 is tantalizingly close to 1.0, and by not constructing arbitrary boundaries for dendritic fields, we can demonstrate the existence of irregular tilings. The rule is simple; if dendrites of neighboring cells intersect and cross one another, their dendritic fields overlap and the cells are not part of a tiling. If branches from neighboring dendritic fields do not cross and yet leave no sizable gaps between one cell and the next, they must be part of a tiling with unity coverage.

The example in Figure 15.6 shows ganglion cells in rabbit retina; there are two dendritic fields belonging to cells of the same type. These ON–OFF cells are bistratified, and the two dendritic branching planes are shown separately. Although the areas covered by dendrites differ in the two planes, the same rule applies in both: Dendrites from the two cells touch, but none of the branches cross one another. Therefore, these dendritic fields are two of the tiles in a tiling. There are two independent tilings—one in the OFF portion (Figure 15.6*a*) and

(a) OFF sublayer branching

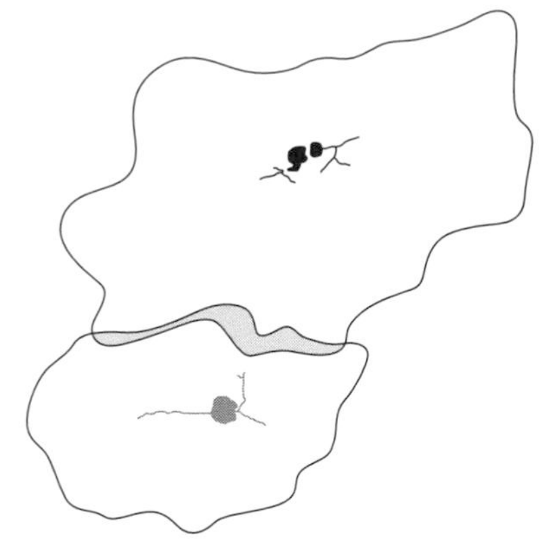

(b) ON sublayer branching

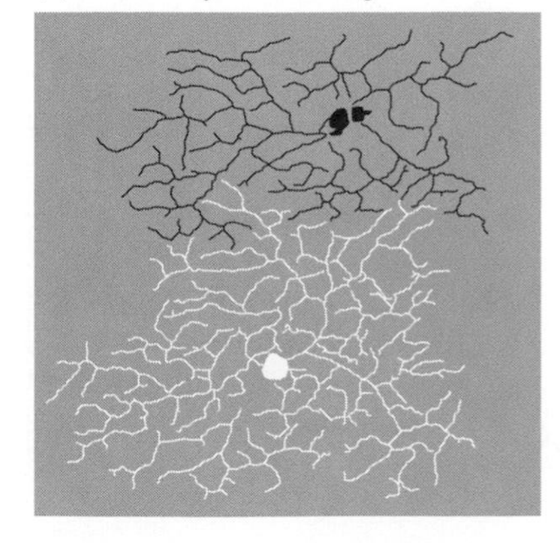

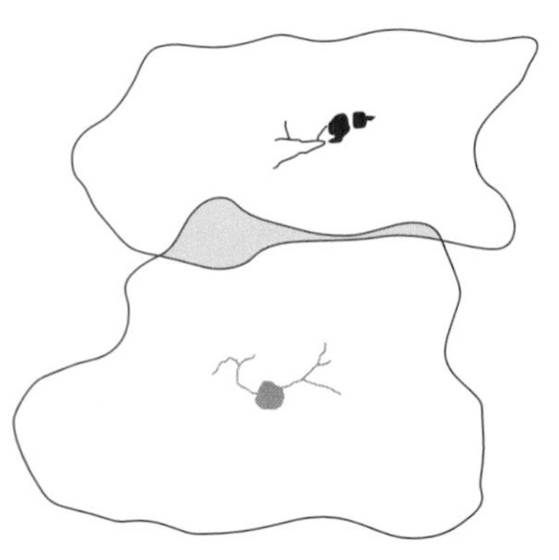

Figure 15.6
Tilings by Ganglion Cell Dendrites
These dendritic fields are from two neighboring direction-selective ganglion cells in rabbit retina. The cells are of the same type, both preferring stimuli moving from left to right. Because these cells are bistratified, they have dendrites ramifying in both the OFF (*a*) and ON (*b*) portions of the inner plexiform layer. In both sublayers, dendrites touch but do not cross. Thus these dendritic fields are tiles in a tiling, but outlines of the dendritic territories (right-hand drawings) overlap because dendrites from the two cells interdigitate. (After Amthor and Oyster 1995.)

one in the ON portion (Figure 15.6*b*) of the inner plexiform layer. The tilings extend completely across the retina such that every point in the retinal plane falls within one of the tiles in each of the tilings.

When boundaries for these dendritic fields are constructed, some overlap appears (shaded areas), which indicates why artificial boundaries are inappropriate. These dendritic tiles are quite irregular, and each tile is probably unique, differing from all others in size and shape. Thus the tiling cannot be specified in mathematical terms (although some of its descriptive statistics may be), and tilings by cells of the same type will differ in detail from one retina to the next. The important point, however, is that dendritic fields or axon terminal arborizations are elements in tilings. Whatever information is being transmitted or sampled by cells of a particular type, the existence of a tiling means that the information about the visual world is being sampled and transmitted without breaks or discontinuities. Every point in visual space is represented by one element in the cell type that forms a tiling.

Coverings, in which the coverage factor is greater than 1, also exist in the retina; not all types of neurons form tilings, in other words. The neurons that do not tile are the horizontal and amacrine cells in the lateral pathways.

Tilings do not need to be regular, and tiles do not have to be the same size

Although the tiling formed by the cone inner segments is close to being a tiling of hexagons, it is not perfectly regular. It becomes less regular away from the foveal center, particularly as rods enter as elements in the tiling, but perfect regularity is not necessary and some irregularity may be beneficial.

Perfectly regular tilings are susceptible to undersampling and aliasing if the image contains frequencies higher than the Nyquist limit for the array. For example, Figure 15.7*a* shows two dot patterns, one regular (left) and one irregular (right). When a grating pattern is superimposed on the dot arrays, aliasing can be seen as low-frequency stripes within the regular dot pattern. The irregular dot pattern does not show the effect strongly, however.

Although the foveal cone array is nearly regular, aliasing is not a problem, because the retinal image does not contain frequencies significantly larger than the Nyquist frequency for the foveal cones. But as cones become more widely spaced away from the fovea, their Nyquist frequency decreases below the highest spatial frequency in the retinal image. Thus aliasing could be a significant problem if high spatial frequencies are present at high contrast; the result, as we have just seen, would be the appearance of low-frequency components in the retinal image and distortion of the relationship between the retinal image and the retinal sketch.

In fact, the high frequencies in extrafoveal images are not high-contrast components, but even if they were, irregularity in the extrafoveal cone array works to offset aliasing. Irregular cone arrays lack strong aliasing because there is no specific Nyquist frequency. Differences in size and spacing of the cones mean that the value of d (cone separation) used to calculate the Nyquist frequency does not have a single value. Instead, d is a variable with a mean and standard deviation, and the calculated Nyquist limit is no longer a single frequency, but a *range* of frequencies varying around a mean value. As the variation of d increases, the Nyquist limit becomes less well defined, and there is no sharp transition from oversampling below a fixed frequency to undersampling at higher frequencies. Only when the standard deviation of d is small relative to the mean value of d, which is the case for foveal cones, is the Nyquist limit a hard boundary between oversampling—with no aliasing—and undersampling *with* aliasing.

Although the irregularity of the extrafoveal cone arrays and the low contrast of high spatial frequencies in the retinal image minimize aliasing under normal

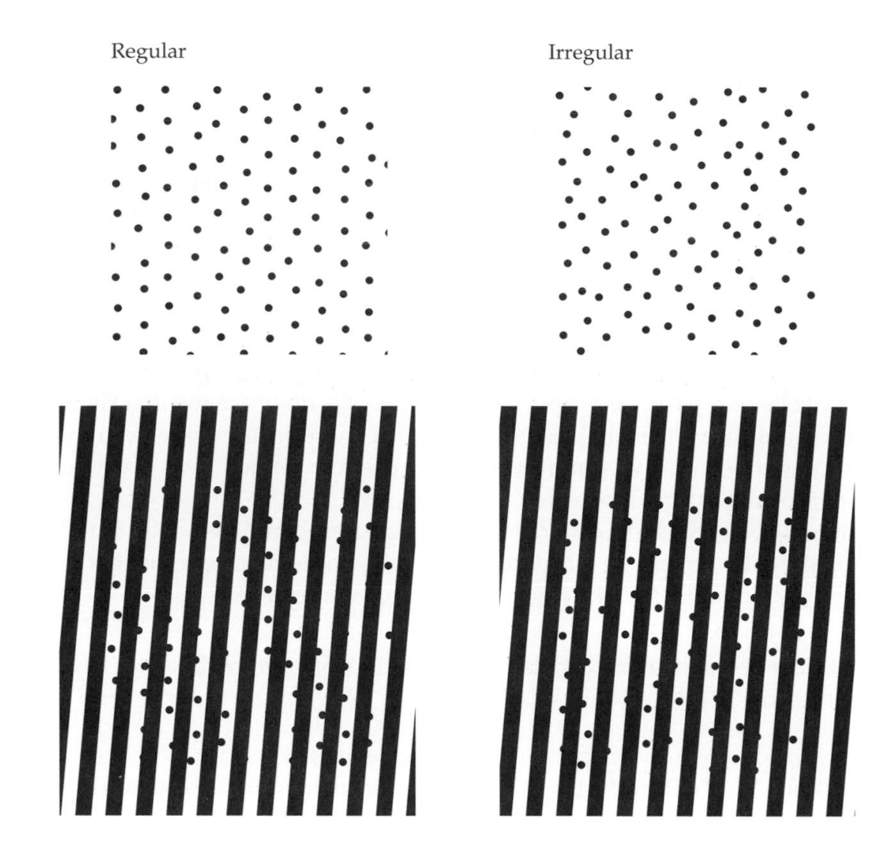

Figure 15.7
Regular Arrays Show More Aliasing Than Irregular Arrays
A grating pattern superimposed on a regular array of dots (left) produces a pattern of oblique stripes at a low spatial frequency, due to aliasing by the regular dot array. Aliasing is reduced and the low-frequency components are less obvious when the dot array is irregular (right). (Dot patterns based on cone arrays in Curcio and Sloan 1992.)

conditions, aliasing does occur when high-contrast, high spatial frequency interference fringes are generated on the retina. (One can see low spatial frequency components under these conditions, but they are not as regular as the stripes in Figure 15.7, because of the imprecise Nyquist limit.) This aliasing can be used to estimate the Nyquist frequency and therefore the cone separation at different locations on the retina. The strategy harks back to Figure 15.1, which shows that the signals by the photoreceptor array are the same for a pair of frequencies, one below the Nyquist frequency and the other above it. If a subject viewed interference fringes whose frequency was gradually increasing, the perception would be of increasing grating frequency—the grating stripes would appear smaller and smaller—until the Nyquist frequency was crossed. Then the subject would see grating stripes becoming wider (spatial frequency decreasing) even as the true frequency in the interference fringes continued to increase. The perceptual change from increasing to decreasing frequency marks the Nyquist frequency. In experiments of this sort, the cone separations predicted by the Nyquist frequencies at different retinal locations agree with direct anatomical measurements. Thus we have a method to study the anatomical arrangement of cones in the living eye, which is a satisfying result.

Although the cone tiling can approach regularity, tilings formed by axonal or dendritic arbors in the retina are invariably irregular. As with extrafoveal cones, any calculations of Nyquist limits for these tilings are subject to the qualification made earlier: The calculated Nyquist limit is an average value with a certain amount of inherent variation. It gives a measure of the spatial resolution limit of a tiling that is useful in comparing one irregular tiling to another, but not a precise one that specifies exact frequencies at which over- or undersampling will occur. But wherever dendritic field sizes limit spatial resolution, the irregularity should minimize aliasing.

Tilings formed by axonal or dendritic arbors at different levels of the retina need not match precisely

The inner segments of the foveal cones form a tiling that samples light in the retinal image without gaps. Assuming that the pedicles of these cones also tile in the outer synaptic layer, the pedicles represent an unbroken sheet of information generated by the cones that is to be passed along to multiple possible dendritic tilings formed by the different types of cone bipolar cells. The axon terminals of the cone bipolar cells then pass information along to the tiled dendritic fields of ganglion cells.

It is important that information contained in one tiling, like that of the cone pedicles, be passed along in such a way that the array is completely sampled; complete sampling keeps the sheet of information about the retinal image from developing gaps or discontinuities. If the recipient tiling is formed by bipolar cell dendrites, for example, the dendritic tiles do *not* have to match the photoreceptor tiles in size and shape to keep the information sheet continuous, but departures from matching have other significant consequences.

Figure 15.8*a* shows two overlaid tilings, one representing cone pedicles and the other representing the areas occupied by bipolar cell dendrites. On average, the bipolar cell tiles are several times larger than the photoreceptor tiles. But this is just another way to describe convergent wiring, as shown by looking from the side at a section through the two tilings (the right-hand portion of Figure 15.8); each bipolar cell is receiving inputs from several cones. One consequence of this tiling mismatch is loss of spatial resolution; the Nyquist frequency for the bipolar cell tiling will be less than that of the cone tiling. The other, less obvious result is a loss of specificity about cone type. The red, green, and blue cones have not been labeled in the cone pedicle tiling, but wherever they are, the tiling drawn to represent bipolar cell dendrites makes no distinction among them. Thus, this partic-

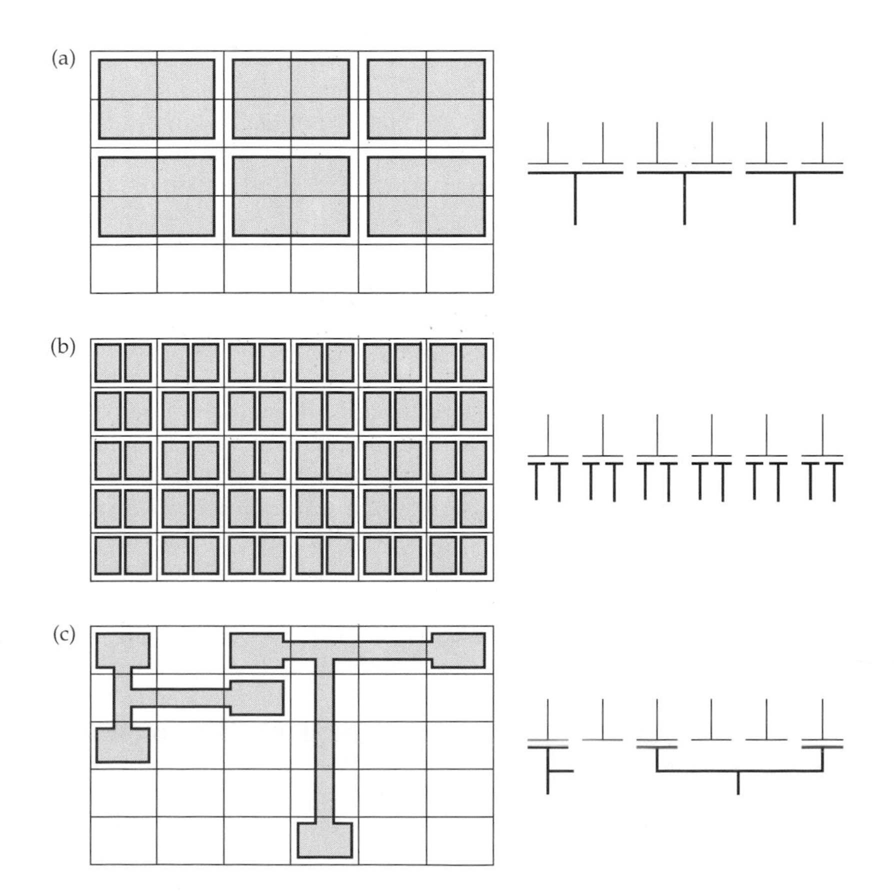

Figure 15.8
Sampling by Overlaid Tilings
The relationship between axon terminals and the dendritic fields they contact can be thought of as overlaid tilings. Here, the input tiling is square and drawn with fine lines. The recipient tiling may contain tiles that are larger (*a*), smaller (*b*), or irregular (*c*). Sectional views (at right) through the overlaid tilings show convergence (*a*), as from cones to diffuse bipolar cells; divergence (*b*), as from cones to midget bipolar cells; and selectivity (*c*), as from blue cones to blue cone bipolar cells.

ular tiling mismatch produces a loss of high spatial frequency information and a lack of specificity to stimulus wavelength. But it is still an unbroken representation of the retinal image.

The example in Figure 15.8*b* shows tiles formed by bipolar cell dendrites that are *smaller* than the cone tiles; there are two bipolar cell tiles for every cone tile. Wavelength specificity is not a problem here; each cone tile will pass on its wavelength properties to each of its attendant bipolar cell tiles. As for the Nyquist frequency of the bipolar cell tiling, there is no change; each bipolar cell tile sees whatever its input cone sees. This is an example of divergent wiring, like that between cones and their ON and OFF midget bipolar cells. We have taken one tiling and made a two-element tiling out of it.

The final example (Figure 15.8*c*) is quite different from the other two. The elements of the bipolar cell tiling do not touch one another; they have gaps between them, and they do not contact all the tiles in the photoreceptor array. And given what has been said already about tilings, this arrangement of elements doesn't seem to fit the definition of a tiling; the elements don't overlap, and the gaps are obvious.

These elements represent the dendritic fields of blue cone bipolar cells; they contact only blue cone pedicles. By skipping over most of the tiles in the cone array, they sample the cone tiling selectively and discontinuously. My justification for calling this a tiling is the absence of overlap among elements and the complete sampling *of the blue cones*. The bipolar cells sample all of the blue cone tiles with unity coverage, and in that sense they represent as good a tiling as either of the previous examples. Gaps in the blue cone bipolar cell tiling simply reflect gaps between the blue cones themselves. And to anticipate a bit, the blue cone bipolar cell terminals in the inner plexiform layer do constitute a well-defined tiling. The main consequence of the selectivity of the blue cone bipolar cell sampling is loss of spatial resolution because the size of the "tiles" implies a low Nyquist frequency.

Although the examples in Figure 15.8 refer to the interface between photoreceptor terminals and bipolar cells, they could also be used to illustrate the meeting of bipolar cell axon terminals and ganglion cell dendrites in the inner plexiform layer. Dendritic tilings by ganglion cells do not have to match the tilings of bipolar cell terminals to maintain a seamless representation of the retinal image, but mismatching has implications for spatial resolution and information specificity.

Spatial Organization of the Retina

The fovea is a depression in the retina where the inner retinal layers are absent

A schematic section through the center of the human fovea is shown in Figure 15.9. At the bottom of the pit, the center of the fovea, the retina is about 100 μm thick and consists only of the pigment epithelium, the photoreceptor layer, external limiting membrane, outer nuclear layer, a bit of outer plexiform layer, and the inner limiting membrane. Moving out from the center to either side along the slope of the pit, we begin to see the other layers appear in succession; outer synaptic layer, inner nuclear layer, inner plexiform layer, and, finally, near the rim of the depression, the ganglion cell and nerve fiber layers. From rim to rim, the diameter of the pit is around 1500 μm, corresponding to about 5° of visual angle (1° of visual angle is approximately equal to 300 μm of linear extent on the retinal surface).

The photoreceptors at the center of the fovea are all cones, with thin inner segments and greatly elongated outer segments. Since the bipolar cells to which the foveal cones connect have been pushed to the sides, the photoreceptor axons run

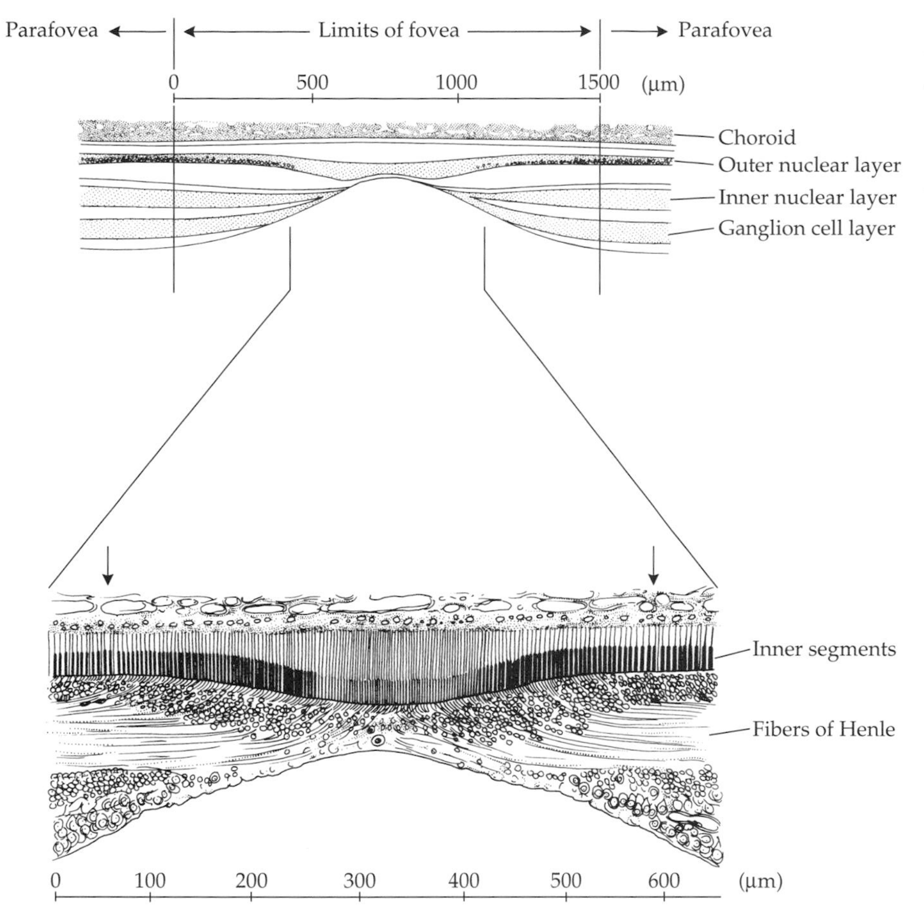

Figure 15.9
The Fovea
The human fovea is about 1500 μm (5°) in diameter and has only cones at its center. Only one of the nuclear layers is present in the center of the fovea. Locations of rods closest to the foveal center are indicated with arrows (lower panel). (After Polyak 1941.)

obliquely out from the center to their pedicles. Collectively, these oblique cone axons are called **Henle's fiber layer**.

Humans and other primates are the only mammals to have foveas in their retinas, but foveas are present in a variety of nonmammalian species. (Most mammalian retinas have an area centralis with elevated photoreceptor density, but lack the foveal pit.) Foveas are common among birds and are particularly prominent in raptorial birds, which sometimes have two foveas at different locations in the retina; foveas are also found in retinas of diurnal lizards and snakes, and a few species of fish. Foveate retinas usually have cones within the fovea, but there is one notable exception: The retina of the *Tuatara* (sphenodon), an endangered and strictly nocturnal New Zealand reptile, has a rod-dominated fovea.

Although the deep, well-defined foveas in bird retinas may have a significant optical function, the relatively broad and shallow human fovea probably does not, but moving the inner retinal layers aside should reduce light scatter and therefore reduce some image degradation that might otherwise occur. The most important feature is the absence of retinal capillaries; they have been pushed aside along with the retinal layers they supply, so there is no blood in front of the foveal cones to absorb and scatter light (the photoreceptors' blood supply is the choriocapillaris in the choroid).

The spatial distribution of a pigment in and around the fovea is responsible for entoptic images associated with the fovea

Entoptic phenomena are visual perceptions produced by something intrinsic to the eye, not to the visual stimulus. Among the simplest, most commonly

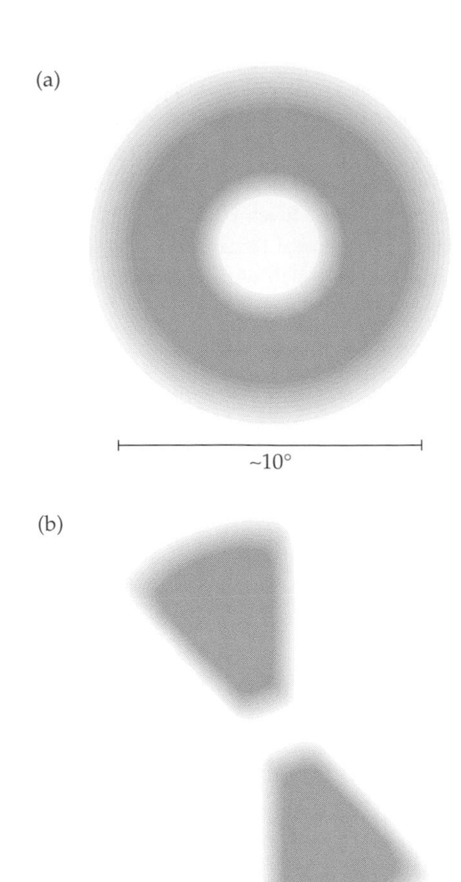

Figure 15.10
Entoptic Images Produced by the Macular Pigment
(*a*) Maxwell's spot appears to me as a dark ring or annulus with indistinct margins. The fixation point is at the center of the annulus. (*b*) Haidinger's brush also appears slighly darker than the background, but it has the shape of an hourglass.The waist of the hourglass marks the point of fixation.

observed examples are the small dark spots or blobs that sometimes drift through the visual field when we view a bright, uniformly illuminated surface. They are shadows cast by small clumps of debris adrift in the vitreous, and because they flit and jitter around as the eye moves, they are called **muscae volitantes** (flying flies). The Purkinje tree, which is the shadow cast by the retinal blood vessels, is another familiar entoptic image.

When the human retina is removed from an enucleated eye, the region containing the fovea acquires an intense lemon-yellow color; this is the **macula lutea** (yellow spot), described by Samuel Soemmering in 1795. The color is produced by xanthophyll pigments that are distributed throughout the retinal layers, but most densely within the outer plexiform layer, in a region about 2 mm in diameter centered on the fovea. The macula lutea is characteristic of diurnal primate retinas, but is not found in any other species. (This is but one of several types of intraocular filters, however; colored oil droplets within photoreceptors of fish, bird, and turtle retinas and the yellow pigment in the lenses of squirrel eyes are other examples. Although a macula lutea may be exceptional, intraocular filters of one sort or another are common in diurnal species.)

The function of the macular pigment is a matter for speculation, but it absorbs short wavelength light very strongly and may filter out potentially harmful ultraviolet light not already absorbed by the ocular media. The pigment may also mitigate the image-degrading effects of chromatic aberration by absorbing out-of-focus blue light. But whatever its function, one effect of the macular pigment is an entoptic image called **Maxwell's spot**, after the great physicist and investigator of color vision, James Clerk Maxwell (1831–1879).

Maxwell's spot can be seen by viewing a uniformly illuminated field through a filter that transmits short wavelengths (a filter that appears violet in color is particularly effective). The strong absorption by the macular pigment reduces the amount of light reaching the photoreceptors, so that the region containing the pigment will appear darker than the rest of the field; the result is an annular region of darkening around the point of fixation, as shown in Figure 15.10*a*. Although the macular pigment within the outer plexiform layer extends though the foveal center, the pigment density is less than it is elsewhere. Therefore, the region corresponding to the foveal center is lighter than the surrounding annulus. (And not everyone has dense macular pigment—Maxwell's wife, for example, could not see Maxwell's spot.)

The macular pigment is within the retinal cells, including the photoreceptor axons running radially out from the center of the fovea, which means that the pigment is not distributed uniformly but instead is restricted to radial bundles (see Figure 15.19). This radial geometry, combined with the fact that the pigment is dichroic (its transmission varies with the plane of polarization of light), produces another entoptic phenomenon, first reported in 1844, called **Haidinger's brushes**.

When a polarizing filter is placed in front of the eye observing a uniformly illuminated field, the brush appears as a dark hourglass-shaped pattern with the narrow neck of the hourglass at the fixation point (Figure 15.10*b*); the darkening is due to the reduced transmission perpendicular to the plane of polarization. Rotating the polaroid filter continually changes the polarization angle, and the brush will rotate around the fixation point like an airplane propeller.

Both Maxwell's spot and Haidinger's brushes can be used clinically as approximate markers for the position of the fovea in situations where it is important to ensure foveal fixation.

Clinical usage often treats the words "fovea" and "macula" as if they were synonyms, but they are not. Fovea refers to the pit, the region of thinning of the retina, while macula refers to the region of retina containing the xanthophyll pigment. Although the fovea lies near the center of the macula, the macula is larger, including the fovea and much of the parafovea. A person lacking macular pigment will have a fovea but will not have a macula.

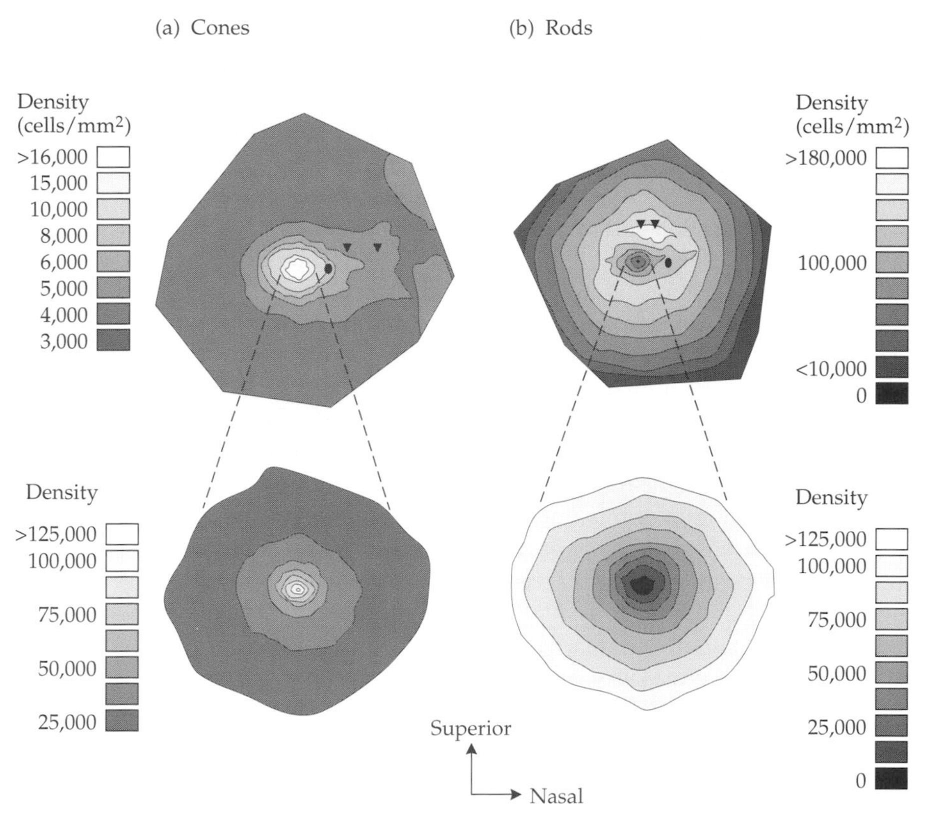

Figure 15.11
Density Maps of Human Cones and Rods
Cone (*a*) and rod (*b*) densities are shown for the entire retina and for the central 10° (bottom panel)—that is, the fovea and parafovea. The lines in these maps are iso-density contours drawn through points with the same density. Density changes most rapidly where the contour lines are close together, and lighter shading means higher density. Arrows indicate a "visual streak" on the cone map and the region of maximum rod density on the rod map. (After Curcio et al. 1990.)

Photoreceptor densities vary with respect to the center of the fovea, where cones have their maximum density and rods are absent

Although the fovea is neither at the geometric center of the retina nor situated on the eye's anatomical axis of symmetry, it is normally the functional reference point for the retina and for vision generally.* Much of the fovea's importance is directly related to the different distributions of the photoreceptors, which make the fovea the region in which we obtain our best spatial resolution and the more peripheral retina the site of highest sensitivity.

Measuring photoreceptor densities is difficult and tedious. One must count all photoreceptors in each of many small sampling fields judiciously placed at different retinal locations; there may be hundreds of sampling fields, and their placement is critical to prevent undersampling of the cell population. And if the sampling is to represent the whole retina and not just its principal meridians, there is a tricky problem in geometry; the retina in vivo is a hemisphere, but to study it under a microscope, it must be cut and flattened. The spatial relationship between locations on the flattened tissue and locations in the intact eye must be retained. Also, distinguishing between rods and cones near the fovea is not easy. But these problems have been solved, and the results are shown in Figure 15.11.

These representations of rod and cone densities are contour maps, like those used to map the terrain of Earth's surface, but the contour lines here connect

*There are some abnormalities for which this is not true. Individuals with **eccentric fixation** do not use the center of the fovea when looking at a target with one eye. Their fixational eye movements are larger than normal and centered around a point outside the fovea. Individuals with **anomalous correspondence** do not use the foveas together when attempting to fixate a target with both eyes. Instead, the fovea in one eye is associated with an extrafoveal point in the other eye. Anomalous correspondence is always accompanied by strabismus (see Chapter 4).

points having the same cell density rather than points having the same altitude. The rod and cone maps are very different. Cone density is relatively low over most of the retina, as indicated by the wide spacing of the isodensity contour lines, and increases sharply at the fovea (Figure 15.11*a*). This abrupt increase in cone density is more obvious in the map that includes only the fovea and its immediate surroundings. Rod density is almost the opposite: relatively high over most of the retina and disappearing, as if down into the crater of a volcano, into the depths at the fovea (Figure 15.11*b*). The only similarity in the maps is an elliptical hole at the location of the optic nerve head.

The density contours are irregular, and would be even more so if the data had not been smoothed somewhat, but a few of the bumps in the maps appear to be characteristic of all human retinas. In the cone map, for example, the arrows point to a region nasal to the fovea in which cell density is slightly more elevated relative to surrounding regions. This asymmetry is like a very diminished visual streak. A bulge on the rod map (arrow) superior and slightly nasal to the fovea appears to be the site of maximum rod density in the human retina. Because of this high rod density, this superior retinal region should be the most sensitive part of the dark-adapted retina; its location may be a reminder that not so long ago the darkest part of our nighttime environment was below the horizon, imaged on the superior retina. This and other sorts of differences between superior and inferior retinas are common in other species.

Figure 15.12 shows details of the photoreceptor density changes along the retina's horizontal meridian. The photomicrographs above the graph are the raw data from which the densities are calculated; they are optical sections through the photoreceptors at the level of the inner segments near the external limiting membrane. The central panel (0.0), from the foveal center, contains only cones, packed tightly together; in the flanking panels (1.35) the cones are noticeably larger and the rods that have appeared are about the same diameter as the foveal cones. In the more peripheral panels (5.0, 8.0, 16.0), the cones have about the same diameter and density, but rods are much more numerous and they become larger at the extreme peripheral retina (16.0).

Quantitative values for rod and cone densities corresponding to the photomicrographs are indicated on the curves for densities along the horizontal meridian. The maximum cone density is about 200,000 cones/mm^2; it falls by a factor of 10 at the edge of the fovea and continues to decline slowly to a minimum of around 3000 cones/mm^2 at the retinal periphery. Over most of the retina, cone density is less than 5000 cones/mm^2. Maximum rod density is around 150,000 rods/mm^2, about 20° from the foveal center, falling to just under 100,000 rods/mm^2 in the periphery. Rod density falls to zero in the center of the fovea. The **rod-free area** is about 300 μm (1° of visual angle) in diameter. Note that the rod-free area is considerably smaller than the fovea; the occasional statement that the fovea contains no rods is incorrect.

The human retina varies from center to periphery in terms of the spatial detail in the retinal sketch

The human retina is organized around the point of maximum cone density, and most measures of visual performance vary relative to the fovea. Figure 15.13 shows how visual acuity, plotted on a relative scale, declines with retinal eccentricity—that is, with distance (or degrees of visual angle) from the maximum at the center of the fovea. Visual acuity is almost halved at 1° from the foveal center (which is still within the fovea) and halved again at 5° from the foveal center. In Snellen notation, if the maximum acuity were 20/20, it would fall to about 20/40 at 1° of eccentricity and to 20/80 at 5° eccentricity. Beyond 20° eccentricity, acuity is below 20/200, which is a common legal standard for blindness.

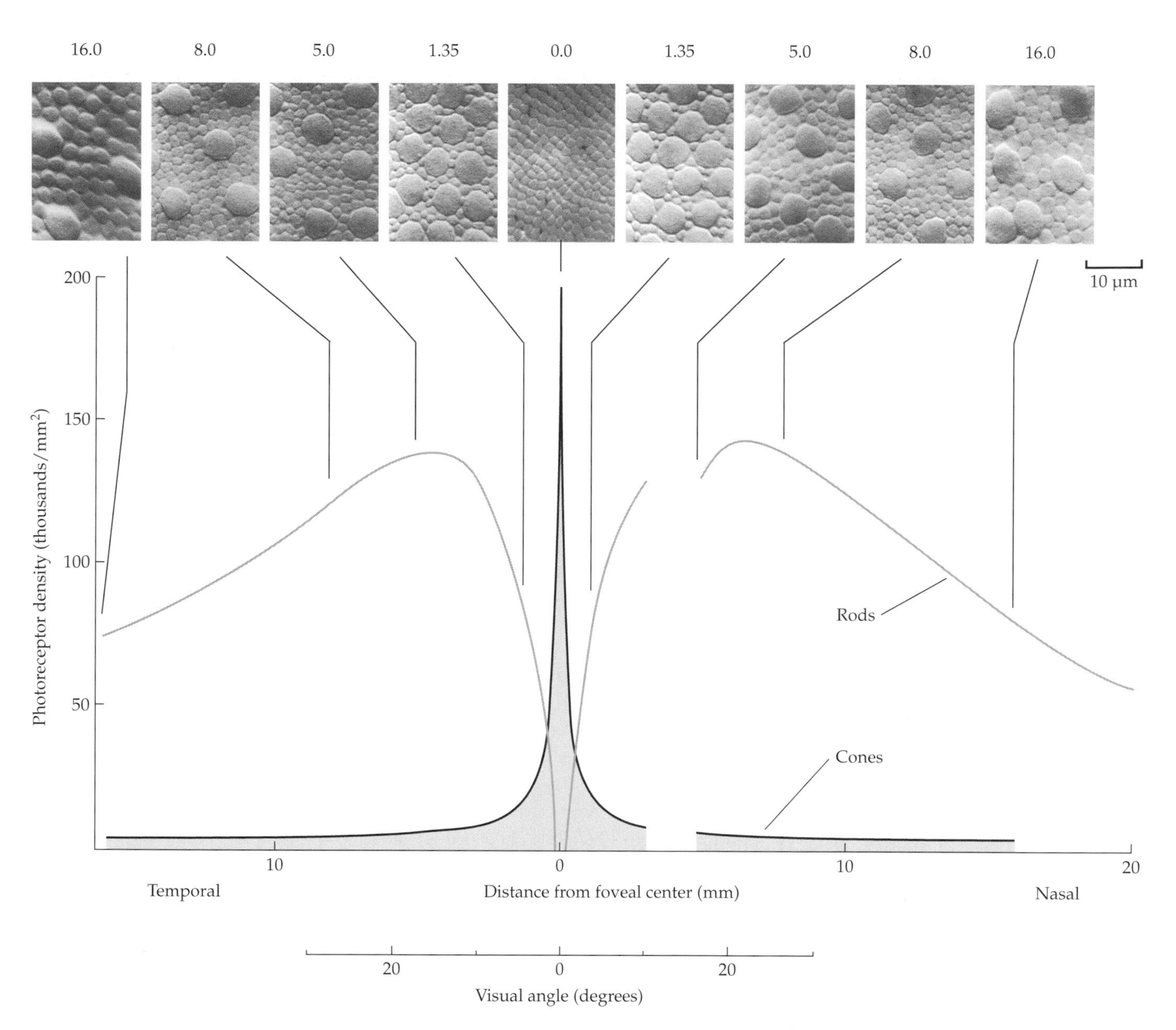

Figure 15.12
Photoreceptor Densities as Functions of Eccentricity along the Horizontal Meridian
The photographs are optical sections through the photoreceptor inner segments. They show photoreceptors at different distances from the center of the fovea (0.0); the distances are given in millimeters above each photograph, and their locations are correlated with the rod density curve below. Except for the central panel (0.0), which contains only cones, the larger cells are cones and the smaller ones are rods. The break on the nasal side of the graph marks the location of the optic nerve head. (Photographs and data [replotted] from Curcio et al. 1990.)

Another measure of performance shown in Figure 15.13 is one of the hyperacuities, tasks whose thresholds are smaller than the diameter of a single cone. (I hesitate to mention this because a spatial threshold smaller than a cone diameter seems to invalidate the earlier discussion about cone spacing as a limitation on spatial resolution. Hyperacuities are not resolution tasks, however; they involve judgments of relative location, which is a different problem from sampling spatial frequency in an image.) The relative hyperacuity sensitivity shown here relates to detection of the displacement of a line. It also declines significantly as a function of retinal eccentricity, though not at the same rate as resolution acuity does.

Other measures of visual performance not shown here, such as color discrimination, also decline in the peripheral retina. There are, however, some performance measures that improve in the periphery, because they are more dependent on the rod pathways. The eye's absolute sensitivity to light is the most obvious example, but temporal discrimination also increases with eccentricity from the fovea.

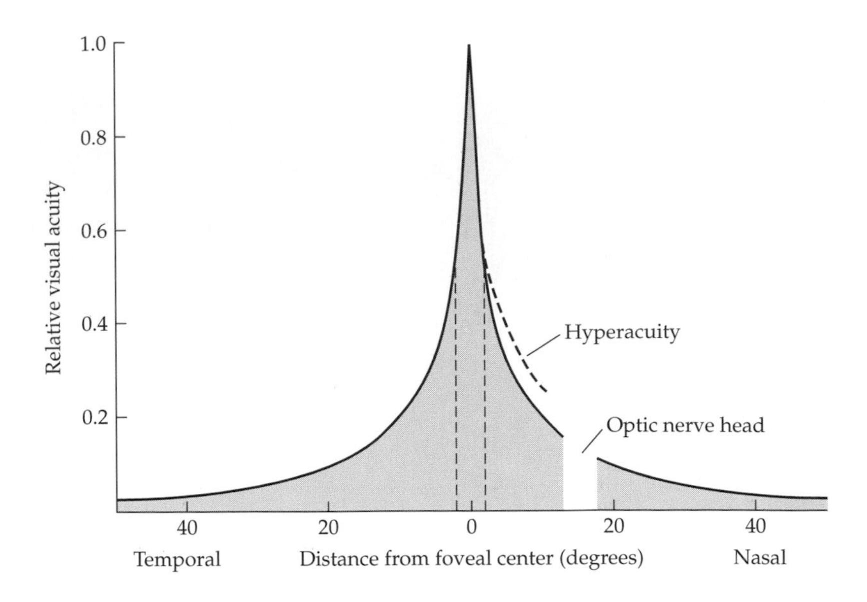

Figure 15.13
Decline in Spatial Resolution with Retinal Eccentricity
Dashed vertical lines show the limits of the fovea. The dashed curve shows relative hyperacuity sensitivity to target displacement (a line). (Replotted from Wertheim 1894 [relative visual acuity] and Westheimer 1979 [relative hyperacuity].)

Maximum cone densities vary among different retinas by a factor of three

Figure 15.14 shows cone densities from six retinas plotted as functions of eccentricity from the foveal center out to near the edge of the fovea. The maximum cone densities at the center of the fovea vary by a factor of three, ranging from just over 100,000 cones/mm^2 to about 350,000 cones/mm^2. At the edge of the fovea, however, the cone density is almost identical for all retinas.

Assuming that the observed variation is not artifactual, the results imply substantial differences in the spatial resolution limit set by the cone spacing. For these retinas, the Nyquist frequencies should range from just over 50 cycles/degree to almost 90 cycles/degree, or, in terms of Snellen acuities, from about 20/14 to 20/7 (one should not make too much of the actual numbers; it is the twofold variation in estimated best acuity that is important). The notion that there is substantial variation in best acuity among individuals will be no surprise to clinicians, but the fact that the variation may be a direct consequence of differences in peak cone density is interesting news.

Rod densities also vary among retinas, but less widely than cone densities do. Although rod density is most variable near the fovea, where cone density also varies significantly, the fovea is where the absolute rod density is relatively low and large percentage variations do not create large numerical differences. In the regions of maximum rod density, the variation among retinas is only about ±10%, a striking contrast to the roughly ±50% variation in peak cone densities.

The human retina has about 4.5 million cones and 91 million rods

No one has ever counted all the photoreceptors in the human retina, but estimates of the numbers can be made from density maps like those in Figure 15.11. A mean density is assigned to each region bounded by isodensity contours; it is usually the average of the densities of the boundary contours. The area of the bounded region can be measured, and this area, multiplied by the mean density, gives an estimate of the total number of cells in the region; that is, density (cells/mm^2) × area (mm^2) = number of cells. When this calculation is done for all the different regions and they are added together, the sum is an estimate of the total number of cells in the retina. The more regions there are—that is, the closer the isodensity contours are spaced—the better the estimate will be (also, the more sampling fields that are used in the original counts, the more accurate are the isodensity contours and the areas that are measured).

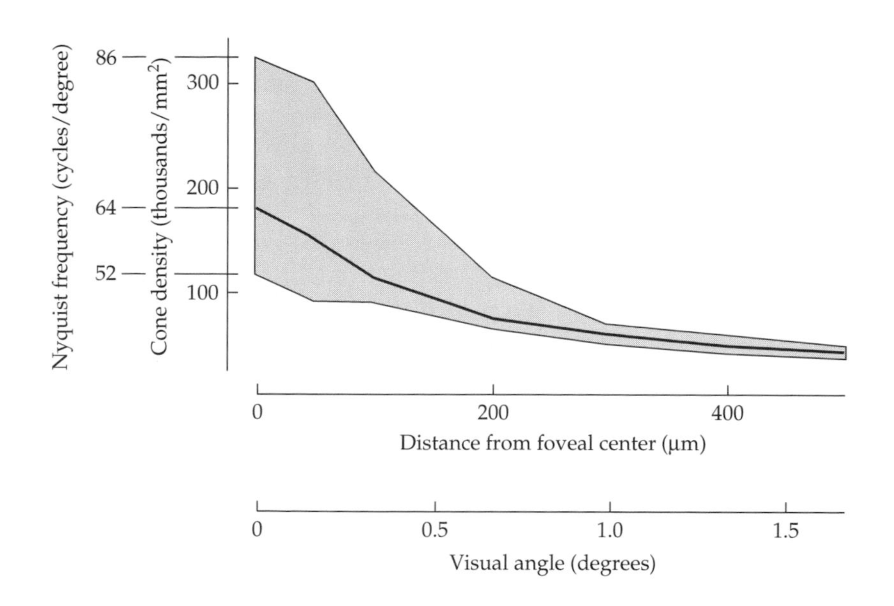

Figure 15.14
Individual Variation in Foveal Cone Density
The shaded region encompasses data from six retinas showing cone density at the foveal center (point of maximum density) and at other locations out to 500 µm from the center. Median cone density for the six retinas is indicated by the dark line within the shaded region. (Replotted from Curcio et al. 1990.)

Table 15.1 shows some of the quantitative results from the six retinas in Figure 15.14; these retinas had a median area of just over 1000 mm^2, with median values of 4.5 million cones and 91 million rods. Higher values usually cited in the literature (6.2 million cones, 110 million rods) are from one retina studied by Østerberg, who reported lower peak densities for both rods and cones but higher total numbers for each because of undersampling near the fovea.

The number and density of photoreceptors change with age, a point that will considered in more detail in the Epilogue. But to summarize, however, the number of cones does not change much throughout life, while the number of rods decreases steadily and dramatically.

Blue cones have a different distribution than red and green cones have: the center of the fovea is dichromatic

Normal human color vision is trichromatic, requiring mixtures of red, green, and blue primaries to match a test color. Dichromats are insensitive to one of these three primaries and require only two primaries to match a test color; **protanopia** is red "blindness," **deuteranopia** is green "blindness," and **tritanopia** is blue "blindness." Protanopes and deuteranopes have difficulty distinguishing

Table 15.1

Numbers and Maximum Densities of Human Rods and Cones

		Cones		Rods	
Eye	Area of retina (mm^2)	Max. density (cones/mm^2)	Total cones (millions)	Max. density (rods/mm^2)	Total rods (millions)
H1	1105	311,000	4.7	189,000	—
H3	856	120,000	4.1	158,000	77.9
H4	964	182,000	5.3	181,000	107.3
H5	1104	178,000	4.4	167,000	90.7
H6	976	324,000	4.5	—	—
H7	1106	181,000	—	—	—
Median	1040	181,500	4.5	174,000	90.7

Source: From Curcio et al. 1990, rounded one decimal point.

between reds and greens, while tritanopes confuse blues and greens. Protanopia and deuteranopia are much more common than tritanopia, and the milder forms—**protanomaly** and **deuteranomaly**—are even more prevalent. Roughly 10% of males and 1% of females exhibit some deficiency in color discrimination. But psychophysical studies have for many years indicated that the normal retina is not trichromatic everywhere; specifically, the center of the fovea is dichromatic, exhibiting a condition called **foveal tritanopia**.

We now know that foveal tritanopia is due to a lack of blue cones in the center of the fovea. There are a variety of methods by which blue cones can be labeled selectively and recognized in anatomical preparations. One technique uses a monoclonal antibody (see Box 15.1) that is specific for the blue cone photopigment. The blue cones can be located and mapped in retinas treated with this antibody, as in Figure 15.15*a*, which shows a small region near the center of the fovea. The center of the fovea has no blue cones; this blue cone–free area lies within, but is smaller than, the rod-free area.

The density of blue cones reaches its maximum (about 1800 cells/mm^2) along the foveal slope about halfway between the center and edge of the fovea. Density declines in the parafovea to around 500 cells/mm^2. It is not shown here, but the rise and fall of blue cone density around the fovea is asymmetric; density falls off most rapidly nasal and inferior to the fovea.

Blue cones are very much a minority cell population. In terms of cell density, the maximum blue cone density is less than 2000 cells/mm^2 in a part of the fovea where the total cone density is more than 50,000 cells/mm^2 (Figure 15.15*c*); this is barely 4% of the total. At most, blue cones never constitute more than about 8% of the cone population at any retinal location (Figure 15.15*d*). The data are not complete enough to specify a number for all blue cones, but they are unlikely to constitute much more than about 5% of all cones, and their total number is probably less than 250,000 cells.

Since the blue cones have a low density and wide separation in comparison to the more numerous red and green cones, one would expect visual acuity for blue or violet stimuli that stimulate only blue cones to be relatively poor. This has been shown to be the case, and the maximum acuity for blue targets changes with retinal eccentricity in parallel with the density of blue cones.

There are more red cones than green cones, and more green cones than blue cones

At the moment, we do not have selective labels for human red or green cones that allow their specific retinal distributions to be studied. But some methods that are meant to isolate blue cones also produce two different shades of staining in the remaining cones; assuming that this difference corresponds to the red–green cone dichotomy, one group appears to outnumber the other by about twofold. Indirect evidence from psychophysical studies using stimuli designed to isolate one or the other of the red and green cone systems indicates that the more numerous group is the red cones, as does microspectrophotometry on small samples of cones in a limited retinal region (the ratio is less than 2:1, however). If we use the 2:1 relationship for red and green cones, the total cone population in human retina breaks down as follows: red cones, 63% (about 2.9 million); green cones, 32% (about 1.4 million); blue cones, 5% (about 0.2 million).

The distribution of different types of cones is neither regular nor random

Collectively, the photoreceptors tile the retina, but the individual types—the rods and the three cone types—do not. Each photoreceptor type is distributed within the overall tiling, and the nature of the distributions has implications for

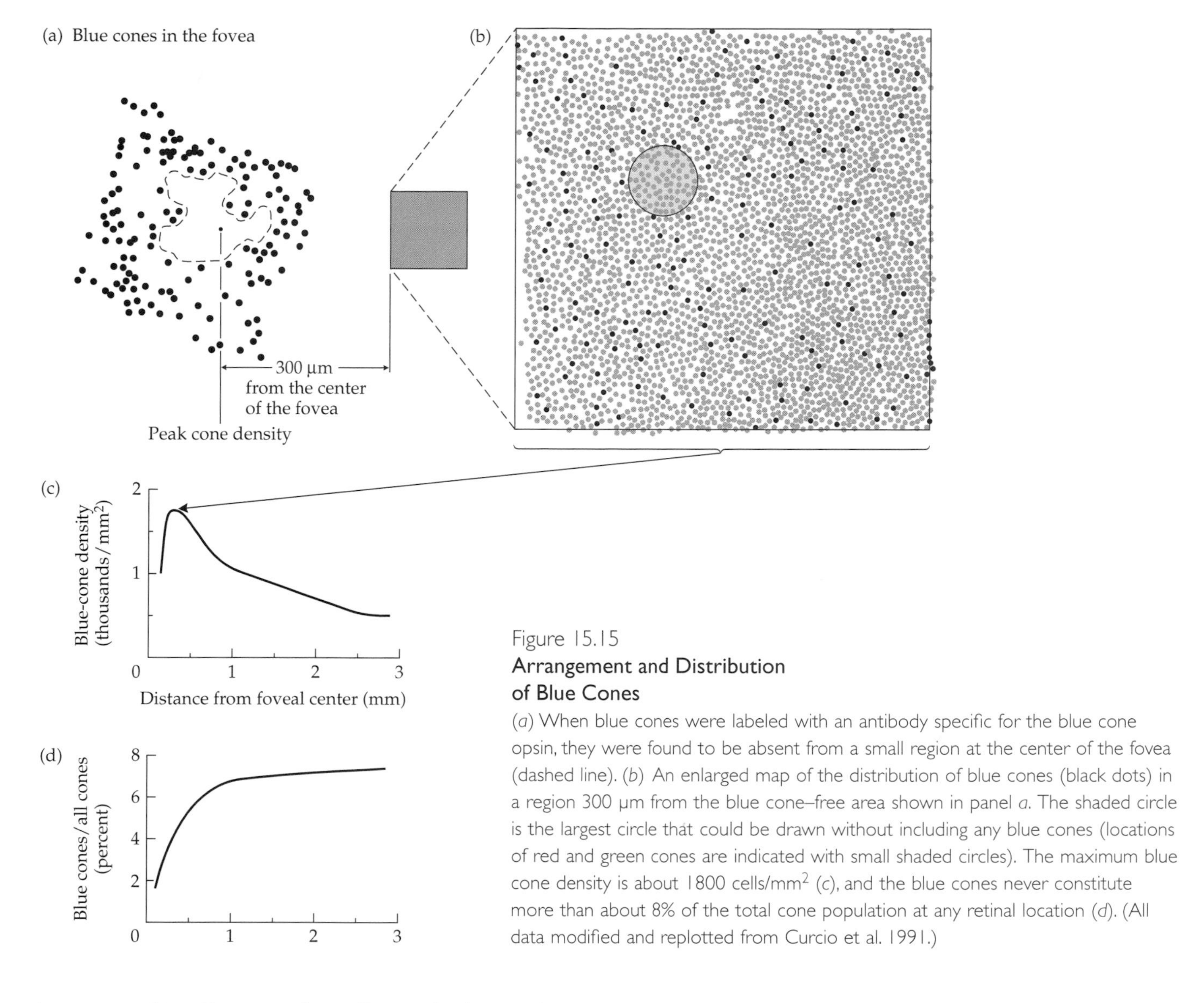

Figure 15.15
Arrangement and Distribution of Blue Cones
(*a*) When blue cones were labeled with an antibody specific for the blue cone opsin, they were found to be absent from a small region at the center of the fovea (dashed line). (*b*) An enlarged map of the distribution of blue cones (black dots) in a region 300 μm from the blue cone–free area shown in panel *a*. The shaded circle is the largest circle that could be drawn without including any blue cones (locations of red and green cones are indicated with small shaded circles). The maximum blue cone density is about 1800 cells/mm^2 (*c*), and the blue cones never constitute more than about 8% of the total cone population at any retinal location (*d*). (All data modified and replotted from Curcio et al. 1991.)

image sampling. For example, will a patch of retina large enough to include 20 cones always have 12 red, 7 green, and 1 blue, or will the proportions vary to the extent that blue or green cones might be absent in some locations and present in high proportions in others? Are the cone types distributed regularly, which could make aliasing an issue, or are they irregular?

The blue cone mosaic is the best understood; Figure 15.15*b* is a map showing locations of blue cones within the fovea, along with the undifferentiated red and green cones. Although some blue cones occasionally form the vertices of equilateral triangles, suggesting an underlying triangular lattice for the blue cone mosaic, this rigidity is exceptional; the blue cones are not all equally spaced. The shaded circle drawn on the mosaic shows a region lacking blue cones; it could not have been drawn that large if the array were perfectly regular. On the other hand, the blue cones are not randomly distributed; if they were, one should occasionally find places where two blue cones are side by side, and there are none.* It therefore seems that any sizable patch of retina containing blue cones will have

*The defining property of random distributions is independence of locations. Nearest neighbors are sometimes side by side and sometimes widely separated, because there is no rule that specifies a minimum distance between neighbors. Nonrandom distributions always have some constraints on locations, which means that cells are not distributed independently of one another.

[Box 15.1] Locating Species of Molecules: Immunohistochemistry

In addition to the many molecular species that cells have in common, cell specialization demands some molecular specialization. As a result, cells differ in molecular composition. Muscle cells contain contractile proteins not found in other cell groups, for example, and neurons differ from one another in their amino acids, specialized proteins at receptor sites, synthetic enzymes, and so on. These differences can be used to label and visualize cells that contain specific molecular species. If the target molecule is distributed throughout the cells, the entire cellular structure can be seen.

The methods for locating molecular species in cells are collectively called **immunohistochemistry** because they exploit the capacity of the immune system to recognize foreign molecules and generate **antibodies** that preferentially bind to them. The reaction of the immune system is an effort to neutralize and render the invading substances ineffective, but it can be controlled and used to advantage.

For example, the enzyme tyrosine hydroxylase participates in the synthesis of dopamine, which is a neuromodulator used in the central nervous system and the retina. Molecules of tyrosine hydroxylase injected into the body of a mouse elicit the production of antibodies that are specific for tyrosine hydroxylase; once produced, the antibodies can be extracted from the mouse's serum. When a piece of neural tissue is exposed to the antibody-containing serum, the antibodies will bind to the **antigen** that elicted their production in the first place—in this case to molecules of tyrosine hydroxylase. If other radioactive, fluorescent, or oxidant molecules (like horseradish peroxidase) are coupled to the antibodies, their aggregation in the tissue can be visualized microscopically; the elements of the procedure at the molecular level are shown in Figure 1.

Antibodies can be made for the pure form of any antigen; although the purity requirement is important, it is not unduly restrictive and many antibodies directed against purified antigens are commercially available. Antibodies produced in this way are **polyclonal** antibodies; they are usually mixtures of antibodies specific for different parts (**epitopes**) of large antigens, like proteins, but they may also be directed with high specificity to small molecules, like amino acids.

Monoclonal antibodies bind to a single part of an unknown antigen. After an antigen source, such as some homogenized retina, is injected into a mouse, the animal's lymphocytes will produce antibodies to many antigens in the retinal tissue. These antibody-producing lymphocytes are combined with a mutant strain of lymphocytes that grow readily as tumors, producing fused cells called **hybridomas**. The hybridomas will survive and grow only in a special culture medium. Tests of the medium for immunoreactivity with the target tissue—the retina, in this case—can identify specific antibody-producing hybridomas so that they can be cloned to provide a continuous source of the antibody. None of this process is particularly easy (the people who figured it out received a Nobel prize for it), but the difficulty of producing monoclonal antibodies is outweighed by the ability to generate a highly specific antibody without having a purified antigen. As a result, we now have monoclonal antibodies that bind to different photopigment opsins, to different neurotransmitter receptor sites, or even to different parts of an acetylcholine receptor.

Examples of immunohistochemical labeling in Figure 2 include an amacrine cell in human retina labeled with a tyrosine hydoxylase antibody, primate bipolar cells labeled with an antibody to GLT-1, which is a glutamate transporter protein, blue cones labeled with an antibody to the blue cone opsin, and a section of monkey retina treated with an antibody to glutamate. The glutamate labeling shows the amount of glutamate present; the darker the reaction product, the higher the glutamate concentration. In this case, the glutamate labeling includes photoreceptors, most bipolar cells, some amacrine

Figure 1
Method for Indirect Immunolabeling
Primary antibodies carry a label for the species in which they were produced—for example, rabbit. These primary antibodies will bind to their target antigen when they are added to the antigen-containing tissue. Here, two antibodies bind to different sites on the target antigen. Secondary antibodies specific for the "rabbit" component of the primary antibody are then added, and they bind to the primary antibodies. The secondary antibodies have markers or labels attached to them. The markers, such as horseradish peroxidase, fluorescent dyes, or gold particles, become visible after appropriate histological processing of the tissue. Since several secondary antibodies bind to each primary antibody, there is a certain amount of amplification; in this case, each antigen molecle is represented visually by six secondary antibodies.

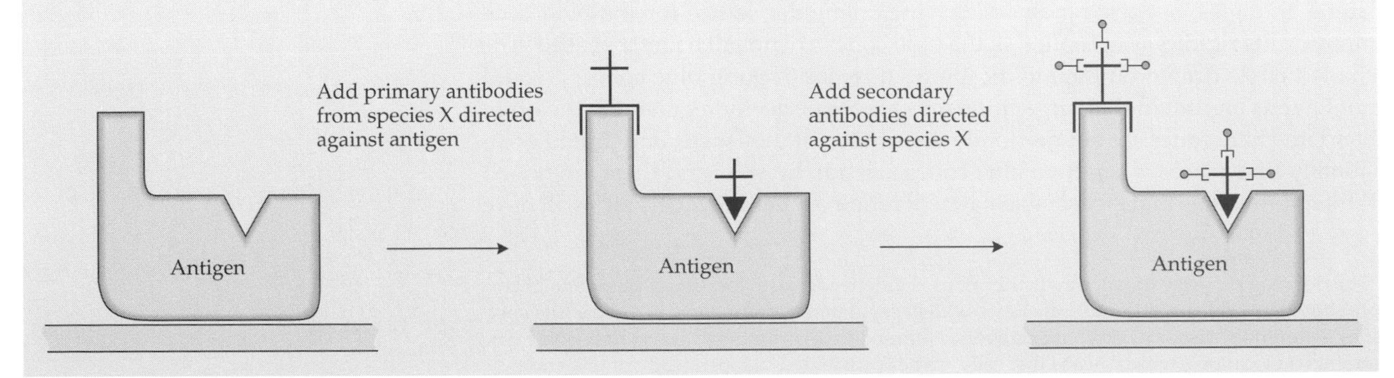

cells, and most ganglion cells, indicating that these retinal neurons contain glutamate.

Immunohistochemical methods are important because they allow us to isolate and study specific cell groups from mixed populations of cells, but the methods are not foolproof and do not invariably produce unequivocal answers to experimental questions. In general, there are two broad categories of problems with immunohistochemical labeling. False positives result when something other than the intended target of the antibody shows labeling. At the other extreme, failure to label the antigen even when it is present is a false negative.

False positives can occur when the antibody is specific for a very short sequence of amino acids in a target protein; if the same sequence exists in another protein, it will be labeled along with the intended target. Also, the antigen may be distributed too generally; tyrosine hydroxylase, for example, is in the synthetic pathway for all catecholamines, not just dopamine, and tyrosine hydroxylase labeling does not automatically mean that a cell contains dopamine. Investigators routinely compare results from different antibodies targeting the same antigen to recognize and control for false positives, but residual uncertainties often require qualified descriptions of results. Phrases like "tyrosine hydroxylase–like immunoreactivity" are awkward, but honest.

False negatives can be difficult to recognize. Does the failure to label mean that the antigen is not present, which would be a *true* negative, or has the labeling failed for a procedural reason? If other evidence indicates that a particular antigen should be present, investigators can modify their protocols until they get positive results. But this is not always possible; some negative results have been reported that were later found to be *false* negatives.

In spite of the possible interpretive problems, immunohistochemical methods employed with proper controls are very powerful. Most of the specific types of retinal neurons discussed in the text were visualized with specific antibodies. And the potential has yet to be fully realized. Any cell in the body, for example, has several thousand different proteins that are potential targets for antibodies, and most of these proteins have yet to be targeted. Also, the chemical messengers for induction, guidance, and other cellular interactions during development are not well understood, but their temporal and spatial sequences of expression will probably become known by means of immunohistochemical labeling.

(a)

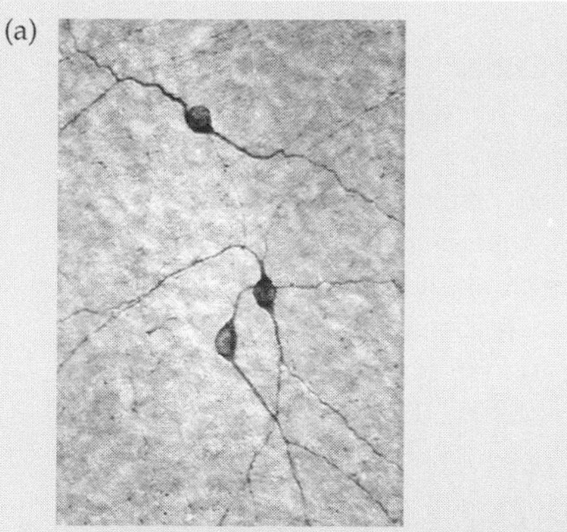

(b)

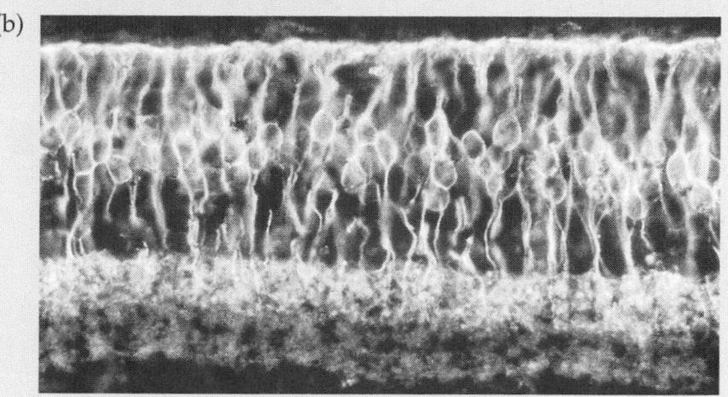

(c)

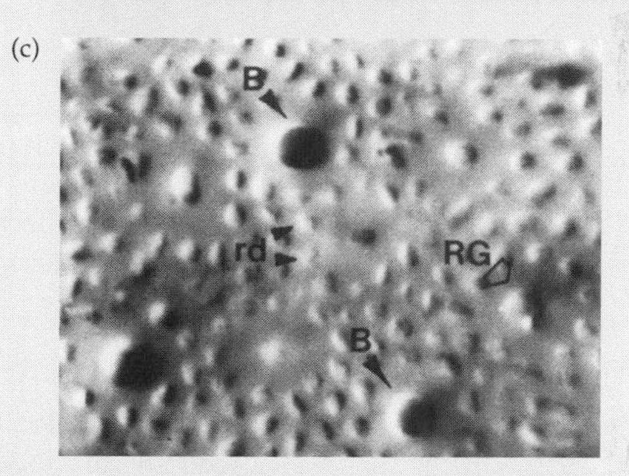

(d)

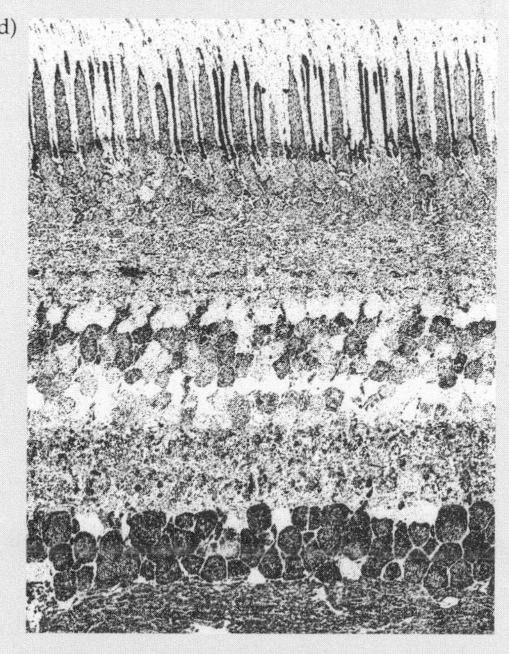

Figure 2
Immunolabeling in the Retina
(*a*) Amacrine cells in flat-mounted rabbit retina labeled with antibodies directed against tyrosine hydroxylase; these are presumed to be dopamine-containing cell cells. (*b*) Diffuse cone bipolar cells in sectioned monkey retina labeled with antibodies directed against GLT-1, which is a glutamate transporter protein. (c) Blue cones (B) in flat-mounted human retina labeled with a monoclonal antibody directed against the blue cone opsin. (*d*) Neurons in a section through monkey retina labeled with an antibody to glutamate. The density of the label is proportional to the glutamate concentration in the cells. (*b* courtesy of Thomas Rauen and Heinz Wässle; *c* from Curcio et al. 1991; *d* courtesy of Michael Kalloniatis and Robert Marc.)

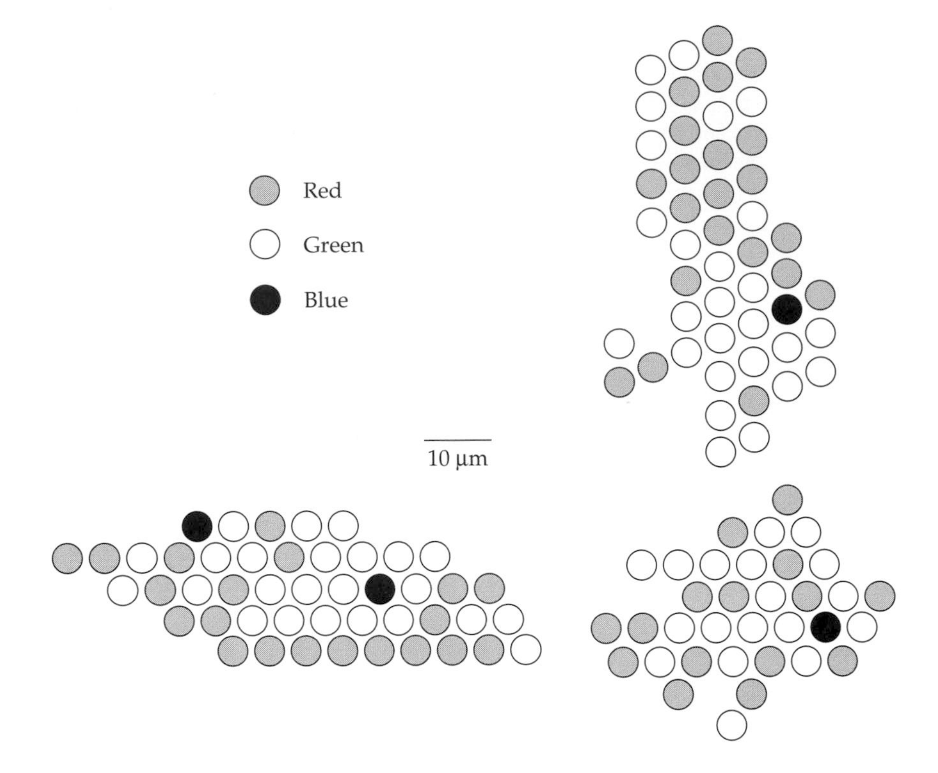

Figure 15.16
Arrangement of Foveal Cones
Locations of cones identified as red, green, or blue by microspectrophotometry in three small patches of retina within the monkey fovea are indicated by circles; the circles are drawn in a triangular lattice that is more regular than the actual cone locations. Both red and green cones tend to form clusters. (After Mollon and Bowmaker 1992.)

a fairly uniform sampling for "blueness" but that the irregularity in the blue cone mosaic will reduce aliasing.

The arrangements of the red and green cones are not known to the same level of detail. Psychophysical experiments, however, suggest that their arrays are not perfectly regular. But at the other extreme, they are probably not distributed randomly. In other primates, in which the spectral properties of each cone have been determined for small patches of retina, the red and green cones seem to cluster, forming small clumps of five or six cones of a particular type within the overall cone mosaic (Figure 15.16). Clustering is a form of irregularity, but it is not characteristic of random distributions. In theory, there would be no color discrimination if an image fell within one of these red or green cone clusters, but the likelihood that this would happen is extremely small, particularly since fixational eye movements normally move the retinal image across areas larger than this.

There are two implications of the "neither regular nor random" arrangements of the different cone types. Outside the fovea, where the spatial frequencies in the retinal image may exceed the Nyquist frequency of the individual cone arrays, the irregularity should reduce aliasing. Yet the degree of order in the arrays should give reasonably uniform sampling by the different cone types at all retinal locations.

Cone pedicles probably tile the retina in and near the fovea, but rod spherules probably never form a single-layered tiling

Cone pedicles form a nearly continuous sheet in the outer synaptic layer near the fovea (Figure 15.17); the gaps between pedicles are due primarily to Müller's cells. Thus the array of pedicles is a tiling, or something very close to a tiling. Rod spherules, which are not shown here, are piled up several layers deep just outside the layer of cone pedicles. This is not a tiling, in the sense of covering a plane with a single layer of tiles, but proximity of neighboring rods should be retained in the layers of spherules. Their arrangement is like beginning with a line of people standing shoulder to shoulder, having every second

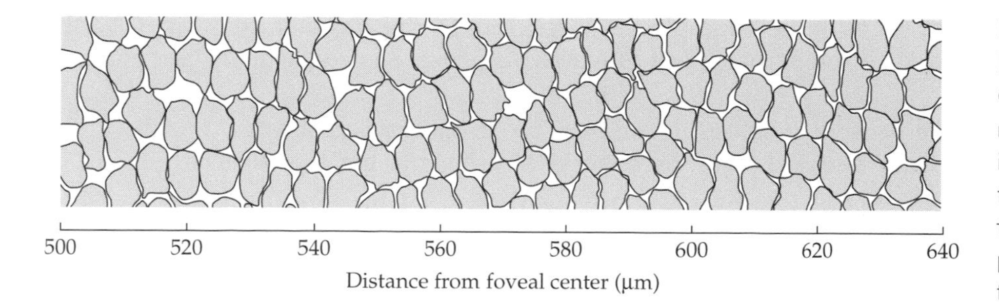

Figure 15.17
Distribution of Cone Pedicles
Cone pedicles in the outer synaptic layer near the fovea form a nearly complete, irregular tiling. These pedicles lie on the foveal slope between 500 and 640 μm from the foveal center and are just outside the pedicles corresponding to the cones at the foveal center. (After Tsukamoto et al. 1992.)

person take one step back to form a second row, and then closing up the spaces again. Spatial neighborliness is then preserved in columns of two.

As cone density declines away from the fovea, gaps open between cone pedicles and the pedicles do not maintain a complete tiling of the outer synaptic layer. These gaps in the array of cone pedicles, however, correspond to gaps in the array of cone inner segments to which the pedicles are connected. That is, the cones do not form a complete tiling away from the fovea, and their pedicles don't either. The gaps in the outer synaptic layer array are probably filled by rod spherules, much as gaps in the cone inner segment array are filled by rod inner segments. Like the tiling by the photoreceptor inner segments, the tiling of the outer synaptic layer is a tiling by *all* photoreceptors; none of the different elements in the array ever tiles completely on its own.

The progressive breakdown of the cone pedicle tiling and the intrusion of rod spherules into the array creates no problem for passing signals from the photoreceptors to bipolar cells in the central retina (Figure 15.18). Diffuse cone bipolar cells still receive inputs from all red and green cones within reach of their dendrites, and midget bipolar cells still contact just one cone, either red or green. Rod bipolar cells extend their dendrites past the cone pedicles to contact rod spherules in both the layers of spherules. And the blue cone bipolar cells send their sparsely branched dendrites selectively to several neighboring blue cones. Thus all the elements in the photoreceptor array are sampled, either selectively or indiscriminately, to form multiple continuous samplings of the photoreceptor signals.

There are no significant changes in the underlying rules as the cone pedicles become more and more widely spaced. Diffuse bipolar cells become more diffuse in appearance, but they still contact all red and green cones within reach of their dendrites. Midget bipolar cells still receive input from just one cone throughout the central retina, but in the periphery they receive inputs from sev-

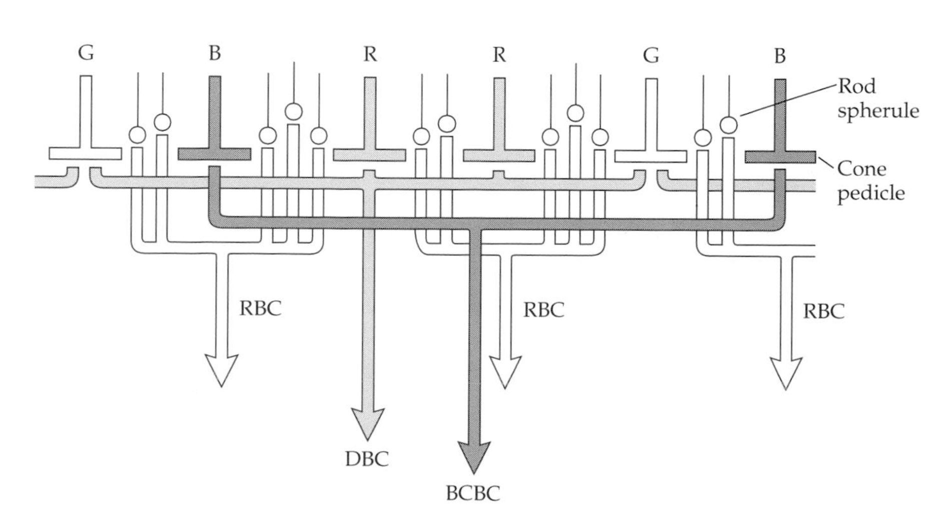

Figure 15.18
Connections from Photoreceptor to Bipolar Cells outside the Fovea
Although cone pedicles have gaps between them and do not form a complete tiling at locations outside the fovea, they are still completely sampled by all of the different types of cone bipolar cells, thus maintaining an uninterrupted functional sampling of the cones. For clarity, only one type of diffuse bipolar cell is shown, and the midget bipolar cells are omitted altogether. Rod bipolar cell also make a complete sampling of the rods. R, red cone; G, green cone; B, blue cone; RBP, rod bipolar cell; DBC, diffuse bipolar cell; BCBC, blue cone bipolar cell.

eral cones of the same type (red or green). Peripheral blue cone bipolar cells, like their counterparts near the fovea, recognize and contact only blue cones.

The pedicles of cones in and near the fovea are displaced radially outward from the cone inner segments, but spatial order is preserved

Sweeping the inner retinal layers aside at the fovea means that the pedicles of foveal cones do not line up along the radial axis of the cone inner and outer segments; the pedicles are also swept aside. But as we have already seen, the foveal cones form a regular tiling of the retinal image, and this arrangement should be preserved in the cone pedicles that are passing signals along to the bipolar cells. The way order is maintained is shown in Figure 15.19.

The radial distance of a cone pedicle from the foveal center is related to the location of its parent inner segment. Pedicles belonging to cones at the foveal center lie along the foveal slope about 300 μm or so from the cone inner segments. The next pedicles outward from the first belong to the next cones outward from the first, and so on, with each successive pedicle corresponding to the next successive cone. Since the same thing is happening along many radii from the foveal center, the pedicles of the central cone cluster form a circle lying 300 μm from the center. Thus the cone axons—the fibers of Henle—run radially from the center, with Haidinger's brushes as the visible consequence.

Because of this arrangement, two neighboring cones near the center of the fovea may have their pedicles on opposite sides of the fovea, separated by more than 600 μm. And because the axon terminals of the midget bipolar cells are also swept outward from their dendrites contacting the cone pedicles, the corresponding midget ganglion cells for the two central cones will be even farther

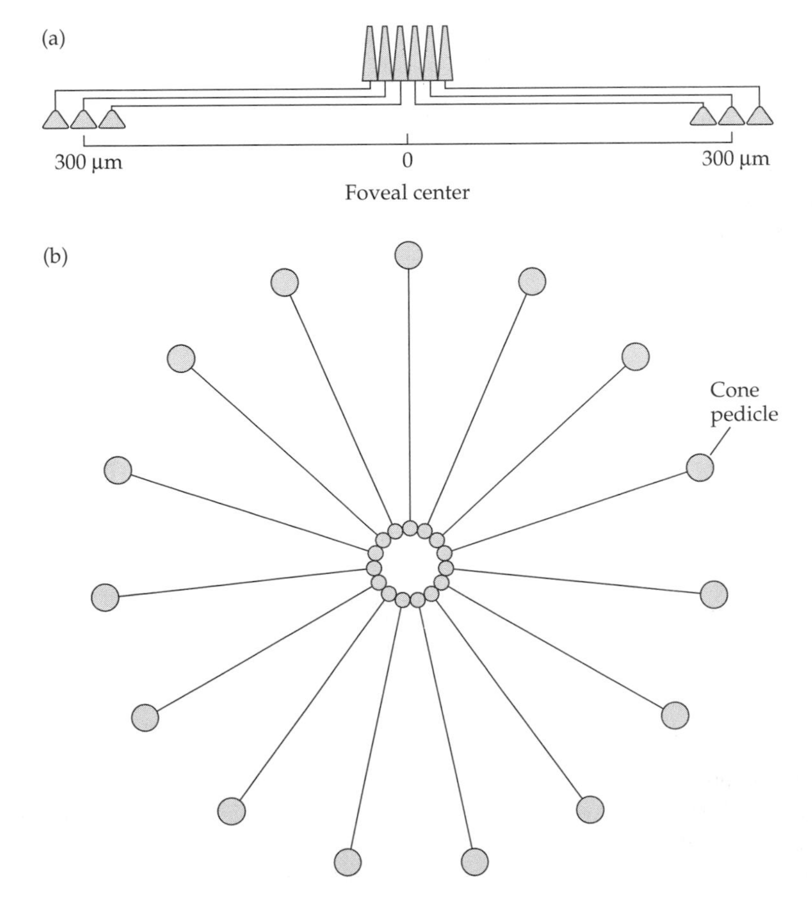

Figure 15.19
Displacement and Arrangement of Pedicles from Foveal Cones
(*a*) The pedicles of the central cones are displaced outward about 300 μm from the cone inner segments. Spatial order is maintained because neighboring pedicles correspond to neighboring inner segments.
(*b*) Viewed from above, a circle of cones around the center of the fovea can be seen to have axons running radially out to their pedicles. The gaps between the pedicles from neighboring cones are artifacts of the scaling; many more cone inner segments will lie in the circle around the foveal center than are shown here. (After Tsukamoto et al. 1992.)

apart. Thus signals from cones that are neighbors and are responding to stimuli very close together in visual space end up in two widely separated vertical pathways.

This separation does not create a sizable discontinuity in the mapping of visual space, because the axons from the two separate ganglion cells end up side by side when they terminate in the lateral geniculate nucleus. The vertical pathways for neighboring cones have been separated by the fovea, but when the pathways come together again, it is as if the fovea does not exist; the continuous mapping of visual space is maintained. Some of the mechanisms underlying the organization of ganglion cell projections to the brain during development were considered in Chapter 5.

The density of horizontal cells is highest near the fovea and declines in parallel with cone density

The distribution of horizontal cells in the human retina is not known, but it is likely to be similar to the distribution in monkey retina (Figure 15.20). No horizontal cell bodies appear in the center of the fovea where the inner nuclear layer is absent. They begin to appear along the foveal slope and quickly rise to their maximum density (about 20,000 cells/mm^2) around the edge of the fovea. From this maximum, the density decreases more or less exponentially out to the retinal periphery; the decline is less rapid nasal to the fovea, apparently in keeping with the less rapid decline in cone density there. At the retinal periphery, the horizontal cell density is about 20 times less than the maximum.

Were it not for the absence of horizontal cells in the foveal center, the distribution of horizontal cells would be similar to the plots of cone distribution. But since the cones in the center of the fovea have their pedicles pushed radially outward, the direct comparison is a bit misleading; a plot of the distribution of cone *pedicles* would also go to zero in the center of the fovea and would therefore be

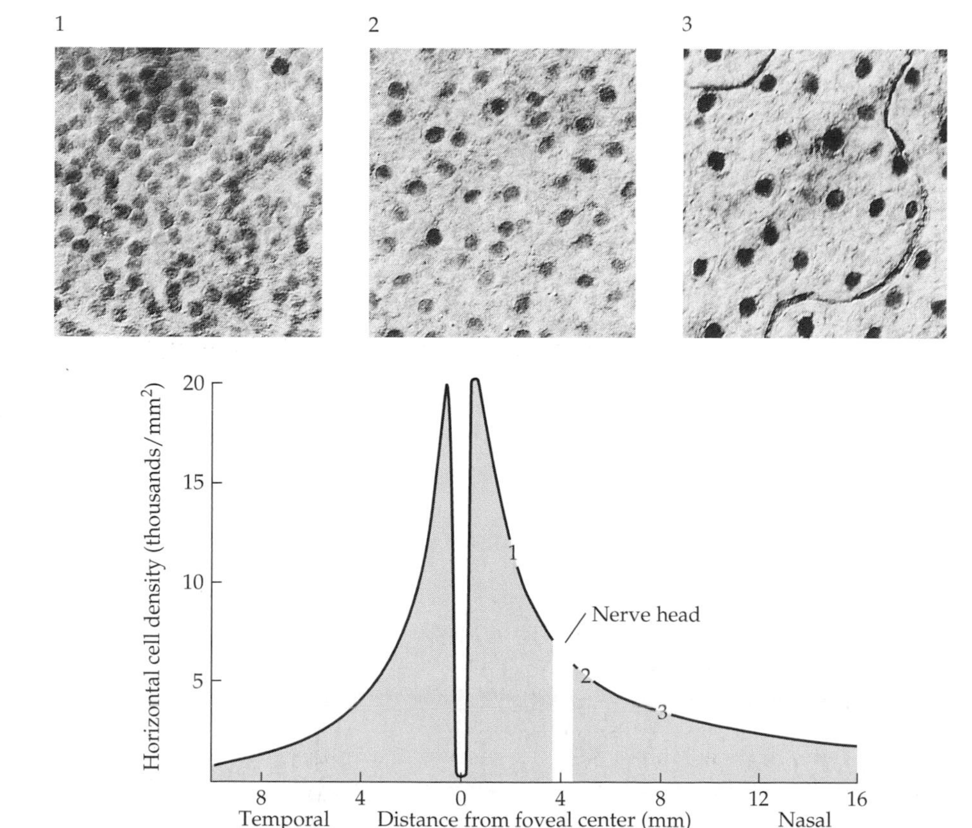

Figure 15.20
Horizontal Cell Density as a Function of Eccentricity
Horizontal cells in monkey retina were labeled with antibodies directed toward a calcium-binding protein (parvalbumin). The numbered inset photographs show labeled cells from corresponding locations in extrafoveal retina. Horizontal cell density in human retina is likely to be generally similar, though different in numerical detail. (Photographs and data [replotted] from Röhrenbeck et al. 1989.)

very similar to the horizontal cell distribution. In other words, there seems to be a strong association between cone and horizontal cell densities.

We lack solid estimates of the number of horizontal cells in the retina, but since their density parallels that of the cones and is everywhere about ten times less, there are likely to be around 500,000 horizontal cells in the human retina.

Neither H1 nor H2 horizontal cells form tilings

The relative numbers and distributions of the different horizontal cell types are not known for human or monkey retinas; there are as yet no immunohistochemical markers for the different cell types. But because of the gap junctions between horizontal cells, it is possible to inject one cell with Neurobiotin, which will pass through the gap junctions, and see a small group of labeled horizontal cells. This technique labels two different populations of horizontal cells (Figure 15.21), indicating that gap junctions are made only between cells of the same type, that is H1 with H1 cells and H2 with H2 cells. The H1 horizontal cells outnumber H2 cells by about 2 to 1. (This discussion assumes that there are just two types of horizontal cells, which is still an open question.)

Branches cross everywhere in these pictures because both types of horizontal cell dendrites overlap extensively (the fine branches running through the pictures are the axons of the horizontal cells; no terminals are shown here). Quantitative estimates of coverage factors are not available, but they are likely to be at least three- or fourfold coverages, meaning that each cone has an opportunity to provide output to several horizontal cells and, in turn, to be influenced by several horizontal cells. The difference, as mentioned earlier, is that the H1 cells avoid blue cones.

These meshworks of crisscrossing branches are unlike anything we have seen or will see for the vertical pathway neurons, but these pictures will be duplicated with even more dense coverings at the amacrine cell level. With the possible exception of some small-field amacrine cells, lateral pathway neurons form coverings, not tilings.

Coverings by horizontal cells are sensible in the context of our earlier discussion about horizontal cell function. Operations like contrast enhancement or sensitivity adjustment should be applied symmetrically, so that the effects of a single photoreceptor radiate outward in all directions from it and lateral interactions impinging on the photoreceptor converge from all directions. It is difficult to imagine a reason for biasing the network so that adaptational effects are propagated upward but not downward, for example. Horizontal cell coverings appear to provide the underlying radial symmetry. Any small region is crossed by branches from several cells, and because the cells have different locations around the region, the branches run in multiple radial directions. There is no obvious bias for any particular direction.

We will come back to this point again in discussing amacrine cells, but most lateral pathway neurons form coverings, not tilings, and radial symmetry of their interactions appears to be a direct consequence.

All types of cone and rod bipolar cells are distributed like their photoreceptor types

Variations in cone bipolar cell density for monkey retina (Figure 15.22*a*) are generally similar to the variation in horizontal cell density, except that the bipolar cell densities are greater. The maximum density is just outside the fovea, where the density of cone pedicles is at its maximum. We do not know exactly how the various types of cone bipolar cells are represented in these data, but there are undoubtedly significant differences; midget bipolar cells, for example, should constitute more of the total centrally than peripherally. And the overall distribu-

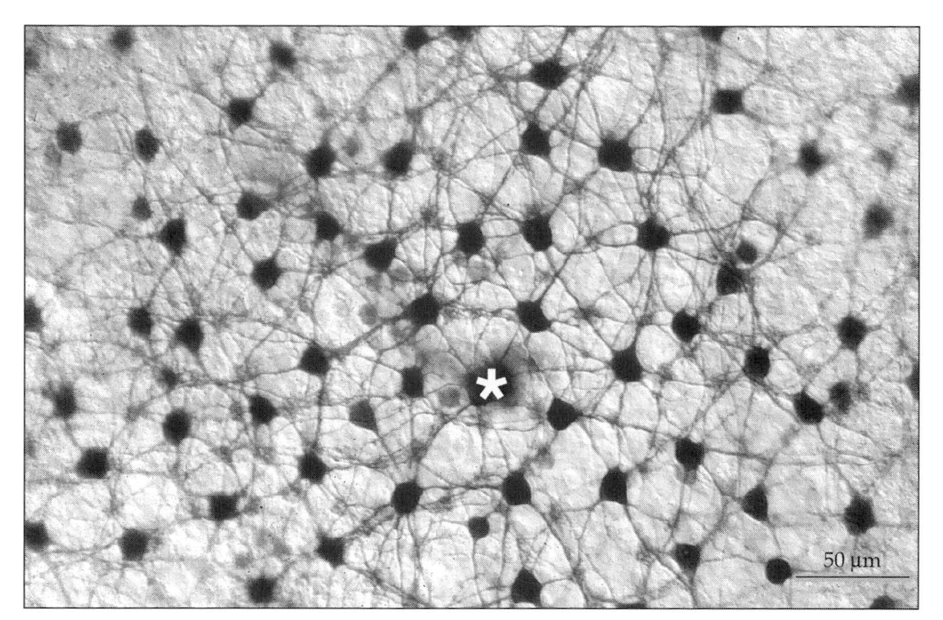

H1 horizontal cells

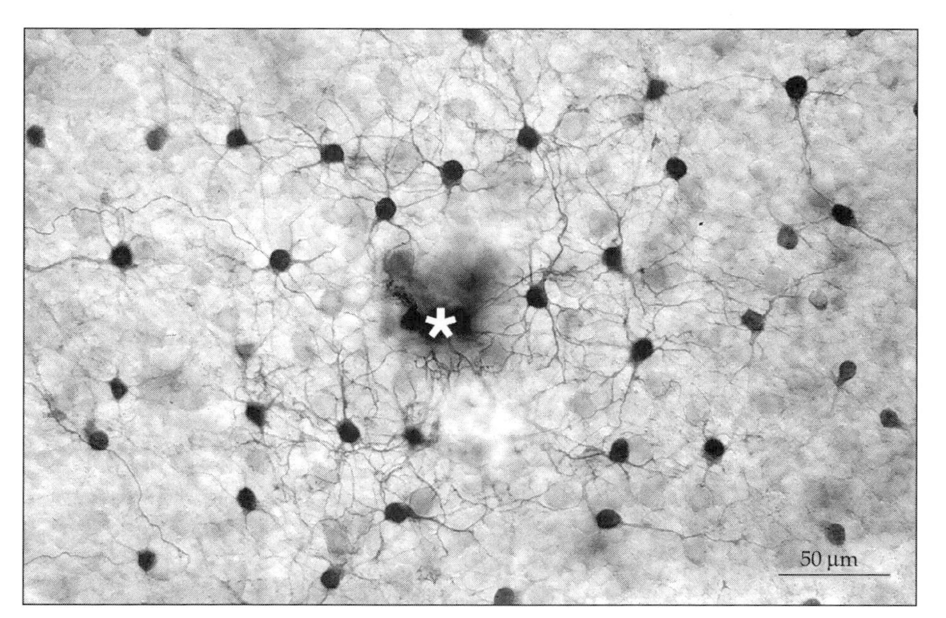

H2 horizontal cells

Figure 15.21
Arrays and Coverings of H1 and H2 Horizontal Cells
Groups of H1 or H2 horizontal cells labeled by injection of Neurobiotin into one cell (marked with asterisks in each panel) form dense coverings with considerable overlap of processes from neighboring cells. H1 cells are more numerous, and their coverage factor is higher. (From Dacey et al. 1996.)

tion should be about equally divided between ON and OFF cone bipolar cells.

The density plot for one type of diffuse bipolar cell (DB4) has the same shape and peak density location as the overall bipolar cell density plot, but with a much lower peak density, as one would expect—the peak is around 7000 cells/mm^2. The graph in Figure 15.22*a* compares the density of diffuse bipolar cells to the density of ganglion cells, which will be similar to the density of cone pedicles, showing that the peak densities coincide near the edge of the fovea. Ganglion cell density declines more rapidly with retinal eccentricity, however, reflecting the increasing convergence of bipolar cells onto a single ganglion cell in peripheral retina.

Blue cone bipolar cells, which are not included here, have a distribution similar to the blue cone population, but with lower density (there are three or four blue cones for each blue cone bipolar cell).

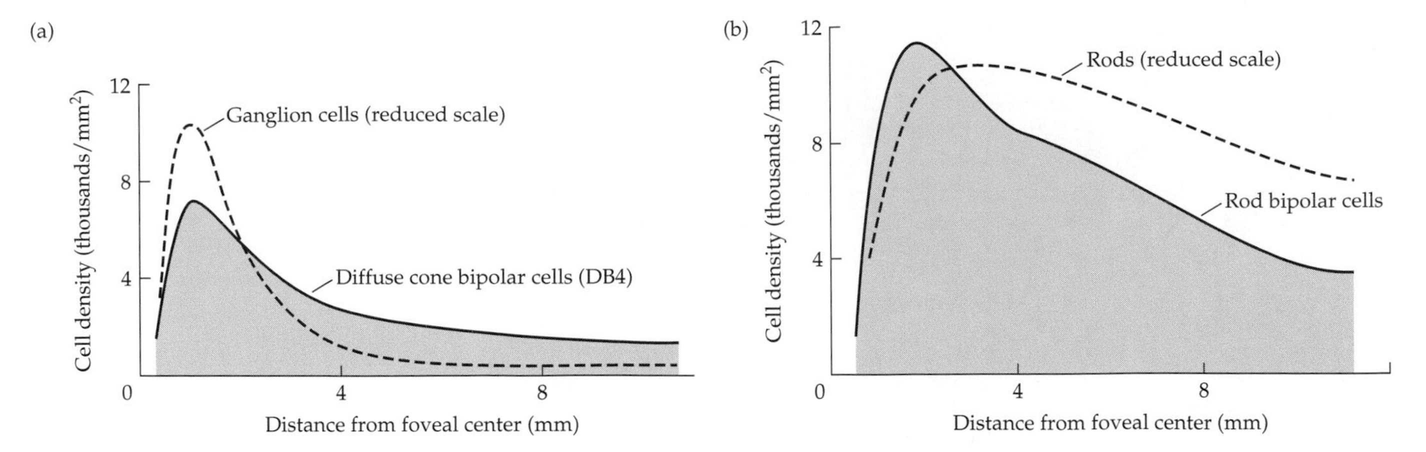

Figure 15.22
Cone and Rod Bipolar Cell Densities as Functions of Eccentricity
Both cone and rod bipolar cells in monkey retina were labeled with specific antibodies. The density for one type of diffuse bipolar cell varies much as ganglion cell density does (*a*). Since ganglion cell density is much like the density of cone pedicles, it follows that cone bipolar cell densities are like cone pedicle densities. Rod bipolar cell density parallels rod density. (Replotted from Grünert et al. 1994 and Martin and Grünert 1992.)

Comparing the number of cones to the number of cone bipolar cells favors the bipolar cells: On average, there are about four cone bipolar cells for every cone. Since there are about 4.5 million cones in the retina, the total number of cone bipolar cells should be about 18 million, which is twice the number of rod bipolar cells. The fact that there are more cone bipolar cells than cones is consistent with the number of types of cone bipolar cells in primate retina. Divergence in the cone pathways (each cone provides inputs to eight cone bipolar cells) outweighs convergence in the cone pathways outside the fovea (diffuse bipolar cells and peripheral midget bipolar cells receive inputs from several cones). This is quite different from the rod system, in which convergence always dominates.

Rod bipolar cells in many species contain relatively high levels of the enzyme protein kinase C (PKC), and antibodies to PKC constitute a specific marker for rod bipolar cells, allowing them to be seen and counted in isolation. The distribution of rod bipolar cells based on PKC labeling in monkey retina is quite similar to the plot of rod density (Figure 15.22*b*). The maximum rod bipolar cell density is in the parafovea, not far from the region of maximum rod density, and clearly farther from the fovea than the peak cone bipolar cell densities (see Figure 15.22*a*). The number of rods converging onto a single rod bipolar cell increases dramatically from fovea to periphery, which makes the peak rod bipolar cell density lie closer to the fovea than the peak rod density does (a given number of rods need fewer rod bipolar cells in the periphery). Each rod appears to make contact with two rod bipolar cells, on average, implying at least a twofold redundancy in the sampling of rod output.

Rod bipolar cell density is always about one-tenth the rod density. Taking the number of rods in the human retina to be just over 90 million, the total rod bipolar cell population should be on the order of 9 million cells.

The different types of bipolar cells provide different amounts of coverage with their dendrites

What little we know about the arrangement of diffuse bipolar cell dendrites comes from fortuitous staining of neighboring diffuse bipolar cells of the same type, as in Figure 15.23. In places, the two bipolar cells contact the same cones, but their dendrites never cross one another to reach a cone within the other bipolar cell's dendritic field. If this is typical behavior for the rest of the diffuse bipolar cells of this type, the dendritic arbors will form an irregular tiling. And if the other types of diffuse bipolar cells behave in the same way, the dendrites of the diffuse bipolar cells will form six overlaid tilings that sample the array of cone pedicles. All visual space sampled by the cones will have been remapped six times, without gaps or overlap, into the vertical pathways represented by the diffuse bipolar cells.

Since each cone pedicle contacts two midget bipolar cells, one ON and one OFF, the total population of midget bipolar cell dendrites will form a two-element tiling wherever the cone pedicles tile, with an occasional gap where there is a blue cone pedicle. (Actually, it is a four-element tiling: red-ON, red-OFF, green-ON, and green-OFF.) In peripheral retina, where the cone pedicles become separated and do not tile the outer synaptic layer, the organization of midget bipolar cell dendrites will change. Because the peripheral midget bipolar cells receive inputs from several cones of the same type (red or green), their dendritic fields will be larger and there may be overlap among fields from the different types of midget bipolar cells. In this respect, the organization will become more like that of the diffuse bipolar cells, in which there are overlaid tilings. At the same time, however, gaps will appear between dendritic fields of the same type because the input cones become more widely spaced.

Blue cone bipolar cell dendrites in monkey retina form a sparse but complete coverage of the plane in which the blue cones terminate. There are often large gaps between the dendrites of neighboring bipolar cells, but this is mainly because the blue cones themselves can be rather widely separated; all blue cones are contacted, however, and the dendrites of the blue cone bipolar cells do not overlap. In effect, the blue cone bipolar cell "tiling" is like the tilings made by red and green midget bipolar cells in peripheral retina.

Since there is only one type of rod bipolar cell, and since dendrites from neighboring cells overlap to some extent, it follows that the rod bipolar cell dendrites do not tile in the outer plexiform layer. Their coverage factor is greater than 1.0, less than 3.0, and probably around 2.0. This small amount of divergence means that rod bipolar cells sample the rods completely, with some redundancy.

All bipolar cell terminals form tilings at different levels in the inner plexiform layer

Although the arrangements of bipolar cell dendrites in the outer synaptic layer require some fudging to fit them into the notion of tilings, matters are better

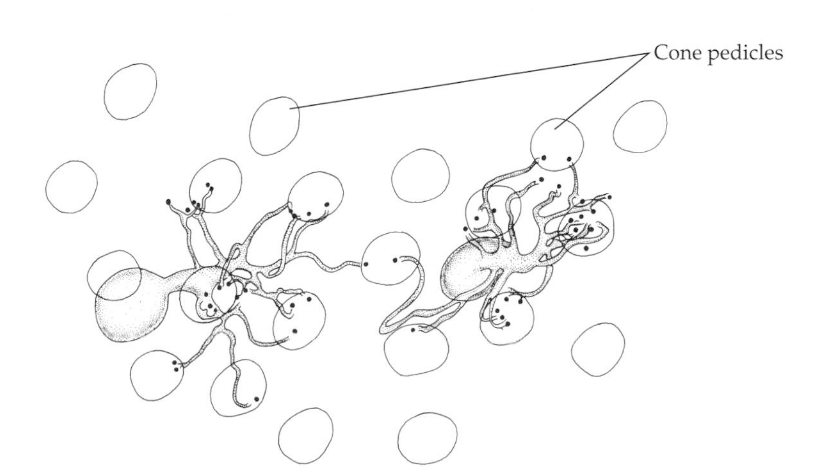

Figure 15.23
Arrangement of Diffuse Bipolar Cell Dendrites
The dendrites of two diffuse bipolar cells of the same type (DB1) neither touch nor overlap, although they share one cone as a source of input. If other diffuse bipolar cells have a similar arrangement, the dendritic arbors constitute tilings. (After Boycott and Wässle 1991.)

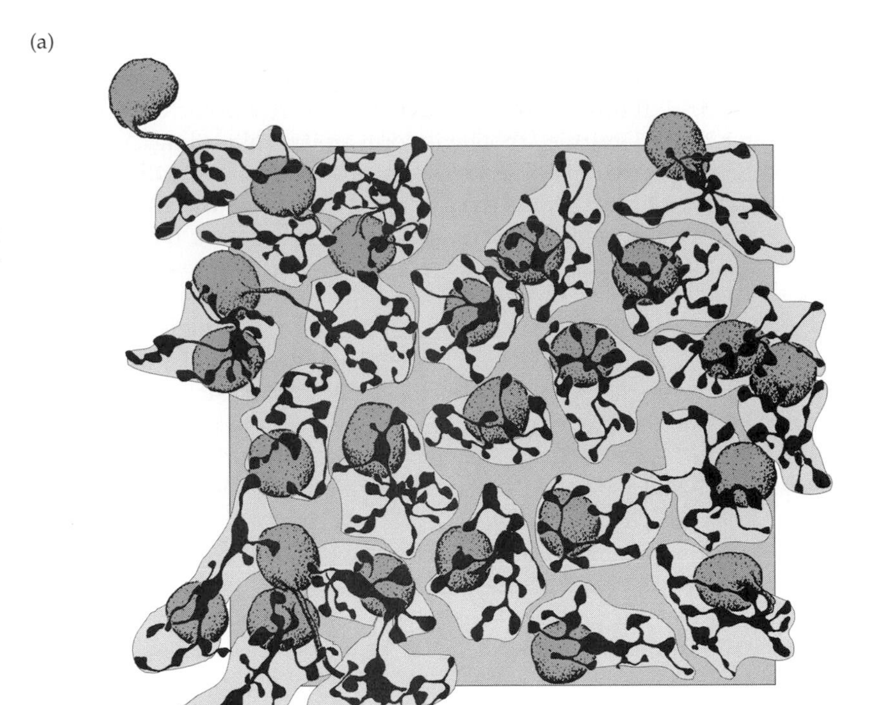

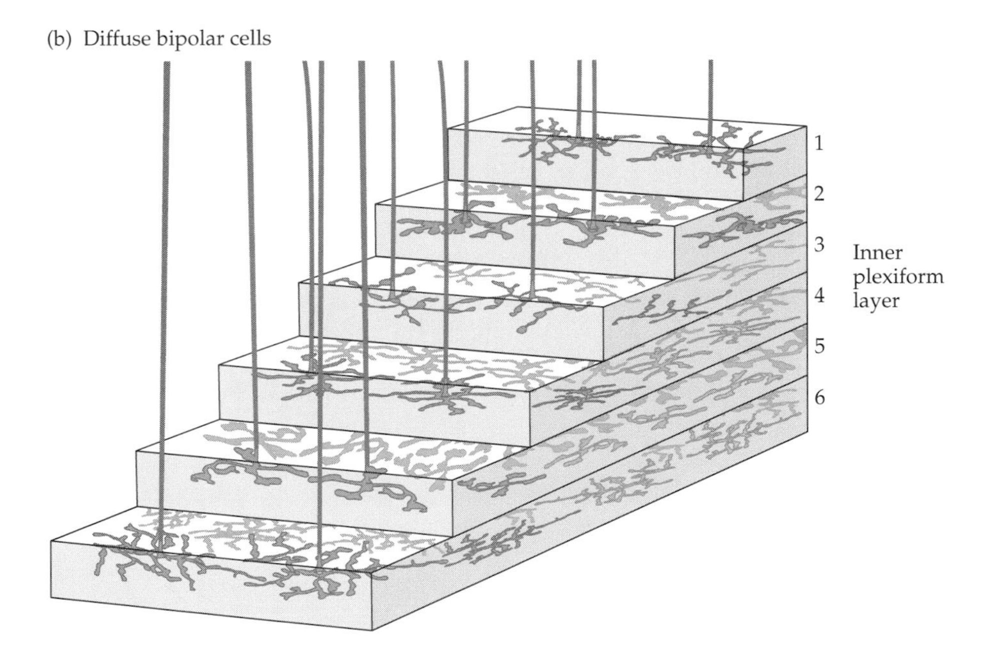

Figure 15.24
Tiling by Diffuse Bipolar Cell Terminals
(*a*) Neighboring terminals from one type of diffuse bipolar cells are shown here as viewed from below. Footprints drawn to include the diffuse bipolar cells form an irregular tiling with small gaps between tiles. (*b*) All six types of diffuse bipolar cell terminals tile their respective sublayers in the inner plexiform layer. (*a* from Rodieck 1988, after Milam, Dacey, and Dizhoor 1993; *b* after Boycott and Wässle 1991.)

defined in the inner plexiform layer. And this is good, because it is here that the tiling concept becomes more critical.

Axon terminals of the diffuse bipolar cells form six tilings at different levels of the inner plexiform layer. Cells belonging to one subset of diffuse bipolar cells are shown in Figure 15.24*a,* along with their footprints—their irregular tiles—in one plane of the inner plexiform layer (Figure 15.24*b*). Gaps between the tiles may indicate that the terminals were not completely stained, but even if they are genuine gaps, they are quite small relative to the size of the tiles. (Some small gaps are inevitable because processes from bipolar, amacrine, and Müller's cells run through the plane in which the terminals lie.) Each bipolar cell terminal clearly has its own territory, and the set of tiles formed by the terminal arbors covers the plane with no significant breaks in coverage. The situation for all the diffuse bipolar cells is illustrated schematically in Figure 15.24*b*, in which the inner plexiform layer is reconstructed in three dimensions so that the different bipolar tilings can be seen individually in their respective sublayers.

Comparable data for the midget bipolar cells are not available, but their terminals must form tilings, because the midget ganglion cells to which they connect are tiled, as we'll see shortly. Blue cone bipolar cell and rod bipolar cell terminals are known to tile, but with sizable gaps. Again, the gaps appear because there are several elements in the overall tiling at this level of the inner plexiform layer. These tilings are shown in Figure 15.25 as they relate to the six levels defined by the diffuse bipolar cells, whose terminals have been omitted for clarity.

The different levels of termination for the bipolar cells within a sublayer cannot be specified precisely, but each cell type may have its own terminal plane; in sublayer 6, for example, the terminals of DB6 cells, blue cone bipolar cells, and rod bipolar cells may each have a slightly different termination level. Thus they will be single-element tilings. The one exception seems to be the midget ganglion cells, where red and green bipolar cell terminals intermingle in the OFF (2) and ON (5) sublayers. These will be two-element tilings.

These pictures of the inner plexiform layer are applicable to all parts of the retina, from fovea to periphery. What change are the relative proportions of the different bipolar types and the size of the tiles—their terminal endings—in each sublayer. Not surprisingly, the axon terminals of all cone bipolar cells become larger away from the fovea, corresponding to the decline of cone density in the peripheral retina. Rod bipolar cell terminals probably enlarge with eccentricity as well.

All amacrine cells tile the retina, varying in density as ganglion cells do

The AII amacrine cells are the only member of the group of small-field amacrine cells that we know much about. They form a loose tiling at two levels of the inner plexiform layer (one of which is shown in Figure 15.26). Their density varies with eccentricity, peaking near the fovea where the ganglion cell density peaks,

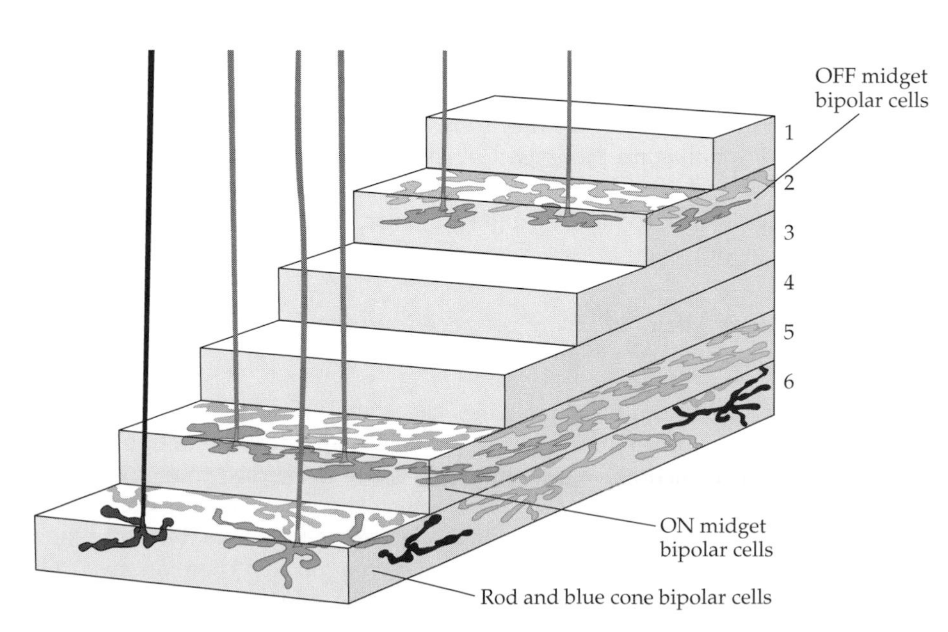

Figure 15.25
Tilings by the Terminals of Midget, Blue Cone, and Rod Bipolar Cells
The correspondence between the layers of diffuse bipolar cell terminals and the other bipolar cell types is shown here. OFF and ON midget bipolar cell terminals are in sublayers 2 and 5, respectively, while the terminals of blue cone and rod bipolar cells intermingle in sublayer 6. Midget bipolar cell terminals form tilings, and the terminals of blue cone and rod bipolar cells may also tile.

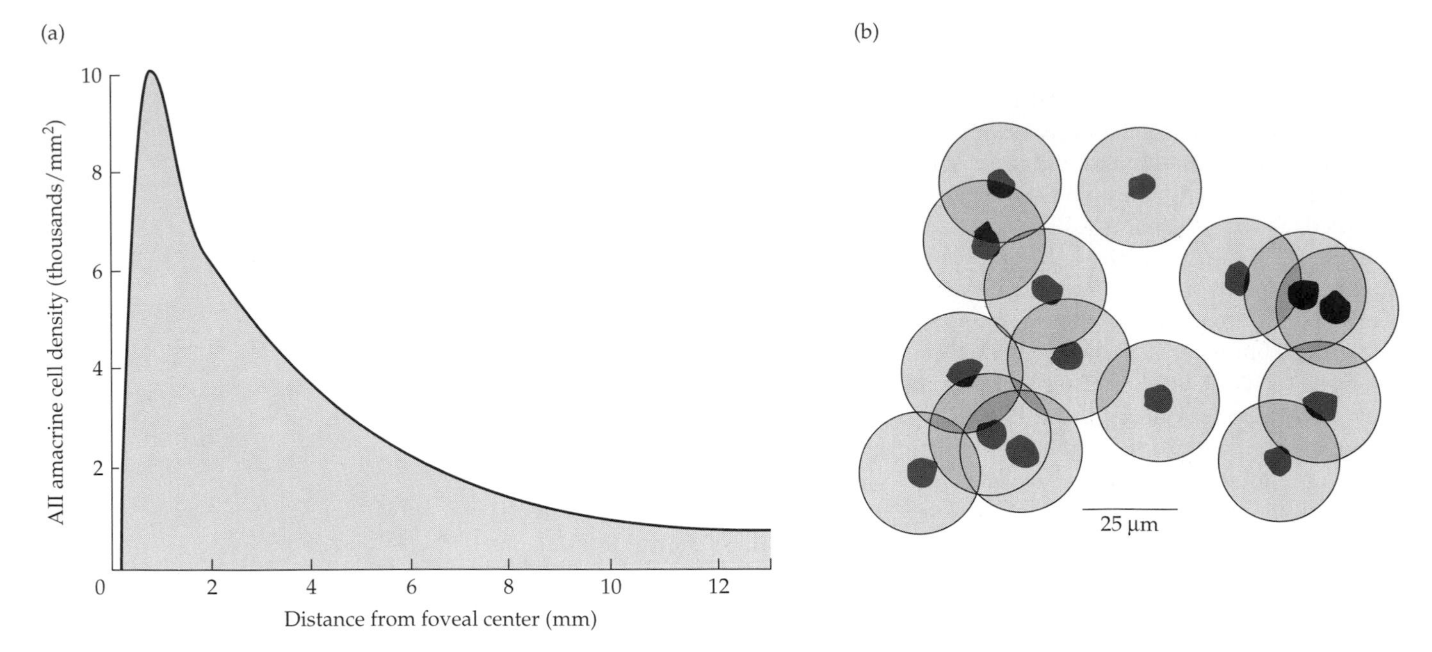

Figure 15.26
Density and Tiling of AII Amacrine Cells
The density of AII amacrine cells in monkey retina peaks outside the fovea and declines in peripheral retina. The density variation is more like that of the ganglion cells than of the rods. The processes from the AII amacrine cells form loose tilings at two levels in the inner plexiform layer, one of which is constructed here with shaded circles indicating the spread of processes around the cell bodies. As discussed in the text, using regular circles will exaggerate overlap and create unrealistically large gaps. (After Wässle et al. 1995.)

and declining out to the retinal periphery. Because of their small size and the fact that they tile, they must be quite numerous if they are to cover the entire retinal plane; their average area divided into the retinal area gives an estimate of at least 1.5 million cells.

We don't know if AII amacrine cells are typical of small-field amacrine cells, but it is tempting to generalize. Because of their position in connecting the rod vertical pathway into the cone pathways, they are as much vertical pathway elements as they are lateral pathway elements. This verticality makes them very different from larger amacrine cells. Like the bipolar cells, AII amacrine cells are in the business of maintaining an orderly, unbroken representation of the retinal image as seen by the rods. Thus, they form a tiling. Other small-field amacrine cells may be in similar positions with respect to cone pathways, particularly those associated with the midget bipolar cells, and if this is the case, they may also form tilings. (They may also be quite numerous, despite the paucity of information about them.)

Medium- and large-field amacrine cells are low-density populations whose processes generate high coverage factors

The best-known medium-field amacrine cell is the "starburst" amacrine cell. Figure 15.27 shows both an individual starburst cell and data for a population of starburst cells. This distinctive morphology has been observed in many species and is invariably associated with immunohistochemical labeling for acetylcholine and GABA (they were often called cholinergic amacrine cells until their co-localization of GABA and acetylcholine was recognized). There are two mirror-image subpopulations of starburst amacrine cells, one consisting of ordinary

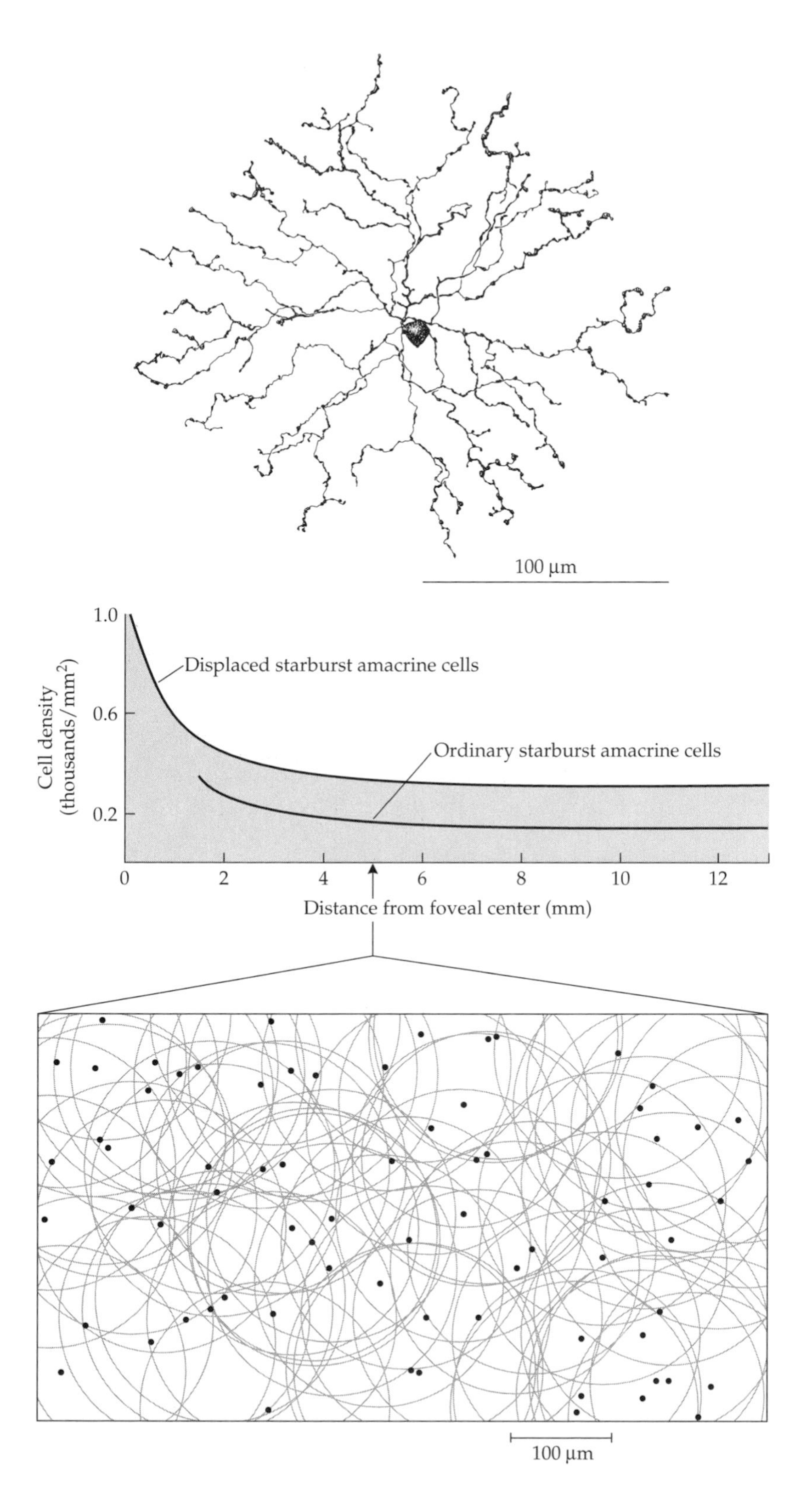

Figure 15.27
Density and Coverage by a Medium-Field Amacrine Cell: The Starburst Amacrine Cell
Starburst amacrine cells have a distinctive morphology, with numerous processes radiating from the cell body. There are two types of starburst amacrine cells (ordinary and displaced); both are low-density cell populations, with peak densities of 1000 cells/mm^2 or less near the fovea. Densities in peripheral retina are a few hundred cells per square millimeter. Because the cell dendrites spread widely and neighboring cells overlap extensively, however, the coverage factors are always high. (After Rodieck and Marshak 1992 and Rodieck 1989.)

amacrine cells with processes ramifying in sublayer 2, and another consisting of displaced amacrine cells whose processes ramify in sublayer 5.

In keeping with their larger size, starburst amacrine cells are not as numerous as the small AII amacrine cells. Their maximum density, around 1000 cells/mm^2 near the fovea, declines to 300 to 400 cells/mm^2 in peripheral retina (see Figure 15.27); in total, there may be 0.5 million or so starburst amacrine cells in human

(a) Coverage by GABA–dopamine amacrine cell processes

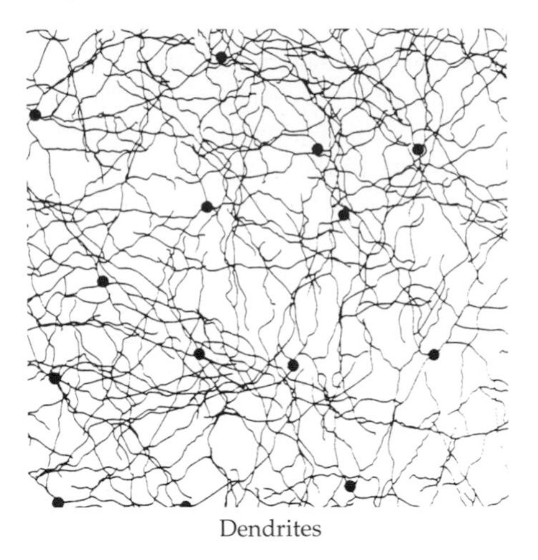

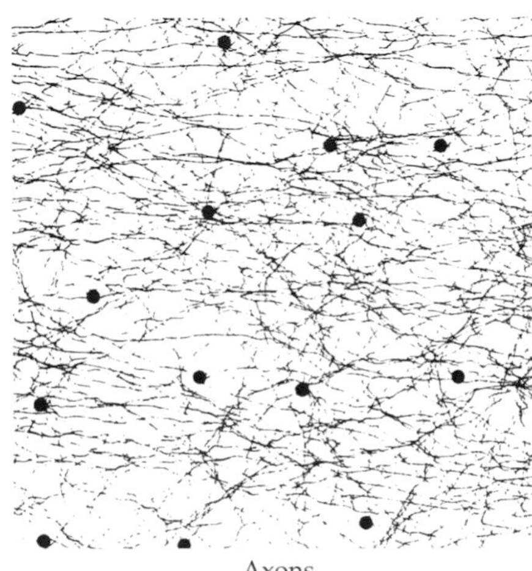

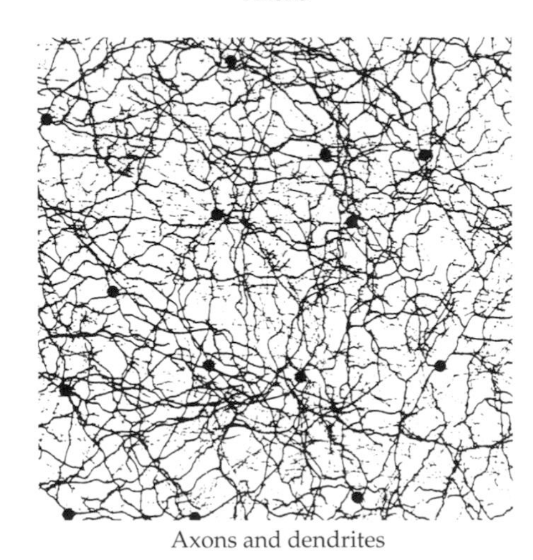

(b) GABA–dopamine amacrine cell density

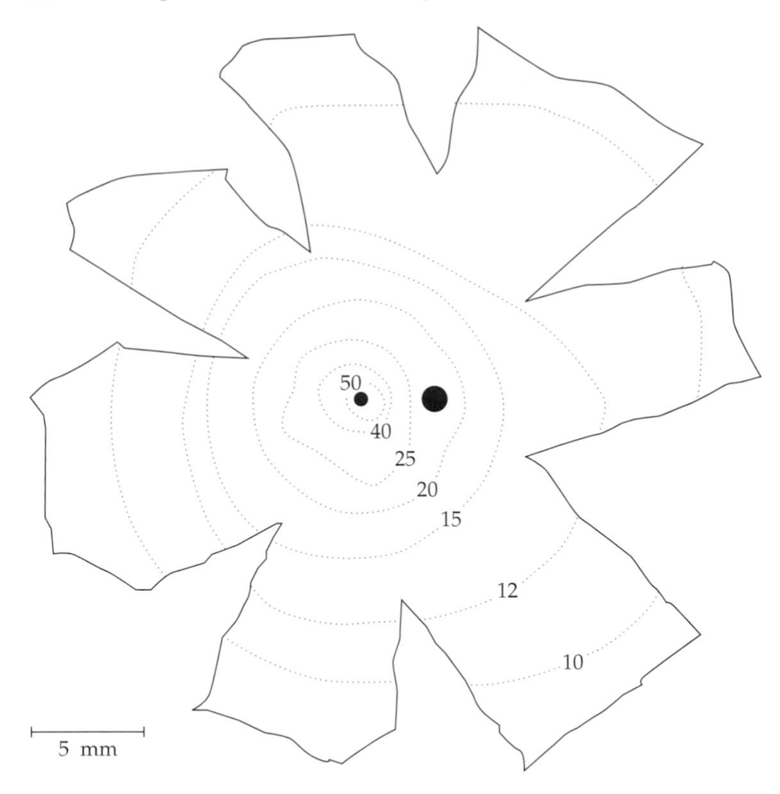

Figure 15.28
Density and Coverage by a Large-Field Amacrine Cell: The GABA–Dopamine Amacrine Cell
(*a*) Illustrations of coverage by GABA–dopmine amacrine cells, simulated by reproducing all the processes from one of four well-stained cells at one of the locations of stained cell bodies (indicated by the black dots). The cell was reproduced at a given location and its orientation varied randomly. Both the dendrites and the axons of these cells overlap extensively. Coverage by dendrites and axons in combination (they both ramify in layer 1) is very dense. (*b*) Map of monkey retina with isodensity contour lines for the GABA–dopamine amacrine cells. The maximum density near the fovea is only about 50 cells/mm^2. Although the cells are few in number and sparsely distributed, they have a very high coverage factor. (After Dacey 1990.)

retina. But since the processes from each starburst amacrine cell extend 100 μm or so from the cell body, the processes from neighboring cells overlap extensively. Each of the starburst amacrine cell types has a coverage factor on the order of 10 or more; that is, every point in sublayers 2 and 5 is within the dendritic fields of ten or more starburst amacrine cells.

Several features of a type of large-field amacrine cell are shown in Figure 15.28. This is the so-called dopaminergic amacrine cell that is commonly labeled with tyrosine hydroxylase antibodies; it is now known to be a GABA amacrine cell that co-localizes dopamine. Although the cell's processes are stratified, usually confined to sublayer 1 of the inner plexiform layer, the processes are not the same; some are dendrites on which the cell receives inputs, and the other, thinner processes are axons. The axons typically extend much farther from the cell body than do the dendrites (Figure 15.28*a*).

Like other types of large-field amacrine cells, GABA–dopamine cells form a population of very low density, consisting of perhaps 10,000 cells in the entire retina (Figure 15.28*b*). But the extensive spread of processes around each cell body creates much the same picture of coverage that we saw for the starburst amacrine cells. The processes from neighboring cells overlap considerably, generating a coverage factor approaching 10. As discussed in Chapter 14, these GABA–dopamine amacrine cells interact with the AII amacrine cells; the coverage factor means that each AII amacrine cell can be influenced by ten or so large-field amacrine cells that receive inputs throughout a large area around the AII amacrine cell.

The dense coverage factors of the starburst and GABA–dopamine amacrine cells are likely to be characteristic of medium- and large-field amacrine cells, and it is worth considering what this coverage means in terms of what amacrine cells do (with the qualification that we don't understand amacrine cell operations in much detail).

Consider what might happen to a bipolar cell input at some point in either of the meshworks of crisscrossing branches illustrated in Figure 15.28. If the bipolar cell contacted all the branches within reach of its terminal arbor, its effect would be conducted outward in all directions from the point of contact, for the simple reason that the amacrine cell processes are crossing in all directions. Or to look at the situation from the perspective of a cell receiving input *from the meshwork* of amacrine cell processes at a given point, the receiving cell—an AII amacrine cell or a ganglion cell dendrite, for example—would see potential signal sources converging on it from all directions. Again, this radial convergence is a consequence of the fact that the numerous amacrine cell processes run every which way.

Sending an input to one of these meshworks is like dropping a stone in a pool of water: The ripples spread out concentrically from the local splash, diminishing in amplitude as they go. How far and how quickly ripples spread in an amacrine cell meshwork probably depends on the particular cell type. And if we had videotaped the stone dropping into the water, running the tape backward would illustrate the radial convergence of influences at the stone's point of entry.

Whatever we think amacrine cells are doing—contrast enhancement, sensitivity adjustment, or temporal sharpening of responses—their activities should have a concentric organization that is a direct consequence of their multiple coverage factors. Thus ganglion cells receiving significant amounts of input from amacrine cells should exhibit an underlying concentric organization—which they do.*

Ganglion cell density declines steadily from the parafovea to the periphery of the retina

Counting ganglion cells is difficult because of displaced amacrine cells in the ganglion cell layer. They may account for about 3% of the cells in the ganglion cell layer near the fovea and as much as 80% in the far peripheral retina. In human retina, ganglion cells and displaced amacrine cells are generally distinguished by size: Displaced amacrine cell bodies are usually smaller in absolute terms, and their size relative to their nuclei is also smaller. (A monoclonal antibody specific for ganglion cells in some species has not worked well in human retina.) Thus, when we talk about ganglion cell densities in human retina, the underlying assumption is that displaced amacrine cells have been recognized and eliminated from the counts.

Figure 15.29 shows ganglion cell density in human retina as maps for the whole retina and for the fovea and parafovea, and as a graph of cell density along the retina's horizontal meridian. There are no ganglion cells in the center

*The difficult job is explaining *asymmetry* in ganglion cell responses, such as the strong directional preferences for image movement exhibited by ganglion cells in some species. Amacrine cells have something to do with these response asymmetries, but no one has figured out how they could be generated by symmetrical amacrine cell networks. The answer may lie with a small-field amacrine cell that has yet to be studied in detail.

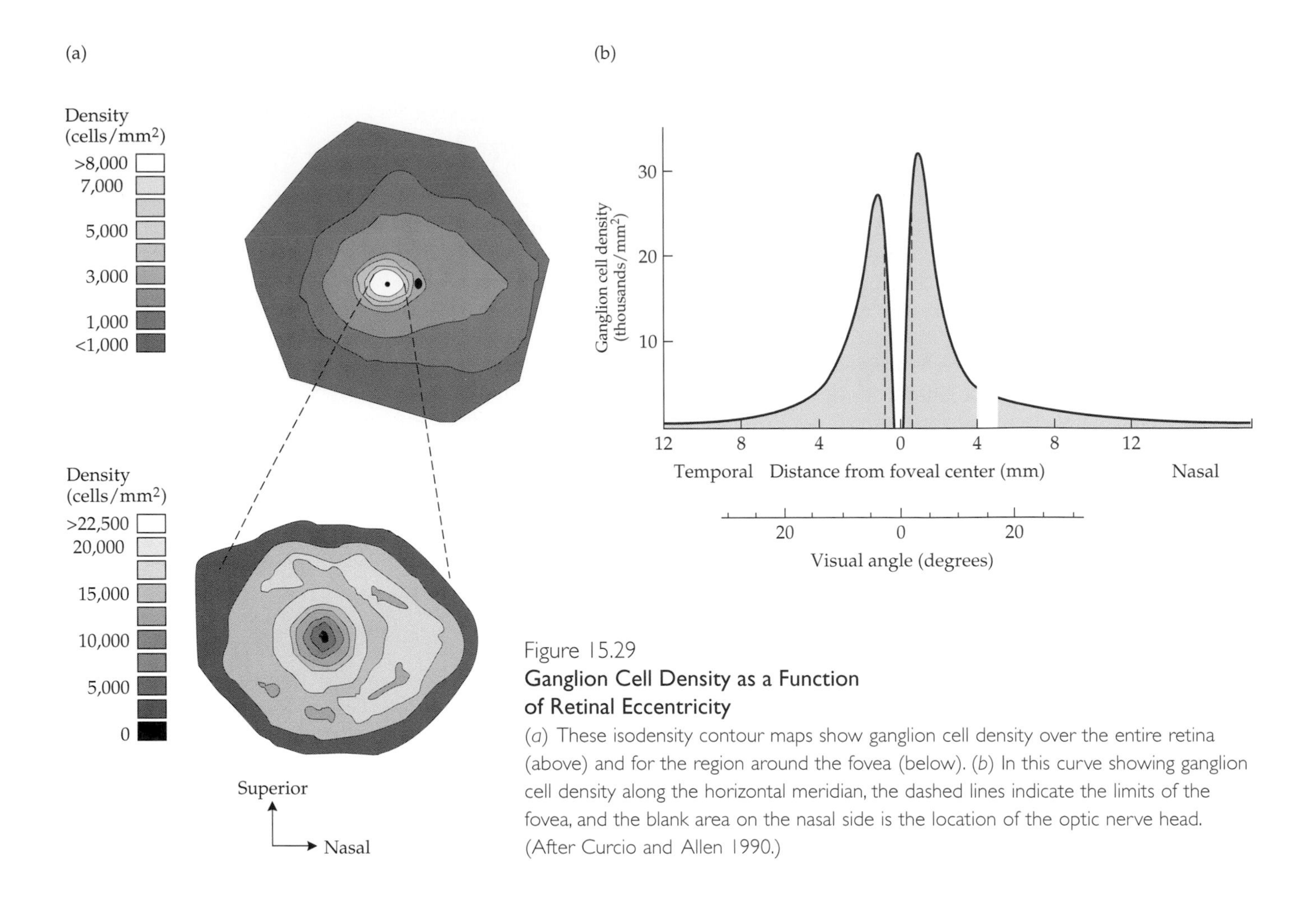

Figure 15.29
Ganglion Cell Density as a Function of Retinal Eccentricity
(*a*) These isodensity contour maps show ganglion cell density over the entire retina (above) and for the region around the fovea (below). (*b*) In this curve showing ganglion cell density along the horizontal meridian, the dashed lines indicate the limits of the fovea, and the blank area on the nasal side is the location of the optic nerve head. (After Curcio and Allen 1990.)

of the fovea, of course; they first appear anywhere from 100 to 500 μm from the foveal center, depending on the meridian, and their density rises rapidly to a peak of 35,000 cells/mm^2 about 1.5 mm from the center. Density declines from this maximum in all directions out to the retinal periphery, where there are only 100 or so cells per square millimeter. The decline in cell density is not symmetric around the fovea; in general, density falls off more rapidly vertically than it does horizontally. Also, the ganglion cell–free zone in the fovea is not circular, and ganglion cells appear first along the nasal and temporal sides of the fovea.

Like the densities and numbers of photoreceptors, numbers of ganglion cells vary when different retinas are compared. On the basis of seven human retinas, the median value for maximum ganglion cell density is 34,550 ganglion cells/mm^2, and the median number of ganglion cells in the retina is 1.12 million. Peak ganglion cell density varies from 31,600 to 37,800 cells/mm^2, which is about ±9%, but estimates of total ganglion cell number range from 0.71 million to 1.54 million; the large variation in ganglion cell number appears to be due to variation in peripheral rather than foveal ganglion cell densities.

Within the overall ganglion cell population, the distributions of midget, parasol, and small bistratified ganglion cell types have been estimated, and their densities all decrease from center to periphery, much like the density of the total ganglion cell population does (Figure 15.30). The curves are not parallel, however, indicating differences among the three cell types. Midget ganglion cells are the most numerous at all locations, and at the fovea they account for 95% or so of the ganglion cell population. The difference between the total ganglion cell density curve and the midget ganglion cells increases away from the

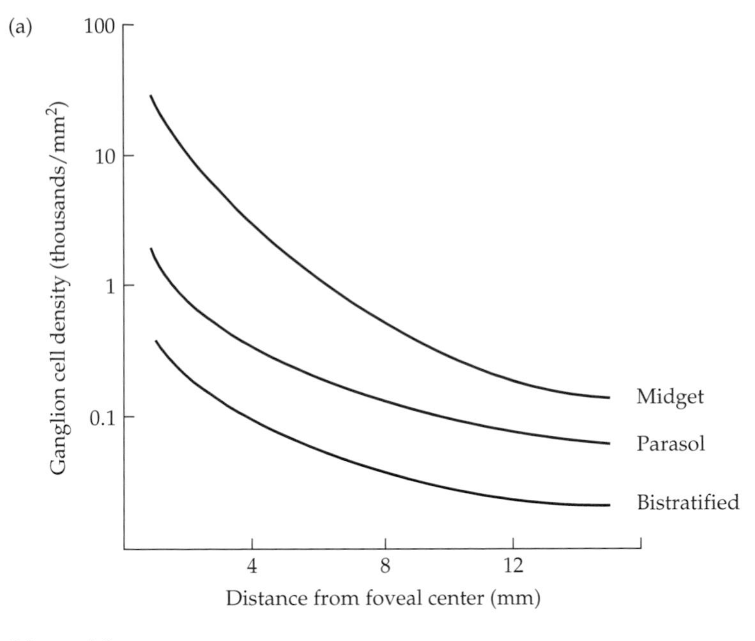

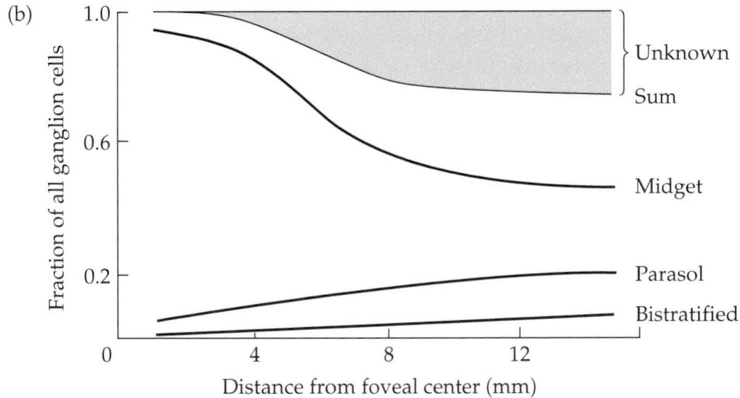

Figure 15.30
Estimated Densities of Midget, Parasol, and Small Bistratified Ganglion Cells
(*a*) Ganglion cell densities for the different cell types have been plotted on a logarithmic scale that compresses the density range; the peak density for midget ganglion cells is almost 100 times greater than the peak bistratified ganglion cell density. (*b*) When the cell densities are compared to the density for all ganglion cells as fraction of total cell density, the sum of midget, parasol, and bistratified ganglion cells falls short of unity outside the parafovea, leaving almost 20% of peripheral ganglion cells unaccounted for (shaded region). (*a* replotted from Dacey 1994.)

fovea, however, indicating that the midget ganglion cells are less dominant in the extrafoveal retina. The parasol and small bistratified ganglion cells are lower-density populations, and there are always about twice as many parasol cells as small bistratified cells. Decreasing cell density is accompanied by an increase in dendritic field size for all cell types (see Figure 14.21).

Replotting the data as a fraction of the total ganglion cell population, which is set equal to 1.0, shows a decline of the midget ganglion cell fraction away from the fovea and an increase in the relative proportions of both parasol and small bistratified ganglion cells. Note that the curve in Figure 15.30*b* labeled "sum," which represents the midget, parasol, and bistratified ganglion cells combined, falls well below unity outside the fovea. The discrepancy, shown here as a shaded region, represents the fraction of ganglion cells not accounted for by the midget, parasol, and small bistratified cell groups. About 20% of the ganglion cell population is not accounted for in peripheral retina, and around 1 to 2% is missing from the foveal ganglion cells.

Midget and parasol ganglion cell dendrites tile at different levels in the inner plexiform layer

We have arrived once again at the critical intersection of bipolar cell terminals, amacrine cell networks, and ganglion cell dendrites. The ganglion cell dendrites are making the final samples of the vertical and lateral pathways, and the result of

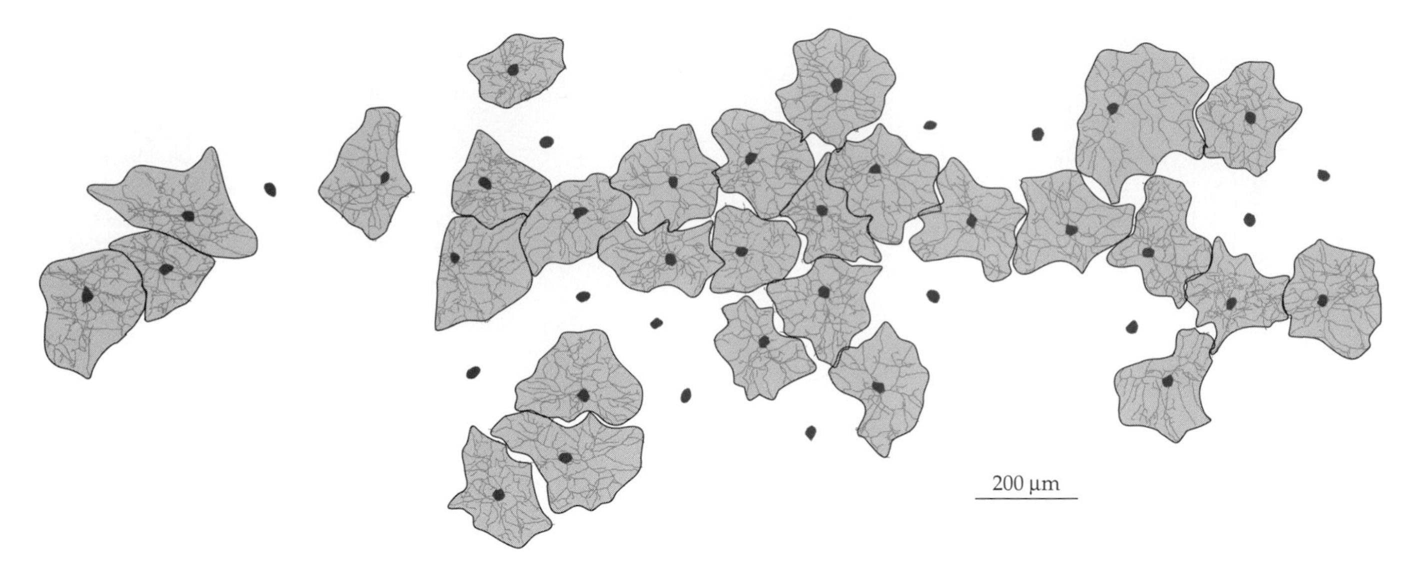

Figure 15.31
Tiling by Midget Ganglion Cell Dendrites
Midget ganglion cells in the peripheral retina were injected with Neurobiotin that labeled the cell bodies and the dendrites. Neighboring cells have dendrites that touch or approach one another closely, but they do not overlap. The dendritic areas for these cells fit together like pieces of a jigsaw puzzle, thereby showing the tiling made by these ON midget ganglion cells. OFF midget ganglion cells not shown here also form a tiling with their dendrites. (From Dacey 1993.)

this sampling will go to the brain. And ganglion cell dendritic fields tile the levels of the inner plexiform layer from which they sample; that's how they manage to make their dots of activity a complete mapping of the retinal image.

When neighboring midget ganglion cells are injected with a visible marker that fills and stains their dendrites (see Box 14.1), a picture of the flat-mounted retina like that in Figure 15.31 emerges; the cell bodies and their dendrites are shown, with borders around the dendritic territories. The dendritic territories are quite irregular in shape, but they fit together like pieces of a jigsaw puzzle, forming a tiling of the plane (the plane in this case is sublayer 5 in the inner plexiform layer, and these are ON midget ganglion cells). Finally, although not shown here OFF midget cells form tilings with their dendritic arbors; at any given eccentricity, the tiles for the ON (inner) midget cells are slightly larger than those for their OFF (outer) counterparts.

On the foveal slope, the dendritic areas are tiny, corresponding to the size of a midget bipolar cell terminal, and the midget ganglion cell density is quite high (see Figure 15.30). Farther out in the retina, the dendritic areas become larger as their density declines; the inverse relationship between dendritic area (size of the tiles) and density is perfectly matched so that the tiles of the midget cells always maintain a unity coverage of their branching planes.

But a problem we have seen before comes up again: Both the ON and the OFF midget ganglion cell tilings contain tiles in different colors, red and green. How can we have an unbroken representation of the retinal image seen by the cones if it is viewed by just the green midget ganglion cell tiles, for example? There will be gaps in the tiling, just as there are spaces between green cones, and the representation will be incomplete.

Although we put different labels on the midget ganglion cells or midget bipolar cells according to the cone type to which they are connected, the difference between a red cone pathway and a green cone pathway is small; it is not easy to devise a stimulus that will activate one pathway and not the other. As discussed in Chapter 14, the wavelength information is embedded in the midget cell pathways, but it may not be highly explicit; the red and green cone pathways are not mutually exclusive. Most of the time—probably *all* of the time under everyday conditions—the red and green pathways are operating in concert and never in isolation. Later comparisons of the small but significant differences in their signals will extract more specific wavelength information. Thus the ON and OFF tilings of midget ganglion cells are complete representations of the retinal image, even though each tiling has two elements within it.

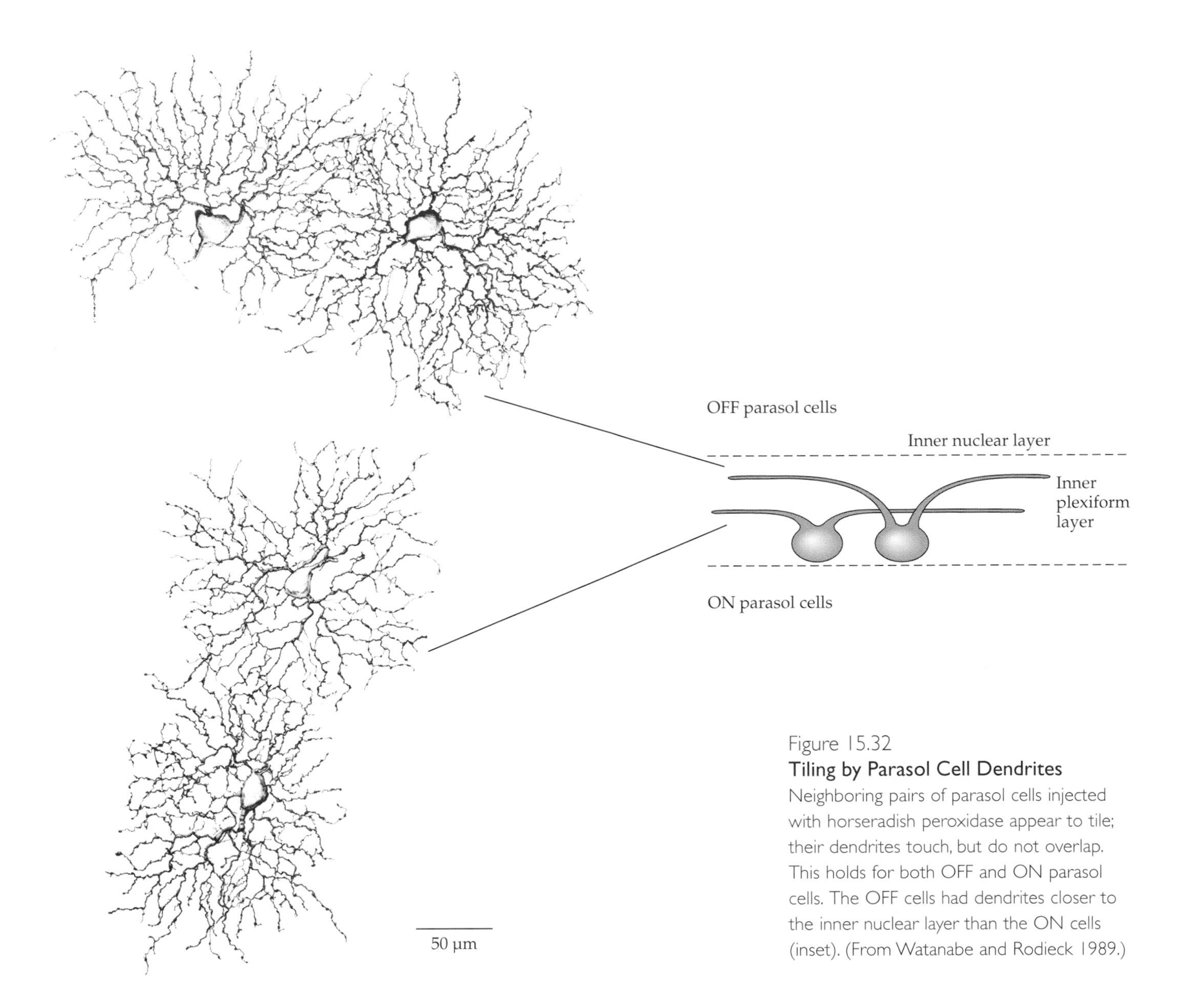

Figure 15.32
Tiling by Parasol Cell Dendrites
Neighboring pairs of parasol cells injected with horseradish peroxidase appear to tile; their dendrites touch, but do not overlap. This holds for both OFF and ON parasol cells. The OFF cells had dendrites closer to the inner nuclear layer than the ON cells (inset). (From Watanabe and Rodieck 1989.)

In peripheral retina, the difference between red and green midget ganglion cell tiles disappears, because the midget ganglion cells receive inputs from both red and green midget bipolar cells. The dendritic fields of the midget ganglion cells take on irregular shapes as the dendrites reach out for bipolar cell terminals that are separated by gaps and the oddly shaped tiles interdigitate, but the ON and OFF tilings are maintained.

Pairs of neighboring ON or OFF parasol cells labeled by horseradish peroxidase injections are shown in Figure 15.32. In each pair, dendrites from neighboring cells touch but rarely cross, indicating that the dendritic fields are tiles in a tiling. Sampling by parasol cells, like that of the midget ganglion cell tilings, will extract a complete, unbroken representation of the retinal image from both the ON and the OFF layers. Parasol cells are, of course, sampling from different sets of bipolar cell terminals than the midget ganglion cells, and their sampling is done at a coarser scale because of their relatively large dendritic fields.

The blue-ON ganglion cells are different from the midget and parasol cells in having dendrites that branch at two levels of the inner plexiform layer; the major branching is in the inner (ON) region, with a subsidiary, less extensive set of dendrites in the outer (OFF) region. Although the presence of bistratified ganglion cells in primate retinas has been known for some time, the fact that they carry information from the blue cone pathway is a recent discovery, and there is as yet no information about their tiling or lack thereof. Using the available data to calculate their coverage factor based on density and average dendritic field sizes, however, gives a value only a little less than unity, which hints that a tiling may be present.

Spatial resolution is limited by cone spacing in the fovea and parafovea and by midget ganglion cells elsewhere in the retina

Figure 15.33 compares visual acuity, the cone Nyquist frequency, and the midget ganglion cell Nyquist frequency, all plotted as functions of retinal eccentricity. Within about 8° of the foveal center, the decline in visual acuity with eccentricity parallels the cone Nyquist frequency, thus reinforcing what was said earlier: Cone spacing is the limiting factor for spatial resolution. For eccentricities greater than 8°, however, visual acuity is well below the cone Nyquist frequency and is almost identical to the Nyquist frequency calculated from the center-to-center distance between midget ganglion cell dendritic fields. Why does this change occur?

Wherever the wiring from cones to midget ganglion cells is one to one, cone spacing is the critical factor. If the midget ganglion cells to which the cones connected had dendritic fields as small as the cones themselves, the cone and midget ganglion cell Nyquist frequencies would be identical. The dendritic arbors of the midget ganglion cells are larger, however, so their Nyquist frequency is lower than the Nyquist frequency of the cones feeding into them. This

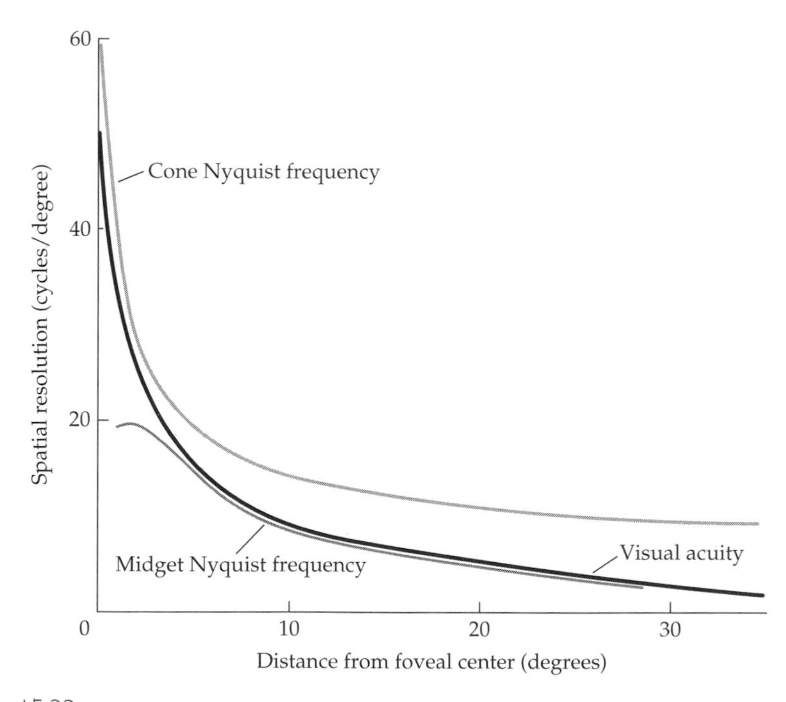

Figure 15.33
Nyquist Frequencies for Cones and Midget Ganglion Cells as Functions of Eccentricity
Cone spacing sets the acuity limit in the central retina, but in peripheral retina, the size of midget ganglion cell dendritic fields is the limiting factor. (Cone Nyquist frequencies calculated from data in Curcio et al. 1990; data for midget ganglion cells replotted from Dacey 1993; visual acuity data from Wertheim 1894.)

is the situation within the central 16° of the retina (the circle with an 8° radius from the foveal center).

Outside this central area, convergence begins and midget ganglion cells receive signals from more than one cone. Resolution is therefore determined not by the spacing between individual cones, but by the spacing between the cone clusters that converge on the midget ganglion cells. Since the size of the cone clusters will be similar to the dendritic field size of the midget ganglion cells, the spacing of the midget ganglion cells becomes the limiting factor for spatial resolution outside the central 16° of the retina.

A Final Look at Three Small Pieces of Retina: Dots for the Retinal Sketches

A sampling unit is the smallest retinal region containing at least one representative from each type of ganglion cell

The term "sampling unit" refers to the smallest retinal region within which all of the parallel information streams in the ganglion cell output are fully represented. In a sense, a sampling unit is a module or basic element from which the retina is constructed. To see what this means in anatomical terms, consider a case in which there are only two types of ganglion cells, both of which tile the plane and have unity coverage, but where one tiling is significantly larger than the other. As Figure 15.34 shows, the tilings are independent of one another, but several smaller tiles always fall within the area enclosed by one of the larger tiles. In this simplified case, each of the larger tiles represents a sampling unit because this is the smallest area within which we can be certain that the retinal image, as transformed into photoreceptor signals, is being sampled by both of the information pathways represented by the ganglion cells.

A sampling unit in this example includes one large tile and a half dozen or so small tiles, but the ratio of small to large tiles need not be a constant throughout the retina. For example, if Figure 15.34*a* represents the fovea, Figure 15.34*b* might represent the central retina outside the fovea; the large tiles have increased in size, meaning that a sampling unit is also larger, but the small tiles have increased even more in size and fewer of them fit within a large tile. And in Figure 15.34*c*, which might be peripheral retina, both sets of tiles have increased in size again, thus making a sampling unit quite large, but the increase has been proportional for both large and small tiles and their ratio has not changed.

Although a sampling unit is defined in terms of information outflow from the retina, it has a structural basis because elements of information outflow are discrete, identifiable anatomical units—namely, the ganglion cells. The ganglion cell type with the largest tiles (dendritic areas) and the coarsest mosaic (lowest density) will set the size and number of the retina's sampling units.

But what about convergence in the wiring from photoreceptors to ganglion cells, particularly in nonfoveal regions of the retina? Won't the area occupied by the photoreceptors be larger than the dendritic area of the ganglion cell on which their signals converge? And what about the lateral horizontal cell and amacrine cell pathways? Won't they spread the area of influence even more?

Convergence will certainly make a converging group of photoreceptors extend over an area larger than the ganglion cell's dendritic field, and lateral interactions will bring in influences from an even larger area. But these facts say merely that a ganglion cell's receptive field (whose limits are set by all the photoreceptors that influence it through vertical or lateral pathways) can be larger than its dendritic field. That is, receptive fields overlap even if dendritic fields do not. And neighboring ganglion cells of the same type will have some elements of their input circuitry in common.

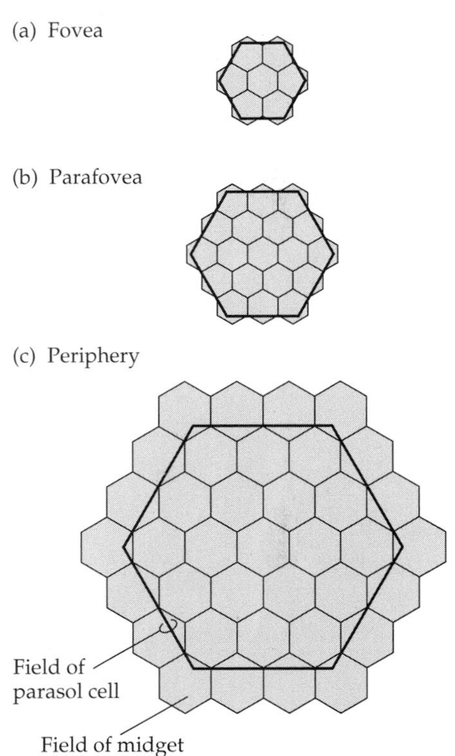

Figure 15.34
Sampling Units at Different Retinal Locations
The larger hexagons represent the dendritic field of the larger of two types of ganglion cells; they define the limits of a sampling unit, which contains at least one representative of each ganglion cell type. The size of sampling units changes with retinal location as the larger dendritic fields increase in size, and the number of the smaller fields included in a sampling unit may also change.

Thinking about retinal modules in terms of information sampling by the ganglion cell dendritic arbors circumvents these apparent complications; whatever information the ganglion cell conveys and however much it may depend on activity in other parts of the retina, all other retinal neurons exert their influences at or before the ganglion cell's dendrites. The dendritic arbors are sampling the combined effects of the vertical and lateral pathways, and all these influences will be contained in the ganglion cells' signals. Although the sizes of receptive fields or the strengths of lateral interactions can and do change with adaptation levels, the sizes of the sampling elements—the tiles formed by the dendritic arbors of the ganglion cells—are invariant. Within the area encompassed by the largest of these arbors, there will always be at least one tile for each of the different types of ganglion cells and therefore at least one dot of activity for each of the retinal sketches.

In this conception, the retina is a large collection of sampling units packed together with no gaps between them. In other words, the retina has a modular construction and the sampling units are the modules. Differences between one part of the retina and another are largely differences in the size of the modules and the relative proportions of the neurons within them. And with this idea in mind, we will look at three small pieces of the retina in detail, each piece containing a sampling unit that generates several or many dots of activity for different retinal sketches. The three retinal pieces are from the fovea, the parafovea, and the perifovea near the region of highest rod density.

Sampling units are smallest at the foveal center and are dominated by cone signals

Figure 15.35 represents a cluster of cones at the center of the fovea connected via the long fibers of Henle and the intervening bipolar cells to ganglion cells lying along the foveal slope; the ganglion cells are about 400 μm away from the cones. This sampling unit is about 30 μm across, which is the diameter of the smallest parasol cell dendritic fields that have been reported for human retina.

The smallest dendritic fields of the midget ganglion cells on the foveal slope are around 5 μm in diameter, so the dendritic field area of a parasol cell is large enough to contain around 35 midget ganglion cell dendritic fields. This is the highest ratio of midget to parasol cells in the retina. This sampling unit corresponding to the central cluster of cones will therefore contain 2 parasol cells (1 ON and 1 OFF) and about 70 midget ganglion cells (35 ON and 35 OFF). (None of the midget ganglion cells have been included in the illustration.)

Since each midget ganglion cell receives a direct input from a single midget bipolar cell, their numbers are equal and there are 70 midget bipolar cells. But since each cone is presynaptic to *two* midget bipolar cells, one ON and one OFF, the number of cones will be half the number of midget bipolar cells—that is, 35 cones, of which 22 to 25 will be red cones and 10 to 13 will be green cones. (There are no blue cones here.) These cones in the photoreceptor layer are packed into a smaller area than the size of the sampling unit at the ganglion cell level; if these cones represent the peak cone density of around 200,000 cells/mm^2, they will occupy a region about 15 μm in diameter. As the cone density drops going away from the foveal center, the area occupied by a sampling unit's photoreceptors will become more nearly the same as the ganglion cell dendritic area that delimits the sampling unit.

We have little quantitative data on the diffuse cone bipolar cells here, but assuming that their axon terminal branching is about 15 μm in diameter, three or four of them should terminate in each of the six levels in the inner plexiform layer, giving a total of 18 to 24 diffuse bipolar cells. Thus the ratio of midget to diffuse bipolar cells is roughly 3:1.

There are, of course, lateral pathways, but they are not shown in Figure 15.35. There are likely to be a half dozen or so horizontal cells contacting the central

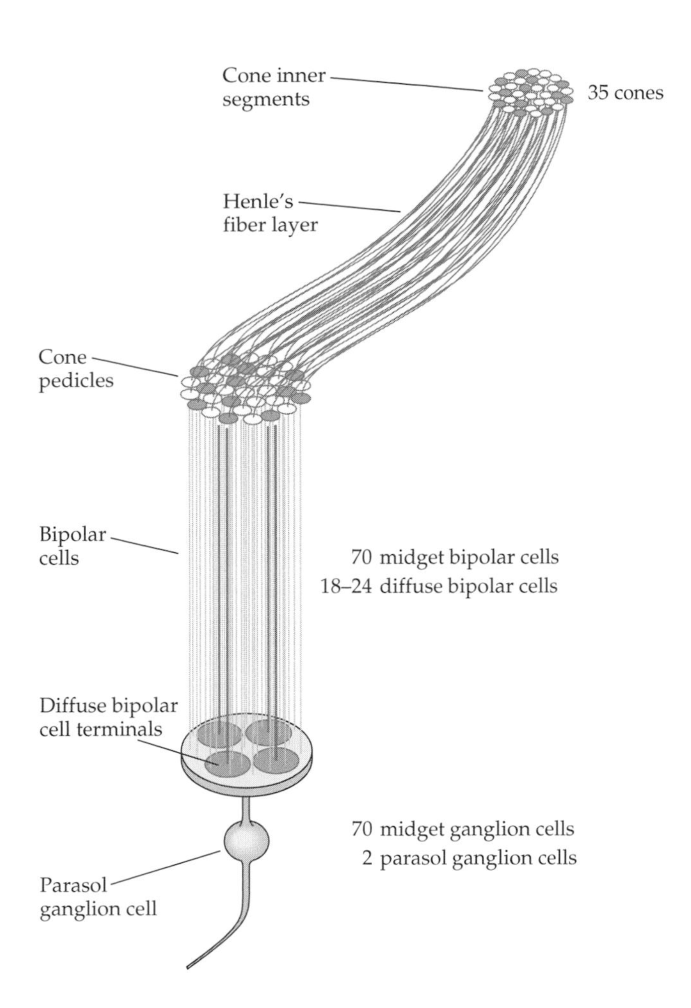

Figure 15.35
A Foveal Sampling Unit
Vertical pathways for a small cluster of 35 cones near the center of the fovea make up a sampling unit with 2 parasol cells and 70 midget ganglion cells. (Here only one parasol cell is shown, and no midget ganglion cells.) Sampling units like this provide high spatial resolution, but the absence of rod pathways make these units quite insensitive at low light levels.

cluster of cones, making them interact with one another and with some of the neighboring cones not included in the illustration. The fact that the horizontal cell processes extend beyond the central cluster of cone pedicles makes the anatomy of the sampling unit rather fuzzy at this level; it means, as already suggested, that interactions occur not only within but also between sampling units. Neighboring sampling units invariably have some cone inputs in common.

Amacrine cells will also generate interactions between sampling units, but it is hard to know which amacrine cells to include here and how many of them there are. There should be some small-field amacrine cells that connect midget bipolar cell terminals in the ON and OFF regions along the lines of the circuits shown in Figure 14.25; there might be a dozen or so amacrine cells operating in each of the ON and OFF regions. Because the amacrine cells are small, their interactions with neighboring sampling units are quite limited. More extensive interactions are indicated by the processes from some larger amacrine cells whose morphology and extent are deliberately vague; they are undoubtedly present, and probably interacting with the diffuse bipolar cells and the parasol ganglion cells, but the details are not known.

This sampling unit represents the rod-free portion of the fovea. There should be around 500 sampling units in this part of the fovea, each examining a very small bit of the retinal image corresponding to a visual angle about 3′ of arc in diameter. Since neither rod nor blue cone pathways are present, all the information conveyed through these sampling units originates in red and green cones. Even so, each sampling unit contains at least six different sets of ganglion cell signals: two coarse-grained signals from the ON and OFF parasol cells and four sets of signals from individual cones carried by the midget ganglion cells (red and green ON-center, red and green OFF-center). And these are the sampling units that provide our most detailed spatial information, because the cones are the most densely packed. These 500 foveal sampling units, with about 36,000 ganglion cells and half that many cones, are so informative that in many respects they are the main focus of attention for the eye and visual system as a whole.

Rods and blue cones become significant in the parafoveal sampling units

The piece of parafoveal retina in Figure 15.36 is just a bit larger than the dendritic fields of parasol cells, which are on average about 40 μm in diameter, because another ganglion cell type—the small bistratified ganglion cell—is larger still and its dendritic field will delimit the sampling unit. In this figure, it is shown as about 50 μm in diameter.

Midget ganglion cell arbors are considerably smaller, about 8 μm on average, so there are approximately 25 to 30 midget ganglion cells for every parasol cell and somewhat more within the total area defined by the bistratified ganglion cell's dendritic field—40 is a reasonable number. Thus in this small bit of tissue, there will be 1 bistratified (blue–yellow) ganglion cell, 2 complete parasol cells (1 ON and 1 OFF), and about 80 midget ganglion cells (40 ON and 40 OFF). As in Figure 15.35, most of the midget ganglion cells are not shown.

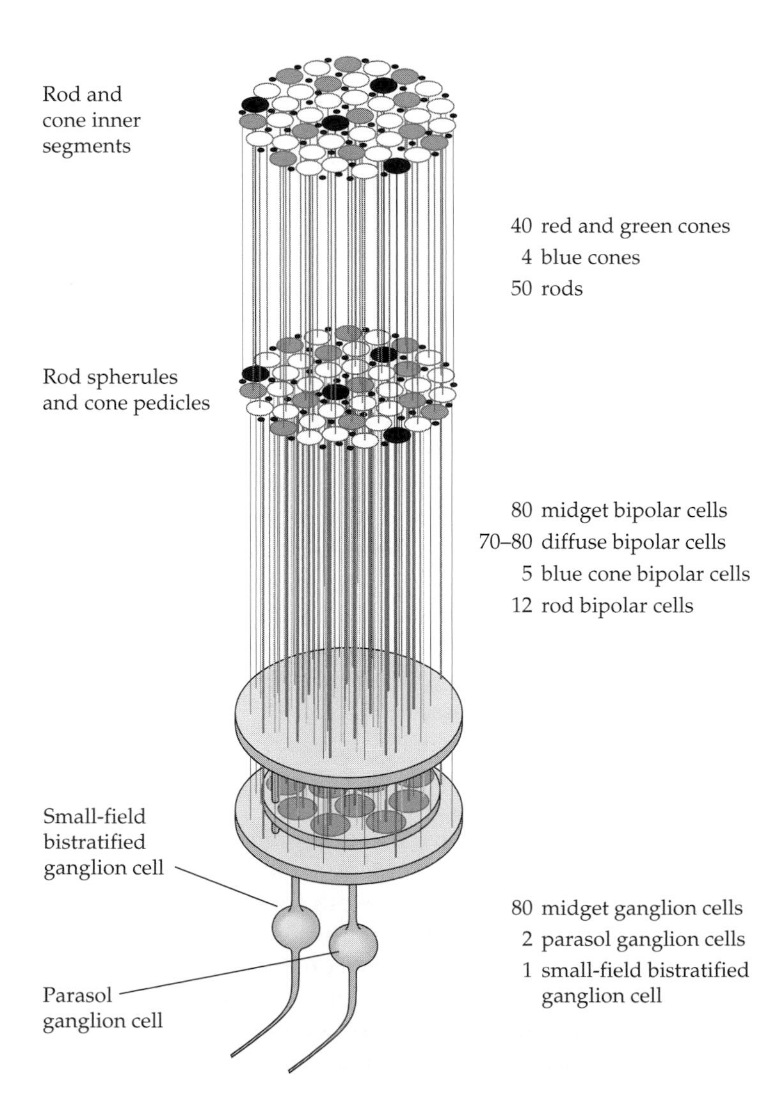

Figure 15.36
A Parafoveal Sampling Unit
The size of a sampling unit just outside the fovea is larger than central foveal sampling units; vertical pathways from about 45 cones and 50 rods end in 1 small bistratified ganglion cell, which sets the size of the sampling unit, 2 complete parasol cells, and 80 midget ganglion cells (not shown). Spatial resolution is less than in the fovea, but these sampling units are trichromatic, because of the blue cones and their pathway, and work well at low light levels because of the rod pathways. Parafoveal sampling units should contain all known cell types in the retina and perform all forms of information processing.

In this part of the retina, midget ganglion cells still receive direct input from only a single midget bipolar cell, so their numbers are equal, as before, and the number of red and green cones will be half that; thus there are about 80 midget bipolar cells and 40 red and green cones. Roughly 8% of the total cone population here will be blue cones, which means there will be three or four of them in this cone cluster (the total cone density is just over 20,000 cones/mm^2).

The number of diffuse bipolar cells can be estimated as it was for the foveal sampling unit. Using the value of 15 μm diameter for axon terminal areas that was applied to the foveal cone cluster leads to 12 or 13 diffuse bipolar cells at each of six levels, or between 70 and 80 in all. The numbers of midget and diffuse bipolar cells are about equal in the parafoveal sampling unit, whereas in the foveal sampling unit, midget bipolar cells outnumbered diffuse bipolar cells by three to one.

The entry of blue cones into the cone population means that some blue cone bipolar cells must be present, but it is difficult to estimate a number. There won't be many, however. Since blue cone bipolar cells invariably receive input from more than one blue cone, and most blue cones appear to make contact with at least two blue cone bipolar cells, the presence of three or four blue cones implies the presence of about four to six blue cone bipolar cells. If the blue cone bipolar cell terminals tile in the inner plexiform layer, the size of their terminal arbors, which are appreciably larger than those of the diffuse bipolar cells, also leads to a number in this range—let's say five blue cone bipolar cells.

Rod density in this part of the retina is somewhat higher than cone density, so there should be about 50 rods interspersed among the cones. Each rod bipolar cell receives inputs from several rods, and the rod bipolar cell dendrites overlap somewhat; this sampling unit should contain about 12 of them, with terminals at level 6 along with the blue cone bipolar cell terminals. The rod bipolar cells have outputs onto AII amacrine cells, and some convergence should occur here also; the signals from the 12 rod bipolar cells are probably handled by convergence on four to six AII amacrine cells, each of which has outputs to several cone bipolar cells, both ON and OFF.

Lateral pathways in the outer synaptic layer differ from those in the foveal sampling unit only by the addition of H1 horizontal cell terminals that are integrating rod signals collected at the spherules. With this exception, however, the number of horizontal cells is roughly the same, that is, six or so that contact cone pedicles.

Amacrine cell pathways include the small-field amacrine cells that interconnect the midget bipolar cell terminals and the AII amacrine cells that have appeared in the rod pathways. The presence of the rod pathways also implies the presence of large-field amacrine cell processes at two levels: the GABA–dopamine amacrine cells in level 1 and the A17 amacrine cells in level 6 (see Figure 14.14). As in the fovea, some medium- or large-field amacrine cells are probably making connections with the diffuse bipolar cells and the parasol ganglion cells, but the type and connectivity have again been left unspecified.

This parafoveal sampling unit differs from its foveal counterpart in several ways, the most obvious being the inclusion of rod and blue cone pathways, with their attendant bipolar cells. Color vision is now trichromatic, and the operating range of the retina has been extended down to much lower light levels than before. The addition of the rod pathways has also brought in more types of amacrine cells, particularly the large-field amacrine cells that may be controlling the rods' access to the cone pathways.

The dominance of the midget cell pathways in the fovea has been reduced along with the decline in cone density, and the spatial grain represented by the Nyquist frequency has coarsened considerably; visual acuity here should be about four times less than at the center of the fovea—that is, about 20/80. The Nyquist frequency for the parasol ganglion cells is also less than in the fovea, but the decline is not as striking as it is in the midget cell system.

These parafoveal sampling units are more numerous and occupy a larger area than the foveal sampling units, which suggests that their limited spatial resolution does not mean they are unimportant. In fact, this is perhaps the most versatile part of the retina. The parafoveal sampling units represent a region of transition between the strict diurnality of the central fovea and the overwhelming nocturnality of the peripheral retina that we will see in the next example. At dusk and dawn, the parafoveal retina is the part of the eye with which we see best.

Rods and rod pathways dominate in peripheral sampling units

Figure 15.37 shows a sampling unit 15 to 20° from the foveal center. As in the parafovea, the largest dendritic fields (of those we know about) will be those of the small bistratified ganglion cells, shown here as being about 250 μm in diameter. Dendritic fields of parasol cells will be about 220 μm in diameter, and those of midget ganglion cells will be about 60 μm in diameter. Relative to the parafoveal sampling unit, everything has been scaled up appreciably, by a factor of five for the bistratified ganglion cell, by almost six for the parasol cells, and by almost eight for the midget ganglion cells. Spatial resolution is clearly not an important feature here, and visual acuity will be quite low.

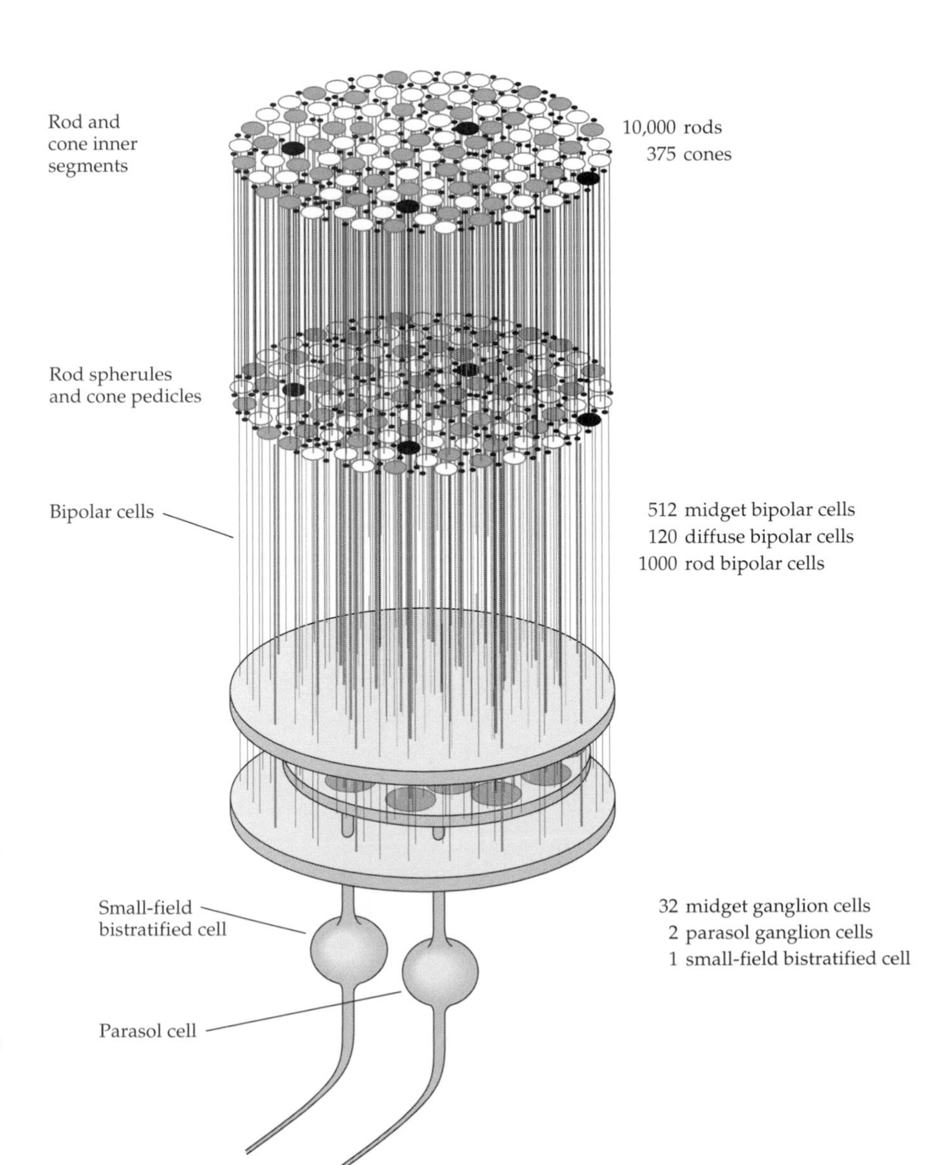

Figure 15.37
A Peripheral Sampling Unit
The dramatic increase in the sizes of ganglion cell dendritic fields makes peripheral sampling units very large. This one contains about 10,000 rods and 375 cones. The sampling unit is defined by 1 small bistratified ganglion cell that encloses 2 complete parasol ganglion cells and 32 midget ganglion cells (not shown). The enormous convergence of photoreceptors onto ganglion cells makes this rod-dominated sampling unit exquisitely sensitive to low light levels.

Rod density is near its maximum (150,000 cells/mm^2), and this sampling unit should contain almost 10,000 of them—for obvious reasons, Figure 15.37 merely hints at the true number. Cone density will be about 6000 cells/mm^2, and there will be about 375 cones scattered among all the rods; about 30 of them will be blue cones. This enormous increase in the number of photoreceptors in the sampling unit has not changed things much at the level of the ganglion cells. By definition of the sampling unit, there are 1 bistratified ganglion cell, 2 complete parasol cells (1 ON and 1 OFF), and about 32 midget ganglion cells (16 ON and 16 OFF), based on the size of their dendritic fields. The convergence from photoreceptors to ganglion cells is obviously very large.

Convergence increases in the vertical pathways at both synaptic levels; there are more photoreceptors per bipolar cell and more bipolar cells per ganglion cell. In the midget cell pathways, the size of the midget bipolar cell terminals (about 15 μm) indicates that each midget ganglion cell may receive inputs from 16 midget bipolar cells, which would put the number of midget bipolar cells at 512 (16×32), equally divided between ON and OFF midget bipolar cells (256 of each). Thus each ON or OFF midget bipolar cell is receiving inputs from more than one cone, but as far as is known, the specificity for cone type—red or green—is preserved. This change in the midget bipolar cell pathways is a dramatic difference from either the foveal or parafoveal sampling units. In the fovea and the parafovea, there was no convergence, and each midget ganglion cell reflected activity in a single cone (ignoring lateral connections). In the peripheral fovea, each midget ganglion cell looks at activity from 25 cones or more.

Convergence from cones to diffuse bipolar cells is estimated to double relative to the parafovea, suggesting that each diffuse bipolar cell receives inputs from about 15 cones. This gives 18 to 20 diffuse bipolar cells for each of the six levels of the inner plexiform layer, for a total of about 120 diffuse bipolar cells.

Data on the density of rod bipolar cells give values around 15,000 cells/mm^2 in this part of the retina, which translates to about 1000 rod bipolar cells in this sampling unit. The convergence from rod to rod bipolar cell is therefore about 10 rods per bipolar cell. As a guess, these rod bipolar cells contact about 200 AII amacrine cells (a convergence of five to one), through which signals are conveyed to the cone bipolar cells and ganglion cells. All these numbers are consistent with the clear rod dominance in this sampling unit.

There should be no significant changes in the lateral pathways, except for increases in the sizes of the horizontal cell and amacrine cell arbors. The main point is that amacrine cell interactions are more extensive, whatever they are doing, in keeping with the expanded size of the peripheral sampling units.

Most of the retina is filled with sampling units like this; they cover more than 90% of the retina, with more than 15,000 sampling units containing somewhat more than half of the 1 million ganglion cells. (The 500 sampling units in the central fovea have 36,000 ganglion cells sampling only 0.01% of the retinal area; the remaining 3000 or so foveal and parafoveal sampling units cover about 9% of the retina, using several hundred thousand ganglion cells.) The peripheral sampling units are dominated by rods and rod pathways, they have low spatial resolution, and they have potentially high sensitivity at very low light levels.

For most people in urban, industrialized societies, the obvious feature of the peripheral retina is its limited performance during the day; except to find the bathroom in the middle of the night, we rarely experience the immense capabilities of the peripheral retina when it has just moonlight and starlight with which to work. Although we are not specialized creatures of the night, the sampling units in our peripheral retinas can detect and signal as few as five photon absorptions occurring within a second of one another. Very few species can do much better than this. (Our real limitation at night is the lack of auditory and olfactory specializations that permit other animals to get around when there is no light. When we can't see, they can't see either, but they have other senses they can employ to compensate.)

The problem of understanding how the retina works can be reduced to the problem of understanding its sampling units

We understand a lot about the retina, which explains in part why this story extends through several chapters. But the length of the story also stems from some critical obscurities that get in the way of concise expression. (It's difficult to be clear about things dimly seen.) More data and more ideas will come along, which is one reason that I have tried to bring the retina down to a manageable size, to reduce it to a relatively simple structural element that can be modified as new information becomes available.

To that end, consider the real parafoveal sampling unit represented in Figure 15.36 as the retina in miniature; it contains all the cell types in the retina, all of the possible interactions between cells, and all of the multiple information streams that create the dots in the retinal sketches. Although it is connected to and interacts with neighboring parts of the retina, its outputs—the dots of activity from its ganglion cells—are potentially unique, reflecting aspects of the visual world that it alone can sample. As such, it represents the thousands of other sampling units that, in combination, constitute the retina.

Yet a parafoveal sampling unit contains only 300 or 400 cells, a far cry from the 130 million or so cells in the retina as a whole. These several hundred cells incorporate, explicitly or implicitly, everything we have considered about the retina to this point, and in this sense they represent a summary of our understanding of the structure and function of the human retina.

The central representation of a sampling unit depends on the number of ganglion cells it contains

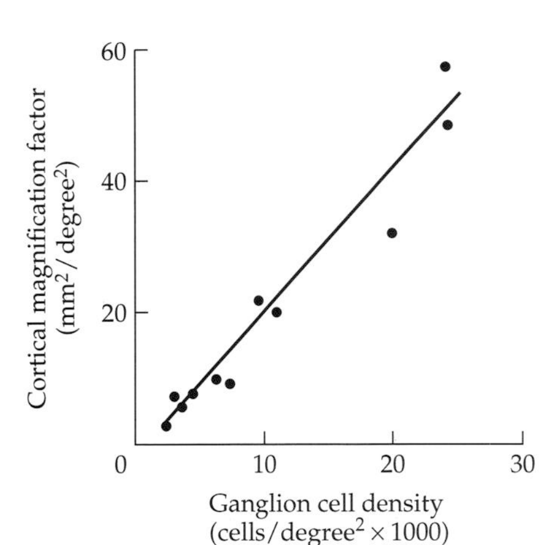

Figure 15.38
Cortical Magnification Factor as a Function of Ganglion Cell Density
The cortical magnification factor is given in square millimeters of visual cortex divided by degrees squared of visual angle for the central 8° of the macaque monkey's visual field. It increases linearly with ganglion cell density (expressed here in cells/degree^2 of visual angle). The value of the constant slope indicates that each ganglion cell is allotted 0.002 mm^2 of cortex, or that each square millimeter of cortex receives inputs derived from about 500 ganglion cells. (After Wässle et al. 1990.)

Although the fovea occupies less than 0.2% of the retina's area, its representation in the primary visual cortex takes up almost 40% of the cortical surface area. This enormous expansion is called cortical magnification. The cortical magnification factor is the numerical relationship between a unit area of cortex and the area of visual space (or of the retina) to which it corresponds. The smaller the region of space or retina seen by a unit of cortex, the larger the magnification factor will be, and the cortical magnification factor for the fovea is several thousand times that of the retinal periphery. If we compare two small pieces of cortex of equal size, one receiving inputs from the fovea and the other from far peripheral retina, the peripheral piece of cortex will see several thousand times more of the visual field area than will the foveal piece of cortex. Thus a large part of the cortex is required to represent the fovea and only a small part need be devoted to the part of the world seen by the peripheral retina.

This expansion of a small retinal region like the fovea to a large area of cortex is not done by the relay at the lateral geniculate nucleus. The relationship between retinal ganglion cell axons and geniculate cells projecting to the visual cortex is one to one, so for the purpose of this discussion, we can talk about retinal ganglion cell axons *as if* they went directly to the cortex. (This does *not* mean that nothing of importance happens in the lateral geniculate nucleus, however.) But if divergence in the lateral geniculate nucleus is not an issue, how is cortical space allotted to inputs from different parts of the retina? And does the fovea receive preferential treatment, receiving more cortical space than it deserves?

The answers turn out to be simple,* and a key summary result is shown in Figure 15.38.Within the central 8° of the monkey's visual field, the cortical magnification factor increases as a linear function of ganglion cell density; that is, the

**Getting* to the answers is not simple, however, and I have omitted all of the underlying data and logic. Moreover, not everyone would agree with my answers, which are based on Wässle et al. 1990. For example, Azzopardi and Cowey (*Nature* 361: 719–721, 1993) conclude that the fovea does receive preferential treatment in the allotment of cortical space.

ganglion cell density determines how much cortical space each bit of retina occupies. This is the result one would expect if each ganglion cell claimed the same amount of cortical space. (If ganglion cells from one part of the retina took up more space than others, the relationship between magnification factor and cell density would not be linear.) The space allotted to each ganglion cell is the slope of the line, which is about 0.002 mm^2 of cortex per ganglion cell. The reciprocal of the slope is the number of ganglion cells per square millimeter of cortex, and this value is about 500.

With each ganglion cell allotted the same amount of space, the parts of the retina that have high ganglion cell density necessarily take up the most cortical area. Thus the fovea is represented to a greater extent, because it has more ganglion cells per unit of retinal area. And since the same rule—cortical space depends on ganglion cell density—applies everywhere in the retina, the fovea is not overrepresented; it gets what it deserves on the basis of the number of ganglion cells it has.

There are around 500 sampling units in the rod-free area of the fovea, and about 36,000 ganglion cells convey signals from about 18,000 cones. These 500 sampling units occupy only 0.03% of the retina's area, but they have 0.4% of the cones and 3.6% of the ganglion cells. As a result, their representation in the primary visual cortex should occupy 3.6% of the available cortical space. By representing such a tiny part of visual space in this much cortex, the brain has the opportunity to examine and compare these dots from the retinal sketch in great detail.

A peripheral sampling unit, in contrast, has 35 ganglion cells, but there are about 15,000 of these sampling units, occupying about 90% of the retina's area and using half the complement of ganglion cells. Thus the retina outside the central 10° or so should take up half the cortical space. But where the foveal representation was expanded, this representation is compressed; most of the visual world claims only half of the primary visual cortex. Although each dot from the retinal sketch may get the same attention from cortical circuitry here as do the dots in the foveal representation, the peripheral dots correspond to larger parts of retina and visual space; the brain's ability to discover what is where in this part of the visual world is therefore limited.

Thus the spatial organization of the retina dictates the spatial organization of the primary visual pathway and the primary visual cortex.

References and Additional Reading

Making Retinal Sketches out of Dots: Limits and Strategies

Amthor FR and Oyster CW. 1995. Spatial organization of retinal information about the direction of image motion. *Proc. Natl. Acad. Sci. USA* 92: 4002–4005.

DeVries SH and Baylor DA. 1997. Mosaic arrangement of ganglion cell receptive fields in rabbit retina. *J. Neurophysiol.* 78: 2048–2060.

Hirsch J and Curcio CA. 1989. The spatial resolution capacity of the human fovea. *Vision Res.* 29: 1095–1101.

Hirsch J and Hylton R. 1984. Quality of the primate photoreceptor lattice and limits of spatial vision. *Vision Res.* 24: 347–355.

Hughes A. 1977. The topography of vision in mammals of contrasting life style: Comparative optics and retinal organization. Chapter 11, pp. 613–756, in *Handbook of Sensory Physiology,* Vol. VII/5, *The Visual System in Vertebrates,* Cresitelli F, ed. Springer-Verlag, Berlin. (See, in particular, Section D, "Retinal Topography," pp. 687–716.)

Vaney DI. 1994. Territorial organization of direction-selective ganglion cells in rabbit retina. *J. Neurosci.* 14: 6301–6316.

Wässle H, Grünert U, Röhrenbeck J, and Boycott BB. 1989. Cortical magnification factor and the ganglion cell density of the primate retina. *Nature* 341: 643–646.

Wässle H, Peichl L, and Boycott BB. 1981. Dendritic territories of cat retinal ganglion cells. *Nature* 292: 344–345.

Wertheim T. 1894. Über die indirekte Sehschärfe. *Zeitschr. Psychol. Physiol. Sinnesorgane* 7: 172–187. Translation by Dunsky IL. 1980. Peripheral visual acuity. *Am. J. Optom. Physiol. Optics* 57: 915–924.

Westheimer G. 1979. The spatial sense of the eye. *Invest. Ophthalmol. Vis. Sci.* 18: 893–912.

Williams DR. 1986. Seeing through the photoreceptor mosaic. *Trends Neurosci.* 9: 193–198.

Williams DR. 1988. Topography of the foveal cone mosaic in the living human eye. *Vision Res.* 28: 433–454.

Williams DR and Collier R. 1983. Consequences of spatial sampling by a human photoreceptor mosaic. *Science* 221: 385–387.

Yellott JI. 1982. Spectral analysis of spatial sampling by photoreceptors: Topological disorder prevents aliasing. *Vision Res.* 22: 1205–1210.

Spatial Organization of the Retina

Boycott BB and Wässle H. 1991. Morphological classification of bipolar cells of the primate retina. *Eur. J. Neurosci.* 3: 1069–1088.

Cicerone CM and Nerger JL. 1989. The relative numbers of long-wavelength-sensitive to middle-wavelength-sensitive cones in the human fovea centralis. *Vision Res.* 29: 115–128.

Curcio CA and Allen KA. 1990. Topography of ganglion cells in human retina. *J. Comp. Neurol.* 300: 5–25.

Curcio CA, Allen KA, Sloan KR, Lerea CL, Hurley JB, Klock IB, and Milam AB. 1991. Distribution and morphology of human cone photoreceptors stained with anti-blue opsin. *J. Comp. Neurol.* 312: 610–624.

Curcio CA and Sloan KR. 1992. Packing geometry of human cone photoreceptors: Variation with eccentricity and evidence for local anisotropy. *Visual Neurosci.* 9: 169–180.

Curcio CA, Sloan KR, Kalina RE, and Hendrickson AE. 1990. Human photoreceptor topography. *J. Comp. Neurol.* 292: 497–523.

Curcio CA, Sloan KR, Packer O, Hendrickson AE, and Kalina RE. 1987. Distribution of cones in human and monkey retina: Individual variability and radial asymmetry. *Science* 236: 579–582.

Dacey DM. 1990. The dopaminergic amacrine cell. *J. Comp. Neurol.* 301: 461–489.

Dacey DM. 1993. The mosaic of midget ganglion cells in the human retina. *J. Neurosci.* 13: 5334–5355.

Dacey DM. 1994. Physiology, morphology, and spatial densities of identified ganglion cell types in primate retina. In *Higher-order Processing in the Visual System* (Ciba Foundation Symposium 184), pp. 12–34. Wiley, Chichester.

Dacey DM and Brace S. 1992. A coupled network for parasol but not midget ganglion cells in the primate retina. *Vis. Neurosci.* 9: 279–290.

Dacey DM, Lee BB, Stafford DK, Pokorny J, and Smith VC. 1996. Horizontal cells of the primate retina: Cone specificity without spectral opponency. *Science* 271: 656–659.

Grünert U and Martin PR. 1991. Rod bipolar cells in the macaque monkey retina: Immunoreactivity and connectivity. *J. Neurosci.* 11: 2742–2758.

Grünert U, Martin PR, and Wässle. 1994. Immunocytochemical analysis of bipolar cells in the macaque monkey retina. *J. Comp. Neurol.* 348: 601–627.

Hart WM. 1992. Entoptic imagery. Chapter 15, pp. 485–501, in *Adler's Physiology of the Eye*, 9th Ed. Hart WM, ed. Mosby Year Book, St. Louis.

Kouyama N and Marshak DW. 1992. Bipolar cells specific for blue cones in the macaque retina. *J. Neurosci.* 12: 1233–1252.

Martin PR and Grünert U. 1992. Spatial density and immunoreactivity of bipolar cells in the macaque monkey retina. *J. Comp. Neurol.* 323: 269–287.

Milam AH, Dacey DM, and Dizhoor AM. 1993. Recoverin immunoreactivity in mammalian cone bipolar cells. *Vis. Neurosci.* 10: 1–12.

Mollon JD and Bowmaker JK. 1992. The spatial arrangements of cones in the primate fovea. *Nature* 360: 677–679.

Østerberg GA. 1935. Topography of the layer of rods and cones in the human retina. *Acta Ophthalmol.* 13: 1–97.

Polyak SL. 1941. *The Retina*. University of Chicago Press, Chicago.

Rodieck RW. 1998. Retinal organization. Chapter 9, pp. 196–209, in *The First Steps in Seeing*. Sinauer Associates, Sunderland, MA.

Rodieck RW and Marshak DW. 1992. Spatial density and distribution of choline acetyltransferase immunoreactive cells in human, macaque, and baboon retinas. *J. Comp. Neurol.* 321: 46–64.

Röhrenbeck J, Wässle H, and Boycott BB. 1989. Horizontal cells in the monkey retina: Immunocytochemical staining with antibodies against calcium binding proteins. *Eur. J. Neurosci.* 1: 407–420.

Schein SJ. 1988. Anatomy of macaque fovea and spatial densities of neurons in foveal representation. *J. Comp. Neurol.* 269: 479–505.

Tsukamoto Y, Masarachia P, Schein SJ, and Sterling P. 1992. Gap junctions between the pedicles of macaque foveal cones. *Vision Res.* 32: 1809–1815.

Wässle H, Grünert U, Chun M-Y, and Boycott BB. 1995. The rod pathway of the macaque monkey retina: Identification of AII-amacrine cells with antibodies against calretinin. *J. Comp. Neurol.* 361: 537–551.

Wässle H, Grünert U, Martin PR, and Boycott BB. 1994. Immunocytochemical characterization and spatial distribution of midget bipolar cells in the macaque monkey retina. *Vision Res.* 34: 561–579.

Wässle H, Grünert U, Röhrenbeck J, and Boycott BB. 1990. Retinal ganglion cell density and cortical magnification factor in the primate. *Vision Res.* 30: 1897–1911.

Watanabe M and Rodieck RW. 1989. Parasol and midget ganglion cells of the primate retina. *J. Comp. Neurol.* 289: 434–454.

Chapter 16

The Retina In Vivo and the Optic Nerve

A small trough of clay or paraffin was constructed round the margin of the orbit, so as to contain a quantity of dilute salt solution, when the body was placed horizontally and the head properly secured. Into this solution the terminal of a non-polarisable electrode was introduced, and in order to complete the circuit the other electrode was connected with a large gutta-percha trough containing salt solution, into which one of the hands was inserted. . . . it is possible to . . . detect an electrical variation similar to what is seen in other animals. This method, however, is too exhausting and uncertain to permit of quantitative observations being made.

■ James Dewar, "The Physiological Action of Light"

The living retina was first made accessible by Helmholtz's ophthalmoscope (see Vignette 11.1). For the first time, the retina, its blood vessels, and the optic nerve head where the ganglion cell axons collect to form the optic nerve were all visible. No other part of the brain can be seen this easily, and detailed examination of the retina for signs of local and systemic disease is now routine.

Another step in making the retina open to inspection was made about 20 years after the invention of the ophthalmoscope. The opening quotation at right, from an article by James Dewar,* was the first report that an electrical signal could be recorded from the human eye in vivo when it was illuminated. Similar electrical activity produced by illuminating an animal's eye had already been shown to come from the retina. Although the early method was "exhausting and uncertain," more than a century of improvement has made such recordings a way to study both normal and impaired retinal function.

Much of the visual inspection of the in vivo retina concerns the blood vessels and their integrity, but the blood vessels inevitably lead to the optic nerve head, from which the arteries emanate, and it too is the object of considerable scrutiny. The nerve head is important because all the retina's signals conveyed by the ganglion cell axons and all of its inner vascular systems converge in this small region. Small lesions here can cause widespread, serious problems for the retina and vision. In addition, the optic nerve head is inherently weaker than the surrounding sclera because of its structure; this weakness makes it vulnerable to structural damage that may underlie visual field loss in glaucoma.

The question of whether the retina is functioning normally is often best answered by the retina itself—that is, by determining if central visual acuity is normal (with any refractive error corrected), if color discrimination is normal, and if visual field limits are normal. These evaluations are among the standard battery of subjective tests that are part of routine clinical examinations, and deficits in any of them should trigger a search for an explanation with other test procedures. In numerous pathological situations, however, early and treatable stages rarely produce visual deficits that the patient notices or that are detectable

*Nature 15: 433–434, 1877. James Dewar (1842–1923) is best known as a physicist and chemist who developed methods to liquefy gases and study their properties at extremely low temperatures. He also invented the vacuum flask, variants of which we know as thermos bottles.

with standard subjective evaluations of visual function. Recognizing that a problem exists may depend on seeing evidence of an abnormality by the methods that Helmholtz and Dewar discovered.

Electrical Signals and Assessment of Retinal Function

A difference in electrical potential exists between the vitreal and choroidal surfaces of the retina and between the front and back of the eye

Even when retinal illumination is constant, a considerable amount of current flows as ions move from neurons to the extracellular space and back again. As noted in Chapter 13, photoreceptors in particular show a steady flow of current from inner to outer segments. This current flow for a single photoreceptor is tiny, but the currents from the millions of photoreceptors add together because the photoreceptors all have the same radial orientation; the total is significant and measurable. Thus the photoreceptor currents and those of other radially oriented elements in the retina sum to produce a net flow of current between the inner and outer surfaces of the retina. And since the electrical resistance between the vitreal surface of the retina and the choroid is quite large (much of the resistance is at the level of the pigment epithelium), the current flow produces a significant voltage difference between the inner surface and the choroid; the retina is rather like a small battery.

A battery doesn't do anything until it is connected into a circuit through which current can flow; in this case the retinal battery is part of a circuit formed by the extraocular tissues and the intraocular media. Current is continually flowing through the retina, through the choroid and sclera, around the eye, and back through the cornea and intraocular fluids. Since the various parts of this current pathway have their own electrical resistances, it is possible to measure voltage differences across them; to take one in particular, there is a voltage difference between the front and the back of the eye.

The steady voltage difference across the eye is used in a measure of eye position called the **electro-oculogram** (**EOG**). Electrodes are placed at the medial and lateral canthi, and the difference in electrical potential between the electrode pairs is measured as the eye moves. This technique works because the steady potential at the corneal apex is positive relative to the surrounding tissues, so as the eye moves laterally, for example, the lateral electrode becomes more positive relative to its medial companion. In clinical practice, the EOG is the amount of potential change as the patient looks alternately at targets placed about 30° apart, under both light- and dark-adapted conditions (the change in potential is normally smaller when the eye is dark-adapted). The EOG is abnormal for various disease conditions, but it is not clear that there are any conditions for which the EOG is the best diagnostic indicator; it is often used to confirm what was already suspected. As a measure of eye position, it is too imprecise and artifact-prone to be of much value for research purposes.

The electroretinogram measures a complex change in voltage in response to retinal illumination

When an electrode is placed on the corneal surface and voltage is measured relative to a reference electrode on the skin of the face or scalp, illumination of the eye with a brief light flash produces voltage changes like those in Figure 16.1. This voltage change is the **electroretinogram** (**ERG**), and it represents the summed activity of many retinal neurons.

In the first recording (Figure 16.1*a*), the electrical response is a large upward deflection with a latency of about 200 ms; it peaks about 500 ms after the stimu-

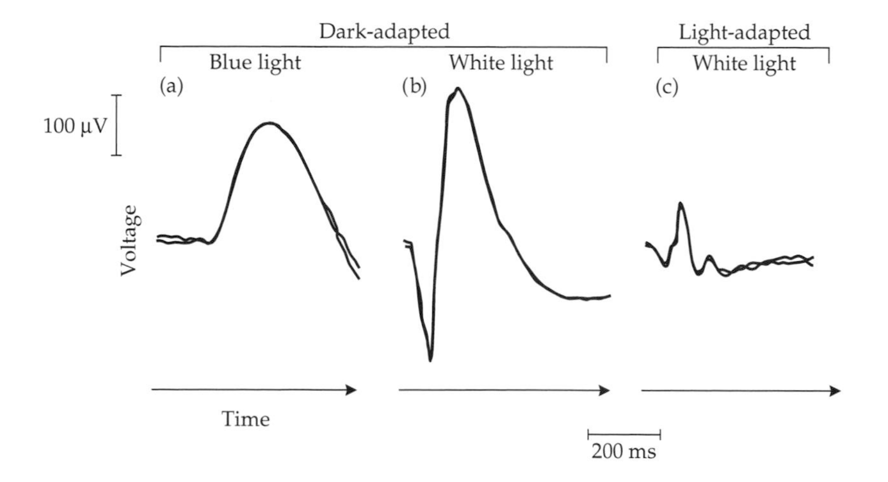

Figure 16.1
The Human Electroretinogram under Different Conditions
In these recordings, the eye was illuminated by a very brief flash of light at the point where each trace begins, producing responses that lasted for about 0.5 s. (*a*) The response to blue light under dark-adapted conditions is mediated almost entirely by rods; it is the simplest waveform. (*b*) White light adds a cone component to the response. (*c*) In light-adapted conditions, the cone-mediated response is both smaller and more complex. (After Berson 1992.)

lus and then decays. Because the light stimulus was blue and the eye was dark-adapted, this ERG should reflect only rod stimulation. Figure 16.1*b* shows the response to a white light stimulus that should activate cones under the dark-adapted condition; this response has a very short latency, with an initial negative deflection prior to the upward rise to the peak. Under light-adapted conditions (Figure 16.1*c*), the overall response is smaller and has two notches in the waveform that were not obvious in the other recordings; this should be a cone-dominated ERG.

Although the ERG obviously changes dramatically with different illumination and stimulus conditions, several components and inflections on the waveform are always present. Those of particular interest are labeled in Figure 16.2.

The small downward deflection just after onset of the light stimulus is the **a-wave**. It is inflected midway to its minimum; the initial phase is due to a cone signal, and the rest of the later part of the a-wave is dominated by a rod signal. The a-wave is followed by a much larger upward deflection, which is the **b-wave**. The b-wave is due mostly to rods, but a small cone inflection is shown on the rising phase. Other components not shown here include a long, slow voltage change called the c-wave; it is more evident with stimuli of longer duration, but the a- and b-waves are nearly complete by the time the c-wave begins. Cone-dominated ERGs under light-adapted conditions exhibit a deflection when a stimulus of long duration is turned off (the off effect), and repetitive stimuli produce a series of small oscillatory potentials in the ERG.

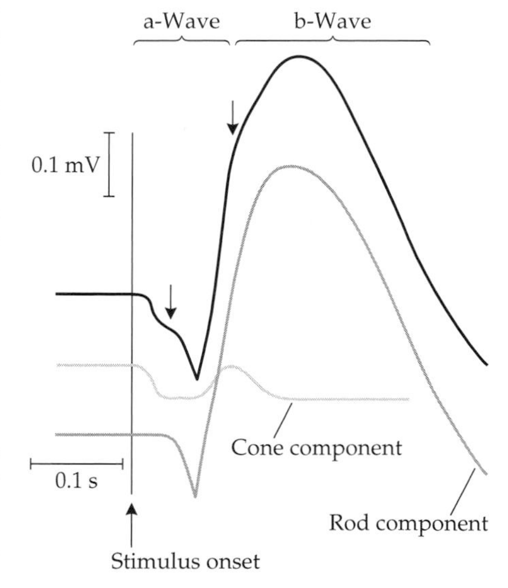

Figure 16.2
Major Components of the ERG
The a- and b-waves of the ERG are inflected (arrows) because both rod and cone components with different amplitudes and time courses contribute to the overall waveform. When the retina is dark-adapted, as it is here, the rod contribution dominates. (After Armington, Johnson, and Riggs 1952.)

The a-wave and off effect are generated by the photoreceptors, the c-wave by the pigment epithelium

Attempts to identify the cellular sources of the ERG components have been going on for many years. The general strategies have been to eliminate one or more populations of cells and observe any change in the ERG waveform and, more recently, to correlate the intracellular responses of individual cells with particular ERG components. The main conclusions of this work are summarized here, omitting all of the steps along the way.

The a-wave is the leading edge of photoreceptor responses; it persists under conditions that eliminate the b-wave (such as cutting off the blood supply to the inner retina), and under conditions that eliminate the c-wave (application of certain metabolic poisons or removal of the retina from the eye cup and pigment epithelium). Since the photoreceptors are the only neurons unaffected by these conditions, they must be responsible for the persistent a-wave (and part of the off-transient as well). The cone contribution to the a-wave is faster, but smaller, than that of the rods.

The c-wave is associated with the pigment epithelium, as implied by the observation that it can be eliminated from the ERG by removing the retina from the eye cup and pigment epithelium. (The isolated retina still produces an ERG.) It may seem odd that an electrical signal produced by illumination of the retina comes from cells that are not neurons and do not respond to light stimuli with electrical potentials. But the pigment epithelial cells are intimately related to the photoreceptors, and they will be caught up in the ion fluxes that occur as the photoreceptors respond. The action spectrum of the c-wave—that is, its amplitude as a function of wavelength—is like the rhodopsin absorption spectrum. In short, the c-wave is a secondary change in electrical potential generated across the pigment epithelium as a result of changing ion fluxes around the photoreceptors.

The b-wave is either a direct reflection of ON bipolar cell activity or is indirectly related to their activity by a secondary potential arising from Müller's cells

Müller's cells have been mentioned several times in the preceding chapters. The cells, as Müller originally described them, are radial elements in the retina extending from the inner segments of the photoreceptors to the vitreal surface of the retina (Figure 16.3). The expanded vitreal ends of Müller's cells interleave to

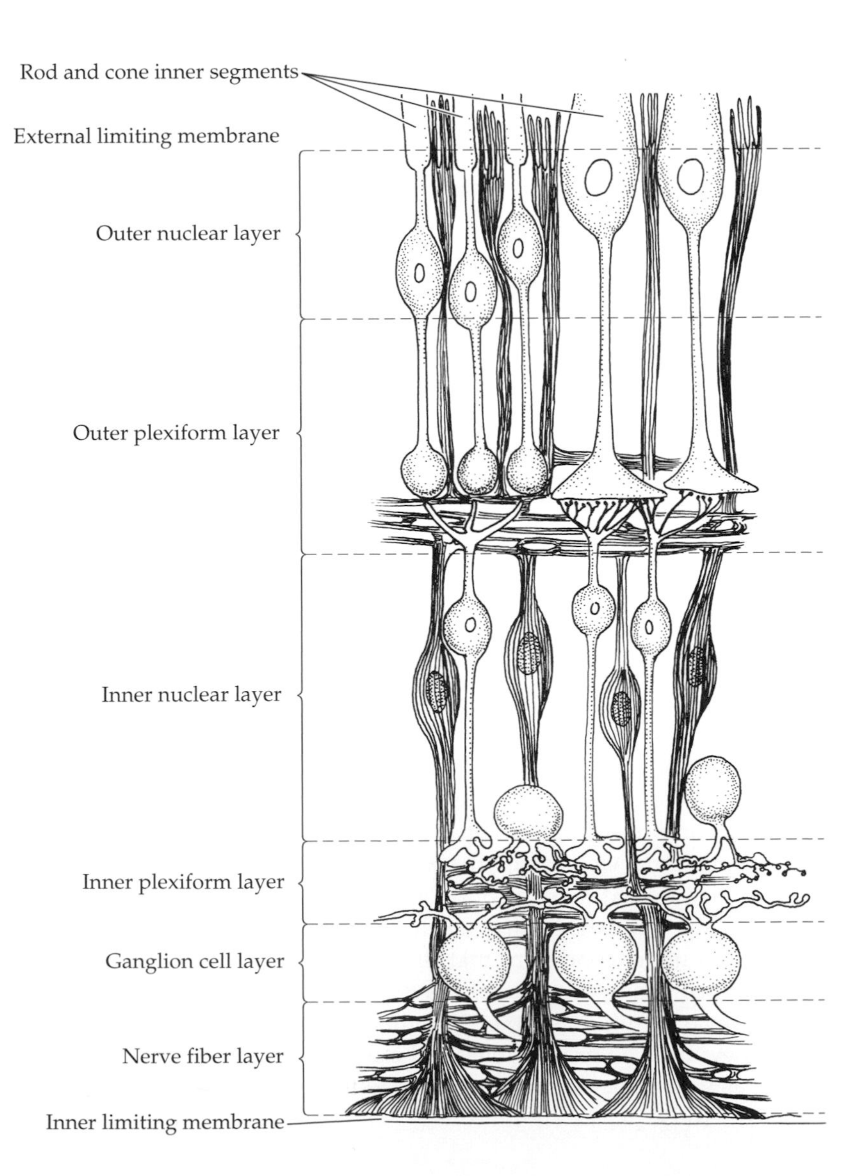

Figure 16.3
Structure of Müller's Cells
In this schematic cross section through the retina, the cytoplasm of Müller's cells has been shown darkly shaded in comparison to that of other cells. Müller's cells extend from the inner surface of the retina, where their expanded feet form the inner limiting membrane, to the photoreceptor layer, where terminal processes extend up between the inner segments of the photoreceptors. Müller's cells attach to photoreceptors with adhering junctions at the external limiting membrane. Their nuclei are in the inner layer, and the cells extend lateral processes into both plexiform layers and the nerve fiber layer. (After Hogan, Alvarado, and Weddell 1971.)

form the retina's inner limiting membrane, and their junctions with photoreceptor inner segments form the external limiting membrane. Their nuclei lie in the inner nuclear layer, and their lateral branches spread out at all levels, but most extensively in the plexiform layers.

The detailed morphology of Müller's cells is variable and not known to be of special importance; since they are not neurons, the levels and extent of their branching are not critical to their function. It is important, however, that Müller's cells are very large relative to the retinal neurons and that they are numerous and ubiquitous. Their size and numbers indicate that they constitute a large part of the total cellular volume of the retina. In effect, the cytoplasm of Müller's cells occupies all the space not taken up by neuronal processes and somata. (Estimates of Müller's cell numbers and their contribution to the retina's volume are difficult to find. Generalizing from a few studies of rat retina, however, the human retina might contain 0.5 to 1 million Müller's cells, occupying perhaps 15% of the retina's volume.)

Because Müller's cells surround everything else, the extracellular space of a retinal neuron—the surrounding pool of ions and small molecules with which it interacts—is Müller's cell cytoplasm. As a result, the ionic currents resulting from activity in retinal neurons are reflected in large, secondary changes in electrical potential across Müller's cell membranes. Potassium, for example, flows into Müller's cells in the outer and inner plexiform layers and flows back out in the nerve fiber layer. As Figure 16.4 indicates, K^+ is released from cells that have been activated by light stimuli.

Until recently, Müller's cells appeared to be the only cells in the retina from which an electrical potential similar to that of a b-wave could be recorded with

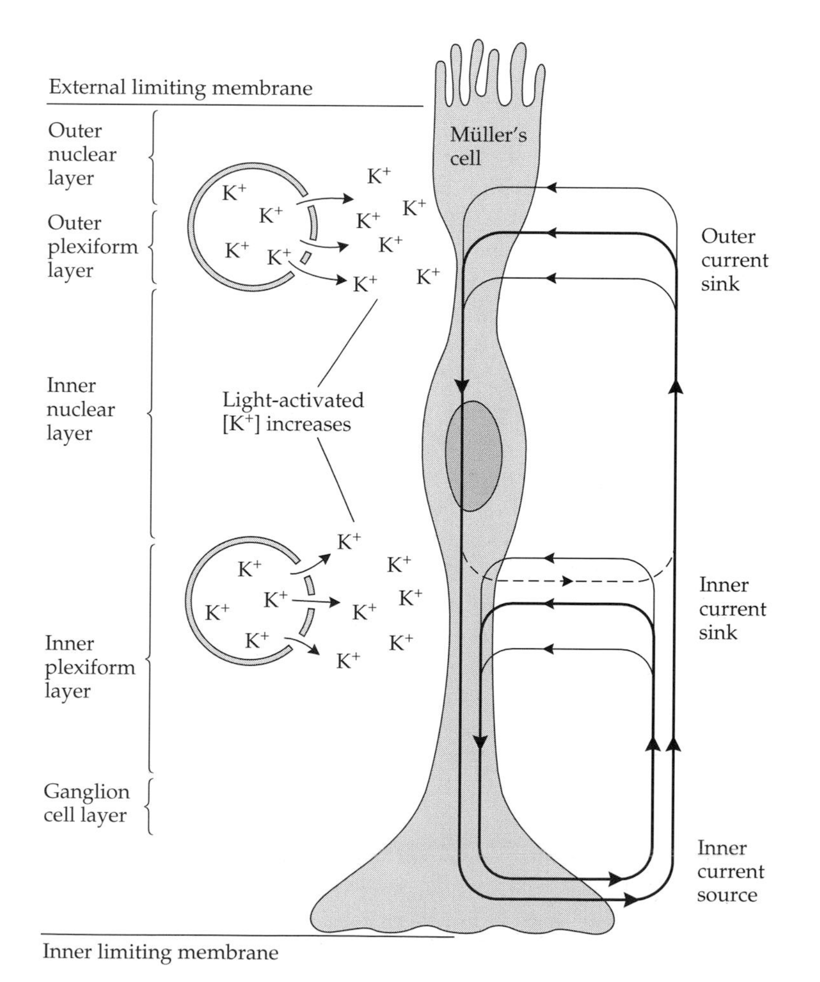

Figure 16.4
K^+ Current Flows in and around Müller's Cells
A steady K^+ current flows through and around Müller's cells; most of the current flows out of the Müller's cell at its base near the vitreous and flows back into the cell in both the inner and outer plexiform layers. K^+ levels are affected by the activity of retinal neurons, resulting in variations in the K^+ current flow that contribute to the ERG's b-wave. (After Newman 1980.)

intracellular electrodes. The size, time course, variation with stimulus intensity, and so on are similar for the ERG and Müller's cell potentials. The obvious source of current generating the response from Müller's cells is the flow of K^+ ions associated with bipolar cell activity.

We can now be more specific about the class of bipolar cells that contribute to the b-wave, for several reasons. One is the observation that the drug APB (2-amino-4-phosphonobutyric acid), which blocks the metabotropic glutamate receptors on ON bipolar cells (see Chapter 14), affects the ERG from amphibian retina in parallel with its effect on the ON bipolar cell response. As the response of the ON bipolar cell decreases and disappears under the influence of APB, the b-wave of the ERG also declines and disappears. In human, ERGs from which the photoreceptor component has been subtracted are very similar in waveform, timing, and variation with intensity to the responses of ON bipolar cells from other species.

Thus the b-wave of the ERG appears to be due to activity in the ON bipolar cells, presumably the various types of ON cone bipolar cells, as well as the rod bipolar cells that also depolarize in response to light. The question that has been raised, but not yet answered, concerns the relative magnitudes of any direct contribution by the ON bipolar cells to the b-wave and the indirect contribution reflected in the responses of Müller's cells.

The ERG is useful as a gross indicator of photoreceptor function

Since ERGs can be easily recorded in human subjects, the method is not invasive, and it requires no subjective responses from the patient, the ERG could be a valuable tool in the diagnosis of retinal pathology, permitting clinicians to detect pathology before it becomes apparent by direct observation. This hope has been realized, but in a modest way; in most instances, ERGs confirm what has been detected by other means and assist in differential diagnoses.

One significant exception is retinal degenerative disease, whose primary target is the photoreceptors; the classic example is **retinitis pigmentosa**, which is a hereditary form of progressive photoreceptor degeneration. A markedly reduced ERG amplitude often identifies this condition long before there is any detectable visual loss or observable change in appearance of the fundus.

Typically, the rods are more prone to degeneration than the cones, or they begin to degenerate earlier, and the course of the disease generally begins with night blindness and peripheral visual field loss. Sometimes, however, cones seem to be affected first, and comparisons of light- and dark-adapted ERGs may assist in distinguishing these forms of retinitis pigmentosa; in the more common form, the dark-adapted ERG is initially much more affected than the light-adapted ERG. The degeneration of the photoreceptors is very slow, usually proceeding from the periphery inward toward the fovea. Affected individuals may be completely blind by 60 or 70 years of age, but not all patients lose their vision completely. Unfortunately, there is no known therapy to halt the progressive degeneration, and management of these cases consists primarily of counseling, good refractive correction, routine monitoring of the disease progression, and special optical aids as they become necessary.

Since ERGs can be elicited with diffuse illumination, even significant opacities in the ocular media, like cataracts, have little effect. Being able to produce an ERG through a cataract so dense that the fundus cannot be seen well implies some degree of normal retinal function and suggests that cataract surgery will produce the desired improvement in vision. Unfortunately, the presence of an ERG under these circumstances does not necessarily mean that the foveal retina is normal; it is quite possible to have degeneration in and around the fovea (**macular degeneration**) with a normal diffuse-light ERG.

Multifocal ERGs provide assessments of retinal function within small areas of the retina

Because the diffuse-light, or full-field, ERG is relatively insensitive to local pathology, particularly pathology confined to the fovea, various types of focal, or small-field, stimuli have been used to make the ERG more sensitive to local changes. Until recently, technical difficulties made this multifocal ERG approach only moderately successful, but it is now possible to determine the local ERG from 200 small areas of the retina simultaneously.

The stimulus for a multifocal ERG is an array of distorted hexagons, each of which can be illuminated or turned off independently (Figure 16.5*a*). The size and shape of the hexagons have been adjusted to produce local ERGs of nearly uniform size. A patient confronted with this stimulus array will see an enlarged version of "snow" on a television set—that is, a meaningless array of small lights flickering on and off with no apparent pattern. The ERG is recorded during 30 s intervals of viewing this stimulus, and 8 to 16 minutes of total recording are used to compute the multiple local responses (Figure 16.5*b*).

What makes this technique work is a specific pattern in which each hexagon is turned on and off. The pattern is the same for all hexagons, but the start of the sequence begins at different times for each hexagon so that what is happening to one hexagon is independent of events in all the others. The ERG recorded in response to this stimulus array looks nothing like the response to a diffuse-light ERG. Because of all the hexagonal lights flickering on and off, the ERG continually varies up and down by different amounts, with no apparent pattern to the variation. But the computer controlling all this knows when each hexagon turns on, and it can record what the ERG is doing at that moment and at various times thereafter. When enough of these comparisons have been made (mathematically, it is a cross-correlation), a picture is built up that shows what a given hexagonal stimulus has contributed to the ERG. That contribution is the local ERG for the part of the retina illuminated by the hexagon. With 200 hexagons in the array and comparisons made at millisecond intervals, the computer makes millions of calculations per second.

The local responses are similar to flash ERGs (see Figure 16.5*b*), but they are not identical. The computed local response is simpler, without the obvious inflections seen on the a- and b-waves of a single-flash ERG. Comparisons of the two types of ERG, however, indicate that the initial downward deflection in the

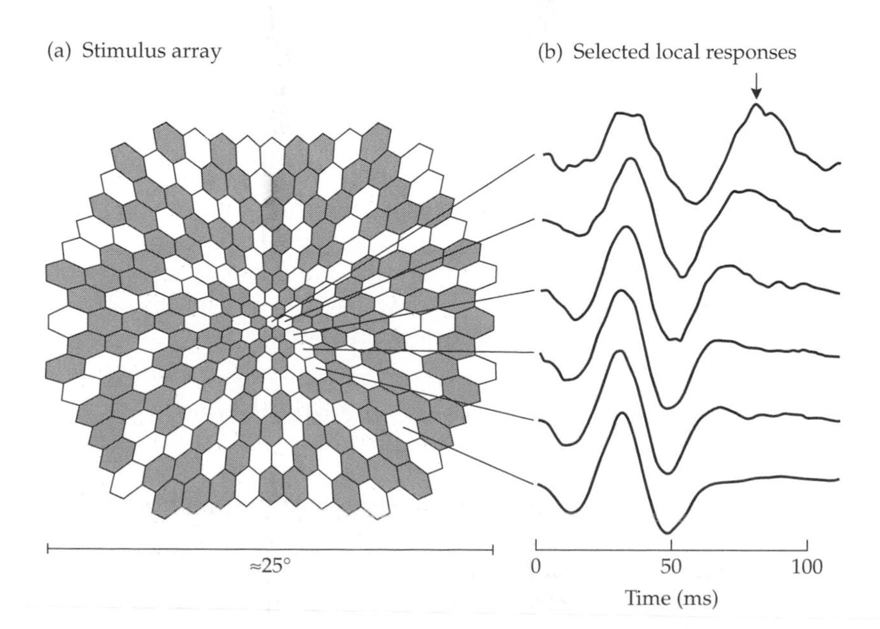

Figure 16.5
Stimulus Array for Multifocal ERGs
(*a*) The stimulus array for a multifocal ERG consists of about 200 hexagons that increase in size from the center to the periphery of the array. The hexagons can be made bright or dark independently of one another. The fixation point is the smallest hexagon at the center of the array. (*b*) Averages of local responses computed from the electrical response as hexagons flicker on and off look somewhat like diffuse-field ERGs, but they have a variable late component (arrow) that is not seen in conventional ERGs. (After Sutter and Tran 1992.)

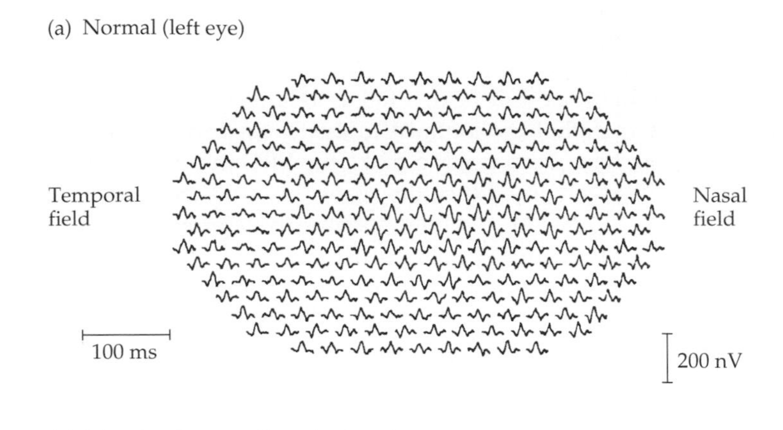

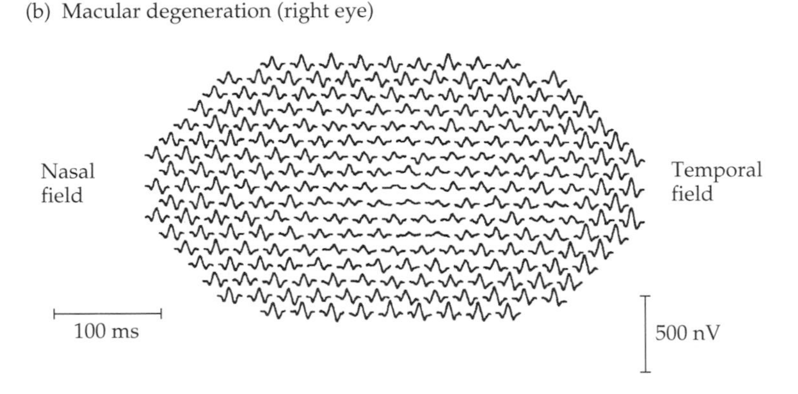

Figure 16.6
Normal and Abnormal Multifocal ERGs
(*a*) The local responses in a normal eye vary somewhat in size, but only a few responses near the temporal edge of the field are clearly smaller than the others. The small responses are due to the optic nerve head. (*b*) Central responses are reduced in an eye with age-related macular degeneration; less than 10° of the central field is affected. Small responses near the temporal side are due to the nerve head. (After Bearse and Sutter 1996.)

computed local response is coming from photoreceptors, as in the a-wave, and the large upward peak comes from the same source as that of the b-wave in the single-flash ERG.

A multifocal ERG in its entirety is the array of all the local responses; Figure 16.6*a* shows a normal multifocal ERG. The individual responses are similar in waveform and differ mainly in amplitude, those near the center being larger. The amplitude differences are due mainly to the size of the hexagonal stimuli and the inability to calibrate them perfectly to record local responses of equal magnitude. Near the temporal side of the array, however, several responses are quite small; they reveal the location of the optic nerve head.

The potential power of this method is indicated by the multifocal ERG in Figure 16.6*b*. Some of the responses near the center are clearly smaller than normal. (The small responses near the temporal edge of the field are due to the nerve head.) The central responses are reduced because of an age-related macular degeneration, the extent of which is less than the central 10° of the visual field. Results like these offer reason to hope that this method will be able to show signs of altered retinal function in some pathologies before they appear ophthalmoscopically. If that proves to be the case, and the response components are clearly related to the activity of photoreceptors and ON bipolar cells, evaluation of the retina's operation in vivo will be much advanced.

The Retinal Vessels and Assessment of Retinal Health

The retina in vivo is invisible

The total thickness of the neural layers of the human retina is never more than about 250 μm. Not only is the retina a thin, delicate tissue, but it is also meant to be transparent so that light reaches the photoreceptors without much attenuation. And the difference in refractive index at the retina–vitreous interface is

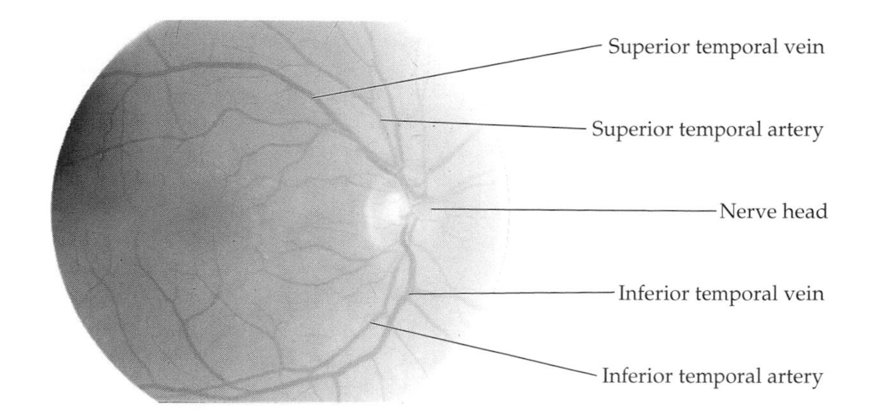

Figure 16.7
The Fundus of the Human Eye
In this ophthalmoscopic view, the background is uniform in color and density, and the branches of the central retinal artery and vein are seen diverging from the nerve head. The fovea is not visible, but the blood vessels temporal to the nerve head arch toward and around the macula, which appears as a dark region. This is a normal fundus, with no visible pathology. (Photograph by Kim Washington.)

small, so there is little specular reflection that might allow one to see the retinal surface. In short, all of the various cellular elements, the retinal layers, and the intricate circuitry discussed in the previous chapters are invisible to the clinician. Only the larger branches of the retinal blood vessels and the optic nerve head can be seen against a background of light that has passed through the retina and has been reflected back by the choroid and sclera. If the retina is healthy, you can't see it—you can see only elements of its vascular systems.

Fragments of the vascular branching pattern can be visualized as an entoptic image by moving a small, bright light source back and forth along the lateral side of the eye. Dark shadows of the blood vessels move as the light source moves, and they look much like the branches of a tree seen against a red-orange background. This **Purkinje tree** is named after the Czech physiologist and anatomist Jan Purkinje (1787–1869), who first described it. The fact that it moves means that the blood vessels and the photoreceptors are some distance apart (see Vignette 14.1).

In a low-magnification ophthalmoscopic view of the fundus,* the visible structures are the optic nerve head and the branches of the central retinal artery and vein that radiate from it (Figure 16.7; see also Box 16.1). The location of the fovea is given by the pattern of the blood vessels as they arc toward and around it and by the darkening in the region due to the macular pigment; there is sometimes a small, bright reflection from the center of the fovea, but that is the only direct sign of its location. The background is normally orange-red in Caucasian eyes (grayer in other races that have more pigmentation), with slight mottling due to the large vessels in the choroid and differences in choroidal pigment density. Therefore, the anatomical evaluation of the fundus is concerned mainly with three things: the uniformity of the background, the characteristics of the arterial and venous branches, and the appearance of the optic nerve head.

The invisibility of the retina's neural elements is not much of a handicap for the clinician. Other than conditions that affect the photoreceptors and their interface with the pigment epithelium, no pathological conditions can be attributed directly to failures of retinal circuitry and neurotransmission (a possible exception is associated with Parkinson's disease, in which the dopamine depletion seems to occur not only in the brain but also in the retina, leading to some moderate visual loss). Many pathologies that threaten retinal function are vascular problems, commonly resulting from systemic disease, and since the retinal vasculature is accessible to direct inspection, clinical examination of the retina concentrates on the blood vessels and their surroundings.

*The word *fundus* is Latin for "bottom." As an anatomical term, it refers to the back surface of an organ with an opening into a hollow interior; it's like the bottom of a well. Thus, the back of the eye visible with an ophthalmoscope is the ocular fundus.

[Box 16.1] Fluorescein Angiography and the Adequacy of Circulation

PARTS OF THE OCULAR CIRCULATION can be observed directly, in particular the superficial branches of the anterior ciliary arteries, the episcleral veins, and the central retinal artery and vein. The ability to see these structures is important diagnostically because a great deal of ocular pathology is directly or indirectly related to vascular problems. But being able to see the larger vessels does not always reveal how well the invisible capillary beds are supplied or drained. **Fluorescein angiography**, however, shows how blood is flowing through the vessels and through the capillary beds.

Sodium fluorescein is a fluorescent dye. When illuminated with ultraviolet light, it absorbs the high-energy photons and fluoresces—that is, reemits lower-energy photons in the yellow-green region of the spectrum. Although sodium fluorescein is a small molecule, the barrier-type capillaries in the retina will not allow much of it to escape, so normally it remains in the vessels instead of moving into the retinal tissue.

A solution of fluorescein injected into a vein in the arm will reach the eye in 10 to 12 seconds; the dye will have been diluted by mixing with the blood in the heart, but the concentration is sufficient to fluoresce strongly when the retina is illuminated with ultraviolet light. An injection of 5 ml of 10% fluorescein produces detectable fluorescence in the retinal vessels for an hour or so after the injection (and discoloration of the skin and urine for a few days thereafter because the capillaries in the skin and the kidneys are more permeable). In examinations of the retinal circulation, it is customary to take a series of photographs, first at short intervals to document the pattern of filling of the arteries, capillaries, and veins, and then at longer intervals to observe the diffusion of fluid into the tissue. More recently, high-resolution video cameras have been developed that can provide a continuous record of the development and waning of the vascular fluorescence.

The photographs in Figure 1 illustrate the normal progression of fluorescence through the retinal vessels: Fluorescence appears in the major arterial branches (Figure 1*a*); then the veins begin to show laminar fluorescence, with the core of the vein still dark, and the choriocapillaris begins to fluoresce (Figure 1*b*); the veins fill completely as the arteries begin to fade (Figure 1*c*); and finally, arteries and veins both fade against the background fluorescence from the choroid (Figure 1*d*). Figure 2 shows the peak intensity of arterial filling at higher magnification. The foveal region appears relatively dark because of the absence of retinal capillaries in this area.

Places in the retinal circulation where the capillary density is abnormally high or the capillaries are leaking to produce abnormal

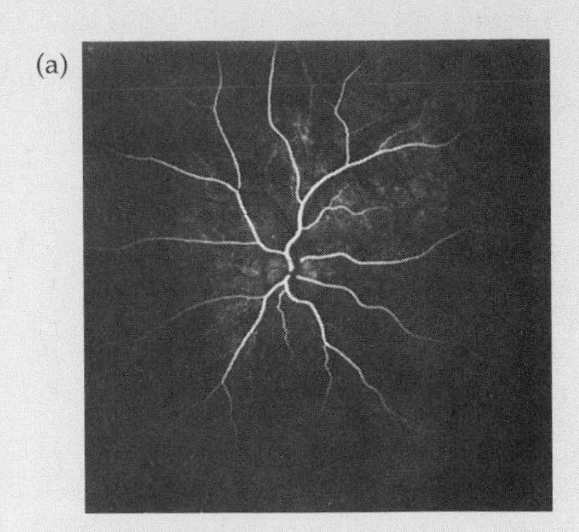

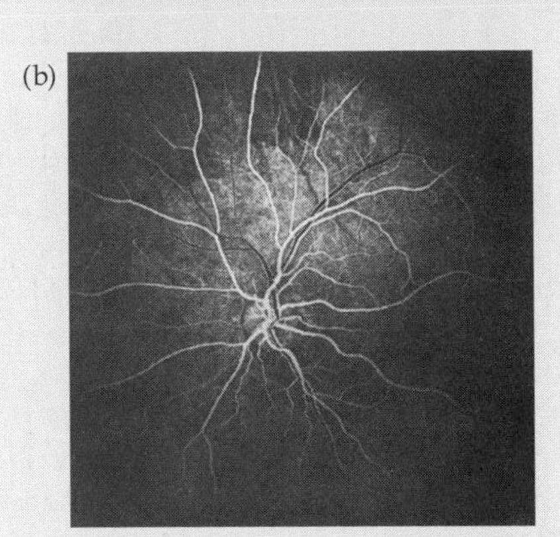

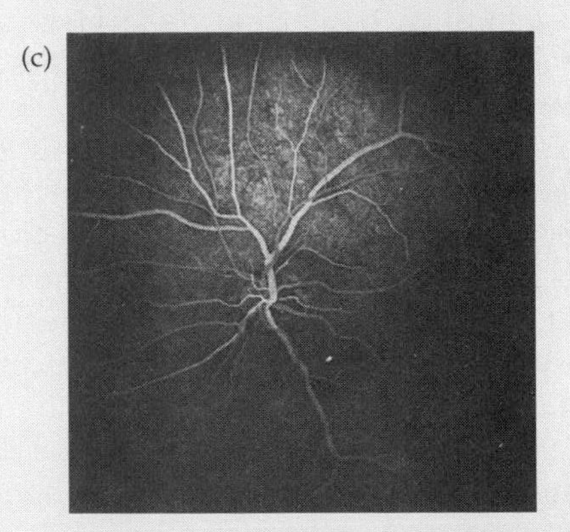

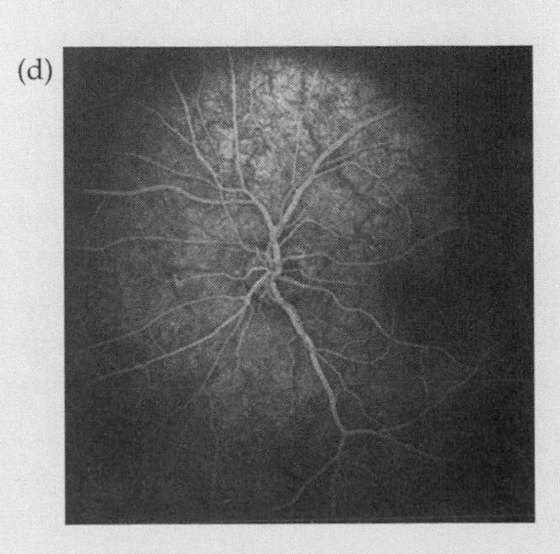

Figure 1
Sequence of Arterial and Venous Filling
A temporal sequence of photographs during fluorescein angiography shows the initial filling of the arteries (*a*), the beginning of venous filling with lamination in the veins (*b*), the height of venous filling with the arteries fading (*c*), and the fading retinal vessels against background fluorescence from the choriocapillaris in the choroid (*d*). (From Alm 1992.)

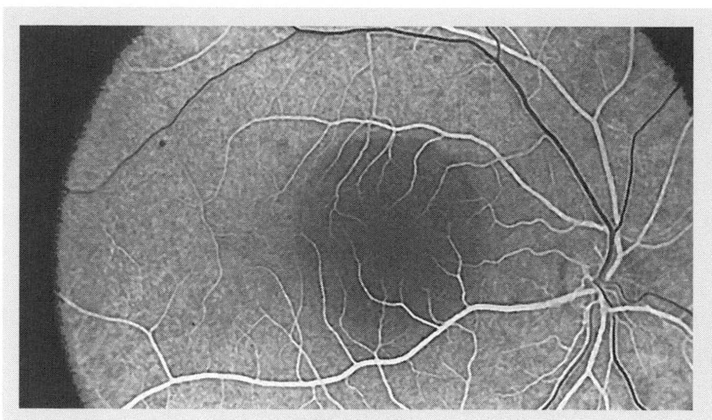

Figure 2

A Normal Fluorescein Angiogram at the Peak of Arterial Filling

The major branches of the central retinal artery and vein mimic the pattern of ganglion cell axons as they run from the nerve head out over the retina. The vessels arc above and below the fovea on the temporal side and follow more radial paths nasally to the nerve head. Small arterial and venous branches run radially in toward the fovea, which lies in the center of the dark region in this photograph. The veins are not completely filled and show a laminar flow of dye. (From Alm 1992.)

fluid movement into the tissue will have a higher concentration of fluorescein and therefore brighter fluorescence. In Figure 3*a*, for example, small bright spots that are not present in any of the normal pictures can be seen. These bright spots are microaneurysms, some of which are leaking fluid. Microaneurysms may occur for various reasons, but they are often the first stage of diabetic retinopathy and may be very difficult to see with normal ophthalmoscopy.

The opposite situation, in which there is less fluorescence than normally expected, is shown in Figure 3*b;* the abnormally dark region is the result of occlusion of an arterial branch. One would expect ophthalmoscopic signs of a problem in this region of the retina, but the fluorescein angiogram eliminates other possible causes by showing the arterial occlusion directly. In so doing, it also confirms the end-arterial nature of the retinal circulation; there is no alternative route for blood to reach the area normally supplied by the blocked artery.

(a)

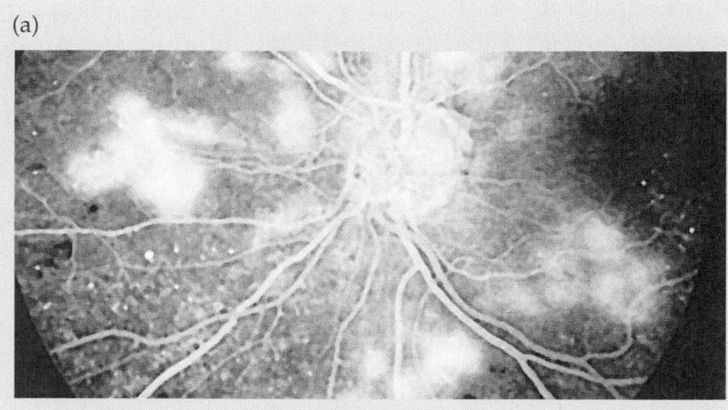

(b)

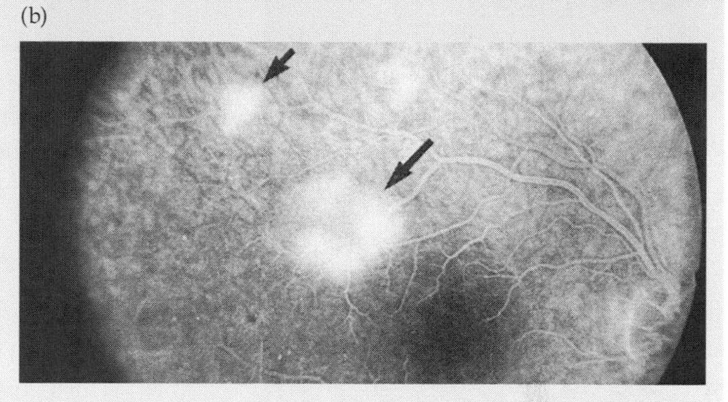

Figure 3

Abnormalities Revealed by Fluorescein Angiography

(*a*) Diabetic retinopathy. The large white areas are patches of edema and the small glowing dots scattered over the fundus are microaneurysms. Although these features would be visible with ordinary illumination, they would be much less obvious. (b) Edema associated with areas of neovascularization (arrows). (From Alexander 1994.)

The use of fluorescein angiography is confined to vascular systems that can be viewed directly, like those in the retina and iris, but other angiographic techniques are used to visualize the ophthalmic artery system within the orbit in situ. The most common procedure is an X-ray method in which a radiopaque substance has been injected into the blood, but specialized forms of magnetic resonance imaging can also be used.

Since the choroidal circulation is usually not directly visible, irregularities and nonuniformities on the fundus are commonly indicators of pathology

Absorptive pigments (melanin and hemoglobin) in the choroid create a fairly uniform and diffuse reflecting surface, with little visible detail associated with the choroidal blood vessels. A striking exception, shown in Figure 16.8*a*, is the appearance of the fundus in **ocular albinism**; here, the absence of choroidal melanin permits one to see some of the larger branches of the long and short posterior ciliary arteries as they run through the choroid. Sometimes the vessels can be seen as blotches and streaks in a normal variation referred to as a **tessellated fundus**. The density and distribution of melanin pigment in the choroid apparently differs from the typical pattern, thus permitting some of the choroidal vasculature to show through.

(a) Ocular albinism

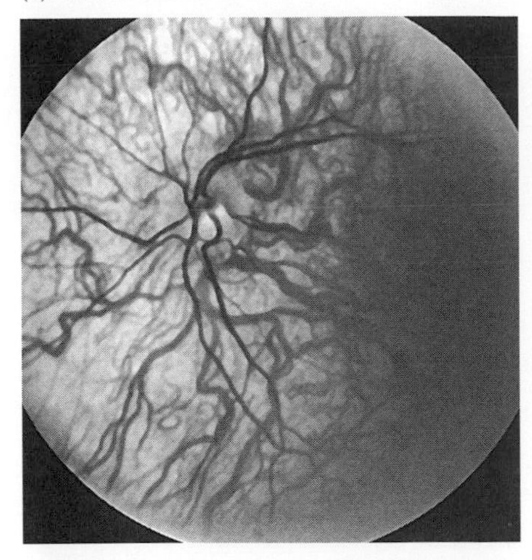

(b) Diabetic retinopathy

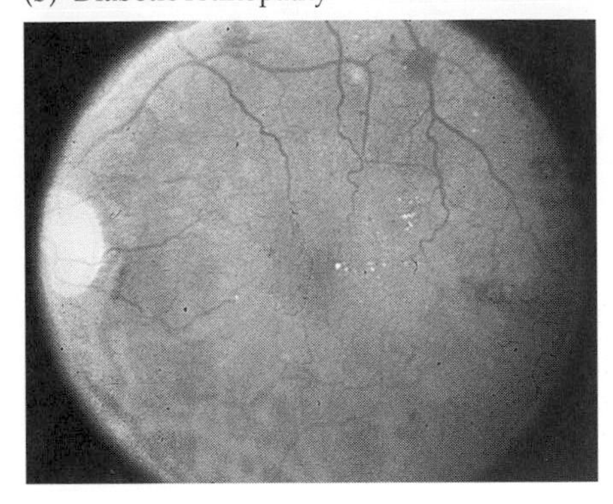

Figure 16.8
Variations in the Appearance of the Fundus
(*a*) Ocular albinism. (*b*) Diabetic retinopathy. (*a* from Alexander 1994; *b* courtesy of Richard Feist.)

The list of spots, patches, streaks, and pigmentary changes that constitute significant and meaningful nonuniformities in the fundus is beyond the proper scope of this book, but the example in Figure 16.8*b* makes the point that nonuniformities are key indicators of abnormality and that they are often vascular in origin. There are two unusual areas—one small whitish spot near the center and another darkened area that is a region with enlarged tortuous capillaries; they are associated with **diabetic retinopathy**, which is a common cause of vision loss and blindness. The fundus signs in diabetic retinopathy all seem to be secondary to changes in the retinal and choroidal blood vessels whose walls have weakened and become leaky. The first signs may be a few microaneurysms, but more developed stages include different forms of exudates, hemorrhages of various sizes, and zones of neovascularization. The most effective treatment for diabetic retinopathy is careful management of the diabetes itself; if the retinopathy is not too advanced, the fundus signs may resolve without significant vision loss. Unfortunately, the retinopathy may progress, becoming **proliferative diabetic retinopathy**, which includes neovascularization, edema, scarring, and vitreal involvement. In spite of recent advances in surgical treatment, about half of the cases of proliferative diabetic retinopathy progress to complete blindness.

The central retinal artery is an end-arterial system

Except for a small anastomosis with the pial plexus surrounding the optic nerve, the central retinal artery has no connections with any other arterial system. Moreover, there are no anastomoses between any of the major or minor branches of the central retinal artery, which means that each precapillary arteriole is the sole source of supply for its capillary bed. Obstruction of any arterial branch, large or small, will deprive some part of the inner retina of its blood supply. The larger the obstructed branch, the greater will be the area of retina affected.

Arterial branch occlusion is devastating; the clinical picture (Figure 16.9) shows large white areas of retina that have begun to degenerate because the blood supply has been cut off. Degeneration of the inner retina following occlusion of an arterial branch is rapid; the blockage must be removed within a few hours if the retina is to be saved. Arterial branch occlusion is often a sign of serious cardiovascular disease that requires treatment.

The distribution of central retinal artery and vein branches shows why arterial shunts do not occur. The vessels are spread out in a plane, and arterial branches of a given size never lie next to one another; a vein always lies between them. In effect, the arrangement of arteries and veins is like the interlaced fingers of one's hands, where the fingers of the right hand (the veins) come between and separate those of the left (the arteries). This alternation of arteries and veins shows very clearly in the vessels around the fovea (Figure 16.10*a*). The arterioles run radially in toward the fovea, and the veins that drain the region run radially outward. The alternating pattern of artery–vein–artery is quite strict; except for the central capillary ring, there is no opportunity for the arterioles to interconnect. Thus each capillary bed is supplied by one arteriole and drained by one venule.

Capillary connections between arteries and veins are like the rungs of a ladder whose sides are an artery and a vein (Figure 16.10*b*). Thus, the field supplied by an arteriole is a strip of retina on either side of the arteriole, and the drainage fields of the venules are similar. If one of the precapillary arterioles is blocked, there is no alternative source of blood supply to the deprived strip of retina, because arterial blood in the adjacent capillary beds cannot flow across the venules into the deprived area. Any blockage inevitably means degeneration of the retinal area that has been deprived of blood.

The only possible exception to the lack of alternate routes of supply is the capillary ring that encircles the central fovea (see Figure 6.12*a*). Here, several small arterioles feed into the capillary circle (and several venules drain it). This region, corresponding to the inner slope of the foveal depression, may be somewhat protected against the effect of arterial branch occlusion. In addition, the presence in some eyes of a cilioretinal artery derived from the short posterior ciliary system (see Chapter 6) may also offset blockage in branches of the central retinal artery that supply the retina nasal to the fovea.

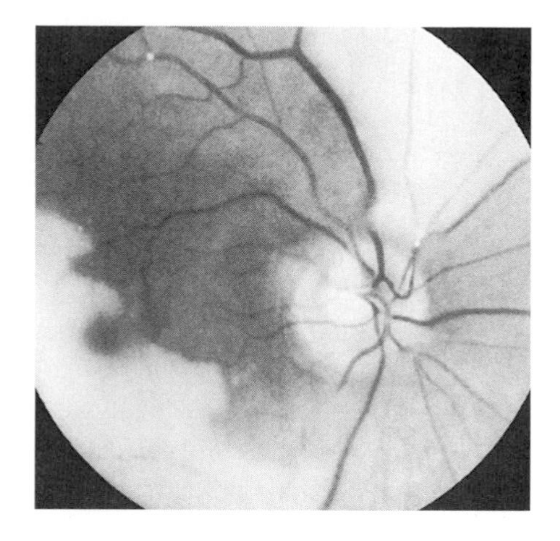

Figure 16.9
Central Retinal Artery Branch Occlusion
The large white areas are parts of the retina that have degenerated because of the loss of blood supply. (From Alexander 1994.)

The capillaries supplied by the central retinal artery ramify in the inner two-thirds of the retina

The precapillary arterioles running over the retinal surface give rise to two sets of capillaries; one set runs laterally in the nerve fiber layer to nearby venules, while the other set penetrates radially into the retina, running toward the scleral side. The capillaries that penetrate the retina never go more than about two-thirds of the way toward the scleral surface; that is, they never go to the outer layers of the retina that contain the photoreceptors. These radial capillaries give off side branches at several levels in the inner retina, however, and these lateral side branches form several flat capillary beds that run parallel to the retinal surface (Figure 16.11).

Shallow capillary beds line both sides of the ganglion cell layer, and deeper capillary beds line both sides of the inner nuclear layer. Thus there can be up to four parallel sheets of capillaries, separated in retinal depth by no more than 50 μm or so. This four-layered arrangement of the capillary beds is most prominent where the retina is thickest, in the region just around the fovea. Except for the vicinity of the optic nerve head, the retina thins from fovea to periphery, and the arrangement of the capillary beds changes as well; the capillary density decreases, and the four layers of capillaries collapse into just two layers, or perhaps a single layer at the extreme periphery of the retina. The decreasing number of capillaries and the reorganization into fewer layers are consistent with the reduced number of cells in peripheral retina and the smaller distance over which diffusion must occur.

Since the capillaries never pass the photoreceptor terminals in the outer plexiform layer (see Figure 16.11), they are farther from the photoreceptor outer segments than the choriocapillaris in the choroid is. The capillaries from the central retinal artery therefore have little to do with supplying the photoreceptors. (Capillaries normally do not go where the local oxygen tension is already high, which explains why the retinal capillaries do not grow into the photoreceptor layer. It also explains why there are capillary-free zones along large central reti-

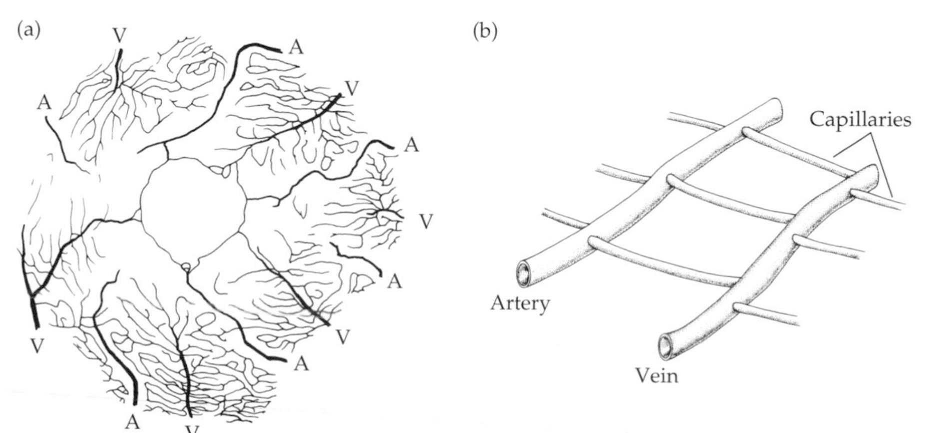

Figure 16.10
Distribution of Central Retinal Artery and Vein Branches
(*a*) The arteries (A) and veins (V) radiating in toward the fovea in monkey retina alternate so that arteries always have a vein running between them. Most of the capillaries have been omitted for clarity. (*b*) The alternation of arteries and veins means that the capillaries are arranged like rungs in a ladder, connecting an artery to its neighboring veins. Unlike this schematic version, however, capillaries do not run at the same level as the arteries and veins on the retinal surface. (*a* after Snodderly, Weinhaus, and Choi 1992.)

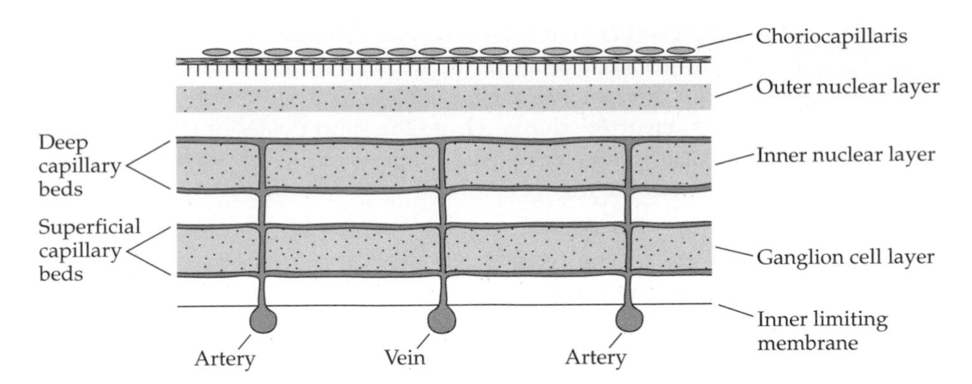

Figure 16.11
Layering of Retinal Capillary Beds
Branches from arteries and veins on the retinal surface run radially into the retina, where the radial branches are interconnected by capillaries at four levels. Superficial capillary beds line the two sides of the ganglion cell layer, and deep capillary beds are on either side of the inner nuclear layer. This schematic section represents the parafovea; other parts of the retina may not have all four capillary layers. The capillary branching pattern is not as rigidly geometric as shown here.

nal artery branches; enough oxygen diffuses from the larger vessels to supply nearby tissue and inhibit capillary formation.)

Experimental obstructions of either the choroidal or the retinal circulation clearly show the duality of the retinal supply. Tying off the central retinal artery causes most cells to degenerate; the photoreceptors are an exception. Conversely, the photoreceptors degenerate when the choroidal circulation is interrupted, while other retinal cells are unaffected.

Retinal detachment separates photoreceptors from their blood supply

Under normal circumstances, the lack of anchoring junctions between the photoreceptors and the pigment epithelium is not important; the intraocular pressure and the viscous interphotoreceptor matrix (see Chapter 13) keep the sensory retina firmly apposed to the pigment epithelium with the outer segments in their proper places. But abnormal conditions can separate the sensory retina from the pigment epithelium, creating a **retinal detachment**. One of the more obvious causes of retinal detachment is trauma; a hard blow to the eye, even if the eyeball remains intact, will deform the eye transiently and produce large, rapid changes in intraocular pressure that may tear the retina and cause part of it to pull away from the pigment epithelium.

Edema may also produce a retinal detachment, particularly when the fluid is leaking from the choriocapillaris. One of the spaces the fluid can penetrate is the extracellular space between the photoreceptor outer segments and the pigment epithelium. In total, this extracellular space is thought to have a very small volume, around 10 μl, so the amount of extra fluid required to fill the space and displace the photoreceptors is also small; the more fluid that accumulates in the space, the greater will be the separation of the epithelium and the photoreceptors. If the retina does not tear, this situation may not be serious, since the retina can drop back into place when the edema subsides. If the retina does tear, however, it may detach over a large area.

Retinal detachment can also be a consequence of shrinkage in the gel component of the vitreous. Attachments between the vitreous cortex and the inner limiting membrane of the retina can pull on the retina as the vitreous volume decreases (see Chapter 12). If the attachments break, the vitreous will separate from the retina (a vitreal detachment), but if they do not, the tension can pull the sensory retina away from the epithelium.

Since the nutritional needs of the photoreceptors are met by the pigment epithelium acting as the agent of exchange between the photoreceptors and the choriocapillaris, separation from the pigment epithelium is serious. Not only are the photoreceptors deprived of their blood supply, but they are also unable to shed their outer segment tips as they normally do. As a result, both rod and cone outer segments begin to degenerate within a day or two of detachment, which is the reason that the retina should be reattached as soon as possible.*

Experimental studies in monkey eyes indicate that reattachment of the retina a week after detachment permits the photoreceptors to regenerate to their normal condition in 6 months to a year. If the time before reattachment is too long, however, the photoreceptors will not regenerate to any significant degree. The length of this window of opportunity is not known precisely, but it is unlikely to exceed a few weeks.

The effect of the retina's dual blood supply shows clearly in retinal detachment. Unless some branches of the central retinal artery and vein have been damaged by a retinal tear, the inner retina is not affected and will function normally when the retina is restored to its proper place in contact with the pigment epithelium.

The foveal center lacks capillaries

Figure 16.12*a* shows the arrangement of the capillary beds in and near the fovea for the cynomolgus monkey. There are no capillaries within a central zone about 600 μm in diameter (the capillary-free zone in human retina is somewhat smaller, around 500 μm). This central region is delineated by a complete capillary circle that represents the deepest of the four capillary planes near the photoreceptor terminals. Moving radially out from the capillary circle, more capillaries are added as more of the inner retinal layers are built up (Figure 16.12*b*). Once the retina has its full complement of layers at the margin of the fovea, all four capillary beds are present.

Note that the fovea is *not* avascular; capillaries are absent only in the central 500 μm (only about one-third of the foveal diameter). Although the absence of capillaries in the central fovea means that nothing intervenes between incoming light and the foveal cones, this optical effect is not a reason for their absence. The foveal cones are adequately supplied by the choriocapillaris behind them, and there is no *nutritional* requirement for additional capillaries from the central retinal artery supply. Hence, there are no capillaries. Capillaries appear along the foveal slope because they are needed to supply the inner layers of the retina.

Retinal capillaries are specialized to create a blood–retina barrier

The walls of the retinal capillaries restrict the movement of molecules from inside to outside the capillaries and create a blood–retina barrier. In this respect, the retinal capillaries are like capillaries in the brain that contribute to a blood–brain barrier or those in the iris that form a blood–aqueous barrier (see Chapter 10). A

*There are several methods for repositioning the retina and anchoring it in place. Detachment of the superior retina can be repositioned by injecting a bubble of inert gas into the vitreous chamber; the bubble pushes the retina back into place. Scars produced by a cryoprobe applied to the outside of the eye and small laser burns to the retina "tack" the retina in place so that it cannot detach again. This procedure (called pneumatic retinopexy) is much simpler than scleral buckling, in which the eye is distorted (buckled) by a band around it that pushes the sclera inward to meet the detached retina.

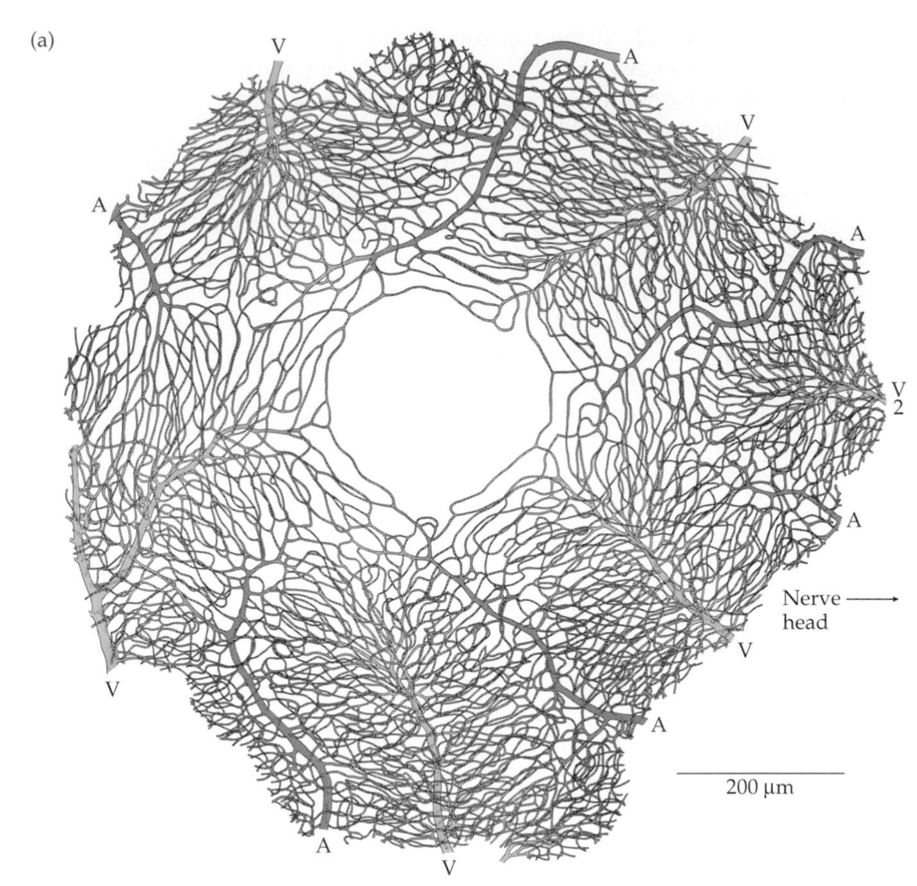

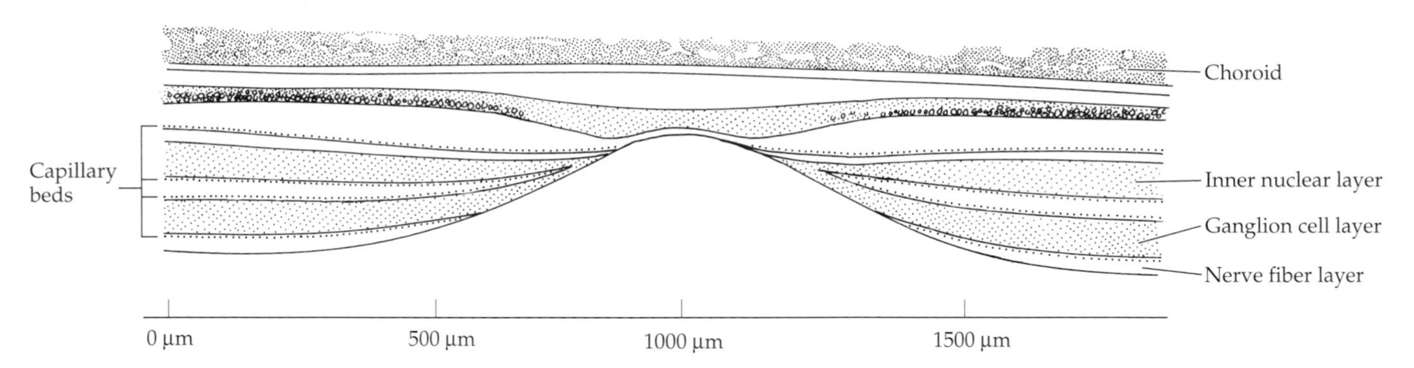

Figure 16.12
Distribution of Capillaries in the Fovea
(*a*) Small arteries (A) and veins (V) radiate in toward the foveal center in alternation: artery–vein–artery, and so on. Capillaries at several levels of depth in the retina run between neighboring arteries and veins. The center of the fovea has no capillaries. The capillary-free zone is encircled by a continuous capillary ring. (*b*) The cellular layers are shaded in this schematic section through the fovea, and the capillary beds are indicated by small dots. The deepest capillaries—those on the scleral side of the inner nuclear layer—extend farthest into the fovea. The most superficial capillaries, in the nerve fiber layer, penetrate the fovea least. (*a* courtesy of R. W. Rodieck, after Snodderly, Weinhaus, and Choi 1992.)

blood–brain barrier is necessary because some molecules circulating normally in the blood are either toxic to neural tissue or interfere with its normal function.

A cross section through a retinal capillary at high magnification shows the basic structural features that contribute to the blood–retina barrier (Figure 16.13). The endothelial cells overlap extensively, thereby creating long and tortuous paths between cells for fluid and molecules to follow if they are moving extracellularly between the inside and outside of the capillary. And these extensive cellular interfaces are interspersed with numerous tight junctions between adjacent cells. These junctions not only impede the movement of molecules in the extracellular space, but also maintain the structural integrity of the capillary wall, holding adjacent cells tightly in apposition and preventing the spaces between cells from opening up. The presence of astrocytes and their interaction with vascular endothelium may induce the development of these anatomical specializations in capillaries of the brain, the retina, and the optic nerve head.

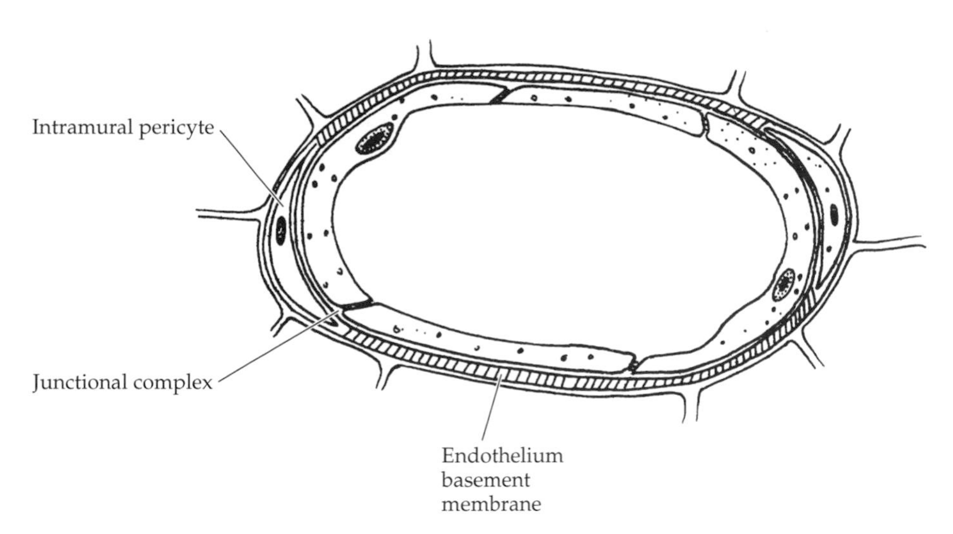

Figure 16.13
Structure of Retinal Capillaries
Retinal capillaries are constructed with overlapping endothelial cells and tight, occluding junctions between cells. This construction and the paucity of mitochondria and pinocytotic vesicles that might be used in membrane transport systems contribute to the creation of a blood–retina barrier.

Blood–brain or blood–retina barriers are not wholly the result of anatomical specialization, however. Peripheral capillaries, like those in the uveal tract, have several mechanisms for transport of substances across the endothelial cells, and the numerous mitochondria and pinocytotic vesicles are the observable signs of these transport systems. Capillaries in the brain are known to lack these general transport systems, and judging from the relative paucity of mitochondria and few signs of pinocytosis, retinal capillaries appear to lack them as well. Thus, while dissolved gases like oxygen or carbon dioxide can diffuse freely, most transport in and out of the retinal capillaries is highly selective.

Retinal blood flow is autoregulated

Various types of nerve endings have been shown to be present on the choroidal arterioles, but none have been found on branches of the central retinal artery. Structurally, however, the central retinal artery and its branches are said to be typical in terms of the presence of smooth muscle in the arterial walls; the muscle is responsive to vasoactive drugs and changes in oxygen tension. The obvious implication is that retinal blood flow is not controlled centrally, but is autoregulated locally (see Chapter 6).

The arterial and venous branches on the retinal surface can be distinguished ophthalmoscopically

Retinal veins are typically somewhat larger than the arterial branches to which they run parallel; they are also darker and less reflective, and they lack the prominent bright streak of reflected light that runs along the center of the arteries (see Figure 16.21). Under normal circumstances, there is no problem distinguishing between arterial and venous branches.

Although the central retinal artery and vein are functionally separated by the intervening capillary beds, they are physically close together, their branches often cross, and they cannot be regarded as mechanically independent of one another. Where arterial and venous branches cross, the arterial branch is often more superficial, so the vein or venule passes between the artery and the retinal tissue. This arrangement, illustrated schematically in Figure 16.14, can lead to constriction of the vein by the crossing arterial branch; it is a result of unusually high arterial pressure and increased rigidity of the arterial wall, often as a consequence of systemic arteriosclerosis and hypertension. The venous constriction (called "nicking") reduces venous drainage from the part of the retina drained by the vein's tributary branches because there is no alternative route for drainage.

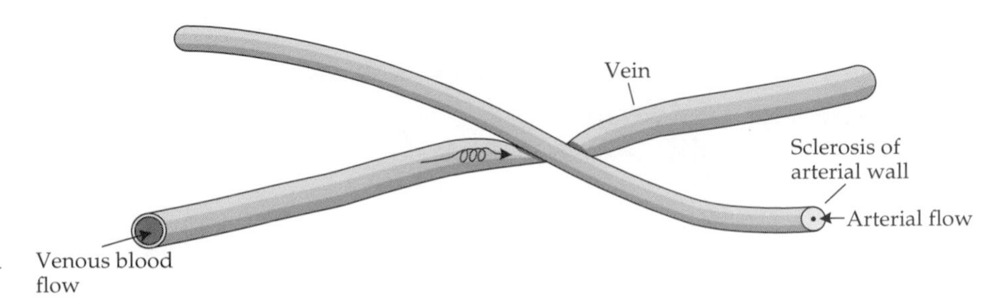

Figure 16.14
Veins May Be Compressed Where Arteries Cross over Them
When branches of the retinal vessels cross, the arterial branch is typically the more superficial. Sclerosis of the arterial wall and increased arterial pressure can partially or completely obstruct blood flow in the underlying vein.

Drainage of the inner retina is segmental

The interlaced pattern of central retinal artery and vein branches and the ladderlike arrangement of the capillaries (see Figure 16.10) mean that a given part of the retina has not only a single source of arterial supply but also a single avenue for venous drainage. The central retinal vein, like the central retinal artery, drains segments of the retina, and the size of the segment depends on the size of the vein; a large vein receives blood from smaller branches, and thus is responsible for a large area of the retina. And branches of the central retinal vein, like branches of the central retinal artery, lack anastomoses at any level; there are no alternative routes for blood to move around a site of blockage.

A retinal venous branch occlusion (Figure 16.15) is associated with edema, hemorrhage, exudates, neovascularization, and severe vision loss in the affected region. It can be a result of trauma, high intraocular pressure, diabetes, or a variety of other systemic factors. Occlusion of venous branches draining the foveal region are particularly threatening. Venous branch occlusions may be located near or within the optic nerve head, blocking either the superior or the inferior branches of the central retinal vein, with all of the associated problems, extending throughout half of the retina. Worse yet, the entire central retinal vein may be occluded and the entire retina compromised.

For these reasons, central retinal vein branch occlusion is no less critical than arterial branch occlusion. Although an arterial occlusion deprives a retinal area of blood, a venous occlusion prevents its egress. If the resulting edema persists long enough, cells will degenerate and vision will be lost from the affected part of the retina.

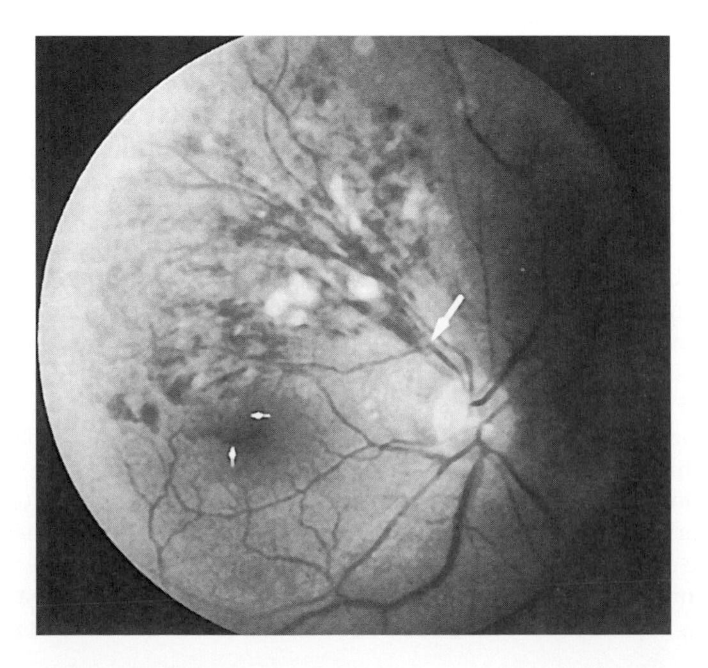

Figure 16.15
Central Retinal Vein Branch Occlusion
The occlusion site is indicated by the large arrow. Most of the superior-temporal quadrant shows hemorrhages and areas of edema (small arrows). (From Alexander 1994.)

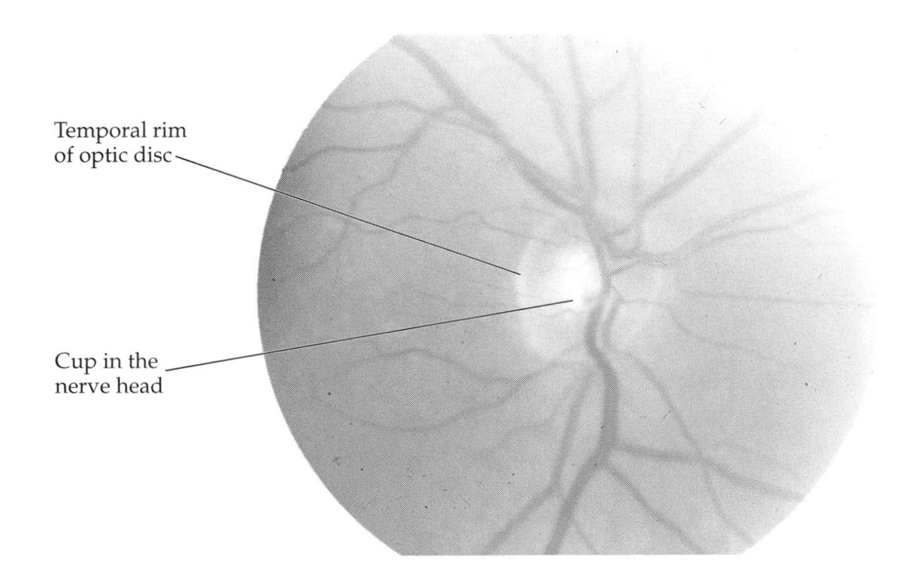

Figure 16.16
Normal Ophthalmoscopic Appearance of the Optic Nerve Head
The nerve head is elliptical, with the long axis vertical. The depression in the center of the disc—the cup—is pale, with a diameter just over a third of the diameter of the disc itself. (Photograph by Kim Washington.)

The Optic Nerve

All ganglion cell axons and all branches of the central retinal artery and vein converge at the optic nerve head

Ganglion cell axons run in a stereotypical pattern from their cell bodies to the optic nerve head, which is their site of exit from the eye (see Figure 16.19). There are about a million axons, each with an average diameter of about 0.5 μm; together they make a bundle about 2 mm in diameter. Seen ophthalmoscopically (Figure 16.16), the nerve head is slightly elliptical, elongated vertically; the horizontal diameter is about 1.5 mm (the same as the foveal diameter) and therefore subtends about 5° of visual angle. There are typically four arterial and four venous branches from the nerve head, creating two superior and two inferior branches of both the central retinal artery and the central retinal vein, but this pattern varies considerably. The branching normally occurs within the nerve head, just below the vitreal surface, which raises the point that the nerve head is a three-dimensional structure about 1 mm thick. The ophthalmoscopic view shows just the surface adjacent to the vitreous; this surface is the **optic disc**.

Since the retina is essentially a plane with substantial area over which the blood vessels ramify, most retinal pathology is local in its early stages and cannot easily affect the entire retina. The situation is very different at the nerve head; all the information outflow from the retina and all the blood vessels for the inner retina are confined to a relatively small space, so even small lesions in this location can have very widespread and profound effects. For this reason, the optic disc receives much attention when the fundus is being examined; its size, shape, color, and surface contour are all significant for diagnostic purposes.

The nerve head and the optic nerve consist primarily of axon bundles separated by sheaths of glial cells and connective tissue

The nerve head can be divided into three structurally different parts: the laminar portion, which is defined by scleral fibers that intersect the axons as they exit the eye (the **lamina cribrosa**); the prelaminar portion, which is the nerve head proper, lying between the lamina cribrosa and the vitreous; and the postlaminar portion, which is the first millimeter or so of the optic nerve just behind the eye (Figure 16.17).

The basic organization of the nerve head is the same in all three regions, in that the ganglion cell axons have been collected in bundles of several thousand

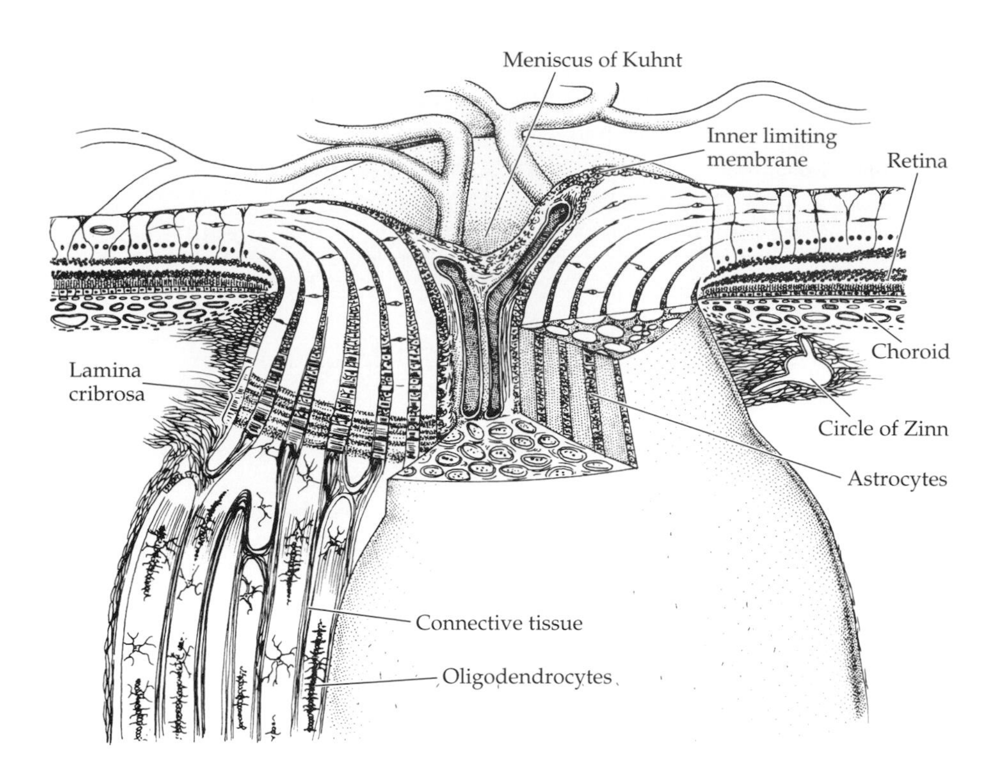

Figure 16.17

Structure of the Optic Nerve Head

The nerve head is divided by the lamina cribrosa into pre- and postlaminar portions that differ in their detailed structure. The bundles of axons in the prelaminar portion are separated by sheaths of astrocytes; in the postlaminar portion they are sheathed by connective tissue and oligodendrocytes. The astrocytes separate the nerve head from the surrounding retina and choroid and form the inner limiting membrane (of Elschnig) on the vitreal surface. Most of the axons are myelinated in the postlaminar portion. (After Hogan, Alvarado, and Weddell 1971.)

axons each that are surrounded by other tissue, but the ensheathing tissue differs from region to region. In the prelaminar portion, most of the tissue around the axon bundles consists of glial cells (specifically, astrocytes), while the postlaminar sheaths contain more connective tissue, fewer astrocytes, and more oligodendrocytes, which produce the myelin around the individual axons. The transition from glial to connective tissue sheaths occurs in the laminar region, where the axon bundles are also separated by the strands of scleral collagen that form the lamina cribrosa.

Two features are confined to the postlaminar region. The axons are myelinated from the lamina cribrosa back to the terminal nuclei in the brain, but the myelination normally does not go past the lamina cribrosa into the eye. In addition, the meninges surrounding the brain—pia mater innermost, arachnoid, and dura mater outermost—also surround the optic nerve up to the back of the eye (see Figure 16.18). A layer of glial cells lies between the pia mater and the outermost axon bundles.

The prelaminar portion of the nerve head is separated from the retina and the choroid by a sheath of glial cells (the *intermediary tissue of Kuhnt* next to the retina and the *border tissue of Jacoby* next to the choroid). The vitreal surface of the nerve head is also covered with a thin layer of astrocytes that are like a continuation of the inner limiting membrane of the retina; this glial layer is the **internal limiting membrane of Elschnig**, and the thickest part of it, at the center of the nerve head, is the **central meniscus of Kuhnt**. The surface of the optic disc is the internal limiting membrane of Elschnig, but because it is thin and transparent, it is not normally visible with an ophthalmoscope.

The significance of the high concentration of glial cells in the prelaminar portion of the nerve is not obvious, although the astrocytes may be required to interact with the oligodendrocytes and stop axon myelination at the lamina cribrosa. Tumors involving the glial cells (astrocytomas) are certainly found predominantly in the prelaminar part of the nerve head, but they are rare.

A congenital condition that is confined to the prelaminar part of the nerve head and may be related to the cellular components of this region is called **buried drusen of the nerve head**. Drusen of the nerve head have nothing in common with choroidal drusen (see Chapter 11), except that they are small and lumpy and usually increase in size and number with advancing age. Drusen of the nerve head are a potential threat to vision; as they increase in size they can press on adjacent nerve axons, probably disrupting axoplasmic flow and preventing the conduction of action potentials. Obstruction of blood vessels can also produce hemorrhages and edema that compound the problem, and the condition can progress to complete blindness. Unfortunately, the etiology of nerve head drusen is unknown, and there is no way to remove them or halt their progression.

The blood supply and drainage differ between the pre- and postlaminar portions of the nerve head

Figure 16.18 shows the vascularization of the nerve head and optic nerve (see also Chapter 6). The postlaminar portion of the nerve is supplied by two different sets of arteries: the central core of the nerve by branches from the central retinal artery, and the outer portion of the nerve by branches from arteries in the pia mater surrounding the nerve. For the most part, the pial arteries are supplied by branches from the short posterior ciliary arteries.

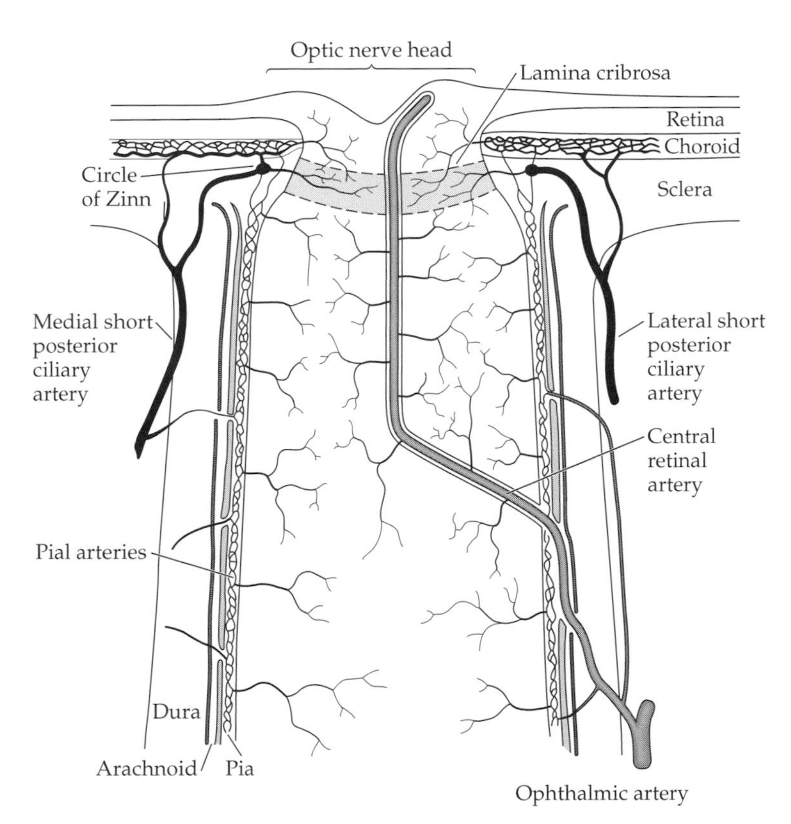

Figure 16.18
Blood Supply to the Optic Nerve and Nerve Head
Behind the lamina cribrosa, the optic nerve has two sources of blood supply: (1) branches from the central retinal artery to the central core of the nerve and (2) branches from the plexus in the pia mater, which is derived from the short posterior ciliary arteries. The laminar and prelaminar regions are supplied by branches from the circle of Zinn and from the choroid, with little or no contribution by central retinal artery branches. The central retinal vein (not shown) runs parallel to the artery and drains both the nerve and the nerve head. Behind the lamina cribrosa, the optic nerve is sheathed by pia mater, arachnoid, and dura mater throughout its orbital course. (After Hogan, Alvarado, and Weddell 1971.)

The laminar and prelaminar parts of the nerve head are supplied by branches from the circle of Zinn and from the choroid, both of which are derived from the short posterior ciliary artery system. It is likely the prelaminar portion depends mostly on the choroidal supply and the laminar portion mostly on the circle of Zinn, as indicated in Figure 16.18, but there is considerable variation and the division is probably not very rigid. Figure 16.18 shows that the central retinal artery does *not* contribute to the supply in the laminar and prelaminar portions of the nerve head, but the absence of a contribution from the central retinal artery is more convincing for the prelaminar portion than for the lamina cribrosa itself.

Whatever the source of the arterioles supplying the nerve head, the capillary beds have the structure of capillaries in the retina (see Figure 16.13), with overlapping endothelial cells and occluding junctions that contribute to a blood–nerve barrier. And even if they arise from the choroidal circulation that receives central innervation, these vessels in the nerve head are also like central retinal artery branches in that they lack autonomic innervation. The lack of innervation implies that the arterial supply of the nerve head is autoregulated, as is the supply to the inner retina.

The venous drainage of the nerve head and nerve roughly parallels the arterial supply, except that there seems to be no well-developed drainage around the nerve that corresponds to the pial artery system. Thus most of the drainage is by branches of the central retinal vein.

The arterial supply to the nerve and nerve head has a fair degree of redundancy, and it is less susceptible to disruption than the end-arterial supply to the inner retina is. The venous drainage is another matter, however; obstruction of the central retinal vein within the nerve cannot be compensated for by other routes of drainage.

Ganglion cell axons form a stereotyped pattern as they cross the retina to the optic nerve head

If the axons from retinal ganglion cells took the most direct line from their somata to the optic nerve head, the distribution of axons would be simply a series of lines radiating in all directions from the nerve head. And in species that lack a fovea in their retinas, this is the basic pattern. The fovea alters things, however, because axons do not cross it, and a strictly radiate, straight-line pattern for the axons running to the nerve head is impossible. Some axons must detour around the fovea, thereby producing the pattern in Figure 16.19.

The presence of the fovea does not affect ganglion cells lying nasal, superior, or inferior to the nerve head; axons from these parts of the retina run directly to the nerve head. The same is true of the ganglion cells just nasal to the fovea, whose axons form the **papillomacular bundle** that runs directly to the nerve head and crosses its temporal margin. The axons most affected by the fovea are those from ganglion cells near the fovea on its temporal side; they cannot run across the fovea to join the papillomacular bundle and are instead diverted in gentle arcs above or below the fovea. They cross the disc margin on its temporal side, above or below the papillomacular bundle. And like the ripples from a pebble dropped in a pool, the diversion of the axons just temporal to the fovea has an effect that spreads throughout the temporal retina. Even axons from ganglion cells well out in the temporal periphery take arcuate paths to the nerve head.

Axons from ganglion cells temporal to the fovea seem not to have a free choice about passing above or below the fovea. Instead, there is a fairly sharp dividing line extending temporally out from the fovea, separating cells whose axons pass above the fovea from those whose axons pass below it. The existence of this **horizontal raphe** means that ganglion cells quite close to one another on the retina, one just above the raphe and the other just below it, send their axons to widely separated points along the margin of the nerve head. (The raphe is an

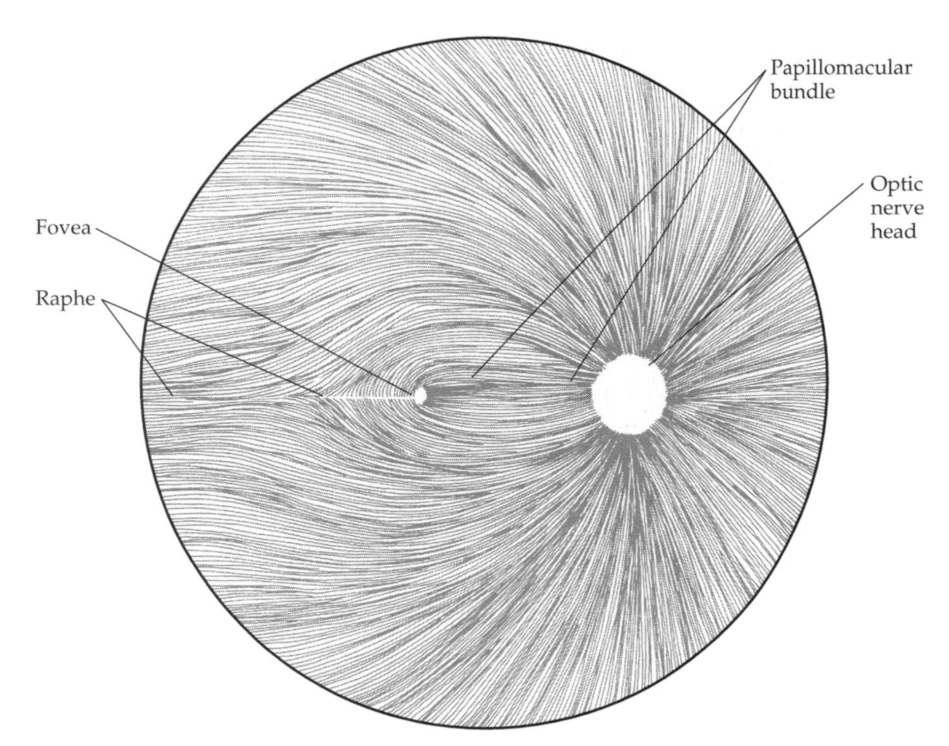

Figure 16.19
Paths Followed by Ganglion Cell Axons to the Optic Nerve Head
Axons from ganglion cells at retinal locations nasal, superior, and inferior to the optic nerve head follow direct, almost straight-line paths to the nerve. Ganglion cells on the nasal side of the fovea also have axons running directly to the nerve, as the papillomacular bundle. Axons from all other ganglion cells follow arcuate paths around the fovea. Temporal to the fovea, the horizontal raphe divides the ganglion cells whose axons run below the fovea from those whose axons run above it. (After Polyak 1941 and Hogan, Alvarado, and Weddell 1971.)

invisible line, seen only by special methods for axonal staining or, as will be illustrated shortly, in certain visual field defects.)

Axons from many widely separated ganglion cells are collected in bundles in the nerve fiber layer

The nerve fiber layer is thickest just around the nerve head. On the temporal side, where the axons in the papillomacular bundle are running, the layer is about 65 μm thick, increasing to almost 200 μm just superior and inferior to the nerve head. A section cut perpendicular to the papillomacular bundle—that is, along a vertical meridian—shows that the axons are bundled in groups separated by glial tissue, much as they are in the nerve head (Figure 16.20). In this case, most of the glial elements are Müller's cells. Among other things, this means that the papillomacular bundle is really a collection of smaller bundles of axons running from the ganglion cells nasal to the fovea. A similar picture can be obtained with any tangential section around the nerve head; the main differences are the thickness of the nerve fiber layer and the size of the axon bundles.

When the retina is viewed ophthalmoscopically with blue-green ("red-free") light, a pattern of striations around the nerve head is often visible (Figure 16.21*a*). These striations are likely to be the axon bundles, since comparisons of the numbers of striations in a small retinal region observed under high magnification are numerically similar to the bundles found in a histological section through the same region. There are numerous reports of abnormalities in this striation pattern, of which Figure 16.21*b* is one example. These defects in the nerve fiber layer are common in glaucomatous eyes, where they are taken as evidence for axonal atrophy (also called *nerve fiber layer dropout*). The visibility of the striations is affected both by the general pigmentation of the fundus and the clarity of the ocular media, however, which means that they cannot be used in all patients as a diagnostic criterion.

The axons in a bundle come not from neighboring ganglion cells in a small part of the retina, but from the entire strip of retina over which the bundle passes. To take a geometrically simple case, an axon bundle approaching the nasal edge

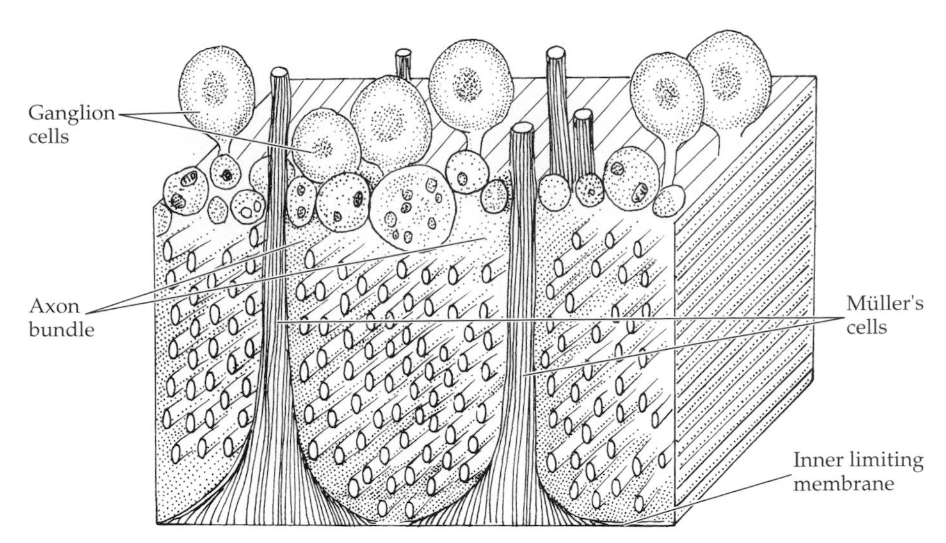

Figure 16.20
Axon Bundles in the Nerve Fiber Layer
Sections cut perpendicular to the path of the ganglion cell axons—that is, cut tangentially to the nerve head—show that the axons are collected into bundles before reaching the nerve head. The bundles are separated by columns of Müller's cell processes. (After Radius and Anderson 1979.)

(a) Normal fundus

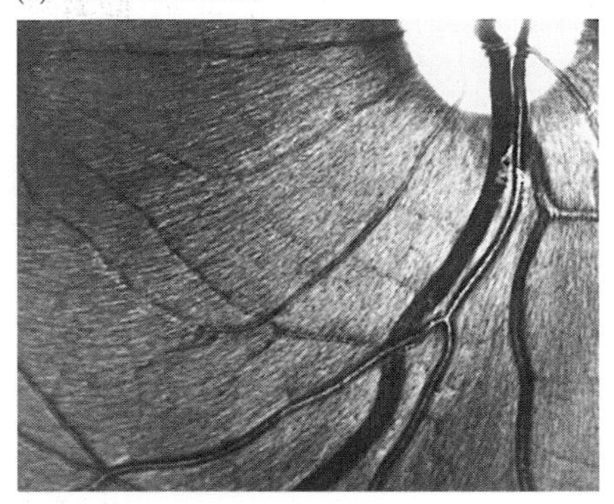

(b) Glaucomatous fundus

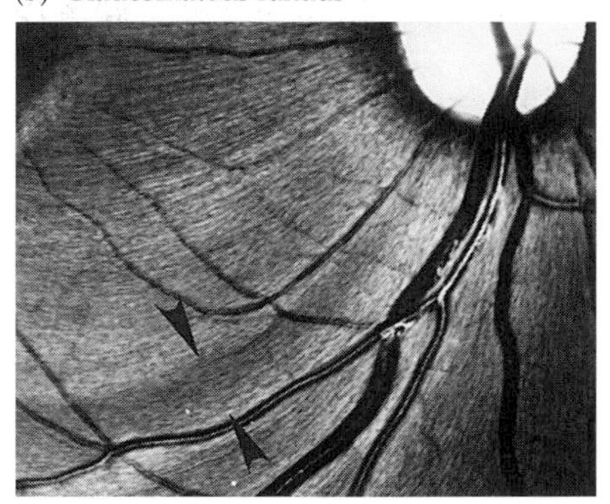

Figure 16.21
Striations in the Nerve Fiber Layer
(*a*) In red-free light, the fundus around the nerve head exhibits a characteristic pattern of bright streaks where the nerve fiber layer is thickest; the streaks are thought to correspond to individual axon bundles in the layer. (*b*) A sector defect in the striation pattern (arrows), produced by experimental glaucoma, is thought to represent axon degeneration and loss. The same rhesus monkey eye is shown in both photographs. (From Quigley et al. 1984.)

of the nerve head has axons from a strip of ganglion cells whose locations extend from near the nerve head all the way out to the peripheral nasal retina.

Studies of ordering or arrangement of the axons in the bundles have produced two distinctly different conclusions. In Figure 16.22*a*, axons from the most peripherally located ganglion cells run closest to the ganglion cell layer, while axons from ganglion cells closer to the nerve head pass perpendicularly through the nerve fiber layer to run along its vitreal surface. Axons from the periphery that are closest to the ganglion cell layer turn into the margin of the nerve head, while those from more central ganglion cells running along the vitreal surface enter near the center of the nerve head.

Figure 16.22*b* reverses the vertical ordering: The most peripheral ganglion cells have the most superficial axons, while those more centrally located have axons nearest the ganglion cell layer. This scheme implies a good deal of reorganization, with much crossing of axons in the nerve head so that the more peripheral axons running along the vitreal surface can turn into the outer part of the nerve head, while the central axons occupy the core, as shown. The experimental observation of numerous axon crossings in the nerve head is a key piece of evidence in favor of this scheme. And since the axons from the more central ganglion cells are the first to reach the nerve head during development, it seems reasonable that the later-developing peripheral axons would overlie them.

Both schemes agree on one important point: Centrally located ganglion cells have axons near the center of the nerve head, while peripheral ganglion cells have axons near the edge of the nerve head.

Axon bundles have an orderly arrangement in the nerve head

The axon bundles in the nerve head preserve a considerable degree of spatial order, such that those around the perimeter of the nerve head are coming from ganglion cells in the far peripheral retina while those in the center of the nerve head are from the central, parafoveal retina.

Figure 16.23, in which the nerve head has been divided into sectors and annuli, shows that the outer annulus contains the axons from ganglion cells that are most distant from the nerve head. In the temporal sector receiving axons in the papillomacular bundle, for example, the axons from ganglion cells at the fovea are outermost and those from ganglion cells closer to the nerve head are at the apex of the wedge near the center of the nerve head. The other sectors are also organized with peripheral ganglion cell axons in the outermost annuli and axons from ganglion cells near the nerve head at the apex of the sector. This

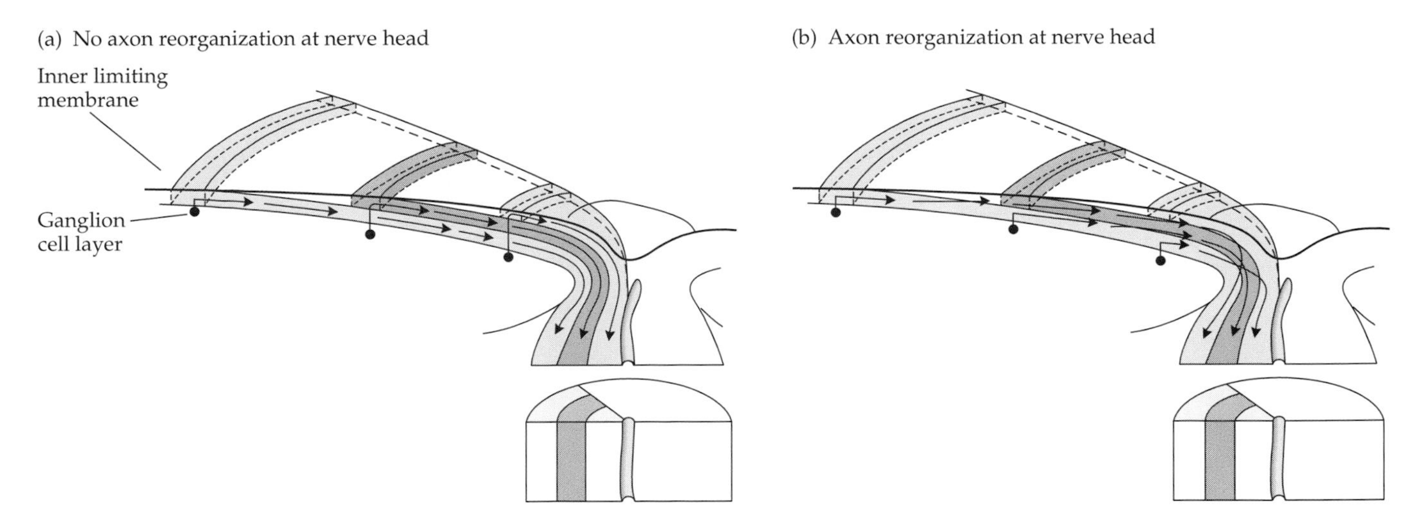

Figure 16.22
Two Schemes for Peripheral versus Central Axon Ordering in the Nerve Fiber Layer at the Nerve Head
(*a*) Axons from ganglion cells very distant from the nerve head run close to the ganglion cell layer, while axons from nearby cells are more superficial, next to the inner limiting membrane. No axons cross in the nerve head. (*b*) Axons from very distant ganglion cells are superficial in the nerve fiber layer, and those from nearby cells run close to the ganglion cell layer. Axons must cross in the nerve head to assume their characteristic locations within the nerve. (*a* after Minckler 1980; *b* after Ogden 1983.)

arrangement holds until the axons pass through the lamina cribrosa into the optic nerve, where significant reordering occurs as the axons approach the optic chiasm (see Chapter 5).

The organization of axon bundles in the nerve head is *chronotopic* ordering, which depends on the timing of axon outgrowth during development. Axons from ganglion cells at the future fovea reach the optic stalk first and occupy a cen-

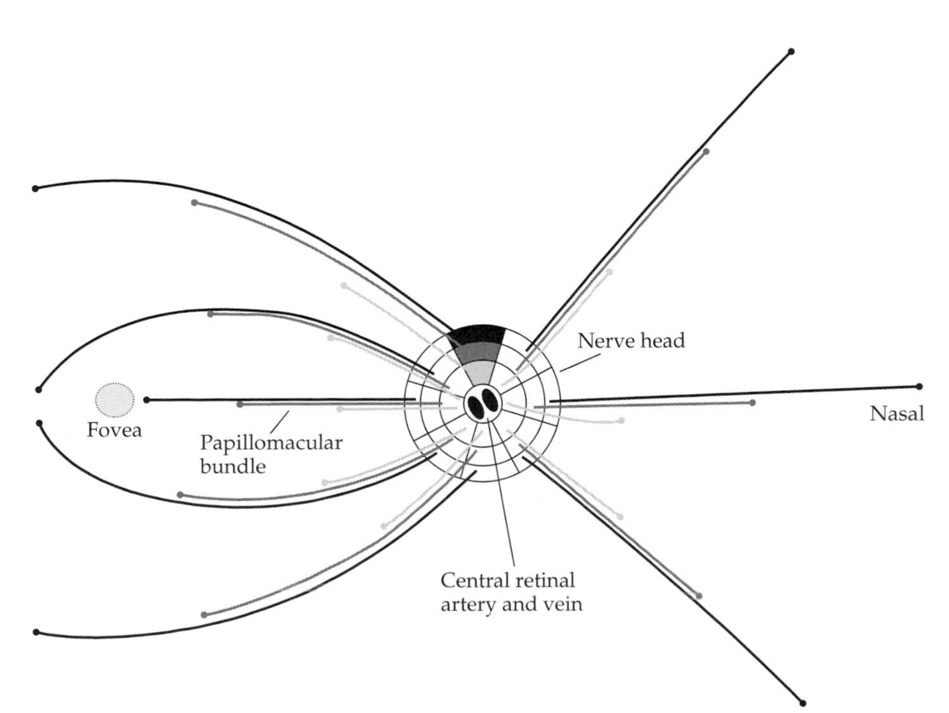

Figure 16.23
Axon Position in the Nerve Head as a Function of Retinal Location
The nerve head has been divided into eight sectors and three concentric annuli. Axons from ganglion cells farthest from the nerve head enter the outermost annulus; axons from cells nearby enter the innermost annulus. Each sector contains cells from different parts of the retina, such that axons from cells in the superior retina are in the superior portion of the nerve head, axons from cells in the nasal retina (relative to the nerve head) are in the nasal portion of the nerve head, and so on. As discussed in the text, the ordering is chronotopic, reflecting the time of arrival of developing axons at the nerve head.

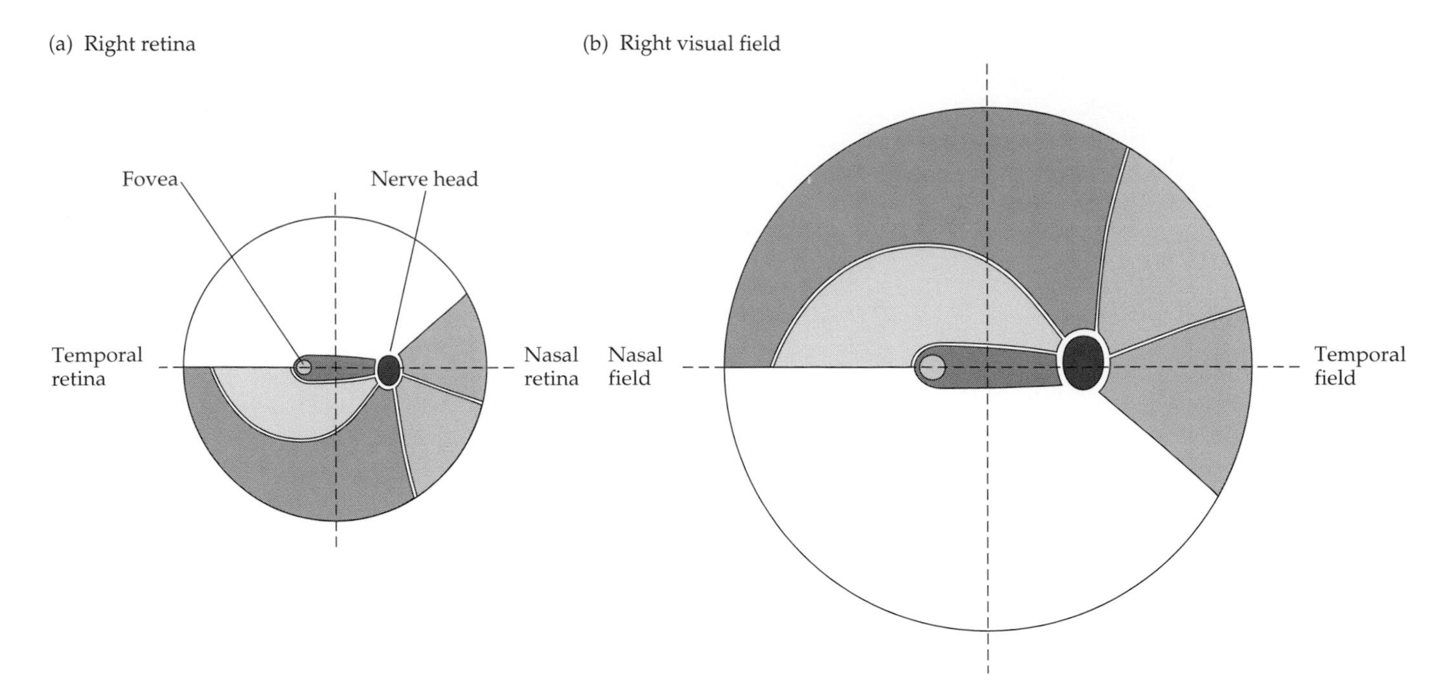

Figure 16.24
Lesions around the Optic Nerve Head Produce Segmental Field Defects with Characteristic Shapes and Locations
(*a*) Severing specific groups of axons at different sites around the circumference of the nerve head affects particular parts of the retina, in keeping with the pattern by which the axons run across the retina to reach the nerve head; the axon bundles along the superior-temporal and inferior-temporal edges of the nerve head come from large arc-shaped regions of the superior or inferior retina, respectively. (*b*) The visual field defects produced by the lesions around the nerve head have similar shapes, but with an up–down inversions.

tral-lateral position, while later-arriving axons from progressively more peripheral ganglion cells fill in around them. Although the result is a center-to-periphery gradient within the nerve head, it is based on time of axon arrival, and only indirectly correlated with retinal location.

Because of the correspondence between the location of axons around the circumference of the nerve head and the position of the parent ganglion cells on the retinal surface, discrete lesions at different points on the disc margin should produce visual field defects in characteristic locations with specific shapes. In Figure 16.24*a*, the disc margin has been arbitrarily divided into small segments, each of which can be thought of as a line along which a small incision could be made to cut all the ganglion cell axons crossing the line as they run to the nerve head. The variously shaded regions extending out from each of the eight disc segments are the retinal regions containing the ganglion cells from which the axons arise.

On the nasal side of the nerve head, the lesions should produce wedge-shaped regions of retina from which transmission has been cut off; a cut along the inferior-nasal part of the disc, for example, affects all the ganglion cells in a segment of the inferior-nasal retina. Lesions on the temporal side of the nerve head affect somewhat larger regions of the retina, and the regions are large curved segments flaring out from the nerve head. Only the lesion cutting the papillomacular bundle affects a narrow strip of retina between the fovea and the nerve head.

In Figure 16.24*b*, the retinal regions affected by lesions around the nerve head are transformed into their corresponding visual field defects (the transformation is simply an up–down inversion). In the visual field, the counterpart of the nerve

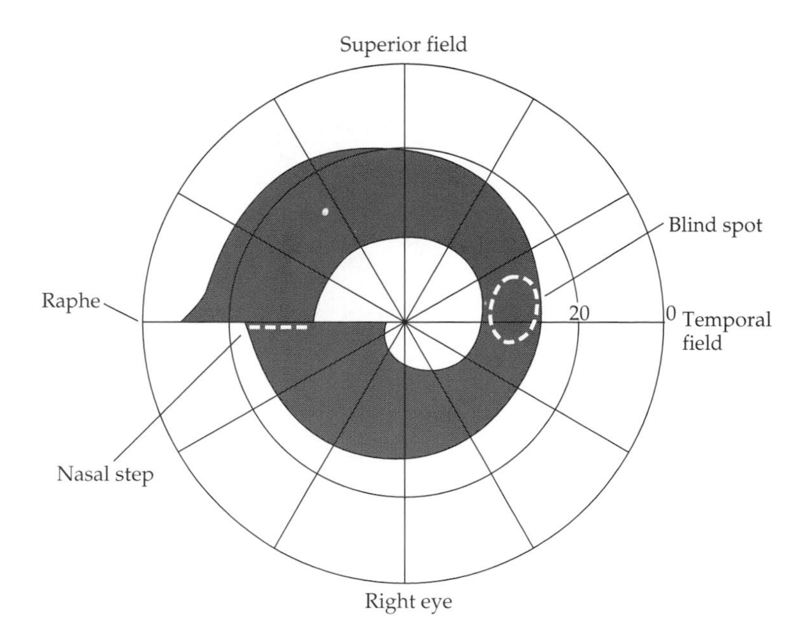

Figure 16.25
Visual Field Defects in Glaucoma
The dark regions in the visual field are absolute scotomas. The arcuate scotomas radiate from the blind spot above and below the horizon into the nasal visual field. The arcuate scotomas are offset where they meet along the raphe, creating a characteristic "nasal step." (After Harrington and Drake 1990.)

head, the **blind spot**, is temporal to the point of fixation (the fovea), so a lesion on the inferior-nasal aspect of the disc margin will produce a wedge-shaped scotoma in the superior-temporal visual field. Lesions on the temporal edge of the disc produce characteristic arcuate scotomas above or below the fixation point, with a sharp termination along a line corresponding to the horizontal raphe.

Scotomas observed in advanced stages of glaucoma correspond to those produced by lesions along the superior and inferior temporal margin of the nerve head

All glaucomas are associated with a characteristic visual field defect if the disease progresses far enough. In a very important sense, the visual field deficit defines glaucoma, since the deficit may occur with high, normal, or low intraocular pressure or with either open or closed anterior chamber angles (see Chapter 9).

The first visual field change in glaucoma is often a vertical enlargement of the blind spot. Later, the scotoma expands into the nasal visual field to produce arcuate defects like those in Figure 16.25, in which both superior and inferior fiber bundles are affected. If the course of the disease is not halted, more and more of the visual field will be affected, and the fovea will be involved. The end stage is complete blindness. (This description of the typical appearance and progression of field defects in glaucoma varies from patient to patient, and with the sensitivity of the testing procedures.)

The point is that visual field defects extending from the blind spot into the peripheral retina direct one's attention to some form of lesion at the nerve head; this is where axons from large parts of the retina converge and it is here that they can be affected by localized damage. For this reason, the problem underlying the visual field deficits in glaucoma must involve structural changes in the nerve head or in its vasculature that can preferentially affect ganglion cell axons crossing the nerve head margin along its superior and inferior temporal aspects.

The lamina cribrosa is weaker than the rest of the sclera

When the collagen fibers of the lamina cribrosa are isolated by digesting away all the surrounding tissue, its sievelike structure can be seen (Figure 16.26). The holes are channels through which the bundles of ganglion cell axons pass (there

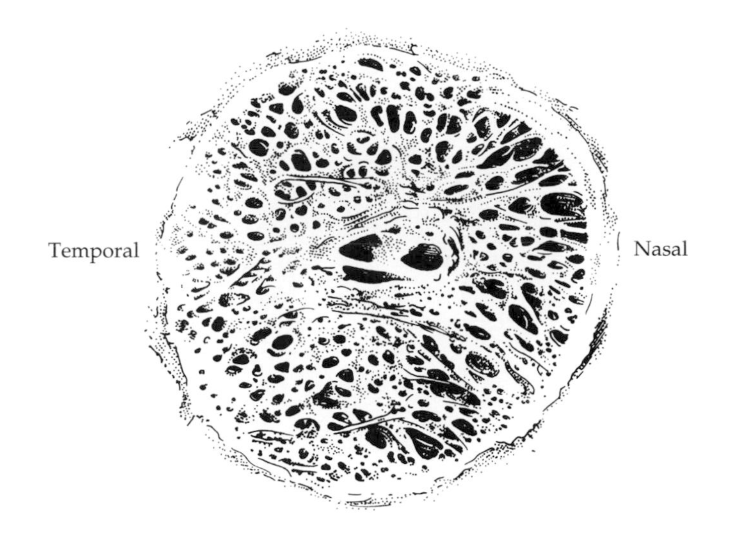

Figure 16.26
Structure of the Lamina Cribrosa
The lamina cribrosa is seen here from the anterior side with all noncollagenous tissue removed. Bundles of axons must pass through several hundred openings as they go into the optic nerve. The larger holes near the center of the lamina are for the central retinal artery and vein. (After Quigley et al. 1990.)

are about 300 of them). Although the lamina cribrosa is continuous with the rest of the sclera, it must be relatively weak, since it is more open space than it is tissue (contrast it with the dense mat of collagen fibers in the sclera proper and in the cornea shown in Chapter 8). If subjected to abnormally high intraocular pressure, the normal lamina cribrosa may give way in time and bulge outward. Similarly, a weaker-than-normal lamina cribrosa may give way in the presence of what would otherwise be regarded as normal intraocular pressure.

The holes in the lamina cribrosa vary in size and appear to be larger in the superior- and inferior-temporal regions where the arcuate axons from the superior- and inferior-temporal retina pass through (see Figure 16.26). Not only are the holes larger, but their meshwork of connective tissue seems thinner than elsewhere. This combination of more open spaces and less connective tissue suggests that these regions are the weakest parts of the lamina cribrosa and therefore the least able to resist deformation resulting from the intraocular pressure. But this suggestion requires that the regional differences in the structure of the lamina cribrosa be fairly consistent from eye to eye—a consistency that may not exist.

First, the channels through the lamina cribrosa often divide from front to back so that the posterior face of the lamina cribrosa has more holes (around 500) than the anterior face (around 350). Thus, what may be large holes on the anterior surface often become two smaller holes on the posterior surface. Moreover, when one compares the amount of open space (the summed areas of the holes) from one region of the lamina cribrosa to the next, the only consistent and statistically significant differences are relatively large holes in the inferior quadrant and relatively small holes in the temporal quadrant through which the axons in the papillomacular bundle pass. So while the superior and inferior quadrants of the lamina cribrosa may be particularly susceptible to deformation, the structural differences that underlie this weakness are not clear.

Field defects in glaucoma may be due to blockage of axonal transport secondary to deformation of the lamina cribrosa

There are two general hypotheses about the origin of field defects in glaucoma—one involving vascular defects at the nerve head, the other invoking structural change in the lamina cribrosa. The vascular hypothesis appears to be based on the notion of a highly segmented, nonredundant blood supply to the optic nerve head in which certain segments of the supply are consistently more prone to disruption than others. But given what is known about the supply to the nerve head, the basic premise is unlikely to be correct. Moreover, variability in the detailed vascular pattern from one person to the next makes it difficult to explain

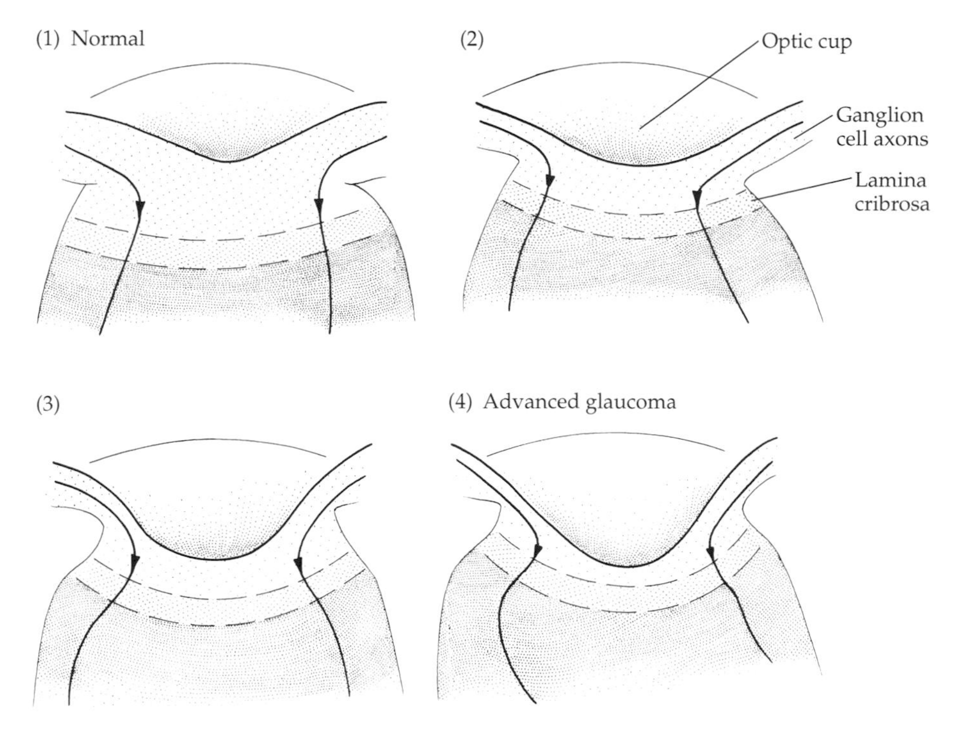

Figure 16.27
Deformation of the Lamina Cribrosa in Glaucoma
These longitudinal sections through the optic nerve head show the progressive deepening of the cup in the nerve head from normal (1) to advanced glaucoma (4). As the cup deepens and the lamina cribrosa becomes more curved, axons passing through the lamina are subject to kinking and pressure as they make their way through the lamina. (After Quigley et al. 1984.)

the consistency of the defects. There is no question that the vascular system at the nerve head can be severely compromised in glaucoma, but when it happens, it is probably secondary to other changes and serves mainly to make a bad situation worse.

The structural-change hypothesis is summarized in Figure 16.27, which shows how the nerve head changes when it is subjected to higher-than-normal intraocular pressure (as mentioned earlier, pressure that is normal at one nerve head may be excessive at another). The surface of the nerve head normally has a shallow depression, or cup, where the tissue overlying the lamina cribrosa is thinnest, but the lamina cribrosa itself is a smooth continuation of the surrounding sclera. Most of the bundles of axons pass perpendicularly through the tunnels through the lamina, with the exception of axons near the margin of the lamina cribrosa, where their courses become slightly oblique as the lamina cribrosa widens from front to back.

A moderate amount of deformation of the lamina may correspond to a stage of glaucoma at which some slight defects in the nerve fiber layer or the visual field would be apparent. The depression in the surface of the nerve head is deeper than normal and the lamina cribrosa has bowed back, forming a distinct bulge in relation to the surrounding sclera. The change in curvature from sclera to the lamina cribrosa is most pronounced at the margin of the nerve head, and this change creates a small kink in the axon bundles at the posterior surface of the lamina.

At a more advanced stage, the cupping in the surface of the nerve head is very pronounced, the lamina cribrosa is severely distorted, and all but the most central axon bundles are markedly kinked as they pass through the lamina.

The kinking of the axons where they pass through the rigid tunnels in the lamina is expected to put pressure on them, interfering with axoplasmic flow and impulse conduction. Blocking the axoplasmic transport systems eventually produces cell death and degeneration.

The main points of evidence supporting this scenario are, first, pictures of the distorted lamina cribrosa associated with severe glaucoma and, second, experimental results showing interference with axoplasmic flow in experimental glaucoma. It is still not clear why the superior and inferior axon bundles should be affected first, however, nor is it known how much deformation is required to produce significant restriction of axoplasmic flow.

Ganglion cell loss in experimental glaucoma does not appear to be selective by cell type or axon diameter

Ganglion cell axons have different diameters that are fairly well correlated with cell body size; larger cells tend to have larger axons. Figure 16.28 compares a relative frequency histogram of axon diameters from a normal monkey optic nerve to a relative frequency histogram in a case of experimental glaucoma. Almost half the normal number of axons had atrophied in the glaucoma case.

The main difference between the two histograms is the shift of the peak of the glaucoma population to a smaller axon diameter than in the normal population. There are proportionally more small axons in the glaucoma case. There is also a reduction *below* normal in the middle of the diameter range. There is little difference in the proportion of largest axons; the largest axon diameter represented in the normal group is also present in the glaucoma group. The simplest interpretation of the differences between the two samples is axon loss throughout the range of axon diameters; the greatest impact falls on the most common axon sizes—that is, those near the *middle* of the normal diameter range.*

Although the original data and those from a related experiment were interpreted as showing a disproportionate loss of large ganglion cells with large axons, that is clearly not the case. But the hypothesis of selective cell loss has been influential because it prompted a search for psychophysical evidence for impaired function due to the large cell loss. The large ganglion cells in primate retina are the parasol cells and small-field bistratified (blue–yellow) cells, and their response properties differ from those of the dominant population of midget cells. Thus deficits in detection of flicker and movement velocity, associated with the parasol cells, and deficits in color discrimination could be better and earlier indications of glaucoma than the standard visual field criteria are.

Some psychophysical studies *do* show elevated flicker and motion thresholds in patients with primary open-angle glaucoma, even in cases of moderate severity. Similar deficits are also observed in patients with ocular hypertension—that is, those with higher-than-normal intraocular pressure but no visual field defects. The occurrence of deficits among the ocular hypertensive group, a substantial number of which can be expected to develop glaucoma, offers hope that measures of this type can provide early warning of glaucoma (although the elevated intraocular pressure is already a danger signal). Unfortunately, no test yet reliably distinguishes among glaucoma patients, ocular hypertensives, and normals, and at least part of the reason is likely to be the underlying lack of selectivity in the effects on optic nerve axons. In short, the experimental rationale is faulty.

The important thing is that the search for a screening test continues even though the original premise was wrong. A subtle visual deficit may show up quite unexpectedly that can be incorporated in a psychophysical test as an early and sensitive predictor of glaucoma yet to come. In the meantime, however, clinicians will have to rely mostly on their standard battery of tests: measurement of intraocular pressure, gonioscopy, determination of visual fields, and ophthalmoscopy of the nerve head.

*Readers familiar with this literature will realize that my interpretation of the results is quite different from that of the investigators who did the work. Replotting the original data in the more appropriate forms of relative frequency histograms or cumulative frequency distributions makes all the difference. A corollary experiment on ganglion cell loss in retinas from glaucomatous eyes has a similar flaw; when reevaluated, it shows little evidence for selective cell loss.

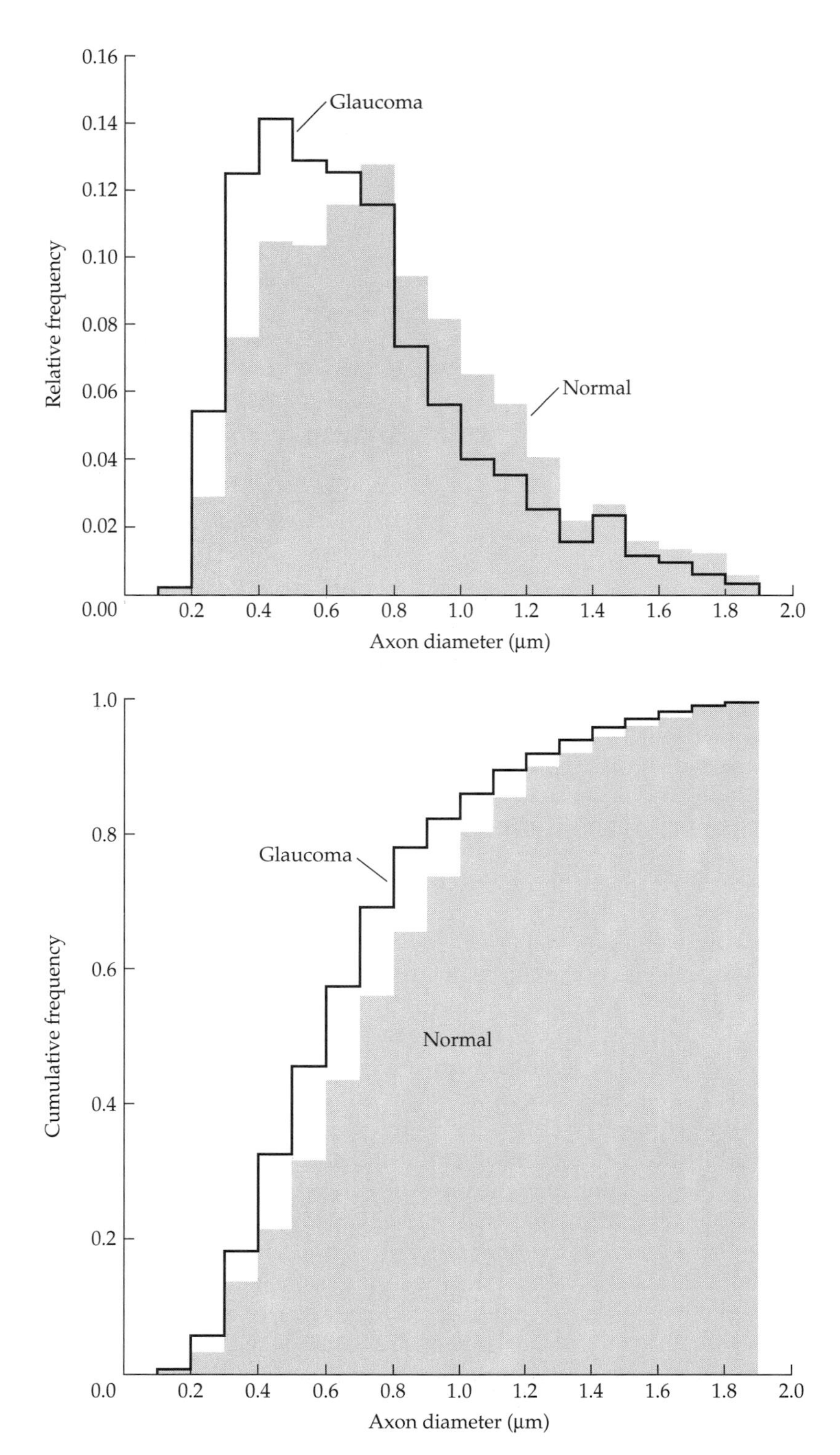

Figure 16.28
Axon Loss in Glaucoma as a Function of Axon Diameter
A sample of optic nerve fiber diameters in normal monkey optic nerve is compared to a sample from an optic nerve in an eye with experimental glaucoma as relative frequency (upper) or cumulative frequency distributions (lower). The optic nerve in glaucoma has a larger proportion of small axons and a smaller proportion of medium-sized axons than the normal optic nerve. The proportions of large axons are almost identical. The cumulative frequency distributions show that the samples differ most widely in the middle of the axon diameter range. (Replotted form Quigley et al. 1987.)

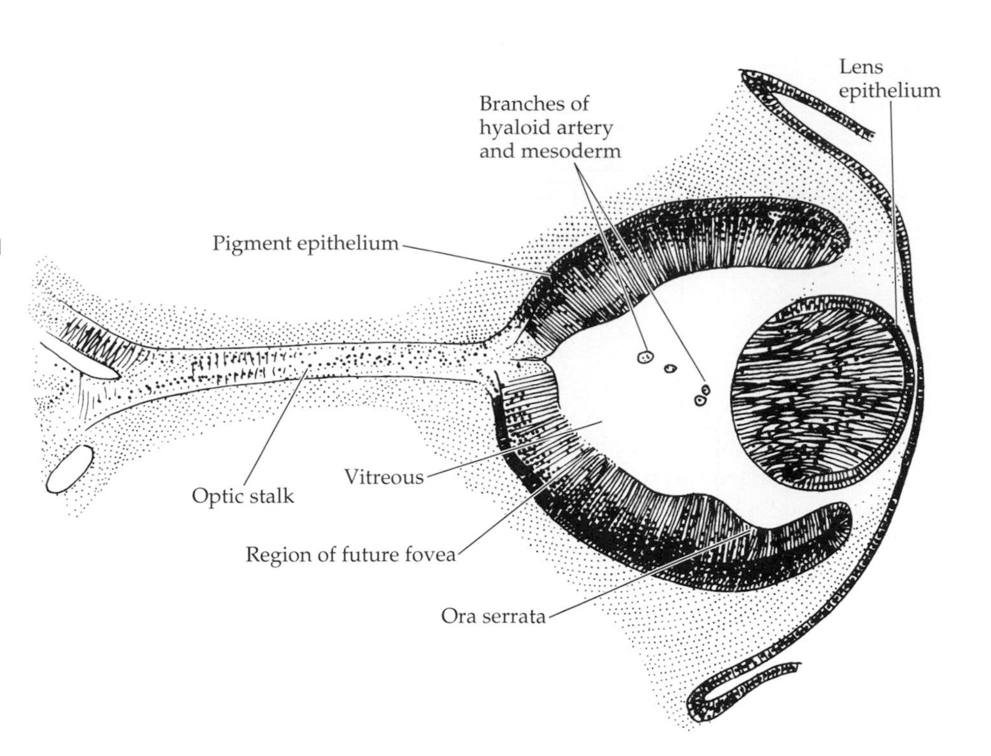

Figure 16.29
Early Differentiation of Cells in the Developing Retina
This horizontal section through the eye and optic stalk at 6 to 7 weeks of gestation shows that the outer layer of the optic cup has begun to synthesize melanin pigment and that some cells in the thick inner layer have begun to move toward the inner surface at the site of the future fovea.

Development of the Retina and Optic Nerve

The retina develops from the two layers of the optic cup

About the time the choroidal fissure closes to make the optic cup complete, the cells in the two layers of the cup begin to differentiate. Differentiation begins first in the outer layer of the cup, as indicated by the appearance of melanin pigment within the cells that will become the pigment epithelium. Pigment appears first near the back of the cup, gradually spreading forward toward the cup's rim; this is the first of several indications that the retina has a center-to-periphery gradient in its developmental course. The pigment-forming cells in the outer layer of the cup gradually lose their markedly columnar shape, becoming much more flattened while retaining their identity as a single confluent layer.

The progression of the pigment epithelium toward its adult appearance is very much in advance of the sensory retina, although some of its details, particularly the sheathing of the photoreceptor outer segments, cannot be made final until the photoreceptors are in place, which may be late in development. In general, however, the story of the pigment epithelium is relatively straightforward, involving proliferation, differentiation, change of cell shape, and cell migration of only one type of cell in a single layer.

Matters are more complex for the rest of the retina. When the optic cup has been completed, the future sensory retina consists of largely undifferentiated cells, whose nuclei form a dense layer throughout most of the thickness of the future retina (Figure 16.29). These undifferentiated cells seem to be totipotent, having the capacity to become any one of the mature cell types in the retina.*

*If embryonic rat retina is labeled with a fragment of a retrovirus that is permanently incorporated into the DNA of a cell when it undergoes mitosis, parent cells (stem cells) and all their progeny will be labeled and can be identified in the mature retina. In this case all classes of mature retinal cells are labeled, including Müller's cells, suggesting that the stem cells are totipotent and that their offspring can differentiate in a variety of directions. (See Turner and Cepko 1987.)

Thus, the story of the sensory retina's development concerns the temporal sequence and spatial gradient of transformation from a layer of undifferentiated stem cells to multiple layers of interconnected neurons of numerous types.

Retinal development proceeds from the site of the future fovea to the periphery

The earliest sign of differentiation in the sensory retina is a thickening of the layer of stem cells and a migration of some cells toward the inner surface of the optic cup. This change occurs first near the posterior pole of the eye, and it marks the site of the future fovea (see Figure 16.29). Cell migration to form two neuroblastic layers in the retina progresses in time from the posterior pole to the periphery; by the time this phase of development reaches the retinal periphery, other changes have taken place at the posterior pole. Thus the organization of the mature retina relative to the fovea is established from the outset, and the fovea is the site where developmental events begin.

Retinal neurons have identifiable birthdays

A cell is born when it represents the last stage of its mitotic lineage—that is, when it drops out of the cell cycle, differentiates, and begins to do the things characteristic of its type. The cell's birthday can be specified if we can determine when the mitotis occurred that produced the cell.

We can make this determination by attaching a radioactive label to the nucleoside thymidine (the label is tritium, and tritiated thymidine is referred to as 3HT) and injecting it into an embryonic eye, where it will be incorporated into the DNA of cells when they divide. If there is no further cell division, the level of radioactivity in the cells will remain high and can be detected with photographic emulsions coated onto thin sections of the mature tissue, even when the sections are cut weeks, months, or years after the 3HT injection. If there is more cell division following the incorporation of 3HT into a cell, its concentration will be halved at each division and the level of radioactivity will quickly become too low to detect. Therefore, any cells showing high radioactivity were born on the day 3HT was injected into the embryonic eye.

The birth date studies most relevant to human retinogenesis are those done on macaque monkeys, but the time from conception to birth in monkeys is around 165 to 170 days, compared to almost 280 days in humans. Making the assumption that the difference in developmental rate is uniform throughout gestation, the date on which events occur in monkey can be transformed to a human date by the ratio of the gestational times: $280/170 = 1.65$. Thus, the monkey's fetal day 30 is *roughly* equivalent to fetal day 50 in human ($30 \times 1.65 = 49.5$).

Ganglion cells, horizontal cells, and cones are the first cells in the retina to be born

Injections of tritiated thymidine in monkey eyes prior to fetal day 33 (about 8 fetal weeks in human) do not label any neurons, meaning that differentiation has yet to begin. Injections within the next week, however, produce heavy labeling of ganglion cells, horizontal cells, and cones in the region of the future fovea. The first amacrine cells in the fovea do not appear for another week or so, along with the first bipolar cells, Müller's cells, and rods.

The cells are born within restricted periods of time (Figure 16.30*a*). Ganglion cells, for example, first appear on fetal day 35 in monkey (around 58 days in human). Each day thereafter, new ganglion cells are labeled and add to those labeled earlier to produce an increase in the cumulative percentage of labeled ganglion cells. All ganglion cells are present by fetal day 70, giving a duration (a

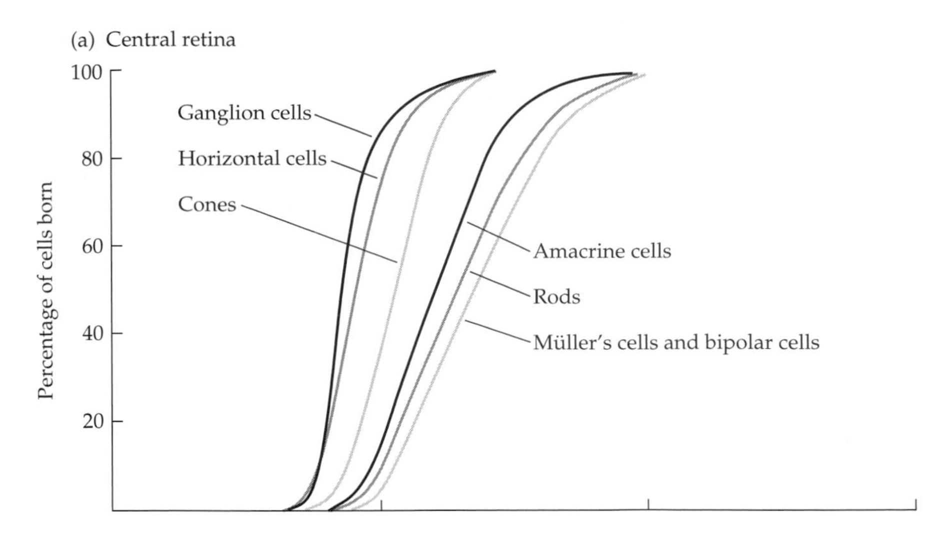

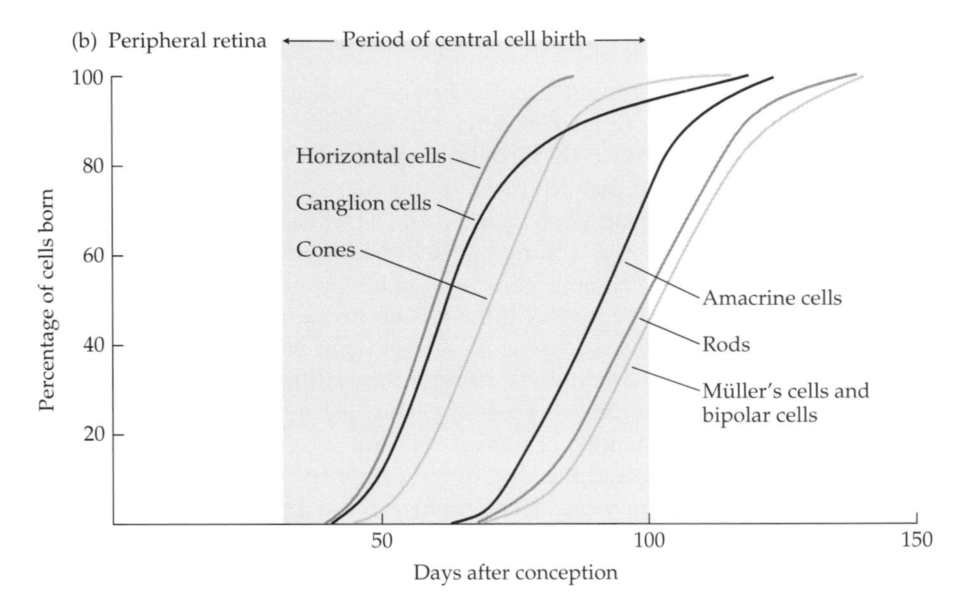

Figure 16.30
Different Classes of Neurons Are Born at Different Times—Central Cells Are Born before Peripheral Cells
(*a*) There are two waves of cell birth in the central retina near the site of the future fovea. The first wave begins at day 35 (monkey) with a cohort including ganglion cells, horizontal cells, and cones; a second cohort (amacrine and bipolar cells, rods, and Müller's cells) appears a few days later. All cell birth in the central retina is complete by day 100. (*b*) The same waves of cell birth are apparent in peripheral retina, but they begin and end later than the corresponding waves of central cell birth. In the periphery, the firstborn cohort appears at about day 40, and the lastborn cells appear just before birth; in humans, the last stages of cell birth in peripheral retina extend into the postnatal period. (Replotted from LaVail, Rapaport, and Rakic 1991.)

window) of about 35 days within which all central ganglion cells are born. The period during which horizontal cells are born has about the same duration as the birth window for ganglion cells, but it begins and ends slightly later. The curve for cone birth is somewhat later still.

These three cell classes—ganglion cells, horizontal cells, and cones—are slightly separated in time, but they are closer to one another than to any of the other cell types and constitute the earliest phase of cell birth. Most ganglion cells have been born before the onset of the second phase, which is led by amacrine cells and followed, in order, by Müller's cells, bipolar cells, and rods. The cells in the second phase are not only delayed with respect to the firstborn group, but their birth windows are longer; whereas 90% of the ganglion cells are born within a 2-week period, rods require more than 100 days between the first and the last birth.

Although some aspects of retinal development show a gradient of retinogenesis from inner to outer retina—that is, from the ganglion cells to the photoreceptors—the gradient is not evident in the cell birth data; in fact, the cells connecting the inner and outer retinal neurons are born well after the cells they will connect are born. The meaning of the sequence we observe is not clear; perhaps

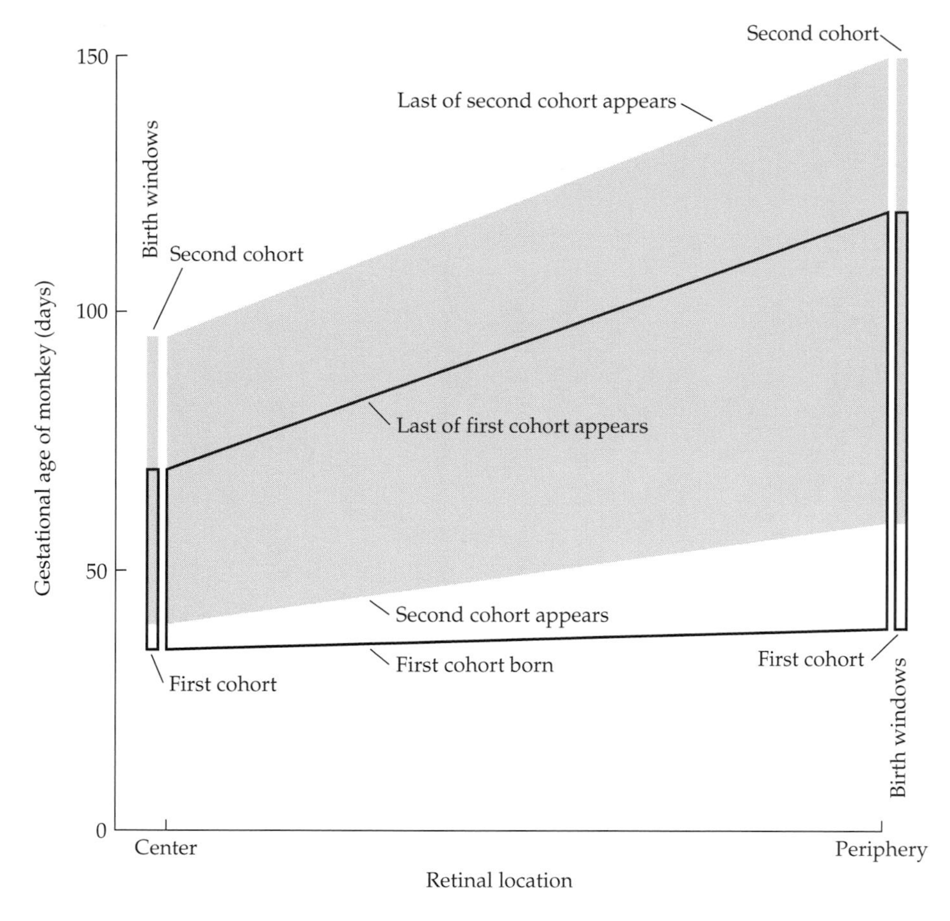

Figure 16.31

Birth Windows and the Center-to-Periphery Gradient of Cell Birth

The periods during which cells are born (the birth windows) are shorter in the central region than in the peripheral region for both the first and the second cell cohorts. In both cases, cell birth begins later in the periphery and the birth windows are longer. The time interval between the appearance of first and second cohort cells is also longer in the periphery than in the central retina. (After LaVail, Rapaport, and Rakic 1991.)

the inner and outer retinal neurons are first organized independently of one another, becoming linked and dependent later, when they are connected by the bipolar cells. We will return to this point when we consider the sequence of synapse formation.

As distance from the fovea increases, the firstborn cells appear at progressively later dates

A center-to-periphery developmental gradient is quite apparent in the dates at which particular cell types are born. The firstborn cohort, for example, appears in the foveal region at about 8 weeks, but they do not appear in the far peripheral retina until around week 24 (Figure 16.30*b*).

Although cell birth in the peripheral retina is quite delayed relative to the foveal region, the sequence in which cell classes are born remains the same; peripheral cell birth begins with ganglion cells and ends with rods. The cell birth windows for the different cell classes are longer in the peripheral retina than in the central retina, however—so much longer that the last rods, Müller's cells, and bipolar cells in human retina are born a few months *post*natally.

If we assume that the delays in birth dates and the increases in the birth windows change linearly from center to periphery, the data can be summarized as in Figure 16.31. Ganglion cells, horizontal cells, and cones have been combined into one group, called the first cohort, while the remaining cell classes constitute the later-born group, the second cohort. All cells in the first cohort are born by fetal day 120, which is about three-fourths of the monkey's gestational period. At this time, the retina has all the cones, horizontal cells, and ganglion cells that it will ever possess; if their numbers change, they can only go down. The more

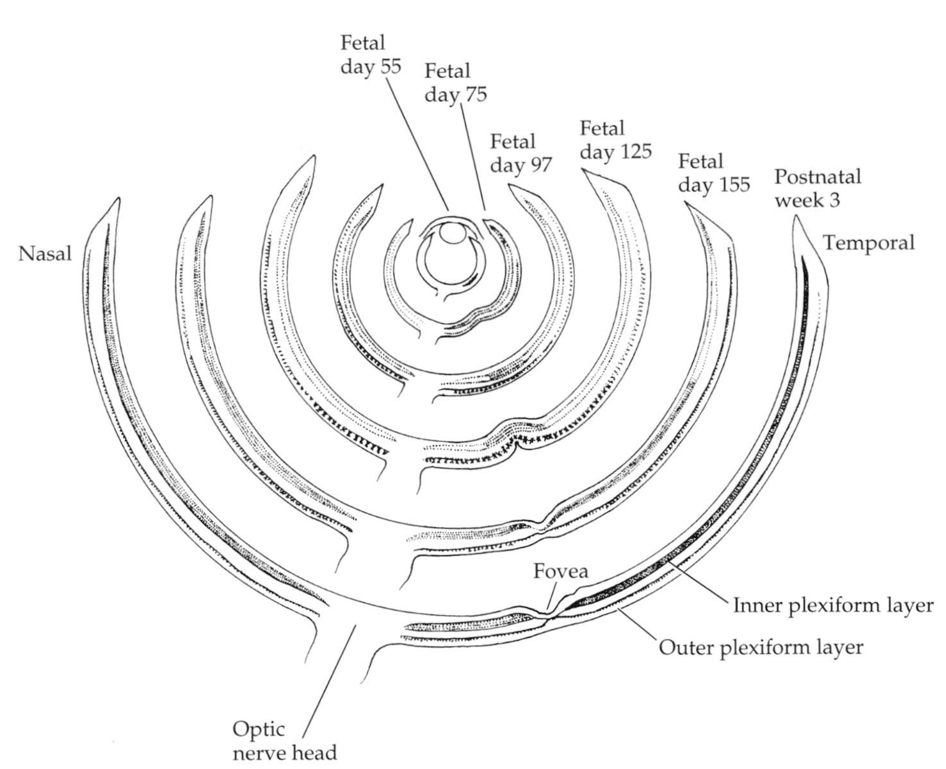

Figure 16.32
Gradients of Synaptogenesis in the Retina
Schematic cross sections through the horizontal meridian of the monkey eye at different gestational ages. In each section, the regions shaded with small dots show where synapses are forming; they appear first at the region of the future fovea in the inner retina (day 55). At progressively later gestational ages, synapses spread from the center to the periphery and from the inner to the outer retina. These data indicate the presence of two center-to-periphery gradients for synaptogenesis, one in the inner retina and one in the outer retina, with the inner gradient somewhat in advance of the outer gradient. (After Okada, Erickson, and Hendrickson 1994.)

leisurely rate of birth in peripheral retina of the second cohort extends cell birth and development of the peripheral retina into the postnatal period.

Synapse formation has a center-to-periphery gradient superimposed on a gradient from inner retina to outer retina

Synapse formation can be studied with a monoclonal antibody called SV2, which recognizes a membrane glycoprotein unique to synaptic vesicles. When SV2 is applied to the developing retina, labeling indicates the presence of vesicle-containing synapses at the labeled site. Lack of a label means no vesicles and no synapses (with a few caveats about false negatives and the possibility of non-vesicular synapses).

SV2 labeling in the developing monkey retina shows two gradients during development (Figure 16.32). Not only do cells differentiate first at the site of the future fovea, but also the first synaptic vesicles are found here at fetal day 50 to 55 (monkey); by this time, the birth of the ganglion cells is complete and the production of bipolar and amacrine cells has begun. At later gestational ages, the synaptic label spreads outward from the foveal region, reaching the retinal periphery around fetal day 100 (monkey). This progression is consistent with the center-to-periphery gradient of cell birth, and the relative times of birth and synapse formation are also consistent; synaptogenesis begins a week or so after the first cells in a given region are born.

The SV2 label always appears first in the inner plexiform layer. In the foveal region, labeling in the outer plexiform layer occurs about a week after labeling in the inner plexiform layer; in the peripheral retina, the outer plexiform layer lags by 2 to 3 weeks. This difference suggests that the establishment of retinal connectivity proceeds from the inner to the outer retina at all parts of the retina. Electron microscopy indicates that most of the early synapses in the inner plexiform layer are associated with the ribbons in the bipolar cell terminals (there seem to be few amacrine synapses early on), and those in the outer plexiform layer are associated with ribbons in photoreceptor terminals, most of which are probably cone terminals.

The location of the future fovea is specified very early; the pit is created by cell migration

The excavation of the fovea—the beginning of a pit in the retina—does not occur until almost the fourth month of gestation. Before this time, there is no pit. In fact, it is just the opposite; the future fovea is the thickest part of the developing retina. By 21 or 22 weeks of gestation (fetal day 90 in monkey), all the nuclear and plexiform layers of the adult retina are present (Figure 16.33*a*). All of the first cohort cells are present (cones, horizontal cells, and ganglion cells), the rest are in process, and synapses have begun to form in both the inner and outer plexiform layers. So by the time a pit begins to appear at 26 to 28 weeks (fetal day 105 in monkey; Figure 16.33*b*), this part of the retina has acquired considerable maturity, with all of its cell complement in place and its connectivity well under way.

The foveal pit forms not by excavation—by death of some of the existing cells—but by migration, movement of the cells of the inner retina away from the center of the future fovea. The evidence for migration is a steady change in the density of cells in the ganglion cell layer, with density *decreasing* at the center of the future fovea (Figure 16.33*d*) and *increasing* in an annular region around it; these changes suggest that the cells are moving radially outward from a central point. In addition, the dendrites of the ganglion cells change. Initially, they run straight up into the inner plexiform layer, but later their trajectories become more and more oblique, as if the cell bodies were shifting away from the foveal center, trailing their dendrites behind them.

Somewhat later, in keeping with the inner-to-outer gradient of development, a second wave of migration begins in the inner nuclear layer, moving these cells away from the center of the fovea as well. This migration includes the bipolar cells and is in part responsible for the oblique course of the cone axons (the fiber layer of Henle) whose terminals must run radially outward to remain in contact with the migrating bipolar cells.

Finally, a wave of migration of cones *into* the fovea produces a steady increase in the central cone density (this migration also contributes to the obliquity of the cone axons in the outer plexiform layer). Since the increase in cone density occurs well after all the cones have been born, it cannot be due to cell birth and must be due to migration.

Foveal cones are incomplete at birth

The infant fovea (Figure 16.33*e*) should be compared to the adult structure (Figure 16.33*f*). Although all the basic elements of the adult fovea are present at birth, the neonate's fovea has more change ahead of it. At this level of detail, the infant fovea is considerably smaller than the adult fovea, and it lacks the flat central floor of the adult foveal center. In addition, the thickest layer in the adult fovea corresponds to the photoreceptor layer that contains the cone inner and outer segments, but this layer is almost nonexistent in the neonate's fovea.

The comparison of neonatal and adult foveas indicates that much of the fovea's development occurs postnatally. Much of the cell migration discussed in the previous section is still in process at birth and will continue for some years; adult cone densities in monkey fovea are not acquired for about 4 years (probably about 6 years in human). In gross morphology, however, the human fovea appears to be adultlike by about 4 years of age.

The virtually complete absence of the photoreceptor layer in the newborn suggests that foveal cones are very immature at birth, and they are (Figure 16.34). The cone inner segments are hardly more than a thick bulge outward

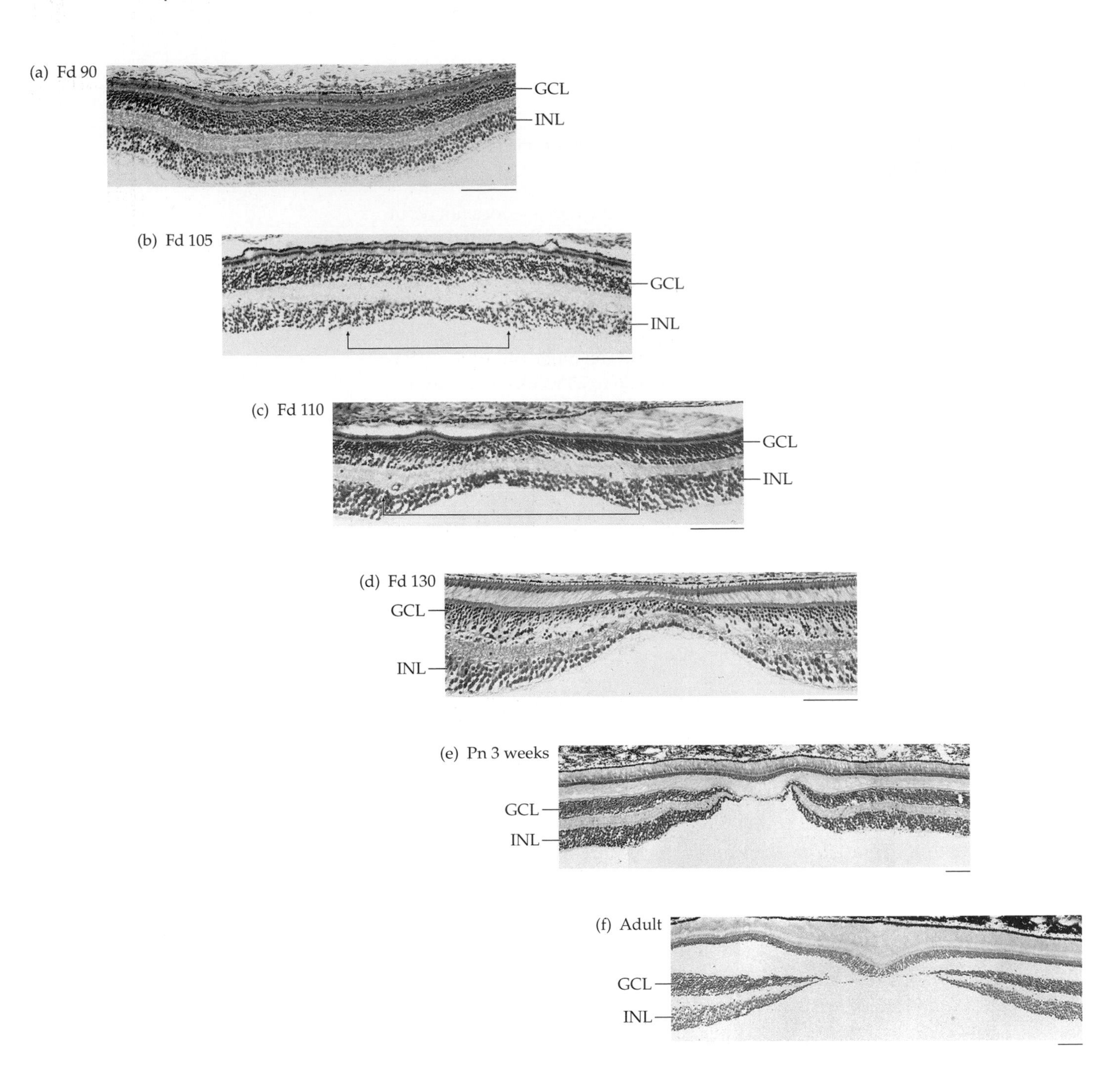

Figure 16.33
Morphological Development of the Primate Fovea
The sequence of photos shows the development of the fovea in the rhesus monkey retina. Birth for monkeys is normally at fetal day (Fd) 165 to 170. (*a*) At fetal day 90, all retinal layers are present in the region of the future fovea and there is no pit in the retina; the ganglion cell layer (GCL) and inner nuclear layer (INL) are very thick. (*b*) About two weeks later, the fovea has begun to form, as indicated by slight depression where the ganglion cell layer has begun to thin. (*c*) By fetal day 110, the depression has widened and deepened and the inner nuclear layer has begun to thin. (*d*) Five weeks before birth, at fetal day 130, most ganglion cells have been eliminated from the center of the fovea, but the outer nuclear layer and the photoreceptor layer are still quite thin. (e) Three weeks after birth, the characteristic flat floor of the monkey's central fovea is obvious, the outer nuclear layer is better defined, and the photoreceptor layer has thickened considerably. As shown by comparison to the adult fovea (*f*), however, there must be continued photoreceptor elongation and migration into the central fovea. The scale bar in all sections represents 100 μm. (Courtesy of Jan Provis and Anita Hendrickson.)

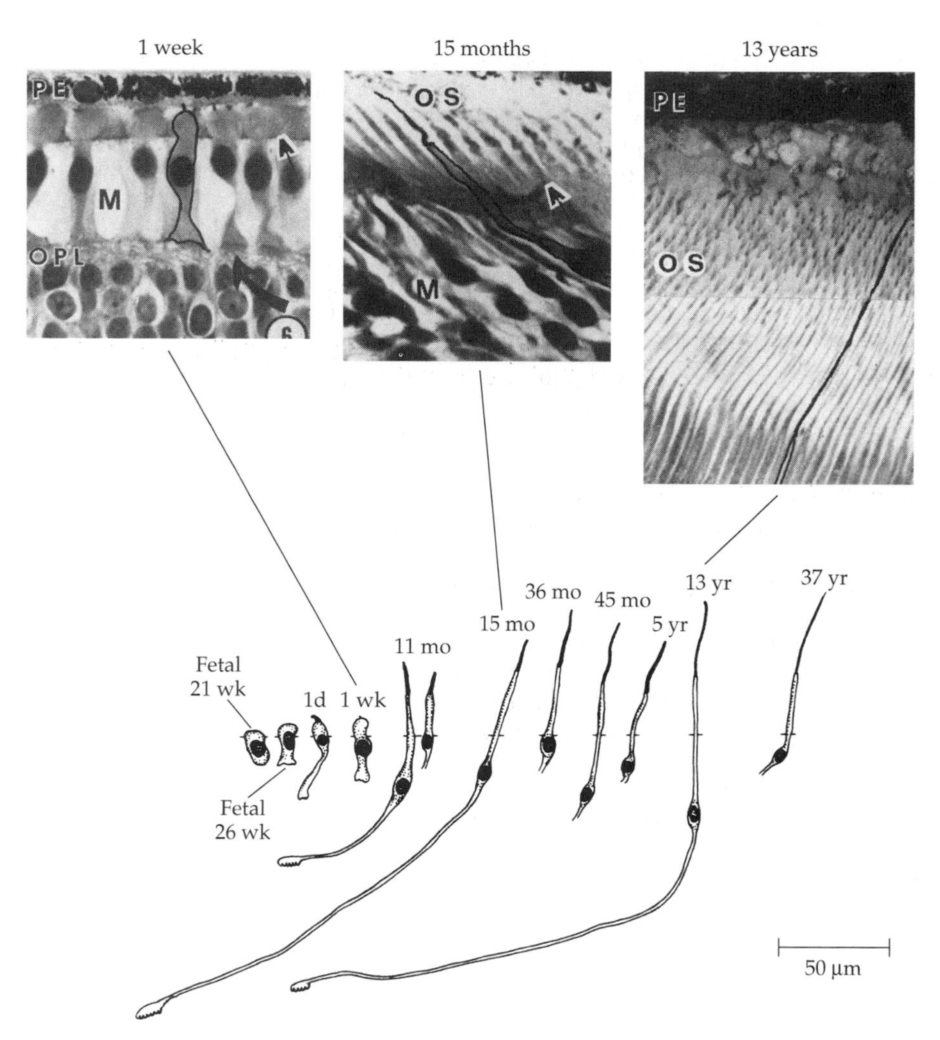

Figure 16.34
Pre- and Postnatal Development of Foveal Cones
The photomicrographs at the top are retinal sections at postnatal ages from 1 week to 13 years of age in which a foveal cone has been outlined; at birth, the inner segment is quite thick and the outer segment is extremely small in comparison to the adult form (13 years). The intermediate photo (15 months) shows a thinner cone, but with incomplete morphology. A more complete sequence of cones is drawn across the bottom. At 26 weeks of gestation, the inner segment has just begun to appear, and it is still quite small at birth. The inner segment achieves its adult length by 36 months of age, but the outer segment is still somewhat immature at 5 years. PE, pigment epithelium; OPL, outer plexiform layer; M, Müller glial processes; OS, outer segments. Arrows indicate outer limiting membrane. (Photographs from Yuodelis and Hendrickson, 1986.)

from the nucleus, and there is only a hint of outer segments. Most of the features associated with adult foveal cones—the long, thin inner and outer segments and the long obliquely running axons in the outer plexiform layer—are absent at birth. More than a year must pass before the cone outer segments acquire half their adult length, with the remainder being added during the next several years. Although the foveal cones are very tightly packed together at birth, their thick inner segments prevent them from acquiring a high density. As the cones mature and elongate, they become thinner, allowing more of them to be packed into a given area, and the density gradually increases to adult levels.

The immaturity of the fovea at birth has implications for vision in infants (see the Epilogue). All of the functions we normally associate with the adult fovea—high spatial resolution, color discrimination, pursuit eye movements, accurate saccades, accurate accommodation—are less than optimal in very young infants. All these functions require time, from one to several years, to reach adult levels of performance; the last to mature is spatial resolution, which may not reach adult levels for 6 years or so, in parallel with the amount of time required to achieve adult cone density in the central fovea.

Photoreceptor densities are shaped by cell migration and retinal expansion

From the time cones are first recognizable in the developing retina, their density is higher in the region of the future fovea than elsewhere. The difference is not

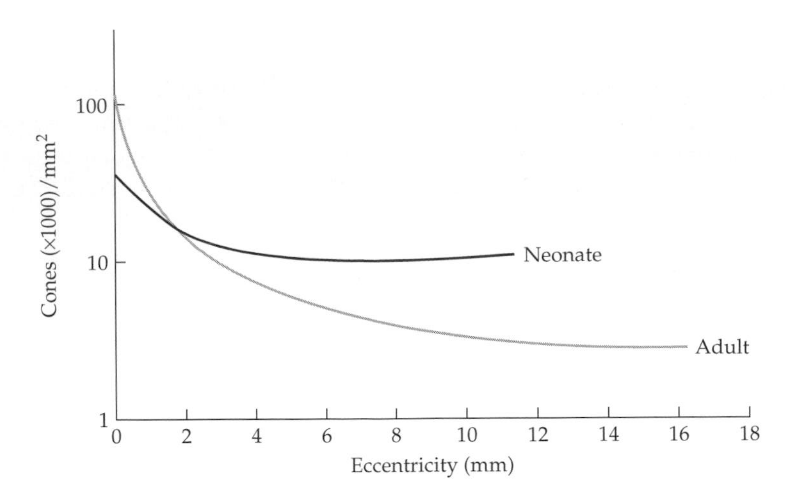

Figure 16.35
Postnatal Development of Maximum Cone Density
Comparison of cone density as a function of retinal eccentricity for the newborn and adult monkey retina. Foveal cone density is very low in the neonate (lower than that of the adult by a factor of 5 or so) but higher in the peripheral retina. Since no new cones are added postnatally and the retina is enlarging to adult size, the acquisition of high cone density in the adult retina must involve migration of cones into the fovea from surrounding regions. (Replotted from Curcio and Hendrickson 1991.)

large, certainly nowhere near the difference between foveal and peripheral cone densities in the adult, but the general decline in cone density from center to periphery is present very early. It takes years for the adult pattern of cone distribution to form, however; how is this done?

One mechanism we have already considered is the inward migration of photoreceptors in the foveal region, a change that increases density by packing the cones more tightly. Another change, however, has the potential to *decrease* photoreceptor density, and that is the expansion of the retina as the eye enlarges. This expansion occurs at a fairly constant rate during gestation and slows thereafter, but much of the enlargement of the retinal area takes place after the majority of photoreceptors, particularly the cones, have been born. If a constant number of cells are distributed across the retinal surface, it follows that their density will decrease as the area increases. In some species, changes in cell densities can be explained by a uniform balloonlike expansion of the retina that maintains preexisting density differences while producing a decrease in mean cell density. This explanation does not work for primate retinas, however, in which central densities *increase* in the face of continual retinal expansion.

In monkey retina, foveal cone density in the immature retina is lower and peripheral cone density higher than in the adult (Figure 16.35), and the immature foveal and peripheral densities differ by barely a factor of 3; in adult retinas, the difference is well over 20-fold. If the immature retina were to expand uniformly, peripheral cone densities might decrease to adult levels, but foveal density would also decrease and keep the density difference at three to one. This is one reason for invoking migration of cones toward the foveal center to account for the increase in foveal cone density in the face of retinal expansion.

Density increases not just in the fovea, but all over a central region several millimeters in diameter. This broader increase in density is more difficult to justify as migration offsetting or overcoming the effect of uniform expansion, which leads to the notion that the retina does not expand uniformly. The retina expands much less in the center than it does peripherally, thereby reducing peripheral cell density relatively more than central density. Combined with photoreceptor migration toward the foveal center, nonuniform expansion acts to exaggerate the difference between central and peripheral density, sending the former up and the latter down toward adult levels.

Ganglion cell density is shaped by migration, retinal expansion, and cell death

Migration out from the fovea may account for the clearing of ganglion cells from the foveal center and for the increase in the parafoveal ganglion cell density, but this story is complicated by compelling evidence for massive death within the ganglion cell population. The evidence for ganglion cell death is simply a matter of numbers; estimates of the total number of ganglion cells and optic nerve axons in midgestation are on the order of 2 to 3 million, while the comparable numbers at birth (and in the adult) are just over 1 million. It is difficult to escape the conclusion that somewhere between 50 and 70% of the fetal ganglion cells and their axons die before birth. And during the period when these numbers are declining, ganglion cells in histological sections often have changes in their nuclei that are commonly associated with death and degeneration.

Cell death seems to have less effect in the central region than it does in the periphery. The thinning of the ganglion cells at the foveal center begins prior to the onset of significant cell death anywhere in the retina, and when cell degeneration is common in the retina later, it is rarely seen in the foveal region. These observations suggest that the central ganglion cell density is mostly the result of cell migration, while the decline in density peripherally is a combination of deceases from nonuniform retinal expansion and cell death. The final shaping of the ganglion cell distribution is certainly *not* a product of cell death, however, because cell death ceases at birth while retinal expansion and cell density changes continue for some years.

This discussion has centered on the cones and the ganglion cells because they are the cell groups for which we have the most information, but among the numerous issues that are waiting to be addressed are the relative distributions of the different types of cones or ganglion cells and the other major groups of retinal neurons. How, for example, is the high-density ring of rods established in the perifovea? Why does the distribution of blue cones differ from those of the red and green cones? Are midget ganglion cells more, or less, susceptible to cell death than parasol cells? Are foveal ganglion cell axons less prone to die because their axons reach the central targets first and have a better chance to establish secure synapses? These and other related questions await the development of very specific labels for different cell types—labels that will mark the cells as soon as they differentiate.

The spatial organization of the retina may depend on specific cell–cell interactions and modifications of cell morphology during development

The tilings formed by ganglion cell dendrites, bipolar cell dendrites and axons, and cone terminals (see Chapter 15) have implications for the developmental mechanisms; there are just a few ways in which these tilings could be generated. In essence, the existence of a tiling means that neighboring cells have some way of establishing mutually acceptable territories for their dendritic or axonal arborizations. Agreement requires communication, and the tilings cannot be established without communication of some sort. The mode of communication is not known, but during development, neighboring cells of like type must have some way of recognizing one another and some means of responding to the presence of processes from the neighbor.

One form of recognition might be molecular tags specific to a particular cell type, a kind of labeling that may be involved in recognition and acquisition of particular synaptic inputs. Here, the tags might be used by cells of the same type to recognize one another during development so that they can slow or cease growth and avoid encroaching on the dendritic territory of a neighbor of like type. Or a similar result might be achieved by competition for synaptic space, much like that of ganglion cell axons jockeying for position in the lateral geniculate nucleus (see Chapter 5). A growing ganglion cell dendrite might recognize and acquire appropriate synaptic inputs as it grows, and stop growing when it can't find any more. And it might not find more because a neighboring cell of the

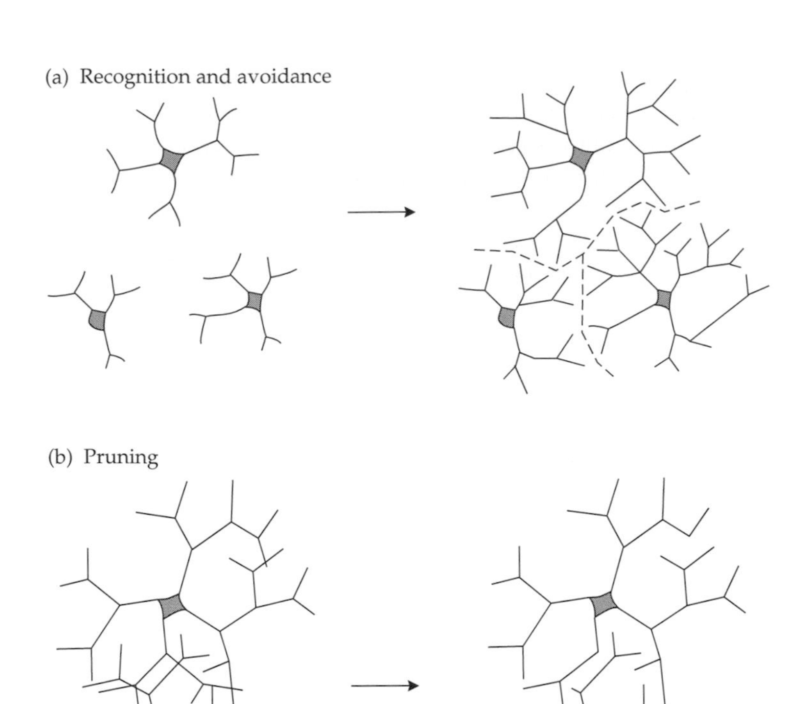

Figure 16.36
Cell Interactions and Shaping during Development
(*a*) If cells recognize and respond to their neighbors of the same type, they may have mutual limitations on their early growth that establishes their nonoverlapping dendritic territories. (*b*) Final cell morphology and territory establishment may also be accomplished by pruning of extra dendritic branches.

same type has already found them. This mechanism would also produce a tiling of the dendrites, not because of their direct interaction, but because they were competing for the same class of input neurons.

Whether directly or indirectly recognizing each other's presence, however, neighboring cells modify the behavior of their dendritic growth cones such that approaching processes may touch that of the neighbor, but do not cross it (Figure 16.36*a*). Once a cell is completely hemmed in by its neighbors, it will no longer grow at the ends of its processes; all future growth will consist of elongation of existing branches (interstitial growth), which will maintain the tiling while allowing the tiles to expand as the retina enlarges.

Another possibility is that like-type neighbors grow for a certain amount of time or grow to a certain size, at which point they recognize that their territories are overlapping, perhaps because of insufficient synaptic space for both of them. Then they "prune" their branches until they no longer overlap (Figure 16.36*b*). Once pruning is complete and the tiling has been established, future growth will again be interstitial.

Some evidence suggests that the dendritic territories of ganglion cells can be established either by cessation of dendritic growth or by pruning of excess branching. Pruning of extra branches is the best-documented phenomenon; immature forms of ganglion cells of a specific type often have many more dendrites than the mature cells have.

The main reason for thinking that dendritic growth is inhibited by dendrites from neighboring cells is what happens when neighbors are removed. When a patch of ganglion cells is experimentally deleted, undamaged cells start growing again and send their dendrites into the vacated region (the new growth is quite selective for the vacated region; there is no growth where neighbors are still present). It looks as if the possibility of further growth is present, but inhibited; removing the neighboring cells removes the inhibition, and dendritic growth resumes.

Whatever the details may be, these mechanisms are extremely important for retinal development because they allow the retina to organize itself as it develops without a great many specific a priori (genetic) instructions. For example, the locations of cells of a particular type do not need to be specified precisely for them to establish a uniform coverage with their dendrites; they simply grow until they recognize a signal to stop growing. The dendritic territories will vary in exact size and shape between one retina and the next, but the plane will be tiled. Cells that are widely spaced will have large territories; cells that are densely packed will have small territories. Deletions by normally occurring cell death can be compensated for without knowing in advance which cells will die.

If the cells in the neuroblastic layer of the early retina are genuinely totipotent, the ultimate fate and identity of a particular cell is probably a matter of chance, all else depending on where the cell happens to be when the signal—whatever it is—arrives to say that it is time for the inner cells to become ganglion cells or for the outer cells to become cones. And the specific cell type may also be based on probabilities; a newly born cone, for example, may have a 63% chance of becoming a red cone, a 32% chance of becoming a green cone, and only a 5% chance of becoming a blue cone. This would mean that the initial arrangement of the different cone types would be random (except for the future fovea, where the probability of becoming "blue" is essentially zero), but interactions and migration could be invoked to account for nonrandomness in the final arrangement. A similar argument may apply to all other cell types as well.

This kind of thinking also implies that cells of the same type—say, midget ganglion cells—will not be morphologically identical, indistinguishable clones; they will differ in detail and be recognizable as individuals because of their unique responses to the contingent events in their developmental environments. No ON-center midget ganglion cell will be exactly like any of its neighbors, but they will share the fact that they receive input from ON midget bipolars. How these connections are recognized, established, and maintained are among the most important of the outstanding questions about the retina's organization of itself during development.

Retinal blood vessels develop relatively late

Shortly after the initial wave of cell birth in the central retina, cells appear in the vicinity of the optic nerve head that are thought to be vascular precursor cells; they have no obvious association with the hyaloid vessels and their origin is unknown. During the next several weeks, however, distinct blood vessels form along the inner surface of the retina, each new vessel apparently preceded by a cord of precursor cells. Unlike the vessel formation by budding discussed in Chapter 6, a process called *angiogenesis,* this de novo formation of new vessels from vascular precursor cells is called **vasculogenesis**; it appears to be the mechanism by which the large vessels of the retina are formed. The intraretinal capillary beds are angiogenic in origin, as is pathological neovascularization (see Chapter 6).

Vasculogenesis follows a distinct center-to-periphery gradient, but unlike the gradient in retinal maturation, the focus of development is the optic nerve head and the hyaloid artery, not the fovea. Thus with increasing gestational age, the new vessels spread farther from the nerve head toward the periphery, carefully avoiding the region of the fovea and the horizontal raphe (Figure 16.37). The raphe is vascularized and the surface completely covered out to the ora serrata by about the fifth month of gestation.

Blood initially appears to be supplied to the developing retina by diffusion from these large vessels, because the intraretinal capillary beds do not begin to appear until the surface vessels have almost reached the ora serrata. The capillary beds appear first around the optic nerve head, formed by budding from the

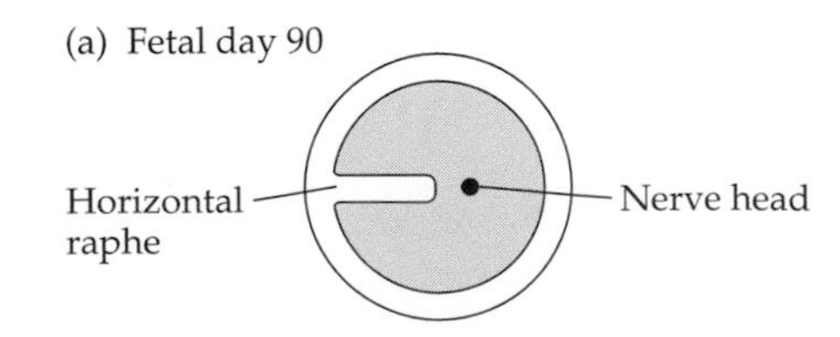

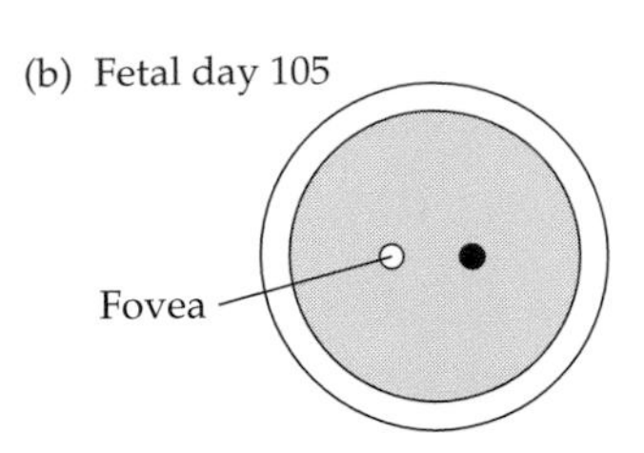

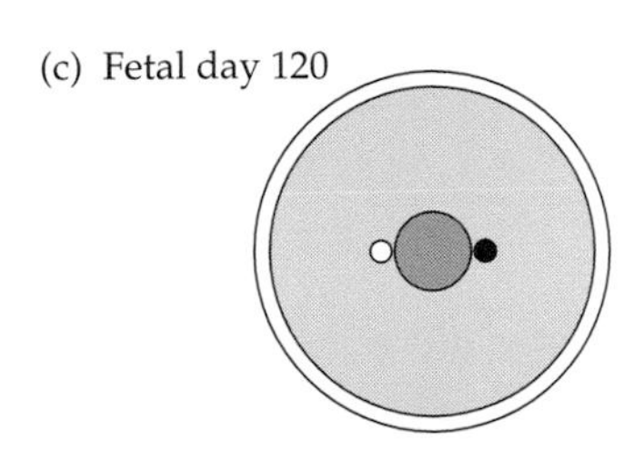

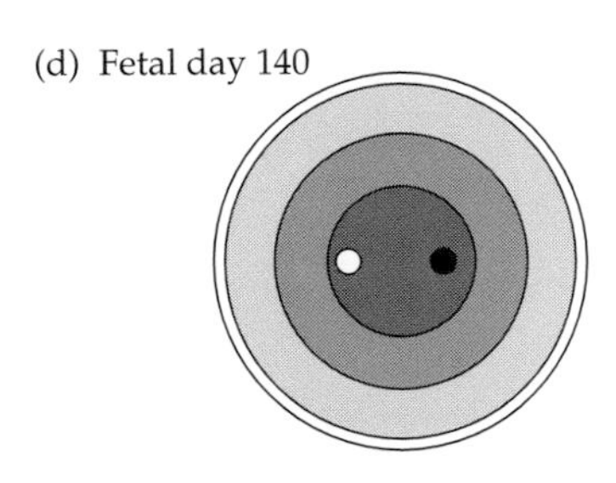

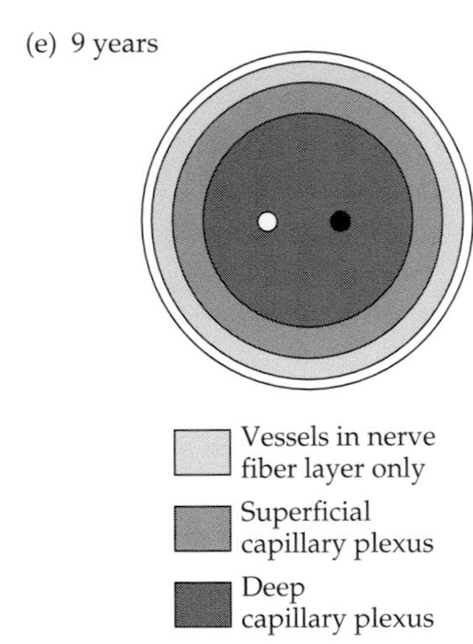

Figure 16.37
Development of the Retinal Blood Vessels
In monkey retina, new blood vessels on the surface of the nerve fiber layer spread outward from the nerve head. (*a*) By gestation day 90 they cover about two-thirds of the retina, except for the region of the future fovea and the temporal raphe. (*b*) The raphe is vascularized by gestational day 105, but the foveal region remains avascular. (*c*) Superficial capillary beds in the inner plexiform layer appear around the nerve head by gestational day 120, and the deeper capillary beds form somewhat later. (*d*) By day 140, the surface vessels are complete out to the periphery, but the superficial and deep capillary beds have not reached their adult extents (e.g., 9 years). (e) In humans, neither the superficial nor the deep capillary beds are complete at birth. (After Gariano, Iruela-Arispe, and Hendrickson 1994.)

surface vessels, and spread peripherally, in accordance with the center-to-periphery development strategy.

As discussed earlier, capillaries ramify at several levels in the retina; the shallow capillary beds lie on either side of the ganglion cell layer, the deep capillaries on either side of the inner nuclear layer. Angiogenesis begins with the surface vessels, first generating the shallow capillary beds, which in turn send branches deeper into the retina to form the deep capillary beds. As a result, capillary formation has an inner-to-outer gradient similar to that for neuronal differentiation and maturation.

With increasing gestational age, the capillary beds can be found farther and farther from the nerve head, with the shallow capillaries leading the deeper ones outward (see Figure 16.37). The foveal center is never vascularized, however, and the extreme peripheral retina never contains anything other than the most superficial of the capillary beds in the nerve fiber layer. At birth, vascularization of the human retina is incomplete and the last of the peripheral capillary beds are added in the first few postnatal weeks.

The fact that even the first large surface vessels avoid the fovea is sometimes invoked as evidence of the fovea's almost mystic power to be special, but the avoidance is not surprising. By the time vasculogenesis extends as far as the fovea, the migration of ganglion cells outward from the foveal center has already begun, clearing the region of cell bodies and their axons. Since the nerve fiber layer appears to be the requisite substrate for vessel growth, it follows that they will pass around the fovea. Similarly, the vessels may avoid the horizontal raphe because of a discontinuity in the nerve fiber layer that is initially more profound than it later becomes. Later, capillary beds do not enter the foveal center, because the cell layers around which the capillaries normally form are not continuous across the fovea.

The use of axons in the nerve fiber layer as a substrate for vasculogenesis probably accounts for the general similarity between the paths followed by both axons and blood vessels as they course over the retinal surface.

Developing vessels are inhibited by too much oxygen

The extension of vascular development in peripheral retina into the early postnatal period is partly responsible for **retinopathy of prematurity**. The problem arises because premature infants (those born at 36 weeks gestation or less) often need supplemental oxygen, which increases oxygen tension within the eye and *inhibits* the formation of new vessels in the peripheral retina. Returning the infant to a normal air supply produces a rebound in vascular proliferation because the oxygen tension is now lower than the retina would experience in utero. Unfortunately, this vascular proliferative phase is too intense as the system tries to compensate for vessels that should have formed earlier but didn't; new vessels may grow into the vitreous, there may be hemorrhages and edema, connective tissue can invade, and the retina may be put under enough traction to produce a detachment. Unchecked, this usually bilateral condition can produce severe vision loss.

It is not always possible to avoid this kind of stress on the retina's developing vascular system, particularly with extremely premature infants whose survival may depend on supplemental oxygen. With careful monitoring of the fundus appearance, it may be possible to regulate oxygen exposure to keep the neovascularization within bounds such that it will spontaneously resolve in time with no significant visual deficit. Alternatively, photocoagulation of leaking vessels or vitrectomy (see Chapter 12) may be required to prevent excessive damage.

The optic nerve forms as tissue in the optic stalk is replaced with developing ganglion cell axons and glial cells

The optic stalk is tubular—basically a hollow cylinder of neuroepithelial cells. Like the invaginating optic cup, the optic stalk folds inward, becoming effectively a double-walled tube with an open groove along the bottom (Figure 16.38*a*); the groove is part of the choroidal fissure in the optic cup. The hyaloid artery enters the groove in the optic stalk and grows forward toward the lens within the eye. Closure of the choroidal fissure traps the artery, placing it at the core of the developing optic nerve (Figure 16.38*b*).

At this stage, the future nerve has three concentric parts: the central core formed by the hyaloid artery, as well as intermediate and outer sheaths of relatively undifferentiated neuroepithelium. The intermediate cell sheath will differentiate to become the glial cells—both astrocytes and oligodendrocytes—within the optic nerve; the outer sheath will become the outermost glial sheath of the nerve, separating it from surrounding mesodermal tissues during development and from the pia mater after development is complete.

When the ganglion cells first begin to differentiate in the region of the future fovea, the fissure along the optic stalk has closed, giving the stalk its solid structure as a blood vessel core surrounded by two layers of differentiating neuroepithelial cells. The firstborn ganglion cells almost immediately sprout axons that grow to the optic stalk, which at this stage is a fraction of a millimeter away. The approach of these pioneering axons coincides with atrophy and differentiation in the cells around the hyaloid artery, and this region of atrophy serves as a rough guide to the path the axons should follow. It is not clear whether axonal approach induces the cell atrophy, or whether atrophy is the primary event to which the axons are attracted.

Since the firstborn ganglion cells are those in the region of the future fovea, the first axons to reach the optic stalk are the beginnings of the papillomacular bundle; they run through the center of the optic stalk next to the artery, creating the first element of topographic order in the nerve, whereby axons from central ganglion cells are central in the nerve and axons from peripheral ganglion cells are in the nerve's outer annulus. Thus the order in the optic nerve is a direct consequence of the center-to-periphery gradient of differentiation in the retina; axons of the firstborn central ganglion cells occupy the core of the optic stalk, forcing later-born and later-arriving axons from more peripheral ganglion cells to fill in around them at progressively more centrifugal locations in the stalk until it is completely filled (Figure 16.38*c*). Or to put it another way, timing, not retinal location, is the main determinant of axonal position in the optic nerve; the ordering is chronotopic rather than retinotopic.

As the axons grow through the optic stalk toward the developing brain, optic stalk cells differentiating around them form the beginnings of the glial sheaths that surround axon bundles in the adult nerve (the connective tissue component of the sheaths comes from mesodermal cells associated with the hyaloid artery). At this time, the myelin-producing oligodendrocytes have yet to appear, so the

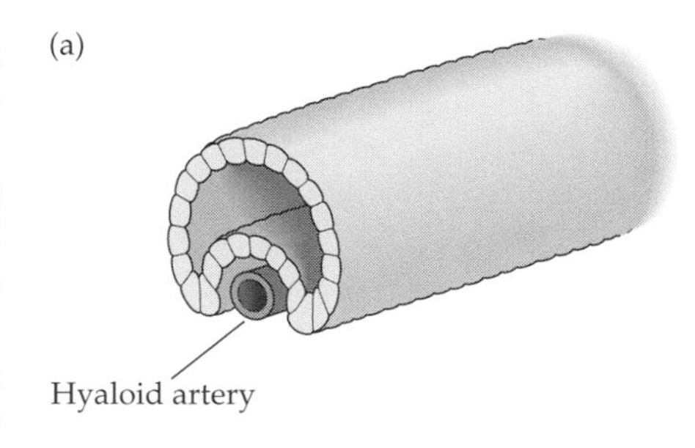

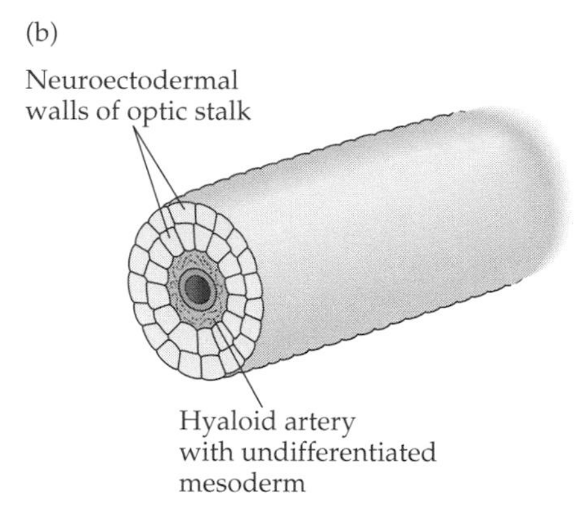

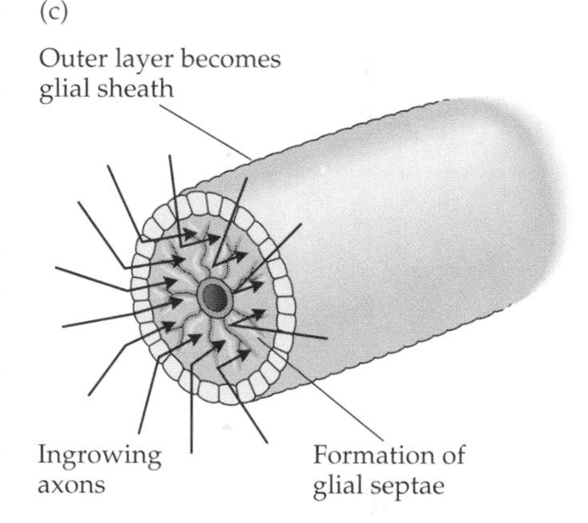

Figure 16.38
Formation of the Optic Nerve
(*a*) A section though the optic stalk just behind the optic cup shows a tubular structure with a groove on the underside (the choroidal fissure) in which the hyaloid artery runs. (*b*) After closure of the fissure, the optic stalk is a tube with several layers of neuroepithelial cells on the outside surrounding a cellular core containing the hyaloid artery. (*c*) As axons from developing retinal ganglion cells reach the nerve, the inner cells of the nerve's core either degenerate or differentiate into glial cells, and the nerve is filled by the axons growing to the brain.

axons have no myelin sheaths, and the lamina cribrosa has yet to form, except perhaps as a glial cell lamina. Thus the axons do not grow through a meshwork of scleral tissue; rather the sclera, represented by the lamina cribrosa, must grow through the optic nerve.

Fusion of the optic stalks produces the optic chiasm, where pioneering axons must choose the ipsilateral or contralateral path

The optic stalks are initially quite separate, diametrically opposite one another on the two sides of the closing neural tube. The explosive growth of the dorsal part of the neural tube tends to push the optic stalks together as the stalks elongate and the optic cups form, so that they eventually touch and merge at a single site. This merger occurs just before the arrival of axons growing in from the eyes.

Since there is such a tremendous overproduction of ganglion cells, it is tempting to think of decussation as we see it in the adult as a product of cell death, wherein axons do not make a choice at the chiasm, but simply go one way or the other at random and later die if the direction was inappropriate. In this scenario, a temporal retinal axon that crossed the midline to the contralateral side would find itself in inhospitable territory and die. This would mean, however, that the axonal projections to the two sides of the brain would be initially quite disorganized and diffuse, with little distinction between regions of ipsilateral and contralateral input. And this does happen to some extent; ipsilateral axons in particular are more widespread early than they are later, but random assignment to the ipsilateral or contralateral side with subsequent refinement through cell death is far from the whole story. Most axons appear to make the correct choice at the optic chiasm.

The signal that tells an axon from the temporal retina to remain on the same side and a nasal retinal axon to cross the midline is not known, but there are several possibilities. In the mature retina, ganglion cells appear to have a retinotopic marker, presumably a membrane protein, that specifies their relative location. If a marker of this type were expressed early in development and were incorporated in the developing axons, they might be recognized at the chiasm as originating in either the nasal or temporal retina. But even so, something needs to be doing the recognizing, perhaps erecting a sign telling the nasal axons to cross and the temporal axons to stay in the outside lane. There is evidence in ferrets (a favorite animal for developmental studies) that the chiasm contains a transient glial cell structure at the midline of the chiasm; it appears for a time, then disappears, and its presence coincides with the early stages of axon growth through the chiasm. It is possible, though not proven, that this transient structure is involved in producing a chemical message that gives axons their proper direction in the chiasm.

In albinism, something goes awry at the chiasm; albinos of all mammalian species have anomalous decussation patterns involving more crossing by temporal retinal axons than normally occurs (i.e., the ipsilateral projections are reduced). This phenomenon suggests a link between melanin in the developing eye and the choice of axon direction at the chiasm, but the nature of the link is by no means obvious. One speculative possibility is that melanin in the pigment epithelium (or its precusors or the by-products of melanin breakdown) are involved in the process of "marking" developing ganglion cells as temporal or nasal.

The last stages of development in the optic nerve are axon loss and myelination

The growth of axons into the optic nerve reaches its peak around fetal week 16, when the number of axons in a section of nerve just behind the eye is around 3.5 million, roughly three times the adult number. From this peak, the number

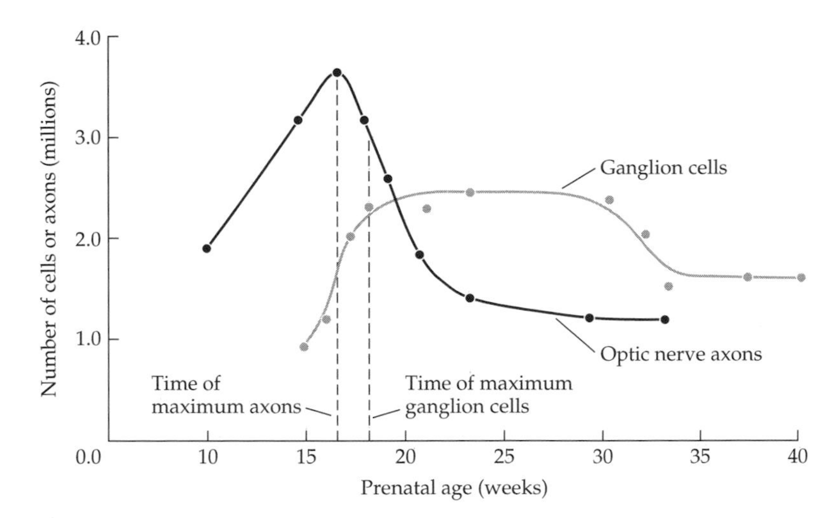

Figure 16.39
Prenatal Death of Ganglion Cells and Their Axons
Estimates of the total number of retinal ganglion cells and axons in the optic nerve as functions of gestational age (human). There are many more ganglion cells in the retina and axons in the optic nerve prenatally than there are at birth; the extra cells and axons are removed by a wave of prenatal cell death. The elimination of axons from the nerve appears to precede the elimination of ganglion cell bodies by several months (see text). (After Provis et al. 1985*b*.)

declines during the next 5 to 6 weeks, reaching the normal adult value about 4 months before birth. Although the time at which the number of axons in the optic nerve is highest coincides roughly with the peak number of ganglion cells in the retina, as one would expect, the loss of axons from the nerve appears to precede ganglion cell death in the retina by about 2 months (Figure 16.39). Note also that the estimates of axon numbers are significantly higher than the estimates of ganglion cell numbers.

These discrepancies in timing and number are probably associated with some unavoidable limitations on the accuracy of the estimates for ganglion cell numbers. One is that a ganglion cell body may be dying, with its axon already gone, but still be physically present for staining and counting; this could explain, in part, why ganglion cell body loss lags axon loss. In addition, the ganglion cell layer contains other cell groups, particularly displaced amacrine cells, that are often difficult to recognize as something other than ganglion cells; if displaced amacrine cells were moving into the ganglion cell layer at the appropriate time, their unrecognized presence could effectively mask some of the ganglion cell loss. As for the discrepancy in peak numbers of axons and ganglion cells, it could mean that immature ganglion cells have axon collaterals that arise very early in their course (giving the appearance of more axons than cell bodies), but a more likely explanation is a systematic underestimate of ganglion cell numbers because of relatively coarse sampling of the population. All told, the data on axon numbers are probably the better indicator of ganglion cell overproduction and the timing of their functional death.

About the time of cell death and axon loss, the oligodendrocytes in the nerve begin to form myelin sheaths around the axons (the myelin-forming cells in peripheral nerves are Schwann cells). Myelination begins centrally near the axon terminals and spreads peripherally toward the eye, reaching the lamina cribrosa just before birth. Clearly something in the normal lamina cribrosa and prelaminar nerve head stops the myelination. It may be something as simple as the absence of oligodendrocytes in the prelaminar region, or it may be a more complicated interaction between oligodendrocytes and astrocytes, which are known to be

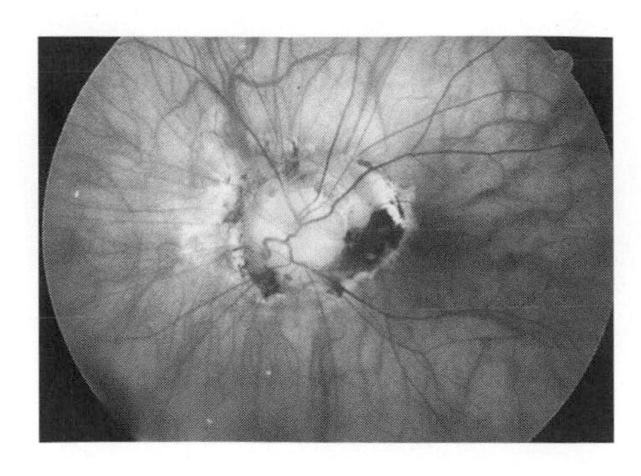

Figure 16.40
Optic Nerve Coloboma
The large white area is the coloboma where tissue did not develop normally, because the choroidal fissure failed to close properly. (Courtesy of Richard Feist.)

involved in regulating myelin production by the oligodendrocytes. Whatever the mechanism, its failure allows myelin production to extend into the nerve fiber layer of the retina, producing an apparent, irregular expansion of the nerve head.

The inner retina seems relatively immune to congenital anomalies

With the exception of gross malformations like coloboma (see the next section), the inner retina and its vasculature exhibit no abnormalities that we can confidently attribute to developmental failures. This apparent paucity of congenital defects may simply be due to our inability to recognize the perhaps subtle effects of improper formation of, say, a diffuse cone bipolar pathway, one of the lateral pathways, or even one of the neurotransmitter systems. Given the enormous developmental flexibility and contingent mechanisms operating during the development of these systems, however, it is more likely that the apparent lack of anomalies is real. In fact, the kinds of serious things one can imagine, such as a failure of the retinal GABA systems, would almost certainly be early failures, widespread in the developing brain and devastating to the embryo.

The most common developmental anomalies are failures to complete embryonic structures or eliminate transient structures

As mentioned in Chapter 1, in some situations the eye fails to grow to its proper size at birth (microphthalmos); the retinas in such eyes contain differentiated tissue that looks as if the development proceeded normally for a time and then stopped, for reasons unknown. Not surprisingly, the retina in a microphthalmic eye is not functional.

Incomplete closure of the choroidal fissure and failure of the optic cup to fuse properly is manifested as a typical *coloboma,* in which there are incomplete regions in one or more of the ocular tissues. The typical coloboma shown in Figure 1.14*b* affects both the retina and the choroid; a white wedge appears in the fundus where the sclera can be seen without the normal obscuration by overlying pigment in the pigment epithelium and the choroid. The region of the coloboma in this instance corresponds with a defect in the visual field.

Since the cleft of the choroidal fissure originally extends back along the bottom of the developing optic nerve, there is also a possibility of failed closure in this region, and some colobomas involve the optic nerve either along with or separately from the retina. As Figure 16.40 shows, the appearance of a large optic nerve coloboma can be quite dramatic; the nerve head seems very large and pale, and is surrounded by dark hyperplasia from the pigment epithelium, and some of the blood vessels emanating from the bit of unaffected nerve head (at the upper left) are abnormally small in diameter. Depending on the size and location of the coloboma, the effects on vision and the amount of visual field loss can range from modest to severe.

The hyaloid artery with its vascular tunic around the developing lens is one of the more obvious embryonic structures that is transient; normally, all the intraocular components of this system atrophy and have no counterpart in the mature eye. Atrophy is sometimes incomplete, however, and various remnants may remain on the lens, within the vitreous, or on the surface of the nerve head (see Chapter 12).

The last stages of clearing away hyaloid artery tissue and excavating the cup in the nerve head occur fairly late in gestation. As a result, more than 90% of eyes in premature infants exhibit hyaloid artery remnants, while less than 5% of full-term infants have them.

References and Additional Reading

Electrical Signals and Assessment of Retinal Function

Armington JC, Johnson EP, and Riggs LA. 1952. The scotopic *A*-wave in the electrical response of the human retina. *J. Physiol.* 118: 289–298.

Bearse MA and Sutter EE. 1996. Imaging localized retinal dysfunction with the multifocal electroretinogram. *J. Opt. Soc. Am. (A)* 13: 634–640.

Berson EL. 1992. Electrical phenomena in the retina. Chapter 21, pp. 641–707, in *Adler's Physiology of the Eye,* Hart WM, ed. Mosby Year Book, St. Louis.

Breton ME, Schueller AW, Lamb TD, and Pugh EN. 1994. Analysis of the ERG a-wave amplification and kinetics in terms of the G-protein cascade of phototransduction. *Invest. Ophthalmol. Vis. Sci.* 35: 295–309.

Carr RE. 1997. Electroretinography. Chapter 103 in *Duane's Foundations of Clinical Ophthalmology,* Vol. 2. Tasman W and Jaeger EA, eds. JB Lippincott, Philadelphia.

Dowling JE. 1987. The electroretinogram and glial responses. Chapter 6, pp. 164–186, in *The Retina: An Approachable Part of the Brain.* Harvard University Press, Cambridge, MA.

Hood DC and Birch DG. 1992. A computational model of the amplitude and implicit time of the *b*-wave of the human ERG. *Vis. Neurosci.* 8: 107–126.

Hood DC and Birch DG. 1996. *b* Wave of the scotopic (rod) electroretinogram as a measure of the activity of human on-bipolar cells. *J. Opt. Soc. Am. (A)* 13: 623–633.

Hood DC, Seiple W, Holopigian K, and Greenstein V. 1997. A comparison of the components of the multifocal and full-field ERGs. *Vis. Neurosci.* 14: 533–544.

Newman E. 1980. Current source-density analysis of the b-wave of frog retina. *J. Neurophysiol.* 43: 1355–1366.

Newman E and Reichenbach A. 1996. The Müller cell: A functional element of the retina. *Trends Neurosci.* 19: 307–312.

Odom JV and Weinstein GW. 1993. Clinical visual electrophysiology. Chapter 5 in *Duane's Clinical Ophthalmology,* Vol. 3, *Diseases of the Retina,* Tasman W and Jaeger EA, eds. Lippincott-Raven, Philadelphia.

Sutter EE and Tran D. 1992. The field topography of ERG components in man—I. The photopic luminance. *Vision Res.* 32: 433–446.

Tian N and Slaughter MM. 1995. Correlation of dynamic responses in the ON bipolar neuron and the *b*-wave of the electroretinogram. *Vision Res.* 35: 1359–1364.

The Retinal Vessels and Assessment of Retinal Health

Ahmed J, Braun RD, Dunn R, and Linsenmeir RA. 1993. Oxygen distribution in the macaque retina. *Invest. Ophthalmol. Vis. Sci.* 34: 516–521.

Alexander LJ. 1994. *Primary Care of the Posterior Segment,* 2nd Ed. Appleton & Lange, Norwalk, CT.

Alm A. 1992. Ocular circulation. Chapter 6, pp. 198–227, in *Adler's Physiology of the Eye,* Hart WM, ed. Mosby Year Book, St. Louis.

Chase J. 1982. The evolution of retinal vascularization in mammals: A comparison of vascular and avascular retinae. *Ophthalmology* 89: 1518–1525.

Federman JL and Maguire JI. 1989. Intravenous fluorescein angiography. Chapter 4 in *Duane's Clinical Ophthalmology,* Vol. 3, *Diseases of the Retina,* Tasman W and Jaeger EA, eds. Lippincott-Raven, Philadelphia.

Guérin CJ, Lewis GP, Fisher SK, and Anderson DH. 1993. Recovery of photoreceptor outer segment length and analysis of membrane assembly rates in regenerating primate photoreceptor outer segments. *Invest. Ophthalmol. Vis. Sci.* 34: 175–183.

Iwasaki M and Inomata H. 1986. Relation between superficial capillaries and foveal structures in the human retina. *Invest. Ophthalmol. Vis. Sci.* 27: 1698–1705.

Janzer RC and Raff MC. 1987. Astrocytes induce blood-brain barrier properties in endothelial cells. *Nature* 325: 253–257.

Sagaties MJ, Raviola G, Schaeffer S, and Miller C. 1987. The structural basis of the inner blood-retina barrier in the eye of *Macaca mulatta. Invest. Ophthalmol. Vis. Sci.* 28: 2000–2014.

Snodderly DM, Weinhaus RS, and Choi JC. 1992. Neural-vascular relationships in central retina of macaque monkeys (*Macaca fasicularis*). *J. Neurosci.* 12: 1169–1193.

Wise GN, Dollery CT, and Henkind P. 1971. *The Retinal Circulation.* Harper & Row, New York.

The Optic Nerve

Anderson DR and Quigley HA. 1992. The optic nerve. Chapter 20, pp. 616–640, in *Adler's Physiology of the Eye,* Hart WM, ed. Mosby Year Book, St. Louis.

Caprioli J. 1989. Correlation of visual function with optic nerve and nerve fiber layer structure in glaucoma. *Surv. Ophthalmol.* 33: 319–330.

Cioffi GA and Van Buskirk EM. 1996. Vasculature of the anterior optic nerve and peripapillary choroid. Chapter 8, pp. 177–188, in *The Glaucomas*, Vol. 1. *Basic Sciences*, 2nd Ed., Ritch R, Shields MB, and Krupin T, eds. Mosby Year Book, St. Louis.

Elkington AR, Inman CBE, Steart PV, and Weller RO. 1990. The structure of the lamina cribrosa of the human eye: An immunocytochemical and electron microscopical study. *Eye* 4: 42–57.

Fechtner RD and Weinreb RN. 1994. Mechanisms of optic nerve damage in primary open angle glaucoma. *Surv. Ophthalmol.* 39: 23–42.

Goldberg I. 1994. Therapeutic possibilities for calcium channel blockers in glaucoma. *Prog. Ret. Eye Res.* 13: 593–604.

Harrington DO and Drake MV. 1990. Glaucoma. Chapter 12, pp. 179–218, in *The Visual Fields; Text and Atlas of Clinical Perimetry,* 6th Ed. Mosby, St. Louis.

Hogan MJ, Alvarado JA, and Weddell J. 1971. Optic nerve. Chapter 10, pp. 523–606, in *Histology of the Human Eye.* WB Saunders, Philadelphia.

Johnson CA and Samuels SJ. 1997. Screening for glaucomatous visual field loss with frequency-doubling perimetry. *Invest. Ophthalmol. Vis. Sci.* 38: 413–425.

Jonas JB, Müller-Bergh JA, Schlötzer-Schrehardt UM, and Naumann GOH. 1990. Histomorphometry of the human optic nerve. *Invest. Ophthalmol. Vis. Sci.* 31: 736–744.

Jonas JB, Schmidt AM, Müller-Bergh JA, Schlötzer-Schrehardt UM, and Naumann GOH. 1992. Human optic nerve fiber count and optic disc size. *Invest. Ophthalmol. Vis. Sci.* 33: 2012–2018.

Minckler DS. 1980. The organization of nerve fiber bundles in the primate optic nerve head. *Arch. Ophthalmol.* 98: 1630–1636.

Minckler DS and Milam AH. 1994. Optic nerve axonal transport. Chapter 27 in *Duane's Foundations of Clinical Ophthalmology,* Vol. 1, Tasman W and Jaeger EA, eds. Lippincott-Raven, Philadelphia.

Ogden TE. 1983. Nerve fiber layer of the macaque retina: Retinotopic organization. *Invest. Ophthalmol. Vis. Sci.* 24: 85–98.

Ogden TE, Duggan J, Danley K, Wilcox M, and Minckler DS. 1988. Morphometry of nerve fiber bundle pores in the optic nerve head of the human. *Exp. Eye Res.* 46: 559–568.

Polyak SL. 1941. *The Retina.* University of Chicago Press, Chicago.

Quigley HA, Brown AE, Morrison JD, and Drance SM. 1990. The size and shape of the optic disc in normal human eyes. *Arch. Ophthalmol.* 108: 51–57.

Quigley HA, Dunkelberger GR, and Green WR. 1988. Chronic human glaucoma causing selectively greater loss of large optic nerve fibers. *Ophthalmology* 95: 357–363.

Quigley HA, Hohman RM, Addicks EM, and Green WR. 1984. Blood vessels and the glaucomatous optic disc in experimental primate and human eyes. *Invest. Ophthalmol. Vis. Sci.* 25: 918–931.

Quigley HA, Sanchez RM, Dunkelberger GR, l'Hernault NL, and Baginski TA. 1987. Chronic glaucoma selectively damages large optic nerve fibers. *Invest. Ophthalmol. Vis. Sci.* 28: 913–920.

Radius RH and Anderson DR. 1979. The histology of retinal nerve fiber layer bundles and bundle defects. *Arch. Ophthalmol.* 97: 948–950.

Silverman SE, Trick GL, and Hart WM. 1990. Motion perception is abnormal in primary open-angle glaucoma and ocular hypertension. *Invest. Ophthalmol. Vis. Sci.* 31: 722–729.

Varma R and Minckler DS. 1996. Anatomy and pathophysiology of the retina and optic nerve. Chapter 7, pp. 139–175, in *The Glaucomas,* Vol. 1, *Basic Sciences,* 2nd Ed., Ritch R, Shields MB, and Krupin T, eds. Mosby Year-Book, St. Louis.

Weber AJ, Kaufman PL, and Hubbard WC. 1998. Morphology of single ganglion cells in the glaucomatous primate retina. *Invest. Ophthalmol. Vis. Sci.* 39: 2304–2320.

Development of the Retina and Optic Nerve

Curcio CA and Hendrickson AE. 1991. Organization and development of the primate photoreceptor mosaic. *Prog. Ret. Res.* 10: 89–120.

Diaz-Araya C and Provis JM. 1992. Evidence of photoreceptor migration during early foveal development: A quantitative analysis of human fetal retina. *Vis. Neurosci.* 8: 505–514.

Dunlop S, Fraley S, and Beazley L. 1993. The morphology of developing and regenerating retinal ganglion cells. Chapter 12, pp. 148–161, in *Formation and Regeneration of Nerve Connections,* Sharma SC and Fawcett JW, eds. Birkhäuser, Boston.

Foos RY. 1987. Retinopathy of prematurity: Pathologic correlation of clinical stages. *Retina* 7: 260–270.

Gariano RF, Iruela-Arispe ML, and Hendrickson AE. 1994. Vascular development in primate retina: Comparison of laminar plexus formation in monkey and human. *Invest. Ophthalmol. Vis. Sci.* 35: 3442–3455.

Gariano RF, Kalina RE, and Hendrickson AE. 1996. Normal and pathological mechanisms in retinal vascular development. *Surv. Ophthalmol.* 40: 481–490.

Gariano RF, Sage EH, Kaplan HJ, and Hendrickson AE. 1996. Development of astrocytes and their relation to blood vessels in fetal monkey retina. *Invest. Ophthalmol. Vis. Sci.* 37: 2367–2375.

Grierson I, Hiscott P, Hogg P, Robey H, Mazure A, and Larkin G. 1994. Development, repair and regeneration of the retinal pigment epithelium. *Eye* 8: 255–262.

Hendrickson AE. 1994. Primate foveal development: A microcosm of current questions in neurobiology. *Invest. Ophthalmol. Vis. Sci.* 35: 3129–3133.

Hendrickson AE and Yuodelis C. 1984. The morphological development of the human fovea. *Ophthalmology* 91: 603–612.

Jeffery G. 1997. The albino retina: An abnormality that provides insight into normal retinal development. *Trends Neurosci.* 20: 165–169.

LaVail MM, Rapaport DL, and Rakic P. 1991. Cytogenesis in the monkey retina. *J. Comp. Neurol.* 309: 86–114.

Linberg KA and Fisher SK. 1990. A burst of differentiation in the outer posterior retina of the eleven-week human fetus: An ultrastructural study. *Vis. Neurosci.* 5: 43–60.

Magoon EH and Robb RM. 1981. Development of myelin in human optic nerve and tract. A light and electron microscopic study. *Arch. Ophthalmol.* 99: 655–664.

Okada M, Erickson A, and Hendrickson A. 1994. Light and electron microscopic analysis of synaptic development in *Macaca* monkey retina as detected by immunocytochemical labeling for the synaptic vesicle protein, SV2. *J. Comp. Neurol.* 339: 535–558.

Provis JM, Diaz CM, and Dreher B. 1998. Ontogeny of the primate fovea: A central issue in retinal development. *Prog. Neurobiol.* 54: 549–581.

Provis JM, van Driel D, Billson FA, and Russell P. 1985a. Development of the human retina: Patterns of cell distribution and redistribution in the ganglion cell layer. *J. Comp. Neurol.* 233: 429–451.

Provis JM, van Driel D, Billson FA, and Russell P. 1985b. Human fetal optic nerve: Overproduction and elimination of retinal axons during development. *J. Comp. Neurol.* 238: 92–100.

Ramoa AS, Campbell G, and Schatz CJ. 1988. Dendritic growth and remodeling of cat retinal ganglion cells during fetal and postnatal development. *J. Neurosci.* 8: 4239–4261.

Reese BE, Maynard TM, and Hocking DR. 1994. Glial domains and axonal reordering in the chiasmatic region of the developing ferret. *J. Comp. Neurol.* 349: 303–324.

Reichenbach A and Robinson SR. 1995. Phylogenetic constraints on retinal organization and development. *Prog. Ret. Eye Res.* 15: 139–171.

Rhodes RH. 1990. Development of the optic nerve. Chapter 25 in *Duane's Foundations of Clinical Ophthalmology,* Vol. 1, Tasman W and Jaeger EA, eds. Lippincott-Raven, Philadelphia.

Streeten BW. 1969. Development of the human retinal pigment epithelium and the posterior segment. *Arch. Ophthalmol.* 81: 383–394.

Trisler D. 1982. Are molecular markers of cell position involved in the formation of neural circuits? *Trends Neurosci.* 5: 306–310.

Turner DL and Cepko CL. 1987. A common progenitor for neurons and glia persists in rat retina late in development. *Nature* 328: 131–136.

Van Driel D, Provis JM, and Billson FA. 1990. Early differentiation of ganglion, amacrine, bipolar and Müller cells in the developing fovea of human retina. *J. Comp. Neurol.* 291: 203–219.

Wilson HR, Metz MB, Nagy SE, and Dressel AB. 1988. Albino spatial vision as an instance of arrested visual development. *Vision Res.* 29: 979–990.

Wong ROL. 1990. Differential growth and remodeling of ganglion cell dendrites in the postnatal rabbit retina. *J. Comp. Neurol.* 294: 109–132.

Yuodelis C and Hendrickson AE. 1986. A quantitative and qualitative analysis of the human fovea during development. *Vision Res.* 26: 847–856.

Epilogue

Time and Change

Nothing I cared in the lamb white days, that time would take me
Up to the swallow thronged loft by the shadow of my hand,
In the moon that is always rising,
Nor that riding to sleep
I should hear him fly with the high fields
And wake to the farm forever fled from the childless land.
Oh as I was young and easy in the mercy of his means,
Time held me green and dying
Though I sang in my chains like the sea.

■ Dylan Thomas, "Fern Hill"

The life of the eye includes birth, growth, maturation, and senescence. And as the structure of the eye changes throughout its lifetime, its functional quality also changes. The newborn eye is structurally and functionally immature and must develop for a dozen years or so. Development is followed by several decades of relative stability, after which structural changes make the aging eye a less effective window on the world than it once was. As the poet said (right), youth is blissfully unaware of its mortality, and most of the normal changes in our eyes are so slow that we are unaware of them, but our eyes are always changing. And the changes, however slow, are inexorable.

Among other things, the changes in the eye with time mean that what we think of as the "normal" anatomy of the eye depends on the age of the eye. For the most part, the anatomy described in this book depicts the eye in its early maturity, at roughly 25 to 30 years of age. Thus our final bit of business is to consider briefly the expected changes in the eye throughout life, and by so doing, to indicate both what happened to the eye in youth and what will become of it later in life.

Postnatal Growth and Development

The newborn eye increases in overall size for the next 15 years

Figure 1*a* compares cross sections of eyes from infants, children, and adults. All parts of the newborn eye are smaller than their adult counterparts. The axial length of the newborn eye is about 17 mm, which is about 70% of the 24 mm axial length of the typical adult eye. In volume, however, the newborn eye is only 35% as large as the adult eye (6.5 ml). Most of the difference will be made up by growth of the sclera and an increase in the volume of the vitreous chamber.

The lens in the neonatal eye has a smaller equatorial diameter than the adult lens has (5.8 mm, compared to 10 mm), and it is thinner along its axial dimension (about 3.8 mm, compared to 5.2 mm at age 80). The infant's lens is relatively more spherical, so the lens curvatures are higher than in the adult and the neonatal lens has more dioptric power by calculation based on a Gullstrand schematic eye (about 24 D compared to the adult's 20 D).

The infant's cornea is also more powerful (about 50 D, compared to 43 D in adults). The higher corneal power, like the higher lens power, is in keeping with the shorter axial length of the infant's eye. Although the newborn's cornea is more curved, it is thicker than it will become later, and it is disproportionately large in diameter (the diameter is 9.5 to 10 mm, which is 83 to 87% of the

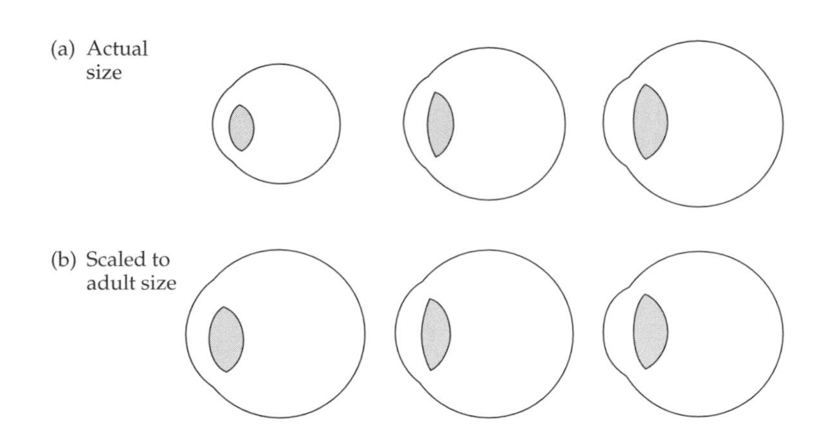

Figure 1
Eye Size in Infants, Children, and Adults
(*a*) The eyes of a newborn infant, a 6-year-old child, and an adult, drawn to their actual dimensions. The axial lengths are 17, 22.5, and 24 mm, respectively. (*b*) The dimensions in the eyes of the infant and child have been scaled up by factors of 1.41 and 1.07, respectively, so that their scleral diameters are the same as that of the adult eye. The relatively large cornea and more curved lens in the infant eye are now obvious. The child's eye is notable for its relatively thin lens.

adult 11.5 mm). At birth, the cornea occupies about 20% of the external surface area of the eye; this value will decrease to just under 16% by the time the eye ceases to grow.

The depth of the anterior chamber is at its minimum in the newborn eye (about 2.5 mm). This small separation between the cornea and the lens adds to the combined optical power of the system, which is about 90 D at birth.

An infant's eye is not just a scaled-down version of the adult eye, as shown in Figure 1*b*, where the eyes of infants and children have been enlarged to match the axial length of the adult eye. The relatively large size of the infant's cornea, the more spherical lens, and the shallower anterior chamber are now quite obvious. The child's eye is more adultlike, as one would expect, but it is still growing and changing.

Different parts of the eye grow at different rates and cease growth at different ages (Figure 2). Beginning at 17 mm, axial length increases rapidly during the first year and less rapidly thereafter, reaching its adult value sometime between the ages of 12 and 16 (Figure 2*a*). The anterior chamber depth also shows a rapid early increase, but reaches its final value between the ages of 8 and 12 (Figure 2*b*). Data on the cornea are measures of corneal power, rather than size, but the acquisition of adult dioptric power by age 3 suggests that the cornea has ceased to grow by this age, which is about 10 years before scleral growth ceases and the eye stops enlarging (Figure 2*c*). Early cessation of corneal growth is consistent with the nearly adult size of the cornea at birth.

The lens data are for the anterior–posterior thickness of the lens; thickness declines from its value at birth (about 3.8 mm) to about 3.5 mm at age 10 (Figure 2*d*). (It will increase after age 10; see Figure 12.10). The early thinning of the lens in combination with an increase in its equatorial diameter means that it is becoming less spherical, with flatter surface curvatures and reduced optical power (assuming a constant index of refraction).

Refractive error is quite variable among newborn infants, but the variation decreases with growth

About 28% of infants are myopic at birth,* and the median refractive error is 1.5 D of hyperopia (Figure 3). The large proportion of infants with significant amounts of hyperopia is primarily responsible for variability of refractive errors

*These percentages of myopia are taken from Figure 3 as the percentile at which a cumulative frequency distribution crosses the vertical at zero refractive error; thus any refractive error more negative than zero is myopia. Other percentages of myopia in the literature are lower because they do not include –0.50 D or less as myopia; these small amounts of myopia do not require a refractive correction and, in that sense, don't really count as myopia. In short, using –0.50 D rather than 0.0 D as the criterion at which more negative errors are myopia reduces the percentages dramatically.

Figure 2
Growth of the Eye's Optical Elements
The shaded regions enclose data from various sources on measures of axial length (*a*), anterior chamber depth (*b*), corneal power (*c*), and lens thickness (*d*) from birth to age 14. In all cases, change is most rapid during the first year and slows thereafter. Axial length and anterior chamber depth increase for more than 10 years, but corneal growth is complete by age 3. The apparently biphasic nature of the decrease in lens thickness is probably an artifact from combining several data sets. After the decrease in lens thickness shown here, lens thickness will increase throughout life. (Replotted from Mutti and Zadnick 1997, including data from Larsen 1971a, b, c; Sorsby, Benjamin, and Sheridan 1961; and Zadnick et al. 1993.)

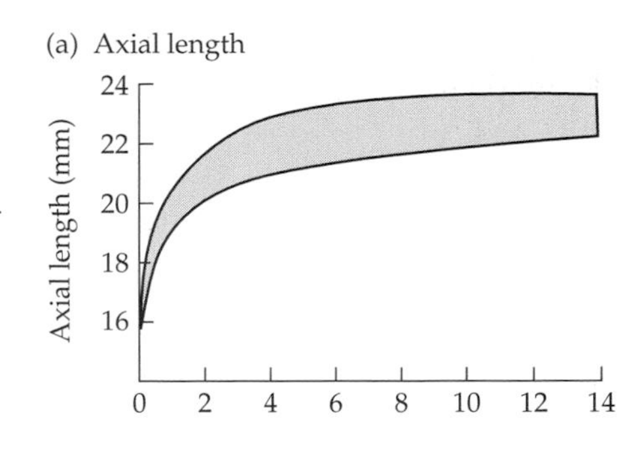

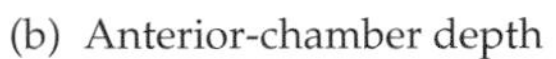

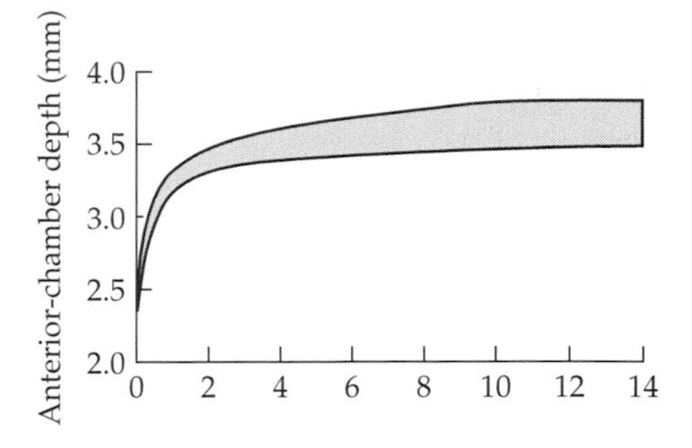

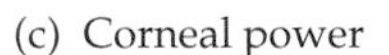

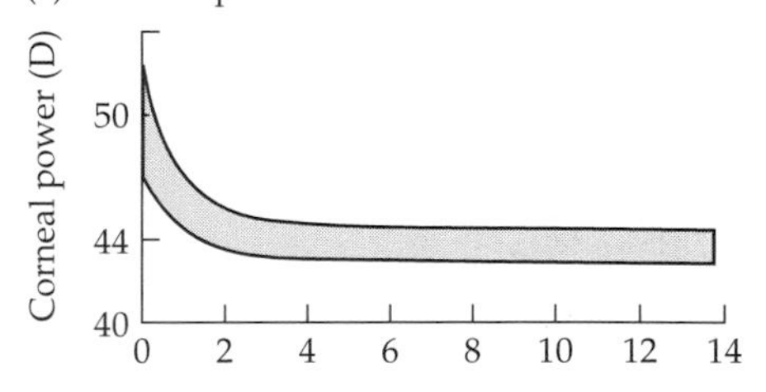

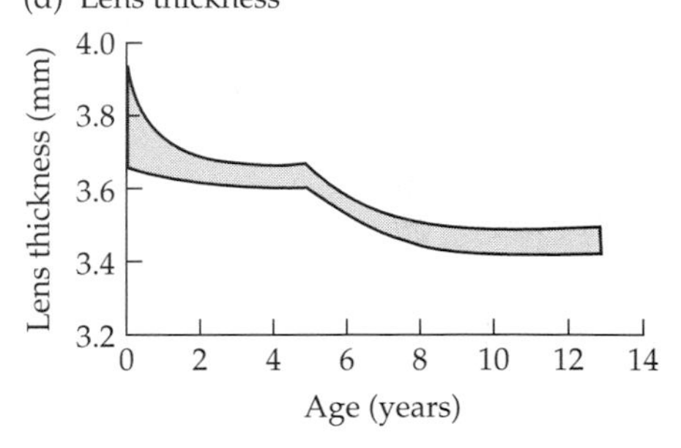

at birth; the range from the tenth to the ninetieth percentiles (i.e., for the middle 80% of infants) is 7 D, from –2 to +5 D.

As the eye grows, the distribution of refractive errors changes dramatically. At primary school age, about 30% of children are myopic, but less than 10% have more than 1 D of myopia. The median refractive error is 0.5 D of hyperopia, and the range from the tenth to the ninetieth percentiles has been reduced to 2 D (–0.7 to +1.3 D). In other words, almost 90% of children 6 or 7 have no significant refractive error (accommodation should compensate for all but the highest hyperopias).

This remarkable reduction in the variability and magnitude of refractive error during the first 6 years of life occurs because infants with hyperopia have become less hyperopic and infants with myopia have become less myopic. The phenomenon, called *emmetropization,* is due to correlated growth of the eye's optical elements. As the optical elements—corneal curvature, lens curvatures, anterior chamber depth—decrease in power with growth, the axial length of the eye keeps pace, elongating at a rate that produces and maintains in-focus images of distant objects.* The rate of axial length elongation required to establish emmetropia must be more rapid in hyperopic eyes than in myopic ones.

Growth of the eye from age 6 until the mid-teens shifts the distribution of refractive errors further so that the median refractive error is zero (i.e., emmetropic) in adult eyes. In so doing, however, about 50% of eyes have some myopia and 20% have more than 1 D of myopia. Most of this new myopia appears during the years of primary school, as I can attest. I was slightly hyperopic in both eyes at age 6 (my father was an optometrist, and I have all the examination records from my youth); by age 11, my left eye had become significantly myopic, and I began to wear spectacles at that time. By age 16, I had significant astigmatism in both eyes. The data in Figure 3 do not show astigmatic changes, but they are much like the changes in the spherical component of the refractive error. Infants exhibit a high prevalence of against-the-rule astigmatism (greater power in the horizontal meridian) that largely disappears by age 6 and then reappears later as with-the-rule astigmatism (greater power in the vertical meridian).

Studies of experimental myopia in primates and chickens indicate that the retina controls growth of the sclera and the rate at which the eye's axial length increases. To exert this control, the retina seems to detect image blur, to distinguish blur produced by hyperopia from that produced by myopia, and to generate a signal that increases or decreases the rate of scleral growth to compensate for the refractive error. How the retina does these things is largely unknown. Its apparent ability to distinguish hyperopic blur from myopic blur is a particularly neat trick; we can't *perceive* any difference.

*The issue is stated as axial length keeping pace with changes in the eye's focal length produced by the optical system rather than the other way around because the axial length is easily altered by changes in the visual environment that have no effect on growth of the cornea and the lens.

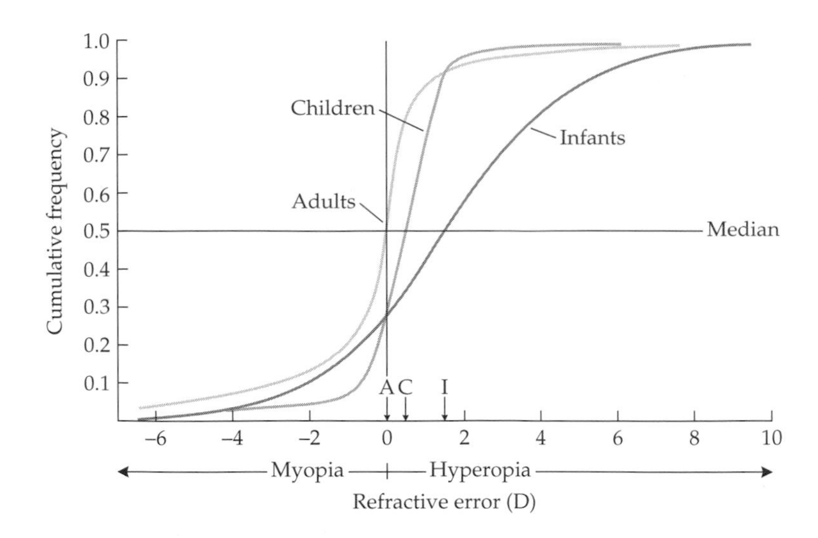

Figure 3
Refractive Errors in Infants, Children, and Adults
The cumulative frequency distribution for refractive error in infants shows much variability around a median refractive error of 1.5 D of hyperopia (arrow labeled "I"). In children aged 6 to 7, the variability is drastically reduced and the median refractive error is about 0.5 D of hyperopia ("C"). The median refractive error for adults is essentially zero ("A"), but the variability has increased somewhat, including a long myopic tail extending beyond the limits plotted here. (Replotted from Cook and Glasscock 1951 [infants]; from Mutti and Zadnick 1997 [children]; and from Stenstöm 1946 [adults].)

The appearance of myopia in juveniles may represent some degree of failure in the generation or the effectiveness of the control signals produced by the retina, and understanding the process of emmetropization might lead to a drug that can prevent the development of myopia. (Attempts to halt the progression of myopia with spectacle lenses meant to minimize accommodative effort or with contact lenses meant to permanently flatten the cornea have not been notably successful.) At present, however, we cannot predict which hyperopic children will develop myopia later, and thus we cannot know which children would be candidates for therapy.

Visual functions mature at different rates during the first 6 years of life

Reliable behavioral measures of an infant's visual perception can be made within a few weeks of birth. Not surprisingly, newborn infants don't see very well in comparison to adults; it takes time for their eyes and visual system to develop their full capability. Grating acuity measured in cycles per degree (see Chapter 2) for an infant 2 weeks old is about 60 times poorer than it is for an adult (1.2 ver-

Newborn

1 month

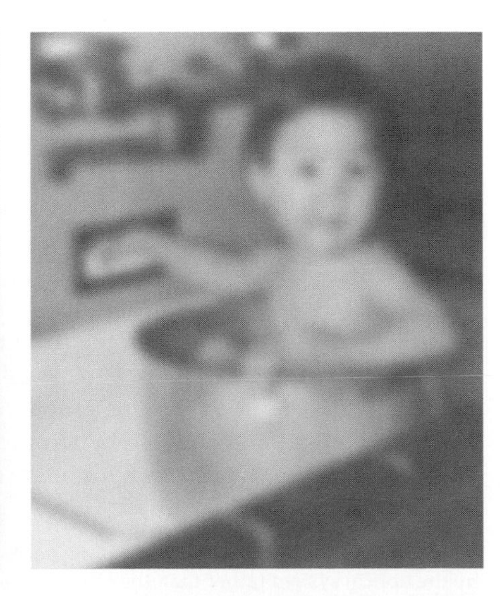

3 months

Figure 4
The Appearance of a Visual Scene to Infants of Different Ages
Newborn infants have very poor visual acuity. Acuity improves to about 20/200 (6/60) at 3 months of age and to about 20/80 (6/24) at 6 months. At 3 years of age, acuity will be about 20/40 (6/12) and will require another 2 or 3 years to reach the adult level. (After Teller 1997; photograph by Henry M. Takahashi, images by Tony Young, © 1998.)

sus 60 cycles/degree). Figure 4 incorporates data on the development of visual acuity to show how a scene would appear to infants and toddlers from 1 month to 3 years of age. For the newborn infant, the world is indistinct and shadowy, but vision improves sufficiently over the next 6 months that faces should be recognizably different, even at several meters distance. But adult visual acuity requires years to acquire. By age 3, acuity is about half the adult value; it reaches the adult level around age 5 or 6.

The experimental data support a rule of thumb for the development of visual acuity: Acuity in cycles per degree equals age in months. Since the adult level is around 60 cycles/degree, it should be attained at age 60 months, or 5 years.

The lengthy developmental period for visual acuity is not due to refractive error or to any other deficiency in the eye's optical system. Most of the refractive errors are hyperopia, and they can be offset by accommodation, which appears to be functioning after the third postnatal month. In addition, a full-term infant's optical media are fully transparent, so the retinal image quality should permit good spatial resolution.

Immaturity of the fovea at birth underlies the infant's initially poor acuity and its subsequent slow development. At birth, foveal cones are morphologically immature, and their density (18,000 cells/mm^2) is about ten times lower than the adult value (181,500 cells/mm^2); by 4 years of age, density is more than 100,000 cells/mm^2 and still increasing. The age at which adult cone density is reached is not known precisely, but it is probably between 5 and 7 years, the age range within which adult acuity levels are attained. The increasing cone density in the fovea during the early years of life is a result of cone migration into the fovea and nonuniform expansion of the retina as the eye grows (see Chapter 16).

Wavelength discrimination is poor at birth, and this deficiency is also likely to be related to the immaturity of the cones. It improves rapidly, however, and infants can make reliable discriminations between red and green at 3 months of age. Discriminations involving the blue cones appear somewhat later, but by 1 year of age, the infant's color vision appears to be like that of adults.

Temporal sensitivity, as measured by the ability to detect rapidly flickering stimuli, is near adult values by 1 month of age and fully adult a month later. This early maturation probably reflects the relative maturity of the extrafoveal retina in comparison to the fovea. Rods outside the fovea are mature at birth; the infant's scotopic sensitivity function (see Chapter 13) is adultlike at 1 month of age and probably is so from birth.

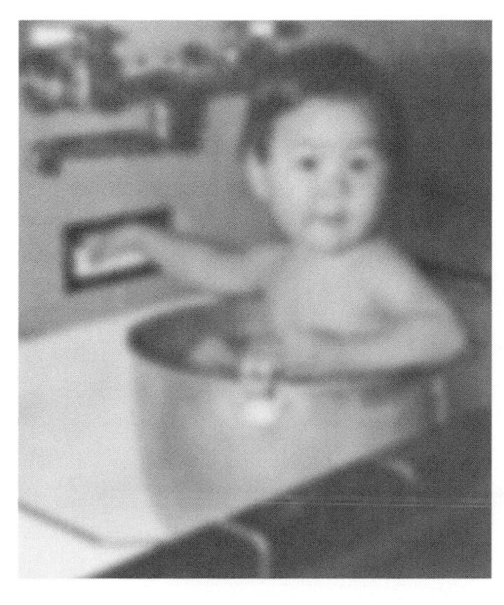

6 months

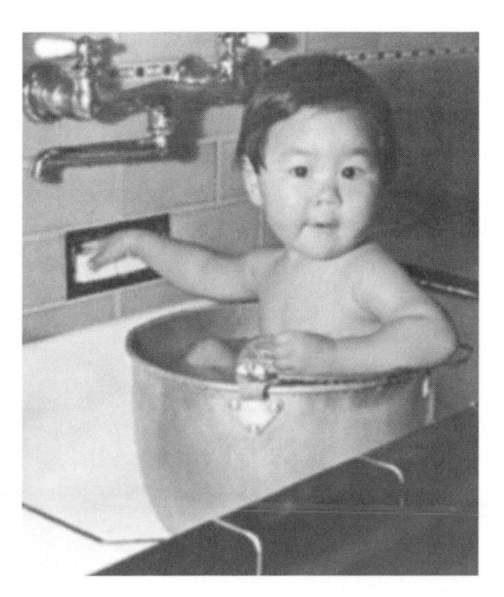

3 years

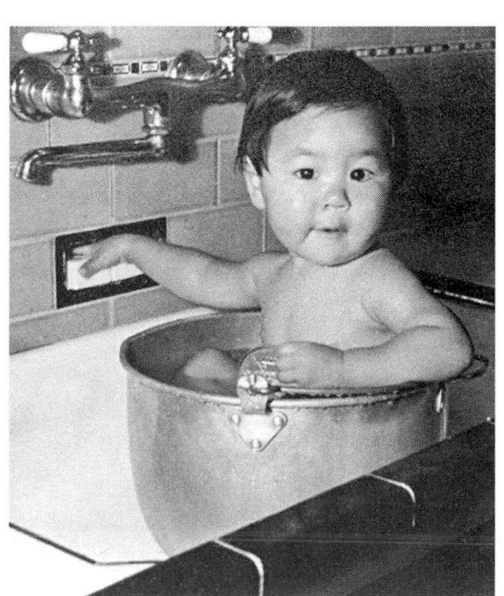

Adult

The various patterns of eye movements develop during the first year of life. Optokinetic nystagmus can be elicited at birth, and newborn infants make obvious, but poorly directed saccades. Saccades are directed to objects of interest with reasonable accuracy by 3 months of age and with accuracy approaching that of adults by 1 year of age. Newborn infants cannot make smooth pursuit movements unless the object of interest is large and moving very slowly, and the pursuit movements are often accompanied by head movements. By 1 year of age, pursuit movements are nearly like those of adults. Vergences and foveal fixation of a target with both eyes begin to appear during the second month and become consistently accurate by 6 months of age.

Most of the numerous cues the visual system uses to judge the relative distances of objects in space are monocular, and some may be used by infants. But stereopsis, which is the more accurate binocular judgment of relative depth, is not present until the third month. Stereo acuity increases rapidly, however, reaching 80% of the adult level by 6 months of age.

The visual system is relatively plastic as it is developing in infancy, and it is susceptible to abnormalities in the visual environment. The most common abnormalities are substantial differences in the images in the two eyes produced by unilateral congenital cataracts, strabismus, or very different refractive errors in the two eyes (anisometropia). Visual functions such as spatial resolution or stereopsis have **critical periods**, which are the times during development at which a visual function can be adversely affected by an altered visual environment. The onset and duration of critical periods for different visual functions in humans are not known precisely, except that some begin by 6 months of age and some may last 6 or 7 years. In the face of this uncertainty, the usual strategy is correction of the abnormality as soon as possible. Thus, surgery for congenital cataract or for strabismus not related to accommodation may be done on infants 1 year old or less. Surgery early in the critical period is meant to forestall the development of amblyopia (reduced spatial resolution in one eye).

In general, the plasticity underlying critical periods for binocular visual functions resides in the central visual pathways, not in the eyes themselves. A critical period for the development of myopia, on the other hand, depends very much on events in the eye because of the retina's role in the control of ocular growth. Amblyopia probably involves effects in both the retina and the central visual pathways.

Changes in the lens and vitreous that begin in infancy continue throughout life

The lens grows rapidly for several years after birth, then at a more leisurely pace for a lifetime (see Figures 12.10 and 12.12). No other component of the eye does this. The continual growth of the lens is accompanied by continual change; its ability to change its shape in response to changing tension imposed by the zonule declines almost from the earliest age at which it can be measured (see Figure 11.23). Less obviously, proteins aggregate to produce a steady, lifelong increase in the optical density of the lens (Figure 5). As we will see shortly, the cumulative effect of this change affects vision adversely in almost all older eyes; an exaggeration of the normal density increase will be cataracts in some eyes.

Lifelong change is also characteristic of the vitreous. Although it probably has a very small liquid component at birth, the liquid fraction is measurable at age 3 and increases steadily thereafter (see Figure 12.32), eventually making up just over half of the vitreous volume. As the liquid fraction increases, the collagen in the gel apparently aggregates. This aggregation, begun in infancy and continued thereafter, eventually leads to vitreal detachment in most older eyes.

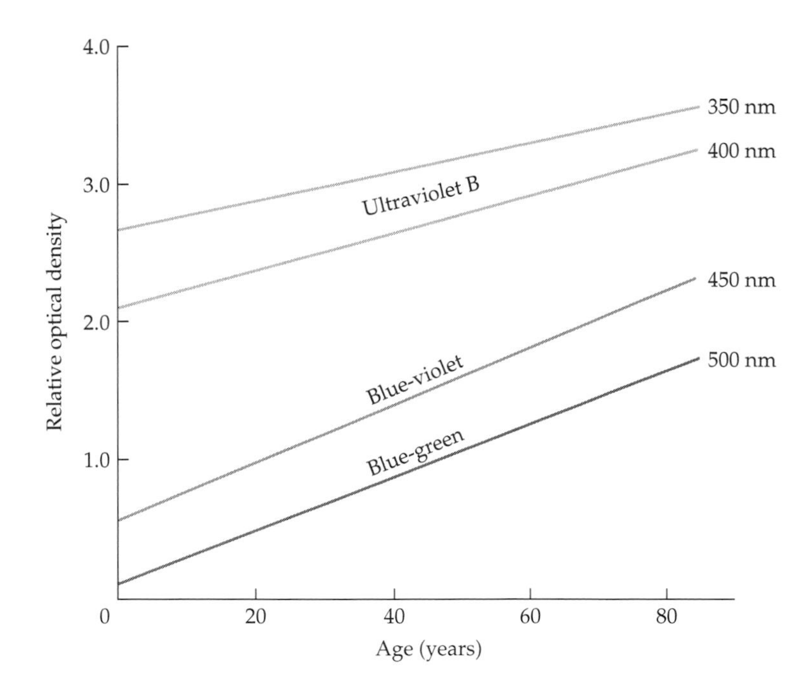

Figure 5
Optical Density of the Lens as a Function of Age
The lines are regression lines calculated from absorbance measurements of the lens at different wavelengths. (Data points have been omitted.) Optical density increases significantly with age in both the ultraviolet and the visible parts of the spectrum. The slopes of the regression lines vary somewhat, but the increase in optical density is roughly 1.0 density unit in 40 years. (Replotted from Weale 1988.)

Maturation and Senescence

The average refractive error is stable from ages 20 to 50, but the eye becomes more hyperopic and then more myopic later in life

Figure 3 showed how the mean hyperopia in newborn eyes lessens with eye growth to become more nearly emmetropic. That reduction of early hyperopia can also be seen in Figure 6, by comparing the refractive error at birth (+2.3 D) with that at age 20 (+1.2 D).* The ages between 20 and 50 represent an ocular interlude, a period of relative stability between the early years of growth and the senescent changes in the aging eye. From age 20 to age 50, the average spherical refractive error varies by little more than 0.5 D. Visual acuity is at its maximum and stable throughout this period. For most eyes, this 30-year interval is their time of prime maturity, during which they reach and maintain a broad plateau of functional excellence.

But stability in mean refractive error and visual acuity notwithstanding, the lens has been changing constantly since birth, as evidenced by the onset of presbyopia at about age 45. (Since refractive error is defined in terms of viewing distant objects, it is unaffected by the loss of accommodative amplitude, no matter how dramatic and significant presbyopia seems to the person affected.) The 50-year-old lens is about ten times less transparent than an infant's lens, and the refractive index may also be changing. In the absence of other ways to explain the increase in hyperopia after age 50, a decrease in the refractive index of the lens seems to be responsible (corneal curvature, anterior chamber depth, and axial length do not change significantly). Visual acuity begins to decline after age 55.

The hyperopic change peaks at about age 65, after which the average change is in the direction of myopia. For individuals over 90 years of age, the mean refractive error is on the myopic side (–0.4 D). The myopic shift, like the hyperopic shift that preceded it, is apparently due to a refractive index change in the

*The increase in hyperopia to a maximum at age 7, followed by a steady decline, however, was not evident in the earlier data (see Figure 3), and it is probably not accurate; the maximum occurs at birth.

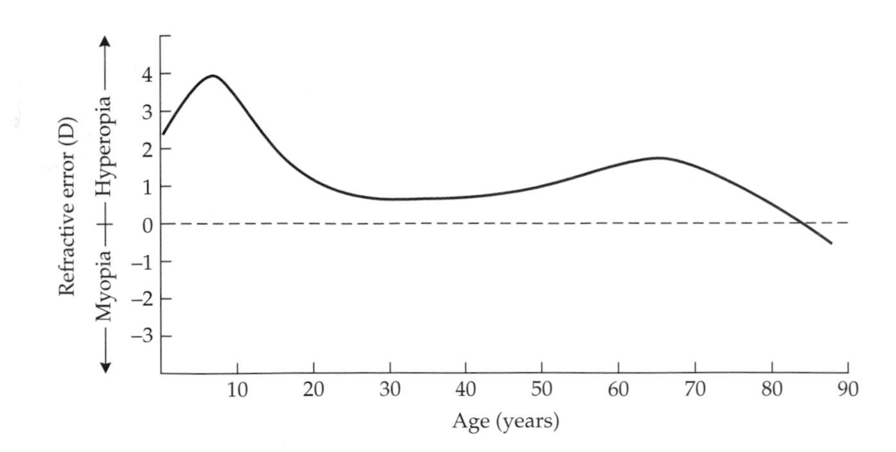

Figure 6
Variation of Mean Refractive Error from Birth to Old Age
Mean refractive error is the "spherical equivalent" refractive error; in the presence of astigmatism, it is the mean of the errors in the meridians of highest and lowest power. (Plotted from data in Slataper 1950.)

lens, but in this case it is an increase. This index change is accompanied by a continuing loss of transparency and a steady decline in visual acuity. The transparency loss is a normal change that is not large enough to be classified as a cataract, although it may become a cataract eventually. In other words, most lenses become optically more dense and therefore less transparent with age. Some become much more dense; these are the cataractous lenses. Fortunately, the variability works both ways. Some people in their 80s have eyes that are childlike in their clarity.

Although the gross structure of the eye is stable after the age of 20, tissues and membranes are constantly changing

Some of the cellular tissues in the eye consist of cells that have relatively short lives. As a result, the tissues are maintained by steady replacement of cells that die in the normal process of apoptosis. The corneal epithelium is the best-known example; cells have a lifetime of 10 days or so, and they are replaced by mitosis and by migration into the cornea from the stem cell pool that resides in the limbal conjunctiva (see Figure 8.38). Because of the continual replacement and renewal of the tissue, the corneal epithelium and its cells do not vary in appearance or size with age. The vascular endothelium in the blood vessels of the eye is also replaced on a regular basis, every month or so.

Other cells have much longer lives. Individual fibroblasts in the cornea and sclera probably survive for many years, for example, but it is not known if they are replaced after they die. The increased cell death in the anterior corneal fibroblasts trigged by epithelial damage (see Chapter 8) appears to be offset by migration from the surviving population rather than by mitosis. Thus the population of fibroblasts is likely to become smaller with age. Cell loss with age is characteristic of the corneal endothelium because cells lost through normal cell death are not replaced; the survivors expand and migrate to plug the gaps in the endothelial mosaic. The number of endothelial cells declines by about half from birth to age 90. The lifetimes of the trabecular endothelial cells and the epithelia on the iris and ciliary body are not known, but they are probably quite long; their longevity would account for their morphological deterioration with advancing age. The morphological changes in the ciliary body epithelium are accompanied by a decrease in the rate of aqueous production, but it is not known if these phenomena are causally linked.

Lens cells are permanent structures, but they change with age. The older cells in the lens nucleus have lost their nuclei, their membranes have become irregular and distorted, and their biochemical activity has slowed; they are dead, or nearly so. The stem cells in the lens epithelium, however, seem to be perennially youthful.

Most retinal cells are also permanent. The number of pigment epithelial cells is stable throughout life, although there are density changes from one region to the next as the eye enlarges during youth and adolescence. After the prenatal wave of ganglion cell death has subsided, there is no convincing evidence for any further loss of ganglion cells over an individual's lifetime. Except for the photoreceptors, the other retinal neurons are assumed to be permanent as well. Photoreceptors differ from other neurons by continually renewing their outer segments and replacing all of the membranes and photopigment every few weeks (see Chapter 13). The photoreceptor inner segment, nucleus, axon, and terminal are long-lived, but not necessarily permanent. The rod population decreases steadily after age 30 (see Figure 9*b*).

Bruch's membrane in the choroid thickens throughout life, as does Descemet's membrane in the cornea, and the characteristic five-layered structure of Bruch's membrane becomes less well defined as collagen and material from outside are deposited in its inner layers. Locally thick deposits in Bruch's membrane, called drusen, are common in older eyes. The most important change, however, is the development of breaks or discontinuities in Bruch's membrane; edema is often the result.

All of the intraocular muscles acquire more connective tissue with age. Suggestions that this additional connective tissue is related to a decreased contractile ability in the ciliary muscle (accounting for presbyopia) and the dilator (accounting for the age-related miosis to be considered in the next section) are unlikely to be correct; both the loss of accommodative amplitude and the age-related decrease in pupil size begin long before increased connective tissue is apparent in the muscles.

Retinal illuminance and visual sensitivity decrease with age

The normal pupil size for a given level of ambient illumination decreases with age (Figure 7). Where the average pupil diameter at full dark adaptation is about 7 mm in a 20-year-old eye, in an 80-year-old eye it is only about 4.2 mm. This proportional difference (about 60%) holds for other illumination conditions and adaptation levels because reflex pupil constriction does not change much with age. Thus the minimum pupil diameter at age 80 is about 1.2 mm (compared to 2 mm at age 20).

A pupil with a diameter 60% that of a pupil in a 20-year-old eye will have just over one-third the area of the young pupil. Since retinal illumination is the prod-

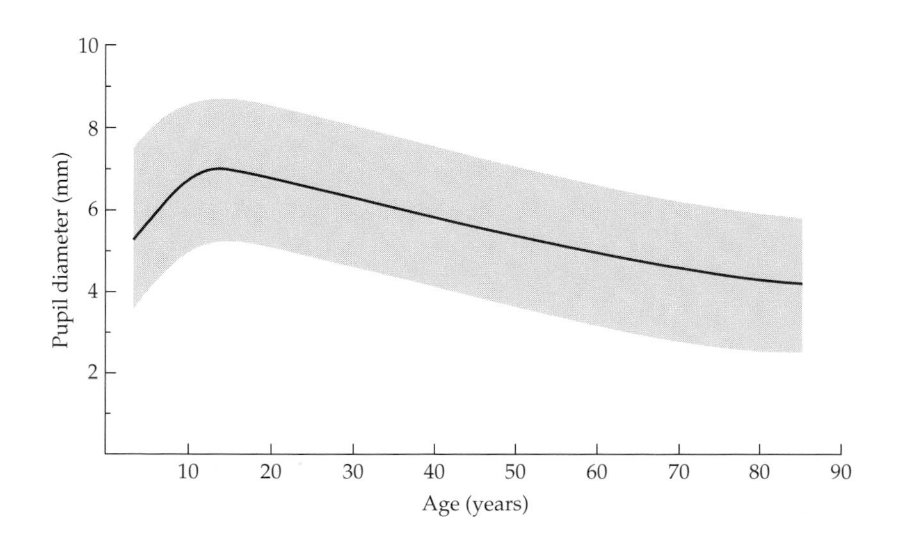

Figure 7
Pupil Diameter as a Function of Age
The shaded region encloses all but a few outlying data points for pupil diameter in darkness from 1263 individuals. At any age, pupil diameter varies in the population by about 4 mm. The line through the center of the region connects the means for different age groups. After an initial increase in pupil size from birth to about age 15, pupil size decreases slowly with advancing age. The mean pupil diameter decreases by about 0.4 mm/decade after age 15. (Replotted from Loewenfeld 1979.)

uct of luminance and pupil area (see Chapter 10), it follows that the retinal illumination will be reduced threefold in the elderly eye. Add to this a substantial increase in the optical density of the lens (around tenfold) and we have at least a 30-fold reduction of retinal illumination in the 80-year-old eye. The effect of reduced retinal illumination is significant; although the combination of decreased pupil size and increased lenticular density in my eyes is unlikely to produce a 30-fold reduction of retinal illumination, I have been using more and higher-wattage bulbs for illumination in my home since I was 50.

Older persons see less well at night than younger persons. For fully dark-adapted eyes, the absolute threshold at age 80 is 1.5 to 2.0 log units higher than at age 20 (Figure 8). Although changes in the pupil and the lens can account for most of the change in sensitivity, they may not be the only factors affecting sensitivity. In particular, a loss of rods may contribute.

Figure 9 shows the total number of cones within the fovea and the total number of rods within 4 mm of the foveal center, both as functions of age. The cone population is stable from ages 30 through 90 (Figure 9*a*), but the rod population decreases dramatically during this age range (Figure 9*b*). Sightly more than 2 million rods, about one-third of the total in this part of the retina, are lost during this time. We do not know what is happening at younger ages, but from age 30 on, the central 15° of the retina loses 100 rods each day. The remaining rods compensate for these lost cells by enlarging their inner segments slightly, thereby maintaining their tiling.

In the peripheral retina near the equator, however, the situation is different. Cone density declines modestly but significantly, in a statistical sense, and the rods show little change with age. Since the cone density in this part of the retina is quite low (see Figure 15.11), the small decrease in cone density is probably of no functional consequence.

Although the large loss of rods from the retina would seem to be an obvious cause of some sensitivity loss, the issue depends more on the rhodopsin density than on rod density. Assuming that the expansion of the rod inner segments to compensate for rod losses is accompanied by outer segment expansion and a larger complement of rhodopsin in each rod, the rhodopsin density may remain high enough to to keep the retina's sensitivity constant with age. We do not know how good the compensation is, however, so we also do not know the extent to which rod loss contributes to sensitivity loss.

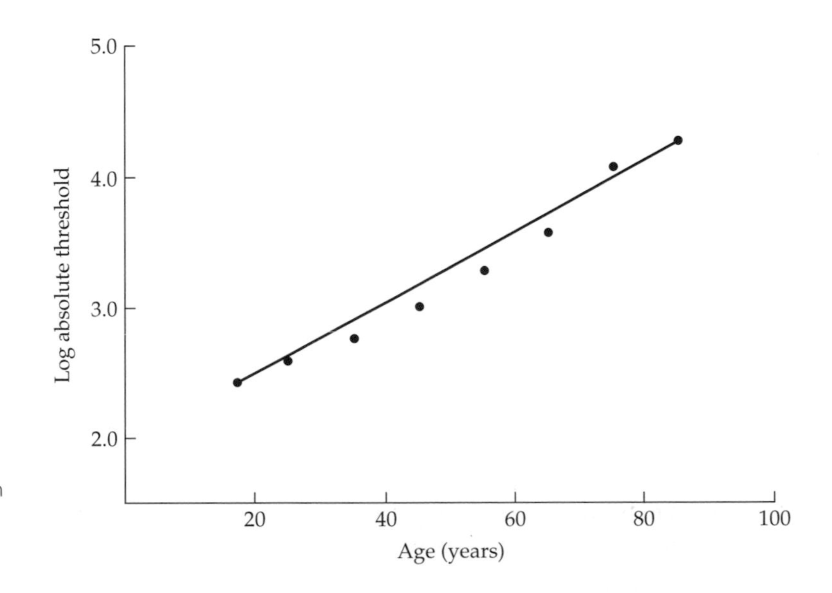

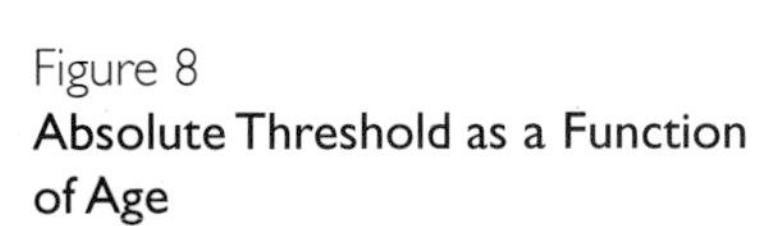

Figure 8
Absolute Threshold as a Function of Age
The data points are the thresholds after 40 minutes of dark adaptation; the straight line through them was fitted by eye. Sensitivity declines (threshold increases) about tenfold (1 log unit) in 30 years. (Plotted from data in Domey, McFarland, and Chadwick 1960.)

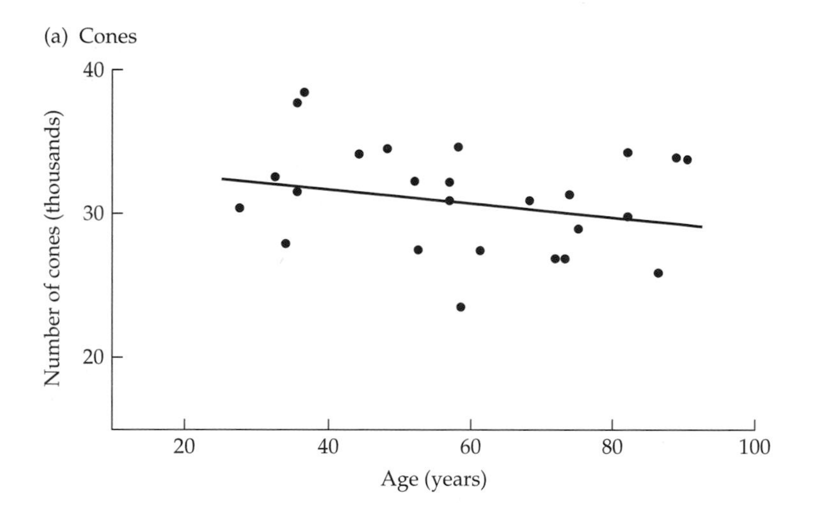

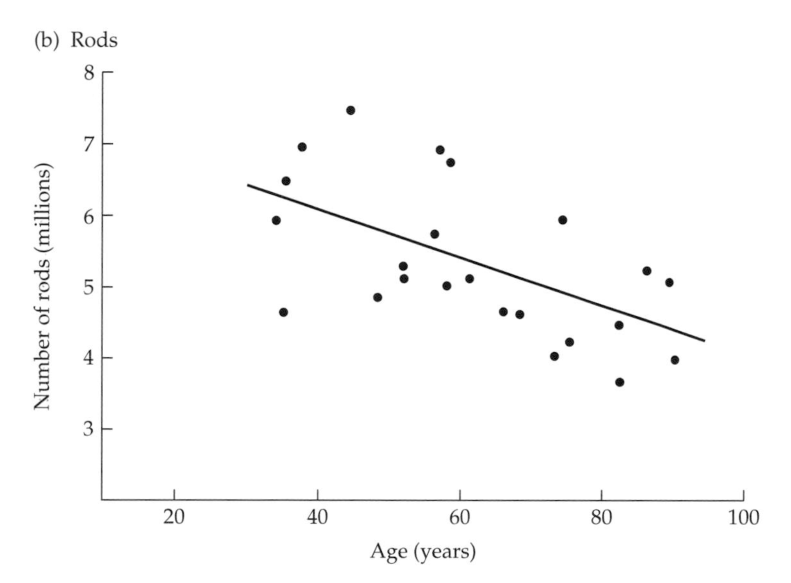

Figure 9
Number of Cones and Rods in Central Retina as a Function of Age
(*a*) The number of cones in the center of the fovea does not change significantly with age. (The slope of the regression line is not significantly different from zero.) (*b*) The number of rods within a region 8 mm in diameter centered on the fovea decreases markedly with age, however. The slope of the regression line is –0.034, which means that 34,000 rods are lost each year from this region. (Replotted from Curcio et al. 1993.)

Visual acuity declines after age 50, largely because of optical factors

Even after refractive errors have been corrected, visual acuity begins to decline slowly in the mid-50s, then declines more rapidly with increasing age (Figure 10). Although the data include individuals with cataracts, macular degeneration, and other pathologies, removing these individuals from the sample does not make much difference; acuity declines with age, pathology or no. Half of the people 80 years of age will have visual acuities less than 20/40 (6/12). Ten percent will be legally blind. By age 85, half will have acuities less than 20/80 (6/24).

Viewed more positively, half of the people 80 years of age will have acuities better than 20/40 and 10% better than 20/25. In fact, the tabulated data on which Figure 10 is based indicate that 1.5% of people over the age of 80 have visual acuities of 20/15 (6/4.5).

Much of the decline of visual acuity with age can be attributed to transmission losses in the eye's optical media. Contrast sensitivity functions (see Chapter 2) for older eyes are typically reduced in comparison to those for young eyes. But when sinusoidal interference fringes are generated on the retina, a technique that eliminates contrast reduction by the eye's optical system, old and young subjects exhibit similar contrast sensitivity functions. The latter observation means, first, that neural impairment plays no role (if it did, the sensitivity functions for interference fringes would not be the same in young and old eyes) and, second, that

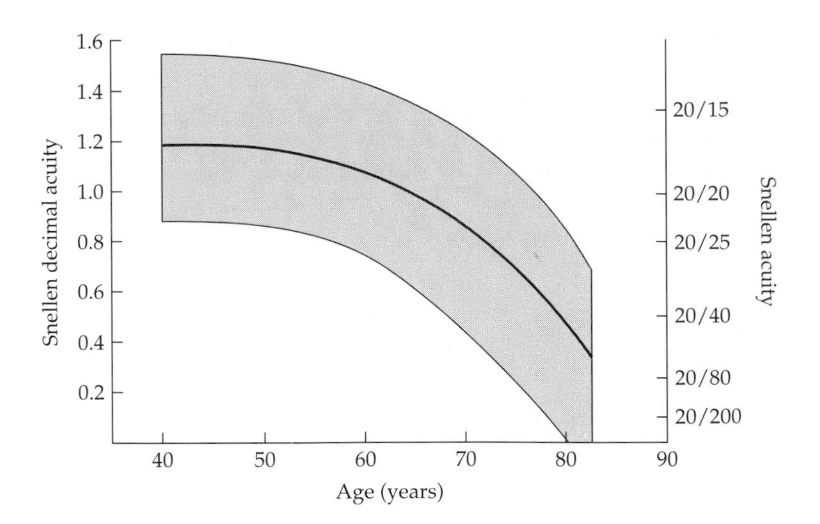

Figure 10
Visual Acuity after Age 40
Acuity is given as the Snellen decimal, which is the numerator of the Snellen fraction divided by the denominator. Thus 20/20 (6/6) becomes 1.00, 20/40 (6/12) becomes 0.50, and so on. (Some Snellen fractions are indicated along the right-hand edge of the graph.) The shaded region contains the central 80% of eyes; that is, it is bounded by the tenth and ninetieth percentiles of the acuity distributions at each age. The heavy, dark line in the middle of the shaded region traces the median acuity at each age. (Replotted from Weymouth 1960.)

the reduced contrast sensitivity for grating targets in object space is largely an optical effect. The optical effect in question is the loss of transmission (increase in optical density) in the lens. The reduced retinal illumination produced by the smaller pupils in older eyes has little effect when the mean luminance is high (photopic), and it may improve contrast sensitivity slightly by reducing the effects of aberrations.

Contrast sensitivity is reduced in young and old eyes at lower luminance levels, but the old eyes are more affected by the luminance reduction. The heavier impact of reduced luminance on the contrast sensitivity function in older eyes is difficult to explain by optical effects. Increased optical density in the lens of older eyes, for example, should have a constant effect, regardless of luminance level. By default, a neural deficit seems to be contributing to the exaggerated contrast sensitivity loss at low luminance levels.

The nature of the neural deficit is not clear. Even at low luminance levels, the contrast sensitivity functions are the result of cone activation, and the number of foveal cones does not change appreciably with age. Another possibility might be a loss of ganglion cells with age, but one would expect its effect to be obvious when the optics are bypassed by using interference fringes, and there is no effect. In any case, the most recent study of axon numbers in the optic nerve from eyes of different ages shows no significant age-related loss. At present, it seems that any neural deficit affecting spatial resolution must reside in the central visual pathways, but this story is by no means complete.

Decreased visual acuity in the elderly raises a practical concern in our automobile-dominated society. If elderly people are healthy and active, they want to drive because it allows them tc lead independent lives. But they have the highest rate of accidents per mile driven, and some part of this increased accident rate is likely to be related to their diminished visual capabilities. Visual acuity may may not be the most important visual function in driving an automobile, but it typifies and calls attention to the decline in other visual functions. Visual processing speed, visual acuity for moving targets, depth perception, and visual search procedures are some of the more complex functions that also exhibit age-related impairment. Although many people recognize these diminished abilities in themselves and voluntarily limit their driving or stop driving altogether, others do not. In short, the decrease in visual functions with age is not just a problem for the individuals affected; it is a problem for society.

Figure 11 shows the scene from Figure 4 as it would appear to adults from ages 60 to 90 who have the median value of visual acuity for their age. The views through eyes 80 years old and 3 years old (see Figure 4) are regarded as equivalent, as are the views through eyes 85 years old and 6 months of age. The next step is not known for certain, but I have equated the eye at age 90 with an eye approximately 3 months of age, in which the acuity is about 20/200 (6/60). This is legally blind, but as Figure 11 shows, it is good enough for a pair of elderly eyes to see their newest great-grandchild (and for the infant to see the great-grandparent). The hope is that a lifetime of seeing has been rewarded with insight that more than compensates for a lack of optical clarity.

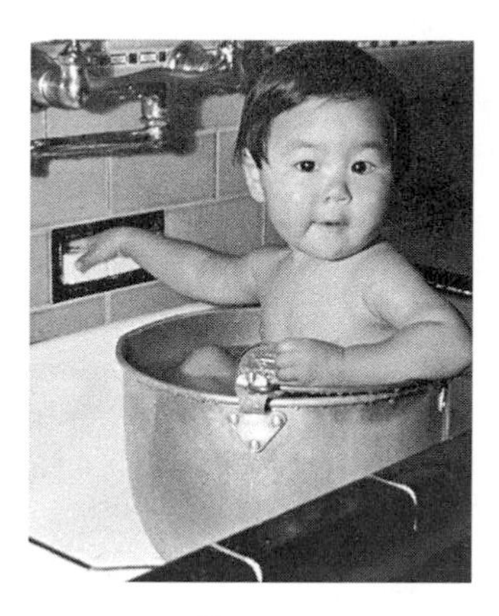

60 years

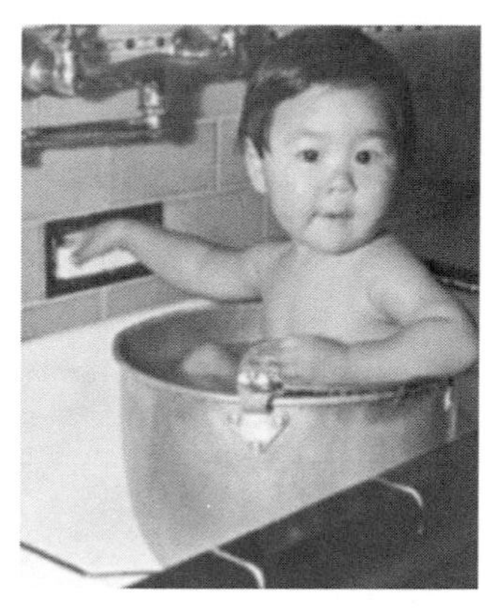

80 years

85 years

90 years

Figure 11
The Appearance of a Visual Scene to Older Adults of Different Ages
Compared to the average visual acuity at age 60, which is better than 20/20, acuities at later ages are reduced. The views for ages 80, 85, and 90 correspond roughly to acuities of 20/40 (6/12), 20/80 (6/24), and 20/200 (6/60), respectively. These views represent only the effects of deterioration in the eye and take no account of any perceptual compensation or of any deterioration in the perceptual apparatus.

References and Additional Reading

Postnatal Growth and Development

Cook RC and Glasscock RE. 1951. Refractive and ocular findings in the newborn. *Am. J. Ophthalmol.* 34: 1407–1413.

Curcio CA and Hendrickson AE. 1991. Organization and development of the primate photoreceptor mosaic. *Prog. Ret. Res.* 10: 89–120.

Daw NW. 1995. *Visual Development*. Plenum, New York.

Eustis HS. 1995. Postnatal development. Chapter 2, pp. 45–59, in *Pediatric Ophthalmology and Strabismus*, Wright KW, ed. Mosby, St. Louis.

Gwiazda J, Mohindra I, Brill S, and Held R. 1985. The development of visual acuity in infant astigmats. *Invest. Ophthalmol. Vis. Sci.* 26: 1717–1723.

Hung L-F, Crawford MJL, and Smith EL. 1995. Spectacle lenses alter eye growth and the refractive status of young monkeys. *Nature Med.* 1: 761–765.

Larsen JS. 1971a. The sagittal growth of the eye. I. Ultrasonic measurement of the depth of the anterior chamber from birth to puberty. *Acta Ophthalmol. (Copenh.)* 49: 239–262.

Larsen JS. 1971b. The sagittal growth of the eye. II. Ultrasonic measurement of the axial diameter of the lens and the anterior segment from birth to puberty. *Acta Ophthalmol. (Copenh.)* 49: 427–440.

Larsen JS. 1971c. The sagittal growth of the eye. IV. Ultrasonic measurement of the axial length of the eye from birth to puberty. *Acta Ophthalmol. (Copenh.)* 49: 873–886.

Lind GJ, Chew SJ, Marzani D, and Wallman J. 1998. Muscarinic acetylcholine receptor antagonists inhibit chick scleral chondrocytes. *Invest. Ophthalmol. Vis. Sci.* 39: 2217–2231.

Mutti DO and Zadnick K. 1997. Biometry of the eye in infancy and childhood. Chapter 4, pp. 31–46, in *Eye Care for Infants and Young Children*, Moore BD, ed. Butterworth-Heinemann, Boston.

Mutti DO, Zadnick K, and Adams AJ. 1996. Myopia: The nature versus nurture debate goes on. *Invest. Ophthalmol. Vis. Sci.* 37: 952–957.

Simons K, ed. 1993. *Early Visual Development, Normal and Abnormal*. Oxford University Press, New York.

Sorsby A, Benjamin B, and Sheridan M. 1961. *Refraction and Its Components during the Growth of the Eye from the Age of Three* (Medical Research Council special report series, no. 301). Her Majesty's Stationery Office, London.

Stenström S. 1946. Untersuchungen über die Variation and Kovariation der optische Elemente des menschlichen Auges. *Acta Ophthalmol. (Copenh.)* Suppl. 26.

Teller DY. 1979. The forced-choice preferential looking procedure: A psychophysical technique for use with human infants. *Infant Behav. Dev.* 2: 135–153.

Teller DY. 1997. First glances: The vision of infants. *Invest. Ophthalmol. Vis. Sci.* 38: 2183–2203.

Wallman J. 1993. Retinal control of eye growth and refraction. *Prog. Ret. Res.* 12: 133–153.

Wallman J and McFadden S. 1995. Monkey eyes grow into focus. *Nature Med.* 1: 737–739.

Weymouth FW. 1963. Visual acuity of children. Chapter 4, pp. 119–143, in *Vision of Children*, Hirsch MJ and Wick BE, eds. Chilton, Philadelphia.

Zadnick K, Mutti DO, Friedman NE, and Adams AJ. 1993. Initial cross-sectional results from the Orinda longitudinal study of myopia. *Optom. Vis. Sci.* 70: 750–758.

Maturation and Senescence

Adams AJ, Wong LS, Wong L, and Gould B. 1988. Visual acuity changes with age: Some new perspectives. *Am. J. Optom. Physiol. Optics* 65: 403–406.

Artal P, Ferro M, Miranda I, and Navarro R. 1993. Effects of aging in retinal image quality. *J. Opt. Soc. Am. (A)* 10: 1656–1662.

Burton KB, Owsley C, and Sloane ME. 1993. Aging and neural contrast sensitivity: Photopic vision. *Vision Res.* 33: 939–946.

Curcio CA, Millican CL, Allen KA, and Kalina RE. 1993. Aging of the human photoreceptor mosaic: Evidence for selective vulnerability of rods in central retina. *Invest. Ophthalmol. Vis. Sci.* 34: 3278–3296.

Domey RG, McFarland RA, and Chadwick E. 1960. Threshold and rate of dark adaptation as functions of age and time. *Hum. Factors* 2: 109–119.

Feeney-Burns L and Ellersieck MR. 1985. Age-related changes in the ultrastructure of Bruch's membrane. *Am. J. Ophthalmol.* 100: 686–697.

Harman AM, Fleming PA, Hoskins RV, and Moore SR. 1997. Development and aging of cell topography in the human retinal pigment epithelium. *Invest. Ophthalmol. Vis. Sci.* 38: 2016–2026.

Loewenfeld IE. 1979. Pupillary changes related to age. Chapter 2.2, pp. 124–150, in *Topics in Neuro-Ophthalmology*, Thompson HS, ed. Williams and Wilkins, Baltimore.

Morgan MW. 1988. Vision through my aging eyes. *J. Am. Optom. Assn.* 59: 278–280.

Owsley C, Ball K, McGwin G, Sloane ME, Roenker DL, White MF, and Overley ET. 1998. Visual processing impairment and risk of motor vehicle crash among older adults. *JAMA* 279: 1083–1088.

Ramrattan RS, van der Schaft TL, Mooy CM, de Bruijn WC, Mulder PGH, and de Jong PTVM. 1994. Morphometric analysis of Bruch's membrane, the choriocapillaris, and the choroid in aging. *Invest. Ophthalmol. Vis. Sci.* 35: 2857–2864.

Repka MX and Quigley HA. 1989. The effect of age on normal human optic nerve fiber number and diameter. *Ophthalmology* 96: 26–31.

Shipp MD and Penchansky R. 1995. Vision testing and the elderly driver: Is there a problem meriting policy change? *J. Am. Optom. Assn.* 66: 343–351.

Slataper FJ. 1950. Age norms of refraction and vision. *Arch. Ophthalmol.* 43: 466–481.

Sloane ME, Owsley C, and Alvarez SL. 1988. Aging, senile miosis and spatial contrast sensitivity at low luminance. *Vision Res.* 28: 1235–1246.

Weale RA. 1988. Age and the transmittance of the human crystalline lens. *J. Physiol.* 395: 577–587.

Weale RA. 1992. *The Senescence of Human Vision*. Oxford University Press, Oxford, England.

Werner JS, Peterzell DH, and Scheetz AJ. 1990. Light, vision, and aging. *Optom. Vis. Sci.* 67: 214–229.

Weymouth FW. 1960. Effect of age on visual acuity. Chapter 4, pp. 37–62, in *Vision of the Aging Patient*, Hirsch MJ and Wick RE, eds. Chilton, Philadelphia.

Yee RW, Matsuda M, Schultz RO, and Edelhauser HF. 1985. Changes in the normal corneal endothelial cell pattern as a function of age. *Curr. Eye Res.* 4: 671–678.

Glossary

"But 'glory' doesn't mean 'a nice knock-down argument,' " Alice objected.

"When I use a word," Humpty Dumpty said, in rather a scornful tone, "it means just what I choose it to mean—neither more nor less."

"The question is," said Alice, "whether you can make words mean so many different things."

"The question is," said Humpty Dumpty, "which is to be master—that's all."

■ Lewis Carroll, *Through the Looking-Glass*

A-scan The basic mode of ultrasonography, in which echoes from ultrasound beams are displayed as pulses along a time axis.

a-wave A component of the electroretinogram produced by summed signals from photoreceptors.

AII amacrine cell A small amacrine cell that receives inputs from rod bipolar cells and passes those signals to both ON and OFF cone bipolar cells.

abducens nerve Cranial nerve VI, which innervates the ipsilateral lateral rectus muscle.

abduction Horizontal rotation of the eye temporally (away from the nose). Contrast *adduction.*

absorbance spectrum A measure of the amount of light absorbed by a photopigment as a function of the wavelength of light.

AC/A ratio The amount of accommodative convergence (AC) that accompanies 1 D of accommodation (A).

accessory lacrimal glands The glands of Krause and Wolfring, which are distributed along the fornix of the upper eyelid.

accessory optic system The medial, lateral, and dorsal terminal nuclei, along with the bundles of retinal ganglion cell axons that terminate in the nuclei.

accommodation The increased optical power of the lens that occurs when the eye shifts its gaze from a distant to a near object.

accommodative vergence Horizontal movement of the eyes in opposite directions as accommodation changes.

acetylcholine An excitatory neurotransmitter employed by the parasympathetic nervous system, by motor nerves, and by some neurons in the central nervous system, including the retina.

actin A cytoskeletal protein found in many cells, used in conjunction with myosin for cell movement and muscle contraction.

action potential A brief (1 msec) change in electrical potential across a nerve cell membrane that is propagated along the cell's axon to the axon terminal.

adduction Horizontal rotation of the eye nasally (toward the nose). Contrast *abduction.*

Adie's syndrome A condition in which the pupil is dilated and pupillary light reflexes are reduced or absent. The cause is unknown.

aesthesiometer A device for measuring the cornea's sensitivity to touch.

afocal apposition eye A type of compound eye in which the independent optical elements do not form images (hence, are afocal) but instead send parallel rays of light into the photosensitive elements. Contrast *focal apposition eye.*

albinism A congenital lack of melanin pigment in skin, hair, and eyes, among other things.

aliasing A consequence of undersampling by discrete elements, such as photoreceptors, in which low-frequency components are visible in the presence of high-frequency stimuli.

alternate cover test A procedure in which an occluder is moved from one eye to the other, not permitting binocular viewing. It is used to measure the angle of strabismus or to establish the presence of a heterophoria.

alternating strabismus Strabismus in which one eye may be the fixating eye at some times and the deviating eye at others. Contrast *unilateral strabismus.*

amacrine cells A class of retinal interneurons that make connections in the inner plexiform layer among bipolar cells, ganglion cells, and other amacrine cells. There are about 30 types of amacrine cell.

amblyopia A reduction in visual acuity that is not due to uncorrected refractive error or to pathology. It is often associated with strabismus or anisometropia.

ametropia Refractive error; distant objects are not focused on the retina. Contrast *emmetropia.*

ampulla (pl. ampullae) In general, a flask-shaped dilation in a tubelike structure. In the eye, an expansion in the canaliculi of the lacrimal drainage system a few millimeters from the puncta.

angiogenesis The formation of new blood vessels by budding from existing vessels.

angle kappa The angle of intersection in the plane of the pupil between the pupillary axis and the line of sight; it is typically about 5°. Technically, this is angle lambda.

angle of strabismus The amount, in degrees or prism diopters, by which the deviating eye misses bifixation.

angle of the anterior chamber See *anterior chamber angle.*

aniridia Absence of the iris.

anisocoria A difference in pupil diameter between the two eyes of more than 0.5 mm.

anisometropia A difference in refractive error between the two eyes.

anterior border layer The discontinuous anterior surface of the iris made of fibroblasts and melanocytes.

anterior chamber The fluid-filled space between the anterior surface of the iris and the posterior surface of the cornea.

anterior chamber angle The angle formed by the intersection of the trabecular meshwork in the limbus with the iris and ciliary body.

anterior chamber depth The distance between the posterior surface of the central cornea and the anterior surface of the lens; it is about 3.5 mm in the adult eye.

anterior ciliary arteries The extensions of the muscular arteries after they pass through the rectus muscles.

anterior hyaloid membrane A thin layer of dense collagen that forms the anterior surface of the vitreous.

anterior segment The ciliary body and all other parts of the eye anterior to it (cornea, iris, limbus, etc.).

anterior synechia (pl. synechiae) An attachment of the iris to the corneal endothelium.

antigen Any substance that is recognized as foreign by the body's immune system and provokes the production of antibodies.

aperture stop An aperture, such as the eye's pupil, located in the optical pathway where it affects the amount of light passing through the system without affecting the field of view. Contrast *field stop.*

aphakia Absence of the lens.

apoptosis Normally occurring, programmed cell death.

apposition eye A compound eye in which the numerous optical elements are optically independent of one another. Apposition eyes are less sensitive than superposition eyes.

aqueous See *aqueous humor.*

aqueous flare Light scatter in the aqueous, usually due to the abnormal presence of proteins and cellular debris.

aqueous humor The fluid produced by the ciliary body epithelium that fills the anterior chamber; the adjective "aqueous" is often used alone as a noun.

aqueous vein A small vein that comes directly from the canal of Schlemm and therefore contains aqueous, not blood.

area centralis A small punctate region of the retina in which photoreceptor density reaches its maximum.

Argyll Robertson pupil A condition pupil in which both the direct and consensual light reflexes of the pupil are reduced or absent but pupil constriction to accommodative stimuli is normal.

arrestin A molecule that attaches to activated rhodopsin, terminates the activated state, and promotes dissociation of the chromophore from the opsin.

astigmatism Refractive error in which the optical power of the eye is not the same in the vertical and horizontal meridians.

autoregulation Regulation of blood flow through a tissue that is exerted by local influences on the precapillary arterioles; it is not centrally controlled.

axon The long, thin process that extends from the cell body or major dendrite of a neuron, along which action potentials are conducted to the axon terminal.

axon collateral An axon branch whose destination is different from that of the main axon.

axoplasmic transport The normal flow of molecules from the cell body to the axon terminal (orthograde transport) or in the opposite direction (retrograde transport).

B-scan A mode of ultrasonography in which the echoes from ultrasound beams passed through the eye in different directions are combined to show an acoustic section through the eye.

b-wave A component of the electroretinogram produced by changes in electrical potential changes in Müller's cells secondary to changes in ON bipolar cells.

barrier capillaries Capillaries with endothelium constructed to restrict the movement of molecules between blood and tissue.

basal cells Columnar cells that form the bottom layer of the corneal epithelium.

Bell's phenomenon Upward rotation of the eyes during voluntary lid clusure.

Bergmeister's papilla (pl. papillae) A small mass of glial cells and remnants of the hyaloid artery that protrudes into the vitreous from the surface of the optic nerve head.

biplexiform cell A type of retinal neuron that is thought to have dendrites extending to the outer plexiform layer, where it receives direct inputs from rods. It may be a displaced horizontal cell.

bipolar cells A class of retinal neurons that receive inputs from photoreceptors in the outer plexiform layer and have outputs to amacrine and ganglion cells in the inner plexiform layer. There are at least a dozen types of bipolar cell in human retina.

blastocoel The cavity within the blastocyst.

blastocyst The embryo at an early stage, a few hundred cells; it will implant in the uterine wall.

bleaching The breakdown of photopigment that results from absorption of light.

blepharospasm Involuntary, forcible lid closure.

blind spot The small scotoma in the temporal visual field of each eye corresponding to the optic nerve head.

blowout fracture A fracture of one of the orbital plates, usually the orbital floor, without an associated fracture of the orbital rim. Contrast *comminution fracture.*

blue cone bipolar cell A bipolar cell that receives inputs only from blue cones.

Bowman's layer The thin anterior layer of the corneal stroma in which fibroblasts are absent.

brachium of the superior colliculus A bundle of retinal ganglion cell axons that leaves the optic tract en route to the superior colliculus.

Brown's syndrome An eye movement disorder characterized by an inability to elevate the adducted eye above the horizontal plane. Some cases may be due to abnormal connective tissue sheaths around the superior oblique tendon.

Bruch's membrane The thin membrane (3 μm) that lines the entire choroid between the choriocapillaris and the retinal pigment epithelium; it extends forward into the pars plana of the ciliary body.

brunescent cataract A cataract with a brownish color.

bulbar conjunctiva See *conjunctiva.*

buried drusen of the nerve head An aggregation of small dense masses in the optic nerve head, sometimes mimicking the appearance of papilledema.

c-wave A long-duration component of the electroretinogram arising that arises from the retinal pigment epithelium.

canal of Petit An annular space or potential space around the lens between the posterior part of the zonule and the anterior hyaloid membrane of the vitreous.

canal of Clouquet The primary vitreous and the site of the hyaloid artery. In the adult eye, it is the normally invisible column of vitreous between the posterior surface of the lens and the optic nerve head.

canal of Schlemm The annular vessel in the limbus into which most of the aqueous is drained.

canaliculi (s. canaliculus) The small tubes extending between the puncta on the nasal eyelid margins to the lacrimal sac.

canals of Sondermann The internal collecting channels that extend from the canal of Schlemm into the trabecular meshwork.

canthus (pl. canthi) The angle where the upper and lower eyelids meet. There are two canthi: an inner, or medial, canthus and an outer, or lateral, canthus.

carotid plexus A network of interweaving sympathetic axons around the carotid artery at the level of the cavernous sinus. The axons arise primarily from cells in the superior cervical ganglion, and they enter the orbit as the sympathetic root of the ciliary ganglion.

cartilage Connective tissue in which the collagen is associated with a very extensive proteoglycan matrix.

caruncle The small mound of tissue on the plica semilunaris in the inner canthus of the eye.

cataract An opacity in the lens.

cavernous sinus One of the paired venous sinuses beside the sphenoid bone where the superior and inferior ophthalmic veins and other veins join; it drains indirectly to the internal jugular vein.

central meniscus of Kuhnt A thickened region in the glial cell membrane overlying the optic nerve head.

central retinal artery A branch of the ophthalmic artery that runs through the center of the optic nerve to supply the inner two-thirds of the retina.

chalazion (pl. chalazia) A swelling produced by a chronic, subacute inflammation of one or more tarsal glands.

channel A specialized region in cell membranes through which one or more species of ions can pass. A channel may open when a particular molecule binds to it (ligand-gated channel) or in response to a change in the electrical potential across the cell membrane.

check ligament A sheet of connective tissue extending from one of the rectus muscles to the periorbita. The medial and lateral recti typically have check ligaments.

chondrocyte A specialized fibroblast found in cartilage.

choriocapillaris The dense single layer of interconnected capillaries that lines the inner choroid. The photoreceptors are supplied by the choriocapillaris.

chorion The part of the blastocyst that implants into and fuses with the placenta.

choroid The heavily vascular part of the uveal tract between the sclera and the retina.

choroidal fissure The cleft along the underside of the optic cup that is the last part of the cup to be completed.

chromophore The photolabile part of the photopigments; it is 11-cis *retinal* in rhodopsin and the cone pigments.

cilia (s. cilium) The eyelashes. Also, tiny hairlike processes consisting of ten pairs of microtubules extending from basal bodies in a cell's membrane; they are used for cell motility or to move fluid or mucus over the cell surface.

ciliary body The portion of the uveal tract between the iris anteriorly and the choroid posteriorly; it contains the ciliary muscle and the ciliary processes.

ciliary ganglion A small parasympathetic ganglion within the orbit; it lies beside the optic nerve about 1 cm behind the eye.

ciliary muscle The ring of smooth muscle in the ciliary body that is responsible for accommodation.

ciliary plexus The network of interweaving small veins in the ciliary muscle.

ciliary process A ridge on the inner surface of the ciliary body that contains numerous large capillaries.

cilioretinal arteries Small arteries derived from the choroidal circulation that supply the inner retina between the fovea and the optic nerve head; about 30% of eyes have them.

circadian rhythm A variation in a physiological function or process that has an approximately 24-hour cycle.

circle of Zinn An arterial circle within the sclera around the optic nerve; it is formed by anastomoses among branches of the short posterior ciliary arteries. Also called circle of Haller or circle of Zinn–Haller.

cleavage The earliest form of cell division in the embryo, in which cells become smaller at each division.

Cloquet's canal See *canal of Cloquet.*

clump cell A phagocytic cell in the iris that ingests melanin.

collarette The thickest portion of the iris, containing the minor arterial circle.

collateral circulation Alternative or redundant sources of blood supply to a tissue.

coloboma A sector, typically in the inferior part of the eye, in which tissue has failed to develop properly. Colobomas may be found in the optic nerve, retina, choroid, ciliary body, or iris, singly or in combination.

co-localization The presence of two neuroactive substances in a single neuron.

comitant strabismus Strabismus in which the angle of deviation does not change with direction of gaze. Also called concomitant strabismus. Contrast *noncomitant strabismus.*

common tendon of Zinn The annular tendon at the back of the orbit from which the rectus muscles originate.

compound eye An eye with multiple optical systems, commonly associated with arthropods. Contrast *simple eye.*

computed tomography (CT) An X-ray method that produces sections through the body.

cone A photoreceptor that contains one of three photopigments maximally sensitive to red, green, or blue light, used primarily for vision in daylight. Contrast *rod.*

cone bipolar cell One of several types of bipolar cell that receive inputs only from cones. See *blue cone bipolar cell*, *diffuse bipolar cell*, and *midget bipolar cell*. Contrast *rod bipolar cell.*

confrontation test Generally, a test for limits or ranges, usually of the vield of view. It is also a test for normal ranges of eye movement; the patient is instructed to look at a penlight or other fixation target as it is moved in an H pattern. Also called an excursion test or H-test.

conjunctiva (pl. conjunctivae) Tissue consisting of an epithelium and an underlying connective tissue stroma. The conjunctiva overlying the eye (bulbar conjunctiva) is folded above and below the eye at the fornix to form the palpebral conjunctiva, which lines the posterior surfaces of the eyelids.

connecting cilium A short thin tube containing cilium-like microfilaments that joins the inner and outer segments of the photoreceptors.

consensual light reflex The reflex pupil constriction of one eye when light is shined in the other eye. Contrast *direct light reflex.*

constant strabismus Strabismus present at all times.

contrast The mean difference between maximum and minimum luminances divided by the mean luminance: that is, $(L_{max} - L_{min})/(L_{max} + L_{min})$.

contrast sensitivity function The visual system's sensitivity to sinusoidal gratings as a function of spatial frequency of the grating.

convergence Simultaneous adduction of the two eyes that brings the intersection of the visual axes closer. Contrast *divergence.*

convergent evolution The independent evolution of similar structures in different phyla.

corneal arcades Radially oriented loops of capillaries in the limbus around the margin of the cornea.

corona ciliaris A collective term for all of the ciliary processes.

cortex (pl. cortices) In general, the outer part of a structure. In the eye, the outer part of the adult lens or the outer part of the vitreous.

cortical cataract An opacity confined to the outer part of the lens.

cortical magnification factor The ratio of a unit area of visual cortex to its corresponding retinal area or area of visual space. The cortical magnification factor for the fovea is very high because only small areas of retina are associated with each cortical unit.

couching The ancient method of cataract surgery in which the lens was displaced into the inferior part of the vitreous.

coverage factor Cell density multiplied by mean dendritic area. The coverage factor for retinal neurons indicates the number of cells of a particular type that overlap a single retinal point.

covering In general, the way in which many single elements cover a surface. Coverings may have gaps between the elements, and the elements may overlap.

craniofacial dysostosis A disorder characterized by anomalous ossification of facial bones during development.

critical period The period of time during development in which structure or function are is plastic and can be affected by an abnormal visual environment.

cryptophthalmos An abnormality in which the eyelids are absent and the skin of the face is continuous over the eye.

crypts of Fuchs Holes in the anterior border layer of the iris.

crystallins The family of lens proteins, rarely found elsewhere in the body.

CT See *computed tomography.*

cul-de-sac See *fornix.*

cutoff frequency The highest spatial frequency that an optical system will transmit. In a modulation transfer function, it is the frequency at which the ratio of image to object contrast goes to zero.

decussation A crossing over from one side of the midline to the other. The trochlear nerves decussate completely, but the decussation of the optic nerves is partial.

degenerative anophthalmos The absence of an eye because of a developmental regression.

dendrite A branch from a neuron that receives inputs from other neurons.
depolarization A more positive (less negative) displacement of a cell's membrane potential from its resting level. Contrast *hyperpolarization.*
depression Rotation of the eye such that the visual axis moves downward. Contrast *elevation.*
depth of field The dioptric equivalent of the distance in external space along which a target can be moved forward and backward while appearing to remain in good focus. It is the counterpart, in object space, to the depth of focus.
depth of focus The dioptric equivalent of the distance along which an image can be moved forward and backward relative to the retina while appearing to remain in good focus. It is the counterpart, in image space, to the depth of field.
dermal bones Bones formed from foci of osteoclasts without being preformed in cartilage. Most of the orbital bones are of this type.
dermatochalasis Drooping of the folds of skin on the upper eyelid, which may overhang the lid margin.
descemetocoele A blisterlike bulge in the cornea following erosion or removal of all tissue anterior to Descemet's membrane; herniation of Descemet's membrane.
Descemet's membrane A membrane made of regularly arranged collagen fibers lying between the corneal stroma and the corneal endothelium that secretes the membrane.
desmosome An intercellular anchoring junction made of adjoining dense plaques of protein into which intermediate filaments (cadherins) insert.
determination The commitment by an embryonic cell to a particular line of development.
deuteranomaly See *deuteranopia.*
deuteranopia A lack of the green photopigment that results in a confusion between reds and greens. The less severe form is called deuteranomaly.
diabetic retinopathy A condition characterized by retinal changes, including hemorrhages and exudates, secondary to vascular leakage produced by diabetes.
difference of Gaussians A sombrero-shaped function produced by the difference between two Gaussian functions (normal distributions) with different variances. It is used to describe the variation in sensitivity across the receptive fields of some retinal ganglion cells.
differentiation The explicit result of determination, in which an embryonic cell undergoes changes that characterize its developmental path.
diffuse bipolar cell A bipolar cell that receives inputs from several red and green cones.
dilator muscle A sheet of radially oriented myoepithelial cells in the iris that dilate the pupil when the cells contract.
diplopia Double vision.
direct light reflex Constriction of the pupil when light is shone in the eye. Contrast *consensual light reflex.*
disparity The situation in which the images of a single object do not fall on corresponding points on the two retinas; images on noncorresponding points do not seem to lie in the same direction from the observer. Disparity is a stimulus for vergence eye movements and stereopsis, and it may produce diplopia if large enough.
divergence Simultaneous abduction of the two eyes that moves the intersection of the visual axes farther away. Contrast *convergence.*
drusen (s. and pl.) Local thickenings in Bruch's membrane.
dry eye A syndrome in which tear production is inadequate or tear composition is inappropriate to properly wet the cornea and conjunctiva.
Duane's retraction syndrome A condition characterized by the inability to abduct the eye past the midline and retraction of the eye into the orbit during adduction.
duction Any movement of one eye when the other is covered.
dyad synapse A characteristic synapse made by bipolar cells that is marked by a synaptic ribbon and two postsynaptic elements. Contrast *triad synapse.*

eccentric fixation A monocular condition in which the fixating eye does not use the center of the fovea for fixation.
ectoderm The outermost of the three germ tissues in the early embryo; most of the eye is derived from ectoderm.
ectopia lentis Congenital displacement of the lens away from its normal location.
ectopic pupil A pupil that is not centered in the iris.
ectropion An eversion, or outward turning, of an eyelid margin. Contrast *entropion.*
Edinger–Westphal nucleus A collection of preganglionic parasympathetic neurons lying near the oculomotor nerve nucleus. The axons travel via the oculomotor nerve to the ciliary ganglion.
efference copy Information about the innervational commands going to the extraocular muscles that is fed back to the motor control centers.
electro-oculogram (EOG) A measure of eye position that exploits the difference in electrical potential between the front and the back of the eye.
electroretinogram (ERG) A measure of the gross change in electrical potential across the retina in response to a visual stimulus.
elevation Rotation of the eye such that the visual axis moves upward. Contrast *depression.*
embryonal nucleus The central portion of the adult lens; it forms during the first 3 months of gestation. Contrast *fetal nucleus.*
emmetropia Lack of refractive error; distant objects are in focus on the retina. Contrast *ametropia.*
emmetropization The process by which the eyes of children exhibit less refractive error as they age.
en grappe ending A cluster of small buttonlike terminals on a muscle fiber from a single motor nerve axon.
encyclovergence Simultaneous intorsion of the two eyes. Contrast *excyclovergence.*
end-arterial Describing an artery, like the central retinal artery, whose branches supply separate regions with no overlap or alternative sources of supply.

endoderm The innermost of the three germ tissues in the early embryo; it makes no direct contribution to the eye. Also called entoderm.
enophthalmos A condition in which the eye is placed farther back in the orbit than normal. Contrast *exophthalmos.*
enteroceptor A sensory receptor that monitors the internal conditions of the body. Muscle spindles, tendon organs, and arterial pressure sensors (baroreceptors) are all enteroceptors.
entrance pupil The image of the pupil and iris formed by the cornea; it is what we see when we look at another person's eye.
entropion An inversion, or inward turning, of an eyelid margin. Contrast *ectropion.*
EOG See *electro-oculogram.*
epiblast The portion of the inner cell mass in the blastocyst that becomes the embryo.
epicanthus (pl. epicanthi) A fold of skin over the inner canthus.
epimysium (pl. epimysia) The connective tissue sheath around a muscle.
epiphora Overflow of tears onto the face.
episclera A thin layer of connective tissue and small blood vessels that overlies the scleral stroma.
episcleral arterial circle An arterial circle formed by anastomoses among episcleral arteries.
episcleral arteries Branches of the anterior ciliary arteries that course through the episclera.
epithelial plug Corneal epithelium that has grown into a radial keratotomy incision.
equator The imaginary circle produced by a vertical plane as it intersects the eye (or the lens) midway between the anterior and posterior poles.
ERG See *electroretinogram.*
esotropia Strabismus in which one eye deviates nasally from bifixation. Contrast *exotropia.*
ethmoidal foramina (s. foramen) The openings in the medial wall of the orbit through which the ethmoidal arteries and nerves pass.
excitation Increased depolarization of a neuron's membrane potential. Contrast *inhibition.*
excyclovergence Simultaneous extorsion of the two eyes. Contrast *encyclovergence.*
exophthalmos A condition in which the eye protrudes farther forward from its normal position in the orbit. Also called proptosis. Contrast *enophthalmos.*
exotropia Strabismus in which one eye deviates temporally from bifixation. Contrast *esotropia.*
external hordeolum See *hordeolum.*
external limiting membrane The array of junctions between Müller's cells and photoreceptors along the base of the photoreceptor inner segments. Also called the outer limiting membrane.
external ophthalmoplegia Paralysis of all the extraocular muscles.
extinction coefficient A measure of the absorbance of a photopigment as a function of wavelength of the incident light; it is an inherent property of the pigment's molecular structure.
extorsion Rotation of the eye around its anterior–posterior axis (torsion) such that the 12 o'clock position moves away from the nose. Contrast *intorsion.*
extrafusal fiber A muscle fiber in the muscle proper, not in one of the muscle spindles. Contrast *intrafusal fiber.*

facial nerve Cranial nerve VII; several of its branches innervate the orbicularis oculi muscle.
farsightedness See *hyperopia.*
fast-twitch fiber A muscle fiber that has a very rapid all-or-nothing mode of contraction. Contrast *slow-twitch fiber.*
fenestrated capillary A capillary in which the endothelium has small thin spots (fenestrae, or windows) covered only by the cell membranes. Fenestrated capillaries in the choroid and ciliary processes are highly permeable.
fetal nucleus All but the center of the adult lens nucleus; it forms between the third month of gestation and birth. Contrast *embryonal nucleus.*
field stop An aperture placed in an optical system so that it limits the field of view. Contrast *aperture stop.*
fixation reflex A reflex that operates to keep a selected image at the center of the fovea.
fluorescein A particular dye that emits yellow-green light when it absorbs ultraviolet light.
fluorescein angiography A technique for evaluating the integrity of retinal or iridial vessels by observing the pattern of fluorescence under ultraviolet illumination.
focal (or simple) apposition eye A compound eye in which images formed by multiple optical elements are combined on a single photosensitive element. Also called simple apposition eye. Contrast *afocal apposition eye.*
fornix (pl. fornices) The region of the conjunctiva where the bulbar conjunctiva folds to cover the back of the eyelids. Also called cul-de-sac.
frequency-of-seeing curve For visual acuity, a plot showing the percentage of correct responses as a function of increasing test letter size.
frontal nerve The branch of the ophthalmic nerve that innervates the skin of the forehead.
fusional vergence Vergence eye movement produced by retinal image disparity.

G protein–linked receptors A family of membrane receptors with that have seven transmembrane regions; they activate G proteins after binding a particular ligand. See also *metabotropic receptor.*
G proteins A large family of proteins that interact with characteristic membrane receptors and are involved in second-messenger pathways.
GABA The amino acid gamma-aminobutyric acid; an inhibitory neurotransmitter in the retina and central nervous system.
GAG See *glycosaminoglycan.*
ganglion (pl. ganglia) A collection of nerve cell bodies in some a location outside the brain.
ganglion cell layer The innermost cellular layer of the retina; it contains the cell bodies of ganglion cells and displaced amacrine cells.

ganglion cells In general, cells located in a ganglion. In the retina, a class of neurons whose axons go to the central nervous system via the optic nerve and optic tracts.
gap junction An intercellular communicating junction that allows ions and small molecules to pass from one cell to the next. The permeability of gap junctions can often be varied by neuromodulators or hormones.
gasserian ganglion The trigeminal nerve ganglion; it lies next to the brainstem. It has three main peripheral branches (ophthalmic, maxillary, and mandibular) and two central roots (sensory and motor). Also called semilunar ganglion.
gastrula (pl. gastrulae) The embryonic stage following the blastula and preceding formation of the neural tube.
germ tissues Cellular components of the gastrula after an initial stage of differentiation; they are ectoderm, mesoderm, and endoderm.
giant vacuolevacuoles A fluid-filled blister in the endothelial cell lining the canal of Schlemm; aqueous is believed to enter the canal when the giant vacuoles open and empty their contents.
glands of Moll Modified sweat glands that empty onto the eyelid margins.
glands of Zeis Small glands associated with the eyelash follicles.
glutamate The amino acid that is the primary excitatory neurotransmitter in the retina; it is used by photoreceptors, bipolar cells, and ganglion cells.
glycine An amino acid and an inhibitory neurotransmitter found in cells throughout the central nervous system. In retina, it is used primarily by AII amacrine cells.
glycosaminoglycan (GAG) One of various long polysaccharides that link to core proteins to form proteoglycans.
goblet cells Unicellular mucus-secreting glands in the fornix and bulbar conjunctiva.
Golgi tendon organ A receptor in tendons where they join the muscle; the Golgi tendon organs are sensitive to and transmit signals about muscle tension.
gonioscopy An optical technique that permits a direct view of the anterior chamber angle.
gradient index lens A lens whose refractive index varies smoothly and continuously from center to surface. Ocular lenses in most species are gradient index lenses.
gradu The gritty material that collects at the inner canthi while one is asleep.
growth cone The specialized tip of a growing axon that moves, pulling the elongating axon along behind it.
guttata Locally thickened regions in Descemet's membrane of the cornea. They are called Hassall–Henle bodies when they are located near the margin of the cornea.

Hasner's valve See *valve of Hasner.*
Hassall–Henle bodies See *guttata.*
hemianopia A visual field defect (scotoma) that affects half of the visual field; blindness in one half of the visual field. Contrast *quadrantanopia.*
hemidesmosome An anchoring junction made by an epithelial cell with its basement membrane.
Henle's fiber layer The outer plexiform layer of the retina in the vicinity of the fovea where the axons of the photoreceptors run radially outward to the terminals.
Hering diagram A plot of the trajectories that the visual axis of the eye would follow if each of the extraocular muscles were to contract in isolation; that is, a plot that purports to show the actions of the extraocular muscles.
heterochromia An abnormality in which the irises of the two eyes have significantly different colors.
heterophoria A deviation of one eye from bifixation when binocular vision is interrupted.
heterotropia See *strabismus.*
hippus Small, normal fluctuations in pupil size.
hordeolum (pl. hordeola) Swelling of the eyelid produced by infection. Also called stye. An internal hordeolum is swelling on the inner side of the eyelid that is produced by an inflammatory infection of a tarsal gland. An external hordeolum is swelling on the outer side of the eyelid produced by inflammatory infection of Zeis's glands.
horizontal cells A class of retinal interneurons that receive inputs from photoreceptors and have inhibitory outputs back onto photoreceptors. H1 horizontal cells have few inputs from blue cones, but blue cones are a major souce of input to H2 horizontal cells.
horizontal raphe In general, a seam or a line dividing something in half. In the retina, it is a line extending temporally from the fovea, above which ganglion cell axons run upward, and below which axons run downward on their way to the optic nerve head.
Horner's muscle The lacrimal portion of the orbicularis oculi muscle, which attaches to the posterior lacrimal crest.
Horner's syndrome A condition characterized by a combination of ptosis, miosis, and anhidrosis produced by a lesion in the preganglionic portion of the sympathetic pathway to the superior cervical ganglion.
horseradish peroxidase (HRP) A plant enzyme that is transported throughout a neuron after it has been injected into the cell or has been taken up by a cut axon. Several chemical reactions cause the enzyme to precipitate as a dark reaction product, thereby making the cell visible.
HRP See *horseradish peroxidase.*
hyalocytes Cells in the vitreous that probably produce the major glycosaminoglycan (hyaluronic acid) in the vitreous.
hyaloid artery A transient artery in the fetal eye that runs through the vitreous chamber to supply the developing lens. Its intraocular portion atrophies and reorganizes to become the central retinal artery supplying the retina.
hyaloideocapsular ligament See *Wieger's ligament.*
hyaluronic acid The primary glycosaminoglycan in the vitreous.
hyperopia A refractive error in which the focal point of the optical system lies behind the retina. It can be compensated for by accommodation or by the use of lenses

with positive power. Also called farsightedness. Contrast *myopia.*

hyperpolarization A more negative displacement of a cell's membrane potential from its resting level. Contrast *depolarization.*

hypertropia Strabismus in which one eye is deviated vertically.

hyphema An accumulation of blood along the inferior anterior chamber angle.

hypoblast Tissue that becomes the lining of the yolk sac surrounding the embryo.

hypopyon An accumulation of pus along the inferior anterior chamber angle.

immunohistochemistry A collection of methods employing labeled antibodies that bind to specific molecules or parts of molecules (antigens) in cells. The label makes the targeted cells visible.

induction The use of chemical signals by cells during embryogenesis to influence the development of nearby cells and tissues.

inferior orbital fissure The opening in the orbit between the posterior part of the orbital floor and the lateral wall.

infraorbital foramen (pl. foramina) The opening in the maxillary bone below the orbital rim through which the infraorbital artery and nerve pass.

infraorbital groove The groove in the orbital floor that becomes the infraorbital canal leading to the infraorbital foramen.

infraorbital suture The line above the infraorbital canal where different developmental foci of the maxillary bone came into contact. It is often the site of blowout fracture.

inhibition Increased hyperpolarization of a neuron's membrane potential. Contrast *excitation.*

inner cell mass The tissue from which the embryo forms.

inner limiting membrane The inner lining of the retina next to the vitreous. It is formed by the expanded feet of Müller's cells.

inner nuclear layer The retinal layer that contains the cell bodies of the horizontal cells, bipolar cells, most amacrine cells, and Müller's cells. It lies between the outer and inner plexiform layers.

inner segment The part of a photoreceptor between the outer segment and the cell nucleus. It is the photoreceptor's metabolic center and the site of ion pumps responsible for the flow of electric current in the photoreceptor.

insertion spiral of Tillaux An imaginary line connecting the insertions of the rectus muscles that spirals outward as it is traced clockwise from the medial rectus insertion (closest to the cornea) to the superior rectus insertion (farthest from the cornea).

intermittent strabismus Strabismus that is present at some fixation distances or fields of gaze but absent in others. Contrast *constant strabismus.*

internal carotid nerve A bundle of axons from cells in the superior cervical ganglion that run upward along the internal carotid artery. Also called nerve to the internal carotid artery.

internal hordeolum See *hordeolum.*

internal limiting membrane of Elschnig The layer of astrocytes covering the optic nerve head next to the vitreous.

internuclear ophthalmoplegia A condition characterized by impaired eye movements due to a lesion in the medial longitudinal fasciculus that carries axons from abducens internuclear neurons. Signs vary, but adduction weakness on the ipsilateral side during a version to the contralateral side is common.

interphotoreceptor matrix (pl. matrices) A gel containing proteins, glycoproteins, and glycosaminoglycans that fills the extracellular space between photoreceptor outer segments and the retinal pigment epithelium.

interplexiform cell A retinal neuron that receives inputs from bipolar cells in the inner plexiform layer and has outputs to horizontal cells and bipolar cells in the outer plexiform layer.

interstitial growth Growth by addition of membrane to the middle of an existing axon or dendrite rather than at a growth cone.

interstitial retinoid-binding protein (IRBP) A protein that transports retinol and *retinal* across the interphotoreceptor matrix.

intorsion Rotation of the eye around its anterior–posterior axis (torsion) such that the 12 o'clock position moves toward the nose. Contrast *extorsion.*

intrafusal fiber A small muscle fiber within a muscle spindle. Contrast *extrafusal fiber.*

intramuscular arterial circle An incomplete circle of arteries within the ciliary muscle formed by anastomoses among branches of the major perforating arteries from the anterior ciliary arteries.

ionotropic receptor A receptor in which binding of the appropriate molecule (ligand) affects an ion channel directly because the receptor and channel are parts of the same structure. Contrast *metabotropic receptor.*

IRBP See *interstitial retinoid-binding protein.*

iris The anteriormost part of the uveal tract; it contains the pupil and is visible through the cornea.

iris bombé A condition characterized by forward bulging of the iris produced by a buildup of aqueous trapped behind the iris. It is the result of posterior synechiae that prevent the normal flow of aqueous out of the posterior chamber through the pupil.

iris processes Columns of cells from the anterior border layer of the iris that extend up to the trabecular meshwork.

iris root The junction of the iris with the ciliary body. It is the thinnest part of the iris.

irregular astigmatism Astigmatism in which the meridians of highest and lowest optical power are not perpendicular.

juxtacanalicular tissue The tissue between the trabecular meshwork and the canal of Schlemm.

keratoconjunctivitis sicca A form of dry eye associated with a decreased aqueous component in the tears.

keratoconus A condition in which the cornea has a region of abnormally high curvature and astigmatism.

keratocytes Corneal fibroblasts.

keratoplasty A corneal graft or transplant.
Krukenberg's spindle An hourglass-shaped deposit of pigment on the posterior surface of the cornea.

lacrimal bone The small bone on the medial wall of the orbit between the nasal process of the maxillary bone and the ethmoid bone.
lacrimal crest A ridge along the edge of the nasolacrimal fossa. The anterior lacrimal crest is on the maxillary bone, the posterior lacrimal crest on the lacrimal bone.
lacrimal fossa (pl. fossae) A shallow depression in the orbital plate of the frontal bone marking the site of the lacrimal gland.
lacrimal gland The main source of tears. The gland is on the upper lateral side of the eye just behind the orbital margin.
lacrimal nerve A branch of the ophthalmic nerve that is primarily a sensory nerve for the skin above and lateral to the orbit. Parasympathetic fibers going to the lacrimal gland travel in the lacrimal nerve for a short distance.
lacrimal sac The hollow sac in the nasolacrimal fossa into which tears drain via the canaliculi.
lacrimal tubercle A small bump on the nasal process of the maxillary bone to which the lower edge of the medial palpebral ligament attaches.
lamina cribrosa The meshwork of scleral fibers through which the optic nerve passes.
lamina fusca The innermost layer of the sclera, in which the scleral stroma contains melanocytes.
laminated vein An episcleral vein downstream from the mouth of an aqueous vein. The streams of blood and aqueous do not mix immediately, creating the laminated appearance.
laser in situ keratomileusis (LASIK) Corneal refractive surgery in which the optic zone is undercut and folded back before the underlying stroma is reshaped by laser ablation.
LASIK See *laser in situ keratomileusis.*
lateral geniculate nuclei Large six-layered nuclei in the midbrain to which most retinal ganglion cells project. Lateral geniculate neurons send axons to the primary visual cortex via the optic radiations.
lateral orbital tubercle A small bump on the zygomatic bone that marks the site of attachment of the lateral palpebral ligament. Also called Whitnall's tubercle.
lens capsule The membrane surrounding the lens that is secreted by the lens epithelium.
lens paradox The phenomenon in which lens curvatures increase with age but lens power does not increase.
lens placode The thickened plate of surface ectoderm from which the lens develops.
lens sutures The irregular branched lines along which lens fibers join and interdigitate.
lens vesicle The developing lens before its interior was filled with cells.
levator palpebrae superioris The muscle primarily responsible for raising the upper eyelid. Also called simply levator.
ligand-gated channel See *channel.*
limbus In general, a border. In the eye, an annulus of tissue about 1.5 mm wide that encircles the cornea; it contains the canal of Schlemm and other structures for aqueous drainage.
line of sight The line between a point of fixation and the center of the entrance pupil.
long ciliary nerves Branches of the nasociliary nerve that carry sensory fibers from the cornea and conjunctiva.
luminance A measure of the quantity of light emitted by a source or reflected from a surface.
lysozyme A bacteriolytic enzyme that is a component of the tears.

macula (pl. maculae) The region of central retina, including the fovea, that contains xanthophyll pigments.
magnetic resonance imaging (MRI) A method for creating sectional images through the body. Magnetic fields are used to align hydrogen atoms, from which radio waves are generated as the atoms return to alignment after being perturbed. The emitted radio waves can be detected and used to compute the spatial locations of the radio sources.
magnocellular layers The bottom layers (1 and 2) in the lateral geniculate nuclei, which contain neurons with large cell bodies. Contrast *parvocellular layers.*
major arterial circle An incomplete circle of arteries in the ciliary body formed mainly by anastomoses among branches of the long posterior ciliary arteries.
major perforating arteries Branches from the anterior ciliary arteries that pass through the sclera to the ciliary body, where they form the intramuscular arterial circle.
mandibular nerve One of the three branches of the trigeminal nerve, designated V_3. It innervates the lower jaw.
Marcus Gunn sign Reduced direct and consensual pupil constriction to light shone in one eye as compared to the other. It indicates a defect in the afferent pathway.
Marcus Gunn's syndrome A condition characterized by fluttering of the eyelids as the lower jaw is moved from side to side.
marginal myotomy A surgical technique for making an extraocular muscle less effective. Several cuts are made partway through the muscle, perpendicular to its long axis, after which the muscle elongates. Contrast *myectomy* and *tucking.*
maxillary nerve One of the three branches of the trigeminal nerve; it is designated V_2. It innervates the upper jaw.
mean The arithmetic average.
Meckel's ganglion See *pterygopalatine ganglion.*
medial longitudinal fasciculus (pl. fasciculi) A tract running between the abducens nucleus and the oculomotor nucleus, carrying axons from abducens internuclear neurons responsible for coordinating horizontal version eye movements.
megalocornea A congenitally enlarged cornea. Contrast *microcornea.*
meibomian gland See *tarsal gland.*
mesoderm One of the three germ tissues in the gastrula; the sclera, extraocular muscles, and much of the vascular system are derived from mesoderm.
metabotropic receptor A receptor that affects ion channels indirectly after it binds an appropriate molecule

(ligand); the indirect action is often mediated by a G protein. See also G protein–linked receptors. Contrast *ionotropic receptor.*

microcornea A congenitally smaller-than-normal cornea. Contrast *megalocornea.*

micronystagmus The constant tremor of the eye that occurs even during steady fixation; its amplitude is about one foveal cone diameter (2 μm or about 0.5′ arc).

microphthalmos A very small eye whose development was interrupted.

midget bipolar cell A type of cone bipolar cell that receives input from either red or green cones; in central retina, the input is from a single cone.

midget ganglion cell A type of ganglion cell that receives excitatory inputs from midget bipolar cells; in central retina, the input is from a single midget bipolar cell.

minor arterial circle A circle of arteries in the iris near the sphincter; it represents anastomoses among radial arteries running into the iris from the intramuscular and major arterial circles in the ciliary body.

miosis Pupil constriction. Contrast *mydriasis.*

Mittendorf dot A small remnant of the hyaloid artery on the posterior lens capsule.

mode In a relative frequency histogram, the value at which the peak occurs—that is, the value most frequently observed.

modulation transfer function The ratio of image to object contrast for an optical system as a function of spatial frequency. It is a measure of the quality of the optical system that is mathematically related to the light spread in the image of a point light source.

molecular chaperones Proteins that assist other proteins in folding to their desired configuration. α-Crystallins in the lens may serve this role.

Moll's glands See *glands of Moll.*

monoclonal antibodies Antibodies raised from a clone of lymphocytes; unlike polyclonal antibodies, they may target a single antigenic site on a molecule. Contrast *polyclonal antibodies.*

monocular diplopia Double vision under monocular conditions.

morula (pl. morulae) The small cell cluster produced by the initial cleavages of the zygote; it will become the blastocyst.

motility In embryogenesis, the ability of cells to move around and to change shape.

motor end plate The broad expanded terminal of a motor nerve axon that makes synaptic contact with a muscle fiber.

motor root of the ciliary ganglion A bundle of parasympathetic axons from the Edinger–Westphal nucleus that run from the inferior division of the oculomotor nerve to the ciliary ganglion.

motor unit A motor nerve axon and the muscle fibers it innervates. Alternatively, all the muscle fibers innervated by a single motor nerve axon.

MRI See *magnetic resonance imaging.*

Müller's cells Large radially oriented retinal glial cells; they form the inner and outer limiting membranes of the retina and are responsible for neurotransmitter recycling and ion storage.

muscle spindle A collection of specialized muscle fibers enclosed in a capsule that is sensitive to changes in muscle length.

muscle stars Fingerlike expansions at the ends of ciliary muscle fibers, particularly those ending in the choroid.

myasthenia gravis An autoimmune disease in which acetylcholine receptors are attacked and destroyed.

mydriasis Pupil dilation. Contrast *miosis.*

myectomy A surgical technique for making an extraocular muscle more effective; the muscle is made shorter by cutting a piece out of it and rejoining the cut ends. Contrast *marginal myotomy* and *tucking.*

myofibrils Long bundles of actin, myosin, and other proteins; they are the contractile elements in muscle fibers.

myogenic response Changes in the contraction of vascular smooth muscle that result from changes in pressure within the blood vessels. The myogenic response tends to keep blood flow constant.

myopia A refractive error in which the focal point of the eye's optical system lies in front of the retina. It is corrected by lenses with negative power. Also called nearsightedness. Contrast *hyperopia.*

myosin A large protein that is organized in a long chain; in conjunction with actin, it is the major protein used for muscle fiber contraction.

nasociliary nerve One of the three major branches of the ophthalmic nerve; it conveys all of the somatosensory innervation from the eye.

nasolacrimal canal The tunnel that leads from the lacrimal fossa to the nasal cavity.

nasolacrimal duct The tube extending down through the nasolacrimal canal from the lacrimal sac that drains tears into the nasal cavity.

nasolacrimal fossa (pl. fossae) The depression at the junction of the lacrimal and maxillary bones in which the lacrimal sac lies.

nasotemporal overlap The narrow vertical strip passing through the center of the fovea, within which ganglion cell axons may project ipsilaterally or contralaterally; that is, the region in which nasal, contralaterally projecting retina overlaps temporal, ipsilaterally projecting retina.

near triad Simultaneous accommodation, convergence, and pupil constriction.

nearsightedness See *myopia.*

neovascularization The formation of new capillary beds from existing blood vessels.

nerve fiber layer The layer of retinal ganglion cell axons lying next to the vitreous.

nerve growth factor A protein that interacts supportively with growing axons.

nerve loop of Axenfeld A loop of the long ciliary nerve out to the conjunctiva in company with a major perforating artery.

neural crest A collection of embryonic cells from the roof of the neural tube. They migrate out to form peripheral nerve ganglia, as well as some tissues in the eye, such as the corneal stroma and endothelium, trabecular meshwork cells, iris stroma, and ciliary muscle.

neural ectoderm The ectoderm that separates from the surface of the embryo to become the neural groove, neural tube, and neural crest.

neural groove The line of depression along the embryo's long axis that deepens and invaginates to become the neural tube.

neural superposition eye An optical apposition eye in which a particular pattern of neural connectivity is used to sum signals from several optical elements.

neural tube The tube of neural ectoderm that is the forerunner of the brain and spinal cord.

neuromodulator A substance that alters the sensitivity of neurons to a neurotransmitter.

neurotransmitter A substance released at synapses by axon terminals; it interacts with receptors on the postsynaptic cell, causing specific ion channels to open or close.

neurulation The process of forming the neural tube and the primitive nervous system.

nociceptors Receptors sensitive to noxious, injurious stimuli; pain receptors.

noncomitant strabismus Strabismus in which the angle of deviation varies with distance, gaze position, or fixating eye. Contrast *comitant strabismus.*

norepinephrine An excitatory neurotransmitter used by postganglionic sympathetic neurons.

notochord A transient column of mesodermal cells lying below the neural tube.

nuclear cataract An opacity in the lens nucleus.

nucleus (pl. nuclei) In general, the innermost part of something, such as the inner part of the lens. In the central nervous system, an aggregation of nerve cell bodies.

nucleus of the optic tract One of the pretectal nuclei. Its neurons may be involved in the control of reflex eye movements.

Nyquist frequency The maximum frequency that can be sampled faithfully by an array of discrete sampling elements, such as cones. It is twice the frequency at which the elements are spaced.

nystagmus Alternating rotation of the eyes in one direction and then in the opposite direction.

oblique astigmatism Astigmatism in which the axes of maximum and minimum power are not aligned with the vertical or the horizontal meridians of the eye.

ocular hypertension Intraocular pressure greater than the ninety-fifth percentile of normal intraocular pressures—that is, above 21 mm Hg. Individuals with ocular hypertension may be at greater risk for glaucoma.

oculocardiac reflex Slowing or stopping of the heart in response to pulling on the extraocular muscles and associated tissue.

oculomotor nerve Cranial nerve III; it is the motor nerve for the medial rectus, superior rectus, inferior rectus, inferior oblique, and levator muscles.

OFF bipolar cell A cone bipolar cell that depolarizes in response to decreasing light intensity (OFF) and terminates in the outer half of the inner plexiform layer. Contrast *ON bipolar cell.*

ommatidium (pl. ommatidia) One of the individual optical elements in compound eyes. An ommatidium usually contains a cornea, lens or lens cone, a photosensitive element (rhabdom), and screening pigment.

ON bipolar cell A cone bipolar cell that depolarizes in response to increasing light intensity (ON) and terminates in the inner half of the inner plexiform layer. Contrast *OFF bipolar cell.*

ophthalmic arteries The arteries branching from the internal carotid arteries that run into the orbits through the optic foramina to supply the eye and surrounding structures.

ophthalmic nerve One of the three branches of the trigeminal nerve, designated V_1. It has three branches itself: the frontal, lacrimal, and nasociliary nerves. They innervate the forehead, brow, and the eye.

ophthalmic veins The veins, superior and inferior, that drain the eye and surrounding areas and convey blood to the cavernous sinuses.

opsin One of several transmembrane proteins to which 11-cis *retinal* attaches to form a photopigment. Its structure resembles that of other G protein–linked receptor proteins.

optic canal The tunnel in the sphenoid bone through which the optic nerve and ophthalmic artery pass.

optic chiasm In general, a chiasm is a crossing or decussation. The optic chiasm is the site of decussation for axons from the nasal retinas; it lies just above the pituitary gland.

optic cup The form taken by the developing eye as the optic vesicle invaginates.

optic disc The optic nerve head viewed ophthalmoscopically.

optic foramen (pl. foramina) The opening to the optic canal in the sphenoid bone through which the optic nerve and the ophthalmic artery pass.

optic nerve Cranial nerve II, which carries retinal ganglion cell axons running from the eye to the brain.

optic neuritis Inflammation of the optic nerve.

optic pit A bulge on the side of the neural tube that is the primordial eye.

optic stalk The hollow tube connecting the optic vesicle to the developing brain; it will become the optic nerve.

optic tract The bundle of retinal ganglion cell axons as they run from the optic chiasm to the lateral geniculate nucleus.

optic vesicle The balloonlike expansion at the end of the optic stalk that will inveaginate invaginate to become the optic cup.

optical superposition eye A compound eye in which the individual optical elements are not independent but act as groups to contribute to the image. Superposition eyes are more sensitive than apposition eyes.

optical zone The central part of the cornea; it is about 4 mm in diameter.

optokinetic nystagmus Nystagmus in which slow eye movements that track extended, slowly moving targets are interrupted by periodic rapid eye movements in the opposite direction.

ora serrata The anterior limit of the retina next to the pars plana of the ciliary body.

orbicularis oculi The large muscle around the orbit, extending into the eyelids, that is responsible for lid closure. Also called simply orbicularis.
orbit The roughly pyramid-shaped bony structure from which the eye looks out. Also, the space enclosed by the orbital bones.
orbital cellulitis Inflammation of connective tissue in the orbit.
orbital septum (pl. septa) The sheet of connective tissue that is attached to the orbital margin and runs into the eyelids. Also called septum orbitale.
orthograde transport See *axoplasmic transport.*
osteoblasts Specialized fibroblasts that produce the matrix in developing bone.
osteoclasts Specialized fibroblasts in mature bone that are responsible for absorption and removal of bone.
osteocytes Osteoblasts resident within cavities (lacunae) in the bone matrix.
outer nuclear layer The retinal layer that contains the photoreceptor nuclei.
outer plexiform layer The retinal layer that contains photoreceptor axons leading to the photoreceptor terminals that make synaptic contact with horizontal cells and bipolar cells.
outer segment The part of a photoreceptor that contains flattened discs with photopigment molecules embedded in their membranes.
outer synaptic layer The part of the outer plexiform layer that contains the cone pedicles and rod spherules.

pachometer A device for measuring corneal thickness. Some pachometers are optical devices; others use ultrasound.
palatine bone A bone lying below the orbital floor at the back of the orbit that is a small part of the orbital floor.
palisades of Vogt Radially oriented channels in the limbal stroma through which the corneal arcade vessels run.
palpebral arcades Small arteries that run horizontally within the eyelids; they are anastomoses between the medial and lateral palpebral arteries.
palpebral conjunctiva See *conjunctiva.*
palpebral furrows Creases in the skin of the eyelids, roughly along the outer margins of the tarsal plates.
palpebral ligaments The ligaments, medial and lateral, that attach the orbicularis muscle to the orbital bones.
palpebral muscles of Müller See *tarsal muscles.*
palpebral raphe A horizontal line in the orbicularis muscle along the center of the lateral palpebral ligament.
papilledema Bulging of the optic nerve head because of fluid accumulation.
papillomacular bundle The bundle of retinal ganglion cell axons between the fovea and the temporal side of the optic nerve head.
parabolic superposition eye An optical superposition eye in which both refraction and reflection are used to combine light from neighboring optical elements.
parasol ganglion cell A unistratified ganglion cell that receives excitatory inputs from diffuse bipolar cells.
pars plana The part of the ciliary body that lacks ciliary processes; it is smooth.
pars plicata The part of the ciliary body that has ciliary processes.
parvocellular layers The four layers (3 through 6) of the lateral geniculate nucleus that contain relatively small cell bodies. Contrast *magnocellular layers.*
past-pointing In some individuals with strabismus, an overrotation of the hand used to indicate the direction of gaze of the normally deviated eye when it is forced to be the fixating eye.
patellar fossa (pl. fossae) The depression on the anterior surface of the vitreous in which the lens resides.
pecten A darkly pigmented, heavily vascularized structure that projects into the vitreous in avian eyes.
pedicle The broad synaptic terminal of a cone photoreceptor. Contrast *spherule.*
perimysium (pl. perimysia) The connective tissue sheath around a bundle of muscle fibers within a muscle.
periorbita The periosteum that lines the orbital bones.
periosteum (pl. periostea) A thin layer of connective tissue that adheres to the surfaces of bones.
persistent hyaloid artery A residual portion of the embryonic hyaloid artery due to incomplete atrophy.
persistent hyperplastic primary vitreous (PHPV) An abnormal retention of part of the anterior hyaloid artery, tunica vasculosa lentis, and primary vitreous. It is accompanied by other anomalies, including microphthalmos and lens opacities.
persistent pupillary membrane A residual portion of the pupillary membrane, seen as strands of iris tissue crossing the pupil, due to incomplete atrophy.
Peter's anomaly Improper development of the anterior segment with corneal opacity, anterior synechiae, and variable defects in the cornea, limbus, and iris.
Petit's canal See *canal of Petit.*
phagosome An inclusion in retinal pigment epithelial cells resulting from the cells' engulfing of outer segment tips shed from the photoreceptors.
photoreceptor layer The retinal layer that contains the inner and outer segments of the photoreceptors.
photorefractive keratectomy (PRK) A method of refractive surgery in which the corneal epithelium is removed and the underlying stroma is reshaped by ablating tissue with pulses from an ultraviolet-emitting laser.
phototransduction The sequence of events in which the absorption of a photon by a photopigment molecule produces a change in the photoreceptor's membrane potential.
PHPV See *persistent hyperplastic primary vitreous.*
physiological insertion The point on the eye where the force of a contracting muscle is located; it will be some distance from the anatomical insertion.
pigment epithelium of the retina (RPE) The single layer of melanin-containing cells that lie between the photoreceptors and Bruch's membrane of the choroid.
pigmentary dispersion glaucoma Glaucoma resulting from blockage of the trabecular meshwork by pigment shed from the iris.
pinhole eye A type of simple eye in which the aperture is the principal optical element.
plica lacrimalis See *valve of Hasner.*

plica semilunaris The crescent moon–shaped fold of modified conjunctiva in the inner canthus.
polyclonal antibodies Antibodies produced by different lymphocytes in response to a specific antigen. Contrast *monoclonal antibodies.*
posterior chamber The fluid-filled space between the anterior surface of the lens and the posterior surface of the iris.
posterior ciliary arteries Branches, lateral and medial, of the ophthalmic artery that supply blood to the eye.
posterior segment All of the eye behind the ciliary body—that is, the lens, vitreous, choroid, retina—and the surrounding sclera.
posterior subcapsular cataract An opacity of the lens at the posterior pole.
posterior synechia (pl. synechiae) An attachment of the lens capsule to the posterior surface of the iris.
posterior vitreal detachment (PVD) Separation of the vitreous cortex from the retina.
presbyopia Loss of accommodation.
pretectal nuclei A complex of five nuclei in the midbrain just anterior to the superior colliculi; they are the nucleus of the optic tract, the pretectal olivary nucleus, and the anterior, medial, and posterior pretectal nuclei. Several of them receive inputs from retinal ganglion cell axons.
pretectal olivary nucleus One of the pretectal nuclei; it receives retinal inputs involved in the pathway for controlling pupil size.
primary position of gaze Roughly speaking, the straightforward position of gaze when the head is erect.
primitive dorsal ophthalmic artery A transient artery in the embryo; it will become the ophthalmic artery.
primitive streak A line along the long axis of the early embryo that defines the future midline.
PRK See *photorefractive keratectomy.*
proliferation Increase in cell number by mitosis.
proliferative diabetic retinopathy Diabetic retinopathy with neovascularization into the vitreous.
proliferative vitreoretinopathy (PVR) Formation of abnormal collections of vitreous cells, connective tissue, and abnormal vascularization at the boundary between retina and vitreous.
proptosis See *exophthalmos.*
protanomaly See *protanopia.*
protanopia A lack of the red cone photopigment, resulting in a confusion between reds and greens. The less severe form is called protanomaly.
proteoglycan A core protein and its associated glycosaminoglycans.
pterygopalatine ganglion A parasympathetic ganglion in the pterygopalatine fossa below the orbital floor; some of the neurons innervate the lacrimal gland. Also called sphenopalatine ganglion or Meckel's ganglion.
ptosis Drooping of the upper eyelid.
puncta (s. punctum) The small openings to the canaliculi on the lid margins near the inner canthus.
pupil The circular hole in the center of the iris.
pupillary axis The imaginary line perpendicular to the cornea at its apex that passes through the center of the entrance pupil.
pupillary membrane The part of the fetal iris that will atrophy to become the pupil.
pupillary ruff The epithelial layers of the iris that line the margin of the pupil. Also called pupillar frill.
Purkinje shift The change in the peak of the eye's spectral sensitivity function between light and dark adaptation.
Purkinje tree The visible shadow of the retinal blood vessels produced by diffuse illumination of the retina.
PVD See *posterior vitreal detachment.*
PVR See *proliferative vitreoretinopathy.*

quadrantanopia A visual field defect (scotoma) that affects about one-fourth of the visual field. Contrast *hemianopia.*

radial keratotomy (RK) A method of refractive surgery for myopia in which six or more radial cuts are made in the cornea outside the optical zone; as the incisions gape, the central cornea flattens, thus reducing its optical power.
rami oculares Several small bundles of postganglionic parasympathetic axons that penetrate the posterior sclera en route to the choroid and the choroidal vessels.
ramus communicans See *sensory root of the ciliary ganglion.*
Rayleigh criterion The minimum spacing at which two point images may be seen as two images. It is proportional to the wavelength of light and inversely proportional to pupil diameter.
receptive field The region in visual space within which visual stimuli will affect the activity of a visual neuron. The definition can be generalized to other sensory modalities.
receptor A membrane molecule that binds other molecules (ligands) that have a particular configuration. It affects the status of ion channels either directly (ionotropic receptor) or indirectly (metabotropic receptor).
recession A method for making an extraocular muscle less effective by moving its insertion.
reciprocal innervation Innervational yoking in agonist–antagonist muscle pairs in which one muscle is inhibited when the other is activated.
reciprocal synapse A synaptic configuration in which synapses in opposite directions between two cells lie close to one another.
reflecting superposition eye An optical superposition eye in which reflection is the primary method for combining light from neighboring optical elements. Contrast *refracting superposition eye.*
refracting superposition eye An optical superposition eye in which refraction is the primary method for combining light from neighboring optical elements. Contrast *reflecting superposition eye.*
regeneration In general, regrowth, as in regeneration of a nerve fiber. In photoreceptors, it is the process that restores bleached photopigment to its active form.

resection A surgical method for increasing the effectivity of a muscle by shortening it.
retina An array of photoreceptors and their accompanying neural circuitry.
retinal An aldehyde of retinol (vitamin A) that is the active element in photopigments.
retinal detachment Separation of the photoreceptors from the pigment epithelium.
retinal pigment epithelium See *pigment epithelium of the retina.*
retinitis pigmentosa A hereditary form of photoreceptor degeneration.
retinopathy of prematurity Neovascularization, hemorrhage, and scar formation in the vitreous resulting from the use of supplemental oxygen for premature infants.
retinotopic organization The pattern of projection of axons such that neighboring areas on the retina remain neighbors in the target nuclei.
retrobulbar block Anesthesia of the eye produced by injection of the anesthetic into the center of the muscle cone.
retrograde transport See *axoplasmic transport.*
retrolental space of Berger The space or potential space between the posterior surface of the lens and the anterior surface of the vitreous in the patellar fossa.
rhabdom The light-sensitive structure in the ommatidium of a compound eye.
rhodopsin The visual pigment of rods.
rhubarb harvest The gathering of a useless, unpalatable vegetable, therefore meaningless discourse. Committee meetings and most political utterances are examples.
RK See *radial keratotomy.*
rod A photoreceptor that contains rhodopsin and used primarily for vision at night. Contrast *cone.*
rod bipolar cell A bipolar cell that receives inputs only from rods; the outputs are to amacrine cells. Rod bipolar cells depolarize to increasing light intensity. Contrast *cone bipolar cell.*
rod-free area The region in the center of the fovea that contains only cones.
RPE See *pigment epithelium of the retina.*

sarcolemma The cell membrane of a muscle fiber.
sarcoplasmic reticulum The membranous structure within muscle fibers that acts as a reservoir of the calcium used in initiating muscle fiber contraction.
Scheimpflug imaging A method for photographing a slit-lamp image without the parallax normally associated with an oblique angle of view.
Schlemm's canal See *canal of Schlemm.*
Schwalbe's ring The annulus of limbal tissue between the margin of the cornea and the anterior extent of the trabecular meshwork.
scleral spur A thickened ridge of sclera running around the limbus just behind the canal of Schlemm; it is the site of attachment for ciliary muscle fibers and for cords of tissue in the trabecular meshwork.
sclerocornea A congenital condition in which the cornea is much like the sclera in structure and opacity.
scotoma A region of blindness or reduced sensitivity in the visual field.
secondary positions of gaze All eye positions along the vertical or the horizontal lines that intersect at the primary position of gaze.
sella turcica The depression in the upper part of the sphenoid bone in which the pituitary gland is located.
semilunar ganglion See *gasserian ganglion.*
sensory root of the ciliary ganglion The bundle of sensory axons running between the ciliary ganglion and the nasociliary nerve. Also called ramus communicans.
septum orbitale See *orbital septum.*
shedding The process by which the tips of photoreceptor outer segments are discarded at regular intervals.
short ciliary nerves The small bundles of nerve axons from the ciliary ganglion that penetrate the sclera around the optic nerve. They are mixed nerves carrying sympathetic, parasympathetic, and sensory axons.
sign-conserving synapse A synapse in which the membrane potentials of the pre- and postsynaptic cells always change in the same direction. Usually an excitatory synapse. Contrast *sign-inverting synapse.*
sign-inverting synapse A synapse in which the membrane potentials of the pre- and postsynaptic cells change in opposite directions. Usually an inhibitory synapse. Contrast *sign-conserving synapse.*
simple apposition eye See *focal apposition eye.*
simple eye An eye with a single optical system and multiple photoreceptors. Contrast *compound eye.*
simple reflecting eye A simple eye in which the primary optical element is a reflecting surface.
simple refracting eye A simple eye in which the primary optical elements refract light.
slit-lamp A combination of biomicroscope and variable-configuration light source used to examine ocular structure in vivo.
slow-twitch fiber A muscle fiber with a relatively slow all-or-nothing mode of contraction. Contrast *fast-twitch fiber.*
small bistratified ganglion cell A retinal ganglion cell with dendrites ramifying at two levels of the inner plexiform layer; it carries blue-ON, yellow-OFF signals.
smooth pursuit movement An eye movement made to track a small moving object, thereby keeping its image on the fovea.
sphenoid bone The bone lying on the midline behind the orbits that forms the apices of the orbits and contributes to their lateral and medial walls.
sphenopalatine ganglion See *pterygopalatine ganglion.*
spherical aberration The difference in focal points for light rays passing through a lens close to its axis and those for rays passing through the periphery of the lens.
spherule The synaptic terminal of a rod photoreceptor. Contrast *pedicle.*
sphincter In general, any muscle so arranged that it closes an aperture or opening when it contracts. The sphincter of the iris is responsible for pupillary constriction; the ciliary muscle acts as if it were a sphincter.

spina trochlearis A small ossified cone on the frontal bone at the location of the trochlea for the superior oblique muscle.

squamous cells In general, flattened cells. In the cornea, cells in the outer two or three layers of the corneal epithelium.

staphyloma An abnormal, localized outward bulge of the sclera.

stem cells Cells in mature tissues from which new members of a particular cell type can be produced.

stereoisomer One of the different three-dimensional configurations that a molecule may assume.

strabismus A deviation of one eye from bifixation under binocular viewing conditions. Also called heterotropia.

straightforward position of gaze The position of the eyes when the visual axes are perpendicular to the plane of the face when the head is erect.

stye See *hordeolum.*

subluxated lens A partial dislocation of the lens.

superior cervical ganglion One of the uppermost ganglia in the two chains of sympathetic ganglia lying alongside the vertebral column. It sends axons to the dilator muscle in the iris, to the tarsal muscles, and to the choroidal blood vessels.

superior colliculus (pl. colliculi) One of the two paired nuclei on the surface of the midbrain that is a secondary target for retinal ganglion cell axons.

superior orbital fissure The opening between the greater and lesser wings of the sphenoid bone at the back of the orbit.

superior transverse ligament A sling of connective tissue attached to the orbital ceiling through which the levator muscle passes. Also called Whitnall's ligament.

superposition eye See *optical superposition eye.*

suprachiasmatic nucleus A small nucleus in the hypothalamus, just above the optic chiasm. It receives retinal inputs and may play a role in regulating a biological clock.

suprachoroidea The outermost part of the choroid next to the sclera.

supraorbital notch The notch, sometimes a foramen, near the middle of the superior margin of the orbit. It contains the supraorbital artery and nerve.

surface ectoderm The germ tissue that forms the outer surface of the early embryo. The corneal and conjunctival epithelia are derived from it, as are the lens, the lacrimal and other glands in the eyelids, and the lacrimal drainage system.

suspensory ligament of the lens See *zonule.*

sympathetic root of the ciliary ganglion The bundle of sympathetic axons that run from the carotid plexus to the ciliary ganglion and into the short ciliary nerves.

synapse The site at which neurotransmitter is released by one cell to interact with receptors in the adjacent membrane of another cell.

synaptic ribbon An elongated ribbonlike structure associated with synaptic terminals of photoreceptors and bipolar cells.

T tubules Small membrane-lined tunnels through muscle fibers; they effectively bring the extracellular space close to all parts of the cytoplasm.

tapetum (pl. tapeta) A reflecting layer in the choroid of many species; it is not an image-forming element.

tarsal gland One of the oil-secreting glands embedded in the tarsal plates. Also called meibomian gland.

tarsal muscles The smooth muscles, inferior and superior, in the eyelids that are innervated by the sympathetic system. Also called the palpebral muscles of Müller.

tarsal plate A layer of dense connective tissue within the eyelid, in which the tarsal glands are located.

telodendria (s. telodendrion) Small fingerlike extensions from cone pedicles, often containing gap junctions with other cone pedicles.

Tenon's capsule A sheet of connective tissue that lines all of the eye out to the limbus and ensheathes the extraocular muscles.

tight junction An occluding junction that restricts the movement of substances through the extracellular space between cells; tight junctions vary in tightness.

tiling An array of discrete elements that covers a surface without gaps or overlap.

tonic fiber A muscle fiber whose contraction is prolonged and can be graded depending on the strength of the innervation. Smooth muscle fibers are typically tonic fibers, as are some of the extraocular muscle fibers.

tonometer A device for measuring intraocular pressure indirectly.

torsion Rotation of the eye around its anterior–posterior axis.

trabecular meshwork The collection of tissue cords and open spaces that separates the canal of Schlemm from the anterior chamber angle.

trabeculectomy A surgical procedure to increase aqueous drainage by the creation of a drainage channel through the limbus and sclera from the anterior chamber.

trabeculoplasty A surgical procedure to increase aqueous drainage by the creation of openings from the anterior chamber into the canal of Schlemm. It is commonly done with a laser.

transneuronal atrophy Atrophy of cells in a nucleus after their input axons degenerate.

transposition Strabismus surgery in which muscle insertions are interchanged or moved out of their normal line of action.

triad synapse The synapse within the invaginations in photoreceptor terminals. There are often, but not always, three postsynaptic elements. Contrast *dyad synapse.*

tritanomaly See *tritanopia.*

tritanopia Confusion between blues and greens because of the absence of blue cone pigment. The less severe form is tritanomaly.

trochlea (pl. trochleae) The U-shaped pulley through which the tension of the superior oblique muscle passes; it is on the upper medial wall of the orbit, just behind the orbital rim.

trochlear fossa (pl. fossae) The small depression in the frontal bone at the site of the trochlea.

trochlear nerve Cranial nerve IV, which innervates the superior oblique muscle.

trophoblast The part of the blastocyst that is responsible for implantation in the uterine wall; it becomes the embryonic part of the placenta.

tucking A surgical technique for making an extraocular muscle shorter and more effective by suturing a Z-fold in the muscle. Contrast *marginal myotomy* and *myectomy.*

tunica vasculosa lentis A transient, embryonic network of blood vessels around the developing lens.

ultrafiltration Movement of fluid along a hydrostatic pressure gradient; it may be the mechanism by which water moves into the vitreous from the pars plana of the ciliary body.

ultrasonography The use of ultrasound echoes from acoustic reflecting surfaces to image internal structures.

undersampling The condition in which the discrete elements in a sampling array are too far apart to faithfully capture the high frequencies in the signal.

unilateral strabismus Strabismus in which one eye is always the deviated eye.

unistratified Having neuronal branching, usually dendritic branching, in a plane.

uveal tract The vascular middle coat of the eye, consisting of the iris, ciliary body, and choroid.

uveoscleral flow An ill-defined path of aqueous outflow from the anterior chamber into the ciliary body; it is a secondary pathway to the normal outflow route through the canal of Schlemm.

valve of Hasner A flap of tissue at the end of the nasolacrimal sac in the nasal cavity. It has also been known as the plica lacrimalis.

vasa hyaloidea propria A complex of small blood vessels around the hyaloid artery in the primary vitreous; the vessels atrophy with the hyaloid artery.

vascular casting The technique for making a plastic replica of the eye's vascular system (a vascular cast) by injecting resin into the blood vessels.

vasculogenesis The formation of new blood vessels from precursor cells.

vergence Movement of the eyes in opposite directions (one right and one left, or one up and one down). Contrast *version.*

version Movement of the eyes in the same direction (both right or left, or both up or down). Contrast *vergence.*

vestibulo-ocular reflex The reflex movement of the eyes in response to stimulation of receptors in the vestibular system.

visual axis The imaginary line from the object of regard to the primary nodal point.

visual streak A line of high photoreceptor density extending horizontally across the retina.

vitrectomy Microsurgery to remove abnormal tissue from the vitreous and to remove some of the vitreous itself.

vitreous See *vitreous humor.*

vitreous base The site of attachment of vitreous collagen to the pars plana of the ciliary body.

vitreous chamber The space bounded by the retina, ciliary body, and lens that contains the vitreous.

vitreous cortex The outermost part of the vitreous, where most of the collagen is located.

vitreous humor The transparent gel that fills the vitreous chamber; the adjective "vitreous" is usually used alone as a noun.

vortex veins The large veins, typically four, that drain all of the choroid and parts of the ciliary body and iris.

Whitnall's ligament See *superior transverse ligament.*

Whitnall's tubercle See *lateral orbital tubercle.*

Wieger's ligament A ring of apparent attachment of the posterior lens capsule to the anterior vitreous. Also called hyaloideocapsular ligament.

wing cells Cells in the corneal epithelium in the layers just superficial to the basal cells.

Zeis's glands See *glands of Zeis.*

zonule (of Zinn) The collection of small fibers that run between the ciliary body and the lens capsule and hold the lens in place. Also called suspensory ligament of the lens.

zygomatic nerve A branch of the maxillary nerve (V_2) in which axons from the pterygopalatine ganglion run en route to the lacrimal gland.

Historical References and Additional Reading

Questions like "What is light?" and "How do I see the world around me?" must extend back into the night of time.

■ David Park, *The Fire within the Eye*

Primary or Authoritative Sources

Adams F. 1886. *The Genuine Works of Hippocrates.* William Wood & Company, New York.

Bartisch G. 1583. ΟΦΘΑΛΜΟΔΟΥΛΕΙΑ *Das ist, Augendienst.* Gedruckt zu Dressden durch Mathes Stöckel. (Rey)*

Bowman W. 1892. Lectures on the parts concerned in the operations on the eye and on the structure of the retina, delivered at the Royal London Ophthalmic Hospital, Moorfields, June, 1847. In *The Collected Papers of Sir W. Bowman, Bart., F.R.S.*, Vol. II, J Burdon-Sanderson and JW Hulke, eds. Harrison and Sons, London. (Rey)

Darwin C. 1870. *On the Origin of Species by Means of Natural Selection, or the Preservation of Favored Races in the Struggle for Life,* 5th Ed. D Appleton, New York. (Rey)

Descartes R. 1637. *Discourse de la Methode pour Bien Conduire Sa Raison, & Chercher la Verité dans les Sciences. Plus la Dioptrique. Les Meteores. Et la Geometrie. Qui Sont des Essais de Cete Methode.* A Lyde, de I'mprimerie de I. Maire (Rey)

Descartes R. 1677. *Tractatus de Homine, et de Formatione Foetus.* Danielen Elsevirium, Amstelodami. (Rey)

Dogiel AS. 1891. Ueber die nervösen Element in der Retina des Menschen. *Archiv für Mikroskopische Anatomie,* 38: 317–344.

Eustachii B. 1722. *Tabulae Anatomicae Clarissum Viri Bartholomaie Eustachii Quas è Tenebris Tandem Vindicatus et Clementis XI Pont. Max.* R & G Wetstenios, Amstelaedami. (Rey)

Fabricius ab Aquapendente H. 1600. *De Visione Voca Auditu.* Venetiis per Franciscum bolzettam. (Rey)

Falloppio G. 1561. *Observationes Anatomicae.* Venetiis. Apud Marcum Antonium Vlmum. (Rey)

Harvey W. 1628. *De Motu Cordis & Sanguinis in Animalibus.* Francofurti, sumptibus Guilielmi Fitzeri. (Rey)

Helmholtz H von. 1924. *Helmholtz's Treatise on Physiological Optics*, JPC Southall, ed. Optical Society of America, Rochester, NY. Reprinted by Dover, New York, 1962.

Hooke R. 1665. *Micrographia: Or Some Physiological Descriptions of Minute Bodies Made by Magnifying Glasses.* J Martyn and J Allestry, London. (Rey)

Hoole S. 1800–1807. *The Select Works of Antony van Leeuwenhoek, Containing His Microscopical Discoveries in Many Works of Nature,* 2 vols. G Sidney (Vol. 1) and the Philanthropic Society (Vol. 2), London. (Rey)

Leonardo da Vinci. 1911–1916. *Quaderni d'Anatomica,* 6 vols. (Folgi della Royal Library di Windsor). J Dybwad, Christiania [Oslo, Norway]. (Rey)

*Reynolds Historical Library, University of Alabama at Birmingham.

Lind LR. 1949. *The Epitome of Andreas Vesalius*. Macmillan, New York.

Mann IC. 1928. *The Development of the Human Eye*. University Press, Cambridge.

Müller H. 1856. Anatomisch-physiologische Untersuchungen über die Retina des Menschen und der Werbelthiere. *Zeitschrift für wissenschaftliche Zoologie*, 8: 1–128.

Newton I. 1730. *Opticks: Or, a Treatise of the Reflections, Refractions, Inflections, and Colours of Light*, 4th Ed. Reprinted by Dover, New York, 1952.

O'Malley CD and Saunders JB de CM. 1952. *Leonardo da Vinci on the Human Body: The Anatomical, Physiological, and Embryological Drawings of Leonardo da Vinci*. H Schuman, New York. Reprinted by Dover, New York, 1983.

Paley W. 1847. *Paley's Natural Theology*, E Bartlett, ed. Harper & Brothers, New York.

Polyak SL. 1941. *The Retina*. University of Chicago Press, Chicago.

Ramón y Cajal S. 1892. La rétine des vertébrés. *La Cellule* 9: 17–257. Translated in Rodieck RW, *The Vertebrate Retina: Principles of Structure and Function*, pp. 781–904, WH Freeman & Co., San Francisco, 1973, and in Thorpe SA and Glickstein M, *The Structure of the Retina*, Charles C Thomas, Springfield, IL, 1972.

Scheiner C. 1619. *Oculus Hoc Est: Fundamentum Opticum*. Londini: Excudebat J. Flesher, & prostant apud Cornelium Bee. (Rey)

Sherrington CS. 1906. *The Integrative Action of the Nervous System*. Yale University Press, New Haven. (Rey)

Siegel RE. 1970. *Galen on Sense Perception: His Doctrines, Observations and Experiments on Vision, Hearing, Smell, Taste, Touch and Pain, and Their Historical Sources*. S Karger, Basel.

Soemmerring ST von. 1795. *Abbildungen des menschlichen Auges*. Frankfurt am Main: Varrentrapp und Wenner. (Rey)

Tartuferi F. 1887. Sull'anatomia della retina. *International Monatsschrift Anatomie Physiologie*, 4: 421–441.

Taylor J. 1738. *Le Mechanisme, ou le Nouveau Traité de l'Anatomie du Globe de l'Oeil*. Michel-Estienne David, Paris. (Rey)

Vesalius A. 1543. *De Humani Corporis Fabrica Libri Septum*. Ioannis Oporini, Basel. (Rey)

Walls GL. 1942. *The Vertebrate Eye and Its Adaptive Radiation*, Cranbrook Institute of Science, Bloomfield Hills, MI. Reprinted by Hafner, New York, 1963.

Withering W. *Anatomical and Physiological Observations Collected from Lectures of Doct^r Hunter and Doct^r Monro by William Withering: Anno 1765*. Holograph notes in ink on paper. (Rey)

Zinn JG. 1780. *Descriptio Anatomica Oculi Humanis Iconibus Illustrata, Nunc Altera Vice Edita, et Necessario Supplemento, Novisque Tabulis Aucta*. Abrami Vandenhoeck, Goettingae. (Rey)

Biography

Albert DM. 1980. Allvar Gullstrand, 1862–1930: "A lonesome giant of optical theory." *Trends Neurosci*. 3(30): IV–VI.

Albert DM. 1993. *Men of Vision: Lives of Notable Figures in Ophthalmology*. WB Saunders, Philadelphia.*

Bowlby J. 1990. *Charles Darwin: A New Life*. WW Norton, New York.

Bramly S. 1991. *Leonardo: Discovering the Life of Leonardo da Vinci*. S Reynolds, trans. HarperCollins, New York.

Calder R. 1970. *Leonardo & the Age of the Eye*. Simon and Schuster, New York.

Christianson GE. 1984. *In the Presence of the Creator: Isaac Newton and His Times*. Free Press, New York.

Cohen of Birkenhead, Lord. 1958. *Sherrington: Physiologist, Philosopher, Poet*. Liverpool University Press, Liverpool.

Craigie EH and Gibson WC. 1968. *The World of Ramon y Cajal, with Selections from His Nonscientific Writings*. Charles C Thomas, Springfield, IL.

Crescitelli F. 1963. Obituary: Gordon Lynn Walls (1905–1962). *Vision Res*. 3: 1–7.

Eccles JC and Gibson WC. 1979. *Sherrington: His Life and Thought*. Springer International, Berlin.

Gillispie CC, ed. 1970. *Dictionary of Scientific Biography*. Charles Scribner's Sons, New York.†

Keele KD. 1965. *William Harvey: The Man, the Physician, and the Scientist*. Nelson and Sons, London.

Keele KD. 1983. *Leonardo da Vinci's Elements of the Science of Man*. Academic Press, New York.

Keynes G. 1966. *The Life of William Harvey*. Clarendon Press, Oxford.

Konigsberger L. 1906. *Hermann von Helmholtz*. FA Welby, trans. Clarendon Press, Oxford.

O'Malley CD. 1965. *Andreas Vesalius of Brussels 1514–1564*. University of California Press, Berkeley.

Piccolino M. 1988. Cajal and the Retina: A 100-Year Retrospective. *Trends Neurosci*. 11: 521–525.

Ramón y Cajal S. 1937. *Recollections of My Life*. EH Craigie, trans. American Philosophical Society, Philadelphia.

Snyder C. 1967. *Our Ophthalmic Heritage*. Little, Brown, Boston.††

Vasari G. 1568. *The Lives of the Artists*. Translation by JC Bondanella and P Bondanella, 1991. Oxford University Press, Oxford.

Williams H. 1954. *Don Quixote of the Microscope: An Interpretation of the Spanish Savant, Santiago Ramon y Cajal (1852–1934)*. Cape, London.

Yates T, Constable I, and Lowe R. 1984. Obituary: Dame Ida Caroline Mann, 1893–1983. *Austral. J. Ophthalmol*. 12: 95–96.

History of Science

Boorstin DJ. 1983. *The Discoverers*. Random House, New York.

Boorstin DJ. 1992. *The Creators*. Random House, New York.

Bracegirdle B. 1987. *A History of Microtechnique*, 2nd Ed. Science Heritage, Lincolnwood, IL.

Bradbury S. 1968. *The Microscope: Past and Present*. Pergamon Press, Oxford.

Canfora L. 1989. *The Vanished Library*. M Ryle, trans. University of California Press, Berkeley.

Choulant L, Frank M, Garrison FH, and Streeter EC. 1920. *History and Bibliography of Anatomic Illustration in Its Relation to Anatomic Science and the Graphic Arts*. University of Chicago Press, Chicago.

Hirschberg J. 1982. *The History of Ophthalmology*. FC Blodi, trans. JP Wayenborgh, Bonn.

Kronfeld PC. 1943. *The Human Eye in Anatomical Transparencies*, with a historical appendix by SL Polyak. Bausch & Lomb Press, Rochester, NY.

Lassek AM. 1958. *Human Dissection: Its Drama and Struggle*. Charles C Thomas, Springfield, IL.

Lindberg DC. 1976. *Theories of Vision from al-Kindi to Kepler*. University of Chicago Press, Chicago.

Lindberg DC. 1992. *The Beginnings of Western Science*. University of Chicago Press, Chicago.

Park D. 1997. *The Fire within the Eye: A Historical Essay on the Nature and Meaning of Light*. Princeton University Press, Princeton, NJ.

Polyak SL. 1957. History of investigations of the structure and function of the eye and of the visual pathways and centers of the brain. Part I, pp. 9–203, in *The Vertebrate Visual System*, H Kluver, ed. University of Chicago Press, Chicago.

Roberts KB and Tomlinson JDW. 1992. *The Fabric of the Body: European Traditions of Anatomical Illustration*. Clarendon Press, Oxford.

Ronchi V. 1970. *The Nature of Light: An Historical Survey*. V Barocas, trans. Harvard University Press, Cambridge, MA.

Silver BL. 1998. *The Ascent of Science*. Oxford University Press, New York.

Singer CJ. 1926. *The Evolution of Anatomy*. AA Knopf, New York.

*Sir William Bowman, pp. 300–316, and Hermann von Helmholtz, pp. 345–360, by P Henkind. Sir Isaac Newton, pp. 21–32, and John Taylor, pp. 41–49, by DM Albert.

†This multivolume work contains biographies for the following persons mentioned in the vignettes or in the text: Alhazen (Ibn al-Haytham): Vol. VI, pp. 189–210; William Bowman: Vol. II, pp. 375–377; Charles Darwin: Vol. III, pp. 565–577; René Descartes: Vol. IV, pp. 51–65; Eustachius (Bartolommeo Eustachio): Vol. IV, pp. 486–488; Fabricius ab Aquapendente (Girolamo Fabrici): Vol. IV, pp. 507–512; Fallopius (Gabriele Falloppio): Vol. IV, pp. 519–521; Galen: Vol. V, pp. 227–237; Johann von Gudden: Vol. V, pp. 569–572; Alvar Gullstrand: Vol. V, pp. 590–591; William Harvey: Vol. VI, pp. 150–162; Hermann von Helmholtz: Vol. VI, pp. 241–253; Herophilus: Vol. VI, pp. 316–319; Robert Hooke: Vol. VI, pp. 481–488; Hunain ibn Ishak: Vol. XV, pp. 230–249; Leonardo da Vinci: Vol. VIII, pp. 192–245; Antoni van Leeuwenhoek: Vol. VIII, pp. 126–130; Isaac Newton: Vol. X, pp. 42–92; William Paley: Vol. X, pp. 277–279; Jan Purkinje (Purkyne): Vol. XI, pp. 213–217; Santiago Ramón y Cajal: Vol. XI, pp. 273–276; Frederick Ruysch: Vol. XII, pp. 39–42; Christoph Scheiner: Vol. XII, pp. 151–152; Max Schultz: Vol. XII, pp. 230–233; Charles Scott Sherrington: Vol. XII, pp. 395–403; Samuel Soemmerring: Vol XII, pp. 509–511; Gottfried Treviranus: Vol. XIII, pp. 460–462; Andreas Vesalius: Vol. XIV, pp. 3–12; William Withering: Vol. XIV, pp. 463–465.

††"Charles Babbage and his rejected ophthalmoscope," pp. 1–3; "Helmholtz at Columbia University," pp. 4–7; "Johann Sebastian Bach and Chevalier Taylor," pp. 77–79; "Doctor Franklin's double spectacles," pp. 89–92; "Allvar Gullstrand, Nobel Laureate," pp. 149–152.

Index

B

C

D

E

F

G

H

I

Q

R

S

T

U

V

W

X

Y

Z

About the Book

Editor: Peter Farley

Project Editor: Paula Noonan

Copy Editor: Stephanie Hiebert

Production Manager: Christopher Small

Illustrations: Nancy Haver

Electronic Art: Michele Ruschhaupt and Precision Graphics

Book Design: Jean Hammond

Cover Design: Jefferson Johnson

Composition: Wendy Beck

Book Manufacturer: Courier Companies, Inc.

P9-DBV-382
ROMA
ITALY

THIS BOOK BELONGS TO

ROME

The Adventures of Bella & Harry

Let's Visit Rome!

Written By

Lisa Manzione

Illustrated By

Kristine Lucco

Bella & Harry, LLC

www.BellaAndHarry.com
email: BellaAndHarryGo@aol.com

"I am a gladiator!
I am a gladiator!"

"Harry,
who are you talking to?"
"I am practicing, Bella.
I am a 'Roman Gladiator'!"

"**Harry**, gladiators have not been around for thousands of years. A long time ago, gladiators were a part of competitions that were held in large areas, usually with a crowd watching the competition. Today Harry, gladiators can only be found posing for pictures outside the Roman Colosseum or in history books."

"**Where** is the Roman Colosseum, Bella?"

"The Roman Colosseum is in Rome, Italy. Remember, Harry, we visited Venice, Italy with our family earlier this year. Italy is in Europe. A lot of people think Italy looks like a boot!

Let's look at the map before we get started. See Harry, Rome is here."

"Yes, Bella, I see Rome and I think Italy still looks like a boot!"

"I agree Harry, but Italy is a very fancy boot!"

BELGIUM
GERMANY
AUSTRIA
SWITZERLAND
SLOVENIA
CROATIA
ITALY
Venice
Bosnia-
Herzegovina
FRANCE
Adriatic
Sea
Rome
Corsica
Mediterranean
Sea
Sardinia
Tyrrhenian
Sea
Sicily

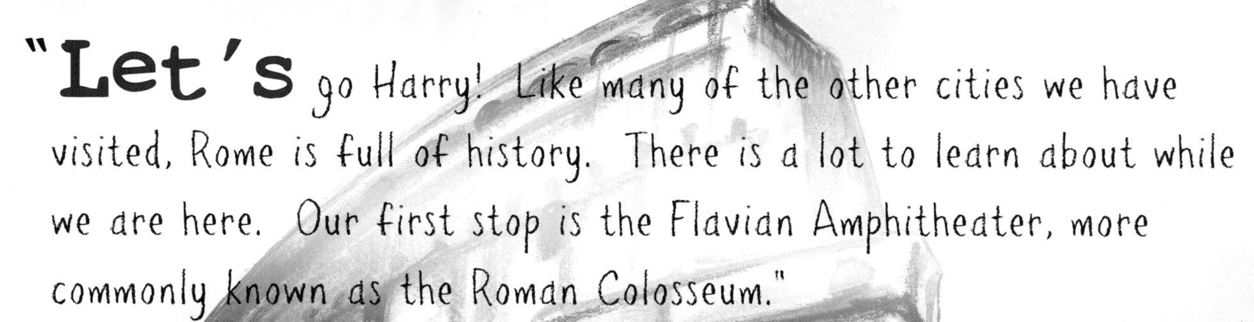

"**Let's** go Harry! Like many of the other cities we have visited, Rome is full of history. There is a lot to learn about while we are here. Our first stop is the Flavian Amphitheater, more commonly known as the Roman Colosseum."

"**Today**, the Colosseum is one of the most visited sites in Rome. The Colosseum is in ruins now, mostly because of earthquakes and stone robbers. There is not much left of the original floor, but you can still see the 'hypogeum' from the inside of the Colosseum."

"**Bella**, what is a 'hypogeum'?"

"A 'hypogeum' is an area underground. If you were a gladiator Harry, this is where you would have entered the stadium for your contest."

"**Most** people believe the Colosseum is one of the best buildings ever built by the Romans. The building is about 2,000 years old. It's an 'amphitheater', which means the building is usually oval or round, and there is no roof covering the center area of the theater."

"Look up there! That area is called Palatine Hill. Palatine Hill is one of the oldest areas in Rome. It is also one of the seven hills that Rome was built on. If you look below the hill, you will see the Roman Forum."

"We are going to the Roman Forum next. Come on, let's go!"

"**First**, the Roman Forum was a marketplace. Later Harry, many other buildings were built in and around the Roman Forum area during ancient times."

"**There** were all sorts of meetings held at the Roman Forum for both work and fun. Today, the Roman Forum is only ruins, but it too is one of the oldest areas of Rome."

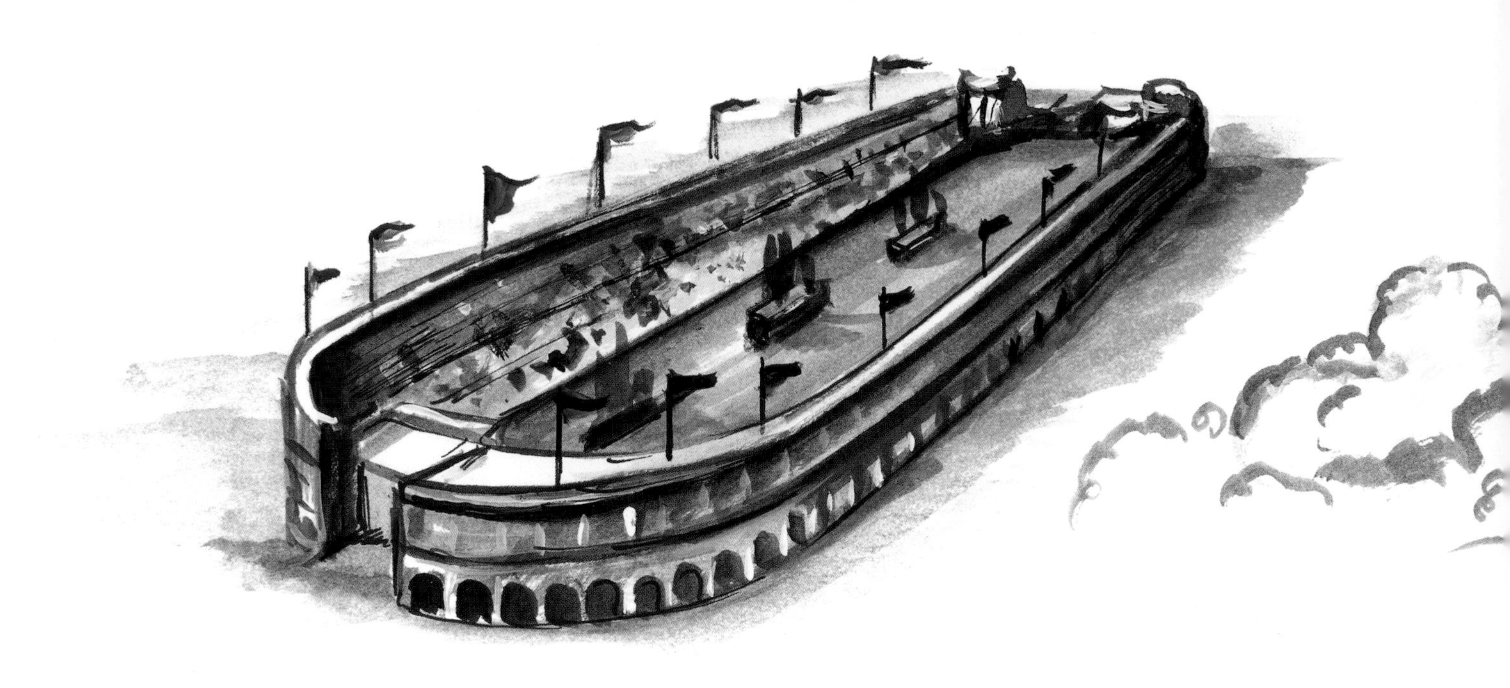

"**Next** stop, Circus Maximus!"

"Yay! I love the circus!"

"**Harry**, Circus Maximus is not a circus with animals or games. Circus Maximus was a place for chariot races long ago. It was the first and largest stadium in old Rome. It measures about 2,050 feet long ... or about 342 average lions ... standing tail to tail. It's now a public park."

"**Lunch** time! We are going to have lunch at Piazza Navona today. The piazza, or open square, has a big fountain, which has one obelisk (a four sided, tall stone) in the middle of the fountain."

"**There** are two more fountains at each end of the piazza. Also, there are a lot of places in the piazza that serve lunch and dinner."

"It looks like the antipasto (or first course) includes cheeses, meats, and olives... you know, all of your favorites!"

"Yummy!"

"Let's go Harry! It's time to see a famous fountain and toss a coin in the water!"

"**Harry**, this is the Trevi Fountain, or Fontana di Trevi. It is the largest and most famous Baroque (a type of design) fountain in Rome. I am going to turn around and toss this coin in the water.
Harry? Harry?"

"Harry, what are you doing? Get out of the water!"

"**Bella**, look at all of the coins I found!"

"Harry, no! People toss coins in the fountain because legend says if you toss a coin in the water, you will be sure to return to Rome. We must come back to Rome, so leave the coins in the water. There is so much more to see in this fun city! I am sure everyone who tossed a coin in the water wants to come back for another visit too!"

"Okay, Bella."

"**Harry**, it's time to start walking back to our hotel.

Let's take the scenic route."

"**Harry**, we are now in an area called the 'Trident', which has lots of shopping and restaurants. Most of the streets in this area start with the name 'via', which means 'by way of' in our language."

"**Last** stop... the Spanish Steps! The Spanish Steps is a favorite location of both visitors and locals because of its beauty."

"The Spanish Steps make up the longest and widest staircase in Europe and they connect the lower piazza (Piazza di Spagna) with the upper piazza (Piazza Trinità dei Monti). Our hotel is located in the upper piazza, next to a beautiful old church. Let's race to the top!"

Whew! What a race! We are at our hotel at the top of the Spanish Steps. Harry and I are going to rest for a while after our fun tour of Rome. We can't wait for our next adventure but for now it's good-bye or "arrivederci" from Bella Boo and Harry too!

Our Adventure to Rome

Bella and Harry at the Pantheon.

Harry and Bella with the Vatican's Swiss Guard.

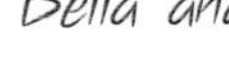

Bella and Harry enjoying spaghetti and meatballs.

Bella and Harry at the Vatican.

Fun Italian Words and Phrases

Yes – Si

No – No

Family – Famiglia

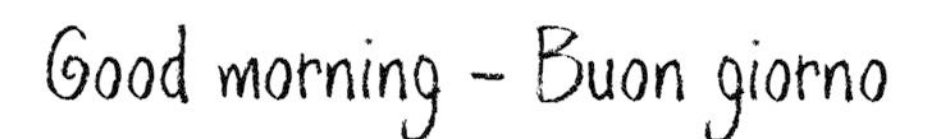

Good morning – Buon giorno

Good evening – Buona sera

Good night – Buona notte

Publisher's note: This book is a work of fiction. Names, characters, places and incidents either are the product of the author's imagination or are used fictitiously, and any resemblance to actual persons or animals living or dead, events, or locales is entirely coincidental.

Requests for permission to make copies of any part of the work should be directed to BellaAndHarryGo@aol.com or 855-235-5211.

Library of Congress Cataloging-in-Publications Data is available

Manzione, Lisa

The Adventures of Bella & Harry: Let's Visit Rome!

ISBN: 978-1-937616-08-3

First Edition

Book Eight of Bella & Harry Series

For further information please visit:

www.BellaAndHarry.com

or

Email: BellaAndHarryGo@aol.com

CPSIA Section 103 (a) Compliant

www.beaconstar.com/ consumer

ID: L0118329. Tracking No.: MR210171-1-10823

Printed in China